# CARBON NANOMATERIALS SOURCEBOOK

Nanoparticles, Nanocapsules, Nanofibers,
Nanoporous Structures, and Nanocomposites

# CARBON NANOMATERIALS SOURCEBOOK

## VOLUME II

Nanoparticles, Nanocapsules, Nanofibers, Nanoporous Structures, and Nanocomposites

*Edited by*

## Klaus D. Sattler

University of Hawaii at Manoa, Honolulu, USA

**CRC** CRC Press
Taylor & Francis Group
Boca Raton  London  New York

CRC Press is an imprint of the
Taylor & Francis Group, an **informa** business

CRC Press
Taylor & Francis Group
6000 Broken Sound Parkway NW, Suite 300
Boca Raton, FL 33487-2742

First issued in paperback 2020

© 2016 by Taylor & Francis Group, LLC
CRC Press is an imprint of Taylor & Francis Group, an Informa business

No claim to original U.S. Government works

ISBN-13: 978-1-4822-5270-5 (hbk)
ISBN-13: 978-0-367-78307-5 (pbk)

---

**Library of Congress Cataloging-in-Publication Data**

Names: Sattler, Klaus D.
Title: Carbon nanomaterials sourcebook. Nanoparticles, nanocapsules, nanofibers, nanoporous structures, and nanocomposites / [edited by] Klaus D. Sattler.
Description: Boca Raton : Taylor & Francis Group, 2016. | "A CRC title." | Includes bibliographical references and index.
Identifiers: LCCN 2016002145 | ISBN 9781482252705 (alk. paper)
Subjects: LCSH: Nanostructured materials. | Nanoparticles. | Nanocapsules. | Nanofibers. | Nanocomposites (Materials)
Classification: LCC TA418.9.N35 C34227 2016 | DDC 620.1/15--dc23
LC record available at http://lccn.loc.gov/2016002145

---

Visit the Taylor & Francis Web site at
http://www.taylorandfrancis.com

and the CRC Press Web site at
http://www.crcpress.com

# Contents

## SECTION III   Nanocapsules

## SECTION IV   Porous Materials

## SECTION V   Hybrids/Composites

# Preface

In the last three decades, zero-dimensional, one-dimensional, and two-dimensional carbon nanomaterials have attracted significant attention due to their unique electronic, optical, thermal, mechanical, and chemical properties. There is a need for understanding the science of carbon nanomaterials, the production methods, and the applications, in order to use these exceptional properties. In particular, in the last two decades, carbon nanotubes, fullerenes, nanodiamonds, and mesoporous carbon nanostructures became a new class of nanomaterials. Research on graphene, a single sheet of graphite, is one of the fastest-growing fields today and holds the promise of someday replacing silicon in computers and electronic devices. Many scientists and engineers are redirecting their work toward carbon nanomaterials. It is in this context that the editor of this book determined that a new resource concentrating on this subject would be a unique and useful source of information and learning.

*Carbon Nanomaterials Sourcebook* is the most comprehensive reference covering the field of carbon nanomaterials, reflecting its interdisciplinary nature that brings together physics, chemistry, materials science, molecular biology, engineering, and medicine. The scope of the two volumes spans from fundamental properties, growth mechanisms, and processing of nanocarbons to their functionalization and electronic device, energy conversion and storage, and biomedical and environmental applications. It encompasses a wide range of areas from science to engineering. In addition to addressing the latest advances, the *Sourcebook* presents core knowledge with basic mathematical equations, tables, and graphs in order to provide the reader with the tools necessary to understand current and future technology developments.

The contents are made up of 54 chapters organized into nine subject areas, with each chapter covering one type of carbon nanomaterial. Materials have been selected to showcase exceptional properties, good synthesis and large-scale production methods, and strong current and future application prospects. Every chapter covers the three main areas: formation, properties, and applications. This setup makes the book a unique source where a reader can easily navigate to find the information about a particular material. The chapters will be written in tutorial style where basic equations and fundamentals are included in an extended introduction.

# Editor

**Klaus D. Sattler** pursued his undergraduate and master's courses at the University of Karlsruhe in Germany. He received his PhD under the guidance of Professors G. Busch and H. C. Siegmann at the Swiss Federal Institute of Technology (ETH) in Zurich, where he was among the first to study spin-polarized photoelectron emission. In 1976, he began a group for atomic cluster research at the University of Konstanz in Germany, where he built the first source for atomic clusters, and led his team to pioneering discoveries such as "magic numbers" and "Coulomb explosion." He was at the University of California, Berkeley, for three years as a Heisenberg fellow, where he initiated the first studies of atomic clusters on surfaces with a scanning tunneling microscope.

Dr. Sattler accepted a position as a professor of physics at the University of Hawaii, Honolulu, in 1988. There, he initiated a research group for nanophysics, which, using scanning probe microscopy, obtained the first atomic-scale images of carbon nanotubes directly confirming the graphene network. In 1994, his group produced the first carbon nanocones. He has also studied the formation of polycyclic aromatic hydrocarbons and nanoparticles in hydrocarbon flames in collaboration with ETH Zurich. Other researches have involved the nanopatterning of nanoparticle films, charge density waves on rotated graphene sheets, bandgap studies of quantum dots, and graphene folds. His current work focuses on novel nanomaterials and solar photocatalysis with nanoparticles for the purification of water.

Dr. Sattler is the editor of the seven-volume *Handbook of Nanophysics* (CRC Press, 2011) and the *Fundamentals of Picoscience* (CRC Press, 2013). Among his many other accomplishments, Dr. Sattler was awarded the prestigious Walter Schottky Prize from the German Physical Society in 1983. At the University of Hawaii, he teaches courses in general physics, solid-state physics, and quantum mechanics.

# Contributors

**Ludwik Adamowicz**
Department of Chemistry and Biochemistry
University of Arizona
Tucson, Arizona

**Noemí Aguiló-Aguayo**
Institute for Textile Chemistry and Physics
Leopold-Franzens-University Innsbruck
Dornbirn, Austria

**Juan Alcañiz-Monge**
Departamento de Química Inorgánica
Universidad de Alicante
Alicante, Spain

**Marco Aurilia**
Cytec Industries Italia
Naples, Italy

**Yakup Aykut**
Department of Textile Engineering
Uludag University
Bursa, Turkey

**Nicolas Batisse**
Institut de Chimie de Clermont-Ferrand
Clermont Université, Université Blaise Pascal
Clermont-Ferrand, France

and

CNRS, UMR 6296
Institut de Chimie de Clermont-Ferrand
Aubière, France

**Pierre Bonnet**
Institut de Chimie de Clermont-Ferrand
Clermont Université, Université Blaise Pascal
Clermont-Ferrand, France

and

CNRS, UMR 6296
Institut de Chimie de Clermont-Ferrand
Aubière, France

**Diego Cazorla-Amorós**
Departamento de Química Inorgánica
Universidad de Alicante
Alicante, Spain

**Alexey S. Cherevan**
Vienna University of Technology
Institute of Materials Chemistry
Vienna, Austria

**Samuel Chigome**
Botswana Institute for Technology Research
and Innovation
Gaborone, Botswana

**Steven W. Cranford**
Laboratory of Nanotechnology in Civil
Engineering
Department of Civil and Environmental
Engineering
Northeastern University
Boston, Massachusetts

**Romildo Dias Toledo Filho**
Civil Engineering Department
Núcleo de Materiais e Tecnologias Sustentáveis
    (NUMATS)
Universidade Federal do Rio de Janeiro
Rio de Janeiro, Brazil

**Marc Dubois**
Institut de Chimie de Clermont-Ferrand
Clermont Université, Université Blaise Pascal
Clermont-Ferrand, France

and

CNRS, UMR 6296
Institut de Chimie de Clermont-Ferrand
Aubière, France

**Florin Dumitrache**
National Institute for Lasers, Plasma
    and Radiation Physics
Bucharest-Măgurele, Romania

**Hai M. Duong**
Department of Mechanical Engineering
National University of Singapore
Singapore

**Dominik Eder**
Vienna University of Technology
Institute of Materials Chemistry
Vienna, Austria

**Zeng Fan**
Department of Mechanical Engineering
National University of Singapore
Singapore

**Daniele Fazzi**
Max-Planck-Institut für Kohlenforschung
Mülheim an der Ruhr, Germany

**George Filoti**
National Institute of Materials Physics
Bucharest-Măgurele, Romania

**Claudiu Fleaca**
National Institute for Lasers, Plasma
    and Radiation Physics
Bucharest-Măgurele, Romania

**Sylwester Furmaniak**
Physicochemistry of Carbon Materials
    Research Group
Faculty of Chemistry
Nicolaus Copernicus University in Toruń
Toruń, Poland

**Bhaskar Garg**
Department of Chemistry
National Tsing Hua University
Hsinchu, Taiwan

**Piotr A. Gauden**
Physicochemistry of Carbon Materials
    Research Group
Faculty of Chemistry
Nicolaus Copernicus University in Toruń
Toruń, Poland

**Paul Gebhardt**
Vienna University of Technology
Institute of Materials Chemistry
Vienna, Austria

**Katia Guérin**
Institut de Chimie de Clermont-Ferrand
Clermont Université, Université Blaise Pascal
Clermont-Ferrand, France

and

CNRS, UMR 6296
Institut de Chimie de Clermont-Ferrand
Aubière, France

**Kaori Hirahara**
Department of Mechanical Engineering/Center
    for Atomic and Molecular Technologies
Osaka University
Osaka, Japan

**Patrick C. H. Hsieh**
Institute of Biomedical Sciences
Academia Sinica
Taipei, Taiwan

**Andrzej Huczko**
Laboratory of Nanomaterials Physics
and Chemistry
Department of Chemistry
Warsaw University
Warsaw, Poland

**Gan-Lin Hwang**
Nano-Powder and Thin Film Technology
Center
Industrial Technology Research Institute
Tainan, Taiwan

**Branka Kaludjerović**
Vinča Institute of Nuclear Sciences
Materials Science Department
University of Belgrade
Belgrade, Serbia

**Saad A. Khan**
Department of Chemical and Biomolecular
Engineering
North Carolina State University
Raleigh, North Carolina

**Tokushi Kizuka**
Division of Materials Science
Faculty of Pure and Applied Sciences
University of Tsukuba
Ibaraki, Japan

**Ashley Kocsis**
Laboratory of Nanotechnology in Civil
Engineering
Department of Civil and Environmental
Engineering
Northeastern University
Boston, Massachusetts

**Piotr Kowalczyk**
School of Engineering and Information
Technology
Murdoch University
Western Australia, Australia

**Victor Kuncser**
National Institute of Materials Physics
Bucharest-Măgurele, Romania

**Meng Li**
Faculty of Engineering
Department of Materials Science and Engineering
National University of Singapore
Singapore, Singapore

**Angel Linares-Solano**
Departamento de Química Inorgánica
Universidad de Alicante
Alicante, Spain

**Yong-Chien Ling**
Department of Chemistry
National Tsing Hua University
Hsinchu, Taiwan

**Zhenyu Liu**
Kennametal Inc.
Latrobe, Pennsylvania

**Dolores Lozano-Castelló**
Departamento de Química Inorgánica
Universidad de Alicante
Alicante, Spain

**Manuel Macias-Montero**
Nanotechnology and Integrated Bio-Engineering
Centre
University of Ulster
Newtownabbey, United Kingdom

**Davide Mariotti**
Nanotechnology and Integrated Bio-Engineering
Centre
University of Ulster
Newtownabbey, United Kingdom

**Moein Mehdipour**
Department of Textile Engineering
University of Guilan
Rasht, Iran

**Oscar Aurelio Mendoza Reales**
Civil Engineering Department
Núcleo de Materiais e Tecnologias Sustentáveis
(NUMATS)
Universidade Federal do Rio de Janeiro
Rio de Janeiro, Brazil

**Somak Mitra**
Nanotechnology and Integrated Bio-Engineering
    Centre
University of Ulster
Newtownabbey, United Kingdom

**Somen Mondal**
Department of Chemical Sciences
Indian Institute of Science Education
    and Research Kolkata
West Bengal, India

**Ion Morjan**
National Institute for Lasers, Plasma
    and Radiation Physics
Bucharest-Măgurele, Romania

**Takashi Nakamura**
Institute of Multidisciplinary Research
    for Advanced Materials
Tohoku University
Miyagi, Japan

**Son T. Nguyen**
Department of Mechanical Engineering
National University of Singapore
Singapore

**Babak Noroozi**
Department of Textile Engineering
University of Guilan
Rasht, Iran

**Kinga Ostrowska**
Department of Organic Chemistry
Faculty of Pharmacy
Medical University of Warsaw
Warsaw, Poland

**Petru Palade**
National Institute of Materials Physics
Bucharest-Măgurele, Romania

**Jong Deok Park**
Department of Energy Systems Engineering
Daegu Gyeongbuk Institute of Science
    and Technology (DGIST)
Daegu, South Korea

**Drew Parsons**
School of Engineering and Information
    Technology
Murdoch University
Western Australia, Australia

**Sunil A. Patil**
Laboratory of Microbial Ecology and Technology
Faculty of Bioscience Engineering
Ghent University
Ghent, Belgium

**Georgios Chr. Psarras**
Department of Materials Science
School of Natural Sciences
University of Patras
Patras, Greece

**Pradipta Purkayastha**
Department of Chemical Sciences
Indian Institute of Science Education
    and Research Kolkata
West Bengal, India

**Fatemeh Razmjooei**
Department of Energy Systems Engineering
Daegu Gyeongbuk Institute of Science
    and Technology (DGIST)
Daegu, South Korea

**Conor Rocks**
Nanotechnology and Integrated Bio-Engineering
    Centre
University of Ulster
Newtownabbey, United Kingdom

**Monica Scarisoreanu**
National Institute for Lasers, Plasma
    and Radiation Physics
Bucharest-Măgurele, Romania

**Gabriel Schinteie**
National Institute of Materials Physics
Bucharest-Măgurele, Romania

**Ruslan Sergiienko**
Physics-Technological Institute of Metals and
    Alloys of NASU
Kyiv, Ukraine

**Cameron J. Shearer**
Flinders University
School of Chemical and Physical Sciences
Adelaide, Australia

**Etsuro Shibata**
Institute of Multidisciplinary Research for
    Advanced Materials, Tohoku University
Miyagi, Japan

**Kiran Pal Singh**
Department of Energy Systems Engineering
Daegu Gyeongbuk Institute of
    Science and Technology (DGIST)
Daegu, South Korea

**Luigi Sorrentino**
Istituto per i Polimeri, Compositi e Biomateriali
Consiglio Nazionale delle Ricerche
Naples, Italy

**Michał Soszyński**
Laboratory of Nanomaterials Physics
    and Chemistry
Department of Chemistry
Warsaw University
Warsaw, Poland

**Fabian Suárez-García**
Instituto Nacional del Carbón–Consejo Superior
    de Investigaciones Científicas
Oviedo, Spain

**Vladimir Svrcek**
Research Center for Photovoltaics
Advanced Industrial Science and Technology
Tsukuba, Japan

**Alan C. L. Tang**
Aulisa Medical Technologies
Taipei, Taiwan

**Artur P. Terzyk**
Physicochemistry of Carbon Materials
    Research Group
Faculty of Chemistry
Nicolaus Copernicus University in Toruń
Toruń, Poland

**Bartosz Trzaskowski**
Centre of New Technologies
University of Warsaw
Warsaw, Poland

**Caterina Vozzi**
Istituto di Fotonica e Nanotecnologie, CNR-IFN
Milano, Italy

**Karolina Werengowska-Ciećwierz**
Physicochemistry of Carbon Materials
    Research Group
Faculty of Chemistry
Nicolaus Copernicus University in Toruń
Toruń, Poland

**Marek Wiśniewski**
Physicochemistry of Carbon Materials
    Research Group
Faculty of Chemistry
Nicolaus Copernicus University in Toruń
and
Invest-Tech R&D Center
Toruń, Poland

**Junmin Xue**
Faculty of Engineering
Department of Materials Science
    and Engineering
National University of Singapore
Singapore, Singapore

**Jong-Sung Yu**
Department of Energy Systems Engineering
Daegu Gyeongbuk Institute of Science
    and Technology (DGIST)
Daegu, South Korea

**Bing Zhang**
School of Petrochemical Engineering
Shenyang University of Technology
Liaoyang, People's Republic of China

**Zhian Zhang**
School of Metallurgy and Environment
Central South University
Hunan, People's Republic of China

# Allotropes

I

# 1

# Carbyne: A One-Dimensional Carbon Allotrope

Ashley Kocsis

Steven W. Cranford

**Abstract**

One of the newest potential carbon allotropes being investigated exhibits the bare minimum possible molecular structure—a one-dimensional monoatomistic chain of atoms known as carbyne. Theoretically, carbyne may take a cumulene ([=C=]) repeating double bond form, or a polyyne form with alternating single and triple bonds (–[C≡C]–), which has been determined to be more energetically favorable. Carbyne is known to have a relatively high axial tensile stiffness and, due to its monoatomistic cross section, incredible specific strength and high surface area per given mass. Carbyne's practical applications include energy storage devices and nanoscale electronic devices. Due to its tunable electrical properties, it is also possible that carbyne could act as a smart connection between electronic or mechanical components at the atomistic scale. Due to its chemical stability, mechanical properties, and the natural abundance of carbon, carbyne applications could lead the way in nanotechnology applications. This molecular rod or cable has potential as an effective wire or even structural element at the nanoscale.

## 1.1 Introduction

The remarkable emergence of nanotechnology has made new possibilities flourish for the characterization of nanomaterials. These small structures have at least one dimension at the nanoscale (<100 nm),

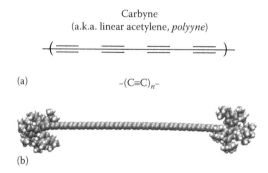

Carbyne
(a.k.a. linear acetylene, *polyyne*)

(a)          $-(C\equiv C)_n-$

(b)

**FIGURE 1.1** (a) Chemical structure of carbyne, a linear monoatomistic chain of carbon atoms with a repeating single–triple (polyyne) bond structure (also known as linear acetylene); (b) full atomistic model of carbyne chain with $n = 44$ with tris(3,5-di-t-butylphenyl)methyl group end-caps for stability. (From Chalifoux, W. A. & Tykwinski, R. R., *Nature Chemistry*, 2, 11, 2010.)

and can be categorized according to their dimensionality as zero-dimensional (0D), one-dimensional (1D), two-dimensional (2D), and three-dimensional (3D) materials. In particular, the precise control and flexibility of carbon at the nanoscale has resulted in all-carbon based representative systems of limited dimension, including 0D fullerenes, 1D carbon nanotubes (CNTs), 2D graphene, and 3D diamond (Baughman et al. 2002, Geim & Novoselov 2007, Hirsch 2010). While CNTs are classified as 1D due to their length far exceeding their cross-sectional dimensions (e.g., length $\gg$ radius), there is nominal width and, thus, material understanding must take into account the entire 3D structure of the tube (e.g., the chirality and curvature of the carbon atoms). Intuitively, one can easily discern the difference between 2D and 3D objects: restrict the size of a material volume to its length and width alone and reduce its height to zero. The transition from 3D to 1D is similar by zeroing both height and width, leaving *only* the length. From a materials science or engineering perspective, a reduction in dimensions necessarily changes how a system is both described and characterized. Yet most parameters/behaviors are accustomed to 3D understanding—what consequences occur when a material is constrained to one dimension poses intriguing questions.

Theoretically, this would involve a linear chain of bonded atoms with a cross section defined by atomistic orbitals alone, and all bonds confined to a single direction/axis. Is such a theoretical 1D molecule possible? If so, can such a molecule be composed of carbon? Herein, we focus on such a minimal 1D nanostructure—a monoatomistic chain of carbon atoms known as *carbyne* (Figure 1.1).

## 1.1.1 Carbyne by Any Other Name

As the very existence of carbyne has been disputed (see next section), it is without surprise that there are some deviations in the nomenclature of monoatomistic chains of carbon. In organic chemistry, *carbyne* is often a general term for any compound whose molecular structure includes an electrically neutral carbon atom with three nonbonded electrons, connected to another atom by a single bond (McNaught et al. 1997). Here, the suffix -*yne* refers to the acetylenic or alkyne nature of the bond structure of carbon atoms (carbon + *yne* = carb*yne*). As such, carbyne is also called *linear acetylenic carbon*—an allotrope of carbon that has the chemical structure $(-C\equiv C-)_n$ as a repeating chain, with alternating single and triple bonds (Heimann et al. 1999, Baughman 2006). It would thus be a member of the polyyne family. The simplest polyyne example is diacetylene or buta-1,3-diyne, or $H-C\equiv C-C\equiv C-H$. These compounds have also been called *oligoynes* or *carbinoids* (Heimann et al. 1999). Regardless, we adopt the name *carbyne* to refer to a repeating and linear $(-C\equiv C-)_n$ structure of arbitrary length $n$.

## 1.1.2 Controversial History of Carbyne

Both CNTs and graphene had been theoretically predicted but debated for decades before they were ultimately isolated (in 1991 [Iijima 1991] and 2004 [Novoselov et al. 2004], respectively), and carbyne has a similar history. Until recently, carbyne was considered only a hypothetical allotrope of carbon that would be the ultimate example of a polyyne structure (Chalifoux & Tykwinski 2009). The classification of carbyne and its possible existence in condensed phases has been contested mainly due to claims that the chains would cross-link exothermically (and perhaps explosively) if they approached each other. As early as 1885, Adolf von Baeyer attempted to create the third state of carbon (in addition to diamond and graphitic) through a "stepwise sequence of transformations," but deemed it impossible after having difficulty synthesizing lower polyynes (Chuan et al. 2005). A continuous, 1D polymer of carbon, carbyne, was not known to exist.

The synthesis of this substance has been claimed several times since the 1960s, but those reports have been disputed (Kroto 2010). Indeed, the substances identified as short chains of "carbyne" in many early organic synthesis attempts (Akagi et al. 1987) would be called polyynes today. In 1960, research by Russian scientists regarding Glaser coupling led to the discovery of carbyne, and 8 years later, it was found as a natural mineral in the Ries crater in Bavaria (West Germany) (Goresy & Donnay 1968). This mineral form of carbyne was named *chaoite*, after American scientist E. C. T. Chao (Heimann et al. 1999). However, this discovery did not end the carbyne controversy. Researchers and scientists continued to argue its existence throughout the 1980s. At the forefront of the argument were Smith and Buseck (1982, 1985), continually challenging any cases that pushed for the existence of carbyne. In 1982, Smith and Buseck claimed that the reflections of the material found in the Ries crater better agree with quartz rather than chaoite, and that the X-ray reflections may actually be nontronite and quartz contaminants. Smith and Buseck (1982) defined carbyne as a "poorly ordered graphite" and indicated that silicate materials may be mistaken for carbyne due to the similar interplanar spacings. Whittaker later argued against Smith and Buseck's article by reasoning that they did not make use of all the diffraction data available and utilized an unreliable analytical method. Smith and Buseck (1985) continued to disagree, publishing a direct reply to Whittaker's article restating the support of their previous work and stating that Whittaker had an "incomplete understanding of diffraction theory." It appeared that the existence of carbyne would be a never-ending controversy, but in 1995, Richard Lagow published findings of carbyne using laser-assisted synthesis (Lagow et al. 1995). Interestingly enough, Whittaker had already correctly discovered carbyne utilizing the same method more than 20 years earlier (Heimann et al. 1999). Carbyne has been observed naturally in shock-compressed graphite, interstellar dust, and meteorites. More recently, carbyne has been created by gas-phase deposition, epitaxial growth, electrochemical synthesis, or pulling the atomic chains from graphene or CNTs (Liu et al. 2013). The possibility of 1D carbon chains would not disappear.

## 1.1.3 Current Synthetic Approaches

While progress is being made, long polyyne carbyne-like chains are said to be inherently unstable in bulk because they can cross-link with each other in an exothermal (indeed explosive) reaction (Kroto 2010) and explosions are a real hazard in this area of research (Baughman 2006). Carbynes, however, can be fairly stable, even against moisture and oxygen, if they are end-capped with suitable inert groups (such as tert-butyl or trifluoromethyl) rather than hydrogen atoms (Lagow et al. 1995), especially bulky ones that can keep the chains apart (Gibner et al. 2002, Chalifoux & Tykwinski 2009, 2010). Indeed, whether or not carbyne "exists" seems to be a matter of defining the chain length in which a polyyne can be considered carbyne. A polyyne compound with 10 acetylenic units (20 atoms), with the ends capped by Fréchet-type aromatic polyether dendrimers, was isolated and characterized in 2002 (Gibner et al. 2002). As of 2010, the polyyne with the longest chain yet isolated had 22 acetylenic units (44 atoms), end-capped with tris(3,5-di-t-butylphenyl) methyl groups (Chalifoux & Tykwinski 2010). For current achievements in the synthesis of such carbynes, with $n = 2-22$, the reader is referred to the recent works of Chalifoux and Tykwinski (2009, 2010), Chalifoux et al. (2009, 2012), Tykwinski et al. (2010), Lucotti et al. (2012), and Januszewski and Tykwinski (2014).

Carbyne chains have also been seen bridging between graphene systems, offering a different means to potentially isolate individual chains (Chuvilin et al. 2009). Monoatomistic carbon chains can be formed from graphene using electron irradiation (Figure 1.2). During this continuous process, holes in the graphene grow, and graphene constrictions arise and then shrink where there are neighboring holes. These constrictions become carbon chains before separating from the graphene membrane (Chuvilin et al. 2009). A single chain with a typical length of 10–15 carbon atoms is seen in more than 50% of the cases as the final product of bridge thinning. It has been concluded that atomic rearrangements must occur during electron irradiation in order for the result to be 1D carbon chains. The formed carbon chains may take either a cumulene or polyyne form depending whether there is an even or an odd number of atoms (Chuvilin et al. 2009). Typically, linear chains with an odd number of carbon atoms have a closed-shell

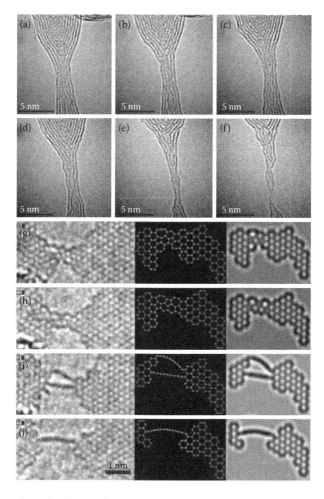

**FIGURE 1.2** Carbyne chains bridging carbon nanotubes and graphene. (a–f) HRTEM images of irradiation of MWCNT to produce carbon chain. (Reprinted from *Carbon*, 66, Casillas, G. et al., New insights into the properties and interactions of carbon chains as revealed by HRTEM and DFT analysis, pp. 436–441, Copyright 2014, with permission from Elsevier.) (g–j) Transition from a graphene ribbon to a single carbon chain. (From Chuvilin, A. et al., *New Journal of Physics*, 11, 2009.)

structure and are said to be more stable with a lower energy than chains with an even number of atoms (Heimann et al. 1999).

1D carbon chains have also been experimentally formed from CNTs, but lack stability for transmission electron microscopy (TEM) imaging. A carbon chain-like structure has been identified when a CNT is thinned out to a diameter of 1 nm and below, and a bridge that is formed between the ends of the CNT appears. In a study by Yuzvinsky et al. (2006), monoatomistic carbon chain properties are used to predict the behavior of this bridge nanotube junction. Additional evidence was attained by in situ TEM radiation experiments by Casillas et al. (2014), wherein holes were formed in the multiwalled carbon nanotubes (MWCNTs) and a bridge was formed between the holes. Again, as MWCNTs broke a carbon chain was formed. Moreover, every time an inner CNT broke, a carbon chain was formed, demonstrating a greater stability inside a CNT. Occasionally, the last CNT of the MWCNT would break and form two carbon chains.

Clearly, today, carbynes of finite length are achievable, and the door is open for characterization. The natural abundance and the chemical stability of carbon, along with its promising mechanical and electrical properties, demonstrate its applicability in a range of fields (Nair et al. 2011). Now, carbyne is at the forefront of research in the nanotechnology field. Some of its applications include use as a molecular wire, terahertz detector, biosensor, energy storage device, nanomechanical resonators, and reinforcing agents in nanocomposites or nanoelectronics (Hu 2011, Cretu et al. 2013, Liu et al. 2013). Since experimental manipulation of carbon chains is extremely hard, several theoretical works have been published regarding the properties of this structure; the properties that have been predicted include a Young's modulus comparable to CNTs (Nair et al. 2011, Castelli et al. 2012), spin polarized electronic transport (Zanolli et al. 2010), magnetic states (Xu et al. 2010), axial torsion effects (Ravagnan et al. 2009), and negative differential resistance (Khoo et al. 2008), among others (Hobi et al. 2010, Shen et al. 2010, Topsakal & Ciraci 2010, Zeng et al. 2010, Ataca & Ciraci 2011, dos Santos et al. 2011, Erdogan et al. 2011, Lin & Ning 2011, Van Wesep et al. 2011). Herein we focus on computational characterization of the exemplary mechanical and transport properties of carbyne. Arising from controversial history, along with its brethren fullerenes and graphene, carbyne has a promising future.

## 1.2 Mechanical Properties

Due to the all-carbon structure, carbyne demonstrates excellent mechanical properties including tensile stiffness, specific stiffness, and specific strength exceeding those of CNTs and graphene. Presently, these properties have been determined using computational materials science approaches, without supporting experimental validation. Primarily, carbyne has been investigated using density functional theory (DFT) and/or full atomistic molecular dynamics (MD) approaches, which have been proven to be relatively precise for other all-carbon based structures (e.g., CNTs and graphene), and can thus provide an accurate indication of the mechanical response of carbyne. Each of these methods, however, has advantages and disadvantages, and prudence is required when interpreting results. For instance, MD simulations can be costly for experiments on a large timescale, and thus typically utilize relatively fast loading rates. Moreover, fundamentally, MD cannot incorporate orbital interactions like DFT (Lee et al. 2012). On the other hand, it has been shown that DFT can underestimate the bandgap, causing a weakened Peierls instability (Liu et al. 2013), and cannot accurately reflect thermal effects (Cranford 2013) or environmental conditions (Mirzaeifar et al. 2014). Regardless of the method employed, upon the first indications of exceptional mechanical properties, it became clear that carbyne is a revolutionary material that can potentially be used in a variety of applications and in a number of fields.

### 1.2.1 Bond Geometry and Stability

Carbyne is a minimal 1D monoatomistic chain of carbon atoms consisting of repeating $sp$-hybridized alkyne groups (Heimann et al. 1999, Zimmerman et al. 2004a). In theory, repeating carbons may take either the cumulene $[=C=]_n$ form, consisting of repeated double bonds, or the polyyne $-[C{\equiv}C]_n-$ form, consisting of alternating single and triple bonds. The polyyne form has been shown using ab initio methods to be more stable and energetically favorable, and the cumulene form may undergo Peierls transition into the polyyne form (Kertesz et al. 1978, Liu et al. 2013). Moreover, recent TEM images suggest that polyyne forms may be further aided by the tensile strain involved in the synthesis of such linear systems (Casillas 2014), which facilitates the Peierls transition. Regardless, along with cumulenes, polyynes are distinguished from other organic chains by their rigidity, which makes them promising for molecular nanotechnology. Of interest is that as the number of atoms increases in a polyyne carbon chain, the average total energy increases (Heimann et al. 1999), degrading chain stability. Thermally, carbyne may be stable at temperatures up to ~3000 K (Liu et al. 2013). However, naked carbyne chains terminated by hydrogen atoms or delocalized electrons are stable only at low temperatures. The free energy of carbynes is higher than that of both graphite (and associated graphene or CNTs) and diamond, and can transition to more stable configurations under high temperature and pressure conditions (Luo & Windl 2009). Thus, carbyne chains at high temperatures require bulky end groups for stability. Carbyne chains may also be stabilized using metal ions.

The spacing of the carbon atoms has a unit cell length of approximately 2.565 Å (Lee et al. 2008, Liu et al. 2013), with a single bond on the order of 1.4 Å and a triple bond on the order of 1.2 Å. As carbyne is essentially a 1D conjugated system (connected/overlapped orbitals with delocalized electrons), a more accurate description of lengths is moot, dependent on the degree of bond length alternation (BLA) (the difference in the bond length between the central single and triple bonds) in the carbyne structure (Kertesz et al. 2005), which itself is dependent on the number of carbon atoms and boundary conditions (e.g., chemical end-cap [Chalifoux et al. 2009]) and depicts variation (Cahangirov et al. 2010).

### 1.2.2 Tensile Properties

Recent investigations of carbyne utilizing molecular dynamics with first-principles-based ReaxFF (van Duin et al. 2001, Chenoweth et al. 2008) yielded several tensile properties including the effective stress–strain curve, tensile Young's modulus or stiffness, and maximum failure stress (Nair et al. 2011, Kocsis et al. 2014). Nair et al. (2011) conducted a uniaxial tension test of varying length carbyne chains in which both ends of the chain were displaced in opposite directions until the chains failed. This test outputted stress and strain values measured using a virial stress (Zimmerman et al. 2004b) formulation (see Figure 1.3a). By utilizing MD simulations instead of first-principles DFT, longer carbyne chains were

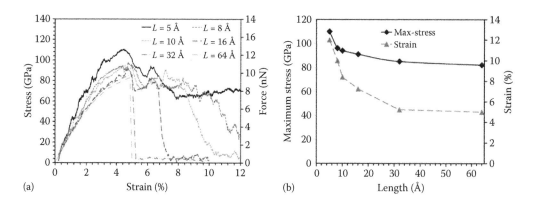

**FIGURE 1.3** (a) Stress–strain curves for carbyne chains of lengths $L$ = 5, 8, 10, 16, 32, and 64 Å with 4, 6, 8, 12, 24, and 48 atoms, respectively. (b) Maximum stress and strain of carbyne chains with different lengths. The data show that carbyne chains become stronger and stiffer and more extensible at shorter lengths. The values of maximum stress and strain reach an asymptotic value as the chain length approaches ≈60 Å. (Modified from Nair, A. K., Cranford, S. W. & Buehler, M. J., *EPL Europhysics Letters*, 95, 1, 2011.)

able to be tested due to the fact that ReaxFF allows for slower mechanical testing (Nair et al. 2011). First-principles methods are not suitable due to the length and timescales required.

Nair et al. calculated the total stress from the virial stress components of the macroscopic stress tensor $S_{ij}$ over a volume $\Omega$, where typically $\Omega = AL$, and $A$ is defined as the effective cross-sectional area, and $L$ is defined as the carbyne chain length. The total stress can be expressed as

$$\sigma_{ij} = S_{ij}\Omega^{-1}. \tag{1.1}$$

Assuming a cross-sectional area of $3.35 \times 3.35$ Å$^2$ (e.g., the interlayer spacing for graphene allowing comparison), the peak stress was determined to be approximately 80–115 GPa, depending on chain length. These magnitudes are approximately on the order of graphene or CNTs. The estimated Young's modulus from the stress–strain curves can be approximated as 3 TPa, which is about three times that of graphene (Nair et al. 2011). For both these calculations, the assumed cross section is quite large. Alternatively, based on the electron orbital radius of 0.39 Å (Liu et al. 2013), the ultimate stresses are then on the order of 6–8 TPa—a ridiculously high strength! Clearly, it is difficult to ascertain the "correct" cross section, especially considering carbyne's propensity to cross-link when grouped (Casari et al. 2004, Belenkov & Mavrinsky 2008). Thus, a definitive chain spacing and effective cross-sectional area is currently unknown.

We note that the derived modulus and ultimate stress assume a cross-sectional area on which the stress acts. But what is the cross section of a purely 1D material? Can such an area be quantified? The same ambiguity has arisen regarding the thickness of 2D materials such as graphene (Huang et al. 2006, Roman & Cranford 2014), and it has been suggested that the stress and elastic moduli of monolayer 2D systems be reported in force per unit length (newtons/meter) rather than force per unit area (newtons/square meter or pascal). Using this logic, if both the height and the width of the effective cross section are lost, 1D stress should merely be reported in terms of the force—stress has no meaning on a dimensionless cross section. As a result, the 1D stress, or the force $F_{ii}$, can also be independently computed from the virial stress components and represented as

$$\sigma_{1D} = F_{ii} = S_{ii}L^{-1}. \tag{1.2}$$

Note that the 1D stress or modulus can be derived for any 1D-like system, by scaling the traditional modulus by the cross-sectional area of the system, where $\sigma_{1D} = \sigma A = F$ and $E_{1D} = EA$. From these simple relations, Nair et al. (2011) report a 1D elastic modulus on the order of 323 nN and strength of 9–13 nN (e.g., traditional units of force rather than stress). These values are independent of cross section, and thus easier to compare to other 1D systems.

Per Figure 1.3b, it is clear that the maximum stress and thus the maximum force of a carbyne chain are dependent on the length of the chain. Not only can this be attributed to the high axial strain energy of the shorter chains, but also to the fact that longer chains have more bonds that could trigger failure (Nair et al. 2011). This was supported by a recent theoretical formulation (Mirzaeifar et al. 2014) relying on the classical Bell model, which gives the probability of failure $P_{fail}$ of a chemical bond under load, where

$$P_{fail} = \omega \exp\left[-\frac{E_b - F \cdot x}{k_B T}\right], \tag{1.3}$$

where $\omega$ is the natural vibrational frequency of the bond, $E_b$ is the energy required to break a bond, $F$ is the applied load, $x$ is the displacement of the force, and $k_B T$ is the thermal energy ($k_B$ being Boltzmann's constant and $T$ the temperature). The above formulation results in (a) a logarithmic dependence on strength with respect to length, and (b) a linear dependence on strength with respect to temperature. Both trends have been supported by MD simulations (Nair et al. 2011, Mirzaeifar et al. 2014), and await validation by experimental means. The length dependence can be interpreted simply as a localization of strain intensity—because of thermal phonons, the chain is never at a single value of strain but rather is locally compressed or extended. The applied strain is merely an average of such local values. A longer chain, therefore, has increased odds of experiencing critical strains locally and failure at lower applied loads.

Moreover, from these simulations, it was further concluded that carbyne chains of above a length of approximately 3 nm have a quick and brittle failure, unlike the more ductile failure of the smaller chains. Since all chains rupture at approximately 5% strain, the brittle responses are attributed to the post-fracture bond relaxation, which, for longer chains, results in larger separation of the free carbon ends and thus loss of any interaction that enables "graceful" failure (e.g., 3 nm strained to 5% results in 1.5 Å of relaxation, which is greater than a carbon–carbon bond length).

While MD is a suitable method to determine other tensile properties of carbyne, Liu et al. utilized first-principles calculations to determine carbyne's tensile stiffness and specific strength (Liu et al. 2013). A tensile stiffness of 95.56 eV/Å was defined energetically as

$$C = \frac{1}{a}\frac{\partial^2 E}{\partial \varepsilon^2},$$ (1.4)

where $a$ is the equilibrium unit cell length (here, $a = 2.565$ Å), $E$ is the strain energy per two carbon atoms (or, equivalently, per a single/triple bond pair), and $\varepsilon$ is the strain (Liu et al. 2013). We note that this procedure is equivalent to taking change in linear strain density (rather than volumetric strain density) to arrive at a 1D modulus, in agreement with the approach described above. Indeed, the unit electron volt per angstrom is equivalent to units of force. Accounting for mass, carbyne's specific stiffness trumps CNTs and graphene, by more than a magnitude, and exceeds diamond by almost three times, equaling approximately $10^9$ N m/kg (Liu et al. 2013).

Liu et al. (2013) further determined the ultimate strength by computing carbyne's activation barrier as a function of strain and finding its breaking force. Its breaking force was determined to be 9.3–11.7 nN, in close agreement with MD results (Nair et al. 2011, Kocsis et al. 2014). Accounting for mass, this is equivalent to a specific strength of $(6.0–7.5) \times 10^7$ N m/kg. Once again, carbyne exceeds the excellent properties of graphene, CNTs, and diamond (Liu et al. 2013).

## 1.2.3 Serial Bonds ≠ Serial Springs

From a modeling perspective, carbyne presents an interesting challenge. On the one hand, the molecular structure is extremely simple—a repeating sequence of single and triple carbon–carbon bonds. On the other hand, the structure is so simple that the electron orbitals have a large effect on the molecular mechanics. If we ignore the orbitals (for now), from a classical MD approach, one could simply assign the bond stiffness associated with such bonds. Many widely used MD potentials implement simple harmonic functions for bonded interactions (Chemistry at Harvard Macromolecular Mechanics [CHARMM], GROningen MAchine for Chemical Simulations [GROMACS], consistent valence force field [CVFF] [Dauberosguthorpe et al. 1988], etc.), or

$$\phi_{\text{bond}} = \frac{1}{2}k(r - r_0)^2,$$ (1.5)

where $k$ represents a harmonic spring stiffness, and $r_0$ is the equilibrium (stress free) spacing of the atoms. With this perspective, carbyne is simply a serial spring system, consisting of repeating single and triple bonds. For argument's sake, we can take the spring stiffness values from a known potential and—choosing the CVFF potential (Dauberosguthorpe et al. 1988)—say $k_{single}$ = 162 kcal/mol/Å$^2$ and $k_{triple}$ = 400 kcal/mol/Å$^2$. Using the simple relation for springs in series, the spring stiffness for a unit cell of carbyne can be approximated as

$$k_{carbyne} = \left( \frac{1}{k_{single}} + \frac{1}{k_{triple}} \right)^{-1} \approx \left( \frac{1}{162} + \frac{1}{400} \right)^{-1} \approx 115 \text{ kcal/mol}\text{Å}^2. \tag{1.6}$$

If we wish to convert to a scale-free material stiffness, we substitute strain for displacement, where $\varepsilon = (r - r_0)/r_0$, such that

$$\phi_{carbyne} = \frac{1}{2} \left( k r_0^2 \right) \varepsilon^2, \tag{1.7}$$

where the 1D material stiffness (per unit length) can be defined as

$$C = \frac{1}{r_0} \frac{\partial^2}{\partial \varepsilon^2} (\phi_{carbyne}) = k r_0 \approx 314 \text{ kcal/mol/Å}. \tag{1.8}$$

Converting to electron volt per angstrom, this results in approximately 13.5 eV/Å, or about 14% the stiffness determined by ab initio calculations reported by Liu et al. (approx. 96 eV/Å). Simply put, *carbyne cannot be thought of as springs in series!* Accordingly, nonreactive classic MD potentials (CHARMM, GROMACS, CVFF, etc.) are inadequate to reflect the bond conjugation of carbyne systems. *Monitus es!*

## 1.2.4 Tensile Properties in Other Environments

While the previously mentioned tensile properties of carbyne were studied with chains in an ideal vacuum, practical applications would necessitate the understanding of carbyne in various chemical environments. Of particular concern is the effect of water—beneficial due to its potential applications as part of filtration and water purification devices (Mirzaeifar et al. 2014). Due to the requirement of (a) larger system sizes and (b) thermal effects, such simulations are beyond the capabilities of the more accurate first-principles methods such as DFT and require an MD formulation. A recent study considered carbyne chains of various lengths (7–60 Å), temperatures (0–400 K), and solvents (water, acidic, and basic) in uniaxial tension tests (Mirzaeifar et al. 2014). The MD simulations were run at different temperatures and modeled inside both a vacuum and a periodic box filled with water molecules for comparison sake (Figure 1.4a).

Once solvent was introduced, the carbyne chains immediately lost linear configurations due to the amorphous state of the adjacent water molecules. This further affects the deformation mechanism upon extension, as the nonlinear shape must first unfold or straighten before direct bond stretching. Interestingly, the maximum rupture force and ultimate strain of carbyne are marginally improved in a water environment, supported by analysis of the bond order distribution during tensile loading (Mirzaeifar et al. 2014). In simple terms, the carbons prefer to bond to themselves rather than to water.

The same study considered both acidic and basic chemical environments for the same reasoning as above. In the MD studies, the acidic environment replaces some water molecules with carbonic acid ($H_2CO_3$) and the basic environment replaces some water molecules with hydroxide groups ($OH^-$). Per

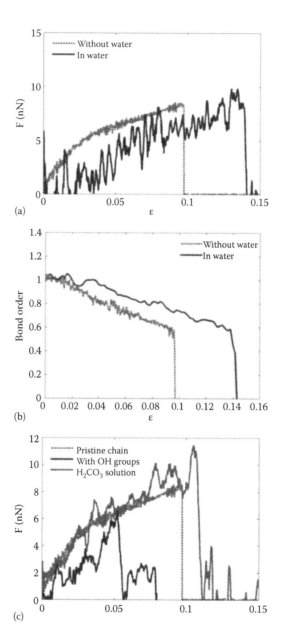

**FIGURE 1.4**  (a) Force extension of carbyne chains with 15 Å length. The force extension curve is compared for the original pristine chains (in vacuum) and chains in water; (b) bond order of the failed single carbon–carbon bond in carbyne chain with and without the interaction of water molecules; and (c) force extension compared for the pristine chains and carbyne chains in acidic and basic solutions. (From Mirzaeifar, R., Qin, Z. & Buehler, M. J., *Nanotechnology*, 25, 37, 2014.)

Figure 1.4c, it is evident that the basic environment decreases the chain strength, while the acidic environment increases chain strength (Mirzaeifar et al. 2014). It was demonstrated that, when the carbyne chain ruptures, the hydroxide group undergoes a reaction with the free carbon end to form a CCOCC chain—an effective oxygen splice. Additional loading results in the rupture of the splice. Such healing behavior is an example of how the unique properties of carbyne may be exploited in nanotechnology applications.

## 1.2.5 Bending Rigidity

Given the extreme tensile stiffness of carbyne, it is interesting to understand how this material performs under bending deformations—Is it rigid like a rod? Or flexible like a polymer? Such behaviors are dependent on the bending stiffness or flexural rigidity of the chain. Of note, Hu has shown, using DFT methods, that the strain energy due to high bending angles of a chain is less than the necessary energy to break the carbon bonds (Hu 2011), making it unlikely that carbyne will fail in flexure. This enables many conformational possibilities, such as looped or knotted structures (Cranford 2013). To determine the bending stiffness of carbyne, one can first determine the change in bending energy due to applied curvature.

Since it is difficult to dynamically bend carbyne chains in simulation, one approach is to model static carbyne rings with constant uniform curvatures to determine the bending energy as a function of ring radii (Liu et al. 2013, Kocsis et al. 2014). The bending stiffness $K$ is defined as

$$K = \frac{1}{a}\frac{\partial^2 E}{\partial q^2},\qquad(1.9)$$

where $q$ is equal to curvature ($1/r$). Small rings (less than eight carbons or so) undergo a transition from constant curvature to constant, near-rigid angles, and thus cannot be used to determine the bending stiffness (Cranford 2013, Kocsis et al. 2014). As the size of the rings increases, the stiffness of the rings converges since it is approaching the limit of infinite straight carbyne and $K = 3.56$ eV Å (Figure 1.5) (Liu et al. 2013). Using similar MD methods with the first-principles-based ReaxFF, a bending stiffness of 1.3 eV Å was determined. The smaller value is attributed to the fact that stiffening due to Jahn–Teller distortions of electron orbitals (which can be considered steric-like effects at the quantum scale) are not captured in ReaxFF.

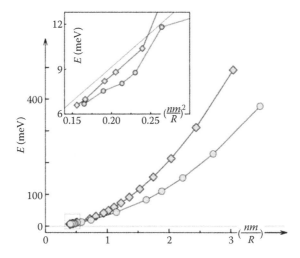

**FIGURE 1.5** Bending energy per unit cell as a function of curvature ($1/R$) calculations used to extract the bending modulus value. (Reprinted with permission from Liu, M. et al., *ACS Nano*, 7(11), pp. 10075–10082. Copyright 2013 American Chemical Society.)

MD approaches, however, can reflect effective bending rigidity in other applications besides static rings. For example, the free vibrational response of a carbyne chain proportional to the bending rigidity, where

$$f = \frac{\alpha^2}{2\pi}\sqrt{\frac{K}{mL^3}},$$   (1.10)

where $K$ is the bending stiffness (equivalent to the continuum value of $EI$), $m$ is the mass of the chain, $L$ is the length, and $\alpha$ is a constant dependent on boundary condition. Via the frequency analysis of a vibrating chain, the fundamental frequency can be used to calculate $K$. Using this method (which is also temperature and length dependent), Nair et al. determined the stiffness of carbyne to be on the order of 8.7–12.4 eV Å (Nair et al. 2011), higher than the first-principles calculation above. The discrepancy could be attributed to higher-order frequencies in the vibrational response.

The bending stiffness can be considered relatively small or large in comparison to other polymers by determining the persistence length of carbyne. For semiflexible molecular chains, the persistence length $l_p$ can be defined as

$$l_p = \frac{K}{k_B T},$$   (1.11)

where $k_B$ is the Boltzmann constant and $T$ is the temperature, effectively relating the thermal energy to the bending rigidity of the molecule. As shown by Equation 1.11, the persistence length is inversely proportional to the temperature—low temperature would induce more rodlike behaviors, while high temperatures would result in more flexible ropelike behavior. This transition can be used to potentially trigger structural transitions (Cranford 2013). The persistence length of carbyne is determined to be ~14 nm when at a temperature of 300 K. In comparison to the polymers and macromolecules in Table 1.1, carbyne is relatively stiff (Liu et al. 2013).

While bending stiffness and persistence length are important properties to quantify carbyne during bending, it is also important to look at how bending may affect carbyne's structure. Utilizing DFT calculations, Hu (2011) utilized two types of carbyne chains (which differ in their end groups), polycumelenic ($C_{20}H_4$) and polyynic ($C_{20}H_2$), with varying arc–chord ratios $\tau$. These ratios could be defined as

$$\tau = \frac{L}{D},$$   (1.12)

**TABLE 1.1**   Persistence Length of Carbyne Compared to Several Important Polymers

| Polymer | $l_p$ (nm) |
|---|---|
| Polyacrylonitrile | 0.4–0.6 |
| Polyacetylene | 1.3 |
| Single-stranded DNA | 1–4 |
| Polyaniline | 9 |
| Double-stranded DNA (dsDNA) | 45–50 |
| Graphene nanoribbons | 10–100 |
| Carbyne (present work) | 14 |

*Source:* Liu, M. et al., *ACS Nano*, 7(11), 10075–10082, 2013.

where $L$ is the chain length and $D$ is the distance between the ends of the chain. An increase in arc–chord ratio implies an increase in curvature of the chain. When these chains are bent, results demonstrate that the curvature causes variations in bond lengths (increase or decrease) depending on the location of the bond. The bond angles are also affected since initially all the bond angles are 180°. When the chain is bent, the bond angles decrease due to the curvature. This is most apparent at the even carbon number carbon atoms versus the odd number carbon atoms, which demonstrates an oscillation behavior with the ordinal number of carbons. It is also concluded that the middle atom is the most pliable since it has the largest change in bond angle (Hu 2011).

### 1.2.6 Torsional Stiffness

At first consideration, twisting of a pure 1D atomistic chain may seem pointless—the atoms rotate about a common axis, and the result is the same structure—e.g., there is no observable angle of twist due to the atomistic symmetry. Indeed, in order to analyze carbyne under torsion, functional group "handles" must be attached to the chain ends to break the cylindrical symmetry of the electron density (Liu et al. 2013). Exploring a variety of end-groups (including methyl [$CH_3$], phenyl [$C_6H_5$], hydroxyl [OH], amine [$NH_2$], and methylene [$CH_2$]), Liu et al. determined that the best functional group to detect torsional stiffness is $=CH_2$. Note that MD methods are not able to reflect torsional stiffness, due to the axial alignment of all bonds—the torsional stiffness arises from orbital interactions alone. While the stiffness relies on the end groups to break the rotational symmetry, from the mechanical perspective, the carbon chain between the handles behaves just like a classical torsion rod where the stiffness $H$ can be defined as

$$H = L\frac{\partial^2 E}{\partial\theta^2},$$ (1.13)

where $\theta$ is the torsion angle. Energy must be determined as a function of angle of twist/rotation. Per DFT calculations, the maximum (end-group dependent) torsional stiffness of carbyne is $H \approx 10.3$ eV Å/rad$^2$ (Liu et al. 2013).

## 1.3 Transport Properties

Beyond the interesting mechanics of this ideal 1D material, there is an obvious potential in applications as a transport medium—e.g., the absolute "mimimal nanowire" for electrical signals, thermal transport, or other phenomena such as spintronics (Zanolli et al. 2010).

The interesting transport properties of carbyne lie in the linear bond conjugation of adjacent carbons. A recent ab initio study (Cahangirov et al. 2010) indicated that the π bands of carbon chains behave as a 1D free-electron system. Interestingly, a local perturbation created by a small displacement of a single atom at the center of a long chain can induce oscillations of charge density, which are carried to long distances over the chain, identified as novel 1D Friedel oscillations showing $1/r$ decay rate along the carbyne axis. This symmetry breaking phenomenon could have interesting applications.

There are two new parameters that can dramatically affect the properties of finite carbon chains—the number of carbon atoms $n$ forming the chain and the end effects (Cahangirov et al. 2010). As both cumulene (consecutive double bonds) and polyyne (alternating single and triple bonds) are possible, the transport characteristics are dependent on both bond types. The symmetric and broken-symmetry forms have very distinct electronic properties. It has been shown that the cumulene form of linear carbon is metallic, due to two degenerate, half-filled $px$ and $py$ bands crossing the Fermi level. However, the cumulene structure is also vulnerable to Peierls instability. Transitioning to polyyne form, a bandgap emerges at the edge of the Brillouin zone, and carbyne behaves as a semiconductor/dielectric. Since this transition is known, of interest is if it can be mechanically controlled.

## 1.3.1 Metal/Insulator Dependence on Strain

As MD approaches cannot account for electric phenomena, DFT calculations are used to estimate the conductance of the carbon chain when subject to a current. As previously stated, the broken-symmetry polyyne form is considered more stable and energetically favorable (Cranford 2013). However, using first-principles approaches, it was demonstrated that by applying tension, the critical temperature of Peierls transition from the symmetric cumulene form and the resulting electron–phonon interaction can be increased (see Figure 1.6). The system can thus be changed from a metal to a dielectric. Moreover, carbyne can change from the conducting cumulene form to the nonconductive polyyne form as the system is subject to additional strain (Artyukhov et al. 2014). In essence, tension and temperature both serve to increase the electron–phonon interaction, whereby rising temperature and strain-induced depression of vibrational levels start to populate the first excited states. At low temperatures, the critical strain is approximately 3%, and it decreases with temperature starting from about 200 K until at 900 K, at which the cumulene form becomes unstable. Of importance is that a theoretical model to explain the transition suggests that the temperature- and strain-dependent transition is valid for arbitrary 1D metals. In addition to strain control, the conductor properties of carbyne may also oscillate depending on the number of its atoms (Cretu et al. 2013), as deviations in the current occur between the even and odd atoms and also decrease for long chains, dependent on strain (Cretu et al. 2013).

## 1.3.2 Bending Effects on Bandgap

While tension can be used to modify the electronic properties of carbyne, other forms of deformation have been considered to influence a response. Under compression, due to the rodlike mechanics of carbyne, buckling would likely occur (Cranford et al. 2012), suggesting *bending* as a primary deformation model subject to compressive axial loads. Using DFT calculations, Hu (2011) determined that bending carbyne chains decreases their stability due to the dependent relationship between strain energy and arc–chord ratio (see Equation 1.12). Clearly, as the arc–chord ratio increases, the chain's strain energy

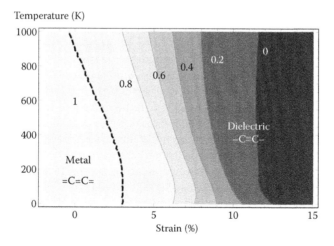

**FIGURE 1.6**  Strain-temperature phase diagram of the polyyne–cumulene transition. The order parameter $R$ indicated measures the shape of the nuclear density distribution, with cumulene at $R = 1$ and polyyne at $R < 1$. (Reprinted with permission from Artyukhov, V. I., Liu, M. & Yakobson, B. I., *Nano Letters*, 2014. 14(8), pp. 4224–4229. Copyright 2014 American Chemical Society.)

increases due to curvature. Hu (2011) demonstrated that when the arc–chord ratio is equal to 4.1 or greater, the strain energy is about three times smaller than the energy required for the chain to fail. From these results, it is concluded that a carbon chain is similar to a soft material since it is rare that it will break due to bending. Thus, bending may be a more promising route to control electronic properties, as application of direct tension may result in failure of the chain at the levels necessary for metal/dielectric transition (e.g., compare maximum strain values of Figure 1.3 with transition strain values of Figure 1.6), depending on the chain length.

Bending has a direct effect on a carbon chain's highest occupied molecular orbital–lowest unoccupied molecular orbital (HOMO–LUMO) gap. This characteristic can define conductance of carbyne when used as a nanowire. A smaller gap is related to a higher conductivity. The bandwidth for an organic metal is 0.5–1 eV; therefore, the HOMO–LUMO energy gap should be less than 0.5 eV. This definition defines a material with an energy gap greater than 5 eV and smaller than 0.5 eV as an insulator and conductor, respectively. A semiconductor is defined with an energy gap between 0.5 and 3.5 eV (Hu 2011).

Again, both cumulene and polyyne structures show distinct differences in electronic behavior while subject to bending. The HOMO–LUMO energy gap for a polycumelenic chain remains virtually unchanged during bending implying constant BLA and no effect on the conductivity of the chain (Figure 1.7a). This can be thought of in terms of the symmetric aligned electron orbitals—subject to

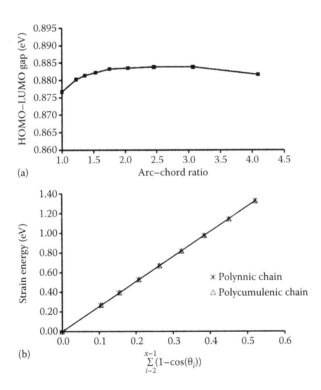

**FIGURE 1.7** Bending to vary electrical properties. (a) HOMO–LUMO energy gap of bent polycumulenic ($C_{20}H_4$) chains versus arc–chord ratios and (b) HOMO–LUMO energy gap of bent polyynic ($C_{20}H_2$) chains versus arc–chord ratios. (Reprinted with permission from Hu, Y. H., *Journal of Physical Chemistry C*, 2011. 115(5), pp. 1843–1850. Copyright 2011, American Chemical Society.)

bending, they are merely compressed together, but the overall structure is unchanged. In contrast, a polyynic chain's energy gap decreased with increasing arc–chord ratio, while the BLA linearly increased, leading to an increase in its conductivity (Figure 1.7b) (Hu 2011). Again, in terms of the orbitals, the non-symmetric are forced to overlap upon bending, changing the electron structure and electron–phonon interaction. Therefore, a polyynic chain may be a more viable option for sensor applications under bending, which is also advantageous due to a more favorable mechanical stability.

### 1.3.3 Thermal Transport

Finally, we consider thermal transport in carbyne. Due to the aligned and linear nature of the carbon bonds, thermal/kinetic vibrations are primarily along a common axis, which suggests efficient thermal transfer between adjacent atoms. Heat removal has become a crucial issue for continuing progress in the electronic industry, and thermal conduction in low-dimensional structures is promising (Balandin 2011). In particular, the room-temperature thermal conductivity of carbon allotropes and their derivatives span an extraordinary large range—of over five orders of magnitude—from the lowest in amorphous carbons to the highest in graphene and CNTs (Balandin 2011), and suggest carbyne may be exceptional due to its limited dimensionality and relative atomistic stiffness. At the same time, there are few studies that have focused on the thermal conductivity of carbyne.

A recent study considered phononic heat transport through ideal molecular chains connecting two thermal reservoirs (Segal et al. 2003). For relatively short molecules at normal temperatures, it was found that heat conduction is dominated by the harmonic part of the molecular force field—e.g., stiffer is better—suggesting that the axial rigidity of carbyne is another virtue. A similar study demonstrated that, at low-temperature limit, the discrete vibrational frequency spectrum of a soft chain may reflect on the thermal conductance by giving rise to a sudden increase due to the appearance of higher-order vibrational responses (Ozpineci & Ciraci 2001). Likewise, at room temperature, the conductance through a stiff chain may oscillate with the number of chain atoms (Ozpineci & Ciraci 2001). The observed quantum thermal effects are comparable with similar effects observed in the electrical conductance. Moreover, due to the limited vibrational modes of a 1D chain (e.g., the molecular vibrational spectral density), an intricate dependence of the heat conduction on molecular length at different temperatures arises (Segal et al. 2003). For example, the heat conduction increases with molecular length for short molecular chains at low temperatures, as the short chains effectively vibrate faster. Such predictions, however, have yet to be confirmed with carbyne.

## 1.4  Case Studies

Once a preliminary understanding of the behavior of carbyne is in place, we can then probe potential applications and more complex responses in devices or molecular systems. Here, we consider two case studies. First, like a cable or rope, while carbyne is strong in tension, it has little mechanical use in compression. At the macroscale, it has been shown that constrained slender rods can support higher compressive loads (Kocsis et al. 2014). We explore the effect of confinement of the effective compressive stiffness of carbyne chains. Second, motivated by the stringlike structure and relatively large extension capacity of carbyne, we explore the effect of tension on the vibrational response—effectively turning carbyne chains into a tunable vibrational signal and/or sensor, similar to a guitar string.

### 1.4.1  Compressive Strength by Confinement

As mentioned earlier, carbyne demonstrates exemplary properties in tension. However, this material is very weak in compression, effectively behaving as a weak entropic spring. Similar to bracing a column, carbyne's compressive stiffness may be controlled in principle by *engineered confinement*. Demonstrated by both first-principles calculations (Zhao et al. 2003, Xu et al. 2009, Kuwahara et al. 2011) and experimental observation (Zhao et al. 2003, Nishide et al. 2006, 2007), carbyne chains are likely stable within the cavity of both single- and multiwalled CNTs. As such, a recent study explored the effect of a carbyne chain within a virtual CNT (Kocsis et al. 2014). By comparing confined carbyne with the confinement of macroscale cables or a braced column, it is evident that both of these techniques have the same goal: to increase compressive strength and help prevent buckling.

Utilizing the first-principles-based ReaxFF potential and MD simulation software, a carbon atom chain was affixed within a virtual tube of various radii. The chain is then subject to regular increments of compressive displacement. The system was then equilibrated at a constant temperature to obtain an average potential energy at each deformed length, which increased due to the stored strain energy (primarily bending) in the carbon chain. In total, 36 simulations were run with varying CNT confinement radii (0.05–2.10 Å), carbyne chain length (approximately 65–120 Å), and the system temperature (100–500 K). By assuming harmonic spring behavior in these simulations, the potential energy output $U$ can be related to the spring constant value $k$, where

$$U = \frac{1}{2}k\Delta^2$$

(1.14)

and $\Delta$ represents the total displacement of the compressive spring.

The results demonstrate that effective stiffness is inversely related to chain length and confinement radius, while an increase in temperature causes a marginal *increase* in effective stiffness. The increase due to temperature is contrary to the typical trend that compliant molecular chains are more flexible as temperature increases, as the confinement limits the possible configuration space. Thus, at higher temperature, more energy is necessary to induce compressive deformation in a reduced volume, similar to an increase in pressure. Of interest, the effective compressive stiffness of confined carbyne could be varied by two orders of magnitude through adjustments of these parameters. Thus, when the wiggle room of the chain decreases by a decrease in either confinement radius or chain length, there is an increase in compressive stiffness (Figure 1.8).

By treating the carbyne chain as a truss element, it is clear that the 1D modulus $E_{1D}$ is proportional to the compressive stiffness or spring constant of carbyne and, hence, also inversely related to chain length $L$ and confinement radius:

$$k_{max} = \frac{E_{1D}}{L}.$$

(1.15)

This control over material properties creates the opportunity for the advancement of carbyne as nanowires and nanorods (Kocsis et al. 2014). This study also demonstrates that confinement might inherently stabilize the chains, creating the opportunity for the synthesis of extraordinarily long chains, something not currently feasible (Zimmerman et al. 2004a,c, Geim & Novoselov 2007, Kocsis et al. 2014).

## 1.4.2 Tuning Carbyne Vibrations

Terahertz and gigahertz sensor technologies are a recent interest in the field of nanotechnology. The possibility of a nanoscale system with tunable vibrational frequencies promises a range of possibilities for potential biosensors (Chartuprayoon et al. 2015). Carbyne, as a 1D nanostructure, demonstrates many of the qualities required and sought after in these sensors including high surface area, reliability, and fast electron transport (Chenoweth et al. 2008, Liu et al. 2013).

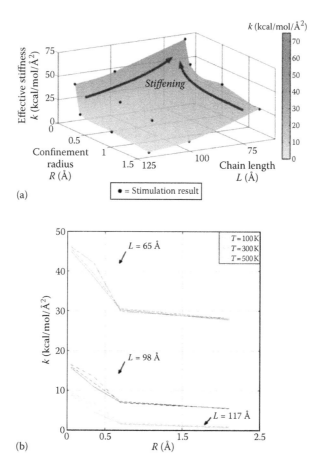

(a)

(b)

**FIGURE 1.8**   (a) Stiffening the compressive stiffness of carbyne by decreasing wiggle room—e.g., variation of confinement and chain length. (b) Plot of the spring constant values versus confinement radii at 100, 300, and 500 K, continuous, dashed, and dash-dot lines respectively, for $L$ = 65, 98, and 117 Å. The results indicate a slight but consistent increase in stiffness with temperature. (Modified from Kocsis, A. J. et al., *Nanotechnology*, 25, 33, 2014.)

By analogously comparing carbyne's features as a nanowire to the tensioned strings of a guitar, the behavior of the vibrational frequency of carbyne can be investigated. The ability to predict and control the frequency of such a material, as well as understand its sensitivity at different lengths, presents many opportunities for sensor applications. To properly examine these properties, it is best to model a single carbyne chain as a nanoscale guitar string.

Before using MD simulations to determine carbyne's frequency range, it must be verified that carbyne may be termed a *string* by meeting the standing wave equation requirements. Due to carbyne's compliance and constant mass per length, the classical 1D wave equation can be applied to the simulation model. However, a limitation on the initial excitation must be applied to avoid higher-order vibrational modes and/or failure. The solution for a standing wave equation is

$$v = c = \sqrt{\frac{T}{\mu}}, \qquad (1.16)$$

where $v$ is the velocity of the string, $T$ is the tensile force in the string, and $\mu$ is the linear density. From this equation and the relationship between velocity and wavelength, the fundamental frequency $f_0$ of a stretched string can be derived to be as follows:

$$f_0 = \frac{1}{2L}\sqrt{\frac{T}{\mu}},$$ (1.17)

where $L$ is the length of the string.

Implementing MD and a large-scale atomic/molecular massively parallel simulator simulation software, carbyne chains of varying length, each fixed at both ends, are initially strained (similar to tuning a guitar) and then the central atom is plucked to induce free vibration (Figure 1.9a). The vertical displacement is recorded and used to determine the fundamental frequency by Fourier analysis (Figure 1.9b and c). Strain values are varied to determine the maximum strain value for each length chain and, hence, the fundamental frequency limit of each chain.

In this simple study, to transcribe the frequency range, the vibrational frequencies are manipulated into an audible language or a musical scale. Using the known frequency of the musical notes, the associating carbyne chain length and strain value can be interpolated (although many scales higher than the human hearing capacity!). Per Figure 1.10, it is evident that by adjusting the chain length, as well as strain value, carbyne can be easily manipulated into different frequency values. Thus, carbyne has a

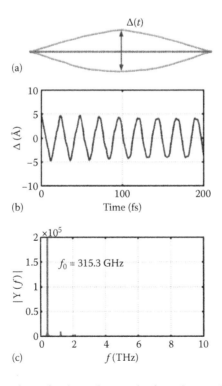

**FIGURE 1.9** (a) Simulation snapshots of carbyne chain under free vibration (fixed ends). The displacement $\Delta$ of a central atom is recorded over time. (b) Plot of vibrational displacement versus time, displaying a repeating sinusoidal-like response. (c) Plot of fast Fourier transform analysis results (amplitude versus frequency spectrum) to determine dominant, fundamental frequency $f_0$. The recorded frequency (315.3 GHz) depicted is equivalent to a D note. (From Kocsis and Cranford, 2015, unpublished.)

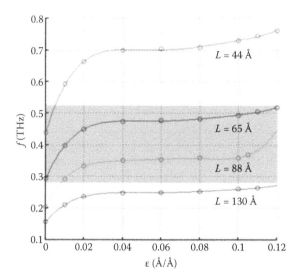

**FIGURE 1.10**  Frequency variation as a function of applied strain ($\varepsilon$) for carbyne chains of varying length ($L = 44, 65, 88,$ and $130$ Å). Frequency response is nonlinear due to the nonlinear stress–strain behavior of carbyne. All chains fail with strain exceeding 12%. The frequencies encompass a range of 156.1–762.8 GHz. Shaded region indicates required frequency range for C major scale. (From Kocsis and Cranford, 2015, unpublished.)

strong frequency sensitivity when strained at different values or fixed to different chain lengths, which presents a favorable quality when linked to the field of biosensors and terahertz technology. Perhaps carbyne-composed nanomusic is in the near future!

## 1.5 Conclusion

Within the last decade, synthetic efforts toward carbynes have enjoyed resurgence, driven in many cases by the success of CNTs and graphene, as well as the possible applications of such molecules as molecular wires and new optical materials. Due to promising mechanical and electrical properties, combined with rapid progress in electrical engineering, carbyne may lead to the development of next-generation miniature electronic platforms that have devices and components as small as theoretically possible—at the single-atom scale. Regardless of the projected application, the study of carbynes is also often motivated by an infatuation with the structural simplicity of these molecules, which are essentially devoid of both steric and conformational effects that might alter the electronic and optical properties. Currently, there is focus on studying the structure and mechanics of carbyne, particularly with regard to the change in transport properties. Experimental efforts to fabricate chains with different lengths are motivated by the tenability of such properties, with high mechanical fidelity. Ultimately, carbyne may pave the way for an ideal nanowire in atomistic scale circuits, due to its novel electronic transport mechanisms and potential to bridge graphene and CNT systems as seamless carbon interconnectors. The potential of this 1D material is undoubtedly multidimensional.

## References

Akagi, K., Nishiguchi, M., Shirakawa, H. et al., "One-dimensional conjugated carbyne—Synthesis and properties," *Synth. Met.* 17 (1987): 557–562.
Artyukhov, V. I., Liu, M. & Yakobson, B. I., "Mechanically induced metal–insulator transition in carbyne," *Nano Lett.* 14 (2014): 4224–4229.

Ataca, C. & Ciraci, S., "Perpendicular growth of carbon chains on graphene from first-principles," *Phys. Rev. B* 83 (2011): 235417.

Balandin, A. A., "Thermal properties of graphene and nanostructured carbon materials," *Nat. Mater.* 10 (2011): 569–581.

Baughman, R. H., "Dangerously seeking linear carbon," *Science* 312 (2006): 1009–1010.

Baughman, R. H., Zakhidov A. A. & de Heer, W. A., "Carbon nanotubes—The route toward applications," *Science* 297 (2002): 787–792.

Belenkov, E. A. & Mavrinsky, V. V., "Crystal structure of a perfect carbyne," *Cryst. Rep.* 531 (2008): 83–87.

Cahangirov, S., Topsakal, M. & Ciraci, S., "Long-range interactions in carbon atomic chains," *Phys. Rev. B* 82 (2010): 195444.

Casari, C. S., Li Bassi, A., Ravagnan, L. et al., "Chemical and thermal stability of carbyne-like structures in cluster-assembled carbon films," *Phys. Rev. B* 69 (2004): 075422.

Casillas, G., Mayoral, A., Liu, M. et al., "New insights into the properties and interactions of carbon chains as revealed by HRTEM and DFT analysis," *Carbon* 66 (2014): 436–441.

Castelli, I. E., Salvestrini, P. & Manini, N., "Mechanical properties of carbynes investigated by ab initio total-energy calculations," *Phys. Rev. B* 85 (2012): 214110.

Chalifoux, W. A., McDonald, R., Ferguson, M. J. et al., "Tert-butyl-end-capped polyynes: Crystallographic evidence of reduced bond-length alternation," *Angew. Chem. Int. Ed.* 48 (2009): 7915–7919.

Chalifoux, W. A., Ferguson, M. J., McDonald, R. et al., "Adamantyl-endcapped polyynes," *J. Phys. Org. Chem.* 25 (2012): 69–76.

Chalifoux, W. A. & Tykwinski, R. R., "Synthesis of extended polyynes: Toward carbyne," *Comptes Rendus Chimie* 12 (2009): 341–358.

Chalifoux, W. A. & Tykwinski, R. R., "Synthesis of polyynes to model the sp-carbon allotrope carbyne," *Nat. Chem.* 2 (2010): 967–971.

Chartuprayoon, N., Zhang, M., Bosze, W. et al., "One-dimensional nanostructures based bio-detection," *Biosens. Bioelectron.* 63 (2015): 432–443.

Chenoweth, K., van Duin, A. C. T. & Goddard, W. A., "ReaxFF reactive force field for molecular dynamics simulations of hydrocarbon oxidation," *J. Phys. Chem. A* 112 (2008): 1040–1053.

Chuan, X.-Y., Wang, T.-K. & Donnet, J.-B., "Stability and existence of carbyne with carbon chains," *New Carbon Mater.* 20 (2005): 83–92.

Chuvilin, A., Meyer, J. C., Algara-Siller, G. et al., "From graphene constrictions to single carbon chains," *New J. Phys.* 11 (2009): 083019.

Cranford, S., "Thermal stability of idealized folded carbyne loops," *Nano Res. Lett.* 8 (2013): 490.

Cranford, S. W., Brommer, D. B. & Buehler, M. J., "Extended graphynes: Simple scaling laws for stiffness, strength and fracture," *Nanoscale* 4 (2012): 7797–7809.

Cretu, O., Botello-Mendez, A. R., Janowska, I. et al., "Electrical transport measured in atomic carbon chains," *Nano Lett.* 13 (2013): 3487–3493.

Dauberosguthorpe, P. Roberts, V. A., Osguthorpe, D. J. et al., "Structure and energetics of ligand-binding to proteins: *Escherichia coli* dihydrofolate reductase trimethoprim, a drug–receptor system," *Proteins Struct. Funct. Genet.* 4 (1988): 31–47.

dos Santos, R. B., Rivelino, R., Mota, F. B. et al., "Effects of N doping on the electronic properties of a small carbon atomic chain with distinct $sp^2$ terminations: A first-principles study," *Phys. Rev. B* 84 (2011): 075417.

Erdogan, E., Popov, I., Rocha, C. G. et al., "Engineering carbon chains from mechanically stretched graphene-based materials," *Phys. Rev. B* 83 (2011): 041401.

Geim, A. K. & Novoselov, K. S., "The rise of graphene," *Nat. Mater.* 6 (2007): 183–191.

Gibtner, T., Hampel, F., Gisselbrecht, J. P. et al., "End-cap stabilized oligoynes: Model compounds for the linear sp carbon allotrope carbyne," *Chem. Eur. J.* 8 (2002): 408–432.

Goresy, A. E. & Donnay, G., "A new allotropic form of carbon from the Ries crater," *Science* 161 (1968): 363–364.

Heimann, R., Evsyukov, S. E. & Kavan L., *Carbyne and Carbynoid Structures* (Springer, New York, 1999).

Hirsch, A., "The era of carbon allotropes," *Nat. Mater.* 9 (2010): 868–871.

Hobi, E., Pontes, R. B., Fazzio, A. E. et al., "Formation of atomic carbon chains from graphene nanorib-bons," *Phys. Rev. B* 81 (2010): 201406.

Hu, Y. H., "Bending effect of sp-hybridized carbon (carbyne) chains on their structures and properties," *J. Phys. Chem. C* 115 (2011): 1843–1850.

Huang, Y., Wu, J. & Hwang, K. C., "Thickness of graphene and single-wall carbon nanotubes," *Phys. Rev. B* 74 (2006): 245413.

Iijima, S., "Helical microtubules of graphitic carbon," *Nature* 354 (1991): 56–58.

Januszewski, J. A. & Tykwinski, R. R., "Synthesis and properties of long [$n$] cumulenes ($n \geq 5$)," *Chem. Soc. Rev.* 43 (2014): 3184–3203.

Kertesz, M., Choi, C. H. & Yang, S. J., "Conjugated polymers and aromaticity," *Chem. Rev.* 105 (2005): 3448–3481.

Kertesz, M., Koller, J. & Azman, A., "Abinitio Hartree–Fock crystal orbital studies: II. Energy bands of an infinite carbon chain," *J. Chem. Phys.* 68 (1978): 2779–2782.

Khoo, K. H., Neaton, J. B., Son, Y. W. et al., "Negative differential resistance in carbon atomic wire-carbon nanotube junctions," *Nano Lett.* 8 (2008): 2900–2905.

Kocsis, A. J., Yedama, N. A. R. & Cranford, S. W., "Confinement and controlling the effective compressive stiffness of carbyne," *Nanotechnology* 25 (2014): 335709.

Kroto, H., "Carbyne and other myths about carbon," *Chem. World* 7 (2010): 37.

Kuwahara, R., Kudo, Y., Morisato, T. et al., "Encapsulation of carbon chain molecules in single-walled carbon nanotubes," *J. Phys. Chem. A* 115 (2011): 5147–5156.

Lagow, R. J., Kampa, J. J., Wei, H. C. et al., "Synthesis of linear acetylenic carbon—The sp carbon allotrope," *Science* 267 (1995): 362–367.

Lee, C., Wei, X., Kysar, J. W. et al., "Measurement of the elastic properties and intrinsic strength of mono-layer graphene," *Science* 321 (2008): 385–388.

Lee, J. W., Nilson, R. H., Templeton, J. A. et al., "Comparison of molecular dynamics with classical density functional and Poisson–Boltzmann theories of the electric double layer in nanochannels," *J. Chem. Theory Comput.* 8 (2012): 2012–2022.

Lin, Z. Z. & Ning, X. J., "Controlling the electronic properties of monatomic carbon chains," *EPL (Europhys. Lett.)* 95 (2011): 47012.

Liu, M., Artyukhov, V. I., Lee, H. et al., "Carbyne from first principles: Chain of C atoms, a nanorod or a nanorope," *ACS Nano* 7 (2013): 10075–10082.

Lucotti, A., Tommasini, M., Chalifoux, W. A. et al., "Bent polyynes: Ring geometry studied by Raman and IR spectroscopy," *J. Raman Spectrosc.* 43 (2012): 95–101.

Luo, W. & Windl, W., "First principles study of the structure and stability of carbynes," *Carbon* 47 (2009): 367–383.

McNaught, A. D., Wilkinson, A. & International Union of Pure and Applied Chemistry, *Compendium of Chemical Terminology: IUPAC Recommendations*, Second edition (Blackwell Science, Oxford, England; Blackwell Science, Malden, 1997), 450 p.

Mirzaeifar, R., Qin, Z. & Buehler, M. J., "Tensile strength of carbyne chains in varied chemical environ-ments and structural lengths," *Nanotechnology* 25 (2014): 371001.

Nair, A. K., Cranford, S.W. & Buehler, M. J., "The minimal nanowire: Mechanical properties of carbyne," *EPL Europhys. Lett.* 95 (2011): 16002.

Nishide, D., Dohi, H., Wakabayashi, T. et al., "Single-wall carbon nanotubes encaging linear chain $C_{10}H_2$ polyyne molecules inside," *Chem. Phys. Lett.* 428 (2006): 356–360.

Nishide, D., Wakabayashi, T., Sugai, T. et al., "Raman spectroscopy of size-selected linear polyyne mol-ecules $C_{2n}H_2$ ($n = 4$–6) encapsulated in single-wall carbon nanotubes," *J. Phys. Chem. C* 111 (2007): 5178–5183.

Novoselov, K. S., Geim, A. K., Morozov, S. V. et al., "Electric field effect in atomically thin carbon films," *Science* 306 (2004): 666–669.

Ozpineci, A. & Ciraci, S., "Quantum effects of thermal conductance through atomic chains," *Phys. Rev. B* 63 (2001): 125415.

Ravagnan, L., Manini, N., Cinquanta, E. et al., "Effect of axial torsion on *sp* carbon atomic wires," *Phys. Rev. Lett.* 102 (2009): 245502.

Roman, R. E. & Cranford, S. W., "Mechanical properties of silicone," *Comput. Mater. Sci.* 82 (2014): 50–55.

Segal, D., Nitzan, A. & Hänggi, P., "Thermal conductance through molecular wires," *J. Chem. Phys.* 119 (2003): 6840–6855.

Shen, L., Zeng, M., Yang, S.-W. et al., "Electron transport properties of atomic carbon nanowires between graphene electrodes," *J. Am. Chem. Soc.* 132 (2010): 11481–11486.

Smith, P. P. K. & Buseck, P. R., "Carbyne forms of carbon: Do they exist?" *Science* 216 (1982): 984–986.

Smith, P. P. K. & Buseck, P. R., "Carbyne forms of carbon: Evidence for their existence," *Science* 229 (1985): 485–487.

Topsakal, M. & Ciraci, S., "Elastic and plastic deformation of graphene, silicene, and boron nitride honeycomb nanoribbons under uniaxial tension: A first-principles density-functional theory study," *Phys. Rev. B* 81 (2010): 024107.

Tykwinski, R. R., Chalifoux, W., Eisler, S. et al., "Toward carbyne: Synthesis and stability of really long polyynes," *Pure Appl. Chem.* 82 (2010): 891–904.

van Duin, A. C. T., Dasgupta, S., Lorant, F. et al., "ReaxFF: A reactive force field for hydrocarbons," *J. Phys. Chem. A* 105 (2001): 9396–9409.

Van Wesep, R. G., Chen, H., Zhu, W. et al., "Communication: Stable carbon nanoarches in the initial stages of epitaxial growth of graphene on Cu(111)," *J. Chem. Phys.* 134 (2011): 171105.

Xu, B., Lin, J. Y., Feng, Y. P. et al., "Structural and electronic properties of finite carbon chains encapsulated into carbon nanotubes," *J. Phys. Chem. C* 113 (2009): 21314–21318.

Xu, B., Lin, J. Y. & Feng, Y. P., "Mechanical control of magnetic states of finite carbon chains encapsulated in single wall carbon nanotubes," *Appl. Phys. Lett.* 96 (2010): 163105.

Yuzvinsky, T. D., Mickelson, W., Aloni, S. et al., "Shrinking a carbon nanotube," *Nano Lett.* 6 (2006): 2718–2722.

Zanolli, Z., Onida, G. & Charlier, J. C., "Quantum spin transport in carbon chains," *ACS Nano* 4 (2010): 5174–5180.

Zhao, X. L., Ando, Y., Liu, Y. et al., "Carbon nanowire made of a long linear carbon chain inserted inside a multiwalled carbon nanotube," *Phys. Rev. Lett.* 90 (2003): 187401.

Zeng, M. G., Shen, L., Cai, Y. Q. et al., "Perfect spin-filter and spin-valve in carbon atomic chains," *Appl. Phys. Lett.* 96 (2010): 042104.

Zimmerman, J. A., Webb, E. B., Hoyt, J. J. et al., "Calculation of stress in atomistic simulation," *Model. Simul. Mater. Sci. Eng.* 12 (2004b): S319–S332.

# 2

# Linear Carbon Chains

Daniele Fazzi

Caterina Vozzi

**Abstract**

Among all allotropic forms that the carbon element presents in nature, $sp$-hybridization is one of the most intriguing and elusive. With respect to $sp^3$ (diamond) and $sp^2$ (graphite, nanotubes, graphene), $sp$-hybridization has been less investigated, and only in the last decade has the scientific community been attracted by what the potential applications of $sp$-carbon-based materials can accomplish. The $sp$-carbon gives rise to atomic 1D wires, whose properties can be tuned and customized by varying the length, the structure, and the functional groups. Linear carbon chains can be classified based on their structure under the name of polyynes and cumulenes, and they represent the building blocks for developing nanostructured, 1D, molecular-based materials for organic electronics, optics, and energy-saving applications. The fabrication methods, the structure–function relationships, and the electronic and optical properties of carbon wires will be reviewed and critically analyzed.

## 2.1 Introduction

The element carbon is present in nature in a variety of allotropes that can be classified based on its hybridization. For instance, $sp^3$-hybridization is responsible for the structure and properties of diamond, while $sp^2$-hybridization characterizes graphite, carbon nanotubes, and graphene, which is a single layer of carbon atoms arranged in a 2D hexagonal lattice. Depending on the atomic hybridization, the macroscopic properties of carbon-based materials can dramatically vary for each allotropic form (Cha et al. 2013). For example, diamond is a good thermal conductor, electrical insulator, and the hardest known naturally occurring material. Graphite, on the contrary, is softer and shows high electrical conductivity. In the last decades, $sp^2$-hybridized carbon allotropes (fullerenes, nanotubes, and graphene) have been widely investigated for their unique features, encompassing thermal and electrical conductivity, and optical and mechanical properties. All these allotropes, featuring $sp^2$- or $sp^3$-hybridization, form 2D and 3D network structures.

The $sp$-hybridization comes next in the series, and the corresponding carbon allotrope is known under the name of *carbyne* which has the particular feature of extending in *one dimension* (Szafert & Gladysz 2003). Because of this peculiarity, carbynes are also named *carbon nanowires*, representing at the nanoscale monodimensional atomic chains. The existence of carbyne was long debated in scientific literature, from both the theoretical and experimental communities (Whittaker & Watts 1980, Smith & Busek 1982, Whittaker 1985). Carbynes were first observed in interstellar dust, meteorites, and as a by-product of shock-fused graphite, as well as in terrestrial plants, fungal, and marine sources (Goresy & Donnay 1968, Hayatsu et al. 1980, Whittaker & Watts 1980, Shi Shun & Tykwinski 2006).

The name *carbyne* defines an entire class of $sp$-carbon-based molecular systems, in principle characterized by an *infinite chain length*. The corresponding *oligomers* (i.e., $sp$-carbon chains with a finite size) can be classified as *polyynes* or *cumulenes* (Januszewski & Tykwinski 2014). Polyynes consist of carbon chains with an alternation of triple and single carbon bonds (…–C≡C–C≡C–C–…); while cumulenes consist of sequences of equalized double bonds (…=C=C=C=C=C=…). The chemical structures of $sp$-chains and their classification are reported in Figure 2.1.

In general, $sp$-carbon chains are more reactive and unstable than $sp^2$- and $sp^3$-carbon-based materials, and polyynes are thermodynamically and chemically more stable than cumulenes. Structural and dynamical properties of polyynes and cumulenes differ because of their different bond topologies and electronic structures. Being molecular nanostructured 1D systems, the properties of carbynes highly depend on a parameter called *bond length alternation* (BLA) (Milani et al. 2006, Yang & Kertesz 2006, Innocenti et al. 2010).

BLA is a geometrical characteristic defined as the difference between averaged adjacent carbon–carbon bond lengths ($\langle R \rangle$). *Hypothetically*, for polyynes BLA = $\langle R^{single} \rangle$ – $\langle R^{triple} \rangle$, while for cumulenes, BLA = $\langle R^{double} \rangle$ – $\langle R^{double} \rangle$, i.e., equal to zero. As explained later, the chain structure of carbynes is far away from being a perfect triple–single bonds alternation (polyynes) or double bonds sequence (cumulenes), resulting in BLA values different from those expected for an ideal hybridization (Innocenti et al. 2010, Januszewski & Tykwinski 2014).

A series of factors, such as the extended π-electron delocalization, the confinement effects due to finite chain lengths, the chemical groups ending the carbon chains, and the quantum confinement effects due to the 1D nature of the systems, induce a complex trend of the BLA that ultimately modulate the electronic, optical, and mechanical properties of carbon chains (Januszewski & Tykwinski 2014).

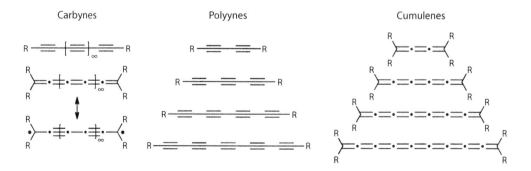

**FIGURE 2.1** Chemical structures of carbon chains: carbynes, polyynes, and cumulenes. (Januszewski, J. A. & Tykwinski, R. R., *Chemical Society Reviews* 43: pp. 3184–3203, 2014. Reproduced by permission of The Royal Society of Chemistry.)

For what concerns the *electronic properties*, polyyne-like carbon chains possess a semiconducting characteristic with a finite energy bandgap ($E_g \geq 0.5$–$1.0$ eV) with BLA values $\geq 0.1$ Å, while cumulenic structures possess a metallic character ($E_g \leq 0.1$ eV), with BLA values $< 0.01$ Å (Milani et al. 2006).

These characteristics make carbynes the ideal candidates for micro- and nanoelectronics applications, in which nanostructured metallic or semimetallic atomic wires can be assembled (Qiu & Liew 2012).

In the last decade, seminal investigations have been carried out for understanding the properties of linear carbon chains and in particular for connecting the chain structure to the electronic, vibrational, and optical properties (Szafert & Gladysz 2003). In this chapter, we will review the relationship between structure and physical properties of linear carbon chains according to the state-of-the-art scientific literature, aiming at illustrating a general panorama regarding the chemical–physical properties of 1D nanostructured carbon materials.

Section 2.2 will summarize the techniques available for synthesizing carbon chains. In Section 2.3, the structural properties of carbynes will be related to the electronic and vibrational properties. Finally, Section 2.4 will be devoted to the discussion of the photophysics of linear carbon chains.

## 2.2 Synthesis and Production of Linear Carbon Chains

A variety of experimental methods is nowadays available to produce and synthesize carbon chains in the laboratory (Cataldo 2005). Most of them are bottom–up approaches, and they can be summarized as (1) *physical* methods, (2) *arc-discharge* methods, and (3) *chemical* methods.

*Physical methods* consist in the vaporization and deposition of carbon clusters, usually in ultrahigh vacuum (UHV) conditions (Milani et al. 2015). The mostly used techniques are (1) pulsed microplasma cluster source (Bogana et al. 2005), where supersonic carbon cluster sources are obtained from arc discharge between graphite electrodes and deposited over a substrate; (2) thermal or laser vaporization of cluster sources (Wakabayashi et al. 2004); (3) ion irradiation of amorphous carbon (Compagnini et al. 2005); and (4) femtosecond laser irradiation of graphite (Hu et al. 2007).

In all these techniques, the final sample consists of a mixture of $sp$–$sp^2$ hybrid amorphous carbon film, in which the content of $sp$-carbon is up to 40%. The $sp$ phase obtained is usually metastable, even in UHV at room temperature conditions. When exposed to oxygen, the carbon network structure rapidly evolves toward an $sp^2$ amorphous phase, with a small residual amount of $sp$ (polyyne-like) aggregates (Casari 2004).

Unfortunately, via physical methods, the stoichiometry of carbynes cannot be controlled, and different phases and chain lengths distributions are always present in the final sample, making a size-dependent characterization difficult.

*Arc-discharge methods* are one-pot processes consisting of an electrical discharge between carbon electrodes (e.g., graphite) (Cataldo 1997, 2005). This procedure has also been used for the production of fullerenes and carbon nanotubes. Arc-discharge methods can be performed in gas phase, under reduced pressure in the presence of a quenching gas (e.g., He), or using two graphite electrodes submerged in a solvent (e.g., acetonitrile). This methodology has been largely adopted by Lucotti et al. (2006), obtaining a mixture of hydrogen-terminated (H–(C≡C)$_N$–H) or cyano-capped (N≡C–(C≡C)$_N$–C≡N) polyynes of different lengths ($N$).

The longest carbon chains obtained through this method consist of 16–20 carbon atoms; however, the most abundant species contains 8 carbon atoms per chain. This method is economic and fast; however, it does not allow to specifically control the size of the carbon chain, and further purification and separation techniques are required, whenever possible, to obtain separated carbynes.

*Chemical methods*, although less accessible, allow to selectively control the length of carbon chains and to obtain stable samples in solution or solid phase with *homogeneous length distribution* (Lagow 1995). Thanks to this unique characteristic, the chemical–physical properties of polyynes or cumulenes can be selectively investigated as a function of the chain length (Eisler 2005).

Over the past years, a variety of synthetic routes has been proposed, and excellent reviews have been published regarding this topic (Januszewski & Tykwinski 2014). Among the most commonly employed chemical routes, we can mention dehydropolycondensation of acetylene, the Glaser reaction based on the oxidative coupling reaction of ethynyl groups by copper salts, polycondensation reactions of halides, and dehydrohalogenation of polymers such as the chemical carbonization of poly(vinylidene halides).

The major problem to overcome with all these methodologies is the stability and isolation of the single carbon chains, due to the high reactivity of the *sp* bonds. Cross-linking mechanisms can in fact occur, yielding to $sp^2$ and $sp^3$ networks, degrading the *sp* chains. To avoid such processes and make the *sp*-carbon chain stable, sterically bulky end-capped groups (*R*) are added. A selection of the mostly used end-capped groups to stabilize and functionalize the carbon chains is reported in Figure 2.2.

Previously, Chalifoux and Tykwinski (2010) were able to lengthen the carbon chain using a modified Eglinton–Galbraith protocol, obtaining solid-state samples (i.e., powder) of polyynic chains (…–C–C≡C–C…) consisting of 44 carbon atoms (22 triple bonds per chain). The result set the world record for the longest stable *sp*-carbon chain ever synthesized, paving the way toward the production of selectively size-controlled carbon chains.

Januszewski et al. (2013) were also able to set another world record: through a careful choice of end-capped groups and synthetic procedures, they synthesized stable, isolated, cumulenic carbon chains, featuring up to 10 carbon atoms per chain. The result represents a step forward not only for the synthesis of highly reactive compounds, but also in the understanding of the physical–chemical properties of carbon chains, allowing the investigation and characterization of isolated 1D carbon systems with a cumulenic-like structure for the first time.

In this frame, end-capped groups are of paramount importance not only for stabilizing the carbon chain, but also for inducing a polyynic or a cumulenic character to the 1D atomic system. Indeed, end-capped groups linked to the carbon chain through a single bond favor a polyynic structure ($R–[C≡C–]_N R$), whereas those connected by a double bond induce a cumulenic pattern ($R=[C=C=]_N R$), as shown in Figure 2.2.

The feature implies a spectacular characteristic of carbon chains as obtained through chemical methods: same chain length (i.e., equal number of carbon atoms per chain) can lead to completely different properties, depending on the terminal groups. For instance, semiconductor-like properties can be

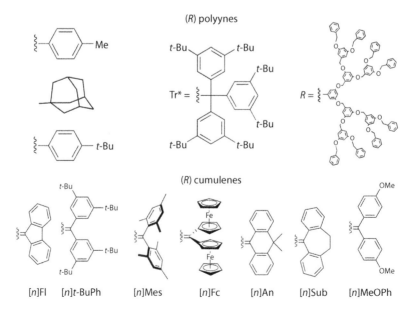

**FIGURE 2.2**   End-capped terminal groups (*R*) for polyynes and cumulenes.

obtained for terminal groups inducing polyyninc structures, while metal-like behaviors can be obtained for those inducing a cumulenic one. The chemical design of the end-capped groups ultimately leads to the design of the functional properties of the *sp*-based materials and their solid state packing.

# 2.3 Structural, Electronic, and Vibrational Properties of Linear Carbon Chains

A unique feature of organic materials is the tight connection between the molecular structure and the functional properties: slight modifications in the intra- and intermolecular structures can lead to significant changes in the macroscopic properties. The understanding and control of the 1D atomic structure allows to design and engineer electronic, optical, chemical, and mechanical properties from the nano- up to the macroscale (Fazzi & Caironi 2015). In the next section, we will discuss the structural properties of carbon chains of infinite and finite size, focusing in particular on the electronic and vibrational properties.

## 2.3.1 Molecular and Electronic Structure

An *sp*-hybridized carbon wire can be ideally modeled as an infinite chain of atoms in two possible geometric arrangements: either a sequence of double equalized bonds (…=C=C=…; cumulene), or a series of alternating triple and single bonds in a dimerized geometry (…–C≡C–C–…; polyyne). The two configurations lead to different electronic structures and properties and the transition from one to the other is driven by the Peierls distortion phenomenon (Milani et al. 2006, 2008). Cumulenes have no (or small) energy bandgap between the valence and conduction bands, featuring a metallic-like behavior, while polyynes show energy bandgap induced by bond length alternation structure, leading to semiconducting properties.

The BLA of polyynes and cumulenes has been the subject of extended experimental and theoretical investigations, with the aim to understand and extrapolate its value up to the infinite limit (carbyne) (Szafert & Gladysz 2003, Lucotti et al. 2006, Yang & Kertesz 2006, Januszewski & Tykwinski 2014).

Figures 2.3 and 2.4 report a survey, taken from literature data, regarding experimental and theoretical evaluations of the bond lengths in isolated polyynes and cumulenes of different chain lengths (Szafert & Gladysz 2003, Innocenti et al. 2010, Januszewski & Tykwinski 2014).

For the simplest polyynic system, ethylene (H–C≡C–H), the triple carbon bond length is evaluated as 1.203(3) Å, while for diacetylene (H–C≡C–C≡C–H) the triple bond lengths range from 1.217 to 1.209 Å, and the single bond is 1.384 Å. As a comparison, the single–single carbon bond for an $sp^3$-hybridization is 1.54 Å, longer than the *sp* one, remarking the high *s* character in the case of carbon chains.

By increasing the number of carbon atoms in the chain, the single–triple bond length alternation is preserved even for very long chains. However, a softening of the inner bonds and a lengthening of the outer ones occur, leading to an equalization effect in the middle of the chain and an alternation at the ends.

A graphical plot of the carbon–carbon bond lengths is reported in Figure 2.4. The BLA has been computed by electronic structure methods for very long hydrogen-terminated carbon chains (e.g., 30 carbon atoms). An increase in the length of the *sp*-carbon chain induces an increase in the π-electron conjugation, corresponding to a decrease in the BLA (Figure 2.4). The longer the carbon chain, the more equalized the bond lengths; however, the bond length alternation pattern in polyynes is preserved up to the infinite limit. Yang and Kertesz (2006) predicted, with a nonlinear extrapolation method, the BLA value for infinitely long hydrogen end-capped polyynes to be around 0.08 Å.

Cumulenes are ideally represented by a sequence of double bonds. The carbon–carbon bond length for the $sp^2$-hybridization is 1.339 Å (ethylene), while for cumulenes it ranges from 1.249 to 1.345 A (see Figure 2.3). Recently, Januszewski (2013) synthesized and isolated stable cumulenes of different chain

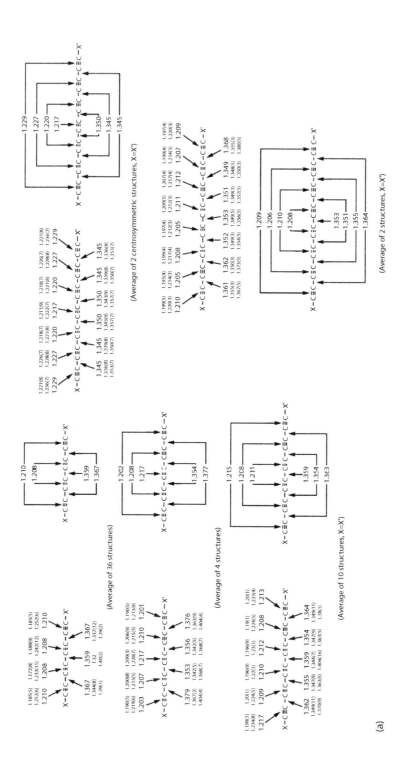

**FIGURE 2.3** Experimental (X-ray data) bond lengths evaluated for (a) polyynes and (b) cumulenes of different chain lengths. (Reprinted with permission from Szafert, S., & Gladysz, J. A., *Chemical Reviews* 103, pp. 4175–4205. Copyright 2003 American Chemical Society; Januszewski, J. A. & Tykwinski, R. R., *Chemical Society Reviews* 43: pp. 3184–3203, 2014. Reproduced by permission of The Royal Society of Chemistry.)

(*Continued*)

**FIGURE 2.3 (CONTINUED)** Experimental (X-ray data) bond lengths evaluated for (a) polyynes and (b) cumulenes of different chain lengths. (Reprinted with permission from Szafert, S. & Gladysz, J. A., *Chemical Reviews* 103, pp. 4175–4205. Copyright 2003 American Chemical Society; Januszewski, J. A. & Tykwinski, R. R., *Chemical Society Reviews* 43: pp. 3184–3203, 2014. Reproduced by permission of The Royal Society of Chemistry.)

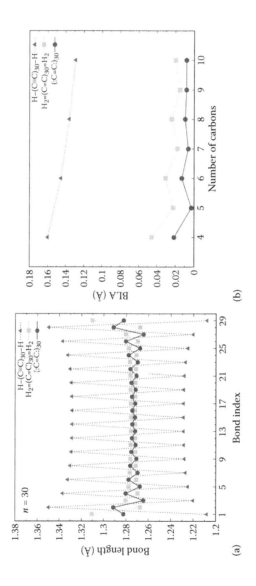

**FIGURE 2.4** (a) Comparison between computed (density functional theory [DFT] data) BLA for hydrogen end-capped polyyne (H–(C≡C)$_N$–H), vynilene end-capped cumulene (H$_2$=(C=C)$_N$=H$_2$), and cumulene (:(C=C:)$_N$) of the same chain length ($N$ = 30). (b) Trend of the BLA for polyyne and cumulenes increasing the number of carbon atoms per chain. (Innocenti, I., Milani, A. & Castiglioni, C.: Can Raman Spectroscopy Detect Cumulenic Structures of Linear Carbon Chains? *Journal of Raman Spectroscopy.* 2010. 41. pp. 226–236. Copyright Wiley-VCH Verlag GmbH & Co. KGaA. Reproduced with permission.)

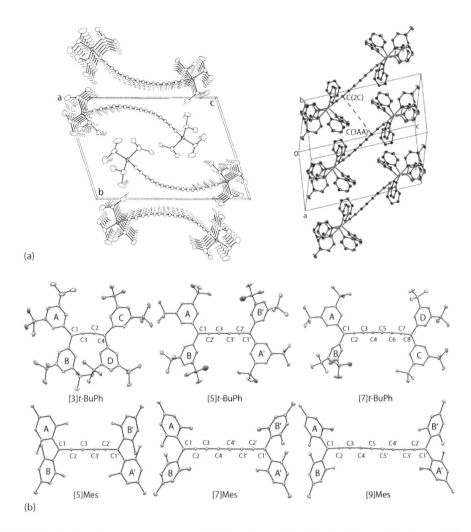

(a)

(b)

**FIGURE 2.5**  (a) Crystal structures and unit cells for polyynes. (Reprinted with permission from Szafert, S. & Gladysz, J. A., *Chemical Reviews* 103, pp. 4175–4205. Copyright 2003 American Chemical Society.) (b) Crystal structure for differently end-capped cumulenes. (Januszewski, J. A. et al.: Synthesis and Structure of Tetraarylcumulenes: Characterization of Bond-Length Alternation versus Molecule Length. *Angewandte Chemie*. 2013. 52. pp. 1817–1821. Copyright Wiley-VCH Verlag GmbH & Co. KGaA. Reproduced with permission.)

lengths. Through X-ray crystallography measurements, they showed that the actual chain structure of cumulenes is linear with a finite value of BLA, although lower than the case of polyynes. This observation confirmed predictions by quantum chemical calculations (Figures 2.3 and 2.4), reporting lower BLA values for cumulenes than for polyynes (Innocenti et al. 2010, Januszewski & Tykwinski 2014). As for polyynes, also in cumulenes, the longer the chain length, the lower the BLA with a softening and equalization of the central bond lengths.

In general, the BLA of carbon wires depends on the length of the atomic chain and on the chemical nature of terminal groups. In this framework, end-capped groups are fundamental to induce a specific BLA pattern, inducing a polyyninc or a cumulenic character. This property paves the way for a functional molecular-based design of new *sp*-carbon chains with desired structural, electronic, and optical properties. It is worth mentioning that the experimental structural data have been collected by X-ray crystallography on solid-state samples (Szafert & Gladysz 2003, Chalifoux & Tykwinski 2010,

Januszewski & Tykwinski 2014). In these conditions, polyynes and cumulenes show packed structures with different structural characteristics depending on the carbon chain length and on the steric hindrance of the terminal groups. In Figure 2.5 are reported some crystal structures of polyynes and cumulenes taken from literature data.

Interestingly, the higher the number of carbon atoms, the longer and more flexible the chain, leading to different conformations, such as linear, symmetric bow, unsymmetric bow, kinked, S-shape, or random, as classified by Szafert and Gladysz (2003). The flexibility of long carbon chains is extremely important because it may explain the possibility to get circular carbon structures, bringing to the generation of fullerene- or nanotube-like patterns. Indeed, from this point of view, $sp$-carbon chains can be seen as a precursor of $sp^2$- and $sp^3$-carbon-based materials.

The molecular structure plays a central role in all organic materials and, in particular, in confined 1D nanosystems like carbon chains, because it directly affects the functional properties. As already mentioned, the BLA is the main parameter for tuning the properties of carbon wires: the modulation of the BLA induces a change in the electronic structure and, in particular, in the energy gap ($E_g$) between the conduction and valence bands, as shown in Figure 2.6.

Considering an infinite carbon chain, different values of BLA lead to semiconducting properties (e.g., polyyninc infinite chain, $E_g \geq 0.5$ eV) or to metallic-like behavior (e.g., cumulenic infinite chain, $E_g \leq 0.05$ eV). For this reason, in 1D carbon systems, there is a univocal connection between the structure and the electronic properties: for a given BLA, a certain $E_g$ is obtained, defining the charge transport and the optical properties of carbon wires. Modulations of the BLA and of $E_g$ can be induced by changing the terminal groups, engineering at the molecular scale the macroscopic functional properties of 1D carbon systems.

In molecular terms, the lower the BLA, the smaller the energy difference between the highest occupied molecular orbital (HOMO) and the lowest unoccupied one (LUMO). Carbon chains with a low HOMO–LUMO gap can be considered as atomic wires featuring high charge mobility that can be exploited, for example, for single molecular junction in molecular electronic applications (Qui et al. 2014).

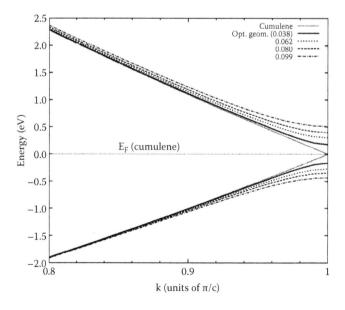

**FIGURE 2.6** Electronic band structure of an infinite linear carbon chain (carbyne), for the case of BLA = 0 (i.e., cumulene) and for different finite values of BLA (polyyne). (Reprinted with permission from Milani, A. et al., *Physical Review B*, 74, pp. 153418, 2006. Copyright 2006 by the American Physical Society.)

## 2.3.2 Nuclear Dynamics and Vibrational Spectroscopy

The length of carbon chain, and consequently the BLA, affects the nuclear dynamics and vibrational properties as well, leading to huge changes by altering the bond topology and the electronic structure.

For ideal carbynes of infinite length, phonon dispersion curves can be computed by adopting the 1D monoatomic chain model, as reported in classical solid-state physics textbooks (Ashcroft & Mermin 2000). For the case of an infinite cumulenic chain (…C=C=C=C…), the periodic unit consists of one atom and the solution of the dynamical problem in periodic boundary conditions leads to one acoustic phonon branch. This means that the system does not present any optical activity. Differently, in the case of an infinite polyynic chain (…–C–C≡C–C…), two atoms per cell have to be considered, leading to an acoustic and an optical phonon branch. For this reason, infinite polyynic structures show optical activity, detectable by vibrational spectroscopies like infrared (IR) and Raman (Ashcroft & Mermin 2000, Milani et al. 2006, Januszewski & Tykwinski 2014).

Figure 2.7a shows the calculated dispersion curves for the longitudinal optical (LO) branch of an infinite carbon chain as a function of the BLA value, modulating the structure from a polyyninc (single–triple bonds) to a cumulenic (double bonds) character.

In real *sp*-carbon chains, finite-size effects break the 1D periodicity relaxing the vibrational and optical selection rules, conferring also to cumulenes IR and Raman activity. For this reason, vibrational spectroscopies turn to be effective and powerful tools to investigate and characterize the structural, electronic, and vibrational properties of real carbon chains (Lucotti et al. 2009a, Milani et al. 2015).

As previously reported for the case of the energy bandgap, BLA also uniquely defines the vibrational frequency and activity of the normal modes. From a polyynic toward a cumulenic structure, the LO branch undergoes huge frequency dispersion, diverging to infinite (negative) values at wave vector $q = 0$ for cumulenic-like structures. The phenomenon is known in condensed matter physics under the name of *Kohn anomaly* (Lucotti et al. 2006), and it has been predicted and observed also for other carbon-based systems such as graphite and carbon nanotubes (Piscanec et al. 2004, 2007).

For finite carbon chains (oligomers), IR and Raman active frequencies can be attributed and placed over the LO branch dispersion curve, depending on the BLA and the length of the carbon chain. Through an *oligomeric approach*, vibrational dispersion curves (typical for infinite, periodic systems) can be rebuilt starting from the finite-size oligomers. This procedure has been effectively applied in the

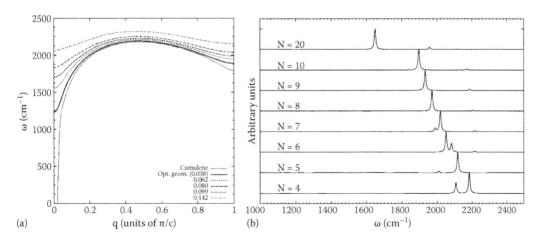

**FIGURE 2.7** (a) DFT calculation of the longitudinal optical phonon branch (LO) for an infinite carbon chain with different values of BLA. (Reprinted with permission from Milani, A. et al., *Physical Review B*, 74, pp. 153418 2006. Copyright 2006 by American Physical Society.) (b) DFT Raman spectra of hydrogen terminated polyynes (H–(C≡C)$_N$–H) of different chain lengths. (Reprinted with permission from Tommasini, M. et al., pp. 11645–11651. Copyright 2007 American Chemical Society.)

last years for both polyynes and cumulenes in order to experimentally derive the dispersion curves for 1D carbon chains (Lucotti et al. 2012, Januszewski & Tykwinski 2014).

The dispersion and modulations of the LO branch frequency suggest that *sp*-chains of finite size have specific active normal modes whose vibrational frequencies shift according to the chain length and the bond length alternation. In fact, polyynes and cumulenes with different chain lengths (i.e., different BLA and π-electron conjugations) are characterized by IR and Raman active normal modes whose frequencies shift accordingly to their size.

An example is reported in Figure 2.7b, where DFT Raman spectra of polyynes end-capped with hydrogen are computed for different lengths (H–(C≡C)$_N$–H, with $2 \leq N \leq 10$). Each polyyne chain has a very selective Raman spectrum, showing an active mode in the 2200–1700 cm$^{-1}$ spectral region, typical of –C≡C– stretching oscillations. The normal mode is assigned to the oscillation of the so-called *effective conjugation coordinate* (ECC) (Castiglioni et al. 1988, 2004). ECC consists of a collective simultaneous in-phase –C≡C– stretching and C–C shrinking of adjacent bonds along the *sp*-carbon chain. ECC mode is characterized by a high Raman scattering cross section, and it is extremely sensitive to the π-electron delocalization (Fazzi et al. 2010). The higher the delocalization of the π-electron density, the softer the force constants between nuclei, leading to lower oscillation frequencies (Castiglioni 2004).

For the case of polyynes, ECC mode undergoes a significant redshift (from 2200 to 1650 cm$^{-1}$) as the chain length increases (Tommasini et al. 2007, Januszewski & Tykwinski 2014). This feature can be directly correlated with the dispersion of the LO branch for the infinite polyynic case, and it makes Raman spectroscopy a unique and selective tool to detect and characterize *sp*-carbon chains. For instance, Raman spectroscopy can be used to investigate the chain lengths distribution in inhomogeneous samples, as those obtained from physical or arc-discharged methods, since each polyyne has one intense Raman active mode whose frequency depends on the chain size (Lucotti et al. 2006, 2009a, Milani 2011). Through spectral decomposition techniques, it is possible to derive the lengths distribution and the concentration profile of a heterogeneous batch of polyynes. In this frame, Raman spectroscopy can be used as a quantitative characterization tool for *sp*-systems.

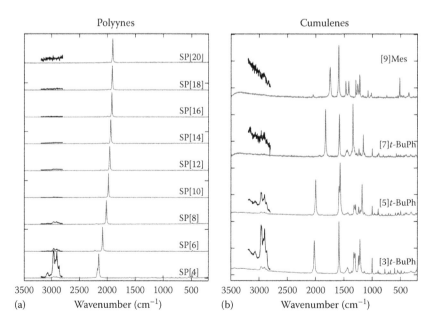

**FIGURE 2.8** Comparison between experimental Raman spectra of (a) polyynes and (b) cumulenes of different chain lengths. (Reprinted with permission from Tommasini, M. et al. 2014, pp. 26415–26425. Copyright 2014 American Chemical Society.)

The possibility shown by Januszewski (2013) of synthesizing stable cumulenes opened the way for their experimental characterization, allowing a comparison with the parent polyynic systems. Figure 2.8 reports the comparison between polyynes' and cumulenes' Raman spectra as a function of the chain length.

As already discussed for the hydrogen end-capped case, the Raman spectra of polyynes are very selective and minor differences in the 2200–1700 cm$^{-1}$ region are caused by bulky terminal groups. In general, few intense normal modes are present, and the longer the polyyne chain, the more redshifted and intense the ECC mode (Lucotti et al. 2009a, Tommasini et al. 2014).

Cumulenes are characterized by more complicated Raman spectra than polyynes: they are less selective, displaying several active lines of comparable intensities in the 2000–1000 cm$^{-1}$ region. No enhancements in the Raman intensity are reported by increasing the chain length, making cumulenes more challenging to characterize by Raman spectroscopy. A comparison of the intensity of the ECC Raman mode with the ones coming from the end groups (black line in Figure 2.8) shows that the former's order of magnitude is higher than the latter in polyynes, while they are of comparable intensity in cumulenes. Generally, cumulenes show a lower Raman cross section than polyynes in accordance with the low optical activity.

IR spectroscopy is another effective, economic, and noninvasive tool to investigate the dynamical properties of carbon chains. We refer to the scientific literature for a specific description of its application for the case of polyynes and cumulenes (Milani et al. 2008, Lucotti et al. 2009a, 2012, Januszewski & Tykwinski 2014).

It is worth mentioning here an interesting application. For a perfectly linear carbon chain, specific selection rules hold, making IR and Raman mutually excluded. In fact, a linear atomic chain belongs to the $D_{\infty h}$ point group, and the Raman active modes are not IR active and vice versa (*mutual exclusion principle* [Herzberg 1945]). Any evidence of a Raman mode becoming active in the IR spectrum (and vice versa) would represent a breaking of the $D_{\infty h}$ symmetry, which would mean a nonlinear chain structure. From this perspective, Raman and IR spectroscopy can be concomitantly adopted to indirectly probe the linearity of the carbon chains (i.e., the structure). This methodology has been applied for the case of polyynes (adamantyl end-capped) featuring different chain lengths, and the activation of a Raman mode in the IR spectrum has been kept as evidence of nonlinear chain structures (Lucotti et al. 2009a).

Furthermore, vibrational spectroscopy has been used to detect polyynes interacting with silver and gold nanoparticles, in order to investigate chemisorption and physisorption mechanisms and the charge transfer processes occurring at the organic–inorganic interface (Milani et al. 2011). The technique used is called *surface-enhanced Raman spectroscopy* (SERS) (Nie et al. 1997). In the vicinity of nanostructured metal clusters, the Raman scattering cross section is enhanced by orders of magnitude due to the local electric field created by the plasmonic oscillations on the metal nanostructured surface. The enhancement of the Raman intensity allows the detection of compounds either with low-scattering section or present in the sample in a very low concentration. Because of these properties, SERS can be used to detect the presence of *sp*-carbon chains in amorphous *sp–sp$^2$* films, even when the amount of *sp* phase is very small. Examples of SERS applied to carbon chains are reported by Lucotti et al. (2009b) and Milani (2011).

Generally, vibrational spectroscopy is an effective tool to qualitatively and quantitatively characterize carbon chains, getting insights in their composition, chain length, structure, and electronic properties.

## 2.4 Photophysics of Carbon Chains

Recently, photophysical properties of carbon chains have been investigated aiming at the possibility to obtain stable and chain-length-selected *sp*-nanostructured systems. Photophysics plays a key role in understanding energy and charge transfer processes, light-harvesting mechanisms, and excitonic processes (Lanzani 2012). Getting insights into the optical properties of *sp*-carbon chains may open the way for their use in linear and nonlinear optics, lasing and photovoltaic applications (Eisler et al. 2005, Wakabayashi et al. 2007).

In the next sections, we will review very recent steady state and ultrafast transient absorption spectroscopy results for polyynes and cumulenes.

## 2.4.1  Ultraviolet–Visible (UV–Vis) Spectroscopy of Carbon Chains

As we already mentioned, the electronic properties of carbynes strongly depend on the BLA value, leading to a finite bandgap for an alternated structure and to no bandgap for an equalized one (Milani et al. 2006, Jochnowitz & Maier 2008). Moreover, in the case of finite-size carbon chains, the BLA is affected by confinement effects induced by the terminal groups.

Recalling the simple physics behind the particle in a box, the longer the 1D carbon chain, the higher the electron delocalization that lowers the energy gap between the HOMO and the LUMO (electronic gap). This lowering of the electronic gap indirectly affects the optical properties, in terms of absorption and emission spectral ranges. Indeed, the first dipole-allowed singlet excited state ($S_1$) has a strong contribution from the HOMO–LUMO transition; thus, a lowering of the HOMO–LUMO energy difference corresponds to a decrease in the optical gap.

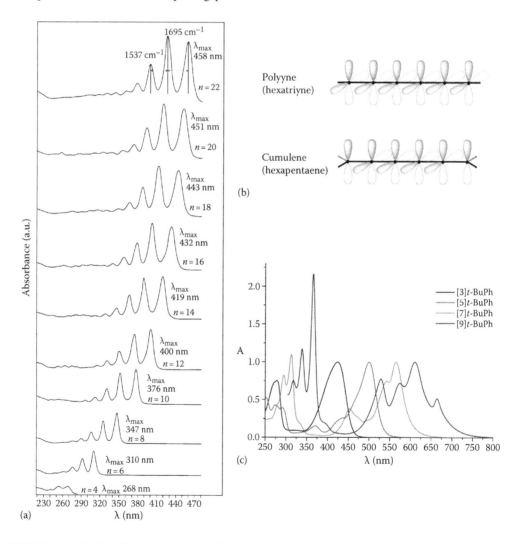

**FIGURE 2.9**  UV–Vis absorption spectra of (a) triethylsilyl end-capped series, end-capped polyynes, and (c) *t*-butyl end-capped cumulenes. (Reprinted by permission from Macmillan Publishers Ltd. *Nature Chemistry*, Chalifoux, W. A. & Tykwinski, R. R., 2: pp. 967–971, copyright (2010).) (b) Polyyne and cumulene π-electron structure. (Januszewski, J. A. & Tykwinski, R. R., *Chemical Society Reviews* 43: pp. 3184–3203, 2014. Reproduced by permission of The Royal Society of Chemistry.)

In Figure 2.9a are the UV–Vis absorption spectra of polyynes with different chain lengths. Each absorption spectrum features a clear vibronic structure, remarking the high electron–phonon coupling characterizing these conjugated systems (Grutter et al. 1998, Wakabayashi et al. 2007). The normal mode coordinate, which couples the excited to the ground state, is the Raman active ECC mode. By increasing the chain length, the vibronic structure is preserved and a remarkable redshift of the main absorption band is observed: $\lambda_{max}$ shifts from 268 to 458 nm as the number of triple bonds per chain increases from 4 to 22.

Attempts to extrapolate the optical gap to the infinite case have been done by Januszewski (2013), as reported in Figure 2.10, and a plausible asymptotic limit has been estimated as $\lambda_{max}$ = 485 nm, for 48 triple bonds (i.e., 96 carbon atoms per chain). Interestingly, the asymptotic limit consists of a finite value, meaning that a finite optical gap is still present for an infinite polyynic chain. The extrapolated value implies that a finite BLA persists for the infinite polyynic case, and these data can be considered as an experimental proof of the theoretical prediction previously reported in literature, which quantified a finite BLA equivalent to 0.08 Å for an infinite polyynic chain (Yang & Kertesz 2006).

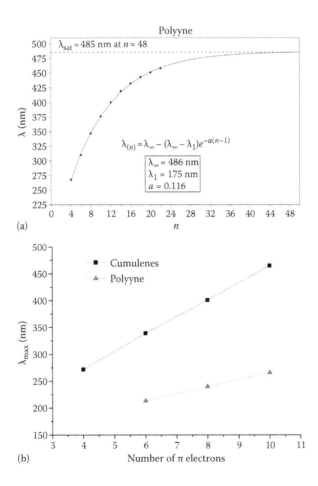

(a)

(b)

**FIGURE 2.10** (a) Evolution of the absorption maxima $\lambda_{max}$ for triethylsilyl end-capped polyynes as recorded in toluene, with the extrapolation up to the asymptotic limit. (Reprinted by permission from Macmillan Publishers Ltd. *Nature Chemistry*, Chalifoux, W. A. & Tykwinski, R. R., 2: pp. 967–971, copyright (2010).) (b) Comparison between $\lambda_{max}$ of polyynes (triangles) and cumulenes (squares) featuring an equivalent number of π-electrons per chain. (Januszewski, J. A. & Tykwinski, R. R., *Chemical Society Reviews* 43: pp. 3184–3203, 2014. Reproduced by permission of The Royal Society of Chemistry.)

The absorption spectra of cumulenes are reported in Figure 2.9b. Since the stability of cumulenic chains is lower than that of polyynes, the maximum length obtained, so far, consists of nine double bonds per chain (i.e., 10 carbon atoms). Generally, the same trend for polyynes is observed: a redshift of the main absorption band with increasing chain length (higher $\pi$-electron delocalization). However, the band structure is more complicated than the one observed for polyynes, and current investigations are ongoing to better understand the electron–phonon coupling effects in cumulenic chains (Januszewski & Tykwinski 2014).

Cumulenes are more $\pi$-electron conjugated than polyynes, and for the same number of carbon atoms per chain, a lower absorption band is observed. For the longest cumulene (10 carbon atoms), $\lambda_{max} = 664$ nm, while for a polyyne with the same number of carbon atoms, $\lambda_{max}$ would be around 300 nm (Januszewski & Tykwinski 2014).

Formally, in order to compare polyynes and cumulenes in a more correct way, the number of $\pi$-electrons present in the system should be considered instead of the number of carbon atoms per chain. This is due to the different valence requirements of the terminal atoms of each chain, i.e., termination with $sp$- versus $sp^2$-carbon. As shown in Figure 2.9c, polyynes possess two degenerate orthogonal $\pi$-systems, while cumulenes have two nondegenerate $\pi$-systems, one that extends over the whole molecule and conjugates to the end groups, and a shorter orthogonal $\pi$-system that cannot conjugate to the end groups. Taking this characteristic into account, it is possible to compare the $\lambda_{max}$ of polyynes and of cumulenes on the same grounds, as reported in Figure 2.10b from Januszewski and Tykwinski (2014). The plot clearly shows that for the same number of $\pi$-electrons, cumulenes have a lower optical gap than polyynes, although we are still far away from predicting the infinite limit value for cumulenes. In principle, this limit would be null; however, Peierls distortion effect and end size effects will limit it to a finite, although small, value.

## 2.4.2 Photogenerated Species and Excited States Dynamics in Carbon Chains

In the last 5 years, the excited states photophysical properties of $sp$-carbon chains have been investigated aiming to get insights into (i) the nuclear structure of the carbon chain in the excited states, (ii) the excited states decay mechanisms and their timescales, and (iii) the intersystem crossing and energy transfer processes. A deep understanding of these properties will pave the way toward a rational design of photoactive devices for optical and photovoltaic applications based on carbon-based materials.

For investigating the molecular structure and nature of photogenerated $sp$-carbon chain species, combined experimental and computational investigations have been recently reported (Yildizhan et al. 2011, Fazzi et al. 2013).

Figure 2.11a shows the comparison between the experimental IR spectra of ground and photoexcited adamantyl end-capped polyyne (Yildizhan et al. 2011) (Figure 2.11b).

In the IR spectra recorded during UV irradiation, a new absorption band appears at 2130 cm⁻¹, representing the vibrational fingerprint of a photogenerated species. Density functional theory (DFT) and time-dependent DFT (TD-DFT) IR spectra of polyyne in ground state ($S_0$), excited state ($S_1$), and triplet state ($T_1$) are also reported in Figure 2.11b. Calculations confirm that the plausible photogenerated species could be related to the first excited singlet or triplet state, both showing an active IR band in the 2100–2050 cm⁻¹.

DFT and TD-DFT methods allow investigating the chain structure in the excited states, as reported in Figure 2.11c. The bond length alternation of the $sp$-chain in both $S_1$ and $T_1$ equilibrium geometries is more equalized than the ground state, leading to softer force constants and, hence, to a redshift of the vibrational frequencies (Tommasini et al. 2007).

The BLA of photoexcited polyynes is comparable to the ground state BLA of cumulenes (see comparison in Figure 2.11c), meaning that polyynes possess a cumulenic character in the excited states.

Temperature-dependent photogenerated IR spectra show the disappearance of the photoinduced absorption band (2130 cm⁻¹) passing from 77 to 223 K, restoring the ground state IR spectrum. This trend

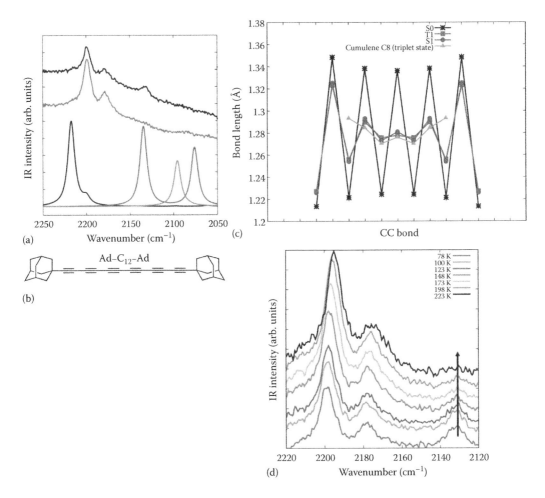

**FIGURE 2.11** (a) Comparison between experimental IR spectra (before and after photoexcitation) and DFT (B3LYP/cc-pVTZ) IR spectra of ground state ($S_0$), first excited state ($S_1$), and triplet state ($T_1$) of Ad–(C≡C)$_6$–Ad. (b) Sketch of an adamantyl polyyne (Ad–(C≡C)$_6$–Ad). (c) DFT-computed bond lengths along the $sp$-chain (Ad–(C≡C)$_6$–Ad) for $S_0$, $S_1$, and $T_1$ states. For comparison, the bond lengths of cumulene C8 (triplet ground state electronic configuration) are also reported. (d) Temperature dependence (78–223 K) of the IR spectra of Ad–(C≡C)$_6$–Ad embedded in polymethyl methacrylate (PMMA) matrix, after the photoirradiation process. (Reprinted with permission from Yildizhan, M. M. et al., *The Journal of Chemical Physics*, 134, 124512. Copyright 2011, American Institute of Physics.)

suggests instability of the photogenerated species by changing the temperature. Since *cumulenic-like* species, as the chain structures predicted in $S_1$ and $T_1$ of polyynes, are known to be particularly unstable and reactive, the observed trends can be rationalized as a destabilization of the photoexcited cumulenic structure (C=C=C), which relaxes from a *metastable* excited state to the ground state when the temperature increases, recovering the more stable alternated (C–C≡C) polyyinic structure (Yildizhan et al. 2011).

Furthermore, time-resolved transient absorption spectroscopy has been recently used for the investigation of the excited states dynamics and deactivation mechanisms occurring after photoexcitation in linear carbon chains.

In Figure 2.12a is the 2D map of the pump–probe signal as recorded on a batch consisting of a mixture of dinaphthylpolyynes (Ar–(C≡C)$_N$–Ar; Ar stands for aryl group) having different chain lengths, being Ar–(C≡C)$_3$–Ar the most abundant species. The sequence of spectra was obtained by exciting the

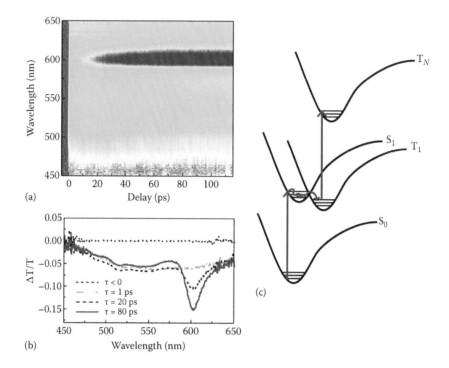

(a)

(b)

(c)

**FIGURE 2.12** (a) 2D map of the pump–probe measurement; contour plot cross sections at different time delays and (b) transient absorption dynamics for two probe wavelengths: 575 nm (solid curve) and 600 nm (dashed curve). (c) Sketch of ground ($S_0$) and excited states singlet ($S_1$) and triplet ($T_1$, $T_N$) involved in the photoinduced process. (Fazzi, D. et al., *Physical Chemistry Chemical Physics: PCCP* 15, pp. 9384–9391, 2013. Reproduced by permission of The Royal Society of Chemistry.)

polyyne mixture with 150 fs pulses centered at 390 nm wavelength. The transient signal was calculated as $\Delta T/T(\lambda,\tau) = [T_{ON}(\lambda,\tau) - T_{OFF}(\lambda)]/T_{OFF}(\lambda)$ (Fazzi et al. 2013).

Over the whole spectral bandwidth, it is possible to observe a broad photoinduced absorption band (450–650 nm), which appears instantaneously upon impulsive photoexcitation. Furthermore, a clear photoinduced absorption signal appears at 600 nm with a time constant of 30 ps and a narrowband spectral shape (around 20 nm).

DFT and TD-DFT calculations revealed that the observed features can be assigned to different excited states transitions. The broad photoinduced absorption band, showing a fast formation time (<1 ps), can be assigned to optical transitions from the excited singlet state ($S_1$) to higher singlet excited states ($S_n$). On the other hand, the narrow feature centered at 600 nm and characterized by 30 ps formation time can be related to an intersystem crossing (ISC) $S_1 \rightarrow T_1$ dynamics that leads to the formation of a triplet state ($T_1$), followed by electronic transitions between triplet states ($T_1 \rightarrow T_n$). In Figure 2.13a and c are the TD-DFT singlet and triplet excited states energies computed for the most abundant species in the batch (Ar–(C≡C)$_3$–Ar), and the frontier molecular orbitals involved in the electronic transitions.

The $T_1 \rightarrow T_8$ dipole active transition, occurring at 1.87 eV, represents the experimental photoabsorption band peaked at 600 nm (2.06 eV). The prominent role of the end-group conformations with respect to the *sp*-chain in facilitating the ISC mechanism clearly appears: distorted conformations, as the (0°, 90°) conformer shown in Figure 2.13b, favor the ISC more than the flat (0°, 0°) conformers.

The data reported by Fazzi et al. (2013) represent the first transient-absorption experimental and computational investigation, shedding light on the dynamics of excited states in linear carbon chains and revealing the character and nature of the photoinduced species.

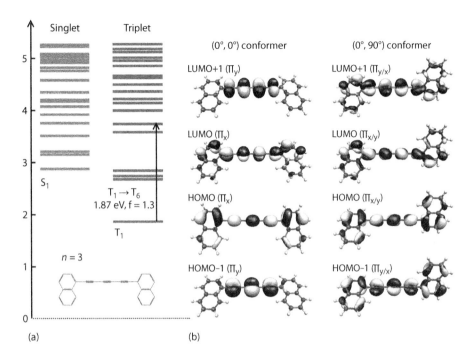

**FIGURE 2.13** (a) TD-DFT (TD-B3LYP/cc-pVTZ) calculated singlet and triplet excited state energies for Ar–$(C\equiv C)_3$–Ar, with the $T_1$-$T_n$ dipole allowed transition also shown. (b) TD-B3LYP/cc-pVTZ frontier molecular orbitals for different conformations of the end-capped groups of Ar–$(C\equiv C)_3$–Ar. (Fazzi, D. et al., *Physical Chemistry Chemical Physics: PCCP* 15, pp. 9384–9391, 2013. Reproduced by permission of The Royal Society of Chemistry.)

In general, singlet–triplet intercrossing mechanisms seem to play a fundamental role in *sp*-systems ruling the dynamics of the excited states and the energy transfer mechanisms. A very recent investigation, carried out by Movsisyan et al. (2014), gets further insights into the photophysics of carbon chains, revealing the role of singlet and triplet excited states in mediating energy transfer processes from the polyynic chain to supramolecular macrocycles. In particular, the structure, excited states dynamics, and singlet–triplet transfer mechanisms have been investigated for a rotaxane consisting of a hexane chain (i.e., polyyne with six triple bonds) threaded through a phenanthroline macrocycle and related compounds containing rhenium(I) chlorocarbonyl complexes. Strong through-space excited states interaction has been detected via time-resolved infrared spectroscopy. Specifically, a quenching of the luminescence of the polyynic chain has been observed as the consequence of a fast (nanosecond timescale) triplet energy migration from the lowest metal-to-ligand charge transfer excited state to the threaded polyynic chain.

Despite excited states properties of *sp*-carbon chains remaining an infant research field where more deep comprehension and understanding are required, the current state-of-the-art investigations suggest polyynes as active compounds to harvest both triplet and singlet energies, exploiting them as photoactive systems for developing nanostructured carbon-based materials for optoelectronics and energy-saving applications.

## 2.5 Conclusions

In this chapter, we reviewed the properties of *sp*-carbon-based molecular systems, such as carbyne, polyyne, and cumulenes, highlighting the link between chemical structure and physical properties. The techniques available for the synthesis of linear carbon chains have been reviewed in Section 2.2. In

Section 2.3, we focused on the structural properties of these materials and how those can be related to the electronic and vibrational properties, while in Section 2.4, we discussed the photophysics of linear carbon chains.

The profound understanding of all these properties still requires further experimental and theoretical investigations, but it would pave the way to the possible use of these materials, e.g., as functional building blocks in nanostructured materials and in supramolecular chemistry or in nonlinear optical applications.

## Acknowledgments

We acknowledge financial support from the European Research Council under the European Community's Seventh Framework Programme (FP7/2007–2013)/ERC Grant Agreement No. 307964-UDYNI and Contract No. 228334 JRA-INREX (Laserlab Europe III) and from the Italian Ministry of Research and Education (ELI project–ESFRI Roadmap). D.F. thanks the Alexander von Humboldt foundation for a Post Doctoral fellowship. The authors thank R. R. Tykwinski for fruitful discussions.

## References

Ashcroft, N. W. & Mermin, N. D., *Solid State Physics* (Cengage Learning Emea, Hampshire, UK, 2000).

Bogana, M., Ravagnan, L., Casari, C. S. et al., "Leaving the fullerene road: Presence and stability of sp-chains in $sp^{2-}$ carbon clusters and cluster-assembled solids," *New J. Phys.* 7 (2005): 81.

Casari, C. S., Li Bassi, A., Ravagnan, L. et al., "Gas exposure and thermal stability of linear carbon chains in nanostructured carbon films investigated by in situ Raman spectroscopy," *Carbon* 42 (2004): 1103–1106.

Castiglioni, C., Navarrete, J. T. L., Zerbi, G. et al., "A simple interpretation of the vibrational spectra of undoped, doped and photoexcited polyacetylene: Amplitude mode theory in the GF formalism," *Solid State Commun.* 65 (1988): 625–630.

Castiglioni, C., Tommasini, M. & Zerbi, G., "Raman spectroscopy of polyconjugated molecules and materials: Confinement effect in one and two dimensions," *Philos. Trans. R. Soc. London A* 362 (2004): 2425.

Cataldo, F., "A study on the structure and electrical properties of the fourth carbon allotrope: Carbyne," *Polym. Int.* 44 (1997): 191–200.

Cataldo, F., *Polyynes, Properties and Applications* (Taylor & Francis, Boca Raton, 2005).

Cha, C., Ryon Shin, S., Annabi, N. et al., "Carbon-based nanomaterials: Multifunctional materials for biomedical engineering," *ACS Nano* 7 (2013): 2891–2897.

Chalifoux, W. A. & Tykwinski, R. R., "Synthesis of polyynes to model the sp-carbon allotrope carbyne," *Nat. Chem.* 2 (2010): 967–971.

Compagnini, G., Battiato, S., Puglisi, O. et al., "Ion irradiation of sp-rich amorphous carbon thin films: A vibrational spectroscopy investigation," *Carbon* 43 (2005): 3025–3028.

Eisler, S., Slepkov, A. D., Elliott, E. et al., "Polyynes as a model for carbyne: Synthesis, physical properties, and nonlinear optical response," *J. Am. Chem. Soc.* 127 (2005): 2666–2676.

Fazzi, D. & Caironi, M., "Multi-length-scale relationships between the polymer molecular structure and charge transport: The case of polynaphthalene diimide bithiophene," *Phys. Chem. Chem. Phys.* 17 (2015): 8573–8590.

Fazzi, D., Canesi, E. V., Negri, F. et al., "Biradicaloid character of thiophene-based heterophenoquinones: The role of electron–phonon coupling," *Chemphyschem* 11 (2010): 3685–3695.

Fazzi, D., Scotognella, F., Milani, A. et al., "Ultrafast spectroscopy of linear carbon chains: The case of dinaphthylpolyynes," *Phys. Chem. Chem. Phys.* 15 (2013): 9384–9391.

Goresy, A. E. & Donnay, G., "A new allotropic form of carbon from the Ries crater," *Science* 161 (1968): 363–364.

Grutter, M., Wyss, M., Fulara, J. et al., "Electronic absorption spectra of the polyacetylene chains $HC_{2n}H$, $HC_{2n}H^-$, and $HC_{2n-1}N^-$ ($n$ = 6–12) in neon matrixes," *J. Phys. Chem. A* 102 (1998): 9785–9790.

Hayatsu, R., Scott, G. R., Studier, M. H. et al., "Carbynes in meteorites: Detection, low-temperature origin, and implications for interstellar molecules," *Science* 209 (1980): 1515–1518.

Herzberg, G., *Molecular Spectra and Molecular Structure*, Vol. 2 (D. Van Nostrand, Princeton, 1945).

Hu, A., Rybachuk, M., Lu, Q. B. et al., "Direct synthesis of sp-bonded carbon chains on graphite surface by femtosecond laser irradiation," *Appl. Phys. Lett.* 91 (2007): 131906.

Innocenti, I., Milani, A. & Castiglioni, C., "Can Raman spectroscopy detect cumulenic structures of linear carbon chains?" *J. Raman Spectrosc.* 41 (2010): 226–236.

Januszewski, J. A. & Tykwinski, R. R., "Synthesis and properties of long [$n$]cumulenes ($n \geq 5$)," *Chem. Soc. Rev.* 43 (2014): 3184–3203.

Januszewski, J. A., Wendinger, D., Methfessel, C. D. et al., "Synthesis and structure of tetraarylcumu-lenes: Characterization of bond-length alternation versus molecule length," *Angew. Chem.* 52 (2013): 1817–1821.

Jochnowitz, E. B. & Maier, J. P., "Electronic spectroscopy of carbon chains," *Ann. Rev. Phys. Chem.* 59 (2008): 519–544.

Lagow, R. J., Kampa, J. J., Wei, H. C. et al., "Synthesis of linear acetylenic carbon: The 'sp' carbon allotrope," *Science* 267 (1995): 362–367.

Lanzani, G., *The Photophysics behind Photovoltaics and Photonics* (Wiley, Hoboken, 2012).

Lucotti, A., Tommasini, M., Del Zoppo, M. et al., "Raman and SERS investigation of isolated sp-carbon chains," *Chem. Phys. Lett.* 417 (2006): 78–82.

Lucotti, A., Tommasini, M., Fazzi, D. et al., "Evidence for solution-state nonlinearity of sp-carbon chains based on IR and Raman spectroscopy: Violation of mutual exclusion," *J. Am. Chem. Soc.* 131 (2009a): 4239–4244.

Lucotti, A., Casari, C. S., Tommasini, M. et al., "sp-carbon chain interaction with silver nanoparticles probed by surface enhanced Raman scattering," *Chem. Phys. Lett.* 478 (2009b): 45–50.

Lucotti, A., Tommasini, M., Chalifoux, W. A. et al., "Bent polyynes: Ring geometry studied by Raman and IR spectroscopy," *J. Raman Spectrosc.* 43 (2012): 95–101.

Milani, A., Tommasini, M., Del Zoppo, M. et al., "Carbon nanowires: Phonon and π-electron confine-ment," *Phys. Rev. B* 74 (2006): 153418.

Milani, A., Tommasini, M., Fazzi, D. et al., "First-principles calculation of the Peierls distortion in an infi-nite linear carbon chain: The contribution of Raman spectroscopy," *J. Raman Spectrosc.* 39 (2008): 164–168.

Milani, A., Lucotti, A., Russo, V. et al., "Charge transfer and vibrational structure of sp-hybridized car-bon atomic wires probed by surface enhanced Raman spectroscopy," *J. Phys. Chem. C* 115 (2011): 12836–12843.

Milani, A., Tommasini, M., Russo, V. et al., "Raman spectroscopy as a tool to investigate the structure and electronic properties of carbon-atom wires," *Beilstein J. Nanotech.* 6 (2015): 480–491.

Movsisyan, L. D., Peeks, M. D., Greetham, G. M. et al., "Photophysics of threaded sp-carbon chains: The polyyne is a sink for singlet and triplet excitation," *J. Am. Chem. Soc.* 136 (2014): 17996–18008.

Nie, S. & Emory, S. R., "Probing single molecules and single nanoparticles by surface-enhanced Raman scattering," *Science* 275 (1997): 1102–1106.

Piscanec, S., Lazzeri, M., Mauri, F. et al., "Kohn anomalies and electron-phonon interactions in graphite," *Phys. Rev. Lett.* 93 (2004): 185503.

Piscanec, S., Lazzeri, M., Robertson, J. et al., "Optical phonons in carbon nanotubes: Kohn anomalies, Peierls distortions, and dynamic effects," *Phys. Rev. B* 75 (2007): 035427.

Qiu, M. & Liew, K. M., "Odd–even effects of electronic transport in carbon-chain-based molecular devices," *J. Phys. Chem. C* 116 (2012): 11709–11713.

Shi Shun, A. L. & Tykwinski, R. R., "Synthesis of naturally occurring polyynes," *Angew. Chem.* 45 (2006): 1034–1057.

Smith, P. & Busek, P. R., "Carbyne forms of carbon: Do they exist?" *Science* 216 (1982): 984–986.

Szafert, S. & Gladysz, J. A., "Carbon in one dimension: Structural analysis of the higher conjugated polyynes," *Chem. Rev.* 103 (2003): 4175–4205.

Tommasini, M., Fazzi, D., Milani, A. et al., "Intramolecular vibrational force fields for linear carbon chains through an adaptive linear scaling scheme," *J. Phys. Chem. A* 111 (2007): 11645–11651.

Tommasini, M., Milani, A., Fazzi, D. et al., "Pi-conjugation and end group effects in long cumulenes: Raman spectroscopy and DFT calculations," *J. Phys. Chem. C* 118 (2014): 26415–264125.

Wakabayashi, T., Ong, A. L., Strelnikov, D. et al., "Flashing carbon on cold surfaces," *J. Phys. Chem. B* 108 (2004): 3686–3690.

Wakabayashi, T., Nagayama, H., Daigoku, K. et al., "Laser-induced emission spectra of polyyne molecules $C_{2n}H_2$ ($n = 5$–8)," *Chem. Phys. Lett.* 446 (2007): 65–70.

Whittaker, A. G., "Carbyne forms of carbon: Evidence for their existence," *Science* 229 (1985): 485–486.

Whittaker, A. G. & Watts, E. J., "Carbynes: Carriers of primordial noble gases in meteorites," *Science* 209 (1980): 1512–1514.

Yang, S. & Kertesz, M., "Bond length alternation and energy band gap of polyyne," *J. Phys. Chem. A* 110 (2006): 9771–9774.

Yildizhan, M. M., Fazzi, D., Milani, A. et al., "Photogenerated cumulenic structure of adamantyl end-capped linear carbon chains: An experimental and computational investigation based on infrared spectroscopy," *J. Chem. Phys.* 134 (2011): 124512.

$3$

# Carbon Nanocoils

Kaori Hirahara

## 3.1 Introduction

Carbon nanocoils (CNCs) are a type of nanocarbon material that consists of fibers of nanometer-sized diameters with a coiled structure (Figure 3.1). The first study on CNCs was reported by Davis et al. (1953); a couple of carbon fibers with a twisted morphology were grown by heterogeneous reaction in the presence of iron (Fe) catalyst. Furthermore, Baker and Chludzinski (1980) and Amelinckx et al. (1994) reported the formation of carbon fibers with coiled morphologies using chemical vapor deposition (CVD). Zhang et al. (2000) reported a large-scale synthesis method for preparing CNCs in the presence of metal composite catalysts. CNCs have found potential applications in various fields owing to their interesting properties related to their unique morphology with carbon-based characteristics. Composites of CNCs have also found potential applications such as electromagnetic wave absorbers (Zhao & Shen 2008). In this chapter, the structural characteristics of CNCs are first discussed together with growth models proposed in past research studies. Synthesis methods for the preparation of CNCs and CNC composites, their resulting properties, and potential applications are introduced.

## 3.2 Structural Properties of CNCs

In general, the geometrical features of a spring coil are defined by the fiber diameter, coil diameter, pitch, and length. In addition to these parameters, the crystallinity and facets can be used to characterize CNCs. The structural parameters of CNCs are determined by the synthetic method employed. However, CNCs of diverse structural features can be obtained within the same synthesis batch as observed in Figure 3.1. The CNCs were synthesized by CVD in the presence of a composite catalyst of iron (Fe) and tin (Sn) as discussed further in Section 3.3. Typically, CNCs produced using such

**FIGURE 3.1**   Carbon nanocoils.

a large-scale CVD-based synthesis method feature fibers with diameters of ~100 nm. Furthermore, CNCs can consist of multiple fibers twisted together resulting in complex facets (Figure 3.2). In contrast, CNCs consisting of single fibers typically feature polygonal facets (Figure 3.2d). The coil diameter typically ranges from several tens of nanometers to several hundred nanometers, and the pitch typically ranges from 10 nm to several hundred nanometers. The length of CNCs depends on the reaction time employed in the CVD process, and can be >100 μm. The tips of individual CNCs can be capped with a catalytic particle (Figure 3.2d).

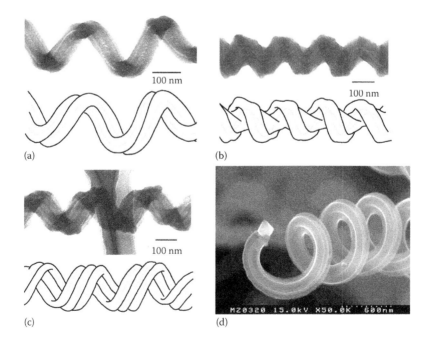

**FIGURE 3.2**   (a–c) Transmission electron microscope (TEM) images of isolated CNCs with various morphologies and corresponding schematics showing facet structures. (d) Scanning electron microscope (SEM) image of a catalytic particle at the tip of a CNC. (From Zhang, M. et al., *Japanese Journal of Applied Physics*, 39, 12 A, 2000.)

The crystallinity of CNCs varies depending on their fiber diameters as well as the synthesis method employed. The fine structures of CNCs are quite different from those of carbon microcoils, of which the diameters are on the micrometer scale (Motojima & Chen 1999). Carbon microcoils typically consist of noncrystalline amorphous fibers, while CNCs have graphitic structures. The fine structure of CNCs with fiber diameters >100 nm is typically similar to the turbostratic-type stacking of graphite. Electron energy loss spectroscopy, however, has shown that the electronic structure of CNCs is similar to that of amorphous carbon rather than highly oriented pyrolytic graphite or CNTs (Chen et al. 2003), owing to the presence of defects in the graphitic structure. We can see fine structures at the thinner regions close to the edge or by electron diffraction (Hirahara et al. 2014). Figure 3.3a and c shows high-resolution images of a CNC, depicting wavy and discontinuous lines corresponding to the graphitic structures. The layered structure can also be confirmed by the Fourier transform of the images, whereby the presence of a couple of spots indicates the periodic arrangement of the graphitic layers, and their broadening nature reflects the low degree of crystallinity. The stacking direction of the graphitic structure generally appears symmetric with respect to the fiber axis because of the cup-stack-type orientation (see schematic in Figure 3.3d). Therefore, some CNCs have interesting surfaces upon assembly of the graphene edges. The center of the fiber consisting of an array of caps of the cup-stacked, however, is not fully packed and the center appears hollow in low-magnification TEM images.

In contrast, CNCs with rather small fibers, i.e., 10 nm in diameter, often feature concentric layered structures, similarly to multiwalled carbon nanotubes (MWCNTs) (Figure 3.4). The five- or seven-membered ring can be introduced into the inner or outer ridgelines to produce a helical structure (Ihara et al. 1993). The CNCs reported by Amelinckx et al. (1994) and Zhang et al. (1994) displayed this type of concentric structure (Akagi et al. 1995), and the selective formation was reported by Xie et al. (2003). Such CNCs feature excellent mechanical properties owing to their high crystallinity, which is similar to that of CNTs.

Materials featuring similar structures to those of CNCs have been reported. For instance, some reports have coined materials such as *coiled CNTs* (Daraio et al. 2006, Wang et al. 2008, etc.). The coiled CNTs consist of an assembly of CNTs with wave features grown vertically from a substrate. The individual

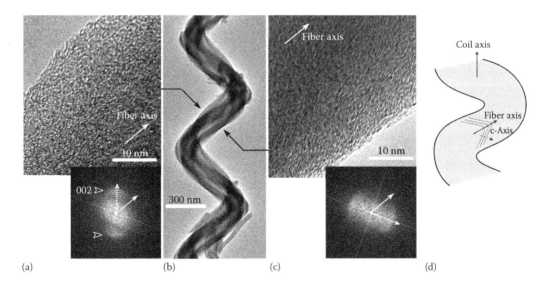

(a) (b) (c) (d)

**FIGURE 3.3** TEM images of (a) a CNC and (b, c) enlarged images with corresponding Fourier transform patterns. The presence of a couple of diffused (002) spots in the patterns indicates the stacking direction of the graphitic layers (c-Axis) with respect to the fiber axis (see schematic in d) and the defective structure of the layers in CNC.

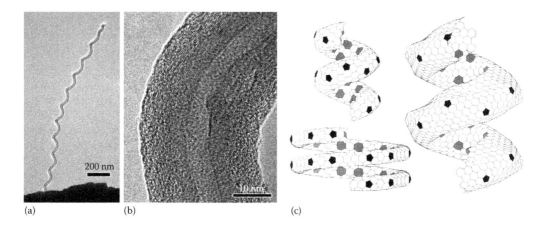

(a)                          (b)                               (c)

**FIGURE 3.4**   TEM images of (a) a concentric-type CNC and (b) a corresponding enlarged image showing the parallel orientation of the graphitic layers relative to the fiber axis, similarly to an MWCNT. (c) Structural models of concentric-type CNCs. The helical cage structures are obtained by introducing five- or seven-membered rings in the hexagonal network of carbon; such structures have the lowest cohesive energy per atom. (Reprinted with permission from Ihara, S., Itoh, S. & Kitakami, J. I., *Physical Review B*, 48, 5643–5647, 1993. Copyright 1993 by the American Physical Society.)

CNTs feature a concentric graphitic structure that resembles that of MWCNTs rather than concentric-type CNCs. Additionally, AuBuchon et al. (2004) reported CNTs with a zigzag morphology via application of an electric field during direct current (DC) arc plasma-enhanced CVD. In such a synthesis, the individual CNTs featured bends with sharp radii of ~25 nm.

## 3.3  Growth Models

The helical coiled morphology of CNCs has attracted much attention. Accordingly, growth models of CNCs have been proposed by many groups. For example, Amelinckx et al. (1994) introduced the concept of hodograph to describe the extrusion of a carbon tubule on a catalytic particle. Fonseca et al. (1996) discussed the growth of perfect tubular connections with coiled morphologies by introducing topological defects in the carbon network. Pan et al. (2002) proposed that the geometrical morphologies of CNCs are strongly related to the structures of the catalysts located at the tips, and furthermore thoroughly investigated the relationship between morphology of a catalytic particle and growth of CNCs (Li et al. 2010a). Another study proposed the contributions of anisotropic activity of the facets (Xia et al. 2008). Detailed crystallographic characterization of individual cementite ($Fe_3C$) catalytic particles by electron diffraction revealed the 3D morphology of the particles (Figure 3.5) and the diffusion of the carbon atoms from a specific facet to another facet. Consideration of these studies reveals that the anisotropic extrusion of carbon at the surface of a catalytic particle is the key to explaining the formation of the coiled morphology. More specifically, crystallographically different structures or different elemental ratios of each facet on a catalytic particle instigate anisotropic catalytic reactions. When a catalytic particle of radius $r$ has different extrusion velocities on the outer and inner sides of CNCs, given as $v_1$ and $v_2$, respectively, a coil forms with an inner diameter $R_i$, which can be calculated as $R_i = 2r/(v_1/v_2 - 1)$. If the $v_1/v_2$ ratio is independent of the catalytic particle size, $R_i$ is proportional to $r$. For example, if $v_1/v_2$ becomes larger, the coil will be rather tightly twisted, resulting in a smaller coil diameter, as illustrated in Figure 3.6. Thus, control over the fiber and coil diameters can be achieved by controlling the size and elemental ratio of the catalytic particles used for growing CNCs (Okazaki et al. 2005). This hypothesis is supported by an empirically established fact that the selective growth of a coiled structure usually requires

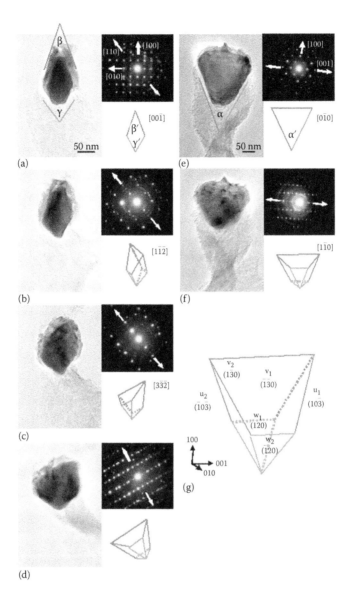

**FIGURE 3.5** (a–f) TEM images and corresponding electron diffraction patterns of an Fe$_3$C catalytic particle recorded in different crystallographic directions, revealing the 3D morphology of the particle schematically depicted in (g). (Reprinted with permission from Xia, J. H., Jiang, X., Jia, C. L. et al. 2008. "Hexahedral Nanocementites Catalyzing the Growth of Carbon Nanohelices." *Applied Physics Letters* 92 (6). Copyright 2008, American Institute of Physics.)

composite catalytic particles consisting of specific combinations of multiple elements. If each facet of the composite crystal displays different composite ratios owing to the different surface plane orientations, differences in the local deposition ratio of carbon may result. Okazaki et al. (2005) suggested that the presence of a large Sn composition gradient in a Fe–Sn catalyst results in the formation of smaller coil diameters. This was confirmed by elemental analysis of the individual catalytic particles located at the tips of CNCs using energy-dispersive X-ray spectroscopy (EDS), as shown in Figure 3.7.

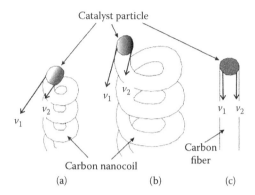

**FIGURE 3.6** Schematic of CNCs with various ratios of extrusion velocities $v_1$ and $v_2$. (a, b) As deduced, larger $v_1/v_2$ ratios produce CNCs with smaller diameters because of the high curvature of the fiber. (c) The fiber can grow straight with uniform rate. (Reprinted with permission from Okazaki, N. et al., pp. 17366–17371. Copyright 2005 American Chemical Society.)

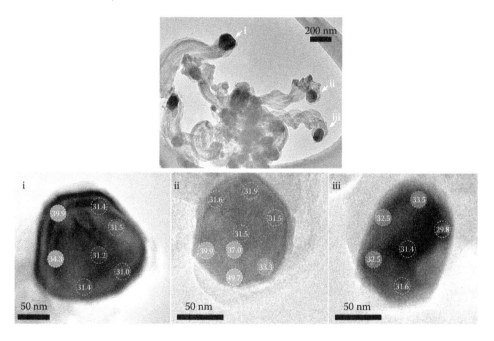

**FIGURE 3.7** Variations in elemental ratio of iron–tin composite catalytic particles at individual tips of three CNCs. Values in particles i–iii indicate the concentration of tin (%) measured at each local area pointed by dotted circle by means of EDS.

The anisotropic deposition of graphitic layers on Fe–Sn catalytic particles subjected to CVD for 5 seconds was also observed by TEM (Figure 3.8).

However, some questions remain unanswered. For instance, how is a complex facet structure formed? Li et al. (2010a) addressed this issue based on an experimental fact that CNCs often consist of multiple twisted fibers, with one catalytic particle typically located at the tip. These fibers grow from the same particle with several active facets on the particle. Another interesting question is how the helical direction (right- or left-handed) of CNCs is determined. To date, no experimental studies have demonstrated the direction-selective growth. Accordingly, macroscale composites involving CNCs feature achiral properties. Thus, achieving control over separation and chirality could offer a broader range of interesting applications.

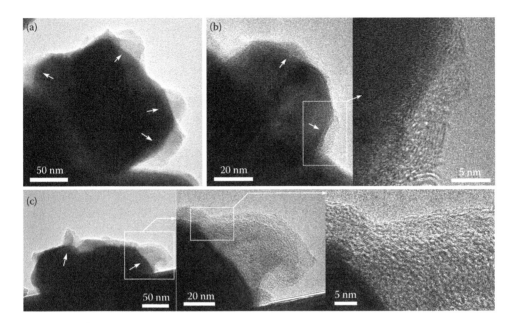

**FIGURE 3.8** (a–c) TEM images of three catalytic particles subjected to CVD for 5 s, showing the earliest stages of CNC growth. The graphitic layers partially deposit on some facets of each particle as indicated by the white arrows, and start coiling as seen in c. The graphitic layers usually stack parallel to the surface plane of the particles at the interface regions.

## 3.4 Synthesis of CNCs

### 3.4.1 Chemical Vapor Deposition

CVD is primarily employed for the synthesis of CNCs. In general, in thermal CVD of CNCs, the catalysts are placed in a furnace in an inert gas atmosphere. Upon temperature increase to the reaction temperature under inert gas flow, the carbon source gas flows to decompose and subsequently deposit carbon at the reaction sites on the catalytic particle. As an example, the procedure for the synthesis of CNCs (Okazaki et al. 2005) shown in Figure 3.1 is as follows: First, an ethanolic solution of hydrates of iron nitrate ($Fe(NO_3)_3 \cdot 9H_2O$), indium chloride ($InCl_3 \cdot 4H_2O$), and tin chloride ($SnCl_4 \cdot 5H_2O$) at an Fe:In:Sn mole ratio of 10:1:1 is prepared as the catalyst precursor solution. Then, the solution is dried on a silica ($SiO_2$) or alumina ($Al_2O_3$) substrate at 100°C, and oxidized at 800°C to produce catalytic Fe–In–Sn–O particles. The catalytic particles are then heated to 700°C in helium (He) atmosphere at a flow rate of 260 standard cm³/minute (sccm) in a quartz tube chamber. A mixed gas of acetylene ($C_2H_2$) and He is then supplied to the chamber at respective flow rates of 30 and 230 sccm for 15 min to grow the CNCs. After the atmosphere gas is replaced with pure He, the chamber is cooled to room temperature. Thus, we can obtain CNCs of several tens of micrometers. Many other procedures have also been studied employing different conditions; however, these will not be discussed considering the breadth of the topic. For reference, such methods and associated results have been summarized by Shaikjee and Coville (2012). In the following sections, some general and important information regarding the use of CVD for the synthesis of CNCs is reported.

### 3.4.2 Choice of Catalysts for Growing CNCs

Improvements in the synthesis techniques of CNCs rely on two essential aspects: mass production and structure control. As CNCs are widely employed as composites, the development of large-scale synthesis processes with high yields, short reaction time requirements, and low cost is essential. Furthermore, the application of CNCs is often based on the properties of the CNC components; hence, fabricating uniform or target structures of CNCs is important. The key to achieving both purposes is the choice of catalyst. As mentioned above, not only its elemental ratio and size but also the gradation of concentration is related to the structure of coils.

To optimize the synthetic procedure of CNCs, most studies to date have reported the use of composite catalysts consisting of several elements. Generally, each element in the catalytic particle can be categorized by the role played in either of these processes: (i) deposition of carbon atoms to form fibrous features, (ii) formation of a coiled structure, or (iii) stabilization and increased rate of the catalytic reaction. Most of the literature reported that composite catalysts are composed of a combination of elements that are suited for the different processes listed above. Iron, nickel, cobalt, and copper belong to type i. The first three elements and their alloy have been generally employed as catalysts for growing CNTs, whereas copper has been typically used for growing graphene sheets. Particularly, iron has been the most frequently used in past studies. Additionally, some papers have reported the growth of CNCs using only type i catalysts (Kong & Zhang 2007, Xia et al. 2008). The formation of CNCs under such conditions was attributed to different facets on catalytic particles as mentioned in Section 3.2. Alternatively, the growth of CNCs can proceed upon addition of another element categorized as type ii to an existing catalyst system. For this purpose, tin is the most commonly employed element in combination with iron. In contrast, the use of tin does not result in the formation of fibrous materials. In an early paper (Zhang et al. 2000), iron was supported as a thin film layer on an In–Sn–O substrate, and another study reported the growth of CNCs on polycrystalline $BaSrTiO_3$ using only tin (Sun et al. 2013). Conversely, indium is believed to play a different role from that of iron and tin. More specifically, it is categorized under type iii. Elements categorized under this group are not absolutely necessary for the growth of CNCs. However, the addition of indium considerably improves the yield of CNCs (Zhang et al. 2000, Okazaki et al. 2005). It is believed that indium works as a catalyst toward optimizing the formation of alloy particles of iron and tin rather than toward growing CNCs. The addition of this element may be essential for the development of large-scale synthesis processes. However, the application of indium, which is a rare metal, incurs high cost for large-scale synthesis processes. Accordingly, S. Kugimiya (personal communication, 2009) investigated relatively inexpensive elements as alternative or more effective elements to indium. The author conducted CVD using an Fe–In–Sn–X or Fe–In–X catalyst system incorporating an additional element X (Co, Na, P), and compared the relative mass of carbon contained in the CNC products. Figure 3.9a shows the carbon product-to-catalyst mass ratio obtained when an Fe–In–Sn–X catalyst system was employed for the growth of CNCs. As deduced from Figure 3.9a, the addition of Co, Na, P, Mg, Ce, Mn, Ni, and Mn to the Fe–In–Sn catalyst system effectively increased the relative content of carbon. It is, however, noted that the addition of these elements also influenced the morphologies of the resulting CNCs. Figure 3.9b through f shows the scanning electron microscopy (SEM) images of the products generated by five different catalyst systems. CNCs produced using Fe–In–Sn–Mg or Fe–In–Sn–P catalyst had similar features to those produced in the presence of Fe–In–Sn catalyst. In contrast, in the other catalyst systems, both the coil diameter and the fiber diameter of the products decreased. More specifically, products generated using Fe–In–Sn–Na and Fe–In–Sn–Co systems displayed thin and twisted structures rather than coiled structures. Such differences were believed to be because of the different roles of each element. Magnesium is believed to behave in a similar fashion as indium. It was observed that the Fe–Mg–Sn system exerted catalytic effects toward the growth of CNCs, similar to the Fe–In–Sn system (Kugimiya, personal communication, 2009). This system is a cost-effective alternative to that containing indium. Using this alternative system is expected to reduce the cost of the catalyst by more than 90% owing to the use of the corresponding metal chloride precursors in the wetting process. It has been demonstrated that the Fe–Co–Mg–Sn system shows good catalytic performance toward the growth of CNCs, achieving extremely high purity (Hirahara & Nakayama 2013).

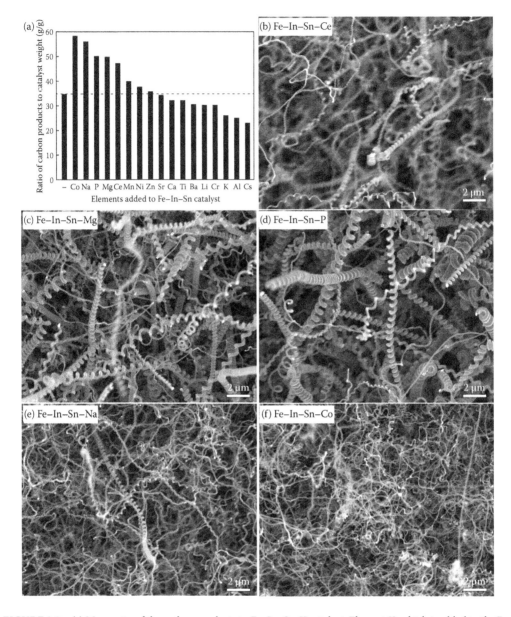

**FIGURE 3.9** (a) Mass ratio of the carbon products to Fe–In–Sn–X catalyst. Element X, which is added to the Fe–In–Sn catalyst, is indicated in the horizontal axis. The initial Fe:In:Sn:X ratio was 10:1:1:1; element X is stated under each bar, and "–" denotes that no additional element was added to the existing Fe–In–Sn system. (b–f) SEM images of the products formed upon addition of different elements X: (b) Ce, (c) Mg, (d) P, (e) Na, and (f) Co. (Courtesy of S. Kugimiya.)

## 3.4.3 Ripening of Catalyst

Another important issue to consider is the provision of uniform catalytic particles, with the required size at optimal concentrations at the start of the reaction. The introduced catalyst changes accordingly to generate particles appropriate for the growth of CNCs during temperature rise under flowing inert gas or mixture in the presence of carbon source gas by grain growth, decomposition of heat-sensitive elements, reduction, alloying, or chemical reaction with substrates. In other words, "ripening" of the catalysts is required prior to the start of the CVD reaction. Previous literature studies reported the use of iron films deposited on an indium–tin–oxide (ITO) coated substrate for the selective growth of CNCs (Zhang et al. 2000), as well as codeposited thin films of iron and tin (Kanada et al. 2008). Upon heating pretreatment, the thin film is converted into nanoparticle aggregates while undergoing concurrent composition changes. In the case of iron film deposited on ITO, X-ray diffraction analysis indicated that indium and iron existed as an oxide independently when the furnace was heated to the reaction temperature of 700°C in inert gas atmosphere; however, no alloys were formed (Zhang et al. 2000). The oxide particles eventually underwent reduction, and alloying occurred only upon introduction of carbon source gas. In contrast, the wetting method usually involves solutions of metal salts as the starting materials for the preparation of the catalysts (Pan et al. 2002, Okazaki et al. 2005, Li et al. 2010b, etc.). This process is more convenient than physical methods because vacuum equipment is not required. Typically, the metal salt solutions are spread, dried, and usually oxidized on the substrate to prepare the particles, which are usually oxides comprising a mixture of single elements and inactive as the catalyst for the growth of CNCs. After introduction of the carbon source gas, such as acetylene, for 1 s, simultaneous formation of Fe–Sn particles alloy and deposition of a carbon layer on the particles occur, as shown in Figure 3.6. In both the physical and wetting methods, the optimal catalytic particles are obtained only upon introduction of the carbon source gas. In recent studies, optimal conditions regarding the elemental ratio, thickness, or total mass have been empirically established. However, the findings vary because of the different degree of structural changes obtained that are dependent on the diverse CVD conditions employed, such as reaction temperature, ratio of the total weight of catalyst to the size of the furnace, and gas flow rate. Particularly, if tin is used in the preparation of the catalysts, appropriate steps must be considered to ensure efficient provision of tin in preparing uniform and optimal particles during the CVD process because of its low melting point. It was confirmed that the external supply of tin acts as the vapor, which means that tin easily decreases from initially installed catalyst under high-temperature conditions. This is an important parameter to consider in the preparation of stable and uniform catalytic particles during the ripening process. Regarding the Fe–Sn system, using an external supply of tin is an effective means to stabilize catalytic particles (Hirahara & Nakayama 2013), resulting in a considerable reduction of by-product layers formation and more than 400% improvement in CVD efficiency (Figure 3.10).

**FIGURE 3.10** SEM images of CVD products obtained using Fe–Mg–Sn–Co catalyst on (a) noncoated and (b) $SnO_2$-coated $Si_3N_4$ substrates. The result indicates that control of the ripening process in catalytic particles upon introduction of $SnO_2$ coating considerably reduces formation of by-products and efficiency improves CNC growth.

### 3.4.4 Controlled Synthesis

Controlling the morphology of individual CNCs is an interesting means to develop applications based on structure-dependent characteristics. Various aspects of CVD conditions have been investigated to obtain specific varieties of CNC structures as defined by the fiber diameter, coil diameter, pitch, and length.

Control over the fiber and/or coil diameters can be achieved by different means. In particular, much effort has been devoted to developing catalytic particles. The fiber diameter can be controlled by selecting the size of the catalytic particle. One of the most common ways for controlling particle size is to control the thickness of the catalytic layers initially deposited on the substrate. Thicker catalytic layers result in CNC products with thicker fibers. For example, catalytic particles with two different diameter ranges of 50–200 and 50–450 nm produced CNCs with corresponding line diameter ranges of 50–150 and 50–200 nm as shown in Figure 3.11 (Hokushin et al. 2007). It should be noted that the initial mixture ratio should be changed according to the thickness of the catalysts, in order to stabilize the alloying environment of catalytic particles to achieve appropriate elemental ratios. Control over the catalytic particle size can be achieved by preannealing prior to CVD, too. Higher annealing temperatures up to 900°C instigate further catalytic particle growth, resulting in CNCs comprising thicker fibers. In the case of wetting methods that employ solutions of metal salts, controlling the solution temperature during coprecipitation is essential. By controlling the temperature during stirring for dissolution and stationary cultivation, the domain size of agglomerates deposited in the solution can be fine-controlled.

The length of CNCs can be controlled by the reaction time employed during CVD. For example, under a CVD condition introduced in Section 3.4.1, the length of CNCs is proportional to the feeding time of the carbon source gas, achieving a growth rate of 300 μm/h. The growth of CNCs reaches saturation with increasing reaction times to several tens of minutes, because the individual catalytic particles gradually deactivate owing to changes in elemental ratio as a result of coverage of the catalytic particles with amorphous carbon layers. Increasing the lifetime of catalytic particles is possible by supplying water or hydrogen gas as reported in the synthesis of millimeter-long CNTs (Chakrabarti et al. 2006). Conversely, the growth rate can be controlled by the flow rate of gas. Generally, growth rate increases as the total flow rate increases; accordingly, longer CNCs can be obtained by using shorter reaction times. However, excess supply of carbon source gas with respect to the content of catalyst causes pyrolysis of the excess gas solely, thereby considerably reducing the purity of the products by generating by-products such as amorphous carbon. Accordingly, a relative carbon source gas concentration of 10%–30% is considered as the optimum concentration range. Furthermore, changes in the flow rate influence the morphologies

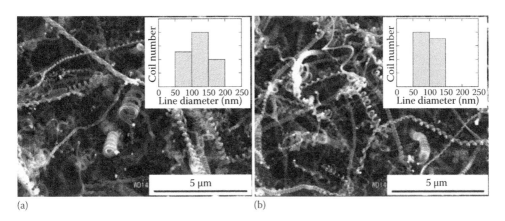

(a)                    (b)

**FIGURE 3.11** SEM images of CNCs synthesized using catalysts prepared from metal carboxylate solutions based on Fe, In, and Sn at different Fe concentrations of (a) 0.3 and (b) 0.15 M. The insets show distributions in fiber diameters. (From Hokushin, S. et al., *Japanese Journal of Applied Physics*, 46, 23 A, 2007.)

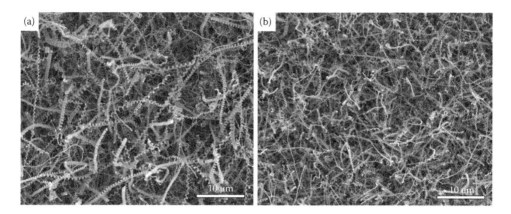

**FIGURE 3.12** SEM images of CNCs synthesized using Fe–In–Mg–Sn catalyst in a mixed gas of nitrogen (580 sccm) and acetylene at different concentrations of (a) 4% and (b) 20%; pyrolysis of acetylene was performed at 700°C for 30 min. (Courtesy of S. Kugimiya.)

of CNCs. The coil and fiber diameters can be changed by tuning either the total flow rate or the concentration of the mixture gas. SEM images in Figure 3.12 show that both the coil and fiber diameters were changed by acetylene concentrations in a mixed gas with nitrogen. In another study, it was observed that the coil diameters decreased with increasing concentrations of acetylene (Xu et al. 2005). Changes in the coil diameter and pitches may indicate variations in the growth rates in the two growth directions along the fiber axis $v_1$ and rotating axis $v_2$ as mentioned in Section 3.3 and the $v_1/v_2$ ratio.

## 3.5 Basic Properties of CNCs and Potential Applications

The properties of CNCs originate from their distinct structures as well as their graphitic nature. Some properties are similar to those of CNTs, but mechanical-spring characteristic, electromagnetic absorbing characteristics, etc., are unique to CNCs featuring coiled structures. In this section, mechanical and electrical properties and dielectrophoretic characteristic are briefly explained, and then potential applications of CNCs expected in various fields are introduced. Regarding electromagnetic wave absorbing characteristic, it is an important issue for composite materials including CNCs, and will be described in the next section.

### 3.5.1 Mechanical Properties

The coiled spring geometry is the most distinctive characteristic of CNCs. Generally, mechanical springs involve the following processes under operation: accumulation and release of energy by deformation and restoration; instigation of deflection upon force application; impact absorption; and absorption/control of mechanical vibration. To employ CNCs as micrometer- or nanometer-sized coiled springs, it is essential to understand their fundamental mechanical properties.

Because individual CNCs have various structures with different pitches and diameters, using manipulators equipped with microscopes is a powerful tool for experimentally investigating the mechanical properties of individual CNCs. Mechanical spring characteristics of individual CNCs have been studied using such a system in SEM (Hayashida et al. 2002). As observed, a single CNC could expand by 200%. The spring constants of the individual CNCs were estimated by measuring the load divided by elongation, and ranged from 0.01 to 0.6 N/m. A related study was reported by Chen et al. (2003). In their case, both ends of a CNC was fixed at the tips of two cantilever probes of an atomic force microscope, and loaded in tension mode (Figure 3.13a). After reaching maximum elongation of 33%, the CNC relaxed to its original length upon slow release of the load. The spring constant of the CNCs was determined

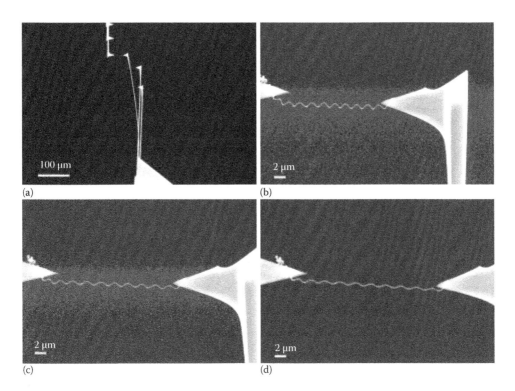

**FIGURE 3.13** A CNC subjected to tensile tests during SEM analysis. (a) A CNC is bridged between two canti-levered probes of an atomic force microscope. (b) The CNC relaxes to its natural length. The CNC is elongated by (c) 20% and (d) 33%. (Reprinted with permission from Chen, X. et al., pp. 1299–1304. Copyright 2003 American Chemical Society.)

as 0.12 N/m at low elongation, and the spring constant increased with increasing relative elongation. Compressive tests using atomic force microscopy also showed nonlinear response when the CNC was subjected to a compressive force using a cantilever probe (Poggi et al. 2004), and a stiffness of 0.7 N/m was determined from the steep linear increase in the force curve at low deflection.

The detailed mechanical properties of individual CNCs are discussed further. The Young and shear moduli of individual CNCs can be calculated using a continuum approximation. The shear modulus can be calculated from the measured spring constant value. The Young modulus can be estimated by the fundamental relationship for continuum materials (Chen et al. 2003). Young moduli of spring constants have been experimentally determined to be ranging from 0.04 to 0.13 TPa (Hayashida et al. 2002). The shear modulus calculated from the measured spring constant by Chen et al. (2003) ranged from 2.1 to 2.5 GPa. These values were considerably lower than those of CNTs, related to the fine structures of CNCs. One factor involves the low crystallinity of CNCs. The wavy and discontinuous graphitic structures result in reduced mechanical strength. Another factor involves the geometry of the orientation of the graphitic structure. CNTs typically consist of rolled-up graphene layers, so that the in-plane mechanical strength of the graphene sheets is directly reflected. In contrast, the stacking direction of most CNCs is usually inclined with respect to the fiber axis because of the cup-stack-type orientation, and each layer is not closed. Such an orientation causes interlayer slipping in the graphitic layer. To obtain CNCs with excellent mechanical properties, it is essential to control the balance between these two factors because the degree of interlayer slipping increases as crystallinity is improved. Such an issue has been discussed by examining the relationship between the shear modulus and the crystallinity of individual CNCs (Hirahara et al. 2014). The crystallinity of the individual CNCs was improved upon annealing treatment under vacuum, and higher annealing temperatures led to the formation of graphitic layers with a more

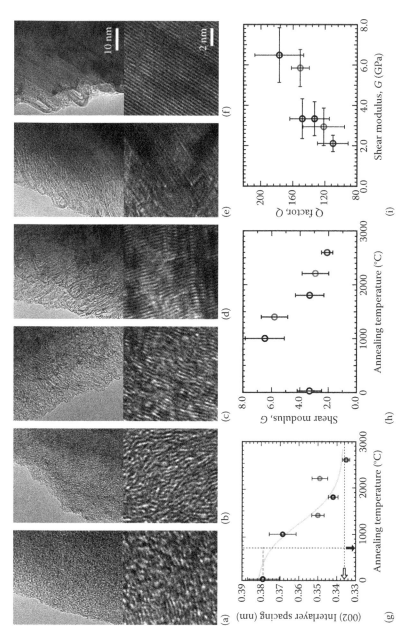

**FIGURE 3.14** Low- and high-magnification TEM images of graphitic layers at edge regions of (a) pristine CNCs and CNCs annealed at (b) 1000°C, (c) 1400°C, (d) 1800°C, (e) 2200°C, and (f) 2600°C. (g) Dependence of the average (002) interlayer spacing in CNCs measured by electron diffraction on annealing temperature. (h) Dependence of the shear modulus on annealing temperature. (i) Correlation between shear modulus and Q factor. (From Hirahara, K. et al., *Materials Science and Engineering A*, 595, 205–212, 2014.)

continuous and straight morphology (Figure 3.14). Electron diffraction of the annealed CNCs showed that the crystallinity was comparable to that of graphite upon annealing at 2600°C (Figure 3.14g). In contrast, the shear modulus can be estimated from the resonant frequency of individual CNCs using the equation on the basis of classical beam theory for oscillation. The resonant oscillation can be observed by SEM upon application of a high-frequency electric field to the cantilevered CNCs. The experimental results revealed the nonlinear correlation between crystallinity and shear modulus. The average value of $G$ improved from 2.6 to 6.5 GPa by annealing at 1000°C, but reduced with further annealing at higher temperatures (Figure 3.14h). Additionally, the Q factor, which indicates the degree of energy dissipation in the mechanical vibration of the cantilevered beams, strongly correlated to the shear modulus (Figure 3.14i). This finding indicates that the mechanical oscillation property of CNCs is dominated by two types of energy dissipation as a result of defects in structures with low crystallinities and interlayer slipping in structures with high crystallinities.

## 3.5.2 Electrical Conductivity

CNCs are ohmic materials when a near-zero bias is applied. Figure 3.15a shows a typical $I$–$V$ curve for a CNC sandwiched between two electrodes and a bias voltage from −3 to 3 V applied in a vacuum of less than $10^{-5}$ Torr at room temperature (Ma et al. 2013). The curve shows an almost-linear correlation for which the current increases with increasing bias voltage. The correlation between resistance and power is shown in Figure 3.15b (Ma et al. 2013). Nonlinearity in the $I$–$V$ curve also appears as a consequence of insufficient contact resistance caused by contamination initially deposited at the contact points. Conductivity values for individual CNCs are in the range of 120–180 S/cm (Hayashida et al. 2002). These values are lower than those of CNTs, but larger than those of microcoils (100 S/cm) consisting of amorphous carbon; the differences are the result of crystallinity. The conductivity also depends on the ambient temperature during measurement (Ma et al. 2014). The temperature dependence of the resistance measured by the four-probe method is shown in Figure 3.15c and d. The former exhibits a linear correlation between $\ln(R)$ and $1/T$, where $R$ and $T$ are the resistance of the CNC and the temperature of a probe used as the electrode, respectively. Despite the highly disordered structure of CNCs, the linearity can be explained by the nearest-neighbor hopping mechanism; the slope of the graph corresponds to the activation energy for electron hopping between the nearest-neighbor localized states.

The crystalline dependence of the resistance of annealed CNCs has also been studied (Ma et al. 2014). As stated above, crystallinity improves by increasing the annealing temperature (Hirahara et al. 2014). Annealing temperature-dependent resistivity measured at room temperature is plotted in Figure 3.15e. The resistivity significantly improves from 1.9 to 7.7 by annealing at 1073 K, which is 100 K higher than the synthesizing temperature, despite the diffraction data showing that the CNCs still have a defective and discontinuous graphitic structure, as seen in Figure 3.14b. Further annealing at more than 2473 K significantly decreases the resistivity to 3.5. This might be due to the improvement not only in the crystallinity of the graphitic layers but also in the interlayer connections at the edges (see TEM images in Figure 3.14), resulting in the formation of additional conductive channels. Indeed, an annealed CNC shows lower resistivity than carbon nanofibers but still higher than the relatively perfect structures of MWCNTs.

## 3.5.3 Assembly of CNCs by Dielectrophoresis

The assembly of CNCs is an important issue to consider for designing applications involving high-performance devices especially as novel mechanical, electromechanical, or electromagnetic elements. As demonstrated, dielectrophoresis is an effective means to align CNCs (Tomokane et al. 2007). When an electric field was applied to a CNC isopropanol-based suspension, the CNCs moved toward the cathode. In the presence of an AC electric field at frequencies higher than 1 kHz, the CNCs aligned, even after drying. The driving force instigating such a phenomenon was ascribed to the dipole moment

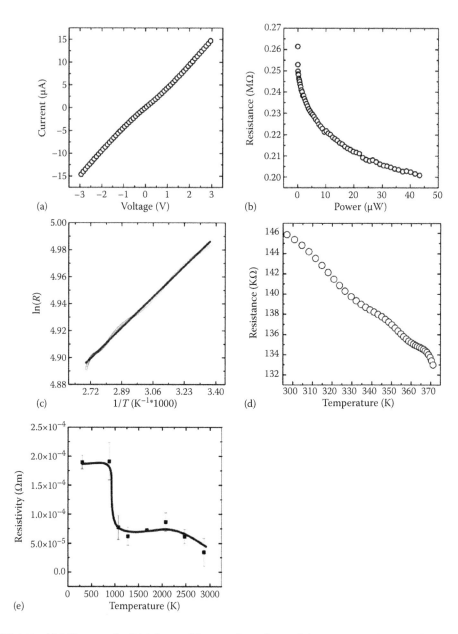

**FIGURE 3.15** (a) *I–V* curve of a CNC device; (b) power dependence of the resistance of a CNC. (Ma, H. et al., *Nanoscale* 5 (3): 1153–1158, 2013. Reproduced by permission of The Royal Society of Chemistry.) (c) ln(*R*) versus 1/*T* of a CNC. (d) Temperature dependence of the resistance; (e) resistance at room temperature of annealed CNCs versus annealing temperature. (From Ma, H. et al., *Carbon*, 73, 71–77, 2014.)

induced by the AC field. Figure 3.16 shows the AC electric field dependence of the CNC alignment at 100 kHz. The SEM images indicate that the orientation degree of the CNCs improved, bridging the gap between the electrodes, with increasing electric fields. Assembly operation technique of CNCs based on dielectrophoresis is also applicable to attach a single CNC to a tip of a probe, which enables single CNC manipulation (Wu et al. 2012).

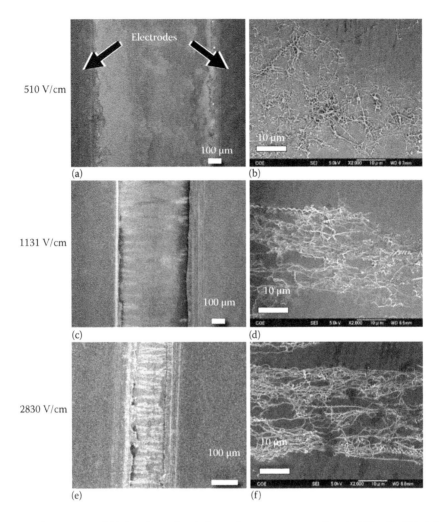

**FIGURE 3.16** SEM images of CNCs subjected to dielectrophoresis at varying electric fields of (a, b) 510, (c, d) 1131, and (e, f) 2830 V/cm, and frequency of 100 kHz. b, d, and f show enlarged images of the gaps between the electrodes in a, c, and e, respectively. (From Tomokane, R. et al., *Japanese Journal of Applied Physics*, 2007.)

## 3.5.4 Structure-Related Potential Applications of CNCs

There are various expected applications of CNCs in many research fields. Two structural types are involved in applications: isolated CNCs and bulk CNCs. Even for the latter, characteristics of individual CNCs are used, so that structural and alignment controls are crucial in their applications. In the following, some of the more attractive applications are listed.

The coiling feature of individual CNCs offers a unique potential in the nanoelectromechanical systems/microelectromechanical systems field as nanomechanical components such as nanomechanical springs (Chen et al. 2003) and self-sensing mechanical resonators (Volodin et al. 2004). Layers of aligned CNCs could provide shock protection; the impact response and nonlinear elastic behavior of coiled CNT forests have already been studied (Daraio et al. 2006, Coluci et al. 2008). Water flow control using a coiled nanotube switch actuated through extension by tensile loading has also been proposed in the field of molecular dynamics (Ju et al. 2013).

Exhibiting high emission current densities and excellent stability (Pan et al. 2001), the field emission property of CNCs makes them suitable as emitters embedded in illumination devices such as displays. CNCs fabricated using Fe–In–Sn catalysts have a unique surface structure that exposes numerous graphene edges. Therefore, they exhibit significantly better field emissions than CNTs because the individual CNCs emit not only from their tips but also from their top surfaces where the electric field is more concentrated (Hokushin et al. 2007).

Energy applications of bulk-quantity CNCs are also promising. Several energy-related applications, e.g., catalyst support for fuel cells (Sevilla et al. 2009, Celerrio et al. 2014), hydrogen storage (Reddy et al. 2011), and electrodes for supercapacitors (Reddy et al. 2011; Barranco et al. 2012), have been developed. In particular, their large surface areas make CNCs suitable as supporting media to catalytic nanoparticles of metals such as platinum (Celerrio et al. 2014). Because numerous dangling bonds can trap individual nanoparticles, they can prevent migration and, hence, promote excess grain growth during electrochemical reactions.

Resin composites that include CNCs as fillers have great practical potential with several attractive features that lead to vibration damping materials, shock absorbers, and electromagnetic wave absorbers. As CNCs are conductive fillers, flexible transparent conductive films can also be made. CNC fillers also harden the resin composites. Because the solubility is good, the composites have good permeability to light. Details of CNC–resin composites are described in the next section.

# 3.6  CNC Composites

Resin-composite-filling CNCs are attractive materials that by exploiting their structural characteristics expand the application of CNCs. CNCs have practical value as nanocarbon fillers because they appear well dispersed in various types of resin matrices. CNC–resin composites show excellent characteristics as electromagnetic wave absorbers over a very wide range of frequencies. In addition, using CNCs as fillers provides a great degree of freedom in the functional design of composites of such electromagnetic wave absorbers because not only the weight ratio of fillers but also the CNC structural parameters such as pitch and coil diameter can be design specific. In the following section, typical fabrication methods and basic properties of CNC composites are described.

## 3.6.1  Fabrication of CNC Composites

Solution kneading is the usual method of fabrication of CNC–resin composites. In this method, both resin and CNCs are dispersed in a resin-soluble solvent. The solution is then dried to evaporate the solvent. There are two main reasons why this method is employed. One is that CNC shows good wettability compared with other graphitic nanocarbon materials such as CNTs. Typical CNCs have numerous dangling bonds present on their surfaces because the surface usually consists of arrays of graphene edges (Section 3.1). This surface structure promotes wettability unlike perfect graphitic layers. The dangling bonds on CNC surfaces are terminated by carboxyl through air exposure; hence, various surface functionalizations are easily attainable. Therefore, CNCs dissolve well in various solvents such as alcohol without needing any surfactants. The other reason to employ the solution method is that melt-kneading, which is a general method used in the fabrication of resin composites, is unsuited to CNC composites. Melt-kneading involves loading high shear forces on fillers that fracture CNCs, particularly in the presence of defects. Moreover, in contrast to straight CNTs, CNCs cannot form bundles because of their coiled geometries. This means that CNCs are difficult to aggregate compared with CNTs; this dispersive feature adds another feature to solvated CNCs.

Here, the typical procedure used in fabricating CNC–resin composites developed for electromagnetic absorbers is introduced. Styrene–ethylene–butadiene–styrene (SEBS) is used as a resin matrix. SEBS is a thermoplastic styrene–ethylene–butylene–styrene copolymer with 0.94 g/cm³ specific weight, and is widely used to make products such as toothbrushes because it is colorless, odorless, and water insoluble, as well as safe for humans. The CNC composite is produced by solution kneading in the following manner. First, CNCs are sonicated with chloroform at rather low power to prevent fracturing the CNCs. SEBS is also solvated with chloroform by stirring at 80°C. These two solutions are then mixed and kneaded by stirring, where the viscosity of the mixture is adjusted by evaporating chloroform so as to have a predetermined viscosity. The CNC/SEBS solution is cast into a mold, and the chloroform is fully evaporated in air. After the resin hardens, it is taken out of the mold. In the platelike resin produced this way, CNCs usually align laterally, as illustrated in Figure 3.17. From measurements of the degree of the orientation by cross-sectional observations, the angular deviation from the direction parallel to the plate surface was within 10°. This orientability of CNCs enables electromagnetic wave effects to be exploited because the area of projection for CNCs is maximal with respect to electromagnetic waves incident on the plate surface. In addition, injection molding is also possible by melt kneading of the pellet-like product, although care must be taken to prevent excessive shear loading. In this case, the CNCs align along the flow of the resin, as seen in injection molding of fibrous fillers.

A SEM cross-sectional image of the CNC–resin composite fabricated by solution kneading is shown in Figure 3.18. Here the surface was prepared by oxygen-plasma etching as the resin is then etched preferentially because of differences in the etching ratio. The concentration of CNCs in the composite in Figure 3.18 is 5 wt% and CNCs are individually embedded in the resin without aggregating. This image also demonstrates the good wettability of CNCs with resin; the resin soaks well into the interior of the coil spring, and voids do not form at the CNC–resin interface.

Furthermore, the aligning of CNCs is essential for producing sophisticated composites. The alignment of CNCs by dielectrophoresis exploits their high aspect ratio (see Section 3.5.3). This method is applicable in producing oriented composites of CNC–resin by employing a UV cure resin as the matrix with a viscosity of about 0.01 Pa s (Fujiyama et al. 2008). This resin can easily solvate CNCs by sonication.

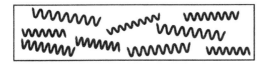

**FIGURE 3.17** Schematic of horizontally oriented CNCs in a composite plate.

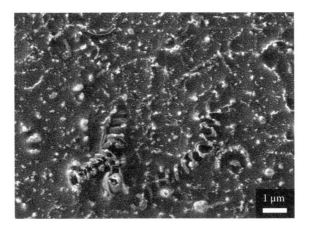

**FIGURE 3.18** High dispersion of CNCs in a CNC–resin composite. (Reproduced from Akita, S., "Design and Development Technology of Radio Wave Absorbant Body Using Carbon Nanocoils." In J. Shibata (ed.), *Kneading/Compound Technologies and Controls of Dispersion/Interface of Composite Materials* (in Japanese), pp. 558–565, Technical Information Institute Co., Ltd., 2013.)

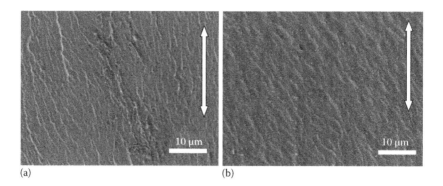

(a)                                                                                     (b)

**FIGURE 3.19**  Cross-sectional images of cleaved CNC composites fabricated (a) with and (b) without applying an AC electric field. The arrow in a shows the direction of electric field. (From *Japanese Journal of Applied Physics*, 2008. With permission.)

The mixture solution is cast into a mold placed between two electrodes separated by 0.5 mm, and by applying an AC electric field; it is then hardened by light irradiation. A cross-sectional image of the composite produced in this way is shown in Figure 3.19. CNCs are all aligned parallel with the electric field; no voids are present. This technique is also suitable in fabricating urethane-based resin composites. In addition, the aligned composites show 23 times better transmittance in the visible region compared with composites produced without applying an electric field.

### 3.6.2 Electroresistivity of CNC Composites

Solvated with various resins, CNCs prove useful as conductive fillers. The surface resistivities of various CNC-resin composites are listed in Table 3.1. For CNC composites, the resistivity significantly decreases to $10^4$ $\Omega/\Upsilon$ for CNC concentrations between 3 and 4 wt%, indicating a gradual increase in percolation. Considering that the effective aspect ratios of CNTs and vapor-grown carbon fibers (VGCFs) are higher than that of CNCs, percolation limits for composites that include them are expected to be lower than those for CNC composites. However, the concentrations of CNTs and VGCFs show a decrease in resistivity at around 5 wt%. This fact reflects the difficulty of isolating monoparticulate dispersions of CNT and VGCF and aggregation in resin, which emphasizes the advantage of CNCs as nanocarbon fillers from the viewpoint of dispersion.

**TABLE 3.1**  Comparison of Volume Resistivities of Nanocarbon Composites

| Weight Ratio of Fillers in the Composites | | 1 wt% | 2 wt% | 3 wt% | 4 wt% | 5 wt% | 20 wt% |
|---|---|---|---|---|---|---|---|
| Resistivity $(\Omega/\Upsilon)$ | CNC | $\gg \sim 10^{13}$ | | $10^5$–$10^{10}$ | | $10^3$–$10^4$ | |
| | CNT | $\gg \sim 10^{13}$ | | | $10^7$–$10^9$ | $10^5$–$10^7$ | |
| | VGCF | | $\gg \sim 10^{13}$ | | $10^{11}$–$10^{13}$ | $10^4$–$10^6$ | |
| | CB | | | | | | $10^5$–$10^7$ |

*Source:*  Akita, S., "Design and Development Technology of Radio Wave Absorbant Body Using Carbon Nanocoils." In J. Shibata (ed.), *Kneading/Compound Technologies and Controls of Dispersion/Interface of Composite Materials* (in Japanese), pp. 558–565, Tokyo: Technical Information Institute Co., Ltd., 2013.

*Note:*  The CNCs used for producing the composite have an average length of 20 μm, an average fiber diameter of 150 nm, an average coil diameter of 440 nm, and an average coil pitch of 560 nm. Those of other composites including commercial CNTs fabricated by CVD, vapor-grown carbon fiber (VGCF), and carbon black are also listed for comparison; these composites were produced by the same method as the CNC composite, specifically solution kneading.

### 3.6.3 Application as Electromagnetic Wave Absorbers

In recent developments of electromagnetic wave absorbers, there are three main types of absorbent materials available: carbon system, ferrite system, and resistive film system. Their loss mechanisms depend on the individual dielectric, magnetic, and conductive characteristics. In regard to carbon systems, electromagnetic absorbers have been developed using mainly carbon black with an amorphous structure. Nevertheless, CNCs are expected to become an effective alternative because of their unique morphologies.

The frequency-dependent complex permittivity for the X bands of CNC–SEBS composites evaluated by the coaxial line method is shown in Figure 3.20 (Akita et al. 2009). It shows frequency dispersion, indicating a decrease in both the real and imaginary parts of the complex permittivity with increasing frequency. Here the composites are not magnetic, because the real and imaginary parts of the complex permeability are almost 1 and 0, respectively. The reflection absorption characteristics calculated from the complex permittivity become large by 40 dB for CNC concentration ratios above 4 wt%. Such large absorbing characteristics are unique to CNC composites; other composites including CNTs, VGCFs, and carbon black exhibit absorption characteristics less than 20 dB. In addition, the concentration value is consistent with values showing a decrease in surface resistivity. The variation in complex permittivity for composite specimens with more than 3 wt% concentration is also consistent with the expected value from percolation theory for a bar-like specimen (Gefen et al. 1983, Lagarkov et al. 1996).

Furthermore, there is another interesting absorbing characteristic. The absorbing frequency of the CNC composite is tunable over its wide X band only by changing its thickness, as seen in Figure 3.21 (Akita et al. 2009). For composites including general fillers, it is difficult to obtain a resonance in such a wide frequency range by changing thickness. This characteristic affords CNC composites an advantage by which design flexibility increases, for example, for multilayered absorbers, and highly functionalized electromagnetic wave absorbers may be realized using an easy procedure. In addition, reflectance losses for the VW bands measured by free-space method are shown in Figure 3.22 (Akita et al. 2009). The loss

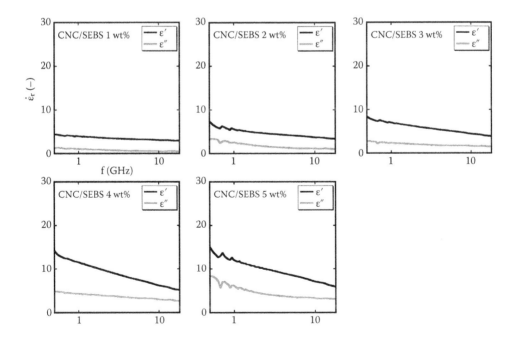

**FIGURE 3.20** Frequency dependence of the complex permittivity of the X bands for CNC–SEBS composites with different CNC concentrations, evaluated by the coaxial line method. (Courtesy of S. Akita.)

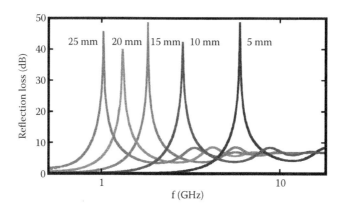

**FIGURE 3.21** Frequency-selective reflection losses in CNC composites with different composite thicknesses. (Courtesy of S. Akita.)

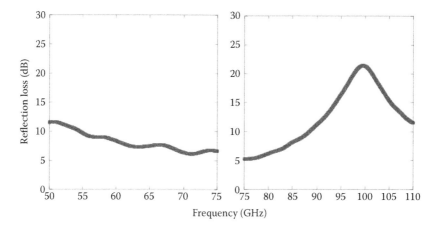

**FIGURE 3.22** Example of reflection loss at V to W bands measured using the free-space method. (Courtesy of S. Akita.)

characteristic of 5 wt% CNC composites, which exhibit the best absorbing characteristics, with 1.6 mm thickness exhibits absorptions of more than 20 dB, even around the high 100 GHz frequency region. Combined with X band measurements using the coaxial line method, the resonance seen in this region is considered to be the second. Accordingly, it was found that 5 wt% CNC composites are suitable materials as electromagnetic wave absorbers not only for the X band but also for these V and W bands. Simply by selecting the appropriate thickness, the CNC composite is also an acceptable frequency-selective absorber over the extremely wide frequency range 0.5100 GHz. In addition, the absorbing characteristics of CNCs are tunable by selecting appropriate values for coil diameter and coil pitch. This heightens CNC's already great potential as an electromagnetic wave absorbent.

# Acknowledgments

Several studies described in this chapter were supported by the Osaka Prefecture Collaboration of Regional Entities for the Advancement of Technological Excellence, Japan Science and Technology Agency (JST), Japan. The author thanks Professor S. Akita of the Osaka Prefecture University for fruitful advice especially in writing Section 3.6.

# References

Akagi, K., Tamura, R., Tsukada, M. et al., "Electronic structure of helically coiled cage of graphitic carbon," *Phys. Rev. Lett.* 74 (1995): 2307–2310. doi:10.1103/PhysRevLett.74.2307.

Akita, S., "Design and development technology of radio wave absorbant body using carbon nanocoils," in Shibata, J., ed., *Kneading/Compound Technologies and Controls of Dispersion/Interface of Composite Materials* (in Japanese) (Tokyo: Technical Information Institute Co., Ltd., 2013), pp. 558–565.

Akita, S., Fujiyama, Y., Higashi, Y. et al., "Electromagnetic wave absorbing sheet," Japan patent WO2009031409 A1 (2009).

Amelinckx, S., Zhang, X. B., Bernaerts, D. et al., "A formation mechanism for catalytically grown helix-shaped graphite nanotubes," *Science (New York, N. Y.)* 265 (1994): 635–639. doi:10.1126/science.265.5172.635.

AuBuchon, J. F., Chen, L.-H., Gapin, A. I. et al., "Multiple sharp bendings of carbon nanotubes during growth to produce zigzag morphology," *Nano Lett.* 4 (2004): 1781–1784. doi:10.1021/nl049121d.

Baker, R. T. K. & Chludzinski, Jr., J. J., "Filamentous carbon growth on nickel–iron surfaces: The effect of various oxide additives," *J. Catal.* 64 (1980): 464–478. doi:10.1016/0021-9517(80)90518-7.

Barranco, V., Celorrio, V., Lazaro, M. J. et al., "Carbon nanocoils as unusual electrode materials for super-capacitors," *J. Electrochem. Soc.* 159 (2012): A464–A469. doi:10.1149/2.083204jes.

Celorrio, V., Florez-Montano, J., Moliner, R. et al., "Fuel cell performance of Pt electrocatalysts supported on carbon nanocoils," *Int. J. Hydrog. Energy* 39 (2014): 5371–5377. doi:10.1016/j.ijhydene.2013.12.198.

Chakrabarti, S., Nagasaka, T., Yoshikawa, Y. et al., "Growth of super long aligned brush-like carbon nanotubes," *Jpn. J. Appl. Phys. Part 2* 45 (2006): 720–722. doi:10.1143/JJAP.45.L720.

Chen, X., Zhang, S., Dikin, D. A. et al., "Mechanics of a carbon nanocoil," *Nano Lett.* 3 (2003): 1299–1304. doi:10.1021/nl034367o.

Coluci, V. R., Fonseca, A. F., Galvão, D. S. et al., "Entanglement and the nonlinear elastic behavior of forests of coiled carbon nanotubes," *Phys. Rev. Lett.* 100 (2008): 086807 1–4. doi:10.1103/PhysRevLett.100.086807.

Daraio, C., Nesterenko, V. F., Jin, S. et al., "Impact response by a foamlike forest of coiled carbon nanotubes," *J. Appl. Phys.* 100 (2006): 064309 1–4. doi:10.1063/1.2345609.

Davis, W. R., Slawson, R. J. & Rigby, G. R., "An unusual form of carbon," *Nature* 171 (1953): 756. doi:10.1038/171756a0.

Fonseca, A., Hernadi, K., Nagy, J. B. et al., "Growth mechanism of coiled carbon nanotubes," *Synth. Met.* 77 (1996): 235–242. doi:10.1016/0379-6779(96)80095-6.

Fujiyama, Y., Tomokane, R., Tanaka, K. et al., "Alignment of carbon nanocoils in polymer matrix using dielectrophoresis," *Jpn. J. Appl. Phys.* 47 (2008): 1991–1993. doi:10.1143/JJAP.47.1991.

Gefen, Y., Aharony, A. & Alexander, S., "Anomalous diffusion on percolating clusters," *Phys. Rev. Lett.* 50 (1983): 77–80. doi:10.1103/PhysRevLett.50.77.

Hayashida, T., Pan, L.-J. & Nakayama, Y., "Mechanical and electrical properties of carbon tubule nanocoils," *Physica B* 323 (2002): 352–353. doi:10.1016/S0921-4526(02)01002-5.

Hirahara, K. & Nakayama, Y., "The effect of a tin oxide buffer layer for the high yield synthesis of carbon nanocoils," *Carbon* 56 (2013): 264–270.

Hirahara, K., Nakata, K. & Nakayama, Y., "Nonlinear annealing effect on correlation between crystallinity and oscillation of carbon nanocoils," *Mat. Sci. Eng. A* 595 (2014): 205–212. doi:10.1016/j.msea.2013.12.018.

Hokushin, S., Pan, L.-J., Konishi, Y. et al., "Field emission properties and structural changes of a standalone carbon nanocoil," *Jpn. J. Appl. Phys.* 46 (2007): L565–L567. doi:10.1143/JJAP.46.L565.

Ihara, S., Itoh, S. & Kitakami, J. I., "Helically coiled cage forms of graphitic carbon," *Phys. Rev. B* 48 (1993): 5643–5647. doi:10.1103/PhysRevB.48.5643.

Ju, S.-P., Lin, J.-S., Hsieh, J.-Y. et al., "Water molecular flow control with a (5,5) nanocoil switch," *J. Nanopart. Res.* 15 (2013): 1889. doi:10.1007/s11051-013-1889-6.

Kanada, R., Pan, L.-J., Akita, S. et al., "Synthesis of multiwalled carbon nanocoils using codeposited thin film of Fe–Sn as catalyst," *Jpn. J. Appl. Phys.* 47 (2008): 1949–1951. doi:10.1143/JJAP.47.1949.

Kong, Q. & Zhang, J., "Synthesis of straight and helical carbon nanotubes from catalytic pyrolysis of polyethylene," *Polym. Degrad. Stabil.* 92 (2007): 2005–2010. doi:10.1016/j.polymdegradstab.2007.08.002.

Lagarkov, A. N. & Sarychev, A. K., "Electromagnetic properties of composites containing elongated conducting inclusions," *Phys. Rev. B* 53 (1996): 6318–6336. doi:10.1103/PhysRevB.53.6318.

Li, D.-W., Pan, L.-J., Liu, D.-P. et al., "Relationship between geometric structures of catalyst particles and growth of carbon nanocoils," *Chem. Vap. Depos.* 16 (2010a): 166–169. doi:10.1002/cvde.200906832.

Li, D.-W., Pan, L.-J., Qian, J. et al., "Highly efficient synthesis of carbon nanocoils by catalyst particles prepared by a sol–gel method," *Carbon* 48 (2010b): 170–175. doi:10.1016/j.carbon.2009.08.045.

Ma, H., Pan, L.-J., Zhao, Q. et al., "Near-infrared response of a single carbon nanocoil," *Nanoscale* 5 (2013): 1153–1158. doi:10.1039/c2nr32641h.

Ma, H., Nakata, K., Pan, L.-J. et al., "Relationship between the structure of carbon nanocoils and their electrical property," *Carbon* 73 (2014): 71–77. doi:10.1016/j.carbon.2014.02.038.

Montazami, R., Liu, S., Liu, Y. et al., "Thickness dependence of curvature, strain, and response time in ionic electroactive polymer actuators fabricated via layer-by-layer assembly," *J. Appl. Phys.* 109 (2011): 104301 1–5. doi:10.1063/1.3590166.

Motojima, S. & Chen, Q., "Three-dimensional growth mechanism of cosmo-mimetic carbon microcoils obtained by chemical vapor deposition," *J. Appl. Phys.* 85 (1999): 3919–3921. doi:10.1063/1.369765.

Okazaki, N., Hosokawa, S., Goto, T. et al., "Synthesis of carbon tubule nanocoils using Fe–In–Sn–O fine particles as catalysts," *J. Phys. Chem. B* 109 (2005): 17366–17371. doi:10.1021/jp050786t.

Pan, L.-J., Hayashida, T., Zhang, M. et al., "Field emission properties of carbon tubule nanocoils," *Jpn. J. Appl. Phys.* 40 (2001): L235–L237. doi:10.1143/JJAP.40.L235.

Pan, L.-J., Hayashida, T. & Nakayama, Y., "Growth and density control of carbon tubule nanocoils using catalyst of iron compounds," *J. Mater. Res.* 17 (2002): 145–148. doi:10.1557/JMR.2002.0022.

Poggi, M. A., Boyles, J. S., Bottomley, L. A. et al., "Measuring the compression of a carbon nanospring," *Nano Lett.* 4 (2004): 1009–1016, Figure 1. doi:10.1021/nl0497023.

Reddy, A. L. M., Jafri, R. I., Jha, N. et al., "Carbon nanocoils for multi-functional energy applications," *J. Mater. Chem.* 21 (2011): 16103–16107. doi:10.1039/c1jm12580j.

Sevilla, M., Sanchis, C., Valdes-Solis, T. et al., "Highly dispersed platinum nanoparticles on carbon nanocoils and their electrocatalytic performance for fuel cell reactions," *Electrochim. Acta* 54 (2009): 2234–2238. doi:10.1016/j.electacta.2008.10.042.

Shaikjee, A. & Coville, N. J., "The synthesis, properties, and uses of carbon materials with helical morphology," *J. Adv. Res.* 3 (2012): 195–223. doi:10.1016/j.jare.2011.05.007.

Sun, J., Koos, A. A., Dillon, F. et al., "Synthesis of carbon nanocoil forests on BaSrTiO₃ substrates with the aid of a Sn catalyst," *Carbon* 60 (2013): 5–15. doi:10.1016/j.carbon.2013.03.027.

Tomokane, R., Fujiyama, Y., Tanaka, T. et al., "Determination of carbon nanocoil orientation by dielectrophoresis," *Jpn. J. App. Phys.* 46 (2007): 1815–1817. doi:10.1143/JJAP.46.1815.

Volodin, A., Buntinx, D., Ahlskog, M. et al., "Coiled carbon nanotubes as self-sensing mechanical resonators," *Nano Lett.* 4 (2004): 1775–1779. doi:10.1021/nl0491576.

Wang, W., Yang, K., Gaillard, J. et al., "Rational synthesis of helically coiled carbon nanowires and nanotubes through the use of tin and indium catalysts," *Adv. Mater.* 20 (2008): 179. doi:10.1002/adma.200701143.

Wu, M.-D., Shih, W.-P., Tsai, Y.-C. et al., "Rapid dielectrophoresis assembly of a single carbon nanocoil on AFM tip apex," *IEEE Trans. Nanotechnol.* 11 (2012): 328–335. doi:10.1109/TNANO.2011.2174253.

Xia, J. H., Jiang, X., Jia, C. L. et al., "Hexahedral nanocementites catalyzing the growth of carbon nanohelices," *Appl. Phys. Lett.* 92 (2008), Article No. 063121. doi:10.1063/1.2842410.

Xie, J., Mukhopadyay, K., Yadev, J. et al., "Catalytic chemical vapor deposition synthesis and electron microscopy observation of coiled carbon nanotubes," *Smart Mater. Struct.* 12 (2003): 744–748. doi:10.1088/0964-1726/12/5/010.

Xu, G. C., Chen, B. B., Shiki, H. et al., "Parametric study on growth of carbon nanocoil by catalytic chemical vapor deposition," *Jpn. J. Appl. Phys.* 44 (2005): 1569–1576. doi:10.1143/JJAP.44.1569.

Zhang, M., Nakayama, Y. & Pan, L.-J., "Synthesis of carbon tubule nanocoils in high yield using iron-coated indium tin oxide as catalyst," *Jpn. J. Appl. Phys.* 39 (2000): L1242–L1244. doi:10.1143/JJAP.39 .L1242.

Zhang, X. B., Zhang, X. F., Bernaerts, D. et al., "The texture of catalytically grown coil-shaped carbon nanotubes." *Euro. Lett. (EPL)* 27 (2). IOP Publishing: 141–146. doi:10.1209/0295-5075/27/2/011.

Zhao, D.-L. & Shen, Z.-M., "Preparation and microwave absorption properties of carbon nanocoils," *Mater. Lett.* 62 (2008): 3704–3706. doi:10.1016/j.matlet.2008.04.032.

# 4
# Carbon Nanohorns

Artur P. Terzyk

Piotr A. Gauden

Sylwester
Furmaniak

Karolina
Werengowska-
Ciećwierz

Piotr Kowalczyk

Marek Wiśniewski

**Abstract**

As it was mentioned by Peter Harris in his fundamental book *Carbon Nanotube Science*, there is uncertainty surrounding the question who actually discovered CNTs, and this probably explains why no Nobel Prize has been awarded in this area. However, in the case of single-walled carbon nanohorn (SWCNH), there are no doubts that they were discovered by Iijima et al. in 1999. From that time, SWCNHs have been subjected to intensive experimental and theoretical studies. In this chapter, we discuss the synthesis and structure of SWCNHs and the morphology and application of those materials. We also discuss the process of nanowindow creation and control, SWCNH filling with, for example, fullerenes and drugs, and recent progress in the creation of realistic atomistic model of SWCNH. Finally, we discuss the most promising applications of SWCNH basing on the reports of different authors as well as on the simulation results reported in our group.

## 4.1  Introduction

One of the newest carbon allotropes are single-walled carbon nanohorns (SWCNHs). They have very interesting and sometimes amazing properties. However, the increasing number of papers reporting the results devoted to SWCNHs is not accompanied by broad review studies. The existing reviews (Delgado et al. 2008, Pagona et al. 2009, Zhu & Xu 2010, Pramoda et al. 2014) are interesting; however, they cover only selected properties of nanohorns. Moreover, they cover the literature from the field published up to the start of 2014. This is why we decided to prepare a more comprehensive study covering the chronology of SWCNH synthesis methods and the properties and applications of these materials. We also report in this chapter some recent results of our studies.

## 4.2 Synthesis

Iijima et al. (1999) reported the discovery of SWCNHs in carbon soot resulting from $CO_2$ laser ablation of graphite. Two different types of carbon nanohorn consisting of globular particles, 80–100 nm in diameter, were observed. The type depends on the experimental conditions and was called dahlia-like and bud-like. In the first type, the SWCNHs protrude from the particle surface, while in the second type the horn-shaped structure appears to develop inside the particle itself (Gattia et al. 2007). From this time, different types of SWCNHs have been discovered (see Figure 4.1).

As it was pointed out by Gattia et al. (2007), among the variety of methods suitable for producing carbon nanostructures, the synthesis routes based on thermal carbon sublimation, like arc discharge and laser ablation, still play a relevant role. These methods are based on carbon atom sublimation at high temperature followed by reaggregation in the solid state in colder regions (Gattia et al. 2007). In the following, we discuss different methods of carbon nanohorn (and related structures) synthesis. To show the progress in this field, the reports are arranged chronologically.

Kasuya et al. (2002) showed that SWCNH aggregates can be produced by $CO_2$ laser vaporization of carbon, and a single aggregate can take either a dahlia-like or a bud-like form. They found that dahlia-like SWCNH aggregates were produced with a yield of 95% when Ar was used as the buffer gas, while

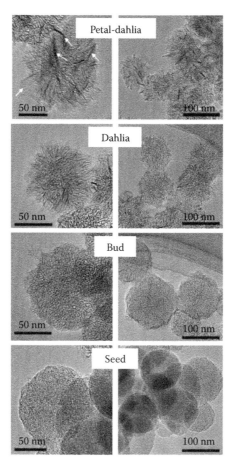

**FIGURE 4.1**   TEM images of different types of SWCNH aggregates. (Reprinted with permission from Azami et al. 2008, pp. 1330–1334. Copyright (2008) American Chemical Society.)

bud-like SWCNH aggregates were formed with a yield of 70% or 80% when either He or $N_2$ was applied (see Figure 4.2). The internal structures of both aggregates were studied by partially burning them in an $O_2$ atmosphere. This burning reveals the presence of graphite inside.

The same year brought a breakthrough in the methods of nanohorn synthesis. Namely, Takikawa et al. (2002) were the first who reported the fabrication of SWCNHs using a torch arc method in open air. A graphite target containing Ni/Y catalyst was used as a counterelectrode (CE) of the welding arc torch. The target was blasted away by the direct current (DC) arc, and soot was deposited on the substrate placed downstream of the arc plasma jet. The deposited soot contained nanotubes and SWCNHs.

Kokai et al. (2002) synthesized aggregates of SWCNHs by $CO_2$ laser vaporization in Ar gas under pressures of 150–760 Torr. At 150 Torr, carbon nanotubes (CNTs) with smaller diameters were synthesized together with amorphous carbon. With an increase in the gas pressure, the yield of nanotubes decreased, the tube diameters increased, and SWCNH aggregates were synthesized. At 760 Torr, most of the products were nanohorn aggregates.

Two years later, Sano (2004) reported a method of synthesizing bulk amounts of SWCNHs on a significant low cost based on the arc-in-water method with $N_2$ injection into the arc plasma. It is elucidated that rapid quenching of the carbon vapor in an inert gas environment is necessary to form the delicate SWCNH structures. To realize this reaction field, the arc plasma between the graphite electrodes was isolated from the surrounding water by a thin graphite wall with an $N_2$ flow that excluded reactive gas species ($H_2O$, CO, and $H_2$) from the arc zone. High concentrations of SWCNHs were found as fine powders floating on the water surface. The particle diameter distribution of the SWCNHs produced was around 70 nm.

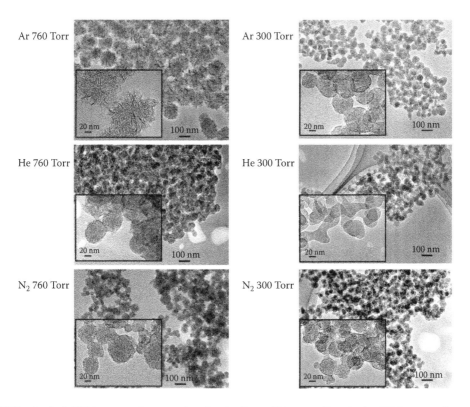

**FIGURE 4.2** Left: TEM images of soot produced in Ar, He, and $N_2$ at 760 Torr. Right: TEM images of soot produced in Ar, He, and $N_2$ at 300 Torr. Each inset is a highly magnified TEM image. (Reprinted with permission from Kasuya et al. 2002, pp. 4947–4951. Copyright (2002) American Chemical Society.)

Also in 2004, Wang et al. was reported as the first team to conduct the procedure of SWCNH synthesis by arc discharge between two graphite electrodes submerged in liquid nitrogen. The product in its powder form was found consisting of spherical aggregates with sizes in the range of 50–100 nm. The internal structure of the aggregates could be described as a mixture of dahlia-like and bud-like. The proposed method is simple and does not require relatively expensive laser and vacuum equipment. As it was stated by the authors, it provides an economical alternative for the synthesis of SWCNHs. Further development of the arc discharge method can be found in the study published by Yamaguchi et al. (2004), who proposed a method of SWCNH preparation by simple pulsed arc discharge between pure carbon rods in the atmospheric pressure of air. Purity of SWCNHs was very high, exceeding 90%. The mean size of SWCNH particles was around 50 nm. It is smaller than those prepared by the $CO_2$ laser method. Preheating of the carbon rod improves the quality of SWCNHs.

Two years later, in 2006, in Japan, Nippon Electric Company (NEC) developed a device producing 1 kg of high-purity SWCNHs a day. It was 100 times larger in capacity than what had been recorded for existing equipment (Fuel Cells Bulletin 2006). NEC proposed a mechanism using a special gas to continuously process carbon nanohorns. It was reported that SWCNHs improve the power generation efficiency of a fuel cell by 20% when used as an electrode. In tests, NEC-made fuel cells with carbon nanohorn electrodes powered a notebook personal computer for at least 10 h (Fuel Cells Bulletin 2006).

Also, the methods of synthesis of metal-loaded nanohorns have been developed. Sano and Ukita (2006) reported a one-step synthesis of Pt-loaded SWCNHs by arc plasma in liquid nitrogen using a Pt-contained graphite anode. The diameter of Pt particles can be controlled by adjusting the concentration of Pt in the graphite anode. In this report, the diameter was smaller than 5 nm, and it depends on the Pt concentration in the anode. To increase the potential application of SWCNHs in nanomedicine, Utsumi et al. (2006) proposed the procedure of synthesis of a SWCNH colloid. Magnetite nanoparticle–anchored SWCNH colloid has the hybrid property of ferrimagnetism and superparamagnetism. It was demonstrated that this colloid dispersed in water by sonication responded to an external magnetic field, gathering toward a magnet.

The state of art in different synthesis methods was reported by Gattia et al. (2007). Moreover, the same authors described a new method of synthesis by a DC-powered electric arc, performed in air and in Ar at atmospheric pressure. The authors reported and designed an experimental device operating in air at atmospheric pressure, in which the yield of nanoparticles is optimized at the expense of the material deposited at the electrodes. Experimental conditions suitable for the synthesis of nanoparticles have been achieved by heating both electrodes above the carbon sublimation temperature with a strategy based on the analysis, even if qualitative, of the heat input and the heat dissipation at the electrode tips (Gattia et al. 2007).

Also, interesting results were published by Kobayashi et al. (2007), who reported the synthesis of SWCNH aggregates hybridized with carbon nanocapsules. The new materials were fabricated with a high yield of around 70%. Carbon was laser-vaporized into an Ar gas atmosphere with one of the following: Fe, Al, Si, Co, Ni, Cu, Ag, $La_2O_3$, $Y_2O_3$, and $Gd_2O_3$. By optimizing the Ar gas pressure and metal content, the authors were able to produce hybridized SWCNH structures for Fe, Co, Ni, Cu, and Ag.

Also, the details of the SWCNHs formation process were studied. For example, Cheng et al. (2007) investigated the formation of carbon nanohorns by laser ablation using a scanning differential mobility analyzer combined with an ultrafine condensation particle counter. They concluded that a longer laser pulse width produces larger SWCNH aggregates. When laser ablation was not used, it was found that carbon nanoparticles with a diameter centered at 20 nm could be produced by thermally desorbing the previously deposited carbon nanoparticles from the reactor wall at temperatures greater than 1300 K. SWCNH growth mechanisms were also studied by Geohegan et al. (2007) and compared for both high- and low-temperature synthesis methods through experiments utilizing time-resolved, in situ imaging and spectroscopy. High-speed videography and pyrometry measured the time frames for growth of single-wall CNTs (SWCNTs) and SWCNHs by laser vaporization at 1423 K, revealing that carbon can self-assemble at high temperatures preferentially into SWCNH structures without catalyst assistance at rates comparable to that for catalyst-assisted SWCNT growth by either laser vaporization or CVD. Laser interferometry and videography reveal the coordinated growth of vertically aligned nanotube arrays by CVD at 823–1173 K. Also, Xing et al. (2007) studied the growth mechanism of carbon nano-materials by arc discharge in water or liquid nitrogen using a special arc discharge apparatus. Results indicate that multiwall CNTs (MWCNTs) and carbon onions can be produced by arc discharge in water and liquid nitrogen, respectively. SWCNHs can be produced by arc discharge in liquid nitrogen, and cobalt-encapsulated carbon nanoparticles can be produced by cobalt-catalyzed carbon arc discharge in water. The liquids acted as quenching walls for the nucleation and growth of carbon nanomaterials, resulting in the formation of different forms of carbon nanomaterials. The authors proposed a model explaining the growth of nanomaterials by arc discharge in liquids.

Azami et al. (2008) tried to optimize the conditions of the laser ablation method to increase the prac-tical daily capacity to produce SWCNHs with high purity by $CO_2$ laser ablation of graphite. The authors proposed a new three-chamber system composed of a target reservoir, a laser ablation chamber, and a collection chamber. The resulting SWCNHs had a purity of 92%–95% under the optimized conditions of laser power density and target rotation speed. A practical production capacity of 1 kg/day was achieved.

Puretzky et al. (2008) studied the experimental conditions for the scaled synthesis of SWCNTs and SWCNHs by laser vaporization at high temperatures. They studied the influence of two laser ablation regimes, namely, *continuous* ablation providing sustained temperatures sufficient for self-organization of the ablated species into SWCNHs and *cumulative* ablation, which is optimal for the growth of nano-tubes with catalyst assistance.

Sano et al. (2008) synthesized SWCNHs by a gas-injected arc-in-water (GI-AIW) method with varied gas flow rate, arc current, and gas component (He, Ne, $N_2$, Ar, $CO_2$, and Kr) to investigate their influ-ence on purity, yield, and structure of the products. The effect of the gas component is important to determine the structure and the yield of the products. The dahlia-like SWCNHs can be produced by use of $CO_2$, although the bud-like SWCNHs were formed when either He, Ne, $N_2$, Ar, $CO_2$, or Kr was used.

Poonjarernsilp et al. (2009) reported the probably cost-effective method to synthesize SWCNHs using GI-AIW. The yield of SWCNHs significantly decreases with the increase in water temperature, although the purity of SWCNHs is not dependent on the temperature change. Li et al. (2009) synthesized multi-walled nanohorns by simple CVD of methane over $Ni/Al_2O_3$. The purity was about 40%–50% and the multiwalled carbon nanohorns (MWCNHs) had well-graphitized multiwalled structures. The mean diameter of tubules was about 20 nm, which is larger than those prepared by the $CO_2$ laser method, and the length of MWCNHs was in the range of 50–500 nm. Also, Li et al. (2010) generated SWCNHs with different morphologies by direct current arc discharge between pure graphite rods in different atmo-spheres, including air, $CO_2$, and CO. In the arc discharge process, the $O_2$ in the air reacted with carbon atoms and was transformed into CO so the formation of SWCNHs was a result of a combination of CO and $N_2$. The effect of different volume ratios of $CO/N_2$ on the formation of SWCNHs was examined and a mechanism for the formation of SWCNHs was additionally proposed. Sano et al. (2011a,b) showed that in a GI-AIW method of SWCNHs synthesis, the yield of the SWCNHs can be significantly increased by increasing the discharge time and the size of the cathode. In using these modified conditions, the horn units in the SWCNH aggregates increased in size, and the thermal stability of SWCNHs in an oxidative environment increased accordingly.

Poonjarernsilp et al. (2011) reported a procedure leading to SWCNHs hybridized with Pd nano-particles. The procedure is a single-step GI-AIW method with a Pd wire inserted inside the anode hole. In the arc zone, carbon and Pd were vaporized simultaneously, leading to the formation of hybrid mate-rial of SWCNHs and Pd nanoparticles due to effective quenching. Pd nanoparticles were found to be embedded inside SWCNH aggregates.

A route to synthesize SWCNHs at a lower than normal temperature was reported by Sano et al. (2011c). SWCNHs were synthesized at 1873 K by a transformation from SWCNTs as the result of the catalytic effect of molten Pd. Nanotubes were first deformed to highly curled shapes, and then were fragmentized into independent horns. Typical structures of SWCNHs were formed by merging these hornlike fragments.

It is also important to search for the procedures of the so-called morphology-selected preparation. The results of Yuge et al. (2012a) show that SWCNHs of seed, bud, dahlia, petal-dahlia, and petal types can be selectively synthesized by changing the buffer gas for $CO_2$ laser ablation on the graphite target. The structure of the SWCNH aggregates strongly depends on the mass number of the buffer gas. The seed, dahlia, and petal types were formed preferentially for He, Ar, and Xe buffer gases, respectively. The aggregates prepared with buffer gas of $N_2$ and Ar are considerably different, although their shapes seem similar.

Sano et al. (2012) reported the synthesis of SWCNHs dispersed with Pd–Ni alloy nanoparticles. The authors used a technique requiring a single step by a GI-AIW method using Pd–Ni–C mixed powders charged in an anode hole. It was found that the Ni/Pd weight ratio in the alloy nanoparticles dispersed in the products depends on the initial Ni/Pd weight ratio.

Sun et al. (2013) reported a method of synthesis of nitrogen-doped SWCNHs (N-SWCNHs). They were synthesized by a flowing nitrogen-assisted arc discharge method at atmospheric pressure in a tubular reactor. N-SWCNHs have a typical spherical structure with a diameter of 40–80 nm. Oxidation treatment suggests the opening of cone-shaped caps of N-SWCNHs. Most of the nitrogen atoms are in N-6, N-5, and triple-bonded CN bonding configuration present at the defect sites or the edges of graphene layers.

In the case of some SWCNHs, applications as sensors or catalysts nanohorns must be first hybridized with metallic nanoparticles. As it was pointed out by Sano et al. (2014a), synthesis of SWCNHs dispersed with metallic nanoparticles (see Figure 4.3) can be performed by one of two means: the first is a two-step method in which SWCNHs are first synthesized and then loaded with metallic nanoparticles, and the other is a one-step method in which as-grown SWCNHs are dispersed with metallic nanoparticles. The syntheses of SWCNHs dispersed with nanoparticles of Pd alloys including nine components, Au, Pt, Cu, Fe, Ni, Ti, Mo, W, or Nb, using the GI-AIW method were investigated by Sano et al. (2014a). To realize these syntheses, wires of Pd and an alloying component were inserted in a hollow graphite anode. It was found that SWCNHs containing Pd-alloy nanoparticles can be synthesized only when the boiling point of the alloying component is below 5000 K. The weight fraction of Pd in the alloy nanoparticles tended to increase when the boiling point of the alloying component was higher, and this boiling point can be used as a threshold to judge whether Pd content is increased or decreased in the products produced by the arc discharge process.

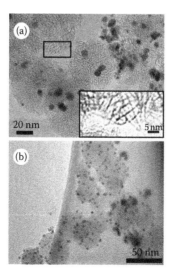

**FIGURE 4.3**   TEM images of (a) Pd–Pt/SWCNHs and (b) Pd–Au/SWCNHs synthesized by the GI-AIW method. The inset of a resolves the horn structures of SWCNHs. (Reprinted with permission from Sano et al. 2014a, pp. 4732–4738. Copyright (2014) American Chemical Society.)

Summing up the reports described above, one can see the evolution of SWCNH synthesis from the discovery of SWCNHs and the synthesis by laser ablation (original method) by the first synthesis by arc discharge method with modifications, and recently the development of the methods of synthesis of hybrid materials containing SWCNHs.

# 4.3 Properties

## 4.3.1 General

SWCNH aggregates, composed of thousands of graphitic tubules (similar in structure to SWCNTs) having wide diameters of 2–5 nm, have a spherical structure with a diameter of 50–100 nm. The agglomerates can be successfully isolated by attaching, for example, gum arabic to achieve a stable suspension through steric stabilization (Fan et al. 2007). Also, an approach to isolating small aggregates of SWCNHs was presented by Zhang et al. (2005) using subsequent centrifuging. This enables the separation of small aggregates from larger aggregates or agglomerations and removal of graphitic particles as the main impurity. The distribution of SWCNHs clusters in solutions was discussed by Torrens and Castellano (2010, 2014).

Zhang et al. (2009) isolated individual nanohorns from the aggregates and claimed that individual SWCNHs have several different shapes such as straight and two-way or three-way branched (see Figure 4.4). The diameters and lengths of the individual SWCNHs or of each branch of the branched SWCNHs ranged from 2 to 10 and 10 to 70 nm, respectively. It was also proved by using neutron diffraction study that the defects are present in the walls of nanohorns (Hawelek et al. 2011). Not only the Stone–Wales defects are observed but also the presence of the mono- and di-vacancies seems to be very probable.

Considering the internal structure of aggregates in 2007, it was claimed that a single C–C bond is present, responsible for the impossibility of dispersion of SWCNHs aggregates (Utsumi et al. 2007). It is also possible that caves occur inside SWCNHs aggregates (Yuge et al. 2009). The presence of

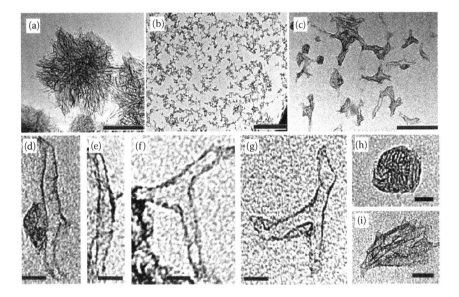

**FIGURE 4.4** Typical TEM images of (a) an oxidized SWCNH (ox-SWCNH) aggregate and (b, c) the particles placed on top of sucrose solutions dispersed in the centrifuge tube, (d, e) straight ox-SWCNHs, (f, g) branched ox-SWCNHs, and (h, i) small aggregates. Scale bars: (a, b) 50, (c) 100, and (d to i) 10 nm. (Reprinted with permission from Zhang et al. 2009, pp. 11184–11186. Copyright (2009) American Chemical Society.)

graphitic carbon inside SWCNHs was first documented by Kasuya et al. (2002) and confirmed by Xu et al. (2011a).

Nuclear magnetic resonance studies confirmed that SWCNHs possess two major components, i.e., nanohorns and internal graphite-like core (Imai et al. 2006). The X-ray powder diffraction analysis indicated the wider spacing: 0.40 nm for the van der Waals distance of the aggregated SWCNHs, as compared with that of the ordinary graphite. This wider distance may be used for the doping space of the donor and acceptor materials (Bandow et al. 2000a). The first extensive experimental analysis on the $sp^2$–$sp^3$ structural changes and electrical conductivities of SWCNHs and ox-SWCNHs was published by Urita et al. (2008). The mechanical treatment weakens the $n$-type and $p$-type nature in SWCNHs and ox-SWCNHs, respectively. It also reduces the pentagons in the cap and bending sites, weakening the $n$-type nature. In contrast, a reduction of the functional groups causes a decrease in the $p$-type nature of ox-SWCNHs.

In 2011, Kumar et al. studied the values of Young's and shear moduli of SWCNHs, showing that similar properties as for CNTs are observed. It is also interesting that the resistivity of pelletized SWCNHs decreases as the pressure for making the pellet increases (Fukunaga et al. 2009).

The reactivity of SWCNHs was discussed by Nakamura et al. (2013). It was shown that carboxylation of thin graphitic sheets is faster than that of carbon nanohorns. Nanofriction properties of SWCNHs were studied by Maharaj et al. (2013), and field emission properties by Bonard et al. (2002). Application of SERS for the study of SWCNHs was described by Fujimori et al. (2008), near-edge X-ray absorption fine structure by Bittencourt et al. (2013), and the effect of neutron bombardment was discussed by Cataldo et al. (2014). Xu et al. (2011b) studied the electrical behavior of SWCNHs under deformation. Raman spectra were presented by Sasaki et al. (2013), electron paramagnetic resonance spectra by Bandow et al. (2001), and electronic absorption spectra by Iglesias-Groth et al. (2014). Thermal properties were discussed by Irie et al. (2010) and thermal transformation by Yudasaka et al. (2003). The full characterization of optical properties of nanofluids consisting of SWCNHs reveals very low scattering albedo. Moreover, the scattering behavior of SWCNHs shows little dependence on the nanohorn morphology (dahlia-like or bud-like), and SWCNH suspensions show good stability properties (Mercatelli et al. 2012).

### 4.3.2 Toxicity

Miyawaki et al. (2008) were the first who studied the toxicity of SWCNHs in vitro as well as in vivo. They stated that present results strongly suggest that as-grown SWCNHs have low acute toxicities. However, later, results confirming this conclusion but also suggesting some possible negative effects of SWCNHs have appeared. Moschino et al. (2014) published a study indicating that SWCNHs may induce sublethal effects in relation to oxidative stress in the digestive system of mussels and lysosomal changes in digestive tubule cells as well as circulating haemocytes, quite remarkable at the highest tested concentrations. The preliminary investigation highlighted the need of focusing future research efforts on possible physiological impairments caused by long-term exposure to SWCNHs in marine species. Tahara et al. (2011) stated that SWCNHs intravenously administered to mice did not show severe toxicity during a 26-week test period, which was confirmed by normal gross appearance, normal weight gain, and the lack of abnormality in the tissues on histological observations of the mice. SWCNHs' biodistribution was influenced by chemical functionalization. Accumulation of SWCNHs in the lungs reduced as SWCNH hydrophilicity increased; however, the most hydrophilic SWCNHs modified with bovine serum albumin (BSA) were most likely to be trapped in the lungs, suggesting that the BSA moiety enhanced macrophage phagocytosis in the lungs. The same group reported the toxicity results obtained for very high SWCNTs doses (Tahara et al. 2012). High uptake was shown to destabilize lysosomal membranes and generate reactive oxygen species that resulted in apoptotic, as well as necrotic, cell death. Despite these dramatic responses, only low levels of cytokines were released.

Horie et al. (2013) examined the induction of oxidative stress by SWCNHs in vitro and in vivo. They stated that although SWCNH did not show any significant cytotoxicity and short-term pulmonary toxicity including oxidative stress, chronic toxicity of SWCNH was still not clear. Therefore, for efficient application of SWCNH, further examinations of the biological influence of SWCNH are necessary. Yang et al. (2014a) studied the role of SWCNHs in the production of reactive oxygen species (ROS). They stated that SWCNHs were accumulated in the lysosomes, where they induced lysosomal membrane permeabilization and the subsequent release of lysosomal proteases, such as cathepsins, which in turn caused mitochondrial dysfunction and triggered the generation of ROS in the mitochondria. Misra and Chaudhari (2013) studied the properties of nylon blended with SWCNHs. The biological functions including cell attachment, viability, proliferation, and morphology, in conjunction with the higher expression level of proteins, suggest that SWCNH is a promising biocompatible material to favorably tune osteoblast functions and cellular interactions of artificial devices.

Zhang et al. (2014a) pointed out that the interactions between SWCNHs and cancerous or normal cells have seldom been studied. To address this problem, they studied the effects of raw SWCNH material on the biological functions of human liver cell lines. The results showed that unmodified SWCNHs inhibited mitotic entry, growth, and proliferation of human liver cell lines and promoted their apoptosis, especially in hepatoma cell lines. Individual spherical SWCNH particles were found inside the nuclei of human hepatoma HepG2 cells and the lysosomes of normal human liver L02 cells, implying that SWCNH particles penetrate human liver cells and the different interaction mechanisms on human normal cell lines compared with those in hepatoma cell lines. The authors pointed out that further research on the mechanisms and application in treatment of hepatocellular carcinoma with SWCNHs is needed. Li et al. (2013) published the results of the study about the direct role of SWCNHs on the growth, proliferation, and apoptosis of microglia cell lines in mice (N9 and BV2) pretreated with or without liposaccharide (LPS). SWCNHs detected on polystyrene surface are individual particles. LPS-induced activation of mice microglia promoted its growth and proliferation and inhibited its apoptosis. SWCNHs inhibit proliferation, delay mitotic entry, and promote apoptosis of mice microglia cells. Individual spherical SWCNH particles smaller than 100 nm in diameter were localized inside lysosomes of mice microglia cells. SWCNHs inhibited mitotic entry, growth, and proliferation of mice microglia cells and promoted their apoptosis, especially in cells pretreated with LPS. SWCNHs inhibited the expression of SIRT3 and the energy metabolism related to SIRT3 in mice microglia cells in a dose-dependent manner, especially in cells pretreated with LPS. The role of SWCNHs on mice microglia was implicating SIRT3 and the energy metabolism associated with it.

Summing up the results presented above, it can be concluded that SWCNHs are rather nontoxic; however, there is still a lack of well-defined procedure of the study of toxic properties of carbon nanomaterials.

# 4.4 Opening, Modifications, and Composites

## 4.4.1 Opening/Closing of SWCNHs

Since SWCNHs are closed, there are different methods proposed to open the internal space. This is done by the creation of so-called nanowindows or nanogates. Murata et al. (2002a) reported the creation of nanowindows by oxidation of SWCNHs with $O_2$. Precise oxidation leads to controllable nanowindow size (Murata et al. 2002a). Bekyarova et al. (2003) described the procedure of SWCNHs opening by the heat treatment with $CO_2$. The porosity of the material can be controlled by the conditions of the process, and the procedure does not introduce surface oxygen groups. Also, light-assisted oxidation of SWCNHs with hydrogen peroxide effectively and rapidly opens holes in SWCNH walls and creates oxygen groups such as carboxylic at the hole edges (Zhang et al. 2007)-Figure 4.5. Miyawaki et al. (2007a) proposed a thermal hole-closing method of controlling nanowindow size. The holes in the sidewalls were difficult to close by thermal annealing, whereas those at the tips were easy to close. Yuge et al. (2007) showed that gadolinium acetate can plug holes of SWCNHs and that unplugging is also possible.

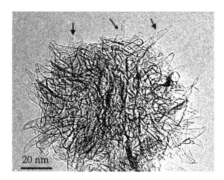

FIGURE 4.5 TEM image of SWCNHs. The arrows indicate the holes in the walls. (Reprinted with permission from Zhang et al. 2007, pp. 265–272. Copyright (2007) American Chemical Society.)

## 4.4.2 Surface Modifications

The major purpose of surface modifications of SWCNHs is to increase the solubility in water and common solvents (Murakami et al. 2006, Matsumura et al. 2009, Mountrichas et al. 2009), and/or to change the electronic properties. Pagona et al. (2008a) improved the solubility of SWCNHs in common organic solvents and water using aryl diazonium functionalization. Shu et al. (2010) described application of a high-speed vibration milling method with acyl peroxide to synthesize water-dispersible SWCNHs. The efficient functionalization of SWCNHs via microwave radiation and combination of two cyclo-addition reactions was reported by Rubio et al. (2009). As a consequence of the covalent attachment, the solubility of the SWCNHs highly increases. Xu et al. (2010) reported the procedure of double SWCNHs' oxidation, to eliminate carbon dust from internal channels of nanohorns and to increase the number of surface oxygen groups. Xu et al. (2010), aiming at improving the dispersion state of SWCNHs in a highly salted aqueous environment for potential biological applications, compared the dispersion ability of different polyethylene glycol (PEG)-based dispersants. They claim that ceramide-conjugated PEG-dispersed SWCNHs are better than phospholipid-conjugated PEG in both water and phosphate buffer saline. Aryee et al. (2014) studied the chemical modification of SWCNHs using $HNO_3$ under reflux. They observed an increase in surface area under optimum functionalization conditions. The pore volume of acid-treated samples also increased.

Yoshida and Sano (2006) studied the minimum irradiances necessary to chemically modify carbon nanohorns by microwave radiation. They showed that only 150 W min, equivalent to operating a home-use microwave oven for less than 10 s, is sufficient for ox-SWCNHs, with the result that nearly all materials used for this study become dispersible in polar solvents. Tagmatarchis et al. (2006) reported the modification of SWCNHs with azomethine ylides. The introduction of a repetitive centrifugation–filtration–solubilization cyclic treatment on the 1,3-dipolar cycloaddition reaction of azomethine ylides with SWCNH was crucial for obtaining functionalized nanohorns of the highest purity, which were soluble in organic solvents and even in water. Cioffi et al. (2006) proposed a procedure of the functionalization of SWCNHs using 1,3-dipolar cycloaddition and concluded that SWCNH derivatives present good solubility in common organic solvents. Pagona et al. (2006a) reported for the first time the functionalization procedure of SWCNH based on the opening of their conical end and the attachment of various organic amines, alcohols, and thiols (see Figure 4.6). In this way, soluble amide, ester, and thioester materials are easily produced.

The same group optimized the conditions of SWCNHs opening with gaseous oxygen (Fan et al. 2006), and reported the addition of porphyrines to SWCNHs (Pagona et al. 2007a). Also for the purpose of covalent SWCNHs functionalization, anionic polymerization with the "grafting-to" approach was proposed (Mountrichas et al. 2007). Polyisoprene homopolymers, as well as polyisoprene-b-polystyrene block copolymers, have been attached on the surface of carbon nanohorns. Functionalized nanostructures

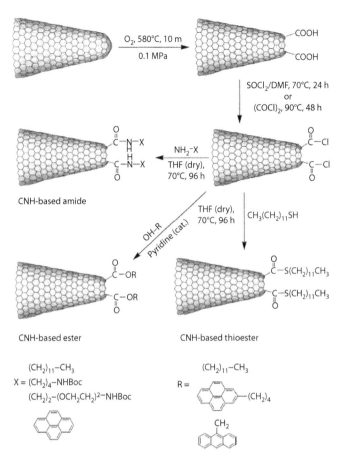

**FIGURE 4.6** Oxidative removal of the cone-shaped tip of SWCNH and covalent functionalization with amines, alcohols, and thiols for the generation of SWCNH-based amide, ester, and thioester materials, respectively. (Reprinted with permission from Pagona et al. 2006a, pp. 3918–3920. Copyright (2006) American Chemical Society.)

showed enhanced solubility in common organic solvents. Also, in the field of functionalization, the modification with ethylene glycol chains and porphyrins was reported (Cioffi et al. 2007). Jadhav et al. (2008) showed very strong hydrophobic properties of a film of glass substrates when covered with BSA dispersed in SWCNHs solution. SWCNHs covalently functionalized at the conical tips with porphyrin moieties were used to construct photoelectrochemical solar cells—see Figure 4.7 (Pagona et al. 2008b).

Imidazolium-modified carbon nanohorn-based hybrid materials have been prepared and characterized by Karousis et al. (2010): the covalent conical tip functionalization of carbon nanohorns with a positively charged imidazolium-based moiety bearing bromide as counteranion. This functional group, as an ionic liquid-based component, has allowed anion exchange, thus switching the solubility of the new hybrid material between water and organic solvents and vice versa. The newly prepared hybrid materials have been utilized as suitable substrates to immobilize Pd and Pt nanoparticles. The covalent attachment of a custom functionalized coumarin derivative onto the sidewalls of SWCNHs was reported by Pagona et al. (2011). In their next study (Pagona et al. 2012), the covalent grafting through a rigid ester bond of a dimeric porphyrin and carbon SWCNHs was accomplished. A new material was adsorbed on nanostructured SnO$_2$ electrode, to construct a photoactive electrode, which reveals photocurrent and photovoltage responses with an incident photon-to-current conversion efficiency value as large as 9.6%, without the application of any bias voltage.

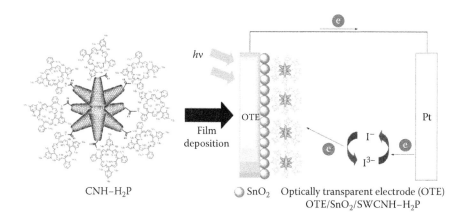

CNH–H₂P

SnO₂   Optically transparent electrode (OTE)
OTE/SnO₂/SWCNH–H₂P

**FIGURE 4.7** Schematic illustrations of SWCNH–H₂P structure and a photoelectrochemical solar cell of optically transparent electrode (OTE)/SnO₂/SWCNH–H₂P. (Reprinted with permission from Pagona et al. 2008b, pp. 15735–15741. Copyright (2008) American Chemical Society.)

In the report of Karousis et al. (2012), SWCNHs were functionalized with aryl units, possessing ethylene glycol chains terminated to amine groups, and were condensed with modified $C_{60}$ carrying a free carboxylic acid group. Chronopoulos et al. (2013) reported benzyne cycloaddition onto SWCNHs, via either conventional conditions or microwave irradiation. In this way, they introduced fused rings onto the graphitic skeleton of SWCNHs, resulting in soluble hybrid materials. Oakes et al. (2013) reported the method of electrophoretic deposition of SWCNHs on different substrates. Efficient deposition on nearly any material, including nonconductive substrates, was proved. Zhu et al. (2003a) proposed a strategy on SWCNH surface modification. The method relied on distinct molecular recognition properties of different functional groups toward the carbon graphitic structure. Hattori et al. (2004) reported the procedure of SWCNHs' fluorination. They concluded that with the rise in fluorination temperature, the nature of C–F bond changes from ionic to covalent; moreover, at 473 K, nanowindows are created, without a change in the shape of each horn and aggregate. Harada et al. (2008) reported the procedure of boron-doped SWCNHs synthesis, and reported the systematic study that the electrical feature was carried out by varying the boron concentration. Amano et al. (2010) discussed the chemical state of nitrogen chemisorbed on SWCNH structures. They concluded that nitrogen is chemisorbed at the graphitic carbon edges and forms a pentagonal or a hexagonal ring with the edge carbon atoms in the zigzag or armchair arrangements. N-SWCNHs were prepared by Yuge et al. (2014a) using $CO_2$ laser ablation of a graphite target under nitrogen atmosphere. Doped nitrogen quantities were about 1.7 and 1.1 atomic percent under nitrogen and nitrogen/Ar mixed atmospheres, respectively. The nitrogen atoms were implicated as the pyridine-like and threefold coordinated $sp^2$ bonding in graphene lattice of SWCNHs, and they were $p$-type dopants.

The surface electronic properties of SWCNHs can be easily modified by physical adsorption of $I_2$. The DC electrical conductivity of SWCNH film prepared by a dip-coating method increased with the iodine adsorption amount by a factor of almost 10 (Khoerunnisa et al. 2011). Chemical doping of SWCNHs with Li, K, and Br reveals an anomalously small charge transfer occurring in saturation-doped SWCNHs using these reagents (Bandow et al. 2000b). Also, microwave-assisted functionalization of SWCNHs was performed (Economopoulos et al. 2009).

### 4.4.3 Hybrids and Composites

Nanohorn–nanotube hybrids were also recently synthesized (Zobir et al. 2013). Palm olein was shown to be a cheap carbon source for the production. Fiber-like polyaniline (PANI)/carbon nanohorn composites were prepared by Maiti and Khatua (2014) as electrode materials for supercapacitors by a simple method that involves in situ polymerization of aniline in the presence of SWCNH in an acidic medium with noteworthy electrochemical performances. The prepared fiber-like PANI/SWCNH composites (PACNs) showed high specific capacitance.

Zhu et al. (2003b) reported the synthesis of a new class of binary nanomaterials from SWCNHs and nanoparticles by utilizing a bifunctional molecule as the bridging interconnector. Also, the studies were performed on the synthesis of nanohybrids for photoinduced electron-transfer processes (Sandanayaka et al. 2007, 2009a, Pagona et al. 2007b).

## 4.4.4 Encapsulation and Metal-Containing Hybrids

Reporting the encapsulation of different fullerenes inside SWCNHs, Kobayashi et al. (2014) concluded that the polarity of the functional group of the $C_{60}$ derivatives is the main factor in determining whether they can be encapsulated in SWCNHs. Li et al. (2003) reported the CVD method of encapsulation of Fe inside SWCNHs. They stated that the formation of Fe@SWCNHs is strongly influenced by the morphology of the catalyst applied during the encapsulation process. Bekyarova et al. (2005) described a direct route for the deposition of Pd nanoclusters on SWCNHs in a one-step reaction involving chemical reduction of metal ions in the presence of a polymer-stabilizer (see Figure 4.8). The applied strategy provides small Pd nanoclusters with an average diameter of 2.3 nm robustly attached to the nanotubular carbon.

The large diameter of SWCNHs allows various molecules to be easily incorporated in the hollow nanospaces (Miyawaki et al. 2006a). Gd–acetate clusters inside SWCNHs were transformed into ultrafine $Gd_2O_3$ nanoparticles with their particle size retained even after heat treatment at 973 K. Nanoparticles of $Gd_2O_3$ must be 2.3 nm in average diameter to be actually useful for magnetic resonance imaging. Rotas et al. (2008) described the preparation of new metallo-nanocomplex and show that the new materials are photoactive. Battiston et al. (2009) reported the coating of SWCNHs with anatase titanium oxide thin films by metalorganic CVD with titanium tetraisopropoxide $Ti(OiPr)_4$ as the precursor, concluding that the deposition process did not alter the typical structures of the SWCNHs. The hydrophilic properties of the titanium oxide coating allowed a stable dispersion of SWCNHs/$TiO_2$ in water, opening new perspectives for water-based nanofluids, biological sensing, or drug delivery systems. Three composites of SWCNH aggregates hybridized with carbon nanocapsules containing $Fe_3C$, Co, or Ag were produced by Kokai et al. (2009) by application of laser vaporization of graphite mixed with Fe, Co, or Ag in Ar gas. The average diameters of composites were 96, 90, and 85 nm, respectively; their SWCNH layers had similar thicknesses (17–18 nm on average). The authors reported a strong correlation

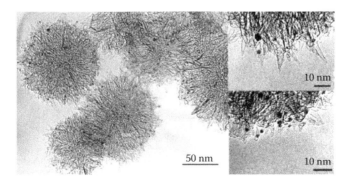

**FIGURE 4.8** TEM micrographs showing the distribution of Pd nanoclusters in Pd/o-SWCNHs (opened nanohorns). (Reprinted with permission from Bekyarova et al. 2005, pp. 3711–3714. Copyright (2005) American Chemical Society.)

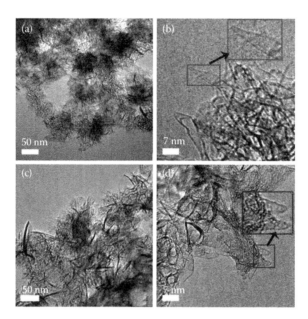

**FIGURE 4.9** Representative TEM images of (a, b) pristine SWCNHs and (c, d) SWCNH–TSCuPc nanohybrids. Insets of b and d show magnified views of the regions indicated. (Reprinted with permission from Jiang et al. 2014, pp. 18008–18017. Copyright (2014) American Chemical Society.)

between the diameters of the hybrids and the thicknesses of the SWCNH layers for the three types, suggesting that the formation mechanism of the three structures is based on the assembly of SWCNHs around a molten metal–carbon particle with certain ranges of lengths and diameters, respectively. Also, metallofullerenes have been encapsulated inside SWCNHs (Zhang et al. 2010a). The resulting material was conjugated with quantum dots, providing a dual diagnostic platform for in vitro and in vivo biomedical applications of the new carbonaceous materials.

Vizuete et al. (2010) described the preparation by simple self-assembly procedures of a new supramolecular assembly of SWCNH and a zinc porphyrin for light-sensitive nanohybrids. A novel and facile one-step method for the preparation of water-soluble SWCNHs and metal phthalocyanines hybrid for photothermal and photodynamic therapies was presented by Jiang et al. (2014)—see Figure 4.9.

## 4.5 Applications

Among the literature reports in major fields, the following applications of SWCNHs can be found: adsorption, catalysis, components applied for preparation of electrodes (and generally application in power sources), nanomedicine and sensors.

### 4.5.1 Catalysis

Nisha et al. (2000) studied the adsorption of liquid ethanol on SWCNHs. They stated that ethanol is adsorbed effectively. The amount of liquid ethanol adsorbed by SWCNHs is about 3.5 times larger than that of a super high surface area carbon. It was also found that SWCNHs acted as a catalyst for the oxidation of ethanol into acetaldehyde. Itoh et al. (2008a) reported the application of palladium nanoparticle-tailored SWCNHs or ox-SWCNHs as catalysts in $H_2/O_2$ gas phase reactions and liquid phase reactions such as Miyaura–Suzuki coupling. They observed increasing catalytic activity of Pd if it is applied with SWCNHs, especially for Pd–ox-SWCNHs. These high catalyst performances might be derived from the specific potential field of the intrahorn nanospaces and the support effects of ox-SWCNHs and Pd nanoparticles, in which the Pd(0) is expected to exhibit a different behavior from molecular Pd(0) species and bulk Pd(0). Thus, Pd-tailored SWCNH is a promising catalyst for both gas phase reactions and liquid phase organic reactions.

One year later, the same group reported a new SWCNH-based catalyst for efficient production of $H_2$ from $CH_4$ without $CO_2$ emission (Aoki et al. 2009). The authors prepared highly dispersed Pd nanoparticles on SWCNH and on ox-SWCNH without an antiaggregation agent. The Pd nanoparticle sizes on SWCNH and ox-SWCNH were around 2.5 and 2.7 nm, respectively. Each sample efficiently provides $H_2$ and hollow carbon nanofibers through $CH_4$ decomposition. The $H_2$ release over the Pd-dispersed SWCNH samples starts from ca. 820 K and is quite a large amount compared with that for a commercial Pd-activated carbon.

Since SWCNHs possess excellent catalytic properties, high purity, and low toxicity, they are suitable for bioelectrochemical application. Wen et al. (2010) reported for the first time the preparation of biofuel cell anode by using SWCNHs as the support for redox mediator and biocatalyst. SWCNHs promoted the electropolymerization of methylene blue (MB) and the resulting nanocomposite (poly–MB–SWCNHs) exhibited prominent catalytic ability to the oxidation of nicotinamide adenine dinucleotide. Glucose dehydrogenase was next immobilized on the poly–MB–SWCNHs–modified electrode for the oxidation of glucose. Employing Pt nanoparticles supported on functionalized $TiO_2$ colloidal spheres with nanoporous surface as cathode catalyst, the as-assembled glucose/$O_2$ biofuel cell operates at the physiological condition with good performance.

Recently sulfonated SWCNHs were shown to be effective catalysts for application in a biodiesel production (Poonjarernsilp et al. 2014). Methyl palmitate, a kind of biodiesel, was produced by the esterification of palmitic acid using sulfonated SWCNHs as catalysts. The catalytic activity was significantly higher than that using a homogeneous sulfuric acid catalyst. Also recently, the applications of PtRu nanoparticles dispersed on N-SWCNHs were reported (Zhang et al. 2014b). It was shown that this catalyst exhibits an obvious enhancement in the tolerance to carbonaceous intermediates and the electocatalytic activity for methanol oxidation reaction in comparison with commercial catalyst.

Additionally, SWCNH was shown as an effective bed inducing the growth of titanium dioxide nanopetals (Battiston et al. 2010), and as an effective support for the catalyst in the synthesis of hydrogen from methane and water (Murata et al. 2006). Kosaka et al. (2009) described the high-power direct methanol fuel cells (DMFCs) using SWCNHs to support the Pt catalyst—see Figure 4.10. The unique structure of the SWCNH aggregate is advantageous to support the fine Pt catalysts. To prevent the growth of Pt particles even under high–Pt content conditions, defects were intentionally created on the surface of SWCNHs by using $H_2O_2$ oxidation. This process reduced the mean diameter of the Pt particles supported on the SWCNHs to 2.9 nm, which is roughly two thirds that of the Pt particles on as-grown SWCNHs, for 60 wt% Pt content. The process was improved by the use of a membrane electrode assembly by immersing the catalyst electrodes in 50 vol% MeOH before the hot pressing process. This increased the active surface area of the Pt catalysts to nearly 100% of the Pt surface and strengthened the joint between the electrodes and the membrane. These improvements enabled us to obtain a power density of 76 mW/cm² at 0.4 V for a passive-type DMFC operated at 313 K (Kosaka et al. 2009). SWCNHs were also applied for the modification of glassy carbon electrode (GCE). Compared with a bare electrode, a modified electrode has stronger electrocatalytic activity for the oxidation of dihydroxybenzenes (Zhu et al. 2014a).

## 4.5.2 Nanomedicine and Nanopharmacology

As pointed out by Wang et al. (2014), the biggest challenge to the application of SWCNHs in medicine is the high retention ratio and inability to degrade in biological systems. These issues may be overcome by adding appropriate surface modifications or decreasing the size of the SWCNH aggregate by optimizing the conditions of synthesis or postsynthetic separation.

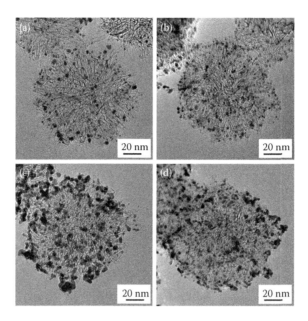

**FIGURE 4.10** TEM images of the supported catalyst: (a) 20 wt% Pt on as-SWCNHs, (b) 20 wt% Pt on treated-SWCNHs, (c) 60 wt% Pt on as-SWCNHs, and (d) 60 wt% Pt on treated-SWCNHs. (Reprinted with permission from Kosaka et al. 2009, pp. 8660–8667. Copyright (2009) American Chemical Society.)

SWCNHs were applied in a new image analysis method called the *spatial phantom evaluation of cellular thermal response in layers* (Sarkar et al. 2011, Whitney et al. 2013). Also, water-dispersed SWCNHs with antitumor activity are widely used as a potential drug carrier for local cancer chemotherapy (Figure 4.11) (Ajima et al. 2005, Miyawaki et al. 2006b, Sawdon et al. 2013, Chen et al. 2014, Guerra et al. 2014), and generally as drug carriers (Xu et al. 2008, Nakamura et al. 2011, Hood et al. 2013) and drug delivery systems for controlled release (Mendes et al. 2013, Jain et al. 2014, Ma et al. 2014).

It is interesting and important to note that some drugs, if dosed from the internal space of SWCNHs, are more effective in vivo than typically administrated drugs (Ajima et al. 2008). Here the nanowindow sizes (Ajima et al. 2006), as well as the chemical composition of nanowindows edges, is very important and strongly influences the kinetics of drug release. It is also very important to note that the results of recent studies show that orally administered SWCNHs were not absorbed into the body from the gastrointestinal tract (Nakamura et al. 2014). They accumulate in the liver and spleen and did not reveal obvious toxicological lesions (Zhang et al. 2013).

The problem of SWCNH distribution in cells is also widely studied. To do this, it is important to synthesize nanohorns/quantum dots nanocomposite, which can be applied in cell labeling and in vivo imaging (Li et al. 2014). Miyawaki et al. (2009) were the first who studied the biodistribution and ultrastructural localization of SWCNHs in mice. They stated that 80% of the material is accumulated in the liver in Kupffer cells but not in hepatocytes. Matsumura et al. (2014) studied the location of SWCNHs in the tumor tissue. Nanohorns were found in macrophages and endothelial cells within the blood vessels. Few SWCNHs were found in tumor cells or in the region away from blood vessels, suggesting that, under these study conditions, the enhanced permeability of tumor blood vessels was not effective for the movement of SWCNHs through the vessel walls. SWCNHs in normal skin tissue were similarly observed. Recently, Zimmermann et al. (2014) also studied the cellular distribution of SWCNHs.

Some new SWCNHs containing composites can be also applied as promising drug delivery systems—for example, SWCNH-supported liposomes (Huang et al. 2011). They are also promising candidates for the application in gene therapy (Guerra et al. 2014) also in the functionalized (with polyamidoamine

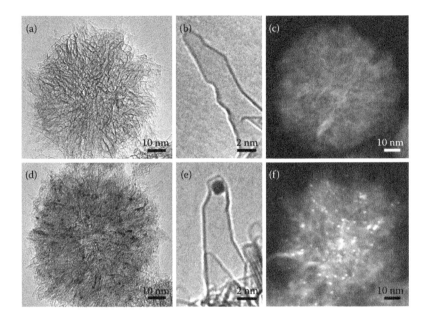

**FIGURE 4.11**   (a, b) HRTEM images of ox-SWCNH (scale bars of 10 and 2 nm, respectively). (c) Z-contrast image of ox-SWCNH aggregate (10 nm). (d, e) HRTEM images of cisplatin(CDDP)@ox-SWCNH (10 and 2 nm) in which black spots are cisplatin clusters. (f) Z-contrast image of CDDP@ox-SWCNH in which bright spots are cisplatin clusters (10 nm). (Reprinted with permission from Ajima et al. 2005, pp. 475–480. Copyright (2005) American Chemical Society.)

dendrimers) forms (Guerra et al. 2012). Also, zinc phthalocyanine-loaded SWCNHs are used in double photodynamic and hyperthermic cancer phototherapy (Zhang et al. 2008).

Another field is the application of SWCNHs in regenerative therapy. It was shown that SWCNHs are useful for bone formation (Kasai et al. 2011), and for application as bone tissue replacements (Stankova et al. 2014).

Near-infrared laser-triggered carbon nanohorns are also effective for selective elimination of microbes (Miyako et al. 2007). Also, near-infrared (NIR) laser-driven functional SWCNH complexes could open the way to a new range of antiviral materials (Miyako et al. 2008a). SWCNHs are additionally promising materials if applied as thermal enhancers when excited by near-infrared light for tumor cell destruction (Whitney et al. 2011).

### 4.5.3 Sensors

The attractive electrochemical and structural properties of SWCNHs suggest potential application of SWCNHs for electrocatalysis and biosensors (Liu et al. 2008). Also, the changes in electrical conductivity upon gas adsorption and nanowindow creation can be potentially applied in sensors (Urita et al. 2006). Sensors using SWCNHs can be fabricated by applying different more or less advanced techniques. For example, Suehiro et al. (2006) described an electrokinetic fabrication method for a gas sensor composed of SWCNHs using dielectrophoresis. Sano and Ohtsuki (2007) described the SWCNH-based sensor for detection of ozone in water. This sensor was based on the phenomenon that the electric resistance of the SWCNH film decreased with the adsorption of ozone molecules. Liu et al. (2008) utilized SWCNHs for the first time to fabricate glucose biosensor. This biosensor based on the Nafion–SWCNHs nanocomposite has high sensitivity, low detection limit, and good selectivity. Dai et al. (2013) described a simple and sensitive electrochemical sensor constructed with alkoxy silane and

SWCNHs. It was used to detect some foodborne contaminants, including malachite green, sudans, triclosan, and bisphenol A (Xu et al. 2013). SWCNHs have been used to construct an aptasensor for the first time by Zhu et al. (2011a). This novel sensor was successfully used for the detection of thrombin with high sensitivity, excellent selectivity, and detection limit of as low as 100 pM. Also, the application of SWCNHs in immunosensors has been reported (Ojeda et al. 2014). SWCNH–hollow Pt nanospheres were successfully applied as procalcitonin (a common clinical biomarker of septicemia) sensor. Higher sensitivity, improved stability, and ideal selectivity of the novel immunosensor compared with a sensor prepared using traditional methods were demonstrated (Liu et al. 2014a). In the study of Liu et al. (2010), SWCNHs were noncovalently functionalized with poly(sodium 4-styrenesulfonate). Heme protein myoglobin was adsorbed onto the surface of the functionalized SWCNHs to prepare the electrochemical biosensor. Also, other biosensors have been widely studied (Shi et al. 2009, Zhu et al. 2009a, 2011b, 2012, Barsan et al. 2014, Liu et al. 2014b). A sensitive electrochemical immunosensor was proposed by Zhang et al. (2010b) by functionalizing SWCNHs with analyte for microcystin-LR detection. Also, other immunosensors using SWCNHs are proposed and studied (Zhao et al. 2013, Yang et al. 2014b). Lu et al. (2012) proposed noncovalent functionalization of SWCNHs with Schiff-base cobalt(II) (Co-salen). The hybrid material provides a good electrochemical sensing platform and has potential applications in environment analysis and biosensors.

## 4.5.4 SWCNHs in Electrochemistry and Power Sources

SWCNHs have large surface area, high pore volume, good electric conductivity, and semiconductor character. Moreover, the application of SWCNHs in electrochemistry and catalysis has recently increased mainly by their metal-free synthesis and relatively easy suspensibility in organic solvents (Zhu & Xu 2010, Guldi & Sgobba 2011, Su & Centi 2013, Hussein 2015).

SWCNH hybrid materials can be used as catalysts (Bandow et al. 2000a, Yoshitake et al. 2002, Bekyarova et al. 2005, Itoh et al. 2008a,b, Kosaka et al. 2009, Zhu & Xu 2010). The application requires uniform nanoparticles attached to the surface in order to improve the durability of catalysts. The deposition of small Pd nanoclusters on ox-SWCNHs has been reported (Bekyarova et al. 2005, Itoh et al. 2008a,b). SWCNHs were also employed as a platinum catalyst support in electrodes of polymer electrolyte fuel cells (Yoshitake et al. 2002). Pt nanoparticles that could act as a catalyst in a fuel cell were incorporated into the nanospaces of SWCNHs (Yuge et al. 2004). Obtained Pt particle sizes were about 1.5–2.5 nm in the interior of SWCNHs. Recently, high-power DMFCs were made by using SWCNHs to support the Pt catalyst (Kosaka et al. 2009, Brandão et al. 2011). Moreover, production of hydrogen using methane and water at low temperature, mediated by EuPt supported on SWCNHs, was also reported (Murata et al. 2006). A highly durable and active nanocomposite cathode catalyst (Pt/N-SWCNH) was assembled with "unprotected" Pt nanoclusters and carbon material as building blocks (Zhang et al. 2012). The present Pt/N-SWCNH catalyst exhibited higher catalytic activity for oxygen reduction reaction and strikingly enhanced stability compared with a commercial Pt/C catalyst. A variety of materials has been examined as alternatives to Pt, but most of these had low activity and their performance deteriorated even further in use. Unni et al. (2015) published studies on an alternate electrocatalyst obtained by a simple surface modification of SWCNHs by simultaneous doping with Fe and N at 1173 K. The synthesized catalyst showed greater activity toward oxygen reduction reactions, both in terms of onset potential and half-wave potential, than commercial 40 wt% Pt/C. Moreover, the electrocatalytic activity of the catalyst was found to increase during the cycling of potential, as indicated by the positive shift in the onset potential of oxygen reduction. Under similar experimental conditions, Pt, on the other hand, was found to be susceptible to degradation.

SWCNHs have also been used as a good carbon matrix to support metal oxides used as anode materials for lithium ion batteries (LIBs) (Zhao et al. 2011a,b). Different SWCNH composites (i.e., with $Fe_2O_3$ [Zhao et al. 2011a], $Fe_3O_4$ [Deshmukh & Shelke 2013], $SnO_2$ [Zhao et al. 2011b], and $MnO_2$ [Wang et al. 2011, Lai et al. 2012]) can be prepared as an anode material for LIBs. The metal oxide/SWCNHs material shows high capacity and better cyclic stability than most simple metal oxide composites. The SWCNH matrix and nanosized $SnO_2$ delivered a high capacity of 530 mAh/g even after 180 cycles under a current density of 500 mA/g, which is better than most $SnO_2$ composites (Zhao 2011b). $Fe_2O_3$/SWCNHs composite shows excellent rate performance and cycle stability, even at a high current density of 1000 mA/g (Zhao 2011a). The $Fe_3O_4$/SWCNH nanocomposite exhibited a high specific capacitance of 377 F/g and delivered a stable discharge capacitance at a current density of 1 A/g over 1000 cycles between 0 and 1.2 V (Deshmukh & Shelke 2013). Moreover, a specific capacitance measured for the materials is significantly higher in comparison with composite electrodes of nanostructured carbon forms (CNT, graphene, carbon nanofiber, porous carbon, etc.) with $Fe_3O_4$ nanoparticles—the range of 72–135 F/g. The $MnO_2$/ SWCNH composite exhibits excellent electrochemical properties with long-cycling capacity and high-rate performance. The large capacity of 565 mAh/g for the $MnO_2$/SWCNH composite measured at a high current density of 450 mA/g after 60 cycles can be obtained (Lai et al. 2012). Interesting properties of $MnO_2$/SWCNH were also investigated by Wang et al. (2011)—see Figure 4.12. The maximum specific capacitances of 357 F/g for total electrode at 1 A/g were achieved in 0.1 M $Na_2SO_4$ aqueous solution. A novel SWCNH–sulfur composite (with high sulfur content up to 76%) was synthesized via a straightforward melt infusion strategy by Wu et al. (2014). The composite exhibited excellent electrochemical performance with a high capacity of 693 mAh/g retained after 100 cycles at a high rate of 1.6 A/g. Yuge et al. (2014b) tried to prepare novel composites of graphite, vapor-grown carbon fibers (VGCFs), and carbon nanohorns covered by carbon films (C-graphite/VGCF/SWCNH). Obtained materials were fixed and unified by carbon film. The C-rate properties of half-cell for C-graphite/VGCF/SWCNH were superior to those for the mixture of graphite, VGCF, and SWCNHs (C-graphite/VGCF/SWCNH [discharge]: 3C/0.1C, 85%; graphite/VGCF/SWCNH [discharge]: 3C/0.1C, 50%), accelerating a promising application for quick charge–discharge of LIBs.

Also, a potential application of SWCNHs to electrodes for supercapacitors has been recently reported (Noguchi et al. 2007a, Yang et al. 2007, Wei et al. 2010, Hiralal et al. 2011, Izadi-Najafabadi et al. 2011, Yuge et al. 2012b, Jung et al. 2013, Meyyappan 2013, Maiti & Khatua 2014). A double layer-based hybrid

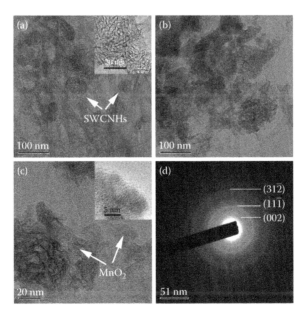

**FIGURE 4.12** (a) TEM images of the SWCNH/SWCNT thin film electrode; (b, c) TEM images and (d) selected-area electron diffraction of the $MnO_2$/SWCNH/SWCNT thin film electrode. (Reprinted with permission from Wang et al. 2011, pp. 4185–4189. Copyright (2011) American Chemical Society.)

capacitor electrode, based on a hierarchical combination of SWCNHs and SWCNTs (Noguchi et al. 2007a, Wei et al. 2010, Maiti & Khatua 2014) or polyaniline nanobrushes (Wei et al. 2010, Maiti & Khatua 2014), was also investigated. The SWCNHs can be considered as a promising candidate for electrode material of supercapacitors, owing to their porous structure and novel physical properties. Their potential as a supercapacitor electrode has been proven by initial research showing superior performance compared with activated carbon when operated in aqueous electrolyte (Yang et al. 2007). Yuge et al. (2012b) investigated the relationship between the specific surface area of ox-SWCNH and electrochemical capacitive properties. The authors found that an electrochemical device with ox-SWCNH showed an excellent specific capacitance of about 100 F/g, significantly higher in comparison with raw material (~60 F/g). Noguchi et al. (2007a) presented that the capacitance depended on the opening of SWCNH and charging time for organic electrolyte, i.e., $(C_2H_5)_4NBF_4$ in propyrene carbonate. Jung et al. (2013) studied the relationship between electrical conductivity of SWCNHs and accessibility of the electrolyte ions in the SWCNH-based supercapacitor. After heat treatment of SWCNHs, the ratio of $sp^2/sp^3$ carbons dramatically increased, suggesting an enhanced electrical conductivity of the SWCNHs. Even though the specific surface area slightly decreased by 16% as a result of heat treatment, the specific capacitance per surface area of the SWCNH electrode remarkably increased from 22 to 47 $\mu F/cm^2$. Such a result indicated an explicit increase in accessible effective surface area by electrolyte ions. Composites made with SWCNT (20%)/SWCNH (80%) showed exceptionally high power density of 990 kW/kg and a high durability with only 6.5% decline in capacitance over $10^5$ cycles (Izadi-Najafabadi et al. 2011). The electrodes with 100% SWCNTs, although having higher surface area, showed lower power density at 400 kW/kg. PACN composites were prepared by Maiti and Khatua (2014) as electrode materials for supercapacitors by a simple method that involves in situ polymerization of aniline in the presence of SWCNH in acidic (HCl) medium with noteworthy electrochemical performances. The prepared composites showed a high specific capacitance value of ~834 F/g at 5 mV/s scan rate compared with ~231 F/g for pure PANI and nanohorns (~145 F/g) at the same scan rate of 5 mV/s. In addition, the composites indicated high electrical conductivity on the order of ~$6.7 \times 10^{-2}$ S/cm, which responded to the formation of continuous interconnected conducting network path in the PACN composites. Recently, SWCNHs were applied for the modification of glassy carbon electrode. This modified electrode exhibited high electrocatalytic activities toward the oxidation of uric acid, dopamine, and ascorbic acid by significantly decreasing their oxidation overpotentials (Zhu et al. 2009a). Zhu et al. (2014b) also reported a simple method for the direct and quantitative determination of L-tryptophan (Trp) and L-tyrosine (Tyr) using a GCE modified with SWCNHs. The SWCNH-modified GCE exhibited high electrocatalytic activity toward the oxidation of both Trp and Tyr. In addition, the modified GCE displayed good selectivity and good sensitivity, thus making it suitable for the determination of Trp and Tyr in spiked serum samples. SWCNH paste electrode was used for amperometric determination of concentrated hydrogen peroxide ($H_2O_2$) (Zhu et al. 2008). In another work, a new approach based on the direct growth of SWCNHs on carbon microfiber substrates was developed for the straightforward fabrication of freestanding (binderless) electrodes for electrochemical applications (Aïssa et al. 2009). Preliminary studies demonstrated their possible application in LIBs.

It is striking that, despite excellent properties of SWCNHs, only a few groups have reported their implementation into solar energy devices, such as photoelectrochemical cells (Vizuete et al. 2010, Mercatelli et al. 2011, Pagona et al. 2012, Costa et al. 2013, Cruz et al. 2013, Casillas et al. 2014, Hussein 2015, Lodermeyer et al. 2015). A dimeric porphyrin and carbon nanohorns hybrid was adsorbed on nanostructured OTE/$SnO_2$ to construct a photoactive electrode by Pagona et al. (2012). They obtained materials revealing photocurrent and photovoltage responses with an incident photon-to-current conversion efficiency value as large as 9.6%, without the application of any bias voltage. Vizuete et al. (2010) prepared a new supramolecular assembly of SWCNH and a zinc porphyrin for light-sensitive nanohybrids by relatively simple self-assembly procedures. Photoelectrochemical cells of this material were fabricated onto OTE/$SnO_2$ electrodes. This electrode exhibited a maximum incident photon-to-current conversion efficiency of 9%. Hussein et al. (2015) and Mercatelli et al. (2011) investigated the scattering and absorption properties of nanofluids consisting of aqueous suspensions of SWCNHs. The characteristics of these nanofluids were evaluated in order to use them as direct sunlight absorber fluids in solar devices. The differences in optical properties induced by carbon nanoparticles compared with those of pure water led to considerably higher sunlight absorption with respect to the pure base fluid. The authors concluded that the carbon nanohorns could efficiently be used for increasing the overall efficiency of the sunlight-exploiting device. SWCNHs and their respective oxidized analogs have been used to fabricate novel doped $TiO_2$ photolectrodes for dye-sensitized solar cells (DSSCs) (Costa et al. 2013). The results published by Costa et al. (2013) clearly demonstrated that all of the nanocarbons (graphene, partially reduced graphene oxide, SWCNTs, ox-SWCNTs, SWCNHs, and ox-SWCNHs) significantly enhance the $TiO_2$ electrode characteristics and, subsequently, the photoresponse of the device with respect to standard $TiO_2$ electrodes. The most outstanding finding of the work is that SWCNH derivatives are also a valuable dopant for fabricating highly efficient DSSCs. The catalytic activity of SWCNHs as counterelectrodes (CEs) of DSSCs was studied for the iodide/triiodide redox reaction by Cruz et al. (2013). A half-cell configuration was used, and CE performances were compared with the performance of CEs made of carbon black and Pt. A CE assembled with high surface SWCNH and mixed with 10 wt% of hydroxyethyl cellulose was found to have the highest electrocatalytic activity among all the carbon-based CEs tested when annealed at 453 K (charge-transfer resistances equal to 141 $\Omega$ cm$^2$). Carbon nanohorns were implemented into solid-state electrolytes for highly efficient solid-state and quasi-solid-state DSSCs (ssDSSCs) by Lodermeyer et al. (2015). They featured an effective catalytic behavior toward the reduction of $I_3^-$ and enhance the $I_3^-$ diffusivity in the electrolyte. Thereby, ssDSSCs without SWCNHs exhibited rather poor performance of 0.42%, which was improved up to 2.09% upon the addition of SWCNHs. In a final device, solar cells with 7.84% efficiency at room temperature were achieved. Casillas et al. (2014) demonstrated the impact that interlayers of pristine SWCNHs exert on the charge transfer across the fluorine-doped tin oxide FTO/$TiO_2$ interfaces and the charge recombination at FTO/electrolyte interface. The implementation of SWCNH interlayers with thicknesses ranging from 5 to 15 nm that were placed in between FTO and $TiO_2$ provided the same or better performance as the standard $TiCl_4$ pretreatment. This procedure assisted in establishing an easy and environmentally friendly process to eliminate the $TiCl_4$ oxidation step from the DSSC fabrication process.

## 4.5.5 SWCNH as Adsorbents

Low-temperature $N_2$ adsorption is one of the techniques commonly used to detect and characterize the porosity of different adsorbents (Sing 2001, 2004). Therefore, $N_2$ adsorption measurements have also been widely used to study the structure of SWCNHs (Murata et al. 2000, 2001a,b, 2002b, Bekyarova et al. 2002a,b,c, 2003, Yang et al. 2004, 2005, Kowalczyk et al. 2005, Tanaka et al. 2005a, Utsumi et al. 2005, Matsumura et al. 2007, Comisso et al. 2010, Hashimoto et al. 2011, Ohba et al. 2011, 2012a, 2012b, Sano et al. 2014b). For example, Bekyarova et al. (2003) indicated analyzing the changes in $N_2$ adsorption isotherms that SWCNHs' oxidation by $CO_2$ provided a sufficient nanohorn opening. Kowalczyk et al. (2005) determined the distribution of diameters of internal cylindrical channels of SWCNHs' sample. They found (using local isotherms obtained from Monte Carlo simulations and Derjaguin–Broekhoff–de Boer theory) that the distribution of SWCNH sizes was broad (internal pore radii varied in the range of 1.0–3.6 nm with the maximum at 1.3 nm). A similar conclusion was also provided by Furmaniak et al. (2006), who analyzed the same experimental $N_2$ adsorption isotherm using simple analytical model.

The structure of SWCNHs has also been detected by measurements of noble gases adsorption (Zambano et al. 2002, Krungleviciute et al. 2008, 2009, 2013). Zambano et al. (2002) studied xenon adsorption on aggregates of SWCNHs. The binding energy value for Xe on the SWCNHs was intermediate between those for Xe on SWCNTs and that for Xe on graphite. Krungleviciute et al. (2008) explored the adsorption behavior of Ne and $CF_4$ on closed dahlia-like SWCNHs aggregates. The isotherms for both gases showed the presence of two substeps before monolayer completion. The authors concluded that this universal behavior was the result of adsorption in interstitial spaces of varying dimension in the aggregates. The same group (Krungleviciute et al. 2009) also measured Ne adsorption isotherms on a sample of closed, mostly bud-like, aggregates of SWCNHs and compared the results with the previous one (on dahlia-like SWCNHs). While for the closed dahlia-like aggregates, the data showed that there were two groups of sites with different binding energies present at relatively low loadings, only one group of adsorption energies was present for the closed bud-like aggregates at comparable loadings. They also stated that, as for Xe adsorption, the binding energy of the sites present on the bud-like aggregates has a value intermediate between that obtained for adsorption on planar graphite and that measured on the grooves of close-ended bundles on nanotubes for the same adsorbate. Krungleviciute et al. (2013) also reported the results of a thermodynamics and kinetics study of the adsorption of neon and carbon dioxide on aggregates of chemically opened SWCNHs. The adsorption characteristics differed for these two adsorbates—for neon, the adsorption isotherms displayed two steps before reaching the saturated vapor pressure, in contrast to carbon dioxide, which did not reveal any substeps in the adsorption isotherms. The observed difference was explained in terms of differences in the relative strengths of adsorbate–adsorbate to adsorbate–sorbent interaction for these species.

SWCNHs have also been tested as adsorbents for hydrogen storage (Murata et al. 2002b,c, Tanaka et al. 2005a,b, Fernandez-Alonso et al. 2007, Noguchi et al. 2007b, Comisso et al. 2010, Liu et al. 2012, Sano et al. 2014b). Murata et al. (2002b) studied the adsorption of supercritical hydrogen on SWCNHs. They observed the presence of two types of physical adsorption sites of interstitial and internal spaces on SWCNH assemblies. $H_2$ density was higher in interstitial spaces (especially for ambient temperature) due to the additional stabilization by the strong fluid–fluid interaction and the cluster formation. However, $H_2$ molecules adsorbed in the interstitial spaces could not form the stable cluster owing to the space limitation. Comisso et al. (2010) investigated the effect of thermal oxidation on the $H_2$ storage properties of SWCNHs. They demonstrated that oxidation of the material was connected with more than three times higher adsorption (at low temperature). Fernandez-Alonso et al. (2007) studied the effects of confining molecular hydrogen within SWCNHs via high-resolution quasielastic and inelastic neutron spectroscopies. The results showed that $H_2$ interacted far more strongly with SWCNHs than it did with CNTs and suggested that nanohorns and related nanostructures may offer significantly better prospects as lightweight media for hydrogen storage applications. Tanaka et al. (2005a,b) measured $H_2$ and $D_2$ adsorption on SWCNHs. They observed the difference between hydrogen isotopes' adsorption, which increased as pressure decreased. The results confirmed that quantum effects on $H_2$ adsorption depended on pore structures and were very important even at 77 K. Sano et al. (2014b) prepared SWCNHs containing Pd–Ni alloy nanoparticles (Pd–Ni/SWCNHs) and investigated high-pressure $H_2$ absorption by these nanoparticles. The amount of $H_2$ absorbed by Pd–Ni/SWCNHs was larger than the predicted combined absorption contributed by Pd–Ni alloy and pure-carbon SWCNHs. The observed synergetic $H_2$ absorption was induced by the combination of Pd–Ni alloy nanoparticles and SWCNHs and occurred because of a spillover effect. Liu et al. (2012) examined metal-assisted hydrogen storage, studying $H_2$ adsorption on SWCNHs decorated with Pt nanoparticles. The performed measurements clearly proved additional excess storage measured at low temperatures induced by metal-assisted activated processes at room temperature.

Adsorption of other gases important for industry and environment, like $CH_4$ (Murata et al. 2005, Noguchi et al. 2007b, Hashimoto et al. 2011, Ohba et al. 2011, 2012a) and $CO_2$ (Ohba et al. 2011, Krungleviciute et al. 2012, 2013), has also been studied on SWCNHs. These investigations confirmed unique properties of these materials. For example, Hashimoto et al. (2011) examined vibrational-rotational properties of $CH_4$ adsorbed on the nanopores of SWCNHs using IR spectroscopy. The obtained results indicated that the rotation of $CH_4$ confined in the nanospaces of SWCNHs was highly restricted, resulting in a rigid assembly structure, which was an anomaly in contrast to that in the bulk liquid phase. Krungleviciute et al. (2012) showed that the behaviors of $CO_2$ adsorbed on SWCNHs were opposite to what was observed for adsorption of other simple gases (like argon, methane, or neon). Ohba et al. (2011)

studied, among others, the adsorption of $CH_4$ and $CO_2$ on SWCNHs with superuniform nanosized gates donated by controlled oxidation. The performed modification conferred a high selectivity for $CO_2$ and a large adsorption capacity. In addition, Ohba et al. (2011) also stated that highly selective separation of other molecules could also be possible by further controlling the nanogate size.

Bekyarova et al. (2002a) reported experimental data of water adsorption on SWCNHs. The analysis of the data supported a cluster-mediated model for filling the interstitial and intratube nanospaces in contrast to the monolayer formation observed for simple nonpolar molecules. The enthalpies of water immersion of SWCNHs showed a very weak interaction between the water molecules and the hydrophobic carbon nanotubular structure and were lower than the values for graphite. Ohba et al. (2012b) studied the water penetration mechanism through zero-dimensional nanogates of SWCNHs. They observed that $H_2O$ vapor adsorption via the nanogates was delayed in the initial adsorption stage but then proceeded at a certain rate. The mechanism water cluster–chain–cluster transformation via the nanogates was proposed, i.e., the growth of water clusters in internal nanospaces facilitated water penetration into these nanospaces. Fan et al. (2004) analyzed the influence of water on desorption of benzene adsorbed on SWCNHs. They found that the rate of benzene desorption that was touched by water droplets or exposed to water vapor in advance became faster in dry air, but slowed in air saturated with water vapor.

Some studies on SWCNHs have also been based on magnetic properties of adsorbed molecules (Bandow et al. 2005, Matsumura et al. 2007). Bandow et al. (2005) showed that magnetic susceptibilities for $O_2$ adsorbed on heat-treated SWCNHs showed anomalous features as compared with those on as-grown SWCNHs. Magnetism for the latter case was similar to that for bulky solid $O_2$ in contrast to $O_2$ adsorbed on heat-treated SWCNHs, which may get into the interior of SWCNH and solidify. Matsumura et al. (2007) investigated the magnetism of organic radical molecules confined in nanospace of ox-SWCNHs. The work showed that studied molecules adsorbed on SWCNHs did not form any ordered structures, but the formation of disordered clusters was suggested.

Ajima et al. (2004) studied the mechanism of materials storage in the inner space of SWCNHs. They tried to visualize this process by adsorption of $C_{60}$ fullerenes. They concluded that the mechanism of storage was not regular and fullerene molecules were heterogeneously distributed and mainly localized near the nanohorn tips—$C_{60}$ molecules behaved in a different way than what was observed for CNTs, where they were packed regularly. Fan et al. (2008a) investigated how the hole size affected the incorporation of $C_{60}$ in ox-SWCNHs and their release. They observed that the release rate of $C_{60}$ from inside SWCNH was slower than that of $C_{60}$ and were not influenced by the hole sizes. Miyako et al. (2008b) demonstrated that $C_{60}$ molecules encapsulated in SWCNHs may be released by NIR laser irradiation, suggesting that SWCNHs are potentially useful for a photocontrolled release of drugs. Miyawaki et al. (2007b) found that ethanol vapor forced $C_{60}$ encapsulated inside SWCNHs to exit from them and, in reverse, toluene vapor forced the ejected $C_{60}$ to reenter.

SWCNHs have also been used as adsorbents from solutions (Ohkubo et al. 2005, Pagona et al. 2006b, Yuge et al. 2008, Zhu et al. 2009b, Shinke et al. 2010, Jiménez-Soto et al. 2012a,b, Yamazaki et al. 2013, Fresco-Cala et al. 2014). Yuge et al. (2008) studied the interaction of tetracyano-p-quinodimethane (TCNQ) with SWCNHs. They noticed that the amounts of TCNQ adsorbed on the inside and outside of SWCNH corresponded to about 20% and 60% of respective monolayer coverage, which did not increase even when the TCNQ was overdosed. Furthermore, Raman spectra suggested that the quinoid rings of TCNQ developed benzene-like character—charge was transferred from SWCNH to TCNQ. Pagona et al. (2006b) investigated porphyrin adsorption on SWCNHs. The obtained results suggested charge separation from the photoexcited tetracationic porphyrin to SWCNH. Shinke et al. (2010) and Yamazaki et al. (2013) tested SWCNHs as adsorbent in plasma apheresis on example of bilirubin adsorption. They stated that bilirubin was selectively adsorbed against albumin, especially by SWCNHs with high oxidation levels. Ohkubo et al. (2005) elucidated (using X-ray absorption fine structure technique) the structure of RbBr aqueous solution confined in the interstitial nanospaces of SWCNH assemblies. The performed analysis indicated that the highly distorted and dehydrated structure around ions of aqueous solution confined in the nanospace formed quasi-1D structure in the interstitial nanospaces of SWCNH. This hydration structure was not only unique and quite different from that of the bulk aqueous solution, but also confined in the slit-shaped nanospaces. SWCNHs have also been tested as adsorbents for solid-phase extraction (Zhu et al. 2009b, Jiménez-Soto et al. 2012a,b, Fresco-Cala et al. 2014).

### 4.5.6 Other Applications

SWCNHs have also been applied as pseudostationary and stationary phases for electrophoresis (Jiménez-Soto et al. 2010), in microsolid-phase extraction (Jiménez-Soto et al. 2013, Roldán-Pijuán et al. 2014), and in photodynamic therapy (Sandanayaka et al. 2009b).

## 4.6 Theoretical Studies on SWCNHs

Experimental investigations on SWCNHs have been supported by theoretical studies. Computer simulations and quantum chemical calculations have been used both for SWCNH structure modeling and for examining of their properties.

A lot of theoretical studies on adsorption on SWCNHs have been based on open-ended conical carbonaceous structures (Majidi & Tabrizi 2010, Gotzias et al. 2011) or on an artificial model constructed from two parts: CNT-like and conical one (Ohba et al. 2005, 2012b, Tanaka et al. 2005b). In the second case, the connection between both parts has been neglected. In some model studies, properties of SWCNH assemblies have been tested using simplified models constructed only from CNTs (Ohba et al. 2001, 2012a, Tanaka et al. 2005a, Hashimoto et al. 2011). In contrast, Furmaniak et al. (2013a) generated (using Monte Carlo simulations and realistic potential of carbon–carbon interactions) a series of fully atomistic energetically stable models of SWCNHs having different geometric parameters (apex angle, diameter, and length)—see Figure 4.13. The model structures consisted of conical and tubular parts but their bottoms were also closed by graphene sheets. Using these model SWCNHs during simulations of Ar adsorption, we showed that for the calculation of the pore size distribution curve for experimental systems, one can use the approximation of local isotherms by the isotherms simulated for SWCNTs. However, for the modeling of applications of SWCNHs as nanocontainers or nanoreaction chambers, the application of the atomistic model is necessary, since the values of the adsorption energy suggested a very large influence of the tips on the energetics of the adsorption process.

Hawelek et al. (2013) modeled the structure of dahlia-type SWCNH aggregates using classical MD simulations. They constructed the structure consisting of an outer part formed from SWCNHs with diameters of 2–5 nm and a length of 40 nm and an inner turbostratic graphite-like core with a diameter

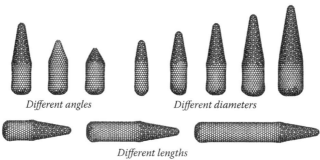

*Different angles*     *Different diameters*

*Different lengths*

**FIGURE 4.13** Schematic representation of considered model nanohorns having different angles, lengths, and diameters.

of 13 nm. They showed that the diffracted intensity and the pair correlation function computed for such a constructed model were in good agreement with the neutron diffraction experimental data, which confirmed the model validity. Tsai and Jeng (2014), using MD simulations, investigated the heat treatment-induced coalescence of SWCNHs with different conical angles. They found the formation of a metastable venturi-like structure from two individual nanohorns connected with successive Stone–Wales transformations following initial junctions. The simulation results revealed that the coalescence temperature and the melting temperature were both dependent on the conical angle of the nanostructure (these temperatures were lower in the nanohorn junctions with sharp nanohorns). Fan et al. (2008b) used tight-binding MD simulations to elucidate the diameter dependence of the hole closing. They stated that in the case of small holes, the low-frequency vibration squashed the tube structure, and as soon as the opposite edges came close together, they connected. When the holes were larger, the edges were further apart, so they came close together less frequently, and the hole closing took a longer time.

Molecular simulations have also been used to study adsorption properties of SWCNHs (Ohba et al. 2001, 2005, 2012a,b, Tanaka et al. 2005a,b, Majidi & Tabrizi 2010, Gotzias et al. 2011, Terzyk et al. 2012, Furmaniak et al. 2013a,b, 2014, Kowalczyk et al. 2014). Majidi and Tabrizi (2010) investigated Ne adsorption on single-walled carbon nanocones (SWCNCs) using MD simulations. They found that temperature had a considerable effect on saturation coverage and saturation pressure. The adsorption coverages on SWCNCs were greater than those on CNTs, while the binding energy was smaller. Ohba et al. (2001, 2005) simulated $N_2$ adsorption on SWCNHs using the grand canonical Monte Carlo (GCMC) technique. The detailed comparison of the simulated adsorption isotherm with the experimental isotherm in the internal nanospaces provided 2.9 nm of the average pore width of the internal nanospaces. They also stated that the interstitial nanopores of SWCNH assemblies can be regarded as quasi-1D pores due to the partial orientation of the SWCNH particles (the average pore width of the interstitial pores was 0.6 nm), which was confirmed by a good agreement between the GCMC simulation and an experimental adsorption isotherm in a low-pressure region.

$H_2$ adsorption modeling has also attracted a great interest (Tanaka et al. 2005a,b, Gotzias et al. 2011, Kowalczyk et al. 2014). For example, Tanaka et al. (2005a,b) simulated $H_2$ and $D_2$ adsorption on SWCNHs. They observed a good agreement between the simulated and experimentally measured results, which suggested that the hydrogen isotope adsorption can be successfully explained with the use of the effective Feynman–Hibbs potential. The hydrogen isotopes were preferentially adsorbed in the conical part of the SWCNH. Because of a strong potential field and quantum effects, the density of adsorbed $H_2$ inside the SWCNH was smaller than that of $D_2$. The difference between $H_2$ and $D_2$ adsorption increased as pressure decreased because the quantum spreading of $H_2$, which was wider than that of $D_2$, was fairly effective at the narrow conical part of the SWCNH model. This indicated that quantum effects on hydrogen adsorption depended on pore structures and were very important even at 77 K. Kowalczyk et al. (2014) presented the first in silico modeling of the Pd–H–SWCNH nanocomposites. The temperature-quench Monte Carlo simulations were applied to generate the most stable morphologies of Pd clusters deposited inside the SWCNH. The optimized nanocomposites were next used in calculating the hydrogen binding energy distributions. We concluded that small Pd nanoclusters stabilized inside ox-SWCNHs were responsible for the observed experimentally enhanced $H_2$ storage.

Ohba et al. (2012a) investigated $CH_4$ storage in SWCNHs using, among others, GCMC simulations. They considered the adsorption mechanism in the internal and external nanopores of SWCNHs. The adsorption of larger amounts of $CH_4$ in the narrow nanopores at pressures lower than 3 MPa was the result of strong adsorption potential fields. In the wider nanopores (where higher density was observed above 3 MPa), $CH_4$ molecules were stabilized by trimer formation, which compensated for the weak potential fields and enabled high-density adsorption and adsorption of large amounts of $CH_4$. Furmaniak et al. (2013b, 2014) examined $CO_2/CH_4$ mixture adsorption and separation inside the above-mentioned model SWCNHs. Our comparison of the equilibrium separation factor computed for SWCNHs and their CNT counterparts clearly showed that the horn-shaped tip of SWCNHs pronounced effects on the $CO_2/CH_4$

separation efficiency. We stated that key structural properties responsible for the excellent separation performance of SWCNHs were a large surface-to-volume ratio and unique defective morphology.

Ohba et al. (2012b) performed MD simulations to examine the mechanism of water penetration through zero-dimensional nanogates of a SWCNH. Water vapor adsorption via the nanogates was delayed in the initial adsorption stage but then proceeded at a certain rate. These simulation results confirmed a water cluster–chain–cluster transformation via the nanogates mechanism; i.e., the growth of water clusters in internal nanospaces facilitated water penetration into these nanospaces. Terzyk et al. (2012) presented MD simulation results of $C_{60}$ fullerenes' adsorption inside a SWCNH. We demonstrated a relatively good agreement between simulation and experiment. We also showed that the confinement of water and ethanol inside a carbon nanohorn strongly changed the properties of confined liquids leading to a decrease in the number of hydrogen bonds, and diffusion coefficients in comparison to bulk and the appearance of $C_{60}$ molecules inside the SWCNH led to further decrease in diffusion coefficients of confined solvents.

The studies of the electronic structure of nanohorns, fullerenes, nanotubes, and objects of more complicated geometry is not easy, and yet a few interesting experimental and theoretical results were obtained in this area (Kolesnikov & Osipov 2009). The role of topological defects and heteroatoms should be particularly emphasized.

The electronic structures of five SWCNH isomers ($C_{115}H_{25}$, $C_{158}H_{26}$, $C_{135}H_{21}$, $C_{102}H_{16}$, and $C_{118}H_{16}$) were calculated by the discrete variational X$\alpha$ method (Amano & Muramatsu 2009). Comparing the densities of state (DOSs) of the pentagonal and hexagonal rings among the SWCNH isomer models indicated that there is little difference in DOS among the isomers. This suggested that the electronic structures of the pentagonal and hexagonal rings near pentagonal rings can be determined by their local structure, but not the arrangement of pentagonal rings in the hexagonal carbon layer. Comparing the DOSs among the pentagonal and hexagonal rings in SWCNH to the hexagonal rings in graphite suggests that the pentagonal rings have less covalent character than hexagonal rings in graphite. In addition, the hexagonal rings near the pentagonal rings in SWCNH have an intermediate character. The next interesting aspect of the stability of nanohorns/nanocones is the inclusion of noncarbon atoms into the network of carbon nanostructures. Of course, this procedure modifies the electronic and chemical properties due to variations in electronic structure. A density functional theory (DFT) study was carried out to predict the relative reactivities of different sites on the external surface of pristine and nitrogen-doped carbon nanocones (Esrafili & Mahdavinia 2013). These authors showed that the presence of N indeed improves the surface reactivity of the carbon structures.

In the field of theoretical chemistry, there is a great lack of studies regarding inclusion molecules on carbon nanocomposites, which can be explained by the relatively large size of these structures and also by the inherent difficulties on describing noncovalent interactions (Basiuk & Irle 2008). Stability of these systems can be elucidated using, for example, the DFT approach. Molecular orbital calculations and charge population were also performed, seeking an understanding of the nature of intermolecular interactions existing in the complexes studied. The carbon nanohorn due to severe curvature could bind hydrogen molecule through enhanced binding at the top section adjacent to its closed top end (Chen et al. 2010). The storage capacity limited by the room at the top end section is only 1.8 wt% for hydrogen to be captured inside the studied nanohorn (8.0 wt% on the outer one). An extensive search for configurations of the Li atoms adsorbed on carbon nanohorn was also carried out. The Li atoms were found to be distributed separately on the sidewalls of the nanohorn rather than aggregate at low Li atom content. The adsorbed Li atom could enhance the binding strength of the hydrogen molecules. The binding energy of hydrogen molecules adsorbed around the Li atoms would not be altered much if both sidewalls of nanohorn were capped by Li atoms. Chen et al. (2010) concluded that the nanohorn studied in their paper may store 8.6 wt% hydrogen by adsorbing $H_2$ on both sidewalls. The interactions of a gold atom with a nanohorn conic tip and their complexes with a CO molecule were studied by Khongpracha et al. (2008) using the DFT method. The authors identified five sites on both the SWCNT tip and the nanohorn where attachment of a gold atom leads to a stable complex. Moreover, the electronic characteristics of Au/SWCNH are very sensitive upon adsorption of the CO molecules. This result suggested that nano-horns could be one of the promising candidates for the development of high-specificity nanosensors. DFT calculations indicated a lack of one preferred binding site if chemically bonded H occurred in SWCNH samples (Liu et al. 2012). This is not unexpected due to the great variety of curvatures inherent in the different adsorption sites on SWCNHs. Additionally, Liu et al. (2012) could not identify the exact nature of the binding of the additional hydrogen in these computer experiments, confirming the loss of molecular hydrogen and significant metal-assisted hydrogen storage on Pt–SWCNHs that is activated at $T > 150$ K (experimental measurements). Petsalakis et al. (2007) studied the 217 carbon atom model of a nanohorn and a fused pyrrolidine ring at different sites in an effort to resolve questions raised on the validity of the quantum chemistry calculations. Nanohybrids consisting of a carbon nanohorn and different electron donor substituent groups at the fused pyrrolidine ring have been calculated by DFT and time-dependent DFT methods. Significant binding is found at sites closer to the conical tip. DFT calculations have been carried out on SWCNH nanohybrids with photo- or electroactive substituents, porphyrin, pyrene, and tetrathiafulvalene, of interest for photoinduced charge transfer. It is found that the relevant excitations in the individual substituents retain their characteristics in the nanohybrid systems. A paper written by De Souza et al. (2013) reported a quantum chemical investigation of the inclusion complex formation between a carbon nanohorn structure and cisplatin molecule, using the DFT. The published results indicated that the SWCNH and cisplatin can indeed form a stable inclusion complex, with the calculated $^1$H NMR and $^{15}$N NMR chemical shifts for cisplatin atoms revealing very substantial changes due to complex formation.

## 4.7 Perspectives

Future possible applications of SWCNHs were given by Zhu and Xu (2010). They pointed out that the controllable creation of holes (nanowindows) can be used for gas separation. Also, there are many per-spectives in the application of SWCNHs as drugs and drug carriers. Also, catalytic properties are very promising. There are also some possible applications in chromatography and microextraction. In our opinion, SWCNHs, due to their interesting properties, become the most important carbon nanomaterial.

## References

Aïssa, B., Hamoudi, Z., Takahashi, H. et al., "Carbon nanohorns-coated microfibers for use as free-standing electrodes for electrochemical power sources," *Electrochem. Commun.* 11 (2009): 862–866.

Ajima, K., Yudasaka, M., Suenaga, K. et al., "Material storage mechanism in porous nanocarbon," *Adv. Mater.* 16 (2004): 397–401.

Ajima, K., Yudasaka, M., Murakami, T. et al., "Carbon nanohorns as anticancer drug carriers," *Molec. Pharm.* 2 (2005): 475–480.

Ajima, K., Maigné, A., Yudasaka, M. et al., "Optimum hole-opening condition for cisplatin incorporation in single-wall carbon nanohorns and its release," *J. Phys. Chem.* B 110 (2006): 19097–19099.

Ajima, K., Murakami, T., Mizoguchi, Y. et al., "Enhancement of in vivo anticancer effects of cisplatin by incorporation inside single-wall carbon nanohorns," *ACS Nano* 2 (2008): 2057–2064.

Amano, T. & Muramatsu, Y., "Electronic structure calculations of carbon nanohorns for their chemical state analysis using soft X-ray spectroscopy," *Int. J. Quantum Chem.* 109 (2009): 2728–2733.

Amano, T., Muramatsu, Y., Sano, N. et al., "Adsorption structure analysis of entrapped nitrogen in carbon-nanohorns by soft X-ray emission and absorption spectroscopy," *J. Electron Spectrosc. Rel. Phenom.* 181 (2010): 186–188.

Aoki, Y., Urita, K., Noguchi, D. et al., "Efficient production of $H_2$ and carbon nanotube from $CH_4$ over single wall carbon nanohorn," *Chem. Phys. Lett.* 482 (2009): 269–273.

Aryee, E., Dalai, A. K. & Adjaye, J., "Functionalization and characterization of carbon nanohorns (CNHs) for hydrotreating of gas oils," *Top. Catal.* 57 (2014): 796–805.

Azami, T., Kasuya, D., Yuge, R. et al. "Large-scale production of single-wall carbon nanohorns with high purity," *J. Phys. Chem. C* 112 (2008): 1330–1334.

Bandow, S., Kokai, F., Takahashi, K. et al., "Interlayer spacing anomaly of single-wall carbon nanohorn aggregate," *Chem. Phys. Lett.* 321 (2000a): 514–519.

Bandow, S., Rao, A. M., Sumanasekera, G. U. et al., "Evidence for anomalously small charge transfer in doped single-wall carbon nanohorn aggregates with Li, K and Br," *Appl. Phys. A* 71 (2000b): 561–564.

Bandow, S., Kokai, F., Takahashi, K. et al., "Unique magnetism observed in single-wall carbon nanohorns," *Appl. Phys. A* 73 (2001): 281–285.

Bandow, S., Yamaguchi, T. & Iijima, S., "Magnetism of adsorbed oxygen on carbon nanohorns," *Chem. Phys. Lett.* 401 (2005): 380–384.

Barsan, M. M., Prathish, K. P., Sun, X. et al., "Nitrogen doped graphene and its derivatives as sensors and efficient direct electron transfer platform for enzyme biosensors," *Sens. Actuators B* 203 (2014): 579–587.

Basiuk, V. & Irle, S., eds., *DFT Calculations on Fullerenes and Carbon Nanotubes* (Research Signpost, Kerala, 2008).

Battiston, S., Bolzan, M., Fiameni, S. et al., "Single wall carbon nanohorns coated with anatase titanium oxide," *Carbon* 47 (2009): 1321–1326.

Battiston S., Minella, M., Gerbasi, R. et al., "Growth of titanium dioxide nanopetals induced by single wall carbon nanohorns," *Carbon* 48 (2010): 2470–2477.

Bekyarova, E., Hanzawa, Y., Kaneko, K. et al., "Cluster-mediated filling of water vapor in intratube and interstitial nanospaces of single-wall carbon nanohorns," *Chem. Phys. Lett.* 366 (2002a): 463–468.

Bekyarova, E., Kaneko, K., Yudasaka, M. et al., "Micropore development and structure of single-wall carbon nanohorn assemblies by compression," *Adv. Mater.* 14 (2002b): 973–975.

Bekyarova, E., Kaneko, K., Kasuya, D. et al., "Oxidation and porosity evaluation of budlike single-wall carbon nanohorn aggregates," *Langmuir* 18 (2002c): 4138–4141.

Bekyarova, E., Kaneko, K., Yudasaka, M. et al. "Controlled opening of single-wall carbon nanohorns by heat treatment in carbon dioxide," *J. Phys. Chem. B* 107 (2003): 4479–4484.

Bekyarova, E., Hashimoto, A., Yudasaka, M. et al., "Palladium nanoclusters deposited on single-walled carbon nanohorns," *J. Phys. Chem. B* 109 (2005): 3711–3714.

Bittencourt, C., Ke, X., Van Tendeloo, G. et al., "NEXAFS spectromicroscopy of suspended carbon nanohorns," *Chem. Phys. Lett.* 587 (2013): 85–87.

Bonard, J.-M., Gaál, R., Garaj, S. et al., "Field emission properties of carbon nanohorn films," *J. Appl. Phys.* 91 (2002): 10107–10109.

Brandão, L., Passeira, C., Gattia D. M. et al., "Use of single wall carbon nanohorns in polymeric electrolyte fuel cells," *J. Mater. Sci.* 46 (2011): 7198–7205.

Casillas, R., Lodermeyer, F., Costa, R. D. et al., "Substituting $TiCl_4$–carbon nanohorn interfaces for dye-sensitized solar cells," *Adv. Energy Mater.* 4 (2014): 1301577.

Cataldo, F., Iglesias-Groth, S., Hafez, Y. et al., "Neutron bombardment of single wall carbon nanohorn (SWCNH): DSC determination of the stored Wigner–Szilard energy," *J. Radioanal. Chem.* 299 (2014): 1955–1963.

Chen, G., Peng, Q., Mizuseki, H. et al., "Theoretical investigation of hydrogen storage ability of a carbon nanohorn," *Comput. Mater. Sci.* 49 (2010): S378–S382.

Chen, D., Wang, C., Jiang, F. et al., "In vitro and in vivo photothermally enhanced chemotherapy by single-walled carbon nanohorns as a drug delivery system," *J. Mater. Chem. B* 2 (2014): 4726–4732.

Cheng, M.-D., Lee, D.-W., Zhao, B. et al., "Formation studies and controlled production of carbon nanohorns using continuous in situ characterization techniques," *Nanotechnology* 18 (2007): 185604.

Chronopoulos, D., Karousis, N., Ichihashi, T. et al., "Benzyne cycloaddition onto carbon nanohorns," *Nanoscale* 5 (2013): 6388–6394.

Cioffi, C., Campidelli, S., Brunetti, F. G. et al., "Functionalisation of carbon nanohorns," *Chem. Commun.* 20 (2006): 2129–2131.

Cioffi, C., Campidelli, S., Sooambar, C. et al., "Synthesis, characterization, and photoinduced electron transfer in functionalized single wall carbon nanohorns," *J. Am. Chem. Soc.* 129 (2007): 3938–3945.

Comisso, N., Berlouis, L. E. A., Morrow, J. et al., "Changes in hydrogen storage properties of carbon nanohorns submitted to thermal oxidation," *Int. J. Hydrog. Energy* 35 (2010): 9070–9081.

Costa, R. D., Feihl, S., Kahnt, A. et al., "Carbon nanohorns as integrative materials for efficient dye-sensitized solar cells," *Adv.Mater.* 25 (2013): 6513–6518.

Cruz, R., Brandão, L. & Mendes, A., "Use of single-wall carbon nanohorns as counter electrodes in dye-sensitized solar cells," *Int. J. Energy Res.* 37 (2013): 1498–1508.

Dai, H., Gong, L., Xu, G. et al., "An electrochemical sensing platform structured with carbon nanohorns for detecting some food borne contaminants," *Electrochim. Acta* 111 (2013): 57–63.

De Souza, L. A., Nogueira, C. A. S., Lopes, J. F. et al., "DFT study of cisplatin@carbon nanohorns complexes," *J. Inorg. Biochem.* 129 (2013): 71–83.

Delgado, J. L., Herranz, M. Á. & Martín, N., "The nano-forms of carbon," *J. Mater. Chem.* 18 (2008): 1417–1426.

Deshmukh, A. B. & Shelke, M. V., "Synthesis and electrochemical performance of a single walled carbon nanohorn–$Fe_3O_4$ nanocomposite supercapacitor electrode," *RSC Adv.* 3 (2013): 21390–21393.

Economopoulos, S. P., Pagona, G., Yudasaka, M. et al., "Solvent-free microwave-assisted Bingel reaction in carbon nanohorns," *J. Mater. Chem.* 19 (2009): 7326–7331.

Esrafili, M. D. & Mahdavinia, G., "Nitrogen-doping improves surface reactivity of carbon nanocone," *Superlattices Microstruct.* 62 (2013): 140–148.

Fan, J., Yudasaka, M., Kasuya, Y. et al., "Influence of water on desorption rates of benzene adsorbed within single-wall carbon nanohorns," *Chem. Phys. Lett.* 397 (2004): 5–10.

Fan, J., Yudasaka, M., Miyawaki, J. et al., "Control of hole opening in single-wall carbon nanotubes and single-wall carbon nanohorns using oxygen," *J. Phys. Chem. B* 110 (2006): 1587–1591.

Fan, X., Tan, J., Zhang, G. et al., "Isolation of carbon nanohorn assemblies and their potential for intracellular delivery," *Nanotechnology* 18 (2007): 195103.

Fan, J., Yuge, R., Maigne, A. et al., "Effect of hole size on the incorporation of $C_{60}$ molecules inside single-wall carbon nanohorns and their release," *Carbon* 46 (2008a): 1792–1828.

Fan, J., Yuge, R., Miyawaki, J. et al., "Close-open-close evolution of holes at the tips of conical graphenes of single-wall carbon nanohorns," *J. Phys. Chem. B* 112 (2008b): 8600–8603.

Fernandez-Alonso, F., Bermejo, F. J., Cabrillo, C. et al., "Nature of the bound states of molecular hydrogen in carbon nanohorns," *Phys. Rev. Lett.* 98 (2007): 215503.

Fresco-Cala, B., Jimenez-Soto, J. M., Cardenas, S. et al., "Single-walled carbon nanohorns immobilized on a microporous hollow polypropylene fiber as a sorbent for the extraction of volatile organic compounds from water samples," *Microchim. Acta* 181 (2014): 1117–1124.

Fuel Cells Bulletin, "NEC boosts carbon nanohorn output," *Fuel Cells Bull.* 2 (2006): 9.

Fujimori, T., Urita, K., Aoki, Y. et al., "Fine nanostructure analysis of single-wall carbon nanohorns by surface-enhanced Raman scattering," *J. Phys. Chem. C* 112 (2008): 7552–7556.

Fukunaga, Y., Harada, M., Bandow, S. et al., "Variable range hopping conduction and percolation networks in the pellets formed from pristine and boron-doped carbon nanohorn particles," *Appl. Phys. A* 94 (2009): 5–9.

Furmaniak, S., Terzyk, A. P., Gauden, P. A. et al., "Simple models of adsorption in nanotubes," *J. Colloid Interface Sci.* 295 (2006): 310–317.

Furmaniak, S., Terzyk, A. P., Kaneko, K. et al., "The first atomistic modelling-aided reproduction of morphologically defective single walled carbon nanohorns," *Phys. Chem. Chem. Phys.* 15 (2013a): 1232–1240.

Furmaniak, S., Terzyk, A. P., Kowalczyk, P. et al., "Separation of $CO_2$–$CH_4$ mixtures on defective single walled carbon nanohorns—Tip does matter," *Phys. Chem. Chem. Phys.* 15 (2013b): 16468–16476.

Furmaniak, S., Terzyk, A. P., Kaneko, K. et al., "Surface to volume ratio of carbon nanohorn—A crucial factor in $CO_2/CH_4$ mixture separation," *Chem. Phys. Lett.* 595–596 (2014): 67–72.

Gattia, D. M., Antisari, M. V. & Marazzi, R., "AC arc discharge synthesis of single-walled nanohorns and highly convoluted graphene sheets," *Nanotechnology* 18 (2007): 255604.

Geohegan, D. B., Puretzky, A. A., Styers-Barnett, D. et al., "In situ time-resolved measurements of carbon nanotube and nanohorn growth," *Phys. Status Solidi (B)* 244 (2007): 3944–3949.

Gotzias, A., Heiberg-Andersen, H., Kainourgiakis, M. et al., "A grand canonical Monte Carlo study of hydrogen adsorption in carbon nanohorns and nanocones at 77 K," *Carbon* 49 (2011): 2715–2724.

Guerra, J., Herrero, M. A., Carrión, B. et al., "Carbon nanohorns functionalized with polyamidoamine dendrimers as efficient biocarrier materials for gene therapy," *Carbon* 50 (2012): 2832–2844.

Guerra, J., Herrero, M. A. & Vázquez, E., "Carbon nanohorns as alternative gene delivery vectors," *RSC Adv.* 4 (2014): 27315–27321.

Guldi, D. M. & Sgobba, V., "Carbon nanostructures for solar energy conversion schemes," *Chem. Commun.* 47 (2011): 606–610.

Harada, M., Inagaki, T., Bandow, S. et al., "Effects of boron-doping and heat-treatment on the electrical resistivity of carbon nanohorn-aggregates," *Carbon* 46 (2008): 766–772.

Hashimoto, S., Fujimori, T., Tanaka, H. et al., "Anomaly of $CH_4$ molecular assembly confined in single-wall carbon nanohorn spaces," *J. Am. Chem. Soc.* 133 (2011): 2022–2024.

Hattori, Y., Kanoh, H., Okino, F. et al., "Direct thermal fluorination of single wall carbon nanohorns," *J. Phys. Chem. B* 108 (2004): 9614–9618.

Hawelek, L., Wrzalik, W., Brodka, A. et al., "A pulsed neutron diffraction study of the topological defects presence in carbon nanohorns," *Chem. Phys. Lett.* 502 (2011): 87–91.

Hawelek, L., Brodka, A., Dore, J. C. et al., "Structural modeling of dahlia-type single-walled carbon nanohorn aggregates by molecular dynamics," *J. Phys. Chem. A* 117 (2013): 9057–9061.

Hiralal, P., Wang, H., Unalan, H. E. et al., "Enhanced supercapacitors from hierarchical carbon nanotube and nanohorn architectures," *J. Mater. Chem.* 21 (2011): 17810–17815.

Hood, R. L., Carswell, W. F., Rodgers, A. et al., "Spatially controlled photothermal heating of bladder tissue through single-walled carbon nanohorns delivered with a fiberoptic microneedle device," *Lasers Med. Sci.* 28 (2013):1143–1150.

Horie, M., Komaba, L. K., Fukui, H. et al., "Evaluation of the biological influence of a stable carbon nanohorn dispersion," *Carbon* 54 (2013): 155–167.

Huang, W., Zhang, J., Dorn, H. C. et al., "Assembly of single-walled carbon nanohorn supported liposome particles," *Bioconjug. Chem.* 22 (2011): 1012–1016.

Hussein, A. K., "Applications of nanotechnology in renewable energies—A comprehensive overview and understanding," *Renew. Sustain. Energy Rev.* 42 (2015): 460–476.

Iglesias-Groth, S., Cataldo, F., Angelini, G. et al., "Single-walled carbon nanohorn: Electronic absorption spectra in neutral and oxidized state," *Nanotubes Carbon Nanostruct.* 22 (2014): 938–948.

Iijima, S., Yudasaka, M., Yamada, R. et al., "Nano-aggregates of single-walled graphitic carbon nanohorns," *Chem. Phys. Lett.* 309 (1999): 165–170.

Imai, H., Babu, P. K., Oldfield, E. et al., "$^{13}C$ NMR spectroscopy of carbon nanohorns," *Phys. Rev. B* 73 (2006): 125405.

Irie, M., Nakamura, M., Zhang, M. et al., "Quantification of thin graphene sheets contained in spherical aggregates of single-walled carbon nanohorns," *Chem. Phys. Lett.* 500 (2010): 96–99.

Itoh, T., Danjo, H., Sasaki, W. et al., "Catalytic activities of Pd-tailored single wall carbon nanohorns," *Carbon* 46 (2008a): 172–175.

Itoh, T., Urita, K., Bekyarova, E. et al., "Nanoporosities and catalytic activities of Pd-tailored single wall carbon nanohorns," *J. Colloid Interface Sci.* 322 (2008b): 209–214.

Izadi-Najafabadi, A., Yamada, T., Futaba, D. N. et al., "High-power supercapacitor electrodes from single-walled carbon nanohorn/nanotube composite," *ACS Nano* 5 (2011): 811–819.

Jadhav, A. D., Ogale, S. B. & Prasad, B. L. V., "Carbon nanohorn and bovine serum albumin hierarchical composite: Towards bio-friendly superhydrophobic protein film surfaces," *J. Mater. Chem.* 18 (2008): 3422–3425.

Jain, K., Mehra, N. K. & Jain, N. K., "Potentials and emerging trends in nanopharmacology," *Curr. Opin. Pharm.* 15 (2014): 97–106.

Jiang, B.-P., Hu, L.-F., Shen, X.-C. et al., "One-step preparation of a water-soluble carbon nanohorn/phthalocyanine hybrid for dual-modality photothermal and photodynamic therapy," *ACS Appl. Mater. Interfaces* 6 (2014): 18008–18017.

Jiménez-Soto, J. M., Moliner-Martínez, Y., Cárdenas, S. et al., "Evaluation of the performance of single-walled carbon nanohorns in capillary electrophoresis," *Electrophoresis* 31 (2010): 1681–1688.

Jiménez-Soto, J. M., Cárdenas, S. & Valcárcel, M., "Dispersive micro solid-phase extraction of triazines from waters using oxidized single-walled carbon nanohorns as sorbent," *J. Chromatogr. A* 1245 (2012a): 17–23.

Jiménez-Soto, J. M., Cárdenas, S. & Valcárcel, M., "Evaluation of single-walled carbon nanohorns as sorbent in dispersive micro solid-phase extraction," *Anal. Chim. Acta* 714 (2012b): 76–81.

Jiménez-Soto, J. M., Cárdenas, S. & Valcárcel, M., "Oxidized single-walled carbon nanohorns as sorbent for porous hollow fiber direct immersion solid-phase microextraction for the determination of triazines in waters," *Anal. Bioanal. Chem.* 405 (2013): 2661–2669.

Jung, H. J., Kim, Y.-J., Han, J. H. et al., "Thermal-treatment-induced enhancement in effective surface area of single-walled carbon nanohorns for supercapacitor application," *J. Phys. Chem. C* 117 (2013): 25877–25883.

Karousis, N., Ichihashi, T., Chen, S. et al., "Imidazolium modified carbon nanohorns: Switchable solubility and stabilization of metal nanoparticles," *J. Mater. Chem.* 20 (2010): 2959–2964.

Karousis, N., Sato, Y., Suenaga, K. et al., "Direct evidence for covalent functionalization of carbon nanohorns by high-resolution electron microscopy imaging of $C_{60}$ conjugated onto their skeleton," *Carbon* 50 (2012): 3909–3914.

Kasai, T., Matsumura, S., Iizuka, T. et al., "Carbon nanohorns accelerate bone regeneration in rat calvarial bone defect," *Nanotechnology* 22 (2011): 065102.

Kasuya, D., Yudasaka, M., Takahashi, K. et al., "Selective production of single-wall carbon nanohorn aggregates and their formation mechanism," *J. Phys. Chem. B* 106 (2002): 4947–4951.

Khoerunnisa, F., Fujimori, T., Itoh, T. et al., "Electronically modified single wall carbon nanohorns with iodine adsorption," *Chem. Phys. Lett.* 501 (2011): 485–490.

Khongpracha, P., Probst, M. & Limtrakul, J., "The interaction of a gold atom with carbon nanohorn and carbon nanotube tips and their complexes with a CO molecule: A first principle calculation," *Eur. Phys. J. D* 48 (2008): 211–219.

Kobayashi, K., Shimazu, T., Yamada, Y. et al., "Formation of single-wall carbon nanohorn aggregates hybridized with carbon nanocapsules by laser vaporization," *Appl. Phys. A* 89 (2007): 121–126.

Kobayashi, K., Ueno, H., Kokubo, K. et al., "Effect of functional group polarity on the encapsulation of $C_{60}$ derivatives in the inner space of carbon nanohorns," *Carbon* 68 (2014): 346–351.

Kokai, F., Takahashi, K., Kasuya, D. et al., "Growth dynamics of single-wall carbon nanotubes and nanohorn aggregates by $CO_2$ laser vaporization at room temperature," *Appl. Surf. Sci.* 197–198 (2002): 650–655.

Kokai, F., Tachi, N., Kobayashi, K. et al., "Structural characterization of single-wall carbon nanohorn aggregates hybridized with carbon nanocapsules and their formation mechanism," *Appl. Surf. Sci.* 255 (2009): 9622–9625.

Kolesnikov, D. V. & Osipov, V. A., "Field-theoretical approach to the description of electronic properties of carbon nanostructures," *Phys. Part. Nucl.* 40 (2009): 502–524.

Kosaka, M., Kuroshima, S., Kobayashi, K. et al., "Single-wall carbon nanohorns supporting Pt catalyst in direct methanol fuel cells," *J. Phys. Chem. C* 113 (2009): 8660–8667.

Kowalczyk, P., Hołyst, R., Tanaka, H. et al., "Distribution of carbon nanotube sizes from adsorption measurements and computer simulation," *J. Phys. Chem. B* 109 (2005): 14659–14666.

Kowalczyk, P., Terzyk, A. P., Gauden, P. A. et al., "Toward in silico modeling of palladium–hydrogen–carbon nanohorn nanocomposites," *Phys. Chem. Chem. Phys.* 16 (2014): 11763–11769.

Krungleviciute, V., Calbi, M. M., Wagner, J. A. et al., "Probing the structure of carbon nanohorn aggregates by adsorbing gases of different sizes," *J. Phys. Chem. C* 112 (2008): 5742–5746.

Krungleviciute, V., Migone, A. D. & Pepka, M., "Characterization of single-walled carbon nanohorns using neon adsorption isotherms," *Carbon* 47 (2009): 769–774.

Krungleviciute, V., Migone, A. D., Yudasaka, M. et al., "$CO_2$ adsorption on dahlia-like carbon nanohorns: Isosteric heat and surface area measurements," *J. Phys. Chem. C* 116 (2012): 306–310.

Krungleviciute, V., Ziegler, C. A., Banjara, S. R. et al., "Neon and $CO_2$ adsorption on open carbon nanohorns," *Langmuir* 29 (2013): 9388–9397.

Kumar, D., Verma, V., Bhatti, H. S. et al., "Elastic moduli of carbon nanohorns," *J. Nanomater.* 2011 (2011): 127952.

Lai, H., Li, J., Chen, Z. et al., "Carbon nanohorns as a high-performance carrier for $MnO_2$ anode in lithium-ion batteries," *ACS Appl. Mater. Interfaces* 4 (2012): 2325–2328.

Li, X., Lei, Z., Ren, R. et al., "Characterization of carbon nanohorn encapsulated Fe particles," *Carbon* 41 (2003): 3068–3072.

Li, H., Zhao, N., Wang, L. et al., "Synthesis of carbon nanohorns by the simple catalytic method," *J. Alloys Cmpd.* 473 (2009): 288–292.

Li, N., Wang, Z., Zhao, K. et al., "Synthesis of single-wall carbon nanohorns by arc-discharge in air and their formation mechanism," *Carbon* 48 (2010): 1580–1585.

Li, L., Zhang, J., Yang, Y. et al., "Single-wall carbon nanohorns inhibited activation of microglia induced by lipopolysaccharide through blocking of SIRT3," *Nanoscale Res. Lett.* 8 (2013): 100.

Li, J., He, Z., Guo, C. et al., "Synthesis of carbon nanohorns/chitosan/quantum dots nanocomposite and its applications in cells labeling and in vivo imaging," *J. Lumin.* 145 (2014):74–80.

Liu, X., Shi, L., Niu, W. et al., "Amperometric glucose biosensor based on single-walled carbon nanohorns," *Biosens. Bioelectron.* 23 (2008): 1887–1890.

Liu, X., Li, H., Wang, F. et al., "Functionalized single-walled carbon nanohorns for electrochemical biosensing," *Biosens. Bioelectron.* 25 (2010): 2194–2199.

Liu, Y., Brown, C. M. & Neumann, D. A., "Metal-assisted hydrogen storage on Pt-decorated single-walled carbon nanohorns," *Carbon* 50 (2012): 4953–4964.

Liu, F., Xiang, G., Chen, X. et al., "A novel strategy of procalcitonin detection based on multi-nanomaterials of single-walled carbon nanohorns–hollow Pt nanospheres/PAMAM as signal tags," *RSC Adv.* 4 (2014a): 13934–19340.

Liu, F., Xiang, G., Yuan, R. et al., "Procalcitonin sensitive detection based on graphene–gold nanocomposite film sensor platform and single-walled carbon nanohorns/hollow Pt chains complex as signal tags," *Biosens. Bioelectron.* 60 (2014b): 210–217.

Lodermeyer, F., Costa, R. D., Casillas, R. et al., "Carbon nanohorn-based electrolyte for dye-sensitized solar cells," *Energy Environ. Sci.* 8 (2015): 241–246.

Lu, B., Zhang, Z., Hao, J. et al., "Electrochemical sensing platform based on Schiff-base cobalt(II)/single-walled carbon nanohorns complexes system," *Anal. Methods* 4 (2012): 3580–3585.

Ma, X., Shu, C., Guo, J. et al., "Targeted cancer therapy based on single-wall carbon nanohorns with doxorubicin in vitro and in vivo," *J. Nanopart. Res.* 16 (2014): 2497.

Maharaj, D., Bhushan, B. & Iijima, S., "Effect of carbon nanohorns on nanofriction and wear reduction in dry and liquid environments," *J. Colloid Interface Sci.* 400 (2013): 147–160.

Maiti, S. & Khatua, B. B., "Polyaniline integrated carbon nanohorn: A superior electrode materials for advanced energy storage," *eXPRESS Polym. Lett.* 8 (2014): 895–907.

Majidi, R. & Tabrizi, K. G., "Study of neon adsorption on carbon nanocones using molecular dynamics simulation," *Phys. B* 405 (2010): 2144–2148.

Matsumura, T., Tanaka, H., Kaneko, K. et al., "Magnetism of organic radical molecules confined in nanospace of single-wall carbon nanohorn," *J. Phys. Chem. C* 111 (2007): 10213–10216.

Matsumura, S., Sato, S., Yudasaka, M. et al., "Prevention of carbon nanohorn agglomeration using a conjugate composed of comb-shaped polyethylene glycol and a peptide aptamer," *Molec. Pharm.* 6 (2009): 441–447.

Matsumura, S., Yuge, R., Sato, S. et al., "Ultrastructural localization of intravenously injected carbon nanohorns in tumor," *Int. J. Nanomed.* 9 (2014): 3499–3508.

Mendes, R. G., Bachmatiuk, A., Büchner, B. et al., "Carbon nanostructures as multi-functional drug delivery platforms," *J. Mater. Chem. B* 1 (2013): 401–428.

Mercatelli, L., Sani, E., Zaccanti, G. et al., "Absorption and scattering properties of carbon nanohorn-based nanofluids for direct sunlight absorbers," *Nanoscale Res. Lett.* 6 (2011): 282.

Mercatelli, L., Sani, E., Giannini, A. et al., "Carbon nanohorn-based nanofluids: Characterization of the spectral scattering albedo," *Nanoscale Res. Lett.* 7 (2012): 6.

Meyyappan, M., "Nanostructured materials for supercapacitors," *J. Vac. Sci. Technol. A* 31 (2013): 050803.

Misra, R. D. K. & Chaudhari, P. M., "Osteoblasts response to nylon 6,6 blended with single-walled carbon nanohorn," *J. Biomed. Mater. Res. A* 101A (2013): 1059–1068.

Miyako, E., Nagata, H., Hirano, K. et al., "Near-infrared laser-triggered carbon nanohorns for selective elimination of microbes," *Nanotechnology* 18 (2007): 475103.

Miyako, E., Nagata, H., Hirano, K. et al., "Photoinduced antiviral carbon nanohorns," *Nanotechnology* 19 (2008a): 075106.

Miyako, E., Nagata, H., Hirano, K. et al., "Photodynamic release of fullerenes from within carbon nanohorn," *Chem. Phys. Lett.* 456 (2008b): 220–222.

Miyawaki, J., Yudasaka, M., Imai, H. et al., "Synthesis of ultrafine $Gd_2O_3$ nanoparticles inside single-wall carbon nanohorns," *J. Phys. Chem. B* 110 (2006a): 5179–5181.

Miyawaki, J., Yudasaka, M., Imai, H. et al., "In vivo magnetic resonance imaging of single-walled carbon nanohorns by labeling with magnetite nanoparticles," *Adv. Mater.* 18 (2006b): 1010–1014.

Miyawaki, J., Yuge, R., Kawai, T. et al., "Evidence of thermal closing of atomic-vacancy holes in single-wall carbon nanohorns," *J. Phys. Chem. C* 111 (2007a): 1553–1555.

Miyawaki, J., Yudasaka, M., Yuge, R. et al., "Organic-vapor-induced repeatable entrance and exit of $C_{60}$ into/from single-wall carbon nanohorns at room temperature," *J. Phys. Chem. C* 111 (2007b): 9719–9722.

Miyawaki, J., Yudasaka, M., Azami, T. et al., "Toxicity of single-walled carbon nanohorns," *ACS Nano* 2 (2008): 213–226.

Miyawaki, J., Matsumura, S., Yuge, R. et al., "Biodistribution and ultrastructural localization of single-walled carbon nanohorns determined in vivo with embedded $Gd_2O_3$ labels," *ACS Nano* 3 (2009): 1399–1406.

Moschino, V., Nesto, N., Barison, S. et al., "A preliminary investigation on nanohorn toxicity in marine mussels and polychaetes," *Sci. Total Environ.* 468–469 (2014): 111–119.

Mountrichas, G., Pispas, S. & Tagmatarchis, N., "Grafting living polymers onto carbon nanohorns," *Chem. Eur. J.* 13 (2007): 7595–7599.

Mountrichas, G., Ichihashi, T., Pispas, S. et al., "Solubilization of carbon nanohorns by block polyelectrolyte wrapping and templated formation of gold nanoparticles," *J. Phys. Chem. C* 113 (2009): 5444–5449.

Murakami, T., Fan, J., Yudasaka, M. et al., "Solubilization of single-wall carbon nanohorns using a PEG–doxorubicin conjugate," *Molec. Pharm.* 3 (2006): 407–414.

Murata, K., Kaneko, K., Kokai, F. et al., "Pore structure of single-wall carbon nanohorn aggregates," *Chem. Phys. Lett.* 331 (2000): 14–20.

Murata, K., Kaneko, K., Steele, W. A. et al., "Molecular potential structures of heat-treated single-wall carbon nanohorn assemblies," *J. Phys. Chem. B* 105 (2001a): 10210–10216.

Murata, K., Kaneko, K., Steele, W. A. et al., "Porosity evaluation of intrinsic intraparticle nanopores of single wall carbon nanohorn," *Nano Lett.* 1 (2001b): 197–199.

Murata, K., Hirahara, K., Yudasaka, M. et al., "Nanowindow-induced molecular sieving effect in a single-wall carbon nanohorn," *J. Phys. Chem. B* 106 (2002a): 12668–12669.

Murata, K., Kaneko, K., Kanoh, H. et al., "Adsorption mechanism of supercritical hydrogen in internal and interstitial nanospaces of single-wall carbon nanohorn assembly," *J. Phys. Chem. B* 106 (2002b): 11132–11138.

Murata, K., Yudasaka, M., Iijima, S. et al., "Classification of supercritical gas adsorption isotherms based on fluid–fluid interaction," *J. Appl. Phys.* 91 (2002c): 10227–10229.

Murata, K., Miyawaki, J., Yudasaka, M. et al., "High-density of methane confined in internal nanospace of single-wall carbon nanohorns," *Carbon* 43 (2005): 2826–2830.

Murata, K., Yudasaka, M. & Iijima, S., "Hydrogen production from methane and water at low temperature using EuPt supported on single-wall carbon nanohorns," *Carbon* 44 (2006): 818–820.

Nakamura, M., Tahara, Y., Ikehara, Y. et al., "Single-walled carbon nanohorns as drug carriers: Adsorption of prednisolone and anti-inflammatory effects on arthritis," *Nanotechnology* 22 (2011): 465102.

Nakamura, M., Irie, M., Yuge, R. et al., "Carboxylation of thin graphitic sheets is faster than that of carbon nanohorns," *Phys. Chem. Chem. Phys.* 15 (2013): 16672–16675.

Nakamura, M., Tahara, Y., Murakami, T. et al., "Gastrointestinal actions of orally-administered single-walled carbon nanohorns," *Carbon* 69 (2014): 409–416.

Nisha, J. A., Yudasaka, M., Bandow, S. et al., "Adsorption and catalytic properties of single-wall carbon nanohorns," *Chem. Phys. Lett.* 328 (2000): 381–386.

Noguchi, D., Hattori, Y., Yang, C.-M. et al., "Storage function of carbon nanospaces for molecules and ions," *ECS Trans.* 11 (2007a): 63–75.

Noguchi, H., Kondo, A., Noguchi, D. et al., "Adsorptive properties of novel nanoporous materials," *J. Chem. Eng. Jpn.* 40 (2007b): 1159–1165.

Oakes, L., Westover, A., Mahjouri-Samani, M. et al., "Uniform, homogenous coatings of carbon nanohorns on arbitrary substrates from common solvents," *ACS Appl. Mater. Interfaces* 5 (2013): 13153–13160.

Ohba, T., Murata, K., Kaneko, K. et al., "$N_2$ adsorption in an internal nanopore space of single-walled carbon nanohorn: GCMC simulation and experiment," *Nano Lett.* 1 (2001): 371–373.

Ohba, T., Kanoh, H., Yudasaka, M. et al., "Quasi one-dimensional nanopores in single-wall carbon nanohorn colloids using grand canonical Monte Carlo simulation aided adsorption technique," *J. Phys. Chem. B* 109 (2005): 8659–8662.

Ohba, T., Kanoh, H. & Kaneko, K., "Superuniform molecular nanogate fabrication on graphene sheets of single wall carbon nanohorns for selective molecular separation of $CO_2$ and $CH_4$," *Chem. Lett.* 40 (2011): 1089–1091.

Ohba, T., Kaneko, K., Yudasaka, M. et al., "Cooperative adsorption of supercritical $CH_4$ in single-walled carbon nanohorns for compensation of nanopore potential," *J. Phys. Chem. C* 116 (2012a): 21870–21873.

Ohba, T., Kanoh, H. & Kaneko, K., "Facilitation of water penetration through zero-dimensional gates on rolled-up graphene by cluster–chain–cluster transformations," *J. Phys. Chem. C* 116 (2012b): 12339–12345.

Ohkubo, T., Hattori, Y., Kanoh, H. et al., "EXAFS study of electrolytic nanosolution confined in interstitial nanospaces of single-wall carbon nanohorn colloids," *Phys. Scr.* T115 (2005): 685–687.

Ojeda, I., Garcinuño, B., Moreno-Guzmán, M. et al., "Carbon nanohorns as a scaffold for the construction of disposable electrochemical immunosensing platforms: Application to the determination of fibrinogen in human plasma and urine," *Anal. Chem.* 86 (2014): 7749–1156.

Pagona, G., Tagmatarchis, N., Fan, J. et al., "Cone-end functionalization of carbon nanohorns," *Chem. Mater.* 18 (2006a): 3918–3920.

Pagona, G., Sandanayaka, A. S. D., Araki, Y. et al., "Electronic interplay on illuminated aqueous carbon nanohorn–porphyrin ensembles," *J. Phys. Chem. B* 110 (2006b): 20729–20732.

Pagona, G., Sandanayaka, A. S. D., Araki, Y. et al., "Covalent functionalization of carbon nanohorns with porphyrins: Nanohybrid formation and photoinduced electron and energy transfer," *Adv. Funct. Mater.* 17 (2007a): 1705–1711.

Pagona, G., Sandanayaka, A. S. D., Maigné, A. et al., "Photoinduced electron transfer on aqueous carbon nanohorn–pyrene–tetrathiafulvalene architectures," *Chem. Eur. J.* 13 (2007b): 7600–7607.

Pagona, G., Karousis, N. & Tagmatarchis, N. et al., "Aryl diazonium functionalization of carbon nanohorns," *Carbon* 46 (2008a): 604–610.

Pagona, G., Sandanayaka, A. S. D., Hasobe, T. et al., "Characterization and photoelectrochemical properties of nanostructured thin film composed of carbon nanohorns covalently functionalized with porphyrins," *J. Phys. Chem. C* 112 (2008b): 15735–15741.

Pagona, G., Mountrichas, G., Rotas, G. et al., "Properties, applications and functionalisation of carbon nanohorns," *Int. J. Nanotechnol.* 6 (2009): 176–195.

Pagona, G., Katerinopoulos, H. E. & Tagmatarchis, N., "Synthesis, characterization, and photophysical properties of a carbon nanohorn–coumarin hybrid material," *Chem. Phys. Lett.* 516 (2011): 76–81.

Pagona, G., Zervaki, G. E., Sandanayaka, A. S. D. et al., "Carbon nanohorn–porphyrin dimer hybrid material for enhancing light-energy conversion," *J. Phys. Chem. C* 116 (2012): 9439–9449.

Petsalakis, I. D., Pagona, G., Tagmatarchis, N. & Theodorakopoulos, G., "Theoretical study in donor–acceptor carbon nanohorn-based hybrids," *Chem. Phys. Lett.* 448 (2007): 115–120.

Poonjarernsilp, C., Sano, N. & Tamon, H., "Hydrothermally sulfonated single-walled carbon nanohorns for use as solid catalysts in biodiesel production by esterification of palmitic acid," *Appl. Catal. B Environ.* 147 (2014): 726–732.

Poonjarernsilp, C., Sano, N., Tamon, H. et al., "A model of reaction field in gas-injected arc-in-water method to synthesize single-walled carbon nanohorns: Influence of water temperature," *J. Appl. Phys.* 106 (2009): 104315.

Poonjarernsilp, C., Sano, N., Charinpanitkul, T. et al., "Single-step synthesis and characterization of single-walled carbon nanohorns hybridized with Pd nanoparticles using $N_2$ gas-injected arc-in-water method," *Carbon* 49 (2011): 4920–4927.

Pramoda, K., Moses, K., Ikram, M. et al., "Synthesis, characterization and properties of single-walled carbon nanohorns," *J. Clust. Sci.* 25 (2014): 173–188.

Puretzky, A. A., Styers-Barnett, D. J., Rouleau, C. M. et al., "Cumulative and continuous laser vaporization synthesis of single wall carbon nanotubes and nanohorns," *Appl. Phys. A* 93 (2008): 849–855.

Roldán-Pijuán, M., Lucena, R., Cárdenas, S. et al., "Micro-solid phase extraction based on oxidized single-walled carbon nanohorns immobilized on a stir borosilicate disk: Application to the preconcentration of the endocrine disruptor benzophenone-3," *Microchem. J.* 115 (2014): 87–94.

Rotas, G., Sandanayaka, A. S. D., Tagmatarchis, N. et al., "(Terpyridine)copper(II)-carbon nanohorns: Metallo-nanocomplexes for photoinduced charge separation," *J. Am. Chem. Soc.* 130 (2008): 4725–4731.

Rubio, N., Herrero, M. A., Meneghetti, M. et al., "Efficient functionalization of carbon nanohorns via microwave irradiation," *J. Mater. Chem.* 19 (2009): 4407–4413.

Sandanayaka, A. S. D., Pagona, G., Fan, J. et al. "Photoinduced electron-transfer processes of carbon nanohorns with covalently linked pyrene chromophores: Charge-separation and electron-migration systems," *J. Mater. Chem.* 17 (2007): 2540–2546.

Sandanayaka, A. S. D., Ito, O., Tanaka, T. et al., "Photoinduced electron transfer of nanohybrids of carbon nanohorns with amino groups and tetrabenzoic acid porphyrin in aqueous media," *New J. Chem.* 33 (2009a): 2261–2266.

Sandanayaka, A. S. D., Ito, O., Zhang, M. et al., "Photoinduced electron transfer in zinc phthalocyanine loaded on single-walled carbon nanohorns in aqueous solution," *Adv. Mater.* 21 (2009b): 4366–4371.

Sano, N., "Low-cost synthesis of single-walled carbon nanohorns using the arc in water method with gas injection," *J. Phys. D Appl. Phys.* 37 (2004): L17–L20.

Sano, N. & Ohtsuki, F., "Carbon nanohorn sensor to detect ozone in water," *J. Electrost.* 65 (2007): 263–268.

Sano, N. & Ukita, S., "One-step synthesis of Pt-supported carbon nanohorns for fuel cell electrode by arc plasma in liquid nitrogen," *Mater. Chem. Phys.* 99 (2006): 447–450.

Sano, N., Kimura, Y. & Suzuki, T., "Synthesis of carbon nanohorns by a gas-injected arc-in-water method and application to catalyst-support for polymer electrolyte fuel cell electrodes," *J. Mater. Chem.* 18 (2008): 1555–1560.

Sano, N., Akita, Y. & Tamon, H., "Effects of synthesis conditions on the structural features and methane adsorption properties of single-walled carbon nanohorns prepared by a gas-injected arc-in-water method," *J. Appl. Phys.* 109 (2011a): 124305.

Sano, N., Suzuki, T., Hirano, K. et al., "Influence of arc duration time on the synthesis of carbon nano-horns by a gas-injected arc-in-water system: Application to polymer electrolyte fuel cell electrodes," *Plasma Sources Sci. Technol.* 20 (2011b): 034002.

Sano, N., Ishii, T. & Tamon, H., "Transformation from single-walled carbon nanotubes to nanohorns by simple heating with Pd at 1600 °C," *Carbon* 49 (2011c): 3698–3704.

Sano, N., Ishii, T., Mori, H. et al., "One-step synthesis of single-walled carbon nanohorns dispersed with Pd–Ni alloy nanoparticles by gas-injected arc-in-water method and effects of synthesis factors on their hydrogen sensor sensitivity," *J. Appl. Phys.* 112 (2012): 044301.

Sano, N., Suntornlohanakul, T., Poonjarernsilp, C. et al., "Controlled syntheses of various palladium alloy nanoparticles dispersed in single-walled carbon nanohorns by one-step formation using an arc dis-charge method," *Ind. Eng. Chem. Res.* 53 (2014a): 4732–4738.

Sano, N., Taniguchi, K. & Tamon, H., "Hydrogen storage in porous single-walled carbon nanohorns dis-persed with Pd–Ni alloy nanoparticles," *J. Phys. Chem. C* 118 (2014b): 3402–3408.

Sarkar, S., Gurjarpadhye, A. A., Rylander, C. G. et al., "Optical properties of breast tumor phantoms con-taining carbon nanotubes and nanohorns," *J. Biomed. Opt.* 16 (2011): 051304.

Sasaki, K., Sekine, Y., Tateno, K. et al., "Topological Raman band in the carbon nanohorn," *Phys. Rev. Lett.* 111 (2013): 116801.

Sawdon, A., Weydemeyer, E. & Peng, C.-A., "Tumor photothermolysis: Using carbon nanomaterials for cancer therapy," *Eur. J. Nanomed.* 5 (2013): 131–140.

Shi, L., Liu, X., Niu, W. et al., "Hydrogen peroxide biosensor based on direct electrochemistry of soy-bean peroxidase immobilized on single-walled carbon nanohorn modified electrode," *Biosens. Bioelectron.* 24 (2009): 1159–1163.

Shinke, K., Ando, K., Koyama, T. et al., "Properties of various carbon nanomaterial surfaces in bilirubin adsorption," *Colloids Surf. B* 77 (2010): 18–21.

Shu, C., Zhang, J., Ge, J. et al., "A facile high-speed vibration milling method to water-disperse single-walled carbon nanohorns," *Chem. Mater.* 22 (2010): 347–351.

Sing, K. S. W., "The use of nitrogen adsorption for the characterisation of porous materials," *Colloids Surf. A* 187–188 (2001): 3–9.

Sing, K. S. W., "Characterization of porous materials: Past, present and future," *Colloids Surf. A* 241 (2004): 3–7.

Stankova, L., Fraczek-Szczypta, A., Blazewicz, M. et al., "Human osteoblast-like MG 63 cells on polysul-fone modified with carbon nanotubes or carbon nanohorns," *Carbon* 67 (2014): 578–591.

Su, D. S. & Centi, G., "A perspective on carbon materials for future energy application," *J. Energy Chem.* 22 (2013): 151–173.

Suehiro, J., Sano, N., Zhou, G. et al., "Application of dielectrophoresis to fabrication of carbon nanohorn gas sensor," *J. Electrost.* 64 (2006): 408–415.

Sun, L., Wang, C., Zhou, Y. et al., "Flowing nitrogen assisted-arc discharge synthesis of nitrogen-doped single-walled carbon nanohorns," *Appl. Surf. Sci.* 277 (2013): 88–93.

Tagmatarchis, N., Maigné, A., Yudasaka, M. et al., "Functionalization of carbon nanohorns with azo-methine ylides: Towards solubility enhancement and electron-transfer processes," *Small* 2 (2006): 490–494.

Tahara, Y., Miyawaki, J., Zhang, M. et al., "Histological assessments for toxicity and functionalization-dependent biodistribution of carbon nanohorns," *Nanotechnology* 22 (2011): 265106.

Tahara, Y., Nakamura, M., Yang, M. et al., "Lysosomal membrane destabilization induced by high accu-mulation of single-walled carbon nanohorns in murine macrophage RAW 264.7," *Biomaterials* 33 (2012): 2762–2769.

Takikawa, H., Ikeda, M., Hirahara, K. et al., "Fabrication of single-walled carbon nanotubes and nano-horns by means of a torch arc in open air," *Phys. B* 323 (2002): 277–279.

Tanaka, H., Fan, J., Kanoh, H. et al., "Quantum nature of adsorbed hydrogen on single-wall carbon nano-horns," *Molec. Simul.* 31 (2005a): 465–474.

Tanaka, H., Kanoh, H., Yudasaka, M. et al., "Quantum effects on hydrogen isotope adsorption on single-wall carbon nanohorns," *J. Am. Chem. Soc.* 127 (2005b): 7511–7516.

Terzyk, A. P., Gauden, P. A., Furmaniak, S. et al., "Material storage mechanism in porous nano-carbon—Comparison between experiment and simulation," *Comput. Methods Sci. Technol.* 18 (2012): 45–51.

Torrens, F. & Castellano, G., "Cluster nature of the solvent features of single-wall carbon nanohorns," *Int. J. Quantum Chem.* 110 (2010): 563–570.

Torrens, F. & Castellano, G., "Cluster solvation models of carbon nanostructures: Extension to fullerenes, tubes, and buds," *J. Molec. Model.* 20 (2014): 2263.

Tsai, P.-C. & Jeng, Y.-R., "Theoretical investigation of thermally induced coalescence mechanism of single-wall carbon nanohorns and their mechanical properties," *Comput. Mater. Sci.* 88 (2014): 76–80.

Unni, S. M., Ramadas, S., Illathvalappil, R. et al., "Surface-modified single wall carbon nanohorn as an effective electrocatalyst for platinum-free fuel cell cathodes," *J. Mater. Chem.* A 3 (2015): 4361–4367.

Urita, K., Seki, S., Utsumi, S. et al., "Effects of gas adsorption on the electrical conductivity of single-wall carbon nanohorns," *Nano Lett.* 6 (2006): 1325–1328.

Urita, K., Seki, S., Tsuchiya, H. et al., "Mechanochemically induced $sp^3$-bond-associated reconstruction of single-wall carbon nanohorns," *J. Phys. Chem.* C 112 (2008): 8759–8762.

Utsumi, S., Miyawaki, J., Tanaka, H. et al., "Opening mechanism of internal nanoporosity of single-wall carbon nanohorn," *J. Phys. Chem.* B 109 (2005): 14319–14324.

Utsumi, S., Urita, K. & Kanoh, H. et al., "Preparing a magnetically responsive single-wall carbon nano-horn colloid by anchoring magnetite nanoparticles," *J. Phys. Chem.* B 110 (2006): 7165–7170.

Utsumi, S., Honda, H., Hattori, Y. et al., "Direct evidence on C–C single bonding in single-wall carbon nanohorn aggregates," *J. Phys. Chem.* C 111 (2007): 5572–5575.

Vizuete, M., Gómez-Escalonilla, M. J., Fierro, J. L. G. et al., "A carbon nanohorn–porphyrin supramolecular assembly for photoinduced electron-transfer processes," *Chem. Eur. J.* 16 (2010): 10752–10763.

Wang, H., Chhowalla, M., Sano, N. et al., "Large-scale synthesis of single-walled carbon nanohorns by submerged arc," *Nanotechnology* 15 (2004): 546–550.

Wang, N., Wu, Ch., Li, J. et al., "Binder-free manganese oxide/carbon nanomaterials thin film electrode for supercapacitors," *ACS Appl. Mater. Interfaces* 3 (2011): 4185–4189.

Wang, J., Hu, Z., Xu, J. et al., "Therapeutic applications of low-toxicity spherical nanocarbon materials," *NPG Asia Mater.* 6 (2014): e84.

Wei, D., Wang, H., Hiralal, P. et al., "Template-free electrochemical nanofabrication of polyaniline nano-brush and hybrid polyaniline with carbon nanohorns for supercapacitors," *Nanotechnology* 21 (2010): 435702.

Wen, D., Deng, L., Zhou, M. et al., "A biofuel cell with a single-walled carbon nanohorn-based bioanode operating at physiological condition," *Biosens. Bioelectron.* 25 (2010): 1544–1547.

Whitney, J. R., Sarkar, S., Zhang, J. et al., "Single walled carbon nanohorns as photothermal cancer agents," *Lasers Surg. Med.* 43 (2011): 43–51.

Whitney, J., DeWitt, M., Whited, B. M. et al., "3D viability imaging of tumor phantoms treated with single-walled carbon nanohorns and photothermal therapy," *Nanotechnology* 24 (2013): 275102.

Wu, W., Zhao, Y., Wu, C. & Guan, L., "Single-walled carbon nanohorns with unique horn-shaped structures as a scaffold for lithium–sulfur batteries," *RSC Adv.* 4 (2014): 28636–28639.

Xing, G., Jia, S. & Shi, Z., "The production of carbon nano-materials by arc discharge under water or liquid nitrogen," *New Carbon Mater.* 22 (2007): 337–341.

Xu, J., Yudasaka, M., Kouraba, S. et al., "Single wall carbon nanohorn as a drug carrier for controlled release," *Chem. Phys. Lett.* 461 (2008): 189–192.

Xu, J., Zhang, M., Nakamura, M. et al., "Double oxidation with oxygen and hydrogen peroxide for hole-forming in single wall carbon nanohorns," *Appl. Phys.* A 100 (2010): 379–383.

Xu, J., Tomimoto, H. & Nakayama, T., "What is inside carbon nanohorn aggregates?" *Carbon* 49 (2011a): 2074–2078.

Xu, J., Shingaya, Y., Tomimoto, H. et al., "Irreversible and reversible structural deformation and electro-mechanical behavior of carbon nanohorns probed by conductive AFM," *Small* 7 (2011b): 1169–1174.

Xu, G., Gong, L., Dai, H. et al., "Electrochemical bisphenol A sensor based on carbon nanohorns," *Anal. Methods* 5 (2013): 3328–3333.

Yamaguchi, T., Bandow, S. & Iijima, S., "Synthesis of carbon nanohorn particles by simple pulsed arc discharge ignited between preheated carbon rods," *Chem. Phys. Lett.* 389 (2004): 181–185.

Yamazaki, K., Shinke, K. & Ogino, T., "Selective adsorption of bilirubin against albumin to oxidized single-wall carbon nanohorns," *Colloids Surf. B* 112 (2013): 103–107.

Yang, C.-M., Kasuya, D., Yudasaka, M. et al., "Microporosity development of single-wall carbon nanohorn with chemically induced coalescence of the assembly structure," *J. Phys. Chem. B* 108 (2004): 17775–17782.

Yang, C.-M., Naguchi, H., Murata, K. et al., "Highly ultramicroporous single-walled carbon nanohorn assemblies," *Adv. Mater.* 17 (2005): 866–870.

Yang, C.-M., Kim, Y.-J., Endo, M. et al., "Nanowindow-regulated specific capacitance of supercapacitor electrodes of single-wall carbon nanohorns," *J. Am. Chem. Soc.* 129 (2007): 20–21.

Yang, M., Zhang, M., Tahara, Y. et al., "Lysosomal membrane permeabilization: Carbon nanohorn-induced reactive oxygen species generation and toxicity by this neglected mechanism," *Toxic. Appl. Pharm.* 280 (2014a): 117–126.

Yang, F., Han, J., Zhuo, Y. et al., "Highly sensitive impedimetric immunosensor based on single-walled carbon nanohorns as labels and bienzyme biocatalyzed precipitation as enhancer for cancer biomarker detection," *Biosens. Bioelectron.* 55 (2014b): 360–365.

Yoshida, S. & Sano, M., "Microwave-assisted chemical modification of carbon nanohorns: Oxidation and Pt deposition," *Chem. Phys. Lett.* 433 (2006): 97–100.

Yoshitake, T., Shimakawa, Y., Kuroshima, S. et al., "Preparation of fine platinum catalyst supported on single-wall carbon nanohorns for fuel cell application," *Phys. B* 323 (2002): 124–126.

Yudasaka, M., Ichihashi, T., Kasuya, D. et al., "Structure changes of single-wall carbon nanotubes and single-wall carbon nanohorns caused by heat treatment," *Carbon* 41 (2003): 1273–1780.

Yuge, R., Ichihashi, T., Shimakawa, Y. et al., "Preferential deposition of Pt nanoparticles inside single-walled carbon nanohorns," *Adv. Mater.* 16 (2004): 1420–1423.

Yuge, R., Yudasaka, M., Miyawaki, J. et al., "Plugging and unplugging holes of single-wall carbon nanohorns," *J. Phys. Chem. C* 111 (2007): 7348–7351.

Yuge, R., Yudasaka, M., Maigné, A. et al. "Adsorption phenomena of tetracyano-p-quinodimethane on single-wall carbon nanohorns," *J. Phys. Chem. C* 112 (2008): 5416–5422.

Yuge, R., Ichihashi, T., Miyawaki, J. et al., "Hidden caves in an aggregate of single-wall carbon nanohorns found by using $Gd_2O_3$ probes," *J. Phys. Chem. C* 113 (2009): 2741–2744.

Yuge, R., Yudasaka, M., Toyama, K. et al., "Buffer gas optimization in $CO_2$ laser ablation for structure control of single-wall carbon nanohorn aggregates," *Carbon* 50 (2012a): 1925–1933.

Yuge, R., Manako, T., Nakahara, K. et al., "The production of an electrochemical capacitor electrode using holey single-wall carbon nanohorns with high specific surface area," *Carbon* 50 (2012b): 5569–5573.

Yuge, R., Bandow, S., Nakahara, K. et al., "Structure and electronic states of single-wall carbon nanohorns prepared under nitrogen atmosphere," *Carbon* 75 (2014a): 322–326.

Yuge, R., Tamura, N., Manako, T. et al., "High-rate charge/discharge properties of Li-ion battery using carbon-coated composites of graphites, vapor grown carbon fibers, and carbon nanohorns," *J. Power Sour.* 266 (2014b): 471–474.

Zambano, A. J., Talapatra, S., Lafdi, K. et al., "Adsorbate binding energy and adsorption capacity of xenon on carbon nanohorns," *Nanotechnology* 13 (2002): 201–204.

Zhang, M., Yudasaka, M., Miyawaki, J. et al., "Isolating single-wall carbon nanohorns as small aggregates through a dispersion method," *J. Phys. Chem. B* 109 (2005): 22201–22204.

Zhang, M., Yudasaka, M., Ajima, K. et al., "Light-assisted oxidation of single-wall carbon nanohorns for abundant creation of oxygenated groups that enable chemical modifications with proteins to enhance biocompatibility," *ACS Nano* 1 (2007): 265–272.

Zhang, M., Murakami, T., Ajima, K. et al., "Fabrication of ZnPc/protein nanohorns for double photo-dynamic and hyperthermic cancer phototherapy," *Proc. Natl. Acad. Sci.* 105 (2008): 14773–14778.

Zhang, M., Yamaguchi, T., Iijima, S. et al., "Individual single-wall carbon nanohorns separated from aggregates," *J. Phys. Chem. C* 113 (2009): 11184–11186.

Zhang, J., Ge, J., Shultz, M. D. et al., "In vitro and in vivo studies of single-walled carbon nanohorns with encapsulated metallofullerenes and exohedrally functionalized quantum dots," *Nano Lett.* 10 (2010a): 2843–2848.

Zhang, J., Lei, J., Xu, C. et al., "Carbon nanohorn sensitized electrochemical immunosensor for rapid detection of microcystin-LR," *Anal. Chem.* 82 (2010b): 1117–1122.

Zhang, L., Zheng, N., Gao, A. et al., "A robust fuel cell cathode catalyst assembled with nitrogen-doped carbon nanohorn and platinum nanoclusters," *J. Power Sour.* 220 (2012): 449–454.

Zhang, M., Yamaguchi, T., Iijima, S. et al., "Size-dependent biodistribution of carbon nanohorns in vivo," *Nanomed. Nanotechnol. Biol. Med.* 9 (2013): 657–664.

Zhang, J., Sun, Q., Bo, J. et al., "Single-walled carbon nanohorn (SWNH) aggregates inhibited proliferation of human liver cell lines and promoted apoptosis, especially for hepatoma cell lines," *Int. J. Nanomed.* 9 (2014a): 759–773.

Zhang, L., Gao, A., Liu, Y. et al., "PtRu nanoparticles dispersed on nitrogen-doped carbon nanohorns as an efficient electrocatalyst for methanol oxidation reaction," *Electrochim. Acta* 132 (2014b): 416–422.

Zhao, Y., Li, J., Ding, Y. et al., "Single-walled carbon nanohorns coated with $Fe_2O_3$ as a superior anode material for lithium ion batteries," *Chem. Commun.* 47 (2011a): 7416–4718.

Zhao, Y., Li, J., Ding, Y. et al., "A nanocomposite of $SnO_2$ and single-walled carbon nanohorns as a long life and high capacity anode material for lithium ion batteries," *RSC Adv.* 1 (2011b): 852–856.

Zhao, C., Lin, D., Wu, J. et al., "Nanogold-enriched carbon nanohorn label for sensitive electrochemical detection of biomarker on a disposable immunosensor," *Electroanalysis* 25 (2013): 1044–1049.

Zhu, J., Yudasaka, M., Zhang, M. et al., "A surface modification approach to the patterned assembly of single-walled carbon nanomaterials," *Nano Lett.* 3 (2003a): 1239–1243.

Zhu, J., Kase, D., Shiba, K. et al., "Binary nanomaterials based on nanocarbons: A case for probing carbon nanohorns' biorecognition properties," *Nano Lett.* 3 (2003b): 1033–1036.

Zhu, S., Fan, L., Liu, X. et al., "Determination of concentrated hydrogen peroxide at single-walled carbon nanohorn paste electrode," *Electrochem. Commun.* 10 (2008): 695–698.

Zhu, S., Li, H., Niu, W. et al., "Simultaneous electrochemical determination of uric acid, dopamine, and ascorbic acid at single-walled carbon nanohorn modified glassy carbon electrode," *Biosens. Bioelectron.* 25 (2009a): 940–943.

Zhu, S., Niu, W., Li, H. et al., "Single-walled carbon nanohorn as new solid-phase extraction adsorbent for determination of 4-nitrophenol in water sample," *Talanta* 79 (2009b): 1441–1445.

Zhu, S. & Xu, G., "Single-walled carbon nanohorns and their applications," *Nanoscale* 2 (2010): 2538–2549.

Zhu, S., Han, S., Zhang, L. et al., "A novel fluorescent aptasensor based on single-walled carbon nano-horns," *Nanoscale* 3 (2011a): 4589–4592.

Zhu, S., Liu, Z., Zhang, W. et al., "Nucleic acid detection using single-walled carbon nanohorns as a fluo-rescent sensing platform," *Chem. Commun.* 47 (2011b): 6099–6101.

Zhu, S., Liu, Z., Hu, L. et al., "Turn-on fluorescence sensor based on single-walled-carbon-nanohorn-peptide complex for the detection of thrombin," *Chem. Eur. J.* 18 (2012): 16556–16561.

Zhu, S., Gao, W., Zhang, L. et al., "Simultaneous voltammetric determination of dihydroxybenzene-isomers at single-walled carbon nanohorn modified glassy carbon electrode," *Sens. Actuators B* 198 (2014a): 388–394.

Zhu, S., Zhang, J., Zhao, X. et al., "Electrochemical behavior and voltammetric determination of L-tryptophan and L-tyrosine using a glassy carbon electrode modified with single-walled carbon nanohorns," *Microchim. Acta* 181 (2014b): 445–451.

Zimmermann, K. A., Inglefield, Jr., D. L., Zhang, J. et al., "Single-walled carbon nanohorns decorated with semiconductor quantum dots to evaluate intracellular transport," *J. Nanopart. Res.* 16 (2014): 2078.

Zobir, S. A. M., Zainal, Z., Keng, C. S. et al., "Synthesis of carbon nanohorn–carbon nanotube hybrids using palm olein as a precursor," *Carbon* 54 (2013): 492–494.

# 5

# Geodesic Arenes

Bartosz Trzaskowski

Kinga Ostrowska

Ludwik Adamowicz

## Abstract

In 1966, when a successful synthesis of corannulene, the first member of the geodesic arenes class of the polycyclic aromatic hydrocarbons, was performed, its authors immediately noticed that the intrinsic strain of the molecule may result in unusual physical and chemical properties of this system. Today, geodesic arenes are a rapidly growing family of chemical compounds because they offer numerous potential applications, including their use as new electronic and optical organic materials, as components in the synthesis of fullerene isomers, isomerically pure SWCNHs, and many others. Interestingly, the questions concerning their structural strain and aromaticity asked over 50 years ago are still relevant today within the context of the basic understanding of the structure–property relationship. This chapter is intended as a comprehensive guide to geodesic arenes by describing their fundamental properties, synthesis, and potential applications across many scientific disciplines.

## 5.1 Introduction

The 1950s and 1960s were perceived by many as the golden age of annulene chemistry, with many new polycyclic aromatic hydrocarbons (PAHs) synthesized at that time. One of the most interesting systems first obtained in this period was fluoranthene, which consists of naphthalene and benzene units connected by a five-membered ring as first identified in 1940 (Rehner 1940) (Figure 5.1). In the next 20 years, numerous fluoranthene derivatives have been synthesized yielding increasingly more complex planar structures. In 1966, Barth and Lawton published their work on the synthesis of dibenzo[ghi,mno]fluoranthene, the first PAH with an intrinsic structural curvature. Despite the lack of a crystal structure of this species (which was obtained 10 years later by Hanson and Nordman in 1976), the authors immediately noted that the compound exhibits an unusual strain resulting from its geometrical configuration. They also correctly assumed that the strain is distributed in a symmetric manner, although they were not able to predict whether the system in question remains planar or becomes bowl shaped. They also suggested a resonance structure, dibenzo[ghi,mno]fluoranthene, which, in simple terms, explained its aromaticity by satisfying the $4n + 2$ Hückel rule and named the species corannulene. Hence, the family of geodesic arenes was born.

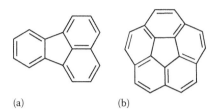

(a)                              (b)

**FIGURE 5.1**   Chemical structures of (a) fluoranthene and (b) corannulene.

Interestingly, despite the obvious uncommonness of the structure of corannulene, the number of studies on this system in the next 20 years was very low; there were a total of 11 published works with the word *corannulene* published between 1967 and 1990. Unfortunately, the difficulty and sheer length of the synthetic process discouraged researchers from undertaking a more extensive investigation of the system. It all changed in the early 1990s, when many scientists were looking for new methods to synthesize fullerenes and its derivatives. The $C_{60}$ fullerene, first synthesized in 1985 (Kroto et al. 1985), as well as other fullerenes, are also members of the geodesic arenes family. Fullerenes will not be discussed here, though, because they fall beyond the scope of this chapter. The enormous interest in fullerenes turned, however, the attention of researchers to PAHs and the development of new more efficient synthetic approaches to make corannulene and its derivatives become a target of numerous studies.

In this chapter, we summarize the most important studies on open geodesic arenes. First, in Section 5.2, we describe the fundamental properties of the systems, showing the differences between geodesic arenes and planar PAHs. In Section 5.3, we describe the most commonly used synthetic protocols of obtaining geodesic arenes. Lastly, in Section 5.4, we talk about their potential applications in various fields.

## 5.2 Fundamental Properties

Planarity is the fundamental geometric characteristic of the planar PAHs, which has a major impact on their chemical and physical properties. However, in corannulene, the arrangement of the benzene rings around the central five-membered ring induces a significant nonplanarity of the conjugated carbon network. The planarity (or the lack of planarity) may be measured using several different parameters; the three most convenient parameters to use are the depth of the bowl, the pyramidalization angle, and the puckering angle (Figure 5.2). The depth of the bowl can be defined as the distance between the plane of the central ring and the plane of the rim of the CH carbon atoms (Sygula 2011). This parameter is a good measure of the global curvature of the system. On the other hand, the pyramidalization angle is a good parameter to describe the local curvature. It is defined based on the $\pi$-orbital axis vector, $\theta_{\sigma\pi}$ (the vector that forms equal angles with the three sigma bonds at a conjugated carbon atom), as $\theta_{pyr} = \theta_{\sigma\pi} - 90$. Finally, the bond-puckering angle can be defined as the divergence of the two six-membered rings connected with the five-membered ring with a single chemical bond from the perfect flat surface of graphene or another ideally flat system. For an ideally flat surface, the puckering angle is 0°

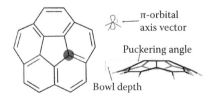

**FIGURE 5.2**   Definitions of the $\pi$-orbital axis vector, depth of the bowl, and bond-puckering angle in corannulene.

(Trzaskowski et al. 2013). The X-ray structure of corannulene (Hanson & Nordman 1976, Petrukhina et al. 2005) allows one to calculate the three parameters for this system; they are equal to 0.87 Å (bowl depth), 8.9° (pyramidalization angle), and 27.4° (puckering angle). One can expect that the intrinsic strain of geodesic arenes, which makes their structures nonplanar, affects the chemical and physical properties of these compounds.

In the case of monocyclic conjugated systems, Hückel's $4n + 2$ rule is fundamental to understanding their stability and reactivity within the aromaticity framework (Hückel 1931). Unfortunately, the rule cannot be applied to polycyclic systems. Among many attempts to extend the rule to larger chemical compounds, Clar's π-sextet rule has been probably the most successful (Clar 1972, Sola 2013). According to this rule, the most important resonance structure for predicting the properties of PAHs is the one with the largest number of disjoint aromatic π-sextets (the π-sextet being defined as a single benzene-like ring separated from the adjacent rings by C–C single bonds). Unfortunately, while this rule gives good results for flat PAHs and systems consisting solely of six-membered rings, it does not give satisfactory results for corannulene and other geodesic arenes.

The annulene-within-an-annulene model, originally proposed by Barth and Lawton (1966) in their work describing the first synthesis of corannulene, is another method to predict aromatic properties of PAHs. Using this approach, we can draw a resonance structure of corannulene as an aromatic 6-electron cyclopentadienyl anion surrounded by an also aromatic 14-electron annulenyl cation. This interesting model has been used for various geodesic arenes, but was later shown to fail to properly describe the chemical properties of [N]-circulenes, including corannulene (Sygula & Rabideau 1995, Steiner et al. 2001, Monaco et al. 2008). More recent studies of PAHs finally led to the realization that a correct description of the aromatic properties may be realized via the measurements of the ring currents (Pople 1956). Although not directly observable, the ring currents are inferred through their manifestation in the nuclear magnetic resonance (NMR) spectra and through measurements of the magnetic anisotropy. Fortunately, they can also be relatively easily calculated by means of the Hückel–London–Pople–McWeeny π-electron ring-current and bond-current calculations (London 1937, McWeeny 1958, Pople 1958, Acocella et al. 2002, Dickens & Mallion 2011).

That being said, corannulene is definitely an aromatic system, even though not all models are able to predict its aromaticity. As the final argument, one can mention Schleyer's NICS index calculations which show that, while the inner five-membered ring of corannulene is antiaromatic, the outer rings are definitely aromatic (Buhl 1998). The value of the NICS index for corannulene is −9 ppm and is very similar to the value obtained for benzene (−9.7 ppm). The aromatic properties of corannulene are also evident from the chemistry of this molecule, which shows similarity to the chemistry of benzene and its derivatives.

Another interesting feature of geodesic arenes is their configurational dynamics; due to their curvature, each geodesic arene may appear in two bowl conformations (Figure 5.3). The conformations are separated by a bowl-to-bowl inversion barrier, which is specific to each system. The inversion requires the molecule to pass through a transition state where the molecule temporarily attains a completely flat conformation, which is energetically favorable from the π-delocalization point of view (Dobrowolski

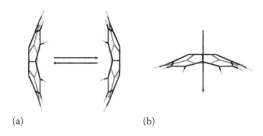

(a)                              (b)

**FIGURE 5.3**   Corannulene: (a) bowl-to-bowl inversion; (b) dipole moment.

et al. 2011), but unfavorable from the strain energy point of view. The inversion barriers of many corannulene derivatives lie in the region of 7–30 kcal/mol and can be measured by variable-temperature NMR study of a suitably derivatized molecule or estimated computationally with good accuracy (Scott et al. 1992b, Sygula et al. 1996). The mentioned range of the energy barriers means that for most small geodesic arenes the inversion is thermally accessible at ambient temperature. Interestingly, as the hydrogen atoms of corannulene in peri positions are replaced by larger moieties, the inversion barriers get smaller; e.g., for the corannulene with all 10 hydrogen atoms replaced by methyl groups, the inversion barrier is only 2.2 kcal/mol and the bowl depth is equal to 0.58 Å, suggesting a much flatter system (Sygula et al. 1996). On the other hand, benzocorannulenes have much larger inversion barriers than corannulene (~20 kcal/mol), although the bowl depths in these systems are similar to the bowl depth in corannulene (Marcinow et al. 2001).

Corannulene also displays interesting electrochemical properties. It was predicted that a closed-shell tetraanion of this species should be stable. Reduction of corannulene with excess metallic lithium at low temperature gives a series of three color changes, first to green, then to purple, and finally to brownish red. Quenching the solution with water results in the formation of tetrahydrocorannulene as the major product, accompanied by smaller amounts of dihydrocorannulene and corannulene in 4:2:1 ratio (Ayalon et al. 1992). The first and second reduction potentials and the first oxidation potential of corannulene can be measured by cyclic voltammetry and were found to depend on the conditions of the experiment, such as the temperature, and the solvent and the electrolytes used (Seiders et al. 1999b).

Another important consequence of the local curvature is the nonzero dipole moments of geodesic arenes. For corannulene, the dipole moment is already substantial and equal to 2.07 D (Lovas et al. 2005), but for larger and more curved arenes, it can attain values of up to 6 D (Baldridge & Siegel 1997). As with all other properties, the dipole moment of corannulene derivatives may be finely tuned to a particular value by attaching specific chemical groups to the core of the molecule.

# 5.3 Synthesis

The pioneering synthesis of corannulene by Barth and Lawton (1971) was definitely a breakthrough but, at the same time, had many disadvantages. Their synthetic approach consisted of 17 steps and the final yield was less than 1%. Only after new synthetic protocols were designed that the systematic studies of geodesic arenes could blossom. Here we present the two most important approaches in the synthesis of geodesic arenes and show a number of examples of the synthetic process starting from corannulene and moving toward larger and more complex corannulene derivatives. The in-depth analysis of all geodesic arenes synthesized up to date is beyond the scope of this chapter and those interested in it should refer to one of the specialized reviews on this topic, which include the studies of Tsefrikas and Lawrence (2006), Wu et al. (2008), and Sygula (2011).

## 5.3.1 Flash Vacuum Pyrolysis

Flash vacuum pyrolysis (FVP) is nowadays one of the most effective methods for synthesis of geodesic polyarenes (Brown 1990, McNab 2004). This method relies on heating the reactants in the gas phase for a very short period of time (milliseconds) to very high temperatures (200°C–1000°C), which allows for breaking aryl halide bonds resulting in radicals and can lead to the formation of carbon–carbon bonds. The typical FVP apparatus consists of a sublimation zone (vacuum), a hot zone (quartz tube), and a cold trap where the final products are collected (Necula & Scott 2000). In the synthesis of large molecules, a stream of nitrogen is commonly used to facilitate the flight of the molecules through the hot zone. Vapors of substrates are heated to temperatures in the range of 200°C–1100°C, allowing the final pressure to fall to the range of 0.1–1.0 mm Hg. While this method is very fast, it has numerous drawbacks, including low control over the synthetic process, lack of regioselectivity, relatively poor yields (~15%–30%), and limited scale of the synthesis (~1 g) in the case of corannulene and its derivatives.

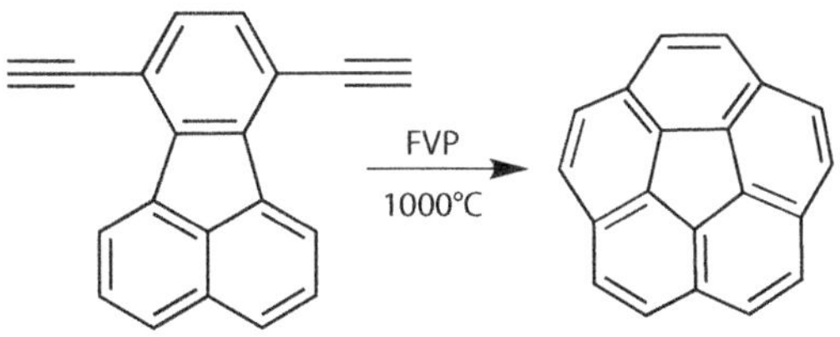

**SCHEME 5.1**  First synthesis of corannulene.

The first synthesis of the geodesic polyarene by FVP was reported in 1991 and involved a synthesis of corannulene ($C_{20}H_{10}$) from 7,10-diethynylfluoranthane using a relatively simple procedure (Scott et al. 1991). Terminal alkynes are good precursors of corannulene because of their ability to isomerize under FVP conditions by 1,2 shifts of hydrogen atoms (Brown et al. 1974a). The resulting vinylidenes can be trapped by insertion into the nearest C–H bonds (Scheme 5.1). The strategy proposed by Scott (1996) involved deformation of the flat diyne with heat and then "catching" the deformed-ring systems by C–C bond-forming reactions at the rim. It was realized at that time that previous numerous attempts to obtain corannulene using the conventional cyclization reaction failed due to insufficient temperature needed to form the new C–C bonds in the rings.

The yield of the FVP synthesis of corannulene from diyne was only 10%; however, the use of other precursors, e.g., tetrabromidefluoranthene or 7,10-bis(2-chlorovinyl)-fluoranthene, allowed to raise the yield to 32% (Liu & Rabideau 1996, Preda unpublished data). The best precursor, in terms of the final yield of the reaction, proved to be bis(1-chlorovinyl)fluoranthene. With this precursor a 35%–40% overall yield was achieved (Cheng 1992, Scott et al. 1992a, 1997).

The FVP route to corannulene may also start from a system different from the preferred fluoranthene nucleus (Scheme 5.2). In 1994, 3-(4H-cyclopenta[def]phenanthrylidene)-1,5-bis-(trimethylsilyl)-1,4-pentadiyne was synthesized and converted by hydropyrolysis into corannulene (Zimmerman et al. 1994). Five years later, a similar approach led to the synthesis of corannulene from a diyne derivative with a 36% yield (Knolker et al. 1999). Finally, it was shown that corranulene may also be synthesized from substituted [4]helicenes using cyclodehydrogenation and involved closure of the five-membered ring, although with a relatively low yield (Mehta & Panda 1997a).

The bowl geometry of corannulene may be effectively locked by the addition of a five-membered ring into its structure. This method is used in the synthesis of acecorannulene ($C_{22}H_{10}$), which was the first corannulene derivative synthesized by the FVP method (Scheme 5.3). The starting material in the synthesis of this system was the 5,8-bis(1-chlorovinyl)-1,2-dihydrocyclopenta[cd]fluoranthene. The pyrolysis of this substrate accompanied by the dehydrogenation of the ethane unit and followed by a partial detachment of the ethane fragment yielded a mixture containing corannulene as the by-product (Abdourazak et al. 1993). Acecorannulene may also be synthesized from 2-chlorovinylcorannulene (Preda unpublished data), 1-chlorovinylcorannulene (Scott 1996), 1-(prop-1-en-2-yl) corannulene (Preda 2001), or from 1-(2,2-dibromovinyl)-corannulene (Mack & Scott unpublished data) under FVP conditions (Preda et al. 2001, Tsefrikas & Lawrence 2006). Unfortunately, in most cases the main product is contaminated by small amounts of corannulene.

Benzo[a]corannulene ($C_{24}H_{12}$) is a corannulene derivative with an extra benzene ring. This compound is synthesized with the FVP method using 7-(2-bromophenyl)-10-(1-chlorovinyl)fluoranthene as a substrate (Scheme 5.4). The synthesis is performed at 1100°C and a 46% yield is achieved (McComas 1996). In this process, the 1-chlorovinyl group is used as a "masked acetylene" to close a ring across one bay region of the fluoranthene nucleus. The same strategy with 9,12-bis(1-chlorovinyl)benzo[e]acephenanthrylene as a substrate gives benzo[a]corannulene with a 20% yield. Other substrates used as benzo[a]corannulene precursors are (Z)-9-(2-chlorovinyl)-4a,4b,10b,14b-tetrahydrobenzo[g]chrysene, 9-(2,2-dibromovinyl)-4a,4b,10b,14b-tetrahydrobenzo[g]chrysene, or 9-ethynyl-4a,4b,10b,14b-tetrahydrobenzo[g]chrysene, but with much lower yields (Mehta & Srirama Sarma 2000).

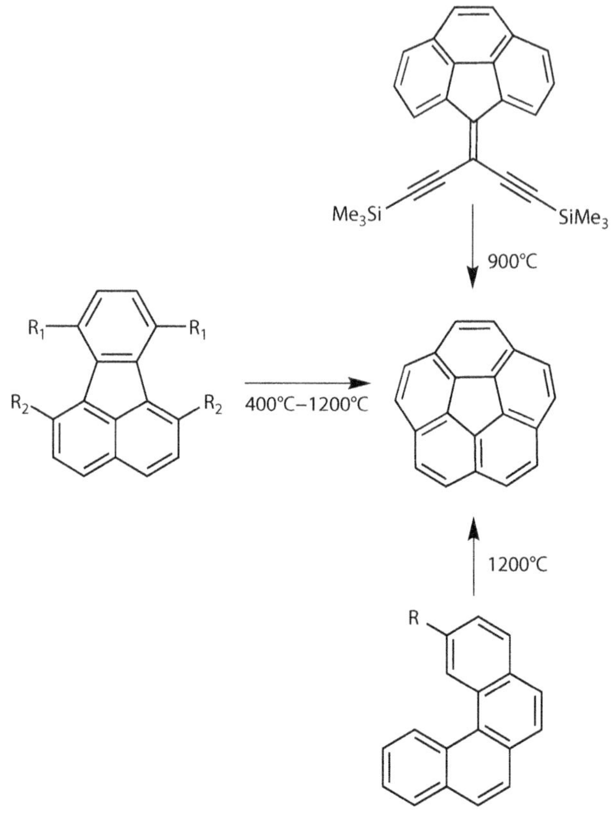

**SCHEME 5.2**   Various synthetic routes to obtain corannulene using FVP.

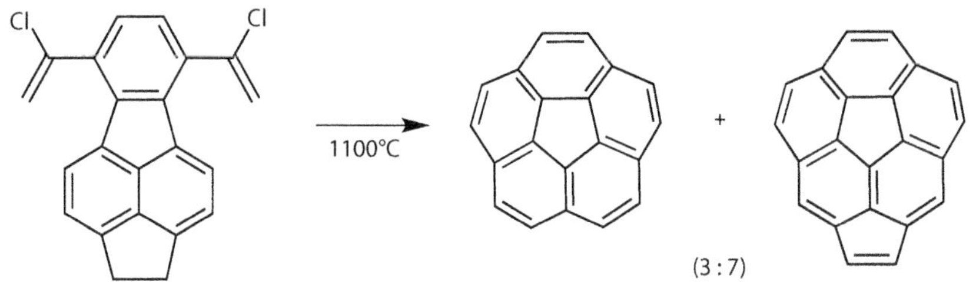

**SCHEME 5.3**   Synthesis of acecorannulene.

*Tetrabenzopyracylene, yin-yang fluoranthenes, indeno[1,2,3-bc]corannulene,* and *benzo[g]-acecor-annulene* are four known isomers of geodesic polyarenes with the same molecular formula $C_{26}H_{12}$. Tetrabenzopyracylene was first prepared from 1,1'-dibromobifluorenylidene; FVP of this compound at 1000°C–1050°C on a 100 mg scale gave the desired product (Bronstein et al. 2002). The use of 2,2'-dibro-mobifluorenylidene in a similar synthetic process results in only 2%–3% yield due to the fact that the major product of the pyrolysis is the singly closed polycyclic aromatic hydrocarbon (Scheme 5.5).

The yin-yang fluoranthene is an interesting system built from two fluoranthene units combined in a shape of the Chinese yin-yang symbol (Scott & Preda 2000). The synthesis proceeds in the FVP condi-tions with 1,6-bis((Z)-2,3-dichlorostyryl)naphthalene as the starting material. The first step of the pro-cess is the cyclization of the substrate followed by the aromatization through HCl loss and, finally, by

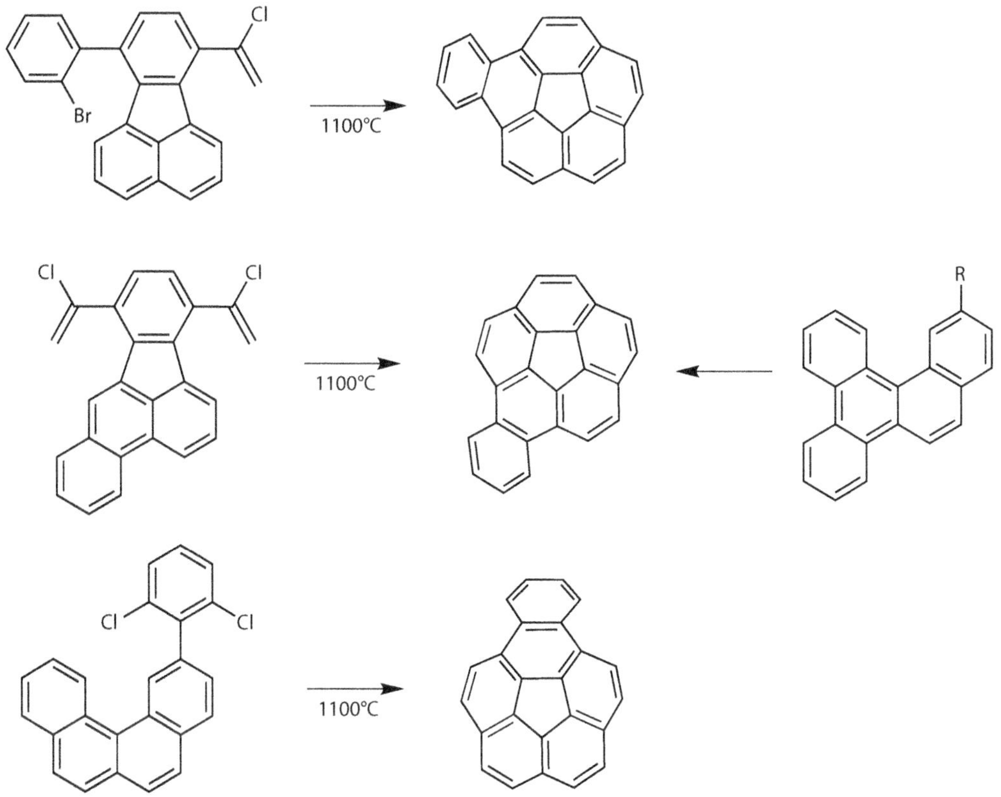

**SCHEME 5.4** Synthesis of benzo[a]corannulene.

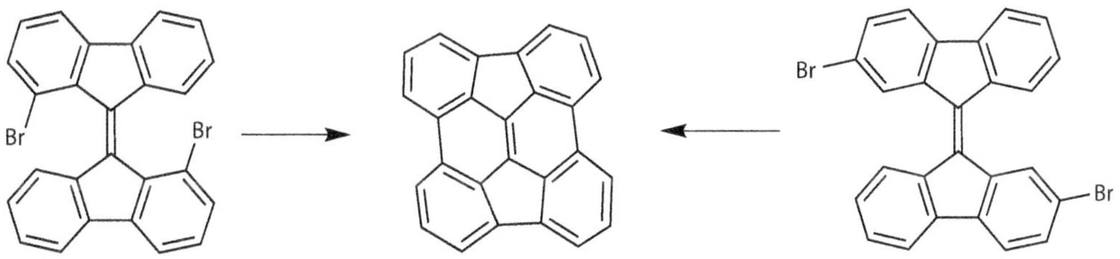

**SCHEME 5.5** Synthesis of tetrabenzopyracylene.

five-membered ring closure. Indenocorannulene, the third $C_{26}H_{12}$ isomer, may be synthesized by direct FVP of phenylcorannulene or (2-bromobenzoyl)corannulene (Scheme 5.6). The yield of the first synthesis is only 5% due to the formation of other undesirable products of cyclodehydrogenation, while the second approach gives a healthy 24% yield. The fourth and final $C_{26}H_{12}$ isomer, benzo[g]acecorannulene, may be synthesized in the FVP conditions using 7,10-bis(2-bromophenyl)acenaphtho[1,2-d] pyridazine as a substrate with 15%–20% yield. Interestingly, a potentially interesting aromatic aza-bowl 1,2-diazadibenzo[*d,m*]corannulene is not formed in this reaction (Tsefrikas et al. 2006).

The next group of isomeric bowls shares the $C_{28}H_{12}$ formula and consists of three compounds: fluoreno[1,9,8-abcd]corannulene, difluoreno[1,9,8,7-cdefg:2′1′9′8′-klmno]-anthracene, and acenaphto[1,2,3-bcd]corannulene. The synthesis of fluoreno[1,9,8-abcd]corannulene starts from 4-[9H-fluorenylidene-(9)]-4H-cyclopenta[def]phenanthrene at 780°C in nitrogen as the carrier gas (Scheme 5.7). First, the electrolitic ring

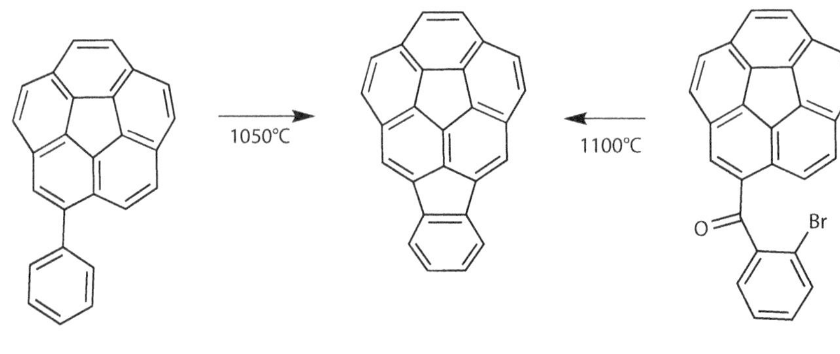

**SCHEME 5.6**   Synthesis of indeno[1,2,3-bc]corannulene.

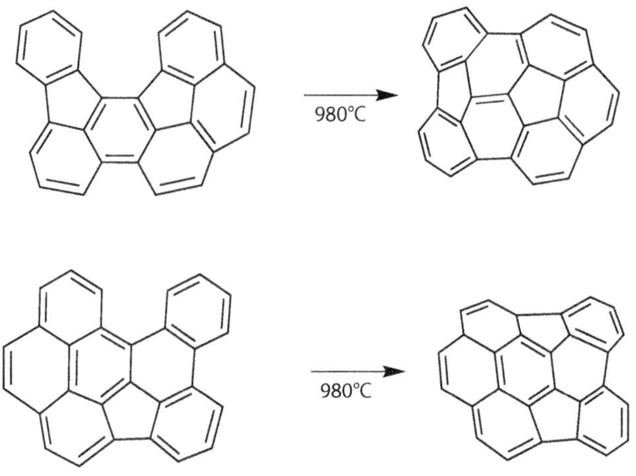

**SCHEME 5.7**   Synthesis of fluoreno[1,9,8-abcd]corannulene (top) and difluoreno[1,9,8,7-cdefg:2′,1′,9′,8′-klmno]-anthracene (bottom).

closure followed by hydrogen loss leads to the synthesis of diindeno[4,3,2,1-cdef;1′,2′,3′-hi]chrysene. In the next step, thermal dehydrocyclization occurs and the final product is obtained, although with a very low yield of only 0.6% (Hagen et al. 1995). The same starting system, 4-[9H-fluorenylidene-(9)]-4H-cyclopenta[def]phenantrene, can also lead to the synthesis of difluoreno[1,9,8,7-cdefg:2′1′9′8′-klmno]-anthracene via a homoallyl–cyclopropyl–carbinyl rearrangement (which can simultaneously occur with the previous reaction). The final yield of this process is even lower at 0.2%. The final $C_{28}H_{12}$ isomer, acenaphto[1,2,3-bcd]corannulene, is created from (Z)-1-(2,3-dichlorostyryl)dibenzo[ghi,mno]fluoranthene at 1080°C in 20% yield (Tsefrikas et al. 2006).

The synthesis of geodesic polyarenes by FVP can also be used to obtain two isomeric $C_{28}H_{14}$ bowls, dibenzo[a,g]corannulene and dibenzo[a,d]corannulene. The FVP of 7,11-diphenyl-8H-cyclopenta[k]fluoranthene-8,10(9H)-dione gives dibenzo[a,g]corannulene as the major product at temperatures in the range of 850°C–1000°C (Tsefrikas et al. 2006). 7,10-Bis(2-bromophenyl)fluoranthene, its corresponding dichloro precursor 7,10-bis(2-chlorophenyl)fluoranthene, and 7,10-bis(3-bromophenyl)fluoranthene also give dibenzo[a,g]corannulene with 38%, 17%, and 3% yields, respectively (Bratcher 1996). Dibenzo[a,d] corannulene is synthesized from 5-(2,2-dibromovinyl)dibenzo[c,p]chrysene at 1150°C with the yield of 5%–7% (Mehta & Srirama Sarma 2000). The synthetic route involves, in this case, cyclodehydrogenation to generate a five-membered ring followed by insertion of vinylidene carbene to form a six-membered ring, exactly as in the benzo[a]corannulene case (Scheme 5.8).

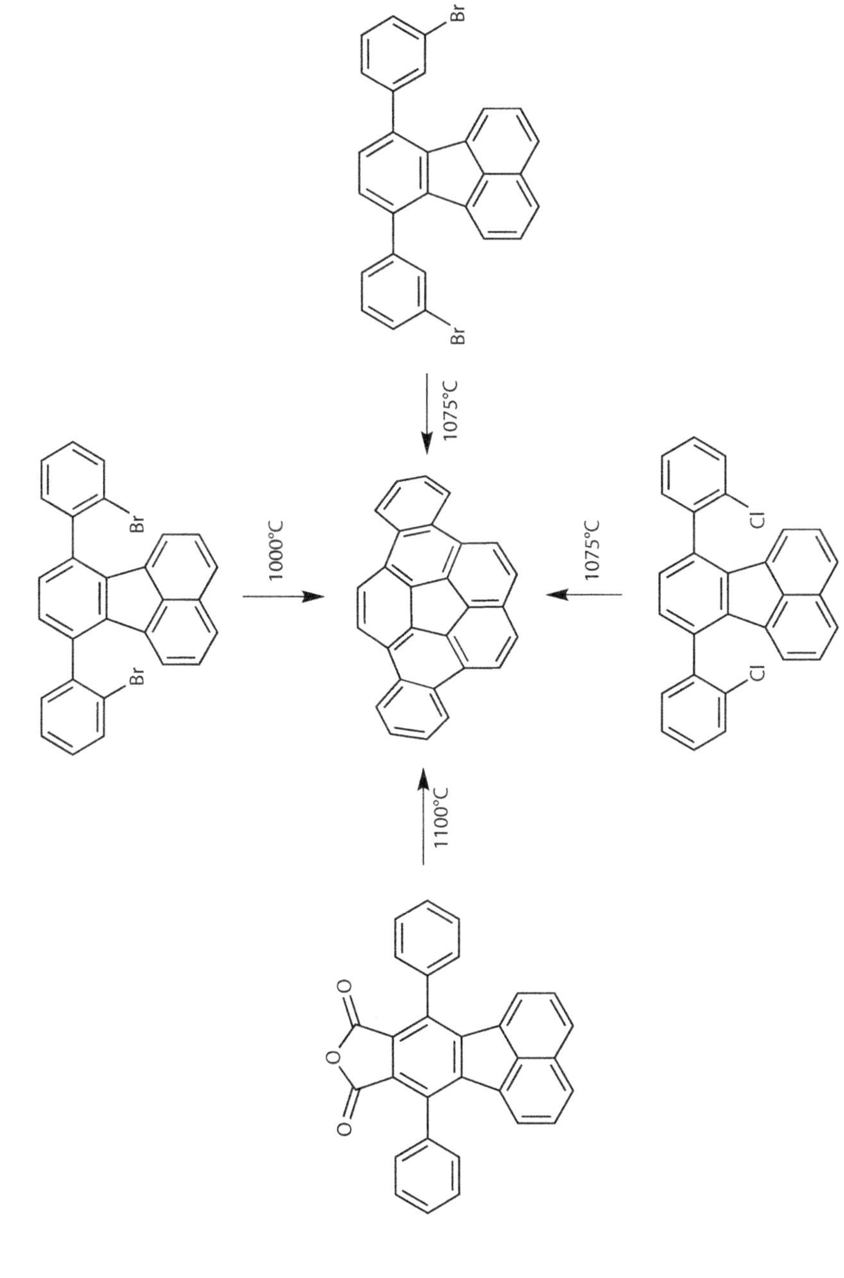

**SCHEME 5.8**  Synthesis of dibenzo[a,g]corannulene.

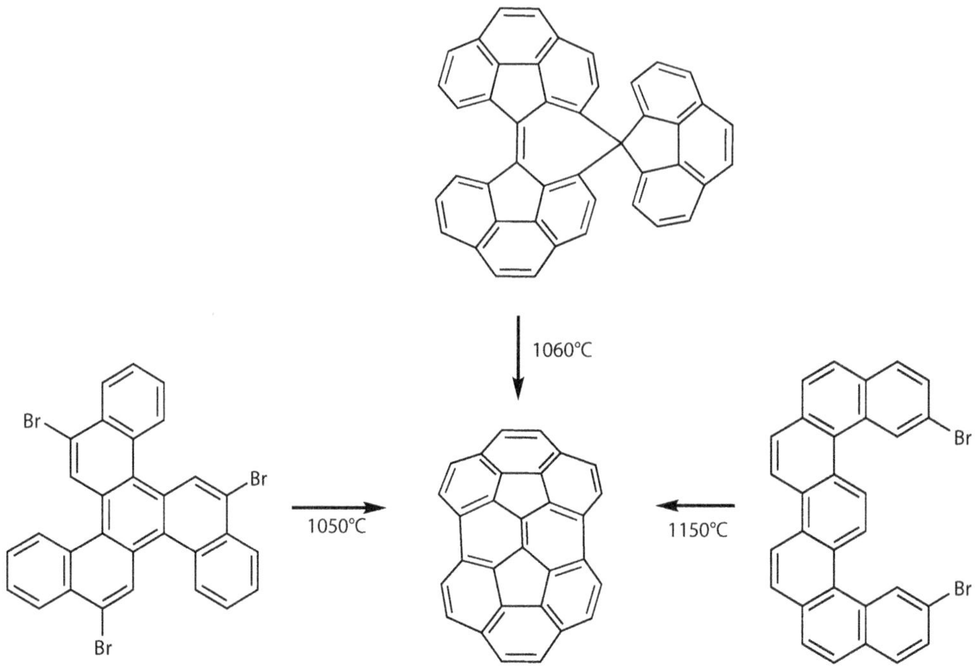

**SCHEME 5.9**    Synthesis of [5,5]circulene.

The method used in the first FVP synthesis of corannulene from bis(1-chlorovinyl)fluoranthene may be useful to prepare more complex systems, like [5,5]circulene ($C_{30}H_{12}$) from 1,4,7,10-tetrakis(1-chlorovinyl)indeno[1,2,3-cd]fluoranthene (Scheme 5.9). The yield of this process at 1000°C is around 5% (Clayton et al. 1996). The same system can be obtained by FVP of a spiro corannulene derivative, although with only 2.1% yield (Hagen et al. 1997). A conceptually different approach toward the synthesis of [5,5]circulene starts from 11,16-dibromodibenzo[a,o]picene and, in a multistage process, yields the final product with 2%–3% yield (Mehta & Panda 1997b).

The $C3$-symmetric "hemifullerene," triindenotriphenylene, is the next arene from the $C_{30}H_{12}$ group (Scheme 5.10). This particular system may be synthesized from chlorinated tris-fulvene using FVP with a 5%–10% yield (Abdourazak et al. 1995). The process involves triple homolysis of the C–Cl bond and cyclization of the resulting methine radical. An alternative synthesis of triindenotriphenylene from 6,12,18-tribromobenzo[c]naphtho[2,1-p]chrysene involves a 1,2-shift of hydrogen atoms in the intermediate aryl radical prior to the C–C bond formation and generation of a reactive aryl radical at the site where ring closure needs to occur (Hagen et al. 1997). A similar strategy may be used in the synthesis of even larger fullerene fragments (see, for example, the synthesis of tetrabenzopyracylene). The third $C_{30}H_{12}$ isomer, [2,1,9,8,7-defghi:2′,1′,9′,8′,7′-mnopqr]naphthacene, may be prepared at very high temperatures (1250°C) under FVP conditions with a 3% yield (Hagen et al. 1997).

Dibenz[f,l]acecorannulene ($C_{30}H_{14}$) contains a central corannulene ring system fused on two sides with additional benzene rings and a second five-membered ring in the form of an etheno bridge (Scheme 5.11). It is an interesting compound because its synthesis is a combination of the syntheses of dibenzo[a,g]corannulene and acecorannulene using 7,10-bis(2-bromophenyl)-3-(1-chlorovinyl)fluoranthene as a substrate (Weitz et al. 1998).

The pyrolysis of 1,2-bis(2-bromophenyl)-4,9-dibromocorannulene leads to the formation of $C_{32}H_{12}$ buckybowl, acenaphtho[3,2,1,8-ijlkm]diindeno[4,3,2,1-cdef:1′,2′,3′,4′-pqra] triphenylene (Marcinow et al. 2002), with a 7% yield (Scheme 5.12). Likewise, the benzannulated derivative of 7,10-bis(2-bromophenyl)fluoranthene (the FVP precursor of dibenzo[a,g]corannulene) forms acenaphth[3,2,1,8-fghij]-as-indaceno-[3,2,1,8,7,6-pqrstuv]

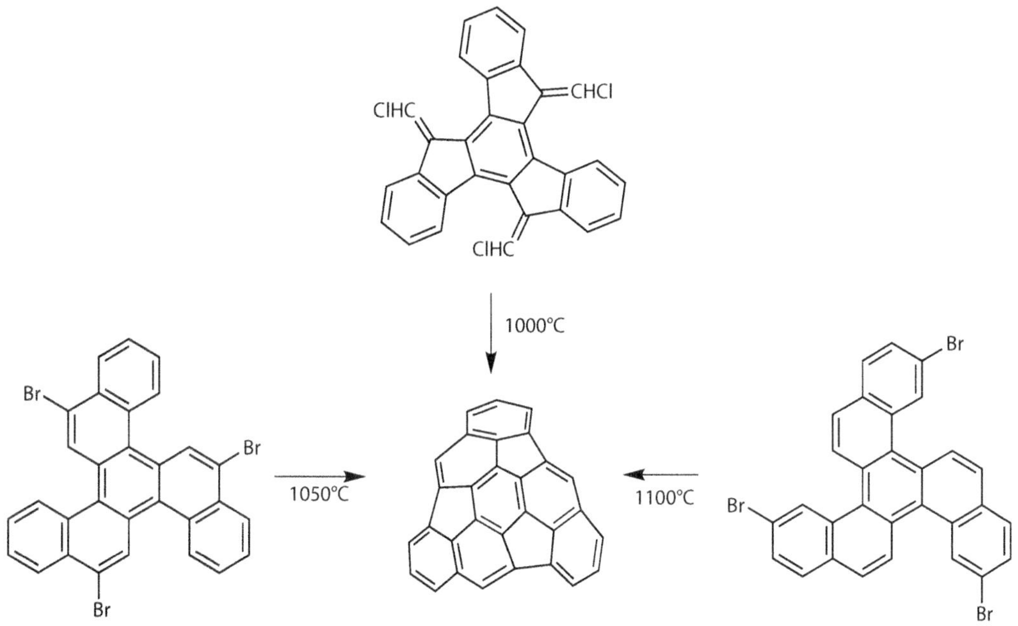

**SCHEME 5.10** Synthesis of triindenotriphenylene.

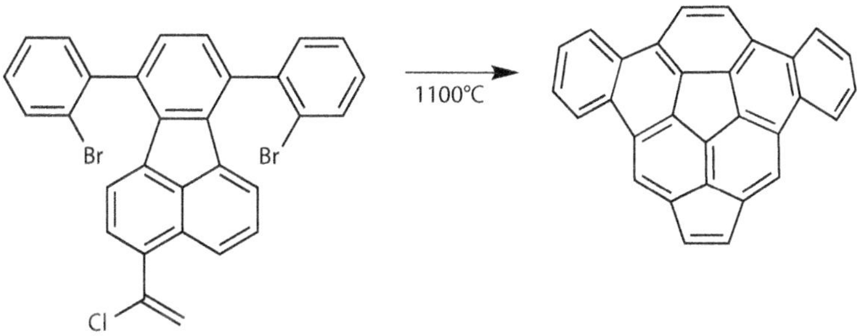

**SCHEME 5.11** Synthesis of dibenz[f,l]acecorannulene.

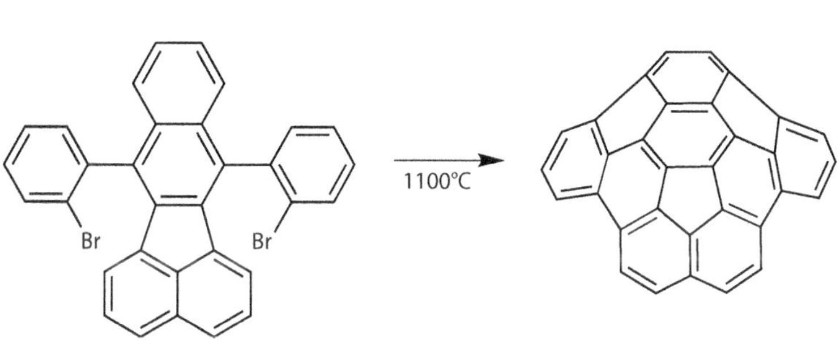

**SCHEME 5.12** Synthesis of acenaphth[3,2,1,8-fghij]-as-indaceno-[3,2,1,8,7,6-pqrstuv]picene.

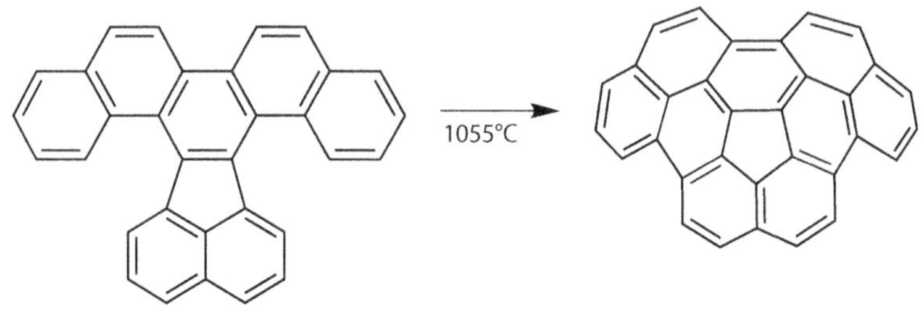

**SCHEME 5.13**   Synthesis of acenaphtho[3,2,1,8-efghi]peropyrene.

picene ($C_{32}H_{12}$), an isomeric bowl of acenaphtho[3,2,1,8-ijlkm]diindeno[4,3,2,1-cdef:1′,2′,3′,4′-pqra]triphenylene. The sequence of the reactions from the substrate to the product in this process involves cyclodehydrogenation to form two sets of rings, followed by loss of bromine to produce aryl radicals, and formation of the second set of rings. In the end, four new C–C bonds form (Clayton & Rabideau 1997). An analogous scheme is used to obtain benzo[a]fluoreno[1,9,8-ghij]corannulene ($C_{32}H_{14}$) from 9,12-bis(2-bromophenyl)benzo[e]acephenanthrylene (Tsefrikas et al. 2006).

The FVP of the final reaction mixture after corannulene double acylation can be used to obtain *o*-diindenocorannulene ($C_{32}H_{14}$) and *p*-diindenocorannulene ($C_{32}H_{14}$) with 5% and 38% yields, respectively (Scheme 5.13). This process is possible because the corannulene acylation with 2-bromobenzoyl chloride gives a 3:2 mixture of the 1,5-isomer (*o*-diindenocorannulene precursor) and 1,8-isomer (*p*-diindenocorannulene precursor). An interesting conclusion from the above results is that cyclizations of 2-benzoyl radicals produce more five-membered-ring ketones than six-membered-ring ketones in the initial ring closure reaction (Aprahamian et al. 2006). A different $C_{32}H_{14}$ isomer, acenaphtho[3,2,1,8-efghi]peropyrene, can be obtained by double cyclodehydrogenation of acenaphtho[1,2-s]picene.

Tribenzo[a,d,j]corannulene is a representative of the $C_{32}H_{16}$ group of compounds. This system may be synthesized under FVP conditions from 9,12-bis(2-bromophenyl)benzo[e]acephenanthrylene with a very high yield of 72% (Scheme 5.14). A different approach to the synthesis of this system is the use of 7-(2,2-dibromovinyl)naphtho[1,2-f]picene as a substrate at 1150°C under FVP conditions (Mehta & Srirama Sarma 2000). As in the previously discussed examples, this bowl-shaped benzocorannulene is formed from appropriate polycyclic aromatic employing FVP as the key step, in which a five- and a six-membered ring are sequentially formed.

The synthesis of larger geodesic arenes is based on the previously discussed examples. Dibenz[*a,j*]indeno[*ef*]corannulene is synthesized analogously to the formation of dibenzocorannulene and tribenzocorannulene at 1100°C, by using either 1,4-bis(2-bromophenyl)indeno[1,2,3-cd]fluoranthene (52% yield) or 7,11-diphenylindeno[1′,2′,3′:3,4]fluorantheno[8,9-c]furan-8,10-dione (10% yield) as substrates (see Scheme 5.11) (Tsefrikas et al. 2006). Triacenaphtho-[3,2,1,8-cdefg:3′,2′,1′,8′-ijklm:3″,2″,1″,8″-opqra]-triphenylene (circumtrindene) ($C_{36}H_{12}$) may be synthesized utilizing the triple cyclodehydrogenation

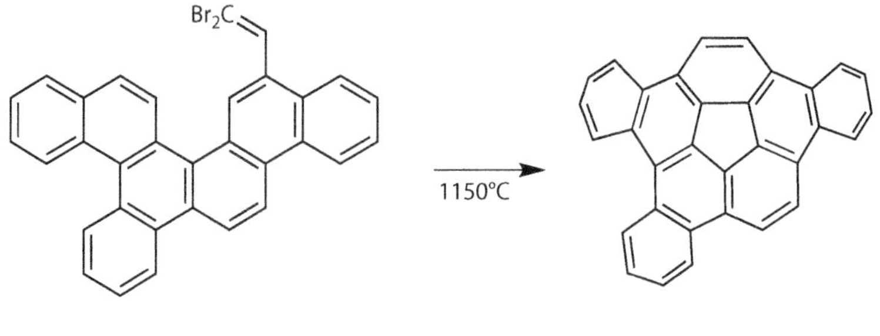

**SCHEME 5.14**   Synthesis of tribenzo[a,d,j]corannulene.

form diacenaphtho[1,2-j:1′,2′-l]fluoranthene (0.6% yield) or 2,9,15-trichlorodiacenaphtho[1,2-j:1′,2′-l]flu-oranthene (25%–27% yield). It has been found that the incorporation of functional groups capable of generating radical centers at the sites of the desired ring closures dramatically improves the formation yields of the geodesic arenes (Ansems & Scott et al. 2000).

Finally, a $C_{40}H_{14}$ diphenanthro[1,10,9,8-abcde:1′,10′,9′,8′-jklmn]tetracyclopentapyrene bowl is obtained as the main product starting from 1,3,6,8-tetrakis(2-chlorophenyl)pyrene as a substrate with a 1.3% yield. Benzo[e]triindeno[1,2,3-cd:1′,2′,3′-fg:1″,2″,3″-jk]pyrene ($C_{34}H_{16}$) is obtained as a secondary product in this process (Tsefrikas et al. 2006). The last compound described here—indeno[1,2,3-bc]circumtrin-dene ($C_{34}H_{16}$)—is formed by FVP at 1100°C from 2-bromophenyl-substituted circumtrindene by loss of the aryl group and is the largest open geodesic polyarene synthesized up to date.

## 5.3.2 Solution-Phase Synthesis

The solution-phase synthesis of geodesic arenes is the second major approach toward obtaining this class of compounds. The main advantage of this methodology is a higher control over the synthetic process allowing access to simple alkyl derivatives of corannulene in a regioselective manner. On the other hand, it often requires optimization of multiple synthetic steps and is generally slower than FVP.

The first successful solution-phase synthesis of corannulene was performed 30 years after the semi-nal work of Barth and Lawton (1966). In their work, diastereomeric mixture of the benzylic bromide of 1,6-bis(bromomethyl)-7,10-bis(1-bromoethyl)fluoranthenes was subjected to internal cyclization by reductive coupling with $TiCl_3/LiAlH_4$ or $VCl_3/LiAlH_4$. Dehydrogenation of the obtained mixture of *cis*- and *trans*-dimethyltetrahydrocorannulene with 2,3-dichloro-5,6-dicyanobenzoquinone produced 2,5-dimethylcorannulene with an 18% total yield (Seiders et al. 1996). It was the first synthesis of geo-desic polyarene which relied on intramolecular reductive coupling of benzylic bromides (Scheme 5.15). Currently, the most common method of the synthesis is an improved version of that approach utilizing low-valent titanium or vanadium coupling of the dibromomethyl groups (Seiders et al. 1999a, Sygula & Rabideau 1999).

Other corannulene derivatives including methylcorannulene, dimethylcorannulene, tetramethylcor-annulenes, acecorannulylene, pentamethylcorannulene, and decamethylcorannulene may be formed using similar low-valent titanium carbon–carbon coupling chemistry and halogen-for-alkyl-exchange chemistry mediated by trimethylaluminum and catalytic nickel salts (Seiders et al. 1999a). The yields of the double ring closure step are usually very good, around the 70%–80% mark.

A large-scale synthesis of the corannulene system can also be accomplished using a mixture of dibromo- and tribromocorannulenes as substrates. The mixture is formed during the hydrolysis of 1,6,7,10-tetrakis(dibromomethyl)fluoranthene in acetone/water solutions containing either sodium carbonate or sodium hydroxide. After treating the mixture with *n*-butyllithium in tetrahydrofuran at –78°C, corannulene is formed and the overall yield is 50%–55% for the two steps combined (Sygula & Rabideau 2000). An important corannulene derivative, dimethyl 1,2-corannulene dicarboxylate, can

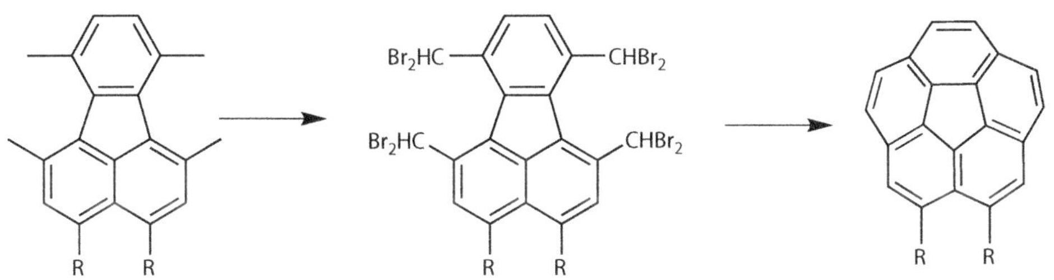

**SCHEME 5.15**   General solution-phase synthesis of corannulene derivatives.

be synthesized by nickel-mediated intramolecular coupling of benzyl and benzylidene bromide groups (Sygula et al. 2002). This process starts with carbinol, which is transformed to fluoranthene. In the next step fluoranthene is brominated, and the obtained hexabromide is transformed to dimethyl 1,2-corannulenedicarboxylate in dimethylformamide with commercial nickel powder with a good yield of ca. 60%. Hexabromide under reflux with dioxane/water (3:1) and sodium hydroxide gives dibromo corannulene dicarboxylic acid.

Nonpyrolytic preparation of larger buckybowls from fluoranthenes and related precursors is also possible. Cyclopentacorannulene may be synthesized by employing low-valent titanium coupling of hexakis(dibromomethyl)fluoranthene with a 20%–30% yield. Acecorannulene ($C_{22}H_{10}$) is produced from 1,10-bis(bromomethyl)dibenzo[ghi,mno]fluoranthene with a 20% yield by either palladium-catalyzed Stille-type coupling or treatment with phenyllithium (Seiders et al. 1999a, 2001) (Scheme 5.16). Semibuckminsterfullerene may be synthesized by nonpyrolytic route from the dodecabromide with a 20% yield (Sygula & Rabideau 1999). Tetrabromosemibuckminsterfullerene is also prepared from dodecabromide with a 26% yield, and can be converted to tetramethylsemibuckminsterfullerene (Sygula et al. 2000).

Dibenzo[a,g]-corannulene and 1,2-dihydrocyclopenta[b,c]dibenzo[g,m]corannulene are formed in palladium-catalyzed intramolecular arylation reactions of 7,10-bis(2-bromophenyl)fluoranthenes with 57% and 36% yields, respectively (Reisch et al. 2000, Marcinow et al. 2001). This process was the first solution-phase synthesis of a geodesic polyarene that relied on palladium-catalyzed intramolecular arylation. The planar ring system of 7,10-di(2-bromophenyl)fluoranthene is converted to the $C_{28}H_{14}$ bowl-shaped fullerene fragment dibenzo-[a,g]corannulene (Scheme 5.17).

Indenocorannulene ($C_{26}H_{12}$) can be synthesized under palladium catalysis by a Suzuki–Heck coupling cascade reaction of pinacol corannuleneborate and *o*-dibromobenzene with a 40% yield (Wegner et al. 2003) (Scheme 5.18). This process consists of two stages—a Suzuki coupling reaction to join two ring systems together and an intramolecular Heck-type arylation to close the five-membered ring.

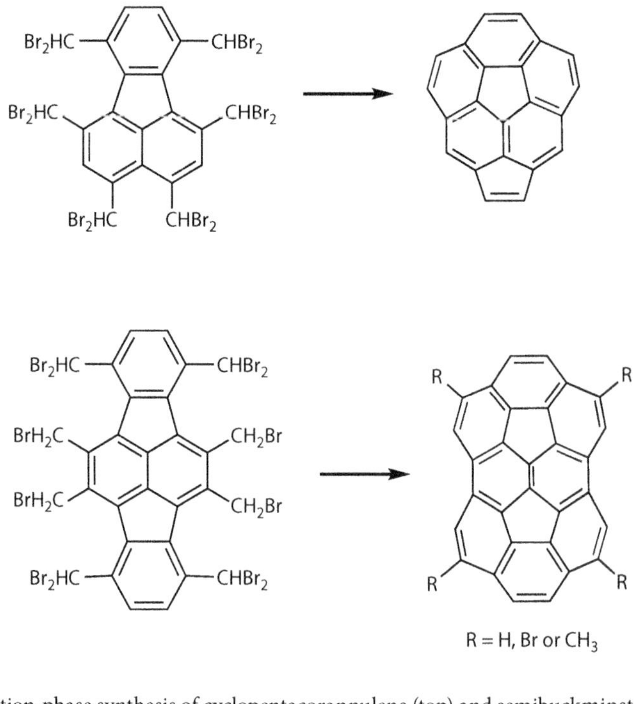

**SCHEME 5.16**  Solution-phase synthesis of cyclopentacorannulene (top) and semibuckminsterfullerenes (bottom).

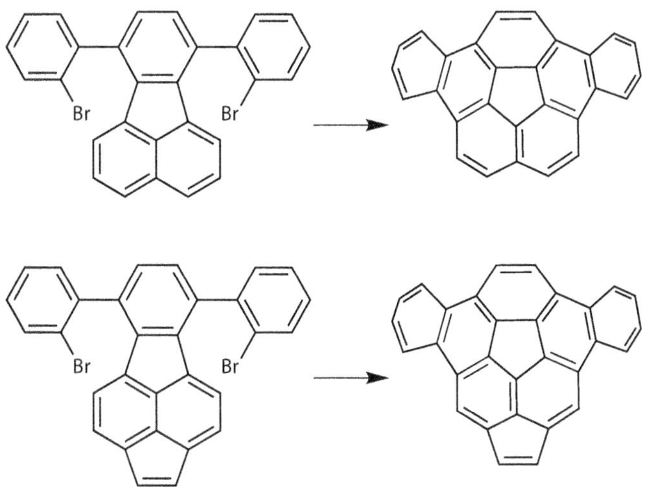

**SCHEME 5.17**  Solution-phase synthesis of dibenzocorannulenes.

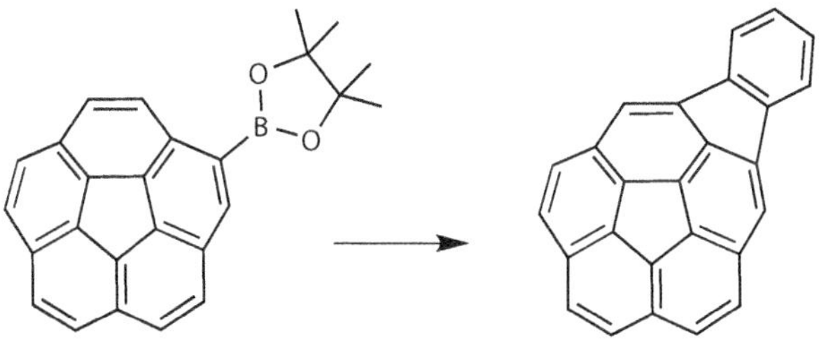

**SCHEME 5.18**  Solution-phase synthesis of indenocorannulene.

Both steps are catalyzed by palladium. Pentaindenocorannulene ($C_{50}H_{20}$) and tetraindenocorannulene ($C_{44}H_{18}$) syntheses start from 1,3,5,7,9-pentachlorocorannulene and 1,2,5,6-tetrabromocorannulene, respectively. The processes include fivefold or fourfold Suzuki–Miyaura coupling of the substrates with 2-chlorophenylboronic acid to give 1,3,5,7,9-pentakis(2-chlorophenyl)corannulene or 1,2,5,6-tetrakis(2-chlorophenyl)corannulene. The subsequent fivefold or fourfold cyclization to pentaindenocorannulene or tetraindenocorannulene by an intramolecular arylation reaction is the last part with a 35% or 13% yield (Jackson et al. 2007). This method was also used in the synthesis of two isomeric diindenocorannulenes ($C_{32}H_{14}$) and two isomers of triindenocorannulene ($C_{38}H_{16}$), but for these systems, ring closures were carried out by means of microwave-assisted palladium-catalyzed intramolecular arylation (Steinberg et al. 2009).

Some of the functionalized precursors can be converted to the desired nanostructures by mild liquid-phase methods such as Pd(0)-catalyzed direct arylation. In the presence of Herrmann's catalyst, the acenaphtho[3,2,1,8-fghij]picene is obtained with a 57% yield, while the 1,2-dihydrocyclopenta[b,c] dibenzo[g,m]corannulene has a 36% yield (Alberico et al. 2007). The synthesis of benz[e]acephenantrylenes and corannulenes using this methodology is also possible (Echavarren et al. 2003). A similar approach leads to the synthesis of diindeno[1,2,3,4-*defg*;1',2',3',4'-*mnop*]chrysene (Chang et al. 2010) and indenocorannulene (Pascual et al. 2007). Finally, pentaindenocorannulene may be obtained with a 35% yield from 1,3,5,7,9-pentakis(2-chlorophenyl)corannulene through a palladium-catalyzed and microwave-induced fivefold intramolecular arylation reaction (Pascual et al. 2007) (Scheme 5.19).

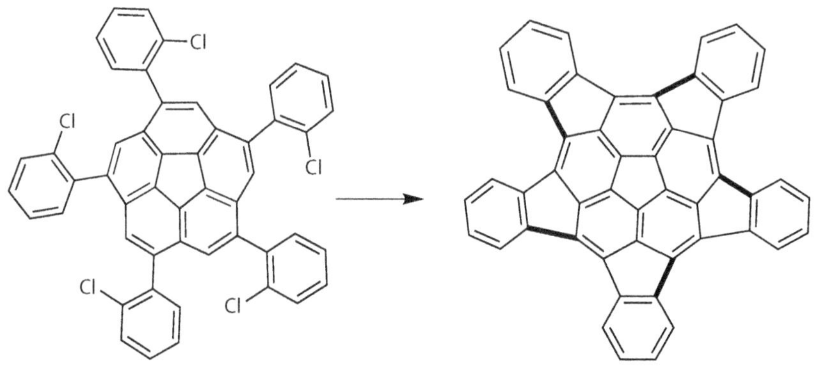

**SCHEME 5.19**   Synthesis of pentaindenocorannulene.

## 5.4 Applications

As mentioned before, the sudden increase in studies of corannulene and other geodesic arenes observed in the early 1990s, was a result of the synthesis of the $C_{60}$ buckminsterfullerene by Kroto et al. in 1985. In the following years, a large number of fullerene derivatives have been synthesized. The enormous interest in fullerenes also turned the attention of the chemical community to PAHs and, in particular, to geodesic arenes, as these systems can be viewed as fragments of fullerenes. In 1991, a remarkably simple procedure of synthesizing corannulene using FVP was published and in the next few years a large number of new geodesic arenes have been synthesized (Scott et al. 1991). To date, this method remains the leading approach to obtaining geodesic arenes (see Section 5.3), but it is also used to synthesize fullerenes and other close geodesic arenes starting from open geodesic arenes.

Historically, the first chemical synthesis of $C_{60}$ fullerene in isolable quantities used FVP in the last step to close the chlorinated polyarene precursor (Scott et al. 2002). In the multistep scheme, all intermediate products were, however, planar polyarenes and the introduction of the curvature occurred only in the last step of fullerene synthesis. Geodesic arenes possess an intrinsic strain, which, in principle, should lead to lower energy being needed during the synthesis of larger geodesic arenes and fullerenes. This is indeed the case.

One of the main strategies of obtaining larger geodesic arenes is the selective halogenation of corannulene or other small geodesic arenes (Xu et al. 2000). The substitution of the H atom bonded to a carbon atom by an F, Cl, Br, or I atom greatly enhances the reactivity of the C atom. This strategy has been used in the synthesis of, e.g., substituted dibenzocorannulenes (starting from tetraiodocorannulene), pentagonal dendrimers based on corannulene core (starting from sym-pentasubstituted corannulene), and other highly substituted corannulenes (Seiders et al. 1999a, Bancu et al. 2004, Pappo et al. 2009). From the many already synthesized corannulene derivatives, one family of systems, indenocorannulenes, warrants a special attention (Wegner et al. 2003). These compounds have a further increased intrinsic curvature and represent an even larger part of the fullerene surface. Many members of the family have been obtained up to date. They may also be considered as potential precursors for the synthesis of carbon nanotubes of uniform diameters.

The direct synthesis of $C_{60}$ fullerene using corannulene as the starting structure was first realized in 1995 by pyrolysis of this system with two fluoranthene derivatives (Crowley et al. 1995). This approach, while giving good yields, is unfortunately not very specific. Since the discovery of fullerenes, scientists have been looking for many years for means of synthesizing these systems using a controlled chemical method to obtain pure products. In general, there are two strategies which can be used to synthesize fullerenes. The first one uses completely planar molecules throughout the synthetic process and utilizes heat in the final step to force molecule bending. Such an approach was shown to be successful in 2002, when the first rational synthesis of $C_{60}$ fullerene was accomplished (Scott et al. 2002). In that seminal work, $C_{60}$ fullerene was synthesized in 12 steps from commercially available compounds. In the last step, a molecular polycyclic aromatic precursor bearing chlorine substituents at key positions formed $C_{60}$ when subjected to FVP at 1100°C. No other fullerenes were formed, although the overall yield was only in the range of 0.1%–1.0%. The second approach uses already curved molecules (like corannulene or other geodesic arene) in the synthesis of even more bent $C_{60}$ fullerene. While this method seems sound in theory, it did not yet find practical use due to technical problems with the synthesis of highly strained arenes and with fusing them together. The most interesting attempt in this regard was the synthesis of two identical hemispherical hydrocarbons resembling two halves of a fullerene (Hagen et al. 1997). Unfortunately, no method currently exists that is capable of assembling these two pieces together to form the fullerene.

Corannulene may be viewed as a structural motif of the $C_{60}$ fullerene, but also as an end-cap of the [5,5]-CNT. CNTs are usually produced by carbon arc discharge or other methods which yield mixtures of different types of CNTs that are virtually impossible to separate (Guldi & Martin 2010). Therefore, new methods with more control over the products of the synthesis of CNTs are clearly needed. One of such approaches is the synthesis of a small CNT part (e.g., the cap) and a slow controlled elongation by the repetitive addition of carbon atoms to annulate new six-membered rings. In the work of Scott et al. (2012), the starting material for such nanotube formations was corannulene, which in a three-step synthesis yielded a short isometrically pure [5,5]-CNT. It is possible that via the use of more complex geodesic arenes, other CNTs can be synthesized using this approach.

Due to their chemical properties, geodesic arenes seem to be perfect candidates for the molecular recognition of fullerenes and other systems with curved surfaces through dispersion-dominated attractive $\pi$–$\pi$ interactions. Early calculations showed, however, that the interaction of a single corannulene molecule with $C_{60}$ fullerene is not strong enough to be observed in a solution (Zhao & Truhlar 2008a). Since the bonding energy of the arene with the fullerene depends on the size of the arene surface, many scientists turned their attention to the synthesis of such supramolecular assemblies. The first successful attempt was made in 2007 when a system composed of two corannulene pincers joined together via a tetrabenzocyclooctatetraene tether was synthesized (Sygula et al. 2007). The crystal structure obtained from a mixture of the system and $C_{60}$ fullerene revealed a complex between the fullerene and corannulene subunits. The system, named *buckycatcher*, is able to adopt four major conformations with very similar energies and relatively small barriers for interconversion between them. As a result, buckycatcher is a very dynamic system in gas-phase or solution, but upon $C_{60}$ complexation adopts one specific conformation to maximize the interactions with the guest molecule. The complexation strength can be experimentally estimated by analyzing shifts in the NMR spectra; a similar approach in the solution led to the determination of the binding constant value $K_{assoc}$, equal to $8600 \pm 500$ M$^{-1}$ in toluene. Interestingly, the same system can also strongly bind ($K_{assoc} = 6800 \pm 500$ M$^{-1}$) $C_{70}$ fullerene, making it nonspecific molecular tweezers (Mück-Lichtenfeld et al. 2010).

It is interesting to note the role of theoretical methods in the search for new fullerene clips, tweezers, pincers, and cages. Nowadays, computational methods based on modern density functionals, including the most popular B3LYP and M06 (Lee et al. 1988, Becke 1993, Zhao & Truhlar 2008b), can accurately predict both the relative energies of geodesic arenes conformations, as well as the interaction energies with other PAHs, fullerenes, or other similar systems. This is possible due to the proper description of the London dispersion forces by some of these methods. The London interactions are the major forces behind the nonbonding attractive van der Waals interactions. Unfortunately, due to the unfavorable scaling of the density functional methods with the number of electrons in the system, such calculations are feasible only for systems up to approximately 120–200 atoms depending on the available computing power. For larger systems, the self-consistent charge density functional tight binding (SCC-DFTB) method is a reasonable choice, as it is known to produce reasonably good results (similar to DFT) at a fraction of the cost of the DFT calculations, particularly for $\pi$–$\pi$ bonded systems (Porezag et al. 1995). By using SCC-DFTB, it is realistic to perform computational studies on systems consisting of up to 1000 atoms.

For the buckycatcher complex with $C_{60}$ fullerene, the best binding energy estimates give 36–40 kcal/mol, suggesting a relatively strong interaction (Grimme et al. 2010, Mück-Lichtenfeld et al. 2010). On the other hand, the experimental $\Delta G$ value for the formation of this complex in toluene is only –5.3 kcal/mol (Sygula et al. 2007), showing that the impact of entropy and solvation effects is immense, and indicating that the propensity to form the complex is much lower than obtained from the calculations which do not include these effects.

Computational studies have been used to propose other systems with similar structures but with much larger fullerene binding energies. There are many potential candidates for fullerene hosts, including calix[n]arenes (Pérez & Martin 2008), cyclic paraphenyleneacetylenes (Kawase et al. 2003), cyclotriveratrylenes (Matsubara et al. 1998), *p*-extended tetrathiafulvalenes (Pérez et al. 2006), and porphyrins (Bonifazi et al. 2004), among others. Corannulene derivatives have, however, one major advantage; their concave surface matches the convex surface of $C_{60}$ fullerene. Recently, we have used a simpler system, indaceno[3,2,1,8,7,6-pqrstuv]picene, as a starting point for designing larger geodesic arenes with high affinity toward $C_{60}$ fullerene (Amsharov et al. 2012, Trzaskowski et al. 2014). Figure 5.4 shows a number of hypothetical geodesic arenes with different topologies and the equilibrium structures obtained from the calculations. One can see that through a controlled synthesis, the number of the cove regions in an arene can be increased and the desired curvature of the cove surface can be achieved. For example, systems 1 and 2 have large curvatures resembling those of buckyballs and may be potentially useful as precursors in the synthesis of fullerenes or nanotubes. All these systems may also be precursors of materials for trapping molecules with specific shapes, e.g., molecular traps for $C_{60}$ and larger fullerenes. Due to the large surface on which the $\pi$-electrons delocalize, these systems should strongly interact with fullerenes, nanotubes, and other similar structures via attractive $\pi$–$\pi$ stacking interactions. We have estimated the energies of these interactions for some of the investigated structures at the DFTB level of theory at 0 K. As expected, for some of the systems, the interaction energies are particularly high. These include system 2 (the interaction energy is –56.0 kcal/mol), system 5 (–89.9 kcal/mol), system 9 (–156.9 kcal/mol), and system 10 (–157.8 kcal/mol). These values are much larger than, e.g., the corresponding interaction energy of the well-known double-concave hydrocarbon buckycatcher mentioned before.

Geodesic arenes may also be used to build new molecular scaffolds by attaching them through covalent bonds to fullerenes or nanotubes. The advantage of using geodesic arenes over PAHs with flat surfaces in such structures lies in their intrinsic curvature, which enhances their chemical reactivity. This phenomenon can be seen, e.g., during the formation of the $C_{60}$–$C_{60}$ dimer, which, due to the high surface curvatures of the reacting fullerenes, occurs with a relatively low energy barrier of approximately 28–36 kcal/mol (Wang et al. 1994, Davydov et al. 2001, Han et al. 2004). The phenomenon is also evident in the study of the impact of the curvature of selected nanotubes on their reactivities (Zheng et al. 2006). Using computational techniques and some model systems, we have shown that the local curvature of

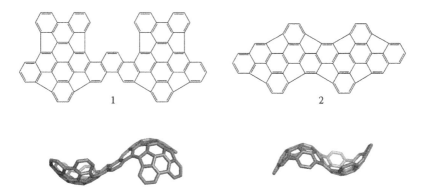

**FIGURE 5.4**   Hypothetical geodesic arenes with large curvature.

graphene-like systems is the main factor affecting the energy barrier in the synthesis of supramolecular systems based on PAHs (Trzaskowski et al. 2013). For geodesic arenes (and other systems with intrinsic curvature, e.g., nanocones), the carbon atoms located at the apex and at the edges of the structure are the most reactive and, in many cases, can be used to selectively attach other carbon-based structures to the systems (Trzaskowski et al. 2007, 2014). However, in the case of a covalent attachment of corannulene to $C_{60}$, the computational studies suggest that the resulting structure is less stable than the corannulene–$C_{60}$ complex interacting via dispersion interactions (Denis 2014). On the other hand, for smaller fullerenes (e.g., $C_{44}$), the system connected with a covalent bond to corannulene is favored over the complex with $\pi$–$\pi$ interactions between the subunits.

The reduction potential of corannulene (mentioned briefly in Section 5.2) gives rise to its electrochemiluminescence (Valenti et al. 2010). A mixture of this system with benzoyl peroxide and arylamine derivatives produces an intense blue light. Interestingly, the emission can be turned on and off by changing the coreactant, which opens the possibility of using this intriguing carbon structure for building efficient blue-light emitting devices. Similarly, many multiethynylcorannulenes display high-quantum-efficiency luminescence and variable emission wavelengths in solution (Wu et al. 2008). Also, the unusual electronic structure of corannulene can stabilize neutral radicals, a phenomenon observed for a number of corannulene derivatives (Morita et al. 2004, 2008). Such neutral radicals are stable as solids in air, as well as in degassed toluene solution for a long period.

Geodesic arenes can also play an important role as scaffolds for catalysts or metal ligands (Bandera et al. 2011). In 1997, the first transition metal corannulene complex, $(\eta^6$-corannulene)Ru(C$_5$Me$_5$)$^+$, has been synthesized (Seiders et al. 1997). Soon after, more complexes of geodesic arene and other PAHs with transition metals (including Pt, Ag, Ir, and Rh) were reported (Shaltout et al. 1998, Álvarez et al. 2012). Interestingly, it has been shown that some metal ions may form complexes with bowl-shaped polyarenes while located on either the exterior or the interior of the convex surface (Filatov et al. 2010). If located in the interior, the metal ions may increase the curvature of the entire hydrocarbon. In some cases, the metal coordination was also shown to be stereoselective, providing a possible platform for studying chiral ligand–metal molecular recognition centers (Bandera et al. 2011).

One final application that warrants mentioning involves the deposition of the geodesic arenes on metal surfaces. This opens the possibility to study these systems using various experimental approaches designed to study surface phenomena. Surface-deposited arenes may lead to designing new materials for photovoltaics and organic light-emitting devices. It was shown that the fivefold symmetric corannulene spontaneously forms enantiomorphous 2D-lattice structures on the rectangular Cu(110) surface (Parschau et al. 2007). A different copper surface, the Cu(111) surface, was also used to study phase transitions of corannulene (Merz et al. 2009). Surface-deposited organic molecules allow one to investigate the phase transitions at the molecular level in detail by scanning tunneling microscopy. In the case of corannulene, this technique enabled the identification of periodic molecular movements in the supercooled crystal phase and showed the coexistence of two different phases at the temperatures in the 236–248 K range.

## 5.5 Conclusions and Outlook

Despite a slow start caused by having to design and implement laborious synthetic processes to produce geodesic arenes, these systems immediately attracted a lot of attention due to their potential applications in various fields. Today, due to advances in FVP and solution-based synthetic methods, corannulene and its derivatives are commonly used in the design and construction of large carbon-based networks with various structures and properties. The structural feature, which gives these systems interesting physical and chemical properties, is their off-plane structural curvature. This curvature gives rise to a higher chemical reactivity in comparison to corresponding systems with flat structures. In addition to the propensity to form covalent bonds with other systems, the corannulenes can also form complexes connected via $\pi$–$\pi$ interactions facilitated by extended delocalized conjugated $\pi$-bond networks. These complexes are usually characterized by strong intermolecular interactions, which can be utilized in processes involving molecular recognition and selective molecular docking. The unusual properties of corranulenes have been studied using various experimental techniques, as well as with the tools of computational modeling and simulation. Theoretical calculations are particularly useful in designing new structures with desired properties prior to their synthesis. The results have demonstrated that the calculations performed with the state-of-the-art methods are reliable enough to provide guidance to the experiment.

# References

Abdourazak, A. H., Sygula, A. & Rabideau, P. W., "'Locking' the bowl-shaped geometry of corannulene: Cyclopentacorannulene," *J. Am. Chem. Soc.* 115 (1993): 3010–3011.

Abdourazak, A. H., Marcinow, Z., Sygula, A. et al., "Buckybowls 2: Toward the total synthesis of buckminsterfullerene (C60); Benz[5,6]-as-indaceno-[3,2,1,8,7-mnopqr]indeno[4,3,2,1-cdef]chrysene," *J. Am. Chem. Soc.* 117 (1995): 6410–6411.

Acocella, A., Havenith, R. W. A., Steiner, E. et al., "A simple model for counter-rotating ring currents in [n] circulenes," *Chem. Phys. Lett.* 363 (2002): 64–72.

Alberico, D., Scott, M. E. & Lautens, M., "Aryl–aryl bond formation by transition-metal-catalyzed direct arylation," *Chem. Rev.* 107 (2007): 174–238.

Álvarez, C. M., Barbero, H., García-Escudero, L. A. et al., "$\eta^6$-Hexahelicene complexes of iridium and ruthenium: Running along the helix," *Inorg. Chem.* 51 (2012): 8103–8181.

Amsharov, K. Yu., Kabdulov, M. A. & Jansen, M., "Facile bucky-bowl synthesis by regiospecific cove-region closure by HF elimination," *Angew. Chem. Int. Ed.* 51 (2012): 4594–4597.

Ansems, R. B. M. & Scott, L. T., "Circumtrindene: A geodesic dome of molecular dimensions; Rational synthesis of 60% of $C_{60}$," *J. Am. Chem. Soc.* 122 (2000): 2719–2724.

Aprahamian, I., Preda, D. V., Bancu, M. et al., "Reduction of bowl-shaped hydrocarbons: Dianions and tetraanions of annelated corannulenes," *J. Org. Chem.* 71 (2006): 290–298.

Ayalon, A., Rabinovitz, M., Cheng, P.-C. et al., "Corannulene tetraanion: A novel species with concentric anionic rings," *Angew. Chem. Int. Ed.* 31 (1992): 1636–1637.

Baldridge, K. K. & Siegel, J. S., "Corannulene-based fullerene fragments $C_{20}H_{10}$–$C_{50}H_{10}$: When does a buckybowl become a buckytube?" *Theor. Chem. Acc.* 97 (1997): 67–71.

Bancu, M., Rai, A. K., Cheng, P.-C. et al., "Corannulene polysulfides: Molecular bowls with multiple arms and flaps," *Synlett* (2004): 173–176.

Bandera, D., Baldridge, K. K., Linden, A. et al., "Stereoselective coordination of $C_5$-symmetric corannulene derivatives with an enantiomerically pure [Rh$^I$(nbd*)] metal complex," *Angew. Chem. Int. Ed.* 50 (2011): 865–867.

Barth, W. E. & Lawton, R. G., "Dibenzo[ghi,mno]fluoranthene," *J. Am. Chem. Soc.* 88 (1966): 380–381.

Barth, W. E. & Lawton, R. G., "The synthesis of corannulene," *J. Am. Chem. Soc.* 93 (1971): 1730–1745.

Becke, A. D., "Density-functional thermochemistry: III. The role of exact exchange," *J. Chem. Phys.* 98 (1993): 5648–5652.

Bonifazi, D., Spillmann, H., Kiebele, A. et al., "Supramolecular patterned surfaces driven by cooperative assembly of $C_{60}$ and porphyrins on metal substrates," *Angew. Chem. Int. Ed.* 43 (2004): 4759–4763.

Bratcher, M. S., PhD dissertation, Boston College, Chestnut Hill (1996).

Bronstein, H. E., Choi, N. & Scott, L. T., "Practical synthesis of an open geodesic polyarene with a fullerene-type 6:6-double bond at the center: Diindeno[1,2,3,4-defg;1′,2′,3′,4′-mnop]chrysene," *J. Am. Chem. Soc.* 124 (2002): 8870–8875.

Brown, R. F. C., "Are flush pyrolytic reactions useful?" *Pure Appl. Chem.* 62 (1990): 1981–1986.

Brown, R. F. C., Eastwood, F. W., Harrington, K. J. et al., "Methyleneketenes and methylenecarbenes III: Pyrolytic synthesis of arylacetylenes and their thermal rearrangements involving arylmethylenecarbenes," *Aust. J. Chem.* 27 (1974): 2393–2402.

Buhl, M., "The relation between endohedral chemical shifts and local aromaticities in fullerenes," *Chem. Eur. J.* 4 (1998): 734–739.

Chang, H.-I., Huang, H.-T., Huang, C.-H. et al., "Diindeno[1,2,3,4-defg;1′,2′,3′,4′-mnop]chrysenes: Solution-phase synthesis and the bowl-to-bowl inversion barrier," *Chem. Commun.* 46 (2010): 7241–7243.

Cheng, P.-C., MS thesis, University of Nevada, Reno (1992).

Clar, E., *The Aromatic Sextet* (Wiley, London, 1972).

Clayton, M. D., Marcinow, Z. & Rabideau, P. W., "Improved synthesis of a $C_{30}H_{12}$ buckybowl via benzylic oxidation with benzeneseleninic anhydride," *J. Org. Chem.* 61 (1996): 6052–6054.

Clayton, M. D. & Rabideau, P. W., "Synthesis of a new $C_{32}H_{12}$ bowl-shaped aromatic hydrocarbon," *Tetrahedron Lett.* 38 (1997): 741–744.

Crowley, C., Kroto, H. W., Taylor, R. et al., "Formation of [60]fullerene by pyrolysis of corannulene,7,10-bis(2,2′-dibromovinyl)fluoranthene, and 11,12-benzofluoranthene," *Tetrahedron Lett.* 36 (1995): 9215–9218.

Davydov, V. A., Kashevarova, L. S., Rakhmanina, A. V. et al., "Pressure-induced dimerization of fullerene $C_{60}$: A kinetic study," *Chem. Phys. Lett.* 333 (2001): 224–229.

Denis, P. A., "A theoretical study on the interaction between well curved conjugated systems and fullerenes smaller than $C_{60}$ or larger than $C_{70}$," *J. Phys. Org. Chem.* 27 (2014): 918–925.

Dickens, T. K. & Mallion, R. B., "Topological ring-current assessment of the 'annulene-within-an-annulene' model in [N]-circulenes and some structures related to kekulene," *Chem. Phys. Lett.* 517 (2011): 98–102.

Dobrowolski, M. A., Ciesielski, A. & Cyrański, M. K., "On the aromatic stabilization of corannulene and coronene," *Phys. Chem. Chem. Phys.* 13 (2011): 20557–20563.

Echavarren, A. M., Gómez-Lor, B., González, J. J. et al., "Palladium-catalyzed intramolecular arylation reaction: Mechanism and application for the synthesis of polyarenes," *Synlett* 5 (2003): 585–597.

Filatov, A. S., Rogachev, A. Y., Jackson, E. A. et al., "Increasing the curvature of a bowl-shaped polyarene by fullerene-like η²-complexation of a transition metal at the interior of the convex surface," *Organometallics* 29 (2010): 1231–1237.

Grimme, S., Antony, J., Ehrlich, S. et al., "A consistent and accurate ab initio parametrization of density functional dispersion correction DFT-D for the 94 elements H–Pu," *J. Chem. Phys.* 132 (2010): 154104.

Guldi, D. M. & Martin, N., *Carbon Nanotubes and Related Structures: Synthesis, Characterization, Functionalization, and Applications* (Wiley-VCH, Weinheim, 2010).

Hagen, S., Bratcher, M. S., Erickson, M. S. et al., "Novel syntheses of three $C_{30}H_{12}$ bowl-shaped polycyclic aromatic hydrocarbons," *Angew. Chem. Int. Ed.* 36 (1997): 406–408.

Hagen, S., Christoph, H. & Zimmerman, G., "High temperature synthesis to bowl-shaped subunits of fullerenes—IV: From 4-[9H-fluorenylidene-(9)]-4H-cyclopenta[def]phenanthrene to fluoreno[1,9,8-abcd]corannulene and difluoreno[1,9,8,7-cdefg;2′,1′,9′,8′-klmno]anthracene," *Tetrahedron* 51 (1995): 6961–6970.

Han, S., Yoon, M., Berber, S. et al., "Microscopic mechanism of fullerene fusion," *Phys. Rev. B* 70 (2004): 113402.

Hanson, J. C. & Nordman, C. E., "The crystal and molecular structure of corannulene, $C_{20}H_{10}$," *Acta Crystallogr. Sect. B* 32 (1976): 1147.

Hückel, E., "Quantentheoretische beiträge zum benzolproblem: I. Die elektronenkonfiguration des benzols und verwandter verbindungen," *Z. Chem.* 70 (1931): 204–286.

Jackson, E. A., Steinberg, B. D., Bancu, M. et al., "Pentaindenocorannulene and tetraindenocorannulene: New aromatic hydrocarbon pi systems with curvatures surpassing that of C60," *J. Am. Chem. Soc.* 129 (2007): 484–485.

Kawase, T., Tanaka, K., Fujiwara, N. et al., "Complexation of a carbon nanoring with fullerenes," *Angew. Chem. Int. Ed.* 42 (2003): 1624–1628.

Knolker, H. J., Braier, A., Brocher, D. J. et al., "Transition metal complexes in organic synthesis: Part 55—Synthesis of corannulene via an iron-mediated [2+2+1] cycloaddition," *Tetrahedron Lett.* 40 (1999): 8075–8078.

Kroto, H. W., Heath, J. R., O'Brien, S. C. et al., "C$_{60}$: Buckminsterfullerene," *Nature* 318 (1985): 162–163.

Lee, C., Yang, W. & Parr, R. G., "Development of the Colle–Salvetti correlation-energy formula into a functional of the electron density," *Phys. Rev. B* 37 (1988): 785–789.

Liu, C. Z. & Rabideau, P. W., "Corannulene synthesis via the pyrolysis of silyl vinyl ethers," *Tetrahedron Lett.* 37 (1996): 3437–3440.

London, F., "Théorie quantique des courants interatomiques dans les combinaisons aromatiques," *J. Phys. Radium* 8 (1937): 397–409.

Lovas, F. J., McMahon, R. J., Grabow, J.-U. et al., "Interstellar chemistry: A strategy for detecting polycyclic aromatic hydrocarbons in space," *J. Am. Chem. Soc.* 127 (2005): 4345–4349.

Marcinow, Z., Grove, D. I. & Rabideau, P. W., "Synthesis of a new C$_{32}$H$_{12}$ bowl-shaped aromatic hydrocarbon: Acenaphtho[3,2,1,8-ijklm]diindeno[4,3,2,1-cdef:1′,2′,3′,4′-pqra]triphenylene," *J. Org. Chem.* 67 (2002): 3537–3539.

Marcinow, Z., Sygula, A., Ellern, A. et al., "Lowering inversion barriers of buckybowls by benzannelation of the rim: Synthesis and crystal and molecular structure of 1,2-dihydrocyclopenta[b,c]dibenzo[g,m]corannulene," *Org. Lett.* 3 (2001): 3527–3529.

Matsubara, H, Hasegawa, A., Shiwaku, K. et al., "Supramolecular inclusion complexes of fullerenes using cyclotriveratrylene derivatives with aromatic pendants," *Chem. Lett.* 9 (1998): 923–924.

McComas, C. C., BS thesis, Boston College, Chestnut Hill (1996).

McNab, H., "Chemistry without reagents: Synthetic application of flash vacuum pyrolysis," *Aldrichim. Acta* 37 (2004): 19–26.

McWeeny, R., "Ring currents and proton magnetic resonance in aromatic molecules," *Molec. Phys.* 1 (1958): 311–321.

Mehta, G. & Panda, G., "A new synthesis of corannulene," *Tetrahedron Lett.* 38 (1997a): 2145–2148.

Mehta, G. & Panda, G., "Buckybowls: A simple, conceptually new synthesis of C$_{2v}$-semibuckminsterfullerene (C$_{30}$H$_{12}$, [5,5]-fulvalene circulene)," *Chem. Commun.* 21 (1997b): 2081–2082.

Mehta, G., Panda, G. & Srirama Sarma, P. V. V., "A short synthesis of 'bucky-bowl' C-3-hemifullerene (triindenotriphenylene)," *Tetrahedron Lett.* 39 (1998): 5835–5836.

Mehta, G. & Srirama Sarma, P. V. V., "Bucky-bowls: A general approach to benzocorannulenes; Synthesis of mono-, di- and tri-benzocorannulenes," *Chem. Commun.* 1 (2000): 19–20.

Merz, L., Parschau, M., Zoppi, L. et al., "Reversible phase transitions in a buckybowl monolayer," *Angew. Chem. Int. Ed.* 48 (2009): 1966–1969.

Monaco, G., Scott, L. T. & Zanasi, R., "Magnetic euripi in corannulene," *J. Phys. Chem. A* 112 (2008): 8136–8147.

Morita, Y., Nishida, S., Kobayashi, T. et al., "The first bowl-shaped stable neutral radical with a corannulene system: Synthesis and characterization of the electronic structure," *Org. Lett.* 6 (2004): 1397–1400.

Morita, Y., Ueda, A., Nishida, S. et al., "Curved aromaticity of a corannulene-based neutral radical: Crystal structure and 3D unbalanced delocalization of spin," *Angew. Chem. Int. Ed.* 47 (2008): 2035–2038.

Mück-Lichtenfeld, C., Grimme, S., Kobryn, L. et al., "Inclusion complexes of buckycatcher with C$_{60}$ and C$_{70}$," *Phys. Chem. Chem. Phys.* 12 (2010): 7091–7097.

Necula, A. & Scott, L. T., "High temperature behavior of alternant and nonalternant polycyclic aromatic hydrocarbons," *J. Anal. Appl. Pyrolysis* 54 (2000): 65–87.

Pappo, D., Mejuch, T., Reany, O. et al., "Diverse functionalization of corannulene: Easy access to pentagonal superstructure," *Org. Lett.* 11 (2009): 1063–1066.

Parschau, M., Fassel, R., Ernst, K. H. et al., "Buckybowls on metal surfaces: Symmetry mismatch and enantiomorphism of corannulene on Cu110," *Angew. Chem. Int. Ed.* 46 (2007): 8258–8261.

Pascual, S., de Mendoza, P. & Echavarren, A. M., "Palladium-catalyzed arylation for the synthesis of polyarenes," *Org. Biomolec. Chem.* 5 (2007): 2727–2734.

Pérez, E. M. & Martin, N., "Curves ahead: Molecular receptors for fullerenes based on concave-convex complementarity," *Chem. Soc. Rev.* 37 (2008): 1512–1519.

Pérez, E. M., Sánchez, L., Fernández, G. et al., "exTTF as a building block for fullerene receptors: Unexpected solvent-dependent positive homotropic cooperativity," *J. Am. Chem. Soc.* 128 (2006): 7172–7173.

Petrukhina, M. A., Andreini, K. W., Mack, J. et al., "X-ray quality geometries of geodesic polyarenes from theoretical calculations: What levels of theory are reliable?" *J. Org. Chem.* 70 (2005): 5713–5716.

Pople, J. A., "Proton magnetic resonance of hydrocarbons," *J. Chem. Phys.* 24 (1956): 1111.

Pople, J. A., "Molecular orbital theory of aromatic ring currents," *Molec. Phys.* 1 (1958): 175–180.

Porezag, D., Frauenheim, T., Kohler, T. et al., "Construction of tight-binding-like potentials on the basis of density-functional theory: Application to carbon," *Phys. Rev. B* 51 (1995): 12947–12957.

Preda, D. V., PhD dissertation, Boston College, Chestnut Hill (2001).

Rehner, J., "The isolation and identification of fluoranthene from carbon black," *J. Am. Chem. Soc.* 62 (1940): 2243–2244.

Reisch, H. A., Bratcher, M. S. & Scott, L. T., "Imposing curvature on a polyarene by intramolecular palladium-catalyzed arylation reactions: A simple synthesis of dibenzo[*a,g*]corannulene," *Org. Lett.* 2 (2000): 1427–1430.

Scott, L. T., "Fragments of fullerenes: Novel syntheses, structures and reactions," *Pure Appl. Chem.* 68 (1996): 291–300.

Scott, L. T., Boorum, M. M., McMahon, B. J. et al., "A rational chemical synthesis of $C_{60}$," *Science* 295 (2002): 1500–1503.

Scott, L. T., Cheng, P.-C. & Bratcher, M. S., Seventh International Symposium on Novel Aromatic Compounds, Victoria, British Columbia, Canada, July 19–24, 1992 (1992a): Abstract 64.

Scott, L. T., Cheng, P.-C., Hashemi, M. M. et al., "Corannulene: A three-step synthesis," *J. Am. Chem. Soc.* 119 (1997): 10963–10968.

Scott L. T., Hashemi, M. M. & Bratcher, M. S., "Corannulene bowl-to-bowl inversion is rapid at room temperature," *J. Am. Chem. Soc.* 114 (1992b): 1920–1921.

Scott, L. T., Hashemi, M. M., Meyer, D. T. et al., "Corannulene: A convenient new synthesis," *J. Am. Chem. Soc.* 113 (1991): 7082–7084.

Scott, L. T., Jackson, E. A., Zhang, Q. et al., "A short, rigid, structurally pure carbon nanotube by stepwise chemical synthesis," *J. Am. Chem. Soc.* 134 (2012): 107–110.

Scott, L. T. & Preda, D. V., "New, simple pyrolytic approach toward benzo[ghi]fluoranthene-type PAHs," in *Abstracts of Papers*, 219th National Meeting of the American Chemical Society, San Francisco, March 2000 (American Chemical Society, Washington, DC, 2000), Abstract ORGN-516.

Seiders, T. J., Baldridge, K. K., O'Connor, J. M. et al., "Hexahapto metal coordination to curved polyaromatic hydrocarbon surfaces: The first transition metal corannulene complex," *J. Am. Chem. Soc.* 119 (1997): 4781–4782.

Seiders, T. J., Baldridge, K. K., Elliott, E. L. et al., "Synthesis and quantum mechanical structure of sym-pentamethylcorannulene and decamethylcorannulene," *J. Am. Chem. Soc.* 121 (1999a): 7439–7440.

Seiders, T. J., Baldridge, K. K., Grube, G. H. et al., "Structure/energy correlation of bowl depth and inversion barrier in corannulene derivatives: Combined experimental and quantum mechanical analysis," *J. Am. Chem. Soc.* 123 (2001): 517–525.

Seiders, T. J., Baldridge, K. K. & Siegel, J. S., "Synthesis and characterization of the first corannulene cyclophane," *J. Am. Chem. Soc.* 118 (1996): 2754–2575.

Seiders, T. J., Elliott, E. L., Grube, G. H. et al., "Synthesis of corannulene and alkyl derivatives of corannulene," *J. Am. Chem. Soc.* 121 (1999b): 7804–7813.

Shaltout, R. M., Sygula, R., Sygula, A. et al., "The first crystallographically characterized transition metal buckybowl compound: $C_{30}H_{12}$ carbon–carbon bond activation by Pt(PPh$_3$)$_2$," *J. Am. Chem. Soc.* 120 (1998): 835–836.

Sola, M., "Forty years of Clar's aromatic π-sextet rule," *Frontiers Chem.* 1 (2013): 1–8.

Steinberg, B. D., Jackson, E. A., Filatov, A. S. et al., "Aromatic π-systems more curved than $C_{60}$: The complete family of all indenocorannulenes synthesized by iterative microwave-assisted intramolecular arylations," *J. Am. Chem. Soc.* 131 (2009): 10537–10545.

Steiner E., Fowler, P. W. & Jenneskens, L. W., "Counter-rotating ring currents in coronene and corannulene," *Angew. Chem. Int. Ed.* 40 (2001): 362–366.

Sygula, A., "Chemistry on a half-shell: Synthesis and derivatization of buckybowls," *Eur. J. Org. Chem.* 2011 (2011): 1611–1625.

Sygula, A., Abdourazak, A. H. & Rabideau, P. W., "Cyclopentacorannulene: π-Facial stereoselective deuterogenation and determination of the bowl-to-bowl inversion barrier for a constrained buckybowl," *J. Am. Chem. Soc.* 118 (1996): 339–343.

Sygula, A., Fronczek, F. R., Sygula, R. et al., "A double concave hydrocarbon buckycatcher," *J. Am. Chem. Soc.* 129 (2007): 3842–3843.

Sygula, A., Karlen, S. D., Sygula, R. et al., "Formation of the corannulene core by nickel-mediated intramolecular coupling of benzyl and benzylidene bromides: A versatile synthesis of dimethyl 1,2-corannulene dicarboxylate," *Org. Lett.* 4 (2002): 3135–3137.

Sygula, A., Marcinow, Z., Fronczek, F. R. et al., "The first crystal structure characterization of a semibuckminsterfullerene, and a novel synthetic route," *Chem. Commun.* 24 (2000): 2439–2440.

Sygula, A. & Rabideau, P. W., "Structure and inversion barriers of corannulene, its dianion and tetraanion: An ab initio study," *J. Molec. Struct. THEOCHEM* 333 (1995): 215–226.

Sygula, A. & Rabideau, P. W., "Non-pyrolytic syntheses of buckybowls: Corannulene, cyclopentacorannulene, and a semibuckminsterfullerene," *J. Am. Chem. Soc.* 121 (1999): 7800–7803.

Sygula, A. & Rabideau, P. W., "A practical, large scale synthesis of the corannulene system," *J. Am. Chem. Soc.* 122 (2000): 6323–6324.

Trzaskowski, B., Adamowicz, L., Beck, W. et al., "Impact of local curvature and structural defects on graphene-$C_{60}$ fullerene fusion reaction barriers," *J. Phys. Chem. C* 117 (2013): 19664–19671.

Trzaskowski, B., Adamowicz, L., Beck, W. et al., "Exploring structures and properties of new geodesic polyarenes," *Chem. Phys. Lett.* 595–596 (2014): 6–12.

Trzaskowski, B., Jalbout, A. F. & Adamowicz, L., "Functionalization of carbon nanocones by free radicals: A theoretical study," *Chem. Phys. Lett.* 444 (2007): 314–318.

Tsefrikas, V. M., Arns, S., Merner, P. M. et al., "Benzo[a]acecorannulene: Surprising formation of a new bowl-shaped aromatic hydrocarbon from an attempted synthesis from 1,2diazadibenzo[d,m]corannulene," *Org. Lett.* 8 (2006): 5195–5198.

Tsefrikas, V. M. & Lawrence, T. S., "Geodesic polyarenes by flash vacuum pyrolysis," *Chem. Rev.* 106 (2006): 4868–4684.

Valenti, G., Bruno, C., Rapino, S. et al., "Intense and tunable electrochemiluminescence of corannulene," *J. Phys. Chem. C* 114 (2010): 19467–19472.

Wang, Y., Holden, J. M., Bi, X. et al., "Thermal decomposition of polymeric $C_{60}$," *Chem. Phys. Lett.* 217 (1994): 413–417.

Wegner, H. A., Scott, L. T. & de Meijere, A., "A new Suzuki-Heck-type coupling cascade: Indeno[1,2,3]-annelation of polycyclic aromatic hydrocarbons," *J. Org. Chem.* 68 (2003): 883–887.

Weitz, A., Shabtai, E., Rabinovitz, M. et al., "Dianions and tetraanions of bowl-shaped fullerene fragments dibenzo[a,g]corannulene and dibenzo[a,g]cyclopenta[kl] corannulene," *Chem. Eur. J.* 4 (1998): 234–239.

Wu, Y. T., Bandera, D., Maag, R. et al., "Multiethynyl corannulenes: Synthesis, structure, and properties," *J. Am. Chem. Soc.* 130 (2008): 10729–10739.

Xu, G., Sygula, A., Marcinow, Z. et al., "Chemistry on the rim of buckybowls: Derivatization of 1,2,5,6-tetrabromocorannulene," *Tetrahedron Lett.* 41 (2000): 9931–9934.

Zhao, Y. & Truhlar, D. G., "Computational characterization and modeling of buckyball tweezers: Density functional study of concave–convex π...π interactions," *Phys. Chem. Chem. Phys.* 10 (2008a): 2813–2818.

Zhao, Y. & Truhlar, D., "The M06 suite of density functionals for main group thermochemistry, thermo-chemical kinetics, noncovalent interactions, excited states, and transition elements: Two new functionals and systematic testing of four M06 functionals and twelve other functionals," *Theor. Chem. Acc.* 120 (2008b): 215–241.

Zheng, G., Wang, Z., Irle, S. et al., "Origin of the linear relationship between $CH_2/NH/O$-SWNT reaction energies and sidewall curvature: Armchair nanotubes," *J. Am. Chem. Soc.* 128 (2006): 15117–15126.

Zimmerman, G., Nuechter, U., Hagen, S. et al., "Synthesis and hydropyrolysis of bis-trimethylsilyl sub-stituted 3-(4h-cyclopenta[def]phenanthrylidene)-1,4-pentadiyne—A new route to corannulene," *Tetrahedron Lett.* 35 (1994): 4747–4750.

# 6

# Cubic Carbon Polymorphs

Piotr Kowalczyk

Drew Parsons

Artur P. Terzyk

Piotr A. Gauden

Sylwester Furmaniak

**Abstract**

This chapter reviews recent theoretical and experimental results on the structure–property relationships of cubic carbon polymorphs—a new class of dense crystalline carbon materials. At the start of the chapter, we briefly describe the evolution of carbon science toward the development of novel exotic carbon nanostructures. Next, we provide details of C–C bonding and intermolecular potential used for modeling of both crystalline and disordered phases of carbon. Then, we introduce four cubic carbon polymorphs investigated recently (fcc-$C_{32}$, bcc-$C_{20}$, fcc-$C_{12}$, and fcc-$C_{10}$). Their basic mechanical and electronic properties are presented and compared with that of a reference material, namely, cubic diamond. Finally, we discuss the impact of the internal pore structure of these cubic carbon polymorphs on the adsorption and transport of selected adsorbates (e.g., He, $H_2$, Ne, Ar, and CO). Some potential applications of the most open fcc-$C_{10}$ carbon polymorph in adsorptive recovery of He from gas mixtures are discussed.

**Key words:** cubic carbon polymorphs, adsorption, molecular simulation, GCMC.

## 6.1 Introduction

Carbon is one of the abundant elements of the earth. Carbon materials, which consist mainly of carbon atoms, have been used since the prehistoric era as charcoal. Inagaki and Feiyu (2006) classified carbon materials into three classes, namely, *classical carbons*, *new carbons*, and *nanocarbons*.

*Classical carbons*, e.g., carbon blacks, artificial graphite blocks, and activated carbons, were established before 1960 (Inagaki & Feiyu 2006). These carbon materials are still the principal products and principal sources of income for carbon industries throughout the world. Carbon blacks are spherical particles with diameters from 10 nm to a few hundred nanometers (Kühner & Voll 1993). They have been primarily used as pigments and reinforcing phases in automobile tires. Artificial graphite blocks include graphite electrodes, which can provide high levels of electrical conductivity and are capable of withstanding extremely high levels of generated heat. These classical carbons have been commonly used in the refinement of steel and similar smelting processes (Inagaki & Feiyu 2006). Activated carbons are porous materials (having narrow pores of irregular shapes) that are commonly produced through the carbonization and subsequent activation of organic precursors (e.g., peach-stones, coals, coconut shells, wood, and lignite) (Bansal & Goyal 2005). Purification of air from potentially harmful molecules (such as toxic industrial chemical and military agents), drinking water treatment, biogas purification, electrochemical storage of energy, drug delivery systems, and hemoperfusion, hydrogen storage are common processes that utilize activated carbons (Bansal & Goyal 2005).

*New carbons*, such as poly(acrylonitrile) (PAN) carbon fibers, pyrolytic carbons, glass-like carbons, have been developed since 1960 (Inagaki & Feiyu 2006). Carbon fibers (PAN or pitch-based) are fascinating materials with extraordinary strength and flexibility (Bahl et al. 1998). Thus, it seems to be not surprising that carbon fibers have been widely used in electronics, automobile, aircrafts, nuclear energy, and other technologies. Highly porous materials with a homogeneous structure of narrow micropores are produced by activation of carbon fibers. Fast kinetics of adsorption–desorption processes and high adsorption capacities of activated carbon fibers have been already utilized in drinking water purification systems and air cleaning technologies (Inagaki & Feiyu 2006). Pyrolytic carbons are produced by a new method known as CVD (Bokros 1969). In contrast to conventional experimental techniques used for production of classical carbons, CVD allows production of a well-developed and well-oriented basal plane of graphite (Bokros 1969). Highly oriented pyrolytic graphite produced by CVD has been used as a monochromator for X-ray and neutrons. Glass-like carbons are very hard and brittle materials (Miyake 2003). These carbonaceous solids have high resistance against various chemicals and corrosion, which facilitates their electrochemical applications (Miyake 2003).

*Nanocarbons*, such as fullerenes, nanotubes, graphene, cubic carbon polymorphs, and others, emerged in the 1980s. In 1985, Kroto et al. (1985) reported a cage (cluster) composed of 60 carbon atoms, known as buckminsterfullerene $C_{60}$. The structure of fullerenes consists of polyhedrons formed by the assembling of a series of pentagonal and hexagonal rings of carbon atoms. Thus, the hybridization of fullerenes is intermediate between $sp^3$ and $sp^2$ (for example, $C_{60}$ hybridization is $sp^{2.28}$) (Kroto et al. 1985). In 1991, Iijima (1991) observed carbon nanotubes, and soon after various tubular carbon molecules were synthesized. These nanocarbons exhibit extraordinary strength and unique electrical, optical, and chemical properties. They can potentially be used for the fabrication of the next generation of energy storage supercapacitors, field emission transistors, high-performance catalysts, photovoltaics, and biomedical devices and implants (Inagaki & Feiyu 2006). Graphene is a crystalline allotrope of carbon with 2D properties. The pioneering research of Novoselov et al. (2004) on the synthesis of graphene was awarded the 2010 Nobel Prize in Physics. A pure one-atom-thick graphene sheet with fascinating strength, flexibility, and electrical conductivity is very promising for novel high-tech applications, including fabrication of a new generation of computers, smartphones, and high-capacity batteries (Novoselov et al. 2005). Graphene doped with nitrogen is a very important 2D material with potential applications in catalysis and optoelectronics (Wang et al. 2012).

Various research groups have recently identified novel dense and superhard nanocarbons, known as *cubic polymorphs of carbon*. For example, body-centered cubic (bcc) superdense carbon, which has a lattice parameter of 0.448 nm, was found in plasma-deposited amorphous carbon films (Strel'niskii et al. 1978). The existence of another carbon allotrope, face-centered cubic carbon (fcc), with a lattice parameter of about 0.356 nm, has been confirmed through electron diffraction (Hirai & Kondo 1991). Furthermore, this fcc carbon was also discovered in natural Mexican crude oil (Santiago et al. 2004) and in the Younger Dryas boundary sediment layer (Kennett et al. 2009). A third controversial cubic carbon, with a lattice parameter of 0.5545 nm, was obtained by experimentally compressing graphite up to 50 GPa (Aust & Drickamer 1963). Pokropivny and Pokropivny (2004) posited a structure for the carbon as a 3D polymer of $C_{24}$ cages (termed as 3D-$C_{24}$). Apart from the above three controversial allotropes, other cubic carbons with uncertain crystallography have been also reported in experiments (Hu et al. 2012).

The discovery of new cubic, dense, transparent, and superhard polymorphs of carbon, in addition to known natural high-pressure polymorphs of carbon (i.e., cubic diamond [space group Fd3m] (Belyanin & Samoilovich 2004) and lonsdaleite [space group P63/mmc]) (Denisov et al. 2011), has stimulated a range of in-depth theoretical investigations. In Section 6.2, we briefly describe the concept of carbon bonding. Here, we also present the basic mathematical structure of the intermolecular potential used for modeling crystalline and disordered phases of carbon (Marks 2000, Kowalczyk et al. 2013). In Section 6.3, we present the optimized crystal structures of four cubic carbon polymorphs investigated by Hu et al. (2012). Mechanical and electronic properties of these materials computed by ab initio methods are presented and compared with those corresponding to a reference material, namely, cubic diamond. In Section 6.4, we briefly present the statistical mechanical methods (i.e., Monte Carlo simulations and transition state theory) used for the investigations of the adsorptive properties of cubic carbon polymorphs, accompanied with extensive simulation results. Finally, in Section 6.5, we present our conclusions with some potential directions for future research on nanocarbons.

## 6.2 Carbon Bonding and Intermolecular Potential

Carbon (an electron configuration of $1s^22s^22p^2$) is a complex element that can bond to other carbon atoms in many different ways via a variety of hybridized states (Inagaki & Feiyu 2006).

The three different orbitals $sp^3$, $sp^2$, and $sp^1$ (constructed from the $s$ and $p$ atomic orbitals, as shown in Figure 6.1) give a variety of combinations of chemical bonds. Therefore, carbon atoms give a large family of organic molecules and phases. Among them, a series of aromatic hydrocarbons with $sp^2$ hybrid orbitals, various aliphatic hydrocarbons with $sp^3$ hybrid orbitals, and $sp^1$-hybridized carbyne chains are the basic building blocks of giant organic molecules, disordered and crystalline phases of carbon (e.g., buckminsterfullerene $C_{60}$, diamond-like disordered carbon, carbon nanotubes, graphene, and diamond).

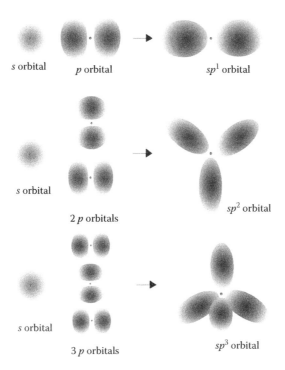

**FIGURE 6.1** An illustration of the combination of $s$ and $p$ orbitals to produce $sp^1$, $sp^2$, and $sp^3$ hybridized orbitals. (Adapted from Opletal 2005.)

The theoretical description of C–C bonding is a challenging problem even in $sp^3$, $sp^2$, and $sp^1$ "pure" phases because of the mathematical complexity of intermolecular potentials and associated computational overhead (Marks 2010, Kowalczyk et al. 2012, Terzyk et al. 2012). The three-body intermolecular environment-dependent interaction potential (EDIP) for carbon has been extensively used for modeling "pure" and "mixed" $sp^3$, $sp^2$, and $sp^1$ orbitals in various carbonaceous materials (Marks 2000, Kowalczyk et al. 2014). A combination of ab initio methods and EDIP can be extended to the investigations of novel carbon polymorphs and exotic carbon nanostructures. In the authors' opinion, such computational modeling is an important step toward smart design and synthesis of novel functional carbonaceous materials.

The EDIP for carbon has the following general form (Marks 2000):

$$U_i = \sum_j U_2(r_{ij}, Z_i) + \sum_{j<k} U_3(r_{ij}, r_{ik}, \theta_{jik}, Z_i), \tag{6.1}$$

where the total energy of an atom $i$ is the sum of the two-body and three-body terms, which in turn depend upon a generalized coordination $Z_i$ of the atom $i$ and where $r_{ij}$ represents the separation between atoms $i$ and $j$, $r_{ik}$ represents the separation between atoms $i$ and $k$, and $\theta_{jik}$ is the angle subtended between the two vectors $i - j$ and $j - k$.

The two-body interaction of the EDIP for carbon is given by the following expression (Marks 2000):

$$U_2(r_{ij}, Z_i) = \varepsilon \left[ \left( \frac{B}{r_{ij}} \right)^4 - \exp\left(-\beta Z_i^2\right) \right] \exp\left( \frac{\sigma}{r_{ij} - a - a'Z_i} \right), \tag{6.2}$$

where the dependence on $Z_i$ ensures that the two-body term has an energy minimum and consequently an equilibrium bond length which varies with the value of $Z_i$. The remaining parameters of EDIP are given elsewhere (Marks 2000).

The three-body term is given by the following expressions (Marks 2000):

$$U_3(r_{ij}, r_{ik}, \theta_{jik}, Z_i) = \lambda(Z_i)g(r_{ij}, Z_i)g(r_{ik}, Z_i)h(\theta_{jik}, Z_i), \tag{6.3}$$

$$\lambda(Z_i) = \lambda_0 \exp[-\lambda'(Z_i - Z_0)^2], \tag{6.4}$$

$$g(r, Z_i) = \exp\left[ \frac{\gamma}{r - a - a'Z_i} \right], \tag{6.5}$$

$$h(\theta_{jik}, Z_i) = 1 - \exp\{-q[\cos \theta_{jik} + \tau(Z_i)]^2\}, \tag{6.6}$$

$$\tau(Z_i) = 1 - \frac{Z_i}{12}\tanh\left[ t_1(Z_i - t_2) \right]. \tag{6.7}$$

The $\tau(Z_i)$ function describes the variation in the ideal bonding angles for various hybridization geometries, namely, 180°, 120°, and 109.5° for coordinations of 2, 3, and 4, respectively. Both two- and three-body terms are dependent upon the generalized coordination $Z_i$ as follows (Marks 2000):

$$Z_i = z_i + \pi_3(z_i)X_i^{\text{dih}} + \pi_3(z_i)X_i^{\text{rep3}} + \pi_2(z_i)X_i^{\text{rep2}}, \tag{6.8}$$

$$\pi_2(z_i) = \begin{cases} \left[ \left( z_i - 2 \right)^2 - 1 \right]^2 & 1 \le z \le 3 \\ 0 & z < 1 \text{ or } z < 3 \end{cases}, \tag{6.9}$$

$$\pi_3(z_i) = \begin{cases} \left[ \left( z_i - 3 \right)^2 - 1 \right]^2 & 2 \le z \le 4 \\ 0 & z < 2 \text{ or } z < 4 \end{cases}, \tag{6.10}$$

$$z_i = \sum_j f(r_{ij}) \tag{6.11a}$$

where

$$f(r_{ij}) = \begin{cases} 1 & r_{ij} < r_{low} \\ \exp\left[ \dfrac{\alpha}{1 - \left( \dfrac{r_{ij} - r_{low}}{r_{high} - r_{low}} \right)^{-3}} \right] & r_{low} \le r_{ij} \le r_{high} \\ 0 & r_{ij} < r_{high} \end{cases} \tag{6.11b}$$

The terms $X_i^{\text{dih}}$, $X_i^{\text{rep3}}$, and $X_i^{\text{rep2}}$ represent various geometrical constraints simulating dihedral rotation and $\pi$ bonding behavior. The spherical coordination contribution is denoted by $z_i$ and the sum in Equation 6.11 is over the neighborhood of atom $i$. It is easy to show that beyond 2 Å, an atom does not contribute to the definition of the coordination of atom $i$, while a separation below 1½ Å represents a bonded neighbor ($z_i = 1$). The $\pi_2$ and $\pi_3$ terms are switching functions which provide a weighting to the geometrical terms based on the coordination of atom $i$.

As pointed out by Marks (2000), EDIP provides a vastly better description of disordered carbon phases compared to the traditional empirical carbon potentials of Tresoff, Brenner, and Stillinger-Weber. Note that cubic carbon polymorphs are dense and superhard carbon materials, similar to diamond and diamond-like disordered carbons (Hu et al. 2012). Therefore, in our computer simulations, the basic units of the cubic carbon polymorphs studied were replicated in space and further optimized in the general utility lattice program (GULP) using EDIP (Gale 1997, Kowalczyk et al. 2013).

## 6.3 Cubic Carbon Polymorphs: Mechanical and Electronic Properties

In 2012, four novel cubic carbon polymorphs (fcc-$C_{10}$, fcc-$C_{12}$, bcc-$C_{20}$, and fcc-$C_{32}$) were discovered and extensively characterized using ab initio calculations at zero temperature and Monte Carlo simulations at finite temperature (Hu et al. 2012, Kowalczyk et al. 2013).

As shown in Figure 6.2, the four carbon structures are 3D polymers consisting of unique, small $C_{10}$, $C_{12}$, $C_{20}$, and $C_{32}$ cages with quite low density. Diamond is a dense crystal (i.e., 3.63 g/cm³), where each carbon atom is $sp^3$-hybridized. Graphite is a more open crystal, as shown by its lower density of

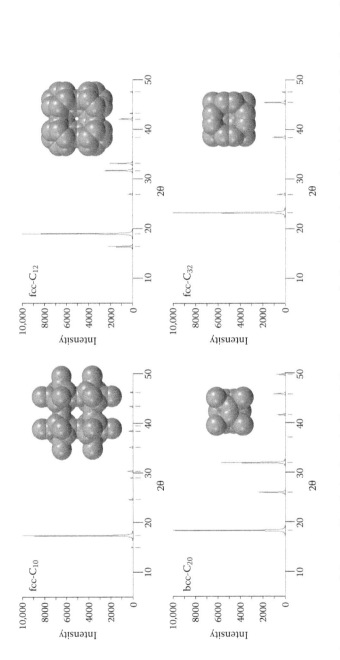

**FIGURE 6.2**  Crystal structures and theoretically calculated powder X-ray diffraction patterns of cubic carbon polymorphs (nm): fcc-C$_{10}$, fcc-C$_{12}$, fcc-C$_{32}$, and bcc-C$_{20}$. (Adapted from Hu et al. 2012.)

**TABLE 6.1** Elastic Constants $C_{ij}$, Bulk Modulus $B$, Shear Modulus $G$, $B/G$ Ratio, and Poisson's Ratio $v$ of the Cubic Carbon Polymorphs and Diamond at Ambient Temperature

| Carbon Crystal | $C_{11}$ (GPa) | $C_{44}$ (GPa) | $C_{12}$ (GPa) | $B$ (GPa) | $G$ (GPa) | $B/G$ | $v$ |
|---|---|---|---|---|---|---|---|
| fcc-$C_{32}$ | 590.45 | 334.64 | 165.85 | 307.38 | 278.83 | 1.10 | 0.15 |
| bcc-$C_{20}$ | 389.51 | 167.70 | 105.34 | 200.06 | 156.83 | 1.27 | 0.19 |
| fcc-$C_{12}$ | 546.67 | 136.81 | 115.75 | 259.39 | 164.24 | 1.58 | 0.24 |
| fcc-$C_{10}$ | 218.79 | 26.62 | 51.30 | 107.13 | 43.04 | 2.49 | 0.32 |
| **Diamond** | **1092.35** | **595.57** | **135.62** | **454.53** | **545.55** | **0.83** | **0.07** |

*Source:* Hu, M., Tian, F., Zhao, Z. et al., *Journal of Physical Chemistry C*, 116, 24233–24238, 2012.

*Note:* The bulk modulus $B$ represents the resistance to material fracture, whereas the shear modulus $G$ represents the resistance to plastic deformation of a material. Hence, high $B/G$ ratio indicates high ductility of crystal, whereas brittle materials are characterized by low $B/G$ ratio.

2.09–2.23 g/cm$^3$. Each carbon atom in graphite is $sp^2$-hybridized. The densities of fcc-$C_{32}$, bcc-$C_{20}$, fcc-$C_{12}$, and fcc-$C_{10}$ cubic carbon polymorphs are 2.75, 2.47, 2.34, and 1.65 g/cm$^3$, respectively (Hu et al. 2012). Interestingly, the fcc-$C_{10}$ and bcc-$C_{20}$ have a mixed $sp^2$ and $sp^3$ hybridization, while fcc-$C_{32}$ and fcc-$C_{12}$ are all $sp^3$-hybridized with variable bond angles (109.5°, 125.3°, and 90° for fcc-$C_{32}$, and 135.0°, 60°, and 90° for fcc-$C_{12}$) (Hu et al. 2012). The calculated elastic constants indicate that these cubic carbon polymorphs are mechanically stable at ambient temperature (Hu et al. 2012). The bulk-to-shear modulus ($B/G$) ratios of fcc-$C_{10}$, fcc-$C_{12}$, bcc-$C_{20}$, and fcc-$C_{32}$ are 2.49, 1.58, 1.27, and 1.10, respectively, which are all higher than that of diamond (0.83) (see Table 6.1). Therefore, Hu et al. (2012) concluded that all cubic carbon crystals have a very ductile nature as opposed to superhard but brittle diamond. The fcc-$C_{10}$ is a ductile material with a $B/G$ value almost triple than that of diamond. The Poisson's ratios of fcc-$C_{10}$, fcc-$C_{12}$, bcc-$C_{20}$, and fcc-$C_{32}$ are 0.32, 0.24, 0.19, and 0.15, respectively (see Table 6.1), which also indicates that the four cubic polymorphs are more ductile than diamond (0.07) (Hu et al. 2012). Thus, we concluded that all the cubic carbon polymorphs studied possess multiple mechanical performances, making them act as cutting tools, coatings, and so on. Electronic properties of the studied cubic carbon polymorphs were reported by Hu et al. (2012). Following DFT calculations, fcc-$C_{32}$, bcc-$C_{20}$, fcc-$C_{12}$, and fcc-$C_{10}$ are all semiconductors with indirect bandgaps of 2.00, 2.26, 2.75, and 1.85 eV, respectively, which are all much smaller than that of diamond (4.20 eV). It is true that larger bandgaps are expected because DFT tends to underestimate the bandgap systematically by about 30%–40% (Hu et al. 2012). Nevertheless, these unique electronic properties make cubic carbon polymorphs interesting materials for potential applications in semiconductor technologies.

# 6.4 Cubic Carbon Polymorphs: Adsorptive Properties

## 6.4.1 Pore Size Distribution

It is expected that cage-like structures of cubic carbon polymorphs can be intercalated by small foreign species, such as alkali metal ions, helium, hydrogen, and argon atoms. Pore size distribution (PSD) is a fundamental measure used for the analysis of the porosity in porous materials (Kowalczyk et al. 2003). The absolute (geometric) PSDs of fcc-$C_{10}$, fcc-$C_{12}$, bcc-$C_{20}$, and fcc-$C_{32}$ cubic crystals were computed using the geometric method proposed by Bhattacharya and Gubbins (2006) called the BG method. For each cubic carbon crystal optimized in GULP using EDIP, a uniform grid of points (100 × 100 × 100) is generated in a cubic box (of ~4 nm in length). Next, for each such point (located in the pore), the largest sphere containing this point (and situated in the pore) is found. The diameter of this sphere is equal to the diameter of the pore, and the collection of histograms of the diameters of pores for each such point of the grid makes it possible to plot the histogram of pore diameters (related to the PSD curve; see Figure 6.3 for a graphical illustration of the BG method). The program, constructed in our group, works in an interactive way (Terzyk et al. 2007).

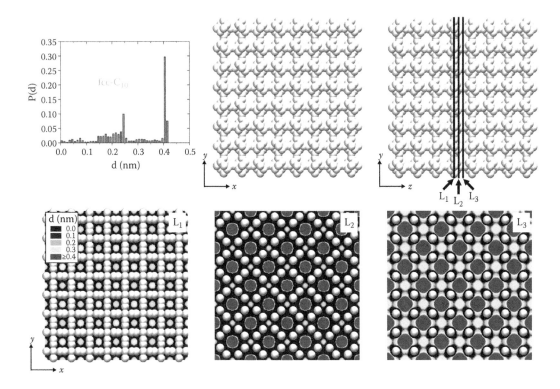

**FIGURE 6.3** Pore size distribution computed from the BG method (Bhattacharya and Gubbins 2006) for fcc-$C_{10}$ polymorph (see upper left panel). In addition, we present pore sizes detected by the BG method for points on arbitrary chosen $xy$ planes inside the box (lower panels, $L_1$–$L_3$ are marked on the upper middle panel). Neighboring carbon cages with pore size of ~0.4 nm (see panel $L_3$) are separated by small carbon cages with pore size of ~0.24–0.26 nm (see panel $L_1$). We can identify narrow windows with free-energy barriers (see panels $L_2$ and $L_3$). Diffusion requires the particles to cross these free-energy barriers so that they can move within the fcc-$C_{10}$ crystal. The VMD program (Humphrey et al. 2006) was used for visualization.

PSD histograms computed for all cubic carbon polymorphs are displayed in Figures 6.3 and 6.4. Taking into account the details of the pore size analysis, it seems clear that the fcc-$C_{12}$, fcc-$C_{32}$, and bcc-$C_{20}$ crystals cannot accommodate light particles, including the smallest He atoms (with effective size of ~0.256 nm).

From the data presented in Figure 6.4, we concluded that carbon cages in fcc-$C_{32}$ and bcc-$C_{20}$ samples are smaller than 0.2 nm. The intercalation of foreign species (such as various atoms, ions, and even molecules) into the graphite interlayer space between hexagonal layer planes of carbon atoms has been well documented (Inagaki & Feiyu 2006). Note, however, that the interlayer spacing between nonintercalated graphite layers is believed to keep that of graphite, i.e., 0.335 nm (Inagaki & Feiyu 2006). Rigid 3D networks of fcc-$C_{32}$ and bcc-$C_{20}$ structures consisting of small carbon cages are inaccessible to foreign species. They can accept foreign atoms by substitution for carbon atoms (for example, N and B), as what is well documented for diamond (Inagaki & Feiyu 2006). Carbon cavities in the fcc-$C_{12}$ crystal are inaccessible to foreign species too. This is because the size of the windows is only about ~0.2 nm and the spaces around the windows are even smaller, as shown in Figure 6.4. Strictly speaking, these carbon cavities should be treated as excluded volumes (Flory 1953). The fcc-$C_{10}$ carbon crystal has a bimodal PSD (see Figure 6.3). Larger carbon cavities with geometrical size of ~0.4 nm are connected through windows with geometrical size of ~0.24–0.26 nm. Note that sizes of the windows are comparable with the size of an He atom. This result is particularly interesting for potential development of thin fcc-$C_{10}$ carbon membranes for He purification from natural gas. Nanoporous membrane produced from fcc-$C_{10}$

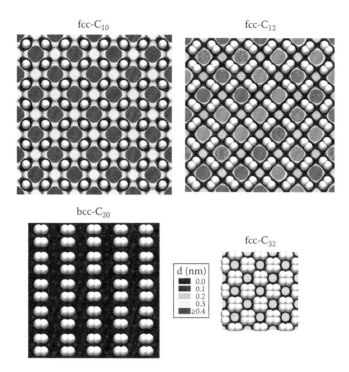

**FIGURE 6.4** Visualization of pore sizes in fcc-$C_{10}$, fcc-$C_{12}$, bcc-$C_{20}$, and fcc-$C_{32}$ polymorphs detected by the BG method (Bhattacharya and Gubbins 2006) for arbitrarily chosen *xy* planes intersecting main cages (other details as on lower panels in Figure 6.3). Note that only cages of fcc-$C_{10}$ crystal are accessible to small foreign particles. The VMD program (Humphrey et al. 2006) was used for visualization.

crystal should be impermeable to typical components of natural gas (e.g., methane, nitrogen, propane, carbon dioxide, butane, and pentane) because their effective sizes are larger than the size of the windows (Kowalczyk et al. 2013). Thus, we expect that He atoms may permeate through a fcc-$C_{10}$ crystal.

## 6.4.2 Thermodynamic Equilibrium: Constant Chemical Potential Simulations

All carbon materials, except highly oriented graphite, contain pores, because they are polycrystalline. The pores in carbonaceous materials are scattered over a wide range of sizes and shapes (Bansal & Goyal 2005). The connectivity of various pores has a significant impact on mass transfer in adsorption and catalytic processes. In most disordered carbonaceous materials, a large amount of pores in various sizes at nanometer scale (i.e., micropores and mesopores), are formed (Bansal & Goyal 2005). The distribution of pore sizes, shapes, and connectivities is very difficult to control during decomposition of organic precursors. Thus, controlling the pore sizes and shapes at nanometer scale by production of ordered carbon crystals is very promising for advancement in carbon nanotechnology.

The fcc-$C_{10}$ polymorph has regular carbon cages connected through narrow windows. Let us first assume that these carbon cages are kinetically open to small particles, including He, $H_2$, Ne, Ar, and CO. To compute the single-component adsorption isotherms of these particles at 298 K, we conducted simulations in the GCMC method (Frenkel & Smit 1996). In these simulations, adsorbed particles are allowed to move in and out of the simulation cell until equilibrium with coexisting fluid is established. For simple spherical atoms (i.e., He, Ne, and Ar), we used the standard GCMC method described elsewhere (Frenkel & Smit 1996). For $H_2$ and CO molecules treated by a fully atomistic rigid model, we implemented the rotational-bias GCMC (RB-GCMC) algorithm of Chávez-Páez et al. (2001). Here we used bias trial insertion/deletion of atomistic rigid molecules into very small fcc-$C_{10}$ nanopores (with pore size comparable with the molecular size of the studied adsorbates). In our implementation, we first inserted the center of the mass of the studied molecule at a random position in the simulation cell. Once inserted, we generated $k$ random orientations of the molecule by using Euler angles. For each molecular orientation, we computed and stored the change in energy $\delta U_i$, where $i = 1, 2, \ldots, k$ (Kowalczyk et al. 2013). The trial insertion of the molecule in the $j$th orientation is then accepted with the following probability (Chávez-Páez et al. 2001):

$$P_{acc} = \min\left[1, \frac{1}{kP_j}\exp\left\{-\beta\left[\delta U_{nm} + \ln\left(\frac{zV}{N+1}\right)\right]\right\}\right],$$ (6.12)

$$P_j = \frac{\exp[-\delta U_j/k_B T]}{\sum_{i=1}^{k}\exp[-\delta U_i/k_B T]},$$ (6.13)

where $U_{nm}$ is the change in the energy of the system due to the insertion of the molecule in that particular orientation, and $z = \exp(\beta\mu)/\Lambda^3$, where $\Lambda$ is the thermal length of the molecule. The trial destruction of a randomly selected molecule is accepted with the following probability (Chávez-Páez et al. 2001):

$$P_{acc} = \min\left[1, kP_1\exp\left\{-\beta\left[\delta U_{nm} + \ln\left(\frac{N}{zV}\right)\right]\right\}\right],$$ (6.14)

$$P_1 = \frac{\exp[-\delta U_1/k_B T]}{\sum_{i=1}^{k}\exp[-\delta U_i/k_B T]},$$ (6.15)

where $(k - 1)$ random orientations, $i = 2, 3, \ldots, k$, are generated by using Euler angles. It has to be pointed out that the RB-GMC method is recommended for simulations of thermodynamic equilibrium in highly confined systems (for example, formation of hydrates in clays and adsorption of particles in small cage-like zeolites and carbon molecular sieves) (Frenkel & Smit 1996, Chávez-Páez et al. 2001).

Ar atoms and CO molecules are strongly adsorbed in the quasi-spherical carbon cages of fcc-$C_{10}$ polymorph at 298 K, as shown in Figure 6.5. The adsorption isotherms of Ar and CO are Langmuir-type with a convex shape with respect to bulk pressure. High values of Ar and CO adsorption enthalpy (~25–27 kJ/mol) are pressure independent, which indicates the dominance of surface forces over fluid–fluid interactions.

Clathrate-formation-mediated adsorption of Ar and CO is clearly seen from the equilibrium RB-GCMC configurations collected at 6 MPa, as displayed in Figure 6.6 (Kowalczyk et al. 2013). The formation of ordered phases of adsorbed Ar and CO is responsible for a large decrease in isothermal compressibility. Interestingly, for the remaining adsorbates, the adsorbed amounts decreased significantly (see linear Henry-type isotherms in Figure 6.5). Isothermal compressibility and microscopic GCMC configurations collected at 6 MPa reveal that the adsorbed phases are gas-like. The enthalpy of $H_2$ and Ne adsorption of ~12 kJ/mol is still significant, which results in partial filling of fcc-$C_{10}$ cages at 6 MPa. Not surprisingly, He atoms occupy the smallest numbers of fcc-$C_{10}$ cages. The interactions between carbon and He atoms are weak compared to the thermal energy at 298 K. Therefore, as far as thermodynamic equilibrium of the mixture in fcc-$C_{10}$ crystal at 298 K is concerned, we expect that the adsorbed phases will be enriched in Ar, CO, Ne, and $H_2$ mixture components, whereas the coexisting

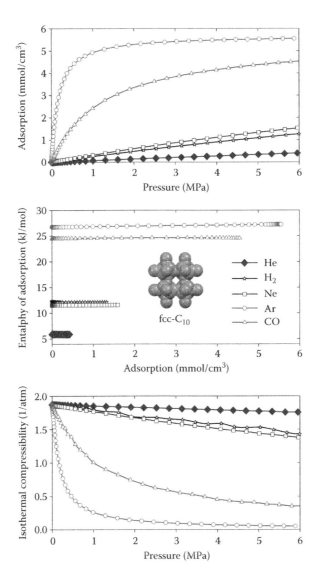

**FIGURE 6.5** Absolute adsorption isotherm (top), enthalpy of adsorption (middle), and isothermal compressibility (bottom) of He, $H_2$, Ne, Ar, and CO at 298 K in fcc-$C_{10}$ polymorph computed by on-the-fly GCMC simulations. (Adapted from Kowalczyk et al. 2013.)

gaseous phases will be enriched in He. Note, however, that the studied fcc-$C_{10}$ crystal is composed of quasi-spherical cages connected by narrow windows (see Figure 6.3). These geometric constrictions represent the free-energy barriers that need to be overcome during the adsorption processes. If they are too high, the fcc-$C_{10}$ carbon cages cannot be accessible to the adsorbed particles within the time scale of the experiment.

## 6.4.3 Kinetics: Self-Diffusion Coefficient

It must be pointed out, as explained above, that the internal structure of cubic carbon polymorphs resembles the internal structure of cage-like zeolites. Consequently, the transport of particles in/out of carbon cages is strongly restricted by the narrow windows. Therefore, calculations of the free-energy

fcc-C$_{10}$

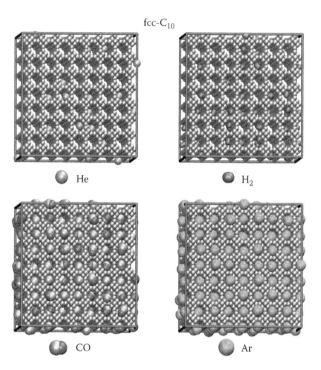

FIGURE 6.6 Grand canonical Monte Carlo snapshot of He, H$_2$, CO, and Ar adsorbed in the fcc-C$_{10}$ polymorph at 6 MPa and 298 K. Under the studied operating conditions, the fcc-C$_{10}$ polymorph adsorbs 0.0017 g/cm$^3$ of He and 0.22 g/cm$^3$ of Ar (volumetric density of liquid He at 4.2 K and liquid Ar at 85.6 K are 0.147 g/cm$^3$ and 1.4 g/cm$^3$, respectively for details see (Kowalczyk et al. 2013). The VMD program (Humphrey et al. 2006) was used for visualization.

barriers associated with the hopping of particles between carbon cages are essential for microscopic understanding of molecular transport in these carbon materials.

We used classical transition state theory (TST) to compute the self-diffusion coefficients of selected particles in the most open fcc-C$_{10}$ crystal (Frenkel & Smit 1996, Kärger et al. 2012). The selection of the TST is justified by the fact that we focused our attention to the limit of infinite dilution of adsorbed phase. At higher pore loadings, TST gives a poor description of transport processes because the contribution from recrossing events can be significant (Kärger et al. 2012).

The transmission rate $k_{A \to B}^{TST}$ corresponding to a spherical particle jumping between cage A and cage B within a rigid cubic carbon crystal is given by Kärger et al. (2012) as

$$k_{A \to B}^{TST} = \sqrt{\frac{k_B T}{2\pi m}} \frac{\exp[-\beta F(q^\star)]}{\int\limits_{\text{cage A}} \exp[-\beta F(q) dq]}, \quad (6.16)$$

where $\beta = 1/k_B T$, $m$ is the mass of particle, and $F(q)$ is the free energy (i.e., potential of mean force) as a function of the reaction coordinate $q$ (Kärger et al. 2012):

$$F(q) = -k_B T \ln[Z(q)], \quad (6.17)$$

where $Z(q)$ is the configurational partition function. For noble gas atoms, $Z(q)$ depends only on the translational degrees of freedom. Rotational degrees of freedom have to be included in $Z(q)$ for the calculation of the H$_2$ and CO free-energy profiles. Here, we assume that, for the studied small quasi-spherical molecules, the rotational degrees of freedom do not play a significant role in the reaction coordinate for elementary jumps through which translational self-diffusion occurs. Therefore, for $H_2$ and CO molecules, we computed $F(q)$ from a preaveraged potential of mean force integrated over all orientational degrees of freedom (Kowalczyk et al. 2013).

In the limit of infinite dilution (i.e., when the correlations between adsorbed particles can be neglected), the self-diffusion coefficient can be computed straightforward from the transmission rate (Kärger et al. 2012) as follows:

$$D_s^{TST} = k_{A \to B}^{TST} \lambda^2 , \tag{6.18}$$

where $\lambda$ denotes the center-to-center lattice distance between two neighboring carbon cages in fcc-$C_{10}$ crystal.

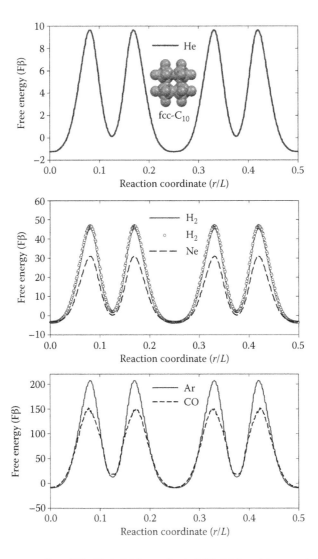

**FIGURE 6.7** Free-energy profiles of He, $H_2$ (solid line: spherical (12,6) LJ model; open circles: five-site atomistic model), Ne, Ar, and CO in fcc-$C_{10}$ polymorph at 298 K and infinite dilution, obtained using Widom's particle insertion method (Widom 1963). (Adapted from Kowalczyk et al. 2013.)

Figure 6.7 presents free-energy profiles of studied adsorbates in fcc-$C_{10}$ crystal computed from Widom's (1963) particle insertion method. Because of cubic symmetry and periodicity, we present the free-energy profiles for three neighboring carbon cages. The free-energy minima corresponding to carbon cages are separated by large free-energy barriers that need to be overcome by self-diffusing particles. For He, the middle minimum has slightly higher free energy compared to neighboring cages. The jumping of He atoms from the left to the middle cavity occurs with a self-diffusion coefficient of ~$1.3 \times 10^{-7}$ cm$^2$/s. The self-diffusion coefficient for jumping of He atoms from the middle cavity to the right cage is higher than $1.3 \times 10^{-6}$ cm$^2$/s. The diffusion coefficients for molecular liquids typically fall in the range of $10^{-5}$–$10^{-4}$ cm$^2$/s (Frenkel & Smit 1996). Moreover, the self-diffusion of quasi-spherical $CH_4$ molecules adsorbed in (10,10) and (12,12) SWCNTs at 300 K and low pressures is even faster (i.e., order of $10^{-3}$ cm$^2$/s) (Skoulidas et al. 2002). Thus, even for the smallest He atoms, there is a significant transport resistance in the fcc-$C_{10}$ polymorph. Because the studied carbon crystal resembles the internal structure of cavity-like zeolites, the computed self-diffusion coefficients fall in the range characteristic of molecular diffusion in these materials (~$10^{-15}$–$10^{-4}$ cm$^2$/s) (Kärger et al. 2012). Not surprisingly, for Ne atoms, the self-diffusion coefficients drop to ~$10^{-18}$–$10^{-16}$ cm$^2$/s (see free-energy profiles displayed in Figure 6.7). Therefore, we conclude that carbon cages in the fcc-$C_{10}$ polymorph are kinetically closed to all gaseous contaminants of He. We would like to point out that thermal vibrations of carbon atoms (i.e., the flexibility of fcc-$C_{10}$ framework) should increase the self-diffusion of He atoms in fcc-$C_{10}$ crystal around ambient temperature. This flexibility of pore windows may promote mass transfer of He atoms, especially at higher temperatures.

## 6.5 Conclusions

During recent years, major progress has been achieved in the synthesis of nanomaterials, leading to discovery of novel exotic carbon nanostructures such as SWCNTs and nanohorns, double-walled carbon nanotubes and MWCNTs, ordered porous carbons, carbon nanofibers and foams, activated carbon fibers, and carbon onions. Cubic carbon polymorphs have been found in nature. The "explosive" synthesis of these novel nanocarbons has also been demonstrated recently. Applications of first-principles calculations and computer simulations not only lead to the reconstruction of basic units of cubic carbon polymorphs, but also provide useful information about their mechanical, electrical, optical, and adsorptive properties. Four cage-like cubic carbons (fcc-$C_{32}$, bcc-$C_{20}$, fcc-$C_{12}$, and fcc-$C_{10}$) with lattice parameters within the range of the experimentally synthesized and naturally existing cubic carbons have been predicted and characterized. Among them, fcc-$C_{10}$ polymorph with a density comparable to $C_{60}$ has the most open structure that may be potentially used for adsorptive separation of He from gaseous contaminates near ambient temperatures via molecular sieving mechanisms. Furthermore, ductile fcc-$C_{10}$ material may be potentially used for coatings and composites, as emphasized by Hu et al. (2012). We predicted that fcc-$C_{32}$, bcc-$C_{20}$, and fcc-$C_{12}$ polymorphs are too dense to accommodate foreign particles inside their cages. However, they have excellent mechanical performances. The fcc-$C_{32}$ crystal is a superhard carbon material with a hardness of 79.9 GPa, indicating its potential application in production of cutting tools (Hu et al. 2012). More experimental work is needed to further advance the structure–property relationships of cubic carbon polymorphs, a fascinating family of novel exotic carbon crystals.

## Acknowledgments

We acknowledge Professor Klaus Sattler for his invitation. P. K. and D. P. acknowledge the financial support from the Research and Development Office, Murdoch University. P. G., A. P. T., and S. F. acknowledge the use of the computer cluster at Poznan Supercomputing and Networking Centre and the Information and Communication Technology Centre of the Nicolaus Copernicus University. We would like to thank Professor D. N. Theodorou (National Technical University of Athens) and Professor J. Gale (Curtin University) for fruitful discussions and comments.

# References

Aust, R. B. & Drickamer, H. G., "Carbon: A new crystalline phase," *Science* 17 (1963): 817–819.

Bahl, O. P., Shen, Z., Lavin, J. G. et al., "Manufacture of carbon fibers," in Donnet, J.-B., Wang, T. K., Rebouillat, S. et al., eds., *Carbon Fibers* (Marcel Dekker, Inc., New York, 1998).

Bansal, R. C. & Goyal, M., *Activated Carbon Adsorption* (CRC Press, Boca Raton, 2005).

Belyanin, A. F. & Samoilovich, M. I., "Nanostructured carbon materials in thin film technology," in Voevodin, A. A., Shtansky, D. M., Levashov, E. A. et al., eds., *Nanostructured Thin Films and Nanodispersion Strengthened Coatings* (Kluwer Academic Publishers, Dordrecht, 2004).

Bhattacharya, S. & Gubbins, K. E., "Fast method for computing pore size distributions of model materials," *Langmuir* 22 (2006): 7726–7731.

Bokros, J. C., "Deposition, structure, and properties of pyrolitic carbon," in Walker, P. L., ed., *Chemistry and Physics of Carbon*, Vol. 5 (Dekker, New York, 1969), pp. 1–118.

Chávez-Páez, M., Van Workum, K., de Pablo, L. et al., "Monte Carlo simulations of Wyoming sodium montmorillonite hydrates," *J. Chem. Phys.* 114 (2001): 1405–1413.

Denisov, V. N., Mavrin, B. N., Serebryanaya, N. R. et al., "First-principles, UV Raman, X-ray diffraction and TEM study of the structure and lattice dynamics of the diamond–lonsdaleite system," *Diam. Relat. Mater.* 20 (2011): 951–953.

Flory, P. J., *Principles of Polymer Chemistry* (Cornell University Press, New York, 1953).

Frenkel, D. & Smit, B., *Understanding Molecular Simulation: From Algorithms to Applications* (Academic Press, San Diego, 1996).

Gale, J., "GULP: A computer program for the symmetry-adapted simulation of solids," *J. Chem. Soc. Faraday Trans.* 93 (1997): 629–637.

Hirai, H. & Kondo, K.-I., "Modified phases of diamond formed under shock compression and rapid quenching," *Science* 253 (1991): 772–774.

Hu, M., Tian, F., Zhao, Z. et al., "Exotic cubic carbon allotropes," *Journal of Physical Chemistry C* 116 (2012): 24233–24238.

Humphrey, W. F., Dalke, A. & Schulten, K., "VMD—Visual molecular dynamics," *J. Molec. Graph.* 14 (1996): 33–38.

Iijima, S., "Helical microtubules of graphitic carbon," *Nature* 354 (1991): 56–58.

Inagaki, M. & Feiyu, K., *Carbon Materials Science and Engineering—From Fundamentals to Applications* (Tsinghua University Press, Beijing, 2006).

Kärger, J., Ruthven, D. M. & Theodorou, D. N., *Diffusion in Nanoporous Materials* (Wiley-VCH Verlag & Co., Weinheim, 2012).

Kennett, D. J., Kennett, J. P., West, A. et al., "Nanodiamonds in the Younger Dryas boundary sediment layer," *Science* 323 (2009): 94.

Kowalczyk, P., Terzyk, A. P., Gauden, P. A. et al., "Estimation of the pore-size distribution function from the nitrogen adsorption isotherm: Comparison of density functional theory and the method of Do and co-workers," *Carbon* 41 (2003): 1113–1125.

Kowalczyk, P., Gauden, P. A. & Terzyk, A. P., "Structural properties of amorphous diamond-like carbon: Percolation, cluster, and pair correlation analysis," *RSC Adv.* 2 (2012): 4292–4298.

Kowalczyk, P., He, J., Hu, M. et al., "To the pore and through the pore: Thermodynamics and kinetics of helium in exotic cubic carbon polymorphs," *Phys. Chem. Chem. Phys.* 15 (2013): 17366–17373.

Kowalczyk, P., Terzyk, A. P., Gauden, P. A. et al., "Carbon molecular sieves: Reconstruction of atomistic structural models with experimental constraints," *J. Phys. Chem. C* 118 (2014): 12996–13007.

Kroto, H. W., Heath, J. R., O'Brien, S. C. et al., "$C_{60}$: Buckminsterfullerene," *Nature* 318 (1985): 162–163.

Kühner, G. & Voll, M., "Manufacture of carbon black," in Donnet, J.-B., Bansal, R. Ch. & Wang, M.-J., eds., *Carbon Black: Science and Technology* (Marcel Dekker, New York, 1993).

Marks, N. A. "Generalizing the environment-dependent interaction potential for carbon," *Phys. Rev. B* 63 (2000): 035401.

Marks, N. A. "Amorphous carbon and related materials," in Colombo, L. & Fasolino, A., eds., *Computer-Based Modeling of Novel Carbon Systems and Their Properties: Beyond Nanotubes* (Springer, Dordrecht, 2010).

Miyake, M., "Electrochemical functions," in Yasuda, E., Inagaki, M., Kaneko, K. et al., eds., *Carbon Alloys: Novel Concepts to Develop Carbon Science and Technology* (Elsevier Science Ltd., Oxford, 2003).

Novoselov, K. S., Geim, A. K., Morozov, S. V. et al., "Electric field effect in atomically thin carbon films," *Science* 306 (2004): 666–669.

Novoselov, K. S., Geim, A. K., Morozov, S. V. et al., "Two-dimensional gas of massless Dirac fermions in graphene," *Nature* 438 (2005): 197–200.

Opletal, G. J., *Structural Simulations Using the Hybrid Reverse Monte Carlo Method* (RMIT University, Melbourne, Australia, 2005).

Pokropivny, V. V. & Pokropivny, A. V., "Structure of 'cubic graphite': Simple cubic fullerite $C_{24}$," *Phys. Solid State* 46 (2004): 392–394.

Santiago, P., Camacho-Bragado, G. A., Martin-Almazo, M. et al., "Diamond polytypes in Mexican crude oil," *Energy Fuels* 18 (2004): 390–395.

Skoulidas, A. I., Ackerman, D. M., Johnson, J. K. et al., "Rapid transport of gases in carbon nanotubes," *Phys. Rev. Lett.* 89 (2002): 185901.

Strel'niskii, V. E., Padalka, V. G. & Vakula, S. I., "Properties of the diamond-like carbon film produced by the condensation of a plasma stream with an RF potential," *Soviet Phys. Tech. Phys.* 23 (1978): 222–224.

Terzyk, A. P., Furmaniak, S., Harris, P. J. F. et al., "How realistic is the pore size distribution calculated from adsorption isotherms if activated carbon is composed of fullerene-like fragments?" *Phys. Chem. Chem. Phys.* 9 (2007): 5919–5927.

Terzyk, A. P., Furmaniak, S., Gauden, P. A. et al., "Virtual porous carbons," in Tascón, J. M. D., ed., *Novel Carbon Adsorbents* (Elsevier Ltd., Oxford, UK, 2012).

Wang, H., Maiyalagan, T. & Wang, X., "Review on recent progress in nitrogen-doped graphene: Synthesis, characterization, and its potential applications," *ACS Catal.* 2 (2012): 781–794.

Widom, B., "Some topics in the theory of fluids," *J. Chem. Phys.* 39 (1963): 2808–2812.

# II

# Nanocarbons

# 7

# High-Energy-Synthesized Carbon-Related Nanomaterials

Michał Soszyński

Andrzej Huczko

## 7.1 Introduction

Following the bottom–up approach of nanomaterials production, we propose the high-energy synthesis of carbon-related nanostructures. To produce gaseous building blocks of nanomaterials (atoms, molecules, radicals, etc.), we use either combustion synthesis (CS) or plasma activation of reactants. Both routes provide temperatures high enough to vaporize any reactants including high-melting carbon (ca. 4000 K). The fast condensation of gaseous carbon-related species results in the formation of novel nanomaterials, i.e., silicon carbide nanowires (SiCNWs), carbon nanotubes (CNTs), and graphene flakes. The different obtained carbon-related nanomaterials were chemically purified and characterized using X-ray powder diffraction (XRD), scanning electron microscopy (SEM), transmission electron microscopy (TEM), elemental analysis, and Raman spectroscopy.

CS, also known as self-propagating high-temperature synthesis (SHS), is a fast, efficient, low-cost, self-sustainable, and one-step method for the production of various industrially useful materials, including novel carbon-related nanostructures. The thermite reaction

$$Al + Fe_2O_3 = Al_2O_3 + Fe$$

is an example of a commercial large-scale application of the CS technique (Yeh 2010). This technique has also been recognized as an attractive alternative to conventional methods for preparing, to date, more than 700 types of advanced materials (Wu et al. 2005). Upon ignition of reacting mixture, the highly exothermic reaction, between a strong reducer and a strong oxidizer, becomes self-sustaining in the form of a combustion wave and yields the final product progressively without requiring additional external heat. Typically, CS is carried out in a modified high-pressure stainless steel calorimetric bomb, modified with an observation port for spectroscopic diagnostics. Recently, this system was also used for the fast synthesis of YAG:Ce$^{3+}$ garnet nanopowders (Huczko et al. 2013).

The thermal plasma synthesis offers a highly efficient way to prepare nanopowders. The attractiveness of plasma processes stems, in general, from high energy densities in the reaction zone resulting in high precursor flow rates and increased temperatures, with both factors causing significant reductions in nanomaterials' growth reaction time. As an example, silicon carbide is an important nonoxide semi-conductor with many diverse industrial applications (Mukasyan et al. 2013). Kong and Pfender (1987) synthesized nanosized β-SiC powders with arc plasma by reacting methane and silicon monoxide. The products showed a bimodal distribution of particle sizes in the characteristic ranges depending on the carbon-to-silicon ratio. There have also been studies on using thermal plasma for silicon carbide nano-coatings, namely, plasma-enhanced CVD (Soum-Glaude et al. 2005, Hafiz et al. 2006, Beaber et al. 2007, Girshick & Hafiz 2007) or thermal plasma physical vapor deposition (Wang et al. 2002, Wang et al. 2003). These techniques were also used for making bulk SiC (Tong & Reddy 2006, Lin et al. 2008, Károly et al. 2011, Vennekamp et al. 2011, Rai et al. 2014). Czosnek et al. (2015) used the DC thermal plasma method for the conversion of selected organic silicon precursors toward SiC-based nanopowders. The DC plasma technique is characterized by relatively short particle residence times and limited solid particles' interactions in the plasma stream. It leads to nanostructured products that are kinetically and diffusion controlled.

1D nanostructures have received steadily growing interest (Xia et al. 2003). 1D nano-SiC exhibits novel properties due to quantum phenomena and much better mechanical characteristics compared to bulk material (Macmillan 1972, Lambrecht et al. 1991, Wong et al. 1997). 1D nano-SiC can now be fabricated using a number of advanced techniques and there are several routes to produce SiC whiskers (Meng et al. 1998):

- Carbothermic reduction of silica (Chrysanthou et al. 1991, Choi & Lee 1995)
- Pyrolysis of silicon-related organic compounds (Addamiano 1982, Czosnek 2004)
- Reactions of silicon halogen derivatives with $CCl_4$ (Setaka & Ajiri 1972)
- Hydrogen reduction of $CH_3SiCl_3$ (Ryan et al. 1967)
- Vapor–liquid–solid catalytic growth (Wagner & Ellis 1964, Milewski et al. 1985)
- Reactions of carbon with SiO at 1700°C (Zhou & Seraphin 1994)

Recently, CVD was used for the efficient preparation of epitaxial $2H$-SiC–$\alpha$-$Al_2O_3$ 1D heterostructures (Sun et al. 2013) and $FeCl_2$-assisted synthesis of β-SiC nanowires, and its photoluminescence was presented (Wang et al. 2015). SiC 1D nanostructures were also integrated into field-effect transistors (Ollivier et al. 2015).

The main drawback of the techniques listed above is related to high-energy consumption and long duration of, mostly solid-state phase, chemical processes. We demonstrated earlier the unconventional fast and efficient one-pot formation of silicon carbide nanofibers (SiCNFs) using SHS (Soszyński et al. 2010, Huczko et al. 2011a).

Graphene, a single-atom-thick sheet of hexagonally arrayed $sp^2$-bonded carbon atoms, is considered to be the most attractive and groundbreaking material for the scientific community, because of its exceptional, mostly electronic and mechanical, properties (Geim et al. 2007). Since the first isolation of single graphene sheet by Scotch tape method (Novoselov et al. 2004), there have been many efforts to develop efficient routes for bulk synthesis of graphene (Avouris & Dimitrakopoulos 2012). There are dozens of possible methods of producing graphene currently being researched (Bonaccorso et al. 2012). Each known approach for graphene synthesis possesses, however, has its own advantages and disadvantages. Regarding prosperous electronic applications, the CVD is one of the mastered techniques to directly grow large surface graphene on metal foils (mostly Cu and Ni) (Yu et al. 2008) to be later transferred to any desired substrate (Reina et al. 2008), even in a continuous mode of operation; but its cost is high. High-quality graphene sheets for electronics can also be produced via epitaxial growth on SiC (Berger et al. 2006). This is, again, a very costly approach due to the specific process environment (Robertson 2009). Physicochemical exfoliation of graphite has been commonly used for the synthesis of bulk graphene (Li et al. 2008), mostly for composites (Ramanthan et al. 2008). It is, however, a very tedious, time-consuming wet chemistry approach, which involves highly hazardous chemicals (strong oxidants and reducers), thus creating severe environmental problems. In the first step, the graphite is oxidized (by wet chemistry) into graphite (or graphene) oxide (GO), which is followed by subsequent reduction into graphene or reduced graphene oxide. Such drastic chemical treatment also introduces different defects in graphene morphology. Thermal exfoliation of GO was also proposed (McAllister et al. 2007), but the product suffered from high oxygen-containing defects and structural defects. We envisage here a one-pot, fast, and energetically autothermal CS route to produce graphene-related nanostructures via thermal reduction of carbon compounds.

## 7.2 High-Energy-Synthesized Carbon-Related Nanostructures

### 7.2.1 Combustion Synthesis

#### 7.2.1.1 Synthesis of Silicon Carbide Nanowires

The unique SHS technique allows for an effective, energetically autothermal formation of different novel materials, including materials that are nanostructural, nonstoichiometric, and bearing new phases, all of this during fast reaction in the mixture of strong oxidant/strong reducer. This all results from specific process characteristics: high temperatures/pressures, short reaction times, and very fast quench of gaseous reaction products during their expansion from the combustion zone toward the cooling zone. Silicon carbide, as so-called special refractory ceramics, possesses very special physical and chemical properties, especially in the case of its nanostructural morphology, including 1D nanofibers.

Silicon carbide is an important material in advanced technology applications due to not only its widely explored excellent mechanical properties and chemical resistivity, but also the fact that it is a wide-bandgap semiconductor. Nanocrystalline powders are commercially available and used mostly for the fabrication of nanocomposites. The hardness and the tensile strength of SiCNWs are greater than in the case of the larger SiC whiskers and, therefore, they may be applied as reinforcement to monolithic ceramics in order to improve fracture tolerance and toughness of the composites (Wang et al. 2006, Rich et al. 2009). Studies on the application of semiconductor properties of nanocrystalline SiC are still at the stage of planning. SiCNFs are perspective materials for applications in electronic devices. Very promising results on the efficient synthesis of SiCNWs and preliminary studies on the optical properties of the synthesized materials formed a basis for presenting this project. SiC has many interesting properties including a very high melting temperature, 2792 K; low thermal expansion coefficient, $4 \times 10^{-6}$ K$^{-1}$; good thermal conductivity, 4.9 W/cm K (better than copper); large Young modulus, 424 GPa; and high fracture toughness, 4.6 MPa m$^{1/2}$. It is a semiconductor, and its large bandgap, 2.39 eV for the cubic SiC, 3.33 eV for the 2H phase, and 3.265 eV for the 4H, makes it an attractive candidate for electronic devices operating at high temperatures. The bandgap in silicon is only 1.1 eV and in porous silicon, it approaches 2.0 eV. It is, therefore, surprising that research on SiCNWs has not attracted attention similar to that of Si nanowires (Figure 7.1).

Because of various excellent properties and specific characteristics due to its nanosize dimensionality, silicon carbide has long been sought as an extremely interesting nanomaterial (Seyller 2006, Wright et al. 2008). Its 1D morphology (nanofibers, nanowires, nanotubes) is even more promising, mostly because of its electronic and mechanical properties (Huczko et al. 2009, Poornima et al. 2010, Schoen et al. 2010). Thus, there have been many attempts to produce 1D SiC (Wong et al. 1999, Zhang et al. 2001, Gao et al. 2002, Saulig-Wenger et al. 2002, Borowiak-Palen et al. 2005, Sundaresan et al. 2007, Lee et al. 2008). For example, Zhang et al. (2001) used CVD, while Wong et al. (1999) obtained SiCNWs on silicon base using the hot filament method. The numerous methods and techniques to produce 1D nano-SiC listed above are, however, multistep and energy-consuming routes, which unambiguously makes them unattractive for scale-up. The goal of our study is to demonstrate that SiCNFs can be produced in a one-step, fast, and simple mode. In addition, the proposed CS route is autothermal and what makes the process economically attractive.

It is crucial to understand the phenomena occurring during the synthesis process and the mechanisms for creating and maximizing the amount of nanofibers synthesized, as well as other various

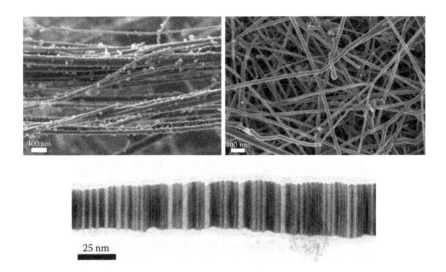

**FIGURE 7.1**    SEM (top) and TEM (bottom) images of SiC fibers obtained by a SHS process.

characteristics of the products obtained in high-energetic conditions. These studies will be carried out using modern methods of experimental processing and diagnostics, combustion theory, kinetics, chemical thermodynamics, and high-temperature physical chemistry. The obtained data will be used to formulate the control means (e.g., effect of reactants and reactor's configuration, effect of atmosphere type and pressure, and effect of ignition) for reaction progress, temperature fields, and degree of conversion, and the accumulated experience will ensure reliable control of the composition, morphology, and structure of the end product.

### 7.2.1.1.1  Experimental System

CS requires a reactor that is resistant to drastic process parameters: temperature and high-pressure gradients and chemically aggressive reaction medium. Used for this purpose is the bomb calorimeter, after appropriate modifications. This reactor is made of stainless steel and is pretested on the instantaneous pressure of up to 10 MPa. In the Laboratory of Physical Chemistry of Nanomaterials (Faculty of Chemistry, Warsaw University, Poland) are three reactors of different volumes: 340, 550, and 2700 $cm^3$ (Figure 7.2).

The structure of a typical modified bomb calorimeter is as follows:

- A sight glass is attached with a window allowing tracking, for example, spectroscopic monitoring of combustion or burning optical wave propagation. The window is made of heat-resistant and pressure-shocked polymer–polycarbonate.
- The system is expanded by a controlled introduction of gas mixtures (working atmosphere during the process of filling the reactor). It consists of an inserter gas at the bottom of the crucible so as not to lose its content. The pressure inside the bomb is measured with the high-pressure transducer connected to a recorder, which allows direct measurement of pressure before and after the reaction, and of the maximum pressure within the reactor during combustion.

The experimental system may be filled with any mixture of gases, such as $O_2$, Ar, CO, $CO_2$, and air. The synthesis reaction is initiated by resistively heating the crucible contents, which are a powder mixture of the reactants of combustion. Carbon tape (eventually Kanthal or tungsten) is used as an "igniter." The tape provides warming-up energy to start allowing a highly exothermic reaction between the oxidant and reductant to occur (Figure 7.3).

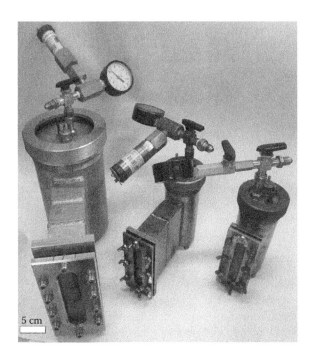

**FIGURE 7.2** Modified bomb calorimeters with volumes of 340, 550, and 2700 cm³.

**FIGURE 7.3** Experimental setup: (1) modified calorimetric bomb ($V \sim 500$ cm²); (2) place for the crucible with substrates—a mixture of Si powder (36 wt%) and PTFE (64 wt%); (3) extremely fast process takes about 1–2 s; (4) solid products contain SiCNFs—50 nm ± 20 nm in diameter, a few/several microns in length.

### *7.2.1.1.2 Combustion Reaction*

The method for the preparation of SiC fibers by CS is a redox reaction between the oxidizer, polytetra-fluoroethylene (PTFE, Teflon®), and the regulator, elemental silicon. The reaction proceeds according to the following general scheme:

$$2Si + -CF_2CF_2- \rightarrow C + SiF_4 \uparrow + SiC.$$

The presented reaction is very highly exothermic (after crossing the activation barrier); its duration is approximately 1 s. This is the time in which the combustion wave passes by the mixture powder substrates. Compared to other methods for obtaining SiC fibers, for example like: carbothermic reduction, plasma synthesis, laser ablation, it is worth noting the very short duration of the synthesis and low energy demand. The initiation of the process is carried out by heating the carbon tape to the appropriate temperature (it takes a few seconds). The temperature reached by the incendiary tape is sufficient to decompose PTFE (melting point: 600 K) and is close to the melting point of silicon (melting point: 1687 K).

The process of combustion may be photographically recorded (Figure 7.4), which enables the accurate determination of the duration of response, and ultimately serves no recognition of the so-called finger-print reaction.

The emitted light in a specified spectral range set on the monochromator can also be recorded on a matrix charge-coupled device camera. Upon completion of combustion, data are sent to a computer, where the digitally recorded spectrum leads to obtaining quantitative information about the mean temperature of combustion, as well as qualitative identification of individuals that are excited radiation sources. From the emission spectrum recorded during the synthesis of the combustion, the temperature is estimated at the average of 1800 ± 200 K (Huczko et al. 2011b).

CS products constitute two fractions: a solid and a gas. The solid contains, inter alia, SiC fibers; unreacted silicon, depending on the atmosphere of silica SiO$_2$; amorphous carbon, and carbon black, as well as other products, depending on the substrates used. The synthesized gaseous products mainly include SiF$_4$. The resulting gaseous compounds give rise to the final pressure. Gases through the scrubber system containing an aqueous solution of KOH are output to the lift. The solid products are subjected to the purification process. In its simplest form, the process of isolating SiCNF can be divided into two stages (Figure 7.5) (Huczko et al. 2005, 2007):

- Boiling in 30% KOH solution for 0.5 hour removes the unreacted Si and SiO$_2$. Silicon carbide does not react with the solutions of the rules.
- Calcining the product at 873 K for 3.5 h in air removes the soot and probably the thinnest and defective nanofibers.
- Reboiling in an aqueous KOH solution is an optional step to remove potentially occluded carbon black and silica in the silicon grains, or to remove the resulting thermal decomposition of the silica ultrafine SiC fibers.

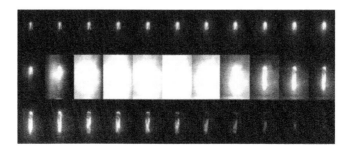

**FIGURE 7.4**   Example photographic registration of combustion. (Courtesy of A. Dąbrowska.)

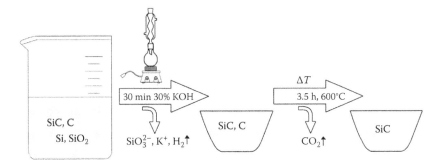

**FIGURE 7.5** Isolation of SiC fiber scheme.

The purification process is very simple and fast and allows obtaining the product with a purity of up to about 95%. Elemental analysis of the carbon content of pure material gives results of about 28–30 wt%. This means that the nanofibers have a purity of about 95% or more.

### 7.2.1.1.3 Characteristics of the Product

Solid products obtained by SHS reaction were collected and subjected to purification using the methods described earlier. The raw solid product is analyzed using the following basic techniques:

- Chemical analysis (determination of unreacted silicon)
- Observation of SEM/TEM/energy-dispersive X-ray spectroscopy (EDX) (morphology and identification of the composition)
- XRD (phase analysis)
- Raman spectroscopy (SiC identification)

Electron microscopy reveals the morphology of the sample, and quantitatively indicates the presence of SiC, as well as enables the identification of the major components of the sample (Figure 7.6). The analysis is also very useful for studying the structure of the product and the presence of various types of defects. XRD analysis provides identification of crystallographic phases of the product. Raman spectroscopy helps identify the main ingredients of the product. Further characterization of SiC material may concern, e.g., the use of scanning tunneling spectroscopy/scanning tunneling microscopy

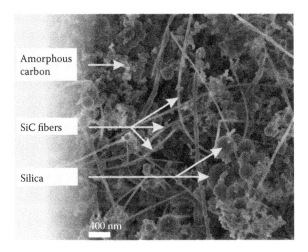

**FIGURE 7.6** SEM image of raw product (CS in the system Si/PTFE).

conductivity of nanomaterials (Busiakiewicz et al. 2008, 2010). EDX technique and X-ray photoelectron spectroscopy (XPS) are used to analyze the surface composition and purity of the product.

Based on SEM images, statistical analysis of received SiCNW diameter distribution was also performed (at a confidence level of 95%) (Soszyński et al. 2010). It was based on the estimated diameter measurements of nanofibers (the number of measured fibers is between 7 and 24 for each set):

- Before cleaning, $\varnothing$ = 60 nm $\pm$ 8
- After boiling for 30 min in 30% KOH solution and firing at 873°C for 3.5 h, $\varnothing$ = 75 $\pm$ 20 nm

TEM and high-resolution transmission electron microscopy (HRTEM) images (Figure 7.1, bottom) reflect that the SiCNW surface is coated with a thin layer of amorphous material (silica), but also surprisingly, it contains a substantial amount of nitrogen (which is responsible for the local conductivity); these tests are continued (Busiakiewicz 2012).

### 7.2.1.1.4 The Influence of Parameters on the Yield of the Synthesis

*7.2.1.1.4.1 Atmosphere of Combustion* The SiC synthesis process requires testing the effects of a number of process parameters on the morphology and performance of the products obtained. One of the most important parameters of the product is changing the morphology of the combustion atmosphere, which also indicates that the reaction is substantially in the gas phase. A series of systematic studies was conducted on the combustion of a powder mixture of Si/PTFE (36/64 wt%) in argon and nitrogen with various additives of oxygen (Huczko 2007). The unexpected result was that the presence of oxygen in the reaction zone is by no means run; the process moves toward the synthesis of silicon oxide only.

In the SEM images shown in Figure 7.7, we find that the presence of oxygen can promote the formation of SiC fibers. Very interesting results were obtained by analyzing the reaction products performed in pure oxygen; here in addition to the expected product—silica—in the strong oxidative conditions 1D nanostructures are also formed from silicon carbide. The graph in Figure 7.8 compares the general degree of conversion of silicon as a function of the composition of the atmosphere of combustion.

Due to the fact that the increasing oxygen content in the gas mixture also increases the silica content of the process, the oxygen content must be optimized in terms of the degree of conversion to SiCNFs. A detailed balance silicon conversion was conducted on combustion synthesis reactions carried out in the Si/PTFE system under an atmosphere of argon, nitrogen, air, and carbon monoxide and dioxide (initial pressure of 1 MPa) (Soszyński 2011). From the mixture containing the solid reaction products, including silica, silicon, and unreacted carbon black, SiC is isolated. Carrying out the material balance of the whole purification process determined the same overall degree of conversion of the starting silicon and silicon conversions to silica and silicon carbide—the data are summarized in Figure 7.9.

Macroscopic morphology of the products varies from loose, black soot mainly located at the bottom of the reactor from tests carried out under argon or nitrogen to gray-brown foam that fills the entire interior of the sealed reactor for testing in an atmosphere comprising an admixture of oxygen. The number emerging from SiC is the highest for the atmosphere of air (the optimum amount of oxygen). Previously, it was found that the formation of SiC fibers favored by the presence of oxygen is extended, so a range of tests was conducted in an atmosphere of carbon monoxide to see whether SiC fibers constitute an additional source of carbon/oxygen in the system. Particularly intriguing is the possibility to use carbon dioxide (a greenhouse gas) as an additional source of both carbon and oxygen, acting as reagent in the transition process of silicon monoxide (SiO).

The CS execution of a mixture of Si/PTFE, therefore, gives the best results when working in carbon monoxide atmosphere; it manifests a high degree of conversion to SiC from Si. This is also confirmed by SEM observations.

*7.2.1.1.4.2 Waste Reagents, Scale-Up Process* One of the aims of the research is to understand the mechanism of parametric tested reaction, and, hence, to increase its efficiency. This will help reduce the cost

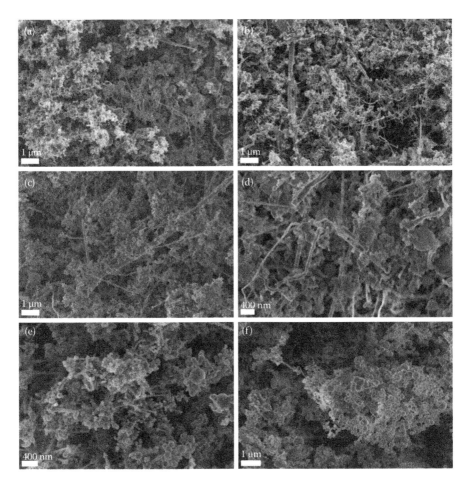

**FIGURE 7.7** The morphology of the products in (a) pure argon, (b) pure nitrogen, (c) a content of 20.0 vol% $O_2$ in Ar, and (d) 19% $O_2$ in $N_2$, (e, f) in pure oxygen.

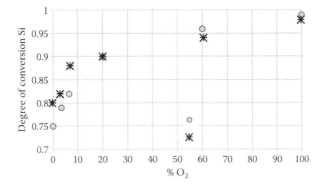

**FIGURE 7.8** The total degree of conversion of the starting silicon according to the reaction atmosphere.

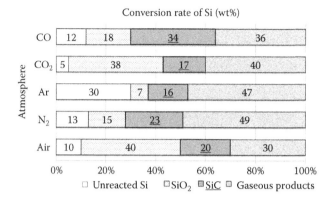

**FIGURE 7.9**  Silicon conversions depending on the gas filling the reactor.

of the process and the isolation of the final product—SiCNFs (Figure 7.10). Lowering the cost of the synthesis of nanomaterials is, however, extremely important for further development of nanotechnology, which is inhibited by the high price of their precursors.

The determination of optimum process conditions, however, will reduce the cost of synthesis. The Laboratory of Nanomaterials Physics and Chemistry (Department of Chemistry, University of Warsaw) investigated a number of systems using the parametric and optimization tests (Soszyński et al. 2010, 2011). The reactants used, which should result in SiCNFs containing a source of carbon and silicon, were usually Teflon and silicon powder (commercially available reagents). The conducted research focuses on the full knowledge of the mechanism of the reaction of the test, which will optimize the synthesis and the ability to maximize its effectiveness. This research, however, will do little to improve the economic characteristics of the reaction, and it is obvious that the cost of nanomaterials is always one of the barriers to the development of nanotechnology. Previously used in the synthesis of pure SiCNFs, commercial reagents (Sigma-Aldrich®) that are relatively expensive will be now replaced by waste PTFE powders (as oxidant) and the silicon elemental from spent solar panels (Figure 7.11). The cost is a few orders of magnitude (!) lower. It will drastically reduce the costs of SiCNF production since the operational costs of energetically autogenic SHS technique are obviously very low. It was found

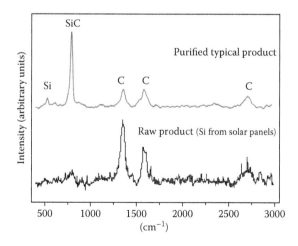

**FIGURE 7.10**  Raman spectra of raw and purified products (combustion in air).

**FIGURE 7.11** SEM images of transverse cross section of a piece of fragmented solar panel, where (a) and (b) are the top part of a solar cell and (c) is the dark, technical part of the cell.

earlier that SiCNFs growth is favored by the presence of the oxygen. Thus, this aspect will be carefully studied within the parametric investigations, with the use of carbon oxide combustion atmosphere, which proved to be superior during the exploratory runs. The exploitation of carbon dioxide (so-called greenhouse gas) looks here especially intriguing, as an additional source of both elemental carbon and of the oxygen which is a component of intermediate compounds (gaseous SiO). The preliminary testing brought very unexpected and encouraging results; even so, the stable thermally compound $CO_2$ can be, under partially nonequlibrium conditions, atomized to elements.

The limited control of nanomaterial production and the lack of the knowledge of its growth mechanism definitely hamper the further progress of nanotechnology. With more than 1000 "nano" products already in the market, the high expectations regarding its development in the 21st century definitely encourage more in-depth studies of nanomaterial formation. Thus, new and fundamental aspects of the nanomaterials growth science are, evidently, worthy of further exploration, especially considering the inherent limitations of existing production techniques. Among them, a search for new procedures of the fabrication of the 1D materials and the exploratory research on its characterization and applications are worth taking.

### 7.2.1.2 Synthesis of Carbon-Related Nanomaterials

The CS system presented previously was used to produce different carbon-based nanostructures. It was shown by Dąbrowska et al. (2011) that solid carbonates can be atomized in the solid phase by using strong reducers with the formation of novel nanocarbons via gas phase condensation. Different starting mixtures of magnesium powder with various carbonates ($Li_2CO_3$, $Na_2CO_3$, $CaCO_3$, $FeCO_3$, $(NH_4)_2CO_3$) were autothermally combusted out under both reactive (air) and neutral atmospheres (argon) with an initial pressure of 1 or 10 atm to yield novel nanomaterials. Under the applied conditions, the presence of crystalline MgO in products (confirmed by XRD analysis), even for the reaction under neutral atmosphere, points to the deep conversion/reduction of carbonates. For producing fibrous products, the $Na_2CO_3$ system proved to be the most promising one (in other tested carbonate systems, except $Li_2CO_3$, the content of fibrous phase was insignificantly lower). SEM images in Figure 7.12 (upper row) show the morphology of the products with some 1D nanostructures resembling CNTs. In fact, it was shown earlier by Szala (2008) that condensing carbon vapors, produced via SHS, can yield CNTs. Also, Bendjemil (2009) produced CNTs by combustion decomposition of carbonates under low pressure. One should mention here that $CO_2$, which is, in fact, an intermediate by-product of the carbonate decomposition under SHS conditions, can also be reduced to CNTs by metallic Li (Lou et al. 2003).

Carbon oxides were also efficiently atomized (Dąbrowska et al. 2012) during the fast, one-step direct CS reduction with the use of strong reducers yielding different solid nanoproducts (oxides, carbonates, carbides, and exfoliated graphite carbon), which were chemically purified and characterized using XRD, SEM, and Raman spectroscopy. To predict whether the reduction will occur spontaneously, the Gibbs free energies, $\Delta G° = \Delta H° - T\Delta S°$, were calculated for several reactions. Since it is well known that oxidation of metals is accompanied by a release of large heat and the temperature of those reactions is usually well above 1000 K, our calculations were performed at 1300 K (in the following, $R$ means "reducer"):

1. Reactions with Al, Fe, and B (formation of trioxides):

$$3CO_2 + 4R \rightarrow 2R_2O_3 + 3C$$

$$3CO + 2R \rightarrow R_2O_3 + 3C$$

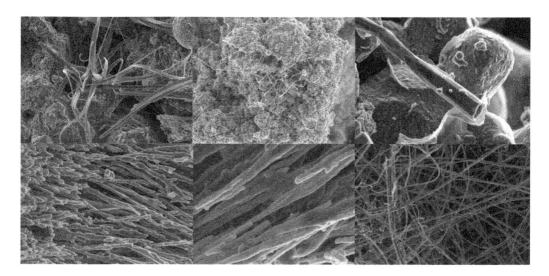

**FIGURE 7.12** Products obtained from carbonate (upper row) and silicon carbide (lower row) systems, from upper left, clockwise: $Na_2CO_3$/Mg/Co, 1 atm, air; $Na_2CO_3$/Mg/PTFE/Fe, 1 atm, air; $FeCO_3$/Mg, 10 atm, air; Si/PTFE/$NaN_3$, 10 atm, air (two pictures with different magnification); Si/PTFE, 10 atm, air.

2. Reactions with Sn, Zr, and Ti (formation of dioxides):

$$CO_2 + R \rightarrow RO_2 + C$$

$$2CO + R \rightarrow RO_2 + 2C$$

3. Reactions with Ca, Mg, and Zn (formation of monoxides):

$$CO_2 + 2R \rightarrow 2RO + C$$

$$CO + R \rightarrow RO + C$$

4. Reaction with Li (formation of suboxides):

$$CO_2 + 4R \rightarrow 2R_2O + C$$

$$CO + 2R \rightarrow R_2O + C$$

The results of the calculations are given in Table 7.1.

From the calculations, one can predict that out of 11 reducers, 8 elements, such as Al, B, Ca, Li, Mg, Si, Ti, and Zr, can be used to reduce carbon oxides. In fact, we observed indeed those self-propagated reactions with both carbon oxides being reduced. Thus, the calculations were in good agreement with the experiments, carried out in the modified calorimetric bomb filled with either gaseous CO or $CO_2$ (initial pressure of 2 MPa). The reduction was initiated by a short resistive heating of powdered reducer (a few grams of sample) using a carbon tape, and terminated within a few seconds. The products were collected and purified via wet chemistry route to remove residual reducer and by-products different from carbon-related materials.

The XRD spectra of raw products showed not only the presence of semiamorphous nanocarbons and reducer oxides, but also other strong reflections which can be attributed to oxalates, carbonates, and carbides (the latter ones in the case of B, Ti, Zr, and Al). The example of XRD data (Figure 7.13; for Mg/ CO raw and purified product) shows the presence of C, Mg, and MgO in a raw product, generally in line

**TABLE 7.1** Calculated Gibbs Free Energy of Reduction of 1 Mole of Carbon Oxide at 1300 K

| Metal | Melting Point (K) | Free Energy of Reaction at 1300 K (kJ mol⁻¹) | |
|---|---|---|---|
| | | With $CO_2$ | With CO |
| Al | 933 | −445.29 | −194.2 |
| Fe | 1811 | +71.85 | +64.67 |
| B | 2348 | −239.99 | −91.54 |
| Si | 1687 | −282.45 | −112.74 |
| Sn | 505 | +85.22 | +142.11 |
| Zr | 2127 | −456.87 | −199.99 |
| Ti | 1943 | −314.36 | −128.21 |
| Ca | 1115 | −601.80 | −272.28 |
| Mg | 923 | −520.13 | −231.62 |
| Zn[a] | 693 | −35.21 | +10.84 |
| Li | 454 | −455.06 | −199.08 |

[a] Data at 1100 K.

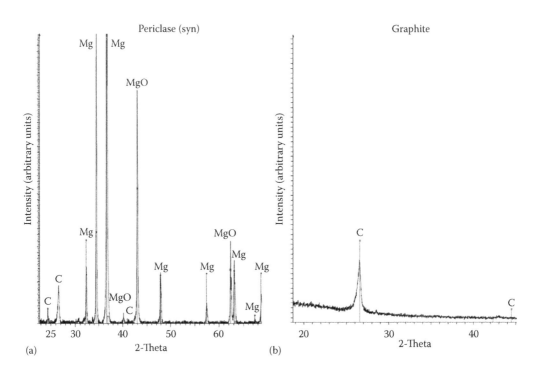

**FIGURE 7.13** XRD data for raw (a) and purified (b) products obtained in Mg/CO system.

with thermodynamic predictions. The high efficiency of purification protocol and isolation of carbon product is clearly confirmed, too.

The change in reducer type had a strong influence on the product's morphology (SEM images; Figure 7.14), leading to nanocarbon 2D flakes (few-layered graphene [FLG], for Mg/CO system), cubes, or porous fibers.

To better evaluate the nature of carbon phase, Raman spectra were measured for all products with representative spectra in Figure 7.15. The peak at around 1600 cm$^{-1}$ (G band) is related to the vibrations of $sp^2$-bonded carbon atoms in a 2D hexagonal lattice, such as in a graphene. The D band at 1350 cm$^{-1}$ is

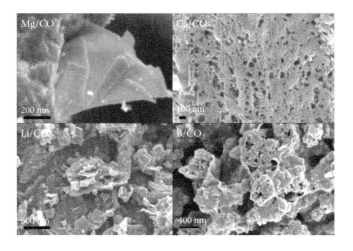

**FIGURE 7.14** SEM images of different carbon-related nanoproducts.

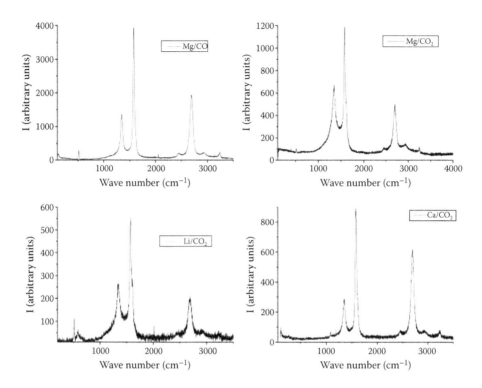

**FIGURE 7.15** Raman spectra ($\lambda = 532$ nm) for selected purified products.

associated with the defects in the graphitic layers and its intensity is high for poorly graphitized materials. Thus, the $I_G/I_D$ ratio serves as a measure of the graphitic ordering. The G/D band ratio definitely differs depending on the starting reactants (Figure 7.15). However, the relatively high ordering of carbon atoms in the produced nanocarbons is evident due to the high temperature of the combustion and product growth.

Thus, the results of the preliminary runs for Me/CO and Me/CO$_2$ combustion systems showed the formation of carbonaceous matter including FLG nanostructures.

Autogeneous CS in the powdered mixtures of a strong reducer (Mg, Si, AlH$_3$, CaH$_2$, NaN$_3$, Li$_3$N, and BH$_3$NH$_3$) and a strong, carbon-bearing oxidizer (C$_{2.78}$F$_2$, Teflon PTFE (CF$_2$)$_n$, and PCV (C$_2$H$_3$Cl)$_n$) yield efficiently accordion-like carbon nanostructures (Huczko et al. 2014). They can be sonicated to produce FLG. The combustions were carried out under both neutral (Ar) and active atmospheres (CO$_2$) at the starting pressure equal to 10 atm. The solid products (layered graphite, nanocarbides) were chemically purified and characterized using XRD, SEM, TEM, and Raman spectroscopy. A mechanism of the reaction (for Me/(CF)$_x$ mixture as an example) can be presented in the following simplified form:

$$x\text{Me} + (\text{CF})_x \rightarrow x\text{MeF} + x\text{C}.$$

All combustions were successful. The raw products were purified via a wet chemistry route to remove the starting reducer and by-products different from carbon-related materials. In the case of the Mg/C$_{2.78}$F$_2$ system, to remove MgF$_2$ from the raw products, the material was boiled (2 h) under reflux in a 3 M solution of nitric acid, suspension filtered; the residue is flushed with distilled water, rinsed with acetone, and air-dried at 70°C until the mass remained constant. This protocol was verified by using reference C/MgF$_2$ compositions. Here we present in more detail only the results of the reaction of Mg with fluorinated graphite (C$_{2.78}$F$_2$), combusted in Ar atmosphere. Different weight ratios of reactants and initial combustion pressures were tried.

XRD patterns of raw and purified products show (Figure 7.16) that they consist of a mixture of carbon and magnesium fluoride, as expected. Interestingly, as the XRD spectra of raw and purified products are similar, it may be concluded that at least some of the produced magnesium fluoride was partially carbon encapsulated.

Figure 7.17 presents SEM images of reaction products. It is clearly seen that during the combustion, the fluorinated graphite is effectively defluorinated with strong reducers and is, at least, partially split. Its structure is profoundly reorganized and the product is dominated by voluminous accordion-like microstructures. The EDX analyses (not presented here) confirmed the qualitative results of XRD analyses.

The products were also characterized by TEM. As shown in Figure 7.18, both raw and purified samples contain planar sheet-like graphite and $MgF_2$ ball-like particles (which were not spotted after purification). The sheets are relatively transparent and were not damaged under electron beam irradiation. The featureless sheets, resembling crumpled silk (Kim et al. 2009), are considered to be FLG sheets entangled with each other, with many ripples and wrinkles on their surface. Some of the wrinkles could be attributed to the collapse of individual graphene sheets after defluorination. However, it is worth noting that

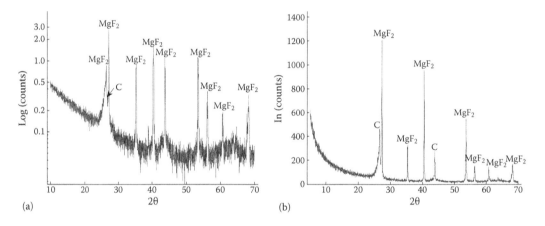

**FIGURE 7.16** XRD patterns of powders obtained by combustion synthesis in a system. (a) Raw product, Ar, 10 atm, $Mg/C_{2.78}F_2 = 1:2$; (b) purified, Ar, 10 atm, $Mg/C_{2.78}F_2 = 1:2$.

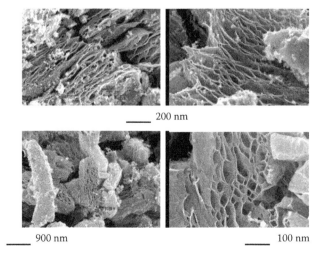

**FIGURE 7.17** SEM images of raw products of combustion: Mg and fluorinated graphite $C_{2.78}F_2$ in argon at initial pressure of 1.5 MPa.

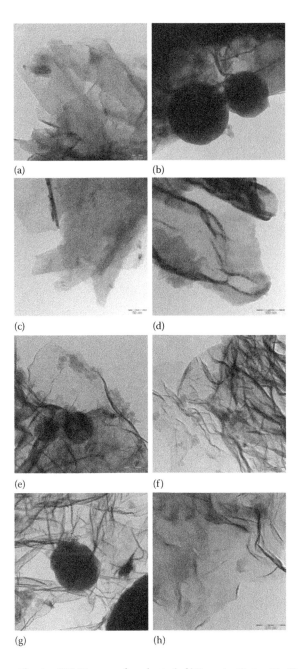

**FIGURE 7.18** High-magnification TEM images of products. (a, b) Raw, Ar, 10 atm, $Mg/C_{2.78}F_2 = 1{:}2$; (c, d) purified, Ar, 10 atm, $Mg/C_{2.78}F_2 = 1{:}2$; (e) raw, Ar, 1 atm, $Mg/C_{2.78}F_2 = 1{:}2$; (f) purified, Ar, 1 atm, $Mg/C_{2.78}F_2 = 1{:}2$; (g) raw, Ar, 1 atm, $Mg/C_{2.78}F_2 = 4{:}5$; (h) purified, Ar, 1 atm, $Mg/C_{2.78}F_2 = 4{:}5$.

there are small smooth regions in these big sheets. The multilayer graphene sheets, of size up to tens of square microns, corrugated together.

To overcome the strong van der Waals forces that still bind graphene layers together, the product from the test with $Mg/C_{2.78}F_2 = 1{:}2$ (Ar, 10 atm) was sonicated in water with sodium dodecylbenzenesulfonate as a surfactant. After sonication, the suspension was centrifuged for 10 minutes (1000 rpm) and separated into two fractions, i.e., the supernatant and the sediment collected from the bottom of the centrifuge vial.

Figure 7.19 presents SEM images of the supernatant and precipitated material. While the accordion-like graphite nanostructures can still be spotted in the precipitate, they were not found in the supernatant. Thus, the sonication evidently helped to further split the loose accordion-like nanocarbons into FLG samples.

Raman spectra were acquired for further evaluation of the nature of the carbon phase. The spectra were measured for the purified and sonicated products obtained from Mg/C$_{2.78}$F$_2$ (1:2; Ar, 10 atm) using a 514.5 nm excitation laser with the spectral resolution of 2 cm$^{-1}$. The spectra comprise five distinct bands (Figure 7.20). The G band, which is typical for all graphitic materials, is located at 1581–1582 cm$^{-1}$. Importantly, the G band has a weak shoulder located at higher wave numbers. This broadening can be ascribed to the so-called D′ band.

The D band (located at 1350–1352 cm$^{-1}$) is associated with defects and its intensity is generally increased in carbon materials with a low degree of graphitization. The G/D integral ratio is a convenient indicator of the degree of graphitization. The G/D ratio values were found to be 1.60, 1.55, and 1.46 for the

FIGURE 7.19 SEM images of the purified product (Mg/C$_{2.78}$F$_2$ = 1:2; Ar, 10 atm) after sonication. (a, b): Supernatant; (c, d): precipitate.

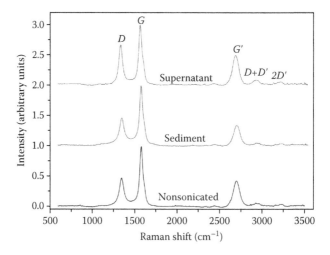

FIGURE 7.20 Raman spectra of nonsonicated, sediment, and supernatant collected after 40 h sonication (product obtained from Mg/C$_{2.78}$F$_2$ = 1:2; Ar, 10 atm).

nonsonicated, sediment, and supernatant samples, respectively. These values do not differ significantly and this observation shows that long-term sonication does not change the overall degree of graphitization and does not induce defects. This observation is in disagreement with other published works. Lotya et al. (2010) studied the sonication of graphite in sodium cholate solutions. They observed that the degree of graphitization decreases even for relatively short sonication times (24 h). All studied samples have a relatively strong feature (G' band), which is located at 2700–2705 cm$^{-1}$. This feature is present in carbon materials having a graphitic structure. Usually, it is used to evaluate the number of layers and their stacking order in graphene and FLG materials (Malard et al. 2009). The G' band was fitted using a single Lorentzian profile and a goodness of fit ($R^2$) larger than 0.97 was obtained. The fitted spectral profile has a width at half maximum between 60 and 70 cm$^{-1}$. The shape and the width of the G' band is consistent with the data published for mechanically exfoliated pyrolytic graphite (Faugeras et al. 2008). The G'/G ratio values were on a similar level and were found to be 0.73, 0.64, and 0.76 for the nonsonicated, sediment, and supernatant samples, respectively. These findings along with the data from electron microscopy clearly show that sonication resulted in the exfoliation of pristine crystallites. However, it did not cause further "delamination" toward thinner flakes. Finally, one should also mention that the variations in the operational parameters of starting mixture combustions did not seem to distinctly influence the morphology and compositions of products.

## 7.2.2 Plasma Synthesis of FLG

The arc discharge method has been extensively used for the production of different nanocarbons for many years. Krätschmer et al. (1990) discovered the carbon arc synthesis of fullerenes, a new allotropic form of carbon. One year later, Iijima (1991) found MWCNTs, the "black diamonds" of the 21st century (Yumura 1999), in the cathode deposit formed during the arc synthesis of fullerenes. Bystrzejewski et al. (2004, 2008, 2009) and Łabędź et al. (2014) have been producing carbon-encapsulated magnetic nanomaterials since 2004 using the carbon arc route. It was also shown by Manning et al. (1999) that the covalently bonded fluorinated graphite CF$_x$ forms exfoliated graphite when processed in an atmospheric pressure induction (27.12 MHz) argon plasma. Recently, Wu et al. (2009) synthesized graphite sheets by hydrogen arc discharge exfoliation from graphite oxide, but the process yield was low (10 mg).

We envisaged a rapid and reactive heating of fluorinated graphite, CF$_x$, by carbon arc discharge in a mixed buffer gas to produce FLG. Figure 7.21 presents the arc plasma route for the reduction/exfoliation of CF$_x$ to bulk amount of FLG.

Fluorographite polymer, CF$_x$ ($x = 0.72$), was used as the starting reactant. The plasma splitting of the starting material was carried out (batch mode of operation) in a carbon arc plasma reactor (Figure 7.22) described in detail elsewhere (Lange et al. 1997). Required amounts of fluorinated graphite CF$_x$ and graphite powder (below 50 μm) was mixed properly and a homogeneous, composite anode (diameter: 8

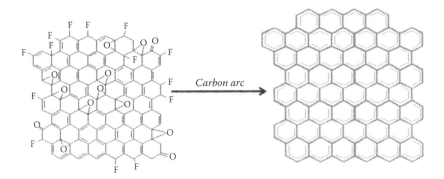

**FIGURE 7.21** Efficient and smooth plasma transformation of CF$_x$ to FLG within minutes.

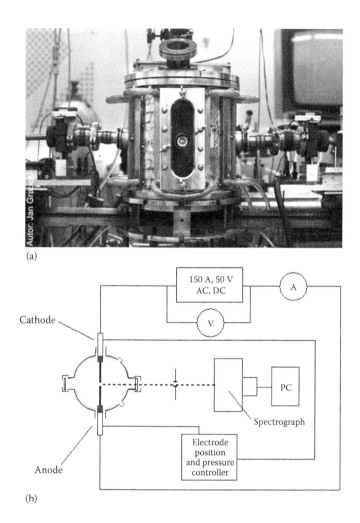

**FIGURE 7.22**  (a) Carbon arc reactor and (b) the scheme of experimental setup.

mm) was manufactured following the procedure presented elsewhere (Bystrzejewski et al. 2013). Drilled and filled heterogeneous anodes were also arced in reference runs. During the arcing, a composed graphite anode was consumed and the fluorine was eliminated from the polymer while the carbon vapors condensed, forming the desired exfoliated graphite. The powders from the reactor walls were characterized using SEM and TEM, XRD diffraction, and Raman spectroscopy. Table 7.2 presents the operational parameters of all runs.

Following the procedure presented elsewhere and taking into account the self-absorption phenomenon (Lange et al. 1996), the $C_2$ emission spectra from the interelectrode gap (with 2 mm distance) were collected and the plasma temperatures were estimated (Figure 7.23). It is evident from the results obtained that the arc plasma temperatures (within 4000–4400 K) are high enough to sublime the carbon anode material.

Figure 7.24 presents the results of SEM observations of the products.

The product morphology depends upon the process operating parameters. It has a loose, accordion-like structure (Figure 7.24a). Most of $CF_x$ were efficiently defluorinated into FLG to form separate ultrathin platelets (Figure 7.24b) or petal-like (Figure 7.24d and f) nanostructures (specifically under hydrogen atmosphere). Evidently, however, the individual sheets in the FLG were agglomerated and overlapped. Therefore, the following sonication is required to obtain single-layer graphene. Surprisingly,

**TABLE 7.2**  Operational Parameters of Carbon Arc Exfoliation of Fluorinated Graphite

| Run No. | Anode Composition | Plasma Gas | Pressure (mbar) | Arc Current (A) | Product |
|---|---|---|---|---|---|
| 1 | Drilled anode, 50 wt% $CF_x$/50 wt% graphite | Ar | 600 | 10–70 | Soot-like, ca. 50 mg |
| 2 | Drilled anode, 50 wt% $CF_x$/50 wt% graphite | He/$H_2$ (1:1) | 600 | 10–70 | Soot-like, ca. 100 mg |
| 3 | Homogeneous anode, 20 wt% $CF_x$ | Ar/$H_2$ (1:1) | 600 | 40–60 | Soot-like, few mg |
| 4 | Homogeneous anode, 60 wt% $CF_x$ | Ar/$H_2$ (1:1) | 600 | 40–60 | Soot-like, few mg |
| 5 | Homogeneous anode, 20 wt% $CF_x$ | He | 600 | 40–60 | Soot-like, several hundreds of milligrams |
| 6 | Homogeneous anode, 40 wt% $CF_x$ | He | 600 | 40–60 | Soot-like, several hundreds of milligrams |

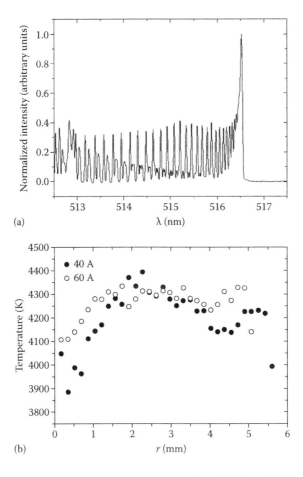

(a)

(b)

**FIGURE 7.23** (a) $C_2$ emission spectrum, homogeneous anode (40 wt% $CF_x$), Ar/$H_2$ (1:1) plasma gas, arc current 40 A; (b) interelectrode gap temperature distribution estimated from $C_2$ emission spectrum, homogeneous anode (40 wt% $CF_x$), Ar/$H_2$ (1:1) plasma gas, arc current 40 A and 60 A.

**FIGURE 7.24**    SEM images of arc discharge-exfoliated and reduced FLG: (a–c) run 1; (d) run 2; (e) run 6; (f) run 4.

CNTs have also been spotted in the raw product (Figure 7.24c). In the case of He atmosphere (Figure 7.24e), only soot-like product was obtained.

TEM was used to further characterize the structure of as-prepared FLG (Figure 7.25).

There is definitely a profound difference between the morphology of starting $CF_x$ (Figure 7.25a and b) and the product (Figure 7.25c through f). The arc discharge exfoliated $CF_x$ looks like a transparent thin sheet with few folds within its plane. The micrographs of as-prepared FLG show randomly aggregated sheets. This is quite different from the graphene sheets obtained via conventional thermal exfoliation,

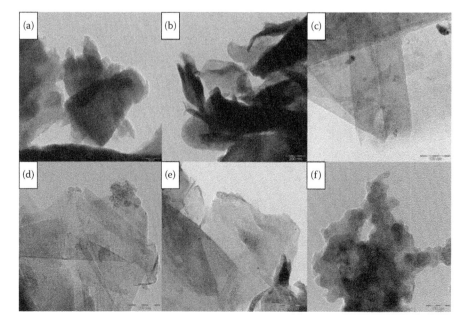

**FIGURE 7.25**    Structural TEM characterization of arc discharge-exfoliated and reduced FLG: (a) starting $CF_x$; (b) starting electrode, test 5; (c) run 1; (d) run 2; (e) run 4; (f) run 6.

which generally look like a wrinkled thin paperlike structure with many folds. This suggests that arc discharge is a more effective route to remove functional groups (than the commonly used thermal exfoliation and postreduction), thus yielding the FLG. Hydrogen atmosphere definitely intensifies the defluorination process (Figure 7.25c through e) compared to He environment (Figure 7.25f).

To further access the nature of the products, XRD analysis was carried out (Figure 7.26).

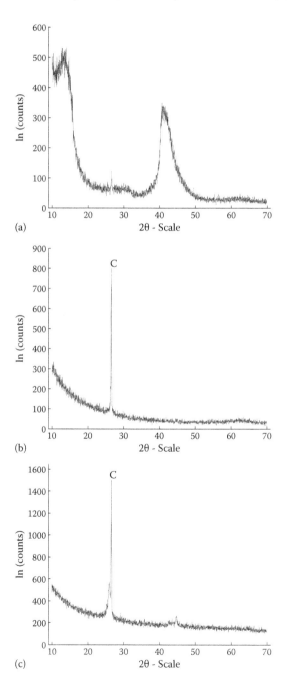

**FIGURE 7.26** XRD patterns of structural characterization of arc discharge-exfoliated and reduced FLG: (a) starting $CF_x$; (b) run 2; (c) run 4.

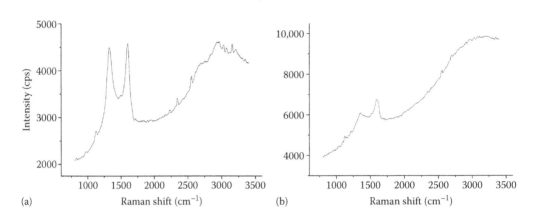

**FIGURE 7.27**   Raman spectra of products obtained in (a) run 1 and (b) run 6.

It is evident from the comparison of XRD spectra between the starting polymer (Figure 7.26a) and the products (Figure 7.26b and c) that a profound transformation of the $CF_x$ into well-graphitized FLG has occurred. The patterns in Figure 7.26b and c are typical for graphitic materials with large ordering in the c direction due to the presence of very well-defined (002) reflection at 26.5°. The (100) and (101) peaks are absent (Figure 7.26b) or have very low intensity (Figure 7.26c). This observation shows that the ordering in the basal planes is weak. Interestingly, the (002) reflection in the sample obtained in Figure 7.26c is composed of two peaks. This finding indicates that this product comprises graphitic phases of two distinct interlayer distances.

Raman spectroscopy is considered to be a fingerprint characterization for graphene and carbon materials (Malard et al. 2009). The representative Raman spectra are shown in Figure 7.27.

Two typical first-order bands appear, i.e., the G band (located at 1590 cm$^{-1}$) and the D band (at 1350 cm$^{-1}$). The first spectral feature is related to stretching vibrations of C–C bonds in hexagonal graphitic lattice. The second band is due to the disorder and defects. More importantly, both spectra are influenced by relatively strong fluorescence. This effect plausibly originates from the presence of polycyclic aromatic hydrocarbons, which could be synthesized during the arc plasma processing of fluorine-containing electrodes. The fluorescence can also be related to fluorine containing small hydrocarbon compounds. In the second-order, 2D (ca. 2700 cm$^{-1}$) peak is absent in each spectrum. This observation along with the moderately low G/D ratio shows that the synthesized carbon materials do not have a high degree of graphitization. Moreover, the spectra in Figure 7.27 do not resemble the typical spectrum of single-layer graphene. These spectra are similar to the spectra reported for low-graphitized FLG (Myriano et al. 2013). In fact, this statement is coherent with electron microscopy images (Figure 7.25).

From the obtained results it is evident that optimization and further study is, obviously, required. The research under way concentrates on the further splitting of the produced FLG into singular graphene sheets.

# References

Addamiano, A., "Preparation and properties of 2H SiC crystals," *J. Cryst. Growth* 58 (1982): 617–622.
Avouris, P. & Dimitrakopoulos, C., "Graphene: Synthesis and applications," *Mater. Today* 15 (2012): 86–90.
Beaber, A. R., Qi, L. J., Hafiz, J. et al., "Nanostructured SiC by chemical vapor deposition and nanoparticle impaction," *Surf. Coat. Technol.* 202 (2007): 871–875.
Bendjemil, B., "Electronic and optical properties of the express purified SWCNTs produced by HiPCO process," *Int. J. Nanoelectron. Mater.* 2 (2009): 173–182.

Berger, C., Song, Z., Li, X. et al., "Electronic confinement and coherence in patterned epitaxial graphene," *Science* 312 (2006): 1191–1196.

Bonaccorso, F. et al., "Graphene synthesis, characteristics and applications," *Mater. Today* 15 (2012): 564–595.

Borowiak-Palen, E., Ruemmeli, M. H., Gemming, T. et al., "Bulk synthesis of carbon-filled silicon carbide nanotubes with a narrow diameter distribution," *J. Appl. Phys.* 97 (2005): 056102.

Busiakiewicz, A., Klusek, Z., Kowalczyk, P. J. et al., "Silicon carbide nanowires studied by scanning tunneling spectroscopy," *Surf. Sci.* 602 (2008): 316–320.

Busiakiewicz, A., Huczko, A., Dudziak, T. et al., "Defects of SiC nanowires studied by STM and STS," *Appl. Surf. Sci.* 256 (2010): 4771–4776.

Busiakiewicz, A., Huczko, A., Soszyński, M. et al., "Surface chemical composition of SiC-cored nanowires investigated at room and elevated temperatures in ultra-high vacuum," *Vacuum* 86 (2012): 1974–1978.

Bystrzejewski, M., Huczko, A., Lange, H. et al., "Arc plasma synthesis of FeNdB containing carbon encapsulates," *Solid State Phenom.* 99–100 (2004): 273–278.

Bystrzejewski, M., Grabias, A., Borysiuk, J. et al., "Mössbauer spectroscopy studies of carbon-encapsulated magnetic nanoparticles obtained by different routes," *J. Appl. Phys.* 104 (2008): 054307.

Bystrzejewski, M., Pyrzyńska, K., Huczko, A. et al., "Carbon-encapsulated magnetic nanoparticles as separable and mobile sorbents of heavy metal ions from aqueous solutions," *Carbon* 47 (2009): 1201–1204.

Bystrzejewski, M., Łabedź, O., Kaszuwara, W. et al., "Controlling the diameter and magnetic properties of carbon-encapsulated iron nanoparticles produced by carbon arc discharge," *Powder Technol.* 246 (2013): 7–15.

Choi, H. J. & Lee, J. G., "Continuous growth of silicon carbide whiskers," *J. Cryst. Growth* 30 (1995): 1982–1986.

Chrysanthou, A., Grieveson, P., Jha, A. et al., "Formation of silicon carbide whiskers and their characteristics," *J. Mater. Sci.* 26 (1991): 3463–3476.

Czosnek, C., Wolszczak, J., Drygaś, M. et al., "Nano-SiC implantation into the structure of carbon/graphite materials made by pyrolysis (carbonization) of the precursor system coal tar pitch/poly(dimethylsiloxane)," *J. Phys. Chem. Solids* 65 (2004): 647–653.

Czosnek, C., Bućko, M. M., Janik, J. F. et al., "Preparation of silicon carbide SiC-based nanopowders by the aerosol-assisted synthesis and the DC thermal plasma synthesis methods," *Mater. Res. Bull.* 63 (2015): 164–172.

Dąbrowska, A., Huczko, A., Soszyński, M. et al., "Ultra-fast efficient synthesis of one-dimensional nanostructures," *Phys. Status Solidi B* (2011): 2704–2707.

Dąbrowska, A., Huczko, A., Dyjak, S. et al., "Fast and efficient combustion synthesis route to produce novel nanocarbons," *Phys. Status Solidi B* (2012): 2373–2377.

Faugeras, C., Nerrìre, A., Potemski, M. et al., "Few-layer graphene on SiC, pyrolitic graphite, and graphene: A Raman scattering study," *Appl. Phys. Lett.* 92 (2008): 011914.

Gao, Y. H., Bando, Y., Kurashima, K. et al., "SiC nanorods prepared from SiO and activated carbon," *J. Mater. Sci.* 37 (2002): 2023–2029.

Geim, A. K., Novoselov, K. S. et al., "The rise of graphene," *Nat. Mater.* 6 (2007): 183–191.

Girshick, S. L. & Hafiz, J., "Thermal plasma synthesis of nanostructured silicon carbide films," *J. Phys. D Appl. Phys.* 40 (2007): 2354–2360.

Hafiz, J., Mukherjee, R., Wang, X. et al., "Analysis of nanostructured coatings synthesized by ballistic impaction of nanoparticles," *Thin Solid Films* 515 (2006): 1147–1151.

Huczko, A., Bystrzejewski, M., Lange, H. et al., "Combustion synthesis as a novel method for production of 1-D SiC nanostructures," *J. Phys. Chem. B* 109 (2005): 16244–16251.

Huczko, A., Osica, M., Rutkowska, A. et al., "A self-assembly SHS approach to form silicon carbide nanofibers," *J. Phys. Condens. Matter* 19 (2007): 395022.

Huczko, A., Dąbrowska, A., Savchyn, V. et al., "Silicon carbide nanowires: Synthesis and cathodoluminescence," *Phys. Status Solidi B* 246 (2009): 2806–2808.

Huczko, A., Dabrowska, A., Soszynski, M. et al., "Ultrafast self-catalytic growth of silicon carbide nanowires," *J. Mater. Res.* 26 (2011a): 3065–3071.

Huczko, A., Dabrowska, A., Soszynski, M. et al., Ultrafast self-catalytic growth of silicon carbide nanowires *J. Mater. Res.* 26 (2011b): 3065–3071.

Huczko, A., Dabrowska, A., Soszynski, M. et al., "Fast combustion synthesis and characterization of YAG:Ce$^{3+}$ garnet nanopowders," *Phys. Status Solidi B* (2013): 2702–2708.

Huczko, A., Dabrowska, A., Soszynski, M. et al., "Facile and fast combustion synthesis and characterization of novel carbon nanostructures," *Phys. Status Solidi B* (2014): 2563–2568.

Iijima, S., "Helical microtubules of graphitic carbon," *Nature* 354 (1991): 56–58.

Károly, Z., Mohai, I., Klébert, S. et al., "Synthesis of SiC powder by RF plasma technique," *Powder Technol.* 214 (2011): 300–305.

Kim, C.-D., Min, B.-K., Jung, W.-S. et al., "Preparation of graphene sheets by the reduction of carbon monoxide," *Carbon* 47 (2009): 1610–1612.

Kong, P. C. & Pfender, E., "Formation of ultrafine b-silicon carbide powders in an argon thermal plasma jet," *Langmuir* 3 (1987): 259–265.

Krätschmer, W., Lamb, L. D., Fostiropoulos, K. et al., "Solid C$_{60}$: A new form of carbon," *Nature* 347 (1990): 354–358.

Łabędź, O., Huczko, A., Gawraczynski, J. et al., "Carbon arc plasma: Characterization and synthesis of nanosized SiC," *J. Phys. Conf. Ser.* 511 (2014): 012068.

Lambrecht, W. R. L., Segall, B., Methfessel, M. et al., "Calculated elastic constants and deformation potentials of cubic SiC," *Phys. Rev. B* 44 (1991): 3685–3688.

Lange, H., Huczko, A., Byszewski, P. et al., "Spectroscopic study of C$_2$ in carbon arc discharge," *Spectrosc. Lett.* 29 (1996): 1215–1228.

Lange, H., Baranowski, P., Huczko, A. et al., "Optoelectronic control of arc gap during formation of fullerenes and carbon nanotubes," *Rev. Sci. Instrum.* 68 (1997): 3723–3727.

Lee, J.-S., Byeun, Y.-K., Lee, S.-H. et al., "In situ growth of SiC nanowires by carbothermal reduction using a mixture of low-purity SiO$_2$ and carbon," *J. Alloys Compd.* 456 (2008): 257–263.

Li, D., Müller, M. B., Gilje, S. et al., "Processable aqueous dispersions of graphene nanosheets," *Nat. Nanotechnol.* 3 (2008): 101–105.

Lin, H. F., Gerbec, J. A., Sushchikh, M. et al., "Synthesis of amorphous silicon carbide nanoparticles in a low temperature low pressure plasma reactor," *Nanotechnology* 19 (2008): 325601–325608.

Lotya, M., King, P. J., Khan, U. et al., "High-concentration, surfactant-stabilized graphene dispersions," *ACS Nano* 4 (2010): 3155–3162.

Lou, Z. et al., "Synthesis of carbon nanotubes by reduction of carbon dioxide with metal lithium," *Carbon* 41 (2003): 3063–3074.

Macmillan, N. H., "Review: The theoretical strength of solids," *J. Mater. Sci.* 7 (1972): 239–254.

Malard, L. M., Pimenta, M. A., Dresselhaus, G. et al., "Raman spectroscopy in graphene," *Phys. Rep.* 473 (2009): 51–87.

Manning, T. J., Mitchell, M., Stach, J. et al., "Synthesis of exfoliated graphite from fluorinated graphite using an atmospheric-pressure argon plasma," *Carbon* 37 (1999): 1159–1164.

McAllister, M. J., Li, J.-L., Adamson, D. H. et al., "Single sheet functionalized graphene by oxidation and thermal expansion of graphite," *Chem. Mater.* 19 (2007): 4396–4404.

Meng, G. W., Zhang, L. D., Mo, C. M. et al., "Preparation of β-SiC nanorods with and without amorphous SiO$_2$ wrapping layers," *J. Mater. Res.* 13 (1998): 2533–2538.

Milewski, J. V., Gac, F. D., Petrovic, J. J. et al., "Growth of beta-silicon carbide whiskers by the VLS process," *J. Mater. Sci.* 20 (1985): 1160–1166.

Mukasyan, A. S., Lin, Y.-C., Rogachev, A. S. et al., "Direct combustion synthesis of silicon carbide nanopowder from the elements," *J. Amer. Ceram. Soc.* 96 (2013): 111–117.

Myriano, H., Oliveira, Jr., Schumann, T. et al., "Mono- and few-layer nanocrystalline graphene grown on Al$_2$O$_3$ (0001) by molecular beam epitaxy," *Carbon* 56 (2013): 339–350.

Novoselov, K. S., Geim, A. K., Morozov, S. V. et al., "Electric field effect in atomically thin carbon films," *Science* 306 (2004): 666–669.

Ollivier, M., Latu-Romain, L., Salem, B. et al., "Integration of SiC-1D nanostructures into nano-field effect transistors," *Mater. Sci. Semicond. Process.* 29 (2015): 218–222.

Poornima, V., Thomas, S. & Huczko, A., "Epoxy resin/SiC nanocomposites: Synthesis and characterization," *Composites* 10 (2010): 11–14.

Rai, P., Park, J.-S., Park, G.-G. et al., "Influence of carbon precursors on thermal plasma assisted synthesis of SiC nanoparticles," *Adv. Powder Technol.* 25 (2014): 640–646.

Ramanthan, T., Abdala, A. A., Stankowich, S. et al., "Functionalized graphene sheets for polymer nanocomposites," *Nat. Nanotechnol.* 3 (2008): 327–331.

Reina, A., Son, H., Jiao, L. et al., "Transferring and identification of single- and few-layer graphene on arbitrary substrates," *J. Phys. Chem. C* 112 (2008): 17741–17745.

Rich, R., Stelmakh, S., Patyk, J. et al., "Bulk modulus of silicon carbide nanowires and nanosize grains," *J. Mater. Sci.* 44 (2009): 3010–3013.

Robertson, J., "Focus on carbon electronics: Graphene—Nanotubes—Diamond," *Phys. Status Solidi RRL* 3 (2009): A77–A78.

Ryan, C. E., Berman, I., Marshall, R. C. et al., "Vapor–liquid–solid and melt growth of silicon carbide," *J. Cryst. Growth* 1 (1967): 255–262.

Saulig-Wenger, K., Cornu, D., Chassagneux, F. et al., "Direct synthesis of β-SiC and h-BN coated β-SiC nanowires," *Solid State Commun.* 124 (2002): 157–161.

Schoen, D. T., Schoen, A. P., Hu, L. et al., "High speed water sterilization using one-dimensional nanostructures," *Nano Lett.* 10 (2010): 3628–3632.

Setaka, N. & Ajiri, K., "Influence of vapor-phase composition on morphology of SiC single-crystals," *J. Am. Ceram. Soc.* 55 (1972): 540–546.

Seyller, T., "Electronic properties of SiC surfaces and interfaces: Some fundamental and technological aspects," *Appl. Phys. A* 85 (2006): 371–385.

Soszyński, M., Dąbrowska, A., Bystrzejewski, M. et al., "Combustion synthesis of one-dimensional nanocrystalline silicon carbide," *Cryst. Res. Technol.* 45 (2010): 1241–1244.

Soszyński, M., Dąbrowska, A. & Huczko, A., "Spontaneous formation and characterization of silicon carbide nanowires produced via thermolysis," *Phys. Status Solidi B* 248 (2011): 2708–2711.

Soum-Glaude, A., Thomas, L., Tomasella, E. et al., "Selective effect of ion/surface interaction in low frequency PACVD of SiC:H films: Part A—Gas phase considerations," *Surf. Coat. Tech.* 200 (2005): 855–858.

Sun, Y., Cui, H., Pang, C. L. et al., "Structural investigation of epitaxial 2H-SiC–α-Al$_2$O$_3$ 1-D heterostructures," *CrystEngComm* 15 (2013): 6477–6482.

Sundaresan, S. G., Davydov, A. V., Vaudin, M. D. et al., "Growth of silicon carbide nanowires by a microwave heating-assisted physical vapor transport process using group VIII metal catalysts," *Chem. Mater.* 19 (2007): 5531–5537.

Szala, M., "SHS production of carbon nanotubes," *Int. J. Self-Propag. High-Temp. Synth.* 17 (2008): 106–111.

Tong, L. R. & Reddy, R. G., "Thermal plasma synthesis of SiC nano-powders/nano-fibers," *Mater. Res. Bull.* 41 (2006): 2303–2310.

Vennekamp, M., Bauer, I., Groh, M. et al., "Formation of SiC nanoparticles in an atmospheric microwave plasma," *Beilstein J. Nanotechnol.* 2 (2011): 665–673.

Wagner, R. S. & Ellis, W. C., "Vapor–liquid–solid mechanism of single crystal growth," *Appl. Phys. Lett.* 4 (1964): 89–90.

Wang, X. H., Eguchi, K., Iwamoto, C. et al., "High-rate deposition of nanostructured SiC films by thermal plasma PVD," *Sci. Technol. Adv. Mater.* 3 (2002): 313–317.

Wang, X. H., Eguchi, K., Iwamoto, C. et al., "Ultrafast thermal plasma physical vapor deposition of thick SiC films," *Sci. Technol. Adv. Mater.* 4 (2003): 159–165.

Wang, Y., Voronin, G. T., Zerda, W. et al., "SiC–CNT nanocomposites: High pressure reaction synthesis and characterization," *J. Physics Cond. Matter* 18 (2006): 275–282.

Wang, F., Qin, X., Zhu, D. et al., "FeCl$_2$-assisted synthesis and photoluminescence of β-SiC nanowires," *Mater. Sci. Semicond. Process.* 29 (2015): 155–160.

Wong, E. W., Sheehan, P. E., Lieber, C. M. et al., "Nanobeam mechanics: Elasticity and toughness of nanorods and nanotubes," *Science* 277 (1997): 1971–1975.

Wong, K. W., Zhou, X. T., Au, F. C. K. et al., "Field-emission characteristics of SiC nanowires prepared by chemical-vapor deposition," *Appl. Phys. Lett.* 75 (1999): 2918–2920.

Wright, N. G., Horsfall, A. B. & Vassilevski, K., "Prospects for SiC electronics and sensors," *Mater. Today* 11 (2008): 16–21.

Wu, K. H., Huang, W. C., Wang, G. P. et al., "Effect of pH on the magnetic and dielectric properties of SiO$_2$/NiZn ferrite nanocomposites," *Mat. Res. Bull.* 40 (2005): 1822–1831.

Wu, Z.-S., Ren, W., Gao, L. et al., "Synthesis of high-quality graphene with a predetermined number of layers," *Carbon* 47 (2009): 493–499.

Xia, Y., Yang, P., Sun, Y. et al., "One-dimensional nanostructures: Synthesis, characterization, and applications," *Adv. Mater.* 15 (2003): 353–359.

Yeh, C.-L., *Encyclopedia of Materials: Science and Technology* (Elsevier, Oxford, UK, 2010).

Yu, Q. K., Lian, Y., Siriponglert, S. et al., "Graphene segregated on Ni surfaces and transferred to insulators," *Appl. Phys. Lett.* 11 (2008): 113103–113106.

Yumura, M., "Methods for preparation of carbon nanotubes," *Kagaku Kogaku* 63 (1999): 321–324.

Zhang, Y., Wang, N., He, R. et al., "Synthesis of SiC nanorods using floating catalyst," *Solid State Commun.* 118 (2001): 595–598.

Zhou, D. & Seraphin, S., "Production of silicon carbide whiskers from carbon nanoclusters," *Chem. Phys. Lett.* 222 (1994): 233–238.

# Activated Carbon Nanogels

Zhian Zhang

This chapter is served to generally introduce activated carbon nanogels, including their preparation and characterization, their chemical and physical activation methods, and the application of activated carbon nanogels.

## 8.1  Introduction of Carbon Gels

Carbon gels are novel porous carbon materials with an interconnected network structure that have attracted considerable attention over the past several decades (Pierre & Pajonk 2002). Because the porous structure of the carbon gel is nanosized, carbon gels are also called *carbon nanogels* in this chapter. Carbon gels have been prepared by drying and carbonizing organic gels synthesized by traditional sol–gel chemistry reaction. Their textural and structural characteristics can be controlled according to the synthesis protocol. As a result, the carbon nanogels keep the unique porous texture which they had in the organic gels. With a high specific surface area (SSA), low apparent density, and tunable continuous porosity, carbon nanogels can be used in many fields, such as an effective absorbent, catalyst support, media for gas storage, and electrode materials for batteries and supercapacitors, etc. (Moreno-Castilla & Maldonado-Hódar 2005).

Carbon aerogels (CAs), carbon xerogels, and carbon cryogels are three kinds of carbon nanogels, mainly depending on the drying method of organic gels (corresponding to supercritical drying, evaporative drying, and freeze-drying) (Elkhatat & Al-Muhtaseb 2011). Among them, CAs are extensively reported, which can be commonly synthesized via the carbonization of organic gels and subsequent extraction of pore solvent or template by a supercritical fluid (carbon dioxide, acetone, or ethanol) where structure collapse can be avoided because the surface tensions are eliminated; thus, no obvious shrinkage occurs. In principle, the dried products are called *carbon aerogels* and retain a majority of the original gel texture. However, supercritical drying with carbon dioxide is extremely expensive. In contrast, carbon xerogels are prepared after organic gels are dried under ambient conditions, giving rise to obvious structural changes due to capillary forces during evaporation of solvent. In consequence, low-density

CAs make porosities of up to 90–98 vol%, whereas carbon xerogels show only 50 vol%. Another way of drying is freeze-drying, which is a less expensive drying process. The dried materials are the so-called carbon cryogels, which have been successfully obtained by the sol–gel polycondensation of resorcinol with formaldehyde followed by a freeze-drying technique after the solvent exchange and pyrolysis in an inert atmosphere. However, the freeze-drying technique always leads to brittle or broken samples, making them particularly difficult to obtain monoliths. Therefore, the methods wherein the organic gels are dried have a significant effect on the properties of the resultant carbon gels. CAs and cryogels have higher pore volumes, whereas capillary forces unavoidably reduce the pore volumes of xerogels. Moreover, CAs are quite costly because the synthesis usually contains supercritical carbon dioxide. Carbon cryogels are comparatively cheaper; nevertheless, their preparation processes need longer periods, commonly at least 5 days. Meanwhile, the pore structure of carbon cryogels is generally coarser.

Specifically, the most common CAs were first developed at the Lawrence Livermore National Laboratory in the 1990s, which are a special kind of porous carbons that display many fascinating properties, including low bulk density, abundant porosities, large Brunauer–Emmett–Teller (BET) surface areas (500–1000 $m^2 g^{-1}$), and large mesopore volume (>0.89 $cm^3 g^{-1}$). Various final forms of CAs with a tunable pore texture have monoliths, beads, disks, microspheres, or thin films according to different synthesis and processing conditions. CAs are typically synthesized via three sequential steps: (i) polymerization of molecular precursors into an organic gel, i.e., gel synthesis; (ii) supercritical drying of the organic gels, where the solvent is eliminated; and (iii) carbonization of the organic gels at an elevated temperature (ca. 1000°C) to produce the final CAs. At the first stage, the sol–gel polymerization takes place in an aqueous solution, wherein molecular reagents combine to transform into cross-linked organic gels. The process parameters (concentration, pH, curing time, temperature, etc.) can affect the nucleation, growth, and interconnectivity of the primary particles and finally determine the structure of the organic gels. At the second stage, supercritical drying can minimize the surface tension and avoid the collapse of the pores and porosity on the account of the inherent weakness of the tenuous links within the gel, therefore retaining the formal skeletal structure of the organic gels. During the process of carbonization, the oxygen and hydrogen functionalities of the CAs are lost, forming a pure carbon structure. An obvious disadvantage of the typical preparation of CAs is that the organic gels usually require supercritical drying prior to the carbonization to obtain the final CAs (Robertson & Mokaya 2013). Often, the CAs are then activated in order to get a high surface area carbon. The most widely reported organic gels are what are synthesized from resorcinol (1,3-dihydroxylbenzene) ($C_6H_4(OH)_2$) (R) and formaldehyde (HCHO) (F) mixtures dissolved in water (W) with either a basic or an acid catalyst (C). The physical and chemical properties of CAs have been widely reported during the past several decades. Distinctive optical, thermal, mechanical, and electrical properties of CAs and the corresponding organic precursors have been extensively explored. The structure and properties of the CAs and the organic gels are mainly determined by three factors: (i) the molar ratio of the monomer building block (resorcinol) to the catalyst (sodium carbonate) (R/C ratio) of the starting solution, (ii) the pyrolysis temperature, and (iii) the activation process. The microstructure of CAs varies with the molar ratio R/C. When the R/C ratio is low, polymeric aerogels will form, which contain small particles in a well-connected configuration. With the increase of R/C ratios, colloidal aerogels consisting of well-defined spherical particles will develop. Carbonization of the dried aerogel is achieved by heating the material in $N_2$ or Ar atmosphere at temperature ranging from 500°C to 2000°C. The porosity of the CAs is influenced by pyrolysis temperature. When the temperature is low, the macropore volume reduces, while the mesopore volume increases, owing to the material shrinkage, and the micropore volume and surface area increase, owing to gas evolution during the process of carbonization. At higher carbonization temperature, all the parameters are inclined to reduce. A partial graphitization of different areas of the CAs occurs at quite a high carbonization temperature (above 2000°C). Moreover, the parameters of activation (activation agent, activation method, activation time, activation time, etc.) will have an effect on the SSA, pore volume, and pore size distribution of the activated carbon nanogels. This will be further explained in the following section of this chapter.

Although CAs and activated carbon aerogels (ACAs) have outstanding porous properties, their wide commercial application is limited, because the preparation technique, including supercritical drying and the starting reactants (i.e., resorcinol), is not a low-cost method. Not only does it require a multi-step procedure, but it also yields a low production rate, resulting in a substantial amount of time and energy consumption. Thus, great efforts have been devoted to bring the high cost down either by using cheaper precursors or by substituting the supercritical drying step or using microwave radiation heating to shorten the process.

## 8.2 Activation of Carbon Nanogels

Compared with other porous carbon materials, including activated carbon (up to 2500 m² g⁻¹), the SSAs of organic gels and carbon nanogels are not very large (about 200 and 1000 m² g⁻¹, respectively); there-fore, the practical application of carbon nanogels has been impeded. To further enlarge surface area values and design abundant pore texture, such as pore volume and pore size distribution, various acti-vation strategies are proposed to generate highly porous carbon materials with new pore and expanded pore size. The activation of carbon materials can be carried out as (i) physical activation, which refers to the carbonization of the carbon precursors at the temperature of 600°C–1200°C with carbon dioxide ($CO_2$), water stream, and oxygen plasma, and (ii) chemical activation, which is involved in impregnating chemical activation agents (such as phosphoric acid, NaOH, or KOH) into a carbon precursor, which is then disposed at high temperature (400°C–900°C) (Wang & Kaskel 2012). A high surface area up to 3000 m² g⁻¹ can be obtained. In general, chemical activation possesses outstanding advantages like lower activation temperature, higher yield, less activation time, and higher SSA. However, there are some disadvantages, such as the corrosiveness of the chemical agents (e.g., KOH) and the necessary washing process to clear the chemical agents. The combination of both physical activation and chemical activation is also proposed. The SSA and pore structure of the activated carbon nanogels can be tailored by varying activation conditions such as activation agents, ratio of agents to raw materials, activation temperature, gas flow rate, heating rate, and activation time. This content will be further explained in the following section.

### 8.2.1 Physically Activated Carbon Nanogels

The activation of carbon nanogels can be performed via physical activation process in the presence of an oxidizing gas, at the temperature of 350°C–1000°C for 1–7 h to develop the pore structure and the SSA. After the activation step, inert gas (e.g., pure $N_2$) may be flowed through the dried materials for 2–5 h to replace the oxidizing gas and cool down the product to the room temperature. Air, $O_2$, $CO_2$, steam ($H_2O$), or their mixtures are most commonly used as the oxidizing gas (Sevilla & Mokaya 2014). When using physical activation with $CO_2$, initially, the tarry pyrolysis off-products trapped within the pores is burned off by the active oxygen in the activating agent, resulting in the opening of some closed pores. The microporous structure is then formed when the oxidizing agent burns the more reactive areas of the carbon skeleton away, producing CO and $CO_2$, with the extent of the burn-off relying on the nature of the employed gas and activation temperature. $CO_2$, air, $O_2$, and steam are employed as activating agents. The chemical and physical properties of physically activated nanogels mainly depend on the precursor, the oxidizing agent used, activation temperature, activation time, and activation degree. Depending on these factors, both activated nanogels from moderate to high porosity, as well as with variable surface chemistry, can be obtained. It is a universal trend that the higher the temperature of activation, the larger the devel-oped porosity is. Previous research shows that the activated carbon nanogels with a surface area of up to 3000 m² g⁻¹ can be obtained after activation with $CO_2$ for several hours. Moreover, the physical activation also affects the surface group. For example, physical activation by air can generate mainly surface pheno-lic and carbonyl groups, whereas oxygen plasma activation, in which carbon gels are exposed to oxygen plasma, can form oxygen groups selectively on the external surface of carbon nanogels.

Job et al. (2005) synthesized carbon xerogels by polycondensation of resorcinol–formaldehyde (RF) in water. Then, the physical activation process using $CO_2$ as the activation agent was applied to the carbon xerogels. As a result, the SSA increased from 600 to 2000 $m^2\,g^{-1}$. Baumann et al. (2008) first synthesized unactivated CAs derived from pyrolysis of resorcinol and formaldehyde gel. At the first stage, the resorcinol and formaldehyde solution was dissolved in water, followed by addition of glacial acetic acid. The obtained organic gels were dried with supercritical $CO_2$. The activated carbon gels were obtained by carbonizing at 1050°C. The activated carbon nanogels with the BET surface area of 3125 $m^2\,g^{-1}$ and the total pore volume of 1.88 ± 0.09 $cm^3\,g^{-1}$ were achieved under flowing $CO_2$ as the agents at 950°C. However, the BET surface area and the total pore volume of the unactivated carbon nanogels are only 463 $m^2\,g^{-1}$ and 0.195 ± 0.01 $cm^3\,g^{-1}$. Chang et al. (2013) also prepared carbon xerogels with the surface area of 3419 $m^2\,g^{-1}$ derived from RF with an R/F ratio of 1:2 and an R/C ratio of 10:1, followed by solvent exchange, pyrolysis, and carbon dioxide activation for 3 h. The intense weight loss of the prepared carbon xerogels after pyrolysis and carbon dioxide activation is 88%–96% compared with that of conventional RF-derived xerogel (50%–60%). Fang et al. (2005) also prepared ACAs from RF and $NaHCO_3$ as the catalyst, followed by dried RF gel under a $CO_2$ flow at 1000°C for 1 h. The results revealed that after the activation by $CO_2$, many micropores (less than 2 nm in diameter) were generated and the SSA of the activated carbon nanogels was approximately twice that without activation.

## 8.2.2 Chemically Activated Carbon Nanogels

Chemically activated carbon nanogels are prepared via heating treatment of a mixture of carbon nanogels and chemical activation agent at a typical temperature between 450°C and 900°C. Many activating agents are found in the literatures. KOH, NaOH, $H_3PO_4$, and $ZnCl_2$ are the most commonly used agents. Strong bases like KOH and NaOH do not commonly serve as dehydrating agents, but as oxidants, whereas $ZnCl_2$ (Lewis acid) and $H_3PO_4$ (Brönsted acid) act as dehydrating agents. Compared with the physical activation, there are many advantages for chemical activation and they are as follows: (i) the process usually relates to only one step, (ii) lower activation temperatures are employed, (iii) much higher carbon yield is achieved, (iv) activated carbon nanogels with very high surface area are obtained, and (v) the microporosity can be well developed, controlled, and tailored to be narrowly distributed. The properties enable the carbon nanogels to be applied in various fields such as gas storage ($H_2$, $CH_4$, or $CO_2$) and power storage (battery or supercapacitor). However, there are disadvantages, including the corrosiveness of the chemical agents and the necessary washing process to remove the chemical agents.

Although KOH chemical activation is proven to be a universal way to develop the pore structure in carbons, the mechanism of chemical activation has not been fully understood due to the process complexity and variable process parameters such as the ratio of KOH/C, activation time, and activation temperature. Both carbon gels and activation process parameters also have significant influence on the pore microstructure and surface chemistry. In general, the reaction of carbon nanogels and KOH begins with solid–solid reactions and undergoes solid–liquid reaction including the reduction of potassium (K) compound to generate metallic K, the oxidation of carbon to carbon oxide and carbonate, and other reactions among various active intermediates. The carbon skeleton is etched to form pores because of the oxidation of the carbon into carbonate ions and intercalation of the resulting potassium compounds, which are to be removed during subsequent washing steps. The production of $CO_2$ from the decomposition of $K_2CO_3$ at temperatures above 700°C can devote to further developing porosity through carbon gasification. Some researches have revealed that the higher the reactivity of the precursor, the lower the temperature required to trigger gasification; the higher the degree of gasification incurred by the $CO_2$ evolved from $K_2CO_3$, the larger the resultant porosity is. Typically, the KOH activation process is as follows: KOH is mixed with the carbon gels either physically or by impregnation with KOH aqueous solution. The KOH/carbon mass ratio is generally 2–5. An extra step with low-temperature preheating process for evaporation of water at low temperature of 300°–500°C is necessary when the impregnation method is employed. The resulting mixture is then carbonized at high temperatures of 650°C–1000°C.

After cooling, the product is completely washed with either aqueous acid solution or just water in order to clear any soluble impurities, and then dried to get the activated carbon nanogels.

The amount of chemical activation agent and the activation temperature greatly affect the textural properties of the carbon nanogels. In general, the SSA, micropore volume, and the total pore volume increase with increasing the mass ratio of KOH/carbon. Boosting the activation temperature has the similar effect as enlarging the KOH amount has. Besides affecting the surface area and pore volume, the carbon burn-off during activation is another important factor to be considered. High carbon burn-off increases cost and gas emissions, which are undesirable from both economic and environmental perspectives. More chemical activation agents and higher activation temperature can enlarge the surface area of mesoporous carbons, but at the cost of increasing carbon burn-off, therefore imposing a trade-off between the desired material properties and economic/environmental impacts (Zhai et al. 2011).

Conceição et al. (2009) first prepared CAs with the ratio of R/F = 0.5 and resorcinol/catalyst (sodium carbonate) ratio of 100, and then synthesized a series of activated carbon nanogels with different $H_3PO_4$/gels ratios of 1, 2, or 3. The porosity structure can be controlled by varying the $H_3PO_4$/gels ratio. Increasing this ratio results in higher pore volume and larger size, so the activated carbon nanogels with micro/mesopore volumes as high as 1.23 $cm^3\,g^{-1}$ with pore widths of up to 7 nm can be achieved.

KOH chemical activation, which is used to tailor the pore structure of carbon nanogels, has been reported in literatures. For example, Wang et al. (2008) prepared RF CAs and obtained activated carbon nanogels via KOH activation processes. The typical process is as follows: about 2 g of CAs was mixed with KOH at a mass ratio of KOH to CAs (K/CA ratio) ranging from 0.5/1 to 7/1, and 5 mL of ethanol was added to dissolve the KOH. The resultant was then dried at 110°C and carbonized in a tubular furnace at 900°C for 3 h with the heating rate of 5°C/min under flowing nitrogen. After the mixture is cooled down to room temperature, the product was taken out and washed with 10% HCl and distilled water. Finally, the activated carbon nanogels were obtained. The SSA of the activated carbon nanogels is enlarged up to 1628 $m^2\,g^{-1}$, while the SSA of the unactivated carbon nanogel is only 607 $m^2\,g^{-1}$. The $S_{micro}$ is increased to 647 from 326 $m^2\,g^{-1}$, the $S_{meso}$ is increased up to 969 from 310 $m^2\,g^{-1}$, and the total pore volume is up to 1.92 from 1.29 $cm^3\,g^{-1}$.

### 8.2.3 Combination of Physical and Chemical Activation

In general, a considerable amount of microporosity is developed by chemical activation (typically KOH), whereas physical activation (typically $CO_2$) maintains high fractions of mesopores and generates a slight increased mesopore size. A small fraction (40%–50%) of the total surface area is linked with micropores. This difference may be caused by the large dimension of $CO_2$ molecule, which restrains the accessibility of $CO_2$ from getting into micropores, thus leading to an "etching" of mesopores and a low microporosity. Two steps of activation processes consisting of a chemical activation step (normally with KOH or $ZnCl_2$) followed by physical activation (usually with $CO_2$) have also been used to further enhance the porosity development and tune the pore size distribution of carbon nanogels.

# 8.3 Applications

## 8.3.1 Electrochemical Applications

### 8.3.1.1 Supercapacitor

Supercapacitors are a special kind of capacitor based on charging and discharging at electrode–electrolyte interface of high surface area materials, also named as *electrochemical capacitors*, *ultracapacitors*, or *electrochemical double layer capacitors* (EDLCs), which possess a long cycle life (>100,000 cycles), high efficiency (90%), long shelf life, fast charge/discharge processes (within seconds), and low cost of maintenance. Compared to conventional electrolytic capacitors, supercapacitors can store more capacitance for a given device size. Supercapacitors also have a much higher power density over conventional batteries, thus stimulating wide interests as an alternative or supplement to batteries in the field of energy storage. Currently, supercapacitors are widely used in consumer electronics, power backup systems, industrial power and energy management, and electrical/hybrid electric vehicles and other devices.

In principle, supercapacitors can be divided into three categories based on their different modes of energy storage and construction: EDLC, in which the capacitance arises from electrostatic attraction between ions and the charged surface of an electrode, such as carbon materials; pseudocapacitor, which is associated with fast and reversible oxidation/reduction (redox) or faradic charge transfer reactions of the electroactive species on the surface of an electrode, such as transition metal oxides and conducting polymers; and a hybrid system incorporating combinations of double layer and pseudocapacitance.

The double layer capacitance at the electrode–electrolyte interface can be described by the Helmholtz equation as

$$C = \varepsilon_r \varepsilon_0 A/d,$$

where $C$ is the capacitance of the EDLC, $\varepsilon_r$ is the dielectric constant of electrolyte, $\varepsilon_0$ is the permittivity of free space, $A$ is the surface area of the electrode, and $d$ is the effective thickness of the electrical double layer (charge separation distance).

For the pseudocapacitor, the charges are strongly related to the electrode potential as

$$C = dQ/dV,$$

where $C$ is the capacitance of the redox supercapacitor, $Q$ is the charge accumulated in the capacitor, and $V$ is the potential. A pseudocapacitor has a much higher capacitance than that of EDLCs, but suffers from lower power density and poorer cyclic stability owing to its poor electrical conductivity and structural collapse during cycling. Despite the wide array of possible materials and device architecture, EDLCs are currently the most developed form of supercapacitors and are extensively used in the commercial application.

Therefore, for the EDLCs, the higher the surface area, the higher the specific capacitance that can be obtained. Carbon materials are currently the most widely utilized active materials in practical EDLCs due to their low cost and various forms such as powders, fibers, composites, monoliths, and tubes. Many reviews have been discussed about both the science and the technology of supercapacitors using carbon-based materials, such as graphene, CNTs, solid activated carbon, and carbide-derived carbons.

Therefore, activated carbon nanogels with abundant porosity from a carbonized organic precursor have attracted much attention in the development of porous carbon materials for supercapacitors, because of their high surface area, electrical conductivity, chemical stability, tunable porosity, etc.

According to the electrolyte used in the EDLCs, there are two kinds of capacitors, aqueous supercapacitor and nonaqueous supercapacitor. The typical specific capacitances for the activated carbon nanogels are 120–250 and 40–130 F g$^{-1}$ in the aqueous electrolyte ($H_2SO_4$, KOH, $Na_2SO_4$) and the nonaqueous electrolyte (1M TEABF$_4$ in propylene carbonate or acetonitrile), respectively.

Previous studies reveal that there is a linear relationship between specific capacitance and SSA for the ACAs with similar pore system. The highest specific capacitance of the ACAs reached is 245 F g$^{-1}$ in KOH electrolytes, 2.3 times that of CAs. Therefore, the activation process can increase the SSA of the carbon material, thus enlarging the specific capacitance of the carbon material. So, the ACA is considered as a promising material for supercapacitors.

For example, Pekala et al. (1998) prepared ACAs and investigated the electrochemical capacitance behavior in 5 M KOH and 0.5 M Et$_4$NBF$_4$/propylene carbonate. After physical activation with $CO_2$ for 3 h, the SSA is enlarged from ~650 to 1200 m$^2$ g$^{-1}$. And the capacitance density is increased from 16 to 26 F cm$^{-3}$. The results reveal that the activation process can also have a significant effect on the specific capacitance of CAs.

Zapata-Benabithe et al. (2013) prepared carbon xerogels microspheres by the sol–gel polymerization reaction of resorcinol (R) with formaldehyde (F) in water using potassium carbonate as catalyst (C). The R/F and R/C molar ratios were 0.5 and 800, respectively. The carbon xerogels were further steam-activated at 840°C using a $N_2$ flow (500 cm$^3$ min$^{-1}$) saturated with water vapor to obtain samples with different burn-off percentages. Electrochemical measurements were also investigated at room temperature using 1 M $H_2SO_4$ as an electrolyte in a typical three-electrode cell.

Chang et al. (2013) reported that a RF condensation reaction catalyzed by acetic acid is used to prepare carbon xerogels for supercapacitor. During the process, the carbon dioxide was introduced for 3 h to activate and obtain the porous carbon with a surface area of 3419 m$^2$ g$^{-1}$. Electrochemical results demonstrate that the specific capacitance is 324.8 F g$^{-1}$ in 0.5 M $H_2SO_4$ aqueous solution at 1 A g$^{-1}$ for a potential window of 0–1 V.

Fang and Binder (2007) synthesized CA derived from pyrolysis of an RF gel. ACA was prepared by carbonization of CA with KOH at 900°C for 4 h. The heating profile was the same as that for the synthesis of CA. The mass ratio of KOH to CA was set as 3:1. After activation, the CA was washed with deionized water until the pH value reached approximately 7. On the basis of the ACAs, surface modification of activated carbon with surfactant vinyltrimethoxysilane can greatly enhance the hydrophobization of carbon materials. After KOH chemical activation, the SSA is improved to 2371 m$^2$ g$^{-1}$ from 592 m$^2$ g$^{-1}$ of CA. It can be seen that the specific capacitance of modified ACAs shows the highest specific capacitance with nonaqueous electrolyte solution (1 M $Et_4NBF_4$/PC). The activated carbon nanogels can deliver better specific capacitance than the activated carbon.

### 8.3.1.2 Lithium Battery

Rechargeable lithium batteries with a high energy density and longer cycle life are urgently needed as the most promising storage technology to satisfy the increasing requirements for portable electronics, electric vehicles, and grid scale stationary storage systems.

#### 8.3.1.2.1 Lithium Ion Batteries (LIBs)

The rechargeable LIB was first introduced as a commercially viable product by the Sony Corporation in 1991 following more than two decades of research in the field. LIBs commonly consist of transition metal oxide ($LiCoO_2$, $LiFePO_4$) as a cathode and carbon as an anode with a nonaqueous electrolyte between ionic conduction. During charging, $Li^+$ is deintercalated from a cathode, passed across the electrolyte, and intercalated between the graphite layers in an anode. The process reverses during discharging. Carbon materials are commonly used as conductive agents in the process of electrode fabrication.

To obtain a network with high electronic and ionic conductivity of $LiFePO_4$ cathode material for its electrochemical reaction, carbon nanogels with abundant porosity as a carbon scaffold are applied in the LIB. First, the carbon nanogels as a good electrical conductor can enhance electronic conductivity for $LiFePO_4$ particles. Second, the mesopores in carbon nanogels can accommodate $LiFePO_4$ particles and at the same time prevent their growth and aggregation. Third, abundant micropores can store sufficient electrolyte for rapid electrochemical reactions.

Zhou et al. (2013) added KOH–ACAs into the preparation process of $LiFePO_4$ materials, in which $LiNO_3 \cdot H_2O$, $Fe(NO_3)_3 \cdot 9H_2O$, and $NH_4H_2PO_4$ served as lithium sources, iron sources, and phosphorus sources, respectively. The KOH activation treatment of CAs is employed to enlarge the micropores that serve as electrolyte reservoirs and contribute to ionic conductivity. The resultant composite electrodes reveal superior rate performance and good cyclic stability owing to the optimized nanostructure.

Graphite has been commonly used as a standard anode material in LIBs because lithium can be inserted/deinserted during discharging and charging, leading to a theoretical specific capacity of 372 mAh g$^{-1}$. Li-storage anode materials, such as alloy and transition metal oxides, have been explored as anode candidates for LIBs due to their high theoretical capacity, but the application of these materials is hampered because of the large volumetric expansion during charge–discharge process and the intrinsic low conductivity. Carbon materials such as CA with a 3D well-connected conductive network structure can serve as a confining buffer to ensure the stability of the structures and to enhance the electrical conductivity during cycling. Transition metal oxides, such as $Fe_2O_3$, $Co_3O_4$, and $ZnMn_2O_4$, have been used to construct oxides/CA hybrid as electrode materials for LIBs.

Liu et al. (2013) prepared $Fe_2O_3$ nanoparticle/CA composite ($Fe_2O_3$/CA) from a CA which is prepared by a sol–gel process by a simple soaking in an $Fe(NO_3)_3$ solution and subsequent heat treatment at 600°C. $Fe_2O_3$/CA-60 exhibited good capacity retentions of 916 and 617 mAh $g^{-1}$ for the 1st and 100th cycles, respectively, which are much better than those of pure $Fe_2O_3$ and CA. The improvement in cycling performance, specific capacity, and rate capability of $Fe_2O_3$/CA is mainly attributed to the synergistic effects of the nanoporous network skeleton of CA and the uniformly dispersed $Fe_2O_3$ nanoparticles.

Similarly, Hao et al. (2013) prepared a $Co_3O_4$/CAs composite, in which $Co_3O_4$ nanoparticle-anchored CA architecture hybrids were used as anode materials for lithium-ion batteries with improved electrochemical properties.

Yin et al. (2014) also designed 3D interconnected spinel $ZnMn_2O_4$/CA hybrids via a low-temperature solvothermal two-step route. The electrochemical performance demonstrated that the developed $ZnMn_2O_4$/CA hybrids displayed considerably outstanding higher coulombic efficiency, better capacity retention, and better rate capability compared with pure $ZnMn_2O_4$ and CA materials. This novel hybrid provided a new route to design and synthesize future electrode materials for applications with high performance LIBs.

It is interesting that CAs with large open pores and high surface area are recently fabricated via pyrolysis of a readily available natural resource, e.g., bacterial nanocellulose (BNC) aerogels. Freeze-drying of the BNC hydrogels is used to preserve the 3D open network structure upon calcination, whereas using Fe(III) improves the yield and H/C ratio. These CAs are explored as anodes in LIB batteries, where it is shown that they deliver superior capacity retention and rate performance compared to other carbon-based materials (Wang et al. 2014b).

### 8.3.1.2.2 *Lithium Sulfur Batteries*

High-energy density rechargeable batteries based on lithium/sulfur (Li/S) system have gained much attention in recent years in promising applications, such as power source for electric vehicles, energy storage devices, and smart grids, due to their high theoretical specific capacity of 1675 mAh $g^{-1}$ and specific energy of 2600 Wh $kg^{-1}$. As a cathode active material, sulfur also has the advantages of nontoxicity and abundance in nature. However, there are still many challenges hindering the practical application of Li/S batteries. Sulfur and its final discharge products ($Li_2S_2$, $Li_2S$) are electrical insulators, which can cause poor electrochemical accessibility, leading to a low utilization of active materials. In addition, polysulfides ($Li_2S_x$, $4 \leq x \leq 8$) produced in discharge/charge processes can dissolve into organic electrolyte and be reduced to lower-order polysulfides at the interface of the lithium anode. These reduced products will migrate back to the cathode in which they may be reoxidized. This process repeatedly takes place, creating polysulfide shuttle, which can cause loss of active materials and low coulombic efficiency of Li/S batteries. To overcome these problems, sulfur should be well dispersed into electric conductive species, so that the polysulfides can be adsorbed on surfaces of the species. In recent years, various porous carbon-based materials, such as active carbon, CNT, mesoporous carbon, graphene, carbon fiber, and carbon sphere, have been applied to enhance the conductivity and confine sulfur or soluble polysulfides. In a typical process, the carbon–sulfur composite can be obtained by heating the mixture of carbon and elemental sulfur up to 155°C for 12 h to facilitate sulfur diffusion into the pore of carbon material.

CAs have been recognized as promising electrode materials owing to their versatile properties, such as relatively large surface area, fine pore size, high porosity, controllable particle size, and especially its outstanding electrical conductivity.

Li et al. (2013) synthesized CAs by ambient pressure drying with the source of formaldehyde, resorcinol, and sodium acetate. By optimizing the ratio of resorcinol and sodium acetate for CAs, the CAs with higher specific area and larger pore volume exhibit superior electrochemical capacitance; the CAs with SSA of 613.8 m$^2$ g$^{-1}$ and pore volume of 0.6124 cm$^3$ g$^{-1}$ is used as a carbon matrix to obtain the sulfur–CA composite, which can deliver the highest initial specific capacity of 1491 mAh g$^{-1}$.

Zhang et al. (2014) prepared CAs based on resorcinol and formaldehyde. Sodium carbonate was used as a catalyst. The obtained CAs with the SSA of 642 m$^2$ g$^{-1}$ and the total pore volume of 1.02 cm$^3$ g$^{-1}$ were mixed with elemental sulfur at 149°C for 6 h in an argon gas flow, and kept at 300°C in various ratios of sulfur/CA (S/CA). The 3D porous S/CA hybrids exhibit significantly improved reversible capacity, high-rate capability, and excellent cycling performance as a cathode electrode for Li/S batteries. The S/CA hybrid with an optimal incorporating content of 27% S shows an excellent reversible capacity of 820 mAh g$^{-1}$ after 50 cycles at a current density of 100 mA g$^{-1}$. Even at a current density of 3.2 C (5280 mA g$^{-1}$), the reversible capacity of 27% S/CA hybrid can still remain at 521 mAh g$^{-1}$ after 50 cycles.

Jiang et al. (2013) synthesized ACAs with high bimodal porosity obtained by KOH at 800°C for 120 min using 4:1 KOH/C mass ratio under argon atmosphere. Before KOH activation, CAs showed a predominant peak centered at 14 nm, while ACAs exhibited a bimodal pore size distribution (one large peak centered at 2 nm and another peak centered at 18 nm) after KOH activation. When the prepared CAs were applied in Li/S battery, the initial discharge capacities for S–ACA composite and S–CA composite are 1493 and 1269 mAh g$^{-1}$ at the rate of 0.2 C, respectively. After 100 cycles, the reversible specific capacitances of 528 and 412 mAh g$^{-1}$ were kept. The additional micropores played a significant role in absorbing the polysulfides and restraining the shuttle phenomenon, consequently improving the electrochemical performance of Li/S batteries. Compared with single-peak CAs, the cathodes based on ACAs, with high bimodal porosity, exhibited better electrochemical reversibility and less severe polysulfide shuttle.

### 8.3.1.2.3 Lithium Air Batteries

As one of the most promising secondary battery technologies, rechargeable lithium air batteries (or lithium O$_2$ [Li–O$_2$] batteries) have attracted extensive attention owing to their high energy density of 5200 Wh kg$^{-1}$ (3212 Wh kg$^{-1}$ as O$_2$ included), which almost equals that of gasoline and is 5 to 10 times higher than those of the state-of-the-art LIBs. Li–O$_2$ batteries consist of Li as anode and the O$_2$ electrode with oxygen as a cathode active material. Generally, the O$_2$ electrode is a carbon electrode having a porous structure in which several electrochemical and transport processes simultaneously take place. As the skeletal material for the cathode of Li–O$_2$ batteries, carbon materials have a significant impact on the cell performance, not only supplying room for the insoluble discharge product (Li$_2$O$_2$) deposition, but also acting as the transport way of oxygen and Li$^+$ during the charge–discharge process, which is related to the formation or decomposition of Li$_2$O$_2$. Therefore, the pore structure design and optimum of the carbon materials is very important, and a large pore volume is essential for an excellent carbon material used in Li–O$_2$ batteries.

Mirzaeian and Hall (2009) synthesized ACA by polycondensation of resorcinol (R) and formaldehyde catalyzed by sodium carbonate followed by carbonization of the resultant aerogels in an inert atmosphere and further activation with CO$_2$ agent. With increasing the activation temperature, the burning-off, the surface area, and the total and mesopore volumes of ACAs are increased (as shown in Table 8.1).

**TABLE 8.1** Porous Parameters of CRF003 Carbon Samples Activated at Different Degrees of Burn-Off

| Sample | Time (min) | Burn-Off (%) | $S_{BET}$ (m$^2$/g) | $V_{total}$ (cm$^3$/g) | $V_{micro}$ (cm$^3$/g) | $V_{meso}$ (cm$^3$/g) | % $V_{micro}$ | % $V_{meso}$ | $D_{avg}$ (nm) |
|---|---|---|---|---|---|---|---|---|---|
| CRF003 | – | 0 | 647 | 1.2245 | 0.2101 | 1.0144 | 17 | 83 | 21.65 |
| ACRF003-1073K | 150 | 9.7 | 732 | 1.200 | 0.253 | 0.947 | 21 | 79 | 16.77 |
| ACRF003-1123K | 125 | 21 | 891 | 1.336 | 0.313 | 1.023 | 24 | 76 | 15.93 |
| ACRF003-1173K | 120 | 54 | 1497 | 1.942 | 0.448 | 1.494 | 23 | 77 | 15.43 |
| ACRF003-1223K | 60 | 58 | 1506 | 1.965 | 0.434 | 1.531 | 22 | 78 | 15.01 |
| ACRF003-1273K | 30 | 65 | 1585 | 2.075 | 0.488 | 1.587 | 23 | 77 | 14.76 |
| ACRF003-1323K | 20 | 67 | 1687 | 2.195 | 0.463 | 1.732 | 21 | 79 | 14.23 |

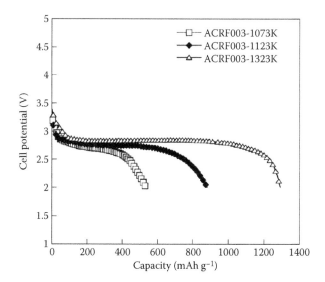

**FIGURE 8.1** Discharge capacities of ACRF003 carbons discharged at 20 mA/g.

It is found that the cell performance (i.e., discharge capacity and discharge voltage) depended on the morphology of carbon, and a combined effect of pore volume, pore size, and surface area of carbon affected the storage capacity. An $Li/O_2$ cell using the carbon with the largest pore volume (2.195 $cm^3$ $g^{-1}$) and a wide pore size (14.23 nm) showed a specific capacity of 1290 mAh $g^{-1}$ when the activation temperature is increased to 1323 K and duration time is 20 min, as demonstrated in Figure 8.1.

Wang et al. (2014a) synthesized CAs by RF polycondensation and carbonization method and prepared a novel highly conductive dual-pore CA-based air cathode which was fabricated to construct a Li–$O_2$ battery exhibiting 18 to 525 cycles in the Li bis(trifluoromethylsulfonyl)imide/sulfolane electrolyte at a current density varying from 1.00 to 0.05 mA $cm^{-2}$, accompanied by a high energy efficiency of 78.32%.

## 8.3.2 Hydrogen Storage

Hydrogen is an ideal clean energy carrier since $H_2$ is both carbon-free and pollution-free and its only burning product is water. Hydrogen is one of the most abundant natural resources in the universe and makes up 75% of normal matters by mass, which provides rich sources for $H_2$ generation. However, storing $H_2$ in an efficient, economical, and safe way is one of the main challenges. In general, hydrogen can be stored either as liquid $H_2$ via liquefaction, in compressed gas cylinders, or in solid state as chemical hydrides or absorbed in porous materials. Among the various approaches to $H_2$ storage, various materials have been explored as the candidates for $H_2$ storage including metal hydrides, metal-organic frameworks, porous polymer, and porous carbon due to the fast kinetics, excellent cyclability, and high adsorption capacity.

Since carbon material was first applied to hydrogen storage in the 1980s, various carbon materials, including activated carbon, CA, CNT, template carbon, carbon fiber, and even graphene have been extensively explored as hydrogen storage media. In particular, carbon nanogel-based porous materials have been widely studied as potential sorbents for hydrogen stores due to their high surface area, large pore volume, light weight, good chemical stability, and the ease with which their porosity can be tailored. High effective hydrogen sorbents should satisfy the requirements as follows: (1) enough large SSA that exposes a large number of sorption sites for adatom or admolecule interaction and (2) sufficiently deep potential wells so that the storage material can be utilized at reasonable operating temperatures. In general, the $H_2$ storage capacity of porous carbon material is proportional to the SSA and micropore volume. Thus, a higher SSA and a well-developed micropore size distribution of the activated carbon nanogels used as sorbents are superiorly advantageous for enhancing $H_2$ adsorption capacity.

In general, the amount of surface excess hydrogen adsorbed on porous carbons at 77 K and ~3.5 MPa varies linearly with BET surface area, and the gravimetric uptake is ~1 wt% $H_2$ per 500 $m^2$ $g^{-1}$ of surface area. Moreover, the size and the shape of the pores in hydrogen physisorbents play a critical role in hydrogen uptake. Activation is commonly used to tailor the pore structure and increase the surface area of the carbon gels, especially creating new micropores (pores smaller than 2 nm) in the structure.

Tian et al. (2009) prepared KOH-catalyzed CAs derived from resorcinol and furfural, displaying a hydrogen uptake of 5.2 wt% and a micropore volume of $1.06 \pm 0.02$ $cm^3$ $g^{-1}$. More recently, Tian et al. (2011b) prepared organic aerogel from acetic acid-catalyzed resorcinol and furfural, which were then further carbonized in nitrogen and activated in $CO_2$. Hydrogen sorption measurements were performed on a Sieverts manometric apparatus at 77 K with a maximum pressure of 6.5 MPa. The enhanced hydrogen storage capacity of $5.4 \pm 0.3$ wt% under 4.6 MPa, surface area of 2617 $m^2$ $g^{-1}$, and micropore volume of 0.92 $m^2$ $g^{-1}$ were obtained.

Robertson and Mokaya (2013) prepared CA from resorcinol and formaldehyde, then obtained ACA by mixing KOH and CA at KOH/carbon ratio of 4/1. The surface areas of the CA and the ACA are 508 and 1980 $m^2$ $g^{-1}$, respectively. The pore volume is enlarged from 0.68 to 1.92 $cm^3$ $g^{-1}$. The KOH activation generates abundant micropores; thus, the proportion of microporosity is up to 87%. The obtained ACAs exhibit 3.5–4.3 wt% hydrogen storage at −196°C and 20 bar. The hydrogen storage density of the ACAs is up to 16.2 $\mu$mol $H^2$ $m^{-2}$.

However, hydrogen binding energy between the absorbent and the hydrogen molecules is low, typically on the order of 3–6 kJ $mol^{-1}$, in the absence of relatively strong polarizing centers. The performance of these modified carbon gels with the incorporation of metal nanoparticles into the carbon framework as the next-generation hydrogen storage materials is evaluated. An increased enthalpy due to doping can benefit the hydrogen adsorption properties in a near-room-temperature physisorption process.

Tian et al. (2011a) also prepared cobalt-doped CAs with high surface area (~1667 $m^2$ $g^{-1}$) and large micropore volume (~0.6 $cm^3$ $g^{-1}$), exhibiting 4.38 wt% $H_2$ uptake under 4.6 MPa at −196°C.

Substitutional doping of carbon with nitrogen or other light elements has also been presented as a promising route toward increasing hydrogen binding energy in these sorbent materials. Kang et al. (2009) synthesized nitrogen-doped carbon xerogels from carbon carbonization of resorcinol–formaldehyde polymer in an ammonia atmosphere. The maximum hydrogen uptake of 3.2 wt% was obtained at −196°C for the nitrogen-doped carbon xerogels (4.5 wt%) with high surface area of 1602 $m^2$ $g^{-1}$.

Beyond their use as hydrogen sorbents, CA scaffolds used for complex hydride confinements for reversible hydrogen storages have been recently investigated and have been shown to improve the rates of both hydrogenation and dehydrogenation for these materials. Previous results revealed that several groups have proved the promise of the CA scaffolding approach with a number of different hydrogen storage materials. For example, Gosalawit-Utke et al. (2013) proposed nanoconfined 2LiBH$_4$–MgH$_2$–TiCl$_3$ in CA scaffold for reversible hydrogen storage, exhibiting good hydrogen storage performance of the material.

### 8.3.3 Catalyst

Carbonaceous materials present some advantages in catalysis such as stability in acid or basic media, inertness, high electrochemical conductivity, hydrophobic character, low cost, mechanical robustness, and easy recuperation of the active phase from spent catalysts. Nanostructured carbon gels are prepared by sol–gel polymerization of resorcinol–formaldehyde. The flexibility of the sol–gel process permits the control of the morphology, porosity, and surface chemistry of carbon gels and allows carbon gels to be fitted as catalyst or catalyst support. Therefore, the carbon gels have attracted much attention in the catalyst field. Carbon nanogels have been used not only as catalyst supports in fuel cells, using metals like Pt and Ru, but also as catalysts in advanced oxidation processes, such as wet air oxidation, and in fine chemical applications.

There are four critical steps in the synthesis: preparation of the initial solution (monomers, concentration, catalyst, pH, etc.), polymerization and curing, drying, and, finally, carbonization and/or activation (temperature, time, activating agents, etc.). By controlling different process parameters, the physicochemical properties of pure carbon gel as supports or doped-carbon are tailored.

Mager et al. (2014) presented functional oxidized carbon xerogels treated by $HNO_3$ or air to increase the number of acidic functions on the surface. Oxygen groups on the surface of carbon xerogels were used by grafting and activation of palladium to yield bifunctional catalysts with redox and acidic sites. The results demonstrate that the palladium-supported catalyst is a promising catalyst applied in sugar transformations.

Park et al. (2013) described the synthesis of CAs by sol–gel polymerization of resorcinol and formaldehyde and various ACAs were prepared by chemical activation using $H_3PO_4$ as activation agent. Palladium catalysts were then supported on ACAs by an incipient wetness impregnation method for use in the decomposition of 4-phenoxyphenol to aromatics, which is the promising candidate for lignin production.

Hardjono et al. (2011) successfully synthesized Co oxide-doped CA catalyst in heterogeneous oxidation of phenol in water. Machado et al. (2010) also reported carbon xerogel-supported noble metal catalyst (Pt, Ir, and Ru) for fine chemical application.

Zhang et al. (2013) synthesized nanosized carbides supported on CA composites by an incipient wetness impregnation method for use in the decomposition of 4-phenoxyphenol to aromatics polycondensation method of RF in the presence of sodium tungstate and sodium molybdate. The study found that the Pd nanoparticles supported on binary carbide and CA composites ($Pd/WC–Mo_2C/C$) can deliver the lowest activation free energy for ethylene glycol (EG) oxidation on $Pd/WC–Mo_2C/C$ electrocatalyst and will be the novel catalyst support in the direct alcohol fuel cells. Fort et al. (2013) prepared iron-doped CA, which was used for $H_2O_2$ electrocatalytic reduction. The finding showed that Fe–CA is a stable and sensitive electrode material for $H_2O_2$ reduction. Previous research results have proved that incorporation of catalysis particle into the carbon nanogels' structure to add catalytic activity is a promising method for the field of catalysis.

### 8.3.4 Dye Adsorption

In the past decades, the pollution of water resources has been paid great attention due to the indiscriminate disposal of toxic dyes from some industries, such as textile, paper, leather, plastic, food processing, and ink production. Removal of the contaminants from wastewater is a big challenge and becomes environmentally important. Various strategies like physical, chemical, and biological treatment technologies are developed for the removal of dyes from the wastewater. Among some existing technologies, removal of dyes by adsorption using porous carbon has been found to be one of the most economic, effective, and widely used methods. Many studies have proved that the porosity structure of porous carbon can play a key role in dye adsorption performance.

CAs with many unique properties, such as high surface areas, controllable porosities, low mass densities, and different surface chemistry, are potential porous carbon absorbent for the dye industry.

Wu et al. (2007) reported CAs as a textile dye, reactive brilliant red X-3B dye (RBRX) adsorbent (Figure 8.2). Firstly, CAs and their organic aerogels (AGs) were synthesized by sol–gel polymerization method from resorcinol, formaldehyde, and cetyltrimethyl ammonium bromide (CTAB) surfactant.

The SSA, pore volume, and pore size distribution of the CA and AGs according to $N_2$ adsorption and desorption isotherms are shown in Table 8.2 and Figure 8.3, revealing that the BET SSA of the CA is larger than that of its precursory AG. Kinetic studies indicated that the adsorption of CA followed pseudo-first- and pseudo-second-order kinetic models, but the adsorption of AG followed only pseudo-first-order kinetic models. The adsorption capacity of CA was higher than that of AG due to high surface areas of CA.

In Wu's (2012) group work, the CA and the ACA were used as fixed bed adsorptions for dye removal. The CA was first heated in $N_2$ flow of 800 $cm^3$ $min^{-1}$ to 900°C, and then activated at 900°C in mixed gas

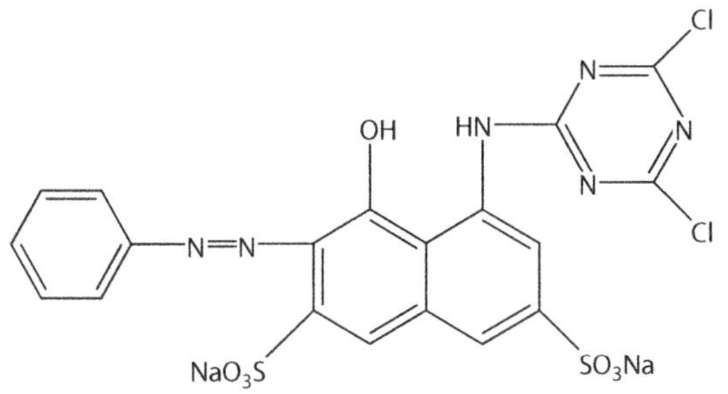

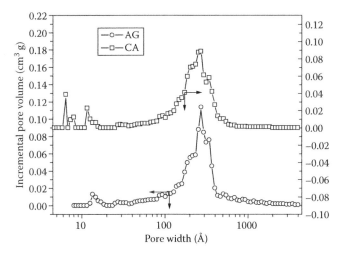

**FIGURE 8.2**   Structure of RBRX (chemical formula: $C_{19}H_{10}O_7N_6Cl_2S_2Na_2$; molecular weight: 615).

**TABLE 8.2**   Textual Characteristics of AG and CA

| Sample | AG-125 | CA-125 |
|---|---|---|
| $S_{BET}$ (m$^2$ g$^{-1}$) | 296.1 | 582.7 |
| $S_{micro}$ (m$^2$ g$^{-1}$) | 37.8 | 312.8 |
| $S_{BJH}$ (m$^2$ g$^{-1}$) | 291.3 | 302.1 |
| $V_{micro}$ (cm$^3$ g$^{-1}$) | 0.0136 | 0.145 |
| $V_{BJH}$ (cm$^3$ g$^{-1}$) | 1.216 | 1.150 |
| dp (4V/A by BJH) (Å) | 167.0 | 152.3 |
| dp (4V/A by BET) (Å) | 162.8 | 87.1 |

**FIGURE 8.3**   Pore size distributions of AG and CA.

flow of $CO_2$ (1000 cm$^3$ min$^{-1}$) and $N_2$ (400 cm$^3$ min$^{-1}$) to obtain the ACA. For the textile dye RBRX, the results revealed that the adsorption ability of the CA could be controlled by adjusting the molar ratio of resorcinol to surfactant and carbonization condition. Compared with the CA, the adsorption capacity of the ACA with higher surface area and larger pore volume is increased from 0.78 to 1.14 mol g$^{-1}$.

Ling et al. (2011) also prepared organic CAs by a sol–gel method from polymerization of resorcinol, furfural, and hexamethylenetetramine catalyzed by KOH at around pH 9 using ambient pressure drying. Activated carbon at different temperatures (900°C and 1000°C) using $CO_2$ as activation agent were obtained. The CAs with $S_{BET}$ of 2796 m$^2$ g$^{-1}$ and $V_{total}$ of 1.89 cm$^3$ g$^{-1}$ exhibited higher adsorption for two different types of dyes, including basic blue 9 (color index [CI] 52015; $C_{16}H_{18}N_3ClS$) and acid red 183 (CI 18800; $C_{16}H_{11}ClN_4Na_2O_8S_2$). The CAs form stable porosity and develop larger pores during $CO_2$ physical activation, which provide more active sites for adsorption of dye molecules. Adsorption tests indicated that the CAs were effective for both basic and acid dye adsorption and the adsorption increased by increasing surface area and pore volume.

# 8.4 Conclusions

Activated carbon nanogels with high SSA, low apparent density, and tunable continuous porosity are widely used in a variety of fields such as energy storage, hydrogen storage, absorbent, and catalyst. This chapter presents a brief overview on the synthesis, performance, and applications of activated carbon nanogels. While tremendous progress has been made in the syntheses of carbon nanogels and activated carbon nanogels, the SSA and pore structure of the activated carbon nanogels can be tailored by varying activation methods and activation process parameters such as the ratio of activation agents to raw materials, activation temperature, gas flow rate, heating rate, and activation time, which play an important role in the modification of the surface characteristics of the produced carbon nanogels. For different applications, the requirements for porosity structure of the carbon nanogels are different.

Meanwhile, compared with other porous carbon materials, it is still difficult for activated carbon nanogels to compete due to the high cost of carbon nanogels, especially in the large-scale applications. To reduce their cost is a great challenge. Surface fictionalization such as surface chemical modification to form a group to optimize their properties for some specific applications needs to be further developed.

# References

Baumann, T. F., Worsley, M. A., Han, T. Y.-J. et al., "High surface area carbon aerogel monoliths with hierarchical porosity," *J. Non-Cryst. Solids* 354 (2008): 3513–3515.

Chang, Y. M., Wu, C.-Y. & Wu, P.-W., "Synthesis of large surface area carbon xerogels for electrochemical double layer capacitors," *J. Power Sources* 223 (2013): 147–154.

Conceição, F. L., Carrott, P. J. M. & Ribeiro Carrott, M. M. L., "New carbon materials with high porosity in the 1–7 nm range obtained by chemical activation with phosphoric acid of resorcinol–formaldehyde aerogels," *Carbon* 47 (2009): 1867–1885.

Elkhatat, A. E. & Al-Muhtaseb, S. A., "Advances in tailoring resorcinol–formaldehyde organic and carbon gels," *Adv. Mater.* 23 (2011): 2887–2903.

Fang, B. & Binder, L., "Enhanced surface hydrophobisation for improved performance of carbon aerogel electrochemical capacitor," *Electrochim. Acta* 52 (2007): 6916–6921.

Fang, B., Wei, Y.-Z., Maruyama, K. et al., "High capacity supercapacitors based on modified activated carbon aerogel," *J. Appl. Electrochem.* 35 (2005): 229–233.

Fort, C. I., Cotet, L. C., Danciu, V. et al., "Iron doped carbon aerogel—New electrode material for electrocatalytic reduction of $H_2O_2$," *Mater. Chem. Phys.* 138 (2013): 893–898.

Gosalawit-Utke, R., Milanese, C., Javadian, P. et al., "Nanoconfined $2LiBH_4$–$MgH_2$–$TiCl_3$ in carbon aerogel scaffold for reversible hydrogen storage," *Int. J. Hydrog. Energy* 38 (2013): 3275–3282.

Hao, F., Zhang, Z. & Yin, L., "$Co_3O_4$/carbon aerogel hybrids as anode materials for lithium-ion batteries with enhanced electrochemical properties," *ACS Appl. Mater. Interfaces* 5 (2013): 8337–8344.

Hardjono, Y., Sun, H., Tian, H. et al., "Synthesis of Co oxide doped carbon aerogel catalyst and catalytic performance in heterogeneous oxidation of phenol in water," *Chem. Eng. J.* 174 (2011): 376–382.

Jiang, S., Zhang, Z., Wang, X. et al., "Synthesis of sulfur/activated carbon aerogels composite with a novel homogeneous precipitation method as cathode materials for lithium–sulfur batteries," *RSC Adv.* 3 (2013): 16318–16321.

Job, N., Théry, A., Pirard, R. et al., "Carbon aerogels, cryogels and xerogels: Influence of the drying method on the textural properties of porous carbon materials," *Carbon* 43 (2005): 2481–2494.

Kang, K. Y., Lee, B. I. & Lee, J. S., "Hydrogen adsorption on nitrogen-doped carbon xerogels," *Carbon* 47 (2009): 1171–1180.

Li, L., Guo, X., Zhong, B. et al., "Preparation of carbon aerogel by ambient pressure drying and its application in lithium/sulfur battery," *J. Appl. Electrochem.* 43 (2013): 65–72.

Ling, S. K., Tian, H. Y., Wang, S. et al., "KOH catalysed preparation of activated carbon aerogels for dye adsorption," *J. Colloid Interface Sci.* 357 (2011): 157–162.

Liu, N., Shen, J. & Liu, D., "A $Fe_2O_3$ nanoparticle/carbon aerogel composite for use as an anode material for lithium ion batteries," *Electrochim. Acta* 97 (2013): 271–277.

Machado, B. F., Gomes, H. T., Serp, P. et al., "Carbon xerogel supported noble metal catalysts for fine chemical applications," *Catal. Today* 149 (2010): 358–364.

Mager, N., Meyer, N. & Léonard, A. F., "Functionalization of carbon xerogels for the preparation of palladium supported catalysts applied in sugar transformations," *Appl. Catal. B Environ.* 148–149 (2014): 424–435.

Mirzaeian, M. & Hall, P. J., "Preparation of controlled porosity carbon aerogels for energy storage in rechargeable lithium oxygen batteries," *Electrochim. Acta* 54 (2009): 7444–7451.

Moreno-Castilla, C. & Maldonado-Hódar, F. J., "Carbon aerogels for catalysis applications: An overview," *Carbon* 43 (2005): 455–465.

Park, H. W., Kim, J. K., Hong, U. G. et al., "Catalytic decomposition of 1,3-diphenoxybenzene to monomeric cyclic compounds over palladium catalysts supported on acidic activated carbon aerogels," *Appl. Catal. A Gen.* 456 (2013): 59–66.

Pekala, R. W., Farmer, J. C., Alviso, J. C. et al., "Carbon aerogels for electrochemical applications," *J. Non-Cryst. Solids* 225 (1998): 74–80.

Pierre, A. C. & Pajonk, G. M., "Chemistry of aerogels and their applications," *Chem. Rev.* 102 (2002): 4243–4265.

Robertson, C. & Mokaya, R., "Microporous activated carbon aerogels via a simple subcritical drying route for $CO_2$ capture and hydrogen storage," *Microporous Mesoporous Mater.* 179 (2013): 151–156.

Sevilla, M. & Mokaya, R., "Energy storage applications of activated carbons: Supercapacitors and hydrogen storage," *Energy Environ. Sci.* 7 (2014): 1250–1280.

Tian, H. Y., Buckley, C. E., Wang, S. B. et al., "Enhanced hydrogen storage capacity in carbon aerogels treated with KOH," *Carbon* 47 (2009): 2112–2142.

Tian, H. Y., Buckley, C. E. & Paskevicius, M., "Nanoscale cobalt doped carbon aerogel: Microstructure and isosteric heat of hydrogen adsorption," *Int. J. Hydrog. Energy* 36 (2011a): 10855–10860.

Tian, H. Y., Buckley, C. E., Paskevicius, M. et al., "Acetic acid catalysed carbon xerogels derived from resorcinol-furfural for hydrogen storage," *Int. J. Hydrog. Energy* 36 (2011b): 671–679.

Wang, J. & Kaskel, S., "KOH activation of carbon-based materials for energy storage," *J. Mater. Chem.* 22 (2012): 23710–23725.

Wang, J., Yang, X., Wu, D. et al., "The porous structures of activated carbon aerogels and their effects on electrochemical performance," *J. Power Sources* 185 (2008): 589–594.

Wang, F., Xu, Y.-H., Luo, Z.-K. et al., "A dual pore carbon aerogel based air cathode for a highly rechargeable lithium-air battery," *J. Power Sources* 272 (2014a): 1061–1071.

Wang, L., Schütz, C., Salazar-Alvarez, G. et al., "Carbon aerogels from bacterial nanocellulose as anodes for lithium ion batteries," *RSC Adv.* 4 (2014b): 17549–17554.

Wu, X., Wu, D. & Fu, R., "Studies on the adsorption of reactive brilliant red X-3B dye on organic and carbon aerogels," *J. Hazard. Mater.* 147 (2007): 1028–1036.

Wu, X., Wu, D., Fu, R. et al., "Preparation of carbon aerogels with different pore structures and their fixed bed adsorption properties for dye removal," *Dyes Pigments* 95 (2012): 689–694.

Yin, L., Zhang, Z., Li, Z. et al., "Spinel $ZnMn_2O_4$ nanocrystal-anchored 3D hierarchical carbon aerogel hybrids as anode materials for lithium ion batteries," *Adv. Funct. Mater.* 24 (2014): 4176–4185.

Zapata-Benabithe, Z., Carrasco-Marín, F., de Vicente, J. et al., "Carbon xerogel microspheres and monoliths from resorcinol–formaldehyde mixtures with varying dilution ratios: Preparation, surface characteristics, and electrochemical double-layer capacitances," *Langmuir* 29 (2013): 6166–6173.

Zhai, Y., Dou, Y., Zhao, D. et al., "Carbon materials for chemical capacitive energy storage," *Adv. Mater.* 23 (2011): 4828–4850.

Zhang, X., Tian, Z. & Shen, P. K., "Composite of nanosized carbides and carbon aerogel and its supported Pd electrocatalyst for synergistic oxidation of ethylene glycol," *Electrochem. Commun.* 28 (2013): 9–12.

Zhang, Z., Li, Z., Hao, F. et al., "3D interconnected porous carbon aerogels as sulfur immobilizers for sulfur impregnation for lithium-sulfur batteries with high rate capability and cycling stability," *Adv. Funct. Mater.* 24 (2014): 2500–2509.

Zhou, J., Liu, B. H. & Li, Z. P., "Nanostructure optimization of $LiFePO_4$/carbon aerogel composites for performance enhancement," *Solid State Ionics* 244 (2013): 23–29.

# 9

# Activated Carbon Nanoadsorbents

Babak Noroozi

Moein Mehdipour

**Abstract**

Because activated carbon nanostructures have a large specific surface area and porous and fine-layered structures, they have been considered as proficient adsorbents for different metal ions and for elimination of inorganic and organic molecules from polluted water and wastewater streams. The characteristic which governs their efficiency at low concentrations is micropore size distribution and, more specifically, the ratio between the diameter of the molecule to be eliminated and the micropore size of the adsorbent. On the other hand, at high adsorbate concentrations, the entire micropore system is involved in the adsorption process and the adsorption capacity is directly related to the total pore volume. Thus, because of the widely spread spectra of the types and concentrations of pollutants, which should be removed from the environment, designing adsorbents with specific micro- and mesopore volume and surface area is very important. This chapter attempts to highlight activated carbon nanoadsorbents bearing special pore structure and surface chemistry to emphasize the role of these features as a key factor affecting the adsorption process as a realistic method to sieve a diverse range of pollutant molecules.

**Key words:** activated carbon, nanoadsorbent, pore size, tailoring.

## 9.1 Introduction

Activated carbon (AC) has been the most effective adsorbent for the removal of a wide range of contaminants from an aqueous or gaseous environment. It is a widely used adsorbent in the treatment of wastewater because of its exceptionally high surface area, which ranges from 500 to 1500 $m^2$ $g^{-1}$, and well-developed internal microporosity structure (Bandosz 2006). AC can be used for removing taste and odor compounds, synthetic organic chemicals, and dissolved natural organic matter from water (Muniz et al. 1998). The overall adsorption behavior depends on the surface area, the pore size distribution

(PSD), and the oxygen content of adsorbents as well as the hydrophobicity and the molecular structure of adsorbates (Apul et al. 2013). The adsorption capacity depends on the accessibility of the organic molecules to microporosity, which is dependent on their size (Yin et al. 2007). Although the effectiveness of ACs to act as adsorbents for a wide range of pollutant materials is well noted (Marsh & Reinoso 2006), it is necessary to understand the various key factors that influence the adsorption capacity of AC by tailoring its specific physical and chemical properties, including specific surface area, PSD, pore volume, and the presence of surface functional groups (Bach 2007).

The unusual structural and electronic properties of carbon nanostructures make them appropriate as electrode materials of electric double-layer capacitors (EDLCs) and batteries (Lee et al. 2006). Also, they may possess extraordinary physical and chemical properties. Thus, significant progress has been made in preparing nanocarbon-based adsorbents and a novel adsorption process throughout the recent years. On the other hand, pore size control is an important factor influencing AC or carbon nanostructures' adsorption in drinking water treatment or other adsorbent application. Therefore, certain engineered pore sizes in carbon materials are required for their diverse applications as molecular sieving carbons (MSCs). MSCs have a smaller pore size with a sharper distribution in the range of micropores in comparison with other ACs for gas- and liquid-phase adsorbents. They have been used for adsorbing and eliminating of pollutants at very low concentrations.

An important application of MSCs was developed in gas-separation systems (Miura et al. 1991). It has been shown that the adsorption rate of gas molecules, such as nitrogen, oxygen, hydrogen, and ethylene, depends strongly on the pore size of the MSC; the adsorption rate of a gas becomes slower for an MSC with smaller pore size. The temperature also governs the rate of adsorption of a gas because of activated diffusion of adsorbate molecules in micropores: the higher the temperature, the faster the adsorption (Verma & Walker 1992, Kawabuchi et al. 1998). From the columns of MSC, therefore, one adsorbate-rich stream comes out in the adsorption process. By using more than two columns of MSC and repeating these adsorption/desorption processes, one selected adsorbate can be isolated from the mixed flow. This adsorption method provides separation techniques with advantages such as low energy cost, room temperature operation, and compact equipment (Rao & Sircar 1996, Katsaros et al. 1997).

## 9.2 Selective Adsorption by Pore Control

Pore sizes and their distributions in adsorbents have to fulfill the requirements of different applications. Accordingly, relatively small pores are needed for gas adsorption and relatively large pores for liquid adsorption, and a very narrow PSD is required for molecular sieving applications (Zhu et al. 2011). Figure 9.1 shows the schematic features of this type of adsorption. A wide range of distribution of pore sizes and shapes is usually obtained in carbon materials. Actually, PSD, especially at the nanoscale, can be controlled during preparation of carbon-based adsorbents, by selecting a proper precursor, the condition of carbonization, and the conditions of activation. This control of pore structure is essential to compete in adsorption performance with porous inorganic materials such as silica gels and zeolites and to use the advantages of carbon materials such as high chemical stability, high temperature resistance, and low weight. For applications in modern technology fields, not only high surface area and large pore volume but also a sharp PSD at a definite size and control of surface nature of pore walls are strongly required (Bandosz 2006, Marsh & Reinoso 2006). Figure 9.2 shows the PSDs for AC and two types of activated carbon fibers (ACFs). It can be found from the figure that the two ACFs have narrow PSDs and F400 has a wide PSD, which can impact on their adsorption efficiency for certain molecules. It is found that the PSD of the adsorbent plays a major role in the oligomerization of the adsorbates. ACFs with narrow PSDs are more effective in obstructing oligomerization compared to granular activated carbon, which has a wide PSD (Lu & Sorial 2004).

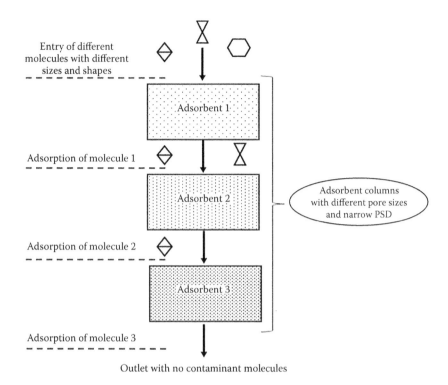

**FIGURE 9.1** Schematic of selective adsorption by different adsorbents using altered pore structures.

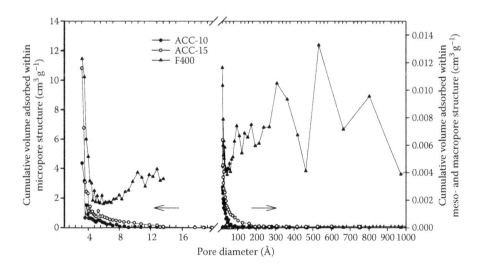

**FIGURE 9.2** PSDs of AC (F400) and two kinds of ACFs (ACC-10 and ACC-15). (From Lu, Q., and G. A. Sorial, *Carbon*, 42 (15), 3133–3142, 2004.)

# 9.3 Specific Composite Carbon Nanostructures as Adsorbents

It has been speculated that because of the unique structure of carbon nanomaterials, they may possess extraordinary physical and chemical properties. Because of their small size and their extraordinary physicochemical properties, much attention has been paid to fibrous carbons, including carbon nanotubes (CNTs) (Poole & Owens 2003). CNTs are the most fascinating materials playing an important role in nanotechnology nowadays. Their unique mechanical, electronic, and other properties are expected to result in revolutionary new materials and devices. To completely consume these properties, several CNT/polymer composites have been produced and investigated, especially as novel adsorbents.

Advanced continuous fibers produced a revolution in the field of structural materials and composites in the last few decades as a result of their high strength, stiffness, and continuity, which, in turn, meant processing and alignment that were economically feasible (Dzenis 2004). Fiber mechanical properties are known to substantially improve with a decrease in the fiber diameter. Hence, there is a considerable interest in the development of advanced continuous fibers with nanoscale diameters (Reneker & Chun 1996, Ramakrishna et al. 2005).

With the development of nanotechnology in fiber fields, carbon nanofibers (CNFs) gradually attracted much attention after the discovery of CNTs. CNFs, like other 1D nanostructures such as nanowires, nanotubes, and molecular wires, are receiving increasing attention because of their large length to diameter ratio (Gu et al. 2005). They are currently widely used in many fields, such as reinforcement materials, gas adsorption/desorption, rechargeable batteries, templates for nanotubes, high-temperature filters, supports for high-temperature catalysis, nanoelectronics, and supercapacitors, because of their high aspect ratio, large specific surface area, high-temperature resistance, and good electrical/thermal conductivities.

CNFs with controllable nanoporous structures can be prepared via different ways. The conventional preparation methods for CNFs, including the substrate method, spraying method, vapor growth method, and plasma-enhanced chemical vapor deposition (CVD) method, are known to be very complicated and costly (Park & Keane 2001). A simple and inexpensive electrospinning process, first patented by Cooley in 1902, has been increasingly utilized as the optimum method for fabricating continuous CNFs during the last decade (Liu et al. 2009). Continuous CNFs were successfully prepared via electrospinning of precursors such as poly(acrylonitrile) (PAN) solution (Gu et al. 2005, Zussman et al. 2005, Kim et al. 2008, Wu et al. 2008, Zhang & Hsieh 2009). As we know, electrospinning is a unique method for producing nanofibers or ultrafine fibers, and uses an electromagnetic field with a voltage sufficient to overcome the surface tension. Electrospun fibers have the characteristics of high specific surface area, high aspect ratio, dimensional stability, etc. (Oh et al. 2008b).

It is mentioned (Barranco et al. 2010) that certain graphitic CNFs could be obtained by procedures such as the following:

1. From blends of polymers in which a polymer acts as the carbon precursor and the other gives way to porosity, the latter being removed along the carbonization process
2. By electrospinning of precursors, which gives way to webs of CNFs after carbonization
3. By using an anodic alumina as a template, which leads to CNFs with obvious mesoporosity

The low crystallinity of the CNFs could be an advantage for subsequent activation; thus, activated CNFs with a larger specific surface area and a higher specific capacitance can be obtained.

In the following, some examples of CNT–carbon composites are presented.

## 9.3.1 Carbon Nanotube–Activated Carbon Composite

Carbon is an important support material in heterogeneous catalysis, in particular for liquid-phase catalysis (Schwandt et al. 2010).

Specific surface area, pore volume, and PSD are the important factors affecting the performance of adsorbents. Studies showed that the PSD of CNT-AC changes by the pores in bundles of CNTs and from the pores generated by the stacking of CNTs, and the size of pores slightly increases because of the presence of AC particles in the composite. Also, the micropore volume of the AC is slightly diminished by the presence of CNTs; however, the mesoporous volume increases and the $V_{meso}/V_{micro}$ ratio shows a significant increase (Chou et al. 2010, Xu et al. 2011).

## 9.3.2 Carbon Nanotube–Carbon Nanofiber Composite

In principle, any polymer with a carbon backbone such as PAN, poly(vinyl alcohol) (PVA), polyimides, polybenzimidazol, poly(vinylidene fluoride), phenolic resin, and lignin can be used to convert electrospun polymer nanofibers to carbon nanofibers by carbonization (Huang et al. 2010).

To combine CNTs with carbon fibers, most studies report the direct synthesis of CNTs on carbon fibers by CVD, with special attention paid to the control of CNT length, diameter, and density as well as arrangement and anchorage on carbon fibers. The approach which was mostly investigated consists of impregnating carbon fibers with a liquid solution of catalyst precursors (nickel, iron or cobalt nitrates, iron chloride, etc.), followed by CNT growth from carbonaceous gaseous precursors such as ethylene and methane. Carbon fibers can also be covered by catalyst particles (iron, stainless steel) using sputtering or evaporation techniques or even electrochemical deposition before introducing a gaseous carbon source (methane, acetylene) for CNT growth (Cao & Wu 2005, Chou et al. 2010).

Nanocomposites made of carbon nanofibers have been characterized and investigated as a novel structure for engineered processes. Since CNFs have a much larger functionalized surface area compared to that of CNTs, the ratio of surface-active groups to volume of these materials is much larger than that of the glass-like surface of CNTs (Huang et al. 2003, Saeed & Park 2010, Yusof & Ismail 2012). Single-wall carbon nanotubes (SWCNTs), double-wall carbon nanotubes, and multiwall carbon nanotubes (MWCNTs) can be assembled into nanofibers by spinning to increase the electrical property, the strength, and the toughness of CNFs. It was found that the effectiveness of CNT in reinforcing precursor is vastly dependent on the dispersion and arrangement of CNT. Alignment is achieved by controlling the spinning parameters. The molecular orientation and the microstructure of precursors are strongly affected by incorporation of CNTs (Minet et al. 2010), which enhances many properties such as tensile strength, modulus, and chemical resistance and also reduces thermal shrinkage. Reinforcement is one logical application of mechanically strong electrospun composite fibers. The presence of CNTs may also enhance the growth of carbon crystals during carbonization (Jagannathan et al. 2008).

Since the number and the type of functional groups on the outer surface of CNFs can be controlled well, specific materials are expected to be produced for the selective immobilization and stabilization of biomolecules such as proteins, enzymes, and DNA and also for hydrogen storage in fuel cells (Huang et al. 2003, Saeed & Park 2010, Yusof & Ismail 2012). Structural and conformational changes due to the addition of CNTs are mostly greater in gel-spun fibers than in solution-spun fibers, mainly because of a decrease in PAN interchain spacing. Using CNTs as fillers would offer property improvement in the derived CNFs (Chronakis 2005, Inagaki et al. 2012).

## 9.3.3 Carbon Nanotube–Activated Carbon Fiber Composite

To resolve the aggregation and dispersion problem of CNTs, preparation of CNT/ACF composite and its application has been investigated for the removal of specific molecules by solution, melting, and electrospinning methods (Vilatela et al. 2012, Gupta et al. 2013). PAN/CNT, polyaniline/CNT, and CNT/ PVA are important composites prepared by solution spinning. Polyurethane/CNT, polypropylene (PP)/ CNT, and polycarbonate/CNT are significant composites that are prepared by melt spinning. PAN/ CNT, poly(methyl methacrylate)/CNT, and CNT/poly(butylene terephthalate) are main composites that are prepared by electrospinning.

Most composite fibers processed to date contain less than 10 wt% CNTs, although there are a few studies where the CNT content is 60 wt% or higher. In most cases, the addition of CNT results in increased tensile and compressive properties, enhanced fatigue resistance, and increased solvent resistance, as well as increased glass transition temperature. CNTs act as a template for polymer orientation and a nucleating agent for polymer crystallization. This ability of CNTs is expected to have a profound impact on polymer and fiber processing, as well as on the resulting morphology and properties (Vilatela et al. 2012).

The decoration of CNTs tends to lower the porosity of the ACF. However, as-produced CNTs offer more attractive sites, including grooves between adjacent tubes on the perimeter of the bundles, accessible interstitial channels, and external nanotube walls. Therefore, the appearance of CNTs plays a positive role in the following:

- Facilitating pore accessibility to adsorbates
- Providing more adsorptive sites for the liquid-phase adsorption

This reveals that CNTs/ACF may contain a large number of mesopore channels, thus preventing the pore blockage from the diffusion path of micropores for adsorbates to penetrate (Gupta & Saleh 2013).

Understanding the nature of polymer–CNT interaction will lead to a more efficient use of CNT in composites as well as better processing methods, which may in turn contribute to lower cost of these materials, more widespread use, and improvement in the properties of composite fibers, as well as the production of functional polymer/CNT composite fibers tailored for specific environmental remediation applications (Min et al. 2009).

### 9.3.4 Carbon Nanotube–Activated Carbon Nanofiber (ACNF) Composite

The ACNF is the physically or chemically activated CNF, which has been, in many investigations, practically applied in EDLCs, organic vapor recovery, catalyst support, hydrogen storage, and so on. Commonly, when CNFs are highly activated, the surface area and the pore volume increase and, consequently, the adsorption capability is enhanced. In practice, the physical activation method involves carbonizing the carbon precursors at high temperatures and then activating CNF in an oxidizing atmosphere such as carbon dioxide or steam (Liu & Adanur 2010). Higher adsorption capacity of ACNF can be related to the high porous structure providing physical trapping of the adsorbate molecules.

The steam activation also makes it possible to tune the nitrogen contents on some kinds of ACNFs, which play a dominant role in adsorption. For increasing the surface area and the pore volume while maintaining the amount of nitrogen-containing functional groups, mounting of narrow micropores on carbon fiber with a small diameter must be applied (Lee et al. 2010).

The chemical activation method involves chemically activating agents such as alkali, alkaline earth metals, some bases such as potassium hydroxide (KOH) and sodium hydroxide, zinc chloride, and phosphoric acid ($H_3PO_4$) (Korovchenko et al. 2005).

The superficial and homogeneous microporous structure is attained by controlling the carbonization and steam activation conditions (Liu et al. 2011). The activation process generates micropores, which contribute to a large surface area and a large micropore volume. It is shown that activated carbon nanofibers have many more micropores than spun ones, increasing their prospective applications in adsorption. Based on a novel solvent-free coextrusion and melt spinning of polypropylene/(phenol formaldehyde polyethylene)-based core/sheath polymer blends, a series of ACNFs has also been prepared (Nataraj et al. 2012). Figure 9.3 shows electron microscope images of ACNF prepared by chemical and physical activation (Blackman et al. 2006). Figure 9.4 shows the graphical scheme of ACNF production from a PAN precursor (Rafiei et al. 2014).

From an application point of view, some of the best recent applications of ACNFs include their uses as anodes in lithium-ion battery (Ji & Zhang 2009), in removal of organic pollutants from wastewater (Lee et al. 2006, Wang et al. 2012), as cathode catalysts or as anodes for microbial fuel cells (MFCs)

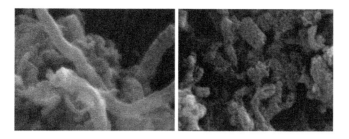

**FIGURE 9.3** SEM images of ACNFs prepared by carbon dioxide activation (left) and potassium hydroxide activation (right). (From Blackman, J. M. et al., *Carbon*, 44 (8), 1376–1385, 2006.)

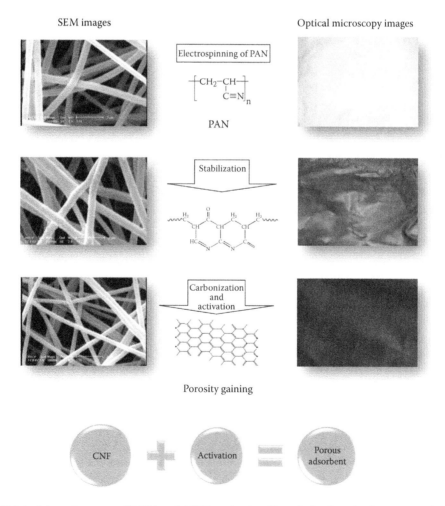

**FIGURE 9.4** Schematic process of PAN-based ACNF production. (From Rafiei, S. et al., *Chinese Journal of Polymer Science*, 32 (4), 449–457, 2014.)

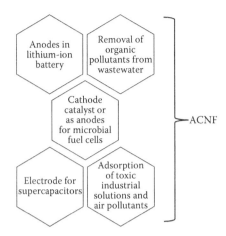

**FIGURE 9.5**   Some applications of ACNFs.

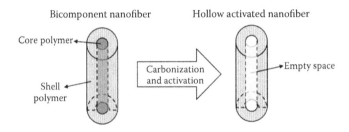

**FIGURE 9.6**   Schematic process of production of hollow-type ACNF.

(Karra et al. 2013), as an electrode for supercapacitors (Kim 2005, Seo & Park 2009, Jung & Ferraris 2012, Jung et al. 2013), in adsorption of some toxic industrial solutions and air pollutants (Lee et al. 2010, Dadvar et al. 2013), and counterelectrodes for dye-sensitized solar cells (Park et al. 2013b). Figure 9.5 shows some examples of ACNF applications. MFC technologies are an emerging approach to wastewater treatment. MFCs are capable of recovering the potential energy present in wastewater and converting it directly to electricity. Compared with other AC materials, the unique features of ACNF are its mesoporous structure, excellent porous interconnectivity, and high bioavailable surface area for biofilm growth and electron transfer (Karra et al. 2013).

Some researchers explore the probability of the production of electrospun hollow activated carbon nanofibers with highly mesoporous characteristics (Park et al. 2013a). In the proposed procedure, the inner polymer will be the object of the core hollow part of the ACNF after the activation process. Figure 9.6 shows the schematic of the process.

## 9.4  Pore Formation and Control in Carbon Adsorbents

The pore foundation in most carbon materials has been known since they are polycrystalline and result from thermal decomposition of organic precursors. During pyrolysis and the carbonization process, a large amount of decomposed gases is formed over a wide range of temperatures depending strongly on the precursors. On the other hand, the gas evolution behavior from organic matter is strongly dependent on the heating rate and pressure. Hence, the pores in carbon materials spread over a wide range of sizes and shapes, which can be classified as shown in Table 9.1 (Bandosz 2006, Marsh & Reinoso 2006, Bach 2007).

**TABLE 9.1** Different Classes of Pore Formation in Carbon Materials

| Classification | Type of Pores | |
|---|---|---|
| Based on origin | Intraparticle pores | Intrinsic intraparticle pores |
| | | Extrinsic intraparticle pores |
| | Interparticle pores | Rigid interparticle pores |
| | | Flexible interparticle pores |
| Based on size | Ultramicropores | <0.7 nm |
| | Supermicropores | 0.7–2 nm |
| | Micropores | <2 nm |
| | Mesopores | 2–50 nm |
| | Macropores | >50 nm |
| Based on state | Open pores | |
| | Closed pores (latent pores) | |

The classification of pores based on pore sizes was proposed by the International Union of Pure and Applied Chemistry.

Pores in carbon materials have been identified by different techniques depending mostly on their sizes. Pores with nanometer sizes, i.e., micropores and mesopores, are identified by the analysis of gas adsorption isotherms. The basic theories, equipment, measurement practices, analysis procedures, and many results obtained by gas adsorption have been reviewed in different publications.

The most important pore characterization methods include scanning tunneling microscopy, atomic force microscopy, transmission electron microscopy, gas adsorption, calorimetric methods, small-angle X-ray scattering, small-angle neutron scattering, positron annihilation lifetime spectroscopy, SEM, optical microscopy, mercury porosimetry, and molecular resolution porosimetry (Lastoskie et al. 1993a, Dandekar et al. 1998, Bach 2007).

As mentioned before, the important parameters that greatly affect the adsorption performance of a porous carbonaceous adsorbent are porosity and pore structure. The presence of micropores is critical in the adsorption of small molecules; nevertheless, when the adsorbate is a large molecule, mesopores are the only motivation of the adsorption process. The importance of pores type can determine the use of AC-based adsorbent for technical applications. Accordingly, control and determination of PSD of carbon nanostructures adsorbents is of particular interest (Inagaki 2009, Okhovat et al. 2012). Different methods have been applied for pore structure control, including activation and template methods (Table 9.2).

Macropores and mesopores are formed during the activation process; the pore development in carbon materials on the nanometric scale by air oxidation has been studied in detail, which is an activation process with mild thermal conditions and is also energy- and resource-saving among different activation processes. Oxidation is supposed to start only on the physical surface of carbons and proceed to the inside by forming macropores, mesopores, and micropores. It has been reported that at the beginning of oxidation, the main process is the formation of ultramicropores. Beyond 10 h to approximately 60 h, the relative amount of ultramicropores decreases; nevertheless, supermicropores increase with increasing oxidation time. Beyond 65 h, both ultramicropores and supermicropores decrease rapidly but mesopores increase slightly, which results in decrease in surface areas. This change in pore volumes might suggest the gradual enlargement of pore size from ultramicropore through supermicropore to mesopore and possibly to macropores by prolonged oxidation above 100 h. Activation processes may have some drawbacks. The mesopores can usually be created by the enlargement of micropores; in fact, a certain part of carbon atoms has to be gasified to CO and/or $CO_2$ during the activation process, causing the final yield of activation to be low. These disadvantages were one of the aims in developing some new preparation processes of porous carbon materials. For example, activation in high temperature for a

**TABLE 9.2**  Summary of Models for Controlling Pore Size in Carbon Nanoadsorbents

| Carbon Material Type | Applied Model and Simulation Methods | References |
| --- | --- | --- |
| Carbon structures | $N_2$ adsorption at 77 K and NLDFT model | Lastoskie et al. 1993b |
| Carbon structures | $N_2$ adsorption at 77 K and alpha-plot | Kaneko 1994 |
| Carbon structures | ASA algorithm DFT, ND | Kowalczyk et al. 2003 |
| Carbon structures | Ar adsorption at 87 K and IAST | Tanaike et al. 2003 |
| Carbon structures | NLDFT | Zhang et al. 2013 |
| AC | GCMC and DFT | Olivier 1998 |
| AC | $N_2$ adsorption and DR equation–BJH | Khalili et al. 2000 |
| AC | Ar adsorption at 87 K and BET | Daud et al. 2000 |
| AC | IAST, BET, BJH | Ebie et al. 2001 |
| AC | DA, ND, alpha-plot, GCMC | Terzyk et al. 2002 |
| AC | SIE equation, DR | Cao et al. 2002 |
| AC | BJH | Py et al. 2003 |
| AC | ND, HK, DA | Gauden et al. 2004 |
| AC | $N_2$ adsorption at 77 K, DR | Tseng & Tseng 2005 |
| AC | Monte Carlo simulations | Kowalczyk et al. 2008 |
| AC | DFT–$N_2$ adsorption at 77 K | Gong et al. 2009 |
| AC | NLDFT | Jagiello & Olivier 2013 |
| AC | DS-HK-IHK | Okhovat et al. 2012 |
| ACF | Alpha-plot–$N_2$ adsorption at 77 K | Kawabuchi et al. 1998 |
| ACF | BET | Pelekani & Snoeyink 2000 |
| ACF | Ar adsorption at 87 K and alpha-plot | Pelekani & Snoeyink 2001 |
| ACF | *t*-plot, DFT | Mangun et al. 2001 |
| ACF | $H_2S$ adsorption | Feng et al. 2005 |
| ACF | BET–$N_2$ adsorption at 77 K | Purewal et al. 2009 |
| ACNF | BET | Wang et al. 2012 |
| ACF and AAPFs | DR equation | Mangun et al. 1998 |
| AC, ACF, MSC | $N_2$ adsorption at 77 K and *t*-plot, alpha-plot | Kyotani 2000 |
| MSCs-ACFs | DR–$N_2$ adsorption at 77 K | Kakei et al. 1990 |
| MSC, AC | $N_2$ adsorption at 77 K, BET, BJH | Moreira et al. 2001 |
| P-ACS | DR, *t*-plot, BJH | Yang et al. 2002 |
| C-aerogels | $N_2$ adsorption at 77 K and alpha-plot, BET | Lin et al. 1999 |
| C-ZIFs | $N_2$ adsorption at 77 K and *t*-plot, BET | Banerjee et al. 2009 |
| Carbon black | NLDFT | Ustinov et al. 2006 |
| p-carbon | NLDFT-BJH method | Peter 2007 |
| ACH structure | $N_2$ adsorption at 77 K, DR | Endo et al. 2001 |
| CDCs | Ar adsorption at 87 K, NLDFT | Chmiola et al. 2006 |
| Microporous carbon | DFT | Lastoskie et al. 1993a |
| Microporous carbon | CONTEN, DFT, DA | Kowalczyk et al. 2002 |
| Glassy carbon | Monte Carlo simulation, $N_2$ adsorption at 77 K | Pérez-Mendoza et al. 2006 |
| Micro- and mesopores carbon | ASA algorithm, $N_2$ adsorption at 77 K | Gauden et al. 2007 |
| Micro- and mesoporous carbon | NLDFT, QSDFT | Landers et al. 2013 |
| Microporous carbons | Fullerene-like models for adsorption | Harris 2013 |
| MWCNT | DCV-GCMD | Keil 2010 |

*Note:* AAPFs: air activated phenolic fibers; ACH: activated carbon honeycomb; ASA: adsorption stochastic algorithm; BJH: Barrett-Joyner-Halenda; CDCs: carbide-derived carbons; CONTEN: an alternative method for analyzing the autocorrelation function can be achieved through an inverse Laplace transform developed by Steven Provencher; DA: Dubinin–Astakhov; DCV: dual control volume; DR: Dubinin–Radushkevich; DS: Dubinin-Stoeckli; GDMC: grand canonical Monte Carlo; HK: Horvath and Kawazoe; IAST: ideal adsorbed solution theory; IHK: improved Horvath–Kawazoe; ND: Nguyen and Do; NFLDT: nonlocalized density functional theory; P-ACS: phenolic activated carbon spheres; QSDFT: quench solid density functional theory; SIE: statistics integral equation.

short time, followed by the second step at a low temperature for a long residence time, provides adequate results (Bandosz 2006, Inagaki 2009, Seo & Park 2009, Inagaki et al. 2012).

Microporous carbons, of which the highest surface area and pore volume have been prepared through the carbonization of a carbon precursor in nanochannels of zeolite, are formed using the procedure called the template carbonization technique. Since the size and the shape of the channels in zeolites are strictly defined by their crystal structures and the pores formed in the resultant carbon are inherited from their channels, micropores formed in the resultant carbons are homogeneous in both size and morphology. Mesoporous carbons can be also prepared by the template method using mesoporous silica. By coupling with an activation process, micropores could be easily introduced into these mesoporous carbons (Harris 2013).

A simple heat treatment of thermoplastic precursors, such as PVA, hydroxylpropyl cellulose, polyethylene terephthalate, and pitch, with the coexistence of MgO particles as substrate ceramics, proved that the carbons formed have a higher surface, which is known to be mostly because of mesopore formation (Inagaki 2009).

Porous carbons containing both micropores and mesopores were also prepared by using a nickel hydroxide template (Okhovat et al. 2012).

Carbon aerogels, which have been well known as one of the mesoporous carbons, are prepared from the pyrolysis of organic aerogels of resorcinol and formaldehyde. These carbon aerogels contain predominantly interparticle mesopores formed in a 3D network of interconnected tiny carbon particles, and only a small amount of intraparticle micropores may be formed in primary carbon particles. Carbon aerogels could be activated to increase microporosity. Doping of Ce and Zr into carbon aerogels was found to result in micropore-rich carbon materials. The type of drying process of gels (intermediate product of aerogels) such as freeze-drying, microwave drying, and hot air-drying, is effective in preparing mesoporous carbons. Carbon aerogels can be used as a template for the preparation of highly crystalline adsorbent with uniform mesoporous structure (Dombrowski et al. 2000, Bandosz 2006, Inagaki 2009).

## 9.5 Conclusion

Porous carbon materials have been of interest because of their potential for applications in various fields of science. Introducing nanometer-sized porosity has been shown as an effective strategy to achieve desired technical properties. Nanoporous materials have filled many needs in many areas because of their intrinsically high surfaces, interstructure pores, and engineering versatility. In concept, adding nanoporosity to ultrafine nanofibers should lead to the highest possible specific surface and fibrous materials. Hence, the synthesis of organic polymer or inorganic nanofibers with controlled nanoporous structures might be very useful in a wide range of areas, such as membrane technology, tissue engineering, drug delivery, adsorption materials, filtration and separation, sensors, catalyst supports, and electrode materials.

## References

Apul, O. G., Wang, Q., Zhou, Y. et al., "Adsorption of aromatic organic contaminants by graphene nanosheets: Comparison with carbon nanotubes and activated carbon," *Water Res.* 47 (2013): 1648–1654.

Bach, M. T., *Impact of Surface Chemistry on Adsorption: Tailoring of Activated Carbon* (University of Florida, Gainesville, 2007).

Bandosz, T. J., *Activated Carbon Surfaces in Environmental Remediation*, Vol. 7 (Academic Press, Waltham, 2006).

Banerjee, R., Furukawa, H., Britt, D. et al., (2009). "Control of pore size and functionality in isoreticular zeolitic imidazolate frameworks and their carbon dioxide selective capture properties," *J. Am. Chem. Soc.* 131 (2009): 3875–3877.

Barranco, V., Lillo-Rodenas, M., Linares-Solano, A. et al., "Amorphous carbon nanofibers and their activated carbon nanofibers as supercapacitor electrodes," *J. Phys. Chem. C* 114 (2010): 10302–10307.

Blackman, J. M., Patrick, J. W., Arenillas, A. et al., "Activation of carbon nanofibres for hydrogen storage," *Carbon* 44 (2006): 1376–1385.

Cao, D. & Wu, J., "Modeling the selectivity of activated carbons for efficient separation of hydrogen and carbon dioxide," *Carbon* 43 (2005): 1364–1370.

Cao, D., Wang, W., Shen, Z. et al., "Determination of pore size distribution and adsorption of methane and CCl4 on activated carbon by molecular simulation," *Carbon* 40 (2002): 2359–2365.

Chmiola, J., Yushin, G., Gogotsi, Y. et al., "Anomalous increase in carbon capacitance at pore sizes less than 1 nanometer," *Science* 313 (2006): 1760–1763.

Chou, T.-W., Gao, L., Thostenson, E. T. et al., "An assessment of the science and technology of carbon nanotube-based fibers and composites," *Compos. Sci. Technol.* 70 (2010): 1–19.

Chronakis, I. S., "Novel nanocomposites and nanoceramics based on polymer nanofibers using electrospinning process—A review," *J. Mater. Process. Technol.* 167 (2005): 283–293.

Dadvar, S., Tavanai, H., Morshed, M. et al., "A study on the kinetics of 2-chloroethyl ethyl sulfide adsorption onto nanocomposite activated carbon nanofibers containing metal oxide nanoparticles," *Sep. Purif. Technol.* 114 (2013): 24–30.

Dandekar, A., Baker, R. T. K. & Vannice, M. A., "Characterization of activated carbon, graphitized carbon fibers and synthetic diamond powder using TPD and DRIFTS," *Carbon* 36 (1998): 1821–1831.

Daud, W. M. A. W., Ali, W. S. W. & Sulaiman, M. Z., "The effects of carbonization temperature on pore development in palm-shell-based activated carbon," *Carbon* 38 (2000): 1925–1932.

Dombrowski, R. J., Hyduke, D. R. & Lastoskie, C. M., "Pore size analysis of activated carbons from argon and nitrogen porosimetry using density functional theory," *Langmuir* 16 (2000): 5041–5050.

Dzenis, Y., "Spinning continuous fibers for nanotechnology," *Am. Assoc. Adv. Sci.* 304 (2004): 1917–1919.

Ebie, K., Li, F., Azuma, Y. et al., "Pore distribution effect of activated carbon in adsorbing organic micropollutants from natural water," *Water Res.* 35 (2001): 167–179.

Endo, M., Takeda, T., Kim, Y. et al., "High power electric double layer capacitor (EDLC's); from operating principle to pore size control in advanced activated carbons," *Carbon Sci.* 1 (2001): 117–128.

Feng, W., Kwon, S., Borguet, E. et al., "Adsorption of hydrogen sulfide onto activated carbon fibers: Effect of pore structure and surface chemistry," *Env. Sci. Technol.* 39 (2005): 9744–9749.

Gauden, P. A., Terzyk, A. P., Jaroniec, M. et al., "Bimodal pore size distributions for carbons: Experimental results and computational studies," *J. Colloid Interface Sci.* 310 (2007): 205–216.

Gauden, P. A., Terzyk, A. P., Rychlicki, G. et al., "Estimating the pore size distribution of activated carbons from adsorption data of different adsorbates by various methods," *J. Colloid Interface Sci.* 273 (2004): 39–63.

Gong, G.-Z., Xie, Q., Zheng, Y.-F. et al., "Regulation of pore size distribution in coal-based activated carbon," *New Carbon Mater.* 24 (2009): 141–146.

Gu, S. Y., Ren, J. & Vancso, G. J., "Process optimization and empirical modeling for electrospun polyacrylonitrile (PAN) nanofiber precursor of carbon nanofibers," *Eur. Polym. J.* 41 (2005): 2559–2568.

Gupta, V. K. & Saleh, T. A., "Sorption of pollutants by porous carbon, carbon nanotubes and fullerene—An overview," *Environ. Sci. Pollut. Res.* (2013): 1–16.

Gupta, V. K., Kumar, R., Nayak, A. et al., "Adsorptive removal of dyes from aqueous solution onto carbon nanotubes: A review," *Adv. Colloid Interface Sci.* 193 (2013): 24–34.

Harris, P. J. F., "Fullerene-like models for microporous carbon," *J. Mater. Sci.* 48 (2013): 565–577.

He, J.-H., Wan, Y.-Q. & Yu, J.-Y., "Effect of concentration on electrospun polyacrylonitrile (PAN) nanofibers," *Fibers Polym.* 9 (2008): 140–142.

Huang, Z.-M., Zhang, Y.-Z., Kotaki, M. et al., "A review on polymer nanofibers by electrospinning and their applications in nanocomposites," *Compos. Sci. Technol.* 63 (2003): 2223–2253.

Huang, J., Liu, Y. & You, T., "Carbon nanofiber based electrochemical biosensors: A review," *Anal. Meth.* 2 (2010): 202–211.

Inagaki, M., "Pores in carbon materials—Importance of their control," *New Carbon Mater.* 24 (2009): 193–232.

Inagaki, M., Yang, Y. & Kang, F., "Carbon nanofibers prepared via electrospinning," *Adv. Mater.* 24 (2012): 2547–2566.

Jagannathan, S., Chae, H. G., Jain, R. et al., "Structure and electrochemical properties of activated polyacrylonitrile based carbon fibers containing carbon nanotubes," *J. Power Sources* 185 (2008): 676–684.

Jagiello, J. & Olivier, J. P., "2D-NLDFT adsorption models for carbon slit-shaped pores with surface energetical heterogeneity and geometrical corrugation," *Carbon* 55 (2013): 70–80.

Ji, L. & Zhang, X., "Generation of activated carbon nanofibers from electrospun polyacrylonitrile–zinc chloride composites for use as anodes in lithium-ion batteries," *Electrochem. Commun.* 11 (2009): 684–687.

Jung, K.-H. & Ferraris, J. P., "Preparation and electrochemical properties of carbon nanofibers derived from polybenzimidazole/polyimide precursor blends," *Carbon* 50 (2012): 5309–5315.

Jung, M.-J., Jeong, E., Kim, Y. et al., "Influence of the textual properties of activated carbon nanofibers on the performance of electric double-layer capacitors," *J. Ind. Eng. Chem.* 19 (2013): 1315–1319.

Kakei, K., Ozeki, S., Suzuki, T. et al., "Multi-stage micropore filling mechanism of nitrogen on microporous and micrographitic carbons," *J. Chem. Soc., Faraday Trans.* 86 (1990): 371–376.

Kaneko, K., "Determination of pore size and pore size distribution: 1. Adsorbents and catalysts," *J. Membr. Sci.* 96 (1994): 59–89.

Karra, U., Manickam, S. S., McCutcheon, J. R. et al., "Power generation and organics removal from wastewater using activated carbon nanofiber (ACNF) microbial fuel cells (MFCs)," *Int. J. Hydrog. Energy* 38 (2013): 1588–1597.

Katsaros, F. K., Steriotis, Th. A., Stubos, A. K. et al., "High pressure gas permeability of microporous carbon membranes," *Microporous Mater.* 8 (1997): 171–176.

Kawabuchi, Y., Oka, H., Kawano, S. et al., "The modification of pore size in activated carbon fibers by chemical vapor deposition and its effects on molecular sieve selectivity," *Carbon* 36 (1998): 377–382.

Keil, F. J., "Temperature and pore size effects on diffusion in single-wall carbon nanotubes," *J. Univ. Chem. Technol. Metallurgy* 45 (2010): 161–168.

Khalili, N. R., Campbell, M., Sandi, G. et al., "Production of micro-and mesoporous activated carbon from paper mill sludge: I. Effect of zinc chloride activation," *Carbon* 38 (2000): 1905–1915.

Kim, C., "Electrochemical characterization of electrospun activated carbon nanofibres as an electrode in supercapacitors," *J. Power Sources* 142 (2005): 382–388.

Kim, J. H., Ganapathy, H. S., Hong, S. S. et al., "Preparation of polyacrylonitrile nanofibers as a precursor of carbon nanofibers by supercritical fluid process," *J. Supercrit. Fluids* 47 (2008): 103–107.

Korovchenko, P., Renken, A. & Kiwi-Minsker, L., "Microwave plasma assisted preparation of Pd-nanoparticles with controlled dispersion on woven activated carbon fibres," *Catal. Today* 102 (2005): 133–141.

Kowalczyk, P., Ciach, A. & Neimark, A. V., "Adsorption-induced deformation of microporous carbons: Pore size distribution effect," *Langmuir* 24 (2008): 6603–6608.

Kowalczyk, P., Terzyk, A. P. & Gauden, P. A., "The application of a CONTIN package for the evaluation of micropore size distribution functions," *Langmuir* 18 (2002): 5406–5413.

Kowalczyk, P., Terzyk, A. P., Gauden, P. A. et al., "Estimation of the pore-size distribution function from the nitrogen adsorption isotherm. Comparison of density functional theory and the method of Do and co-workers," *Carbon* 41 (2003): 1113–1125.

Kyotani, T., "Control of pore structure in carbon," *Carbon* 38 (2000): 269–286.

Landers, J., Gor, G. Y. & Neimark, A. V., "Density functional theory methods for characterization of porous materials," *Colloids and Surfaces A: Physicochemical and Engineering Aspects* 437 (2013): 2–32.

Lastoskie, C., Gubbins, K. E. & Quirke, N., "Pore size distribution analysis of microporous carbons: A density functional theory approach," *J. Phys. Chem.* 97 (1993a): 4786–4796.

Lastoskie, C., Gubbins, K. E. & Quirke, N., "Pore size heterogeneity and the carbon slit pore: A density functional theory model," *Langmuir* 9 (10) (1993b): 2693–2702.

Lee, J.-W., Kang, H.-C., Shim, W.-G. et al., "Heterogeneous adsorption of activated carbon nanofibers synthesized by electrospinning polyacrylonitrile solution," *J. Nanosci. Nanotechnol.* 6 (2006): 3577–3582.

Lee, K. J., Shiratori, N., Lee, G. H. et al., "Activated carbon nanofiber produced from electrospun polyacrylonitrile nanofiber as a highly efficient formaldehyde adsorbent," *Carbon* 48 (2010): 4248–4255.

Lin, C., Ritter, J. A. & Popov, B. N., "Correlation of double-layer capacitance with the pore structure of sol-gel derived carbon xerogels," *J. Electro. Soc.* 146 (1999): 3639–3643.

Liu, Ch., Lai, K., Liu, W. et al., "Preparation of carbon nanofibres through electrospinning and thermal treatment," *Soc. Chem. Ind.* 58 (2009): 1341–1349.

Liu, W. & Adanur, S., "Properties of electrospun polyacrylonitrile membranes and chemically-activated carbon nanofibers," *Text. Res. J.* 80 (2010): 124–134.

Liu, Y., Chae, H. G. & Kumar, S., "Gel-spun carbon nanotubes/polyacrylonitrile composite fibers: Part II—Stabilization reaction kinetics and effect of gas environment," *Carbon* 49 (2011): 4477–4486.

Lu, Q. & Sorial, G. A., "The role of adsorbent pore size distribution in multicomponent adsorption on activated carbon," *Carbon* 42 (2004): 3133–3142.

Mangun, C., Daley, M., Braatz, R. et al., "Effect of pore size on adsorption of hydrocarbons in phenolic-based activated carbon fibers," *Carbon* 36 (1998): 123–129.

Mangun, C. L., Benak, K. R., Economy, J. et al., "Surface chemistry, pore sizes and adsorption properties of activated carbon fibers and precursors treated with ammonia," *Carbon* 39 (2001): 1809–1820.

Marsh, H. & Reinoso, F. R., *Activated Carbon* (Elsevier, Amsterdam, Netherlands, 2006).

Min, B. G., Chae, H. G., Minus, M. L. et al., "Polymer/carbon nanotube composite fibers—An overview," *Funct. Composit. Carbon Nanotubes Appl.* (2009): 43–73.

Minet, I., Hevesi, L., Azenha, M. et al., "Preparation of a polyacrylonitrile/multi-walled carbon nanotubes composite by surface-initiated atom transfer radical polymerization on a stainless steel wire for solid-phase microextraction," *J. Chromatogr. A* 1217 (2010): 2758–2767.

Miura, K., Hayashi, J. & Hashimoto, K., "Production of molecular sieving carbon through carbonization of coal modified by organic additives," *Carbon* 29 (1991): 653–660.

Moreira, R., José, H. & Rodrigues, A., "Modification of pore size in activated carbon by polymer deposition and its effects on molecular sieve selectivity," *Carbon* 39 (2001): 2269–2276.

Muniz, J., Herrero, J. E. & Fuertes, A. B., "Treatments to enhance the $SO_2$ capture by activated carbon fibres," *Appl. Catal. B Environ.* 18 (1998): 171–179.

Nataraj, S. K., Yang, K. S. & Aminabhavi, T. M., "Polyacrylonitrile-based nanofibers—A state-of-the-art review," *Prog. Polym. Sci.* 37 (2012): 487–513.

Oh, G.-Y., Ju, Y.-W., Jung, H.-R. et al., "Preparation of the novel manganese-embedded PAN-based activated carbon nanofibers by electrospinning and their toluene adsorption," *J. Anal. Appl. Pyrolysis* 81 (2008a): 211–217.

Oh, G. Y., Ju, Y. W., Kim, M. Y. et al., "Adsorption of toluene on carbon nanofibers prepared by electrospinning," *Sci. Total Environ.* 393 (2008b): 341–347.

Okhovat, A., Ahmadpour, A., Ahmadpour, F. et al., "Pore size distribution analysis of coal-based activated carbons: Investigating the effects of activating agent and chemical ratio," *ISRN Chem. Eng.* (2012). doi:10.5402/2012/352574.

Olivier, J. P., "Improving the models used for calculating the size distribution of micropore volume of activated carbons from adsorption data," *Carbon* 36 (1998): 1469–1472.

Park, C. & Keane, M. A., "Controlled growth of highly ordered carbon nanofibers from Y zeolite supported nickel catalysts," *Langmuir* 17 (2001): 8386–8396.

Park, S.-H., Jung, H.-R. & Lee, W.-J., "Hollow activated carbon nanofibers prepared by electrospinning as counter electrodes for dye-sensitized solar cells," *Electrochim. Acta* 102 (2013a): 423–428.

Park, S.-H., Kim, B.-K. & Lee, W. J., "Electrospun activated carbon nanofibers with hollow core/highly mesoporous shell structure as counter electrodes for dye-sensitized solar cells," *J. Power Sources* 239 (2013b): 122–127.

Pelekani, C. & Snoeyink, V. L., "Competitive adsorption between atrazine and methylene blue on activated carbon: The importance of pore size distribution," *Carbon* 38 (2000): 1423–1436.

Pelekani, C. & Snoeyink, V. L., "A kinetic and equilibrium study of competitive adsorption between atrazine and Congo red dye on activated carbon: The importance of pore size distribution," *Carbon* 39 (2001): 25–37.

Pérez-Mendoza, M., Schumacher, C., Suárez-García, F. et al., "Analysis of the microporous texture of a glassy carbon by adsorption measurements and Monte Carlo simulation. Evolution with chemical and physical activation," *Carbon* 44 (2006): 638–645.

Peter, J., "How realistic is the pore size distribution calculated from adsorption isotherms if activated carbon is composed of fullerene-like fragments?" *Phys. Chem. Chem. Phys.* 9 (2007): 5919–5927.

Poole, C. P. & Owens, F. J., "Introduction to nanotechnology," in Nalwa, H. S., ed., *Nanostructured Materials and Nanotechnology* (John Wiley & Sons, Inc., Hoboken, 2003).

Purewal, J., Kabbour, H., Vajo, J. et al., "Pore size distribution and supercritical hydrogen adsorption in activated carbon fibers," *Nanotechnology* 20 (2009): 204012.

Py, X., Guillot, A. & Cagnon, B., "Activated carbon porosity tailoring by cyclic sorption/decomposition of molecular oxygen," *Carbon* 41 (2003): 1533–1543.

Rafiei, S., Noroozi, B., Arbab, Sh. et al., "Characteristic assessment of stabilized polyacrylonitrile nanowebs for the production of activated carbon nano-sorbents," *Chin. J. Polym. Sci.* 32 (2014): 449–457.

Rao, M. B. & Sircar, S., "Performance and pore characterization of nanoporous carbon membranes for gas separation," *J. Membr. Sci.* 110 (1996): 109–118.

Ramakrishna, S., Fujihara, K., Teo, W. et al., *An Introduction to Electrospinning and Nanofibers* (World Scientific Publishing Co. Pte. Ltd., Singapore, 2005).

Reneker, D. H. & Chun, I., "Nanometer diameter fibers of polymer, produced by electrospinning," *Nanotechnology* 7 (1996): 216.

Saeed, K. & Park, S.-Y., "Preparation and characterization of multiwalled carbon nanotubes/polyacrylonitrile nanofibers," *J. Polym. Res.* 17 (2010): 535–540.

Schwandt, C., Dimitrov, A. T. & Fray, D. J., "The preparation of nano-structured carbon materials by electrolysis of molten lithium chloride at graphite electrodes," *J. Electroanal. Chem.* 647 (2010): 150–158.

Seo, M.-K. & Park, S.-J., "Electrochemical characteristics of activated carbon nanofiber electrodes for supercapacitors," *Mater. Sci. Eng. B* 164 (2009): 106–111.

Tanaike, O., Hatori, H., Yamada, Y. et al., "Preparation and pore control of highly mesoporous carbon from defluorinated PTFE," *Carbon* 41 (2003): 1759–1764.

Terzyk, A. P., Gauden, P. A. & Kowalczyk, P., "What kind of pore size distribution is assumed in the Dubinin–Astakhov adsorption isotherm equation?" *Carbon* 40 (2002): 2879–2886.

Tseng, R.-L. & Tseng, S.-K., "Pore structure and adsorption performance of the KOH-activated carbons prepared from corncob," *J. Colloid Interf. Sci.* 287 (2005): 428–437.

Ustinov, E., Do, D. & Fenelonov, V., "Pore size distribution analysis of activated carbons: Application of density functional theory using nongraphitized carbon black as a reference system," *Carbon* 44 (2006): 653–663.

Verma, S. K. & Walker, P. L., "Preparation of carbon molecular sieves by propylene pyrolysis over microporous carbons," *Carbon* 30 (1992): 829–836.

Vilatela, J. J., Khare, R. & Windle, A. H., "The hierarchical structure and properties of multifunctional carbon nanotube fibre composites," *Carbon* 50 (2012): 1227–1234.

Wang, G., Pan, C., Wang, L. et al., "Activated carbon nanofiber webs made by electrospinning for capacitive deionization," *Electrochim. Acta* 69 (2012): 65–70.

Wu, S., Zhang, F., Yu, Y. et al., "Preparation of PAN-based carbon nanofibers by hot-stretching," *Compos. Interfaces* 15 (2008): 671–677.

Xu, G., Zheng, C., Zhang, Q. et al., "Binder-free activated carbon/carbon nanotube paper electrodes for use in supercapacitors," *Nano Res.* 4 (2011): 870–881.

Yang, J.-B., Ling, L.-C., Liu, L. et al., "Preparation and properties of phenolic resin-based activated carbon spheres with controlled pore size distribution," *Carbon* 40 (2002): 911–916.

Yin, C. Y., Aroua, M. K. & Daud, W. M. A. W., "Review of modifications of activated carbon for enhancing contaminant uptakes from aqueous solutions," *Sep. Purif. Technol.* 52 (2007): 403–415.

Yusof, N. & Ismail, A. F., "Post spinning and pyrolysis processes of polyacrylonitrile (PAN)-based carbon fiber and activated carbon fiber: A review," *J. Anal. Appl. Pyrolysis* 93 (2012): 1–13.

Zhang, L. & Hsieh, Y., "Carbon nanofibers with nanoporosity and hollow channels from binary polyacrylonitrile systems," *Eur. Polym. J.* 45 (2009): 47–56.

Zhang, L., Yang, X., Zhang, F. et al., "Controlling the effective surface area and pore size distribution of sp2 carbon materials and their impact on the capacitance performance of these materials," *J. Am. Chem. Soc.* 135 (2013): 5921–5929.

Zhu, Y., Murali, S., Stoller, M. D. et al., "Carbon-based supercapacitors produced by activation of graphene," *Science* 332 (2011): 1537–1541.

Zussman, E., Chen, X., Ding, W. et al., "Mechanical and structural characterization of electrospun PAN-derived carbon nanofibers," *Carbon* 43 (2005): 2175–2185.

# 10

# Heteroatom-Doped Nanostructured Carbon Materials

Kiran Pal Singh

Fatemeh Razmjooei

Jong Deok Park

Jong-Sung Yu

## 10.1 Introduction

Carbon has always been considered to be a fundamental unit, due to its omnipresence in every aspect of moving and static entities. Due to its valence, it is capable of forming different allotropes, each possessing a unique set of properties and among which diamond and graphite are most widely used. Recently, various new classes of carbon allotropes have also come into existence such as fullerene, graphene, nanotubes, nanobuds, and nanoribbons, which have incited a renewed interest in the scientific community for the use of carbon in various applications (Baughman et al. 2002, Chen et al. 2010, Huang et al. 2011). These allotropes of carbon are found to possess unprecedented physicochemical properties, such as ballistic electron transport in graphene and CNT, extremely high thermal conductivity and Young's modulus (~1 TPa) of graphene, and very high chemical stability of CNTs and graphene. However, these allotropes in their purest form are not suitable for many energy applications, such as electrocatalyst in fuel cell, electrodes in supercapacitor, and Li batteries.

It was found that modifying carbon electronic structure significantly enhances its electrochemical performances. Electronic structure manipulation can be achieved either by charge doping, such as cesium, argon, and iodine, which induces disruption in carbon's bandgap by incorporating electron donor or acceptor level near conduction or valence band (Choi et al. 2008, Tolvanen et al. 2009), or by simply doping with heteroatoms, such as nitrogen (N) (Ayala et al. 2010), boron (B) (Wang et al. 2008), sulfur (S) (Yang et al. 2011), phosphorus (P) (Liu et al. 2011a), or sometimes the halogens iodine (I) (Yao et al. 2012), chlorine (Cl), and bromine (Br) as well (Jeon et al. 2013). The influence of these heteroatoms on the enhancement of the catalytic activity of carbon for various electron transfer reactions has been examined quite commonly (Strelko et al. 2000, Singh et al. 2014). Theoretically, trivalent (B, N, and P) or divalent (N and P/or S) heteroatoms directly influence the bandgap of the semiconducting carbon matrix, which significantly helps in improving the catalytic activity of carbon in various energy applications. The electron donor or acceptor characteristic of carbon is highly susceptible to the amount and type of heteroatom present. For example, as the heteroatom (B, N, and P) content is increased in a conjugated system of carbon matrix, the influence on the electron-donor characteristics produced is in the order P > N > B; for electron-acceptor, the order is reversed, i.e., B > N > P (Strelko et al. 2000).

Apart from heteroatom doping, theoretically for energy storage applications, the SSA and conductivity are also highly desired properties for doped carbon electrode (Singh et al. 2014). However, only the surface of the pores that the ions can access can contribute to the performance enhancement. Hence, mesopores (pore diameter of <2 nm) are found to be more useful than micropores (pore diameter of >2 nm). This implies the importance of designing the carbon electrode materials with pores that are large enough for the electrolyte to access them completely, but small enough to ensure a large surface area. Furthermore, the importance of having sufficiently high conductivity, for the unhindered movement of electrolyte ions and reactants, cannot be neglected and is an important property required for the superior performance of carbon (Bhattacharjya & Yu 2014, Park et al. 2015).

The situation of N and B in a carbon matrix, which can affect the electronic structure, or the size of the second-row elements like S and P, which can induce structural effects on carbon, represents a pool of possibilities to tune carbon materials into a more versatile electrochemical electrode. In the following sections, we will therefore give an overview of highly advanced carbon materials, doped with N, B, P, or S, and the effect of this doping on the performance of carbon for various energy applications. Furthermore, the effect of surface area and the conductivity of the carbon sample on the various applications will also be considered and described.

## 10.2 Heteroatom-Doped Carbons

### 10.2.1 N-Doping in Carbon

N-doped carbon nowadays presents the most researched/studied field in energy material development. The roots of the development of N-doped carbons can be found in the field of fuel cell research and date back to 1964, when Jasinski applied metal–phthalocyanine complexes to electrochemically catalyze the reduction of oxygen in the cathodic half-cell, inspired by redox-active enzymes. The literature regarding the study of properties and applications of N-doped carbon has progressed enormously since then and a great interest has been observed toward improving and controlling its properties through various modeling and synthetic procedures. N is a neighbor of carbon and it is relatively easy to chemically bring both the atoms together, which makes N-doped carbon quite stable and popular for various electrochemical applications. Depending on the type of N bonding within the carbon matrix, nitrogen can share one to two π-electrons with the π-electron system of the carbon. This sharing of electron causes an *n*-type doping if N atoms directly substitute the C atoms in the graphitic lattice. N-doping in carbon commonly manifests itself in three forms as pyrrolic-N, pyridinic-N, and quaternary-N, and each form varies the carbon electronic bandgap differently. Bandgaps of pyrrolic-N, pyridinic-N, and quaternary-N-doped carbon are reported to be 1.20, 1.40, and 1.39 eV, respectively. This signifies that the HOMO–LUMO gap is small in pyrrolic-N, due to which it has low kinetic stability and high chemical reactivity and makes the pyrrolic N-doping in carbon more efficient for various electrochemical reactions.

The synthetic pathways leading to such N-doped carbons are nevertheless incredibly manifold and are thus not limited to a single standard procedure. After Jasinski's report, it took almost a decade to realize that heat treatment of macrocyclic compound can lead to much superior catalyst in terms of both activity and stability (Gupta et al. 1989). Later on, it was realized that not only N-doped carbon derived from macrocycles but also N-doped carbon prepared by the posttreatment of crude carbon shows superior electrochemical performance. Classically, posttreatment of as-prepared carbons with reactive nitrogen sources, such as urea, nitric acid, or especially ammonia, is one way of obtaining N-doped carbons (Wang et al. 1999, Jaouen et al. 2006, Sidik et al. 2006, Park et al. 2015). Another possible way is to pyrolyze the nitrogen and carbon-containing precursors, such as heterocycles or melamine, by which a direct incorporation of the nitrogen atoms into the forming carbon backbone becomes possible (Kim et al. 1991, Wu et al. 1998, Gadiou et al. 2008). Another well-established procedure for deriving N-doped carbon is hydrothermal treatment of carbohydrate-rich biomass (Hu et al. 2010, Titirici & Antonietti 2010). Using nitrogen-containing biomass-related precursors and hydrothermally treating them yield nitrogen-containing carbonaceous materials that offer tremendous possibilities for further treatments and energy applications (Titirici et al. 2007, Xu et al. 2012).

Another approach is based on a study on the thermal stability of ionic liquids from 2006 in which the potential of nitrile-functionalized ionic liquids is indicated, as they do not completely decompose to volatile products under an inert gas (Wooster et al. 2006). This led to profound and detailed studies on how these ionic liquids can be used as a N-doped carbon source, ranging from mechanistic and fundamental points of view to more application-oriented studies, as the derived materials are promising candidates, e.g., for fuel cell applications (Paraknowitsch et al. 2010a,b, Fellinger et al. 2012, Hasché et al. 2012). Also, supramolecular types of ionic liquids can form systems suitable for N-doped carbon synthesis. Meanwhile, also, N-rich organic networks and frameworks have been thermally treated to give N-containing carbon material promising properties, e.g., for supercapacitor applications or as cathodic materials in Li batteries (Hao et al. 2012, Sakaushi et al. 2012).

## 10.2.2 P-Doping in Carbon

In this section, we will discuss the changes in the physical properties of $sp^2$ carbon motifs after the addition of P into their lattice. P, an element of the nitrogen group, has the same number of valence electrons as nitrogen and often shows similar chemical properties. However, it has a larger atomic radius and higher electron-donating ability, which make it an astute choice as a dopant to carbon materials and thus is expected to enhance the electrochemical performance of the carbon as well. Computational studies have predicted the influence of P-doping on the bandgap of graphene, with a more pronounced effect than that calculated for sulfur, while also the incorporation of phosphorus should be energetically more favorable. Furthermore, calculations have also shown that P-doping improves the electron-donor properties of a carbon material, conclusively accompanied by an increased catalytic activity, e.g., in oxygen reduction reaction (ORR). It was found that due to the N-type doping induced by the P-doping, the field emission current for P-doped diamond films increased significantly. Due to the addition of P in the carbon matrix, the density of states near Fermi level was also found to have increased, which further increases with the increase in P-doping level (Chen et al. 2001). On the basis of these theoretical studies, it was concluded that the presence of P can significantly improve the physical and chemical properties of crude carbon.

In 2003, Lee and Radovic adopted a truly synthetic and not computational approach for adding P to a crude carbon. The authors used crude carbon fabrics, intending to increase their stability against oxidation by applying phosphoryl chloride or methyl phosphonic acid as activating agents. It was found that the treated carbon contained up to 6.7 wt% P and showed increased stability against oxidation, which portrayed the possibility of obtaining a highly stable carbon after P-doping. However, the above approach was not related to any energy application and, furthermore, the synthetic approach using a kind of an activating agent rather favors the formation of surface functional groups, but not the direct incorporation of P atoms into the carbonaceous backbone of the materials (Lee & Radovic 2003). More recent approaches have established different synthetic pathways toward phosphorus-doped carbon materials, proving themselves as very promising candidates for energy applications such as ORR catalysts, supercapacitor, and LIBs. In 2011, Liu et al. showed the oxygen reduction properties of P-doped carbon in an alkaline medium (Liu et al. 2011a). They used the mixture of toluene containing 2.5 wt% of triphenylphosphine and pyrolyzed it in a tubular furnace, yielding graphite flakes whose phosphorus content was determined by energy-dispersive X-ray spectroscopy and X-ray photoelectron spectroscopy (XPS). XPS further allowed for an insight into the binding states, proving the true incorporation of the phosphorus atoms into the graphite sheets, besides some P–O binding sites, most likely on the material's surface (Liu et al. 2011a). Liu et al. continued their work on P-doped carbons for ORR. The synthetic approach using toluene and triphenylphosphine as the precursor system was altered by adding metal catalysts aiming at the thermolytic formation of P-doped CNTs. The as-prepared multi-walled P-doped CNTs showed excellent performance as ORR catalyst under alkaline reaction conditions, exceeding the electrocatalytic activity of a conventional Pt@C catalyst used as a reference material (Liu et al. 2011b). Also, in comparison to the previously discussed P-doped graphite, the electrocatalytic activity is increased, which was supposed to be due to the nanostructural motifs of the CNT morphology. Nevertheless, it must also be considered that ferrocene is used as a catalyst during the thermolytic P-doped CNT synthesis, while no details on any acidic rinsing or other metal-removing steps are reported; so the influence of the residual metal may require more profound consideration. To portray the effect of surface area synthesis of P-doped ordered mesoporous carbon (P-OMC), a study was carried out by Yang et al. (2012a). The P-OMCs were synthesized by copyrolyzing a phosphorus-containing source and a carbon source together using SBA-15 ordered mesoporous silica as a template without the use of any metal components, which can intrinsically avoid the involvement of any metal components in electrocatalytic activity, to elucidate only the effect of P in the carbon framework (Yang et al. 2012a). The electrocatalytic activity of the prepared catalyst was found to have increased manifolds, which was reasoned to be due to the superior surface area of the catalyst.

## 10.2.3 S-Doping in Carbon

In comparison to phosphorus, S-doping in carbon materials is hitherto still quite rare and represents an emerging field in carbon material research. While nowadays the high potential of such materials for energy applications such as in fuel cells, supercapacitors, or batteries is continuously discovered and exploited, until only a few years ago, only little was known about such S-doped carbonaceous species. S-doping was found to be quite more difficult than doping with N, taking into account the size and the different binding behavior of S atoms. Nevertheless, it was calculated that S-doping should allow for a targeted tuning of the graphene bandgap, depending on the amount of S atoms incorporated into the sheets. The increase in specific capacitance, in supercapacitors and Li-S batteries, by S-doping is related to the high electron density located on the surface of the carbon material due to the presence of the aromatic sulfide, which increases the electrolyte dielectric constant and facilitates the charge transfer process. Cui et al. (2011) have also reported that the presence of S during the preparation of graphitic CNTs is favorable for the formation of pentagon and heptagon carbon rings, and the rings are supposed to result in the formation of CNTs with distorted sidewalls. This kind of CNTs might supply more active positions for catalyst nanoparticles and can have potential usage in the catalyst support area. It was also found that S addition can enhance the N-doping content of CNTs, resulting from heterocyclic rings, which are favorable for the enhancement of N-doping (Cui et al. 2011). Therefore, the presence of S helps in the formation of disordered carbon; in general, the formed S-doped graphene or CNTs are found to possess more energetically active Stone–Wales defect, which is found to be very beneficial for many energy applications. Zhang et al. (2014) showed that S atoms could be adsorbed on the graphene surface, substitute carbon atoms at the graphene edges in the form of sulfur/sulfur oxide, or connect two graphene sheets by forming S cluster rings. These S-doped graphene clusters with S or sulfur oxide located at the graphene edges show enhanced electrocatalytic activity for ORR. Catalytic active sites distribute at the zigzag edge or on the neighboring carbon atoms of doped sulfur oxide atoms, which possess a large spin or charge density. For those being the active catalytic sites, S atoms with the highest charge density take a two-electron transfer pathway, while the carbon atoms with high spin or charge density follow a four-electron transfer pathway. It was predicted from the reaction energy barriers that the S-doped graphene could show ORR catalytic properties comparable to that of state-of-art-platinum/carbon catalyst (Zhang et al. 2014).

Synthesis concepts establishing doped carbons with S atoms firmly and covalently incorporated into the carbon structures have been developed only since 2011, when Paraknowitsch et al. (2011) and Schmidt et al. (2009) used a microporous polymer network containing thienyl building blocks as a precursor for intrinsically microporous S-doped carbon with variable S contents of ~7 to ~20 wt% depending on the carbonization temperature. Comparable to this concept, an approach was developed by Botger-Hiller et al. (2012) that covalently links thienyl monomers to silica precursors. This dually functionalized precursor system allows for the synthesis of silica/S-doped carbon composites yielding additional porosity after silica removal. The authors further induced mesoporosity within this type of material applying hard templates (Botger-Hiller et al. 2012). Sevilla et al. (2011) acknowledged the potential of the thienyl group as a central tool in the synthesis of S-doped carbon, and polymerized thiophene using an oxidative approach, followed by carbonization under activation of the forming carbon with potassium hydroxide. Thus, a rather classical approach for the preparation of activated carbons has been successfully transferred to the design of modern materials. The as-derived S-doped carbons exhibit microspherical morphology and remarkable microporosities with BET surface areas of up to ~3000 m$^2$ g$^{-1}$ and micropore surface areas of up to ~2600 m$^2$ g$^{-1}$, depending on the polythiophene–KOH ratio and the carbonization temperature applied. The S atoms are found to firmly bound to the carbon backbone, mainly in C–S–C binding motifs, in amounts from ~4 to ~14 wt% depending on the reaction temperature (Sevilla et al. 2011, Sevilla & Fuertes 2012).

## 10.2.4 B-Doping in Carbon

Boron is an element with unique and basically incomparable properties within the periodic table. It is thus a highly interesting candidate for the doping of carbon materials, modifying the properties of pure carbons. Some research groups have started focusing not only on fundamental studies on B-doping, but also on applying the obtained materials and exploiting their beneficial properties in energy-related applications. Owing to its three valence electrons, B is considered as a good dopant. The existence of B–C bonds in the carbon framework can lower the Fermi level of the structure and then tune the properties of oxygen chemisorption and electrochemical redox reactions. In addition, the incorporation of boron will also bring other heteroatom functional groups, such as nitrogen functional groups, together. The experimental results indicate that electron-deficient boron doping can turn carbon material into an efficient electrocatalyst with positively shifted potentials and elevated reduction current. The electrocatalytic ability of B-doped CNTs for ORR stems from the sharing of accumulated electron in the vacant $2p_z$ orbital of B dopant to the $\pi^*$ electrons of the carbon-conjugated system; thereafter, the transfer readily occurs to the chemisorbed O$_2$ molecules with boron as a bridge. The transferred charge weakens the O–O bonds and facilitates the ORR on B-doped carbon materials. A similar study was conducted by Yang et al. (2011). The experimental progress and theoretical understanding of their study pointed out the two key factors for the doped CNTs as metal-free ORR catalysts: (1) breaking the electroneutrality of CNTs to create the charged sites favorable for O$_2$ adsorption with electron-deficient dopant B; and (2) effective utilization of carbon $\pi$-electrons for O$_2$ reduction (Yang et al. 2011).

Early works on B-doped carbons were inspired by the fascination of stoichiometric boron nitride compounds that can form hexagonal patterns enabling $sp^2$-carbon-related structures, such as stacked sheets or nanotubes, as has been reviewed, e.g., by Golberg et al. (2010) or Ma et al. (2004). Early works on B/N/C-materials with graphite-like structure and a composition of B$_{0.35}$C$_{0.3}$N$_{0.35}$ have been reported by Kaner et al. in 1987, who used a CVD approach with boron trichloride, acetylene, and ammonia as the precursor mixture (Kaner et al. 1987). Instead of CVD, Stephan et al. (1994) applied an electric arc discharge base synthetic procedure, in which elemental amorphous B and elemental graphite powder served as precursors, yielding a mixture of different B-containing carbon nanostructures, such as thin graphitic sheets, tubes, and filaments. A similar pathway was followed for B/N/C-nanotubular structures by Redlich et al. (1996). Over the years many boron-doped carbons and, due to the fundamental proximity to the stoichiometric boron nitride materials, B- and N-codoped carbons have been synthesized, reaching from CVD and arc discharge approaches over pyrolytic procedures toward the carbonization of ionic liquid-based precursors. For example, Wang et al. (2011) presented B-doped carbon nanorods derived by a spray pyrolysis CVD process.

## 10.2.5 Multiple Doping

After testing all the possible heteroatom dopings, the scientific community has also started exploring the effect of multiple heteroatom doping on the physical and the electrochemical properties of carbon materials. Theoretically, nitrogen doping within the carbon matrix can induce defects due to the difference in atomic radius, bond length, and electronegativity. The conjugated heteroatom and the surrounding site become uneven in the distribution of electric charge state. As a consequence, the carbon atoms become positively charged due to the charge delocalization. The positively charged carbon atoms efficiently reduce oxygen molecules to water. Boron and phosphorus, on the other hand, have electronegativities lower than that of carbon. In comparison to nitrogen, B and P exhibit a different ORR mechanism. The doped B or P dopants become the active sites instead of the carbon. Due to the different electronic structures and chemical properties, the simultaneous doping of N with B or P might further change the electronic structure of the carbon matrix. Therefore, the multiple heteroatom-doped carbon matrices were expected to possess new active sites corresponding to the ratio and position of various dopants. It was found that after multiple doping, the carbon becomes more disordered and more prone to the chemical environment, which causes it to be more active for various energy applications. For instance, Razmjooei et al. (2014) showed that after ternary doping of graphene, the sheet structure of graphene completely transformed into crumpled morphology (Figure 10.1). The doping of carbon using the heteroatoms with electronegativity higher or lesser than that of carbon can disrupt the electroneutrality of the neighboring carbon, due to which the graphitic plane becomes energetically unstable. Therefore, the graphitic plane starts squeezing itself to gain a more energetically stable configuration. The difference in bond length between carbon and heteroatoms can further increase the wrinkles in the structure in addition to inherent defects. Figure 10.1a through d clearly corroborates that the prepared undoped and doped graphene samples have crumpled sheetlike structure, which increases with the incorporation of heteroatoms (Razmjooei et al. 2014). The effect of single and multiple heteroatom doping on the physical structure of the carbon lattice is shown in Figure 10.2.

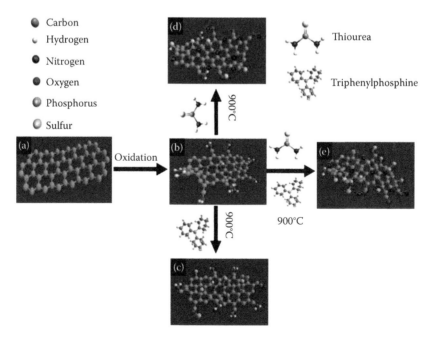

**FIGURE 10.1** Process depicting the formation of wrinkles and defects on the graphene sheets with the introduction of heteroatoms. (a) Graphene sheet, (b) graphene oxide, (c) P-doped reduced graphene oxide (RGO), (d) N- and S-doped RGO, and (e) N-, S-, and P-doped RGO. (From Razmjooei, F. et al., *Carbon*, 78, 257–267, 2014.)

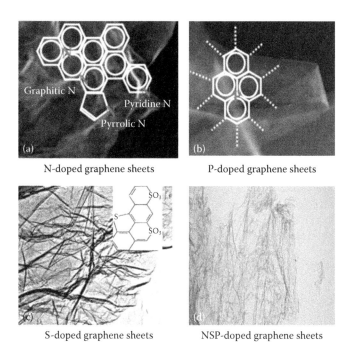

N-doped graphene sheets          P-doped graphene sheets

S-doped graphene sheets          NSP-doped graphene sheets

**FIGURE 10.2** The effect of doping on the graphitic plane of carbon. (a) N-doped graphene, (b) P-doped graphene, (c) S-doped graphene (Reprinted with permission from Yang, Z., Yao, Z., Li, G. et al., *ACS Nano*, 6, 205–211. Copyright 2012 American Chemical Society.), and (d) NSP-doped graphene. (From Razmjooei, F. et al., *Carbon*, 78, 257–267, 2014.)

## 10.3 Applications

### 10.3.1 Fuel Cells

Today with the depletion of natural resources and rapid increase in energy demand, the need for energy is becoming more crucial. In this era, we need an alternative energy source that can replace natural resources and can be environmentally friendly as well. Because of their high energy-conversion efficiency, low pollution, low operating temperature, high power density, and wide range of applications, fuel cells have received a plethora of attention. Among the different kinds of fuel cells, the alkaline fuel cell is drawing much attention due to its lower operating cost, lower operation temperature, higher durability, and higher efficiency than its acidic counterparts. ORR takes place usually on Pt loaded on different carbon supports, but their large-scale commercial application has been precluded mainly by the high cost of the requisite noble metal. Apart from this, several other factors that impede the commercialization of the fuel cells include catalyst layer degradation due to Pt sintering (Pt nanoparticle agglomeration, Pt loss and migration, and Pt active site contamination), carbon support corrosion, and gas diffusion layer degradation.

To overcome all the drawbacks of Pt or Pt-based catalyst, various efforts have been devoted to find its proper substitute. Heteroatom (N, S, P, B, etc.)-doped carbon catalysts, due to their high stability, excellent electrocatalytic performance, and economic viability over the costly Pt-based catalysts, are being pioneered as a suitable alternative (Inamdar et al. 2013, Paraknowitsch & Thomas 2013). Heteroatoms bond covalently with the adjacent carbon in the carbon lattice, and therefore catalytic degradation is much less compared to Pt-based catalysts, which are usually generated through their physical attachment over the carbon support. Furthermore, the difference in electronegativity of the dopants (N: 3.04;

S: 2.58; and P: 2.19) and carbon (2.55) in the carbon matrix can destroy the electroneutrality of the adjacent carbon. Moreover, the C–C bond length of $sp^2$-hybridized carbon changes from 147 to 210 pm with the introduction of N, P, or S. Because of the changes in bond length and electronegativity of the carbon framework due to heteroatom doping, the carbon surface becomes asymmetric in nature and gets more ORR active (Gong et al. 2009, Wu et al. 2012). Zhang et al. (2012) reported that with heteroatom doping, spin density and charge density redistribute on the carbon, which strongly affects the formation of the intermediate molecules during ORR, including OOH or $O_2$ adsorption, O–O bond breaking, and water formation.

As already discussed, after Jasinski's (1964) report, N-doped nonprecious metal– and nitrogen-doped carbon has become famous. However, later on, it was realized that the presence of metal is not necessary and only N-doped carbon can act as an efficient ORR catalyst. In 2006, completely metal-free catalysts were proven to be active ORR catalysts by Matter and Ozkan, finally evidencing the beneficial influence of heteroatoms on the activity of carbon. Also, the possibility that N atoms bound in pyridinic sites at the edges of carbon sheets play an important role was implied. Later, vertically aligned nitrogen-doped carbon nanotube arrays were presented by Gong et al. (2009), remarkably exceeding the activity of conventional Pt@C catalysts while additionally exhibiting an outstanding ORR selectivity avoiding deactivation by crossover effects in an alkaline medium. For the first time, Pt@C catalysts were left behind, at least under alkaline conditions; this was considered as one of the most important breakthroughs in the research on N-doped carbon-based electrocatalysts. The reasons for the electrochemical activity of such materials are meanwhile seen in the charge profile induced in the carbon backbone by the electronegativity of the doping atoms, and also in the influence of pyridinic-N atoms with their free lone pair available for interaction with oxygen.

Similarly, except N, other heteroatom-doped carbons were also tested for ORR; for instance, Yang et al. (2011) directly tested B-doped CNTs for ORR activity. Upon increasing the boron content, an increase in the ORR activity has also been observed, indicating the importance of the boron moieties in the catalytic process (Yang et al. 2011). Also, excellent ORR selectivity and resistance against methanol crossover qualify the B-doped nanotubes as promising ORR candidates. Additional theoretical calculations performed by the authors have closely elucidated the reason for the beneficial influence of the boron doping, which is based on two synergistic effects: on the one hand, boron has a lower electronegativity than carbon, and the positively polarized boron atoms attract the negatively polarized oxygen atoms leading to chemisorption; on the other hand, boron sites can also act as electron donors for the reduction reaction, as the electron density of the graphitic $p$-electron system can be transferred to the free $p_z$ orbital of boron.

After testing N- and B-doped carbon, scientists have started exploring other heteroatoms as well, such as P- and S-doped carbon. In 2012, Yang et al. reported a phosphorus-doped ordered mesoporous carbon material capable of catalyzing the electrochemical reduction of oxygen in an alkaline medium. The material was successfully tested as a catalyst in ORR, in direct comparison to a non-P-doped reference, derived from an analogous synthesis without the addition of triphenylphosphine and to conventional Pt@C-catalyst materials. Rotating disk electrode experiments clearly show how the P-doping significantly boosts the catalytic activity in ORR while nonetheless not totally reaching the values of Pt@C. A similar study was also conducted on the high surface area carbon by Yang et al. (2014a). The P-doped platelet ordered mesoporous carbon (P-pOMC) was synthesized by nanocasting method using platelet ordered mesoporous silica as template and triphenylphospine as P and C source followed by carbonization at various temperatures (700°C, 800°C, 900°C, and 1000°C and by etching of the silica template). It was found that the combination of P-doping and excellent surface properties empowered the P-pOMC catalyst to show high ORR activity, nearly equal to that of state-of-the-art Pt@C catalyst (Figure 10.3), along with superior long-term stability and excellent methanol tolerance. The carbonization temperature was found to be a major factor in determining the catalytic activity of the prepared catalyst and among all the prepared catalysts, p-POMC 900 showed the best ORR performance. The superior activity of this catalyst was reasoned to be due to the unhindered mass transfer in the cathode as well as the availability of higher catalytic active sites.

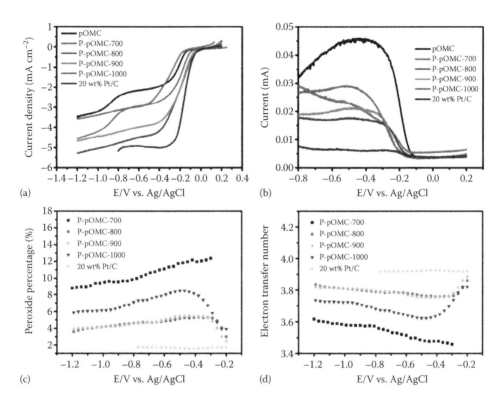

**FIGURE 10.3** Steady-state rotating ring-disk electrode (RRDE) experiments on P-pOMC, pOMC, and 20 wt% Pt/C (E-TEK) catalysts for ORR at 1600 rpm electrode rotation rate and 10 mV s⁻¹ potential scan rate. (a) Disk and (b) ring currents are shown separately for convenient viewing; (c) plot of peroxide percentage formation; and (d) number of electron transfers at different potentials for P-pOMC and 20 wt% Pt/C (E-TEK) catalysts for ORR. (From Yang, D.-S. et al., *Carbon*, 67, 736–743, 2014.)

Similarly, the capability of S-doped graphene to catalyze ORR was also predicted theoretically. The electrocatalytic activity of S-doped graphene synthesized at 1050°C even exceeds the activity of a conventional Pt@C catalyst, while additionally exhibiting higher selectivity for oxygen reduction and thus avoiding crossover effects well (Figure 10.4). Koutecky–Levich plots further revealed that an almost ideal four-electron transfer occurred in the S-doped graphene catalysts, once more accentuating the powerful character of the S-doping method within the field of fuel cell catalyst research. The reason is suggested to be found in the increased spin density in the graphene achieved by S-doping (Yang et al. 2012b). Similarly, Inamdar et al. (2013) showed an interesting way of synthesizing S-doped carbon; they have opted a flame synthesis technique in which S-doped carbon was prepared by directly burning the thiophene and the obtained carbon shoot particles were collected (Figure 10.5) and tested for ORR.

The ORR occurs on a wide variety of carbon electrocatalysts in alkaline electrolytes, like graphite, glassy carbon, active carbon, CNTs, graphene, and other nanostructured carbons. The ORR conducted on carbon electrocatalysts without doping, functionalization, or lattice substitution usually exhibits the two-plateau peroxide pathway rather than the one-plateau 4e⁻ pathway observed on Pt@C, a commercial electrocatalyst. In order to render the high-voltage and high-power performance of alkaline fuel cells or metal–air batteries, the ORR catalysts are usually required to conduct the 4e⁻ pathway that can deliver large current density and high onset potential. However, the two-plateau peroxide pathway of ORR is often associated with small current and low onset potential. Therefore, one major goal of the heterogeneous carbon ORR research is to explore efficient carbon catalysts favorable for the 4e⁻ ORR

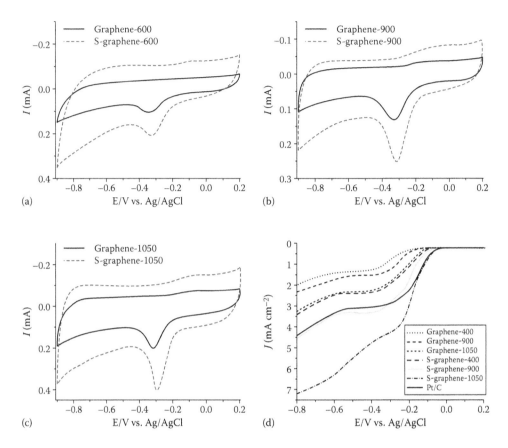

**FIGURE 10.4** Cyclic voltammograms for (a) graphene-600 and S-graphene-600, (b) graphene-900 and S-graphene-900, and (c) graphene-1050 and S-graphene-1050; and (d) linear sweep voltammetry (LSV) curves for various graphenes and a Pt/C catalyst on a glass carbon RDE saturated in $O_2$ at a rotation rate of 1600 rpm. (Reprinted with permission from Yang, Z., Yao, Z., Li, G. et al., *ACS Nano*, 6, 205–211. Copyright 2012 American Chemical Society.)

behavior. Carbon materials can be alternatively used as catalyst supports to provide good electron transfer to less-conducting ORR catalysts. It was reasoned that higher surface area and the conductivity helps in the facile movement of oxygen, electrolyte, and electron toward the active sites, where they form the triple phase boundary (TPB), and the formation of these TPBs is found to be of utmost importance for an efficient ORR catalysis. Therefore, it is highly desirable to modify the interface structure between the carbon support and the catalyst by engineering the carbon surface. Regardless of the carbon-supported hybrid ORR catalysts, enhancing the ORR activity of carbon catalysts through effective surface modifications has attracted tremendous attention. For example, Singh et al. (2014) showed the effect of conductivity on the ORR activity. They studied the effect of iodine treatment on the physiological properties of heteroatom-doped carbon derived from PANI. It was reported that the concentration of iodine present during carbonization had a huge effect on the physical and electrochemical properties of the obtained carbonized polyaniline (CPANI), as shown in Figure 10.6. Such an iodine treatment was found to improve the C/O ratio, the crystallinity, and the electrical conductivity of the prepared samples. Furthermore, the iodine treatment decreases the doping of catalytically active heteroatoms, which is unfavorable for the ORR, but at the same time greatly increases the graphitic nature of the sample, which was found to be beneficial for the ORR (Singh et al. 2014). On the other hand, the importance of the surface area and the porosity on the formation of catalytic active sites was shown by Yang et al. (2014b). They prepared

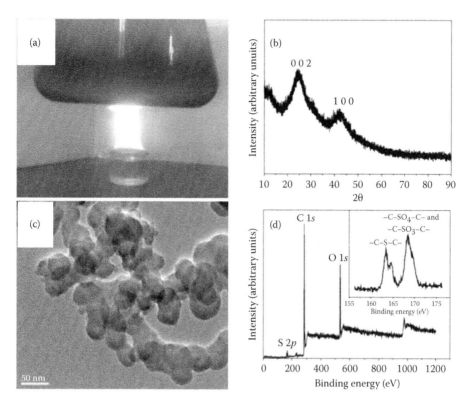

**FIGURE 10.5** (a) Actual photograph of flame pyrolysis setup used to prepare sulfur-containing carbon, (b) XRD pattern, (c) HRTEM image recorded for as-prepared sample S-3, and (d) XPS survey spectrum of S-3 with an inset showing high-resolution S $2p$ spectrum with two distinct peaks. (From Inamdar, S. et al., *Electrochemistry Communications*, 30, 9–12, 2013.)

carbon nanofibers by electrospinning the PAN solution, which was followed by carbonization. Later, the N-containing carbon fibers with high porosity were prepared by cospinning PAN with poly(ethylene oxide) (PEO) as a porogen and the porosity was tuned by changing the concentration of PEO. The variation in ORR performance with porosity as well as the effect of carbon graphitization and the amount of efficient N-doping was also studied (Figure 10.7) (Yang et al. 2014b).

## 10.3.2 Li-Ion Batteries

The depletion of earth's natural resources has been continuously destabilizing and threatening the present fossil fuels-based energy economy. In addition, continuous increase of greenhouse gases has put earth's existence at risk. Therefore, the urgency of energy research requires the development of clean energy sources at a much higher level than that presently in force. Consequently, exploitation of renewable energy resources is increasing worldwide, particularly in the area of wind and solar power energy plants (PEPs). Efficient use of these resources requires high-efficiency energy storage systems. Electrochemical systems, such as batteries and supercapacitors, which can efficiently store and deliver energy on demand in hybrid electric vehicles, plug-in electric vehicles, and portable consumer electronic equipments are playing a crucial role in this field. The effectiveness of batteries in PEPs is directly related to their energy efficiency and lifetime. In recent decades, secondary LIBs have evolved as the foremost energy source for all kinds of consumer electronic products as well as electric/hybrid vehicles due to their high electromotive force and high energy density (Owens et al. 1999, Kang et al. 2006).

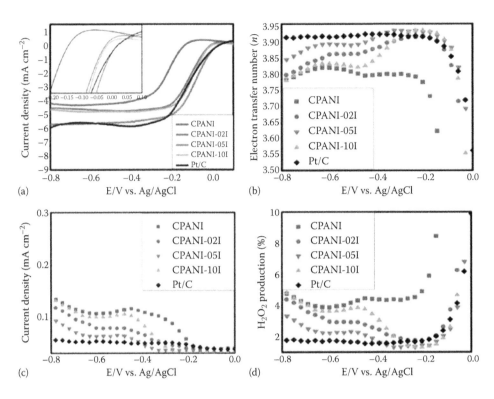

**FIGURE 10.6** The electrochemical activity studied using the RRDE technique at a rotation rate of 1600 rpm in an $O_2$-saturated 0.1 M KOH solution for CPANI, CPANI-02I, CPANI-05I, CPANI-10I, and 20% Pt/C (E-TEK). (a) The effect of the iodine concentration on the LSV curves for the treated and untreated CPANI samples, with the inset showing the expansion of the LSV curves in the potential range of 0.2 and 0.1 V; (b) the number of electrons transferred per oxygen molecule during the ORR reaction; (c) the current produced on the ring electrode; and (d) $HO_2^-$ percentage production during the ORR. (From Singh, K. P. et al., *Journal of Materials Chemistry A*, 2, 18115–18124, 2014.)

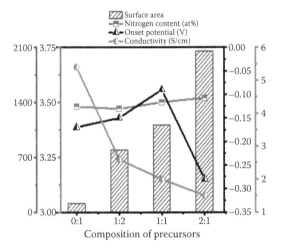

**FIGURE 10.7** Assessment of carbon nanofibers in terms of surface area, electrical conductivity (18 MPa), and nitrogen content along with onset potential as a function of different PEO/PAN precursor ratios, indicating the material properties and their effect on ORR activity. (From Yang, D.-S. et al., *ChemCatChem*, 4, 1236–1244, 2012.)

However, presently many potential electrode materials are experiencing slow Li-ion diffusion and high resistance at the interface of electrode/electrolyte at high charge–discharge rates, and their rate capability, dominated mainly by the diffusion rate of Li ion and the electron transfer in electrode materials, must be improved considerably. Recently, in order to improve the transport of Li ions in electrode, various nanostructured materials with high surface area, nanoscale size, and/or nanoporous structure have been widely investigated. Many nanostructured carbon materials such as ordered mesoporous carbon CMK-3, CNTs, graphene, and hierarchical porous carbons have demonstrated greatly enhanced Li storage capacity, and/or improved rate capability compared with commercial graphite anode (Ergang et al. 2006, Sorensen et al. 2006, Wena & Li 2009).

However, it is still highly desired to explore new materials with further enhanced Li storage capacity, especially at high charge–discharge rates. Another approach to improve the electrochemical performance of the carbon-based anode is to modify its surface functional groups with noncarbon elements such as N, P, S, and B. The presence of heteroatoms on the carbon surface can enhance the reactivity and electric conductivity and, hence, the Li⁺ ion storage capacity. Therefore, for the carbon-based anode materials, the porous nanostructure and the incorporation of heteroatoms are both desirable for Li⁺ ion storage.

An early overview has been provided in 2000 by Endo et al., pointing out difficulties that have been faced with B-doped carbons as lithium battery anodes; as in many cases, the formation of boron carbide or nitride sites in carbonaceous materials has prevented beneficial effects of real B-doping. Later, some examples have shown that B-doping obtained by graphitization of pitches or fossil coals with boron sources at temperatures >2000°C can yield potential anode materials for LIBs, still partially encountering the aforementioned problem (Liu et al. 2010, Tanaka et al. 2010, Rodríguez et al. 2011). A recent study has focused on graphene to be doped with boron for enhanced performance as an anode material in LIBs. Thus, graphene is postfunctionalized at 800°C using boron trichloride as a doping agent, yielding a doping level of 0.88 at% in the material. High electrode capacities of 1549 mAh g⁻¹ at low charge/discharge rates and 235 mAh g⁻¹ at very high rates represent a remarkable result, attributed to numerous effects of the boron doping regarding electrode/electrolyte wetting properties, interlayer distances, electrical conductivities, or heteroatom-induced defect sites, thus leading to enhanced performance of the material concerning its Li⁺ absorption and diffusion properties (Morita & Takami 2004).

There have been many reports on the increase in LIB performance of carbon after doping it with nitrogen. For instance, Qie et al. (2012) reported a facile strategy to synthesize porous carbon that shows a nanostructure with a large surface area and high-level nitrogen doping. Polypyrrole nanofiber webs with a high N content (~16 wt%) were selected as a precursor, and KOH was used as an activating agent to form a porous structure. Benefiting from the unique porous nanostructure and high-level N-doping, the CNFWs exhibit high capacity and excellent rate capability, delivering a reversible capacity as high as 943 mAh g⁻¹ at a current density of 2 A g⁻¹ even after 600 cycles (Qie et al. 2012). Similarly, Bhattacharjya et al. (2014) showed an effect of turbostratic carbon and N-doping on the Li-ion performance. The N-doped turbostratic carbon nanoparticles were prepared using fast single-step flame synthesis by directly burning acetonitrile in air atmosphere. The as-prepared N-doped carbon nanoparticles showed excellent Li-ion storage properties with an initial discharge capacity of 596 mAh g⁻¹, which was 17% more than that shown by the corresponding undoped carbon nanoparticles synthesized by an identical process with acetone as carbon precursor and also much higher than that of commercial graphite anode. Further analysis showed that the charge–discharge process of N-doped carbon was highly stable and reversible not only at high current density but also over 100 cycles, retaining 71% of initial discharge capacity. The authors also performed electrochemical impedance spectroscopy, which proved that N-doped carbon has better conductivity for charge and ions than undoped carbon. The high specific capacity and the very stable cyclic performance of N-doped carbon were attributed to a large number of turbostratic defects and N and associated increased O content in the flame-synthesized N-doped carbon (Bhattacharjya et al. 2014).

## 10.4 Conclusions

Despite the superior chemical and physical properties of carbon and its allotropes, it is highly desirable to dope the carbon surface with heteroatoms to further enhance and improve its properties. In this chapter, we have tried to show various studies performed on heteroatom-doped carbon, doped with various heteroatoms such as N, P, S, and B. It is shown that all the heteroatoms possess varied physical properties, which have varied but beneficial effects on the electrocatalytic and electrochemical energy storage applications of the carbon material. Therefore, N, P, S, and B enable advanced doping concepts for carbon materials. While boron changes the electronic structures of carbon materials in the opposite way, but just as beneficially as nitrogen does, using both dopants at the same time can even yield synergistic effects. Therefore, nonetheless the synthetic procedure must be finely tuned to avoid mutual compensation of the effects that are so beneficial for energy applications. When it comes to S and P, the effects are attributed more to the formation of structural defect sites or favorable spin densities throughout the carbons, as here the electronegativities of carbon and the dopants are closer to each other. Regarding the wide spectrum of numerous synthetic routes and application profiles, it is still difficult to foresee which candidates will have the chance of a long-term breakthrough and will finally enter industrial scales of production. Now, the potential is undoubtful, and it will be highly interesting to follow the further progress in advanced heteroatom-doped carbons, and how such materials will begin to revolutionize the field of materials for energy applications.

## Acknowledgments

This research was supported by a National Research Foundation (NRF) grant (NRF-2010-0029245) and the Global Frontier R&D Program on Center for Multiscale Energy System (NRF-2011-0031571) funded by the Ministry of Education, Science and Technology of Korea. The authors also would like to thank Korea Basic Science Institute (KBSI) at Jeonju, Daejeon, and Chuncheon for SEM, TEM, and XPS measurements.

## References

Ayala, P., Arenal, R., Rümmeli, M. et al., "The doping of carbon nanotubes with nitrogen and their potential applications," *Carbon* 48 (2010): 575–586.

Baughman, R. H., Zakhidov, A. A. & de Heerr, W. A., "Carbon nanotubes: The route toward applications," *Science* 297 (2002): 787–792.

Bhattacharjya, D. & Yu, J.-S., "Activated carbon made from cow dung as electrode material for electrochemical double layer capacitor," *J. Power Sources* 262 (2014): 224–231.

Bhattacharjya, D., Park, H.-Y., Kim, M.-S. et al., "Nitrogen-doped carbon nanoparticles by flame synthesis as anode material for rechargeable lithium-ion batteries," *Langmuir* 30 (2014): 318–324.

Botger-Hiller, F., Mehner, A., Anders, S. et al., "Sulphur-doped porous carbon from a thiophene-based twin monomer," *Chem. Commun.* 48 (2012): 10568–10570.

Chen, C. F., Tsai, C. L. & Lin, C. L., "Electronic properties of phosphorus-doped triode-type diamond field emission arrays," *Mater. Chem. Phys.* 72 (2001): 210–213.

Chen, D., Tang, L. & Li, J., "Graphene-based materials in electrochemistry," *Chem. Soc. Rev.* 39 (2010): 3157–3180.

Choi, W. I., Ihm, J. & Kim, G., "Modification of the electronic structure in a carbon nanotube with the charge dopant encapsulation," *App. Phys. Lett.* 92 (2008): 193110-1–3.

Cui, T., Lv, R., Huang, Z.-h. et al., "Effect of sulfur on enhancing nitrogen-doping and magnetic properties of carbon nanotubes," *Nanoscale Res. Lett.* 6 (2011): 77–83.

Endo, M., Kim, C., Nishimura, K. et al., "Recent development of carbon materials for Li ion batteries," *Carbon* 38 (2000): 183–197.

Ergang, N. S., Lytle, J. C., Lee, K. T. et al., "Photonic crystal structures as a basis for a three-dimensionally interpenetrating electrochemical-cell system," *Adv. Mater.* 18 (2006): 1750–1753.

Fellinger, T. P., Hasché, F., Strasser, P. et al., "Mesoporous nitrogen-doped carbon for the electrocatalytic synthesis of hydrogen peroxide," *J. Am. Chem. Soc.*, 134 (2012): 4072–4075.

Gadiou, R., Didion, A., Ivanov, D. A. et al., "Synthesis and properties of new nitrogen-doped nanostructured carbon materials obtained by templating of mesoporous silicas with amino-sugars," *J. Phys. Chem. Solids* 69 (2008): 1808–1814.

Golberg, D., Bando, Y. & Huang, Y., "Boron nitride nanotubes and nanosheets," *ACS Nano* 4 (2010): 2979–2993.

Gong, K., Du, F., Xia, Z. et al., "Nitrogen-doped carbon nanotube arrays with high electrocatalytic activity for oxygen reduction," *Science* 323 (2009): 760–764.

Gupta, S., Tryk, D., Bae, I. et al., "Heat-treated polyacrylonitrile-based catalysts for oxygen electroreduction," *J. App. Electrochem.* 19 (1989): 19–27.

Hao, L., Luo, B., Li, X. et al., "Terephthalonitrile-derived nitrogen-rich networks for high performance supercapacitors," *Energy Environ. Sci.* 5 (2012): 9747–9751.

Hasché, F., Fellinger, T.-P., Oezaslan, M. et al., "Mesoporous nitrogen doped carbon supported platinum PEM fuel cell electrocatalyst made from ionic liquids," *ChemCatChem* 4 (2012): 479–483.

Hu, B., Wang, K., Wu, L. et al., "Engineering carbon materials from the hydrothermal carbonization process of biomass," *Adv. Mater.* 22 (2010): 813–828.

Huang, X., Yin, Z., Wu, S. et al., "Graphene-based materials: Synthesis, characterization, properties, and applications," *Small* 7 (2011): 1876–1902.

Inamdar, S., Choi, H.-S., Wang, P. et al., "Sulfur-containing carbon by flame synthesis as efficient metal-free electrocatalyst for oxygen reduction reaction," *Electrochem. Commun.* 30 (2013): 9–12.

Jaouen, F., Lefevre, M., Dodelet, J.-P. et al., "Heat-treated Fe/N/C catalysts for $O_2$ electroreduction: Are active sites hosted in micropores?" *J. Phys. Chem. B* 110 (2006): 5553–5558.

Jasinski, R., "A new fuel cell cathode catalyst," *Nature* 201 (1964): 1212–1213.

Jeon, I.-Y., Choi, H.-J., Choi, M. et al., "Facile, scalable synthesis of edge-halogenated graphene nanoplatelets as efficient metal-free eletrocatalysts for oxygen reduction reaction," *Sci. Rep.* 1810 (2013): 1–7.

Kaner, R. B., Kouvetakis, J., Warble, C. E. et al., "Boron–carbon–nitrogen materials of graphite-like structure," *Mater. Res. Bull.* 22 (1987): 399–404.

Kang, K., Meng, Y. S., Bréger, J. et al., "Electrodes with high power and high capacity for rechargeable lithium batteries," *Science* 311 (2006): 977–980.

Kim, D. P., Lin, C. L., Mihalisin, T. et al., "Electronic properties of nitrogen-doped graphite flakes," *Chem. Mater.* 3 (1991): 686–692.

Lee, Y.-J. & Radovic, L. R., "Oxidation inhibition effects of phosphorus and boron in different carbon fabrics," *Carbon* 41 (2003): 1987–1997.

Liu, T., Luo, R., Yoon, S.-H. et al., "Anode performance of boron-doped graphites prepared from shot and sponge cokes," *J. Power Sources* 195 (2010): 1714–1719.

Liu, Z., Peng, F., Wang, H. et al., "Novel phosphorus-doped multiwalled nanotubes with high electrocatalytic activity for $O_2$ reduction in alkaline medium," *Catal. Commun.* 16 (2011b): 35–38.

Liu, Z.-W., Peng, F., Wang, H.-J. et al., "Phosphorus-doped graphite layers with high electrocatalytic activity for the $O_2$ reduction in an alkaline medium," *Angew. Chem. Int. Ed.* 50 (2011a): 3257–3261.

Ma, R., Golberg, D., Bando, Y. et al., "Syntheses and properties of B–C–N and BN nanostructures," *Philos. Trans. R. Soc. A* 362 (2004): 2161–2186.

Matter, P. H. & Ozkan, U. S., "Non-metal catalysts for dioxygen reduction in an acidic electrolyte," *Catal. Lett.* 109 (2006): 115–123.

Morita, T. & Takami, N., "Characterization of oxidized boron-doped carbon fiber anodes for Li-ion batteries by analysis of heat of immersion," *Electrochim. Acta* 49 (2004): 2591–2599.

Owens, B. B., Smyrl, W. H. & Xu, J. J., "R&D on lithium batteries in the USA: High-energy electrode materials," *J. Power Sources* 81–82 (1999): 150–155.

Paraknowitsch, J. P. & Thomas, A., "Doping carbons beyond nitrogen: An overview of advanced hetero-atom doped carbons with boron, sulphur and phosphorus for energy applications," *Energy Environ. Sci.* 6 (2013): 2839–2855.

Paraknowitsch, J. P., Thomas, A. & Antonietti, M., "A detailed view on the polycondensation of ionic liquid monomers towards nitrogen doped carbon materials," *J. Mater. Chem.* 20 (2010a): 6746–6758.

Paraknowitsch, J. P., Zhang, J., Su, D. et al., "Ionic liquids as precursors for nitrogen-doped graphitic carbon," *Adv. Mater.* 22 (2010b): 87–92.

Paraknowitsch, J. P., Thomas, A. & Schmidt, J., "Microporous sulfur-doped carbon from thienyl-based polymer network precursors," *Chem. Commun.* 47 (2011): 8283–8285.

Park, H.-Y., Singh, K. P., Yang, D.-S. et al., "Simple approach to advanced binder-free nitrogen-doped graphene electrode for lithium batteries," *RSC Adv.* 5 (2015): 3881–3887.

Qie, L., Chen, W.-M., Wang, Z.-H. et al., "Nitrogen-doped porous carbon nanofiber webs as anodes for lithium ion batteries with a superhigh capacity and rate capability," *Adv. Mater.* 24 (2012): 2047–2050.

Razmjooei, F., Singh, K. P., Song, M. Y. et al., "Enhanced electrocatalytic activity due to additional phosphorous doping in nitrogen and sulfur-doped graphene: A comprehensive study," *Carbon* 78 (2014): 257–267.

Redlich, Ph., Loeffler, J., Ajayan, P. M. et al., "B–C–N nanotubes and boron doping of carbon nanotubes," *Chem. Phys. Lett.* 260 (1996): 465–470.

Rodríguez, E., Cameán, I., García, R. et al., "Graphitized boron-doped carbon foams: Performance as anodes in lithium-ion batteries," *Electrochim. Acta* 56 (2011): 5090–5094.

Sakaushi, K., Nickerl, G., Wisser, F. M. et al., "An energy storage principle using bipolar porous polymeric frameworks," *Angew. Chem. Int. Ed.* 51 (2012): 7850–7854.

Schmidt, J., Weber, J., Epping, J. D. et al., "Microporous conjugated poly(thienylene arylene) networks," *Adv. Mater.* 21 (2009): 702–705.

Sevilla, M. & Fuertes, A. B., "Highly porous S-doped carbons," *Microporous Mesoporous Mater.* 158 (2012): 318–323.

Sevilla, M., Fuertes, A. B. & Mokaya, R., "Preparation and hydrogen storage capacity of highly porous activated carbon materials derived from polythiophene," *Int. J. Hydrogen Energy* 36 (2011): 15658–15663.

Sidik, R. A., Anderson, A. B., Subramanian, N. P. et al., "$O_2$ reduction on graphite and nitrogen-doped graphite: Experiment and theory," *J. Phys. Chem. B* 110 (2006): 1787–1793.

Singh, K. P., Song, M. Y. & Yu, J.-S., "Iodine-treated heteroatom-doped carbon: Conductivity driven electrocatalytic activity," *J. Mater. Chem. A* 2 (2014): 18115–18124.

Sorensen, E. M., Barry, S. J., Jung, H.-K. et al., "Three-dimensionally ordered macroporous $Li_4Ti_5O_{12}$: Effect of wall structure on electrochemical properties," *Chem. Mater.* 18 (2006): 482–489.

Stephan, O., Ajayan, P. M., Colliex, C. et al., "Doping graphitic and carbon nanotube structures with boron and nitrogen," *Science* 266 (1994): 1683–1685.

Stöhr, B., Boehm, H. P. & Schlögl, R., "Enhancement of the catalytic activity of activated carbons in oxidation reactions by thermal treatment with ammonia or hydrogen cyanide and observation of a superoxide species as a possible intermediate," *Carbon* 29 (1991): 707.

Strelko, V. V., Kuts, V. S. & Thrower, P. A., "On the mechanism of possible influence of heteroatoms of nitrogen, boron and phosphorus in a carbon matrix on the catalytic activity of carbons in electron transfer reactions," *Carbon* 38 (2000): 1499–1524.

Tanaka, U., Sogabe, T., Sakagoshi, H. et al., "Anode property of boron-doped graphite materials for rechargeable lithium-ion batteries," *Carbon* 39 (2001): 931–936.

Titirici, M.-M. & Antonietti, M., "Chemistry and materials options of sustainable carbon materials made by hydrothermal carbonization," *Chem. Soc. Rev.* 39 (2010): 103–116.

Titirici, M.-M., Thomas, A. & Antonietti, M., "Aminated hydrophilic ordered mesoporous carbons," *J. Mater. Chem.* 17 (2007): 3412–3418.

Tolvanen, A., Buchs, G., Ruffieux, P. et al., "Modifying the electronic structure of semiconducting single-walled carbon nanotubes by Ar$^+$ ion irradiation," *Phys. Rev. B* 79 (2009): 125430-1–13.

Wang, H., Cote, R., Faubert, G. et al., "Effect of the pre-treatment of carbon black supports on the activity of Fe-based electrocatalysts for the reduction of oxygen," *J. Phys. Chem. B* 103 (1999): 2042–2049.

Wang, D.-W., Li, F., Chen, Z.-G. et al., "Synthesis and electrochemical property of boron-doped mesoporous carbon in supercapacitor," *Chem. Mater.* 20 (2008): 7195–7200.

Wang, J., Chen, Y., Zhang, Y. et al., "3D boron doped carbon nanorods/carbon–microfiber hybrid composites: Synthesis and applications in a highly stable proton exchange membrane fuel cell," *J. Mater. Chem.* 21 (2011): 18195–18198.

Wena, Z. & Li, J., "Hierarchically structured carbon nanocomposites as electrode materials for electrochemical energy storage, conversion and biosensor systems," *J. Mater. Chem.* 19 (2009): 8707–8713.

Wooster, T. J., Johanson, K. M., Fraser, K. J. et al., "Thermal degradation of cyano containing ionic liquids," *Green Chem.* 8 (2006): 691–696.

Wu, Y., Fang, S. & Jiang, Y., "Carbon anode materials based on melamine resin," *J. Mater. Chem.* 8 (1998): 2223–2227.

Wu, P., Du, P., Zhang, H. et al., "Graphyne as a promising metal-free electrocatalyst for oxygen reduction reactions in acidic fuel cells: A DFT study," *J. Phys. Chem. C* 116 (2012): 20472–20479.

Xu, B., Hou, S., Cao, G. et al., "Sustainable nitrogen-doped porous carbon with high surface areas prepared from gelatin for supercapacitors," *J. Mater. Chem.* 22 (2012): 19088–19093.

Yang, L., Jiang, S., Zhao, Y. et al., "Boron-doped carbon nanotubes as metal-free electrocatalysts for the oxygen reduction reaction," *Angew. Chem. Int. Ed.* 50 (2011): 7132–7135.

Yang, D.-S., Bhattacharjya, D., Inamdar, S. et al., "Phosphorus-doped ordered mesoporous carbons with different lengths as efficient metal-free electrocatalysts for oxygen reduction reaction in alkaline media," *J. Am. Chem. Soc.* 134 (2012a): 16127–16130.

Yang, Z., Yao, Z., Li, G. et al., "Sulfur-doped graphene as an efficient metal-free cathode catalyst for oxygen reduction," *ACS Nano* 6 (2012b): 205–211.

Yang, D.-S., Bhattacharjya, D., Song, M. Y. et al., "Highly efficient metal-free phosphorus-doped platelet ordered mesoporous carbon for electrocatalytic oxygen reduction," *Carbon* 67 (2014a): 736–743.

Yang, D.-S., Chaudhari, S., Rajesh, K. P. et al., "Preparation of nitrogen-doped porous carbon nanofibers and the effect of porosity, electrical conductivity, and nitrogen content on their oxygen reduction performance," *ChemCatChem* 6 (2014b): 1236–1244.

Yao, Z., Nie, H., Yang, Z. et al., "Catalyst-free synthesis of iodine-doped graphene via a facile thermal annealing process and its use for electrocatalytic oxygen reduction in an alkaline medium," *Chem. Commun.* 48 (2012): 1027–1029.

Zhang, L., Niu, J., Dai, L. et al., "Effect of microstructure of nitrogen-doped graphene on oxygen reduction activity in fuel cells," *Langmuir* 28 (2012): 7542–7550.

Zhang, L., Niu, J., Li, M. et al., "Catalytic mechanisms of sulfur-doped graphene as efficient oxygen reduction reaction catalysts for fuel cells," *J. Phys. Chem. C* 118 (2014): 3545–3553.

# 11

# Fluorinated 0D, 1D, and 2D Nanocarbons

Nicolas Batisse

Pierre Bonnet

Katia Guérin

Marc Dubois

**Abstract**

Fluorinated nanocarbons are of great interest for high power and energy densities of primary lithium batteries and as high-temperature solid lubricants. The maintenance of the nano-objects through solid–gas fluorination is not obvious and dedicated fluorination ways should be performed for such purpose. New fluorination ways have been conducted by the $F_2$ team of the Institute of Chemistry of Clermont-Ferrand from Clermont University in order to covalently graft fluorine atoms onto carbons without degradation of the carbon texture into volatile carbon fluorides such as $CF_4$.

The various curvatures of 0D (fullerenes, nanodiamonds, nanographitized carbon blacks), 1D (nanotubes), and 2D (graphene, nanodisks) nanocarbons allow to modulate the C–F bond strength. This bond strength must be deeply characterized in order to master the consecutive properties. Among the numerous available characterization ways, solid-state NMR is of prime interest. Each fluorinated nanocarbon family will be detailed owing to its synthesis way, its C–F bond, and the known properties.

**Key words:** cathode materials, fluorination, fullerene, graphene, nanodiamonds, nanodisks, nanotubes, primary lithium battery, solid lubricant, solid-state NMR.

## 11.1 Introduction

Carbon is the key element to many technological applications from drugs to synthetic materials that have become essential in our daily life and have influenced the world's civilization for centuries. Altering the periodic binding motifs in networks of $sp^3$, $sp^2$, or $sp$-hybridized C atoms represents the conceptual starting point for constructing a wide palette of carbon allotropes. To this end, the past two decades have served as a test bed for measuring the physicochemical properties of carbonaceous nanomaterials in reduced dimensions starting with the advent of 0D fullerenes, followed, in chronological order, by 1D carbon nanotubes (CNTs) and, most recently, by 2D graphene (Guldi & Costa 2013). Those species are now poised for use in wide-ranging applications (Pichler 2007, Imahori & Umeyama 2012, Iurlo et al. 2012, Weiss et al. 2012). Varying the size, the diameter, and the morphology, for example, of these carbon nanostructures changes bandgap energies, electronic conduction, transport features, etc.

Another way to modify those properties is to functionalize the nanocarbons. Two functionalization ways compete: noncovalent and covalent ones. The resulting materials acquire new properties. For example, for the case of functionalized graphene, covalent attachment of organic functionalities to pristine graphene may allow its stable dispersion in common organic solvents to move toward the formation of nanocomposite materials with graphene. In addition, organic functional groups such as chromophores offer new properties that could be combined with the properties of graphene such as conductivity (Georgakilas et al. 2012). On the other hand, noncovalent functionalization and interactions do not disrupt the extended π-conjugation on the graphene surface, unlike covalent functionalization. All the engineering of functionalization lies in the chosen synthesis way, which may then cover from the chemistry in solution to the gas–solid reactivity. One singular chemical element has been the focus of grafting on nanocarbon: fluorine. This can be mainly attributed to the unique nature of the C–F bond, which is characterized by the small size ($r_W = 1.47$ Å) and high electronegativity ($\chi = 4$) of the fluorine atom, resulting in a short, highly polar C–F bond with low polarizability (O'Hagan 2008). These features make the C–F bond the strongest covalent single bond (105 kcal mol$^{-1}$) that carbon forms with any element. In addition to the thermodynamic stability, kinetic issues account for the notorious inertness of C–F bonds, fluorine being neither a good Lewis base nor a good leaving group.

The covalent C–F bond formation is one possibility among many others varying from ionocovalent to ionic bonding depending on the fluorination ways and the nanocarbon precursors (Nakajima & Watanabe 1991). For the case of the fluorine adsorption on the surface of carbonaceous material, these interactions are very weak. In particular, intermediate states are observed in compounds where fluorinated carbon atoms, with $sp^3$ hybridization, and nonfluorinated $sp^2$ ones coexist in the layers (hyperconjugation) (Sato et al. 2004). The C–F bonding depends on the synthesis conditions; for covalent compounds, so-called CF$_x$, namely, graphite fluorides (C$_2$F)$_n$ and (CF)$_n$, prepared with molecular fluorine at 350°C and 600°C, respectively (Ruff & Bretschneider 1934, Rudorff & Rudorff 1947), the carbon skeleton consists of *trans*-linked cyclohexane chairs or *cis–trans*-linked cyclohexane boats with $sp^3$ bonding. For the case of fluorine–graphite intercalation compounds (C$_x$F), obtained at temperature lower than 100°C, the planar configuration of graphite is partially preserved; the nature of the C–F bond evolves from ionic for low fluorine content to weaken its covalent nature for higher fluorine content. The carbon atoms are mainly in $sp^2$ hybridization state. More recently, fluorinated graphites were prepared using a room temperature synthesis in the presence of a gaseous mixture of fluorine, HF, and volatile fluorides (BF$_3$, IF$_5$, ClF$_x$, etc.) (Hamwi et al. 1988, Delabarre et al. 2005, 2006a,b, Giraudet et al. 2006).

However, for many applications such as those which will be detailed in this chapter, i.e., tribological and electrochemical applications, only the covalent C–F bonding is a usable bonding. Indeed, its inertness toward humidity and conventional electrolyte prevents the decomposition of the C–F bonding into the highly toxic HF compound. Only the fluorination ways leading to the formation of such covalent C–F bonding will be addressed and should allow the accessibility to atomic ratio $x =$ F/C varying from 0.1 to 1. Indeed, the aforementioned applications do not need the same level of grafting (Figure 11.1).

For tribological purpose, the development of new lubricant additives whose protective efficiency depends on their reactivity with the sliding surfaces, associated with the occurrence of formation of a protective tribofilm, fluorinated carbonaceous materials are promising for replacing compounds with sulfur and phosphorus contents (Bowden 1950, Deacon & Goodman 1958, Ugarte 1992, Martin & Ohmae 2008). These developments are mainly investigated to overcome the insufficiency of conventional lubricant additives for operation in severe lubrication conditions in engines and environment

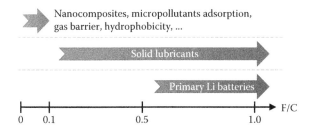

**FIGURE 11.1**  Application of fluorinated nanocarbon as a function of the fluorination level.

protection. Moreover, nanoadditives are subjected to form the protective tribofilm, in the physical–chemical conditions of a sliding boundary contact, without any chemical reactions with the sliding surfaces. Fluorinated carbon nanoparticles made from lamellar structure (carbon nanotubes, onions, nanohorns, peapods, etc.) exhibit good intrinsic friction properties, such as carbon lamellar compounds (graphite) (Fleischauer et al. 1989, Martin & Ohmae 2008). The decrease in nanoparticle surface energies by chemical covalent functionalization with fluorine atoms results in an enhancement of the tribological properties (Thomas et al. 2009, Delbé et al. 2010, Thomas et al. 2011). Moreover, the shape/dimensional factors may be optimized using spherical (0D), cylindrical (1D), and discotic (2D) materials. These good properties are associated with the layered structure of graphite type phases where graphene layers are separated by van der Waals gaps through which interactions between graphene layers are extremely weak (Lancaster 1990).

For electrochemical purpose and more particularly in primary lithium battery, associating gaseous fluorine and lithium in a battery should result in the highest electromotive force of about 6.08 V. However, managing a battery with a gaseous reactive molecule is unusable as far as today. So to benefit from fluorine redox properties, any light compound with no redox properties is an interesting host matrix. Carbon appears as a material of choice. Moreover, when the fluorocarbon lattice is nanostructured, its electrochemical performances may be enhanced because the diffusion length may be decreased by structure control at the nanoscale; as a matter of fact, shorter transport distances for both electrons and lithium/fluorine ions may be achieved as well as a larger electrode/electrolyte contact area (Guo et al. 2008). The nanostructured electrode materials can not only increase the electroactivity of $Li^+$ ions but also improve the flow capacity, i.e., the power densities (Yazami & Hamwi 2007). The curvature of nanostructures nanocarbon fluorides may also be taken advantage of. This curvature is due to the definite amount of $sp^3$ and $sp^2$ carbon atoms. During the fluorination, the C–F bond formation requires the change of the carbon hybridization from $sp^2$ to $sp^3$. The curvature then prevents the formation of pure $sp^3$ hybridization for the carbon atom, since it requires an important local strain. The residual $sp^2$ hybridized orbitals imply a weakening of the overlapping of the hybridized lobes of carbon and the fluorine atomic orbitals. In other terms, this results in the weakening of the C–F bond covalence and, as a consequence, in an increase in the primary battery voltage.

Thus, in this chapter, we will discuss which fluorination way should be performed in order to maintain the nanostructuration of the nanocarbon precursor to get solid fluorinated nanocarbon. Then, we will review the known 0D, 1D, and 2D fluorinated nanocarbons and compare their tribological and electrochemical properties.

## 11.2  Strategies for Nanocarbon Fluorinations

### 11.2.1  Conventional Fluorination Ways

Generally, solid carbon fluorides are prepared by direct reaction of pure fluorine gas with carbonaceous materials (conventional fluorination called direct fluorination). A temperature higher than

350°C is needed if graphite or graphitized carbon materials (for example, petroleum coke heat treated at 2800°C) are used (Watanabe et al. 1988, Nakajima & Watanabe 1991, Nakajima & Groult 2005). The higher the reaction temperature, the higher the fluorination level $x$ ($x$ = F/C, 0.5 < $x$ < 1) of the compounds (then called $CF_x$) and the higher the C–F covalent character, where the carbon atoms are in $sp^3$ hybridization. Nevertheless, because of the decomposition reaction of $CF_x$ (whatever the $x$ value is), the fluorination temperature must not exceed 600°C, at which the $x$ value reaches 1 for graphite as starting material. When amorphous or disordered (less graphitized) carbons (petroleum coke, carbon blacks, activated carbons, etc.) (Morita et al. 1983, Watanabe et al. 1984) are used as starting materials, the fluorination level $x$ sometimes exceeds one fluorine atom per carbon atom, indicating the formation of $CF_2$ and $CF_3$ groups. These latter groups should be less electrochemically active than the C–F groups because C–F bond energies in >$CF_2$ and –$CF_3$ are higher and the specific faradic capacity decreases. For such kind of disordered materials, different processes compete: (i) fluorination with C–F bond formation, (ii) perfluorination with $CF_2$ and $CF_3$ formations on the sheet edges and defects, and even (iii) decomposition, in which volatile carbon fluorides are evolved. Therefore, graphite (natural or artificial) and highly graphitized carbon (coke or other high-temperature heat-treated material) with high amounts of $sp^2$ carbon are more suitable as starting materials because of the easier control of the reaction.

## 11.2.2 Nanocarbon-Dedicated Fluorination Ways

As already mentioned, curvature of nanocarbon sometimes implies high level of $sp^3$ and also high amount of sheet edges which are very sensitive to perfluorination. The fluorination with C–F bond formation appears then as a real challenge. As a consequence, new fluorination ways have been developed especially for nanocarbons such as subfluorination or controlled fluorination, which proceeds by fluorination through the decomposition of a solid fluorinating agent.

The subfluorination process is done by combining a minute control of the fluorination conditions, fluorine gas flow rate, and time (Yazami & Hamwi 2006, 2007, 2008a,b) in order to get fluorinated materials with nanodomains of nonfluorinated carbon atoms in their core. These conductive nonfluorinated domains facilitate the electron flow within the particles. The subfluorination process has been applied on various carbon types and the highest electrochemical densities have first been obtained with carbon nanofibers (CNFs) as a precursor of this new fluorination way.

Another solution for the enhancement of the intrinsic electrical conductivity of $CF_x$ cathode materials consists of favoring a better distribution of the nonfluorinated part in between the fluorinated one. It can be achieved by either using pure molecular fluorine by static process (a defined amount of $F_2$ in a closed reactor for a long duration in order to achieve homogenous distribution of F atoms) or using more diffusive and reactive atomic fluorine F° species formed during the decomposition of a solid fluorinating agent such as $TbF_4$ and $XeF_2$ (Boltalina 2000). This fluorination process was called *controlled fluorination* when applied to nanocarbons. For example, for the case of $TbF_4$ as fluorinating agent, the $TbF_4$ powder was obtained before from $TbF_3$ (Sigma-Aldrich, 99.9%) fluorination in pure $F_2$ gas at 500°C. Its purity (i.e., the absence of residual $TbF_3$) must be systematically checked by X-ray diffraction. The thermogravimetric analysis of freshly prepared $TbF_4$ indicated that exactly 1 mol of F° was released per mole of $TbF_4$ between 300°C and 500°C. For the fluorination by $TbF_4$ of nanocarbon, a closed nickel reactor was used in order to preserve the defined fluorine amount released by the thermal decomposition of $TbF_4$. A two-temperature oven was used: the part containing $TbF_4$ was heated at 450°C whatever the experiment was, whereas nanocarbons were heated at a higher temperature $T_F$ which depended on the nanocarbon reactivity toward fluorine (the samples are denoted C-$T_F$, C for "controlled"). Before the heating, a primary vacuum (~$10^{-2}$ atm) was applied into the reactor. The reactions involved during the fluorination are the following:

$$TbF_4 \xrightarrow{\Delta} TbF_3 + F° \text{ or } \frac{1}{2}F_2,$$

$$C + xF° \text{ or } \frac{x}{2}F_2 \rightarrow CF_x.$$

As the reactive species are different between subfluorination and fluorination, the fluorination mechanisms differ. With $TbF_4$, because the fluorine amount is by the equilibriums $TbF_4(s) \rightleftharpoons TbF_3(s) + F°$ and $2F°(g) \rightleftharpoons F_2(g)$, the fluorination is more progressive and homogeneous, leading to the formation of $(CF)_n$-type phase whatever the fluorination conditions are, i.e., the denser fluorinated phase. On the contrary, the direct fluorination with $F_2$ gas leads to an intermediate $(C_2F)_n$ structural type in a wide temperature range, $(CF)_n$ being reached only at the highest fluorination temperature.

# 11.3 Dimensionality and C–F Bonding

## 11.3.1 0D Fluorinated Nanocarbons

Among the different nanocarbons of zero dimensionality, three have been the focus of functionalization by fluorination: fullerenes, nanodiamonds, and more recently graphitized carbon blacks. Let us begin with the fluorination of $C_{60}$ fullerene.

Due to the large number of addition sites, reaction of fluorine with $C_{60}$ fullerene leads to a wide composition range, from $C_{60}F_2$ to $C_{60}F_{60}$. Owing to the steric hindrance that occurs at the surface of the molecule during the fluorine addition, fluorofullerene with composition of $C_{60}F_{48}$ is considered as the compound with the maximum fluorine content that can be synthesized in pure form through the direct fluorination using pure molecular fluorine (Boltalina 2000, Banerjee et al. 2005). Further fluorination tends to induce internal rupture at the carbon skeleton level. Among many available synthesis routes, the direct fluorination, under gaseous $F_2$, usually results in polydispersed products, as revealed by mass spectrograms. Nevertheless, when the fluorination temperature is increased, the endpoint product composition range can be narrowed, and the maximum fluorine content can be exclusively obtained under $F_2$ gas over a 300°C–350°C temperature range (Gakh et al. 1994, Boltalina et al. 1996, Boltalina 2000, Gakh & Tuinman 2001, Troyanov et al. 2001, Niyogi et al. 2002). $C_{60}F_{48}$ can also be obtained at a high yield (80%) and purity (96%) using the following two-stage method: $C_{60}$ is embedded into an $MnF_2$ matrix and exposed to a flow of fluorine gas allowing the sublimation of volatile products. Subsequent fluorination of the $C_{60}F_{34}$–$C_{60}F_{38}$ mixture thus obtained at elevated pressure resulted in the formation of $C_{60}F_{48}$ final product (Kepman et al. 2006). Direct fluorination can be activated by UV irradiation (Tuinman et al. 1993, Matsuo et al. 1996). This results in higher fluorine content $C_{60}F_{50}$. Moreover, various fluorinating agents were successfully used to synthesize highly fluorinated fullerenes, such as $XeF_2$ ($C_{60}F_{60}$) (Holloway et al. 1991), $KrF_2$ ($C_{60}F_{44}$ and $C_{60}F_{46}$) (Boltalina et al. 1995), and $TbF_4$ ($C_{60}F_{40-44}$) (Boltalina et al. 1998). Treatment with $CeF_4$ ($C_{60}F_{36}$) and $K_2PtF_6$ (mainly $C_{60}F_{18}$) resulted in a lower fluorine content (Niyogi et al. 2002). Using either fluorine gas or fluorinating agent, thermally induced disruption of the fullerene skeleton can occur above 350°C resulting in the formation of $C_{60}F_x$ ($x > 60$) species (hyperfluorination). In a general way, the content of $C_{60}F_{48}$ strongly depends on temperature and time of reaction (Tuinman et al. 1993, Banerjee et al. 2005, Taylor 2006).

The most fluorinated compound, i.e., $C_{60}F_{48}$, is a white powder because the increase in the fluorine content is accompanied with an increase in the depletion of the $\pi/\pi^*$ states and with an increase in the HOMO–LUMO gap, as for isolated fluorofullerene molecules, and therefore with a reduced opportunity of an electronic transition occurrence in the visible range. Thus, over the $C_{60}F_x$ series with $x = 2, 8, 18, 20, 36,$ and 48, the colors brown, red, yellow green, off-white, cream white, and white are distinctly observed (Boltalina et al. 2001, 2002). $C_{60}F_{48}$ samples exhibit high stability in air for very long storage duration (several months) since neither weight uptake nor changes in X-ray diffractograms or $^{19}F$ NMR spectra were observed (Zhang et al. 2010a). $^{13}C$ spectrum recorded using a solid echo sequence shows that the $x$ value in $C_{60}F_x$ of 47.7 is obtained for $C_{60}$ fluorinated at 300°C under pure fluorine gas (Zhang et al. 2010b). The fitting procedure results in a maximal error of 0.05 for $x$. This value is not in accordance with weight uptake ($x = 44$) because the formation of volatile species during fluorination (decomposition) at 300°C results in an underestimation of the value by weight uptake. The increase in fluorination temperature results in the increase in both decomposition and underestimation. The bond length can be obtained by NMR measurements with inverse $^{19}F \rightarrow {}^{13}C$ cross polarization sequence. The extracted averaged C–F bond lengths ($d_{C-F}$) were equal to $0.147 \pm 0.002$ nm (Zhang et al. 2010a). Two reasons could explain why the average C–F bond in highly fluorinated fullerene is significantly longer than for covalent graphite fluorides: (i) Fluorinated $C_{60}$ are independent molecules on which the fluorine atoms stand outside of the fullerene cage. This could lead to steric hindrance between fluorine atoms. By analogy with graphite fluorides and fluorine–graphite intercalation compound, the hyperconjugation results in the weakening of the C–F covalence. (ii) The curvature of the carbon lattice also acts, as discussed in the Section 11.1. It is important to note that the curvature is important for the case of spherical fullerene. A weakening of the C–F bonding covalence occurs, which affects the C–F bond strength and leads to a lack of robustness of the C–F interaction within fluorofullerene clusters. Such an effect can be experimentally observed by their thermally activated decomposition. Whereas a perfluorinated graphite plane remains unaffected, fluorofullerenes readily lose bonded fluorine above 350°C in the absence of a fluorine atmosphere.

A novel carbonaceous material emerges and appears as a very interesting 0D nanocarbon for starting material of fluorination: the graphitized carbon blacks (GCBs). The commercially available product consists of submicrometric aggregates of spherical nanoparticles with 40–50 nm average diameters (Figure 11.2). The main interest in such compounds is their highly ordered (graphitic) structure at the nanoscale and their low cost (Barsukov et al. 2006), which make the final fluorinated derivatives promising compounds for industrial synthesis implementation and uses.

In order to reach different distributions of fluorinated and nonfluorinated regions in samples where the fluorine content $x$ in $CF_x$ is lower than 1, i.e., subfluorinated, two different fluorination ways were applied to GCBs: (i) a direct process using a flux of pure molecular fluorine $F_2$ (conventional dynamic process) and (ii) a recently developed controlled fluorination using thermal decomposition of a solid fluorinating agent ($TbF_4$); both are performed at high temperature in order to achieve C–F covalent bonds (Ahmad et al. 2014). TEM images recorded on initial GCBs and on the different fluorinated derivatives obtained with direct ($F_2$) and controlled ($TbF_4$) fluorinations are presented in Figure 11.2. GCBs appear as faceted nanoparticles resulting from their structure constituted of well-ordered nanodomains joined by faulted regions with a high curvature of the carbon layers subjected to be more reactive than the well-organized and flat domains. The direct fluorination results in a distribution of the fluorine atoms on the first layers of the GCBs, and is therefore essentially a surface fluorination when F/C is low. When the fluorination temperature increases, fluorination progresses toward the inner layers of the GCBs. The highly fluorinated angular portions are opened when the fluorination level increases, creating an opening for the insertion of fluorine atoms. The accommodation of fluorine atoms on the periphery of the particles leads to the destructuration of the fluorocarbon sheets. An opening of the carbon lattice may then occur when an excess of fluorine reacts with the more reactive parts, i.e., where a curvature is present. Controlled fluorination by solid fluorinating agent, here $TbF_4$, allows a better control of the fluorine addition, as mentioned before; this is due both to the progressive generation of atomic

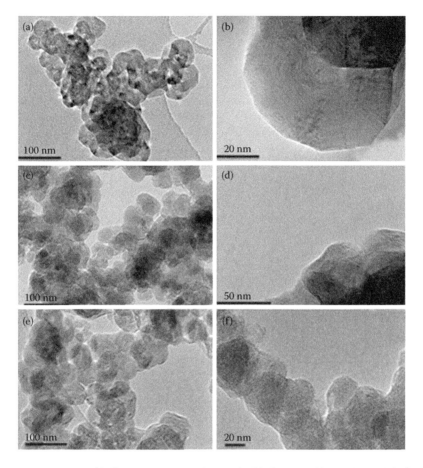

**FIGURE 11.2**    TEM images of (a, b) pristine GCBs and GCBs highly fluorinated by (c, d) $F_2$ and (e, f) $TbF_4$.

fluorine by thermal decomposition of $TbF_4$ and to its high reactivity and diffusion because of the higher temperature used for this process (480°C) compared to the direct one (420°C) for a target composition of around $CF_{0.8}$. Fluorinated particles obtained with a controlled method do not exhibit significant nanotextural differences with the raw material. The higher the fluorination temperature, the higher the kinetic diffusion. This method promotes more homogeneous fluorine distributions in the whole volume of the nanospheres.

A last case of fluorination of 0D particle is that of crystalline diamond nanoparticles or nanodiamonds (NDs), which are extensively investigated because of their unique physical and chemical properties that provide various potential applications of NDs in luminescence biomedical imaging, drug delivery, tribology, quantum engineering, surface coatings, etc. (Krueger & Lang 2012, Mochalin et al. 2012). Because pure NDs consist of pure $sp^3$ hybridization, no fluorination can occur in the same temperature as that for graphite, up to 650°C. Here, fluorination is used as a purification method in order to eliminate $sp^2$ carbon atom remaining from the ND production. On one hand, prehalogenated nanoparticles, especially fluorinated NDs (F-NDs), are considered to be suitable starting materials for the grafting of more complex polyatomic pieces (Krueger & Lang 2012). On the other hand, atomic and electronic structures of the nanoparticles surface depend on the synthesis method of the NDs and the kind of various carbon groups, which are formed on the surface in the growth processes of nanoparticles and stabilize the nanoparticle by terminating the dangling bonds (Mochalin et al. 2012). NDs are commonly synthesized as powders using the graphite–diamond phase transition at high temperature and pressure (by the detonation of a mixture of strong organic explosives with a negative oxygen balance) (Aleksenskii et al. 1997, Baidakova et al. 1998, Gruen 1999, Danilenko 2004, Dolmatov et al. 2004, Dolmatov 2005, 2007, Terranova et al. 2008, Zou et al. 2009) or as films using chemical formation with plasma-assisted CVD techniques (Tang et al. 2003, Li et al. 2012). Along with these techniques, the other ways of the ND synthesis—laser ablation (Spencer et al. 1976, Yang et al. 1998), high-energy ball milling of high-pressure high-temperature (HPHT) diamond microcrystals (Hall 1958, Loshak & Alexandrova 2001), shock compression (SC) of different carbon precursors in a metallic matrix (De Carli 1966, Setaka 1983), autoclave synthesis from supercritical fluids (Gogotsi et al. 1996), etc.—are known and often used.

Different types of NDs have been fluorinated under a flux of pure $F_2$ (1 atm) at 500°C for 12 h:

1. Those provided from the raw detonation (raw-NDs) produced by Gansu Lingyun Nano-Material Co., Ltd., Lanzhou, China, without an additional treatment and size separation. In this case, diamond particles of size 4–5 nm were combined into aggregates with a size up to 100 nm (Krüger et al. 2005).
2. Raw-ND treated by acid treatments and disintegration with large excess of $ZrO_2$ microbeads to reduce the particle size distribution; the resulting sample called ND4 (of particle size about 4 nm) was purchased as NanoAmando° from NanoCarbon Research Institute.
3. Size-separated powdered diamond from bulk synthesized by static HPHT method (ND30) with a median diameter of 30 nm. So-produced NDs were treated with heated solutions of acids to remove impurities (ND30 were provided by Tomei Diamond Co.).
4. ND powder synthesized by SC method; the size-separated particles had an average diameter of 50 nm (these NDs were provided by Nichiyu Co., Japan).

The various 4–50 nm in size diamond nanoparticles prepared by different synthesis methods and their fluorinated derivatives were studied by near-edge X-ray absorption fine structure (NEXAFS), solid-state NMR, and Fourier transform infrared spectroscopy (Ferrari & Robertson 2004, Chung et al. 2007, Kharisov et al. 2010, Dworak et al. 2014). C 1s and F 1s NEXAFS spectra of as-prepared (NDs) and F-NDs were measured in the bulk- and surface-sensitive modes—total electron yield (TEY) and partial electron yield, respectively. The comparative analysis of the C 1s absorption spectra of as-prepared NDs and reference compounds (graphitized carbon nanodisks, phenol and amino acid molecules, graphite oxide) revealed that all the diamond nanoparticles studied have crystalline diamond cores and their surfaces are covered with graphite-like carbon clusters which are partially amorphized and oxidized by forming functional groups C–OH, C=O, or O=C–OH. The properties of the surface shell also depend on the synthesis method of the NDs. An investigation of the fluorination effects by NEXAFS gives direct information about the surface chemistry of the ND precursors and underlines the extent of the $sp^2$ carbon shell, low in ND4, medium in raw-ND and ND30, and large in ND50. C 1s NEXAFS spectra of various F-NDs highlighted that the fluorination of diamond nanoparticles has a purely superficial character: it almost completely cleans the ND particles from the carbon clusters by formation of volatile $CF_4$ and $C_2F_6$ species, and then the fluorine atoms are attached to the near-surface carbon atoms of the diamond nanoparticles. A strong hydrophobic character is then added by the conversion of C–OH into C–F groups. $^1H \rightarrow {}^{13}C$ cross-polarization NMR spectra with spinning speed of 8 kHz (Figure 11.3a) were also recorded to underline the presence of oxygenated/hydrogenated groups (C–OH, CO, or COOH) in the as-prepared NDs. Whatever the sample is, the line with a $^{13}C$ chemical shift close to 50 ppm is observed in addition to the one of diamond carbon (35 ppm); it is typical of aliphatic carbons and it underlines the presence of C–H and $CH_2$ groups. The second common line at 75 ppm is assigned to C–OH groups. The ND50 sample exhibits different characteristics with the presence of additional lines at around 130 and 170 ppm, which are assigned to aromatic carbons (C=C) and carbonyl carbons (C=O), respectively. The presence of such aromatic carbons in an aromatic carbon shell which covers the diamond core was predicted by Raty and Galii (2003) and Raty et al. (2003) and observed using NMR by Panich (1999) for detonation NDs (Shames et al. 2002). Contrary to the NDs investigated by Panich et al. (2009) (also produced by detonation technique, i.e., mixture of TNT and hexogen 60/40), only one line with a chemical shift of 34.9 ppm/TMS is observed in the magic angle spinning (MAS) $^{13}$C spectrum (Dubois et al. 2009). This line is assigned to $sp^3$-hybridized carbons. Panich (1999) reported an additional line at 111 ppm (then close to aromatic carbons), not observed in our case, assigned to fullerene-like (or onion-like) shell that covers the diamond core. The observations underline the low extent of the $sp^2$ C shell for the case of ND4 in agreement with NEXAFS data (about 10% of $sp^2$-hybridized carbon atoms were obtained from NEXAFS TEY spectrum). $^1$H MAS spectra with 30 kHz spinning speed confirm the simultaneous presence of C–OH and C–H groups by the presence of two lines (Figure 11.3b). The C–OH group exhibits a chemical shift of 4.5 ppm and the value of 1.5 ppm is related to the proton bonded to the aliphatic carbon (CH and/or CH$_2$). Both the linewidth and the relative amounts of C–OH differ according to the sample. C–OH groups are in large excess for ND30 and ND50, whereas the relative amount of C–H is larger in raw-ND and ND4 (Figure 11.3b).

$^1$H → $^{13}$C CP-MAS study is in accordance with the NEXAFS data. The spectrum of ND30 sample exhibits several additional lines with low intensities (Figure 11.3a), showing that some impurities are still present; they could be carbonyl, carboxyl, ether-based resin, and other functional groups (Zou et al. 2009). Regarding the $^{19}$F MAS spectra (Figure 11.3c), the line at −164 ppm/CFCl$_3$ is observed whatever the fluorinated ND sample was. This chemical shift significantly differs from the one of CF$_2$ groups, which is located at around −130 ppm. By analogy with (C$_2$F)$_n$-type graphite fluoride, the line at −164 ppm is assigned to $^{19}$F atoms bonded to $sp^3$ carbon atoms (Dubois et al. 2006, Zhang et al. 2010b). For fluorine linked to $sp^3$ carbon, the chemical shift depends on its neighboring: when C–F bonds are present, $\delta_{19F}$ is

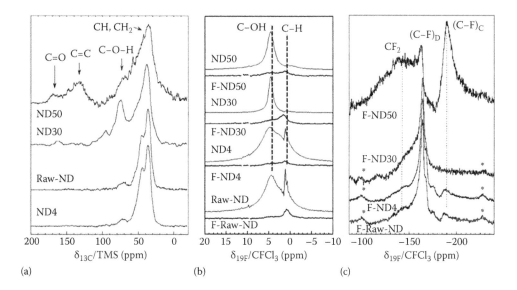

**FIGURE 11.3** Room temperature NMR spectra of the various as-prepared and fluorinated NDs: (a) $^1$H → $^{13}$C CP-(MAS spectra and spinning rate of 8 kHz); (b) $^1$H NMR spectra (500.33 MHz) (MAS with spinning rate of 30 kHz); and (c) room temperature $^{19}$F NMR spectra (30 kHz spinning rate). Solid-state NMR experiments were carried out in room temperature using Tecmag spectrometer with operating frequencies of 500.33, 470.74, and 125.81 MHz for $^1$H, $^{19}$F, and $^{13}$C chemical shifts, respectively. The NMR spectra recorded were the following ones: $^1$H NMR spectra (MAS with spinning rate of 30 kHz), $^{19}$F NMR with spinning rate of 30 kHz, and cross-polarization (CP) spectra $^1$H → $^{13}$C CP-MAS (spinning rate of 8 kHz) as well as $^{19}$F → $^{13}$C MAS (spinning rate of 12 kHz) of the as-prepared and fluorinated NDs. A simple sequence (τ-acquisition) was used with a single π/2 pulse length of 3.5 μs for $^1$H and $^{13}$C nuclei and recycle times of 3 and 5 s. For $^{19}$F MAS spectra, the π/2 pulse duration was of 5.5 μs and the recycle time was equal to 5 s. In addition to the time gain for the acquisition, cross-polarization sequence allows to favor carbon atoms in the neighboring of $^1$H or $^{19}$F nuclei. The measurements were then focused on the hydrogenated surface shell or on the fluorinated layer. The reference for both $^1$H and $^{13}$C chemical shifts in measured NMR spectra was taken as tetramethylsilane (TMS), while $^{19}$F chemical shifts were referenced to CFCl$_3$.

equal to −190 ppm, whereas if the neighboring consists of nonfluorinated carbons, the chemical shift is −164 ppm. This is related to the interaction between fluorine and the neighbor, which weakens the C–F covalence. The lower the C–F covalence, the higher the chemical shift (Krawietz & Haw 1998, Panich 1999, Giraudet et al. 2006). These C–F bonds seem to be mainly formed through the conversion under fluorine gas at 500°C of C–OH rather than CH groups. As a matter of fact, the $^1H$ spectra of as-prepared and fluorinated samples underline that the C–H line at 1.5 ppm is still measured after the fluorination treatment contrary to C–OH line (Figure 11.3b). It is also important to note from Figure 11.3c that the amount of CF groups is very low ($\delta_{19F} = -130$ ppm). Such groups could be formed by the conversion of $CH_2$ groups that indirectly underlines that the initial state contains low amount of $CH_2$. Due to the high content of amorphous carbons in ND50, already highlighted by NEXAFS and $^{13}C$ NMR, the fluorination results in two types of C–F bonds: fluorine bonded on $sp^3$ carbon atoms as for the other NDs (−164 ppm) and C–F bonds in fluorinated amorphous carbons (−190 ppm), denoted as $(C–F)_C$ in Figure 11.3c. These latter C–F bonds exhibit $\delta_{19F}$ similar to covalent graphite fluoride.

## 11.3.2 1D Fluorinated Nanocarbons

The halogenation of the CNTs has been extensively studied in the literature, especially the fluorination. It appears that the iodination and bromination leads to noncovalent functionalization of 1D nanocarbons (Rao et al. 1997, Grigorian et al. 1998, Jin et al. 2000, Do Nascimento et al. 2008, 2009, Bulusheva et al. 2012, Ghosh et al. 2012), except in the case of the iodination of SWCNTs with a modified Hunsdiecker reaction (Coleman et al. 2007) and the microwave-assisted bromination of double-walled carbon nanotubes (DWCNTs) (Colomer et al. 2009), while their fluorination is still covalent (Mickelson et al. 1998, An et al. 2002, Lee et al. 2003b, Unger et al. 2004, Lee 2007, Bittencourt et al. 2009, Zhang et al. 2012). Chlorination is little studied and, most often, it leads to covalent bonds (Abdelkader et al. 2013, 2014, Mercier et al. 2013, 2014).

Fluorination is one of the most studied and important ways for covalent chemistry of CNTs. Not only is fluorination a good starting point for further covalent sidewall modification of CNTs (alkylation, hydroxylation, amino functionalization, etc.) (Bahr & Tour 2002, Zhang et al. 2004, Banerjee et al. 2005, Khabashesku 2011) but also fluorinated carbon nanotubes (F-CNTs) exhibit a large range of possible applications themselves. Fluorination allows dispersion of CNTs in alcoholic solvents (Mickelson et al. 1998) and potential applications of F-CNTs include electric storage as cathodes in lithium batteries (Peng et al. 2001, Root 2002), in supercapacitor electrodes (Lee et al. 2003a), sensors, as solid lubricants (Vander Wal et al. 2005, Thomas et al. 2009, Im et al. 2011), or hydrophobic films (Armentano et al. 2008). Experimentally, a wide variety of conditions and parameters were used to fluorinate CNTs: nature of the CNT (SWCNTs, DWCNTs, few-walled carbon nanotubes [FWCNTs], MWCNTs, CNFs) (Zhang et al. 2010b), temperature (from −191°C to ~500°C) (Chamssedine & Claves 2008, Chamssedine et al. 2011, Zhang et al. 2012), and nature of the fluorinating agent ($F_2$, $CF_4$, $XeF_2$, etc.). Most of CNT fluorination experiments have been performed with molecular fluorine gas ($F_2$) pure or diluted, in dynamic or static mode (Mickelson et al. 1998, Marcoux et al. 2002, Lee et al. 2003b, Pehrsson et al. 2003, Zhang et al. 2012), but several other methods have also been reported including plasma treatments with $CF_4$ (Khare et al. 2004, Plank et al. 2005, Valentini et al. 2005) or $SF_6$ (Hou et al. 2008), bromine trifluoride ($BrF_3$) (Yudanov et al. 2002, Bulusheva et al. 2010), and xenon difluoride ($XeF_2$) (Unger et al. 2004); liquid medium with $HPF_6$ has also been used (Maiti et al. 2012).

The fluorine content $x$ = F/C depends on the conditions of synthesis and the nature of the CNT (Figure 11.4). For SWCNTs treated under $F_2$ at 250°C, the composition can achieve the theoretical limit $x$ = 0.5. In general, fluorination under $F_2$ gives the highest fluorine content, which increases with the temperature. Nevertheless, above 400°C under $F_2$, SWCNTs are destroyed. The stability of MWCNTs increases under fluorine atmosphere due to the reduced curvature. Thus, it is possible to fluorinate MWCNTs under $F_2$ at 520°C and reach the saturation level $x$ = 1. As a general rule, for similar fluorine content, the larger the CNT diameter, the higher the fluorination temperature (Figure 11.4). Nevertheless, it is possible to fluorinate MWCNTs at low temperature. For instance, MWCNTs were fluorinated at room temperature under catalytic conditions (mixture $F_2$, HF, and $IF_5$). To achieve the low-temperature fluorination (<250°C) of MWCNTs with significant fluorine amounts, Wang et al. (2013a) proposed to pretreat MWCNTs by oxidizing treatments, and then, oxygen-related groups provide sites for fluorination. As summarized in Figure 11.4, the fluorine content $x$ depends not only on the fluorination temperature, the nature of CNT, and their diameters, but also on the nature of the fluorinating reagent and the fluorination conditions (static versus dynamic, diluted versus pure gas, exposure time, etc.). The nature of the C–F bond can be very variable from covalent to semi-ionic; it strongly depends on the conditions of fluorination (temperature, reactive species [molecular or atomic fluorine], etc.). For example, fluorination of SWCNTs at low temperature (from −191°C to 25°C) under pure $F_2$ gas leads to C–F bonds significantly weaker than for samples fluorinated at 280°C (Chamssedine et al. 2011). Reversely, if the fluorination is done at low temperature, fluorine atoms can then be removed by moderated heating until 300°C or by vacuum without strong damage. When the C–F bonding is stronger, the defluorination is generally performed by thermal treatment at 400°C or by washing with hydrazine (Mickelson et al. 1998). Because high temperatures of fluorination lead to strong C–F bonds, high-temperature defluorination is needed. Moreover, the higher the curvature, the lower the covalence.

The distribution and homogeneity of fluorine atoms on nanotubes is an important issue. Thus, the fluorination reaction is often an interesting way to activate the surface of the nanotubes for various applications or further functionalizations, but simultaneously it can be wished to keep the inner walls intact (for DWCNT, MWCNT, or CNF) to retain their intrinsic properties (conductivity, etc.). The following paragraphs will show how, by proper selection of the fluorinating agent, it is possible to control, for near fluorine contents, the distribution of fluorine atoms in the shells of the nanotubes.

The fluorination of SWCNTs by $F_2$ and $XeF_2$ is singular. $XeF_2$ resulted in samples with $x$ = F/C fluorine contents of 0.05, 0.10, 0.20, 0.28, and 0.32. For comparison, two samples were synthesized under $F_2$ at 100°C and 200°C with F/C values of 0.04 and 0.37, respectively.

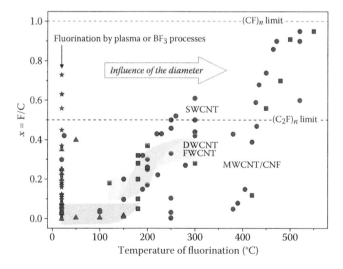

**FIGURE 11.4** Evolution of fluorine content ($x$ = F/C) depending on the temperature of fluorination for 1D nanocarbons (SWCNT in black; DWCNT and FWCNT in red; MWCNT and CNF in blue). Fluorination reagents and processes are symbolized as follows: $F_2$ (spheres, ●); $XeF_2$ and $TbF_4$ (squares, ■); plasma processes (stars, ★) and other types of processes (triangles, ▲). Data extracted from the literature.

The TEM pictures in Figure 11.5 clearly show that for similar fluorine contents of fluorinated SWCNT, the structures are very different according to the fluorination method. The direct method (under $F_2$) appears to be very aggressive, whereas the fluorination with $XeF_2$ is less damaging and more homogeneous. Samples were studied by solid-state NMR. $^{19}F$ MAS NMR spectra were recorded with a spinning speed of 20 kHz. The spectrum for fluorinated SWCNT ($x = 0.32$) prepared by direct fluorination under $F_2$ exhibits an intense line at $-163$ ppm/$CFCl_3$ assigned to covalent C–F bonds (Mallouk et al. 1985, Panich 1999, Giraudet et al. 2004, 2007, Dubois et al. 2006). The spectrum of fluorinated SWCNT prepared with $XeF_2$ ($x = 0.28$), in addition to these covalent bonds, also features a line at $-152$ ppm indicating weaker carbon–fluorine bonding. The $-163$ ppm value for both samples is explained by this main effect and represents high surface density fluorination. For this sample, the spectrum shows a second line at $-152$ ppm. If fluorine atoms are homogenously dispersed over the tube surface, C–F bonds must be separated by nonfluorinated carbon atoms for compositions lower than $CF_{0.5}$. Hyperconjugation may occur, and coupled with mechanical lattice resistance, the formation of fully $sp^3$ C results in a further weakening of the covalence. This explains the $^{19}F$ chemical shift increase from $-163$ to $-152$ ppm. In the case of controlled fluorination (with $XeF_2$), the two types of C–F bonds coexist.

The fluorination of SWCNTs strongly affects the intensity but not the absolute shift of the radial breathing mode (RBM) peaks. For pristine SWCNTs, Raman spectrum in the RBM region (low-frequency range) exhibits a main peak around 250 cm$^{-1}$ (Figure 11.6). The fluorination of SWCNT with $XeF_2$ leads to main contributions to the RBM peaks at around 200 cm$^{-1}$ at low fluorine contents ($x = 0.05$), and 250 cm$^{-1}$ at high concentrations ($x = 0.28$). In the case of fluorination by the direct method ($F_2$), the Raman spectrum at low fluorine contents is very similar to that of pristine nanotubes with intense modes at around 250 cm$^{-1}$. The differences at low fluorine contents for the two methods are attributed to different fluorine atom concentrations on the nanotube surfaces. If the fluorine atoms are homogeneously dispersed on the surface of the shell, the resonance conditions are changed, leading to changes in Raman spectrum in the RBM region; conversely, if a large part of the surface of SWCNT is not functionalized, their Raman signal will be similar to the signal of the pristine SWCNT. At higher concentrations, there are no differences between the two methods because of a large coverage of the nanotube's surface by the fluorine atoms. In conclusion, the choice of the fluorination method allows obtaining fluorinated SWCNT with not only similar fluorine contents but different morphologies, as demonstrated via TEM imaging (Figure 11.7), but also different fluorine concentrations, at low fluorine content. Indeed, the process using atomic fluorine ($XeF_2$) appears as a controlled method compared with direct fluorination ($F_2$). The fluorine atoms are homogeneously dispersed on the walls of the nanotubes, whatever their position in the bundle is, as shown by our NMR and Raman results. On the contrary, the fluorinated parts formed by direct fluorination under $F_2$ flow seem more concentrated. DFT calculations suggest that the reaction with $F_2$ probably starts on the defects of SWCNTs and the fluorination process

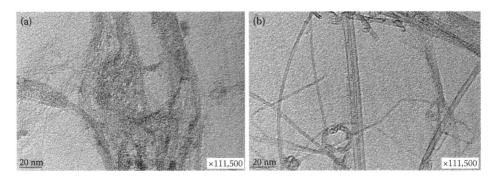

**FIGURE 11.5** TEM pictures: (a) fluorinated SWCNT (F-SWCNT) $x = 0.37$ under $F_2$ and (b) F-SWCNT $x = 0.32$ with $XeF_2$. Nanotubes appear more individualized and regular.

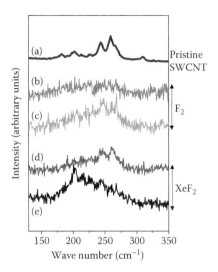

**FIGURE 11.6** Raman spectra in RBM region of (a) pristine SWCNT, SWCNT fluorinated under $F_2$: (b) $x = 0.37$, and (c) $x = 0.04$, and SWCNT fluorinated under $XeF_2$: (d) $x = 0.32$ and (e) $x = 0.05$.

then proceeds only via HF catalysis neighboring these regions, resulting in fluorinated domains less homogeneous both in the bundles and along the tubes.

Another interesting point is the fluorination of FWCNT and DWCNT and the repartition of fluorine atoms onto the surface. The fluorination of DWCNT or FWCNT was performed by $F_2$ or $BrF_3$ (Hayashi et al. 2008, Lavskaya et al. 2009). In each case, the studies showed that only the outer wall was fluorinated, leaving the internal walls intact. This leads to an inhomogeneous fluorination of CNT, but it can be

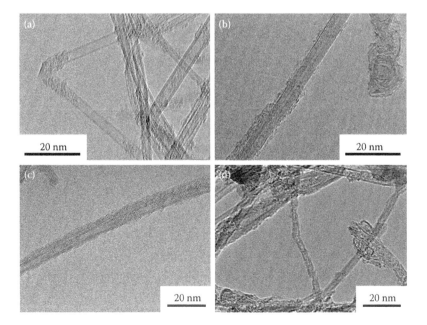

**FIGURE 11.7** TEM pictures of pristine (a) FWCNT, (b) FWCNT fluorinated by $XeF_2$, (c) FWCNT fluorinated by $F_2$, and (d) FWCNT fluorinated by $TbF_4$. The outer walls of FWCNT fluorinated with $F_2$ and $XeF_2$ appear damaged, while FWCNTs fluorinated by $TbF_4$ are totally deteriorated.

changed with other fluorinating agents. Indeed, the fluorination of FWCNT by direct fluorination with $F_2$ and the fluorination with fluorinating agents ($XeF_2$ and $TbF_4$) show noticeable differences in the homogeneity of the fluorine atoms in the structure obtained. Fluorinations were performed between 150°C and 300°C, and fluorine contents $x = F/C$ between 0.18 and 0.28 were achieved. The TEM pictures in Figure 11.7 show pristine FWCNT and fluorinated nanotubes by $F_2$, $XeF_2$, and $TbF_4$. The walls of FWCNT are damaged by fluorination: for the cases of FWCNT fluorinated by $F_2$, and $XeF_2$, only the outer walls seem to be deteriorated, while for the sample fluorinated with $TbF_4$, all the nanotube shells were damaged. Raman scattering spectroscopy was used for structural characterization of FWCNT fluorinated by different ways. In the frequency range of D and G modes (~1300 and ~1600 cm$^{-1}$), an increase in the $I_D/I_G$ ratio after fluorination was noticed; this increase was particularly important in the case of the fluorination by $TbF_4$, while the ratio for samples fluorinated by $F_2$ and $XeF_2$ were similar (Figure 11.8a). Changes in the low-frequency region (RBM modes) occurred (Figure 11.8b); these modes are related to the diameters of the shells in FWCNT and then indicate changes in outer, inner, or intermediate walls. The fluorination of FWCNT using $TbF_4$ decomposition induced the complete disappearance of the RBM modes, indicating that the complete structure of the nanotubes was fluorinated, in good agreement with the TEM images. In parallel, fluorinations with $F_2$ and $XeF_2$ do not affect all the RBM peaks; indeed, Raman spectra reveal that in both cases, the modes at the lower frequencies (~150 cm$^{-1}$) had disappeared, revealing that the outer walls of the CNT were fluorinated, as confirmed by TEM. However, the fluorination with $XeF_2$ resulted in a significant decrease in the intensity of the modes at the higher frequencies, which correspond to the smaller diameters (inner tubes). Generally, fluorination with a fluorinating agent leads to a homogenous fluorination; thus, fluorine atoms are uniformly dispersed within the tubes using $TbF_4$ as the fluorinating agent. In both cases, the fluorinations with $XeF_2$ and $F_2$ lead to fluorinated outer walls, but only the fluorination using $XeF_2$ leads to functionalization of the inner walls. Please note that with the two methods ($F_2$ and $XeF_2$), no modifications of the intermediate walls are observed, unlike with $TbF_4$.

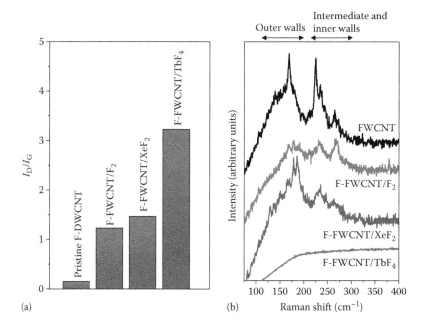

FIGURE 11.8   (a) $I_D/I_G$ ratio evolution for pristine FWCNT and fluorinated FWCNT under $F_2$ (F-FWCNT/$F_2$), with $XeF_2$ (F-FWCNT/$XeF_2$) and with $TbF_4$ (F-FWCNT/$TbF_4$), respectively. (b) Raman spectra in RBM region for FWCNT and fluorinated FWCNT under $F_2$ (F-FWCNT/$F_2$), with $XeF_2$ (F-FWCNT/$XeF_2$) and with $TbF_4$ (F-FWCNT/$TbF_4$), respectively (excitation line at 561 nm). Top double arrows indicate outer, intermediate, and inner walls (nanotube diameter evolves as $d \sim A/w$, where $w$ is the RBM nanotube frequency).

The comparative fluorination by $F_2$ and $TbF_4$ has also been performed on CNT with a higher number of walls as CNF. Because of the important diameter of CNF, a higher fluorination temperature than for SWCNT or FWCNT is needed. CNFs were heated at fluorination temperatures ranging between 420°C and 500°C. The fluorination level increases regularly with the fluorination temperature. When the fluorination temperature is lower than 420°C, $x$ is lower than 0.2, and a highly fluorinated CNF was obtained at 500°C with a corresponding fluorination level of 0.9. The TEM images of the resulting sample, fluorinated CNFs (F-CNFs) (Figure 11.9), underline that the layered structure is maintained even for high temperature and high fluorine level, and clearly show changes in the structural order during the accommodation of fluorine atoms in the graphite layers. In the case of CNF fluorinated with $TbF_4$, whatever the fluorination temperature is, the interlayer distance increases from the one of pristine CNFs (0.34 nm) up to the classical value of $(CF)_n$-type graphite fluoride structure (0.60 nm), as seen in Figure 11.9, which is confirmed by the XRD measurements presented in Figure 11.10. Conversely, for the

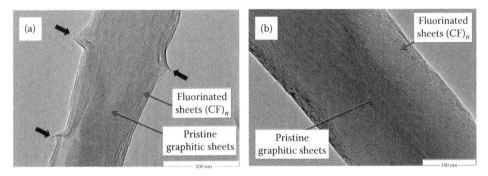

**FIGURE 11.9** TEM pictures of fluorinated CNFs (a) under $F_2$ ($x = 0.86$) (black arrows indicate exfoliated fluorinated layers; red arrows indicate fluorinated layers in $(CF)_n$ phase in outer sheets and pristine sheets in inner part) and (b) under $TbF_4$ ($x = 0.91$); no exfoliated layers are seen (red arrows indicate fluorinated layers in $(CF)_n$ phase in outer sheets and pristine sheets in inner part).

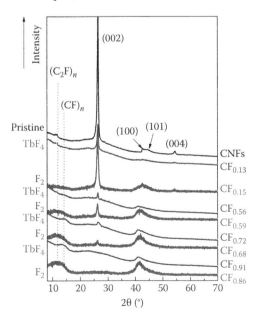

**FIGURE 11.10** XRD patterns of pristine CNF (black line) and fluorinated CNF under $F_2$ (blue lines) and $TbF_4$ (red lines) with different fluorine contents. Identification of $(C_2F)_n$ and $(CF)_n$ phases.

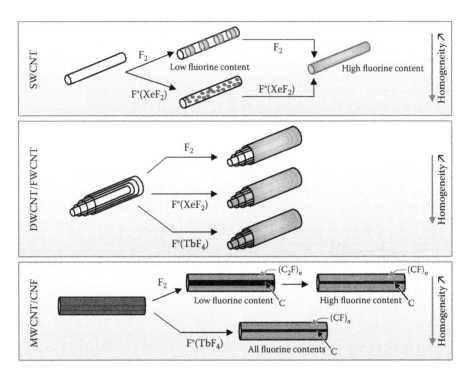

**FIGURE 11.11**  Schematic summary of the fluorination of 1D nanocarbons: cases of SWCNT, FWCNT, and CNF. The influence of the fluorinating reagent ($F_2$ versus $F°$ [$XeF_2$ and $TbF_4$]) on the fluorine atoms' homogeneity on the surface of nanocarbons.

$F_2$ way, a $(C_2F)_n$-type phase is formed first; when the fluorine content increased at the highest fluorination temperatures ($T > 480°C$), a $(CF)_n$-type phase is mainly formed. It is important to note that the fluorination under $F_2$ at high temperatures ($T > 480°C$) induces a partial exfoliation on the surface of the CNF. The higher homogeneity of the fluorinated CNF with $TbF_4$ by comparison with fluorination by $F_2$ can be explained by the diffusion rate and the reactivity of atomic fluorine $F°$ generated by the thermal decomposition of $TbF_4$, which allows a homogeneous dispersion of fluorine atoms and a more progressive fluorination. Such a behavior is common whatever carbonaceous nanomaterials, graphitized carbon blacks, or nanotubes are used.

According to the desired applications, the choice of the fluorination method is strategic to control the distribution of fluorine on the structure of CNTs. For instance, the maintenance of a consequent fraction of residual nonfluorinated tubes in CNFs is interesting in terms of electrochemical performances (maintenance of the electrical conductivity despite fluorination) (Ahmad et al. 2013). As a general conclusion, functionalizations of 1D nanocarbons with atomic fluorine ($F°$) generated by thermal decomposition of a fluorinating agent ($XeF_2$ or $TbF_4$) lead to more homogeneous fluorination than with molecular fluorine ($F_2$), as summarized in Figure 11.11.

## 11.3.3  2D Fluorinated Nanocarbons

Fluorination appears as a promising method to open and tune the bandgap of graphene (Robinson et al. 2010, Chang et al. 2011), resulting in potential applications in optoelectronic and photonic devices. For instance, transistors in a single graphene film may be fabricated (Ho et al. 2014). Fluorination is also the most effective route to introduce localized spins in graphene (Feng et al. 2013). Fluorinated graphene at high fluorine content appears as an excellent insulator with high thermal and chemical stability

(Nair et al. 2010). The covalent grafting of fluorine atoms on graphene drastically increases the friction coefficient between a nanoscale probe tip (silicon) and a surface. Such a fact is explained by the increased corrugation of the interfacial potential due to the strong local charge concentrated in the fluorine sites (Li et al. 2014). Nevertheless, the preparation of fluorinated graphene is still a real challenge because of its reactivity with fluorine gas, even when diluted with an inert gas ($N_2$ or Ar). Graphene strongly reacts with molecular fluorine $F_2$ to form volatile fluorocarbons such as $CF_4$ and $C_2F_6$ species. In other terms, graphene burns in fluorine atmosphere. A possible strategy then is to fluorinate graphite flakes or highly ordered pyrolytic graphite with $F_2$ or $XeF_2$ and to perform a mechanical exfoliation. Xenon difluoride has been used as the fluorinating agent. This method results in a homogeneous dispersion of fluorine atoms into the whole volume of the graphite, contrary to $F_2$ gas, and favors the mechanical exfoliation by cleavage of the fluorocarbon sheets (Jeon et al. 2011); exfoliation by sonication has also been performed in sulfolane (Zbořil et al. 2010). Ionic liquids (1-butyl-3-methylimidazolium bromide) were used as the only chemical to exfoliate commercially available fluorinated graphite into single- and few-layer F-graphene; the ionic liquid molecules can be mostly or fully removed from as-made F-graphene by solvent rinsing and/or thermal annealing (Chang et al. 2011). Rather than graphite fluoride, graphite intercalation compounds with $ClF_3$, with a composition of $C_2F_{0.13}ClF_3$, were mechanically exfoliated. Two different types of bonds, namely, a covalent C–F bond and an ionic C–F bond, were then present in graphene. After selective elimination of ionic C–F bonds by acetone treatment, the sample recovers the highly conductive property of graphene (Lee et al. 2013). The hydrothermal reaction between dispersed graphene oxide (GO) and HF allows fluorine atoms to be grafted onto the basal plane of graphene, and the C/F can be adjusted by controlling the reaction conditions (2.1 < C/F < 6.2); oxygenated groups, $CF_2$ and $CF_3$, are present because of the nature of the starting materials (GO) (Wang et al. 2012). Fluorinating agents may also be used directly. Few-layer fluorinated graphene sheets were prepared by fluorination with $F_2$ gas of GO up to 180°C. The amorphous region containing the oxygenated groups easily reacts with fluorine gas even at low temperature, while the aromatic region is progressively fluorinated at a higher temperature. The oxygen content further decreased with increasing temperature. The highest fluorine content is $C_1F_{1.02}O_{0.09}$ starting from GO ($C_1O_{0.426}$) (Wang et al. 2013b). Reduced GO was also treated with $XeF_2$ for investigation of magnetic properties; it was found that fluorination of those materials could greatly increase magnetization and localized spin magnetic moment (Feng et al. 2013).

$XeF_2$ was directly applied to graphene monolayer (Nair et al. 2010, Robinson et al. 2010, Stine et al. 2013, Li et al. 2014). On graphene monolayer (CVD-grown graphene), $CF_4$ plasma treatment results in fluorine radical chemisorption onto the surface (Ho et al. 2014, Wang et al. 2014). Existing grain boundaries and lattice defects of CVD graphene play an important role in controlling its rate of fluorination and the damage of the sheet. Radio-frequency-plasma with $CF_4$ was applied to dispersed exfoliated GO (Bon et al. 2009). It is important to note that fluorinated graphene is in this case used for the covalent grafting of the amino groups onto the surface for the thorough elimination of the fluorine atoms. $SF_6$ plasma treatment with ion etching system also appears efficient (Yang et al. 2011). Defluorination with hydrazine (Robinson et al. 2010) or an electron beam (Withers et al. 2010) is useful for tailoring the fluorine content and the bandgap. Treatment or irradiation effectively removes fluorine while retaining the carbon skeleton. An original method consists of irradiating fluoropolymer-covered graphene with a laser. This decomposition of the fluoropolymer produces active fluorine radicals under laser irradiation that react with graphene but only in the laser-irradiated regions (according to the laser power and fluoropolymer thickness). Such a way appears then as selective (Lee et al. 2012).

Monolayer graphene is much more susceptible to be fluorinated than thicker graphene (Yang et al. 2011). Larger corrugations of monolayer graphene than those of thicker graphenes increase the reactivity in addition to high SSA. The distribution of fluorine atoms on CVD graphene (treated with $CF_4$ plasma) is highly inhomogeneous, where multilayer islands and structural features such as folds, wrinkles, and ripples are less fluorinated and consequently form a conductive network through which charge transport occurs (Wang et al. 2014).

Thus, the stacking of graphene layers leads to an enhancement of stability under atmosphere. Carbon nanodisks were investigated as a model compound for 2D materials because they consist of stacking of disks along the c-axis. The carbon nanocones and nanodisks, provided by n-Tec Norway (Zhang et al. 2009, Stine et al. 2013), were produced by pyrolysis of heavy oil using the Kvaerner carbon black and hydrogen process (CBH) (Lynum et al. 1996). The CBH is an emission-free industrial process that decomposes hydrocarbons directly into carbon and $H_2$, based on a specially designed plasma torch, with a plasma temperature above 2000°C. The solid output was found to consist of a significant amount of opened carbon nanocones (20 wt%), as well as a large number of flat carbon disks (70 wt%), and the rest being carbon black. Some impurities are also present such as Fe (257 ppm), Si (85 ppm), Ca (65 ppm), and Al (34 ppm). The size of the crystalline domains, i.e., the coherence length along the c-axis ($L_c$) determined by XRD, is around 2 nm, which means that the as-product exhibits a low degree of order. An annealing treatment under argon at 2700°C was then performed in order to decrease the amount of amorphous carbon (the resulting sample is denoted CND). It improves the crystallinity since $L_c$ reaches a much higher value of 39 nm (Zhang et al. 2009). SEM images underline, firstly, the size dispersion of nanodisks and nanocones for the raw material. In order to quantify this dispersion, the frequency of occurrence was estimated as a function of the diameter for disks and cones. Compared to the data provided by n-Tec, no difference was observed between the as-prepared and graphitized samples, indicating that the posttreatment at 2700°C did not significantly change the geometry. Whatever the considered parameters are, the dispersion is rather large with a maximum size giving the average diameter centered at 1.5 µm for nanodisks and at 1.0 µm for nanocones. Moreover, the same average thickness of 35 nm is found for both cones and disks.

Subfluorination was carried out with pure fluorine gas flow in a nickel reactor. The fluorination temperatures $T_F$ were 480°C and 520°C in order to achieve a fluoride level between 0.7 and 1. The duration of each experiment was of 3 h. The resulting samples are denoted D-$T_F$ (D for "direct"). CNDs were fluorinated at temperatures $T_F$ of 500°C and 550°C. A reaction time of 20 h was used. Prior to the heating, a primary vacuum (~$10^{-2}$ atm) was applied into the reactor. The resulting samples are denoted C-$T_F$, C meaning "controlled fluorination." The fluorine contents $x$ in $CF_x$, after fluorination, obtained by weight uptake and confirmed by quantitative $^{19}F$ NMR, are 0.72, 0.95, 0.78, and 0.96 for C-500, C-550, D-480, and D-500, respectively. The fluorination level increases with $T_F$. Among the different samples, the highly fluorinated CNDs were obtained at 550°C using $TbF_4$ and at 500°C using $F_2$ with corresponding fluorination levels of 0.95 and 0.96, respectively. The detailed physical–chemical characterization of all synthesized materials has been fully explained elsewhere (Ahmad et al. 2012). The fluorination using $TbF_4$ decomposition generates the formation of a unique highly fluorinated phase, i.e., $(CF)_n$ type, whatever the fluorination temperature is, the C–F bonding being covalent. This fluorinated phase is always mixed with some residual nonfluorinated CND and a high fluorination level can be obtained ($x = 0.95$) without hyperfluorination, contrary to the $F_2$ process. The higher homogeneity of the materials obtained using this controlled process by comparison with the direct fluorination using fluorine gas can be explained by the low kinetic of decomposition of $TbF_4$ that allows a continuous addition of fluorine into the carbon matrix and a more progressive fluorination.

Comparisons of samples having a similar ratio of fluorination obtained by direct and controlled fluorinations indicate different mechanisms of fluorination. Structural phases $(C_2F)_n$ and $(CF)_n$ formed during the fluorination, and previously determined by XRD, are confirmed by NMR. For the direct method, an intermediate phase of $(C_2F)_n$ is formed, whereas for the controlled process, only the $(CF)_n$ phase is observed. Indeed, the peak at −180 ppm on the $^{19}F$ NMR spectra and at 42 ppm on the $^{13}C$ NMR spectra are indicators of the presence of diamond-like carbons in interaction with neighboring carbon–fluorine bonds, which are the characteristic of the $(C_2F)_n$ phase (Zhang et al. 2008b, Ahmad et al. 2013). The better distribution of atomic fluorine generated by the thermal decomposition of $TbF_4$ and the reactivity of atomic fluorine F° supposed to be higher compared to molecular one $F_2$ favor the formation of the $(CF)_n$ phase (Zhang et al. 2008a,b, 2010b).

SEM images underlined the maintenance of the geometry for both direct and control fluorinations, whatever the reaction temperature is (Ahmad et al. 2012). Nevertheless, as revealed by atomic force microscopy, the swelling due to the accommodation of the fluorine atoms differs according to the fluorination route as underlined by atomic force microscopy, which was performed in the tapping mode. D-500 and C-550 fluorinated nanodisks have been studied due to their high content in fluorine ($x$ = F/C = 0.96 and 0.95, respectively). The aspects of their surfaces are very different. The accommodation of the fluorine atoms occurs by a swelling, which is more pronounced on the edges for D-500. The surface is rougher and without visible cracking. The maximum extent of swelling is observed in the overall perimeter. The edge seems then to be the preferential area for the direct fluorination. On the contrary, the disk surface is smoother and cracks are observed on the whole surface for the case of controlled fluorination. The accommodation of fluorine atoms in the sample is then more homogeneous across the disks, thus leading to a homogeneous nonlocalized swelling. This, therefore, generates cracks on the material surface to accommodate the volume expansion associated with this process. The different locations of fluorine atoms as well as different C–F bonds according to the fluorination change the applicative properties. Considering selected examples, Section 11.4 will underline how the control of the fluorine content and the repartition of fluorine atoms allow the applicative properties to be improved.

## 11.4 Modulating C–F Bond Strength and Consecutive Properties

### 11.4.1 Tribological Properties

0D materials (spherical carbon blacks), 1D nanocarbons (CNF), and 2D ones (carbon nanodisks) with similar crystalline order (graphitized materials) were selected and fluorinated using molecular fluorine $F_2$. The tribological properties of the fluorinated nanocarbons were evaluated using a ball-on-plane tribometer. The experimental details are given by Nomède-Martyr et al. (2012). Whatever the investigated samples are, the C–F bonding is covalent without weakening due to curvature effect or hyperconjugation (see the parts related to GCBs in Section 11.3.1, CNFs in Section 11.3.2, CNDs in Section 11.3.3). This choice has been done both to facilitate the comparison and to focus on the effects of the dimensionality.

The tribological experiments start in the presence of pentane in order to establish a stable and compact film of nanoparticles on the sliding surfaces without significant changes in the nanoparticles' structures and morphologies (Thomas et al. 2009). The evaporation of the liquid allows us to determine the intrinsic friction coefficient of the materials in air. As a typical example of the tribological behavior of the various fluorinated derivatives, Figure 11.12 presents the evolution of the friction coefficient of

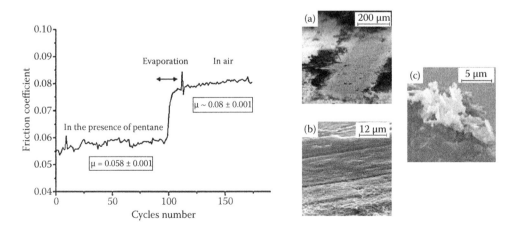

**FIGURE 11.12** Evolution of the friction coefficient of CNF-480 according to cycle numbers and SEM micrographs showing the aspect of the tribofilms for fluorinated nanocarbons: (a) GCBs, (b) CNFs, and (c) CNDs.

fluorinated nanofibers (F/C = 0.72). In the presence of pentane, the friction coefficient was around 0.06 and remained stable as long as the pentane was present. The friction reduction recorded in the presence of pentane is attributed to the simultaneous presence of the solid particles and the organic liquid in the sliding contact. During pentane evaporation, the friction coefficient increased and raised the intrinsic value of the tested compound here to 0.08 after complete evaporation of the liquid. The stabilized friction coefficients obtained for selected studied compounds are reported in Table 11.1. The friction coefficients of the nonfluorinated nanocarbons significantly differ in comparison with the fluorinated ones. For the cases of graphitized carbon blacks ($\mu_{GCBs}$ = 0.12 ± 0.01) and CNFs ($\mu_{CNFs}$ = 0.14 ± 0.01), the friction coefficients are close to those obtained for graphite in air (0.10–0.15), whereas $\mu$ is lower ($\mu_{CNDs}$ = 0.08 ± 0.01) for the graphitized nanodisks. This was attributed to the highly organized structure of CNDs (large graphitic domains) compared to the other nanocarbons (Thomas et al. 2011). The fluorination process leads to an improvement of the friction properties (decrease of the friction coefficient) for CNFs and GCBs (0.065 < $\mu$ < 0.085), whereas the friction coefficient recorded in the case of CNDs remains constant whatever the F/C content is.

At a given fluorination level, the friction coefficients of the graphitized carbon blacks are lower than for CNDs. This difference is clearly visible for the highest fluorine content (F/C = 1) as $\mu_{GCBs-400}$ = 0.070, $\mu_{CNFs-480}$ = 0.080, and $\mu_{CNDs-520}$ = 0.095. Because the fluorinated nanocarbons present similar structures and C–F bonding, the effect of the shape factor of the fluorinated nanocarbons on the tribological properties can be proposed.

Due to their structure and shape factors (diameter/thickness ratio), the CNDs appear as a particular case. The cleavage is facilitated even in the nonfluorinated sample, contrarily to both CNFs and GCBs, for which the presence of fluorine atoms, within the interlayer spacing, significantly enhances the cleavage. Moreover, the nanostructuration with either concentric cylinders of graphite sheets (Russian doll model) or a single graphite sheet rolled in around itself (parchment model) and as onion-like structure may allow the local pressure during friction to be accommodated without strong damage. Nevertheless, differences are noted between fluorinated GCBs and CNFs. The 0D GCBs exhibit at least equal and most of the time lower friction coefficients. The spherical shape of GCBs without edges could act on the friction properties. Raman analyses reveal that the friction process induces a partial deterioration of the CNFs, and SEM studies show the fibrous nature of the tribofilm, inducing that individual CNFs play a role in the friction reduction process (Thomas et al. 2009). The effect of the fluorination may be discussed in two points:

1. The fluorination of nanocarbons results in a lowering of the interparticle interactions by the reduction in the surface free energy. The fluorination of the first external layers allows such effect in the cases of CNFs and GCBs. Above the fluorination level of F/C = 0.15, the nanoparticle interaction stabilizes, resulting in friction coefficients independent of the fluorination rate in both cases. The $CF_{0.15}$ composition corresponds to a shell constituted of some tens of perfluorinated external layers for CNFs (thickness of the perfluorinated shell ≈ 15–20 nm). Both peeling of some fluorinated graphene layers and decrease of the interparticle interaction could occur for nanofibers and graphitized carbon blacks.

**TABLE 11.1** Intrinsic Friction Coefficients of the Fluorinated Nanocarbons Measured in Air

| F/C | Friction Coefficients | | |
| --- | --- | --- | --- |
| | GCBs | CNFs | CNDs |
| 0 | 0.12 ± 0.001 | 0.14 ± 0.02 | 0.08 ± 0.01 |
| 0.15 | 0.075 ± 0.001 | 0.085 ± 0.001 | 0.075 ± 0.015 |
| 0.65 | 0.065 ± 0.001 | 0.075 ± 0.001 | 0.085 ± 0.005 |
| 0.75 | 0.070 ± 0.001 | 0.080 ± 0.001 | 0.080 ± 0.005 |
| 1 | 0.070 ± 0.001 | 0.080 ± 0.001 | 0.095 ± 0.005 |

2. The mechanisms of tribofilm formation must be evoked to explain the differences between tubular and spherical shapes. SEM observation of the tribofilm formed with nanofibers revealed a thickness of 1 μm and a continuous surface which exhibits undulations parallel to the sliding direction (Figure 11.12). Their sizes are of the same order as fluorinated fibers' diameters (Thomas et al. 2009, 2011). An orientation of the fibers occurred during the formation of the tribofilm. In the same way, fluorinated GCBs could disaggregate and easily form a dense tribofilm, as seen on the SEM images (Figure 11.12). The thickness of the film is 0.6 μm. The friction properties differ for GCBs and CNFs probably according to the nature of the tribofilm and suggest different friction mechanisms involving surface effects for the case of CNFs and bulk effects for the case of CNDs. Once again, nanodisks appear as a particular case. A very smooth aspect is observed in the high-pressure zone, whereas the disks are oriented parallel to the sliding direction in the low contact pressure zone (Figure 11.12c). The tribofilm is mainly amorphous with some individual disks embedded in the disordered phase. Such disorganization in the high-pressure zone could be a consequence of severe cleavage of the disks. Also for this case, bulk effects seem to be at the origin of the friction reduction mechanism.

A new class of solid lubricants emerges then from those works: very low friction coefficients were achieved thanks to the incorporation of fluorine atoms into carbonaceous nanomaterials. When the fluorine content is intermediate or high, i.e., from 0.5 to 1, the tribological properties do not change significantly. The situation is different considering the electrochemical properties since conductive residual nonfluorinated carbon atoms act on the discharge in primary lithium batteries. The challenge was to determine the best balance between high fluorine content, to increase the capacity, and the presence of sufficient nonfluorinated carbon atoms, to ensure the conductivity.

## 11.4.2 Electrochemical Properties

Actually, in the Li/CF$_x$ battery system, a high oxidation–reduction potential of the cathode reaction is combined with low weight density of light C and F elements. The use of CF$_x$ materials as the active cathode in primary lithium batteries was first demonstrated by Watanabe and Fukuda (1970, 1972) and then a few years later, the first Li/CF$_1$ batteries were commercialized by Matsushita Electric Co. in Japan (Fukuda & Iijima 1975). Commercial Li/CF$_x$ batteries consist of a coke-based cathode having an F/C molar ratio equal or slightly higher than unity. The main features of the Li/CF$_1$ batteries are high energy density (up to 1560 Wh kg$^{-1}$), high average operating voltage (around 2.4 V versus Li$^+$/Li), long shelf life (higher than 10 years at room temperature), stable operation, and wide operating temperatures (−40°C/170°C), but low power density (around 1400 Wh kg$^{-1}$) and low faradic yield (not more than 75% because of too high amounts of inactive CF$_2$ and CF$_3$ groups and dangling bonds which are considered as structural defects and hinder the lithium diffusion).

The use of fluorinated nanocarbon as cathode materials has permitted to overcome all the conventional performances of commercial Li/CF$_x$. Concerning the increase in the operating voltage, the curvature of the carbon lattice allowed to get weakened covalent C–F bonding as studied by $^{19}$F and $^{13}$C chemical shifts (measured by solid-state NMR) and wave number of the C–F vibration band (measured by IR spectroscopy) applied on nanotubes (single-, double-, and multiwalled), spherical fluorinated fullerene, and compared to planar graphite (Zhang et al. 2010a). Both IR wave numbers and chemical shifts exhibit progressive change as a function of the curvature, indicating that the covalence is weakened when the curvature is increased (Figure 11.13).

The discharge potentials in Li batteries are unambiguously correlated with the diameter of the outer tube, i.e., the curvature of the carbon lattice. This has been done with precautions, i.e., similar fluorine content, exclusive fluorination of the outer tubes, and same electrochemical operating conditions. The higher the curvature, i.e., the lower the diameter, the higher the potential $E_{1/2}$ is. The potential of fluorinated nanofibers and covalent graphite fluoride are close, meaning that the low curvature of this kind of nanotubes does not act on the covalence anymore. This is in accordance with the nondependence

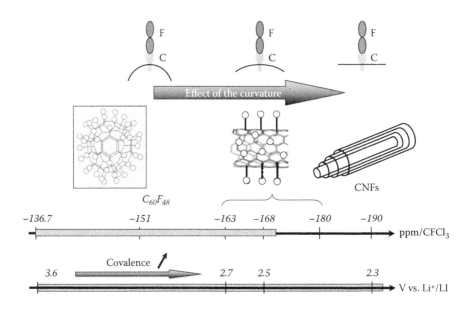

**FIGURE 11.13** Schematic comparison of operating voltage and C–F bonding shift obtained by [19]F NMR modulated by curvature effect.

of the potential according to the fluorine content in fluorinated CNFs; nearly the same potential has been measured whatever the fluorinated wall was, outer wall for low F/C ratio or inner wall when the fluorination progresses toward the core. The quasi linearity is verified in the overall series of fluorinated nanotubes and graphites. Fluorinated $C_{60}$ does not strictly follow the relationship observed for fluorinated nanotubes and graphites. An additional effect is expected to explain such a difference: steric hindrance between fluorine atoms, for the case when the fluorination rate is high, could require different bond lengths, and could then affect the C–F bonding. As fluorinated fullerenes contain several kinds of C–F bonds, depending not only on the position in pentagon or hexagon, but also on their neighbor, this case must be separately discussed. A linear dependence of the potentials with the chemical shift is also observed.

The increase in the energy density, resulting from the increase in the potential because of curvature, is very promising. Nevertheless, another condition is required to achieve high energy densities: to enhance the capacity, as energy density is the multiplication of capacity by operating voltage. The maximum capacity for $CF_1$ carbon fluorides is of 865 mAh $g^{-1}$. This capacity has been widely exceeded by the use of carbon nanodisks partially fluorinated by atomic fluorine released by thermal decomposition of $TbF_4$, for which a capacity of up to 1180 mAh $g^{-1}$ was registered (Araujo Da Silva et al. 2014). These capacities are obtained for the first time, whereas in the past many researchers have devoted their interest in fluorinated nanomaterials in primary lithium batteries. This suggests a strong effect of the nanomaterial nature, its morphology, its shape factor, and the stacking of the graphene layers (in tubes or sheets). The discharge mechanism was investigated using mainly SEM and solid-state NMR in order to understand this "extra capacity." The reinforcement effect of the central disks, coupled with a low amount of structural defects $CF_2$ and $CF_3$, is a key parameter to achieve the extra capacity. This reinforcement effect allows the rebuilding of the carbon lattice during the electrochemical defluorination, and then different sites of insertion of $Li^+$ species are possible in the reformed carbon lattice. Furthermore, the presence of a shell of LiF seems to allow the diffusion of the lithium ions (either $Li^+$ or $Li_2F^+$) through the sheet edges or the surface cracks (disruptions of the sheets); thus, insertion may be an additional phenomenon after the electrochemical defluorination resulting in the extra capacities. The discharge mechanism study underlines that the electrochemical discharge forms LiF particles outside the carbonaceous matrix. An additional lithium insertion could occur in the newly formed carbon by the electrochemical defluorina-tion. Both salt and solvent of the electrolytes were changed to better understand this experimental fact. And PC LiClO$_4$ 1M electrolyte appears as the best electrolyte to favor the formation of Li$_2$F$^+$ ions and extra capacity. Extra capacities were achieved thanks to the reinforcement effect of the central disks. The study is now extended to nanostructured carbonaceous materials where the central parts (disks, tubes, core for 2D, 1D, and 0D aspects, respectively) may act in the same manner. Low structural disorder is also required in the pristine fluorinated nanocarbons in order to achieve efficient diffusion of Li$^+$/F$^-$ ions. The low amount of CF$_2$ and CF$_3$ groups as well as the super hyperfine structure is a good indicator of the lowering of disorder, necessary to rebuild carbon-favoring extra capacities.

The last and more spectacular performances in power density increased thanks to nanostructured carbon fluorides. This can be achieved by enhancing the intrinsic electrical conductivity of CF$_x$, main-taining intimate nonfluorinated conductive carbon into the vicinity of insulating fluorinated carbon. Selective fluorination of the outer part of the nanocarbon without fluorination of the core has been the focus of numerous studies. For example, fluorinated MWCNTs with average diameter larger than 50 nm exhibit high energy density, i.e., 1923 Wh kg$^{-1}$. Moreover, the power density is also improved with a value of 7114.1 Wh kg$^{-1}$, the batteries operating with high discharge rate of up to 5 C (Li et al. 2013). This study follows our previous works on the fluorination of CNFs. An impressive 8057 W kg$^{-1}$ power density associated with a high 1749 Wh kg$^{-1}$ energy density was achieved for subfluorinated CNFs with F/C equal to 0.76 (obtained by both quantitative NMR and gravimetry). The 1D tubular morphology is up to now the better strategy to enhance the power density (Yazami & Hamwi 2006, 2007, 2008a,b, Yazami et al. 2007).

Our study on fluorinated nanofibers underlined that this effect could be very efficient on electro-chemical properties. The strategy that takes profit of the surface also involves research to prepare new fluorinated carbons starting from graphene and porous carbons or postfluorination process such as ball-milling. Some examples underline the potential improvements. Fluorinated graphene appears as very promising. Fluorinated graphene with F/C equal to 0.47, 0.66, and 0.89 exhibited faradic yield of 75%–81% at moderate rates when used as cathode materials for primary lithium batteries (Meduri et al. 2013). Among the samples, the one with CF$_{0.47}$ composition maintained a capacity of 356 mAh g$^{-1}$ at a 5 C rate, then higher than that of traditional fluorinated graphite.

## 11.5 Conclusion

The diversity of carbonaceous nanomaterials in terms of dimensionality (from 0D to 2D), nanosize, stacking, i.e., structural order (which can be tailored using thermal posttreatment), and SSA complicates the works of the fluorine chemists, the final aim being the enhancement of the applicative properties, e.g., for lubrication and lithium batteries. The most difficult challenge was probably the fluorination of a single layer of graphene because of its high surface. As a general rule, the decomposition under fluorine atmosphere must be as low as possible. The purification of nanodiamonds is an opposite example for which the decomposition under F$_2$ gas is useful to remove the $sp^2$ carbon covering. The challenge of the fluorination of a large panel of nanocarbons without large decomposition was overcome thanks to new fluorinated routes, e.g., using fluorinated agent rather than F$_2$ gas, and a minute control of the treatment conditions. The change of scale from micro (graphite and petroleum coke) to nano needed new solutions of fluorination but the benefits were huge as exemplified by the increase in the power densities by a fac-tor of 5 for subfluorinated nanofibers or the 40% extra capacity for fluorinated nanodisks. The panel of reactive species is large enough now, with solid fluorinating agents such as XeF$_2$ and TbF$_4$, in addition to F$_2$ gas, to investigate other highly reactive carbonaceous materials. For instance, the fluorination of highly porous carbons is of great interest for organic micropollutant trapping and the stability improve-ment of the catalyst in fuel cells. The presence of fluorine atoms always favors the hydrophobicity of the surface and decreases the effect of adsorbed water molecules. The hydrophobicity may be useful to increase the interaction with commercial polymers (polyethylene, polypropylene, polystyrene, etc.) and fabricated enhanced nanocomposites. In many cases, functionalization with covalent C–F bonds favors the dispersion in conventional solvents and consequently helps for further uses which require a homogenous dispersion. The solutions which have been developed for nanocarbons may be applied to another reactive material type, polymers, in order to achieve specific properties, such as gas barrier, printability, lubrication, and (super)hydrophobicity. The challenges for fluorine chemists in the field of carbonaceous nanomaterials and polymers are still numerous and stimulating.

# References

Abdelkader, V. K., Scelfo, S., García-Gallarín, C. et al., "Carbon tetrachloride cold plasma for extensive chlorination of carbon nanotubes," *J. Phys. Chem. C* 117 (2013): 16677–16685.

Abdelkader, V. K., Domingo-García, M., Gutiérrez-Valero, M. D. et al., "Sidewall chlorination of carbon nanotubes by iodine trichloride," *J. Phys. Chem. C* 118 (2014): 2641–2649.

Ahmad, Y., Disa, E., Dubois, M. et al., "The synthesis of multilayer graphene materials by the fluorination of carbon nanodiscs/nanocones," *Carbon* 50 (2012): 3897–3908.

Ahmad, Y., Guérin, K., Dubois, M. et al., "Enhanced performances in primary lithium batteries of fluorinated carbon nanofibers through static fluorination," *Electrochim. Acta* 114 (2013): 142–151.

Ahmad, Y., Disa, E., Guérin, K. et al., "Structure control at the nanoscale in fluorinated graphitized carbon blacks through the fluorination route," *J. Fluor. Chem.* 168 (2014): 163–172.

Aleksenskii, A. E., Baidakova, M. V., Vul', A. Y. et al., "Diamond–graphite phase transition in ultradisperse-diamond clusters," *Phys. Solid State* 39 (1997): 1007–1015.

An, K. H., Heo, J. G., Jeon, K. G. et al., "X-ray photoemission spectroscopy study of fluorinated single-walled carbon nanotubes," *Appl. Phys. Lett.* 80 (2002): 4235–4237.

Araujo Da Silva, K., Dubois, M. & Hamwi, A., "Utilisation de nanoobjets en carbone sous fluore comme matériau d'électrode de batteries primaires au lithium de fortes capacités," Patent WO2014091422 (2014).

Armentano, I., Álvarez-Pérez, M. A., Carmona-Rodríguez, B. et al., "Analysis of the biomineralization process on SWNT-COOH and F-SWNT films," *Mater. Sci. Eng. C* 28 (2008): 1522–1529.

Bahr, J. L. & Tour, J. M., "Covalent chemistry of single-wall carbon nanotubes," *J. Mater. Chem.* 12 (2002): 1952–1958.

Baidakova, M. V., Vul', A. Y., Siklitskii, V. I. et al., "Fractal structure of ultradisperse-diamond clusters," *Phys. Solid State* 40 (1998): 715–718.

Banerjee, S., Hemraj-Benny, T. & Wong, S. S., "Covalent surface chemistry of single-walled carbon nanotubes," *Adv. Mater.* 17 (2005): 17–29.

Barsukov, I. V., Gallego, M. A. & Doninger, J. E., "Novel materials for electrochemical power sources: Introduction of PUREBLACK® Carbons," *J. Power Sources* 153 (2006): 288–299.

Bittencourt, C., Van Lier, G., Ke, X. et al., "Spectroscopy and defect identification for fluorinated carbon nanotubes," *ChemPhysChem* 10 (2009): 920–925.

Boltalina, O. V., "Fluorination of fullerenes and their derivatives," *J. Fluor. Chem.* 101 (2000): 273–278.

Boltalina, O. V., Abdul-Sada, A. K. & Taylor, R., "Hyperfluorination of [60]fullerene by krypton difluoride," *J. Chem. Soc. Perkin Trans.* 2 (1995): 981–985.

Boltalina, O. V., Sidorov, L. N., Bagryantsev, V. F. et al., "Formation of $C_{60}F_{48}$ and fluorides of higher fullerenes," *J. Chem. Soc. Perkin Trans.* 2 (1996): 2275–2278.

Boltalina, O. V., Lukonin, A. Y., Pavlovich, V. K. et al., "Reaction of [60]fullerene with terbium(IV) fluoride," *Fuller. Sci. Technol.* 6 (1998): 469–479.

Boltalina, O. V., Markov, V. Y., Troshin, P. A. et al., "$C_{60}F_{20}$: 'Saturnene,' an extraordinary squashed fullerene," *Angew. Chem. Int. Ed.* 40 (2001): 787–789.

Boltalina, O. V., Darwish, A. D., Street, J. M. et al., "Isolation and characterisation of C60F4, C60F6, $C_{60}F_8$, $C_{60}F_7CF_3$ and $C_{60}F_2O$, the smallest oxahomofullerene: The mechanism of fluorine addition to fullerenes," *J. Chem. Soc. Perkin Trans.* 2 (2002): 251–256.

Bon, S. B., Valentini, L., Verdejo, R. et al., "Plasma fluorination of chemically derived graphene sheets and subsequent modification with butylamine," *Chem. Mater.* 21 (2009): 3433–3438.

Bowden, F. P., "Frictional properties of porous metals containing molybdenum disulphide," *Research* 3 (1950): 383–384.

Bulusheva, L. G., Fedoseeva, Y. V., Okotrub, A. V. et al., "Stability of fluorinated double-walled carbon nanotubes produced by different fluorination techniques," *Chem. Mater.* 22 (2010): 4197–4203.

Bulusheva, L. G., Okotrub, A. V., Flahaut, E. et al., "Bromination of double-walled carbon nanotubes," *Chem. Mater.* 24 (2012): 2708–2715.

Chammsedine, F. & Claves, D., "A kinetic, morphological and mechanistic approach of the fluorination of multiwall carbon nanotubes," *Chem. Phys. Lett.* 454 (2008): 252–256.

Chammsedine, F., Guérin, K., Dubois, M. et al., "Fluorination of single walled carbon nanotubes at low temperature: Towards the reversible fluorine storage into carbon nanotubes," *J. Fluor. Chem.* 132 (2011): 1072–1078.

Chang, H., Cheng, J., Liu, X. et al., "Facile synthesis of wide-bandgap fluorinated graphene semiconductors," *Chem. Eur. J.* 17 (2011): 8896–8903.

Chung, P. H., Perevedentseva, E. & Cheng, C. L., "The particle size-dependent photoluminescence of nanodiamonds," *Surf. Sci.* 601 (2007): 3866–3870.

Coleman, K. S., Chakraborty, A. K., Bailey, S. R. et al., "Iodination of single-walled carbon nanotubes," *Chem. Mater.* 19 (2007): 1076–1081.

Colomer, J. F., Marega, R., Traboulsi, H. et al., "Microwave-assisted bromination of double-walled carbon nanotubes," *Chem. Mater.* 21 (2009): 4747–4749.

Danilenko, V. V., "On the history of the discovery of nanodiamond synthesis," *Phys. Solid State* 46 (2004): 595–599.

De Carli, P. S., "Method of making diamond," Google Patents (1966).

Deacon, R. F. & Goodman, J. F., "Lubrication of lamellar solids," *Proc. R. Soc. Lond. Ser. A* 243 (1958): 464–482.

Delabarre, C., Guérin, K., Dubois, M. et al., "Highly fluorinated graphite prepared from graphite fluoride formed using $BF_3$ catalyst," *J. Fluor. Chem.* 126 (2005): 1078–1087.

Delabarre, C., Dubois, M., Giraudet, J. et al., "Electrochemical performance of low temperature fluorinated graphites used as cathode in primary lithium batteries," *Carbon* 44 (2006a): 2543–2548.

Delabarre, C., Dubois, M., Guérin, K. et al., "Room temperature graphite fluorination process using chlorine as catalyst," *J. Phys. Chem. Solids* 67 (2006b): 1157–1161.

Delbé, K., Thomas, P., Himmel, D. et al., "Tribological properties of room temperature fluorinated graphite heat-treated under fluorine atmosphere," *Tribol. Lett.* 37 (2010): 31–41.

Do Nascimento, G. M., Hou, T., Kim, Y. A. et al., "Double-wall carbon nanotubes doped with different $Br_2$ doping levels: A resonance Raman study," *Nano Lett.* 8 (2008): 4168–4172.

Do Nascimento, G. M., Hou, T., Kim, Y. A. et al., "Comparison of the resonance Raman behavior of double-walled carbon nanotubes doped with bromine or iodine vapors," *J. Phys. Chem. C* 113 (2009): 3934–3938.

Dolmatov, V. Y., "Some speculations as to the structure of a cluster of a detonation-produced nanodiamond," *J. Superhard Mater.* 1 (2005): 28–32.

Dolmatov, V. Y., "Detonation synthesis nanodiamonds: Synthesis, structure, properties and applications," *Russ. Chem. Rev.* 76 (2007): 339–360.

Dolmatov, V. Y., Veretennikova, M. V., Marchukov, V. A. et al., "Currently available methods of industrial nanodiamond synthesis," *Phys. Solid State* 46 (2004): 611–615.

Dubois, M., Giraudet, J., Guérin, K. et al., "EPR and solid-state NMR studies of poly(dicarbon monofluoride) $(C_2F)_n$," *J. Phys. Chem. B* 110, 24 (2006): 11800–11808.

Dubois, M., Guérin, K., Petit, E. et al., "Solid-state NMR study of nanodiamonds produced by the detonation technique," *J. Phys. Chem. C* 113, 24 (2009): 10371–10378.

Dworak, N., Wnuk, M., Zebrowski, J. et al., "Genotoxic and mutagenic activity of diamond nanoparticles in human peripheral lymphocytes in vitro," *Carbon* 68 (2014): 763–776.

Feng, Q., Tang, N., Liu, F. et al., "Obtaining high localized spin magnetic moments by fluorination of reduced graphene oxide," *ACS Nano* 7 (2013): 6729–6734.

Ferrari, A. C. & Robertson, J., "Raman spectroscopy of amorphous, nanostructured, diamond-like carbon, and nanodiamond," *Phil. Trans. R. Soc. Lond. A* 362 (2004): 2477–2512.

Fleischauer, P. D., Lince, J. R., Bertrand, P. A. et al., "Electronic structure and lubrication properties of molybdenum disulfide: A qualitative molecular orbital approach," *Langmuir* 5, 4 (1989): 1009–1015.

Fukuda, M. & Iijima, T., "Lithium/poly-carbonmonofluoride cylindrical type batteries" in Collins, D. H. (ed.), *Power Sources* (Academic Press, New York, 1975).

Gakh, A. A. & Tuinman, A. A., "'Fluorine dance' on the fullerene surface," *Tetrahedron Lett.* 42 (2001): 7137–7139.

Gakh, A. A., Tuinman, A. A., Adcock, J. L. et al., "Selective synthesis and structure determination of $C_{60}F_{48}$," *J. Am. Chem. Soc.* 116 (1994): 819–820.

Georgakilas, V., Otyepka, M., Bourlinos, A. B. et al., "Functionalization of graphene: Covalent and non-covalent approaches, derivatives and applications," *Chem. Rev.* 112 (2012): 6156–6214.

Ghosh, S., Yamijala, S. R. K. C. S., Pati, S. K. et al., "The interaction of halogen molecules with SWNTs and graphene," *RSC Adv.* 2 (2012): 1181–1188.

Giraudet, J., Dubois, M., Hamwi, A. et al., "Solid-state NMR ($^{19}F$ and $^{13}C$) study of graphite monofluoride $(CF)_n$: $^{19}F$ spin–lattice magnetic relaxation and $^{19}F/^{13}C$ distance determination by Hartmann–Hahn cross polarization," *J. Phys. Chem. B* 109 (2004): 175–181.

Giraudet, J., Delabarre, C., Guérin, K. et al., "Comparative performances for primary lithium batteries of some covalent and semi-covalent graphite fluorides," *J. Power Sources* 158 (2006): 1365–1372.

Giraudet, J., Dubois, M., Guérin, K. et al., "Solid-state NMR study of the post-fluorination of $(C_{2.5}F)_n$ fluorine–GIC," *J. Phys. Chem. B* 111 (2007): 14143–14151.

Gogotsi, Y. G., Nickel, K. G., Bahloul-Hourlier, D. et al., "Structure of carbon produced by hydrothermal treatment of [small beta]-SiC powder," *J. Mater. Chem.* 6 (1996): 595–604.

Grigorian, L., Williams, K. A., Fang, S. et al., "Reversible intercalation of charged iodine chains into carbon nanotube ropes," *Phys. Rev. Lett.* 80 (1998): 5560–5563.

Gruen, D. M., "Nanocrystalline diamond films," *Annu. Rev.Mater. Sci.* 29 (1999): 211–259.

Guldi, D. M. & Costa, R. D., "Nanocarbon hybrids: The paradigm of nanoscale self-ordering/self-assembling by means of charge transfer/doping interactions," *J. Phys. Chem. Lett.* 4 (2013): 1489–1501.

Guo, Y.-G., Hu, J.-S. & Wan, L.-J., "Nanostructured materials for electrochemical energy conversion and storage devices," *Adv. Mater.* 20 (2008): 2878–2887.

Hall, H. T., "Ultrahigh-pressure research: At ultrahigh pressures new and sometimes unexpected chemical and physical events occur," *Science* 128 (1958): 445–449.

Hamwi, A., Daoud, M. & Cousseins, J. C., "Graphite fluorides prepared at room temperature: 1. Synthesis and characterization," *Synth. Met.* 26 (1988): 89–98.

Hayashi, T., Shimamoto, D., Kim, Y. A. et al., "Selective optical property modification of double-walled carbon nanotubes by fluorination," *ACS Nano* 2 (2008): 485–488.

Ho, K.-I., Liao, J.-H., Huang, C.-H. et al., "One-step formation of a single atomic-layer transistor by the selective fluorination of a graphene film," *Small* 10 (2014): 989–997.

Holloway, J. H., Hope, E. G., Taylor, R. et al., "Fluorination of buckminsterfullerene," *J. Chem. Soc. Chem. Commun.* 14 (1991): 966–969.

Hou, Z., Cai, B., Liu, H. & Xu, D., "Ar, $O_2$, $CHF_3$, and $SF_6$ plasma treatments of screen-printed carbon nanotube films for electrode applications," *Carbon* 46 (2008): 405–413.

Im, J. S., Kang, S. C., Bai, B. C. et al., "Thermal fluorination effects on carbon nanotubes for preparation of a high-performance gas sensor," *Carbon* 49 (2011): 2235–2244.

Imahori, H. & Umeyama, T., "Applications of supramolecular ensembles with fullerenes and CNTs: Solar cells and transistors," in *Supramolecular Chemistry of Fullerenes and Carbon Nanotubes* (Wiley-VCH Verlag GmbH & Co. KGaA, Weinheim, Germany, 2012).

Iurlo, M., Rapino, S., Valenti, G. et al., "Electrochemistry of carbon nanostructures: From pristine materials to functional devices," in *Handbook of Carbon Nano Materials* (World Scientific, Singapore, 2012).

Jeon, K.-J., Lee, Z., Pollak, E. et al., "Fluorographene: A wide bandgap semiconductor with ultraviolet luminescence," *ACS Nano* 5 (2011): 1042–1046.

Jin, Z.-x., Xu, G. Q. & Goh, S. H., "A preferentially ordered accumulation of bromine on multi-wall carbon nanotubes," *Carbon* 38 (2000): 1135–1139.

Kepman, A. V., Sukhoverkhov, V. F., Tressaud, A. et al., "Novel method of synthesis of $C_{60}F_{48}$ with improved yield and selectivity," *J. Fluor. Chem.* 127 (2006): 832–836.

Khabashesku, V. N., "Covalent functionalization of carbon nanotubes: Synthesis, properties and applications of fluorinated derivatives," *Russ.Chem. Rev.* 80 (2011): 705–725.

Khare, B. N., Wilhite, P. & Meyyappan, M., "The fluorination of single wall carbon nanotubes using microwave plasma," *Nanotechnology* 15 (2004): 1650.

Kharisov, B. I., Kharissova, O. V., Valdés, J. J. R. et al., "Coordination and organometallic nanomaterials: A microreview," *Synth. React. Inorg. Metal-Org. Nano-Metal Chem.* 40 (2010): 640–650.

Krawietz, T. R. & Haw, J., "Characterization of poly(carbon monofluoride) by $^{19}F$ and $^{19}F$ to $^{13}C$ cross polarization MAS NMR spectroscopy," *Chem. Commun.* 19 (1998): 2151–2152.

Krueger, A. & Lang, D., "Functionality is key: Recent progress in the surface modification of nanodiamond," *Adv. Funct. Mater.* 22 (2012): 890–906.

Krüger, A., Kataoka, F., Ozawa, M. et al., "Unusually tight aggregation in detonation nanodiamond: Identification and disintegration," *Carbon* 43 (2005): 1722–1730.

Lancaster, J. K., "A review of the influence of environmental humidity and water on frictions," *Tribol. Int.* 23 (1990): 371–389.

Lavskaya, Y. V., Bulusheva, L. G., Okotrub, A. V. et al., "Comparative study of fluorinated single- and few-wall carbon nanotubes by X-ray photoelectron and X-ray absorption spectroscopy," *Carbon* 47 (2009): 1629–1636.

Lee, Y.-S., "Syntheses and properties of fluorinated carbon materials," *J. Fluor. Chem.* 128 (2007): 392–403.

Lee, J. Y., An, K. H., Heo, J. K. et al., "Fabrication of supercapacitor electrodes using fluorinated single-walled carbon nanotubes," *J. Phys. Chem. B* 107 (2003a): 8812–8815.

Lee, Y. S., Cho, T. H., Lee, B. K. et al., "Surface properties of fluorinated single-walled carbon nanotubes," *J. Fluor. Chem.* 120 (2003b): 99–104.

Lee, W. H., Suk, J. W., Chou, H. et al., "Selective-area fluorination of graphene with fluoropolymer and laser irradiation," *Nano Lett.* 12 (2012): 2374–2378.

Lee, J. H., Koon, G. K. W., Shin, D. W. et al., "Property control of graphene by employing 'semi-ionic' liquid fluorination," *Adv. Funct. Mater.* 23 (2013): 3329–3334.

Li, Q., Liu, X.-Z., Kim, S.-P. et al., "Fluorination of graphene enhances friction due to increased corrugation," *Nano Lett.* 14 (2014): 5212–5217.

Li, Y. S., Zhang, C. Z., Ma, H. T. et al., "CVD nanocrystalline diamond coatings on Ti alloy: A synchrotron-assisted interfacial investigation," *Mater. Chem. Phys.* 134 (2012): 145–152.

Li, Y., Feng, Y. & Feng, W., "Deeply fluorinated multi-wall carbon nanotubes for high energy and power densities lithium/carbon fluorides battery," *Electrochim. Acta* 107 (2013): 343–349.

Loshak, M. G. & Alexandrova, L. I., "Rise in the efficiency of the use of cemented carbides as a matrix of diamond-containing studs of rock destruction tool," *Int. J. Refract. Met. Hard Mater.* 19 (2001): 5–9.

Lynum, S., Hox, K. & Hugdahl, J., "Pyrolytic decomposition in plasma torch," US5527518 A, 1996.

Maiti, J., Kakati, N., Lee, S. H. et al., "Fluorination of multiwall carbon nanotubes by a mild fluorinating reagent HPF6," *J. Fluor. Chem.* 135 (2012): 362–366.

Mallouk, T., Hawkins, B. L., Conrad, M. P. et al., "Raman, infrared and n.m.r. studies of the graphite hydrofluorides $C_xF_{1-\delta}(HF)_\delta$ (2< $x$ <5)," *Phil. Trans. R. Soc. Lond. A* 314 (1985): 179.

Marcoux, P. R., Schreiber, J., Batail, P. et al., "A spectroscopic study of the fluorination and defluorination reactions on single-walled carbon nanotubes," *Phys. Chem. Chem. Phys.* 4 (2002): 2278–2285.

Martin, J. M. & Ohmae, N. *Nanolubricants* (Wiley, Hoboken, 2008).

Matsuo, Y., Nakajima, T. & Kasamatsu, S., "Synthesis and spectroscopic study of fluorinated fullerene, $C_{60}$," *J. Fluor. Chem.* 78 (1996): 7–13.

Meduri, P., Chen, H., Xiao, J. et al., "Tunable electrochemical properties of fluorinated graphene," *J. Mater. Chem. A* 1 (2013): 7866–7869.

Mercier, G., Herold, C., Mareche, J.-F. et al., "Selective removal of metal impurities from single walled carbon nanotube samples," *New J. Chem.* 37 (2013): 790–795.

Mickelson, E. T., Huffman, C. B., Rinzler, A. G. et al., "Fluorination of single-wall carbon nanotubes," *Chem. Phys. Lett.* 296 (1998): 188–194.

Mochalin, V. N., Shenderova, O., Ho, D. et al., "The properties and applications of nanodiamonds," *Nat. Nanotechnol.* 7 (2012): 11–23.

Morita, A., Eda, N., Ijima, T. et al., in Thompson, J. (ed.), *Power Sources* (Academic Press, New York, 1983), p. 435.

Nair, R. R., Ren, W., Jalil, R. et al., "Fluorographene: A two-dimensional counterpart of Teflon," *Small* 6 (2010): 2877–2884.

Nakajima, T. & Watanabe, N. *Graphite Fluorides and Carbon—Fluorine Compounds* (CRC Press, Boca Raton, 1991).

Nakajima, T. & Groult, H. *Fluorinated Materials for Energy Conversion* (Elsevier, Amsterdam, Netherlands, 2005).

Niyogi, S., Hamon, M. A., Hu, H. et al., "Chemistry of single-walled carbon nanotubes," *Acc. Chem. Res.* 35 (2002): 1105–1113.

Nomède-Martyr, N., Disa, E., Thomas, P. et al., "Tribological properties of fluorinated nanocarbons with different shape factors," *J. Fluor. Chem.* 144 (2012): 10–16.

O'Hagan, D., "Understanding organofluorine chemistry: An introduction to the C–F bond," *Chem. Soc. Rev.* 37 (2008): 308–319.

Panich, A. M., "Nuclear magnetic resonance study of fluorine–graphite intercalation compounds and graphite fluorides," *Synth. Met.* 100 (1999): 169–185.

Panich, A. M., Vieth, H.-M., Shames, A. I. et al., "Structure and bonding in fluorinated nanodiamond," *J. Phys. Chem. C* 114 (2009): 774–782.

Pehrsson, P. E., Zhao, W., Baldwin, J. W. et al., "Thermal fluorination and annealing of single-wall carbon nanotubes," *J. Phys. Chem. B* 107 (2003): 5690–5695.

Peng, H., Gu, Z., Yang, J. et al., "Fluorotubes as cathodes in lithium electrochemical cells," *Nano Lett.* 1 (2001): 625–629.

Pichler, T., "Molecular nanostructures: Carbon ahead," *Nat. Mater.* 6 (2007): 332–333.

Plank, N. O. V., Forrest, G. A., Cheung, R. et al., "Electronic properties of n-type carbon nanotubes prepared by $CF_4$ plasma fluorination and amino functionalization," *J. Phys. Chem. B* 109 (2005): 22096–22101.

Rao, A. M., Eklund, P. C., Bandow, S. et al., "Evidence for charge transfer in doped carbon nanotube bundles from Raman scattering," *Nature* 388, 6639 (1997): 257–259.

Raty, J.-Y. & Galli, G., "Ultradispersity of diamond at the nanoscale," *Nat. Mater.* 2 (2003): 792–795.

Raty, J.-Y., Galli, G., Bostedt, C. et al., "Quantum confinement and fullerenelike surface reconstructions in nanodiamonds," *Phys. Rev. Lett.* 90 (2003): 037401.

Robinson, J. T., Burgess, J. S., Junkermeier, C. E. et al., "Properties of fluorinated graphene films," *Nano Lett.* 10 (2010): 3001–3005.

Root, M. J., "Comparison of fluorofullerenes with carbon monofluorides and fluorinated carbon single wall nanotubes: Thermodynamics and electrochemistry," *Nano Lett.* 2 (2002): 541–543.

Rudorff, W. & Rudorff, G., "Zur konstitution des kohlenstoff-monofluorids," *Z. Anorg. Allg. Chem.* 293 (1947): 281.

Ruff, O. & Bretschneider, O., "Die Reaktionsprodukte der verschiedenen Kohlenstoffformen mit Fluor II (Kohlenstoff-monofluorid)," *Z. Anorg. Allg. Chem.* 217 (1934): 1–18.

Sato, Y., Itoh, K., Hagiwara, R. et al., "On the so-called 'semi-ionic' C–F bond character in fluorine–GIC," *Carbon* 42 (2004): 3243–3249.

Setaka, N., "Process for producing diamond powder by shock compression," Patent US4377565 (1983).

Shames, A. I., Panich, A. M., Kempiński, W. et al., "Defects and impurities in nanodiamonds: EPR, NMR and TEM study," *J. Phys. Chem. Solids* 63 (2002): 1993–2001.

Spencer, E. G., Schmidt, P. H., Joy, D. C. et al., "Ion-beam-deposited polycrystalline diamondlike films," *Appl. Phys. Lett.* 29 (1976): 118–120.

Stine, R., Lee, W.-K., Whitener, K. E. et al., "Chemical stability of graphene fluoride produced by exposure to XeF$_2$," *Nano Lett.* 13 (2013): 4311–4316.

Tang, Y. H., Zhou, X. T., Hu, Y. F. et al., "A soft X-ray absorption study of nanodiamond films prepared by hot-filament chemical vapor deposition," *Chem. Phys. Lett.* 372 (2003): 320–324.

Taylor, R., "Addition reactions of fullerenes," *C. R. Chim.* 9 (2006): 982–1000.

Terranova, M. L., Orlanducci, S., Tamburri, E. et al., "Polycrystalline diamond on self-assembled detonation nanodiamond: A viable route for fabrication of all-diamond preformed microcomponents," *Nanotechnology* 19 (2008): 415601.

Thomas, P., Himmel, D., Mansot, J. L. et al., "Tribological properties of fluorinated carbon nanofibres," *Tribol. Lett.* 34 (2009): 49–59.

Thomas, P., Himmel, D., Mansot, J. L. et al., "Friction properties of fluorinated carbon nanodiscs and nanocones," *Tribol. Lett.* 41 (2011): 353–362.

Troyanov, S. I., Troshin, P. A., Boltalina, O. V. et al., "Two isomers of C$_{60}$F$_{48}$: An indented fullerene," *Angew. Chem. Int. Ed.* 40 (2001): 2285–2287.

Tuinman, A. A., Gakh, A. A., Adcock, J. L. et al., "Hyperfluorination of buckminsterfullerene: Cracking the sphere," *J. Am. Chem. Soc.* 115 (1993): 5885–5886.

Ugarte, D., "Curling and closure of graphitic networks under electron-beam irradiation," *Nature* 359 (1992): 707–709.

Unger, E., Liebau, M., Duesberg, G. S. et al., "Fluorination of carbon nanotubes with xenon difluoride," *Chem. Phys. Lett.* 399 (2004): 280–283.

Valentini, L., Puglia, D., Armentano, I. et al., "Sidewall functionalization of single-walled carbon nanotubes through CF$_4$ plasma treatment and subsequent reaction with aliphatic amines," *Chem. Phys. Lett.* 403 (2005): 385–389.

Vander Wal, R. L., Miyoshi, K., Street, K. W. et al., "Friction properties of surface-fluorinated carbon nanotubes," *Wear* 259 (2005): 738–743.

Wang, Z., Wang, J., Li, Z. et al., "Synthesis of fluorinated graphene with tunable degree of fluorination," *Carbon* 50 (2012): 5403–5410.

Wang, X., Chen, Y., Dai, Y. et al., "Preparing highly fluorinated multiwall carbon nanotube by direct heating-fluorination during the elimination of oxygen-related groups," *J. Phys. Chem. C* 117 (2013a): 12078–12085.

Wang, X., Dai, Y., Gao, J. et al., "High-yield production of highly fluorinated graphene by direct heating fluorination of graphene-oxide," *ACS Appl. Mater. Interfaces* 5 (2013b): 8294–8299.

Wang, B., Wang, J. & Zhu, J., "Fluorination of graphene: A spectroscopic and microscopic study," *ACS Nano* 8 (2014): 1862–1870.

Watanabe, N. & Fukuda, M., "Primary cell for electric batteries," Patent US3536532 (1970).

Watanabe, N. & Fukuda, M., "High energy density battery," Patent US3700502 (1972).

Watanabe, N., Hagiwara, R. & Nakajima, T., "On the relation between the overpotentials and structures of graphite fluoride electrode in nonaqueous lithium cell," *J. Electrochem. Soc.* 131 (1984): 1980–1984.

Watanabe, N., Nakajima, T. & Touhara, H., *Graphite Fluorides* (Elsevier Science, Amsterdam, Netherlands, 1988).

Weiss, N. O., Zhou, H., Liao, L. et al., "Graphene: An emerging electronic material," *Adv. Mater.* 24 (2012): 5776–5776.

Withers, F., Dubois, M. & Savchenko, A. K., "Electron properties of fluorinated single-layer graphene transistors," *Phys. Rev. B* 82, 7 (2010): 073403.

Yang, G. W., Wang, J. B. & Liu, Q. X., "Preparation of nano-crystalline diamonds using pulsed laser induced reactive quenching," *J. Phys. Condens. Matter* 10 (1998): 7923.

Yang, H., Chen, M., Zhou, H. et al., "Preferential and reversible fluorination of monolayer graphene," *J. Phys. Chem. C* 115 (2011): 16844–16848.

Yazami, R. & Hamwi, A., "Fluoration de nanomatériaux carbonés multicouches," Patent EP1976792 (2006).

Yazami, R. & Hamwi, A., "Électrochimie de sous-fluorures de carbone," Patent WO2007098478 (2007).

Yazami, R. & Hamwi, A., "Électrochimie de sous-fluorures de carbone," Patent EP1999812 (2008a).

Yazami, R. & Hamwi, A., "Fluoration de nanomatériaux carbonés multicouches," Patent WO2007126436 (2008b).

Yazami, R., Hamwi, A., Guérin, K. et al., "Fluorinated carbon nanofibres for high energy and high power densities primary lithium batteries," *Electrochem. Commun.* 9 (2007): 1850–1855.

Yudanov, N. F., Okotrub, A. V., Shubin, Y. V. et al., "Fluorination of arc-produced carbon material containing multiwall nanotubes," *Chem. Mater.* 14 (2002): 1472–1476.

Zbořil, R., Karlický, F., Bourlinos, A. B. et al., "Graphene fluoride: A stable stoichiometric graphene derivative and its chemical conversion to graphene," *Small* 64 (2010): 2885–2891.

Zhang, L., Kiny, V. U., Peng, H. et al., "Sidewall functionalization of single-walled carbon nanotubes with hydroxyl group-terminated moieties," *Chem. Mater.* 16 (2004): 2055–2061.

Zhang, W., Dubois, M., Guérin, K. et al., "Solid-state NMR and EPR study of fluorinated carbon nanofibers," *J. Solid State Chem.* 181 (2008a): 1915–1924.

Zhang, W., Guérin, K., Dubois, M. et al., "Carbon nanofibres fluorinated using $TbF_4$ as fluorinating agent: Part I—Structural properties," *Carbon* 46 (2008b): 1010–1016.

Zhang, W., Dubois, M., Guérin, K. et al., "Effect of graphitization on fluorination of carbon nanocones and nanodiscs," *Carbon* 47 (2009): 2763–2775.

Zhang, W., Dubois, M., Guerin, K. et al., "Effect of curvature on C–F bonding in fluorinated carbons: From fullerene and derivatives to graphite," *Phys. Chem. Chem. Phys.* 12 (2010a): 1388–1398.

Zhang, W., Dubois, M., Guérin, K. et al., "Fluorinated nanocarbons using fluorinating agent: Strategies of fluorination and applications," *Eur. Phys. J. B* 75 (2010b): 133–139.

Zhang, W., Bonnet, P., Dubois, M. et al., "Comparative study of SWCNT fluorination by atomic and molecular fluorine," *Chem. Mater.* 24 (2012): 1744–1751.

Zou, Q., Li, Y. G., Zou, L. H. et al., "Characterization of structures and surface states of the nanodiamond synthesized by detonation," *Mater. Charact.* 60 (2009): 1257–1262.

# 12

# Activated Carbon Nanofibers

Yakup Aykut

Saad A. Khan

## 12.1 Introduction

Carbon nanostructured materials have garnered great interest in recent years since carbon nanostructured materials have many tailored features and enhance device performance to several orders of magnitudes. Carbon nanostructured materials are manufactured in the form of nanoparticles, nanofilms, and nanofibers. Activated carbon (AC) materials have high specific surface as a result of microporous structures and have been widely used for various purposes, including purifications of water, beverages, and chemicals; energy storage as anode layers in batteries and supercapacitors; electron-emitting devices; and gas and molecular adsorbents. AC is a form of carbon with small and low volume of pores on and inside the surface that increase surface area and provide available reachable surface for adsorption, chemical reaction, and interactions on the surface. Due to the pore size and volume, pores are named *meso-*, *micro-*, and *macropores*. When the pore diameters are beyond 500 Å, the pores are called macropores, those below 50 Å are called micropores, those around 2–50 nm are called mesopores, and those above 50 nm are called macropores (Sodhi 2013). These pore sizes in an AC can be controlled due to precursor material, reagents, and carbonization and activation process of carbon.

AC has been widely used in the form of powders or granule forms. In the last decades, several researches have been done for the production of activated nanomaterials, including nanoparticles, nanotubes, and nanofibers. Production in activated form gathers lots of new functional properties because

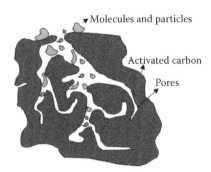

**FIGURE 12.1**   Porous structure of AC and material adsorption and encapsulation on/in its pores.

easily reachable micropore structures on the surface allow capturing and holding of molecules from the surface. As seen in Figure 12.1, there are lots of reachable surface on AC. If we produce this fancy material in the form of nanofibers, the surface-to-volume ratio increases tremendously and the total surface increases to several orders of magnitude compared with micro- and macrofiber forms.

Before production of AC, carbon materials are first produced from precursor materials and then activated in the proper conditions, or simultaneous processes can take place to perform both carbon material production and activation. Two main activation processes of carbon have been reported in the literature, which are called physical and chemical activation of carbon. The physical activation process is generally a process wherein carbon materials are activated with steam or $CO_2$ by heating to develop porous structure. Eliminating particles from carbon to form porous structure is also considered physical activation of carbon. For chemical activation, chemical reagents such as $H_3PO_4$, KOH, $CoCl_2$, and ZnCl are added into the AC precursor materials before the activation process and these materials catalyze activation. In most cases heating leads to catalytic activation of carbon with existence of chemical reagents. Activated carbon nanofibers (ACNFs) are produced via similar ways. Either carbon nanofibers (CNFs) are produced first and then activation takes place, or ACNFs are produced in a one-step simultaneous process. A detailed explanation of these two activation processes for ACNF preparation will be explained in the next sections by giving examples from the literature.

## 12.2  Activation of Carbon Materials (Macro-, Micro-, and Nanoscales)

AC materials, regardless of their scales, are produced either through a one-step process or by subjecting previously prepared carbon to an activation process. Carbon material activations are mainly divided into two groups called physical and chemical activation methods. A general overview on both activations is briefly given in Table 12.1.

### 12.2.1  Physical and Chemical Activation Methods

Physical activation can be either activation of carbon in water steam or $CO_2$ environment by heating or elimination of noncarboneous phase from carbon to obtain porous structure. In contrast, chemical activation is described as activation of carbon by adding chemical reagents such as $H_3PO_4$, KOH, $CoCl_2$, or ZnCl in the precursor material; these materials help pore development during the heat treatment process. For instance, activated carbon fibers (ACFs) from rayon-based precursor produced by Gao et al. (2010) via two-step activation with steam and KOH treatments that have high specific surface area (SSA) and micropore volume have been investigated. Prepared carbon fibers were first activated by steam at 900°C and mixed with KOH in distilled water and subsequently activated at 850°C in nitrogen atmosphere

**TABLE 12.1** Fabrication of ACNFs via Various Ways from Different Resources and Comparison of Activation Mechanisms

| Advantages | Disadvantages | Precursors | Activating Reagents | Process/Method | Applications | References |
|---|---|---|---|---|---|---|
| | | | Physical Activation of Carbon | | | |
| • Eliminates noncarboneous phase from carbon to obtain porous structure<br>• Rate of activating reagents can be controlled during processing<br>• Less residual activating reagents | • High activating temperature<br>• Needs more gas circulation during activation<br>• Long activation time<br>• Lower thermal stability | $C_2H_4$, $C_2H_2$:$NH_3$:$N_2$ Polyacrylonitrile (PAN) Poly(amide imide) Polybenzimidazole (PBI) Organic and recycled materials: polyethylene terephthalate (PET), polyvinyl chloride (PVC), acrylic fabric wastes, pitch, tire, sewage sludge | $CO_2$ gas, $H_2O$ vapor; removes second phase | Nanofibers are produced via<br>• CVD if the precursor is in a gas<br>• Electrospinning if the precursor is in the liquid phase (solution or melt)<br>*Then*<br>Heat treatment is performed in a proper atmospheric condition: carbonization of the raw material and its subsequent activation at high temperature with $CO_2$ or $H_2O$ vapor, or elimination of a phase to form pores. | Gas/chemical adsorption, energy storage, supercapacitors, LIBs, hydrogen storage, power generation, microprocessor cooling, sound absorption, water filtration, reinforcement in composites, electron-emitting devices, sensory systems, biomedical applications. | Aykut 2012<br>Dias et al. 2007<br>Im et al. 2007<br>Im et al. 2009<br>Kim 2005<br>Lee et al. 2010<br>Merino et al. 2005<br>Oh et al. 2008<br>Seo & Park 2009<br>Shim et al. 2006<br>Tao et al. 2006<br>Warhurst et al. 1997<br>Yoon et al. 2004<br>Zhang et al. 2009 |
| | | Chemical Activation of Carbon | | | | |
| • Higher yield<br>• Lower temperature of activation<br>• Less activation time<br>• Higher development of porosity<br>• Good thermal stability | • Activating agents are more expensive<br>• Hydroxides are very corrosive<br>• An additional washing stage might be necessary | The same precursors can be used as those in physical activation method. Additionally, catalytic reagents are used. | KOH, $ZnCl_2$, CoCl, $MnCl$, MgO, $Al_2O_3$, Si, Co, Mn, Ni, Fe, Au, Ni, Cu | The precursor material is previously impregnated with the chemical reagents and then heat treatments are performed. The same nanofiber production methods can be used as those in physical activation methods. | | |

(Gao et al. 2010). Previously prepared CNFs were chemically and physically activated by KOH and $CO_2$ by Ghasemi et al. (2011). They immersed CNFs in the reagents and heat-treated these in argon atmosphere. For physical activation, $CO_2$ is fed into the heating chamber to pass through the sample in the furnace. Activation enhances both the surface area and the catalytic activity of the material in microbial fuel cell. Ghasemi et al. also compared results and reported that chemically activated samples were better than physically activated samples, and the chemically activated samples showed better catalytic activity. Comparison of porous CNFs with MWCNTs for supercapacitors was performed by Huang et al. (2007). They purchased MWCNTs and produced porous CNFs via decomposition of $NiCl_2$ containing polyethylene glycol nanofibers at 600°C in nitrogen atmosphere. Open pores on the surface of CNFs make the surface enriched with oxygen-containing functional groups compared to the basal planes of MWCNTs, providing significant and better pseudocapacitance.

## 12.2.2 AC Materials from Various Resources

To decrease AC material cost and adverse environmental effect, there have been attempts to find alternative precursor materials, such as from natural waste and recycled materials. Uses of various waste materials have been attempted to investigate AC property, including shells and/or stones of fruits such as nuts, peanuts, olives, dates, almonds, apricots, and cherries; walnut shells; coffee bean husks, corncobs; and oak wood, cedar wood, tropical tree wood, and rubber wood sawdusts as agricultural wastes. To reduce the cost, waste materials from natural resources and industrial polymer type waste materials, such as PET, PVC, acrylic fabric waste, pitch, tire, and sewage sludge, are used to obtain AC (Dias et al. 2007). For instance, the carbon-rich property of lignin makes it a good candidate for AC preparation and both physical and chemical activation methods can be applied to obtain lignin-based AC. High-porosity and high–surface area lignin-based AC has been produced and its porosity features were investigated by Brunauer–Emmett–Teller (BET) measurement (Carrott & Carrott 2007). Preparation and potential application of AC from waste husk of *Moringa oleifera* seed by single-step steam pyrolysis was reported by Warhurst et al. (1997) and they demonstrated that the obtained carbon structures show both mesopore and macropore structures. Chemical activation of carbon, including amorphous carbon, graphite, carbon nanospheres, and different types of CNFs (platelet, herringbone, and ribbon), is performed by using KOH as a chemical activating reagent at atmospheric pressure environment by heating in a tube by using $C_2H_4/H_2$ as precursor and $Ni/SiO_2$ as catalyst (Jimenez et al. 2010).

# 12.3 ACNF Preparation via Different Ways

## 12.3.1 ACNF Preparation via Chemical Deposition

CNFs are produced in a chamber by feeding the carbon precursor in gas phase and CNFs are grown in the chamber by adjusting the environment in a wanted condition. Catalytically grown CNFs with high pore content and surface area were prepared by Yoon et al. (2004) via KOH activation at the temperature range 500–1000°C under argon gas atmosphere by using Cu–Ni catalyst powders. According to Yoon et al., some partial gasification and expansion of the interlayer spacing between graphenes in CNFs leads to formation of high–surface area porous structure in CNFs. Local removal, broadening, collapse, and rearrangement of graphene layers enhance the porous structure in CNFS. A similar mechanism was proposed by Merino et al. (2005). They used Fe powders as catalyst and chemical activation was performed with KOH treatment to produce porous ACNFs. The specific surface area (SSA) increased dramatically as a result of activation. In the study, hydrocarbons were used as a carbon source and the selected carbonization temperature was 800°C. Tao et al. (2006) synthesized multibranched porous CNFs via CVD technique on a Cu catalyst doped with Li, Na, or K. In their work, $C_2H_2$:$NH_3$:$N_2$ was fed into the heated reaction chamber, which contained the right amount of catalyst precursor and citric acid dissolved in water. The results demonstrated that multibranched structures of porous CNFs had been obtained. Mesoporous CNFs with tailored nanostructure were prepared by Fang et al. (2011) by using porous anodic aluminum oxide (AAO) membrane and colloidal silica as hard template nanocasting and phenolic resin as a carbon source. They first incorporated phenol and silica in the system and conducted heat treatment in vacuum to obtain silica–phenol resin polymer composite. Then, the composite was treated with HCl at 323 K and AAO was removed. Carbonization was carried out at 973 K under $N_2$ and polymer nanofibers were converted into tailored porous CNF structures.

## 12.3.2 ACNF Preparation via Electrospinning (ES) Methods

### 12.3.2.1 CNF Preparation via ES Process

For shaped carbon material preparation, polymeric materials have been widely used as carbon precursor material. Shape is given when the carbon is in precursor form, and after carbonization a similar shape is transferred to the carbon structure. A polymer or polymer mixtures are dissolved in an appropriate solvent and processed into any wanted shape by using a variety of techniques, including ES, forcespinning, and melt blowing, etc. Since molecular orientations were gathered during the fiber spinning, high tensile strength properties can be also obtained after carbonization process. PAN is a promising material since it can be shaped easily before carbonization and can be blended with other polymers to alter the porous structure and the pore size distribution in final carbon nanomaterials. Polymer-based ACNFs have been produced via ES of a polymer precursor after stabilization, carbonization, and activation processes. In the ES process, first an appropriate ES polymer solution is prepared. Generally, the polymer is dissolved in a solvent, left for some time for relaxation of the dissolved polymer chains, and then electrospun into nanofibrous structure. The ES process is simply demonstrated in Figure 12.2. Briefly, after preparation of an ES solution, a plastic syringe with a metal needle with a proper diameter of the hole is filled with some amount of solution. The syringe is fitted on a micropump system that will provide a controllable flow of ES solution from the needle. A grounded conductive plate is placed in front of the needle at a proper distance. A high-voltage power supplier is attached to the metal needle. When electrical voltage is applied to the metal needle, electron flow occurs from the needle to the grounded collector and an electrical field is created between the needle and the collector. The polymer droplet at the tip of the metal needle is ejected from the needle to the collector. The droplet is elongated and forms a 1D nanofiber structure during its journey from the needle to the collector. During this trip, the surface area of the material is increased tremendously and the solvent rapidly evaporates from the surface until the nanofiber reaches the grounded collector. Thus, the nanofibers are collected in dry form.

Polymers not only can be electrospun into nanofibers from a polymer solution, they also can be melted and electrospun into nanofibers without using any solvent (Lyons et al., 2004). On the other

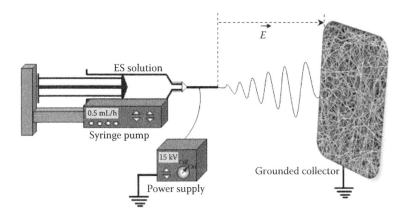

**FIGURE 12.2** Schematic illustration of ES process for nanofiber fabrication.

hand, additives, particles, and any other contaminants can be incorporated to nanofibers by adding these into the ES solution or melt (Aykut 2012, Aykut et al. 2013b).

As-spun polymer nanofibers are stabilized, carbonized, and activated sequentially via heat treatments in appropriate atmospheric conditions. A schematic illustration of nanofibers after each step is shown in Figure 12.3, and chemical structural changes during stabilization and carbonization are given in Figure 12.4.

Chemical carbonization mechanism of PAN has been proposed by several researchers (Daranyi et al. 2010, Nandi et al. 2012). In the mechanism, as-prepared PAN in the form of fiber, film, particle, or any geometrical form is first stabilized in air atmosphere and then carbonized in an inert atmosphere by heating.

Before carbonization, the shaped PAN is heated first in an oxidizing atmosphere at the temperature range 250°C–290°C for proper stabilization. In this stage, C≡N bonds in PAN are converted into C=N bonds and the reaction results in conjugated ladder polymer structure in the stabilized PAN (Daranyi et al. 2010, Nandi et al. 2012). This structure allows polymer to process at high temperatures and the structure obtains resistance against burning out. The main reactions that take place during the stabilization stage are cyclization, dehydrogenation, aromatization, oxidation, and cross-linking (Zhu et al. 2003). Thus, this stage has a kind of exothermic behavior. Typically, the color of the polymer changes from white to yellow or brown after stabilization; the exact reason for this, though, is unknown. It is believed that the color change is a result of formation of the ladder structure. Stabilized shaped PANs are carbonized in an inert atmosphere above 800°C to obtain a carboneous structure. To obtain a more graphitic and crystallized structure, the carbonization temperature can be increased. The presence of oxygen in carbonization atmosphere can burn the material. Volatiles are removed from the fibers and the heating

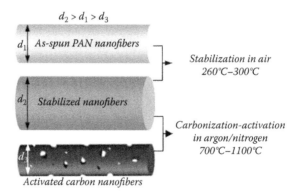

**FIGURE 12.3**  Schematic representation of the nanofibers at stages of ES, stabilization, and carbonization.

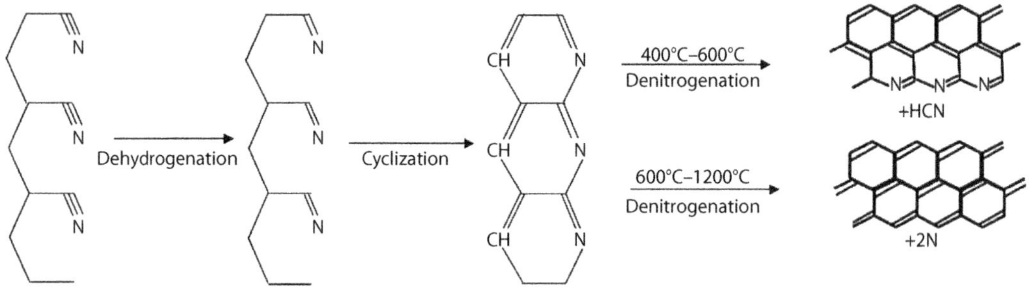

**FIGURE 12.4**  Schematic demonstration of the thermal chemistry of the carbonization of PAN—degradation, cyclization, and denitrogenation—leading to graphitic structures at high temperatures. (Adapted from Daranyi, M. et al., *Process. Appl. Ceram.*, 4, 59–62, 2010; and Nandi, M. et al., *Chem. Commun.*, 48, 10283–10285, 2012.)

chamber. Fibers shrink during carbonization because densification takes place. Also, denitrogenation takes place and the ladder structure gained at the stabilization stage is kept.

In general AC material manufacturing, ACNFs are produced by physical and chemical activation processes. The following explanation in this chapter gives examples from the literature of physical and chemical ACNF preparation via the ES process.

### 12.3.2.2 Physically Activated CNFs via ES Process

#### 12.3.2.2.1 *Feeding the Activating Reagent into the Carbonization Environment*

As mentioned earlier in this chapter, carbon can be activated by feeding the activating reagent such as $CO_2$ and steam gases into the carbonization or activation reaction environment. This approach has been also applied to electrospun CNFs to gather porous carbon nanofibrous structures. For example, Shim et al. (2006) produced ACNFs from electrospun PAN nanofibers after stabilization at 280°C in air and carbonization at 800°C in nitrogen. In their work, physical activation was carried out in 30% steam in nitrogen environment. They also investigated benzene adsorption on their produced porous CNFs at different temperatures and observed that benzene adsorption decreases with increasing adsorption temperature. Electrospun ACNFs with tailored microporosity and high amount of nitrogen-containing functional groups have been produced at 270°C stabilization and at 600°C carbonization temperatures in helium gas with humidity steam feeding in the furnace. ACNFS with improved formaldehyde adsorption that were produced by Lee et al. (2010) are another physically activated electrospun ACNF sample. In another study, the diameter and graphitic structures of electrospun ACNFs were investigated by Wang et al. (2012b) as electrospun PAN nanofibers were first oxidatively stabilized in air atmosphere at 270°C and carbonized at 750°C–900°C in argon. During carbonization, when the activation temperature was reached, $CO_2$ gas was fed into the furnace. The diameter of ACNFs became smaller and more graphitization was seen as the activation temperature increased from 750°C–900°C. In another research study, Wang et al. (2012a) produced hierarchical ACNF webs by incorporation of carbon black into an electrospun polymer solution and carbonization of as-spun nanofibers. As-spun nanofibers stabilized at 280°C in air and carbonized at 700°C–900°C in $CO_2$ flow environment and hierarchical activated CNFs were obtained. This hierarchical structure provides superior capacities as electrode materials in capacitive deionization for desalination to activated CNFs (Wang et al. 2012a). Electrospun ACNFs can also be produced by using other precursor polymeric materials rather than PAN. PBI-based ACNFs were produced by ES of PBI and subsequently activated in steam (Kim 2005). As-spun nanofibers were carbonized by heat treatment between 700°C and 850°C in nitrogen atmosphere and activated in steam flow at given temperature in nitrogen. The prepared ACNFs showed a good electrochemical performance property for supercapacitors. ACNFs were produced from precursor electrospun poly(amide imide) web by Seo and Park (2009). As-spun nanofibers were subsequently stabilized, carbonized, and activated by heating in the proper environments. High SSA and pore size distribution and higher content of surface oxygen functional groups on the surface provide good electrochemical property and make these nanofibers useful electrode materials for supercapacitor applications.

#### 12.3.2.2.2 *Removing the Second Phase*

ACNFs have been produced via the ES process by adding a second phase in the ES solution. Three cases can happen. First, the second phase is removed from the nanofibers after ES and before carbonization by washing or other physical forces. This removed phase leaves voids behind it in nanofibers. Thus, the already existent voids were converted to a porous structure after the carbonization process. Second, since the second phase has not been converted into carbon structure and has not combined with carbon structure, this phase is removed after the carbonization process and leaves a porous structure behind it. Finally, the second phase can be removed from the system during the carbonization process and ACNFs are obtained as a result of this removal. To produce ACNF this way, bi- or multicomponent polymer nanofibers must be produced first.

AC materials in the form of nanofibers and nanoparticles with mesopore structures were obtained from electrospun PAN after physical activation with silica. It is well known that silica is one of the effective reagents for physical activation of AC because silica can be dissolved in hydrofluoric (HF) acid easily. During the process, Im et al. (2007) manufactured silica-doped PAN samples via the ES technique and then performed carbonization of silica-loaded PAN nanofibers. Silica particles were removed from the fibers by immersing the fibers in HF acid and porous carbon nanofibrous structures were obtained. Mesoporous CNFs were produced from phenolic resin precursor by addition of tetraethyl orthosilicate and HCl and combination of sol–gel ES technique (Teng et al. 2012). As-spun and dried nanofibers were pyrolyzed in nitrogen atmosphere at 500°C then carbonized at 900°C. Silicon components in the form of $SiO_2$ in the carbonized nanofibers were removed by washing with aqueous solution, leading to the formation of mesoporous CNFs. As-produced mesoporous CNFs showed good dye adsorption capacity with different molecular sizes of dyes. In another study, porous CNFs with a large surface area were produced from electrospun PAN/$SiO_2$ composite and a subsequent carbonization process (Ji et al. 2009). Fumed $SiO_2$ nanoparticles and PAN solutions with different $SiO_2$ contents in *N,N*-dimethylformamide (DMF) were prepared and electrospun into nanofibrous structures. As-spun nanofibers were first stabilized in air atmosphere at 280°C and carbonized in nitrogen atmosphere at 700°C and 1000°C. Ji et al. (2009) also held as-carbonized $SiO_2$/CNF composite in HF acid for 24 hours and removed $SiO_2$ from the CNFs and washed the fibers with water. They observed that both $SiO_2$ addition and HF acid treatment provide porous CNF structure and increasing carbonization temperature has great influence on both porous structure and electrochemical properties. PAN and polyvinylpyrrolidone (PVP) blend solution was electrospun into a bicomponent nanofibrous system (Zhang et al. 2009). Then, PVP was removed from as-spun composite nanofibers via immersing in water, and the nanofibers were heat-treated in air at 270°C for stabilization and then in $N_2$ at 1000°C for carbonization to obtain porous CNFs. Zhang et al. (2009) reported that porosity can be controlled easily by changing PVP/PAN ratios. Nanoporous, hollow, and U-shaped cross sections were generated in CNFs via phase-separation behavior of polymer additives with PAN. Electrospun binary mixture of PAN with poly(ethylene oxide) (PEO), cellulose acetate (CA), or poly(methyl methacrylate) (PMMA) was produced via ES; removal of one phase and carbonization of remaining PAN phase led to porous CNFs. Different ratios of these additives help to produces a variety of structures. Nanoporous CNFs were generated from the water-treated PAN/PEO precursors. Multichannel hollow fibers were produced from the acetone-treated PAN/CA precursors. Carbon fibers were produced from the chloroform-treated PAN/PMMA precursors and consisted of varied structures, including hollow and U-shaped cross sections (Zhang & Hsieh 2009). Porous CNFs were produced from PAN/poly-l-lactic acid (PLLA) bicomponent nanofibers via ES. As-spun nanofibers were stabilized at 280°C in air, leading to nonplastic cyclic or ladder compounds (PLLA decomposed at this stage), and were then carbonized at 800°C in nitrogen atmosphere, and porous nanofibers were generated. The porous nanofibers show better cyclic performance and can be used as anode in LIBs. Numerous active sites on the surface and short diffusion pathways for lithium ions on porous CNFs provide cyclic performance and can be used as anode in LIBs (Zhang & Ji 2009). ACNFs with hollow core/ highly mesoporous shell structure were prepared by Park et al. (2013) via concentric ES using PMMA as a pyrolytic core precursor with either PAN or PAN/PMMA blended polymer as a carbon shell precursor. As-spun nanofibers were stabilized at 280°C in air and carbonized at 1000°C in nitrogen and then activated at 800°C with steam supply in nitrogen atmosphere. The immiscibility of PAN with PMMA provides this fancy porous structure after the thermal treatment.

### 12.3.2.3 Chemically Activated CNFs via ES process

As explained previously, AC can be obtained by addition of some chemical reagents in the precursor materials. These chemicals play a catalytic role during the carbonization of the precursor and lead to porous carbon structures. In this regard, chemical reagents such as $H_3PO_4$, KOH, $CoCl_2$, and ZnCl have been added into the carbon precursor material and these materials help pore development during the heat-treatment process. Besides forming porous structures during the heat-treatment process, these reagents play a catalytic role and reduce the heat-treatment temperature for carbonization of electrospun nanofibers. Also, more compact graphitic structures are obtained with combination of such reagents (Aykut 2012).

ACNFs with various pore structures via chemical activations have been obtained by our group and other researchers in the literature. In one of our works, we added various amounts of $CoCl_2$ to a PAN/DMF solution and the solution was electrospun into a PAN/$CoCl_2$ nanofibrous structure. After a proper heat-treatment process, we obtained highly porous ACNFs, as seen in Figure 12.5, since the cobaltous phase catalytically activated the nanofibers during the heat-treatment process. The surface area increased with increasing reagent contents in the precursor PAN nanofibers and ranged between 33.56 and 417.3 $m^2 g^{-1}$ and the total pore volume was between 0.005 and 0.340 $cm^3 g^{-1}$ (Aykut 2012). In another work, as-spun PAN/$CoCl_2$ nanofibers were treated with $H_2S$ and then heat-treatment process was performed. Homogeneous macroporosity of ACNFs was observed after the carbonization process under the same heat-treatment conditions (Aykut et al. 2013a).

SEM images of as-spun PAN nanofibers are given in Figure 12.6a and b and of their stabilized and carbonized analogs in Figure 12.6c and d, respectively. TEM images of as-spun PAN and its carbonized nanofiber forms are also given in Figure 12.7a and b. When the cobaltous phase is incorporated into nanofibers, structural and visual changes are observed in each nanofibrous form. SEM images of

**FIGURE 12.5**  SEM images of (a, b) chemically ACNFs and their (c) as-spun and (d) stabilized analogs.

**FIGURE 12.6**  SEM images of (a, b) as-spun net PAN and its (c) stabilized and (d) carbonized nanofiber analogs.

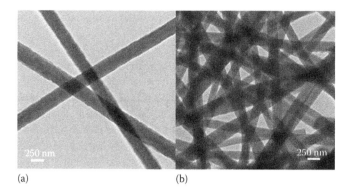

**FIGURE 12.7**    TEM images of (a) as-spun PAN and (b) its carbonized nanofiber analogs.

catalytically ACNFs are shown in Figure 12.5a and b and of their as-spun and stabilized analogs are shown in Figure 12.5c and d, respectively.

TEM images of catalytically activated CNFs are shown in Figure 12.8a and b and of their as-spun and stabilized analogs Figure 12.8c and d, respectively. As seen from the TEM images, cobaltous nanoparticles are observed in the as-spun and carbonized nanofibers, and the nanoparticles become bigger after carbonization process since small particles migrate and come together during carbonization. Some particles leave the nanofibers and leave a porous structure behind them. Oxidized cobaltous phase is formed on the surface of the stabilized nanofibers, as seen in Figure 12.8d and this film layer disappears after the carbonization process, leaving a porous structure. After increasing the reagent (CoCl$_2$) contents in the precursors of the ACNFs, more graphitic structures are observed as well.

In another work, hierarchical porous magnetic crystal Fe$_3$O$_4$–embedded CNFs with an extremely high surface area of 1623 m$^2$ g$^{-1}$ and a pore volume of 1.635 cm$^3$ g$^{-1}$ were prepared via electrospun PAN/polybenzoxazine nanofibers as composite carbon precursor (Ren et al. 2012). In the process, in situ polymerization occurred during stabilization of as-spun benzoxazine (monomer)/PAN/iron acetate nanofibers. Afterward, stabilized nanofibers were immersed in KOH and carbonized in N$_2$ at 850°C, and a hierarchical porous structure of Fe$_3$O$_4$/CNF was obtained. Porous CNFs were synthesized via the self-degradation method by pyrolysis of conducting polymer in argon gas atmosphere as anode material with enhanced storage properties in LIBs (Li et al. 2009). Precursor nanofibers were prepared by addition of methyl orange, pyrrole, and ferric chloride in reaction media and as-prepared nanofibers were annealed in argon at 900°C. ACNFs were obtained by Oh et al. (2008) after stabilization, and after carbonization and activation of manganese (Mn)-dispersed PAN, nanofibers were produced via ES. The activation was carried out by supplying 30 vol% steam for 30 min at 800°C under nitrogen. Pore characteristics show that the ACNFs have a higher SSA and a larger and narrow micropore volume. The SSA and the micropore volume of the CNFs were 853 m$^2$ g$^{-1}$ and 0.280 cm$^3$ g$^{-1}$, respectively, but those

**FIGURE 12.8**    TEM images of (a, b) chemically activated CNFs and their (c) as-spun and (d) stabilized analogs.

of Mn-doped activated CNFs were 1229 $m^2 g^{-1}$ and 0.416 $cm^3 g^{-1}$, respectively. Potassium hydroxide and zinc chloride activations were conducted for activation of electrospun CNFs to increase the SSA and the pore volume of CNFs (Im et al. 2008). Pure electrospun PAN nanofibers were properly stabilized at 523 K in air and carbonized at 1323 in N and then immersed in NaOH and $K_2CO_3$ activating reagents. Immersed CNFs were activated at 1023 K in nitrogen atmosphere and ACNFs with macro-, meso-, and microporous structures were produced. Micropore volume is increased by increasing $K_2CO_3$ content. Micropores are formed with sodium hydroxide activation at low concentrations and then mesopores are formed at high concentrations (Im et al. 2009). CNFs were chemically activated by using nitric acid ($HNO_3$) and then hot-pressed on a carbon cloth support to increase the surface area (Santoro et al. 2013). The activation of the nanofibers (ACNFs) gave an advantage to the cathode performances compared to the raw CNFs. Coal-based ACNFs were prepared via ES by mixing of PAN and acid-treated coal (Zhao et al. 2014). Acid-treated ($HNO_3$ and $H_2SO_4$) coal was added to PAN/DMF solution then the ES process was performed to form the precursor nanofibers. As-spun nanofibers were preoxidized in air at 280°C and carbonized at 800°C in argon and were activated at the same environment by feeding with steam. Chen et al. (2012) used already prepared CNFs to activate nanofibers. In their work, CNF and pyrrole were added to deionized water, sonicated, and stirred. Some amount of aqueous ammonium persulfate solution was also added to the solution. As-prepared solution was dried and heat-treated under nitrogen at temperatures between 500°C and 1000°C. As-prepared CNFs have enhanced porous structure, and their good capacitive performance, good cycling stability, large power capability, high electrochemical capacitance, and good retention capability make them reliable electrode materials for supercapacitors.

## 12.4 Characterizations of ACNFs

### 12.4.1 Morphological Characterizations

Morphological and textural characterizations of ACNFs are conducted with SEM and TEM methods. A detailed imaging information can be obtained via these methods. To get SEM images, the surface of the sample should have a proper conductivity level. Since ACNFs have a level of conductivity, sometimes they do not need conductive coating. But for a deeply focused and detailed morphological characterization of pores on ACNF surface, conductive coating can be necessary. With SEM imaging, the edge and curves of the nanofibers can be imaged (see Figures 12.5 and 12.6). On the other hand, the information inside the nanofibers cannot be gathered with SEM imaging. TEM imaging is an appropriate way to get information inside the nanofibers (see Figures 12.7 and 12.8). Even microstructural characterization can be done via TEM as well. By focusing on the nanofibers under high magnifications, the crystalline structure can be visualized and crystal lattice fringe images can be obtained. As seen in Figure 12.8a through d, crystalline cobaltous nanoparticle phases can be observed. The graphene layers in a graphite structure can be also seen clearly via TEM imaging. Also, installing and mounting an energy-dispersive X-ray spectroscopy detector on an SEM or TEM, elemental mapping and analysis of the nanofibers can be done and the ratio of each element in nanofibers can be screened. Atomic force microscopy (AFM) is another technique for morphological characterization. With this technique only surface analyses can be conducted at the nanoscale. But since AFM characterization takes a long time and the AFM tip breaks very often during measurement, SEM imaging is generally preferable. Because imaging is conducted at the nanoscale, optical imaging methods are not preferred.

### 12.4.2 Surface and Pore Volume Analyses

For a quantitative measurement of surface area and pore characteristics, generally, BET measurement is used. With this measurement, the SSA of the material in square meters per gram is measured by nitrogen multilayer adsorption measurement as a function of pressure. The BET theory gives information on the physical adsorption of gas molecules on a surface. Surface porosity, surface area, and particle size are determined this way in many applications. The Barrett–Joyner–Halenda (BJH) analysis is another theory that refers to the pore area, the specific pore volume, and the pore size distribution of a solid sample. The two theoretical methods (BET and BJH) can be run in the same instrument. Nanolevel high-resolution images of the sample surface and quantitative 3D measurements of ACNFs can be done by AFM as well.

## 12.4.3 Microstructural Analyses

### 12.4.3.1 Raman Characterization

Raman spectroscopy is a technique mostly used to investigate the microstructural property and perfectness of carbon-based materials by utilizing Raman peak position, bandwidth, intensity ratio of the peaks, and total area under the Raman curves. As seen in Figure 12.9, the Raman spectra of carbon-based materials have two distinct broad bands around 1329.6 cm$^{-1}$ (D band) and 1572 cm$^{-1}$ (G band), which correspond to the respective defect-induced vibration mode of turbostratic disordered carbons in graphene layers and ordered graphite phase (Kim et al. 2006, Aykut 2013). The intensity ratios ($R = I_D/I_G$) of the peaks indicate the perfectness of structurally ordered graphite phases in the ACNFs. For instance, increasing activation level leads to increasing G band intensities and decreasing D band intensities when carbon is catalytically activated with cobaltous phase, as seen in Figure 12.10 (Aykut 2012). Increasing the cobaltous phase in the precursor PAN nanofibers leads to more activation and a more graphitic structure is obtained (see Figure 12.10). Sometimes, activation can increase structural disorder and create a more amorphous structure. Thus, by looking at the Raman curves of ACNFs, we can figure out the effect of the activation process, activating agents, temperatures, and any other parameters on the microstructure and carboneous structural perfection of ACNFs.

### 12.4.3.2 XRD Characterization

ACs are mainly composed of three different microstructure phases, which are graphite-like microcrystals, single-reticular-plane carbon, and nonorganized carbon (Lee et al. 2013). XRD is another method for investigating the microstructure of ACNFs. When ACNFs comprise disordered and nonorganized carboneous amorphous phases, they exhibit broad XRD peaks. Sometimes, there are also massive turbostratic graphite microstructures in ACNFs. The two obvious diffraction peaks represent the diffraction-characteristic peaks of the microcrystalline graphitic (002) peak of the ACNF located around 2θ ~ 25° (Ji et al. 2009). From the Bragg equation ($2d \sin \theta = n\lambda$), we can examine the turbostratic

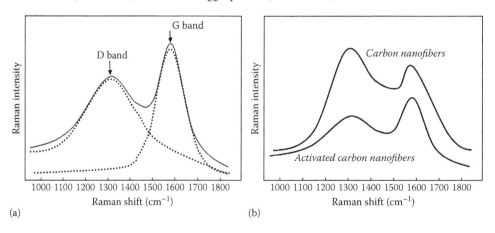

**FIGURE 12.9**  Microstructural investigation of ACNFs with Raman spectroscopy: expressions of (a) D and (b) G bands at ACNFs.

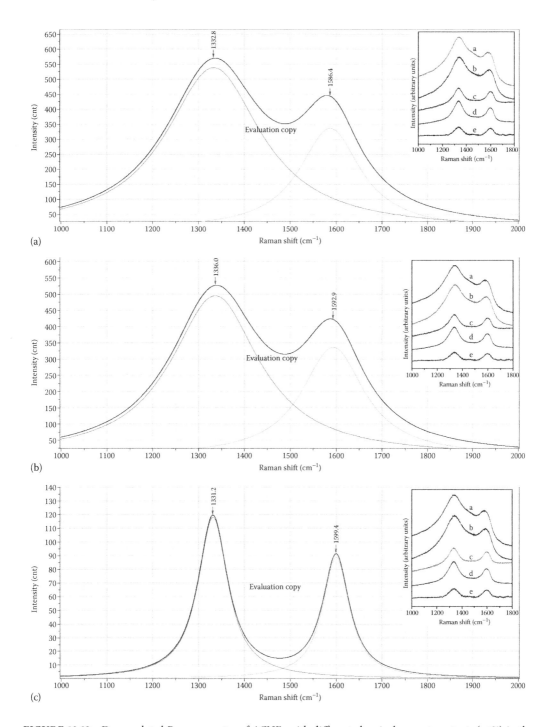

**FIGURE 12.10** Deconvoluted Raman spectra of ACNFs with different chemical reagent contents (wt%) in the PAN precursor nanofibers before carbonization: (a) 0 wt% (pure), (b) 5 wt%, and (c) 10 wt%. (Reprinted with permission from Y. Aykut, *ACS Appl. Mater. Interfaces*, 4(7), pp. 3405–3415. Copyright 2012 American Chemical Society.) *(Continued)*

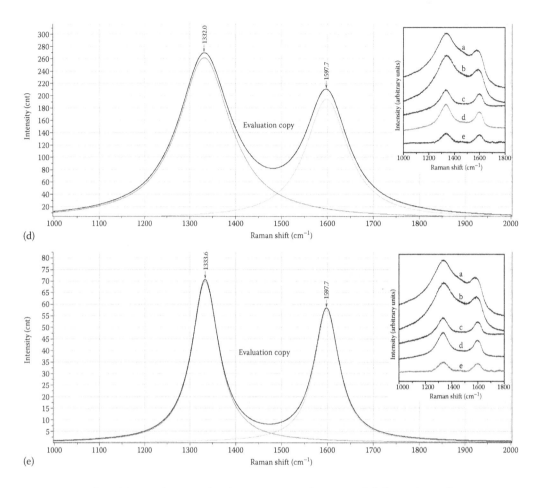

**FIGURE 12.10** **(CONTINUED)** Deconvoluted Raman spectra of ACNFs with different chemical reagent contents (wt%) in the PAN precursor nanofibers before carbonization: (d) 15 wt%, and (e) 30 wt%. (Reprinted with permission from Y. Aykut, *ACS Appl. Mater. Interfaces*, 4(7), pp. 3405–3415. Copyright 2012 American Chemical Society.)

crystallinity of ACNFs by looking at plane spacing *d* (002) of the diffraction crystal face (002) for ACNFs. For instance, Lee et al. (2013) reported that the diffraction angle of the crystal face (002) of ACNFs is smaller than that of CNFs and from the Bragg equation ($2d \sin \theta = n\lambda$), we can see that plane spacing *d* (002) of the diffraction crystal face (002) of ACNFs is relatively larger and indicates the existence of turbostratic structure in ACNFs (Lee et al. 2013). The effect of the activation type, mechanism, precursors used, activating agents, and activating temperatures on microstructure can be estimated via Raman spectroscopy and XRD techniques.

### 12.4.4 Characterization Depending on Use

There are many electrochemical applications of ACNFs, including LIBs, fuel cells, and bio- and chemosensors. Thus, electrochemical characterizations become critical measurements for ACNFs. Charge–discharge characteristics, electrochemical impedance spectra, and electrochemical corrosion measurements have been extensively performed on ACNFs with various properties. Adsorption characteristics of gaseous and small molecules on ACNFs are important for many applications, so the measurement of such adsorption characteristics has a critical importance in the working area. BET instrument and measurement have been used to measure adsorption–desorption characteristics of ACNFs at the surface and pore characteristics are determined via this technique by evaluating adsorption–desorption characteristics of gaseous molecules (generally nitrogen) on the nanofibers. For a specific gas or molecule, just such material must be replaced with nitrogen in the BET instrument. Chemical analyses with Fourier transform infrared spectroscopy and X-ray photoelectron spectroscopy, filtration measurements, electrical conductivity, optical properties, and field electron emission characteristics are other highly wondered properties of ACNFs and have been determined by various research groups working in the area.

# 12.5 Applications and Enhancements of ACNF Devices

## 12.5.1 Energy Applications

ACNFs have widely been used in energy-related devices including LIBs for energy storage and in fuel cells for energy production. To build a better energy storage device, engineered ACNFs with highly porous structures have been used in batteries. ACNFs as negative electrodes in supercapacitors with high energy density were produced from rod-shaped polyaniline by carbonization and activation with KOH by heat treatment at high temperature in a nitrogen environment (Fan et al. 2011). Fan et al. used a graphene/$MnO_2$ composite as the positive electrode and neutral aqueous $Na_2SO_4$ as the electrolyte. The graphene/$MnO_2$/ACNF asymmetric supercapacitor exhibits excellent cycling durability, with 97% specific capacitance retained even after 1000 cycles. The result reveals possible potentially great use of ACNFs in energy storage devices. Park et al. (2013) used ACNFs in dye-sensitized solar cells as counterelectrodes. The high efficiency of 7.21%, which is comparable to that of Pt counterelectrode-based solar cell, was obtained and demonstrated the potential use of ACNFs in solar cell devices. ACNFs are also used in microbial fuel cells for power generation (Karra et al. 2013). Ghasemi et al. (2011) reported the effect of physical and chemical activation of ACNFs on power generation in a microbial fuel cell. The results show that chemically activated (KOH) CNFs show greater power generation than both physically activated CNFs and traditionally used platinum cathode. Thus, these examples show that CNFs are potentially useful in energy application devices with enhanced properties.

## 12.5.2 Gas and Chemical Adsorptions

Adsorption is a process wherein gas or liquid molecules are collected on a solid surface. AC is an extensively porous form of carbon and acts like a sponge to adsorb gases and small chemical molecules on its surface through van de Waals forces. This way, numerous pollutants from air or water can be removed by using AC. When AC is used in the form of nanofibrous structure, the reachable surface is increased marvelously and more gas or chemical is adsorbed by the surface. Shim et al. (2006) produced electrospun ACNFs and investigated benzene adsorption on porous CNFs at different temperatures and observed that, even though energetically and structurally heterogeneous surfaces cause more benzene molecule adsorption, benzene adsorption decreases with increasing adsorption temperature. Highly efficient formaldehyde adsorption was observed by Lee et al. (2010) with PAN-based electrospun ACNFs. Manganese-embedded and chemically activated PAN-based CNFS were used for toluene adsorption and effective adsorption of toluene on ACNFs was reported by Oh et al. (2008). To sum up, as a result of high adsorption characteristics of ACNFs, pore characteristics of nanofibers are determined via BET measurement by evaluation of nitrogen gas adsorption and desorption with the instrument.

## 12.5.3 Other Applications

ACFs were used as catalyst support because of their excellent adsorbent properties. Tzeng et al. (2006) produced CNFs on ACF fabric. Since the porous surface property of ACFs in the fabric catalyst was distributed uniformly on the fibers, most of the fibers were synthesized uniformly and densely on ACF fabrics. Nickel nitrate was used as catalyst and CNFs were grown via CVD by using a hydrocarbon source (Tzeng et al. 2006). ACF film was immersed in Ni–Al catalyst precursor solution. In situ growth of CNFs on the surface of ACFs took place with the flow of acetylene and hydrogen mixture gas into the CVD chamber at 823 K (Zhan et al. 2011). ACNFs were grown on AC surface from $C_2H_4$ source with nickel (Ni) catalyst source via CVD, and the texture of CNTs (micropore structure on the surface) was changed by tuning the CNF growth temperature between 500°C and 850°C (Rinaldi et al. 2009). ACNFs have been used for liquid and gas filtration, treatment, and purifications. Environmentally damaged substances and biogases can be released from the system to reduce potential harmful effects by using ACNFs. Filtration of water and cigarette filters can be given as examples. Air and odor-control filter gas masks, respirators, and wound dressing are personal and collective protection devices. ACNFs are used in these for medical and pharmacological purposes. AC cloth as support for the growth and differentiation of human mesenchymal umbilical-cord stromal stem cells was investigated by Farias et al. (2013). The results demonstrate that the scaffold provides a suitable environment for cell growth, providing advantageous new possibility for bone-tumor engineering and treatment of traumatic and degenerative bone diseases. ACs can also absorb sound, and if these fancy materials are produced and used in nanoscale form such as nanofibers, enhanced sound absorption can be obtained.

## 12.6 Toxicology of ACNFs

Nanomaterials are small substances that can fly and penetrate easily into the human body by smell, physically touch, or transfer from other substrate material. For this reason, the toxicity of nanomaterials, including ACNFs, is an important topic that must be taken into account by the researcher, the producer, and the user of the final devices in which nanomaterials are used.

Because of the high toxicity of carbon nanomaterials, including the immobilization of such materials is desirable. Because of this, production of CNFs via ES is favored since continuous CNFs in a mat have been manufactured. Thus, during both the production and the use of such a structure in the device, no loss of carbon nanostructures in dust form is seen.

ACNFs on AC surface from $C_2H_4$ source with different iron catalyst sources, including iron nitrate, iron acetate, and iron citrate, have been produced by heating between 500°C and 800°C. Iron particles are not readily found when using iron citrate as precursor except for the other iron sources. Iron particles were well dispersed on AC surface and immobilized CNFs have been produced (Chen et al. 2007).

## 12.7 Summary and Outlook

The overview in this chapter demonstrates an essential explanation and a comprehensive literature research on fabrication of ACNFs via different techniques from various resources, including gas, liquid, solid, natural resources, and recycled materials. The mostly used characterization methods include morphological characterizations, such as SSA and pore characteristics, crystalline, and microstructural analyses, and characterizations depending on the application areas such as energy, electrochemistry, optics, biochemistry, filtration, and electron emission gas and molecular adsorption–desorption characteristics of ACNFs have been demonstrated by giving examples from the literature. A brief introduction on ACNFs has been given in Section 12.1. Two main activation methods called physical and chemical activation mechanisms have been explained in detail by giving examples from the literature. The effect of production techniques and source on porosity, microstructure, and morphology and also device enhancement have been discussed in detail. Toxicology of ACNFs is a critical issue as for other nanoscale materials and has been briefly mentioned in this chapter. Study problems given at the end of this chapter induce the readers toward a detailed thinking of ACNF production and characterization methods and application of such porous nanofibers due to their properties, including morphological and microstructural features. Readers will also gather more information and complete their knowledge by discussing the answers to the questions via reading this chapter and novel published researches in the literature.

# References

Aykut, Y., "Enhanced field electron emission from electrospun Co-loaded activated porous carbon nanofibers," *ACS Appl. Mater. Interfaces* 4 (2012): 3405–3415.

Aykut, Y., "Electrospun MgO-loaded carbon nanofibers: Enhanced field electron emission from the fibers in vacuum," *J. Phys. Chem. Solids* 74 (2013): 328–337.

Aykut, Y., Pourdeyhimi, B. & Khan, S. A., "Catalytic graphitization and formation of macroporous activated carbon nanofibers from salt-induced and $H_2S$ treated polyacrylonitrile," *J. Mater. Sci.* 48 (2013a): 7783–7790.

Aykut, Y., Pourdeyhimi, B. & Khan, S. A., "Effects of surfactants on the microstructures of electrospun polyacrylonitrile nanofibers and their carbonized analogs," *J. Appl. Polym. Sci.* 130 (2013b): 3726–3735.

Carrott, S. P. J. M. & Carrott, M. M. L. R., "Lignin—From natural adsorbent to activated carbon: A review," *Bioresour. Technol.* 98 (2007): 2301–2312.

Chen, X. W., Su, D. S. & Hamid, S. B. A., "The morphology, porosity and productivity control of carbon nanofibers or nanotubes on modified activated carbon," *Carbon* 45 (2007): 892–902.

Chen, L. F., Zhang, X. D., Liang, H. W. et al., "Synthesis of nitrogen-doped porous carbon nanofibers as an efficient electrode material for supercapacitors," *ACS Nano* 6 (2012): 7092–7102.

Daranyi, M., Csesznok, T., Sarusi, I. et al., "Beneficial effect of multi-wall carbon nanotubes on the graphitization of polyacrylonitrile (PAN) coating," *Process. Appl. Ceram.* 4 (2010): 59–62.

Dias, J. M., Alvim-Ferraz, M. C. M., Almeida, M. F. et al., "Waste materials for activated carbon preparation and its use in aqueous-phase treatment: A review," *J. Environ. Manag.* 85 (2007): 833–846.

Fan, Z., Yan, J., Wei, T. et al., "Asymmetric supercapacitors based on graphene/$MnO_2$ and activated carbon nanofiber electrodes with high power and energy density," *Adv. Funct. Mater.* 21 (2011): 2366–2375.

Fang, B., Kim, M., Fan, S. Q. et al., "Facile synthesis of open mesoporous carbon nanofibers with tailored nanostructure as a highly efficient counter electrode in CdSe quantum-dot-sensitized solar cells," *J. Mater. Chem.* 21 (2011): 8742.

Farias, V. A., Lopez-Penalver, J., Sires-Campos, J. et al., "The growth and spontaneous differentiation of umbilical-cord stromal stem cells on an activated carbon cloth," *J. Mater. Chem. B* 1 (2013): 3359–3368.

Gao, F., Zhao, D. L., Li, Y. et al., "Preparation and hydrogen storage of activated rayon-based carbon fibers with high specific surface area," *J. Phys. Chem. Solids* 71 (2010): 444–447.

Ghasemi, M., Shahgaldi, S., Ismail, M. et al., "Activated carbon nanofibers as an alternative cathode catalyst to platinum in a two-chamber microbial fuel cell," *Int. J. Hydrog. Energy* 36 (2011): 13746–13752.

Huang, C. W., Wu, Y. T., Hu, C. C. et al., "Textural and electrochemical characterization of porous carbon nanofibers as electrodes for supercapacitors," *J. Power Sources* 172 (2007): 460–467.

Im, J. S., Park, S. J. & Lee, Y. S., "Preparation and characteristics of electrospun activated carbon materials having meso- and macropores," *J. Colloid Interface Sci.* 314 (2007): 32–37.

Im, J. S., Park, S. J., Kim, T. J. et al., "The study of controlling pore size on electrospun carbon nanofibers for hydrogen adsorption," *J. Colloid Interface Sci.* 318 (2008): 42–49.

Im, J. S., Park, S. J. & Lee, Y. S., "Superior prospect of chemically activated electrospun carbon fibers for hydrogen storage," *Mater. Res. Bull.* 44 (2009): 1871–1878.

Ji, L., Lin, Z., Medford, A. J. et al., "Porous carbon nanofibers from electrospun polyacrylonitrile/$SiO_2$ composites as an energy storage material," *Carbon* 47 (2009): 3346–3354.

Jimenez, V., Sanchez, P., Valverde, J. L. et al., "Effect of the nature the carbon precursor on the physicochemical characteristics of the resulting activated carbon materials," *Mater. Chem. Phys.* 124 (2010): 223–233.

Karra, U., Manickam, S. S., McCutcheon, J. R. et al., "Power generation and organics removal from wastewater using activated carbon nanofiber (ACNF) microbial fuel cells (MFCs)," *Int. J. Hydrog. Energy* 38 (2013): 1588–1597.

Kim, C., "Electrochemical characterization of electrospun activated carbon nanofibres as an electrode in supercapacitors," *J. Power Source* 142 (2005): 382–388.

Kim, C., Yang, K. S., Kojima, M. et al., "Fabrication of electrospinning-derived carbon nanofiber webs for the anode material of lithium-ion secondary batteries," *Adv. Funct. Mater.* 16 (2006): 2393–2397.

Lee, K. L., Shiratori, N., Lee, G. H. et al., "Activated carbon nanofiber produced from electrospun polyacrylonitrile nanofiber as a highly efficient formaldehyde adsorbent," *Carbon* 48 (2010): 4248–4255.

Lee, H. M., Kang, H. R., An, K. H. et al., "Comparative studies of porous carbon nanofibers by various activation methods," *Carbon Lett.* 14 (2013): 180–185.

Li, C., Yin, X. & Chen, L., "Porous carbon nanofibers derived from conducting polymer: Synthesis and application in lithium-ion batteries with high-rate capability," *J. Phys. Chem. C* 113 (2009): 13438–13442.

Lyons, J., Li, C. & Ko, F., "Melt-electrospinning part I: Processing parameters and geometric properties," *Polymer* 45 (2004): 7597–7603.

Merino, C., Sato, P., Vilaplana-Ortego, E. et al., "Carbon nanofibres and activated carbon nanofibres as electrodes in supercapacitors," *Carbon* 43 (2005): 551–557.

Nandi, M., Okada, K., Dutta, A. et al., "Unprecedented $CO_2$ uptake over highly porous N-doped activated carbon monoliths prepared by physical activation," *Chem. Commun.* 48 (2012): 10283–10285.

Oh, G. Y., Ju, Y. W., Jung, H. R. et al., "Preparation of the novel manganese-embedded PAN-based activated carbon nanofibers by electrospinning and their toluene adsorption," *J. Anal. Appl. Pyrolysis* 81 (2008): 211–217.

Park, S. H., Kim, B. K. & Lee, W. J., "Electrospun activated carbon nanofibers with hollow core/highly mesoporous shell structure as counter electrodes for dye-sensitized solar cells," *J. Power Sources* 239 (2013): 122–127.

Ren, T., Si, Y., Yang, J. et al., "Polyacrylonitrile/polybenzoxazine-based $Fe_3O_4$@carbon nanofibers: Hierarchical porous structure and magnetic adsorption property," *J. Mater. Chem.* 22 (2012): 15919.

Rinaldi, A., Abdullah, N., Ali, M. et al., "Controlling the yield and structure of carbon nanofibers grown on a nickel/activated carbon catalyst," *Carbon* 47 (2009): 3023–3033.

Santoro, C., Stadlhofer, A. & Hacker, V., "Activated carbon nanofibers (ACNF) as cathode for single chamber microbial fuel cells (SCMFCs)," *J. Power Sources* 243 (2013): 499–507.

Seo, M. K. & Park, S. J., "Electrochemical characteristics of activated carbon nanofiber electrodes for supercapacitors," *Mater. Sci. Eng. B* 164 (2009): 106–111.

Shim, W. G., Kim, C., Lee, J. W. et al., "Adsorption characteristics of benzene on electrospun-derived porous carbon nanofibers," *J. Appl. Polym. Sci.* 102 (2006): 2454–2462.

Sodhi, P., "Activated carbon: Black magic," *Indian J. Res.* 2 (2013): 121–124.

Tao, X. Y., Zhang, X. B., Zhang, L. et al., "Synthesis of multi-branched porous carbon nanofibers and their application in electrochemical double-layer capacitors," *Carbon* 44 (2006): 1425–1428.

Teng, M., Qiao, J., Li, F. et al., "Electrospun mesoporous carbon nanofibers produced from phenolic resin and their use in the adsorption of large dye molecules," *Carbon* 50 (2012): 2877–2886.

Tzeng, S. S., Hung, K. H. & Ko, T. H., "Growth of carbon nanofibers on activated carbon fiber fabrics," *Carbon* 44 (2006): 859–865.

Wang, G., Dong, Q., Ling, Z. et al., "Hierarchical activated carbon nanofiber webs with tuned structure fabricated by electrospinning for capacitive deionization," *J. Mater. Chem.* 22 (2012a): 21819.

Wang, G., Pan, C., Wang, L. et al., "Activated carbon nanofiber webs made by electrospinning for capacitive deionization," *Electrochim. Acta* 69 (2012b): 65–70.

Warhurst, A. M., McConnachie, G. L. & Pollard, S. J. T., "Characterisation and applications of activated carbon produced from *Moringa oleifera* seed husks by single-step steam pyrolysis," *Water Res.* 31 (1997): 759–766.

Yoon, S. H., Lim, S., Song, Y. et al., "KOH activation of carbon nanofibers," *Carbon* 42 (2004): 1723–1729.

Zhan, Y., Nie, C., Li, H. et al., "Enhancement of electrosorption capacity of activated carbon fibers by grafting with carbon nanofibers," *Electrochim. Acta* 56 (2011): 3164–3169.

Zhang, L. & Hsieh, Y. L., "Carbon nanofibers with nanoporosity and hollow channels from binary polyacrylonitrile systems," *Eur. Polym. J.* 45 (2009): 47–56.

Zhang, X. & Ji, L., "Fabrication of porous carbon nanofibers and their application as anode materials for rechargeable lithium-ion batteries," *Nanotechnology* 20 (2009): 155705.

Zhang, Z., Li, X. & Wang, C., "Polyacrylonitrile and carbon nanofibers with controllable nanoporous structures by electrospinning," *Macromolec. Mater. Eng.* 294 (2009): 673–678.

Zhao, H., Wang, L., Jia, D. et al., "Coal based activated carbon nanofibers prepared by electrospinning," *J. Mater. Chem. A* 2 (2014): 9338.

Zhu, D., Koganemaru, A., Xu, C. et al., "Oxidative stabilization of PAN/VGCF composite," *J. Appl. Polym. Sci.* 87 (2003): 2063–2073.

# 13

# Electrospun Carbon Nanofibers

Samuel Chigome

Sunil A. Patil

**Abstract**

The properties and abundance of carbon in its different forms resulted in the wide use of carbon-based materials by diverse sectors of industry, science, and technology. Consequently, emerging and anticipated technological challenges have increased the pressure on researchers to develop advanced carbon-based materials. In addition, there will always be a need to search for alternative sources of carbon as well as innovative ways of using available carbonaceous materials for different applications. To date, materials at the nanoscale have emerged as having the greatest potential of addressing existing and emerging materials science challenges. Needless to say, a better understanding of the fabrication and use of carbon-based materials at the nanoscale level via such approach as electrospinning (ES) has become of paramount importance. Since the last decade, ES of polymeric precursors, such as polyacrylonitrile, has been routinely used as a simple and low-cost method to make continuous nanofibers. The chapter discusses the fabrication and properties of electrospun carbon nanofibers, which have been demonstrated to be useful for several applications ranging from energy storage and conversion, biomedical, and adsorption and separation devices due to their superior physicochemical, electrochemical, and mechanical properties.

## 13.1 Introduction

Carbon nanofibers (CNFs) can be defined as carbonaceous filaments with a diameter of less than 100 nm. The approaches to CNF fabrication can be broadly classified into two on the basis of the precursor state. These are spinning and vapor growth, which rely on solid phase and vapor phase precursors, respectively. The vapor growth approach comprises the decomposition of hydrocarbons in the presence of metal catalysts to produce CNFs (Tibbetts et al. 2007). Vapor-grown CNFs are relatively short, difficult to assemble, and challenging to produce at large scales with acceptable efficiencies at low cost, which collectively limits their applications. The spinning approach comprises fabricating fibers from a polymeric precursor through an extrusion process, with subsequent thermal treatment to produce CNFs. Conventional spinning approaches include wet, dry, melt, and gel spinning. The main advantages of spinning approach are fabrication of long and continuous fibers, possibility to scale up to industrial scale, ability to use a wide range of precursors, and more flexibility in controlling the nanofiber morphology and orientation. Despite the general benefits of spinning over vapor growth, the pressing need for total control of CNF production has resulted in recent research efforts being channeled toward a better understanding of the versatile spinning techniques such as electrospinning (ES) (Greiner & Wendorff 2007, Cavaliere et al. 2011, Zhang et al. 2014). The objective of the chapter is to equip the readers with sufficient knowledge that will enable them to fabricate and employ electrospun carbon nanofibers (ECNFs) for a broad range of applications. Therefore, a discussion on the fundamental fabrication aspects of ECNFs that influence their physical, chemical, electronic, thermal, and mechanical properties is presented.

## 13.2 Electrospinning

ES is a versatile nanofabrication technique that relies on repulsive electrostatic forces to draw a viscoelastic solution into nanofibrous form. The technique is well documented as a lot of research has been conducted over the past decade (Teo & Ramakrishna 2006, Chigome & Torto 2011, Chigome et al. 2011). Although there have been several developments in terms of ES setups, all setups are centered on the three basic components: a high voltage power supply, a mode to deliver a viscoelastic solution, and a means for collecting fibers (Figure 13.1).

When a high voltage is applied to a polymer solution, surface charges accumulate, leading to a deformation that takes a conical shape referred to as a Taylor cone. Progressively, a jet is erupted from the tip of the Taylor cone due to the tensile forces that result from the increased charges (Taylor 1969). As the jet travels toward the collector, increased surface charge repulsion results in a bending motion and it is at this point that polymer chains stretch and orient to form a nanofiber that is deposited on a collector (Doshi & Reneker 1995).

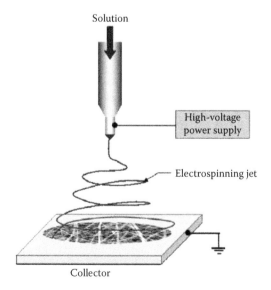

**FIGURE 13.1**   Schematic representation of the basic components of an *ES* setup. (From Teo, W. E. & Ramakrishna, S., *Nanotechnology*, 17, R89–R106, 2006, IOP Publishing.)

The process of ES is still not fully understood; a major challenge being a complete understanding of the exact details of charge motion of the polymer solution jet as it travels toward the collector. Nevertheless, it is believed that charges accumulate at the surface of the polymer jet such that they are essentially static with respect to the moving coordinate systems of the jet (Reneker et al. 2000). On that basis, the ES jet can be thought of as a string of charged elements connected by a viscoelastic medium with one end fixed at the point of origin and the other end free. The free end initially follows a stable trajectory until a point where the electrostatic interactions between the charge elements begin to dominate the ensuing motion, initiating and perpetuating a chaotic motion. Figure 13.2 shows a schematic representation of a simulation of an ES jet based on a numerical modeling proposed by Kowalewski et al. (2005).

If the ES jet is thought of as a moving charged object, electric, magnetic, and mechanical forces can be employed to control the deposition patterns of resulting nanofibers. The ability to easily control the orientation of nanofibers is an important benefit of ES because nanofiber arrangements can have a significant effect on the performance of the subsequent CNF-based device.

Traditionally, ES has been generally viewed as a laboratory scale technique due to the low production rate. However, in the last decade, significant progress has been made toward changing the perception of ES as various strategies have been implemented aimed at scaling up. Initially, the efforts to scale up the ES setup to an industrial production level were biased toward multiplication of the jets using multinozzle constructions (Zhou et al. 2009) (Figure 13.3). However, this approach posed a challenge associated

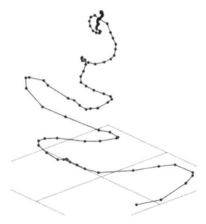

**FIGURE 13.2** Numerical model simulation of an ES jet. (From Kowalewski, T. A. et al., *Bulletin of the Polish Academy of Sciences: Technical Sciences,* 35, 385–394, 2005.)

(a)

(b)

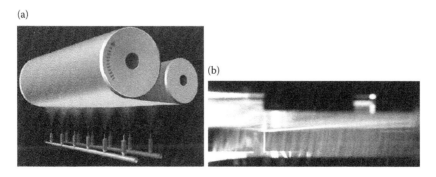

**FIGURE 13.3** (a) Schematic and (b) photograph of a multinozzle spinning head. (Courtesy of TOPTEC, South Korea, www.toptec.co.kr, 2011.)

with the number of jets needed to reach economically acceptable productivity, typically thousands. Furthermore, challenges associated with reliability, quality consistency, and machine maintenance (especially cleaning) also surfaced.

In this context, a welcome development in ES technology was nozzle-less ES reported by Jirsak et al. (2004). The nozzle-less ES solves most of the limitations associated with multiple-needle ES due to its mechanical simplicity. However, the process itself is more complex because of its spontaneous multijet nature. The main difference between nozzle-less ES and conventional ES is that multiple ES jets are ejected from a free liquid surface where the voltage is applied. This means that the number of ES jets is solely dependent on the available liquid surface area unlike conventional ES, which is dependent on the number of needles.

## 13.3  Fabrication of ECNFs

The process of ECNFs fabrication involves two main stages: (i) preparation of a viscoelastic carbonaceous precursor and (ii) heat treatment to achieve CNFs. The initial heat treatment stage can be followed by either further heat treatment at higher temperatures to achieve graphitization or chemical/heat treatment to achieve activation. The atomic structure of ECNFs is similar to that of graphite, consisting of carbon atom layers arranged in a regular hexagonal pattern (Figure 13.4).

Depending upon the precursors and fabrication approach, layer planes in ECNFs may be either turbostratic, graphitic, or a hybrid structure (Fitzer et al. 1990). The basic structural unit of most ECNFs consists of a stack of turbostratic layers. In a turbostratic structure, the parallel graphene sheets are stacked irregularly or haphazardly folded, tilted, or split, while the graphitic structure consists of layered planes stacked parallel to one another in a regular fashion. Figure 13.5 shows a schematic representation of turbostratic and graphite structures.

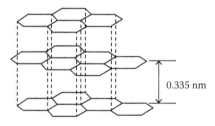

**FIGURE 13.4**  Structure of graphitic crystals. (From Huang, X., *Materials*, 28, 2369–2403, 2006.)

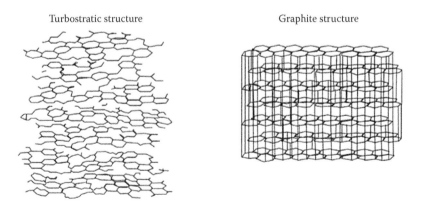

**FIGURE 13.5**  Schematic representation of turbostratic and graphite structures.

## 13.3.1 Precursors

CNFs inherit their properties from their precursors. For example, precursors with high tensile strength and modulus will produce CNFs with similar properties (Jazaeri 2012). Although any material with a carbon backbone can be used as a precursor, ECNFs are generally produced from three polymeric precursors: polyacrylonitrile (PAN), cellulose, and pitch.

### 13.3.1.1 PAN

Of the various precursors for making ECNFs, PAN has been the most commonly used for manufacturing high-performance fibers, owing to a combination of its high carbon, tensile, and compressive properties. The carbon content of acrylonitrile ($CH_2$=CHCN) of 67.9%, together with its chemical structure, results in a very high carbon yield of about 50%–55% in ECNFs. In order to be suitable as a precursor, the acrylic fiber should contain at least 85% of acrylonitrile monomer to provide as high a carbon yield as possible. Additionally, it helps to preserve the fibrous structure of the polymer precursor throughout the carbonization stage (Kuzmenko 2012).

### 13.3.1.2 Pitch

Pitch-based fibers are produced from low-cost by-products of the destructive distillation of coal, crude oil, or asphalt. They can be separated in two groups: (i) isotropic pitch fibers, which have low mechanical properties (Young's modulus is 35–70 GPa) and are relatively cheap; and (ii) mesophase pitch fibers, which have very high modulus (above 230 GPa) but are more expensive. The carbon yield for the pitch-based fibers is even higher than for PAN-based as it can exceed 60% (Kuzmenko 2012).

### 13.3.1.3 Cellulose

Cellulose is one of the oldest polymer precursors used in the production of fibers. Generally, polymers have a lower carbon yield compared to aromatic hydrocarbons since they contain other elements like oxygen, chlorine, or nitrogen that must be removed during pyrolysis. Cellulose gives a carbon yield of about 20% (Kuzmenko 2012). Although cellulose-based fibers are not as strong as PAN-based fibers, they have a huge advantage of being abundant, resulting in a considerable reduction in cost of the resultant ECNFs.

### 13.3.1.4 Other Precursors

Researchers continue to explore other carbon fiber precursors that are low cost, have high carbon content, and are biorenewable. Coal and lignin are seen as the most appropriate of the readily abundant biorenewable materials that can be exploited for ECNF production. Biorenewable lignin has been investigated as a potential precursor material for carbon fibers (Huang 2009, Zhao et al. 2014) as it is the most abundant phenolic compound in nature, and it is produced as the by-product of the pulping and cellulosic ethanol fuel production processes.

Lignin typically forms fibers with a structure that closely resembles that of isotropic pitch, which is presumably related to its relatively poor mechanical properties. For easier electrospinnability, lignin is usually blended with synthetic polymers such as poly(ethylene terephthalate), polypropylene, and poly(ethylene oxide) (PEO) (Kubo & Kadla 2005). Blending with PEO was observed to enhance the spinnability of lignin due to enhanced miscibility between lignin and PEO, which was due to the formation of hydrogen bonds between the hydroxyl groups of lignin and the ether groups in the PEO backbone (Kadla et al. 2002). Therefore, PEO could be a polymer of choice when preparing a lignin precursor for ECNF fabrication. ECNFs are then formed by subsequent heat treatment that is similar to that of pitch-based precursors.

Arc discharge– or arc plasma–assisted chemical vapor deposition methods have been conventionally used to fabricate vapor growth type fibers from coal. In 2014, Zhao et al. were the first to report the use of coal as a carbonaceous precursor for ECNF fabrication. Similar to pitch and lignin, coal requires a

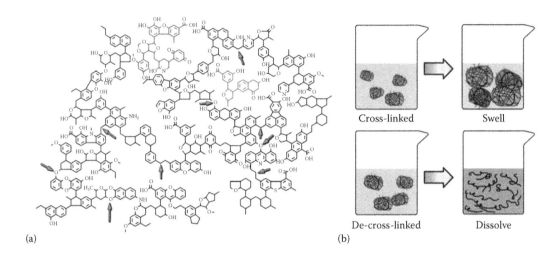

(a) (b)

**FIGURE 13.6** (a) Molecular representation of coal proposed by Shinn (1984). The red and blue molecules represent the low molecular weight parts, and the arrows indicate the bonds that can be broken easily. (Reprinted from *Fuel*, 63, Shinn, J. H., From coal to single-stage and two-stage product, 1187–1196, Copyright (1984), with permission from Elsevier.) (b) The cross-linked polymer network swells in the solvent, while the de-cross-linked polymer chains dissolve in the solvent. (Zhao, H. et al., *Journal of Materials Chemistry A* 2, 9338–9344, 2014. Reproduced by permission of The Royal Society of Chemistry.)

polymer support for easier electrospinnability. Coal consists of cross-linked macromolecular networks and relatively small molecules. The small molecule phase is clathrated in the macromolecular network (Zhao et al. 2014). This explains the swelling of coal in solvents, which is a typical phenomenon for common cross-linked macromolecules. Therefore, the aliphatic or ether bonds within the coal structure are easily broken under harsh reaction conditions (Figure 13.6).

The first stage of preparing a coal-based precursor for ES involves breaking of the aliphatic and ether bonds, which leaves the macromolecule units separated from each other. Harsh solvents like sulfuric acid and nitric acid are suitable for breaking down coal macromolecule units. The separated macromolecule units with polar groups on their edges are much easier to disperse in solvent. Given the polar nature of the molecules, a polar solvent is most suitable for preparing the ES solution as it ensures maximum dispersion. Therefore, PAN dissolved in dimethylformamide (DMF) is a suitable polymer solvent combination for preparing a coal-based precursor.

Generally, polar solvents (such as DMF, dimethyl sulfoxide, dimethyl acetamide among others) are the most preferred for the preparation of CNF precursor solutions and among these, DMF is the most popular due to its high dielectric constant that enhances charge accumulation, favoring better electrospinnability (Kurban et al. 2010, Zhao et al. 2014).

PAN, pitch, and cellulose are the most used precursors; hence, further discussion will primarily focus on these polymers.

### 13.3.2 Carbonization

Carbonization is the conversion of an organic substance into carbon or a carbon-containing residue through a process of heat treatment in a controlled environment. PAN and pitch precursors follow a similar heat treatment pattern as they are first stabilized by slow heating at 200°C–400°C in an $O_2$-containing atmosphere to avoid detrimental chain scission. However, different temperatures are employed at the carbonization stage: 1200°C–1400°C in inert atmosphere for PAN-based fibers, while

800°C–1200°C for pitch-based fibers (Poletto et al. 2011). Given that PAN is the most used backbone polymer for these precursors, the subsequent discussion will focus on the structural changes of PAN that occur during the heat treatment stages.

The oxidative stabilization is considered to be caused by the introduction of thermosetting properties to the PAN fibers and involves the oxygen uptake and subsequent cyclization, dehydrogenation, aromatization, oxidation, and cross-linking (Dalton et al. 1999). The stabilization process is found to play a key role in converting PAN fibers to an infusible stable ladder polymer that converts $C\equiv N$ bonds to $C=N$ bonds and to develop cross-links between PAN molecules, which tend to operate at high temperatures with minimum volatilization of carbonaceous material. The thermal stability of these fibers is attributed to the formation of the ladder structure due to cyclization of the nitrile groups. Setnescu et al. (1999) reported that $CH_2$ and CN groups disappear completely due to elimination, cyclization, and aromatization reactions, and form $C=C$, $C=N$, and $=C-H$ groups. When the temperature rises above 600°C, the cyclized structure undergoes dehydrogenation and links up in a lateral direction, producing a graphite-like layer consisting of three hexagons in the lateral direction and bounded by the nitrogen atom (Fourduex et al. 1971) (Figure 13.7).

Cellulose decomposition during the process of heat treatment is a complex phenomenon comprising a series of reactions occurring in succession or concurrently that eventually leads to mainly three products: (a) water, (b) gases, and (c) char. There have been many studies investigating cellulose decomposition and theories developed explaining the phenomenon (Tang & Bacon 1964, Bacon & Silvaggi 1971, Sekiguchi et al. 1983, Pastor 1999, Plaisantin 2001, Ishida et al. 2004, Sevilla & Fuertes 2009, 2010). However, since cellulose itself has many physical forms in nature, the decomposition mechanism is influenced by its structural configuration. Hence, it is difficult to obtain a general explanation of the decomposition mechanism. Nevertheless, the current consensus can be summarized as follows (Morgan 2005) and as illustrated in Figure 13.8.

- External dehydration (120°C–180°C)

    When cellulose is heated to 120°C, physically absorbed water molecules are released. The hydrogen bonds between water molecules and –OH groups in cellulose are replaced with the hydrogen bonding between –OH groups in cellulose.

- Internal dehydration (180°C–250°C)

    As the temperature increases to around 180°C, the energy is sufficient to excite and break some bonds such as with –OH and –H atoms within the cellulose molecules (Piskorz et al. 1986, Pastorova et al. 1994, Matsuoka et al. 2011, Poletto et al. 2012). This causes dehydration from the cellulose glycosidic rings (internal dehydration) and elimination of water molecules from the rings (Tang & Bacon 1964). At ~250°C, levoglucosan (1-6-anhydro-β-D-glucopyranose) is formed by an intrachain dehydration reaction between an OH group in $CH_2OH$ and the oxygen in 1–4 glycosidic links (Robert et al. 2002, Knill & Kennedy 2003, Collinson & Thielemans 2010). Levoglucosan is volatile and, at higher temperatures, breaks down into tar and some other by-products such as char, $CO_2$, and furan (Lu et al. 2009, Tao et al. 2012). Levoglucosan is also responsible for the flammability of cellulose since its structural breakdown results in the formation of flammable gases. Hence, the prevention of levoglucosan formation will increase the yield of carbon residue (Broido et al. 1973, Broido & Nelson 1975, Broido & Yow 1977). The use of fire retardants such as $ZnCl_2$ has been studied in order to reduce the amount of levoglucosan that was produced during cellulose decomposition (Azzam 1987, Dollimore & Hoath 1987, Amarasekara & Ebede 2009).

- Depolymerization (250°C–400°C)

    Above 250°C, the scission of glycosidic rings occurs that leads to the formation of CO, $CO_2$, $H_2O$, and char (Gaur & Reed 1994, Mamleev et al. 2007, Jin et al. 2008). However, the scission is not a straightforward reaction but a series of other reactions leading to such a phenomenon. On

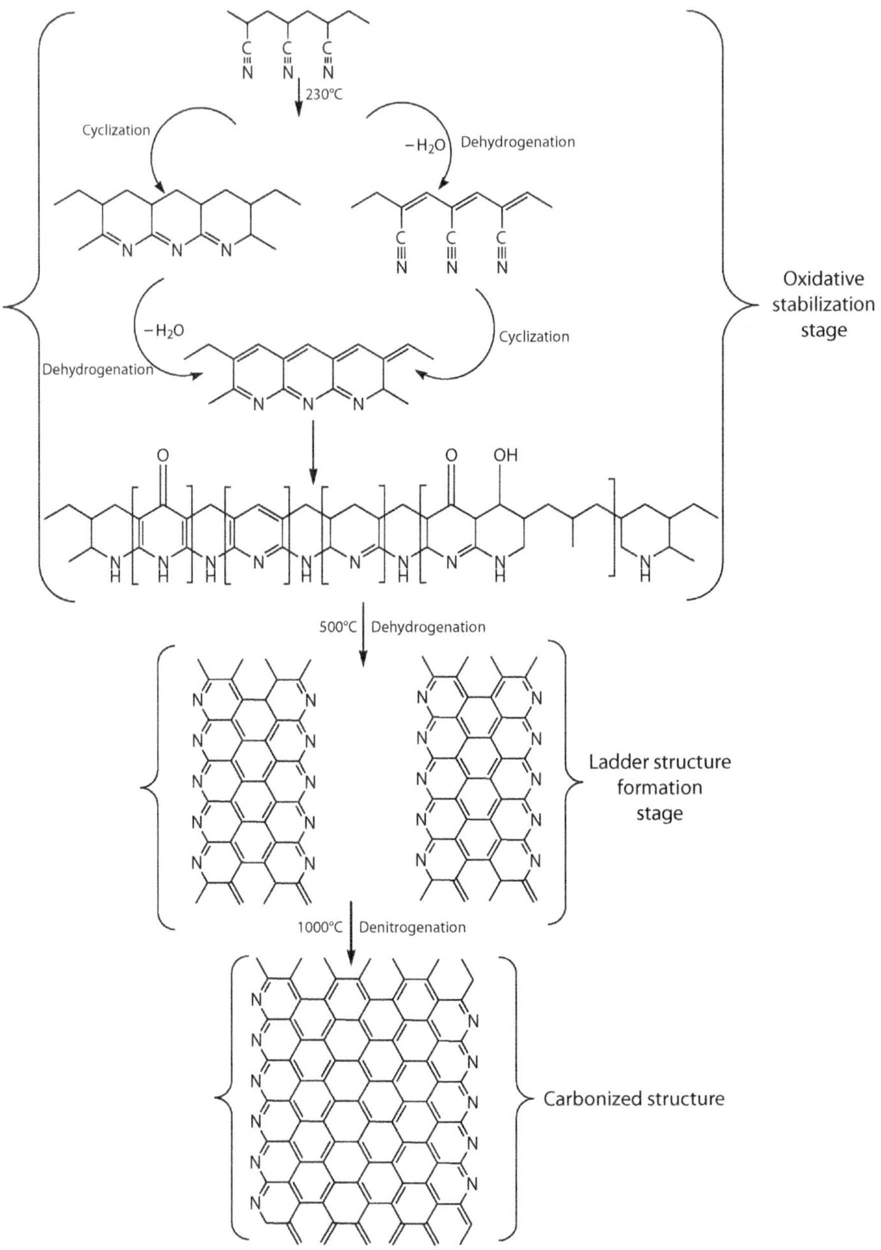

**FIGURE 13.7** Schematic representation of stages involved in the carbonization process. (Reprinted from *Bioresource Technology*, 132, Patil, S. A. et al., Electrospun carbon nanofibers from polyacrylonitrile blended with activated or graphitized carbonaceous materials for improving anodic bioelectrocatalysis, 121–126, Copyright (2013), with permission from Elsevier.)

one hand, the removal of these OH groups from the rings results in the formation of C=C and C=O bonds, while the chair/bed (Zugenmaier 2001, Pérez & Samain 2010) conformation of cellulose is still maintained (Soares et al. 1995, Hajaligol 2002, Paris et al. 2005, Lédé 2012). On the other hand, levoglucosan is also formed at above 250°C. These two reactions are always concurrent and compete with each other (Morgan 2005).

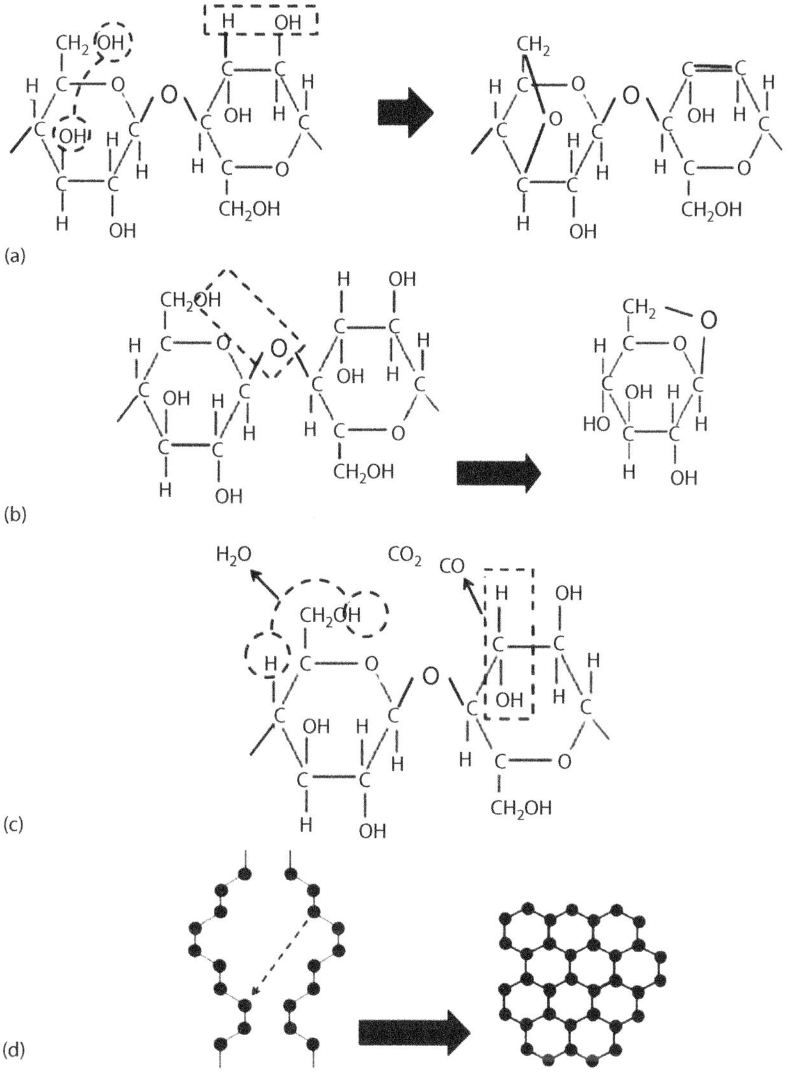

**FIGURE 13.8** Schematic representation of the steps involved in the carbonization of cellulose: (a) external dehydration; (b) internal dehydration; (c) depolymerization; (d) aromatization. (From Jazaeri, E., MEng Thesis, Deakin University, 2013.)

- Aromatization (>400°C)

  Depolymerization is completed at 400°C and further increase in temperature initiates aromatization of CNFs. At the completion of cellulose decomposition, carbon atoms connect with each other and form hexagonal rings similar to graphite. However, the paths in which carbon atoms connect and form rings are not fully understood. It has been suggested that longitudinal polymerization occurs where carbon atoms join to form carbon chains and these chains connect to form rings. However, this model has little experimental support (Konkin 1985).

## 13.3.3 Graphitization

Graphitization is the transformation of carbon structure into graphite structure by heat treatment as well as thermal decomposition at high temperatures (typically 1600°C–3000°C). In other words, graphitization is carbonization at high heating temperatures. At this stage, up to 99% of, for instance, PAN is converted to the carbon structure. Since graphitization starts from the fiber surface, it is expected that the graphitization takes place at lower temperatures in nanofibers than in conventional micron-sized fibers. This means that graphitized ECNFs may be produced with less energy and cost than conventional CNFs (Jazaeri 2012).

## 13.3.4 Activation

Traditionally, activated carbon was manufactured from carbon-rich materials by carbonization followed by activation (heat treatment with an oxidizing agent), or by simultaneous carbonization–activation in the presence of a dehydrating compound (Xu & Chung 2001, Jurewicz et al. 2008). Activation is carried out to remove the disorganized carbon that blocks the pores. Apart from this, it can enlarge the diameters of the pores which are formed during carbonization. As a result, a well-developed and readily accessible pore structure with large internal surface area can be produced (Diefendorf 2000, Xu & Chung 2001, Alam et al. 2009). Broadly, activation approaches can be classified into two: physical and chemical.

### 13.3.4.1 Physical Activation

Physical activation involves heat treatment at 700°C to 1000°C in the presence of oxidizing gases such as carbon dioxide, steam ($H_2O$), air ($O_2$), or their mixtures. To increase the efficiency of activation, the reaction takes place inside the fibers, in contrast to the reaction occurring outside the fibers. Higher carbon removal results in better pore development if the reaction occurs inside the fiber (Linares-Solano & Cazorla-Amorós 2008). The most common activating agents are $CO_2$ and steam, where reactions on carbon are endothermic. There are significant differences observed between these two activating agents on the porous texture and mechanical properties of the activated carbon fibers (Gaur et al. 2006, Linares-Solano & Cazorla-Amorós 2008). For instance, Virginia and Adrian (2012) found that in contrast to steam, oxidizing gases such as $CO_2$ creates final carbon product with larger micropore volume but narrower micropore size distribution. According to Zhang et al. (2004), $CO_2$ is preferably selected and used by many researchers as activation gas. The reasons include cleanliness and easy handling of $CO_2$. Besides, $CO_2$ also helps in controlling the activation process due to the slow reaction rate at the temperature of around 800°C (Bansal & Goyal 2005). Generally, the carbon products are prepared by conventional heating method for activation process. Among modern developments for heating technology, microwave heating is a new approach which has received great attention from many researchers. Microwave activation is still being utilized at a laboratory-scale level for the preparation of high-porosity carbon products. Microwave heating has a high potential to become a feasible alternative to replace conventional heating methods in industrial production due to its number of advantages. These include high yield of products, high heating rate, energy transfer instead of heat transfer, selective heating, better control of the heating process, lower activation temperature, smaller equipment size, and less automation (Minkova et al. 2001, Savova et al. 2001, Sugumaran et al. 2012).

### 13.3.4.2 Chemical Activation

Chemical activation involves treatment of the starting materials with an activating agent such as zinc chloride (Hameed et al. 2009), phosphoric acid (Olivares-Marin et al. 2012), potassium hydroxide (Petrov et al. 2008), sulfuric acid (Li et al. 2010), hydrochloric acid (Sugumaran et al. 2012), sodium hydroxide (Savova et al. 2001), and other dehydrating agents. In this process, the mixture of raw material and activating agent is heat-treated under inert atmosphere at temperatures ranging from 400°C to 700°C. Then, the chemical is removed by exhaustive washing with water, followed by drying and separation from the slurry. The dried fibers are then conditioned according to its application (Xu & Chung 2001).

Essentially, all activation agents are dehydrating agents that inhibit the formation of tar and other by-products, e.g., acetic acid and methanol. It can help in increasing above 30 wt% of yield of carbon in this process compared to thermal activation as reported by Bansal et al. (1988), and Veksha et al. (2009). The impregnation ratio between the mass of activating agent and the mass of raw material is one of the variables posing a great influence on the yield and porosity of the final carbon. In most studies, the impregnation ratios in the range of 0.3–3 (mass of activating agent/mass of raw material) were found to be applied on various raw materials (Yalçýn & Sevinç 2000, Minkova et al. 2001, Nakagawa et al. 2007, Fadhil et al. 2012). The yield of carbon decreases when the impregnation ratio increases. The advantages of chemical activation of CNFs over the physical activation also include one-stage heating in chemical activation, higher yield, and highly microporous final products. Notably, it is more suitable for the materials which yield higher ash content. Apart from these advantages, chemical activation can assist in lowering the temperature for activation, shortening the time for activation, and improving the development of inner porosity to obtain the desired result (Zhang et al. 2004, Gerçel et al. 2007). However, there are also some disadvantages associated with the use of chemical activation. For instance, extensive washing is needed after the heat treatment. Besides, the use of corrosive activating agents is also another drawback in the process (Linares-Solano & Cazorla-Amorós 2008). The use of chemical activation agents in treatment increases the cost of overall processing. Furthermore, the handling of chemical activation agents can be a dangerous and unhealthy practice. Thus, extra precautions are needed to minimize the health and safety risks.

## 13.4 Properties of Electrospun Carbon Nanofibers (ECNFs)

Commercially, ECNFs are produced for three main purposes:

1. General-purpose ECNFs: They are mainly amorphous in structure with low modulus, strength, and production cost.
2. High-performance ECNFs: They are graphitic in structure with higher modulus and strength than general-purpose ECNFs.
3. Activated ECNFs: They are porous in their microstructures, which makes them highly adsorptive to other chemicals (Fridman & Grebennikov 1990).

ECNFs have >92% carbon atoms in their structure which can be amorphous, semicrystalline, or crystalline. Carbon atoms are connected to each other via covalent bonding that leads to the formation of hexagonal rings. Such rings then connect in the cylindrical plane to form graphitic layers. These layers are connected in directions perpendicular to the layer via van der Waals bonding. Such a molecular configuration gives the ECNFs two properties: (a) good electrical and thermal conductivity on the planes and good thermal insulation in the direction perpendicular to the planes and (b) high strength (high modulus of elasticity) on the planes and low strength (low modulus of elasticity) in the direction perpendicular to the planes (Fitzer 1990). In ECNFs, graphitic layers are highly aligned along the fiber axis. This gives the fibers high elastic modulus, large crystallite sizes, high density, high tensile modulus, and high electrical and thermal conductivities along the fiber axis. The highly aligned graphitic layers along the fiber axis also give electrical and thermal insulation properties to the fiber axis (Chung 1994).

Graphitized CNFs consist of long graphitic layers that are oriented along the fiber axis and stacked up perpendicular to the fiber axis (Bennett et al. 1983, Fujimoto 2003, Im et al. 2008). Such structural orientation provides CNFs with very attractive properties such as ultrahigh aspect ratios and high specific surface area (SSA). They exhibit very high tensile strength and modulus which are comparable to those of metals. Superior properties of such CNFs have promised a bright future for many applications where strength combined with flexibility is needed.

During ES, the fibers are also stretched, which gives them higher strength because of the parallel orientation of fiber axes. Stretching of fibers can be performed separately after spinning as well. However, the mechanical properties of ECNFs need to be improved because of the limited crystallinity and orientation during ES (Li et al. 2003). The key aspect that determines the properties of the carbon fiber is the orientation of the polymer chain in the precursors. Usually, the high degree of preferential orientation along the fiber axis of the layer planes is mainly responsible for the extraordinarily high Young's modulus of the fibers. Researchers now realize that understanding and controlling the orientation structure during the precursor formation step is critical if the properties of the ECNFs are to be optimized (Ruland 1990). Therefore, the research on the microstructure of the precursors under different processing conditions is important in the study of the final properties of the carbon fibers (Ko et al. 1989). The influence of the ECNFs fabrication approach on their electrical, mechanical, and thermal properties is discussed in the Sections 10.4.1 through 10.4.3, respectively.

## 13.4.1 Electrical Properties

Electrical conductivity depends greatly on the physical deformation or damage in ECNFs, because any change in the cross section and length of the graphitic layers would reduce the conductivity (Parikh et al. 2009). Although ECNFs are electrically conductive to a certain extent, their conductivity is much lower than that of metals, except for highly graphitic ECNFs. One method to increase the electrical conductivity of ECNFs is intercalation (Donnet & Bansal 1990). In this method, metal particles, such as copper (Oshima et al. 1983), nickel (Gaier et al. 1988, Gall et al. 2000), or bromine (Chiou et al. 1989, Ho & Chung 1990a), are inserted in between graphene layers so as to increase the electrical conductivity up to 70% (Wang & Chung 1997). This method is possible only in graphitic ECNFs which have well-developed graphitic structures (Ho & Chung 1990b,c).

## 13.4.2 Mechanical Properties

Factors such as defects on the surface or within the fibers, handling procedures, and impurities in the nanofiber structure affect the mechanical properties of ECNFs (Dobb et al. 1995). The aspect ratio of nanofibers is another important property that influences the mechanical properties of composites. Since ECNFs have diameters in nanoscale and lengths of up to several meters, their high aspect ratio enables them to evenly distribute the load throughout the matrix with no fibers pulled out of the matrix, and hence increases the reinforcement effects in composites (Naji et al. 2008, Martone et al. 2011).

Tensile strength depends on the orientation of the graphene layers along the fiber axis as well as the interlayer spacing; the higher the alignment and the lower interlayer spacing, the higher the tensile strength. The tensile strength increases as the fiber diameter decreases (Johnson 1990). Pitch-based ECNFs show higher tensile strength than PAN- or cellulose-based ECNFs, because the graphene layers are highly oriented along the fiber axis in these ECNFs. However, an increase in tensile strength results in a decrease in shear modulus, since the higher orientation of graphene layers along the fiber axis makes fibers weaker to shear forces.

## 13.4.3 Thermal Properties

The thermal stability of CNFs depends on the atmosphere in which they are used. ECNFs can be oxidized in air but great mass loss occurs above 400°C (Lin et al. 1994). Graphitized ECNFs show higher stability than ungraphitized ECNFs. However, mass loss still occurs above 600°C. Therefore, if ECNFs are to be used in atmospheric environments, their working temperature should be below 400°C. Thermal stability of ECNFs also depends on the amount of impurities such as Na, Si, Mg, and Ca. Although these impurities may increase thermal conductivity, they lower the stability of ECNFs at high temperatures (Koráb et al. 2002, Wang et al. 2009). Thermal conductivity of highly graphitized ECNFs is generally higher than that of metals such as copper (Katzman et al. 1994, Adams et al. 1998). In addition, increase in tensile strength increases the thermal conductivity due to the more developed graphitic structure (Lavin et al. 1993, Wang et al. 2009).

# 13.5 Applications of ECNFs

Because of their excellent physicochemical, electrochemical, and mechanical properties, ECNFs have found applications in various fields (Doyle et al. 2013, Zhang et al. 2014). The most important applications are illustrated in Figure 13.9 and are discussed further in the following.

## 13.5.1 Energy Conversion and Storage

The need to develop efficient, eco-friendly, and highly durable energy conversion and storage devices has led to the exploration of different kinds of nanomaterials such as continuous nanofibers, nanowires, nanotubes, and nanorods as electrodes or electrolytes in these devices. Since ES allows altering the dimensions, compositional flexibility, design, and other properties, it has become an attractive method for fabricating electrode materials for these devices. The specific properties of electrospun nanofibers such as high surface area and aspect ratios, low density, and high pore volume drive their use in these devices (Cavaliere et al. 2011). In particular, ECNFs are increasingly being investigated for the targeted development of electrode and electrolyte materials for such devices as dye-sensitized solar cells (DSCs), fuel cells, lithium ion batteries (LIBs), and supercapacitors. Several superior properties of ECNF-based electrodes such as ionic conductivity, recyclability, reversibility, low interfacial resistance, electrochemical stability, and mechanical strength make their use appealing in these devices.

In addition to the aforementioned properties, ECNFs represent a promising support material for fuel cells, as these possess increased conductivity over other carbon-based particulate materials. This is mainly due to easier electron transfer along the aligned CNFs, in contrast to carbon particles that possess significant interfaces between the particles, which may add to the resistance of the system. Most studies have focused on PAN-based CNFs as they can be readily electrospun and their graphitization method is well established. Phenolic resin-based CNFs with thin diameters and high conductivity have also been reported (Imaizumi et al. 2009). Examples of CNF-based electrodes in fuel cells include Fe–polyaniline (PANI)–PAN composite cathodes for oxygen reduction reaction (ORR) in proton exchange membrane fuel cells (Zamani et al. 2014), PAN-based ECNFs for ORR in enzymatic biofuel cells (Che et al. 2011),

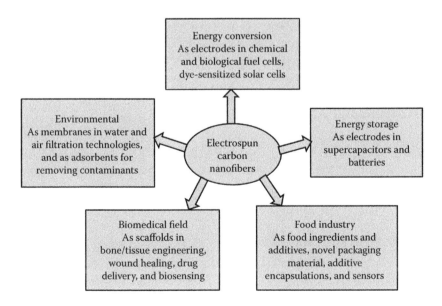

**FIGURE 13.9** An overview of important applications of electrospun carbon nanofibers.

Pt–CNF mat (Li et al. 2008), and PtRu/CNFs composite (Lin et al. 2010) for methanol oxidation in direct methanol fuel cells, and FeCo/CNFs composite for ORR in alkaline fuel cells (Jeong et al. 2010), and as anodes in microbial fuel cells (Chen et al. 2011a,b, He et al. 2011, Patil et al. 2013). CNFs have also been used as templates for the growth of zinc oxide nanowires for use as an anode material for the fabrication of flexible DSCs (Unalan et al. 2008). PAN-based ECNFs, either stand-alone or composites, have been investigated as a low-cost alternative to Pt counterelectrodes in DSCs (Joshi et al. 2010, Poudel et al. 2012). In microbial fuel cells, the excellent bioelectrocatalytic performance of ECNF mats as anodes has been attributed to its fibrous and porous structure that provides a habitat for the growth of microorganisms and channels for the efficient mass transfer. Furthermore, the interconnections between the individual fibers have been proposed to facilitate efficient electron transfer in the ECNF mats (Chen et al. 2011b, Patil et al. 2013).

In LIBs, which are used to power a wide range of electronic devices such as mobile phones, laptops, computers, and hybrid electric automobiles, ECNFs have been tried mainly as anode electrocatalysts. Stand-alone PAN-based (Kumar et al. 2012) and composite PAN + polypyrrole ECNFs (Ji et al. 2010) have been demonstrated to be good anodes for such batteries. In order to overcome the limitations of metal-based anodes and to improve cycling performance of LIBs, several composite PAN-based CNFs with Sn (Yu et al. 2010), $SnO_2$ (Bonino et al. 2011), Ni (Ji et al. 2012), and $TiO_2$ (Yang et al. 2012) have been reported. Composite materials such as $CNFs/LiFePO_4/graphene$ (Toprakci et al. 2012) and CNFs/LiF/Fe (Zhang et al. 2012) have also been tried as cathodes in these batteries. In addition, hybrid or composite ECNFs have been exploited as electrode materials for improving the capacitance of supercapacitors. Examples include CNF–carbon nanotubes (Guo et al. 2009), CNF–PANI (Yan et al. 2011), graphene-integrated CNFs (Kim et al. 2012, Zhou & Wu 2013), and CNFs with boron and nitrogen functional groups (Kim et al. 2013).

## 13.5.2 Biomedical Field/Applications

The main applications of electrospun fibers in biomedical field include bone/tissue engineering, drug delivery, wound healing, and biosensing. For their use in the biomedical field, the precursor polymers have to be biocompatible, safe, nontoxic, nonreactive, and most importantly nonimmunogenic (Saito et al. 2011). The use of ECNFs in biomedical applications such as tissue engineering, implant, and wound healing has been studied and fibers with specific properties suitable for such end uses have been investigated. For example, ECNFs from PAN have been used to prepare CNFs/hydroxyapatite composite to mimic the collagen fiber/hydroxyapatite composite structure in natural bone (Xu et al. 2001). Composite PAN-based CNFs with bioactive glass precursors have also been developed as a substrate for bone regeneration uses (Yang et al. 2013).

## 13.5.3 Environmental Applications

Due to their high SSA and high porosity, ECNFs have found applications in the field of adsorption and separation technologies. For instance, PAN-based CNFs have been demonstrated to be useful for the removal of disinfection by-products such as chloroform and monochloroacetic acid from drinking water (Singh et al. 2010). PAN-based activated ECNFs have also been tried for the adsorption of formaldehyde gas (Lee et al. 2010) and the removal of NO from polluted air (Wang et al. 2011).

## 13.5.4 Applications in Food Industry

The use of nonfood grade polymers limits the use of electrospun nanofibers in the food sector. Nevertheless, their use as new food ingredients, food additives, novel packaging material, food sensors, and additive encapsulations is under investigation (Doyle et al. 2013). In particular, nanofibers produced from natural polymers have potential applications in the development of high-performance packaging for food, food coatings, flavor enhancement, additive encapsulation, and nutraceutical applications due to their biocompatibility and biodegradability. For instance, food-packaging materials constructed from natural polymers can be used to improve the shelf time and retain the flavors in the food (Kriegel et al. 2008, Fabra et al. 2013). Another example of nanofiber application in the food industry is in biosensing. An example includes the detection of phenolic compounds (Arecchi et al. 2010, Xu et al. 2010). It is anticipated that incorporating electrospun nanofibers in the active packaging material could greatly assist regulators and enhance health and safety controls (Doyle et al. 2013).

## 13.6 Conclusions

ES of polymeric carbonaceous precursors offers several advantages such as efficiency, low cost, high yield, high carbon content, and high degree of reproducibility of the obtained ECNF materials. This chapter summarizes the manufacturing of ECNFs by ES of precursors such as PAN, followed by carbonization and activation via graphitization. Several important properties of ECNFs have also been discussed. This class of nanostructured carbon material that possess high SSA and good physicochemical and electrical properties has quickly found applications in several fields ranging from energy conversion and storage, catalysis, sensor and adsorption, to biomedical devices. With continuous ongoing research and development, these carbon-based nanofibrous materials are expected to find many more applications across wide scientific and technological disciplines in the future. With the increase in the number of ES companies over the last decade, ES is expected to progressively move from a laboratory bench process to an industrial-scale process. On the basis of the rapid development of ES companies, large-scale production of ECNFs seems feasible in the near future, thus paving a way for industrial-scale production of ECNF-based devices.

Although ECNFs clearly show a great potential for the future, enhancing mechanical strength by drawing still seems to be a technological challenge, particularly at the commercial ES scale. Therefore, future research efforts could focus on the ways for enhancing mechanical properties that are amenable to industrial scale-up.

## References

Adams, P. M., Katzman, H. A., Rellick, G. S. et al., "Characterization of high thermal conductivity carbon fibers and a self-reinforced graphite panel," *Carbon* 36 (1998): 233–245.

Alam, M. Z., Ameem, E. S., Muyibi, S. A. et al., "The factors affecting the performance of activated carbon prepared from oil palm empty fruit bunches for adsorption of phenol," *Chem. Eng. J.* 155 (2009): 191–198.

Amarasekara, A. S. & Ebede, C. C., "Zinc chloride mediated degradation of cellulose at 200°C and identification of the products," *Bioresourc. Technol.* 100 (2009): 5301–5304.

Arecchi, A., Scampicchio, M., Drusch, S. et al., "Nanofibrous membrane based tyrosinase-biosensor for the detection of phenolic compounds," *Anal. Chim. Acta* 659 (2010): 133–136.

Azzam, A. M., "Saccharification of bagasse cellulose pretreated with $ZnCl_2$ and HCl," *Biomass* 12 (1987): 71–77.

Bacon, R. & Silvaggi, A. F., "Electron microscope study of the microstructure of carbon and graphite fibers from a rayon precursor," *Carbon* 9 (1971): 321–325.

Bansal, R. C. & Goyal, M., *Activated Carbon Adsorption* (Taylor & Francis Group, Boca Raton, 2005).

Bansal, R. P., Donnet, J. B. & Stoecklie, F., *Active Carbon* (Marcel Dekker, New York, 1988).

Bennett, S. C., Johnson, D. J. & Johnson, W., "Strength–structure relationships in PAN-based carbon fibres," *J. Mater. Sci.* 18 (1983): 3337–3347.

Bonino, C. A., Ji, L., Lin, Z. et al., "Electrospun carbon–tin oxide composite nanofibers for use as lithium ion battery anodes," *ACS Appl. Mater. Interfaces* 3 (2011): 2534–2542.

Broido, A., Javierso, A. C., Ouano, A. C. et al., "Molecular-weight decrease in early pyrolysis of crystalline and amorphous cellulose," *J. Appl. Polym. Sci.* 17 (1973): 3627–3635.

Broido, A. & Nelson, M. A., "Char yield on pyrolysis of cellulose," *Combust. Flame* 24 (1975): 263–268.

Broido, A. & Yow, H., "Resolution of molecular-weight distributions in slightly pyrolyzed cellulose using Weibull function," *J. Appl. Polym. Sci.* 21 (1977): 1677–1685.

Cavaliere, S., Subianto, S., Savych, I. et al., "Electrospinning: Designed architectures for energy conversion and storage devices," *Energy Environ. Sci.* 4 (2011): 4761–4785.

Che, A.-F., Germain, V., Cretin, M. et al., "Fabrication of free-standing electrospun carbon nanofibers as efficient electrode materials for bioelectrocatalysis," *New J. Chem.* 35 (2011): 2848–2853.

Chen, S., He, G., Carmona-Martinez, A. A. et al., "Electrospun carbon fiber mat with layered architecture for anode in microbial fuel cells," *Electrochem. Commun.* 13 (2011a): 1026–1029.

Chen, S., Hou, H., Harnisch, F. et al., "Electrospun and solution blown three-dimensional carbon fiber nonwovens for application as electrodes in microbial fuel cells," *Energy Environ. Sci.* 4 (2011b): 1417–1421.

Chigome, S. & Torto, N., "A review of opportunities for electrospun nanofibers in analytical chemistry," *Analytica Chim. Acta* 706 (2011): 25–36.

Chigome, S., Darko, G. & Torto, N., "Electrospun nanofibers as sorbent material for solid phase extraction," *Analyst* 136 (2011): 2879–2899.

Chiou, J. M., Ho, C. T. & Chung, D. D. L., "Effect of bromination on the oxidation resistance of pitch-based carbon fibers," *Carbon* 27 (1989): 227–231.

Chung, D. D. L., *Carbon Fiber Composites* (Butterworth-Heinemann, Newton, 1994), pp. 3–11.

Collinson, S. R. & Thielemans, W., "The catalytic oxidation of biomass to new materials focusing on starch, cellulose and lignin," *Coord. Chem. Rev.* 254 (2010): 1854–1870.

Dalton, S., Heatley, F. & Budd, P. M., "Thermal stabilization of polyacrylonitrile fibres," *Polymer* 40 (1999): 5531–5543.

Diefendorf, R. J., "Pitch precursor carbon fibers," in Chou, T. W., ed., *Comprehensive Materials, Vol 1: Reinforcement Materials and General Theories* (Pergamon Press, Oxford, England, 2000), pp. 1–33.

Dobb, M. G., Guo, H., Johnson, D. J. et al., "Structure-compressional property relations in carbon fibres," *Carbon* 33 (1985): 1553–1559.

Dollimore, D. & Hoath, M. J., "The kinetics of the oxidative degradation of cellulose and cellulose doped with chlorides," *Thermochim. Acta* 121 (1987): 273–282.

Donnet, J. B. & Bansal R. C., "Carbon fibers," in *International Fiber Science and Technology* (Marcel Dekker, New York, 1990), pp. 267–366.

Doshi, J. & Reneker, D. H., "Electrospinning process and application of electrospun fibers," *J. Electrost.* 35 (1995): 151–160.

Doyle, J. J., Choudhari, S., Ramakrishna S. et al., "Electrospun nanomaterials: Biotechnology, food, water, environment, and energy," *Conf. Pap. Mater. Sci.* (2013): 14.

Fabra, M. J., Lopez-Rubio, A. & Lagaron, J. M., "High barrier polyhydroxyalcanoate food packaging film by means of nanostructured electrospun interlayers of zein," *Food Hydrocoll.* 32 (2013): 106–114.

Fadhil, A. B., Dheyab, M. M. & Abdul-Qader, Y., "Purification of biodiesel using activated carbons produced from spent tea waste," *J. Assoc. Arab Univ. Basic Appl. Sci.* 11 (2012): 45.

Fitzer, E., "Carbon fibers—Present state and future expectations," in Figueiredo, J. L., Bernardo, C. A., Baker, R. T. K. et al., eds., *Carbon Fibers Filaments and Composites* (Springer, New York, 1990), pp. 3–41.

Fitzer, E., Edie, D. D. & Johnson, D. J., "Carbon fibers—Present state and future expectation; Pitch and mesophase fibers; Structure and properties of carbon fibers," in Figueiredo, J. L., Bernardo, C. A., Baker, R. T. K. et al., eds., *Carbon Fibers Filaments and Composites* (Springer, New York, 1990), pp. 3–41, 43–72, 119–146.

Fourduex, A., Perret, R. & Ruland, W., "Carbon fibers—Their composites and applications," The Plastic Institute, London (1971): 57.

Fridman, L. I. & Grebennikov, S. F., "Theoretical aspects in the preparation and use of fibrous carbon adsorbents," *Fiber Chem.* 22 (1990): 363–367.

Fujimoto, H., "Theoretical X-ray scattering intensity of carbons with turbostratic stacking and AB stacking structures," *Carbon* 41 (2003): 1585–1592.

Gaier, J. R., Slabe, M. E. & Shaffer, N., "Stability of the electrical resistivity of bromine, iodine monochloride, copper(II) chloride, and nickel(II) chloride intercalated pitch-based graphite fibers," *Carbon* 26 (1988): 381–387.

Gall, N. R., Rut'kov, E. V. & Tontegode, A. Y., "Intercalation of nickel atoms under two dimensional graphene film on (111)Ir," *Carbon* 38 (2000): 663–667.

Gaur, S. & Reed, T. B., "Prediction of cellulose decomposition rates from thermogravimetric data," *Biomass Bioenergy* 7 (1994): 61–67.

Gaur, V., Sharma, A. & Verma, N., "Synthesis and characterization of activated carbon fiber for the control of BTX," *Chem. Eng. Process* 45 (2006): 1–13.

Gerçel, Ö., Özcan, A., Özcan, A. S. et al., "Preparation of activated carbon from a renewable bio-plant of *Euphorbia rigida* by $H_2SO_4$ activation and its adsorption behavior in aqueous solutions," Appl. Surf. Sci. 253 (2007): 4843–4852.

Greiner, A. & Wendorff, J. H., "Electrospinning: A fascinating method for the preparation of ultrathin fibers," *Angew. Chem. Int. Ed.* 46 (2007): 5670–5703.

Guo, Q., Zhou, X., Li, X. et al., "Supercapacitors based on hybrid carbon nanofibers containing multiwalled carbon nanotubes," *J. Mater. Chem.* 19 (2009): 2810–2816.

Hajaligol, M., "Low temperature formation of aromatic hydrocarbon from pyrolysis of cellulosic materials," *Fuel Energy Abstr.* 43 (2002): 252.

Hameed, B. H., Tan, I. A. W. & Ahmad, A. L., "Preparation of oil palm empty fruit bunch-based activated carbon for removal of 2,4,6-trichlorophenol: Optimization using response surface methodology," *J. Hazard. Mater.* 164 (2009): 1316–1324.

He, G., Gu, Y., He, S. et al., "Effect of fiber diameter on the behavior of biofilm and anodic performance of fiber electrodes in microbial fuel cells," *Bioresour. Technol.* 102 (2011): 10763–10766.

Ho, C. T. & Chung, D. D. L., "Carbon fibers brominated by electrochemical intercalation," *Carbon* 28 (1990a): 521–528.

Ho, C. T. & Chung, D. D. L., "Bromination of graphitic pitch-based carbon fibers," *Carbon* 28 (1990b): 831–837.

Ho, C. T. & Chung, D. D. L., "Kinetics of intercalate desorption from carbon fibers intercalated with bromine," *Carbon* 28 (1990c): 825–830.

Huang, X., "Fabrication and properties of carbon fibers," *Materials* 2 (2009): 2369–2403.

Im, J. S., Park, S.-J., Kim, T. J. et al., "The study of controlling pore size on electrospun carbon nanofibers for hydrogen adsorption," *J. Colloid Interface Sci.* 318 (2008): 42–49.

Imaizumi, S., Matsumoto, H., Suzuki, K. et al., "Phenolic resin-based carbon thin fibers prepared by electrospinning: Additive effects of poly(vinyl butyral) and electrolytes," *Polym. J.* 41 (2009): 1124–1128.

Ishida, O., Kim, D. Y., Kuga, S. et al., "Microfibrillar carbon from native cellulose," *Cellulose* 11 (2004): 475–476.

Jazaeri, E., "Fabrication and characterisation of carbon nanofibres from cellulose nanofibres," MEng thesis, Deakin University, Victoria, Australia (2012).

Jeong, B., Uhm, S. & Lee, J., "Iron-cobalt modified electrospun carbon nanofibers as oxygen reduction catalysts in alkaline fuel cells," *ECS Trans.* 33 (2010): 1757–1767.

Ji, L., Yao, Y., Toprakci, O. et al., "Fabrication of carbon nanofiber-driven electrodes from electrospun polyacrylonitrile/polypyrrole bicomponents for high-performance rechargeable lithium-ion batteries," *J. Power Sources* 195 (2010): 2050–2056.

Ji, L., Lin, Z., Alcoutlabi, M. et al., "Electrospun carbon nanofibers decorated with various amounts of electrochemically-inert nickel nanoparticles for use as high-performance energy storage materials," *RSC Adv.* 2 (2012): 192–198.

Jin, S., Zheng, Y., Gao, G. et al., "Effect of polyacrylonitrile (PAN) short fiber on the mechanical properties of PAN/EPDM thermal insulating composites," *Mater. Sci. Eng. A* 483 (2008): 322–324.

Jirsak, O., Sanetrnik, F., Lukas, D. et al., "Process and apparatus for producing nanofibers from polymer solution by electrostatic spinning," Technicka Univerzita v Liberci, Czech Republic (2004): 13.

Johnson, D. J., "Structure and properties of carbon fibres," in Figueiredo, J. L., Bernardo, C. A., Baker, R. T. K. et al., eds., *Carbon Fibers Filaments and Composites* (Springer, New York, 1990), pp. 119–146.

Joshi, P., Zhang, L., Chen, Q. et al., "Electrospun carbon nanofibers as low-cost counter electrode for dye-sensitized solar cells," *ACS Appl. Mater. Interfaces* 2 (2010): 3572–3577.

Jurewicz, K., Pietrzak, R., Nowicki, P. et al., "Capacitance behaviour of brown coal based active carbon modified through chemical reaction with urea," *Electrochim. Acta* 53 (2008): 5469–5475.

Kadla J. F., Kubo, S., Venditti, R. A. et al., "Lignin-based carbon fibers for composite fiber application," *Carbon* 40 (2002): 2913–2920.

Katzman, H. A., Adams, P. M., Le, T. D. et al., "Characterization of low thermal conductivity pan-based carbon fibers," *Carbon* 32 (1994): 379–391.

Kim, S. Y., Kim, B.-H., Yang, K. S. et al., "Supercapacitive properties of porous carbon nanofibers via the electrospinning of metal alkoxide–graphene in polyacrylonitrile," *Mater. Lett.* 87 (2012): 157–161.

Kim, B.-H., Seung, K. Y. & Woo, H.-G., "Boron–nitrogen functional groups on porous nanocarbon fibers for electrochemical supercapacitors," *Mater. Lett.* 93 (2013): 190–193.

Knill, C. J. & Kennedy, J. F., "Degradation of cellulose under alkaline conditions," *Carbohydr. Polym.* 51 (2003): 281–300.

Ko, T. H., Lin, C. H. & Ting, H. Y., "Structure changes and molecular motion of polyacrylonitrile fibers during pyrolysis," *J. Appl. Polym. Sci.* 37 (1989): 553–556.

Konkin, A. A., "Production of cellulose based carbon fibrous materials," in Watt, W. & Perov, V. B., eds., *Strong Fibres* (Elsevier Science Publishers, Amsterdam, Netherlands, 1985), pp. 275–325.

Koráb, J., Štefánik, P., Kavecký, Š. et al., "Thermal conductivity of unidirectional copper matrix carbon fibre composites," *Compos. Part A Appl. Sci. Manuf.* 33 (2002): 577–581.

Kowalewski, T. A., Błoński, S. & Barral, S., "Experiments and modelling of electrospinning process," *Bull. Pol. Acad. Sci. Tech. Sci.* 53 (2005): 385–394.

Kriegel, C., Arrechi, A., Kit, K. et al., "Fabrication, functionalization, and application of electrospun bio-polymer nanofibers," *Crit. Rev. Food Sci. Nutr.* 48 (2008): 775–797.

Kubo, S. & Kadla, J. F., "Lignin-based carbon fibers: Effect of synthetic polymer blending on fiber properties," *J. Poly. Environ.* 13 (2005): 97–105.

Kumar, P. S., Sahay, R., Aravindan, V. et al., "Free-standing electrospun carbon nanofibres—A high performance anode material for lithium-ion batteries," *J. Phys. D Appl. Phys.* 45 (2012): 265302.

Kurban, Z., Lovell, A., Jenkins, D. et al., "Turbostratic graphite nanofibres from electrospun solutions of PAN in dimethylsulphoxide," *Eur. Polym. J.* 46 (2010): 1194–1202.

Kuzmenko, V., "Carbon nanofibers synthesized from electrospun cellulose nanofibers," MS thesis in materials engineering, Chalmers University of Technology, Göteborg, Sweden (2012).

Lavin, J. G., Boyington, D. R., Lahijani, J. et al., "The correlation of thermal conductivity with electrical resistivity in mcsophase pitch-based carbon fiber," *Carbon* 31 (1993): 1001–1002.

Lédé, J., "Cellulose pyrolysis kinetics: An historical review on the existence and role of intermediate active cellulose," *J. Anal. Appl. Pyrolysis* 94 (2012): 17–32.

Lee, K. J., Shiratori, N., Lee, G. H. et al., "Activated carbon nanofiber produced from electrospun polyacrylonitrile nanofiber as a highly efficient formaldehyde adsorbent," *Carbon* 48 (2010): 4248–4255.

Li, D., Wang, Y. & Xia, Y., "Electrospinning of polymeric and ceramic nanofibers as uniaxially aligned arrays," *Nano Lett.* 3 (2003): 1167–1171.

Li, M., Han, G. & Yang, B., "Fabrication of the catalytic electrodes for methanol oxidation on electrospinning-derived carbon fibrous mats," *Electrochem. Commun.* 10 (2008): 880–883.

Li, K., Li, Y. & Zheng, Z., "Kinetics and mechanism studies of p-nitroaniline adsorption on activated carbon fibers prepared from cotton stalk by $NH_4H_2PO_4$ activation and subsequent gasification with steam," *J. Hazard. Mater.* 178 (2010): 553–559.

Lin, C. T., Kao, P. W. & Jen, M. H. R., "Thermal residual strains in carbon fibre-reinforced aluminium laminates," *Composites* 25 (1994): 303–307.

Lin, Z., Ji, L., Krause, W. E. et al., "Synthesis and electrocatalysis of 1-aminopyrene-functionalized carbon nanofiber-supported platinum–ruthenium nanoparticles," *J. Power Sources* 195 (2010): 5520–5526.

Linares-Solano, A. & Cazorla-Amorós, D., "Adsorption on activated carbon fibers," in Bottani, E. J. & Tascón, J. M. D., eds., *Adsorption by Carbons* (Elsevier, Amsterdam, Netherlands, 2008), pp. 431–454.

Lu, Q., Xiong, W. M., Li, W. Z. et al., "Catalytic pyrolysis of cellulose with sulfated metal oxides: A promising method for obtaining high yield of light furan compounds," *Bioresour. Technol.* 100 (2009): 4871–4876.

Mamleev, V., Bourbigot, S. & Yvon, J., "Kinetic analysis of the thermal decomposition of cellulose: The main step of mass loss," *J. Anal. Appl. Pyrolysis* 80 (2007): 151–165.

Martone, A., Faiella, G., Antonucci, V. et al., "The effect of the aspect ratio of carbon nanotubes on their effective reinforcement modulus in an epoxy matrix," *Compos. Sci. Technol.* 71 (2011): 1117–1123.

Matsuoka, S., Kawamoto, H. & Saka, S., "Thermal glycosylation and degradation reactions occurring at the reducing ends of cellulose during low-temperature pyrolysis," *Carbohydr. Res.* 346 (2011): 272–279.

Minkova, V., Razvigorova, M., Bjornbom, E. et al., "Effect of water vapour and biomass nature on the yield and quantity of the pyrolysis products from biomass," *Fuel Process. Technol.* 70 (2001): 53–61.

Morgan, P., "Carbon fiber production using a cellulosic based precursor," in *Carbon Fibers and Their Composites* (CRC Press, Boca Raton, 2005), pp. 269–294.

Naji, H., Zebarjad, S. M. & Sajjadi, S. A., "The effects of volume percent and aspect ratio of carbon fiber on fracture toughness of reinforced aluminum matrix composites," *Mater. Sci. Eng. A* 486 (2008): 413–420.

Nakagawa, Y., Molina-Sabio, M. & Rodriguez-Reinoso, F., "Modification of the porous structure along the preparation of activated carbon monoliths with $H_3PO_4$ and $ZnCl_2$," *Microporous Mesoporous Mater.* 103 (2007): 29–34.

Olivares-Marin, M., Fernández-González, C., Macias-Garcia, A. et al., "Preparation of activated carbon from cherry stones by physical activation in air: Influence of the chemical carbonization with $H_2SO_4$," *J. Anal. Appl. Pyrolysis* 94 (2012): 131–137.

Oshima, H., Woollam, J. A., Yavrouian, A. et al., "Electrical and mechanical properties of copper chloride-intercalated pitch-based carbon fibers," *Synth. Met.* 5 (1983): 113–123.

Parikh, N., Allen, J. & Yassar, R. S., "Effect of deformation on electrical properties of carbon fibers used in gas diffusion layer of proton exchange membrane fuel cells," *J. Power Sources* 193 (2009): 766–768.

Paris, O., Zollfrank, C. & Zickler, G. A., "Decomposition and carbonisation of wood biopolymers—A microstructural study of softwood pyrolysis," *Carbon* 43 (2005): 53–66.

Pastor, A. C., Rodríguez-Reinoso, F., Marsh, H. et al., "Preparation of activated carbon cloths from viscous rayon: Part I—Carbonization procedures," *Carbon* 37 (1999): 1275–1283.

Pastorova, I., Botto, R. E., Arisz, P. W. et al., "Cellulose char structure: A combined analytical Py-GC-MS, FTIR, and NMR study," *Carbohydr. Res.* 262 (1994): 27–47.

Patil, S. A., Chigome, S., Hägerhäll, C. et al., "Electrospun carbon nanofibers from polyacrylonitrile blended with activated or graphitized carbonaceous materials for improving anodic bioelectrocatalysis," *Bioresour. Technol.* 132 (2013): 121–126.

Pérez, S. & Samain, D., "Structure and engineering of celluloses," in Derek, H., ed., *Advances in Carbohydrate Chemistry and Biochemistry* (Academic Press, Waltham, 2010), pp. 25–116.

Petrov, N., Budinova, T., Razvigorova, M. et al., "Conversion of olive wastes to volatiles and carbon adsorbents," *Biomass Bioenergy* 32 (2008): 1303–1310.

Piskorz, J., Radlein, D. & Scott, D. S., "On the mechanism of the rapid pyrolysis of cellulose," *J. Anal. Appl. Pyrolysis* 9 (1986): 121–137.

Plaisantin, H., Pailler, R., Guette, A. et al., "Conversion of cellulosic fibres into carbon fibres: A study of the mechanical properties and correlation with chemical structure," *Compos. Sci. Technol.* 61 (2001): 2063–2068.

Poletto, M., Pistor, V., Zeni, M. et al., "Crystalline properties and decomposition kinetics of cellulose fibers in wood pulp obtained by two pulping processes," *Polym. Degrad. Stab.* 96 (2011): 679–685.

Poletto, M., Zattera, A. J., Forte, M. M. C. et al., "Thermal decomposition of wood: Influence of wood components and cellulose crystallite size," *Bioresour. Technol.* 109 (2012): 148–153.

Poudel, P., Zhang, L., Joshi, P. et al., "Enhanced performance in dye-sensitized solar cells via carbon nano-fibers-platinum composite counter electrodes," *Nanoscale* 4 (2012): 4726–4730.

Reneker, D. H., Yarin, A. L., Fong, H. et al., "Bending instability of electrically charged liquid jets of polymer solutions in electrospinning," *Appl. Phys.* 87 (2000): 4531–4547.

Robert, R., Barbati, S., Ricq, N. et al., "Intermediates in wet oxidation of cellulose: Identification of hydroxyl radical and characterization of hydrogen peroxide," *Water Res.* 36 (2002): 4821–4829.

Ruland, W., "Carbon-fibers," *Adv. Mater.* 2 (1990): 528–536.

Saito, N., Aoki, K., Usui, Y. et al., "Application of carbon fibers to biomaterials: A new era of nano-level control of carbon fibers after 30 years of development," *Chem. Soc. Rev.* 40 (2011): 3824–3834.

Savova, D., Apak, E., Ekinci, E. et al., "Biomass conversion to carbon adsorbents and gas," *Biomass Bioenergy* 21 (2001): 133–142.

Sekiguchi, Y., Frye, J. S. & Shafizadeh, F., "Structure and formation of cellulosic chars," *J. Appl. Polym. Sci.* 28 (1983): 3513–3525.

Setnescu, R., Jipa, S., Setnescu, T. et al., "IR and X-ray characterization of the ferromagnetic phase of pyrolyzed polyacrylonitrile," *Carbon* 37 (1999): 1–6.

Sevilla, M. & Fuertes, A. B., "The production of carbon materials by hydrothermal carbonization of cellulose," *Carbon* 47 (2009): 2281–2289.

Sevilla, M. & Fuertes, A. B., "Graphitic carbon nanostructures from cellulose," *Chem. Phys. Lett.* 490 (2010): 63–68.

Shinn, J. H., "From coal to single-stage and two-stage product," *Fuel* 63 (1984): 1187–1196.

Singh, G., Rana, D., Matsuura, T. et al., "Removal of disinfection byproducts from water by carbonized electrospun nanofibrous membranes," *Sep. Purif. Technol.* 74 (2010): 202–212.

Soares, S., Camino, G. & Levchik, S., "Comparative study of the thermal decomposition of pure cellulose and pulp paper," *Polym. Degrad. Stab.* 49 (1995): 275–283.

Sugumaran, P., Priya Susan, V., Ravicharan, P. et al., "Production and characterization of activated carbon from banana empty fruit bunch and *Delonix regia* fruit pod," *J. Sustain. Energy Environ.* 3 (2012): 125–132.

Tang, M. M. & Bacon, R., "Carbonization of cellulose fibers—I. Low temperature pyrolysis," *Carbon* 2 (1964): 211–214.

Tao, F., Song, H. & Chou, L., "Efficient conversion of cellulose into furans catalyzed by metal ions in ionic liquids," *J. Mol. Catal. A Chem.* 357 (2012): 11–18.

Taylor, G., "Electrically driven jets," *Proc. R. Soc. Lond. A* 313 (1969): 453–475.

Teo, W. E. & Ramakrishna, S., "A review on electrospinning design and nanofibre assemblies," *Nanotechnology* 17 (2006): R89–R106.

Tibbetts, G. G., Lake, M. L., Strong, K. L. et al., "A review of the fabrication and properties of vapor-grown carbon nanofiber/polymer composites," *Compos. Sci. Technol.* 67 (2007): 1709–1718.

Toprakci, O., Toprakci, H. A. K., Ji, L. et al., "LiFePO$_4$ nanoparticles encapsulated in graphene-containing carbon nanofibers for use as energy storage materials," *J. Renew. Sustain. Energy* 4 (2012): 013121.

TOPTEC, Company brochure, www.toptec.co.kr, (2011).

Unalan, H. E., Wei, D., Suzuki, K. et al., "Photoelectrochemical cell using dye sensitized zinc oxide nanowires grown on carbon fibers," *Appl. Phys. Lett.* 93 (2008).

Veksha, A., Sasaoka, E. & Uddin, M. A., "The influence of porosity and surface oxygen groups of peat-based activated carbons on benzene adsorption from dry and humid air," *Carbon* 47 (2009): 2371–2378.

Virginia, H. M. & Adrian, B. P., "Lignocellulosic precursors used in the synthesis of activated carbon—Characterization techniques and applications in the wastewater treatment" (InTech Europe, Croatia, 2012).

Wang, X. & Chung, D. D. L., "Electromechanical behavior of carbon fiber," *Carbon* 35 (1997): 706–709.

Wang, J., Gu, M., Ma, W. G. et al., "Temperature dependence of the thermal conductivity of individual pitch-derived carbon fibers," *Carbon* 47 (2009): 350–354.

Wang, M., Kang, Q. & Pan, N., "Thermal conductivity enhancement of carbon fiber composites," *Appl. Thermal Eng.* 29 (2009): 418–421.

Wang, M.-X., Huang, Z.-H., Shimohara, T. et al., "NO removal by electrospun porous carbon nanofibers at room temperature," *Chem. Eng. J.* 170 (2011): 505–511.

Xu, Y. & Chung, D. D. L., "Silane treated carbon fiber for reinforcing cement," *Carbon* 39 (2001): 1995–2001.

Xu, Q., Yin, X., Wang, M. et al., "Analysis of phthalate migration from plastic containers to packaged cooking oil and mineral water," *J. Agric. Food Chem.* 58 (2010): 11311–11317.

Yalçýn, N. & Sevinç, V., "Studies of the surface area and porosity of activated carbons prepared from rice husks," *Carbon* 38 (2000): 1943–1945.

Yan, X., Tai, Z., Chen, J. et al., "Fabrication of carbon nanofiber–polyaniline composite flexible paper for supercapacitor," *Nanoscale* 3 (2011): 212–216.

Yang, Z., Du, G., Meng, Q. et al., "Synthesis of uniform $TiO_2$@carbon composite nanofibers as anode for lithium ion batteries with enhanced electrochemical performance," *J. Mater. Chem.* 22 (2012): 5848–5854.

Yang, Q., Sui, G., Shi, Y. Z. et al., "Osteocompatibility characterization of polyacrylonitrile carbon nanofibers containing bioactive glass nanoparticles," *Carbon* 56 (2013): 288–295.

Yu, Y., Yang, Q., Teng, D. et al., "Reticular Sn nanoparticle-dispersed PAN-based carbon nanofibers for anode material in rechargeable lithium-ion batteries," *Electrochem. Commun.* 12 (2010): 1187–1190.

Zamani, P., Higgins, D., Hassan, F. et al., "Electrospun iron–polyaniline–polyacrylonitrile derived nanofibers as non-precious oxygen reduction reaction catalysts for PEM fuel cells," *Electrochim. Acta* 139 (2014): 111–116.

Zhang, T., Walawender, W. P., Fan, L. T. et al., "Preparation of activated carbon from forest and agricultural residues through $CO_2$ activation," *Chem. Eng. J.* 10 (2004): 53–59.

Zhang, S., Lu, Y., Xu, G. et al., "LiF/Fe/C nanofibres as a high-capacity cathode material for Li-ion batteries," *J. Phys. D Appl. Phys.* 45 (2012): 395301.

Zhang, L., Aboagye, A., Kelkar, A. et al., "A review: Carbon nanofibers from electrospun polyacrylonitrile and their applications," *J. Mater. Sci.* 49 (2014): 463–480.

Zhao, H., Wang, L., Jia, D. et al., "Coal based activated carbon nanofibers prepared by electrospinning," *J. Mater. Chem. A* 2 (2014): 9338–9344.

Zhou, Z. & Wu, X.-F., "Graphene-beaded carbon nanofibers for use in supercapacitor electrodes: Synthesis and electrochemical characterization," *J. Power Sources* 222 (2013): 410–416.

Zhou, F.-L., Gong, R.-H. & Porat, I., "Three jet electrospinning using a flat spinneret," *J. Mater. Sci.* 44 (2009): 5501–5508.

Zugenmaier, P., "Conformation and packing of various crystalline cellulose fibers," *Prog. Polym. Sci.* 26 (2001): 1341–1417.

# 14

# Carbon-Based Nanomaterials as Nanozymes

Bhaskar Garg

Yong-Chien Ling

## Abstract

Nature is the greatest engineer and the main source of all inspiration throughout. Natural enzymes, the heart of a living organism, are ubiquitous in nature and direct most biological processes as active biocatalysts with high substrate affinity and specificity under mild reaction conditions. Increasing understanding of the general principles as well as the intrinsic drawbacks of natural enzymes, including low operational stability; high costs in preparation, purification, and storage; and sensitivity of catalytic activity toward environmental conditions, has triggered a dynamic field in nanotechnology, biochemistry, and materials science that aims at joining the better of the three worlds by combining concepts adapted from nature with the processability of catalytically active carbon-based nanomaterials (CNMs) as nanozymes. *Nanozyme*, a term coined by Manea et al. (2004), is an exciting and promising topic in biomimetics that has ignited considerable research efforts recently to rationally design and execute functional CNMs as enzyme mimetics for a wide variety of technological applications.

## 14.1 Introduction

Nature is the greatest engineer and the main source of all inspiration throughout. Biologically inspired designs or an adaptation from nature and an attempt to incorporate such ideas into better solutions for technological issues and sustainable living is commonly referred to as biomimetics or bionics, which simply means mimicking the nature or biology. The term *biomimetics* is derived from the Greek word *biomimesis* and was coined by Otto Schmitt (1969), whose work was focused on the development of a physical device that could unequivocally mimic the electrical action of a nerve. However, the word *biomimetics* made its first public appearance only in 1974 in *Webster's Dictionary* and is defined as "the study of the formation, structure, or function of biologically produced substances and materials (as enzymes or silk) and biological mechanisms and processes (as protein synthesis or photosynthesis) especially for the purpose of synthesizing similar products by artificial mechanisms which mimic natural ones."

Natural enzymes, in most cases proteins, are nanometer-sized macromolecules formed by the folding and self-assembly of polypeptides utilizing multivalent interactions. Natural enzymes, exquisite biocatalysts mediating biological processes in living organisms, exhibit highly efficient catalytic activity (up to $10^{19}$ times for specific substrate or reaction), high substrate specificity, and high selectivity under mild conditions such as physiological pH, ambient temperature, and atmospheric pressure (Wolfenden & Snider 2001, Garcia-Viloca et al. 2004). However, the catalytic activity of natural enzymes can be easily affected by acidity, temperature, and inhibitors. In addition, the limited stability of these enzymes due to denaturation and digestion and high costs of preparation, purification, and storage also restrict their widespread applications (Gao et al. 2007, Xie et al. 2012). Consequently, the pursuit of alternative molecules or materials which could achieve most of the above-mentioned challenges has resulted in the introduction of human-made enzymes or artificial enzymes (Bhabak & Mugesh 2010, Friedle et al. 2010, Aiba et al. 2011, Guo & Wang 2011, Hu et al. 2011, Lei & Ju 2012, Wulff & Liu 2012, Xie et al. 2012, He et al. 2013, 2014a, Wei & Wang 2013, Fu 2014, Lin et al. 2014a, 2015, Raynal et al. 2014).

*Artificial enzymes*, a term coined by Breslow (Breslow & Overman 1970) for enzyme mimics, is an exciting and promising topic in biomimetic chemistry that aims to imitate the general and essential principles of natural enzymes by using a variety of alternative materials, including supramolecular catalysts. A conventional route to locate an efficient artificial enzyme is to reproduce the elusive structure of enzyme active site (model enzymes). Alternatively, the functions of an enzyme can be mimicked without copying its structure (enzyme mimic). In the late 1950s, noncovalent interactions were used to recreate these properties with artificial systems (Pauling 1946, 1948). In the same decade, James B. Sumner was awarded the Nobel Prize for his discovery that enzymes can be crystallized (Sumner 1926). In 1965, cyclodextrin inclusion compounds were evaluated as the first hosts to imitate enzyme hydrophobic pocket at a small molecular level (Cramer & Kampe 1965, Hennrich & Cramer 1965). Later on, following the lock-and-key principle of enzymes ([i] a substrate fits perfectly in a host molecule that is also catalytically active, accelerating the rate of a reaction, while spatial factors may induce regioselectivity; [ii] two substrates may bind into a cavity), the early examples of supramolecular catalysis were featured (Breslow & Overman 1970, Breslow 1972, Tabushi et al. 1977, Tabushi 1982). After these pioneering works, a remarkable progress was made exploring calixarene (Cuevas et al. 2000), crown ethers (Behr & Lehn 1978), porphyrin (Bonarlaw & Sanders 1995), metal complexes (Lu et al. 2009), polymers (Takagishi & Klotz 1972), supramolecules (Zeyuan et al. 2011), and a range of structurally diverse biomolecules (Tramontano 1986, Breaker & Joyce 1994) to mimic the structures and functions of natural enzymes using different approaches. Table 14.1 shows the notable developments in the regime of natural enzymes and enzyme mimics.

Over the last decade, enzyme mimics have spurred interest in highly organized biological materials, in particular, from the molecular scale to the nanoscale. The organization of nanomaterials in a hierarchical manner with intricate nanoarchitecture provides a significant impetus to mimicking nature by using nanofabrication techniques (Alberts et al. 2008). Generally, a complex interplay between the surface structures, morphology, and physical and chemical properties results in multifunctional materials with exceptional properties. Although imitating natural enzymes with nanomaterials seems counterintuitive because they appear as different as balls and strikes (most natural enzymes are proteins and thus have well-defined tertiary structures, while most nanomaterials are not anatomically uniform because of their different sizes and shapes; secondly, natural enzymes are not as hard and crystalline as nanomaterials), they do share certain similarities in overall surface charge and size and shape. In addition, the external surfaces of nanomaterials can be functionalized with desirable functional groups (similar to those expressed by natural enzymes), enabling them to be materials of choice for enzyme mimics.

**TABLE 14.1** Timeline of the Development of Natural and Artificial Enzymes

| Year | Events | References |
|---|---|---|
| 1877 | The term *enzyme* was coined by Wilhelm Kühne. | – |
| 1926 | The enzyme urease was crystallized and determined to be a protein by James B. Sumner. | Sumner 1926a,b |
| 1946 | James B. Sumner won the Nobel Prize in Chemistry for his discovery that enzymes can be crystallized. | – |
| 1965 | Cyclodextrin inclusion compounds were used to imitate enzymes. | Cramer & Kampe 1965, Hennrich & Cramer 1965 |
| 1970 | The term *artificial enzyme* was coined by Ronald Breslow. | Breslow & Overman 1970 |
| 1971 | A polymer with enzyme-like activity (synzyme) was reported. | Klotz et al. 1971 |
| 1972 | Molecularly imprinted polymers were invented. | Takagishi & Klotz 1972, Wulff & Sarhan 1972 |
| 1982 | The term *ribozyme* was coined by Thomas R. Cech. | Kruger et al. 1982 |
| 1986 | Catalytic antibodies were invented. | Pollack et al. 1986, Tramontano et al. 1986 |
| 1989 | Sidney Altman and Thomas R. Cech won the Nobel Prize in Chemistry for their discovery of the catalytic properties of RNA. | – |
| 1992 | The first artificial RNAzyme was selected. | Pan & Uhlenbeck 1992 |
| 1993 | DNA cleavage was induced by fullerene derivatives. | Tokuyama et al. 1993 |
| 1994 | The first artificial DNAzyme was selected. | Breaker & Joyce 1994 |
| 1996 | Fullerene derivatives were used as superoxide dismutase (SOD) mimic. | Dugan et al. 1996, 1997 |
| 2004 | Nanogold was used as RNase mimic. The term *nanozyme* was coined. | Manea et al. 2004 |
| 2004 | Nanogold was used as oxidase mimic. | Comotti et al. 2004 |
| 2005 | Nanoceria was used as SOD mimic. | Tarnuzzer et al. 2005 |
| 2007 | Ferromagnetic nanoparticles were used as peroxidase mimic. | Gao et al. 2007, Wei & Wang 2008 |
| 2009 | Nanoceria was used as catalase and oxidase mimic. | Asati et al. 2009, Pirmohamed et al. 2010 |
| 2010 | Carbon nanotubes were used as peroxidase mimic. | Song et al. 2010b |
| 2010 | Graphene oxide was used as peroxidase mimic. | Song et al. 2010a |
| 2012 | Nano–vanadium pentaoxide was used as haloperoxidase mimic. | André et al. 2011 |

*Source:* Wei, H. & Wang, E., *Chem. Soc. Rev.* 42, 6060–6093, 2013. Adapted by permission of The Royal Society of Chemistry.

Nanomaterials are chemical entities at least one dimension smaller than 100 nm. Because of their extremely small size and large surface area, they exhibit characteristic physical, chemical, photochemical, and biological properties that are significantly different from those of the same material in bulk form. For instance, historically, gold has been considered as a highly inert material with little or no reactivity. However, gold nanoparticles (AuNPs) have been evaluated as highly efficient catalysts for a wide range of chemical reactions under different conditions. In a pioneering work, Manea et al. (2004) discovered that surface-functionalized (thiol monolayer-protected) nanogold can catalyze the transphosphorylation of 2-hydroxypropyl *p*-nitrophenyl phosphate, mimicking the nuclease system, RNA. Later on, the "naked" AuNPs could serve as a mimic for glucose oxidase (GOx), an oxidoreductase enzyme (Comotti et al. 2004, 2006). Owing to their outstanding and tunable catalytic properties, high stability against denaturing, and ease of preparation, such nanoparticulates were called *nanozymes* by Manea et al. in analogy to the nomenclature of catalytic polymers (synzymes) (Klotz et al. 1971). Since then, to date, considerable advances have been made in this area because of the exceptional characteristics of nanomaterials and the significant progress in the field of nanotechnology. In this context, various catalytic nanoparticulate artificial enzymes such as AuNPs (Lin et al. 2014c), magnetic nanoparticles (Gao et al. 2007, Park et al. 2011, Chen et al. 2012, Melnikova et al. 2014), vanadium pentaoxide ($V_2O_5$) nanoparticles (Natalio et al. 2012), $VO_2$ nanobelts (Nie et al. 2014), cerium oxide ($CeO_2$)

nanoparticles (Pirmohamed et al. 2010), platinum nanomaterials (Ma et al. 2011), silicon nanomaterials (Chen et al. 2014a), bimetallic nanoparticles (He et al. 2010, Zhang et al. 2011), metal–organic frameworks (Liu et al. 2013b, Zhang et al. 2014b), and, most recently, carbon-based nanomaterials (CNMs), especially 2D graphene and its subtypes (Song et al. 2010a, Shi et al. 2011, Tian et al. 2013, Wang et al. 2013b, 2014a, He et al. 2014, Kim et al. 2014, Lin et al. 2014a), are particularly notable. Because of their own unique characteristics, the as-mentioned nanomaterials have been evaluated as promising candidates for a variety of enzyme mimics (to name but a few, nuclease, peroxidase, SOD, GOx, oxidase, and catalase), displaying their wide applications in numerous fields, including biosensing, environmental protection, immunoassays, and thernostics (diagnostics and therapy). However, since the focus of this chapter strongly lies on CNMs such as fullerenes ($C_{60}$), CNTs, graphene and related compounds, including graphene oxide (GO), reduced graphene oxide (rGO), carbon-based nanodots, and graphitic carbon nitride (g-$C_3N_4$), which have been discovered to have SOD-like and, essentially, peroxidase-like activities for a variety of applications, these topics are chiefly categorized into three sections for ease of understanding:

- "Fullerenes as Nanozymes" (Section 14.3.1)
- "CNTs as Nanozymes" (Section 14.3.2)
- "Graphene-Based Nanomaterials (G-NMs) as Nanozymes" (Section 14.3.3)

## 14.2 An Overview of Selected Natural Enzymes

### 14.2.1 Peroxidases

Peroxidases, a group of oxidoreductases, represent a large family of enzymes (heme proteins containing iron (III) protoporphyrin IX as the prosthetic group) that typically catalyze biological reactions, in which peroxide (hydrogen peroxide [$H_2O_2$] in most cases) is reduced and a redox substrate is oxidized as an electron donor (Scheme 14.1).

Through this catalysis, peroxidases can catalytically scavenge $H_2O_2$ to form water and thus play many functions in biological systems, including detoxification of reactive oxygen species (ROS) and defending against pathogens. The nature of the electron donor is very dependent on the structure of the enzyme. Cytochrome C peroxidase (Cyt C), glutathione peroxidase, nicotinamide adenine dinucleotide peroxidase, haloperoxidase, and horseradish peroxidase (HRP) represent the common examples of peroxidases which exhibit quite different substrate specificities. The last one has been widely used in clinical and bioanalytical chemistry and serves as a peroxidase standard to study peroxidation reactions (the most studied isoform is C). Because of the oxidative nature of peroxidases, they offer a wide range of practical applications in many areas of science, including, but not limited to, biosensing and diagnostics, immunoassay, biofuel production, organic and polymer synthesis, decolorization of synthetic dyes, paper pulp industry, and removal of phenolic contaminants and related compounds (Hamid & Khalilur-Rahman 2009).

In general, the peroxidase activity of, for instance, HRP is the result of a nonsequential mechanism (ping-pong). The ping-pong mechanism, also known as double-displacement reaction, is characterized by the change of the enzyme from an intermediate state to its standard state, just like how a ping-pong ball bounces back and forth. Another key feature of the ping-pong mechanism is that one substrate is converted to the product and dissociates or released before the second substrate binds. Scheme 14.2

$$AH_2 + H_2O_2 \xrightarrow{\text{Peroxidase}} A + 2H_2O$$

$$AH_2 + ROOH \xrightarrow{\text{Peroxidase}} A + ROH + H_2O$$

**SCHEME 14.1** The reactions catalyzed by peroxidase.

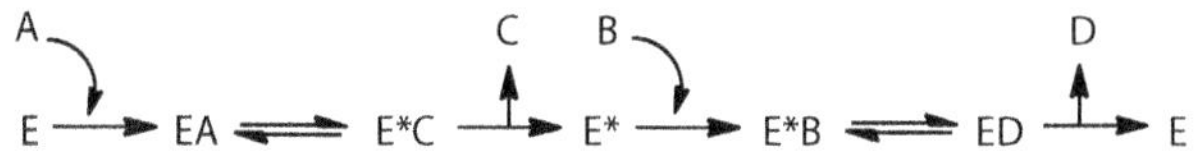

**SCHEME 14.2** A schematic presentation of the ping-pong mechanism.

explains the ping-pong mechanism via an enzymatic action, where A and B are two substrates, C and D are products, and E and E* denote the enzyme in standard and intermediate states, respectively.

In 2007, Gao et al. reported that ferromagnetic metal nanoparticles ($Fe_3O_4$ MNPs) with different sizes (30, 50, and 300 nm) possess intrinsic peroxidase-like activity. In particular, under the optimized conditions (4°C–90°C; pH = 0–12), all sized MNPs oxidized 3,3′,5,5′-tetramethylbenzidine (TMB), *o*-phenylenediamine (OPD), and diazoaminobenzene substrates to their corresponding products in the presence of $H_2O_2$. After this seminal report, many studies have been devoted to the design and development of functional nanomaterials imitating peroxidase catalytic activity (Dai et al. 2009, He et al. 2010, 2013, Zhang et al. 2010, Fan et al. 2011, Liu et al. 2011, Ma et al. 2011, Melnikova et al. 2014).

To detect the peroxidase-like activity of nanoparticulates in a reaction, both the formation of oxidized product and the consumption of $H_2O_2$ are evaluated. The concentration of the latter can be directly measured through a color change of certain chromophores. For instance, the $H_2O_2$ triggered oxidation of 2,2′-azino-bis-(3-ethylbenzothiazoline-6-sulfonic acid) (ABTS), TMB, and OPD, in general, produces green, blue, and yellow colors, respectively (Wei & Wang 2008). Thus, ultraviolet–visible absorption spectroscopy provides a convenient way to determine the enzyme-like activity of nanomaterials. However, if the substrate oxidation is associated with fluorescent markers like rhodamine B and quantum dots, fluorescence spectroscopy can be used (Gao et al. 2011, Jiang et al. 2011). The chromophores 4-aminoantipyrene phenol and *N*,*N*-diethyl-*p*-phenylenediamine sulfate constitute the common examples of other chromophores in peroxidase mimics.

## 14.2.2 Oxidases

Oxidases are yet another subclass of oxidoreductases that use molecular oxygen as an electron acceptor to oxidize the substrate. The molecular oxygen itself is reduced to water or $H_2O_2$ or even superoxide radical in some cases (Scheme 14.3). Similarly, as with peroxidation reactions, oxidase-catalyzed reactions produce a color change (in some cases), making them ideal candidates in biosensing or other detection purposes.

To detect oxidase-like activity in a reaction, both the formation of oxidized product and the decrease in molecular oxygen levels are measured. GOx, nicotinamide adenine dinucleotide phosphate (NADPH) oxidase, Cyt P450 oxidase, xanthine oxidase, monoamine oxidase, and cholesterol oxidase are specific examples of oxidases that catalyze the oxidation of glucose, NADPH, Cyt P450, xanthine, monoamines, and cholesterol, respectively. Owing to its wide applications in clinical settings, biotechnology, and chemical, pharmaceutical, and other industries, GOx is currently receiving much attention (Luo et al. 2010, Ciaurriz et al. 2014).

$$AH + O_2 \xrightarrow{\text{Oxidase}} A + H_2O$$

$$AH + O_2 + H_2O \xrightarrow{\text{Oxidase}} A + H_2O_2$$

$$A_{red} + O_2 \xrightarrow{\text{Oxidase}} A_{ox} + O_2^{\cdot-}$$

**SCHEME 14.3** The reactions catalyzed by oxidase.

$$2O_2^{\cdot-} + 2H^+ \xrightarrow{\text{Superoxide dismutase}} H_2O_2 + O_2$$

$$2H_2O \xrightarrow{\text{Catalase}} 2H_2O + O_2$$

**SCHEME 14.4**   The reactions catalyzed by SOD and catalase.

## 14.2.3 SOD and Catalase

ROS are a class of chemically reactive species that are involved in a wide range of physiological and pathological processes, such as signal transduction, inflammation, carcinogenesis, and neurodegenerative injury (Dickinson et al. 2010, Dickinson & Chang 2011). The mismanagement and accumulation of ROS in mammals leads to a condition broadly referred to as oxidative stress, and this has been linked to aging and a host of diseases, including cancer and neurodegenerative Alzheimer's and Parkinson's diseases (Barnham et al. 2004, Zhang et al. 2004). Among ROS of biological significance, superoxide anion is particularly important and has been known to cause tissue injury and associated inflammation.

SOD is an important enzyme and plays a crucial role in many redox-active processes acting as anti-inflammatory and antioxidant in nearly all cells exposed to oxygen. SOD catalyzes the dismutation of superoxide anions into $H_2O_2$ and molecular oxygen. Alongside SOD, catalase catalyzes the decomposition of $H_2O_2$ into water and molecular oxygen, protecting the cell from oxidative damage by ROS (Scheme 14.4). Indeed, the catalytic functions of SOD and catalase are quite complementary. Whereas SOD generates $H_2O_2$, an ROS, catalase promptly eliminates it. Consequently, the synergistic functioning of these two enzymes has long been shown to effectively scavenge ROS, especially superoxide anion. Owing to the presence of mixed valence ($Ce^{3+}$, $Ce^{4+}$) and oxygen defects, nanoceria ($CeO_2$) has been shown to possess highly catalytic activity for both SOD and catalase mimics (Pirmohamed et al. 2010).

To detect SOD-like activity in a reaction, the formation of $H_2O_2/O_2$ and the consumption of superoxide anion radicals are determined. On the other hand, to detect catalase-like activity, the formation of molecular oxygen is commonly assessed. The latter can be examined with ease by using a dissolved-oxygen meter. This easy setup does not require any expensive instrumentation and can be visualized with the naked eye.

# 14.3 CNMs as Nanozymes and Their Applications

Carbon, symbolized as *C*, is the 15th most abundant element in the earth's crust and the 4th most abundant element by mass after hydrogen, helium, and oxygen. Carbon is abundant in the sun, stars, comets, and in the atmospheres of most planets. This abundance, combined with the exceptional diversity of organic compounds and their unfamiliar polymer-forming ability at Earth's temperature, makes carbon the chemical basis of all known life. The allotropes of carbon, including amorphous carbon or coal and the most familiar ones, graphite and diamond, have found many usages over the centuries. From providing energy, to being the most expensive crystal (*an ideal symbol of an indescribable connection between two human beings*), to supporting a basic need in education, giving us the pencil, a tool for many—carbon has done it all. Indeed, with the added advantage of classical nanofabrication techniques, CNMs, ranging from mesoporous or amorphous carbon to fullerenes, CNTs, and graphene, have shown a great promise in a variety of technological applications, including enzyme mimics (Compton et al. 2010, Song et al. 2010a,b, Singh et al. 2011, Dai et al. 2012, Garg & Ling 2013, Garg et al. 2014a,b).

## 14.3.1 Fullerenes as Nanozymes

Since the initial discovery of 0D fullerenes ($C_{60}$) as an excellent electron acceptor capable of accepting as many as six electrons reversibly, they have received much attention in recent years in nanomaterial science and biotechnology (Badamshina & Gafurova 2012, Yu et al. 2012, Afreen et al. 2015). In early studies, polyhydroxylated fullerenes ($C_{60}(OH)_{12}$, $C_{60}(OH)_nO_m$; $n = 18$–20, $m = 3$–7) were evaluated to have neuroprotective effects reducing the excitotoxic and apoptopic death of cultured cortical neurons (Dugan et al. 1996). However, it was Ali et al. (2004) who first demonstrated that tris-malonic acid derivatives of $C_{60}$ ($C_{60}[C(COOH)]_3$) with $C_3$ and $D_3$ symmetries exhibit SOD-like activity. Owing to its polarity and relatively good penetrating ability toward the cell membrane, the $C_{60}[C(COOH)]_3$ with $C_3$ symmetry ($C_{60}$-$C_3$) exhibits higher activity. It was experimentally evidenced that the SOD-like activity of $C_{60}$-$C_3$ was due to catalytic dismutation, rather than stoichiometric scavenging, of superoxide anions, regenerating both molecular oxygen and $H_2O_2$. Based on electron-distribution map results, it was suggested that $C_{60}$-$C_3$ electrostatically drives superoxide anions toward electron-deficient areas on its surface until a second superoxide anion radical arrives to undergo dismutation with the help of protons from carboxyl groups and/or surrounding water molecules (Figure 14.1). Furthermore, kinetic studies were performed in conjunction with other artificial enzymes. Specifically, the rate constant of $C_{60}$-$C_3$ was calculated to be $2.2 \times 10^6$ $M^{-1}$ $s^{-1}$ under basic conditions (pH = 7.4). This rate was comparable to those of several enzyme mimics but lower than that of native SOD by 100-fold. In Ali et al.'s (2008) subsequent work, the authors further demonstrated that six carboxyfullerenes with different electronic structures (variable number and distribution of –COOH groups) and biophysical characteristics exhibit SOD-like activity. A good correlation was observed between neuroprotection by carboxyfullerenes and their affinity toward superoxide radicals. In yet another work, Quick et al. (2008) reported $C_{60}$-$C_3$–based SOD mimic capable of improving cognition and extending the life span of nontransgenic C57BL/6 mice. Surprisingly, after chronic treatment with $C_{60}$-$C_3$, age-related oxidative damage was significantly reduced in all brain regions as evidenced by fixed brain slice imaging (Figure 14.2). The major applications of fullerene-based SOD mimics include antiaging (Quick et al. 2008) and neuroprotection including against Parkinson's disease (Dugan et al. 1996, 1997, Ali et al. 2008).

Aside from SOD-like activity, recently, $C_{60}$-carboxyfullerene, $C_{60}[C(COOH)_2]_2$, has been proven to function as a peroxidase mimic that can catalyze the reaction of TMB in the presence of $H_2O_2$ to produce a blue-colored reaction (Li et al. 2013b). Based on kinetic studies, it was found that $C_{60}[C(COOH)_2]_2$ exhibits an even higher affinity than that of HRP. By combining GOx with $C_{60}[C(COOH)_2]_2$/TMB system, a colorimetric assay was developed for glucose (limit of detection [LOD] = 0.5 µM) and could also be successfully used for the quantitative determination of glucose in human serum.

## 14.3.2 CNTs as Nanozymes

In recent years, 1D CNTs with a wide spectrum of functional and exploitable properties have received considerable attention in many areas of science, including biomimetics (Ghule et al. 2007, Yu et al. 2008, Araghi & Bokaei 2013, Anirudhan & Alexander 2014, Giraldo et al. 2014, Mao et al. 2014, Rong et al. 2014). In general, depending on the number of carbon layers, CNTS may be categorized as SWCNTs and MWCNTs. In a pioneering work, Song et al. (2010b) demonstrated the peroxidase-like activity of SWCNTs using TMB as a model substrate. As with HRP, the peroxidase-like activity of SWCNTs was found to be dependent on temperature, pH, and $H_2O_2$ concentration. Based on these results, a label-free colorimetric assay was developed for disease-associated single nucleotide polymorphisms in human DNA. In another work, helical CNTs (containing variable Fe content) were evaluated to have intrinsic peroxidase-like activity (Cui et al. 2011). The experimental results indicated that higher Fe content results in better catalytic activity. However, no molecular mechanism was proposed in support of peroxidase-like activity of helical CNTs. Recently, MWCNTs containing iron impurity have been used to interact with the amino functional group of the chitosan polymer, forming a unique complex-like structure (MWCNT–Fe(III/II):$NH_2$–CHIT), similar to that of heme peroxidase (Gayathri & Kumar 2013). The hybrid biomimetic system showed Michaelis–Menten–type reaction kinetics for the $H_2O_2$ reduction reaction with a $K_m$ value of 0.23 mM. Unlike most of the reported systems, MWCNT Fe(III/II):$NH_2$–CHIT displayed a remarkable tolerance to other coexisting interferons (such as cysteine, ascorbic acid, uric acid, nitrate, and nitrite), at a $H_2O_2$ detection potential similar to that of peroxidase. The

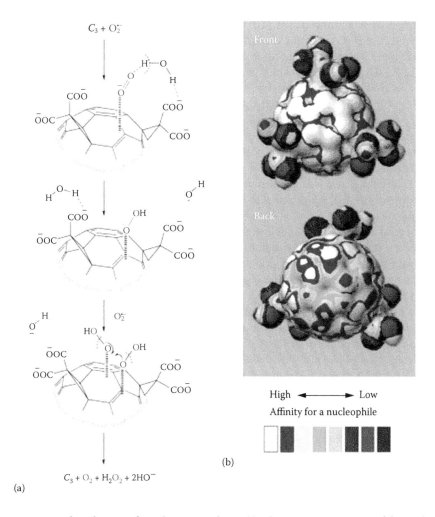

(a)

(b)

**FIGURE 14.1** Proposed mechanism of $O_2^{\cdot-}$ dismutation by $C_3$. (a) Schematic representation of the catalytic interaction of $C_3$ and $O_2^{\cdot-}$. Chemical bonds colored red are associated with electron-deficient areas and were predicted through semiempirical quantum calculations (compare with b). Incoming superoxide ions and oxygen atoms derived from them are colored purple to facilitate visualization of the suggested mechanism. Broken lines represent hydrogen bonding between oxygen and hydrogen atoms and striped lines are used to represent electrostatic attraction between negatively charged oxygen atoms and electron-deficient areas on the $C_{60}$ moiety. In the proposed mechanism, $C_3$ is suggested to electrostatically drive superoxide anions toward electron-deficient areas on its surface until a second $O_2^{\cdot-}$ arrives to undergo dismutation with the help of protons from carboxyl groups and/or surrounding water molecules. (b) Map of the electron density reflecting nucleophilic superdelocalizability. On the resulting isosurface, locations that are susceptible to attack by superoxide are designated by a color scale showing decreasing susceptibility, in the order white through cyan (legend). (Reprinted from *Free Radical Biology and Medicine*, 37, Ali, S. S. et al., A biologically effective fullerene ($C_{60}$) derivative with superoxide dismutase mimetic properties, 1191–1202, Copyright (2004), with permission from Elsevier.)

system was successfully used to detect $H_2O_2$ present in simulated milk and clinical as well as cosmetic samples with appreciable recovery values. In this regime, it would be worthwhile to mention that CNTs can be used as an aid in protein folding and in preventing aggregation (artificial molecular chaperones) as reported previously by us (Ghule et al. 2007). Taking into account the stability of CNTs to withstand harsh physical conditions, the use of CNTs on the industrial scale to recover recombinant proteins from inclusion bodies appears very promising.

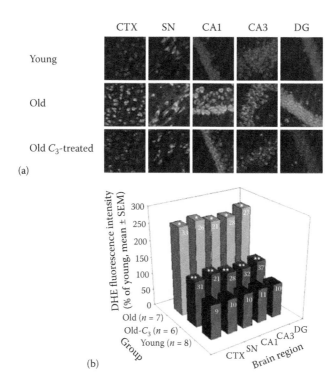

**FIGURE 14.2** Reduction of age-associated brain oxidative stress by $C_3$. (a) Confocal microscopy images of fixed brain slices superoxide-mediated dihydroethidium (DHE) oxidation, shown in cyan. (b) Quantification of DHE oxidation in cortex (CTX), substantia nigra (SN), and CA1, CA3, and dentate gyrus (DG) regions of the hippocampus. Values are mean ± standard error of the mean (SEM); SEM is represented by numbers on the face of each bar. *$p < .05$ by analysis of variance (ANOVA) and Student–Neuman–Keuls (SNK) post hoc analysis for young vs. old. **$p < .05$ by ANOVA and SNK post hoc analysis for old vs. old-$C_3$. (Reprinted from *Neurobiology of Aging*, 29, Quick, K. L. et al., A carboxyfullerene SOD mimetic improves cognition and extends the lifespan of mice, 117–128, Copyright (2008), with permission from Elsevier.)

Intrigued by the potential role of nanogold in peroxidase mimic (Lin et al. 2014b,c), a positively charged AuNP-SWCNT nanohybrid was reported to mimic the peroxidase activity using TMB as a model substrate (Zhang et al. 2013a). The as-prepared nanohybrid with uniformly dispersed (+)AuNPs on the surface of SWCNTs could be used as a label-free colorimetric assay for DNA hybridization detection. The experimental results indicated that in contrast to double-stranded DNA, single-stranded DNA (ssDNA) can resist salt-induced aggregation of (+)AuNPs-SWCNTs. The response to target DNA concentration was linear in the range of 0.025–0.5 mM with a detection limit of 2 nM (3 s).

Very recently, Wang et al. (2014c) have reported a mild, one-pot method for facile filling of MWCNTs with Prussian blue nanoparticles (PBNPs) as peroxidase mimic. As illustrated in Scheme 14.5, ultrasonic treatment of MWCNTs in an acidic aqueous solution of $K_3[Fe(CN)_6]$ and $FeSO_4$ precursors resulted in gradual growth of PBNPs inside and outside the MWCNTs. Fine control of PBNP growth was achieved by modulating the concentration ratio of $K_3[Fe(CN)_6]$ to $FeSO_4$. The most efficient encapsulation was achieved when the diameter of the PBNPs closely matched the internal diameter of the MWCNTs. After removing PBNPs, those that were precipitated on the outer surfaces of MWCNTs as well as dispersed in the solution, the robustly filled PBNPs (24% ± 2%) in the cavities of MWCNTs (MWCNT-PB$_{in}$) were exploited as a peroxidase mimic for the colorimetric assay of $H_2O_2$ in solution by using TMB as a reporter. Furthermore, the unoccupied outer surfaces of MWCNT-PB$_{in}$ were used to anchor GOx to form GOx/MWCNT-PB$_{in}$ bionanocomposites. The cooperation of outer-surface biocatalysis with

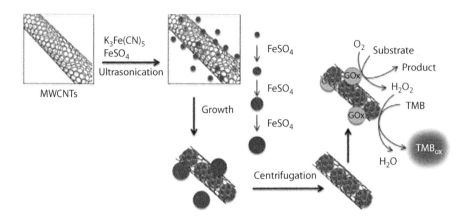

**SCHEME 14.5** Schematic of the preparation of MWCNTs-PB$_{in}$ nanocomposites for chemo/biosensing applications. (Wang, T. & Fu, Y. et al.: Filling Carbon Nanotubes with Prussian Blue Nanoparticles of High Peroxidase-Like Catalytic Activity for Colorimetric Chemo- and Biosensing. *Chemistry—A European Journal*. 2014. 20. 2623–2630. Copyright Wiley-VCH Verlag GmbH & Co. KGaA. Reproduced with permission.)

peroxidase-like catalysis of PB$_{in}$ resulted in a cooperative colorimetric biosensing mode for glucose assay. The uses of MWCNT-PB$_{in}$ and GOx/MWCNT-PB$_{in}$ for colorimetric biosensing of H$_2$O$_2$ and glucose gave linear absorbance responses from 1 µM to 1.5 mM (LOD = 100 nM) and from 1 µM to 1.0 mM (LOD = 200 nM), respectively.

Aside from peroxidases, recently, NiCoBP-doped MWCNT hybrid has been used as an oxidase mimetic system for highly efficient electrochemical immunoassay (Zhang et al. 2014a). The experimental results indicate that the system exhibits good electrochemical responses toward target prostate-specific antigen (PSA; used as a model analyte) with a detection limit of as low as 0.035 ng mL$^{-1}$ (Figure 14.3). The presented protocols may be extended to other multifunctional nanocomposite systems for broad applications in catalysis and biotechnology.

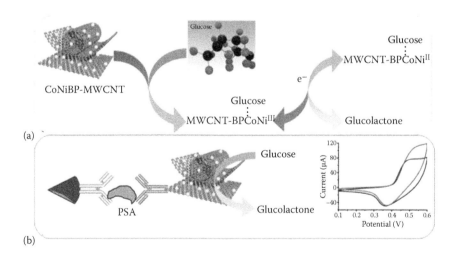

**FIGURE 14.3** Schematic illustration of (a) NiCoBP-MWCNT-based oxidase mimetic system and (b) the enzyme-free electrochemical immunoassay by coupling with the NiCoBP-MWCNT oxidase mimetic system. (Reprinted from *Analytica Chimica Acta*, 851, Zhang, B., He, Y. et al., NiCoBP-doped carbon nanotube hybrid: A novel oxidase mimetic system for highly efficient electrochemical immunoassay, 49–56, Copyright (2014), with permission from Elsevier.)

### 14.3.3 Graphene-Based Nanomaterials (G-NMs) as Nanozymes

Notably, until 2004, the carbon family was dominated only by well-known materials such as graphite, diamond, fullerenes, and CNTs. However, this "aura" lessened with the first isolation of freestanding 2D graphene (Novoselov et al. 2004). In a broader context, graphene could be considered as a potential building block (Figure 14.4) of 3D graphite, 0D fullerenes, and 1D CNTs and possesses many exceptional properties that have surpassed those observed in its predecessor, CNTs (Compton & Nguyen 2010). Not surprisingly, therefore, G-NMs are (and will continue to be) increasingly used for large technological applications in the imminent carbon age (Ganesh & Ling 2012, 2014, Wan et al. 2012, Ganesh et al. 2013, Li et al. 2013a, Hu 2014, Garg et al. 2015a, Wu et al. 2013b, Yang et al. 2013).

In an elegant work, the single-layer, metal-free GO was found to possess peroxidase mimicking activity (Song et al. 2010a). In particular, the reaction of TMB and $H_2O_2$ in the presence of GO-COOH produced a blue-colored solution of 3,3′,5,5′-tetramethylbenzidine diimine (TMBDI), a diagnostic for peroxidases. Because of the very high surface-to-volume ratio of GO as well as its high affinity toward organic substrates, it exhibited higher affinity toward TMB than native HRP. Mechanistically, it was proposed that upon interaction between GO, TMB, and $H_2O_2$, a charge-transfer $n$-type doping of graphene increases the Fermi level and thus the electrochemical potential from the LUMO of $H_2O_2$, accelerating the electron transfer from graphene domain to $H_2O_2$. This provides a high density of catalytically active centers for binding redox substrate. By combining the GO's peroxidase activity with GOx, a simple, cheap, selective, and sensitive colorimetric assay was developed for glucose detection and successfully applied on diluted blood, buffer solution, and two commercial fruit juice samples. Briefly, the mechanism of glucose detection can be understood in two steps. Firstly, in the presence of GOx, glucose is oxidized to gluconic acid with the simultaneous conversion of substrate oxygen to $H_2O_2$. In the second step, GO reduces $H_2O_2$ in the presence of a cosubstrate, TMB, which itself oxidizes into TMBDI (pH = 4.0), producing a blue color (Figure 14.5). Using this method, glucose could be detected as low as $1 \times 10^{-6}$ mol $L^{-1}$ in the linear range from $1 \times 10^{-6}$ to $2 \times 10^{-5}$ mol $L^{-1}$; thus, this method is applicable to real glucose

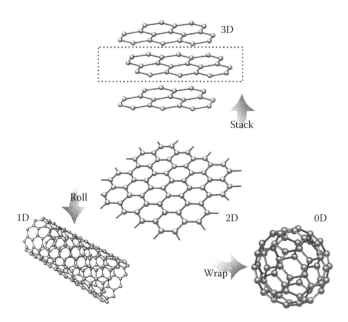

**FIGURE 14.4** Graphene: the basic building block for other carbon allotropes, graphite (3D), fullerene (0D), and CNT (1D). (Reprinted with permission from Wan, X. et al. 2012, 598–607. Copyright 2012 American Chemical Society.)

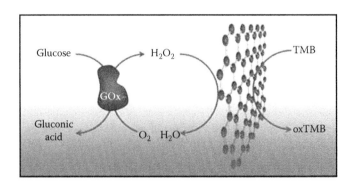

**FIGURE 14.5** Schematic illustration of colorimetric detection of glucose by using GOx- and GO-COOH-catalyzed reactions. (Song, Y., Qu, K. et al.: Graphene Oxide: Intrinsic Peroxidase Catalytic Activity and Its Application to Glucose Detection. *Advanced Materials.* 2010. 22. 2206–2210. Copyright Wiley-VCH Verlag GmbH & Co. KGaA. Reproduced with permission.)

samples like those from healthy (~3–8 mM) and diabetic persons (9–40 mM). Besides glucose detection, the intrinsic peroxidase activity of GO has also been used for colorimetric detection of the cancer bio-marker PSA (Qu et al. 2011). Wang et al. (2013b) demonstrated that few-layer graphene, prepared from chitosan-induced exfoliation of graphite, exhibits a much higher peroxidase catalytic activity than GO and its reduced form, rGO. The excellent catalytic performance of as-prepared graphene could be attrib-uted to the fast electron transfer on the surface of few-layer graphene as evidenced by electrochemical characterization. The few-layer graphene was used to determine $H_2O_2$ in three real water samples with satisfactory results.

The inherent ability for a task-specific functionalization, following covalent and noncovalent strategies on 2D surface of graphene and its derivatives, is one of the most attracting features that have been significantly explored in the last few years (Dreyer et al. 2010, 2014). In this context, the fabrication of graphene with $-SO_3H$ groups affording sulfonated graphene is of crucial importance (Zhao et al. 2011, Garg et al. 2014a,b, Jana et al. 2014, Xiong et al. 2014). Hua et al. (2012) used Cyt C, an electron transfer protein in the respiratory chain, and intercalated it between covalently modi-fied sulfonated graphene nanosheets (GNs) using electrostatic interactions (Figure 14.6). In a con-fined environment, the $G-SO_3H$/Cyt C assembly exhibited an excellent peroxidase-like activity (~8 times higher than that of native Cyt C) in the oxidation of OPD. This was attributed to the higher probability of the substrate colliding with Cyt C. This work sheds light on the fact that the property of proteins in the confined environments may differ significantly from that when they are in free states; thus, the native properties of proteins can be fine-tuned using engineered nanostructures such as graphene.

Aside from the intrinsic enzymatic activity of graphene and its derivatives, they have also been used as excellent supports for a variety of proteins and polymers in enzymatic catalysis because of their large specific surface areas. For instance, functional hybrid nanosheets (assembled through π–π interactions) between graphene and hemin, a natural metalloporphyrin, could be used as a highly effective peroxi-dase mimic using TMB, ABTS, and OPD as substrates (Guo et al. 2011a,b, Xue et al. 2012). Likewise, a folic acid—conjugated graphene—hemin hybrid (GFH) was fabricated by Song et al. (2011), as shown in Figure 14.7. Owing to the large size and the peroxidase-mimicking activity of GFH, a colorimetric method was developed for quantitative and fast detection of as few as 1000 cancer cells. Recently, a GFH material was evaluated as an effective catalyst for selective oxidation of primary C–H bond in toluene (Li et al. 2013c).

Owing to their synergistic properties, graphene–nanoparticle composites are one of the most prom-ising topics in the study of bio–nano interfaces. By using in situ growth of naked AuNPs on graphene

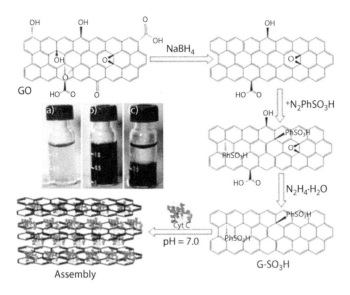

**FIGURE 14.6** Illustration of the synthesis of G–SO₃H and its assembly with Cyt C. Insets: (a) Cyt C solution; (b) fresh mixture solution of Cyt C and G–SO₃H; (c) colorless supernatant and sediment after assembling for 24 h. (Hua, B.-Y. et al., *Chem. Commun.* 48, 2316–2318, 2012. Reproduced by permission of The Royal Society of Chemistry.)

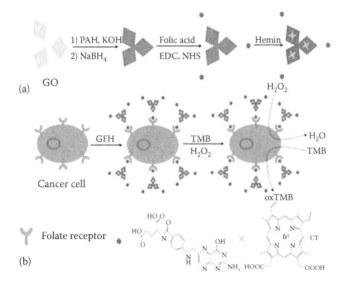

**FIGURE 14.7** Schematic representations of (a) preparation of GFH and (b) cancer cell detection by using target-directed GFH. (Song, Y. J. et al., *Chem. Commun.* 47, 4436–4438, 2011. Reproduced by permission of The Royal Society of Chemistry.)

sheets, a synergistic peroxidase-mimicking catalyst was engineered (Liu et al. 2012a). The catalytic activity of as-prepared AuNPs–graphene hybrid was evaluated using three peroxidase substrates, namely, TMB, OPD, and ABTS. The apparent catalytic activity of AuNPs–graphene hybrid could be fine-tuned reversibly through interface passivation induced by ssDNA due to π-stacking between the nucleobases and AuNPs–graphene (Liu et al. 2012b). When ssDNA was released from the graphene surface by either hybridization with its complementary ssDNA or cleavage, the enzymatic activity of graphene was

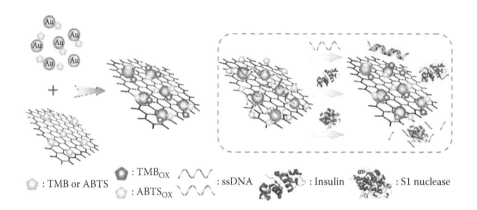

**FIGURE 14.8**  Schematic illustration of a versatile and label-free colorimetric biosensing platform based on the tunable smart interface of catalytic AuNP–graphene hybrid. (Reprinted with permission from Liu, Zhao et al. 2012b, 3142–3151. Copyright 2012 American Chemical Society.)

restored. This unique feature prevented the specific sites of peroxidase substrates from diffusing and binding to the surface of the catalyst, providing a label-free sensing mechanism for DNA or enzyme activity (Figure 14.8). Such a reversible switching interface can be considered as an ideal signal reporter with low background signal because of off–on switching mode and is applicable to a broad range of analytes.

In recent years, because of the greater stability of graphene-based nanoarchitectures (graphene as support) against self-dimerization or oxidative destruction and their relatively large surface areas, which result in a higher reaction turnover rate and binding interactions, they have received considerable attention in enzyme mimics, in particular, peroxidases (Garg et al. 2015b) for a variety of applications, as summarized in Table 14.2. Moreover, in conjunction with graphene and/or its derivatives, the other forms of carbon, such as carbon nanodots (CDs), carbon nitride dots (CNDs), graphene quantum dots (GQDs), and g-$C_3N_4$, have also been explored to mimic peroxidase activity (Table 14.3).

# 14.4  Tuning the Catalytic Activity of CNMs in Enzyme Mimics

The enzyme activity of CNMS can be effectively tuned depending on various factors, including pH, temperature, and size-dependent properties (size, shape, and morphology). For instance, CDs which, owing to their smaller size (<10 nm), provide a higher surface-to-volume ratio and exhibit relatively higher enzyme activity. In Chen et al.'s report (2014b), the shape and the morphology of CNMs have also been found to be critical in tuning CNMs' catalytic activities. Specifically, among the variety of tested nanoarchitectures, PtPd nanodendrites–GNs showed a stronger affinity relative to mono-Pt nanoflowers–GNs, core-shell Pt@Pd nanoflowers–GNs, PtNPs–GNs, and PtPd nanoalloys–GNs. Likewise, compared with $Fe_3O_4$ microspheres–rGO and $Fe_3O_4$ nanopolyhedrons–rGO, the nanocomposite of $Fe_3O_4$ nanospheres–rGO were found to possess the highest peroxidase-like activity (Qian et al. 2014). Aside from this, the construction of nanohybrids allows the enzymatic reactions to be feasible over a broad pH range. Recently, a GO–Au nanocluster hybrid was evaluated as an efficient enzyme mimic that can work even at neutral pH (Tao et al. 2013). The synergistic working action leading to enhanced catalytic activity and stability is an added advantage of nanohybrid systems that cannot be realized by either component alone (Dong et al. 2012, Xie et al. 2013, Xing et al. 2014, Chen et al. 2015). Furthermore, the deposition of metal ions on the surface of nanohybrid systems can stimulate their enzymatic activity (Zhan et al. 2014). Besides metal ions, GO itself can serve as an enzyme activator or modulator to regulate the catalytic activity of a nanohybrid system (Tao et al. 2013).

**TABLE 14.2**  G-NMs as Nanozymes in Peroxidase Mimics

| G-NMs as Nanozymes | Method | Peroxidase Substrate | LOD | Applications | References |
|---|---|---|---|---|---|
| GO | Colorimetric | TMB | 1.0 μM | Glucose detection | Song et al. 2010a |
| GO | Fluorometric | DAB | 5.0 pg/mL | Immunosensing | Lim et al. 2012 |
| G-$Fe_3O_4$ magnetic nanocomposite | Colorimetric | TMB | 0.32, 0.74 μM | $H_2O_2$, glucose detection | Dong et al. 2012 |
| AuNPs-GO hybrid | Colorimetric | TMB, ABTS, OPD | – | DNA detection | Liu et al. 2012a |
| Au–rGO nanocomposite | Colorimetric | Pyrogallol | – | Dye removal | Dutta et al. 2013 |
| QrGO | Colorimetric | TMB | 1.0 μM | Glucose detection | Liu et al. 2013a |
| $Fe_2O_3$–G nanocomposite | Colorimetric | TMB | 0.5 μM | Glucose detection | Xing et al. 2014 |
| Magnetic GO–hemin hybrid | Electrochemical | ABTS | 0.08 nM | GSH detection | Bi et al. 2014 |
| $Co_3O_4$–rGO nanocomposite | Colorimetric | TMB | 0.5, 1.0 μM | $H_2O_2$, glucose detection | Xie et al. 2013 |
| $Fe_3O_4$ NSs–rGO nanocomposite | Colorimetric | TMB | 39 nM | Ach detection | Qian et al. 2014 |
| GO-AuNPs hybrid | Colorimetric | TMB | 0–50 μM | $Hg^{2+}$, $Pb^{2+}$ detection | Chen et al. 2015 |
| MNPs-PtNPs-GO nanocomposite | Colorimetric | TMB | 100 cells | Target cancer cells | Kim et al. 2014 |
| G-FeTMPyP complex | Electrochemical | OPD | – | DNA biosensing | Wang et al. 2013a |
| GO | Voltammetric | TMB | 1.0 nM | $H_2O_2$ detection | Sun et al. 2013 |
| GCNT–$Fe_3O_4$ nanocomposite | Colorimetric, electrochemical | TMB, OPD, DAB, PAP, and HQ | –, 0.022 mM | $H_2O_2$, glucose detection | Wang et al. 2014b |
| AuNPs-GO hybrid | Colorimetric | TMB | 0.04 pg/mL | RSV detection | Zhan et al. 2014 |
| MWCNTs@rGONR heterostructure | Colorimetric | TMB | 10 μM | Cholesterol detection | Qian et al. 2015 |
| FA–GO–Au NCs | Colorimetric | TMB | 1000 cells | Target cancer cells | Tao et al. 2013 |
| Hemin–rGO–AuNPs nanocomposite | Colorimetric | TMB | 5.0 nM | $H_2O_2$ detection | Lv & Weng 2013 |
| CS-GO nanocomposite | Colorimetric | TMB | 0.5 μM | Glucose detection | Wang et al. 2014a |
| $CoFe_2O_4$ ferrite–rGO nanocomposite | Colorimetric | TMB | 0.3 μM | $H_2O_2$ detection | Hao et al. 2013 |
| Few-layer graphene | Electrochemical | TMB | 10 nM | $H_2O_2$ detection | Wang & Weng 2013 |
| PtPd nanodendrites–G nanocomposite | Colorimetric | TMB | 0.1 μM | $H_2O_2$ detection | Chen et al. 2014b |
| AuNPs/Cit-GNs | – | TMB | – | – | Chen et al. 2014c |

*Note:* Ach: acetylcholine; Cit-GNs: citrate-functionalized graphene nanosheets; CS: chitosan; DAB: diaminobenzene; FA: folic acid; FeTMPyP: iron (III)meso-tetrakis(*N*-methylpyridinum-4-yl)porphyrin; G: graphene; GCNT: graphene oxide–dispersed carbon nanotube; GO: graphene oxide; GSH: glutathione; HQ: hydroquinone; MNPs: magnetic nanoparticles; MWCNTs@rGONR: Multiwalled carbon nanotubes@reduced graphene oxide nanoribbon; NCs: nanoclusters; NSs: nanospheres; OPD: *o*-phenylenediamine; PAP: *p*-amino phenol; PtNPs: platinum nanoparticles; QrGO: quinone-mediated reduced graphene oxide; rGO: reduced graphene oxide; RSV: respiratory syncytial virus; TMB: 3,3′,5,5′-tetramethylbenzidine.

**TABLE 14.3**    Carbon-Based Dots as Nanozymes in Peroxidase Mimics

| Carbon-Based Dots | Precursor | Synthetic Method | Peroxidase Substrate | LOD | Applications | References |
|---|---|---|---|---|---|---|
| CDs | Candle soot | $HNO_3$ oxidation | TMB | 0.4 µM | Glucose detection | Shi et al. 2011 |
| CDs | ILs | Microwave oven at 450 W | MR, MO | – | Azo dye degradation | Safavi et al. 2012 |
| GDs | Vulcan XC-72 carbon | $HNO_3$ treatment at 130°C | TMB | 0.5 µM | Biosensing | Zheng et al. 2013 |
| CDs | β-cyclodextrin | $HNO_3$ treatment at reflux | TMB | 0.8, 0.5 µM | $Fe^{3+}$, $Ag^+$ detection | Zhu et al. 2014 |
| CDs | $Na_2EDTA\cdot2H_2O$ | Hydrothermal treatment at 400°C | TMB | 0.3 µM | GSH detection | Shamsipur et al. 2014 |
| CDs | Leaves of *Olea europaea* | Hydrothermal treatment at 200°C | TMB | 0.6, 5.6 µM | $H_2O_2$, glucose detection | Wu et al. 2013a |
| CNDs | Organic amines | Microwave-assisted heating | TMB | 0.4, 0.5 µM | $H_2O_2$, glucose detection | Liu et al. 2012c |
| GQDs–$Fe_3O_4$ NPs | $FeCl_3\cdot6H_2O$ + $FeSO_4\cdot7H_2O$ + GQDs | Co-precipitation method | TMB | – | Phenolic compound removal | Wu et al. 2014 |
| GQDs/Au electrode | GO | Mercury lamp exposure (365 nm, 1000 W) | TMB | 0.7 µM | $H_2O_2$ detection | Zhang et al. 2013b |
| CDs | $Na_2EDTA\cdot2H_2O$ | Hydrothermal treatment at 400°C | TMB | 23 nM | $Hg^+$ detection | Mohammadpour et al. 2014 |
| Fe–g-$C_3N_4$ | Melamine | Pyrolysis at 600°C | TMB | 0.5 µM | Glucose detection | Tian et al. 2013 |
| g-$C_3N_4$ | Melamine | Pyrolysis at 600°C | TMB | 1.0 µM | Glucose detection | Lin et al. 2014a |

*Note:*  CDs: carbon nanodots; CNDs: carbon nitride dots; EDTA: ethylenediaminetetraacetic acid; GD: graphene dot; GO: graphene oxide; GQDs: graphene quantum dots; GSH: glutathione; ILs: ionic liquids; MO: methyl orange; MR: methyl red; TMB: 3,3′,5,5′-tetramethylbenzidine.

# 14.5  Conclusions, Challenges, and Future Perspectives

As with many other technical fields, the research on CNMs as artificial enzymes or nanozymes has seen dramatic advancement recently and is significantly influencing the current biotechnology and biochemistry. Among CNMs of varying dimensions, G-NMs are of particular interest and have been widely studied via integration of AuNPs, magnetic NPs, polymers, metalloproteins, MWCNTs, and a variety of other nanoarchitectures on the surface of 2D graphene or its subtypes. Besides the several distinct advantages shared with other artificial enzymes such as organic catalysts, metal/metal oxide-based nanomaterials, and/or metal–organic frameworks, CNMs as nanozymes are unique because of their exceptional large surface area and ease of functionalization for further modifications (bioconjugation, support), rendering them potentially useful in preparing a library of catalytically active CNMs for enzyme biomimics. Despite this, the challenges or bottlenecks we face are also very significant (Table 14.4) and should be addressed appropriately so that the potential of CNMs as nanozymes could be fully realized. To this end, the following aspects are waiting to be achieved:

1. The types of catalytic reactions using CNMs are limited to only redox-type reactions except a few which are based on SOD. On the other hand, natural enzymes and various types of organic catalysts can catalyze a variety of biochemical reactions. Therefore, further investigations are needed

**TABLE 14.4**  Comparison between Natural Enzymes and CNMs as Nanozymes

| | Key Advantages | Key Disadvantages or Challenges |
|---|---|---|
| Natural enzymes | High catalytic efficiency | Limited stability |
| | High substrate specificity | High cost |
| | High selectivity | Sensitivity of catalytic activity toward environment |
| | Wide range of catalytic types | Hard for mass production |
| | Good biocompatibility | Difficulties in purification, recovery, and recycling |
| | Precise structure | |
| | Tunable activity | |
| CNMs as nanozymes | High operational stability | Limited catalytic reactions (primarily peroxidases) |
| | Relatively low cost | Relatively low efficiency |
| | Easy for mass production | Low substrate selectivity |
| | Relatively low cost | Low substrate specificity |
| | Easy in purification, recovery, and recycling | Difficulties in rational design |
| | Large surface area for further functionalization or support material[a] | Toxicity issues |
| | Self-assembly | Industrial use |
| | Multifunctional properties and applications[a] | |

[a] The properties are unique for CNMs.

to construct highly efficient CNMs for catalyzing other types of reactions, and that, certainly, is going to be the most promising research direction in the near future.

2. Compared with natural enzymes and organic catalysts, CNMs usually exhibit relatively low catalytic activity. In this regard, the development of high-performance CNMs that show excellent activity would be worthwhile for the next generation of carbon-based mimicking systems. A related issue is exploring efficient activators or inhibitors that can tailor the catalytic activities of CNMs. Although some of the research groups have started to address this issue (Tao et al. 2013), but the research progress in this direction is still in its infancy and needs more attention and efforts.

3. The toxicity of nanomaterials is one of the key paradigms currently receiving much attention in the development of new nanotechnologies and materials. Although it may be a subject of intense debate, CNMs are not devoid of possible risks to human health or the environment. For instance, the available experimental results indicate that graphene-based nanocomposites might become a health hazard because of their flat shape and charges, oxidative stress, and damage to cell membrane. The recent findings on the biodegradability of GO as well as the functionalization of graphene and CNTs modulating their toxicity impact are certainly some favorable points in the biomedical domain, but more efforts are needed to translate the leading scientific results for practical applications (Bianco 2013).

4. The industrial applications of natural enzymes are well established, whereas, despite the high operational stability of CNMs, their industrial applications are still awaiting.

5. Alongside the above aspects, the rational design of efficient CNMs as enzyme mimics and their relatively low substrate specificity and selectivity, in most cases, are some common challenges to be tackled.

Despite the fact that there are many unresolved issues and challenges, the structural features and exceptional possibility of on-demand sophisticated surface engineering of CNMs, especially, graphene and its derivatives, indicate that this field will continue to thrive and mature in the near future.

# Acknowledgments

The authors are thankful to the Ministry of Science and Technology (NSC101-2113-M-007-006-MY3, MOST104-2881-M-007-039, MOST104-2113-M-007-008-MY3) and National Tsing Hua University (NTHU) (104N1807E1) of Taiwan for financial support.

# References

Afreen, S., Muthoosamy, K., Manickam, S. et al., "Functionalized fullerene ($C_{60}$) as a potential nanomediator in the fabrication of highly sensitive biosensors," *Biosen. Bioelectron.* 63 (2015): 354–364.

Aiba, Y., Sumaoka, J. & Komiyama, M., "Artificial DNA cutters for DNA manipulation and genome engineering," *Chem. Soc. Rev.* 40 (2011): 5657–5668.

Alberts, B., Johnson, A., Lewis, J. et al., eds., *Molecular Biology of the Cell* (Garland Science, New York, 2008).

Ali, S. S., Hardt, J. I., Quick, K. L. et al., "A biologically effective fullerene ($C_{60}$) derivative with superoxide dismutase mimetic properties," *Free Radical Bio. Med.* 37 (2004): 1191–1202.

Ali, S. S., Hardt, J. I. & Dugan, L. L., "SOD activity of carboxyfullerenes predicts their neuroprotective efficacy: A structure–activity study," *Nanomed. Nanotechnol. Biol. Med.* 4 (2008): 283–294.

André, R., Natálio, F., Humanes, M. et al., "$V_2O_5$ nanowires with an intrinsic peroxidase-like activity," *Adv. Funct. Mater.* 21 (2011): 501–509.

Anirudhan, T. S. & Alexander, S., "Multiwalled carbon nanotube based molecular imprinted polymer for trace determination of 2,4-dichlorophenoxyaceticacid in natural water samples using a potentiometric method," *Appl. Surf. Sci.* 303 (2014): 180–186.

Araghi, M. & Bokaei, F., "Manganese(III) porphyrin supported on multi-wall carbon nanotubes: A highly efficient and reusable biomimetic catalyst for oxidative decarboxylation of a-arylcarboxylic acids and oxidation of alkanes with sodium periodate," *Polyhedron* 53 (2013): 15–19.

Asati, A., Santra, S., Kaittanis, C. et al., "Oxidase-like activity of polymer-coated cerium oxide nanoparticles," *Angew. Chem. Int. Ed.* 48 (2009): 2308–2312.

Badamshina, E. & Gafurova, M., "Polymeric nanocomposites containing non-covalently bonded fullerene $C_{60}$: Properties and applications," *J. Mater. Chem.* 22 (2012): 9427–9438.

Barnham, K. J., Masters, C. L. & Bush, A., "Neurodegenerative diseases and oxidative stress," *Nat. Rev. Drug. Discov.* 3 (2004): 205–214.

Behr, J.-P. & Lehn, J.-M., "Enhanced rates of dihydropyridine to pyridinium hydrogen transfer in complexes of an active macrocyclic receptor molecule," *J. Chem. Soc. Chem. Commun.* (1978): 143–146.

Bhabak, K. P. & Mugesh, G., "Functional mimics of glutathione peroxidase: Bioinspired synthetic antioxidants," *Acc. Chem. Res.* 43 (2010): 1408–1419.

Bi, S., Zhao, T., Jia, X. et al., "Magnetic graphene oxide-supported hemin as peroxidase probe for sensitive detection of thiols in extracts of cancer cells," *Biosens. Bioelectron.* 57 (2014): 110–116.

Bianco, A., "Graphene: Safe or toxic? The two faces of the medal," *Angew. Chem. Int. Ed.* 52 (2013): 4986–4997.

Bonarlaw, R. P. & Sanders, J. K. M., "Polyol recognition by a steroid-capped porphyrin: Enhancement and modulation of misfit guest binding by added water or methanol," *J. Am. Chem. Soc.* 117 (1995): 259–271.

Breaker, R. R. & Joyce, G. F., "A DNA enzyme that cleaves RNA," *Chem. Biol.* 1 (1994): 223–229.

Breslow, R., "Centenary lecture: Biomimetic chemistry," *Chem. Soc. Rev.* 1 (1972): 553–580.

Breslow, R. & Overman, L. E., "Artificial enzyme combing a metal catalytic group and a hydrophobic binding cavity," *J. Am. Chem. Soc.* 92 (1970): 1075–1077.

Chen, Z., Yin, J.-J., Zhou, Y. T. et al., "Dual enzyme-like activities of iron oxide nanoparticles and their implication for diminishing cytotoxicity," *ACS Nano* 6 (2012): 4001–4012.

Chen, Q., Liu, M., Zhao, J. et al., "Water-dispersible silicon dots as a peroxidase mimetic for the highly-sensitive colorimetric detection of glucose," *Chem. Commun.* 50 (2014a): 6771–6774.

Chen, X., Su, B., Cai, Z. et al., "PtPd nanodendrites supported on graphene nanosheets: A peroxidase-like catalyst for colorimetric detection of $H_2O_2$," *Sens. Actuators B Chem.* 201 (2014b): 286–292.

Chen, X., Tian, X., Su, B. et al., "Au nanoparticles on citrate-functionalized graphene nanosheets with a high peroxidase-like performance," *Dalton Trans.* 43 (2014c): 7449–7454.

Chen, X., Zhai, N., Synder, J. H. et al., "Colorimetric detection of $Hg^{2+}$ and $Pb^{2+}$ based on peroxidase-like activity of graphene oxide-gold nanohybrids," *Anal. Methods* (2015): doi: 10.1039/c4ay02801e.

Ciaurriz, P., Bravo, E. & Hamad-Schifferli, K., "Effect of architecture on the activity of glucose oxidase/horseradish peroxidase/carbon nanoparticle conjugates," *J. Colloid Interf. Sci.* 414 (2014): 73–81.

Comotti, M., Della Pina, C., Matarrese, R. et al., "The catalytic activity of 'naked' gold particles," *Angew. Chem. Int. Ed.* 43 (2004): 5812–5815.

Comotti, M., Della Pina, C., Falletta, E. et al., "Aerobic oxidation of glucose with gold catalyst: Hydrogen peroxide as intermediate and reagent," *Adv. Synth. Catal.* 348 (2006): 313–316.

Compton, O. C. & Nguyen, S. T., "Graphene oxide, highly reduced graphene oxide, and graphene: Versatile building blocks for carbon-based materials," *Small* 6 (2010): 711–723.

Cramer, F. & Kampe, W., "Inclusion compounds: XVII. Catalysis of decarboxylation by cyclodextrins; A model reaction for the mechanism of enzymes," *J. Am. Chem. Soc.* 87 (1965): 1115–1120.

Cuevas, F., Di Stefano, S., Magrans, J. O. et al., "Toward an artificial acetylcholinesterase," *Chem. Eur. J.* 6 (2000): 3228–3234.

Cui, R., Han, Z. & Zhu, J. J., "Helical carbon nanotubes: Intrinsic peroxidase catalytic activity and its application for biocatalysis and biosensing," *Chem. Eur. J.* 17 (2011): 9377–9384.

Dai, Z., Liu, S., Bao, J. et al., "Nanostructured FeS as a mimic peroxidase for biocatalysis and biosensing," *Chem. Eur. J.* 15 (2009): 4321–4326.

Dai, L., Chang, D. W., Baek, J.-B. et al., "Carbon nanomaterials for advanced energy conversion and storage," *Small* 8 (2012): 1130–1166.

Dickinson, B. C. & Chang, C. J., "Chemistry and biology of reactive oxygen species in signaling or stress responses," *Nat. Chem. Bio.* 7 (2011): 504–511.

Dickinson, B. C., Srikun, D. & Chang, C. J., "Mitochondrial-targeted fluorescent probes for reactive oxygen species," *Curr. Opin. Chem. Biol.* 14 (2010): 50–56.

Dong, Y.-L., Zhang, H.-G., Rahman, Z. U. et al., "Graphene oxide–$Fe_3O_4$ magnetic nanocomposites with peroxidase-like activity for colorimetric detection of glucose," *Nanoscale* 4 (2012): 3969–3976.

Dreyer, D. R., Park, S., Bielawski, C. W. et al., "The chemistry of graphene oxide," *Chem. Soc. Rev.* 39 (2010): 228–240.

Dreyer, D. R., Todd, A. D. & Bielawski, C. W., "Harnessing the chemistry of graphene oxide," *Chem. Soc. Rev.* 43 (2014): 5288–5301.

Dugan, L., Gabrielsen, J. K., Yu, S. P. et al., "Buckminsterfullerenol free radical scavengers reduce excitotoxic and apoptotic death of cultured cortical neurons," *Neurobiol. Dis.* 3 (1996): 129–135.

Dugan, L. L., Turetsky, D. M., Du, C. et al., "Carboxyfullerenes as neuroprotective agents," *Proc. Natl. Acad. Sci.* 94 (1997): 9434–9439.

Dutta, S., Sarkar, S., Ray, S. et al., "Benzoin derived reduced graphene oxide (rGO) and its nanocomposite: Application in dye removal and peroxidase-like activity," *RSC Adv.* 3 (2013): 21475–21483.

Fan, J., Yin, J.-J., Ning, B. et al., "Direct evidence for catalase and peroxidase activities of ferritin–platinum nanoparticles," *Biomaterials* 32 (2011): 1611–1618.

Friedle, S., Reisner, E. & Lippard, S. J., "Current challenges of modeling diiron enzyme active sites for dioxygen activation by biomimetic synthetic complexes," *Chem. Soc. Rev.* 39 (2010): 2768–2779.

Ganesh, G. & Ling, Y.-C., "Multi-functional graphene for *in vitro* and *in vivo* zebrafish imaging," *Biomaterials* 33 (2012): 2532–2545.

Ganesh, G. & Ling, Y.-C., "Magnetic and fluorescent graphene for dual modal imaging and single light induced photothermal and photodynamic therapy of cancer cells," *Biomaterials* 35 (2014): 4499–4507.

Ganesh, G., Chang, C. C. & Ling, Y.-C., "Facile synthesis of smart magnetic graphene for safe drinking water: Heavy metal removal and disinfection control," *ACS Sustain. Chem. Eng.* 1 (2013): 462–471.

Gao, L., Zhuang, J., Nie, L. et al., "Intrinsic peroxidase-like activity of ferromagnetic nanoparticles," *Nat. Nanotechnol.* 2 (2007): 577–583.

Gao, Y., Wang, G., Huang, H. et al., "Fluorometric method for the determination of hydrogen peroxide and glucose with $Fe_3O_4$ as catalyst," *Talanta* 85 (2011): 1075–1080.

Garcia-Viloca, M., Gao, J., Karplus, M. et al., "How enzymes work: Analysis by modern rate theory and computer simulations," *Science* 303 (2004): 186–195.

Garg, B. & Ling, Y.-C., "Versatilities of graphene-based catalysts in organic transformations," *Green Mater.* 1 (2013): 47–61.

Garg, B., Bisht, T. & Ling, Y.-C., "Graphene-based nanomaterials as heterogeneous acid catalysts: A comprehensive perspective," *Molecules* 19 (2014a): 14582–14614.

Garg, B., Bisht, T. & Ling, Y.-C., "Sulfonated graphene as highly efficient and reusable acid carbocatalyst for the synthesis of ester plasticizers," *RSC Adv.* 4 (2014b): 57297–57307.

Garg, B., Sung, C.-H. & Ling Y.-C., "Graphene-based nanomaterials as molecular imaging agents," *WIREs Nanomed. Nanobiotechnol.* (2015a). doi: 10.1002/wnan.1342.

Garg, B., Bisht, T. & Ling, Y.-C., "Graphene-based nanomaterials as efficient peroxidase mimetic catalysts for biosensing applications: An overview," *Molecules* 20 (2015b): 14155–14190.

Gayathri, P. & Kumar, A. S., "An iron impurity in multiwalled carbon nanotube complexes with chitosan that biomimics the heme-peroxidase function," *Chem. Eur. J.* 19 (2013): 17103–17112.

Ghule, A. V., Kathir, K. M. & Krishnaswamy, T. et al., "Carbon nanotubes prevent 2,2,2 trifluoroethanol induced aggregation of protein," *Carbon* 45 (2007): 1583–1595.

Giraldo, J. P., Landry, M. P., Faltermeier, S. M. et al., "Plant nanobionics approach to augment photosynthesis and biochemical sensing," *Nat. Mater.* 13 (2014): 400–408.

Guo, S. & Wang, E., "Functional micro/nanostructures: Simple synthesis and application in sensors, fuel cells, and gene delivery," *Acc. Chem. Res.* 44 (2011): 491–500.

Guo, Y. J., Deng, L., Li, J., Guo, S. J. et al., "Hemin–graphene hybrid nanosheets with intrinsic peroxidase-like activity for label-free colorimetric detection of single-nucleotide polymorphism," *ACS Nano* 5 (2011a): 1282–1290.

Guo, Y. J., Li, J. & Dong, S. J., "Hemin functionalized graphene nanosheets-based dual biosensor platforms for hydrogen peroxide and glucose," *Sens. Actuators B Chem.* 160 (2011b): 295–300.

Hamid, M. & Khalil-ur-Rahman, "Potential applications of peroxidases," *Food Chem.* 115 (2009): 1177–1186.

Hao, J., Zhang, Z., Yang, W. et al., "In situ controllable growth of $CoFe_2O_4$ ferrite nanocubes on graphene for colorimetric detection of hydrogen peroxide," *J. Mater. Chem. A* 1 (2013): 4352–4357.

He, W., Wu, X., Liu, J. et al., "Design of AgM bimetallic alloy nanostructures (M = Au, Pd, Pt) with tunable morphology and peroxidase-like activity," *Chem. Mater.* 22 (2010): 2988–2994.

He, W., Zhou, Y. T., Wamer, W. G. et al., "Intrinsic catalytic activity of Au nanoparticles with respect to hydrogen peroxide decomposition and superoxide scavenging," *Biomaterials* 34 (2013): 765–773.

He, W., Wamer, W., Xia, Q. et al., "Enzyme-like activity of nanomaterials," *J. Environ. Sci. Health Part C Environ. Carcinogenesis Ecotoxicol. Rev.* 32 (2014a): 186–211.

He, Y., Wang, X., Sun, J. et al., "Fluorescent blood glucose monitor by hemin-functionalized graphene quantum dots based sensing system," *Anal. Chim. Acta* 810 (2014b): 71–78.

Hennrich, N. & Cramer, F., "Inclusion compounds; XVIII. The catalysis of the fission of pyrophosphates by cyclodextrin; A model reaction for the mechanism of enzymes," *J. Am. Chem. Soc.* 87 (1965): 1121–1126.

Hu, Y. H., "The first magnetic-nanoparticle-free carbon-based contrast agent of magnetic-resonance imaging-fluorinated graphene oxide," *Small* 10 (2014): 1451–1452.

Hu, X., Liu, J., Hou, S. et al., "Research progress of nanoparticles as enzyme mimetics," *Sci. China Phys. Mech. Astronomy* 54 (2011): 1749–1756.

Hua, B.-Y., Wang, J., Wang, K. et al., "Greatly improved catalytic activity and direct electron transfer rate of cyctochrome C due to the confinement effect in a layered self-assembly structure," *Chem. Commun.* 48 (2012): 2316–2318.

Jana, M., Khanra, P., Murmu, N. C. et al., "Covalent surface modification of chemically derived graphene and its application as supercapacitor electrode material," *Phys. Chem. Chem. Phys.* 16 (2014): 7618–7626.

Jiang, Z., Kun, L., Ouyang, H. et al., "A simple and sensitive fluorescence quenching method for the determination of $H_2O_2$ using rhodamine B and $Fe_3O_4$ nanocatalyst," *J. Fluoresc.* 21 (2011): 2015–2020.

Kim II, M., Kim, M. S., Woo, M.-A. et al., "Highly efficient colorimetric detection of target cancer cells utilizing superior catalytic activity of graphene-oxide-magnetic-platinum nanohybrids," *Nanoscale* 6 (2014): 1529–1536.

Klotz, I. M., Royer, G. P. & Scarpa, I. S., "Synthetic derivatives of polyethyleneimine with enzyme-like catalytic activity (synzymes)," *Proc. Natl. Acad. Sci.* 68 (1971): 263–264.

Kruger, K., Grabowski, P. J., Zaug, A. J. et al., "Self-splicing RNA: Autoexcision and autocyclization of the ribosomal RNA intervening sequence of tetrahymena," *Cell* 31 (1982): 147–157.

Lei, J. & Ju, H., "Signal amplification using functional nanomaterials for biosensing," *Chem. Soc. Rev.* 41 (2012): 2122–2134.

Li, L., Wu, G., Yang, G. et al., "Focusing on luminescent graphene quantum dots: Current status and future perspectives," *Nanoscale* 5 (2013a): 4015–4039.

Li, R., Zhen, M., Guan, M. et al., "A novel glucose colorimetric sensor based on intrinsic peroxidase-like activity of $C_{60}$-carboxyfullerenes," *Biosens. Bioelectron.* 47 (2013b): 502–507.

Li, Y., Huang, X., Li, Y. et al., "Graphene–hemin hybrid material as effective catalyst for selective oxidation of primary C–H bond in toluene," *Sci. Rep.* 3 (2013c): 1787–1793.

Lim, S. Y., Ahn, J., Lee, J. S. et al., "Graphene-oxide-based immunosensing through fluorescence quenching by peroxidase-catalyzed polymerization," *Small* 8 (2012): 1994–1999.

Lin, T., Zhong, L., Wang, J. et al., "Graphite-like carbon nitrides as peroxidase mimetics and their applications to glucose detection," *Biosen. Bioelectron.* 59 (2014a): 89–93.

Lin, Y., Ren, J. & Qu, X., "Catalytically active nanomaterials: A promising candidate for artificial enzymes," *Acc. Chem. Res.* 47 (2014b): 1097–1105.

Lin, Y., Ren, J. & Qu, X., "Nano-gold as artificial enzymes: Hidden talents," *Adv. Mater.* 26 (2014c): 4200–4217.

Lin, Y., Wu, L., Huang, Y. et al., "Positional assembly of hemin and gold nanoparticles in graphene mesoporous silica nanohybrids for tandem catalysis," *Chem. Sci.* 6 (2015): 1272–1276.

Liu, S., Lu, F., Xing, R. et al., "Structural effects of $Fe_3O_4$ nanocrystals on peroxidase-like activity," *Chem. Eur. J.* 17 (2011): 620–625.

Liu, M., Zhao, H., Chen, S. et al., "Stimuli-responsive peroxidase mimicking at a smart graphene interface," *Chem. Commun.* 48 (2012a): 7055–7057.

Liu, M., Zhao, H., Chen, S. et al., "Interface engineering catalytic graphene for smart colorimetric biosensing," *ACS Nano* 6 (2012b): 3142–3151.

Liu, S., Tian, J., Wang, L. et al., "A general strategy for the production of photoluminescent carbon nitride dots from organic amines and their application as novel peroxidase-like catalysts for colorimetric detection of $H_2O_2$ and glucose," *RSC Adv.* 2 (2012c): 411–413.

Liu, G., Zhang, X., Zhou, J. et al., "Quinone-mediated microbial synthesis of reduced graphene oxide with peroxidase-like activity," *Bioresour. Technol.* 149 (2013a): 503–508.

Liu, Y. L., Zhao, X. J., Yang, X. X. et al., "A nanosized metal–organic framework of Fe–MIL–88$NH_2$ as a novel peroxidase mimic used for colorimetric detection of glucose," *Analyst* 138 (2013b): 4526–4531.

Lu, Y., Yeung, N., Sieracki, N. et al., "Design of functional metalloproteins," *Nature* 460 (2009): 855–862.

Luo, W., Zhu, C., Su, S. et al., "Self-catalyzed, self-limiting growth of glucose oxidase-mimicking gold nanoparticles," *ACS Nano.* 4 (2010): 7451–7458.

Lv, X. & Weng, J., "Ternary composite of hemin, gold nanoparticles and graphene for highly efficient decomposition of hydrogen peroxide," *Sci. Rep.* 3 (2013): 3285–3295.

Ma, M., Zhang, Y. & Gu, N., "Peroxidase-like catalytic activity of cubic Pt nanocrystals," *Colloids Surf. A Physicochem. Eng. Aspects* 373 (2011): 6–10.

Manea, F., Houillon, F. B., Pasquato, L. et al., "Nanozymes: Gold-nanoparticle-based transphosphorylation catalysts," *Angew. Chem. Int. Ed.* 43 (2004): 6165–6169.

Mao, H., Kawazoe, N. & Chen, G., "Cellular uptake of single-walled carbon nanotubes in 3D extracellular matrix-mimetic composite collagen hydrogels," *J. Nanosci. Nanotechnol.* 14 (2014): 2487–2492.

Melnikova, L., Pospiskova, K., Mitroova, Z. et al., "Peroxidase-like activity of magnetoferritin," *Microchim. Acta* 181 (2014): 295–301.

Mohammadpour, Z., Safavi, A. & Shamsipur, M., "A new label free colorimetric chemosensor for detection of mercury ion with tunable dynamic range using carbon nanodots as enzyme mimics," *Chem. Eng. J.* 255 (2014): 1–7.

Natalio, F., Andre, R., Hartog, A. F. et al., "Vanadium pentoxide nanoparticles mimic vanadium haloperoxidases and thwart biofilm formation," *Nat. Nanotechnol.* 7 (2012): 530–535.

Navalon, S., Dhakshinamoorthy, A., Alvaro, M. et al., "Carbocatalysis by graphene-based materials," *Chem. Rev.* 114 (2014): 6179–6212.

Nie, G., Zhang, L., Lei, J. et al., "Monocrystalline $VO_2$ (B) nanobelts: Large-scale synthesis, intrinsic peroxidase-like activity and application in biosensing," *J. Mater. Chem. A* 2 (2014): 2910–2914.

Novoselov, K. S., Geim, A. K., Morozov, S. V. et al., "Electric field effect in atomically thin carbon films," *Science* 306 (2004): 666–669.

Pan, T. & Uhlenbeck, O. C., "In vitro selection of RNAs that undergo autolytic cleavage with $Pb^{2+}$," *Biochemistry* 31 (1992): 3887–3895.

Park, K. S., Kim, M. I., Cho, D.-Y. et al., "Label-free colorimetric detection of nucleic acids based on target-induced shielding against the peroxidase-mimicking activity of magnetic nanoparticles," *Small* 7 (2011): 1521–1525.

Pauling, L., "Molecular architecture and biological reactions," *Chem. Eng. News* 24 (1946): 1375–1377.

Pauling, L., "Nature of forces between large molecules of biological interest," *Nature* 161 (1948): 707–709.

Pirmohamed, T., Dowding, J. M., Singh, S. et al., "Nanoceria exhibit redox state-dependent catalase mimetic activity," *Chem. Commun.* 46 (2010): 2736–2738.

Pollack, S. J., Jacobs, J. W. & Schultz, P. G., "Selective chemical catalysis by an antibody," *Science* 234 (1986): 1570–1573.

Qian, J., Yang, X., Jiang, L. et al., "Facile preparation of $Fe_3O_4$ nanospheres/reduced graphene oxide nanocomposites with high peroxidase-like activity for sensitive and selective colorimetric detection of acetylcholine," *Sens. Actuators B Chem.* 201 (2014): 160–166.

Qian, J., Yang, X., Yang, Z. et al., "Multiwalled carbon nanotube@reduced graphene oxide nanoribbon heterostructure: Synthesis, intrinsic peroxidase-like catalytic activity, and its application in colorimetric biosensing," *J. Mater. Chem. B* (2015): doi: 10.1039/c4tb01702a.

Qu, F., Li, T. & Yang, M., "Colorimetric platform for visual detection of cancer biomarker based on intrinsic peroxidase activity of graphene oxide," *Biosens. Bioelectron.* 26 (2011): 3927–3931.

Quick, K. L., Ali, S. S., Arch, R. et al., "A carboxyfullerene SOD mimetic improves cognition and extends the lifespan of mice," *Neurobiol. Aging* 29 (2008): 117–128.

Raynal, M., Ballester, P., Vidal-Ferran, A. et al., "Supramolecular catalysis: Part 2—Artificial enzyme mimics," *Chem. Soc. Rev.* 43 (2014): 1734–1787.

Rong, Z., Zhou, Y., Chen, B. et al., "Bio-inspired hierarchical polymer fiber–carbon nanotube adhesives," *Adv. Mater.* 26 (2014): 1456–1461.

Safavi, A., Sedaghati, F., Shahbaazi, H. et al., "Facile approach to the synthesis of carbon nanodots and their peroxidase mimetic function in azo dyes degradation," *RSC Adv.* 2 (2012): 7367–7370.

Schmitt, O., "Some interesting and useful biomimetic transform," in *Third International Biophysics Congress* (1969), p. 297.

Shamsipur, M., Safavi, A. & Mohammadpour, Z., "Indirect colorimetric detection of glutathione based on its radical restoration ability using carbon nanodots as nanozymes," *Sens. Actuators B Chem.* 199 (2014): 463–469.

Shi, W., Wang, Q., Long, Y. et al., "Carbon nanodots as peroxidase mimetics and their applications to glucose detection," *Chem. Commun.* 47 (2011): 6695–6697.

Singh, V., Joung, D., Zhai, L. et al., "Graphene based materials: Past, present and future," *Prog. Mater. Sci.* 56 (2011): 1178–1271.

Song, Y., Qu, K., Zhao, C. et al., "Graphene oxide: Intrinsic peroxidase catalytic activity and its application to glucose detection," *Adv. Mater.* 22 (2010a): 2206–2210.

Song, Y., Wang, X., Zhao, C. et al., "Label-free colorimetric detection of single nucleotide polymorphism by using single-walled carbon nanotube intrinsic peroxidase-like activity," *Chem. Eur. J.* 16 (2010b): 3617–3621.

Song, Y. J., Chen, Y., Feng, L. Y. et al., "Selective and quantitative cancer cell detection using target-directed functionalized graphene and its synergistic peroxidase-like activity," *Chem. Commun.* 47 (2011): 4436–4438.

Sumner, J. B., "The isolation and crystallization of the enzyme urease," *J. Biol. Chem.* 69 (1926a): 435–441.

Sumner, J. B., "The recrystallization of urease," *J. Biol. Chem.* 70 (1926b): 97–98.

Sun, W., Ju, X., Zhang, Y. et al., "Application of carboxyl functionalized graphene oxide as mimetic peroxidase for sensitive voltammetric detection of $H_2O_2$ with 3,3′,5,5′-tetramethylbenzidine," *Electrochem. Commun.* 26 (2013): 113–116.

Tabushi, I., "Cyclodextrin catalysis as a model for enzyme action," *Acc. Chem. Res.* 15 (1982): 66–72.

Tabushi, I., Shimizu, N., Sugimoto, T. et al., "Cyclodextrin flexibly capped with metal ion," *J. Am. Chem. Soc.* 99 (1977): 7100–7102.

Takagishi, T. & Klotz, I. M., "Macromolecule–small molecule interactions: Introduction of additional binding sites in polyethyleneimine by disulfide cross-linkages," *Biopolymers* 11 (1972): 483–491.

Tao, Y., Lin, Y., Huang, Z. et al., "Incorporating graphene oxide and gold nanoclusters: A synergistic catalyst with surprisingly high peroxidase-like activity over a broad pH range and its application for cancer cell detection," *Adv. Mater.* 25 (2013): 2594–2599.

Tarnuzzer, R. W., Colon, J., Patil, S. et al., "Vacancy engineered ceria nanostructures for protection from radiation-induced cellular damage," *Nano Lett.* 5 (2005): 2573–2577.

Tian, J., Liu, Q., Asiri, A. M. et al., "Ultrathin graphitic carbon nitride nanosheets: A novel peroxidase mimetic, Fe doping-mediated catalytic performance enhancement and application to rapid, highly sensitive optical detection of glucose," *Nanoscale* 5 (2013): 11604–11609.

Tokuyama, H., Yamago, S. & Nakamura, E., "Photoinduced biochemical activity of fullerene carboxylic acid," *J. Am. Chem. Soc.* 115 (1993): 7918–7919.

Tramontano, A., Janda, K. D. & Lerner, R. A., "Catalytic antibodies," *Science* 234 (1986): 1566–1570.

Wan, X., Huang, Y. & Chen, Y., "Focusing on energy and optoelectronic applications: A journey for graphene and graphene oxide at large scale," *Acc. Chem. Res.* 45 (2012): 598–607.

Wang, Q., Lei, J., Deng, S. et al., "Graphene-supported ferric porphyrin as a peroxidase mimic for electrochemical DNA biosensing," *Chem. Commun.* 49 (2013a): 916–918.

Wang, Z., Lv, X. & Weng, J., "High peroxidase catalytic activity of exfoliated few-layer graphene," *Carbon* 62 (2013b): 51–60.

Wang, G.-L., Xu, X., Wu, X. et al., "Visible-light-stimulated enzyme like activity of graphene oxide and its application for facile glucose sensing," *J. Phys. Chem. C* 118 (2014a): 28109–28117.

Wang, H., Li, S., Si, Y. et al., "Recyclable mimic enzyme of $Fe_3O_4$ nanoparticles loaded on graphene oxide-dispersed carbon nanotubes with enhanced peroxidase-like catalysis and electrocatalysis," *J. Mater. Chem. B* (2014b): doi: 10.1039/C4TB00541D.

Wang, T., Fu, Y., Chai, L. et al., "Filling carbon nanotubes with Prussian blue nanoparticles of high peroxidase-like catalytic activity for colorimetric chemo- and biosensing," *Chem. Eur. J.* 20 (2014c): 2623–2630.

Wei, H. & Wang, E., "$Fe_3O_4$ magnetic nanoparticles as peroxidase mimetics and their applications in $H_2O_2$ and glucose detection," *Anal. Chem.* 80 (2008): 2250–2254.

Wei, H. & Wang, E., "Nanomaterials with enzyme-like characteristics (nanozymes): Next-generation artificial enzymes," *Chem. Soc. Rev.* 42 (2013): 6060–6093.

Wolfenden, R. & Snider, M. J., "The depth of chemical time and the power of enzymes as catalysts," *Acc. Chem. Res.* 34 (2001): 938–945.

Wu, D., Deng, X., Huang, X. et al., "Low-cost preparation of photoluminescent carbon nanodots and application as peroxidase mimetics in colorimetric detection of $H_2O_2$ and glucose," *J. Nanosci. Nanotechnol.* 13 (2013a): 6611–6616.

Wu, M.-C., Deokar, A. R., Liao, J.-H. et al., "Graphene-based photothermal agent for rapid and effective killing of bacteria," *ACS Nano* 7 (2013b): 1281–1290.

Wu, X., Zhang, Y., Han, T. et al., "Composite of graphene quantum dots and $Fe_3O_4$ nanoparticles: Peroxidase activity and application in phenolic compound removal," *RSC Adv.* 4 (2014): 3299–3305.

Wulff, G. & Liu, J., "Design of biomimetic catalysts by molecular imprinting in synthetic polymers: The role of transition state stabilization," *Acc. Chem. Res.* 45 (2012): 239–247.

Wulff, G. & Sarhan, A., "Über die anwendung von enzymanalog gebauten polymeren zur racemattrennung," *Angew. Chem.* 84 (1972): 364–364. doi: 10.1002/ange.19720840838.

Xie, J., Zhang, X., Wang, H. et al., "Analytical and environmental applications of nanoparticles as enzyme mimetics," *TrAC Trends Anal. Chem.* 39 (2012): 114–129.

Xie, J., Cao, H., Jiang, H. et al., "$Co_3O_4$-reduced graphene oxide nanocomposite as an effective peroxidase mimetic and its application in visual biosensing of glucose," *Anal. Chim. Acta* 796 (2013): 92–100.

Xing, Z., Tian, J., Asiri, A. M. et al., "Two-dimensional hybrid mesoporous $Fe_2O_3$-graphene nanostructures: A highly active and reusable peroxidase mimetic toward rapid, highly sensitive optical detection of glucose," *Biosens. Bioelectron.* 52 (2014): 452–457.

Xiong, Z., Gu, T. & Wang, X., "Self-assembled multilayer films of sulfonated graphene and polystyrene-based diazonium salt as photo-cross-linkable supercapacitor electrodes," *Langmuir* 30 (2014): 522–532.

Xue, T., Jiang, S., Qu, Y. et al., "Graphene-supported hemin as a highly active biomimetic oxidation catalyst," *Angew. Chem. Int. Ed.* 51 (2012): 3822–3825.

Yang, K., Feng, L., Shi, X. et al., "Nano-graphene in biomedicine: Theranostic applications," *Chem. Soc. Rev.* 42 (2013): 530–547.

Yu, H., Jin, Y., Li, Z. et al., "Synthesis and characterization of sulfonated single-walled carbon nanotubes and their performance as solid acid catalyst," *J. Solid State Chem.* 181 (2008): 432–438.

Yu, J., Guan, M., Li, F. et al., "Effects of fullerene derivatives on bioluminescence and application for protease detection," *Chem. Commun.* 48 (2012): 11011–11013.

Zeyuan, D., Yongguo, W., Yanzhen, Y. et al., "Supramolecular enzyme mimics by self-assembly," *Curr. Opin. Colloid Interface Sci.* 16 (2011): 451–458.

Zhan, L., Li, C. M., Wu, W. B. et al., "A colorimetric immunoassay for respiratory syncytial virus detection based on gold nanoparticles-graphene oxide hybrids with mercury-enhanced peroxidase-like activity," *Chem. Commun.* 50 (2014): 11526–11528.

Zhang, W., Wang, T. G., Qin, L. Y. et al., "Neuroprotective effect of dextromethorphan in the MPTP Parkinson's disease model: Role of NADPH oxidase," *FASEB J.* 18 (2004): 589–591.

Zhang, X.-Q., Gong, S.-W., Zhang, Y. et al., "Prussian blue modified iron oxide magnetic nanoparticles and their high peroxidase-like activity," *J. Mater. Chem.* 20 (2010): 5110–5116.

Zhang, K., Hu, X., Liu, J. et al., "Formation of PdPt alloy nanodots on gold nanorods: Tuning oxidase-like activities via composition," *Langmuir* 27 (2011): 2796–2803.

Zhang, Y., Xu, C., Li, B. et al., "In situ growth of positively-charged gold nanoparticles on single-walled carbon nanotubes as a highly active peroxidase mimetic and its application in biosensing," *Biosens. Bioelectron.* 43 (2013a): 205–210.

Zhang, Y., Wu, C., Zhou, X. et al., "Graphene quantum dots/gold electrode and its application in living cell $H_2O_2$ detection," *Nanoscale* 5 (2013b): 1816–1819.

Zhang, B., He, Y., Liu, B. et al., "NiCoBP-doped carbon nanotube hybrid: A novel oxidase mimetic system for highly efficient electrochemical immunoassay," *Anal. Chim. Acta* 851 (2014a): 49–56.

Zhang, J.-W., Zhang, H.-T., Du, Z.-Y. et al., "Water-stable metal organic frameworks with intrinsic peroxidase-like catalytic activity as a colorimetric biosensing platform," *Chem. Commun.* 50 (2014b): 1092–1094.

Zhao, G., Jiang, L., He, Y. et al., "Sulfonated graphene for persistent aromatic pollutant management," *Adv. Mater.* 23 (2011): 3959–3963.

Zheng, A.-X., Cong, Z.-X., Wang, J.-R. et al., "Highly-efficient peroxidase-like activity of graphene dots for biosensing," *Biosens. Bioelectron.* 49 (2013): 519–524.

Zhu, W., Zhang, J., Jiang, Z. et al., "High-quality carbon dots: Synthesis, peroxidase-like activity and their application in the detection of $H_2O_2$, $Ag^+$ and $Fe^{3+}$," *RSC Adv.* 4 (2014): 17387–17392.

# III

# Nanocapsules

# 15

# Hollow Carbon Nanocapsules

Tokushi Kizuka

## Abstract

We review the structures and electrical, mechanical, and optical properties of isolated hollow carbon nanocapsules (CNCs). We begin by explaining the methodology used to experimentally study CNCs: synthesis of CNCs and in situ high-resolution transmission electron microscopy (HRTEM) incorporated with nanotip manipulation of the CNCs. Then, we present sample results from the observations of the formation, mechanical deformation, conductance, and electroluminescence of CNCs. A particularly intriguing example is the structure assembled to measure conductance; this structure corresponds to a single nanocapsule junction. In addition to the structures and properties of hollow CNCs, we also describe the encapsulation of metal and metal carbide nanoparticles in CNCs. This chapter includes descriptions of atomic configurations of encapsulated particles and of interfaces between carbon shells and encapsulated particles. CNCs are tough, flexible, and highly conductive; thus, they can be expected to serve as important functional and structural components in many future nanodevices.

## 15.1 Introduction

Recent carbon technology has inspired drastic developments in the synthesis of various nanometer-sized carbon materials that have cage structures. These structures include fullerene $C_{60}$ (Kroto et al. 1985), single-walled and multiwalled nanotubes (Iijima 1991, Bethune et al. 1993, Iijima & Ichihashi 1993), onion-like shells (Iijima 1980, Ugarte 1992, de Heer & Ugarte 1993), cones (Krishnan et al. 1997), nanocages (Vinu et al. 2005, Ariga et al. 2007), nanocases (Yu et al. 2005), and endohedral metallofullerenes (Chai et al. 1991, Ruoff et al. 1993, Saito et al. 1993, 1996, Tomita et al. 1993). However, hollow CNCs were originally synthesized half a century ago by a conventional method: heating carbon soot (Heidenreich et al. 1968). These CNCs are the subject of this chapter. CNCs can be interpreted as giant single-walled and multiwalled fullerenes; thus, the discovery of fullerene materials and the progress in the related technologies are attracting new attention to CNCs. CNCs are expected to exhibit intriguing characteristic features and functions arising from their closed carbon network structures. Examples include graphene-like conduction, stable covering of magnetic nanoparticles, and sustained drug delivery within organisms. In addition to the abovementioned conventional synthesis method, CNCs have been produced by a variety of methods, including arc discharge (Iijima 1980, Saito et al. 1993, 1997, Saito 1995), electron irradiation of mixtures of carbon soot and catalytic metals (Ugarte 1992, Banhart et al. 1998), heating of nanometer-sized diamonds (Andersson et al. 1998), and pyrolysis of carbon soot, *p*-xylenes, and ethanol (Mordkovich et al. 1999, Li et al. 2007, Wang et al. 2007). Also, CNCs can be synthesized from amorphous carbon nanowhiskers by an electrical contact method (Asaka et al. 2006, Kizuka et al. 2009a,b).

To formulate the applications of CNCs to functional and structural devices, their atomic configurations and properties have been investigated. The sizes of CNCs are small, as expressed by the number of atomic wall layers in their shells. However, HRTEM has adequate spatial resolution to evaluate the atomic configurations of CNCs. TEM is based on a transmission-type analysis and elucidates internal hollow shapes, encapsulated particles, defects, and interfaces between graphitic shells and encapsulated particles. Therefore, since the discovery of CNCs, TEM has been used for their structural evaluations (Heidenreich et al. 1968, Saito 1995). Further, by inserting various in situ observation apparatuses with dynamic image recording systems, such as charge-coupled cameras, we can directly observe the formation and deformation processes of CNCs; such apparatus also allows analyses of the differences in electrical, mechanical, and optical properties caused by differences in the structures of individual CNCs (Asaka et al. 2006a,b, 2007a, 2009, Kizuka et al. 2009a,c). Conventional methods using commercial-based instruments can be used to investigate average properties of CNCs in powder and aggregate forms; examples of such average properties include magnetic and optical properties and sustained release of drugs. However, when CNCs are used as elements in electronic devices and nanoelectromechanical systems, evaluating the structure and properties of individual CNCs becomes difficult. This is because boundaries between CNCs affect the electrical and mechanical properties of their aggregates; the boundaries between CNCs increase electric resistivity and degrade toughness. To analyze mechanical properties, manipulation of individual CNCs is required for both deformation and nanonewton-scale force measurements (Asaka et al. 2006c, 2009). For investigating their electrical properties, attaching at least two electrodes at each end of individual CNCs is necessary (Asaka et al. 2007b). Although such operations are challenging even with the use of advanced technologies, some results have been obtained for CNCs by in situ TEM in which conductance and force are measured. In this chapter, the synthesis of CNCs and the in situ TEM are first described, and then, some results are presented.

# 15.2 Methods

## 15.2.1 Synthesis of CNCs

CNCs are synthesized from amorphous carbon nanowhiskers by the electrical contact method (Asaka et al. 2006, Kizuka et al. 2009c). The whiskers were prepared by heating single-crystal $C_{60}$ whiskers with submicrometer diameters and lengths larger than 100 μm (Miyazawa et al. 2001, 2003). Crystal $C_{60}$ nanowhiskers were synthesized by a liquid–liquid interfacial precipitation method using, typically, a toluene solution saturated with fullerene powder and isopropyl alcohol (Miyazawa 2002). X-ray diffraction and TEM revealed that the whiskers were fullerite whiskers with the $C_{60}$ molecules in an ordered crystalline array (Minato & Miyazawa 2005). The crystal $C_{60}$ whiskers were heated in high vacuum at 1273 K for 30 min to transform them into amorphous carbon whiskers.

## 15.2.2 In Situ TEM in the Study of Nanomaterials

The method described here is based on in situ HRTEM combined with subnanonewton force measurements used in atomic force microscopy and electronic conductance measurements used in scanning tunneling microscopy (Figure 15.1) (Kizuka & Tanaka 1994, Kizuka et al. 1997, Kizuka 1998a,b, 1999, 2008, Kizuka & Monna 2009). First, nanometer-sized tips of noble metals are prepared for use as electrodes: a noble metal is deposited, by vapor deposition, on a nanotip probe mounted on a silicon microcantilever; vapor deposition can be achieved using vacuum deposition or sputtering. The cantilever tip is attached to the front of a cylindrical piezoelement on a cantilever holder for TEM. Then, a noble metal plate with a thickness of 50 µm is attached to the second plate holder for TEM. The contact edge of the plate is thinned to 5–20 nm. On the plate, amorphous carbon nanowhiskers, which are sources of CNCs, are dispersed. Both the cantilever and plate holders are then inserted into the in situ TEM, as performed at the University of Tsukuba (Kizuka et al. 1997). The specimen chamber of the microscope is evacuated, first by a turbomolecular pump and then by an ion pump, resulting in a vacuum of $1 \times 10^{-5}$ Pa. Inside the microscope, piezomanipulation is used to contact the cantilever tip with the whiskers on the edge surfaces of the opposing plate. A bias voltage is applied between the nanotip and the plate to heat the whiskers and to synthesize CNCs. After the formation of CNCs, each CNC is manipulated by nanotip operations to sandwich it between the nanotip and the plate. This series of manipulations is performed at room temperature. During this process, structural dynamics is observed in situ by lattice imaging via HRTEM using a television capture system. The time resolution of image observation is 17 ms. Static bright-field and high-resolution images are also recorded using a conventional film shooting system. Two types of forces between the nanotip and the plate can be measured: the force components of deflection and the torsion of the cantilever. These are measured simultaneously by optical detection of cantilever deformation using a quadrant photodiode. The spring constant of the cantilever is typically 1–10 N/m.

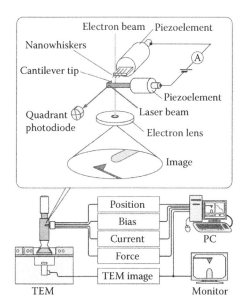

**FIGURE 15.1**  Schematic of in situ TEM for the study of structural dynamics and electrical and mechanical properties of carbon nanowhiskers and CNCs. Two specimen holders (cantilever and plate holders) are manipulated to initiate contact and deform the nanowhiskers and CNCs by two sets of mechanical goniometer stages and piezodriving systems. The structural dynamics of the specimens is directly observed by lattice imaging of TEM and recorded by a charge-coupled device camera. Electrical behaviors, e.g., current, conductance, current–voltage characteristics, conductivity, current density, and electromigration, can be investigated by applying bias voltages. Mechanical tests can be performed to investigate forces for bending, compressing, and buckling. Fracture, strength, elastic limit, elastic constants, such as Young's modulus, and stress–strain relations can also be obtained. By incorporating optical methods into this system, photoluminescence, cathode luminescence, and electroluminescence can be obtained from individual carbon nanowhiskers and CNCs.

Electrical conductance is measured by a two-terminal method. By adding optical attachments to the system for laser illumination and detection, individual spectroscopy, i.e., electroluminescence, photoluminescence, and cathode luminescence of CNCs, can be performed (Kizuka 2011a,b, Kizuka & Oyama 2011). The results from high-resolution imaging and signal detection are simultaneously recorded and analyzed for each image.

By this method, structural dynamics is directly observed. Simultaneously, electrical behaviors, e.g., current, conductance, current–voltage characteristics, and electromigration, are also analyzed. The sizes and shapes obtained from the structural observations allow estimates for conductivity and current density. Mechanical tests provide forces for bending, compressing, buckling, and fracture of CNCs and the elastic limit. Considering the cross section of observed CNCs, strength, elastic constants (such as Young's modulus), and stress–strain relations can be derived. Thus, analyses of the mechanics of materials can be performed on the basis of stress–stain relations. Therefore, the method enables the determination of the mechanics of nanomaterials at atomic resolution. In addition, when optical methods are introduced in the system, individual spectroscopy of nanomaterials can be performed under atomic structural observations.

## 15.3 Formation, Structure, and Properties

### 15.3.1 Formation and Structure

By passing electric current through amorphous carbon nanowhiskers inside the microscope, CNCs are synthesized and aggregated on the whiskers (Figure 15.2) (Asaka et al. 2006, Kizuka et al. 2009b). It was inferred that during current flow, the temperature of the whiskers reaches 3000 K due to Joule heating; this was verified by the melting tests of platinum particles embedded in the whiskers. The surfaces of the initial amorphous carbon whiskers were flat and the thickness was homogeneous along the longer symmetrical axis (Figure 15.2a) (Miyazawa et al. 2001, Miyazawa 2002). After the current flow due to an applied voltage, typically 4 V, the whiskers taper and the whisker surfaces become lumpy (Figure 15.2b);

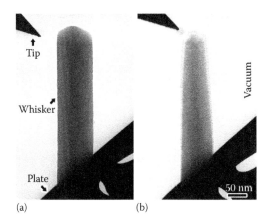

**FIGURE 15.2** In situ TEM observation of the synthesis of CNCs by the electrical contact method. An amorphous carbon nanowhisker, prepared by heating a $C_{60}$ nanowhisker, is fixed on a gold plate. Then, piezomanipulation is used to contact the nanowhisker with the gold nanotip on the microcantilever for atomic force microscopy. Subsequently, a bias voltage is applied between the plate and the nanotip, leading to the formation of CNCs by Joule heating. (a) Low-magnification image of an amorphous carbon nanowhisker prepared by heating a $C_{60}$ nanowhisker. (b) Low-magnification image of the nanowhisker after structural changes due to the applied current. This nanowhisker was used as a filament that emits visible light. (Reprinted from *Carbon*, 20, Kizuka, T., Kato, R., & Miyazawa, K., Structure of hollow carbon nanocapsules synthesized by resistive heating, 105205, Copyright (2009), with permission from Elsevier.)

this signifies the transformation from amorphous to crystalline structures, i.e., CNCs. The CNCs appear on the entire surface of the transformed whiskers (Figure 15.3) (Asaka et al. 2006). The CNCs exhibit hollow single-walled and multiwalled structures; the number of atomic wall layers range from one to five. The most frequent number is two, showing that double-walled CNCs are the most stable. The outer diameter of the CNCs is typically in the range of 3–9 nm with an average of 5 nm. Most CNCs demonstrate polyhedral external shapes. The interlayer spacing of CNCs of five-walled layers is $0.36 \pm 0.01$ nm; however, it reaches $0.42 \pm 0.05$ nm as the number of wall layers decreases to two. Similar stability has been observed in CNTs (Buongiorno et al. 1998, Sugai et al. 2003).

During the growth of CNCs and CNTs, metallic nanoparticles act as nucleating sites (Bethune et al. 1993, Ruoff et al. 1993, Saito et al. 1993). The influence of such catalysts on the formation of CNCs in the electrical contact method has been investigated (Kizuka et al. 2009). One such catalyst, iron nanoparticles embedded in amorphous carbon whiskers, decreases the outer diameter, the number of wall layers, and the sphericity of synthesized CNCs, compared with noncatalyzed synthesis, i.e., using amorphous whiskers of pure carbon with resistive heating (Figure 15.4). The images of such single-walled CNCs appear to resemble larger fullerene molecules and short CNTs (Figure 15.4a). These similarities are not unaccountable. In particular, the reduction in sphericity is seen in CNCs with square-like shapes (Saito & Matsumoto 1998, Kizuka et al. 2009). The nuclei of CNCs are grown on the surfaces of catalytic

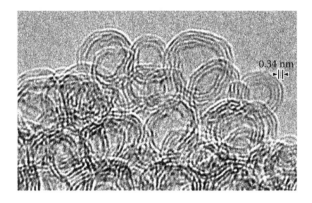

**FIGURE 15.3** High-resolution image of hollow multiwalled CNCs on the surface of the whisker synthesized by the electrical contact method. The sizes of the CNCs are less than 10 nm. (Reprinted with permission from Asaka, K., Kato, R., Maezono, Y. et al., *Applied Physics Letters*, 88, 051914, 2006. Copyright 2006, American Institute of Physics.)

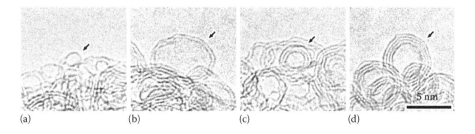

(a)          (b)          (c)          (d)

**FIGURE 15.4** Hollow multiwalled CNCs with various numbers of wall layers synthesized using iron catalysts. (a–d) High-resolution images of CNCs with one, two, three, and four wall layers, as indicated by the arrows, respectively. The scale for a–c is shown in d. (Reprinted from *Diamond and Related Materials*, 18, Kizuka, T., Fujii, J. & Miyazawa, K., Ion catalyzed synthesis of hollow carbon nanocapsules synthesized by resistive heating, 1253–1257, Copyright (2009), with permission from Elsevier.)

nanoparticles and the first layers of CNCs develop around the nanoparticles. Thus, the size and the shape of the nanoparticles are imprinted on the synthesized CNCs, although the nanoparticles dissipate during heating above their melting temperatures. Hence, by using catalysts, controlling the shape and inner diameter (the size of hollow regions) of the CNCs becomes easy. However, the number of layers and the outer diameter (the size of the outermost layers) are determined by growth time, i.e., heating time in this case. Reductions in size and sphericity lead to an increase in the ratio of surface area to weight of CNCs. This will stimulate the activity of various surface reactions, providing possible applications of CNCs to gaseous absorption elements and electrodes of fuel cell batteries. In particular, smaller single-walled CNCs are formed with shorter growth times.

## 15.3.2 Electrical Conduction

After the formation of CNCs by current flow, the cantilever tip is separated from the whisker on the metallic plate and the CNC synthesis stops. Subsequently, one of the CNCs on the whisker is acquired using the cantilever tip and is transferred to a protrusion from the surface of the plate beside the whisker (Figure 15.5) (Asaka et al. 2007a). The CNC is then sandwiched between the surfaces of the gold nanotip and the plate (Figure 15.6). Such assembled structures correspond to single-nanocapsule junctions sandwiched between two metallic electrodes. When a voltage is applied across the electrodes, on/off current flow can be controlled by contacting and separating one of the electrodes from the CNC. This corresponds to a cycle of a working nanometer-sized mechanical relay. When the contact states are maintained and the current is controlled by voltage levels, the current is determined by conductance of the sandwiched CNC. In particular, the conductance through the top graphitic surfaces of the CNC is much higher than that along the layer stacking direction, i.e., the c-axis direction. Notably, the top graphitic surfaces are not graphene sheets because the wall layers of CNCs contain various carbon rings in addition to six-membered rings, e.g., five- and seven-membered rings which occur to form rounded shapes, similar to caps of CNTs. Nevertheless, the CNC junctions cause conductive functions of such pseudographene structures only by sandwiching CNCs between electrodes; this differs from complicated growth processes of graphene on substrates.

Conduction of the CNC junctions depends on the size and shape of the CNCs, as well as on the area, orientational relation, and atomic configuration of the interface between the graphitic layer and the electrodes. In particular, the bonding states of the interfaces are closely related to conductance. For multiwalled CNCs having a few wall layers and several nanometers in diameter, the differential

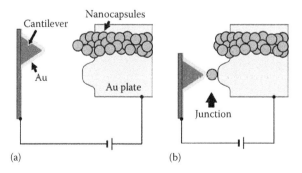

**FIGURE 15.5** Schematic of the assembly of a CNC junction. (a) A cantilever nanotip is used to select one CNC from an aggregation whisker. (b) The selected CNC is transferred to a protrusion on a gold plate beside the aggregation whisker. (Reprinted with permission from Asaka, K., Kato, R., Yoshizaki, R. et al., *Physical Review B*, 76, 113404, 2007. Copyright 2007 by the American Physical Society.)

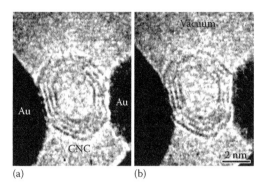

(a)          (b)

**FIGURE 15.6** High-resolution images of a gold/multiwalled CNC/gold junction. (a) Assembled state. Dark regions on left and right sides are the gold nanotips. The CNC is sandwiched between the two nanotips. Bright regions are vacuum. (b) The left nanotip has been shifted to compress the CNC and has caused an increase in interfacial area. At the states observed in a and b, the differential conductances were $0.5G_0$ and $1G_0$, respectively. See also Figure 15.7. (Reprinted with permission from Asaka, K., Kato, R., Yoshizaki, R. et al., *Physical Review B*, 76, 113404, 2007. Copyright 2007 by the American Physical Society.)

conductance of the junction at bias voltages from 0 to 0.20 V corresponds to half of a quantized conductance, i.e., $0.5G_0$, where $G_0 = 2e^2/h$ is the conductance quantum, $e$ is the electron charge, and $h$ is Planck's constant (region a in Figure 15.7). When the interfacial structures between the CNC and the electrode change, the differential conductance increases to $1G_0$ (region b in Figure 15.7). Such conductance values imply that CNC junctions can be used as conductors of high current densities.

When the applied voltage exceeds the critical value for stability of graphitic layers, typically, ca. 0.6 V, the wall layers of the CNCs begin to peel (Figures 15.8 and 15.9) (Kizuka et al. 2009b). For example, as the total number of wall layers of a CNC decreases from four to three, because of the peeling of the outermost layer, the differential conductance decreases from $1G_0$ to $0.5G_0$ and then to $0.2G_0$ (Figure 15.10). Thus, the maximum current and the conductance of CNC junctions can be controlled by selecting the number of wall layers. This is expected to be useful in developing applications of CNCs for nanocarbon technologies.

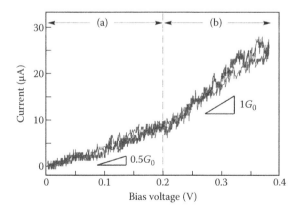

**FIGURE 15.7** Current–voltage curve for the CNC junction in Figure 15.6a and b. The differential conductance at state a in Figure 15.6 was $0.5G_0$ and increased, as the interfacial area increased, to $1G_0$ at state b. (Reprinted with permission from Asaka, K., Kato, R., Yoshizaki, R. et al., *Physical Review B*, 76, 113404, 2007. Copyright 2007 by the American Physical Society.)

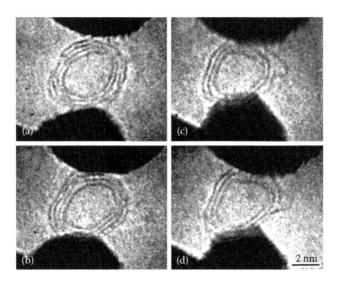

**FIGURE 15.8** Time-sequence series of high-resolution images of peeling of a CNC during conductance measurements. (a) The CNC, constructed of four graphitic wall layers, appears between two gold nanotips in the upper and lower regions. Brighter areas are vacuum. (b–d) Breakdown of the outermost wall results in peeling, i.e., a decrease in the number of wall layers. The transformation process is depicted in Figure 15.9. (Reprinted from Kizuka, T., Kato, R. & Miyazawa, K., "Surface breakdown dynamics of carbon nanocapsules," 20, 105205, 2009, *Nanotechnology*, IOP Publishing.)

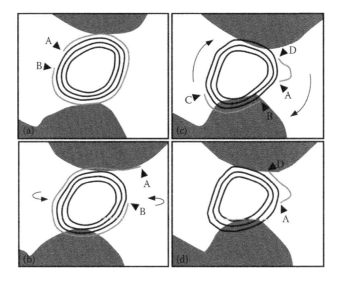

**FIGURE 15.9** Schematic of the high-resolution images of the CNC in Figure 15.8a through d. The edges of the outermost wall are indicated by arrowheads A–D. (a) The region between edges A and B corresponds to the cross section of a hole in the outermost wall layer. The CNC rotates from a to b and from b to c. Simultaneously, (b) the outermost wall turns outward. (c) A new hole with edges C and D forms and expands. (d) A part of the outermost wall remains. (Reprinted from Kizuka, T., Kato, R. & Miyazawa, K., "Surface breakdown dynamics of carbon nanocapsules," 20, 105205, 2009, *Nanotechnology*, IOP Publishing.)

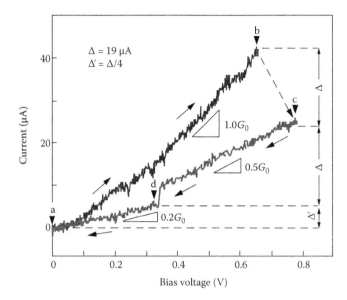

**FIGURE 15.10** Current–voltage curves for the CNC during the peeling shown in Figure 15.8. The states indicated by arrowheads a–d correspond to those observed in Figure 15.8a through d. The differential conductance corresponding to each straight part of the curves is indicated by the triangles. The differences in current between states b and c and states c and d are represented by unit Δ (= 19 μA). The current at d is Δ′ ~ Δ/4. During the voltage increase from 0.63 V at state b, one of the gold nanotips became separated because of the peeling of the outermost layer; then, the gold nanotip was again contacted with the CNC. This separation followed by rejoining is represented by the broken line from arrowhead b to arrowhead c. (Reprinted from Kizuka, T., Kato, R. & Miyazawa, K., "Surface breakdown dynamics of carbon nanocapsules," 47, 105205, 2009, *Nanotechnology*, IOP Publishing.)

## 15.3.3 Toughness

The few carbon walls that shape small hollow CNCs possibly lead us to think that CNCs are structurally weak. The walls of carbon networks joined by covalent bonds might also suggest brittleness even at minimal deformation, unlike the deformation of bulk metals. However, we know that such covalent networks contribute to high stiffness, as evident in CNTs. How do CNCs deform in response to external forces? This question can be addressed by actual mechanical testing. Compressive deformation tests have been applied to investigate the mechanical properties of carbon nanomaterials (Kizuka 1999, Asaka & Kizuka 2005, Asaka et al. 2006b,c, 2007b, Kato et al. 2006, Kizuka et al. 2008, 2012c,d, Saito et al. 2009). Compression is used because strong interfaces between carbon nanomaterials and manipulative chucks (nanotips) cannot be obtained in tensile deformation tests, which are widely used for mechanical tests. Compressive deformation tests show that CNCs are astonishingly flexible and surprisingly tough (Figure 15.11) (Asaka et al. 2006b, 2009). CNCs collapse perfectly under compression, and afterward, the hollow shape is recovered by unloading. The force to perform such compression is on the order of several nanonewtons (Figure 15.12). Because the contact area of the compression tip is several square nanometers, the stress is estimated to be on the order of a gigapascal. This stress is comparable with the ideal strength of defectless metals. CNCs can be expected to be used as nanometer-sized bearing and dispersion-strengthening elements. We also note that CNCs will not break when they are assembled in functional devices and will not be deformed by thermal stresses during current flows and impact shocks during implementation.

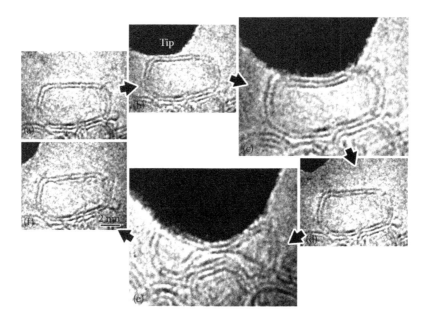

**FIGURE 15.11** Time-sequence series of high-resolution images of a repeated compression process of a double-walled CNC. The dark region at the top is a gold nanotip. (a) The CNC has an elliptical-like cross section. First, (b, c) the CNC was compressed slightly, resulting in a peanut shell-like deformation of the CNC. (d) The CNC recovered after unloading. Subsequently, (e) the CNC was compressed until it was entirely bent. Once the cantilever tip was removed, (f) the NC recovered from its compressed state. For c and e, the images are enlarged to show details. (Reprinted from Asaka, K., Miyazawa, K. & Kizuka, T., "The toughness of multi-wall carbon nanocapsules," 20, 385705, 2009, *Nanotechnology*, IOP Publishing.)

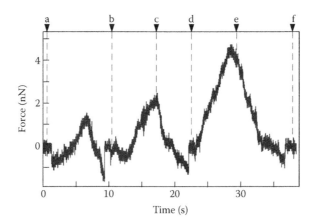

**FIGURE 15.12** Variation in force acting on the CNC during the compression and recovery process presented in Figure 15.11. The times indicated by a to f correspond to those of the high-resolution images in Figure 15.11a through f. Arrowhead a indicates the time at which the nanotip spontaneously jumped into contact with the CNC surface in its first approach to the CNC. The compression–retraction cycle with the nanotip was repeated thrice. At each nanotip–CNC separation, a negative force, i.e., an adhesion force, occurred at the times indicated by arrowheads b, c, and d. (Reprinted from Asaka, K., Miyazawa, K. & Kizuka, T., "The toughness of multi-wall carbon nanocapsules," 20, 385705, 2009, *Nanotechnology*, IOP Publishing.)

### 15.3.4 Electroluminescence

After CNCs are synthesized by applying an electric current, as described in Section 15.3.1, we obtain whisker-shaped CNC aggregates with diameters of 50–100 nm and lengths of ca. 1 μm, i.e., filaments. These CNC filaments emit light at a wavelength of 730 nm (1.7 eV in energy) with several milliwatts (Figures 15.13 and 15.14) (Asaka et al. 2006a). Although each filament is small, the light can be seen by the naked eye at a point distant from the source of 0.3 m. This electroluminescence is stable at an applied voltage of less than 4 V. Thus, the filaments could be used as light sources for illuminating small spaces, such as inside microtubes. Also, expectedly, CNC junctions may emit light, leading to nanometer-sized light-emitting elements.

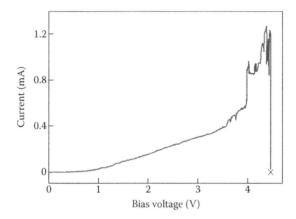

**FIGURE 15.13** Current–voltage curve of the CNC aggregation nanowhisker shown in Figure 15.3. The abrupt increase in current at the bias voltage of 4 V indicates the structural change from the amorphous state to CNC aggregation. The × indicates fracture of the contact. (Reprinted with permission from Asaka, K., Kato, R., Maezono, Y. et al., *Applied Physics Letters*, 88, 051914, 2006. Copyright 2006, American Institute of Physics.)

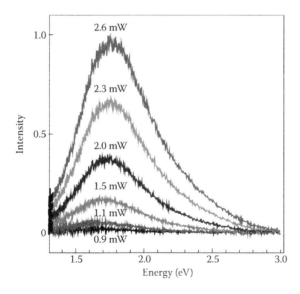

**FIGURE 15.14** Emission spectra of a CNC aggregation nanowhisker at various values of electric power. (Reprinted with permission from Asaka, K., Kato, R., Maezono, Y. et al., *Applied Physics Letters*, 88, 051914, 2006. Copyright 2006, American Institute of Physics.)

## 15.3.5 Encapsulation of Metallic and Carbide Nanoparticles in CNCs

By incorporating pure metals, oxides, and carbides into carbon sources, CNCs can encapsulate various nanoparticles, e.g., iron, cementite, nickel, cobalt and its carbide, vanadium carbide, and niobium carbide (Figures 15.15 and 15.16) (Kizuka et al. 2009a, 2012a,b, Matsuura et al. 2012, Kizuka & Akagawa 2014, Kizuka & Koizumi 2014, Matsuura & Kizuka 2014, Yazaki et al. 2014). Such encapsulation produces various additional effects and functions for both the carbon shells and the inner nanoparticles, such as strengthening, coating, and sustained release of drugs. As described in Section 15.3.3, although the toughness of hollow CNCs is high, the encapsulation of harder nanoparticles, e.g., tungsten carbide and zirconium carbide, further increases toughness. By encapsulating magnetic nanoparticles, aggregation of the nanoparticles can be inhibited, leading to applications in magnetic recording media.

Encapsulation forms interfaces between the graphitic layers and the encapsulated nanoparticles. Because the graphitic layers surround the nanoparticles, interfaces with various orientational relations can be obtained simultaneously (Figure 15.17) (Kizuka & Koizumi 2014; Matsuura & Kizuka 2014, Yazaki et al. 2014). Such observations enable the characterization of various types of interfaces between graphitic layers and different materials. This is useful for the study and development of future applications of graphene interfaces.

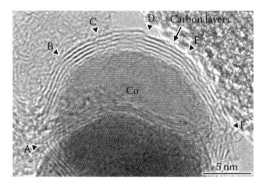

**FIGURE 15.15** High-resolution image of a cobalt-encapsulated CNC. The cobalt nanoparticle shows a face-centered-cubic structure, which is typical for encapsulated materials. (Reprinted from Matsuura, D. & Kizuka, T., *Journal of Nanotechnology*, 2012, 843516, 2012.)

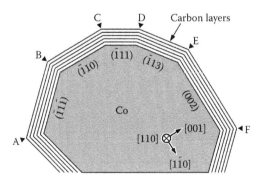

**FIGURE 15.16** Facets of the cobalt-encapsulated CNC shown in Figure 15.15. Interfaces of the graphitic layer and cobalt of various orientational relations can be observed in the CNC. (Reprinted from Matsuura, D. & Kizuka, T., *Journal of Nanotechnology*, 2012, 843516, 2012.)

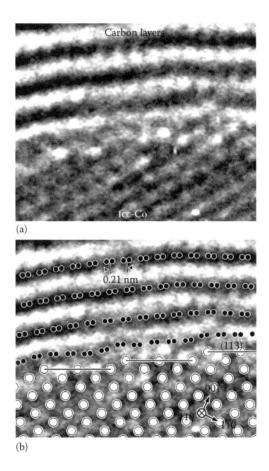

**FIGURE 15.17** Analysis of an interface between the graphitic layer and cobalt. (a) High-resolution image of $(0001)_{\text{graphitic layer}}/(113)_{\text{fcc-Co}}$ interface in the D and E regions of the cobalt-encapsulated CNC shown in Figure 15.15. (b) The atomic configuration overlapping on the image in a. The graphitic layers are observed along $[1\,1\,20]_{\text{graphitic layer}}$. (Reprinted from Matsuura, D. & Kizuka, T., *Journal of Nanotechnology*, 2012, 843516, 2012.)

## 15.4 Conclusions

CNCs can be synthesized in large amounts with easy control of size and structure. They exhibit intriguing functions that originate from their hollow single-walled and multiwalled cage structures. In addition to the low costs of source materials, their lightweight properties, high electrical conduction at quantization unit levels, superior stiffness and toughness, and luminescence characteristics can be applied to a variety of advanced functional and structural nanodevices.

## Acknowledgments

The author sincerely appreciates Professor Koji Asaka and Dr. Kun'ichi Miyazawa for their cooperation with the studies cited in the references. This study was partly supported by Grants-in-Aid from the Ministry of Education, Culture, Sports, Science and Technology, Japan (Nos. 22310065 and 26630032), Cross-Ministerial Strategic Innovation Promotion Program (Materials Integration, Innovative Measurement and Analysis for Structural Materials), and NSK Foundation for the Advancement of Mechatronics.

# References

Andersson, O. E., Prasad, B. L. V., Sato, H. et al., "Structure and electronic properties of graphite nanoparticles," *Phys. Rev. B* 58 (1998): 16387–16395.

Ariga, K., Vinu, A., Miyahara, M. et al., "One-pot separation of tea components through selective adsorption on pore-engineered nanocarbon, carbon nanocage," *J. Am. Chem. Soc.* 129 (2007): 11022–11023.

Asaka, K. & Kizuka, T., "Atomistic dynamics of deformation, fracture, and joining of individual single-walled carbon nanotubes," *Phys. Rev. B* 72 (2005): 115431.

Asaka, K., Kato, R., Maezono, Y. et al., "Light-emitting filaments composed of nanometer-sized carbon hollow capsules," *Appl. Phys. Lett.* 88 (2006a): 051914.

Asaka, K., Kato, R., Miyazawa, K. et al., "Buckling of $C_{60}$ whiskers," *Appl. Phys. Lett.* 89 (2006b): 071912–071913.

Asaka, K., Kato, R., Miyazawa, K. et al., "Deformation of multiwalled nanometer-sized carbon capsules," *Appl. Phys. Lett.* 89 (2006c): 191914.

Asaka, K., Kato, R., Yoshizaki, R. et al., "Conductance of carbon nanocapsule junctions," *Phys. Rev. B* 76 (2007a): 113404.

Asaka, K., Kato, R., Yoshizaki, R. et al., "Fracture surface and correlation of buckling force with aspect ratio of $C_{60}$ crystalline whiskers," *Diam. Relat. Mater.* 16 (2007b): 1936–1939.

Asaka, K., Miyazawa, K. & Kizuka, T., "The toughness of multi-wall carbon nanocapsules," *Nanotechnology* 20 (2009): 385705.

Banhart, F., Redlich, P. & Ajayan, P. M., "Irradiation effects in carbon nanostructures," *Chem. Phys. Lett.* 292 (1998): 554–560.

Bethune, D. S., Klang, C. H., de Vries, M. S. et al., "Cobalt-catalysed growth of carbon nanotubes with single-atomic-layer walls," *Nature* 363 (1993): 605–607.

Buongiorno, N. M., Brabec, C., Maiti, A. et al., "Lip–lip interactions and the growth of multiwalled carbon nanotubes," *Phys. Rev. Lett.* 80 (1998): 313–316.

Chai, Y., Guo, T., Jin, C. et al., "Fullerenes with metals inside," *J. Phys. Chem.* 95 (1991): 7564–7568.

de Heer, W. A. & Ugarte, D., "Carbon onions produced by heat treatment of carbon soot and their relation to the 217.5 nm interstellar absorption feature," *Chem. Phys. Lett.* 207 (1993): 480–486.

Heidenreich, R. D., Hess, W. M. & Ban, L. L., "A test object and criteria for high resolution electron microscopy," *J. Appl. Crystallogr.* 1 (1968): 1–19.

Iijima, S., "Direct observation of the tetrahedral bonding in graphitized carbon black by high resolution electron microscopy," *J. Cryst. Growth* 50 (1980): 675–683.

Iijima, S., "Helical microtubules of graphitic carbon," *Nature* 354 (1991): 56–58.

Iijima, S. & Ichihashi, T., "Single-shell carbon nanotubes of 1-nm diameter," *Nature* 363 (1993): 603–605.

Kato, R., Asaka, K., Miyazawa, K. et al., "In situ high-resolution transmission electron microscopy of elastic deformation and fracture of nanometer-sized fullerene $C_{60}$ whiskers," *Jpn. J. Appl. Phys.* 45 (2006): 8024–8026.

Kizuka, T., "Atomic process of point contacts in gold studied by time-resolved high-resolution transmission electron microscopy," *Phys. Rev. Lett.* 81 (1998a): 4448–44451.

Kizuka, T., "Atomistic visualization of deformation in gold," *Phys. Rev. B* 57 (1998b): 11158–11163.

Kizuka, T., "Direct atomistic observation of deformation in multiwalled carbon nanotubes," *Phys. Rev. B* 59 (1999): 4646–4649.

Kizuka, T., "Atomic configuration and mechanical and electrical properties of stable gold wires of single-atom width," *Phys. Rev. B* 77 (2008): 155401–155411.

Kizuka, T., "Fabrication of silicon oxide nanotips by mechanical contact and elongation methods," *J. Nanosci. Nanotechnol.* 11 (2011a): 1273–1277.

Kizuka, T., "Position-selective emission control of cathodoluminescence using nanotips of optical fibers," *J. Nanosci. Nanotechnol.* 11 (2011b): 5274–5276.

Kizuka, T. & Akagawa, A., "The structure of graphene/nickel interfaces within nickel-encapsulating carbon nanocapsules studied by high-resolution transmission electron microscopy," *J. Nanosci. Nanotechnol.* 2014 (2014): 3176–3180.

Kizuka, T. & Koizumi, H., "Interface structure of niobium carbide-encapsulating carbon nanocapsules studied by high-resolution transmission electron microscopy," *J. Nanosci. Nanotechnol.* 2014 (2014): 3228–3232.

Kizuka, T. & Monna, K., "Atomic configuration, conductance, and tensile force of platinum wires of single-atom width," *Phys. Rev. B* 80 (2009): 205406.

Kizuka, T. & Oyama, M., "Individual cathodoluminescence and photoluminescence spectroscopy of zinc oxide nanoparticles in combination with in situ transmission electron microscopy," *J. Nanosci. Nanotechnol.* 11 (2011): 3278–3283.

Kizuka, T. & Tanaka, N., "Dynamic high-resolution electron microscopy of diffusion bonding between zinc oxide nanocrystallites at ambient temperature," *Philos. Mag. Lett.* 69 (1994): 135–139.

Kizuka, T., Yamada, K., Deguchi, S. et al., "Cross-sectional time-resolved high-resolution transmission electron microscopy of atomic-scale contact and noncontact-type scannings on gold surfaces," *Phys. Rev. B* 55 (1997): R7398–R7401.

Kizuka, T., Saito, K. & Miyazawa, K., "Young's modulus of crystalline $C_{60}$ nanotubes studied by in situ transmission electron microscopy," *Diam. Relat. Mater.* 17 (2008): 972–974.

Kizuka, T., Fujii, J. & Miyazawa, K., "Ion catalyzed synthesis of hollow carbon nanocapsules synthesized by resistive heating," *Diam. Relat. Mater.* 18 (2009a): 1253–1257.

Kizuka, T., Kato, R. & Miyazawa, K., "Surface breakdown dynamics of carbon nanocapsules," *Nanotechnology* 20 (2009b): 105205.

Kizuka, T., Kato, R. & Miyazawa, K., "Structure of hollow carbon nanocapsules synthesized by resistive heating," *Carbon* 47 (2009c): 138–144.

Kizuka, T., Miyazawa, K. & Akagawa, A., "Synthesis of nickel-encapsulated carbon nanocapsules and cup-stacked-type carbon nanotubes via nickel-doped fullerene nanowhiskers," *J. Nanotechnol.* 2012 (2012a): 376160.

Kizuka, T., Miyazawa, K. & Akagawa, A., "Synthesis of carbon nanocapsules and nanotubes using Fe-doped fullerene nanowhiskers," *J. Nanotechnol.* 2012 (2012b): 613746.

Kizuka, T., Miyazawa, K. & Tokumine, T., "Solvation-assisted Young's modulus control of single-crystal fullerene $C_{70}$ nanowhiskers," *J. Nanotechnol.* 2012 (2012c): 583817.

Kizuka, T., Miyazawa, K. & Tokumine, T., "Young's modulus of single-crystal fullerene $C_{70}$ nanotubes," *J. Nanotechnol.* 2012 (2012d): 969357.

Krishnan, A., Dujardin, E., Treacy, M. M. J. et al., "Graphitic cones and the nucleation of curved carbon surfaces," *Nature* 388 (1997): 451–454.

Kroto, H. W., Heath, J. R., O'Brien, S. C. et al., "$C_{60}$: Buckminsterfullerene," *Nature* 318 (1985): 162–163.

Li, Z., Jaroniec, M., Papakonstantinou, P. et al., "Supercritical fluid growth of porous carbon nanocages," *Chem. Mater.* 19 (2007): 3349–3354.

Matsuura, D. & Kizuka, T., "Structures of graphene/cobalt interfaces in cobalt-encapsulated carbon nano-capsules," *J. Nanotechnol.* 2012 (2012): 843516.

Matsuura, D. & Kizuka, T., "Electrical conductivity of single molecular junctions assembled from Co- and $Co_3C$-encapsulating carbon nanocapsules," *J. Nanosci. Nanotechnol.* 2014 (2014): 2441–2445.

Matsuura, D., Miyazawa, K. & Kizuka, T., "Synthesis of cobalt-encapsulated carbon nanocapsules using cobalt-doped fullerene nanowhiskers," *ISRN Nanotechnol.* 2012 (2012): 871208.

Minato, J. & Miyazawa, K., "Solvated structure of $C_{60}$ nanowhiskers," *Carbon* 43 (2005): 2837–2841.

Miyazawa, K., "$C_{70}$ nanowhiskers fabricated by forming liquid/liquid interfaces in the systems of toluene solution of $C_{70}$ and isopropyl alcohol," *J. Am. Ceram. Soc.* 85 (2002): 1297–1299.

Miyazawa, K., Obayashi, A. & Kuwabara, M., "$C_{60}$ nanowhiskers in a mixture of lead zirconate titanate sol–$C_{60}$ toluene solution," *J. Am. Ceram. Soc.* 84 (2001): 3037–3039.

Miyazawa, K., Hamamoto, K., Nagata, S. et al., "Structural investigation of the $C_{60}/C_{70}$ whiskers fabricated by forming liquid–liquid interfaces of toluene with dissolved $C_{60}/C_{70}$ and isopropyl alcohol," *J. Mater. Res.* 18 (2003): 1096–1103.

Mordkovich, V. Z., Umnov, A. G., Inoshita, T. et al., "The observation of multiwall fullerenes in thermally treated laser pyrolysis of carbon blacks," *Carbon* 37 (1999): 1855–1858.

Ruoff, R. S., Lorents, D. C., Chan, B. et al., "Single crystal metals encapsulated in carbon nanoparticles," *Science* 259 (1993): 346–348.

Saito, Y., "Nanoparticles and filled nanocapsules," *Carbon* 33 (1995): 979–988.

Saito, Y. & Matsumoto, T., "Carbon nano-cages created as cubes," *Nature* 392 (1998): 237.

Saito, Y., Yoshikawa, T., Inagaki, M. et al., "Growth and structure graphitic tubules and polyhedral particles in arc-discharge," *Chem. Phys. Lett.* 204 (1993): 277–282.

Saito, Y., Nishikubo, K., Kawabata, K. et al., "Carbon nanocapsules and single-layered nanotubes produced with platinum-group metals (Ru, Rh, Pd, Os, Ir, Pt) by arc discharge," *J. Appl. Phys.* 80 (1996): 3062–3067.

Saito, Y., Masumoto, T. & Nishikubo, K., "Encapsulation of TiC and HFC crystallites within graphite cages by arc discharge," *Carbon* 35 (1997): 1757–1763.

Saito, K., Miyazawa, K. & Kizuka, T., "Bending process and Young's modulus of fullerene $C_{60}$ nanowhiskers," *Jpn. J. Appl. Phys.* 48 (2009): 010217.

Sugai, T., Yoshida, H., Shimada, T. et al., "New synthesis of high-quality double-walled carbon nanotubes by high-temperature pulsed arc discharge," *Nano Lett.* 3 (2003): 769–773.

Tomita, M., Saito, Y. & Hayashi, T., "$LaC_2$ encapsuled in graphite nano-particle," *Jpn. J. Appl. Phys.* 32 (1993): L280–L282.

Ugarte, D., "Curling and closure of graphitic networks under electron-beam irradiation," *Nature* 359 (1992): 707–709.

Vinu, A., Miyahara, M., Sivamurugan, V. et al., "Large pore cage type mesoporous carbon, carbon nanocage: A superior adsorbent for biomaterials," *J. Mater. Chem.* 15 (2005): 5122–5127.

Wang, J. N., Zhang, L., Niu, J. J. et al., "Synthesis of high surface area, water-dispersible graphitic carbon nanocages by an in situ template approach," *Chem. Mater.* 19 (2007): 453–459.

Yazaki, G., Matsuura, D. & Kizuka, T., "The atomic configuration of graphene/VC interfaces in VC-encapsulating carbon nanocapsules," *J. Nanosci. Nanotechnol.* 2014 (2014): 2482–2486.

Yu, J.-S., Yoon, S. B., Lee, Y. J. et al., "Fabrication of bimodal porous silicate with silicalite-1 core/mesoporous shell structures and synthesis of nonspherical carbon and silica nanocases with hollow core/mesoporous shell structures," *J. Phys. Chem. B* 109 (2005): 7040–7045.

# 16

# Hollow Fluorescent Carbon Nanoparticles

Somen Mondal

Pradipta
Purkayastha

## 16.1 Introduction

During the last few decades, a large number of carbon materials, such as carbon nanotubes, fullerenes, graphene, graphene oxide, and graphene quantum dots (QDs), have drawn great attention due to their unexpected physicochemical properties (Jariwala et al. 2013). Besides these, carbon nanomaterials, namely, carbon nanoparticles, which are regularly called carbon nanodots (CNDs), attract considerable scientific attention due to their special properties such as bright photoluminescence (PL), high aqueous solubility, easy functionalization, low toxicity, high photostability, and chemical inertness (Bottini & Mustelin 2007, Baker & Baker 2010, Li et al. 2012, Demchenko & Dekaliuk 2013). In search for a nontoxic alternative to highly fluorescent QD-like nanomaterials, it is recently found that CNDs could be surface-passivated by organic functional groups or biomolecules to induce strong fluorescence in the visible and near-infrared spectral regions (Sun et al. 2006). Although the multifunctional properties of CNDs and their nontoxicity are established, the origin of fluorescence in them is still unclear. A number of research groups have focused on understanding this phenomenon. Recently, Wang et al. (2014) reported that molecule-like states that contain carboxyl groups and carbonyl groups are parts of the so-called edge states in CNDs and such graphene QDs were accounted for their green fluorescence. On a similar aspect, Strauss et al. (2014) reported from theoretical and experimental analyses that the intrinsic luminescence of CNDs could be from $sp^2$-carbon networks. This seems to be a reasonable explanation for the emission at short wavelength, but the other deactivation pathways predominate for luminescence at longer wavelength. Ghosh et al. (2014) found that photoluminescent CNDs behave as electric dipoles, during both absorption and emission of light, and that their emission originates from the recombination of photogenerated charges on defect centers involving strong coupling between the electronic transition and the collective vibrations of the lattice structure. They also showed that each CND contains only one optically active emission center and the structural heterogeneity does not lead to the same heterogeneous behavior in the PL study.

# 16.2 Synthetic Methods of Hollow Fluorescent Carbon Nanoparticles

With their diverse and uniform porosity, porous and hollow nanoparticles attracted huge attention due to their excellent applications in optoelectronics, drug delivery, cell imaging, and photocatalysis (Fang et al. 2012, Choi et al. 2013, Wang et al. 2013, Gong et al. 2014, Zhang et al. 2014). In this ocean of promising potency, hollow carbon nanomaterials with multifunctional capabilities can provide synergistic effects in various fields. Carbon nanomaterial, a superior fluorescent bioimaging agent, possesses low toxicity, stability, and resistance to photobleaching (Fang et al. 2012). Hollow fluorescent carbon nanoparticles (HFCNs) are promising as one of the best biocompatible materials compared with traditional dyes and CdTe QDs.

HFCNs can serve as small containers for applications in catalysis and controlled drug release. They can also be used as a host for various molecular guests, such as small organic molecules, drugs, and genes. Until now, the synthesis of HFCNs has been a challenge and only a limited number of reports appeared in literature. Recently, in a review article, Shen and Fan (2013) described different methods of preparation of porous carbon particles. In another article, a self-assembly approach using hexachlorobenzene and sodium (Na) has been adopted to obtain 50–100 nm hollow nanocarbons (HCNs) (Hu et al. 2002). Through homogenenous contact of the reactant, uniform ball-like carbon particles were synthesized by self-assembly. According to Hu et al., the NaCl generated during the reaction works as a template for the formation of these carbon spheres. Another template-based synthesis of hollow carbon particles was reported by Geng et al. (2007). Carbon-encapsulated ZnSe nanoparticles were synthesized by noncatalytic one-step thermal evaporation method. Porous carbon particles with a shell/core structure were successfully prepared by controlled precipitation from droplets of oil-in-water emulsion, followed by curing and carbonization (Jiang et al. 2007). A synthetic route for HCNs without introducing a template under hydrothermal conditions was reported by Li et al. (2009). Microspheres (100 nm) of HCNs were synthesized using alginate as reagent. In a recent report, Wang et al. (2013) showed a new method to synthesize bright photoluminescent HFCNs from bovine serum albumin (BSA) by solvothermal treatment under mild temperature. BSA was dissolved in 1:1 water–ethanol mixture and sonicated to yield a homogeneous solution, which was heated at 180°C and cooled to yield HFCNs. In this solvothermal reaction, BSA changed its size and degree of carbonization with time. Wang et al. obtained 6.8 nm particles with a pore size of ca. 2 nm. Preparation of green-emitting HFCNs was reported recently by Fang et al. (2012). They reported a carbonization process with self-heating autocatalysis. In this carbonization process all the reagents (acetone, water, and phosphorus pentoxide) played important roles in the self-catalytic reaction to produce the HFCNs. Acetone was particularly selected as the carbon precursor since it acts as an oxygenous carbon resource. These properties induced oxygenous defects and good solubility for the HFCNs. In this self-catalytic reaction, water liberated heat from the system and made the reaction highly exothermic. On the other hand, the dehydrating agent $P_2O_5$ accelerated the carbonization reaction. Thus, water reacted quickly with $P_2O_5$ upon addition of the acetone–water mixture that released heat, promoting the carbonization reaction. $P_2O_5$ reacted with water to produce a highly sticky liquid, polyphosphoric acid (PPA), which captured the hot gas of acetone to suppress the heat loss. Here, both $P_2O_5$ and PPA acted as catalysts. Finally, this self-promoted and self-controlled carbonization reaction led to an "automatic" synthesis of cross-linked HFCNs. In accordance with the synthetic method, they have also discussed the mechanism of the formation of HFCNs and the reactivity of each starting material in the procedure. Wang et al. (2015) synthesized HCNs in noncoordinating solvents in a one-step method. They heated a mixture of a noncoordinating solvent, octadecene, and a capping agent, oleylamine, under argon atmosphere followed by the addition of an aqueous solution of mannose into the hot solution mixture and allowed to react under vigorous stirring. Purification followed by filtration and dialysis against isopropanol yields HCNs.

## 16.3 Characterization of HFCNs

Synthesized HFCNs are generally physically characterized by TEM, atomic force microscopy (AFM), field emission scanning electron microscopy, FTIR, dynamic light scattering (DLS), etc. The hollow core and the vesicular nature of the HFCNs are clearly visible in the AFM image shown in Figure 16.1a and b. A DLS experiment shows the average particle size of the HFCNs (Figure 16.1c) and also demonstrates their monodispersity. Fang et al. (2012) characterized their synthesized HCNs by HRTEM, FTIR, and Raman spectroscopy. HRTEM image clearly shows the void nature of the HCNs and their thin shell. Generally, the shell thickness of HCNs is ~5 nm. The distance between the lattice planes of the HCNs is about 0.33 nm, which is in agreement with the basal spacing of graphite. The surface functional groups of HCNs can be expressed by FTIR spectra.

The negative zeta potential of HFCNs corroborates the presence of residual carboxylic groups on their surface (Mondal et al. 2013). Raman spectra are used to correlate the structural properties of the carbon material. The spectra show three characteristic bands, G, D, and 2D, at around 1586, 1353, and 2700 cm$^{-1}$, respectively. The intensity ratio of the D and G band ($I_D/I_G$) is a measurement of the extent of disorderness and the ratio of $sp^3/sp^2$ carbon. The G and D bands arise due to the vibration of $sp^2$-bonded carbon atoms in a 2D hexagonal lattice and carbon atoms with dangling bonds on the termination plane of disordered graphite, respectively (Fang et al. 2012, Wang et al. 2013, 2015). The intensity ratio ($I_D/I_G$),

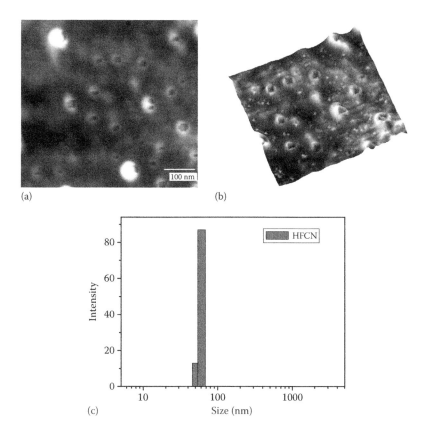

(a)    (b)

(c)

**FIGURE 16.1** (a) 2D and (b) 3D AFM images of HFCNs. (c) Average size distribution of HFCNs in water obtained by DLS analysis. (Reprinted with permission from Mondal et al. 2013, 4260–4267. Copyright 2013 American Chemical Society.)

which was calculated to be larger than 2, reveals that the HFCNs are mainly composed of nanocrystalline graphite. The XRD pattern shows partial graphite-like structure of the HFCNs.

## 16.4 Optical Properties of HFCNs

The relatively large fraction of void space in the hollow structures of HFCNs has been exploited to encapsulate and control the release of drugs, subnanoparticles, and DNA. Moreover, these well-defined voids in the hollow particles modulate their optical properties; the increased active area modulates catalysis as well. HFCNs are practically nontoxic to biological systems and are promising candidates for applications in biological environments. HFCNs can be characterized by UV–visible spectroscopy. The absorption spectrum shows two bands at 245 and 297 nm (Figure 16.2a). The band at 245 nm originates from the formation of multiple polyaromatic chromophores, while the broad peak at 297 nm can be attributed to the $n$–$\pi^*$ transitions in C=O on the HFCNs (Ray et al. 2009, Fang et al. 2012). A carbon source with carboxylic group, such as acetone, introduces oxygenous defects in the HFCNs that impose fluorescence

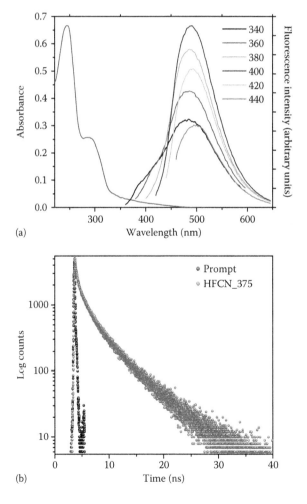

FIGURE 16.2 (a) Absorption spectrum and steady-state fluorescent spectra and (b) time-resolved fluorescence decay of HFCN ($\lambda_{ex}$ = 377 nm; $\lambda_{em}$ = 490 nm). (Reprinted with permission from Mondal et al. 2013, 4260–4267. Copyright 2013 American Chemical Society.)

property. Furthermore, the polyaromatic structure may serve as a fluorescent chromophore that makes HFCN fluoresce. Figure 16.2a shows the emission spectra of HFCNs that arise from excitations at different wavelengths from 340 to 420 nm. The emission maximum has practically no shift, as the excitation wavelength is varied, indicating uniform distribution of size for the HFCNs. The characteristic luminescence decay for the HFCNs is found to be biexponential (Figure 16.2b) with decay times of 2.42 ns (81%) and 360 ps (18%). The reported quantum yield of HFCNs is about 4% in hexane using quinine sulfate in 0.1 M $H_2SO_4$ as the reference (Wang et al. 2015).

## 16.5 Micropolarity of the Interior of HFCNs

The exploration about the prospects of the HFCNs as drug carriers must be preceded by understanding their internal environment. This was tactically performed by Mondal et al. (2013) by selecting specific characteristic small molecules as fluorescence reporters. One such fluorophore is pyrene (Py), a well-known hydrophobic compound and a probe that is extremely sensitive to polarity of the medium (Anthony & Zana 1994, Itoh et al. 1996). Py typically forms excimer in a polar environment and remains as a monomer in lower polarity (Castanheira & Martinho 1991, Ghosh et al. 2012). Anticipating that the interior of the HFCNs can be of lower polarity compared to the bulk aqueous medium, one can measure the parameters of the microenvironment inside the HFCNs by using satellite probes that favor hydrophobic milieu. The absorbance of Py decreases (Figure 16.3a) and the fluorescence emissions from both the Py monomer and the Py excimer get remarkably quenched with an increase in the concentration of HFCNs in the solution (Figure 16.3b). This does not happen for solid fluorescent carbon nanoparticles (SFCNs) (Figure 16.3c). As mentioned earlier that Py tends to exist as a monomer in lower polarity, the conversion of the Py excimer to monomer can be expected if the molecules enter the HFCN cavity. The vibronic structures (peaks I and III) of Py emission are extremely sensitive to the polarity of the environment, and the peak ratio is routinely used to determine the micropolarity of the medium. The peak ratio can be qualitatively taken as a measure of the extent of interaction between the Py system and the solvents of different polarities. Traditionally, a value above 1.5 indicates that Py is experiencing a hydrophilic environment, whereas a value of 0.6 suggests a hydrophobic milieu (Kalyanasundaram & Thomas 1977, Winnik 1993, Kaushlendra & Asha 2012). From the change in the fluorescence spectrum of Py due to the interaction with HFCN, Mondal et al. (2013) found the ratio between the intensities of peak I ($I_1$: ~371 nm) and peak III ($I_3$: ~382 nm) to be 1.18, which indicates ethanol-like polarity of the interior of the HFCNs (Dong & Winnik 1984).

The formation of Py excimer decreases more in the reduced polarity of the interior of the HFCNs. As a result, the $\pi$-electron-rich Py monomers may get docked to the interior $\pi$-electron-rich wall of the HFCN through $\pi$–stacking interaction. This can be verified by steady-state anisotropy measurements monitoring the Py monomer and the excimer emissions with the increase in HFCN concentration. The anisotropy ($r$) increases considerably for the Py monomer and insignificantly for the excimer. Because of the stacking interaction, the internal polarity for HFCN provided by pyrene monomers may not be accurate. This phenomenon, in turn, may lead to the quenching of the fluorescence of the Py monomers due to energy transfer as reflected in Figure 16.3b. In a recent study, Matte et al. (2011) showed that graphene can quench the fluorescence of organic donor molecules, such as pyrenebutanoic acid succinimidyl ester, through photoinduced electron transfer. The fluorescence decay of Py fits a three-component decay (Table 16.1). Photoinduced electron transfer results in a significant decrease in all the three components. The drawbacks in determining the polarity of the interior of the HFCNs by Py can be overcome by using fluorescent probes as oxazine 725 (ox-725). Ox-725 is a cationic dye and, hence, gets easily attracted by the negative surface charge of the HFCNs. The absorption spectral changes in the case of ox-725 are similar to those observed for Py. The absorbance decreased upon addition of HFCNs to the solution (Figure 16.4a) and the fluorescence gets quenched (Figure 16.4b), indicating a change in the environment of the dye molecules. SFCNs cannot change the fluorescence of ox-725 (Figure 16.4c).

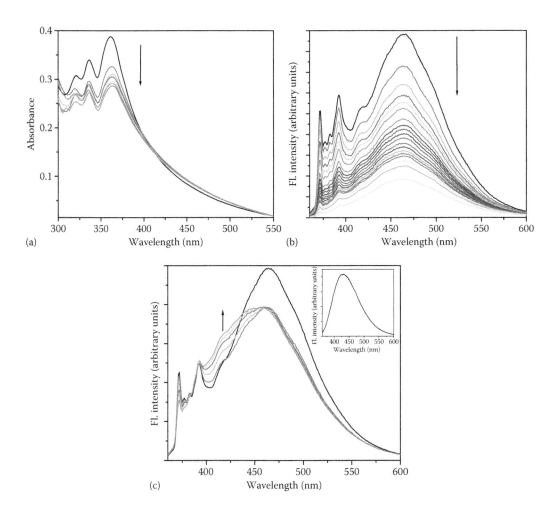

**FIGURE 16.3** (a) Absorption spectra and (b) steady-state fluorescent spectra of Py with increase in HFCN concentration. (c) Fluorescence emission of Py in the presence of gradually enhanced concentration of SFCN ($\lambda_{ex}$ = 340 nm). Inset of c shows the emission spectrum of SFCN in water; the fluorescence spectrum of Py in water is indicated in black. (Reprinted with permission from Mondal et al. 2013, 4260–4267. Copyright 2013 American Chemical Society.)

**TABLE 16.1** Time-Resolved Fluorescence Decay of Py Monomer in Absence and Presence of HFCN

| | $\tau_1$ (ns) | % Contribution | $\tau_2$ (ns) | % Contribution | $\tau_3$ (ps) | % Contribution | $\chi^2$ |
|---|---|---|---|---|---|---|---|
| Pyrene | 3.7 | 20 | 7.11 | 80 | – | – | 1.16 |
| Pyrene with periodic addition of HFCN | 3.7 | 16 | 36.3 | 65 | 50.7 | 19 | 1.01 |
| | 2.1 | 15 | 21.7 | 66 | 75.4 | 19 | 1.09 |
| | 1.8 | 17 | 20.2 | 60 | 94.7 | 23 | 1.18 |
| | 1.8 | 21 | 18.9 | 53 | 110 | 26 | 1.24 |

*Source:* Mondal et al. 2013, 4260–4267. Copyright 2013 American Chemical Society.
*Note:* The sample was excited at 340 nm and monitored at 390 nm.

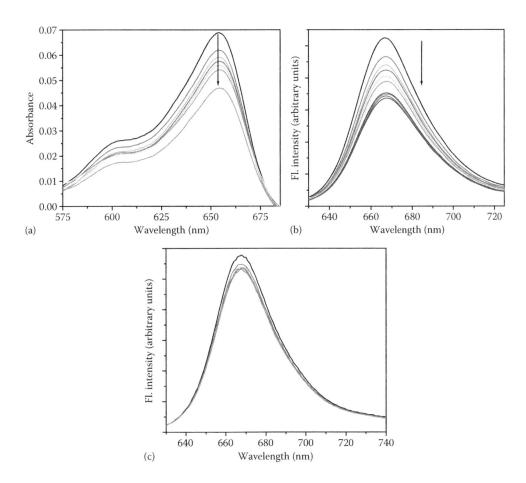

**FIGURE 16.4** (a) Absorption spectra and (b) steady-state fluorescent spectra of ox-725 with increase in HFCN concentration. (c) Fluorescence emission of ox-725 in the presence of gradually enhanced concentration of SFCN ($\lambda_{ex}$ = 630 nm).

Because of the positive charge on the dye molecule, the negatively charged HFCNs attract them, reinforcing the ox-725 molecules to approach the HFCNs and enter the interior. However, the emission maximum of ox-725 does not show any shift due to the polarity of the medium. Thus, to verify the encapsulation of the dye molecules, a fluorescence quencher (iodide, I⁻) may be introduced (Figure 16.5). The Stern–Volmer quenching constants obtained for ox-725 in the absence and presence of the HFCNs are 15.3 and 7.8 M⁻¹, respectively, which confirms the entrapment of ox-725 inside the HFCNs.

The entrapment of ox-725 inside the HFCNs can also be proved by time-resolved fluorescence decay measurements (Table 16.2). The probe shows a single-component decay profile in bulk water environment, whereas in the presence of HFCNs, a second faster component evolves out. The latter picosecond component is evidently due to the HFCN-ox-725 composite. Thus, ox-725 could be suitably used to precisely determine the internal polarity of the HFCNs. The plot of fluorescence energies ($E_F$) against $E_T(30)$ yields the internal polarity of HFCN to be ~43.02, showing that the HFCN interior is even less polar than ethanol/methanol (Figure 16.5) (Deng ct al. 2012).

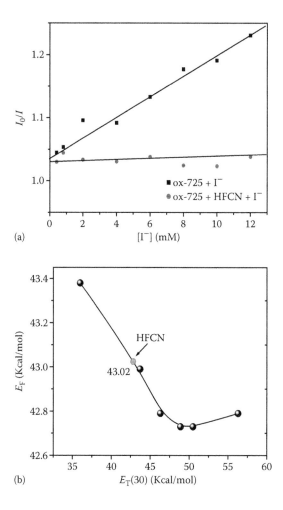

(a)

(b)

**FIGURE 16.5** (a) Plot of relative fluorescence intensity of ox-725 monomer (monitored at 670 nm) against concentration of I⁻. $I_0$ indicates the fluorescence intensity without the quencher ion, and $I$ signifies the intensities at different concentrations of I⁻. (b) Plot of fluorescence energy ($E_F$) obtained from the maxima of the emissions from ox-725 dissolved in dioxane–water mixtures with different ratios of the two solvents against $E_T(30)$ values of the solvent mixtures. The gray dot in the plot shows the $E_T(30)$ value of the interior of the HFCNs.

**TABLE 16.2** Time-Resolved Fluorescence Decay Data for ox-725 in the Absence and Presence of HFCN

| | $\tau_1$ (ps) | % Contribution | $\tau_2$ (ps) | % Contribution | $\chi^2$ |
|---|---|---|---|---|---|
| Ox-725 | 496 | 100 | – | – | 1.18 |
| Ox-725 with periodic addition of HFCN | 446 | 77.0 | 667 | 22.3 | 0.98 |
| | 433 | 75.3 | 716 | 24.6 | 0.98 |
| | 436 | 74.6 | 747 | 25.3 | 1.02 |
| | 440 | 75.3 | 823 | 24.6 | 1.01 |

*Note:* $\chi^2$ signifies the appropriateness of the fit to the raw data. The sample was excited with a 635 nm radiation and monitored at 670 nm.

## 16.6 Possible Applications of HFCNs

### 16.6.1 Drug Delivery

Wang et al. (2013) strategically used HFCNs as a multifunctional vehicle for drug loading and releasing because of their hollow structure and excellent PL. An anticancer drug, doxorubicin (DOX), was chosen as a model for testing HFCNs as a drug delivery system. DOX undergoes π–π stacking through hydrophobic and van der Waals interactions with the HFCNs. DOX loading was confirmed by the change in surface zeta potential from –16.7 to –8.0 mV. About 6% DOX was loaded in HFCNs. Wang et al. also showed the release of DOX by changing the pH of the medium from pH = 7.4 to pH = 5. The drug release study shows that 4% of DOX was released at pH 7.4, while ca. 70% was released at pH 5.0.

### 16.6.2 Watermark Ink and Fluorescent Powder

Fang et al. (2012) applied the HFCNs as watermark ink and fluorescent powder. HFCNs were added to commercial silica gel to produce fluorescent powder of HFCNs. The nonfluorescent silica powder changes into a fluorescent-response powder just after the adsorption of the HFCNs. This fluorescent powder was reported to be stable for more than 3 months.

### 16.6.3 Bioimaging

Low toxicity, stability, and resistance to photobleaching make HFCNs a superior fluorescent marker for living cells. HFCNs are more suitable dye than other traditional dyes and CdTe QDs for long-term cellular imaging. Fang et al. (2012) studied time-dependent fluorescence signals of HEK 293 cells imaged by HFCNs, CdTe QDs, fluorescein isothiocyanate (FITC), and Hoechst. HFCNs-labeled cells are easily noticeable even after 25 min, whereas the cells marked by CdTe QDs, FITC, and Hoechst almost disappear. Choi et al. (2013) used better-quality HFCN fractions (separated by high performance liquid chromatography [HPLC]) to measure the toxicity in MCF-7 cells and as an imaging agent for living cells.

## Acknowledgments

The work was supported by the Department of Science and Technology, Government of India, through Project Number SR/S1/PC-35/2011. SM acknowledges University Grants Commission, Government of India, New Delhi, for his fellowship.

## References

Anthony, O. & Zana, R., "Fluorescence investigation of the binding of pyrene to hydrophobic microdomains in aqueous solutions of polysoaps," *Macromolecules* 27 (1994): 3885–3891.

Baker, S. N. & Baker, G. A., "Luminescent carbon nanodots: Emergent nanolights," *Angew. Chem. Int. Ed.* 49 (2010): 6726–6744.

Bottini, M. & Mustelin, T., "Carbon materials: Nanosynthesis by candlelight," *Nat. Nano* 2 (2007): 599–600.

Castanheira, E. M. S. & Martinho, J. M. G., "Solvatochromic shifts of pyrene excimer fluorescence," *Chem. Phys. Lett.* 185 (1991): 319–323.

Choi, H., Ko, S. J., Choi, Y. et al., "Versatile surface plasmon resonance of carbon-dot-supported silver nanoparticles in polymer optoelectronic devices," *Nat. Photonics* 7 (2013): 732–738.

Demchenko, A. P. & Dekaliuk, M. O., "Novel fluorescent carbonic nanomaterials for sensing and imaging," *Methods Appl. Fluoresc.* 1 (2013): 042001.

Deng, Q., Li, Y., Wu, J. et al., "Highly sensitive fluorescent sensing for water based on poly($m$-aminobenzoic acid)," *Chem. Commun.* 48 (2012): 3009–3011.

Dong, D. C. & Winnik, F. M., "The Py scale of solvent polarities," *Can. J. Chem.* 62 (1984): 2560–2565.

Fang, Y., Guo, S., Li, D. et al., "Easy synthesis and imaging applications of cross-linked green fluorescent hollow carbon nanoparticles," *ACS Nano* 6 (2012): 400–409.

Geng, B. Y., Ma, J. Z., Du, Q. B. et al., "Synthesis of hollow carbon nanospheres through a ZnSe nanoparticle template route," *Mater. Sci. Eng. A* 466 (2007): 96–100.

Ghosh, P., Mandal, S., Das, T. et al., "'Extra stabilisation' of a pyrene based molecular couple by γ-cyclodextrin in the excited electronic state," *Phys. Chem. Chem. Phys.* 14 (2012): 11500–11507.

Ghosh, S., Chizhik, A. M., Karedla, N. et al., "Photoluminescence of carbon nanodots: Dipole emission centers and electron-phonon coupling," *Nano Lett.* 14 (2014): 5656–5661.

Gong, X., Hu, Q., Paau, M. C. et al., "Red-green-blue fluorescent hollow carbon nanoparticles isolated from chromatographic fractions for cellular imaging," *Nanoscale* 6 (2014): 8162–8170.

Hu, G., Ma, D., Cheng, M. et al., "Direct synthesis of uniform hollow carbon spheres by a self-assembly template approach," *Chem. Commun.* 2002, 1948–1949.

Itoh, H., Ishido, S., Nomura, M. et al., "Estimation of the hydrophobicity in microenvironments by pyrene fluorescence measurements: *n*-β-octylglucoside micelles," *J. Phys. Chem.* 100 (1996): 9047–9053.

Jariwala, D., Sangwan, V. K., Lauhon, L. J. et al., "Carbon nanomaterials for electronics, optoelectronics, photovoltaics, and sensing," *Chem. Soc. Rev.* 42 (2013): 2824–2860.

Jiang, Y. B., Wei, L. H., Yu, Y. Z. et al., "Preparation of porous carbon particle with shell/core structure," *eXPRESS Polym. Lett.* 1 (2007): 292–298.

Kalyanasundaram, K. & Thomas, J. K., "Environmental effects on vibronic band intensities in pyrene monomer fluorescence and their application in studies of micellar systems," *J. Am. Chem. Soc.* 99 (1977): 2039–2044.

Kaushlendra, K. & Asha, S. K., "Microstructural reorganization and cargo release in pyrene urethane methacrylate random copolymer hollow capsules," *Langmuir* 28 (2012): 12731–12743.

Li, H., Kang, Z., Liu, Y. et al., "Carbon nanodots: Synthesis, properties and applications," *J. Mater. Chem.* 22 (2012): 24230–24253.

Li, M., Wu, Q., Wen, M. et al., "A novel route for preparation of hollow carbon nanospheres without introducing template," *Nanoscale Res. Lett.* 4 (2009): 1365–1370.

Matte, H. S. S. R., Subrahmanyam, K. S., Rao, K. V. et al., "Quenching of fluorescence of aromatic molecules by graphene due to electron transfer," *Chem. Phys. Lett.* 506 (2011): 260–264.

Mondal, S., Das, T., Ghosh, P. et al., "Exploring the interior of hollow fluorescent carbon nanoparticles," *J. Phys. Chem. C* 117 (2013): 4260–4267.

Ray, S. C., Saha, A., Jana, N. R. et al., "Fluorescent carbon nanoparticles: Synthesis, characterization, and bioimaging application," *J. Phys. Chem. C* 113 (2009): 18546–18551.

Shen, W. & Fan, W., "Nitrogen-containing porous carbons: Synthesis and application," *J. Mater. Chem. A* 1 (2013): 999–1013.

Strauss, V., Margraf, J. T., Dolle, C. et al., "Carbon nanodots: Toward a comprehensive understanding of their photoluminescence," *J. Am. Chem. Soc.* 136 (2014): 17308–17316.

Sun, Y.-P., Zhou, B., Lin, Y. et al., "Quantum-sized carbon dots for bright and colorful photoluminescence," *J. Am. Chem. Soc.* 128 (2006): 7756–7757.

Wang, Q., Huang, X., Long, Y. et al., "Hollow luminescent carbon dots for drug delivery," *Carbon* 59 (2013): 192199.

Wang, L., Zhu, S., Wang, H. et al., "Common origin of green luminescence in carbon nanodots and graphene quantum dots," *ACS Nano* 8 (2014): 2541–2547.

Wang, G., Pan, X., Kumar, J. N. et al., "One-step synthesis of hollow carbon nanospheres in non-coordinating solvent," *Carbon* 83 (2015): 180–182.

Winnik, F. M., "Photophysics of preassociated pyrenes in aqueous polymer solutions and in other organized media," *Chem. Rev.* 93 (1993): 587–614.

Zhang, Z., Xiao, F., Xi, J. et al., "Encapsulating Pd nanoparticles in double-shelled graphene@carbon hollow spheres for excellent chemical catalytic property," *Sci. Rep.* 4 (2014): 4053.

17

# Metal-Filled Carbon Nanocapsules

Takashi Nakamura

Ruslan Sergiienko

Etsuro Shibata

## 17.1 Introduction to Metal-Filled Carbon Nanocapsules

Among various nanostructured materials, metal magnetic nanoparticles are attracting considerable scientific and practical interest because of their size-dependent physical properties. This is reflected in their broad applications, which cover data storage (Reiss & Hütten 2005), catalysis (Herdt et al. 2007), environment protection (Bystrzejewski et al. 2009b), and ferrofluids (Raj et al. 1995, Jeong et al. 2007). Magnetic nanoparticles are one of the most promising candidates for biotechnology applications including magnetic resonance imaging (MRI) contrast agents, targetable drug carriers, hyperthermia-inducing agents, and magnetically controlled media for the sensitive separation and detection of biomolecules (Bystrzejewski et al. 2005, Seo et al. 2006, Lu et al. 2007, Berry 2009, Pankhurst et al. 2009). However, there are significant problems in the use and handling of bare metal nanoparticles. This is a result of their oxidation in air, dissolution in acids, and ease of agglomeration. Oxidation is a significant problem, because it drastically changes (usually deteriorates) the magnetic performance. Thus, an encapsulation route was proposed to preserve the specific properties of the nanoparticles and to overcome the above limitations. Until now, magnetic nanoparticles have been encapsulated by various inorganic materials such as silica (Kobayashi et al. 2003, Wang et al. 2008), noble metals (Carpenter et al. 1999, Sobal et al. 2002), and organic polymers (Luna et al. 2003, Shin & Jang 2007). However, these materials are not completely inert; the silica coating is unstable within a highly basic environment, the organic polymers demonstrate low thermal stability, and the preparation of coatings from noble metals requires expensive and harmful precursors such as metal organic complexes. Alternatively, carbon can be regarded as an ideal coating as it is light and highly inert in extreme chemical and physical environments. Thus, carbon coatings made of graphite layers are resistant to acids, bases, greases, and high temperatures of up to 350°C–400°C under a pure oxygen atmosphere (Bystrzejewski et al. 2006). Moreover, the surface properties of carbon coatings can be modified through their chemical functionalization. Controlled surface modification is therefore possible (Popławska et al. 2010) and carbon coatings suitable for the high capacity binding of biomolecules can be created. Chemical functionalization is of great importance for prospective biomedical applications, and in order to improve the dispersion of carbon-coated particles in a solvent of desired polarity.

Thus, carbon can prevent the rapid oxidization of metal magnetic nanoparticles. Additionally, graphite layers magnetically isolate the nanoparticles from each other. Cobalt, nickel, and iron nanoparticles encapsulated in carbon/or graphite shells are called *carbon nanocapsules* (CNCs) and they have many applications in medicine, e.g., localized radio frequency absorbers in cancer therapy (Xu et al. 2008), for biomedical applications (Bystrzejewski et al. 2005, Abdullaeva et al. 2012), and in drug delivery (Kim et al. 2008). Moreover, CNCs may have applications in the confinement of radioactive waste particles and in the handling of air-sensitive materials.

CNCs were discovered within the carbonaceous deposits formed by the electric arc discharge process. Since 1993, various metal and/or metal carbide-filled CNCs have been synthesized (Figure 17.1). The synthesis of magnetic metal-filled CNCs involves various methods such as carbon arc (Saito 1995), thermal plasma (Bystrzejewski et al. 2007a, 2009a), laser-assisted irradiation (Bi et al. 1993, Park et al. 2008), microwave heating (Jacob et al. 2006), CVD (Wang et al. 2003, Qiu et al. 2006), spray pyrolysis (Wang et al. 2007), explosion (Wu et al. 2003), combustion synthesis (Bystrzejewski et al. 2007b), hydrothermal reactions (Wang et al. 2006), and cocarbonization of various aromatic precursors with transition metal compounds (Huo et al. 2007). However, some of these methods are economically unviable as they require expensive organic precursors and use expensive vacuum equipment. All the above methods result in inhomogeneous products, which contain CNCs in addition to nonencapsulated metallic particles, CNTs, graphite flakes, graphite debris, and amorphous carbon. These by-products cause difficulty with the applications of the CNCs, and costly purification steps have to be performed.

So far, traditional and modified arc discharge methods have been used to synthesize metal-filled CNCs (Table 17.1). Usually, during the arc discharge method, a pure graphite cathode and a metal/metal oxide-loaded graphite anode undergo arc vaporization in a helium gas atmosphere. It is possible to significantly reduce the level of unwanted products, by synthesizing the metal-filled CNCs via the modified arc discharge method. This method uses a tungsten cathode and an anode composed of a metal block placed inside a graphite crucible, where the velocity of the helium gas is controlled (Dravid et al. 1995, Jiao & Seraphin 1998, Bonard et al. 2001). The modified construction decreases the carbon content of the arc within the helium atmosphere. However, neither the traditional nor the modified arc discharge methods can be applied without expensive vacuum equipment. Additionally, the power required by the arc discharge method usually exceeds 1 kW (Table 17.1).

Furthermore, the arc discharge in a liquid media method does not require a vacuum system, but it does require a high electric power input for the generation of the electric plasma discharge. The arc discharge in a gas and liquid media method is initiated by contacting an anode with the cathode, or alternatively, the electrodes must be positioned as near as possible to each other. Without this initiation, a high transient breakdown voltage is necessary to generate the plasma discharge in a neutral medium. For example, the electric breakdown voltages of liquid hydrocarbons are quite high, around several tens of kilovolts, with an electrode separation of 0.5–8 mm (Fuhr et al. 1986, Fuhr & Schmidt 1986). From a physical point of view, the arc discharge in a liquid media method has advantages in comparison to the arc discharge in a gas atmosphere method; these are (1) higher temperature in the arc zone (4000–6500 K) (Lange et al. 2003), and (2) high cooling rates of sputtered nanoparticles by liquid media ($10^6$–$10^{14}$ K/s), which implies a high rate of nucleation and low particle growth (Chuistov & Perekos 1997, Berkowitz et al. 2003).

Note: There are elements in the gray boxes which were encapsulated in carbon shells. A designation of M and C under the chemical symbol represents encapsulated core in the metallic and/or carbide forms, respectively. A number in the bracket means a reference as follows: 1. Ruoff et al. (1993a); 2. Tomita et al. (1993); 3. Saito et al. (1993a); 4. Saito et al. (1994c); 5. Saito et al. (1993b); 6. Saito et al. (1993b); 7. Funasaka et al. (1993); 8. Ruoff et al. (1993b); 9. Hihara et al. (1994); 10. Saito & Yoshikawa (1993); 11. Bandow & Saito (1993); 12. Murakami et al. (1994); 13. Ugarte (1993); 14. Saito et al. (1997b); 15. Li et al. (1998); 16. Babonneau et al. (1998); 17. Jiao & Seraphin (1998); 18. Lamber et al. (1990); 19. Guerret-Piécourt et al. (1994).

| 1 | 2 | 3 | 4 | 5 | 6 | 7 | 8 | 9 | 10 | 11 | 12 | 13 | 14 | 15 | 16 | 17 | 18 |
|---|---|---|---|---|---|---|---|---|---|---|---|---|---|---|---|---|---|
| Li | Be | | | | | | | | | | | B | C | N | O | F | Ne |
| Na | Mg | | | | | | | | | | | Al | Si | P | S | Cl | Ar |
| K | Ca | (3,5) Sc C | (14,17) Ti C | (11) V C | (19) Cr C | Mn | (3,9) Fe m,c | (3) Co m,c | (3,10) Ni m,c | (17) Cu m | (19) Zn M | Ga | Ge | As | Se | Br | Kr |
| Rb | Sr | (4,6) Y C | (11) Zr C | Nb | (19) Mo C | Tc | Ru | Rh | (18,19) Pd m | (16) Ag m | Cd | In | (19) Sn m | Sb | Te | I | Xe |
| Cs | Ba | La | (14) Hf C | (12) Ta C | (15) W C | Re | Os | Ir | Pt | (13) Au m | Hg | Tl | (19) Pb m | (19) Bi m | Po | At | Rn |
| Fr | Ra | Ac | (7) Th C | Pa | (8) U C | Np | Pu | Am | Cm | Bk | Cf | Es | Fm | Md | No | Lr |

| (1,2) La C | (3,4) Ce C | (3,4) Pr C | (3,4) Nd C | Pm | Sm | Eu | (3,4) Gd C | (3,4) Tb C | (3,4) Dy C | (3,4) Ho C | (3,4) Er C | (4) Tm C | (19) Yb C | (3,4) Lu C |
|---|---|---|---|---|---|---|---|---|---|---|---|---|---|---|

**FIGURE 17.1** Metal-filled CNCs.

**TABLE 17.1**   Electric Characteristics of the Different Arc Plasma Discharge Methods

| Environment | Material of Cathode/ (Anode)[a] | Electric Power Input | | | Reference |
|---|---|---|---|---|---|
| | | Current (A) | Voltage (V) | Power (kW) | |
| Helium gas | G/(Ni, NiO, G) | 70 | 25 | 1.75 (DC) | Saito et al. 1994a |
| Helium gas | G/(Co₃O₄, G) | 100 | 30 | 3 (DC) | McHenry et al. 1994 |
| He/CH₄ (4/1) | G/(Fe₂O₃, G) | 150–200 | 20 | 3–4 (DC) | Saraswati et al. 2012 |
| Modified Arc Discharge[b] | | | | | |
| Helium gas | Tungsten/(Ni, G) | 130 | 20–21 | 2.6–2.7 (DC) | Hwang & Dravid 1997 |
| Helium gas He/CH₄ (10/1) | G/(Ni, Co, Cu, Ti, G) | 175 | 25 | 4.375 (DC) | Jiao & Seraphin 1998 |
| Helium gas | G/(Co, G) | 150 | 18 | 2.7 (AC rectified) | Bonard et al. 2001 |
| Helium gas | (Ni, Co, G)/(Ni, Co, G) | 500–650 | 25–30 | 16 (AC) | Ling et al. 2003 |
| Liquid Media | | | | | |
| Water | G/G | 30 | 16–17 | 0.5 (DC) | Sano et al. 2002 |
| Water | G/(0.8 at% Y- or Gd-doped graphite) | 30–40 | 21–22 | 0.8 (DC) | Lange et al. 2003 |
| Water | G/(coal-based carbon rod with 2.5 wt% Ni) | 35–40 | 45–50 | 1.7 (DC) | Qiu et al. 2004 |
| 0.1 M CoSO₄/0.1 M CoNO₃ water solutions | G/G | 100 | 22 | 2.2 (DC) | Hsin et al. 2001 |
| 0.05 M NiSO₄/ CoSO₄/FeSO₄ water solutions | G/G | 30 | 22–26 | 0.72 (DC) | Xu et al. 2006 |
| Ethanol | Pairs electrodes: G, Ni, W, 1010 steel | Pulse generator: peak current 120 A, pulse duration 20 μs, rate 100 Hz | | | Parkansky et al. 2005 |
| Toluene | Pairs of electrodes: G, Mo, Fe, Ni | 1–20 | 15–20 | 0.4 (DC) | Okada et al. 2007 |
| Toluene, water, styrene | G/G, Ti/Ti, Cu/Cu | Pulse generator: 200 V, 5 A, pulse duration 10 μs | | | Omurzak et al. 2007 |
| Ethanol | Pairs of Co-, Ni-doped carbon rods, Ni/Ni, Fe/Fe | Pulse generator: 150–170 V, 1.5–3 A, pulse duration 10 μs, frequency 60 Hz–30 kHz | | | Abdullaeva et al. 2012 |

*Note:* AC: alternating current; DC: direct current; G: graphite.

[a] The metal-loaded graphite anode is prepared with a mixture of metal or metal oxide powder and graphite powder.

[b] In the modified method, a molten metallic pool supported by a graphite crucible is employed as the anode of the material to be encapsulated in carbon shells.

# 17.2  Synthesis of Metal-Filled CNCs by Plasma Discharge Method, Assisted by Ultrasonic Cavitation

Usually in all arc discharge methods, the plasma discharge is initiated by contacting an anode with the cathode. Even at arc powers greater than 1 kW, it is quite difficult to obtain stable and prolonged plasma discharge; therefore, it is necessary to precisely control the electrode gap, in order to run the arc continuously. In some cases (Parkansky et al. 2005, Omurzak et al. 2007), one electrode is mounted onto a vibrator, which brings it into periodic contact with the plate electrode; the arcs are subsequently ignited when the contact between the two electrodes is broken. Liquid hydrocarbons represent a class of almost perfect insulators; therefore, the electric breakdown voltage of organic liquids can be decreased by positioning the electrodes closer together, as demonstrated by Paschen's law (Paschen 1889). This can also be achieved by exposing the electrode gap to an external ion source.

Shibata et al. (2006) developed a method to produce metal-filled CNCs. This method takes advantage of the fact that electric plasma can be generated and maintained in an organic liquid, under ultrasonic irradiation. The ultrasonic cavitation in a liquid media can account for sonoluminescence and sonochemical reactions. In order to explain the origin of sonoluminescence and sonochemical reactions, there are two current theories; these are thermal and electrical. The theory of a hot spot is one of the thermal theories (Fitzgerald et al. 1956, Suslick & Hammerton 1986). Light emission (Suslick & Flint 1987) originates from the excited-state molecules and/or species at a localized high temperature and pressure (>5000 K; >100 MPa) region, where tiny bubbles are collapsing; this region is referred to as a *hot spot*. Several electrical theories have been proposed (Harvey 1939, Frenkel 1940, Degrois & Baldo 1974). The most recent theory was proposed by Margulis (1985) and Margulis and Margulis (2002). It is proposed that the electrical discharge inside the charged cavitation bubble is the source of sonoluminescence and the reason for the sonochemical reactions. Margulis proposed the following theory which describes the electrical charging of fragmented bubbles: An uncompensated electrical charge is localized on the surface of a fragmentation bubble which has been separated from a primary deformed cavitation bubble that is pulsing in the acoustic field. The charging is caused by the shearing of the electrical double layer at the liquid–gas interface.

In the developed method, an ultrasonic cavitation field, with its many activated tiny bubbles, may enhance the electrical conductivity of an organic liquid owing to the high-energy species (radicals, atoms, ions, and free electrons) formed within it. Thus, an electric plasma discharge could be generated at a remarkably low voltage and power of 55 V DC and 87–165 W (Shibata et al. 2004, Sergiienko et al. 2007), respectively. Without ultrasonication, plasma discharge could not take place at such a low level of electric power. Figure 17.2a shows a schematic diagram of the experimental apparatus and Figure 17.2b shows a photograph of the plasma discharge, which is similar to what was previously demonstrated in an experiment for the production of graphite nanosheets (Kim et al. 2010a). The organic liquid and the metal electrode-anode act as precursors for the synthesis of the metal-filled CNCs. In contrast to traditional and modified arc discharge units, the plasma discharge produced under conditions of low electric power with ultrasonication does not require a high electric power input and vacuum equipment (because the plasma discharge occurs in a liquid at atmospheric pressure). Therefore, this process can be more cost-effective and economical. It is important to note that the synthesis products in our method are devoid of any graphite debris and CNTs, which commonly occur in the traditional arc discharge method. The by-products in our method consist of insufficiently encapsulated metal particles and some amounts of amorphous carbon and impure graphite flakes. These by-products can be removed by an acid and oxidation treatment, in addition to magnetic separation. Since the metal anode has high mechanical strength, the consumption of the anode is negligible, and it is not often necessary to replace the electrode. As a result, the new method has the potential to be applied for the continuous synthesis of CNCs.

During the initial experiments (Shibata et al. 2004, Sergiienko et al. 2006a), two separate wire electrodes (anode and cathode) were inserted in a position closely beneath the titanium ultrasonic horn, at a distance of 1 mm apart from each other. Over time, the experimental apparatus was modified (Kim et al. 2008) and one electrode-anode remained, while the ultrasonic horn served as the electrode-cathode, as shown in Figure 17.2. A metal tip composed of the same material as the anode was positioned on top of the ultrasonic horn in order to avoid titanium contamination from the wear of the horn (Figure 17.2). The distance between the electrode-anode and the bottom of the ultrasonic horn was automatically adjusted using a motorized Z-axis precision stage (Sigma Koki Co., Ltd., SGSP80-20ZF) connected to a programmable stage controller (SHOT-602), in order to drive a stepping motor. An ultrasonic homogenizer (Nissei, US-600NCVP) was operated at 600 W and 20 kHz during the experiments to irradiate

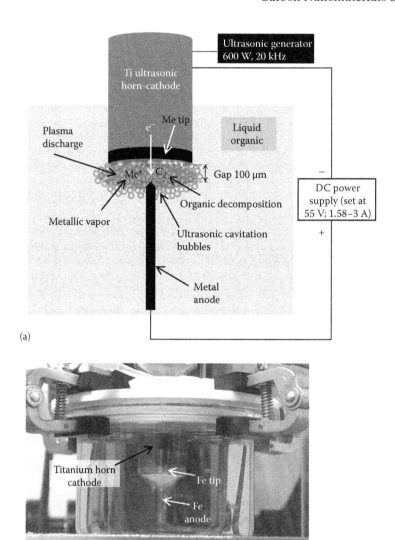

(a)

(b)

**FIGURE 17.2** (a) Schematic view of an experimental apparatus. (From Kim, S. et al., *Mater. Trans.*, 51, 1455–1459, 2010. With permission.) (b) Photograph of the electric plasma discharge in a liquid ethanol (Wako, 99.5% S grade).

100–300 mL of organic liquid. The voltage between the anode and cathode was held at 55 V, and the upper limit of the current on the electrodes was set at 1.58 or 3.0 A throughout the experiment. The plasma discharge occurred between the ultrasonic horn and the electrode-anode at a distance of approximately 100 μm. During the experiments, the electrode-anode was mainly consumed by thermal evaporation. The glass vessel (Figure 17.2b) was cooled in an ice water bath and an argon gas flow was directed into the vessel in order to maintain an inert atmosphere.

In the present work, it was possible to synthesize arbitrary metal-core nanoparticles encapsulated in graphite shells by selecting the material (Ti, Cu, W, Fe, Co, 50Fe–50Pt alloy) of the electrode–anode and metal tip-cathode. The structure, magnetic properties, and formation mechanism of the various metal and/or metal carbide-filled CNCs will subsequently be described and explained. The structure and morphology of the CNCs were evaluated by TEM (JEOL, JEM-3010, JEOL-2010) and field emission scanning electron microscopy (JEOL, JSM-7000 F). The phase constitution of the synthesized carbon powder samples was determined using XRD (Rigaku, RINT2000) with a monochromatic Cu K$\alpha_1$ radiation source. A vibration sample magnetometer (VSM) (TOEI VSM-5-18) operating at room temperature with an applied magnetic field that varied from –1200 to +1200 kA m$^{-1}$ was used to measure the magnetic properties of the CNCs.

## 17.3 Structure Characterization and Magnetic Properties of Metal-Filled CNCs

### 17.3.1 Plasma Discharge in Liquid Benzene: TiC- and Cu-Filled CNCs

To develop the plasma discharge method (assisted by ultrasonic cavitation), liquid benzene was used as a carbon source and copper was selected as the material for the electrodes. Two copper electrodes (2 mm in diameter and with a purity of 99.9%) were inserted into the glass vessel at a distance of 1 mm apart from each other. They were positioned just beneath but very close to the bottom of the ultrasonic horn, as shown in Figure 17.3. When the ultrasonic irradiation was performed, electric plasma was generated in the ultrasonic cavitation field beneath the horn. The tips of the copper electrodes and the bottom of the ultrasonic horn were consumed during the experiment. The positive copper electrode was subjected to a greater amount of wear than the negative copper electrode. During the experiment, which lasted approximately 1 h, very small carbon particles appeared and were dispersed within the liquid benzene. In the static condition, small carbon particles were observed to condense and precipitate in the liquid. Following the experiment, the color of the benzene, which separated from the carbon particles, was observed to be dark yellow. This was due to many soluble aromatic polymers. The gas chromatography–mass spectrometry (GC-MS) spectrum of the liquid benzene sample, following the experiment, is shown in Figure 17.4 and shows the organic compounds present with a molecular weight below 230 Da. The major synthesized compounds were biphenyl ($C_{12}H_{10}$) and 1,4-diethynylbenzene ($C_{10}H_6$) (Figure 17.4a). For reference, the yield of aromatic products from the hexane sample was considerably lower than that yielded from the benzene sample (Figure 17.4b).

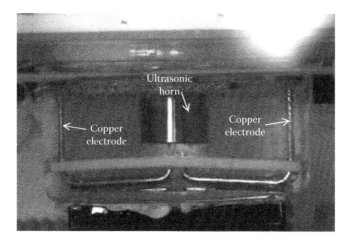

**FIGURE 17.3** Photograph of the electric plasma discharge in the ultrasonic cavitation field of liquid benzene (Wako, S grade). (Reprinted from *Ultrasonics Sonochemistry*, 13, Sergiienko, R., Shibata, E., Suwa, H. et al., Synthesis of amorphous carbon nanoparticles and carbon encapsulated metal nanoparticles in liquid benzene by an electric plasma discharge in ultrasonic cavitation field, 6–12, Copyright (2006), with permission from Elsevier.)

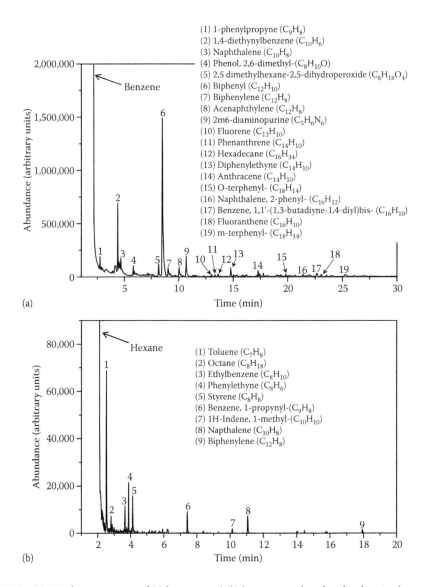

**FIGURE 17.4**  GS-MS chromatograms of (a) benzene and (b) hexane samples after the electric plasma discharge in ultrasonic cavitation field. (Reprinted from *Ultrasonics Sonochemistry*, 13, Sergiienko, R., Shibata, E., Suwa, H. et al., Synthesis of amorphous carbon nanoparticles and carbon encapsulated metal nanoparticles in liquid benzene by an electric plasma discharge in ultrasonic cavitation field, 6–12, Copyright (2006), with permission from Elsevier.)

On the basis of the GC-MS determinations of the compounds shown in Figure 17.4, we speculate that the benzene molecules in the hot zone may have broken up into phenyl, alkane, and H and C radicals. Through quenching by the surrounding liquid, these radicals subsequently evolved into more stable compounds such as carbon nanoparticles and polycyclic aromatic compounds. For example, the production of biphenyl is probably a result of the binding of two phenyl radicals ($C_6H_5$), which were formed in the hot zone by the dissociation of the benzene C–H bonds. The aromatic structure of benzene facilitates the formation of carbonaceous products.

Centrifugation was used to separate the carbonaceous powder produced from the remaining liquid benzene. The presence of a large amount of amorphous carbon was supported by TEM observations and Raman spectroscopy. A selected area electron diffraction (SAED) pattern of carbon nanoparticles

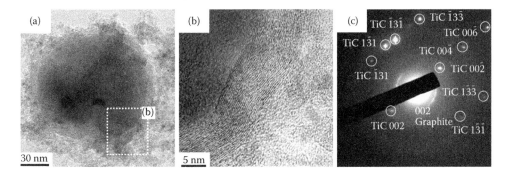

**FIGURE 17.5** (a) TEM image of a TiC particle wrapped in multilayered graphite layers; (b) its corresponding HRTEM image; and (c) selected-area electron diffraction pattern corresponding to the particle in a. (Reprinted from *Ultrasonics Sonochemistry*, 13, Sergiienko, R., Shibata, E., Suwa, H. et al., Synthesis of amorphous carbon nanoparticles and carbon encapsulated metal nanoparticles in liquid benzene by an electric plasma discharge in ultrasonic cavitation field, 6–12, Copyright (2006), with permission from Elsevier.)

consists of blurred rings, indicating that the carbon nanoparticles obtained in this study consisted of one of the amorphous carbons (Shibata et al. 2004). A broad Raman peak centered at approximately 1385 cm$^{-1}$ and a relatively sharp peak at approximately 1580 cm$^{-1}$ were present in the spectra of the experimental powder sample (Sergiienko et al. 2006a); these are commonly referred to as the *D band* and *G band*, respectively. The intensity ratio ($I_D/I_G$) of the two peaks is approximately 0.75. These Raman data suggest that the experimental powder contains carbon nanoparticles with a disordered structure.

In the TEM analysis, we observed spherical crystalline nanoparticles, approximately 50–250 nm in size, dispersed within the amorphous carbon matrix. These particles were wrapped in multilayered shells (CNCs), as shown in Figure 17.5a and b. The spacing of the lattice fringes of the carbon shell in Figure 17.5b is approximately 0.34 nm; this value is close to that of graphite lattice (002) planes. The diffraction spots from the crystalline core of the CNC are superposed over arched semirings originating in the graphite shell (Figure 17.5c). The diffraction spots identify that the crystalline core is an fcc structure of TiC. The TiC particles were probably generated by Ti vapors emitted from the surface of the ultrasonic horn, which reacted with carbon after benzene decomposition, forming the carbide. It is suggested that the TiC-filled CNCs are formed through a mechanism proposed elsewhere (Saito et al. 1993a, Saito 1995). In the plasma discharge zone, small liquid droplets of a titanium–carbon alloy with an excess of carbon are initially formed. Subsequently, graphite layers segregate on the surface of the droplets as the temperature decreases during the quenching from the surrounding liquid benzene. The segregation of the carbon ceases when the composition of the titanium–carbon alloy trapped inside the graphite layers reaches that of the TiC phase. The cores of the large CNCs with thick graphite shells likely consist of TiC compound rather than pure Ti, as the standard free energy of formation of TiC is very low (–43.2 kcal mol$^{-1}$) (Ōya & Ōtani 1979) and TiC can be an effective catalyst for the formation of graphite shells.

We also observed metallic nanoparticles, smaller than 10 nm in size, wrapped in several graphite layers as shown in Figure 17.6a. To clarify the nature of the encapsulated metal nanoparticles, we conducted a fast Fourier transformation (FFT) as shown in Figure 17.6b and c. Figure 17.6b contains the digital diffractogram of the lattice images corresponding to the HRTEM image of Figure 17.6a. Most of the spots in Figure 17.6b can be attributed to copper planes (111) and (200). These nanoparticles were probably generated by vapors emitted from the copper electrodes. The possibility of the formation of Cu-filled CNCs was discussed elsewhere (Jiao & Seraphin 1998). Their particles were synthesized by a modified arc discharge method and showed an average diameter of 11 nm, coated with three to four graphite layers.

The formation mechanisms of TiC- and Cu-filled CNCs are probably different from each other. Ti, as a carbide former, can easily transform to TiC. Therefore, graphite layers can be formed by the

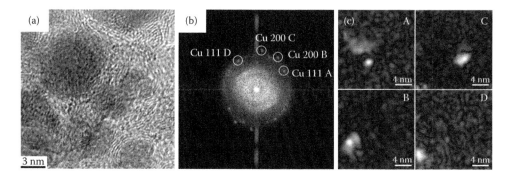

**FIGURE 17.6** (a) HRTEM image of the CNCs formed in a liquid benzene; (b) corresponding digital diffractogram computed by FFT of the image shown in a; (c) inverse FFT images according to each copper spot of digital diffractogram shown in b. (Reprinted from *Ultrasonics Sonochemistry*, 13, Sergiienko, R., Shibata, E., Suwa, H. et al., Synthesis of amorphous carbon nanoparticles and carbon encapsulated metal nanoparticles in liquid benzene by an electric plasma discharge in ultrasonic cavitation field, 6–12, Copyright (2006), with permission from Elsevier.)

segregation of carbon from the saturated titanium–carbon alloy particle. Moreover, the surface of the copper particle must be directly covered by the graphite layers forming the Cu-filled CNC. The carbon shells around the TiC cores were significantly thicker (approximately 10–30 graphite layers), while only a few layers formed around the copper cores.

By using the present method, it is possible to synthesize arbitrary metal nanoparticles encapsulated in graphite shells by selecting the materials of the ultrasonic horn tip and electrode-anode. The benzene decomposition formed a large amount of amorphous carbon; additionally, benzene is a toxic and carcinogenic substance. Therefore, liquid ethanol was chosen as a carbon source in the subsequent series of experiments.

## 17.3.2 Plasma Discharge in Liquid Ethanol: $(WC_{1-x})$-, $W_2C$-, and WC-Filled CNCs

Carbides, particularly those of the transition metals, have a number of valuable properties that make them promising materials for use in various new technology fields. They possess high melting temperatures, great hardness, high chemical resistance, and the electrical and thermal conductivities of a metallic character. They also possess a number of special properties such as the capacity to be transformed to the superconducting state at relatively high temperatures and high emission properties (Lassner & Schubert 1999).

The principal application for carbides is as the major constituent in the so-called cemented carbide. The cemented carbides are combinations of carbides such as WC, TiC, and TaC with binder metals such as cobalt and nickel. These are mass produced as materials for cutting tools and wear-resistant parts. The tungsten carbide–cobalt (WC–Co) composite is the most important for these applications. The increase in the dispersion of the structural components of the composites facilitates the creation of new types of materials which combine high strength and ductility. Recent experiments (Fang et al. 2009, Dvornik 2010) have demonstrated that nanostructured WC–Co composites have superior mechanical properties because of their improved hardness and increased ductility and plasticity. Furthermore, the electrochemical characterization of WC/carbon composites with a high SSA demonstrated that the composite has a potential application as an electrode material for electrochemical capacitors (Liu et al. 2007).

Carbon-encapsulated cubic $WC_{1-x}$ and hexagonal $W_2C$ structure nanoparticles could be produced using arc discharge methods in an inert gaseous atmosphere (Saito et al. 1997a, Li et al. 1998) and also using electrical discharge machining within kerosene fluid (Lin 2005). However, the hexagonal WC phase was not observed when using these methods. The present work proposes an attractive technique for preparing nanocrystalline particles of $WC_{1-x}$, $W_2C$, and WC encapsulated in graphite shells; this uses the electric plasma discharge method in an ultrasonic cavitation field of liquid ethanol, followed by an annealing process.

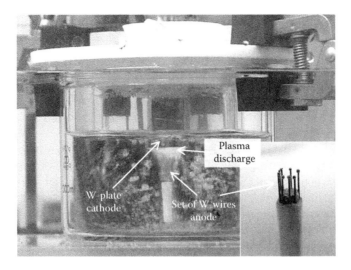

**FIGURE 17.7** Photograph of the electric plasma discharge in the ultrasonic cavitation field of 250 mL of liquid ethanol (Wako, 99.5% S grade). The inset shows the set of tungsten wires attached to the iron rod.

The low current plasma discharge was generated between the tungsten wire anodes (Ø = 0.3–0.5 mm; 99.95% purity), and the bottom of the tungsten plate ($10 \times 10 \times 2$ mm³; 99.95% purity) was mounted on top of a titanium ultrasonic horn, which served as the cathode (Figure 17.7). During the evaporation of the tungsten wires and the decomposition of liquid ethanol in the plasma discharge zone, tungsten carbide nanoparticles were formed and quenched in a liquid organic medium. XRD patterns of the as-synthesized samples which were annealed at 1173 K in an argon atmosphere are presented in Figure 17.8. It can be seen that

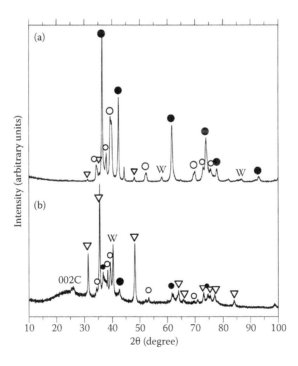

**FIGURE 17.8** XRD patterns obtained from (a) the as-synthesized powder sample and (b) that the powder sample annealed at 1173 K in an argon gas atmosphere. ●: $WC_{1-x}$; ○: $W_2C$; ▽: WC.

the formation of $WC_{1-x}$ and $W_2C$ phases (which are stable at high temperature) occurred as the nanoparticles were quenched in liquid ethanol (Figure 17.8a). The fcc phase, $WC_{1-x}$, is an interstitial solution of carbon in a tungsten matrix. The crystal structure of $W_2C$ can be described as a slightly distorted, hexagonal close packing of tungsten atoms. The carbon atoms occupy half of the octahedral interstices. They may be distributed in an ordered manner; the type and the degree of ordering depend on the temperature. $W_2C$ decomposes at 1523 K into W and WC; $WC_{1-x}$ (high-temperature modification) undergoes eutectoid decomposition at 2803–2808 K into $\beta(W_2C) + \delta(WC)$ (Lassner & Schubert 1999). At room temperature, the phases of $W_2C$ and $WC_{1-x}$ can be only obtained by rapid quenching; this concurs with other researches (Saito et al. 1997b, Li et al. 1998, Lin 2005). The diffraction pattern of the sample that was annealed at 1173 K for 1 h in an argon gas atmosphere (Figure 17.8b) illustrates the increase in the intensity of the diffraction peaks of the WC phase, and the decrease in the intensity of the peaks of the high-temperature modifications ($W_2C$ and $WC_{1-x}$). Thus, at low heating and cooling rates (5–10 K/min) in an argon atmosphere, the hexagonal phase of WC was mainly formed; this phase is stable at room temperature. Under these conditions, the phases of $WC_{1-x}$ and $W_2C$ decomposed. This observation is consistent with the equilibrium phase diagram of "tungsten–carbon" (Lassner & Schubert 1999).

The tungsten carbide nanoparticles encapsulated in graphite shells, which were observed in the as-synthesized and the annealed carbon powder samples, have a wide size distribution (Figures 17.9 and 17.10). The size of the tungsten carbide cores inside the CNCs ranges from approximately 8 to 1000 nm in diameter. For example, it was possible to recognize aggregations of individual nanoparticles with diameters from 20 to 70 nm (Figure 17.9a) by observation with a high resolution SEM. However, the HRTEM image revealed smaller tungsten carbide nanoparticles with an average diameter of approximately 8 nm (Figure 17.9b). In most cases, the aggregated tungsten carbide nanoparticles were separated from each other by graphite layers. One of the CNCs with a typical large size is shown in Figure 17.10a. In this case, the graphite shell (approximately 10 nm thick) completely covers the tungsten carbide core and the shell is almost uniform across the surface of the nanoparticle. The distance between the graphite layers in the shell is approximately 0.34 nm, which corresponds to the distance between the lattice planes (002) of graphite (Figure 17.10b).

Annealing at 1173 K in an argon gas atmosphere intensified the process of graphitization and changed the structure of the tungsten carbides. The carbide cores served as a graphitization catalyst. Some annealed tungsten carbide cores, with a diameter greater than 230 nm, were covered by graphite shells with a thickness of approximately 40 nm (Figure 17.10c and d).

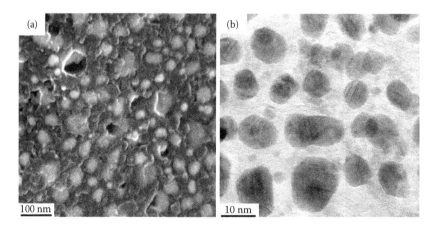

**FIGURE 17.9** High-resolution (a) SEM and (b) TEM images of the as-synthesized nearly spherical tungsten carbide nanoparticles.

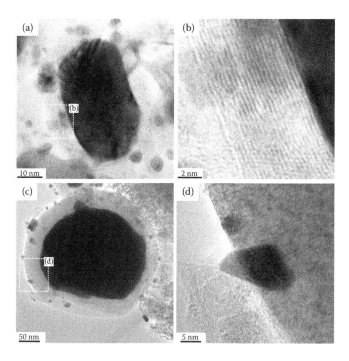

**FIGURE 17.10** TEM images of (a) the as-synthesized and (c) the annealed at 1173 K tungsten carbide nanoparticles encapsulated in graphite shells and (b, d) their corresponding HRTEM images of the enlarged graphite layers.

Thus, on the basis of the present structural investigations, it can be expected that the composite materials reinforced with tungsten carbide nanoparticles and encapsulated in graphite shells should have a high wear resistance and minimal friction coefficient. The graphite shells and the refractory cores of the CNCs should serve as a solid lubricant and a highly rigid reinforcing phase, respectively. For example, it is possible to increase the tribological characteristics of composite materials on the base of aluminum alloys because of complex reinforcement by dispersed refractory particles (WC, SiC) comprising carbon nanostructures (Scheretsky & Zatulovsky 2008). The results of the development of tribological composites, based on a copper powder mixture with CNTs and onion carbon nanostructures, are reported (Kovtun & Pasovets 2006, Kovtun et al. 2008, 2010). It has been determined that the use of such carbon additives facilitates the improvement of the antifriction properties of composite materials, owing to the decreased metal contact area on the friction surface and the inhibition of seizures.

## 17.3.3 Plasma Discharge in Liquid Ethanol: Iron- and Iron Carbide-Filled CNCs

Various magnetic materials have been encapsulated in carbon, i.e., Fe, Co, Ni, FeCo, FePt, $Fe_xO_y$, and Fe–Nd–B phases. Carbon encapsulation of iron nanoparticles is still a great challenge, because Fe has the highest saturation magnetization in comparison to the other ferromagnetic transition metals (Co, Ni, etc.). Our studies have shown that carbon-encapsulated iron nanoparticles could be effectively synthesized by the electric plasma discharge in the ultrasonic cavitation field of liquid ethanol (Sergiienko et al. 2006b). Iron carbide-filled CNCs were synthesized using two schemes of experimental apparatus. In early experiments, the plasma discharge (Sergiienko et al. 2006b) was generated between two iron wire electrodes (Ø = 2 mm; purity 99.7% of both anode and cathode). At that time, the electrodes were supplied manually. In the modified experimental apparatus (Kim et al. 2008, Sergiienko et al. 2009) (Figure 17.2), the plasma discharge was generated between one iron electrode-anode (Ø = 3 mm; purity

99.9%), and the base of the iron tip (Ø = 18 mm; purity 99.9%), which was fixed on the top of a titanium ultrasonic horn, served as a cathode. In the modified apparatus, the distance between the anode and cathode was automatically adjusted by a motorized Z-axis precision stage that significantly stabilized the plasma discharge. There was no difference observed between the structure and morphology of the CNCs synthesized from the two schemes (one or two electrodes). Liquid ethanol (Wako, 99.5% S grade) was used as the organic liquid for the experiments and as a source of carbon. Ethanol was found to be the best carbon source, because the products obtained from this compound contain a much smaller amount of amorphous carbon in comparison with benzene. The products synthesized from aliphatic compounds contain smaller amounts of amorphous carbon than those obtained from aromatic compounds. Thus, previous studies (Bystrzejewski et al. 2009a) on the decomposition of toluene and hexane in Ar plasma demonstrated that the yield of amorphous carbon obtained from toluene was approximately nine times higher in comparison to that obtained from hexane. Ethanol has been demonstrated to produce the highest content of carbon-encapsulated iron nanoparticles (Bystrzejewski et al. 2011). The presence of amorphous carbon is highly undesirable, because this carbon phase deteriorates the overall magnetic performance. It is also very difficult to remove amorphous carbon in comparison to the non-encapsulated metal particles. These are usually removed by leaching in diluted mineral acids.

In this study, the as-synthesized carbon powder sample was separated from the liquid ethanol by centrifugation. It was subsequently subjected to a purification procedure in order to remove the non-encapsulated iron carbide nanoparticles. This was achieved by etching in 4 M HCl at 313 K for 24 h. The powder samples were subsequently washed in water and ethanol and dried in vacuum at 313 K. The powder samples were subsequently annealed in pure Ar at various temperatures (573, 673, 773, 873, 973, 1073, and 1173 K) for 2 h (Sergiienko et al. 2006b).

### 17.3.3.1 Structure and Morphology of the CNCs

The size of the CNCs was measured as the diameter of the outermost graphene layer. The majority of the as-synthesized CNCs (approximately 96.5% of the total) had a very small size (from 5 to 10 nm). The fraction of the CNCs that were larger than 40–100 nm in diameter was approximately 0.01%–0.04%; however, their contribution to the total volume of the carbon powder sample was much greater than that of the small nanoparticles. Following annealing at 1173 K, the number of CNCs with a diameter larger than 100 nm significantly increased owing to sintering. The upper size limit of the CNCs increased from approximately 650 nm in the as-synthesized sample to 1200 nm in the sample that was annealed at 1173 K.

Iron is a strong carbide-forming element, and its vapors in the plasma reacted with carbon from decomposed ethanol molecules to form iron carbides. HRTEM analysis revealed that the iron carbide nanoparticles were covered by graphite shells. The typical morphology of the as-synthesized CNC (of approximately 380 nm in diameter) and its corresponding diffraction pattern are given in Figure 17.11a and c, respectively. The diffraction spots coincide with those expected for the [121] zone-axis pattern from orthorhombic $Fe_3C$; the diffraction arcs originate from the graphitic carbon of the shell. Similar electron diffraction analyses performed for other particles were interpreted as orthorhombic $Fe_3C$ (cementite; Joint Committee on Powder Diffraction Standards [JCPDS] data of Powder Diffraction File [PDF] No. 35-0772) and monoclinic $\chi$-$Fe_{2.5}C$ (the Hägg carbide; JCPDS data of PDF No. 36-1248) crystallites. The graphite layers around the core (Figure 17.11b) appear to be evenly stacked and generally parallel to one another. However, stacking defects are apparent. These defects are of great importance because they introduce curvature and seal the carbon coatings covering the encapsulated nanoparticles. During the TEM analysis, spherical CNCs with amorphous cores were also found in the as-synthesized carbon powder sample (Figure 17.12a). The diffuse ring in the digital diffractogram (Figure 17.12b) suggests that the core is amorphous, but the internal spotty ring originates from the graphitic carbon of the shell. The number of the graphite layers depends on the core size; small cores (less than 15 nm in diameter), similar to that shown in Figure 17.12a and c, were covered by several layers; whereas large particles, similar to that shown in Figure 17.11a, had approximately 25–30 layers of carbon encasing them.

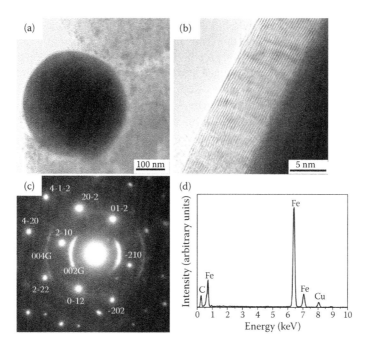

**FIGURE 17.11** (a) TEM image of $Fe_3C$-filled CNC and (b) its corresponding HRTEM image of the graphite layers covering the iron carbide core. (c) The diffraction spots from the crystalline core of the CNC are superimposed over arcs originating from the graphite shell. Miller index indicated by *G* is for the graphitic carbon. (d) Energy-dispersive X-ray spectrum from the CNC shown in a. The copper peak is due to the copper grid supporting the specimen.

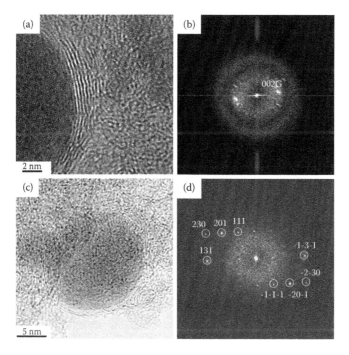

**FIGURE 17.12** HRTEM images of iron carbide-filled CNCs with (a) amorphous core and (c) annealed at 1173 K; (b, d) their corresponding digital diffractograms computed by FFT of images shown in a and c, respectively. Miller index with label *G* is for the graphitic carbon.

Iron carbides are unstable compounds and decompose at well-defined temperatures into α-Fe and/ or γ-Fe iron and graphite. At high annealing temperatures (873–1173 K), the iron carbides were almost completely transformed into α-Fe phase and graphitic carbon, except for the cores of the small nanocapsules (<10–20 nm in diameter), which remained as cementite (Sergiienko et al. 2006b). The digital diffraction pattern with an array of spots (Figure 17.12d), produced the HRTEM image (Figure 17.12c) of an annealed CNC, can be attributed to a single $Fe_3C$ crystallite. It can be assumed that the surface of the small nanoparticles plays a key role in the decomposition of the cementite. This is because as the particle size decreases, the surface energy rises significantly, and this can lead to an increase in the free energy of transformation of cementite nanoparticles, forming α-Fe and/or γ-Fe and graphitic carbon. The cementite stability of the nanoparticles was also confirmed by other researchers (Dong et al. 1998, Zhang et al. 2004), for samples annealed at elevated temperatures.

Figure 17.13 shows the XRD patterns of the samples before and after annealing at various temperatures (673, 873, and 1173 K). The XRD profile (Figure 17.13a) of the as-synthesized carbon powder sample confirmed the presence of iron carbides $Fe_3C$ and $\chi$-$Fe_{2.5}C$. The broad peak over the $2\theta = 35°–50°$ range probably results from some of the as-synthesized nanoparticles being in an amorphous state. These ideas support our interpretation of the TEM results. After annealing at 673 K for 2 h, the Hägg carbide ($\chi$-$Fe_{2.5}C$) was not observed, presumably because this monoclinic structure was converted into cementite (Figure 17.13b). Further heat treatment of the CNCs at elevated temperatures, as high as 873 and 1173 K (Figure 17.13c and d), resulted in iron carbide decomposition, and a large amount of α-Fe was detected in the powder samples. However, traces of cementite that remained intact can be observed from the XRD profile of the CNCs that were annealed at 873 and 1173 K. This concurs with the results of the TEM analysis, where small cores (<10–20 nm in diameter) of CNCs were revealed to be in the cementite phase. It can be assumed that a large number of small nanocapsules retained a cementite structure following annealing, while large carbide cores transformed into α-Fe and graphite.

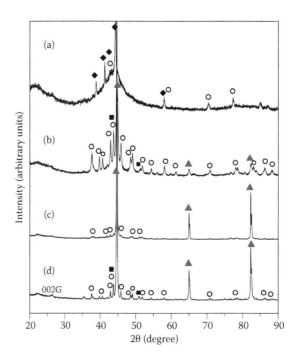

**FIGURE 17.13** XRD patterns of carbon powder samples: (a) the as-synthesized and annealed at (b) 673, (c) 873, and (d) 1173 K. ◆: $\chi$-$Fe_{2.5}C$; ○: $Fe_3C$; ■: γ-Fe; ▲: α-Fe. (Reproduced from Sergiienko, R. et al., *J. Mater. Res.*, 21, 2524–2533, 2006. With permission of the Cambridge University Press.)

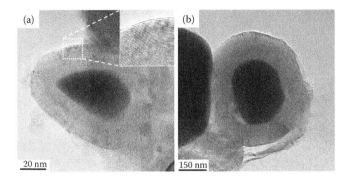

**FIGURE 17.14** TEM images of iron-filled CNCs covered with thick graphite shells after annealing at 1173 K.

Annealing intensifies the graphitization; as the annealing temperature increases, the thickness of the graphite shell increases. Small particles similar to those shown in Figure 17.12c were covered by several layers of graphite (the shell thickness was approximately 1.5–3 nm), regardless of the annealing treatment. However, larger cores had about 40–60 layers (the shell thickness was approximately 14–20 nm) of graphite encasing them (Figure 17.14a). Some of the large CNCs (similar to those shown in Figure 17.14b), with a core diameter greater than 250 nm, were covered with 70–380 graphite layers as thick as 24–130 nm (Sergiienko et al. 2006b).

### 17.3.3.2 Investigation of the Formation of CNCs by Optical Emission Spectroscopy

The atomic emission spectrum collected from the plasma discharge zone is given in Figure 17.15. The dominant lines are determined to be the Balmer line of atomic hydrogen $H_\alpha$ ($\lambda = 656.27$ nm), Swan band system of $C_2$ molecules ($\lambda = 516.52, 558.55$ nm), and atomic Fe I, which result from the cracking of the ethanol molecules and the evaporation of the surface of the ultrasonic iron tip-cathode and the iron electrode-anode. A line of atomic oxygen O I ($\lambda = 777.19$ nm) and peaks of molecular hydrogen $H_2$ are also clearly visible. The spectral lines were identified using the data from references (Striganov & Sventitskii 1968, Pearse & Gaydon 1976).

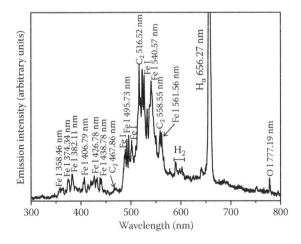

**FIGURE 17.15** Optical emission spectrum from the plasma discharge of liquid ethanol and Fe electrodes. (Reproduced from Sergiienko, R. et al., *J. Mater. Res.*, 21, 2524–2533, 2006. With permission of the Cambridge University Press.)

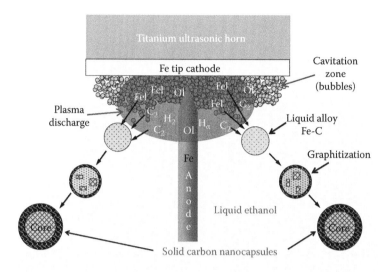

**FIGURE 17.16**  Mechanism of the CNC formation by plasma discharge in liquid ethanol.

Figure 17.16 illustrates the proposed formation mechanism of iron carbide-filled CNCs by the plasma discharge method in an ultrasonic cavitation field of liquid ethanol. Considering the temperature of the plasma discharge zone to be approximately 2500–3000 K, the process of the formation of the iron carbide-filled CNCs can be envisaged as follows: when the iron anode evaporates simultaneously to the decomposition of ethanol molecules by plasma discharge, iron atoms, Fe I, and carbon species, $C_2$, react with each other. The Fe–C alloy particles are subsequently formed in close proximity to the plasma discharge zone. The liquid Fe–C alloy particles begin to solidify out of the plasma discharge zone. As the melting point of carbon is much higher than that of the Fe–C alloy, the supersaturated carbon initially segregates from the liquid alloy particle during quenching. Graphite layers form on the outer surface of the particle; thus, the Fe–C alloy is entrapped in the core. According to the Fe–C binary phase diagram, the segregation of the carbon atoms dissolved in the molten iron–carbon alloy continues, until the composition of the alloy changes to $Fe_3C$ or $\chi$-$Fe_{2.5}C$. This equilibrates with the graphite layers on the surface of the iron carbide core. However, some cores remained in an amorphous state because of rapid cooling. In addition, the appearance of the carbon species, $C_2$, within the plasma spectrum, indicates the formation of solid amorphous and graphitic carbon.

### 17.3.3.3 Magnetic Properties of Iron- and Iron Carbide-Filled CNCs

The magnetic properties of both the as-synthesized and the annealed carbon powder samples were measured at room temperature and are shown by magnetization hysteresis loops in Figure 17.17. The shapes of the hysteresis loops at the various annealing temperatures are similar, except for the larger saturation magnetization that results from an increase in annealing temperature. All the samples demonstrated a soft ferromagnetic behavior, attaining saturation at 1200 kA/m (Figure 17.17). The saturation magnetization ($M_s$) and coercivity ($H_c$) of the reference powders and all the synthesized samples are summarized in Table 17.2, as a function of annealing temperature. $M_s$ directly indicates the maximum magnetic moment per unit mass of the studied sample. The content of the magnetic material in the samples can be derived by comparing the measured $M_s$ with the $M_s$ for the bulk sample (or micron-sized particles). However, this may lead to inaccurate results. The value of the saturation magnetization of the as-synthesized carbon sample (48 A m²/ kg) is much less than the value of 110 A m²/kg reported for the micron-sized iron carbide particles

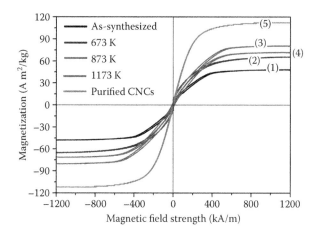

**FIGURE 17.17** Hysteresis loops of the synthesized Fe–C powder samples annealed at different temperatures: 1: as-synthesized; 2: 673 K; 3: 873 K; 4: 1173 K; 5: purified iron carbide-filled CNCs. (Reproduced from Sergiienko, R. et al., *J. Mater. Res.*, 21, 2524–2533, 2006. With permission of the Cambridge University Press.)

**TABLE 17.2**  Magnetic Properties of the Synthesized Carbon Powder Samples

| Samples & Annealing Temperature (K) | Saturation Magnetization $M_s$ (A m²/kg) | Remanent Magnetization $M_r$ (A m²/kg) | Remanence $M_r/M_s$ (%) | Coercivity $H_c$ (kA/m) |
|---|---|---|---|---|
| α-Fe[a] | 204.0 | 6.78 | 3.3 | 2 |
| Fe₃C[b] | 110.0 | 6.10 | 5.5 | 10.42 |
| As-synthesized | 48.0 | 1.80 | 3.75 | 4.16 |
| 573 | 55.6 | 3.52 | 6.33 | 7.54 |
| 673 | 65.5 | 4.82 | 7.36 | 10.90 |
| 773 | 69.3 | 4.63 | 6.68 | 10.85 |
| 873 | 80.6 | 3.9 | 4.84 | 9.36 |
| 973 | 83.4 | 4.31 | 5.17 | 10.41 |
| 1073 | 79.7 | 3.57 | 4.48 | 11.02 |
| 1173 | 71.7 | 2.87 | 4.00 | 10.64 |

*Source:* Sergiienko, R. et al., *J. Mater. Res.*, 21, 2524–2533, 2006. With permission of the Cambridge University Press.
[a] Reference α-Fe (a mean particle size is 45 µm; Wako; purity 99.9%).
[b] Reference cementite (a mean particle size is 130.4 µm; SOEKAWA CHEM.; purity >96%).

(measured at room temperature in a field strength of 1200 kA/m) (Table 17.2). To account for the differences between the saturation magnetization of the as-prepared powder and those of the reference iron carbides, we assume that the proportion of the nonmagnetic phases (amorphous and graphitic carbon) has a detrimental effect on the saturation magnetization. Additionally, the low crystalline or amorphous cores of the iron carbides may exhibit superparamagnetism and reduce the saturation magnetization.

The effect of the annealing temperature on the magnetic properties is attributed to the change in the phase composition of the cores within the CNCs, in addition to their size. In the present samples, the maximum values of saturation magnetization, 80.6 and 83.4 A m²/kg (Table 17.2), were observed in samples with annealing temperatures of 873 and 973 K, respectively. These temperatures are the most favorable for cementite decomposition into α-Fe and graphite (Figure 17.13c). However, as the annealing temperature increased to 1173 K, the saturation magnetization started to decrease (71.7 A m²/kg). It can be assumed that this behavior relates to the thermodynamic stability of cementite at higher temperatures (Figure 17.13d). The CNCs that were annealed at 873 and 973 K have 40% of the saturation magnetization value of the bulk reference $\alpha$-Fe (204 A m$^2$/kg). As mentioned above, CNCs may incorporate nonmagnetic phases (amorphous and graphitic carbon), in addition to remaining cementite particles. The elimination of amorphous carbon from the synthesized samples may increase their saturation magnetization.

Following annealing, the coercivity, which mainly depends on the size of particles and their magnetic anisotropy, was raised in comparison with that of the as-synthesized sample. When the annealing temperatures were set at 673 and 773 K, the coercivity drastically increased to 10.90 and 10.85 kA/m (Table 17.2), respectively. This is because the majority of the CNCs' cores (that were in an amorphous state in the as-prepared sample) had been mainly transformed into crystalline cementite (Figure 17.13b). At elevated temperatures (873–1173 K), the cementite decomposed into $\alpha$-Fe and graphite (Figure 17.13c and d). The value of coercivity of the annealed samples was retained at 10–11 kA/m, which is five times greater than that of the reference bulk $\alpha$-Fe sample (2 kA/m). This is because many cores of the CNCs have critical dimensions close to the size of individual magnetic domains. Quantitatively, the critical sizes amount to 14 and 35 nm for the $\alpha$-Fe and iron carbides, respectively (Leslie-Pelecky & Rieke 1996, Yelsukov et al. 2003).

### 17.3.3.4 Postsynthesis Purification and Size Separation of CNCs

For technical applications, it is necessary to control the size of synthesized CNCs and any inclusions of amorphous and/or graphitic carbon need to be removed. All of the synthesis methods result in inhomogeneous products, which contain CNCs in addition to nonencapsulated metal particles and amorphous and graphitic carbon. These unwanted by-products can be removed using postsynthesis purification routes. Nonencapsulated metal particles are usually removed by leaching in diluted mineral acids. The etching of amorphous carbon has been systematically investigated on CNTs and is commonly realized by using concentrated hydrogen peroxide (Suzuki et al. 2006). The initial reagents should contain oxygen that would selectively etch the amorphous carbon, which has a resistance to oxygen significantly lower than those of other graphite structures. It has been shown (Zhang et al. 2005) that a small addition of molecular oxygen (1 vol% in the buffer gas) substantially reduced the amounts of amorphous carbon, during the growth of CNTs via a plasma-enhanced CVD route. Complete reduction of the amorphous carbon during the continuous synthesis of carbon-encapsulated magnetic nanoparticles was achieved when oxygen-containing compounds (e.g., alcohols: ethanol or methanol) were used as a carbon feedstock in an inductively coupled radio frequency thermal Ar-plasma torch (Bystrzejewski et al. 2009a). The minimal amorphous carbon content could also be achieved when hexane or toluene was processed in Ar–O$_2$ plasma (Bystrzejewski et al. 2009a). However, it has been shown (Bystrzejewski et al. 2009a, 2011) that it is very difficult to remove graphitic carbon forms such as graphite flakes and empty carbon onions.

It is important to be able to control the size of the CNCs, because it may dictate the magnetic response (superparamagnetic or ferromagnetic). The diameter of the CNCs is a key parameter in their applications; e.g., biomedical applications require small nanoparticles in order to minimize flow resistance, following their introduction into the body. There are only a few reports which investigate the size control of CNCs (Dravid et al. 1995, Bonard et al. 2001, He et al. 2006). The size of the CNCs and their production yield could be controlled by an easily adjustable parameter. This parameter is the grain size of the starting iron powder introduced into the Ar or Ar–He plasma torch (Bystrzejewski et al. 2011).

In this study, the CNCs synthesized by the electric plasma discharge method in the ultrasonic cavitation field of liquid ethanol usually include impurities of the nonencapsulated metal particles and other forms of carbon such as amorphous carbon and graphite balls and flakes. Additionally, the electric plasma discharge method synthesizes CNCs with a wide size distribution, from nanometers

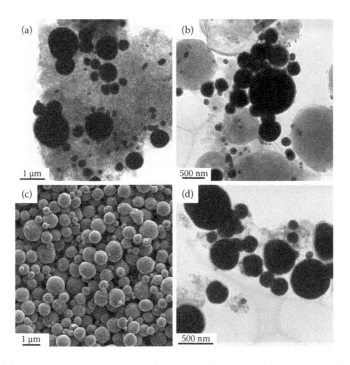

**FIGURE 17.18** (a) TEM image of the as-synthesized Fe–C powder sample; (b) TEM image of CNCs and graphite impurities after selective oxidation and acid treatment; (c, d) SEM and TEM images of CNCs after magnetic separation and centrifugation for 10 min at 1000 rpm.

to micrometers. A selective oxidation was conducted using 25% hydrogen peroxide solution refluxed at 363 K for 24 h (Kim et al. 2008). This is one of the safest, most nontoxic, and effective methods to remove amorphous carbon. Subsequently, the exposed iron carbide particles were removed by etching in 4 M HCl solution. Finally, magnetic separation was conducted in order to remove graphitic carbon impurities, and centrifugation was used to obtain monodispersed CNCs.

Figure 17.18a shows a typical TEM image of the as-synthesized carbon powder sample incorporating CNCs and amorphous carbon particles. A comparison between Figure 17.18a and b shows that the majority of the amorphous carbon was removed after treatment with hydrogen peroxide. The TEM image in Figure 17.18b revealed other impurities such as large graphite balls and flakes. This was because the hydrogen peroxide could not remove graphitic carbon. The results of the TEM and SEM observations indicated that the graphite impurities were easily removed from the sample by the magnetic separator. Thus, Figure 17.18c and d demonstrates the typical morphologies of the magnetically purified CNCs.

Figure 17.19 shows the SEM images of the (almost) monodispersed CNCs separated by different centrifuge speeds. It is obvious that as the centrifuge speed increases, the average size of the CNCs decreases, and their size distribution becomes narrower. Figure 17.20 shows the size distributions of the purified CNCs after each variation in centrifuge speed. The diameters of the CNCs were measured and estimated from the SEM images presented in Figure 17.19. Centrifugation at 1000 rpm resulted in the widest particle size distribution, from 200 to 1200 nm (Figure 17.20a). After centrifuging at 4000 rpm for 10 min, the size of the CNCs was approximately 100–300 nm, and the size distribution became sharper as shown in Figure 17.20d. The (almost) monodispersed CNCs, which have a diameter of approximately 100–300 nm and are covered with graphite shells with a thickness of approximately 5–12 nm, can be considered for use in further medical applications.

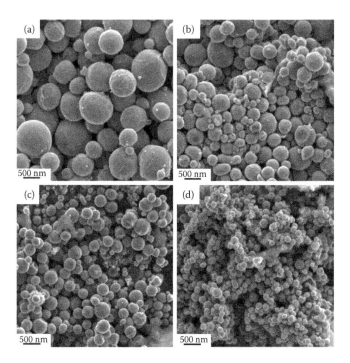

**FIGURE 17.19** SEM images of the purified iron carbide-filled CNCs after centrifugation for 10 min at different speeds: (a) 1000; (b) 2000; (c) 3000; and (d) 4000 rpm. (Reprinted from *Carbon*, 46, Kim, S., Shibata, E., Sergiienko, R. et al., Purification and separation of carbon nanocapsules as a magnetic carrier for drug delivery systems, 1523–1529, Copyright (2008), with permission from Elsevier.)

### 17.3.3.5 Magnetic Properties of Purified Iron Carbide-Filled CNCs

The magnetic properties of the purified CNCs (ranging from 100 to 300 nm in diameter) and the as-synthesized powder sample were measured with the VSM at room temperature. The results are presented in Figure 17.17 (curves 1 and 5) and Table 17.2. The value of the saturation magnetization of 112 A m$^2$/kg and the coercivity of 6 kA/m for the purified CNCs are higher than the respective values of 48 A m$^2$/kg and 4.16 kA/m, as reported for the as-synthesized sample. To account for the differences between the magnetic properties of the purified CNCs and those of the as-synthesized sample, it is clear that the proportion of the nonmagnetic phases (graphite impurities and amorphous carbon) has a detrimental effect on the saturation magnetization and coercivity. The iron carbide cores ($\chi$-Fe$_5$C$_2$, Fe$_3$C) of the purified CNCs have a value of saturation magnetization (112 A m$^2$/kg) that is comparable with the value of 110 A m$^2$/kg for the reference cementite (Fe$_3$C) sample (Table 17.2). This result confirms that the purification of CNCs (from carbon impurities) was satisfactorily conducted.

It was reasonable to compare our results with those for materials produced by different methods. The purified iron carbide-filled CNCs exhibit a significantly higher value of saturation magnetization in comparison with the reference samples (Table 17.3). The only exception is for the controlled size CNCs, with a minimum amount of amorphous carbon, which have the saturation magnetization values of 96 A m$^2$/kg and 118 A m$^2$/kg (Bystrzejewski et al. 2009a, 2011). The higher value of saturation magnetization (112 A m$^2$/kg), as compared with other reference samples (Table 17.3), may be related to a larger amount of iron in our purified material. The presence of carbon impurities could reduce the saturation magnetization of the reference samples (e.g., 56.21 A m$^2$/kg [Jiao et al. 1996], 50 A m$^2$/kg [Zhang et al. 2001], 53 A m$^2$/kg [Borysiuk et al. 2008], and 7.28 A m$^2$/kg [Abdullaeva et al. 2012]), and could cause weaker interaction of the magnetic domains. Moreover, the difference in the saturation magnetization is related

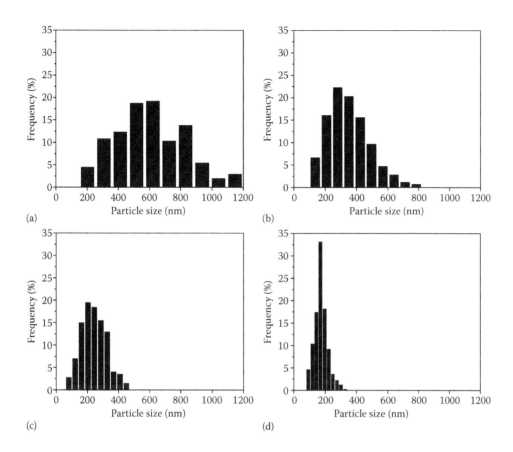

**FIGURE 17.20** The diameter distribution of iron carbide-filled CNCs after centrifugation for 10 min at different speeds: (a) 1000; (b) 2000; (c) 3000; and (d) 4000 rpm. (Reprinted from *Carbon*, 46, Kim, S., Shibata, E., Sergiienko, R. et al., Purification and separation of carbon nanocapsules as a magnetic carrier for drug delivery systems, 1523–1529, Copyright (2008), with permission from Elsevier.)

to a different nanoparticle size. The lower value of coercivity (6 kA/m) may provide evidence that this study created purified CNCs that are larger than the reference CNCs (e.g., Jiao et al. 1996, Si et al. 2003, El-Gendy et al. 2009, and Abdullaeva et al. 2012). The reduced value of coercivity may be related to the fact that the diameter of the iron carbide-filled CNCs is far from the critical size (100–300 nm > $D_c$) (Leslie-Pelecky & Rieke 1996) attributed to a threshold for the single magnetic domain structures and multidomains which exist in the iron carbide cores of the purified CNCs. Nanoparticles with a larger size than the critical size $D_c$ reduce the coercivity. As mentioned above, the critical sizes of $\alpha$-Fe and iron carbide crystallites amount to 14 and 35 nm, respectively (Leslie-Pelecky & Rieke 1996, Yelsukov et al. 2003).

The narrow hysteresis loop (Figure 17.17, curve 5) shows a low coercivity (6 kA/m) and negligible remanent magnetization (2.42 A m$^2$/kg), and thus implies soft ferromagnetic behavior of the purified CNCs. For biomedical applications, the use of iron carbide nanoparticles that possess relatively high saturation magnetization and have soft ferromagnetic behavior at room temperature is preferred. Soft magnetic character during drug delivery is necessary because once the external magnetic field is removed, the magnetization disappears; thus, agglomeration occurs. Hence, the strong magnetic interaction between the particles and the possible embolization of capillary vessels is avoided (Arruebo et al. 2007).

**TABLE 17.3** Comparison of the Magnetic Properties Measured at Room Temperature for the Iron-Filled CNCs Synthesized by Different Methods

| Composition of CNCs (Core/Shell) | $M_s$ (A m$^2$/kg) | $H_c$ (kA/m) | $M_r/M_s$ (%) | Particle Size (nm) | Reference |
|---|---|---|---|---|---|
| As-synthesized | 48 | 4.16 | 3.8 | 5–1000 | This study |
| Purified CNCs | 112 | 6 | 2.2 | 100–300 | This study |
| (α-Fe, γ-Fe/graphite)[a] | 56.21 | 48 | 30 | 32–81 | Jiao et al. 1996 |
| (α-Fe, Fe$_3$C/graphite)[b] | 96 | 19.36 | 15 | 5–80 | Bystrzejewski et al. 2009a |
| (α-Fe, γ-Fe, Fe$_3$C/ graphite)[b] | 118 | 14 | – | 10–100 | Bystrzejewski et al. 2011 |
| (α-Fe, γ-Fe/carbon)[c] | 79 | 35.28 | 20 | 2–58 | El-Gendy et al. 2009 |
| (α-Fe, Fe$_3$C/graphite)[d] | 50 | 14 | 12 | 10–100 | Zhang et al. 2001 |
| (α-Fe, γ-Fe, Fe$_3$C/ graphite)[e] | 85 | 31.76 | 24.5 | 6–40 | Si et al. 2003 |
| (α-Fe, γ-Fe, Fe$_3$C/ graphite)[f] | 53 | 16 | 15 | 10–100 | Borysiuk et al. 2008 |
| (Fe$_7$C$_3$, Fe$_2$C/graphite)[g] | 50 | 2 | – | 20–60 | Kouprine et al. 2006 |
| (α-Fe/graphite)[h] | 7.28 | 38.4 | 33.5 | 2–22 (7.9 average) | Abdullaeva et al. 2012 |

*Source: Materials Chemistry and Physics*, 122, Kim, S., Sergiienko, R., Shibata, E. et al., Iron-included carbon nanocapsules coated with biocompatible poly(ethylene glycol) shells, 164–168, Copyright (2010), with permission from Elsevier.

[a] Arc discharge process in He atmosphere.
[b] Inductively coupled radio frequency thermal Ar-plasma torch with addition of ethanol and C/Fe molar ratio of 0.95 (test A1).
[c] The high-pressure CVD.
[d] Arc discharge process in methane.
[e] Arc discharge in ethanol vapor.
[f] Arc plasma method in Ar–H$_2$ atmosphere.
[g] Capacitively coupled radio frequency plasma discharge.
[h] Pulsed electrical plasma discharge in a liquid ethanol.

Thus, magnetic metal-filled CNCs have the potential for biomedical applications. These include serving as contrast agents for MRI (Uo et al. 2005, Seo et al. 2006), in drug delivery (Bystrzejewski et al. 2005, Taylor et al. 2010, Saraswati et al. 2012), as near-IR light absorbers for selective cancer cell destruction (Seo et al. 2006, Liang et al. 2008), in magnetic hyperthermia (El-Gendy et al. 2009, Abdullaeva et al. 2012), as localized radio frequency absorbers in cancer therapy (Xu et al. 2008), and in magnetic separation of bacteria/antigens/proteins from biosamples.

In biomedical applications, owing to the fact that magnetic CNCs would be used in live environments, a homogenous and fine dispersion of the nanoparticles in liquid (usually water) is required. However, the fabricated particles typically contain elemental carbon, which is strongly hydrophobic and tends to aggregate when introduced into water. Therefore, for further biomedical applications, a CNC should be treated to create graphite surface functionalization. Proper surface modification could enhance the independent dispersion capability and achieve a uniform and stable dispersion in an aqueous medium. The surface treatment of the CNC, using plasma (Saraswati et al. 2012) or acid (Taylor et al. 2010), represents an important step before the biomolecule can be immobilized with various biocompatible hydrophilic polymers such as dextran (Saraswati et al. 2012) or PEG (Zhang et al. 2002). PEG is the most commercially important material for tissue engineering and other biomedical applications. PEG-coated nanoparticles demonstrate excellent solubility and stability in both aqueous solution and physiological saline (Kumagai et al. 2007).

### 17.3.3.6 Iron Carbide-Filled CNCs Coated with PEG

This section describes the structure and morphology of the iron carbide-filled CNCs coated with PEG 3400 (Sigma-Aldrich). The CNCs were purified and centrifuged at 4000 rpm for 10 min. The resultant

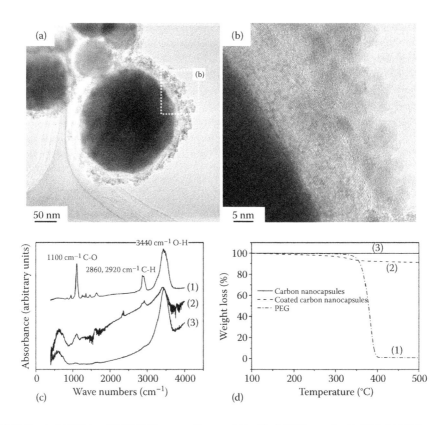

**FIGURE 17.21** (a, b) TEM and HRTEM images of iron carbide-filled CNC coated with PEG; (c) FTIR spectra; and (d) TG curves of (1) a pure PEG compound, (2) the purified CNC sample containing PEG, and (3) the purified CNC sample itself. (Reprinted from *Materials Chemistry and Physics*, 122, Kim, S., Sergiienko, R., Shibata, E. et al., Iron-included carbon nanocapsusles coated with biocompatible poly(ethylene glycol) shells, 164–168, Copyright (2010), with permission from Elsevier.)

CNCs ranged from 100 to 300 nm in diameter and were coated with PEG in a diethyl ether solution (Wako Pure Chemical Industries, Ltd.) (Kim et al. 2008).

The TEM images (Figure 17.21a and b) demonstrate the presence of an additional wispy amorphous coating (with a thickness of approximately 15–20 nm) on the graphite shell of the CNC that was dried from diethyl ether solution. FTIR was conducted to demonstrate the presence of the PEG compound in the purified sample of CNCs. Figure 17.21c presents the infrared spectra for the pure PEG sample (curve 1), the sample of CNCs containing the PEG compound (curve 2), and the sample of pure CNCs (curve 3). The absorption bands of 3435, 2860, 2920, and 1100 cm$^{-1}$ were observed owing to O–H, C–H, and C–O stretching vibrations, respectively, for the pure PEG sample (Figure 17.21c, curve 1). These peaks also existed in the sample of CNCs containing PEG (Figure 17.21c, curve 2). This implies that the amorphous coating observed on the TEM images (Figure 17.21a and b) represents PEG chains on the surface of the CNC. However, the FTIR spectrum (Figure 17.21c) does not provide any strong evidence of covalently bonded PEG chains on the graphite shell surface; thus, further investigation is required to reveal these covalent bonds.

Thermogravimetric (TG) analysis was performed on the CNCs to investigate the content of the PEG compound within the sample. Figure 17.21d shows the TG curves of the purified CNCs (curve 3), the purified CNC sample containing PEG (curve 2), and the pure PEG chemical reagent (curve 1). It is apparent that no weight loss occurs in Figure 17.21d (curve 3), which is ascribed to the purified CNC sample. In contrast, 100% weight loss of the pure PEG sample occurred at approximately 673 K, because of the thermal decomposition of PEG (curve 1). The weight loss of the CNCs with the PEG coating occurred mainly owing to the depolymerization of PEG, which occupied 9.15 mass% of the sample (curve 2).

The saturation magnetization significantly depends on the amount of magnetic material. Therefore, we may expect that the diamagnetic PEG coating will reduce the saturation magnetization of the purified CNCs by approximately 9%, as demonstrated by TG analysis.

Thus, magnetically soft CNCs with the PEG surface coating could possibly be used as biocompatible magnetic nanoparticles within medical applications. Iron carbide-filled CNCs with a PEG coating should stably disperse and provide a longer circulation time in an aqueous biological environment, maximizing the concentration of CNCs in the target tissues for diagnostic or therapeutic uses.

## 17.3.4 Plasma Discharge in Liquid Ethanol: Cobalt- and Cobalt Carbide-Filled CNCs

Cobalt nanoparticles, in addition to iron nanoparticles, are an interesting material for future magnetic applications. Cobalt is a ferromagnetic metal and it has the highest Curie temperature (1394 K) among the known magnetic materials. It also has a higher magnetic susceptibility and anisotropy constant in comparison with iron. The magnetic anisotropy constants of cubic crystal α-Fe are reported to be $K_1 = 4.8 \times 10^4$ J/m$^3$. This is weaker than those of hexagonal α-Co ($K_1 = 4.5 \times 10^5$ J/m$^3$) (Cullity & Graham 2009) and cubic β-Co ($K = 2.7 \times 10^5$ J/m$^3$) (Sucksmith & Thompson 1954). However, the coercivity is proportional to the anisotropy constant; therefore, high-anisotropy materials are attractive candidates for high-coercivity applications. At room temperature, the bulk Co has a saturation magnetization of approximately 162 A m$^2$/kg (Smithells 1998). Below 695 K, Co has a hexagonal close-packed (hcp) structure (α-Co). At higher temperatures, cobalt has an fcc structure (β-Co). There is very little difference between the saturation magnetizations of the α-Co and β-Co phases.

It is well known that the upper limit of the size of the nanostructured scale is considered to be 100 nm, as quantum size effects subsequently appear (Gleiter 1989, Siegel 1993). However, in most cases, a length factor becomes apparent in the range of 10 to 50 nm. Here, the structure, morphology, and magnetic properties of the cobalt- and cobalt carbide-filled CNCs (less than 100 nm in diameter) are reported and discussed.

In a study (Sergiienko et al. 2007), we changed the materials of the ultrasonic horn tip and electrodes from iron to cobalt. Cobalt plate (size 10 × 15 × 3 mm$^3$; purity 99.93%) and cobalt wire (diameter 2 mm; purity 99.8%) were utilized for the production of the ultrasonic tip and electrodes. During the sonication of liquid ethanol (Wako, 99.5% S grade), the voltage on the electrodes was kept at 55 V, using a regulated DC power supply (PAS 60-18). The upper limit of the current of the supply was set at 3 A. Thus, the electric plasma discharge in the liquid ethanol occurred at an electric power input as low as 165 W.

After the experiment, the carbonaceous powder that was dispersed in the liquid ethanol was allowed to stand in a glass bottle. After a few days, larger particles (>100 nm in size) settled to the bottom of the bottle, with smaller particles (<100 nm in size—nanoparticles) remaining in the liquid ethanol dispersion. The black suspension of small particles was poured into another bottle, whereas the large particles were left on the bottom of the first bottle. The nanoparticles were separated from the liquid ethanol by centrifugation, and the resulting carbon powder sample was then etched in 4 M HCl solution at 313 K for 24 h, in order to remove exposed and insufficiently carbon-encapsulated cobalt and cobalt oxide nanoparticles. Following subsequent drying in vacuum at 313 K, the carbon powder samples were annealed in a gas mixture of Ar + 3% H$_2$ at temperatures of 733 and 873 K for 2 h.

### 17.3.4.1 Structure and Morphology of the CNCs

By measuring the approximate diameters of more than 300 nanocapsules using TEM micrographs with various magnifications, the size distribution of the CNCs was obtained for the samples (before and after annealing). The CNC size was measured as the diameter of the outmost graphene layer for the as-synthesized sample, and this ranged from 4 to 40 nm, with a mean size of 8.3 nm (Figure 17.22a). The

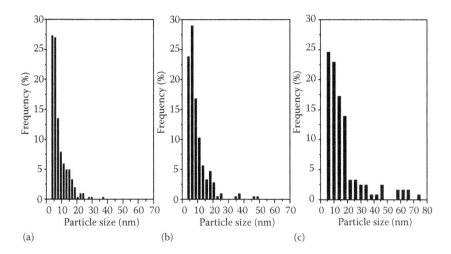

**FIGURE 17.22** The size distribution of cobalt- and cobalt carbide-filled CNCs: (a) as-synthesized sample; (b) after annealing at 733 K; (c) after annealing at 873 K. (Reprinted from *Acta Materialia*, 55, Sergiienko, R., Shibata, E., Zentaro, A. et al., Formation and characterization of graphite-encapsulated cobalt nanoparticles synthesized by electric discharge in an ultrasonic cavitation field of liquid ethanol, 3671–3680, Copyright (2007), with permission from Elsevier.)

diameter of the CNCs for the samples annealed at 733 K increased from 4 to 50 nm, with a mean size of 10.7 nm (Figure 17.22b). For the samples that were annealed at 873 K, the diameter of the CNCs ranged from 5 to 70 nm, with a mean size of 16.75 nm (Figure 17.22c). The increase in the size of the CNCs following annealing was likely owing to their sintering.

Figure 17.23a presents the as-synthesized CNCs; the average diameter of the nanocapsules was measured to be 8.5 nm. The cores of the CNCs were isolated from each other by graphitic carbon layers which were approximately 1–2 nm thick. The HRTEM analysis of the as-synthesized sample indicated weak particle crystallinity; however, there were some well-crystallized nanoparticles and, thus, we

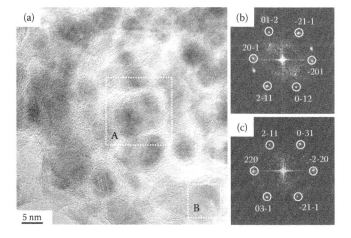

**FIGURE 17.23** (a) HRTEM image of the as-synthesized cobalt carbide-filled CNCs; (b, c) digital diffractograms of the areas surrounded by rectangles A and B in a, respectively. (Reprinted from *Acta Materialia*, 55, Sergiienko, R., Shibata, E., Zentaro, A. et al., Formation and characterization of graphite-encapsulated cobalt nanoparticles synthesized by electric discharge in an ultrasonic cavitation field of liquid ethanol, 3671–3680, Copyright (2007), with permission from Elsevier.)

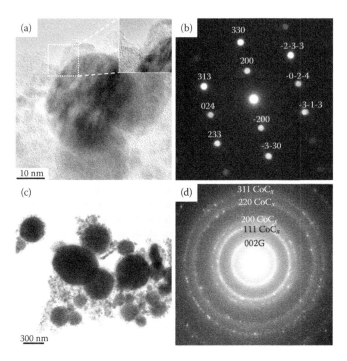

**FIGURE 17.24** (a) HRTEM image of the as-synthesized cobalt carbide-filled CNCs; (b) microbeam electron diffraction of nanocapsule shown in a; (c) TEM image of large precipitated CNCs; (d) SAED pattern of large nanocapsules shown in c. (Reprinted from *Acta Materialia*, 55, Sergiienko, R., Shibata, E., Zentaro, A. et al., Formation and characterization of graphite-encapsulated cobalt nanoparticles synthesized by electric discharge in an ultrasonic cavitation field of liquid ethanol, 3671–3680, Copyright (2007), with permission from Elsevier.)

could apply FFT to clarify the structure of some of the encapsulated nanoparticles. An FFT (Figure 17.23b) of the CNC core (in the area marked by rectangle A in Figure 17.23a) gives an arrangement of spots at spacings of 0.22, 0.21, and 0.205 nm. These are consistent with the (120), (012), and (121) planes, respectively, for the [142] zone-axis pattern of orthorhombic $Co_3C$ (JCPDS data of PDF No. 26-0450). An FFT (Figure 17.23c) of the lattice fringes (at the area marked by rectangle B in Figure 17.23a) gives a hexagonal arrangement of spots at 0.21 and 0.20 Å spacings. These are compatible with the (121) and (103, 022) planes, respectively, for the [–113] zone-axis pattern of cobalt carbide $Co_3C$.

The presence of cobalt carbide $Co_3C$ was also confirmed through microbeam electron diffraction of the larger individual particle (Figure 17.24a). The interlayer spacing derived from the diffraction spots (Figure 17.24b) are 0.22, 0.14, 0.12, 0.115, and 0.11 nm. These respectively correspond to the reported values of 0.2222 nm (200), 0.1392 nm (024), 0.120 nm (313), 0.1142 nm (233), and 0.1106 nm (330) for the orthorhombic $Co_3C$ (JCPDS data of PDF No. 26-0450). The cobalt carbide core of approximately 40 nm in diameter is tightly surrounded by the carbon shell (Figure 17.24a). The carbon shell consists of nine layers, with an interlayer spacing of approximately 0.34 nm, which is close to that of the graphite (002) planes. During the analysis, we also found spherical CNCs with amorphous cores (Sergiienko et al. 2007).

The presence of another cobalt carbide form ($CoC_x$) was verified using diffraction analysis of precipitated large particles similar to that shown in Figure 17.24c. In this study, the large precipitated CNCs (>100 nm in diameter) were considered to be a by-product. Thus, the distinct diffraction rings in Figure 17.24d have lattice spacings of 0.21, 0.18, 0.13, and 0.11 nm. These can respectively be indexed as the (111), (200), (220), and (311) reflections of cobalt carbide $CoC_x$ (a carbon supersaturation into fcc β-Co phase); this is in agreement with the JCPDS data of PDF No. 44-0962. Thus, in the conditions of high-speed quenching, the metastable cobalt carbide phases ($CoC_x$ and $Co_3C$) were formed and subsequently solidified into the cores which were entrapped in the graphite shells of the as-synthesized CNCs.

To clarify the formation mechanism of the synthesized cobalt carbide-filled carbon CNCs, we employed the spectroscopic system, which is described elsewhere (Sergiienko et al. 2007). The optical emission spectrum from the plasma discharge was collected using an optical probe. This was immersed in the ethanol and the transmission was via an optical fiber to a USB 2000 spectrometer, connected to a personal computer. The formation mechanism of the cobalt carbide-filled CNCs was probably similar to the mechanism which was proposed in our previous work (Sergiienko et al. 2006b). That is, when cobalt is evaporated simultaneously to the decomposition of ethanol molecules by the plasma discharge, cobalt atoms and ions (Co I, Co II) and carbon species ($C_2$) react with each other, and the liquid cobalt–carbon alloy particles are formed in close proximity to the plasma discharge zone. The supersaturated carbon segregates from the alloy particles until the composition of the alloy reaches that of the stoichiometric cobalt carbide during the quenching. Graphite layers are formed on the outer surface of the particle, and because cobalt has high carbon content, cobalt carbide is entrapped in the core. The amorphous and low crystalline cobalt carbides have a low saturation magnetization value because of the presence of carbon within the cores of the CNCs. Therefore, they must be annealed at 695 K (or higher) to develop the well-crystallized magnetic phases such as hcp α-Co and fcc β-Co.

The HRTEM observation of the annealed samples revealed that the cobalt nanoparticles had existed in both the fcc β-Co and hcp α-Co forms. For cobalt, the fcc β-Co phase is actually a high-temperature magnetic phase (>695 K), which is characterized by a slightly weaker magnetism (~5%) than the low-temperature hcp phase of α-Co. However, it appears that fcc is a stable phase at room temperature, when the particle size is on the order of a few tens of nanometer (Kitakami et al. 1997).

Figure 17.25a shows an HRTEM image of cobalt nanoparticles encapsulated in graphitic carbon that were annealed at 873 K. The average diameter of the uniformly dispersed cobalt cores was estimated to be approximately 5 nm, with a fairly narrow size distribution (having a standard deviation of 1.1 nm). In size and form, the cores resemble those observed in magnetic films which are synthesized by the codeposition of Co and carbon, using ion-beam sputtering (Hayashi et al. 1996). The cobalt cores are completely isolated from each other by graphitic carbon layers which are approximately 1–2 nm in thickness; the lattice images of carbon with approximately 0.36 nm spacing are clearly seen. The XRD peak at $2\theta \approx 26°$ corresponding to $d_{002} = 0.342$ nm (Figure 17.26) is shown by the annealed and as-synthesized samples. This suggests that a large population of cobalt nanoparticles have graphite shells.

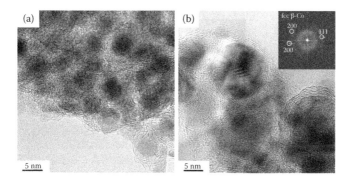

**FIGURE 17.25** (a) HRTEM image of cobalt nanoparticles divided between each other with graphite layers; (b) HRTEM image of cobalt-filled CNCs. Samples annealed at 873 K. Inset in b is the FFT image of CNCs.

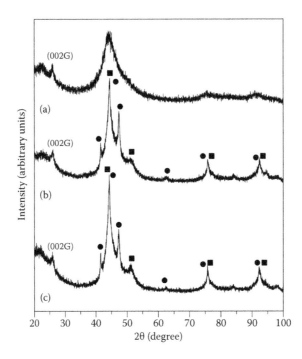

**FIGURE 17.26** XRD patterns of the Co–C powder samples: (a) the as-synthesized sample after etching in 4 M HCl solution; (b) annealed at 733 K; (c) annealed at 873 K. ●: α-Co; ■: β-Co. (Reprinted from *Acta Materialia*, 55, Sergiienko, R., Shibata, E., Zentaro, A. et al., Formation and characterization of graphite-encapsulated cobalt nanoparticles synthesized by electric discharge in an ultrasonic cavitation field of liquid ethanol, 3671–3680, Copyright (2007), with permission from Elsevier.)

The presence of pure cobalt phase in two forms (fcc and hcp) and the cobalt carbide decomposition, which occurred after annealing at elevated temperatures (733 and 873 K), were confirmed by XRD patterns (Figure 17.26b and c) and digital diffractograms of the lattice images of the CNCs' cores (Figure 17.25b). The diffraction spots in the inset of Figure 17.25b can be indexed to pure fcc cobalt. This is in agreement with the JCPDS data of PDF No. 15-0806.

### 17.3.4.2 Magnetic Properties of Cobalt- and Cobalt Carbide–Filled CNCs

The hysteresis loops of the powder samples (both the as-synthesized samples and the samples after annealing at 733 and 873 K) are presented in Figure 17.27. To compare the magnetic properties of the different samples, the data are summarized in Table 17.4. Magnetic measurements revealed that the as-synthesized powder sample was ferromagnetic, with a low ratio of remanent to saturation magnetization, $M_r/M_s = 2.31\%$. The saturation magnetization of the as-synthesized CNCs was approximately 24.22 A m²/kg and the coercivity was 5.17 kA/m. Following annealing at temperatures of 733 and 873 K, the $M_s$ value and the $M_r/M_s$ ratio increased by factors of 2 and 4, respectively, and the $H_c$ tripled in comparison with those of the as-synthesized sample. Upon heating, the saturation magnetization value increased owing to the decomposition of cobalt carbides into α-Co, β-Co, and graphite, and this was confirmed by the XRD and TEM investigations. Following annealing, the coercivity increased. This was likely caused by an increase in the magnetic anisotropy energy with the structure change, in addition to the segregation of carbon from the cores of the CNCs. Nanoparticles with different particle sizes may contribute differently to the mechanism of coercivity (Leslie-Pelecky & Rieke 1996). The ferromagnetic nanoparticles have a maximum coercivity at a critical size, but nanoparticles with a larger or smaller size than the critical size would reduce the coercivity. Studies performed by other researchers on carbon-encapsulated

**TABLE 17.4**　Magnetic Properties of the Synthesized Co–C Powder Samples and Other Reference Cobalt-Filled CNCs Measured at Room Temperature

| Samples (Core/Shell) | $M_s$ (A m²/kg) | $H_c$ (kA/m) | $M_r/M_s$ (%) | Mean Particle Diameter (nm) | Reference |
|---|---|---|---|---|---|
| As-prepared | 24.22 | 5.17 | 2.31 | 8.3 | This study |
| Annealed at 733 K | 49.15 | 15.50 | 8.65 | 10.7 | This study |
| Annealed at 873 K | 51.04 | 15.50 | 9.01 | 16.75 | This study |
| (hcp α-Co/graphite)[a] | 61.8 | 29.46 | 35.60 | 8 | Hayashi et al. 1996 |
| (fcc β-Co/graphite)[b] | 35.7 | 36.63 | 32.21 | 8.5 | Abdullaeva et al. 2012 |
| (fcc β-Co/graphite)[c] | 70 | 26.03 | 21.43 | 10.7 | Host et al. 1998 |
| (fcc β-Co/graphite)[d] | – | 19.11 | 14.50 | 20 | Bonard et al. 2001 |
| (fcc β-Co/graphite)[d] | 114.13 | 15.92 | 28.91 | 40 | Jiao et al. 1996 |
| Bulk (α + β)–Co[e] | 155.64 | 12.46 | 7.31 | 45,000 | Purchased |
| Bulk Co | 152 | 4.38 | 31.70 | 80,000 | Fu et al. 2006 |

[a] Codeposition of Co and carbon by ion-beam sputtering with subsequent film annealing at 623 K.
[b] Pulsed electrical discharge plasma in a liquid ethanol.
[c] The tungsten arc technique where an electric arc coevaporates cobalt and carbon from a liquid metal pool.
[d] An arc-discharge technique modified in the geometry of the anode and the flow pattern of helium gas.
[e] Wako, purity 99.995%.

cobalt nanoparticles, with a morphology and structure comparable to those of ours, yielded higher values of the coercive force (about 26.03–29.5 kA/m), and these reference particles are considered as single magnetic domains (Hayashi et al. 1996, Host et al. 1998). More works (Saito et al. 1997a, Bonard et al. 2001) stated that the cobalt nanoparticles with a mean diameter of less than 20 nm had a single magnetic domain structure. In the majority of this study, we assumed that the annealed cobalt-filled CNCs with diameters of less than 20 nm have a single magnetic domain structure. However, the coercivity of our cobalt nanoparticles was observed to be lower than those of some of the reference samples. This difference can be explained by the fact that the coercivity is dependent on many structural parameters such as the orientation, shape, internal stress, and crystal defects.

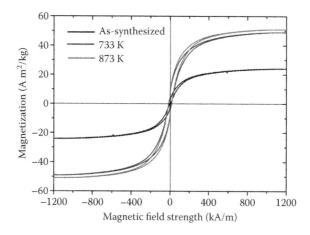

**FIGURE 17.27**　Hysteresis loops of the as-synthesized and annealed cobalt- and cobalt carbide-filled CNCs measured at room temperature. (Reprinted from *Acta Materialia*, 55, Sergiienko, R., Shibata, E., Zentaro, A. et al., Formation and characterization of graphite-encapsulated cobalt nanoparticles synthesized by electric discharge in an ultrasonic cavitation field of liquid ethanol, 3671–3680, Copyright (2007), with permission from Elsevier.)

The difference in the coercivity and saturation magnetization values between the bulk and nanoparticles can be attributed to the small particle size effect. During this study, carbon powder samples annealed at 733 and 873 K reach approximately 32% of the saturation magnetization value of the bulk reference cobalt particles (155.64 and 152 A m$^2$/kg in Table 17.4). A previous research (Jiao et al. 1996) has reported the synthesis of Fe, Ni, and Co nanocrystals produced with a modified arc discharge method with no heat treatment. Table 17.4 demonstrates the data from this previous research and the data from our study. Since the average particle size (40.0 nm) from Jiao et al.'s research is approximately four times that of ours (10.7 nm), they therefore measured larger magnetization (114.13 A m$^2$/kg).

There was little difference between the magnetic properties of the CNCs annealed at temperatures of 733 and 873 K (Table 17.4 and Figure 17.27). This is because it is difficult for carbon to be soluble in both hcp and fcc cobalt cores. Additionally, there was no significant change in the size of the cobalt-filled CNCs.

## 17.4 Summary

This chapter discusses the synthesis of the metal-filled CNCs by the developed plasma processing in the ultrasonic cavitation field of organic liquid. This chapter describes the structure and magnetic properties of the CNCs.

In Section 17.1, we gave a short literary review on the possible biomedical applications and different synthesis methods of magnetic metal-filled CNCs.

In the developed method, electric plasma is discharged in an organic liquid under ultrasonic irradiation. The presence of this ultrasonic cavitation field, with its many activated tiny bubbles, may enhance the electric conductivity owing to the formed radicals, atoms, and ions. The electric plasma discharge is generated at a relatively low power. In contrast to traditional and modified arc discharge units, the developed method has advantages: (1) the plasma processing is produced under conditions of low electric power input (100–200 W); (2) expensive vacuum equipment is not required; (3) graphite debris and CNTs are not present; and (4) high cooling rates of the CNCs are achieved by the organic liquid.

However, all the synthesis methods result in inhomogeneous products. These contain CNCs in addition to nonencapsulated metallic particles and amorphous carbon. For further technical applications of CNCs, the amorphous carbon is an unwanted fraction. This is because this carbon phase diminishes the overall magnetic performance and its removal is more complicated in comparison to that of the nonencapsulated metal particles. The amount of amorphous carbon could be regulated by the use of a wide variety of organic liquids. Ethanol was found to be the best carbon source, because the decomposition of ethanol molecules synthesized a smaller amount of amorphous carbon in comparison with the decomposition of benzene and hexane compounds.

The cores of the as-synthesized CNCs were determined to be carbides in crystalline (TiC, WC$_{1-x}$, W$_2$C, WC, Fe$_3$C, $\chi$-Fe$_{2.5}$C, Co$_3$C, CoC$_x$) and amorphous states. The sizes of the metal carbide-filled CNCs ranged widely from several nanometers to micrometers. After gravitational settling, the diameter of the cobalt carbide-filled CNCs ranged from 4 to 40 nm, with a mean diameter of 8 nm. It was established that the number of graphite layers was dependent on the diameter of the CNC core. For example, iron carbide cores with a diameter of less than 20 nm were covered with several (~3–8) graphite layers, but large cores with a diameter greater than 100 nm had approximately 30–40 graphite layers on their surface.

The optical emission spectra from the plasma discharge in liquid ethanol were used to provide an explanation for the formation mechanism of the CNCs.

For further technical applications, it is very important to control the size and the purity of CNCs. In this study, we used simple techniques for this purpose: (1) a sample treatment with diluted hydrochloric acid and hydrogen peroxide, in order to remove amorphous carbon and nonencapsulated metal particles; (2) magnetic separation of the CNCs from the graphite impurities; and (3) centrifugation to separate the various particle sizes.

The metastable metal carbides had decomposed into stable phases ($\alpha$-Fe, $\alpha$-Co, $\beta$-Co, WC) at annealing temperatures of 733, 873, and 1173 K. Following annealing at elevated temperatures, the nanoparticles became larger because of their sintering.

The effect of the annealing temperature on the magnetic properties of CNCs was examined. It was found that the saturation magnetization and coercivity of the annealed powder samples had been increased several times, in comparison with those of the as-synthesized carbon powder samples. This is because the annealing changed the phase composition, structure, and size of the CNCs. The values of the saturation magnetization of the annealed CNCs were 80.6 and 50 A m²/kg representing 40% and 32% of the saturation magnetizations of the bulk ferromagnetic elements iron and cobalt, respectively. This difference can be explained by the presence of residual nonmagnetic components in the synthesized samples (amorphous carbon and graphite shells) and the small particle size effect. The values of coercivity of the iron ($\alpha$-Fe) and cobalt ($\alpha$-Co, $\beta$-Co) nanoparticles encapsulated in graphite shells that had been annealed at elevated temperatures were enhanced in contrast to that of the bulk reference samples. This observation may be attributed to the nanosize effect.

# References

Abdullaeva, Z., Omurzak, E., Iwamoto, C. et al., "Onion-like carbon-encapsulated Co, Ni, and Fe magnetic nanoparticles with low cytotoxicity synthesized by a pulsed plasma in a liquid," *Carbon* 50 (2012): 1776–1785.

Arruebo, M., Fernandez-Pacheco, R., Ibarra, M. R. et al., "Magnetic nanoparticles for drug delivery," *Nanotoday* 2 (2007): 22–32.

Babonneau, D., Cabioc'h, T., Naudon, A. et al., "Silver nanoparticles encapsulated in carbon cages obtained by co-sputtering of the metal and graphite," *Surf. Sci.* 409 (1998): 358–371.

Bandow, S. & Saito, Y., "Encapsulation of ZrC and $V_4C_3$ in graphite nanoballs via arc burning of metal carbides/graphite composites," *Jpn. J. Appl. Phys.* 32 (1993): L1677–L1680.

Berkowitz, A. E., Hansen, M. F., Parker, F. T. et al., "Amorphous soft magnetic particles produced by spark erosion," *J. Magn. Magn. Mat.* 254–255 (2003): 1–6.

Berry, C. C., "Progress in functionalization of magnetic nanoparticles for applications in biomedicine," *J. Phys. D Appl. Phys.* 42 (2009): 224003–224011.

Bi, X. X., Ganguly, B., Huffman, G. P. et al., "Nanocrystalline $\alpha$-Fe, $Fe_3C$, and $Fe_7C_3$ produced by $CO_2$ laser pyrolysis," *J. Mater. Res.* 8 (1993): 1666–1674.

Bonard, J. M., Seraphin, S., Wegrowe, J. E. et al., "Varying the size and magnetic properties of carbon-encapsulated cobalt particles," *Chem. Phys. Lett.* 343 (2001): 251–257.

Borysiuk, J., Grabias, A., Szczytko, J. et al., "Structure and magnetic properties of carbon encapsulated Fe nanoparticles obtained by arc plasma and combustion synthesis," *Carbon* 46 (2008): 1693–1701.

Bystrzejewski, M., Huczko, A. & Lange, H., "Arc plasma route to carbon-encapsulated magnetic nanoparticles for biomedical applications," *Sensor Actuat. B Chem.* 109 (2005): 81–85.

Bystrzejewski, M., Cudziło, S., Huczko, A. et al., "Thermal stability of carbon-encapsulated Fe–Nd–B nanoparticles," *J. Alloys Compd.* 423 (2006): 74–76.

Bystrzejewski, M., Huczko, A., Lange, H. et al., "Large scale synthesis of carbon encapsulated magnetic nanoparticles," *Nanotechnology* 18 (2007a): 145608 1–9.

Bystrzejewski, M., Huczko, A., Lange, H. et al., "Combustion synthesis route to carbon-encapsulated iron nanoparticles," *Diam. Relat. Mater.* 16 (2007b): 225–228.

Bystrzejewski, M., Karoly, Z., Szepvolgyi, J. et al., "Continuous synthesis of carbon-encapsulated magnetic nanoparticles with a minimum production of amorphous carbon," *Carbon* 47 (2009a): 2040–2048.

Bystrzejewski, M., Pyrzyńska, K., Huczko, A. et al., "Carbon encapsulated magnetic nanoparticles as separable and mobile sorbents of heavy metal ions from aqueous solutions," *Carbon* 47 (2009b): 1201–1204.

Bystrzejewski, M., Karoly, Z., Szépvölgyi, J. et al., "Continuous synthesis of controlled size carbon-encapsulated iron nanoparticles," *Mater. Res. Bull.* 46 (2011): 2408–2417.

Carpenter, E. E., Sangregorio, C. & O'Connor, C. J., "Effects of shell thickness on blocking temperature of nanocomposites of metal particles with gold shells," *IEEE Trans. Magn.* 35 (1999): 3496–3498.

Chuistov, K. V. & Perekos, A. E., "Structure and properties of fine metal particles: I. Phase-structural state and magnetic characteristics," *Metallofiz. Novejš. Tekhnol.* 19 (1997): 36–53 (in Russian).

Cullity, B. D. & Graham, C. D., "Magnetic anisotropy," in *Introduction to Magnetic Materials* (IEEE Press, John Wiley & Sons, Inc., Hoboken, 2009), pp. 197–239.

Degrois, M. & Baldo, P., "A new electrical hypothesis explaining sonoluminescence, chemical actions and other effects produced in gaseous cavitation," *Ultrasonics* 14 (1974): 25–29.

Dong, X. L., Zhang, Z. D., Xiao, Q. F. et al., "Characterization of ultrafine $\gamma$-Fe(C), $\alpha$-Fe(C) and $Fe_3C$ particles synthesized by arc-discharge in methane," *J. Mat. Sci.* 33 (1998): 1915–1919.

Dravid, V. P., Host, J. J., Teng, M. H. et al., "Controlled-size nanocapsules," *Nature* 374 (1995): 602.

Dvornik, M. I., "Nanostructured WC–Co particles produced by carbonization of spark eroded powder: Synthesis and characterization," *Int. J. Refract. Met. Hard Mater.* 28 (2010): 523–528.

El-Gendy, A. A., Ibrahim, E. M. M., Khavrus, V. O. et al., "The synthesis of carbon coated Fe, Co and Ni nanoparticles and an examination of their magnetic properties," *Carbon* 47 (2009): 2821–2828.

Fang, Z. Z., Wang, X., Ryu, T. et al., "Synthesis, sintering, and mechanical properties of nanocrystalline cemented tungsten carbide—A review," *Int. J. Refract. Met. Hard Mater.* 27 (2009): 288–299.

Fitzgerald, M. E., Griffing, V. & Sullivan, J., "Chemical effects of ultrasonics—'Hot spot' chemistry," *J. Chem. Phys.* 25 (1956): 926–933.

Frenkel, Y. I., "Electrical phenomena connected with cavitation caused by ultrasonic oscillations in a liquid," *Russ. J. Phys. Chem.* 14 (1940): 305–308.

Fu, W., Yang, H., Hari-Bala, Liu, S. et al., "Preparation and characteristics of core-shell structure cobalt/silica nanoparticles," *Mater. Chem. Phys.* 100 (2006): 246–250.

Fuhr, J. & Schmidt, W. F., "Spark breakdown of liquid hydrocarbons: II. Temporal development of the electric spark resistance in *n*-pentane, *n*-hexane, 2,2 dimethylbutane, and *n*-decane," *J. Appl. Phys.* 59 (1986): 3702–3708.

Fuhr, J., Schmidt, W. F. & Sato, S., "Spark breakdown of liquid hydrocarbons: I. Fast current and voltage measurements of the spark breakdown in liquid *n*-hexane," *J. Appl. Phys.* 59 (1986): 3694–3701.

Funasaka, H., Sugiyama, K., Yamamoto, K. et al., Paper presented at 1993 Fall Meeting of Mat. Res. Soc., Boston, November–December 1993 (1993).

Gleiter, H., "Nanocrystalline materials," *Prog. Mater. Sci.* 33 (1989): 223–315.

Guerret-Piécourt, C., Le Bouar, Y., Lolseau, A. et al., "Relation between metal electronic structure and morphology of metal compounds inside carbon nanotubes," *Nature* 372 (1994): 761–765.

Harvey, E. N., "Sonoluminescence and sonic chemiluminescence," *J. Am. Chem. Soc.* 61 (1939): 2392–2398.

Hayashi, T., Hirono, S., Tomita, M. et al., "Magnetic thin films of cobalt nanocrystals encapsulated in graphite-like carbon," *Nature* 381 (1996): 772–774.

He, C. N., Du, X. W., Ding, J. et al., "Low-temperature CVD synthesis of carbon-encapsulated magnetic Ni nanoparticles with a narrow distribution of diameters," *Carbon* 44 (2006): 2330–2333.

Herdt, A. R, Kim, B. S. & Taton, T. A., "Encapsulated magnetic nanoparticles as supports for proteins and recyclable biocatalysts," *Bioconjugate Chem.* 18 (2007): 183–189.

Hihara, T., Onodera, H., Sumiyama, K. et al., "Magnetic properties of iron in nanocapsules," *Jpn. J. Appl. Phys.* 33 (1994): L24–L25.

Host, J. J., Block, J. A., Parvin, K. et al., "Effect of the annealing on the structure and magnetic properties of graphite encapsulated nickel and cobalt nanocrystals," *J. Appl. Phys.* 83 (1998): 793–801.

Hsin, Y. L., Hwang, K. C., Chen, F. R. et al., "Production and in-situ metal filling of carbon nanotubes in water," *Adv. Mater.* 13 (2001): 830–833.

Huo, J., Song, H., Chen, X. et al., "Structural transformation of carbon-encapsulated iron nanoparticles during heat treatment at 1000°C," *Mater. Chem. Phys.* 101 (2007): 221–227.

Hwang, J. H. & Dravid, V. P., "Magnetic properties of graphitically encapsulated nickel nanocrystals," *J. Mater. Res.* 12 (1997): 1076–1082.

Jacob, D. S., Genish, I., Klein, L. et al., "Carbon-coated core shell structured copper and nickel nanoparticles synthesized in an ionic liquid," *J. Phys. Chem. B* 110 (2006): 17711–17714.

Jeong, U., Teng, X., Wang, Y. et al., "Superparamagnetic colloids: Controlled synthesis and nice applications," *Adv. Mater.* 19 (2007): 33–60.

Jiao, J. & Seraphin, S., "Carbon encapsulated nanoparticles of Ni, Co, Cu, and Ti," *J. Appl. Phys.* 83 (1998): 2442–2448.

Jiao, J., Seraphin, S., Wang, X. et al., "Preparation and properties of ferromagnetic carbon-coated Fe, Co, and Ni nanoparticles," *J. Appl. Phys.* 80 (1996): 103–108.

Kim, S., Shibata, E., Sergiienko, R. et al., "Purification and separation of carbon nanocapsules as a magnetic carrier for drug delivery systems," *Carbon* 46 (2008): 1523–1529.

Kim, S., Sergiienko, R., Shibata, E. et al., "Production of graphite nanosheets by low-current plasma discharge in liquid ethanol," *Mater. Trans.* 51 (2010a): 1455–1459.

Kim, S., Sergiienko, R., Shibata, E. et al., "Iron-included carbon nanocapsules coated with biocompatible poly(ethylene glycol) shells," *Mater. Chem. Phys.* 122 (2010b): 164–168.

Kitakami, O., Sato, H., Shimada, Y. et al., "Size effect on the crystal phase of cobalt fine particles," *Phys. Rev. B* 56 (1997): 13849–13854.

Kobayashi, Y., Horie, M., Konno, M. et al., "Preparation and properties of silica-coated cobalt nanoparticles," *J. Phys. Chem. B* 107 (2003): 7420–7425.

Kouprine, A., Gitzhofer, F., Boulos, M. et al., "Synthesis of ferromagnetic nanopowders from iron pentacarbonyl in capacitively coupled RF plasma," *Carbon* 44 (2006): 2593–2601.

Kovtun, V. A. & Pasovets, V. N., "Carbon nanostructures: Properties and trends of application in powder composite materials for friction units," *J. Frict. Wear (Trenie Iznos)* 27 (2006): 206–215 (In Russian).

Kovtun, V. A., Pasovets, V. N. & Kharlamov, A. I., "Tribological characteristics of composite materials based on copper–carbon nanotube powder system," *J. Frict. Wear* 29 (2008): 335–339.

Kovtun, V. A., Pasovets, V. N., Mikhovsky, M. et al., "Highly wear-resistant composites based on copper powder mixture with onion carbon nanostructures for self-lubricating friction units," *J. Frict. Wear* 31 (2010): 128–132.

Kumagai, M., Imai, Y., Nakamura, T. et al., "Iron hydroxide nanoparticles coated with poly(ethylene glycol)–poly(aspartic acid) block copolymer as novel magnetic resonance contrast agents for in vivo cancer imaging," *Colloids Surf. B Biointerfaces* 56 (2007): 174–181.

Lamber, R., Jaeger, N. & Schulz-Ekloff, G., "Electron microscopy study of the interaction of Ni, Pd and Pt with carbon: II Interaction of palladium with amorphous carbon," *Surf. Sci.* 227 (1990): 15–23.

Lange, H., Sioda, M., Huczko, A. et al., "Nanocarbon production by arc discharge in water," *Carbon* 41 (2003): 1617–1623.

Lassner, E. & Schubert, W. D., "Tungsten compounds and their application," in *Tungsten: Properties, Chemistry, Technology of the Element, Alloys, and Chemical Compounds* (Kluwer Academic/Plenum Publishers, New York, 1999), pp. 133–178.

Leslie-Pelecky, D. L. & Rieke, R. D., "Magnetic properties of nanostructured materials," *Chem. Mater.* 8 (1996): 1770–1783.

Li, Z. Q., Zhang, H. F., Zhang, X. B. et al., "Nanocrystalline tungsten carbide encapsulated in carbon shells," *Nanostruct. Mater.* 10 (1998): 179–184.

Liang, Y. C., Hwang, K. C. & Lo, S. C., "Solid-state microwave-arcing-induced formation and surface functionalization of core/shell metal/carbon nanoparticles," *Small* 4 (2008): 405–409.

Lin, M. H., "Synthesis of nanophase tungsten carbide by electrical discharge machining," *Ceram. Int.* 31 (2005): 1109–1115.

Ling, J., Liu, Y., Hao, G. et al., "Preparation of carbon-coated Co and Ni nanocrystallites by a modified AC arc discharge method," *Mater. Eng. B* 100 (2003): 186–190.

Liu, W., Soneda, Y., Kodama, M. et al., "Low-temperature preparation and electrochemical capacitance of WC/carbon composites with high specific surface area," *Carbon* 45 (2007): 2759–2767.

Lu, A. H., Salabas, E. L. & Schuth, F., "Magnetic nanoparticles: Synthesis, protection, functionalization, and application," *Angew. Chem. Int. Ed.* 46 (2007): 1222–1244.

Luna, C., Morales, M. P., Serna, C. J. et al., "Effects of surfactants on the particle morphology and self-organization of Co nanocrystals," *Mater. Sci. Eng. C* 23 (2003): 1129–1132.

Margulis, M. A., "Sonoluminescence and sonochemical reactions in cavitation fields: A review," *Ultrasonics* 23 (1985): 157–169.

Margulis, M. A. & Margulis, I. M., "Contemporary review on nature of sonoluminescence and sonochemical reactions," *Ultrason. Sonochem.* 9 (2002): 1–10.

McHenry, M. E., Majetich, S. A., Artman, J. O. et al., "Superparamagnetism in carbon-coated Co particles produced by the Kratschmer carbon arc process," *Phys. Rev. B* 49 (1994): 11358–11363.

Murakami, Y., Shibata, T., Okuyama, T. et al., "Structural, magnetic and superconducting properties of graphite nanotubes and their encapsulation compounds," *J. Phys. Chem. Solids* 54 (1994): 1861–1870.

Okada, T., Kaneko, T. & Hatakeyama, R., "Conversion of toluene into carbon nanotubes using arc discharge plasmas in solution," *Thin Solid Films* 515 (2007): 4262–4265.

Omurzak, E., Jasnakunov, J., Mairykova, N. et al., "Synthesis method of nanomaterials by pulsed plasma in liquid," *J. Nanosci. Nanotechnol.* 7 (2007): 3157–3159.

Ōya, A. & Ōtani, S., "Catalytic graphitization of carbons by various metals," *Carbon* 17 (1979): 131–137.

Pankhurst, Q. A., Thanh, N. K. T., Jones, S. K. et al., "Progress in applications of magnetic nanoparticles in biomedicine," *J. Phys. D Appl. Phys.* 42 (2009): 224001–224015.

Park, J. B., Jeong, S. H., Jeong, M. S. et al., "Synthesis of carbon-encapsulated magnetic nanoparticles by pulsed laser irradiation of solution," *Carbon* 46 (2008): 1369–1377.

Parkansky, N., Alterkop, B., Boxman, R. L. et al., "Pulsed discharge production of nano- and microparticles in ethanol and their characterization," *Powder Technol.* 150 (2005): 36–41.

Paschen, F., "Über die zum Funkenübergang in Luft, Wasserstoff und Kohlensäure bei verschiedenen Drücken erforderliche Potentialdifferenz," *Weid. Ann. Phys.* 37 (1889): 69.

Pearse, R. W. B. & Gaydon, A. G., *The Identification of Molecular Spectra*, Fourth edition (Chapman and Hall, London, UK, 1976).

Popławska, M., Żukowska, G. Z., Cudziło, S. et al., "Chemical functionalization of carbon-encapsulated magnetic nanoparticles by 1,3-dipolar cycloaddition of nitrile oxide," *Carbon* 48 (2010) 1318–1320.

Qiu, J., Li, Y., Wang, Y. et al., "Synthesis of carbon-encapsulated nickel nanocrystals by arc-discharge of coal-based carbons in water," *Fuel* 83 (2004): 615–617.

Qiu, J., Li, Q., Wang, Z. et al., "CVD synthesis of coal-gas-derived carbon nanotubes and nanocapsules containing magnetic iron carbide and oxide," *Carbon* 44 (2006): 2565–2568.

Raj, K., Moskowitz, B. & Casciari, R., "Advances in ferrofluid technology," *J. Magn. Magn. Mater.* 149 (1995): 174–180.

Reiss, G. & Hütten, A., "Magnetic nanoparticles—Applications beyond data storage," *Nat. Mater.* 4 (2005): 725–726.

Ruoff, R. S., Lorents, D. C., Chan, B. et al., "Single crystals metals encapsulated in carbon nanoparticles," *Science* 259 (1993a): 346–348.

Ruoff, R. S., Subramoney, S., Lorents, D. et al., Paper presented at the 184th Meeting of the Electrochemical Society, New Orleans, October 1993 (1993b).

Saito, Y., "Nanoparticles and filled nanocapsules," *Carbon* 33 (1995): 979–988.

Saito, Y. & Yoshikawa, T., "Bamboo-shaped carbon tube filled partially with nickel," *J. Cryst. Growth* 134 (1993): 154–156.

Saito, Y., Yoshikawa, T., Okuda, M. et al., "Carbon nanocapsules encaging metals and carbides," *J. Phys. Chem. Solids* 54 (1993a): 1849–1860.

Saito, Y., Yoshikawa, T., Okuda, M. et al., "Synthesis and electron-beam incision of carbon nanocapsules encaging $YC_2$," *Chem. Phys. Lett.* 209 (1993b): 72–76.

Saito, Y., Okuda, M., Fujimoto, N. et al., "Single-wall carbon nanotubes growing radially from Ni fine particles formed by arc evaporation," *Jpn. J. Appl. Phys.* 33 (1994a): L526–L529.

Saito, Y., Okuda, M., Yoshikawa, T. et al., "Synthesis of $Sc_{15}C_{19}$ crystallites encapsulated in carbon nanocapsules by arc evaporation of Sc–C composite," *Jpn. J. Appl. Phys.* 33 (1994b): L186–L189.

Saito, Y., Okuda, M., Yoshikawa, T. et al., "Correlation between volatility of rare-earth metals and encapsulation of their carbides in carbon nanocapsules," *J. Phys. Chem.* 98 (1994c): 6696–6698.

Saito, Y., Ma, J., Nakashima, J. et al., "Synthesis, crystal structure and magnetic properties of Co particles encapsulated in carbon nanocapsules," *Z. Phys. D At. Mol. Clust.* 40 (1997a): 170–172.

Saito, Y., Matsumoto, T. & Nishikubo, K., "Encapsulation of carbides of chromium, molybdenum and tungsten in carbon nanocapsules by arc discharge," *J. Cryst. Growth* 172 (1997b): 163–170.

Sano, N., Wang, H., Alexandrou, I. et al., "Properties of carbon onions produced by an arc discharge in water," *J. Appl. Phys.* 92 (2002): 2783–2788.

Saraswati, T. E., Ogino, A. & Nagatsu, M., "Plasma-activated immobilization of biomolecules onto graphite-encapsulated magnetic nanoparticles," *Carbon* 50 (2012): 1253–1261.

Seo, W. S, Lee, J. H., Sun, X. et al., "FeCo/graphitic-shell nanocrystals as advanced magnetic-resonance-imaging and near-infrared agents," *Nat. Mater.* 5 (2006): 971–976.

Sergiienko, R., Shibata, E., Suwa, H. et al., "Synthesis of amorphous carbon nanoparticles and carbon encapsulated metal nanoparticles in liquid benzene by an electric plasma discharge in ultrasonic cavitation field," *Ultrason. Sonochem.* 13 (2006a): 6–12.

Sergiienko, R., Shibata, E., Zentaro, A. et al., "Synthesis of Fe-filled carbon nanocapsules by an electric plasma discharge in an ultrasonic cavitation field of liquid ethanol," *J. Mater. Res.* 21 (2006b): 2524–2533.

Sergiienko, R., Shibata, E., Zentaro, A. et al., "Formation and characterization of graphite-encapsulated cobalt nanoparticles synthesized by electric discharge in an ultrasonic cavitation field of liquid ethanol," *Acta Mater.* 55 (2007): 3671–3680.

Sergiienko, R., Shibata, E., Kim, S. et al., "Nanographite structures formed during annealing of disordered carbon containing finely-dispersed carbon nanocapsules with iron carbide cores," *Carbon* 47 (2009): 1056–1065.

Scheretsky, V. A. & Zatulovsky, S. S., "Tribotechnical characteristics of alumomatrix composites with hybrid fillers comprising nano carbon structures," *Liteinoye Proizv. (Foundry Technol. Equip.)* 11 (2008): 11–13 (In Russian).

Shibata, E., Sergiienko, R., Suwa, H. et al., "Synthesis of amorphous carbon particles by an electric arc in the ultrasonic cavitation field of liquid benzene," *Carbon* 42 (2004): 885–901.

Shibata, E., Sergiienko, R. & Nakamura, T., "Method for producing nanocarbon material, producing device, and nanocarbon material," Patent JP2006273707 (A) (2006, October 12).

Shin, S. & Jang, J., "Thiol containing polymer encapsulated magnetic nanoparticles as reusable and efficiently separable adsorbent for heavy metal ions," *Chem. Commun.* 41 (2007): 4230–4232.

Si, P. Z., Zhang, Z. D., Geng, D. Y. et al., "Synthesis and characteristics of carbon-coated iron and nickel nanocapsules produced by arc discharge in ethanol vapor," *Carbon* 41 (2003): 247.

Siegel, R. W., "Nanostructured materials—Mind over matter," *Nanostruct. Mater.* 3 (1993): 1–18.

Smithells, C. J., "Magnetic materials and their properties," in Brandes, E. A. & Brook, G. B., eds., *Smithells Metals Reference Book*, Seventh edition (Butterworth-Heinemann Linacre House, Oxford, England, 1998), pp. 1170–1192.

Sobal, N. S., Hilgendorff, M., Möhwald, H. et al., "Synthesis and structure of colloidal bimetallic nanocrystals: The non-alloying system Ag/Co," *Nano Lett.* 2 (2002): 621–624.

Striganov, A. R. & Sventitskii, N. S., *Tables of Spectral Lines of Neutral and Ionized Atoms* (IFI/Plenum, New York, 1968).

Sucksmith, W. & Thompson, J., "The magnetic anisotropy of cobalt," *Proc. R. Soc. Lond. A* 225 (1954): 362–375.

Suslick, K. S. & Flint, E. B., "Sonoluminescence from non-aqueous liquids," *Nature* 330 (1987): 553–555.

Suslick, K. S. & Hammerton, D. A., "The sites of sonochemical reactions," *IEEE Trans. Ultrason. Ferroelectr. Freq. Control* 33 (1986): 143–147.

Suzuki, T., Suhama, K., Zhao, X. et al., "Purification of single-wall carbon nanotubes produced by arc plasma jet method," *Diam. Relat. Mater.* 16 (2006): 1116–1120.

Taylor, A., Krupskaya, Y., Costa, S. et al., "Functionalization of carbon encapsulated iron nanoparticles," *J. Nanopart. Res.* 12 (2010): 513–519.

Tomita, M., Saito, Y. & Hayashi, T., "$LaC_2$ encapsulated in graphite nano-particle," *Jpn. J. Appl. Phys.* 32 (1993): L280–L282.

Ugarte, D., "How to fill or empty a graphitic onion," *Chem. Phys. Lett.* 209 (1993): 99–103.

Uo, M., Tamura, K., Sato, Y. et al., "The cytotoxicity of metal-encapsulating carbon nanocapsules," *Small* 1 (2005): 816–819.

Wang, Z. H., Zhang, Z. D., Choi, C. J. et al., "Structure and magnetic properties of Fe(C) and Co(C) nanocapsules prepared by chemical vapor-condensation," *J. Alloys Compd.* 361 (2003): 289–293.

Wang, Z., Xiao, P. & He, N., "Synthesis and characteristics of carbon encapsulated magnetic nanoparticles produced by a hydrothermal reaction," *Carbon* 44 (2006): 3277–3284.

Wang, J. N., Zhang, L., Yu, F. et al., "Synthesis of carbon encapsulated magnetic nanoparticles with giant coercivity by a spray pyrolysis approach," *J. Phys. Chem. B* 111 (2007): 2119–2124.

Wang, S., Cao, H., Gu, F. et al., "Synthesis and magnetic properties of iron/silica core/shell nanostructures," *J. Alloys Compd.* 457 (2008): 560–564.

Wu, W., Zhu, Z., Liu, Z. et al., "Preparation of carbon-encapsulated iron carbide nanoparticles by an explosion method," *Carbon* 41 (2003): 317–321.

Xu, B., Guo, J., Wang, X. et al., "Synthesis of carbon nanocapsules containing Fe, Ni or Co by arc discharge in aqueous solution," *Carbon* 44 (2006): 2631–2634.

Xu, Y., Mahmood, M., Li, Z. et al., "Cobalt nanoparticles coated with graphitic shells as localized radio frequency absorbers for cancer therapy," *Nanotechnology* 19 (2008): 435102–435109.

Yelsukov, E. P., Ul'yanov, A. I., Zagainov, A. V. et al., "Hysteresis magnetic properties of the $Fe(100_{-x})C_{(x)}$; $x = 5$–25 at.% nanocomposites as-mechanically alloyed and after annealing," *J. Magn. Magn. Mater.* 258–259 (2003): 513–515.

Zhang, Z. D., Zheng, J. G., Skorvanek, I. et al., "Synthesis, characterization, and magnetic properties of carbon- and boron-oxide-encapsulated iron nanocapsules," *J. Nanosci. Nanotechnol.* 1 (2001): 153–158.

Zhang, Y., Kohler, N. & Zhang, M., "Surface modification of superparamagnetic magnetite nanoparticles and their intracellular uptake," *Biomaterials* 23 (2002): 1553–1561.

Zhang, J., Schneider, A. & Inden, G., "Cementite decomposition and coke gasification in He and $H_2$–He gas mixtures," *Corros. Sci.* 46 (2004): 667–679.

Zhang, G., Mann, D., Zhang, L. et al., "Ultra-high-yield growth of vertical single-walled carbon nanotubes: Hidden roles of hydrogen and oxygen." *Proc. Natl. Acad. Sci. USA* 102 (2005): 16141–16145.

# Carbon-Coated Nanoparticles

Noemí
Aguiló-Aguayo

Zhenyu Liu

## 18.1 Types of Carbon-Coated Nanoparticles

Since the development of thin film fabrication equipment, carbon coating has been used as a common and feasible way to improve the physical and chemical properties of several materials depending on their applications. The general applications of the carbon coating are meant to enhance the chemical stability of the inner material in order to avoid oxidation and increase the stability in harsh conditions, such as acidic or basic environments, or at high operating temperatures and pressures. As a result, the nanoparticles will form a typical core–shell structure of metal/ceramic cores surrounded by multilayers of carbon shells that have an average interplanar distance of 0.34 nm corresponding to the graphite structure. As a matter of fact, graphite starts to oxidize at temperatures of up to 430°C and fullerene supports higher pressures of over 300 atm (Zhang et al. 2009). The carbon coating also protects the cores against agglomeration and its biocompatibility allows the use of the nanoparticles for biomedical applications. For electrochemical applications, it buffers the volume change of the cores and improves the electrical conductivity of the nanoparticles and their cycling performance. For mechanical applications, the carbon-coated nanoparticles function to improve thermal stability, wear resistance, and reduction of friction.

The carbon coating can be found in a crystalline or amorphous form, or a mixture. The crystalline phase can be graphite, diamond, or a family of fullerenes (Figure 18.1). It also exists as an amorphous or quasi-amorphous form of carbon, called diamond-like carbon (DLC), which presents characteristics close to those of diamond, a semiconductor with mechanical hardness, chemical inertness, and optical transparency (Robertson 2002); but in contrast to diamond, it does not require high temperatures or

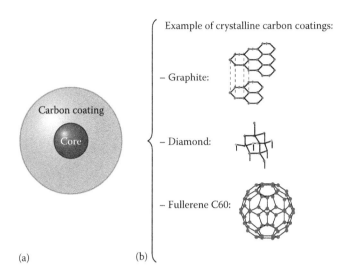

**FIGURE 18.1** Representative images of a carbon-coated nanoparticle (a) and the different allotropes for crystalline carbon coating (b).

special substrates to be deposited and avoids the need for finishing ends. Porous carbon coatings have also been investigated for electrochemical applications since they facilitate the transportation of ions by reducing the resistance and diffusion pathways. In addition, they are also studied in drug transport and delivery even though it is difficult to produce nanoparticles lower than 100 nm with short hierarchical pores (Song et al. 2009). Table 18.1 shows the characteristics of different types of carbon.

Some carbon coatings are prepared separately in an additional stage rather than during the formations of the nanoparticles. For example, carbon black is a type of amorphous carbon, which is cheap and easy to produce, that demonstrates similar properties as graphite. It is mainly used as a coating to improve the conductivity of the core. The porous carbon coating is another example of coating first synthesized with the aid of silica templates and then added to the nanoparticle solution to obtain the final product by hydrothermal method (Liu et al. 2014a). DLC is also deposited a posteriori in order to

**TABLE 18.1**  Characteristics of Different Types of Carbon Materials

| Material | Resistivity ($\rho$) ($\Omega$ m) | Melting Point (°C) | Thermal Conductivity (Room Temperature) (W/mK) | Young's Modulus (E) (GPa) |
|---|---|---|---|---|
| Graphite | $0.04–150 \times 10^{-5}$ (anisotropic conductor)[a] | 3652 (sublimation) | <10 and >1000 (for perpendicular and basal planes) | 4.1–27.6 |
| Graphene | | 3652 | | 1050 |
| Fullerene $C_{60}$ | | | ~0.4 | 53–110 |
| Diamond | ~$10^{12}$ (semiconductor) | 3550 | 600–2000 | 1050–1210 |
| Amorphous | | 3652 (sublimation) | | 100–500 |

*Source:* With kind permission from Springer Science+Business Media: *Appl. Phys. A,* 56, 1993, 219–225, Tea, N. H., Yu, R. C., Salamon, M. B. et al.; Delhaes P., *Graphite and Precursors,* 2000, CRC Press, Boca Raton; Marinescu, I. D., Hitchiner, M. P., Uhlmann, E. et al., *Handbook of Machining with Grinding Wheels,* 2006, CRC Press, Boca Raton; Compton, R. G., and Wadhawan, J. D., *Electrochemistry: Volumen 11—Nanosystems.* 2012. Reproduced by permission of The Royal Society of Chemistry.

[a] Graphite shows an anisotropic conductivity: lower resistivity and higher thermal conductivity across the stacked graphene sheets than on the plane of graphene sheets (first and second values, respectively).

add other properties to the nanoparticles due to its unique characteristics, such as high electrical resistance, optical transparency in infrared and light radiation, and high mechanical hardness. DLC can also enhance the plasmon resonance excitation of Ag nanoparticles and $TiO_2$ photocatalytic activity (Liu et al. 2010a).

We have already mentioned most of the different carbon coatings that one can find in literature. As it is known, the most common carbon coating employed is based on graphite, graphene, or fullerene. These carbon-coated nanoparticles can be divided into three main classes according to their core material: magnetic, metal, and ceramic cores.

### 18.1.1 Magnetic Cores

The nanoparticles have a magnetic core based on transition metals Fe, Co, Ni, and Mn or corresponding alloys FeCo and FeNi, among others (Erokhin et al. 2014). The cores are very reactive and in order to avoid their oxidation, the carbon coating is produced favorably in situ during the formations of the cores. The carbon protects the cores from reactions with the environment and retains their native composition and structure and thus their magnetic properties. The carbon coating also serves to avoid core agglomeration and offers biocompatibility as it is easily functionalized. Magnetic cores that are not oxidized present higher saturation magnetization close to the bulk material. The saturation magnetization at room temperature ($\sigma_S$) for some of the magnetic bulk materials is $\sigma_S(Fe) = 218$ A m$^2$/kg; $\sigma_S(Ni) = 55$ A m$^2$/kg; and $\sigma_S(Co) = 161$ A m$^2$/kg (Cullity & Graham 2008). The coercivity also depends on the core size. The smaller the core diameter, the smaller the coercive field. When the coercivity is close to zero, nanoparticles show the so-called superparamagnetic behavior, which is interesting for biomedical applications (Xu et al. 2014) such as cancer therapy (Xu et al. 2012), drug delivery, diagnosis (Hermann et al. 2010), or resonance imaging enhancement (Bae et al. 2012). The nanoparticles with magnetic cores are also applicable for military applications and in the aerospace industry due to their effectiveness in electromagnetic shielding (Kumar et al. 2014). The shielding mechanism is attributed to the core–shell microstructure, where the dielectric shell and the magnetic cores provide the match between the dielectric and magnetic losses (Zhang et al. 2014a). The magnetic nanoparticles can also behave as ferrofluids and be employed as sealants in harsh environments, or as lubricants to improve the heat transport in bearings and dampers in cars and other machines or in electrical devices (such as speakers) to avoid thermal failures (Gubin 2009).

### 18.1.2 Metal Cores

The carbon coating is also used in noble metal nanoparticles made of Au, Ag, Pd, or Pt (Barone et al. 2014) or other metals such as Cu, Si, Sb, or Ru (He et al. 2011, Chaukulkar et al. 2014, Ma et al. 2014, Zhao et al. 2014). Noble metal nanoparticles, typically containing Au or Ag, exhibit a surface plasmon resonance (SPR) when a polarized electromagnetic field interacts at the interface of the two surfaces metal/dielectric or metal/vacuum. When these nanoparticles are coated with a carbon layer, they exhibit interesting properties, such as strong photoluminescence in the visible region as well as hydrophobicity for bioimaging and photonic applications (Barberio et al. 2013, Choi et al. 2013). The carbon coating also prevents agglomeration among the metal cores, avoiding coalescence and thus obtaining smaller nanoparticles (Jun et al. 2011, Asoro et al. 2014). Carbon-coated nanoparticles containing Pd material are employed as catalysts for energy applications due to high catalytic stability, chemical inertness, and mechanical stability that carbon coating provides (Kim et al. 2008, Zhang et al. 2014b). One of the most common methods to produce this kind of nanoparticles is by laser ablation.

Concerning carbon-coated Cu nanoparticles, the effect of the carbon shells is to improve the SPR band and fluorescence emission (Ma et al. 2014). The carbon coating not only protects the Cu cores against the humid air, but also acts as a highly conductive material. The carbon-coated Cu nanoparticles are applied as aqueous inkjet for printing conductive patterns and catalysts for versatile chemical reactions (Magdassi et al. 2010). The most common method to produce this kind of nanoparticles in a single step is by spray pyrolysis. In addition, the carbon coating is applied in the process of the synthesis of nanoparticles containing Sb, such as InSb, $Cu_2Sb$, SnSb, and Si/C nanoparticles. These types of nanoparticles are used in the anodes of LIBs. The carbon coating improves the cycling performance, by means of controlling the change in volume as well as avoiding the aggregation among the cores. Si/C nanoparticles are produced by plasma, CVD, or spray pyrolysis (Guo et al. 2010, Chaukulkar et al. 2014), whereas Sb/C nanoparticles are mainly produced by chemical synthesis (Wang et al. 2007a, Luo et al. 2014). The carbon coating is also applied in Al–Si nanoparticles in order to improve their mechanical and thermal properties for aerospace and automotive industries (Tulugan et al. 2013).

### 18.1.3 Ceramic Cores

Carbon-coated ceramic nanoparticles consist of cores mostly based on oxides, nitrides, or carbides. Carbon-coated $TiO_2$ nanoparticles can find applications in wastewater treatment. The carbon coating adsorbs the pollutants and favors the interfacial charge transfer process with the semiconductor $TiO_2$ cores. This enhances the process of photo-degradation of organic contaminants on $TiO_2$ particles. Carbon coating also offers higher chemical and thermal stability than other coatings like silica. Simultaneously, it stabilizes the anatase phase of $TiO_2$, enhancing the reaction activity (Lee et al. 2014). The thickness of the carbon coating is a critical factor for this application, since thick coatings benefit the absorption but make it difficult for the UV light to reach the $TiO_2$ core (Nawi & Nawawi 2013). These nanoparticles are prepared by chemical synthesis (hydrothermal method) or by flame spray pyrolysis (Olurode et al. 2012, Anjum et al. 2013).

Carbon-coated $LiFePO_4$ nanoparticles are used as cathodes in LIBs. The carbon coating approach is the most straightforward route to improving the conductivity of the nanoparticles. These particles are prepared by chemical synthesis through solvothermal or hydrothermal processes (Murugan et al. 2008, Yu et al. 2013). Carbides such as WC or $Mo_2C$ are also coated with carbon and applied as electrodes in capacitors. The carbon coating improves the conductivity and suppresses the growth of carbide grains during their formations and the agglomeration of the tungsten and molybdenum hydroxides during the charge and discharge cycles (Morishita et al. 2007).

Transition metal oxides coated with carbon also present enhanced properties for alkaline batteries and fuel cells. The carbon coating not only provides higher electrical conductivity, but also improves the oxygen reduction catalytic activity of the nanoparticles (Malkhandi et al. 2013).

## 18.2 Particle Formation and Growth

In general, the carbon-coated nanoparticle formation process involves several steps: supersaturation, nucleation, particle growth (crystallization), particle coagulation, and coalescence (Figure 18.2).

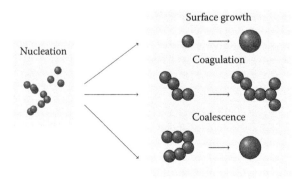

**FIGURE 18.2**   Main steps present in the formation of nanoparticles.

Reaching the supersaturation state is essential to form a nanoparticle. The main process to produce carbon-coated nanoparticles is presented as a gas-to-solid condensation process, where the source material is vaporized. When the supersaturation state is reached, the vapors condensate into nanoparticles. These conditions are established when the pressure of the vaporized atoms is greater than the vapor pressure of the cloud at a certain temperature. This occurs by cooling the atoms/clusters using a cooler gas, for example, in plasma reactors or by promoting chemical reactions by increasing the temperature, for example, in laser or spray pyrolysis reactors (Kruis et al. 1998).

After reaching the supersaturation state, the nucleation process takes place. Metastable nuclei are formed by the collisions of the vaporized atoms. Nucleation can be homogeneous when the reactant and surroundings are in the same phase. Heterogeneous nucleation can also occur, when nuclei are formed in preferable sites such as impurities or surfaces, where less energy is needed to overcome the barrier of particle formation.

Here, we would like to focus on the homogeneous nucleation following the classic theory proposed by Becker and Döring (1935) that neglects the translational and rotational energies of the droplet during its movement through the system and simplifies the model (Lothe & Pound 1962, Moody & Attard 2002). The classical nucleation theory is suitable to predict nucleation rates and critical cluster sizes (Vehkamäki 2006).

The classical expression for the homogeneous nucleation rate depends on the supersaturation ratio $S$, defined as the ratio between the partial pressure of the gaseous reactant $p_i$ and the corresponding saturated vapor pressure $p_S(T)$:

$$S = \frac{p_i}{p_S(T)}. \tag{18.1}$$

When $S$ is larger than unity, the nucleation process may occur (Chazelas et al. 2006).

The free energy required for the creation of a new spherical nucleus of $j$ atoms and radius $R_j$ depends on two terms (Equation 18.2) (Pomogailo & Kestelman 2005). $\Delta G_{Vj}$ is related to the creation of volume and is given by the difference between the chemical potential of the new nucleus ($n$) and the environment phase ($e$), $\Delta \mu = k_B T \ln S$, and $\Delta G_{Sj}$ refers to the energy due to the formation of a new surface (Equations 18.3 and 18.4):

$$\Delta G_j = \Delta G_{Vj} + \Delta G_{Sj} = -\frac{4\pi}{3V_a} R_j^3 \Delta\mu + 4\pi R_j^2 \gamma, \tag{18.2}$$

$$\Delta G_{Vj} = -j\Delta(\mu_e - \mu_n) = -j\Delta\mu = -\frac{4\pi}{3V_a} R_j^3 k_B T \ln S, \tag{18.3}$$

$$\Delta G_{Sj} = 4\pi R_j^2 \gamma, \tag{18.4}$$

where $k_B$ is the Boltzmann constant, $T$ is the temperature in the environment phase, $S$ is the supersaturation ratio, $V_a$ is the atomic volume of the nucleus, and $\gamma$ is the specific free energy.

After nucleation, the particle growth (crystallization) occurs. When the critical energy barrier $\Delta G_{cr}$ is overcome, the nuclei formed during the nucleation process become stable. The value of the critical size for a stable nucleus $R_{cr}$ is determined by $d(\Delta G)/dR = 0$:

$$R_{cr} = \frac{2\gamma V_a}{\Delta\mu} = \frac{2\gamma V_a}{k_B T \ln S}. \tag{18.5}$$

**TABLE 18.2**  Several Metal Examples of Critical Nucleus $R_{cr}$ and the Corresponding Number of Constituent Atoms $j_{cr}$ during Metal Crystallization from the Melt

| Metal | $\gamma \times 10^7$ (J cm$^{-2}$) | $V_a \times 10^{-23}$ (cm$^3$) | $R_{cr}$ (nm) | $j_{cr}$ |
|---|---|---|---|---|
| Fe | 204 | 1.21 | 1.17 | 553 |
| Co | 234 | 1.12 | 1.07 | 457 |
| Ni | 255 | 1.13 | 1.07 | 453 |
| Pd | 209 | 1.51 | 1.36 | 696 |
| Pt | 240 | 1.54 | 1.15 | 418 |
| Cu | 177 | 1.21 | 1.21 | 512 |
| Ag | 126 | 1.75 | 1.27 | 489 |
| Au | 126 | 1.75 | 1.73 | 483 |
| Al | 93 | 1.65 | 1.23 | 466 |

*Source:* With kind permission from Springer Science+Business Media: *Springer Series in Materials Science*, Metallopolymer nanocomposites, 81, 2005, Pomogailo, A. D., and Kestelman, V. N.

Table 18.2 shows some examples of critical radius $R_{cr}$ to form stable metal crystallites (Pomogailo & Kestelman 2005).

The profile of the free Gibbs energy for a homogeneous nucleation and particle growth is depicted in Figure 18.3. The thermodynamic contribution from both volume and surface terms defines a threshold to a critical size, which decides the minimum size of the nanoparticle.

Particle coagulation is the mechanism that occurs at high particle concentrations, when particles are randomly distributed and collide due to Brownian motion. The collision promotes the aggregation of nanoparticles attributed to strong adhesive forces or chemical bonds, among others (Gustch et al. 2002). In plasma reactors, the residence time, which is the time that nanoparticles are located in the plasma region, determines the level of aggregation. To avoid aggregation, it is important to maintain the residence time as short as possible (Kruis et al. 1998).

Coalescence is another kind of agglomeration but, in this case, implies nanoparticles with similar sizes that merge and form a bigger particle. There is also another mechanism, called Ostwald ripening, which is observed during the formation of nanoparticles when there is an ensemble of nanoparticles with slightly different sizes (Lifshitz & Slyozov 1961). This mechanism favors the formation of larger

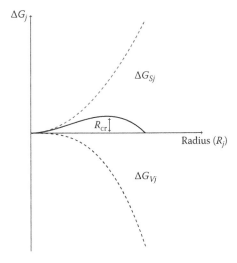

**FIGURE 18.3**  Schematic representation of the free energy as a function of nanoparticle size for a homogeneous nucleation and growth process.

nanoparticles by the exchange matter produced by the atom transfer from smaller crystals to bigger ones driven by the reduction in surface energy. The Ostwald ripening can occur in solid, liquid, or gas phase (Mokari et al. 2005, Hornyak et al. 2008).

## 18.2.1 Carbon-Coating Formation Mechanisms

There are several theories relating to the formations of carbon-coated nanoparticles. Many of these theories are developed from the suggested mechanisms about the formations of CNTs, since they typically use metal nanoparticles as catalysts for their growth (Ding et al. 2004). In the following, we will focus on the mechanisms that take place during the formations of carbon-coated nanoparticles in a single step. The nanoparticle core is formed following the theories explained previously and the carbon coating is generated afterward. When the carbon coating is generated in situ, it is usually composed of graphite, graphene, fullerenes, or amorphous carbon. The mechanism explanations depend on the fabrication technique (Laurent et al. 1998, Kumar et al. 2011). We will discuss more details of the carbon-coated nanoparticle formation mechanism based on the two main widely used approaches: CVD and the arc-discharge plasma method (ADP).

In the case of the CVD method, the most popular theory for the explanation of carbon filaments and/or CNT growth and, hence, carbon-coated nanoparticle formations, is depicted by the vapor–liquid–solid model proposed by Baker (1989) and Baker et al. (1972). According to this model, a nanoparticle acts as a catalyst for the decomposition of hydrocarbons (carbon source) at a lower temperature than that required for the spontaneous reaction. The carbon atoms diffuse into the nanoparticle (in the liquid state) and form a carbon supersaturation, and then carbon precipitates on the substrate and induces the formation and growth of carbon filaments due to the carbon supersaturation and temperature gradient caused by the exothermic decomposition. Oberlin et al. (1976) proposed another model where the carbon diffuses on the surface of the catalytic particle and dissociates at the contact angle between the particle and the surface to form the tubes. Ding and Bolton (2006) showed by thermodynamic analysis that a metal particle only needs to be highly carbon supersaturated to nucleate/initiate carbon coatings.

Regarding the formation of carbon-coated nanoparticles by ADP methods, there are several hypotheses. Saito (1995) explained that there is a correlation between the vapor pressure of metal and the graphite encapsulation. A wide range of composite materials was studied and it was found that volatile elements are not encapsulated. Based on those observations, it was proposed that nanoparticles grow at the cathode with a liquid core coated by a graphite shell, which then segregates and solidifies. For this reason, volatile elements are not encapsulated since they hardly condense on the cathode and diffuse far away from where carbon condenses.

Majetich et al. (1994) and Scott et al. (1995) suggested that, firstly, the core material and the carbon are atomized in the plasma. They collide and nucleate into clusters in the supersaturated vapor until they grow to their characteristic sizes. Finally, particles are deposited on surfaces within the reactor, and by phase segregation, the carbon coating is formed on the outer surface of the core particle (Figure 18.4). Moreover, they speculated that size distribution of the samples depends on the steepness of the cooling profile that the nanoparticles follow during their growth.

Seraphin et al. (1996) studied 20 elements as metal sources in order to investigate the carbon shell formations. They separated the elements into four categories depending on their tendency to form carbides. It was observed that the enthalpy of the carbide formation has an important role in the carbon encapsulation. If the carbide is not formed, then the encapsulation is not observed; in contrast, if the formation of carbide is too aggressive, then there are no formations of carbon shell either. However, there were some exceptions for some element behaviors.

Bystrzejewski et al. (1997, 2005) proposed that it was improbable to form $\alpha$-Fe from the vapor phase under a dominating presence of carbon. They proposed that the carbon-coated nanoparticles emerge from liquid droplets and then solidify in the cold zone. Finally, the core–shell–structured carbon-coated nanoparticles come into being in the phase.

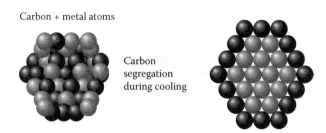

Carbon + metal atoms

Carbon
segregation
during cooling

**FIGURE 18.4** Formation of the carbon coating of carbon-coated nanoparticles by phase segregation during the cooling of the liquid metal nanoparticles. (Reprinted from *Mater. Sci. Eng. A Struct. Mater. Prop. Microstructure Process.*, 204, Majetich, S., Brunsman, E., Scott, J. et al., Formation of nanoparticles in a carbon arc, 29–34, Figure 2, Copyright (1994), with permission from Elsevier.)

Despite all the aforementioned suggested models, the detailed pathways of the formation of these novel forms of materials remain unclear, such as the state of the catalytic particle (vapor, liquid, or solid) and the path of carbon diffusion (volume or surface). More systematic research emphasizing in situ techniques is required for a clear understanding of the mechanisms.

### 18.2.2 Thermally Induced Structural Changes

Core–shell nanoclusters have received considerable attention owing to their physical and chemical properties that are strongly dependent on the structure of the core, shell, and interface. This structure dependence opens a possibility for tuning properties by controlling their chemical composition and relative size of the core and shell. In most cases, carbon-coated nanoparticles present chemical and environmental stability at lower temperature. However, when induced by thermal treatment and other radiation conditions, such as electron beam, both the core and the shell can experience structural changes, such as phase transformation and crystallization. Banhart et al. (1998) observed the behavior of carbon-coated metal nanoparticles by in situ TEM. At high temperature, the curved carbon shell would shrink and contract due to probable carbon vacancy defect and displace the inside metal nanoparticle atoms. Further, the core metal may gradually migrate through the carbon shell by atomic diffusion, and electron beam radiation can enhance the structure rearrangement on both carbon shell and core metal nanocrystals. It is suggested that annealing provides additional thermal energy that makes structural rearrangement possible long after the initial deposition process was terminated. The core size and metal mobility can dominate the migration dynamics. It is revealed that there is an interaction between carbon shell and encapsulated metal nanocrystal. The strong interaction can induce carbon lattice spacing change and the nucleation of diamond inside the carbon onion (Banhart & Ajayan 1996, Huang 2007). That suggested that carbon shells can also be applied as high-pressure microreactor.

Despite intensive research on the encapsulation of metal nanoparticles into carbon clusters, the detailed pathways of the formation of these novel forms of materials remain unclear. The growth of a rich variety of morphologies is not well understood either. Postdeposition annealing or thermal treatment can be introduced as a process that induces structural rearrangements, and thus enables changes in morphologies and further structure modification. More observations may provide insights into the rich variety of morphologies of the deposits obtained at different processes and even in different locations of the reaction chamber. The advantages and disadvantages of the different methods for the preparation of carbon-coated nanoparticles are shown in Table 18.3.

## 18.3 Fabrication Techniques

There are two kinds of fabrication methods to produce nanostructured materials: the top–down and the bottom–up manufacturing processes. The top–down method consists of creating new nanomaterials

**TABLE 18.3** Advantages and Disadvantages of the Different Methods for the Preparation of Carbon-Coated Nanoparticles

| Method | Advantages | Disadvantages |
|---|---|---|
| ADP | Simple and economic technique<br>High-purity samples<br>Very small nanoparticles can be achieved | Higher operating temperatures<br>Difficulty to control the size dispersion<br>Sometimes vacuum is required<br>Low yield |
| CVD | Simple and economic technique<br>Multicomponent nanoparticles | Difficult to control morphology and composition of the nanoparticle<br>Sometimes vacuum is required |
| Laser ablation | Continuous operation<br>Simple procedure<br>Low running costs<br>Wide range of materials can be used<br>Stable handling of colloids | High initial costs because of the lasers<br>Limitation in the size control<br>Limitation in productivity |
| Spray pyrolysis | Low processing temperature<br>High homogeneity and purity of nanoparticles<br>Continuous process | Low control of morphology and composition<br>Difficulty in obtaining unagglomerated nanoparticles |
| Chemical synthesis | Easy to control the size distribution<br>High yields<br>Good reproducibility | Waste of chemicals<br>Not suitable for the preparation of cores with high purity or with accurate stoichiometry<br>For hydrothermal method: high reaction temperatures, high pressure |

from bulk materials that are reduced until the desired nanostructure is achieved. The bottom–up method is characterized by the synthesis of nanomaterials from smaller starting subunits that react with chemical or physical forces and form the desired nanostructures. There is also the hybrid fabrication that is a combination of both the top–down and the bottom–up processes.

Most of the techniques for the production of carbon-coated nanoparticles came from the first techniques used to generate carbon nanostructures, such as MWCNTs, SWCNTs, or fullerenes. The most common methods used nowadays are ADP, CVD, laser ablation, spray pyrolysis, and chemical synthesis.

These methods are based on the gas phase synthesis, which allows the generation of core–shell nanostructures in a single step. Thus, the oxidation and chemical degradation of the cores can be effectively avoided. This is an advantage in comparison to other techniques (sol–gel, precipitation methods) commonly used to produce inorganic (silica, gold) or organic (polymers) coatings, which require several steps or even the combination of different methods to generate the core first and the corresponding encapsulation afterward (Tartaj et al. 2003). In addition, these techniques require less chemical species and process steps, obtaining a high-quality product with high purity and that is environment friendly. However, in comparison with other methods based on liquid-phase synthesis, some nanoparticle aggregation can occur and the yield is lower, thus increasing the cost of production.

## 18.3.1 ADP

The ADP method is the first method applied for the production of CNTs and is widely used for the synthesis of carbon-encapsulated nanoparticles as well (Bystrzejewski et al. 2005, Bera et al. 2006, Aguiló-Aguayo et al. 2009, Wei et al. 2011). This method consists of forming a high-temperature plasma between two electrodes facing each other and connected to a DC, AC, or radio frequency power source (Figure 18.5). A high-energy spark is generated between the close electrodes and plasma is formed due to the ionization of a gas supply consisting mainly of inert gases (He, Ar). One of the electrodes provides the metal source and due to the high temperature of the plasma (about 3000 K or higher), the vaporization of precursors is achieved. The carbon source can be supplied by the electrodes made of graphite or the carbon generated from the

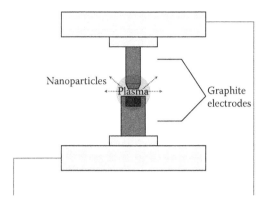

**FIGURE 18.5**   Schematic drawing of an ADP reactor.

decomposition of introduced reactive gases. In this technique, the inert atmosphere, the high temperatures used, and the high cooling rates assure the production of high-purity nanoparticles with good surface activity and of very small sizes due to the very short residence times of the nuclei formed in the plasma. However, the stability and gradient of the cooling rates could deteriorate the uniformity of the nanoparticles.

## 18.3.2  CVD

The synthesis of nanoparticles by CVD commonly consists of two steps. First, the precursor is introduced in the controlled-temperature furnace in a liquid, gas, or solid state and decomposes. Then, a gas mixture of the carrier gas (Ar, $N_2$) and the carbon gas supply (xylene, acetylene) is introduced into the furnace at temperatures in the range 600°C–1000°C in a continuous flow of about 50–800 sccm. The reaction chamber is kept at pressures in the range of 40–100 Pa. The nanoparticles are collected from the walls of the reaction chamber. CVD produces highly dense and pure materials; the deposition temperatures are relatively low, which implies a reasonable processing cost, and a wide range of chemical precursors is available (Liu et al. 2001, Choy 2003, Sarno et al. 2014).

## 18.3.3  Laser Ablation Method

The laser ablation process was first developed at Rice University for the production of MWCNTs (Guo et al. 1995). The apparatus consisted of a quartz tube heated in a temperature-controlled furnace of about 1200°C (Figure 18.6a). A graphite target was placed in the furnace and heated to outgas the target. A neodymium-doped yttrium aluminum garnet laser irradiated the graphite sample in a uniform way providing 10 ns of 250 mJ pulses at 10 Hz. The resulting product was dragged out of the furnace by a carrier inert gas (Ar).

   In order to produce nanoparticles with various compositions, besides the use of laser ablation in a flow reactor (Dumitrache et al. 2004), the method was also modified to be used in a liquid medium (Yang 2007, Park 2008, Kwong et al. 2010, Zeng et al. 2012). A pulsed laser irradiates a bulk target located inside a container filled with the liquid carbon source (i.e., toluene) or irradiates a solution based on metalocene powder (ferrocene, nickelocene, cobaltocene) in the liquid carbon source (i.e., xylene). Nanoparticles can be collected using a magnet or directly obtained in a colloidal solution (Figure 18.6b). The shape and evolution of the plasma plume controls the fabrication of nanoparticles. Complex nanoparticles can be obtained due to the chemical reactions of the target atoms with the environmental molecules. In addition, laser ablation in liquid avoids agglomeration of nanoparticles. However, nanoparticles are produced in a range of 10–100 nm; larger nanoparticles are not possible to fabricate, and it is a costly technique, which requires expensive lasers and has a high power requirement.

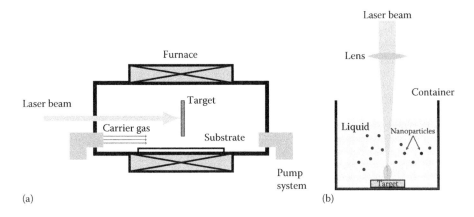

**FIGURE 18.6** (a) Schematic diagram of a pulsed laser ablation system and (b) the same system in a liquid medium.

## 18.3.4 Spray Pyrolysis

The spray pyrolysis technique is also known as aerosol thermolysis, evaporative decomposition, plasma vaporization of solutions, or aerosol decomposition (Jackson & Hargreaves 2008). This method consists of atomizing a reaction solution by means of a spray nozzle containing a capillary tube that has an inner diameter less than 0.4 mm. The solution supplies the carbon and the core material sources, for instance, ethanol ($C_2H_6O$) as carbon source and iron pentacarbonyl ($Fe(CO)_5$) as iron precursor. The solution precursor is carried by a gas into the reactor, where its decomposition occurs in the hot zone of an electrical furnace heated at temperatures of around 500°C–900°C (Figure 18.7). Nanoparticles are collected at the bottom of the reactor. This method offers the possibility to use different types of precursors in gas or liquid phase, and the amount of the resulting nanoparticles can be determined by controlling the flow of the precursor (Yu et al. 2005). However, hollow structures or fractured particles can be obtained and there is some difficulty of scaling-up the production due to the high amount of the required solvents (Wang et al. 2007b, Jung et al.

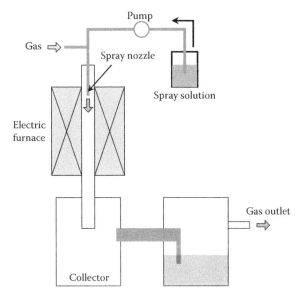

**FIGURE 18.7** Experimental setup of a spray pyrolysis technique.

2010, Atkinson et al. 2011). We should mention that other techniques employ a combination of the methods described above to produce the carbon-coated nanoparticles, such as laser-assisted CVD (Westerberg et al. 1992, Widenkvist et al. 2011) or arc-plasma assisted CVD (Li et al. 2009) or laser-assisted spray pyrolysis (Jager et al. 2006).

### 18.3.5 Chemical Synthesis

One of the classical techniques to produce nanoparticles by chemical route is by precipitation. The chemical synthesis approach allows the formations of large amounts of nanoparticles, although the control of their size is delicate. Many different types of nanoparticles (metal, alloys, ceramic) can be easily synthesized. The metal precursor is dissolved in an aqueous solution and the particles precipitate by means of precipitating agents. The controlled growth depends on the temperature, pH, concentration, and type of precipitating agents (Willard et al. 2004). Another method to prepare carbon-coated nanoparticles is by hydrothermal synthesis, where nanoparticles are produced in aqueous solution at high temperature (above 100°C) and at high pressure (above 1 atm) (Wang et al. 2006, Yu et al. 2013). Sucrose, glucose, and dextrose are the most common compounds used as a carbon source.

# 18.4 Properties of Carbon-Encapsulated Nanoparticles

Carbon-encapsulated nanoparticles exhibit different properties based on the core material. Magnetic and noble metal cores demonstrate superparamagnetism and SPR, respectively, when they present nanometric sizes. In general, ceramic cores are characterized by their electrochemical behavior and carbides or other metal cores are characterized by their thermal properties.

### 18.4.1 Magnetic Properties: Superparamagnetism

Superparamagnetism occurs when ferromagnetic or ferrimagnetic materials (usually based on Fe, Ni, Co) are reduced under a critical diameter and the energy required to form magnetic domain walls is larger than the magnetostatic volume energy associated to a single-domain nanoparticle (Gubin 2009). A schematic drawing of the magnetic domains of (a) superparamagnetic and (b) ferromagnetic nanoparticles is shown in Figure 18.8.

In single-domain nanoparticles, the magnetization randomly reverses due to the effect of temperature (under the Curie temperature) following easy axis directions in which the anisotropy energy is minimal. For uniaxial nanoparticles, the energy barrier required for the magnetization to flip is given

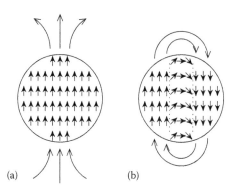

(a)                                             (b)

**FIGURE 18.8**   (a) Single-domain nanoparticle without magnetic domains and showing large demagnetizing field and (b) ferromagnetic nanoparticle with magnetic domains reducing magnetostatic energy (low demagnetizing fields).

by $E_b = KV$ with no magnetic field, while the energy barrier under a magnetic field is decreased to $E_b = KV(1 - HM_S/2K)^2$, where $K$ is the magnetocrystalline anisotropy energy, $V$ is the particle volume, $H$ is the magnetic field, and $M_S$ is the saturation magnetization (Figure 18.9).

The Néel relaxation time is the time between two flips (Néel 1949) and is expressed by

$$\tau = \tau_0 \exp\left(\frac{E_b}{k_B T}\right), \tag{18.6}$$

where $\tau_0$ is an intrinsic time characteristic of the material, $10^{-9}$–$10^{-12}$ s, inversely proportional to the jump attempt frequency of the magnetic moment; $T$ is the temperature; and $k_B$ is the Boltzmann constant. When the window time of the experiments is much longer than the Néel relaxation time, the total magnetization (with the absence of magnetic field) appears to be zero. This is called superparamagnetic state; in the opposite state, when the magnetization is fixed at a certain direction, nanoparticles retain the so-called blocked state (Knobel et al. 2008). For superparamagnetic nanoparticles, the coercivity tends to be zero, whereas for blocked nanoparticles, it reaches a maximum. The coercivity as a function of the diameter is depicted in Figure 18.10.

From the relaxation time $\tau$, it is possible to make an estimation of the critical diameter $D_p$ that separates blocked and superparamagnetic states. If we consider a time reversal of about 100 s, the critical volume $V_p$ is directly obtained from

$$10^{-2} = 10^9 \exp\left(\frac{KV_p}{k_B T}\right), \tag{18.7}$$

$$V_p = \frac{25 k_B T}{K}. \tag{18.8}$$

If we assume that all the spins rotate coherently (Stoner–Wohlfarth mode) and neglect the effect of the surface spin canting, we can then consider that a superparamagnetic nanoparticle acts as a large magnetic dipole, similar to a paramagnetic material. Under the presence of a magnetic field $H$, the

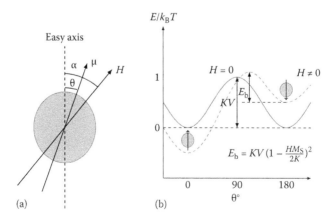

**FIGURE 18.9** Schematic drawing of the easy axis, magnetic moment, and magnetic field orientations for a uniaxial nanoparticle (a), and the energy barriers depending on the orientation of the magnetic moment with the easy axis, with or without an external magnetic field aligned with the easy axis (b).

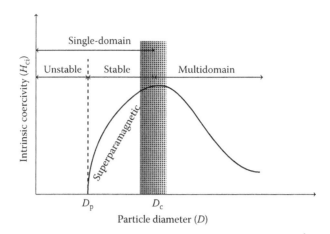

**FIGURE 18.10** Schematic drawing of the intrinsic coercivity dependence on the particle diameter. $D_c$ corresponds to the single-domain particle critical diameter and $D_p$ to the superparamagnetic critical diameter.

magnetic moments orientate with respect to $H$ following a Boltzmann distribution. The total magnetization $M(H)$ is described by the Langevin function (Equation 18.9) (Bean & Livingston 1959); whereas in a paramagnetic material, the magnetic moment $\mu$ corresponds to the magnetic moment per atom, and in superparamagnetic nanoparticles, $\mu$ is described by the magnetic moment per nanoparticle:

$$M(H) = M_S\left[\coth\left(\frac{\mu H}{k_B T}\right) - \frac{k_B T}{\mu H}\right] = M_S L\left(\frac{\mu H}{k_B T}\right). \tag{18.9}$$

In real systems, the size of the nanoparticles is usually not uniform and follows a lognormal distribution. The magnetic moment is given by $\mu = M_S V$, where $\mu$ follows a lognormal distribution due to the volume distribution, and the total magnetization follows (Vargas et al. 2005, Chen et al. 2009):

$$\frac{M(H,T)}{M_S(T)} = \int_0^\infty L\left(\frac{\mu H}{k_B T}\right) f(\mu)d\mu, \tag{18.10}$$

where $f(\mu)$ is the normalized magnetic moment distribution expressed by

$$f(\mu) = \frac{1}{\mu \omega_\mu \sqrt{2\pi}} \exp\left[-\frac{\ln^2(\mu/\mu_0)}{2\omega_\mu^2}\right], \tag{18.11}$$

where $\mu_0$ is the mean geometric magnetic moment and $\omega_\mu$ is the geometric standard deviation of the magnetic moment.

The magnetic behavior of nanoparticles can also be characterized by the zero-field-cooled (ZFC) and field-cooled (FC) curves, which provide information about the history dependence of the magnetization with temperature. The ZFC curve is obtained by annealing the sample with no remnant magnetization and under no magnetic field. Afterward, a magnetic field is applied at a certain temperature and the sample is cooled down to the lowest temperature, and the magnetization is recorded as a function of the temperature to obtain an FC curve. The maximum of the ZFC curve is assumed to be around the blocking temperature $T_B$, above which the superparamagnetic behavior is observed.

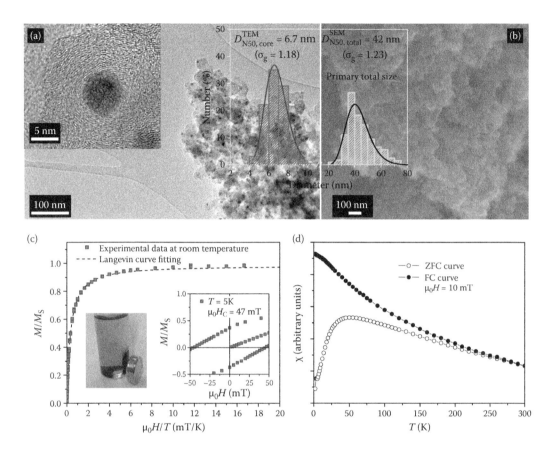

**FIGURE 18.11** (a) TEM image of a representative α-Fe core of about 6.7 nm coated by several graphitic layers. The SAED pattern (inset) confirms the bcc structure of the ferrite cores. (b) SEM images of spherical primary nanoparticles of 42 nm in diameter with a geometric standard deviation $\sigma_g = 1.23$. (c) Hysteresis response at room temperature (300 K) followed a Langevin function corresponding to superparamagnetic behavior. (d) ZFC-FC curves confirmed $T_B$ lower than room temperature and α-Fe composition of the nanoparticles. (Aguiló-Aguayo, N., Maurizi, L., Galmarini, S. et al., *Dalton Trans.* 43, 13764–13775, 2014. Reproduced by permission of The Royal Society of Chemistry.)

Figure 18.11a demonstrates superparamagnetic carbon-coated iron nanoparticles obtained by the ADP method. The crystalline Fe cores present monodispersed sizes of 6.7 nm in diameter coated by several graphitic layers showing a primary total particle diameter of 42 nm. The magnetic moment per particle µ was obtained by fitting experimental data to the typical Langevin function (Equation 18.9) and corresponds to a saturation magnetization of about 60 A m² kg⁻¹, smaller than the bulk α-Fe crystal (51 mT). In Figure 18.11b, a Gittleman model (Gittleman et al. 1974) was fitted to the ZFC curve considering a magnetic diameter of 6.7 nm. The average $T_B$ found was 27.8 K and the effective anisotropy energy $K_{eff}$ was $5.66 \times 10^4$ J m⁻³, in agreement with the bulk α-Fe anisotropy energy $K_1 = 4.8 \times 10^4$ J m⁻³.

Superparamagnetic nanoparticles do not aggregate due to magnetic interactions and, hence, they are very suitable for biomedical applications such as drug delivery, hyperthermia, magnetofection, MRI, and ferrofluids.

## 18.4.2 Optical Properties: Surface Plasmons

Metal nanoparticles, such as Ag, Au, or Cu, are widely studied because of their optical properties, in particular, the SPR. When an electromagnetic field interacts with the nanoparticle surface (interface between metal and dielectric), the electron cloud oscillates coherently in relation to the phonon lattice following the electromagnetic field, thus resulting in a surface charge distribution. A coulombic restoring force between the electron cloud and nuclei is induced and a resonance condition appears when the frequency of these oscillations is very similar to the characteristic frequency of the surface plasmon oscillations. The SPR usually provides an absorption peak in the visible range. In a colloidal solution of nanoparticles, the extinction cross section $C_{ext}$ of a very small particle with radius $R$ and a frequency-dependent dielectric function $\varepsilon = \varepsilon' + i\varepsilon''$ immersed in a medium with a dielectric function $\varepsilon_m$ is given by (Mulvaney 1996):

$$C_{ext} = \frac{24\pi^2 R^3 \varepsilon_m^{3/2}}{\lambda} \frac{\varepsilon''}{(\varepsilon' + 2\varepsilon_m)^2 + \varepsilon''^2},$$

(18.12)

where $\lambda$ is the wavelength of the incident radiation. The frequency of resonance can be modified by changing the type of metal, the nanoparticle size or shape, or the dielectric environment. It is observed that carbon-coated nanoparticles exhibit a redshift of the SPR band of the bare nanoparticles due to the increase in the effective dielectric constant of the carbon coating. For example, Cu nanoparticles present an SPR band in the range of 560–590 nm, whereas carbon-coated Cu nanoparticles' SPR bands are shifted to 608 nm, and carbon-coated Co nanoparticles were redshifted from 270 to 424 nm (Li et al. 2011). The carbon coating protects plasmonic nanoparticles against agglomeration and improves their chemical stability, ensuring the optical properties. Although thick coatings can lead to an exponential decay of the electric field away from the plasmonic nanoparticle surface, Liu et al. (2014b) observed that a carbon coating in Al nanoparticles provides a better confinement of the light due to its higher refractive index in comparison with other coatings such as $Al_2O_3$ or $SiO_2$. A SEM image of the carbon-coated Ag nanoparticles is shown in Figure 18.12a, and the transmittance spectra for the carbon-coated and uncoated Ag nanoparticles are shown in Figure 18.12b.

Plasmonic nanoparticles are very interesting due to their ability to concentrate light into nanoscale volumes and to create a highly localized field for optical trapping, data storage, or photolithography. This ability allows manipulation of light below the diffraction limit, and plasmonic nanoparticles can also be used to develop photodetectors and modulators. In addition, they can be used as catalysts to control chemical reactions or as heat generators, which is very interesting for biomedical applications such as cancer treatment or biosensors (Schuller et al. 2010).

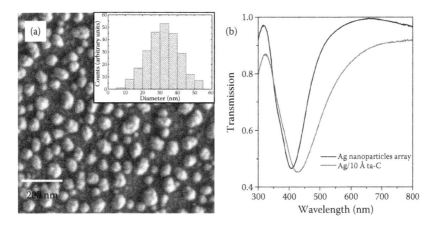

**FIGURE 18.12** (a) SEM image of the Ag nanoparticle film and (b) transmittance spectra of the 1 nm ta-C film-coated Ag nanoparticles and uncoated substrates. (Reprinted by permission from Macmillan Publishers Ltd., Liu, F. X., Tang, C. J., Zhan, P. et al., *Sci. Rep.* 4: 1–6, Copyright (2014).)

### 18.4.3 Electrical, Thermal, and Mechanical Properties

Carbon-based materials are formed by $sp^2$- and/or $sp^3$-hybridized carbon atoms, and the property of good electrical conductivity is related to the extension of the $sp^2$-bonded carbons. For the $sp^2$ hybridization, only three of the four valance electrons of each C atom are used in the sigma bond formation. The other electron is in the $p$ orbital and can overlap with neighboring $p$ orbitals and moves freely within the delocalized $\pi$ orbitals. This is the reason why diamond, which only presents $sp^3$ hybridization (four valance electrons are used in sigma bonding), is a very good insulator; whereas graphite is a very good electrical conductor in the planar orientation, where $sp^2$ hybridization of the carbon atoms leads to the formation of flat layers linked by Van der Waal's forces. Carbon allotropes show anisotropic electrical conductivity. As we have mentioned, graphite presents high conductivity in its planar orientation, CNTs show high conductivity along their length (although it also depends on the chirality of their structure), and fullerenes are insulators except if there is a formation of CNTs. Marinho et al. (2012) studied the DC electrical conductivity of compact MWCNTs, graphene, carbon black, and graphite under high pressures (5 MPa). They observed that graphite exhibits higher conductivities, ~$10^3$ S/m, whereas the rest showed lower values, ~$10^2$ S/m. As expected, the conductivities depended on the preparation process, and a plane orientation was preferred for CNT and graphene measurements. Due to good electrical conductivity, the generation of a carbon coating is an easy route to enhance the electrical conductivity of insulator or semiconductor materials that require improvement of their conductivity for specific applications. This is the case of nanoparticles used in LIBs, such as $LiFePO_4$, $LiCoPO_4$, Si, SnSb, or $SnO_2$, among others (Kumar et al. 2011, Zhou et al. 2013, Fan et al. 2014, Li et al. 2014). For example, in $LiCoPO_4$ nanoparticles, the carbon coating is known to enhance the electrical conductivity more than three orders of magnitude (Kumar et al. 2011).

Concerning thermal properties, carbon coatings like graphene or graphite are particularly interesting since they protect the core from the environment under high-temperature conditions due to their high thermal conductivity (Table 18.1). Microscale TG analysis and conventional TG analysis (TGA) are the most common techniques employed to study the thermal stability of the nanoparticles (Mansfield et al. 2014). Luechinger et al. (2008a,b) investigated graphene-coated Cu nanoparticles and observed that the deposition of two or three layers of graphene (1–2 nm thick) on Cu nanoparticles was enough to protect them from microstructure degradation up to 473 K under humid air. The graphene shell sustains a chemical gradient of over 1 GV/m, offering the nanoparticles a high chemical stability. Similar thermal stability was found in graphite-coated magnetic nanoparticles (Liu et al. 2010b, Sunny et al. 2010); TGA measurements show a core chemical stability up to 500 K.

The thermal stability of the nanoparticles depends on the crystallinity degree of the carbon coating, which in turn depends on the conditions and the method used to produce the nanoparticles (Ou et al. 2008, Aguiló-Aguayo et al. 2013). The crystallinity degree of carbon-based nanomaterials can be determined by Raman spectroscopy. Graphitic materials show two characteristic peaks located around 1560 cm$^{-1}$ (G band) and 1360 cm$^{-1}$ (D band) for visible excitation (Ferrari 2007). The G band is attributed to the Eg$^2$ vibrational mode related to the bond stretching of all the $sp^2$ atoms in both rings and chains and is associated with perfect graphite. The D band is related to the Ag$^1$ mode and is sensitive to disorders in carbon materials, and it is forbidden in perfect graphite. The intensity ratio $I_D/I_G$ provides information of the crystallinity degree. The higher the ratio, the lower the crystallinity.

In order to better understand the behavior of the carbon coating and cores during thermal treatments, the in situ observation by means of the TEM under a controlled annealing is also desired. Aguiló-Aguayo et al. (2013) studied the thermal-induced structural evolution of carbon-coated Fe nanoparticles obtained by two different methods, CVD and ADP. No structural changes were observed up to 600 K. From this temperature, two different thermal-induced growth mechanisms, depending on the size of the iron cores, were identified. Cores larger than 10 nm in diameter migrated outside the carbon shell and coalesce into larger iron clusters, whereas Fe cores smaller than 10 nm required higher operating temperatures (above 973 K) to start diffusing outside the carbon shell via Ostwald ripening. In addition, differences in the graphitic shell were also observed during annealing. The graphitic shells from samples obtained by ADP presented higher structural stability than the CVD samples. The differences are attributed to the characteristics of each method.

Graphitic-based materials present the property of cleavage, which describes the tendency of a crystal to split along weakness planes. For graphite, cleavage occurs on the {001} crystal lattice plane. This property allows graphitic materials to be flexible in other directions and provides antifriction properties. Graphitic materials are flexible but inelastic, or in other words they are capable of bending easily and staying bent after the pressure is removed and resisting fracture under repeated bending. In addition, they present high stiffness, about 1060 GPa for graphene layers, and high strength showing an interlayer shear strength of 0.14 GPa (Falvo et al. 1997, Gu et al. 2002, Stankovich et al. 2006, Liu et al. 2012, Cheng & Liu 2013). Graphitic materials are very suitable for applications where high stiffness and low weight are required, such as spacecraft, automotive, or specific machinery components.

In addition, graphitic materials show self-lubricating and dry lubricating properties. This attribute comes from the cleavage property as well as the capability to adsorb fluids (air, water) between layers. It has been reported that the wear rate and friction can be exceptionally reduced in composites containing carbon-based nanoparticles (Zhang et al. 2014c). Moreover, the high corrosion resistance makes carbon-based nanoparticles, for instance carbon-coated Cu nanoparticles, particularly interesting for the improvement of the tribological properties of oil additives (Tu et al. 2001, Xu et al. 2008).

# 18.5 Applications

Carbon-coated nanoparticles can be found in a wide range of applications from biomedical therapies and diagnostics to electronic or electrochemical devices. Apart from the properties of the core, the carbon coating enhances the core functionalities by providing additional physical and chemical attributes, high electrical and thermal conductivity, chemical stability, high lubricity, and biocompatibility; in addition, it is an inexpensive material with high availability and light weight (Inagaki 2012).

## 18.5.1 Biomedical Applications

The main roles of the carbon coating in nanoparticles used for biomedical applications are to avoid the aggregation among the cores, to protect the core against the external environment in order to ensure the stability of the core magnetic or optical properties, to enhance the NIR absorption of the nanoparticles, to ensure biocompatibility and nontoxicity, and to open the possibility of surface functionalization without damaging the core.

For the use of nanoparticles in biomedical applications, it is necessary to have a good colloidal stability in aqueous solutions. Carbon-based materials present hydrophobic surfaces and require additional treatment to ensure the stability. Different routes have been actively explored (Jiang et al. 2003, Cha et al. 2013, Yang et al. 2013), for instance, the functionalization by means of chemical treatments or mechanical dispersion methods, but these strategies can cause damages to the crystal structure of the carbon, thus changing properties of the nanoparticles. Another route is the use of dispersants that can increase the surface charge density of the nanoparticles through the adsorption of polymers or nonpolar surfactants. Among these, the polyvinyl alcohol polymer is the most accepted for biomedical applications because of its nontoxicity, biocompatibility, water solubility, and biodegradability (Aguiló-Aguayo et al. 2014).

Cancer treatment is one of the biomedical applications of carbon-coated nanoparticles. Due to the selective sensitivity of malignant cells at temperatures above 40°C, one of the strategies for cancer cell killing is to promote heat generation by means of irradiation. The NIR absorption of carbon-based materials makes them very suitable for photothermal therapy (Yang et al. 2010, Markovic et al. 2011), which consists of the electromagnetic radiation in the 700–1100 nm range in the tumor region and the subsequent conversion of the radiation into vibrational energy (heat). The selective absorption of the carbon-based nanoparticles avoids the competition of light absorption by intrinsic chromophores in native tissues. Carbon-coated magnetic nanoparticles can be used as potential photothermal agents, they can be effectively localized to the tumor region by the application of an external magnetic field (Yang et al. 2012a, Gu et al. 2014, Lee et al. 2015). Carbon-coated noble metal nanoparticles with plasmon resonances in the NIR range can also be used as photothermal agents (Wang et al. 2012a). The combination of the SPR effect and the carbon coating NIR absorption remarkably improves the tumor cell-killing efficacy.

Magnetic hyperthermia is another cancer treatment that has gained attention in the last 10 years. It consists of generating heat due to the magnetic losses of magnetic nanoparticles located in the tumor region when an external alternating magnetic field is applied with a specific frequency. The magnetic losses of ferromagnetic (multidomain) nanoparticles come from the heating caused by hysteresis loops, whereas in superparamagnetic nanoparticles the heating is due to the Brown or Néel relaxation losses. Brown relaxation consists of the physical rotation of particles in the dispersed medium hindered by the viscosity of the medium and Néel relaxation is based on the changes in the direction of the magnetic moments hindered by the anisotropy energy that tends to turn the magnetic moments depending on the crystal lattice structure. Superparamagnetic nanoparticles are especially interesting for this application because they present extraordinary specific absorption rates (watts/gram) in comparison with ferromagnetic nanoparticles (Jordan et al. 1999). Carbon-coated magnetic nanoparticles have been investigated to act as potential hyperthermia agents (Taylor et al. 2008, Krupskaya et al. 2009, Verma et al. 2014).

Carbon surfaces can be functionalized for being used for bioimaging, biosensing, and drug and gene delivery as well as for diagnosis. Carbon nanomaterials can be covalently linked to visible-wavelength fluorophores or functionalized with fluorescent amino groups to trace the interaction of carbon nanomaterials with tissues, cells, and organisms using confocal microscope (Wang et al. 2010, Yang et al. 2012b).

A bioimaging technique that allows the noninvasive high spatial resolution of the body is the MRI. Carbon-coated superparamagnetic nanoparticles are used as contrast agents. The contrast of the images is based on the magnetic relaxation process of water protons and is used for tumor detection (Figuerola et al. 2010). Superparamagnetic nanoparticles enhance the contrast between healthy and diseased tissues. There are two types of contrast agents, the positive and the negative contrast agents. The positive contrast (bright signal) agents based on gadolinium-based complexes reduce the proton longitudinal relaxation times, whereas superparamagnetic nanoparticles act as negative contrast (dark signal) agents by shortening the proton transverse relaxation times (Na et al. 2009). Wang et al. (2012c) used carbon-coated iron oxide nanoparticles for lymph nodes mapping (LM) in order to combine the properties of the carbon coating used as a dye for many surgeries when performing LM and the superparamagnetic properties to use nanoparticles as MRI contrast agents. Seo et al. (2006) used FeCo nanoparticles coated by graphitic shells as a contrast agent for MRI. They observed a positive contrast enhancement due to the unusually high longitudinal relaxivity of FeCo/GC nanocrystals (Figure 18.13).

## 18.5.2 Electronics

A critical issue in the field of electronic devices, especially when the tendency of the physical dimensions is scaling down, is to control the heat dissipation generated during their operation that may cause temperature overshoots, thermal stresses, or warping (Zhang et al. 2013). Carbon-based materials present excellent thermal conductivities. At 25°C, graphite shows 100–400 W/mK on plane; carbon black, 6–174 W/mK; CNTs, 2000–6000 W/mK; and diamond, 2000 W/mK (Ebadi-Dehaghani & Nazempour 2012). The addition of carbon-based materials to composites or heat transfer fluids can improve the thermal conductivity to 40%–50% (Leong et al. 2006, Zhang et al. 2010a). In addition, the carbon coating reduces the thermal expansion of the composites and the light weight of carbon is an additional quality (Huang et al. 2009). Similar requirements are applicable to the automobile and construction industries and the handling of carbon-coated nanoparticles is also very convenient (Zhang et al. 2010b).

For instance, carbon-coated Cu nanoparticles are promising substitutes for metal nanoparticles used as conductive inks in the photovoltaic industry or membrane switches. They present potential

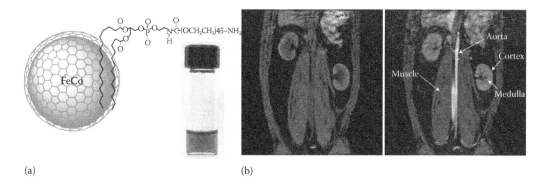

(a)                                                        (b)

**FIGURE 18.13**    (a) Schematic diagram of a graphitic-coated FeCo nanocrystal structure of the phospholipid molecule used for functionalization and a photograph of a phosphate buffer saline (PBS) suspension of functionalized graphitic-coated FeCo nanocrystals taken after heating to 8°C for 1 h and (b) T1-weighted MR images of a rabbit before (left) and 30 min after (right) initial injection of a solution of ~4 nm graphitic-coated FeCo nanocrystals. (Reprinted by permission from Macmillan Publishers Ltd., Seo, W. S., Lee, J. H., Sun, X. et al., *Nat. Mater.* 5: 971–976, Copyright (2006).)

advantages such as low-cost production, protection against oxidation, enhancement in the conductivity, and antistatic properties (Magdassi et al. 2010, Wang et al. 2012b).

Carbon-coated iron nanoparticles are also used to obtain reinforced magnetic epoxy resin polymer nanocomposites that can be integrated into electronic devices. These nanoparticles add magnetic behavior, improve mechanical properties by increasing the tensile strength up to 60%, enhance electrical conductivity, and reduce flammability (Zhang et al. 2013).

In addition, carbon-based materials provide excellent mechanical flexibility and stretchability, which is very interesting for flexible electronic devices, flexible displays, wearable electronics, sensors, or large circuits on curved objects (Yuan et al. 2011, 2012, Sun et al. 2013).

### 18.5.3  Mechanical Applications

Discontinuously reinforced composites have received increasing attention for both academics and manufacturing industries, primarily to meet the demand for lower weight and higher strength; they have been used for applications in aerospace, naval, and automotive structures (Cöcen & Önel 2002). Considerable progress in the field of composite materials has been made, driven by the relentless need for advanced materials with improved performance. Inspired by the dispersion strengthening mechanism, nanoscale-reinforcing materials have been directly used in novel nanocomposite synthesis. A high-strength dispersed phase can be introduced into the metal, ceramic, or polymer matrix. Many recent studies have employed powder metallurgy in the fabrication of Al alloy because the fabrication of the net-shaped parts can be achieved at a low cost. Aluminum alloys not only are lightweight materials, but also have excellent thermal conductivity, electrical conductivity, corrosion resistance, and workability. Therefore, aluminum is widely used as structural components, electrical conductors, and consumer products. Various additives are usually used to modify industrial aluminum alloys. It is demonstrated that carbon-coated silver nanoparticles can be dispersed into 2024 aluminum alloy by mechanical milling. The carbon-coated silver nanoparticles are chemically and environmentally stable and easy to handle. The carbon layers can maintain encapsulated silver nanoparticles intact and they are not dissolved into the Al matrix. With the increase in carbon-coated silver nanoparticle contents, the hardness increases to 30 Brinell units for the case of 2.5 wt% of carbon-coated silver nanoparticles (Carreno-Gallardo et al. 2009).

Nanomaterials offer unique properties that vary with changes in particle size. Metal-based nanoparticles have been used as industrial chemicals, composite reinforcement, catalysts, optical media, magnetic storage materials, and electrode materials. However, metallic nanoparticles without a protective coating often have a high propensity to oxidize or undergo other chemical reactions. Carbon-coated metal nanoparticles can broaden the applications of metallic nanoparticles. Carbon-coated Cu nanoparticle can be applied as new additive, coupling agents, or strengthening filler for polymer composites with improved toughness while maintaining or improving other composite mechanical properties (Lian & Wu 2014). The metal core in such nanoparticles is protected against chemical reactions by the carbon shell and provides the mechanical reinforcement of composites. The carbon shells can be functionalized to enhance the interaction between the carbon-coated nanoparticles and the polymer matrix.

Carbon-coated nanoparticles can be added to the rubber composition to enhance rubber mechanical properties of vehicle tires (tensile strength, 600%; tear strength, 250%; and hardness, 70% of styrene–butadiene rubbers) (Mohseni et al. 2012). Moreover, adding nanoparticles to the rubber composition enhances the durability and fuel efficiency, improves the strain vibration at high speed, leads to a superior traction in wet roads, and reduces inner friction of the rubber matrix.

## 18.5.4 Energy and Environment Applications

The carbon coating in nanoparticles is widely employed for energy applications, especially in LIBs. It is used in electrode materials, both cathodes such as olive-structure orthophosphates $LiMPO_4$ ($M$ = Mn, Fe, Co, and Ni) or lithium metal oxides ($LiMO_2$, $M$ = Co, Ni, Mn), and anodes such as semiconductors Sb, Ge, or Si, among others. The carbon coating improves the battery performance by reaching higher and reversible capacities, long cycle lives, and higher rate capabilities. For example, carbon-coated $LiFePO_4$ nanoparticles can reach high capacity values up to 160 mAh/g ($LiFePO_4$ particles without carbon coating present values of 80 mAh/g at a discharge rate of 60 C) and excellent cycling performance, less than 5% discharge capacity loss over 1100 cycles (Wang et al. 2008). The carbon coating offers the active material good electrical conductivity, improves the ion diffusion, helps the electrode penetration, avoids agglomeration among particles, and offers resistance to corrosion and chemical stability (Liu & Zhou 2012). It can work as an elastic shell to accommodate volume changes and to suppress the grain growth of the active materials during lithiation and delithiation processes. One of the most common carbon coatings used for these applications is the carbon black, in particular acetylene black. Although graphite presents higher electrical conductivity than acetylene black (carbon black consists of a mixture of $sp^2$- and $sp^3$-hybridized carbon atoms), the latter can contain up to three times as much electrolyte (Claus & Besenhard 2011).

Carbon-coated nanoparticles are also being investigated in solar cell architectures, especially in dye-sensitized solar cells, as an alternative to reduce costs, improve stability, and increase power conversion efficiency. The carbon coating offers high surface area due to their porous morphology, corrosion resistance, and enhances electrocatalytic abilities (Kim et al. 2014, Wu et al. 2014, Yeh et al. 2014). For the same main reasons, the enhancement of the photocatalytic effect and the avoidance of particle aggregation, the carbon coating is used in photocatalytic nanoparticles (Pt, $TiO_2$, CdS, or ZnO) for environmental applications, such as water purification or air cleaning. Carbon-coated nanoparticles present the ability to degrade pollutants, such as phenolic compounds, under UV–visible radiation by means of the generation of hydroxyl radicals, one of the most oxidants produced in photocatalytic reactions (Yi et al. 2009, Hu et al. 2010). Carbon black is the most commonly used carbon coating for nanoparticles due to its simplicity in the process formation, high porosity for the easy diffusion of the pollutants, and good electrical conductivity, leading to a reduction in the speed of electron–hole recombination during photocatalytic process (Soltani et al. 2013). For $TiO_2$ nanoparticles, the carbon coating also suppresses grain growth and acts as a stabilizer for the higher catalytic crystalline phase (anatase phase), avoiding other undesired phases like the rutile phase (Inagaki 2012).

Another interest in carbon-based nanomaterials in energy applications is focused on their use as hydrogen reservoirs. Carbon nanostructures can store $H_2$ due to the physisorption on graphenic surfaces and the hydrogen uptake capacity is proportional to the SSA, which in the case of graphene, the theoretical value is 2630 $m^2$/g (Dimitrakakis et al. 2008, Srinivas et al. 2010).

# References

Aguiló-Aguayo, N., Inestrosa-Izurieta, M. J., García-Céspedes, J. et al., "Morphological and magnetic properties of superparamagnetic carbon-coated Fe nanoparticles produced by arc discharge," *J. Nanosci. Nanotechnol.* 9 (2009): 1–4.

Aguiló-Aguayo, N., Liu, Z. Y., Bertran, E. et al., "Thermal-induced structural evolution of carbon-encapsulated iron nanoparticles generated by two different methods," *J. Phys. Chem. C* 117 (2013): 19167–19174.

Aguiló-Aguayo, N., Maurizi, L., Galmarini, S. et al., "Aqueous stabilisation of carbon-encapsulated superparamagnetic alpha-iron nanoparticles for biomedical applications," *Dalton Trans.* 43 (2014): 13764–13775.

Anjum, D. H., Memon, N. K. & Chung, S. H., "Investigating the growth mechanism and optical properties of carbon-coated titanium dioxide nanoparticles," *Mater. Lett.* 108 (2013): 134–138.

Asoro, M. A., Kovar, D. & Ferreira, P. J., "Effect of surface carbon coating on sintering of silver nanoparticles: In situ TEM observations," *Chem. Commun.* 50 (2014): 4835–4838.

Atkinson, J. D., Fortunato, M. E., Dastgheib, S. A. et al., "Synthesis and characterization of iron-impregnated porous carbon spheres prepared by ultrasonic spray pyrolysis," *Carbon* 49 (2011): 587–598.

Bae, H., Ahmad, T., Rhee, I. et al., "Carbon-coated iron oxide nanoparticles as contrast agents in magnetic resonance imaging," *Nanoscale Res. Lett.* 7 (2012): 1–5.

Baker, R., "Catalytic growth of carbon filaments," *Carbon* 27 (1989): 315–23.

Baker, R., Barber, M., Harries, P. et al., "Nucleation and growth of carbon deposits from the nickel catalysed decomposition of acetylene," *J. Catal.* 26 (1972): 51–62.

Banhart, F. & Ajayan, P. M., "Carbon onions as nanoscopic pressure cells for diamond formation," *Nature* 382 (1996): 433–435.

Banhart, F., Redlich, Ph. & Ajayan, P. M., "The migration of metal atoms through carbon onions," *Chem. Phys. Lett.* 292 (1998): 554–560.

Barberio, M., Barone, P., Stranges, F. et al., "Carbon nanotubes/metal nanoparticle based nanocomposites: Improvements in visible photoluminescence emission and hydrophobicity," *Opt. Photonics J.* 3 (2013): 34–40.

Barone, P., Barberio, M., Stranges, F. et al., "Study of coating geometries and photoluminescence properties of metal nanoparticles/graphite composites," *J. Chem.* 2014 (2014): 1–6.

Bean, C. & Livingston, J., "Superparamagnetism," *J. Appl. Phys.* 30 (1959): 120–129.

Becker, R. & Döring, W., "Kinetische Behandlung der keimbildung in übersättigten dämpfen," *Annalen Phys.* 416 (1935): 719–752.

Bera, D., Johnston, G., Heinrich, H. et al., "A parametric study on the synthesis of carbon nanotubes through arc-discharge in water," *Nanotechnology* 17 (2006): 1722–1730.

Bystrzejewski, M., Lange, H., Huczko, A. et al., "Fullerene and nanotube synthesis: Plasma spectroscopy studies," *J. Phys. Chem. Solids* 58 (1997): 1679–1683.

Bystrzejewski, M., Huczko, A. & Lange, H., "Arc plasma route to carbon-encapsulated magnetic nanoparticles for biomedical applications," *Sens. Actuators B Chem.* 109 (2005): 81–85.

Carreno-Gallardo, C., Estrada-Guela, I., Neriet, M. A. et al., "Carbon-coated silver nanoparticles dispersed in a 2024 aluminum alloy produced by mechanical milling," *J. Alloys Compd.* 483 (2009): 355–358.

Cha, C. Y., Shin, S. R., Annabi, N. et al., "Carbon-based nanomaterials: Multifunctional materials for biomedical engineering," *ACS Nano* 7 (2013): 2891–2897.

Chaukulkar, R. P., de Peuter, K., Stradins, P. et al., "Single-step plasma synthesis of carbon-coated silicon nanoparticles," *ACS Appl. Mater. Interfaces* 6 (2014): 19026–19034.

Chazelas, C., Coudert, J., Jarrige, J. et al., "Synthesis of ultra fine particles by plasma transferred arc: Influence of anode material on particle properties," *J. Eur. Ceram. Soc.* 26 (2006): 3499–3507.

Chen, C., Sanchez, A., Taboada, E. et al., "Size determination of superparamagnetic nanoparticles from magnetization curve," *J. Appl. Phys.* 105 (2009): 1–6.

Cheng, Y. & Liu, J., "Carbon nanomaterials for flexible energy storage," *Mater. Res. Lett.* 1 (2013): 175–192.

Choi, H., Ko, S., Choi, Y. et al., "Versatile surface plasmon resonance of carbon-dot-supported silver nanoparticles in polymer optoelectronic devices," *Nat. Photonics* 7 (2013): 732–738.

Choy, K. L., "Chemical vapour deposition of coatings," *Prog. Mater. Sci.* 48 (2003): 57–170.

Claus, D. & Besenhard, J. O., "Carbons," in *Handbook of Battery Materials*, Second edition (Wiley-VCH, Weinheim, Germany, 2011).

Cöcen, U. & Önel, K., "Ductility and strength of extruded SiCp/aluminium–alloy composites," *Compos. Sci. Technol.* 62 (2002): 275–282.

Compton, R. G. & Wadhawan, J. D., *Electrochemistry: Volume 11—Nanosystems* (Royal Society of Chemistry, Cambridge, England, 2012).

Cullity, B. D. & Graham, C. D., *Introduction to Magnetic Minerals*, Second edition (Wiley-IEEE Press, Piscataway, 2008).

Delhaes, P., *Graphite and Precursors* (CRC Press, Boca Raton, 2000).

Dimitrakakis, G. K., Tylianakis, E. & Froudakis, G. E., "Pillared graphene: A new 3-D network nanostructure for enhanced hydrogen storage," *Nano Lett.* 8 (2008): 3166–3170.

Ding, F. & Bolton, K., "The importance of supersaturated carbon concentration and its distribution in catalytic particles for single-walled carbon nanotube nucleation," *Nanotechnology* 17 (2006): 543–548.

Ding, F., Bolton, K. & Rosen, A., "Iron-carbide cluster thermal dynamics for catalysed carbon nanotube growth," *J. Vac. Sci. Technol. A Vac. Surf. Films* 22 (2004): 1471–1476.

Dumitrache, F., Morjan, I., Alexandrescu, R. et al., "Nearly monodispersed carbon coated iron nanoparticles for the catalytic growth of nanotubes/nanofibers," *Diam. Relat. Mater.* 13 (2004): 362–370.

Ebadi-Dehaghani, H. & Nazempour, M., "Thermal conductivity of nanoparticles filled polymers," in Hashim, A., ed., *Nanoparticles Filled Polymers, Smart Nanoparticles Technology* (InTech, Morn Hill, UK, 2012).

Erokhin, A. V., Lokteva, E. S., Yermakov, A. Y. et al., "Phenylacetylene hydrogenation on Fe@C and Ni@C core–shell nanoparticles: About intrinsic activity of graphene-like carbon layer in $H_2$ activation; Selective graphene covering of monodispersed magnetic nanoparticles," *Carbon* 74 (2014): 291–301.

Fan, X. L., Shao, J., Xiao, X. Z. et al., "In situ synthesis of $SnO_2$ nanoparticles encapsulated in micro/mesoporous carbon foam as a high-performance anode material for lithium ion batteries," *J. Mater. Chem. A* 2 (2014): 18367–18374.

Falvo, M. R., Clary, G. J., Taylor, R. M. et al., "Bending and buckling of carbon nanotubes under large strain," *Nature* 389 (1997): 582–584.

Ferrari, A. C., "Raman spectroscopy of graphene and graphite: Disorder, electron–phonon coupling, doping and nonadiabatic effects," *Solid State Commun.* 143 (2007): 47–57.

Figuerola, A., Di Corato, R., Manna, L. et al., "From iron oxide nanoparticles towards advanced iron-based inorganic materials designed for biomedical applications," *Pharmacol. Res.* 62 (2010): 126–143.

Gittleman, J. I., Abeles, B. & Bozowski, S., "Superparamagnetism and relaxation effects in granular Ni–$SiO_2$ and Ni–$Al_2O_3$ films," *Phys. Rev. B* 9 (1974): 3891–3897.

Gu, J. L., Leng, Y., Gao, Y. et al., "Fracture mechanism of flexible graphite sheets," *Carbon* 40 (2002): 2169–2176.

Gu, L., Koymen, A. R. & Mohanty, S. K., "Crystalline magnetic carbon nanoparticle assisted photothermal delivery into cells using CW near-infrared laser beam," *Sci. Rep.* 4 (2014): 1–10.

Gubin, S. P., *Magnetic Nanoparticles* (Wiley-VCH, Weinheim, Germany, 2009).

Guo, J. C., Chen, X. L. & Wang, C. S., "Carbon scaffold structured silicon anodes for lithium-ion batteries," *J. Mater. Chem.* 20 (2010): 5035–5040.

Guo, T., Nikolaev, P., Thess, A. et al., "Catalytic growth of single-walled nanotubes by laser vaporization," *Chem. Phys. Lett.* 243 (1995): 49–54.

Gustch, A., Krämer, M., Michael, G. et al., "Gas-phase production of nanoparticles," *KONA Powder Part. J.* 20 (2002): 24–37.

He, Y., Huang, Li., Li, X. et al., "Facile synthesis of hollow $Cu_2Sb@C$ core-shell nanoparticles as a superior anode material for lithium ion batteries," *J. Mater. Chem.* 21 (2011): 18517–18519.

Hermann, I., Urner, M., Koehler, F. M. et al., "Blood purification using functionalized core/shell nano-magnets," *Small* 6 (2010): 1388–1392.

Hornyak, G. L., Dutta, J., Tibbals, H. F. et al., *Introduction to Nanoscience* (CRC Press, Boca Raton, 2008).

Hu, Y., Liu, Y., Qian, H. et al., "Coating colloidal carbon spheres with CdS nanoparticles: Microwave assisted synthesis and enhanced photocatalytic activity," *Langmuir* 26 (2010): 18570–18575.

Huang, J. Y., "In situ observation of quasimelting of diamond and reversible graphite–diamond phase transformations," *Nano Lett.* 7 (2007): 2235–2240.

Huang, C. H., Wang, H. P., Chang, J. E. et al., "Synthesis of nanosize-controllable copper and its alloys in carbon shells," *Chem. Commun.* 31 (2009): 4663–4665.

Inagaki, M., "Carbon coating for enhancing the functionalities of materials," *Carbon* 50 (2012): 3247–3266.

Jackson, S. D. & Hargreaves, J. S. J., *Metal Oxide Catalysis* (Wiley-VCH, Weinheim, Germany, 2008).

Jager, C., Mutschke, H., Huisken, F. et al., "Iron–carbon nanoparticles prepared by $CO_2$ laser pyrolysis of toluene and iron pentacarbonyl," *Appl. Phys. A Mater. Sci. Process.* 85 (2006): 53–62.

Jiang, L. Q., Gao, L. & Sun, L. "Production of aqueous colloidal dispersions of carbon nanotubes," *J. Colloid Interface Sci.* 260 (2003): 89–94.

Jordan, A., Scholz, R., Wust, P. et al., "Magnetic fluid hyperthermia (MFH): Cancer treatment with AC magnetic field induced excitation of biocompatible superparamagnetic nanoparticles," *J. Magn. Magn. Mater.* 201 (1999): 413–419.

Jun, S., Uhm, Y. & Rhee, K., "Preparation of carbon-encapsulated Ag nanoparticles for dispersion in $La_{0.6}Sr_{0.4}Co_{0.3}Fe_{0.7}O_{3-\delta}$," *J. Korean Phys. Soc.* 59 (2011): 3648–3651.

Jung, D., Bin Park, S. & Kang, Y. C., "Design of particles by spray pyrolysis and recent progress in its application," *Korean J. Chem. Eng.* 27 (2010): 1621–1645.

Kim, S., Oh, S. D., Lee, S. et al., "Radiolytic synthesis of Pd-M (M = Ag, Ni, and Cu)/C catalyst and their use in Suzuki-type and Heck-type reaction," *J. Ind. Eng. Chem.* 14 (2008): 449–456.

Kim, J., Lim, J., Kim, M. et al., "Fabrication of carbon-coated silicon nanowires and their application in dye-sensitized solar cells," *ACS Appl. Mater. Interfaces* 6 (2014): 18788–18794.

Knobel, M., Nunes, W., Socolovsky, L. et al. "Superparamagnetism and other magnetic features in granular materials: A review on ideal and real systems," *J. Nanosci. Nanotechnol.* 8 (2008): 2836–2857.

Kruis, F. E., Fissan, H. & Peled, A., "Synthesis of nanoparticles in the gas phase for electronic, optical and magnetic applications—A review," *J. Aerosol Sci.* 29 (1998): 511–535.

Krupskaya, Y., Mahn, C., Parameswaran, A. et al., "Magnetic study of iron-containing carbon nanotubes: Feasibility for magnetic hyperthermia," *J. Magn. Magn. Mater.* 321 (2009): 4067–4071.

Kumar, P. R., Venkateswarlu, M., Misra, M. et al., "Synthesis, characterization and electrical properties of carbon coated $LiCoPO_4$ nanoparticles," *J. Nanosci. Nanotechnol.* 11 (2011): 3314–3322.

Kumar, R., Kumari, S. & Dhakate, S. R., "Nickel nanoparticles embedded in carbon foam for improving electromagnetic shielding effectiveness," *Appl. Nanosci.* 5 (2014): 553–561.

Kwong, H. Y., Wong, M. H., Leung, C. W. et al., "Formation of core/shell structured cobalt/carbon nanoparticles by pulsed laser ablation in toluene," *J. Appl. Phys.* 108 (2010): 1–5.

Laurent, C., Flahaut, E., Peigney, A. et al. "Metal nanoparticles for the catalytic synthesis of carbon nano-tubes," *New J. Chem.* 22 (1998): 1229–1237.

Lee, S., Kang, Y. I., Ha, S. J. et al., "Carbon-deposited $TiO_2$ nanoparticle balls for high-performance visible photocatalysis," *RSC Adv.* 4 (2014): 55371–55376.

Lee, H. J., Sanetuntikul, J., Choi, E. S. et al., "Photothermal cancer therapy using graphitic carbon-coated magnetic particles prepared by one-pot synthesis," *Int. J. Nanomedicine* 10 (2015): 271–282.

Leong, C. K., Aoyagi, Y. & Chung, D. D. L., "Carbon black pastes as coatings for improving thermal gap-filling materials," *Carbon* 44 (2006): 435–440.

Li, Z. T., Hu, C., Yu, C. et al., "Synthesis and characterization of carbon-encapsulated magnetic nanopar-ticles via arc-plasma assisted CVD," *J. Nanosci. Nanotechnol.* 9 (2009): 7473–7476.

Li, J., Liu, C. Y. & Xie, Z., "Synthesis and surface plasmon resonance properties of carbon-coated Cu and Co nanoparticles," *Mater. Res. Bull.* 46 (2011): 743–747.

Li, L., Seng, K. H., Li, D. et al., "SnSb@carbon nanocable anchored on graphene sheets for sodium ion batteries," *Nano Res.* 7 (2014): 1466–1476.

Lian, K. & Wu, Q., "Carbon-encased metal nanoparticles and sponges as wood/plant preservatives or strengthening fillers," Patent US8828485 (2014).

Lifshitz, I. & Slyozov, V., "The kinetics of precipitation from supersaturated solid solutions," *J. Phys. Chem. Solids* 19 (1961): 35–50.

Liu, H. Q. & Zhou, H. S., "Enhancing the performances of Li-ion batteries by carbon-coating: Present and future," *Chem. Commun.* 48 (2012): 1201–1217.

Liu, Z., Yuan, Z. Y., Zhou, W. et al. "Controlled synthesis of carbon-encapsulated Co nanoparticles by CVD," *Chem. Vapor Depos.* 7 (2001): 248–251.

Liu, F., Cao, Z., Tang, C. et al., "Ultrathin diamond-like carbon film coated silver nanoparticles-based substrates for surface-enhanced Raman spectroscopy," *ACS Nano* 4 (2010a): 2643–2648.

Liu, X. G., Ou, Z. Q., Geng, D. Y. et al., "Influence of a graphite shell on the thermal and electromagnetic characteristics of FeNi nanoparticles," *Carbon* 48 (2010b): 891–897.

Liu, Z., Zhang, S. M., Yang, J. R. et al., "Interlayer shear strength of single crystalline graphite," *Acta Mech. Sin.* 28 (2012): 978–982.

Liu, F. X., Tang, C. J., Zhan, P. et al., "Released plasmonic electric field of ultrathin tetrahedral-amorphous-carbon films coated Ag nanoparticles for SERS," *Sci. Rep.* 4 (2014a): 1–6.

Liu, L., Zhao, C., Zhao, H. et al., "ZnO–CoO nanoparticles encapsulated in 3D porous carbon microsphere for high-performance lithium-ion battery anodes," *Electrochim. Acta* 135 (2014b): 224–231.

Lothe, J. & Pound, G. M., "Reconsiderations of nucleation theory," *J. Chem. Phys.* 36 (1962): 2080–2085.

Luechinger, N. A., Athanassiou, E. K. & Stark, W. J., "Graphene-stabilized copper nanoparticles as an air-stable substitute for silver and gold in low-cost ink-jet printable electronics," *Nanotechnology* 19 (2008a): 1–6.

Luechinger, N. A., Booth, N., Heness, G. et al., "Surfactant-free, melt-processable metal–polymer hybrid materials: Use of graphene as a dispersing agent," *Adv. Mater.* 20 (2008b): 3044–3049.

Luo, W., Lorger, S., Wang, B. et al., "Facile synthesis of one-dimensional peapod-like Sb@C submicron-structures," *Chem. Commun.* 50 (2014): 5435–5437.

Ma, L., Yu, B. & Wang, S., "Controlled synthesis and optical properties of Cu/C core/shell nanoparticles," *J. Nanoparticle Res.* 16 (2014): 613–634.

Magdassi, S., Grouchko, M. & Kamyshny, A., "Copper nanoparticles for printed electronics: Routes towards achieving oxidation stability," *Materials* 3 (2010): 4626–4638.

Majetich, S., Brunsman, E., Scott, J. et al., "Formation of nanoparticles in a carbon arc," *Mater. Sci. Eng. A Struct. Mater. Prop. Microstructure Process.* 204 (1994): 29–34.

Malkhandi, S., Trinh, P., Manohar, A. K. et al., "Electrocatalytic activity of transition metal oxide–carbon composites for oxygen reduction in alkaline batteries and fuel cells," *J. Electrochem. Soc.* 160 (2013): F943–F952.

Mansfield, E., Tyner, K. M., Poling, C. M. et al., "Determination of nanoparticle surface coatings and nanoparticle purity using microscale thermogravimetric analysis," *Anal. Chem.* 86 (2014): 1478–1484.

Marinescu, I. D., Hitchiner, M. P., Uhlmann, E. et al., *Handbook of Machining with Grinding Wheels* (CRC Press, Boca Raton, 2006).

Marinho, B., Ghislandi, M., Tkalya, E. et al., "Electrical conductivity of compacts of graphene, multi-wall carbon nanotubes, carbon black, and graphite powder," *Powder Technol.* 221 (2012): 351–358.

Markovic, Z. M., Harhaji-Trajkovic, L. M., Todorovic-Markovic, B. M. et al., "In vitro comparison of the photothermal anticancer activity of graphene nanoparticles and carbon nanotubes," *Biomaterials* 32 (2011): 1121–1129.

Mohseni, M., Ramezanzadeh, B., Yari, H. et al., "The role of nanotechnology in automotive industries," in Carmo, J., ed., *New Advances in Vehicular Technology and Automotive Engineering* (InTech, Morn Hill, UK, 2012a).

Mokari, T., Sztrum, C., Salant, A. et al., "Formation of asymmetric one-sided metal-tipped semiconductor nanocrystal dots and rods," *Nat. Mater.* 4 (2005): 855–8563.

Moody, M. P. & Attard, P., "Homogeneous nucleation of droplets from a supersaturated vapour phase," *J. Chem. Phys.* 117 (2002): 6705–6714.

Morishita, T., Soneda, Y., Hatori, H. et al., "Carbon-coated tungsten and molybdenum carbides for electrode of electrochemical capacitor," *Electrochim. Acta* 52 (2007): 2478–2484.

Mulvaney, P., "Surface plasmon spectroscopy of nanosized metal particle," *Langmuir* 12 (1996): 788–800.

Murugan, A. V., Muraliganth, T. & Manthiram, A., "Comparison of microwave assisted solvothermal and hydrothermal syntheses of LiFePO$_4$/C nanocomposite cathodes for lithium ion batteries," *J. Phys. Chem. C* 112 (2008): 14665–14671.

Na, H. B., Song, I. C. & Hyeon, T., "Inorganic nanoparticles for MRI contrast agents," *Adv. Mater.* 21 (2009): 2133–2148.

Nawi, M. A. & Nawawi, I., "Preparation and characterization of TiO$_2$ coated with a thin carbon layer for enhanced photocatalytic activity under fluorescent lamp and solar light irradiations," *Appl. Catal. A General* 453 (2013): 80–91.

Néel, L., "Théorie du traînage magnétique des ferromagnétiques en grains fins avec application aux terres cuites," *Ann. Géophys.* 5 (1949): 99–136.

Oberlin, A., Endo, M. & Koyama, T., "Filamentous growth of carbon through benzene decomposition," *J. Cryst. Growth* 32 (1976): 335–349.

Olurode, K., Neelgund, G. M., Oki, A. et al., "A facile hydrothermal approach for construction of carbon coating on TiO$_2$ nanoparticles," *Spectrochim. Acta A Mol. Biomol. Spectrosc.* 89 (2012): 333–336.

Ou, Q., Tanaka, T., Mesko, A. et al., "Characteristics of graphene-layer encapsulated nanoparticles fabricated using laser ablation method," *Diam. Rel. Mater.* 17 (2008): 664–668.

Park, J., "Structural characterisation of hard materials by transmission electron microscopy (TEM): Diamond–silicon carbide composites and yttria-stabilized zirconia," PhD dissertation, Stanford University (2008).

Pomogailo, A. D. & Kestelman, V. N., "Metallopolymer nanocomposites," *Springer Series in Materials Science*, 81. (Springer, Berlin, Germany, 2005).

Robertson, J., "Diamond-like amorphous carbon," *Mater. Sci. Eng. R* 37 (2002): 129–281.

Saito, Y., "Nanoparticles and filled nanocapsules," *Carbon* 37 (1995): 979–988.

Sarno, M., Cirillo, C. & Ciambelli, P., "Selective graphene covering of monodispersed magnetic nanoparticles," *Chem. Eng. J.* 246 (2014): 27–38.

Schuller, J. A., Barnard, E. S., Cai, W. S. et al., "Plasmonics for extreme light concentration and manipulation," *Nat. Mater.* 9 (2010): 193–204.

Scott, J., Majetich, S., Pyrzynska, K. et al., "Morphology, structure and growth of nanoparticles produced in a carbon arc," *Phys. Rev. B* 52 (1995): 564–571.

Seo, W. S., Lee, J. H., Sun, X. et al., "FeCo/graphitic-shell nanocrystals as advanced magnetic-resonance-imaging and near-infrared agents," *Nat. Mater.* 5 (2006): 971–976.

Seraphin, S., Zhou, D. & Jiao, D., "Filling the carbon nanocages," *J. Appl. Phys.* 80 (1996): 2097–2104.

Soltani, R. D. C., Rezaee, A. & Khataee, A., "Combination of carbon black–ZnO/UV process with an electrochemical process equipped with a carbon black–PTFE–coated gas-diffusion cathode for removal of a textile dye," *Ind. Eng. Chem. Res.* 52 (2013): 14133–14142.

Song, C., Du, J., Zhao, J. et al., "Hierarchical porous core-shell carbon nanoparticles," *Chem. Mater.* 21 (2009): 1525–1530.

Srinivas, G., Zhu, Y. W., Piner, R. et al., "Synthesis of graphene-like nanosheets and their hydrogen adsorption capacity," *Carbon* 48 (2010): 630–635.

Stankovich, S., Dikin, D. A., Dommett, G. H. B. et al., "Graphene-based composite materials," *Nature* 442 (2006): 282–286.

Sun, D. M., Timmermans, M. Y., Kaskela, A. et al., "Mouldable all-carbon integrated circuits," *Nat. Commun.* 4 (2013): 1–8.

Sunny, V., Kumar, D. S., Yoshida, Y. et al., "Synthesis and properties of highly stable nickel/carbon core/shell nanostructures," *Carbon* 48 (2010): 1643–1651.

Tartaj, P., del Puerto Morales, M., Veintemillas-Verdaguer, S. et al., "The preparation of magnetic nanoparticles for applications in biomedicine," *J. Phys. D Appl. Phys.* 36 (2003): R182–R197.

Taylor, A., Kraemer, K., Hampel, S. et al., "Carbon coated nanomagnets as potential hyperthermia agents," *J. Urol.* 179 (2008): 392–393.

Tea, N. H., Yu, R. C., Salamon, M. B. et al., "Thermal conductivity of $C_{60}$ and $C_{70}$ crystals," *Appl. Phys. A* 56 (1993): 219–225.

Tu, J. P., Yang, Y. Z., Wang, L. Y. et al. "Tribological properties of carbon-nanotube-reinforced copper composites," *Tribol. Lett.* 10 (2001): 225–228.

Tulugan, K., Kim, H., Park, W. et al., "Aluminum–silicon and aluminum–silicon/carbon nanoparticles with core–shell structure synthesized by arc discharge method," *J. Alloys Compd.* 574 (2013): 529–532.

Vargas, J., Nunes, W., Socolovsky, M. et al., "Effect of dipolar interaction observed in iron-based nanoparticles," *Phys. Rev. B* 72 (2005): 1–6.

Vehkamäki, H., *Classical Nucleation Theory in Multicomponent Systems* (Springer, Heidelberg, Germany, 2006).

Verma, J., Lal, S., Van Noorden, C. J. F. et al., "Nanoparticles for hyperthermic therapy: Synthesis strategies and applications in glioblastoma," *Int. J. Nanomed.* 9 (2014): 2863–2877.

Wang, Z. F., Guo, H. S., Yu, Y. L. et al. "Synthesis and characterization of a novel magnetic carrier with its composition of $Fe_3O_4$/carbon using hydrothermal reaction," *J. Magn. Magn. Mater.* 302 (2006): 397–404.

Wang, Z., Tian, W. H., Liu, X. H. et al., "Synthesis and electrochemical performances of amorphous carbon-coated Sn–Sb particles as anode material for lithium-ion batteries," *J. Solid State Chem.* 180 (2007a): 3360–3365.

Wang, C., Wang, J. & Sheng, Z., "Solid-phase synthesis of carbon-encapsulated magnetic nanoparticles," *J. Phys. Chem. C* 111 (2007b): 6303–6307.

Wang, Y. G., Wang, Y. R., Hosono, E. J. et al., "The design of a $LiFePO_4$/carbon nanocomposite with a core–shell structure and its synthesis by an in situ polymerization restriction method," *Angew. Chem. Int. Ed.* 47 (2008): 7461–7465.

Wang, X., Cao, L., Yang, S. T. et al., "Bandgap-like strong fluorescence in functionalized carbon nanoparticles," *Angew. Chem. Int. Ed.* 49 (2010): 5310–5314.

Wang, S. L., Huang, X. L., He, Y. H. et al., "Synthesis, growth mechanism and thermal stability of copper nanoparticles encapsulated by multi-layer graphene," *Carbon* 50 (2012a): 2119–2125.

Wang, X. J., Wang, C., Cheng, L. et al., "Noble metal coated single-walled carbon nanotubes for applications in surface enhanced Raman scattering imaging and photothermal therapy," *J. Am. Chem. Soc.* 134 (2012b): 7414–7422.

Wang, Y. X., Wang, D. W., Zhu, X. M. et al., "Carbon coated superparamagnetic iron oxide nanoparticles for sentinel lymph nodes mapping," *Quant. Imaging Med. Surg.* 2 (2012c): 53–56.

Wei, Z. Q., Liu, L. G., Yang, H. et al., "Characterization of carbon encapsulated Fe-nanoparticles prepared by confined arc plasma," *Trans. Nonferrous Met. Soc. China* 21 (2011): 2026–2030.

Westerberg, H., Boman, M., Norekrans, A. S. et al. "Carbon growth by thermal laser-assisted chemical vapour deposition," *Thin Solid Films* 215 (1992): 126–133.

Widenkvist, E., Alm, O., Boman, M. et al., "Functionalization and area-selective deposition of magnetic carbon-coated iron nanoparticles from solution," *J. Nanotechnol.* 2011 (2011): 1–4.

Willard, M. A., Kurihara, L. K., Carpenter, E. E. et al., "Chemically prepared magnetic nanoparticles," *Int. Mater. Rev.* 49 (2004): 125–170.

Wu, H., Geng, J., Wang, Y. H. et al., "Bias-free, solar-charged electric double-layer capacitors," *Nanoscale* 6 (2014): 15316–15320.

Xu, Y., Yu, H. L., Xu, B. S. et al., "Preparation and tribological properties of surface-coated nano-copper additives," Paper presented at the Fifth International Conference on Surface Engineering, Dalian, China, July 2008 (2008).

Xu, Y., Kamakar, A., Heberlein, W. E. et al., "Multifunctional magnetic nanoparticles for synergistic enhancement of cancer treatment by combinatorial radio frequency thermolysis and drug delivery," *Adv. Healthc. Mater.* 1 (2012): 493–501.

Xu, J., Sun J., Wang, Y. et al., "Application of iron magnetic nanoparticles in protein immobilization," *Molecules* 19 (2014): 11465–11486.

Yang, G. W., "Laser ablation in liquids: Applications in the synthesis of nanocrystals," *Prog. Mater. Sci.* 52 (2007): 648–698.

Yang, K., Feng, L. Z., Hong, H. et al., "Preparation and functionalization of graphene nanocomposites for biomedical applications," *Nat. Protoc.* 8 (2013): 2392–2403.

Yang, K., Zhang, S. A., Zhang, G. X. et al., "Graphene in mice: Ultrahigh in vivo tumor uptake and efficient photothermal therapy," *Nano Lett.* 10 (2010): 3318–3323.

Yang, Y. H., Cui, J. H., Zheng, M. T. et al., "One-step synthesis of amino-functionalized fluorescent carbon nanoparticles by hydrothermal carbonization of chitosan," *Chem. Commun.* 48 (2012a): 380–382.

Yang, K., Hu, L. L., Ma, X. X. et al., "Multimodal imaging guided photothermal therapy using functionalized graphene nanosheets anchored with magnetic nanoparticles," *Adv. Mater.* 24 (2012b): 1868–1872.

Yeh, M. H., Lin, L. Y., Su, J. S. et al., "Nanocomposite graphene/Pt electrocatalyst as economical counter electrode for dye-sensitized solar cells," *ChemElectroChem* 1 (2014): 416–425.

Yi, X., Sunghwan, H., Seunghwa, Y. et al., "Synthesis and photocatalytic activity of anatase $TiO_2$ nanoparticles-coated carbon nanotubes," *Nanoscale Res. Lett.* 5 (2009): 603–607.

Yu, F., Wang, J. N., Sheng, Z. M. et al., "Synthesis of carbon encapsulated magnetic nanoparticles by spray pyrolysis of iron carbonyl and ethanol," *Carbon* 43 (2005): 3018–3021.

Yu, L. H., Cai, D. D., Wang, H. H. et al., "Synthesis of microspherical $LiFePO_4$–carbon composites for lithium-ion batteries," *Nanomaterials* 3 (2013): 443–452.

Yuan, L. Y., Tao, Y. T., Chen, J. et al., "Carbon nanoparticles on carbon fabric for flexible and high-performance field emitters," *Adv. Funct. Mater.* 21 (2011): 2150–2154.

Yuan, L. Y., Lu, X. H., Xiao, X. et al., "Flexible solid-state supercapacitors based on carbon nanoparticles/ $MnO_2$ nanorods hybrid structure," *ACS Nano* 6 (2012): 656–661.

Zeng, H., Du, X. W., Singh, S. et al., "Nanomaterials via laser ablation/irradiation in liquid: A review," *Adv. Funct. Mater.* 22 (2012): 1333–1353.

Zhang, T. C., Surampalli, R. Y., Lai, K. C. K. et al., *Nanotechnologies for Water Environment Applications* (American Society of Civil Engineers, Reston, 2009).

Zhang, H., Wu, Q. G., Lin, J. et al., "Thermal conductivity of polyethylene glycol nanofluids containing carbon coated metal nanoparticles," *J. Appl. Phys.* 108 (2010a): 1–6.

Zhang, H. Y., Lin, J. & Hong, H. Q., "Thermal conductive composite prepared by carbon coated aluminum nanoparticles filled silicone rubber," Paper presented at the International Conference on Advances in Materials and Manufacturing Processes, Shenzhen, China, November 2010 (2010b).

Zhang, X., Alloul, O., Zhu, J. H. et al., "Iron-core carbon-shell nanoparticles reinforced electrically conductive magnetic epoxy resin nanocomposites with reduced flammability," *RSC Adv.* 3 (2013): 9453–9464.

Zhang, D., Cao, X., Peng, Z. et al., "Simulation and experiment for microwave absorption of carbon-coated nickel nanoparticles composites," *Inf. Technol. J.* 13 (2014a): 1329–1334.

Zhang, Z., Xiao, F., Xi, J. et al., "Encapsulating Pd nanoparticles in double-shelled graphene@carbon hollow spheres for excellent chemical catalytic property," *Sci. Rep.* 4 (2014b): 1–5.

Zhang, Z. J., Simionesie, D. & Schaschke, C., "Graphite and hybrid nanomaterials as lubricant additives," *Lubricants* 2 (2014c): 44–65.

Zhao, Z., Meng, C., Li, P. et al., "Carbon coated face-centered cubic Ru–C nanoalloys," *Nanoscale* 6 (2014): 10370–10376.

Zhou, M., Cai, T. W., Pu, F. et al., "Graphene/carbon-coated Si nanoparticle hybrids as high-performance anode materials for Li-Ion batteries," *ACS Appl. Mater. Interfaces* 5 (2013): 3449–3455.

# 19

# Conjugated Carbon Nanocapsules

Gan-Lin Hwang

Alan C. L. Tang

Patrick C. H. Hsieh

## 19.1 Introduction to Carbon Nanocapsules (CNCs) and Derivative Materials

Since the discoveries of fullerenes and CNTs in 1985 and 1991, respectively, both have garnered considerable interest worldwide. CNCs (Iijima 1991, Saito & Matsumoto 1998, Hwang 2005, 2007), another family of carbon nanomaterials which were originally found as an impurity during the production of CNTs, can now be produced as the major product at high purity. CNCs show properties distinct from those of other carbon family materials, opening them up to a wide variety of interesting new applications.

### 19.1.1 Structure and Morphology of CNCs and Their Derivative Materials

CNCs were observed in the carbon deposit on the cathode by the arc evaporation method. A CNC consists of closed multilayers of graphitic sheets with a nanoscale cavity in the center (Figure 19.1). The closed polyhedral morphology of CNCs exhibits flat graphitic layers except for the corners and edges,

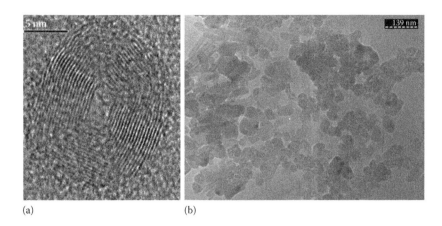

(a)                                                    (b)

**FIGURE 19.1** Morphology of CNCs: (a) HRTEM image of a hollow CNC; (b) low-magnification TEM image showing high-quality hollow CNCs.

where pentagons are located. Various isomers exist in different shapes depending on the cross section, owing to the numerous possibilities of pentagon distribution. These polyhedral morphologies are quite similar to the tip shapes of MWCNTs, but full domes rather than half domes exist at both ends in the tubules. The average diameter of the CNCs ranges from a few to several tens of nanometers, and the aspect ratio is about 1 to 2, looking like a very short MWCNT.

In addition to hollow-structured CNCs, nanocapsules filled with transition metals were fortuitously discovered in the arc process when metal-loaded graphite is evaporated. A variety of metals can be encapsulated in the cavity of CNCs, including not only lanthanide elements but also $3d$ transition metals such as iron, cobalt, and nickel. Metal-filled carbon nanocapsules (M@CNCs) may exhibit magnetism, microwave absorption, or radioactivity depending on the properties of the metals inside, potentially lending M@CNCs to a wide range of applications. The closed graphite structure of CNCs not only offers a protected environment for internal nanometals but is also used to assist dispersion in solvent by means of chemical modification. When functional groups or polymer chains are bonded on CNCs, a wide range of derivative materials can be synthesized to improve the applicability and value of CNCs.

## 19.1.2 Properties of CNCs

The unique properties of CNC are summarized in Table 19.1 (Su 2006). Recently, the free radical quenching ability of CNCs was found to be better than those of 5,5′-dimethyl pyrrolidine $N$-oxide (DMPO)

**TABLE 19.1** Properties of CNCs

| Property | Nanocapsules | $C_{60}$ | Nanotubes |
|---|---|---|---|
| Structure | Multigraphene layers | Single layer | Single layer or multiple layers |
| Size | $d = 10$–$60$ nm; aspect ratio = $1$–$2$ | $d = 1.1$ nm | $d = 2$–$100$ nm, $L =$ several micrometers |
| Thermal stability ($O_2$) | $>600°C$ | $\sim400°C$ | $400°C$–$650°C$ |
| Dispersion | $40$ mg/mL | $\sim2$ mg/mL | Poor |
| Radical quenching rate for $-$(OH) $(g/L)^{-1} s^{-1}$ | $1.16 \times 10^8$ | $6.3 \times 10^7$ | NA |
| Electric conductivity | $10^2$–$10^3$ S/cm$^2$ | Poor | $\sim10^3$ S/cm$^2$ |
| Thermal conductivity | $\sim1600$ W/m K | Poor | $3000$ W/m K |

*Source:* Su, T. T., "Commercialization of nanotechnology—Taiwan experiences," Paper presented at IEEE Conference: Emerging Technologies—Nanoelectronics, Singapore, January 10–13, 2006, 25–28.

and $C_{60}$. By combining this property with good electric conductivity, an anti-inflammatory conducting polymer was developed. Another unique property of CNCs is the incorporation of metal in the core. By using a magnetic core, the anticancer drug cisplatin could be obtained in high yield.

## 19.1.3 Theoretical Prediction

In nature, hollow carbon nanocapsules (HCNCs) have many characteristics that are similar to multi-layer nanotubes. From a structural viewpoint, the multilayer graphite structure of HCNCs is the same as the structure of MWCNTs except that the diameter-to-length ratio is far less than that of nanotubes. Thus, HCNCs are considered to have the same structure as the top ends of MWCNTs. By theoretical calculation, it is predicted that the corner or inflection of nanotubes has a five- or seven-member ring. Noticeably, graphite shows junction phenomena in inflection and has semiconductor properties. Because HCNCs have the same structure as the ends of nanotubes, but no long graphite, semiconductor properties and unique fluorescence are thus manifested.

The outer layer of CNCs resembles the structure of MWCNTs. Furthermore, nanocapsules have an inner space of nanometer scale, which offers new opportunities in the science of atomic clusters and fine particles.

## 19.1.4 Experimental Results

CNCs potentially lend themselves to many interesting applications. For example, CNCs are thermally and electrically conducting materials. The easy dispersion of CNCs without using thermally resistant surfactants makes CNCs good candidates for heat-transfer nanofluid. In addition, nanometallic particles can be protected by closed graphene layers in M@CNCs. The protective layers help prevent the nanosize metals from oxidation and aggregation.

The electromagnetic properties of various CNCs filled with metals have been analyzed by a superconducting quantum interference device. It has been discovered that there is nearly no magnetic hysteresis for Fe@CNCs, in which magnetization increases with increasing external magnetic field $H$, which may be recovered to nonmagnetic particles because of the thermally excited magnetic moment at limited temperature without an external magnetic field. Such properties make Fe@CNCs applicable to the separation of magnetic tubular columns, to the recycling of noble metal catalysts (Figure 19.2) (Hwang & Chang 2005), and as drug synthesis carriers (Hwang et al. 2013a). As reported in the literature, the

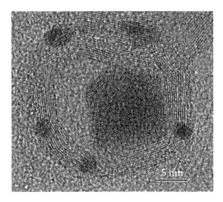

**FIGURE 19.2** Morphology of a magnetic CNC. Shown is a magnetic CNC carrying platinum catalysts, with particle diameter of 2 nm loaded on the corner portion of the shell. (From Hwang, G. L., Jiang, T. S., Yeh, S. L. et al., "Method for preparing a pharmaceutical compound by way of magnetic carbon nanocapsules," U.S. Patent 8,465,764, 2013.)

magnetic field might facilitate the catalysis of spin-exchange reactions (e.g., hydrogenation). It is desirable to enhance the catalytic performance of catalysts or replace expensive platinum catalysts by means of controlling internal magnetic metals.

The Fe@CNCs were used as a new magnetic photocatalyst by growing $TiO_2$ crystal on the surface of Fe@CNC using a simple sol–gel method (Huang et al. 2006a). According to the XRD and energy-dispersive spectrometer results, the anatase phase appears on the surface of the $TiO_2$ crystal. In comparison with pure $TiO_2$, the $TiO_2$ catalyst carried by magnetic Fe@CNCs achieves good performance in removing $NO_x$ gas under UV irradiation. Owing to its easy recycling and good photocatalytic efficiency, this new magnetic photocatalyst is potentially applicable to pollution prevention and photocatalysis.

When CNCs filled with magnetic metals are used as a carrier for a homogeneous catalyst, the nature of the nanoparticles allows the natural maintenance of the reaction system. However, the catalyst can be easily separated from the products by magnetic separation. The separation issue for homogeneous catalysis may be greatly reduced by modifying the surfaces of CNCs with various functional groups to which a variety of molecules can be attached. The size of CNCs can be controlled to the range of several tens of nanometers, a size suitable for a nanocarrier for long circulation in blood vessels. These characteristics make CNCs potentially useful as nanomaterials for biomedical applications such as drug delivery and diagnosis.

# 19.2 Production and Purification of CNCs

Different arc plasma methods are widely used for the production of a series of fullerene-related materials, such as $C_{60}$, CNTs, CNCs, and graphene. The high temperature of the arc process and the presence of an electric field play important roles in the formation of HCNCs and M@CNCs.

High-quality CNCs have been fabricated using a modified arc discharge technique. However, the low solubility of pristine CNCs in most solvents has hindered their chemical manipulation, quantitative characterization, and extensive applications. Strong van der Waals forces exist between CNCs and CNTs. To produce high-purity CNCs, after the collection step, the deposit is usually dispersed in a solution using a surfactant. Next, the CNC main product and the CNT by-product are separated using column chromatography. Finally, the surfactant is removed from the CNCs using rotational centrifugation.

## 19.2.1 Growth Mechanisms of HCNCs

HCNCs were prepared by a carbon arc reaction under pulse current of about 1.2 atm of Ar gas (Hwang 2007). Two graphite rods with a diameter of 6 mm were used as cathode and anode. The arc chamber was cooled by flowing water. A carbon arc reaction was performed under the following conditions: a pulse current of about 60–100 Hz, voltage of about 20 V, and an electric current of about 100–120 A. The carbon arc reaction was allowed to run for about 30 minutes and then stopped. A deposit was formed on the cathode. The deposit was about 34 cm long with the same diameter as the graphitic cathode. The deposit was cut and a black powdery crude product was obtained in the core portion of the deposit. Finally, the deposit formed on the cathode and the anode was collected. The deposit includes an HCNC main product and a short CNT by-product.

It is now well established that arc discharge can produce large quantities of either nanotubes or nanocapsules by changing the conditions of the discharge. The temperature of the interelectrode gas is close to the melting and the boiling temperature of graphite (~4000 K) in an arc discharge. The graphite begins to evaporate, producing a thin layer of saturated carbon vapors near the surface. After the initial expansion, the carbon vapor becomes carbon neutral and ions ($C^+$) deposit and condense to form small carbon clusters on the surface of the cathode. The carbon ions ($C^+$) move to the gap between the positive space charge and the cathode. A smooth current is involved in the formation process of elongated structures (i.e., MWCNTs); in other words, a pulse current resulting in carbon clusters grows to form particle structures. The high structural fluidity of carbon clusters has been observed under intense electron beam irradiation. Since the cooling proceeds from the surface to the center of the carbon cluster, the graphitization starts from the external surface of the cluster and progresses toward the center. The flat planes of the cluster consist of the networks of six-member rings, while five-member rings are located at the corners of the polyhedron. When the solidification of the CNC occurs from the outermost layer, a void is left in the center (such voids are always observed in hollow nanocapsules).

HCNCs may be considered to consist of two ends of an MWCNT without tubules in between. This results in a full dome structure of a CNC rather than the half domes at the two ends of a CNT. Studies on how to control the carbon vapor density during electric arc discharge helped increase understanding of the growth mechanism of CNTs and CNCs. This is significant academically. In the future, such knowledge can be used to manufacture CNTs with different lengths or other types of nanocarbon materials.

## 19.2.2 Growth Mechanisms of M@CNCs

A variety of metals can be encapsulated in CNCs. The preparation of M@CNCs (e.g., Fe@CNCs) is similar to that of HCNCs except for the following (Hwang 2005): The composite graphite electrode was made by mixing iron powder and graphite at a molar ratio of 5:100 (20 wt%). High electrical voltage was applied to induce a pulse current across the cathode and the anode to generate a carbon arc reaction and the deposit formed on the cathode was collected. The carbonaceous products collected from the arc discharge chamber include about 65% Fe@CNCs, ~30% HCNCs/CNTs, and a trace of Fe particles without coating of graphite layers. The Fe-filled CNCs were extracted by magnetic attraction. The surfactant and residue Fe particles were washed away from the Fe@CNCs by 0.1 N HCl solution. The average diameter of Fe@CNCs was about 40 nm. The Fe@CNCs obtained had higher than 95% purity.

## 19.2.3 Purification of CNCs

The crude product contained about 70% HCNCs. The main by-product was short CNTs of 200–1000 nm in length. The crude product was dispersed in water, using cetyltrimethylammonium bromide as a surfactant. Then, the dispersion solution was subjected to column chromatography to separate the HCNCs and CNTs. The column had a filter film of about 0.2 μm pores in the front. Finally, the surfactant was removed from the HCNCs using high-speed centrifugation to obtain the purified product (Hwang 2007).

# 19.3 Functionalization of CNCs

A CNC is a polyhedral carbon cluster with a closed graphitic sheet structure. From a structural perspective, each carbon atom on a CNC has an $sp^2$-electron configuration. Carbon atoms of the flat part of the graphite layers are arranged in a hexagonal network (six-member ring) structure, while those at the corners of the graphite shells are arranged in a pentagonal (five-member ring) structure. Because the closed multilayer graphitic sheet structure can protect the metal within the M@CNCs from oxidation and aggregation, the magnetism of the metal can be preserved in a stable condition. Moreover, the surface of the CNCs can be modified through chemical modification to increase its affinity and, therefore, the CNCs can be easily dispersed in solutions. In this section, functionalization and conjugation modification methods are described. Prospective applications of these properties are also discussed.

## 19.3.1 Functionalization Methods and Subsequent Conjugation of CNCs

The functionalization methods for CNCs are similar to those for preparing fullerene. However, owing to the relatively larger size of CNCs, the nanodispersing technique is important for the chemical modification of CNCs. CNCs can be functionalized by redox reaction, cycloaddition reaction, or radical addition reaction. Specifically, the functional groups are bonded thereon and uniformly distributed over the surface. In addition, CNCs have electrical and magnetic properties different from those of fullerene and CNTs; thus, functionalized CNCs have distinctive applications.

### 19.3.1.1 Dispersion of CNCs

Thus far, mainly the properties of the shells of CNCs have been discussed; however, the inner hollow and the outer surface of CNCs also have interesting features. There is a strong van der Waals force between CNCs, meaning that it is not easy to engineer CNCs for further modification. To improve the dispersion of CNCs in the matrix, CNCs can be chemically functionalized for wetting properties; i.e., the outside of the CNCs can be coated.

Attempts have been made to improve the dispersion behavior of CNCs by chemical modification and physisorption. Chemical functionalization includes strong oxidation, radical addition reaction, atom transfer radical polymerization, etc. (Hwang 2012). The mechanism of chemical modification involves surface modification with reactive functional groups, followed by grafting with organic long chains. Physical adsorption may be carried out by ultrasonication or grinding with solvent, which involves adsorption of small organic molecules, surfactants, polymers, or proteins only through weak intermolecular interactions such as $\pi-\pi$ stacking force and electrostatic force.

The dispersion behavior of CNCs was observed visually and determined by measuring the absorption at the wavelength of 550 nm using an UV–visible spectrometer. The measurements, wherein a higher absorption represents better dispersion behavior, indicate that the dispersion ability of CNCs increased with the aspect ratio. And the absorption was proportional to the concentration of the dispersion, obeying Beer–Lambert's law. At the critical point of the UV–visible absorption spectrum, the attraction force between the CNC particles was almost completely eliminated, leading to a good dispersion state.

### 19.3.1.2 Functionalization Methods

Chemical functionalization of CNCs includes strong oxidation, cycloaddition reaction, radical addition reaction, etc. For example, HCNCs have been chemically functionalized and hydrolyzed to improve the dispersion and interfacial adhesion of HCNCs to the matrix. In a typical experiment, dilute nitric acid was added drop by drop into a beaker containing sodium nitrite ($NaNO_2$). In such way, $NO_2$ gas was in situ generated and then guided into an HCNC–toluene solution. The $NO_2$ gas is chemically reactive toward C=C double bonds of HCNCs and reacts with the graphene layers of HCNCs. After a few tenths of minute of reaction, the solvent was removed by a rotary evaporator, and the nitro-HCNCs were hydrolyzed by concentrated sodium hydroxide solution to convert the nitro group to hydroxyl groups. $HCNC(OH)_n$ was then incorporated into the silicone rubber composites.

In another example, concerning the free radical addition (or scavenging) ability of CNCs, electron spin resonance (ESR) spectroscopy was used to measure the reaction rates of HCNCs with hydroxyl radicals (Figure 19.3). The hydroxyl radicals can be produced by the Fenton reaction, i.e., the reaction of ferrous ion ($Fe^{2+}$) with hydrogen peroxide. DMPO is a well-known free radical trapping reagent; it can react with hydroxyl radicals at a rate constant of $1.8 \times 10^9 \ M^{-1} \ s^{-1}$. The reaction rate of $HCNC(OH)_n$ with hydroxyl radicals can be measured by competition with DMPO. From the magnitudes of ESR signals (see Figure 19.3) in the absence and presence of a given amount of $HCNC(OH)_n$ and the known reaction rates of DMPO with these free radicals, the reaction rate of $HCNC(OH)_n$ with hydroxyl radicals could be obtained as $1.16 \times 10^8 \ (g/L)^{-1} \ s^{-1}$. This result indicates that HCNCs are very good scavengers of hydroxyl radicals (Hwang 2005). In other words, CNCs should be able to be easily functionalized by radical addition reaction.

### 19.3.1.3 Conjugation and Chelator Modification

The surfaces of CNCs can be modified with various functional groups for conjugation to which specific chelating molecules can be attached. The size of CNCs can be controlled to within the range of several tens of nanometers, a size that is suitable for their use as nanocarriers for long-term circulation in blood vessels, making CNCs nanomaterials with potential biomedical applications such as drug delivery and

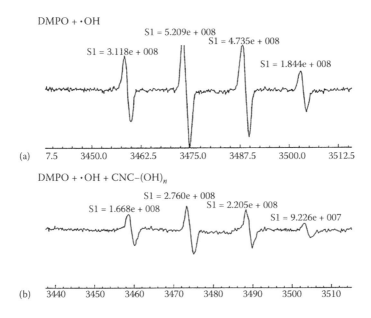

**FIGURE 19.3**  ESR spectra of hydroxylation of HCNCs: the ESR spectra of DMPO-OH adduct in the absence (a) and presence (b) of HCNC(OH)$_n$. The hydroxyl radicals were generated by the Fenton reaction (i.e., reaction of ferrous ions with hydrogen peroxide). The vertical scales of the two ESR spectra are the same. The ESR signal decreases in the presence of HCNC(OH)$_n$.

diagnosis. For example, a unique synthesis method of cisplatin (*cis*-dichlorodiammineplatinum) was achieved by utilizing magnetic metal (Fe)-filled CNCs (Fe@CNCs) as support architectures (Hwang 2006, Hwang et al. 2013a). Cisplatin is one of the most widely used anticancer drugs; however, traditional cisplatin preparation methods have been complicated by low product yield. When high-quality Fe@CNCs were used as carriers with only the *cis* form product located on the surface, with effective collection under magnetic field, a high yield (86%) of cisplatin was obtained. This method was achieved by forming a stable complex medium product through bonding the paired [C(–COOH)$_2$] functional group of the magnetic Fe@CNCs with the coordinated bond of the platinum ion first; then, Cl of the complex Fe@CNC–[C(COO–)$_2$PtCl$_2$]$_n$ was replaced by NH$_2$; finally, the platinum part of the complex was removed from the complex to obtain Pt(NH$_2$)$_2$Cl$_2$ as the product; this was subsequently processed by removing and collecting the complex from the Fe@CNCs.

Figure 19.4 illustrates the flowchart used to synthesize cisplatin utilizing the Fe@CNCs as support architectures. Briefly, the cyclopropanation of Fe@CNCs is based on the Bingel reaction. This was followed by treatment with HCl, transforming the ester group into the malonic acid derivative of Fe@CNCs. A subsequent reaction step utilizing the malonic acid of Fe@CNCs as ligand was used to coordinate with platinum. According to the inherent acid–base strength, carboxylic acid reacts more readily with platinum than chloride does. Two chlorides in potassium tetrachloroplatinate (K$_2$PtCl$_4$) can be substituted with the adjacent carboxyl groups in Fe@CNCs, providing the platinum with anchored sites. After blocking the *cis* site of K$_2$PtCl$_4$, nucleophilic substitution of ammonia occurred in *cis*-dichloroplatinum, generating *cis*-diamine-1,1-cyclobutanedicarboxylatoplatinum (carboplatin) analogue in the support architectures of Fe@CNCs. This carboplatin analogue structure was comprised of two firmly bound ligands of amine and readily displaceable "leaving group"-type carboxylate ligands, which could be cleaved from Fe@CNCs by hydrogen chloride. Finally, the cisplatin anticancer drug could be synthesized from the support architectures of Fe@CNCs, which could be collected by the magnetic field effect and be reused in the Fe@CNC recycle system.

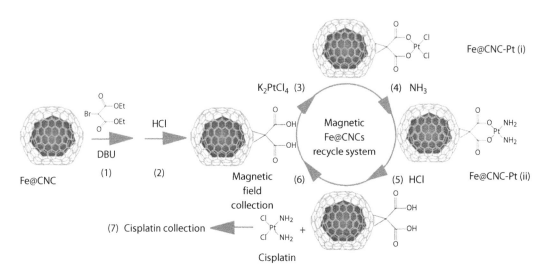

**FIGURE 19.4** Schematic summary of cisplatin synthesis using iron-filled CNCs as support architectures. The following flowchart shows the experimental process: Steps 1 and 2: Modification of the surface of the Fe@CNC bonds with [C(−COOH)$_2$] groups. Step 3: Formation of a complex Fe@CNC–[C(COO−)$_2$PtCl$_2$]$_n$ made of platinum salt and Fe@CNCs. Step 4: Removal of the Cl of the complex prepared in step 3 by NH$_2$ to form a complex of Fe@CNC–[C(COO−)$_2$Pt(NH$_3$)$_2$]$_n$. Step 5: Hydrolysis—removal of the platinum part by hydrolysis, obtaining the product *cis*-Pt(NH$_2$)$_2$(Cl)$_2$. Step 6: Collection of Fe@CNCs by magnetic field and their reuse in the recycle system. Step 7: Collection of cisplatin.

## 19.3.2 Polymer Chain–Grafted CNCs

Polymer chain–grafted CNCs have a CNC in the core and polymer chains bonded on the surface of CNCs to form star-polymer-like particles (Hwang & Abraham 2007). The polymer chains can be thermal plastic polymers, conductive polymers, liquid crystal polymers, biopolymers, etc., thereby expanding the application of CNCs. By grafting polymer chains onto CNCs, the CNCs are easily dispersed in the solvent, bringing out new electro-optical properties and can be further prepared as homogenous composite materials.

For example, in the synthesis of polystyrene (PS)–grafted CNCs, *sec*-butyllithium was used to introduce negative charges on CNCs, and these CNC carbanions then acted as initiators for anionic polymerization of styrene (Huang et al. 2007). The PS content in the PS-grafted CNC sample was approximately 20%, and the molecular weight of the grafted PS on the surface of CNCs was calculated as 1200 g mol$^{-1}$. The PS-grafted CNCs had good dispersion in toluene, tetrahydrofuran, cyclohexane, and other common organic solvents in which PS is dissolvable, thus indicating good compatibility when further blended with other styrenic polymers.

Furthermore, the PS-grafted CNCs were used as lubricant additives applied to retain the fluid viscosity of lubricant at high temperature and effective reduction in the friction coefficient at the contact interface by the rolling tribological mechanism (Hwang et al. 2013b, Jeng et al. 2014). The friction coefficient reduced, giving rise to improved wear resistance and a higher load-bearing capacity. In addition, using CNCs as additives in lube oil, the natural heat convection was increased to four to five times that of the base oil. The thermal dissipation property is a key criterion in assessing lifetime reliability of lubricant systems, to enhance the design quality of future endurance lubricants.

## 19.3.3 Synthesis of Photoluminescent CNCs

The energy-transfer process of the graphite shell allows CNCs to have completely different physical, chemical, and photoelectric properties. For example, CNCs can be oxidized to obtain CNC–(COOH)$_n$,

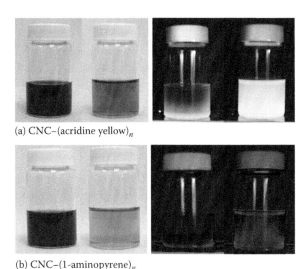

(a) CNC–(acridine yellow)$_n$

(b) CNC–(1-aminopyrene)$_n$

**FIGURE 19.5** Images of photoluminescent iron-filled CNCs: (a) the fluorescent molecules of acridine yellow and (b) 1-aminopyrene are bonded on the surface of CNCs. They are purified and configured to have concentrations of 500 and 50 ppm after 24 h of reaction. The left side shows the fluorescent lamp condition, while the right side shows the UV lamp condition.

which is used for the preparation of photoluminescent CNC/polycyanate composites. The inherent H-bond interactions between the carboxylic acid in CNC–(COOH)$_n$ and the s-triazine ring in polycyanate (PCn) enhance the mutual miscibility between the two components. The energy-transfer process is due to $\pi$–$\pi$ electron overlapping of the aromatic ring between the PCn matrix and the CNC–(COOH)$_n$, resulting in an increase in the $\pi$-conjugation lengths and the generation of redshift (Su et al. 2006).

In addition, polyaniline-grafted HCNCs in homogeneous organic solutions are strongly luminescent. By covalently attaching multiple polyaniline chains onto a HCNC, this HCNC-derived starburst conjugated polymer may exhibit improved inter- and intramolecular solid-state properties and is useful as a thermoelectric material. Furthermore, several photoluminescent modified Fe@CNCs have been effectively synthesized (Figure 19.5). Those photoluminescent modified Fe@CNCs have been used to purify and identify cells to create convenient and rapid cell separation techniques and disease examination methods.

## 19.4 Biocompatibility of Functionalized and Nonmodified CNCs

CNCs, originally discovered as a by-product in the synthesis of carbon nanorods, have characteristics very similar to those of CNTs and most graphene nanomaterials. These capsules conduct heat and electricity well and have high mechanical strength. These properties make graphene-based materials such as nanotubes, capsules, diamonds, fibers, and other composites excellent candidates for circuits, solar applications, batteries, sensors, and other electrical and mechanical applications. As described in the previous section, CNCs are spheres of graphene sheets with aspect ratios between 1 and 2. The structure of these capsules is significantly different from those of nanotubes and rods, which have length-to-width ratios that range from 2 to more than 100. This physical property considerably differentiates the toxicity and effects of using these materials in a biological system. The following section discusses the scope of biocompatibility of nanocapsules in biomedical applications.

Materials used in biological systems are termed *biomaterials*. CNCs have been studied as a drug delivery system and have potential for use in biosensors and biological microelectromechanical systems (bio-MEMSs). However, the safety of using these CNCs in biological systems must first be addressed and overcome.

### 19.4.1 Introduction to Biocompatibility

The biocompatibility of a biomaterial is the extent to which the material elicits an immune response and reaction from the host tissue and its ability to maintain its proper function as a medical device. Medical devices and drugs must be thoroughly studied to determine their safety and biocompatibility during application. Medical devices that use materials such as CNCs and other synthetic materials cannot be digested through biological processes. Therefore, host response and ensuring that the biomaterial does not harm the body and that the immune response does not limit the proper function of the device become important factors when designing and applying carbon-based materials in medical applications.

Biocompatibility of foreign objects in biological systems is determined by several factors, including physical size, surface characteristics, mechanical properties, geometry, and composition. An acute biological response is immediately activated through inflammation by the host. After a prolonged period, a chronic response is initiated by fibroblasts that form an enclosure around the foreign body.

The immune system of a biological system naturally puts up a defense when anything from the outside enters the body. These reactions vary significantly depending on the type of object that enters the system. Because CNTs and nanocapsules are not degradable by a biological system, the immune system reacts to segregate these carbon nanomaterials from the biological system. This reaction against foreign bodies and objects that are nonbiodegradable is known as the foreign body reaction (FBR). This response by the biological tissue occurs against most biomaterials used in medical devices.

### 19.4.2 Toxicity of Nanomaterials and Nanoparticles

Another aspect of safety is the toxicity of the material being used. Toxicity comes in different forms; whether it is chemical, biological, or physical, any type of toxicity drastically affects different aspects of a biological system, ranging from cells and tissues to the entire biological system. Chemical and biological toxicities are determined by the surface chemistry and sterility of the biomaterial being used. Physical toxicity is constrained to the shape, the size, and the morphology of the biomaterial.

Surface chemistry greatly influences how the host cells interact with the surface of the biomaterial. Certain chemical structures are extremely toxic to the body at the cellular or tissue level. Dispersing agents, such as those often used with carbon nanomaterials, are also a source of toxicity. The very function of dispersing agents is harmful and toxic to cells and biological systems, including the circulatory and respiratory systems. Biomaterials with surfaces that are not sterile may contain biologics, bacteria, and viruses that may severely compromise the biological system through disease when the biomaterials are applied invasively.

The basic property of size is also an enormous determining factor of toxicity. More specifically, nanotoxicity is an extensively researched area of the biosafety of carbon nanomaterials. Nanosized particles have different issues when injected into the body or used on the surface of a medical device. Large particles, or small particles that aggregate, may occlude a blood vessel, causing myocardial infarction or stroke. On the other hand, nanosized particles that are able to travel freely in the blood vessels may escape into other parts of the body because of blood vessel dilation. The escaped particles may end up in the lung, causing respiratory issues.

### 19.4.3 Biocompatibility and Toxicity of CNCs

Unlike CNTs that have a large aspect ratio, CNCs are spherical. This rounded morphology assists the dispersion of CNCs, a key aspect in the toxicity of nanomaterials. Functionalized CNCs have even higher solubility because of their hydrophilic properties, paving the way to better performance in different applications including medical applications.

CNCs, when used as small individual nanosized drug carriers, do not induce foreign body responses upon injection into the bloodstream. Typical sizes of CNCs range from 10 to 200 nm, depending on fabrication methods and processes. At this size, cells are able to interact with the capsules without the eventual need for collagen encapsulation by fibroblasts. Similar to fullerenes and nanodiamonds, the extremely low aspect ratio of nanocapsules assists in their dispersion without a dispersing agent. Not only does the high dispersion rate decrease aggregation and avoid blood vessel occlusion, but also the lack of a dispersing agent helps avoid chemical toxicity.

On the other hand, if CNCs are coated onto the surface of an implanted device, aggregation and dispersing agents do not come into play. Instead, the surface morphology and composition largely determines the toxicity and biocompatibility of the biomaterial. Small particles coated onto surfaces significantly increase the surface area–to–volume ratio. An increase in this ratio signifies that more protein and extracellular matrix can adhere to the surface of the medical implant at the same volume. Whether this property affects the function depends on the design and purpose of the implant. However, increased immune response usually negatively impacts the biological system itself through immune response processes such as acute or chronic systemic inflammation.

While CNCs are a relatively novel material in the field of nanomaterials, especially as a biomaterial, their safety and biocompatibility have been studied in several research experiments. Several comprehensive toxicity studies of carbon-based nanomaterials have been conducted with similar conclusions that the high aspect ratios and aggregating properties of CNTs are highly toxic (Lacerda et al. 2006, Kostarelos 2008). As CNCs represent a recent advancement in carbon-based nanomaterial synthesis, they have not yet been extensively studied. However, one study has shown that CNCs were less toxic to mice compared with other nanomaterials including fullerene $C_{60}$ and CNTs (Tang et al. 2002). As expected, most of the carbon-based nanomaterials aggregated in the lungs, especially CNTs. While other studies have attempted to increase the biocompatibility of CNTs, this study shows a relatively safe approach to carbon-based nanomaterials for intravenous drug delivery systems. To use CNCs as a biomaterial, they must be modified for biological use. Several methods to increase nanomaterial biocompatibility are well established, the underlying principle being to decrease the immune response and interaction of the biological system with the foreign body.

## 19.4.4 Methods to Enhance Biocompatibility of CNCs

General methods include local injections to decrease systemic toxicity as well as coinjection with noninflammatory drugs to reduce acute response of the FBR. Advanced methods to reduce immune responses include surface modifications and conjugations of CNCs. By modifying the surface chemistry to become hydrophilic, the nanocapsules are masked from the immune system.

Current advancements in surface modification largely take advantage of the double bonds in functional groups such as carbonyl in carboxyl groups to conjugate polymers, biomolecules, proteins, and other particles to the nanocapsules (Chu et al. 2005). Polyethylene glycol (PEG) is a popular candidate among scientists and engineers when developing drugs and biomaterials for in vivo applications. PEGylation, the conjugation of PEG polymers to the surface of a material, is useful for prolonging the circulating time, increasing solubility and dispersion, decreasing protein adhesion and immunogenicity of nanocapsules. However, the downside of PEGylation is that the increased circulation time of the conjugated material decreases drug efficiency.

Not only has conjugation of nanocapsules have the advantage of increasing the biocompatibility of the particle, but also the functional groups allow therapeutics and drugs to be further attached to capsules through heterobifunctional PEGs. PEGs that have two different functional ends attach two different molecules or particles on both ends. Further methods have been studied to polymerize CNCs by converting the highly strained $sp^2$ atom to $sp^3$ and carbon anions were introduced onto the surface (Chu 2005, Wang et al. 2008). Polymers were then attached onto CNCs through anionic growth and living free radical polymerization processes. Self-polymerization of CNCs has even been demonstrated through radical polymerization of π-bonds (Huang et al. 2006b).

Recent advancements in Y-shaped and comb PEG ligands change the behavior of the particles and further reduce the immune response from previous generations of PEG ligands. Unlike PEG ligands that mask nanoparticles, targeting ligands such as peptides and proteins allow nanomaterials to hone in on specific cells or tissues. As such, the injected nanomaterial is specific and avoids side effects on nontargeted cells or tissues. Other studies have also revealed insights into how size plays a role in nanoparticle distribution and toxicity. Furthermore, surface charge also strongly influences the biocompatibility of nanoparticles. These characteristics, size, surface charge, and other surface modifications can be controlled. Because of surface functionalization, the possibilities are endless for increasing the biocompatibility through surface conjugations on the CNCs. Peptides, proteins, and other ligands can be conjugated onto nanocapsules from functional group activation to direct the characteristics of the nanoparticles.

### 19.4.5 Future Direction of CNC Conjugation and Safety

CNCs have been envisioned to provide groundbreaking advancements in biosensors, bio-MEMS, imaging applications, and even as drug delivery agents by taking advantage of their superior properties and ability for transformation through surface modifications. The basis for applying CNCs into the development and fabrication of biosensors and bio-MEMS is similar to that for creating circuits and sensors with CNTs and other graphene materials. Encapsulation and surface modifications allow CNCs to be truly versatile in various applications, including acting as contrast agents and drug delivery agents.

As new peptides, antibodies, and other biological molecules are being discovered, more and more options are available for CNC surface modification. Peptides are continually being sequenced for specific binding, which helps reduce biological side effects of peptide-conjugated CNCs.

## 19.5 Applications of Functionalized CNCs

Nanocapsules are being developed as a frontier novel material for applications in various fields. As previously mentioned, CNCs conduct heat and electricity very well and exhibit very high strengths. These characteristics lend CNCs to useful applications in various fields. Their high thermal conductivity property makes CNCs suitable for heat transfer applications where heat dissipation is required. In addition, CNCs have the ability to act as radical scavengers and fire-retardant materials. Combining the geometric shape, high thermal conductivity, and high strength of CNCs, cutting fluids (lubricants and coolants) can benefit greatly from the addition of this nanomaterial. The high strength of CNCs also allows its use as a support structure in chemical synthesis or a reinforced composite material.

### 19.5.1 Medical Applications

Combining methods of oxidation for functionalization and conjugation of biomolecules through surface modification, heparin-conjugated CNCs have been demonstrated to act as a drug delivery system in a diseased model of thromboembolism in mice (Tang et al. 2012). Heparin is used clinically as an anticoagulant used during surgery, or as a pharmaceutical agent against cardiovascular diseases. However, because of the previously mentioned circulation and clearance time of chemicals in the body, heparin and many other chemical agents clear the body very quickly. By combining heparin with CNCs, the effect of heparin was extended. The clinical significance of this achievement is that a lower dose of drug can be used to achieve a higher therapeutic effect. Scientists are constantly looking for ways to reduce the dosage for patients, making drug usage safer and, at the same time, reducing costs. The heparin-conjugated CNC is a great example of a medical application of CNCs through multiple modifications of a carbon nanomaterial. Implantable devices can also be coated with CNCs as substrates for further protein or biomolecule immobilization to enhance biocompatibility properties or drug-releasing effects.

As with nanotubes, the center of nanocapsules is hollow. However, nanotubes are long and open ended, making encapsulation difficult. Just like nanodiamonds and nanorods, nanocapsules are fully enclosed. During synthesis, modifications can be made to the process to enclose molecules into the center of the nanocapsules. This is useful in making nanocapsules an imaging agent while leaving the surface free for drug conjugation. Therefore, nanocapsules can act as both an imaging contrast agent and a drug carrier. A possible application is to insert a contrast chemical into CNCs and conjugate binding biomolecules specific to interested cells or other biomolecules. This effectively turns the conjugated CNCs into a homing beacon. Possible medical applications may include diagnostic purposes and even photothermal therapy. Contrast agents help doctors identify malignant cells in cancer patients (Hwang 2010, Yang 2014). Coupled with a photothermal therapy agent like phycocyanin, these center-filled functionalized CNCs can detect cancer and display toxic effects specific to those malignant cells through photothermal therapy with reduced side effects compared to systemic treatment options.

The active site on CNCs through functionalization allows a high degree of manipulation. Whether it is molecule immobilization or polymerization on the surface of the CNCs or acting as a connecting ligand for further more advanced modifications, these functionalized nanoparticles have the ability to act as many different types of tools ranging from antibiotic and antiviral drugs, to immunological agents such as vaccines, and to effective antioxidants and molecular sensors (Chen et al. 2014, Han et al. 2014, Yang 2014).

## 19.5.2 Electrical Applications

Beyond conjugation of a fluorescent molecule to observe the nanomaterial, CNCs can be modified to have photoluminescent properties by using bridging ligands such as dicyanate or polycyanate resins to form carboxylated CNC composites. These CNC composites exhibit photoluminescent properties because of the stacking of the aromatic rings, causing overlapping $\pi-\pi$ bonding. Photoluminescent CNCs can act as biosensors in molecule detection (Su et al. 2006). With some modification, these photoluminescent biosensors can also act as a means by which to purify proteins or cells using the principles of flow cytometry sorting methods. The same principle can even be extended to cell screening and disease diagnosis.

CNCs with high electrical conductivity coupled with high surface area act as a great electrode material, able to catalyze oxygen reduction in fuel cells, batteries, and even supercapacitors effectively (Wu et al. 2007, Hwang 2010). Traditional methods of molecule intercalation into batteries suffer from poor performance in energy and power density. The issue lies in the underperforming efficiency of ion conductivity of sulfur. CNCs not only have the ability to act as a great electron and Li ion conductor but also are a great sulfur host for higher coulombic efficiency (Hu et al. 2014). Titanium oxides immobilized onto the surfaces of CNCs significantly promote the quenching efficiency of light emissions (Huang 2006a). Solar cells can take advantage of the photoluminescence quenching properties of modified CNCs. Titanium oxide-immobilized CNCs also display photocatalytic effects through the removal of nitric oxide in UV environments, acting as a filter for organic pollutants (Huang 2006b).

The transmission efficiency, the interference, and the cost of optical transceivers are continually being improved (Cheng et al. 2004). Conductive CNC and plastic composites were produced for optical transceiver modules and have been shown to efficiently shield against electromagnetic interference without affecting transmission rates.

Because CNCs display superior electrochemical properties, they are ideal for use in biosensor and bio-MEMS production. Biomolecules immobilized onto CNCs act as biosensors for both in vitro and in vivo applications. The superb bioreactivity of CNCs coupled with the large surface area–to–volume ratio is ideal for substrate layering in bio-MEMS production. Similarly, the sensitivity and detection limits of electrochemical immunoassays can be enhanced by CNCs through molecule immobilization on the functional site of CNCs (Hwang 2010). Coupling the electromechanical properties of CNCs with biosensors and bioactuators based on the bio-MEMS principle is possible for various applications including cell sorting, protein purification, and probing, among many other uses.

These CNCs with superior electrical properties have a wide range of possible applications in many fields including optics, fuel cells, actuators, capacitors, and sensors. Combined with biomolecule immobilization, CNCs have a wide array of utilization beyond batteries and circuitry.

### 19.5.3 Other Applications

Many carbon-based nanomaterials exhibit high thermal conductivity and CNCs are no exception. Research has shown through thermal and fluid dynamics of CNCs that addition of CNCs by 0.1% is followed by an increase of 8% in thermal conductivity (Hwang et al. 2005). Because CNCs are naturally stable at high temperatures and enjoy a high surface area–to–volume ratio, dispersion is not an issue like for many other heat-dissipating nanomaterials. Furthermore, CNCs can be functionalized with functional groups to achieve an even higher dispersion rate. The well-dispersed property and the seamless fluid dynamics between the nanocapsules and the solvent further increase the heat dissipation allowed by CNCs. Thus, CNCs are great for various heat and thermal dissipation applications.

Traditional graphite materials lose function easily over time because of the breakdown of the material at boundaries that are chemically active, ending up binding the metal surfaces. Conversely, spherical CNCs do not have unstable edges and are further stabilized by chemical functionalization. Furthermore, the surface energy of CNCs, wet-chemical grinding, and chemical modification by stearic acid and other methods greatly reduce friction between the nanomaterials themselves and the fluid. This high-dispersion low–friction coefficient material makes a great cutting fluid or mineral oil lubricant that is stable in harsh and high-temperature environments (Lan & Cheng 2012, Hwang et al. 2013b, Jeng et al. 2014). Additionally, CNC particles are small enough to fill surface defects caused by scratches and other wear and tear of metal surfaces, facilitating smoother movements and sliding of metals.

CNCs are processed as hollow structures that are fully enclosed, leading to high versatility of manufacturing variations. For example, CNCs can be manufactured to encapsulate magnetic particles. These magnetic CNCs have been demonstrated to act as support architecture in the manufacturing process of the chemotherapy drug cisplatin (Hwang et al. 2006). The synthetic process of cisplatin is both complicated and inefficient. With the use of CNCs as a support architecture, the synthesis of cisplatin yield is increased significantly. A malonic acid derivative of magnetic CNCs coordinated with platinum through cyclopropanation yields cisplatin at a rate higher than those of traditional methods. The magnetic CNCs are collectable through magnetic fields for reuse. These nanoparticles also have the advantage of high yields through magnetic enrichment (Han et al. 2014).

The mechanical and thermal properties of CNCs have much potential in industrial applications ranging from acting as manufacture substrates for drugs to use in cutting fluid and heat-dissipating agents (Wu et al. 2012, Jeng et al. 2014). CNCs also utilize their inherent magnetic properties in the enrichment process of large-scale industrial production of nanoparticles. Magnetic enrichment gives way to higher purity and easier production methods of CNCs.

## 19.6 Conclusion

The challenges in production of CNCs are production purity, production cost, and complicated production methods. Magnetic enrichment is one method through which production purity can be increased. The difficulties and challenges of production methods vary with different types of CNCs, including M@CNCs, functionalized CNCs, and size-specific CNCs; however, ongoing research continues to simplify production methods and reduce costs.

CNCs can be seen as a tool. The real innovation is in how to use this tool for different applications. In this chapter, modifications to CNCs through polymerization, conjugation, functionalization, and encapsulation have been discussed. CNCs can act as supporting substrates for crystallization and drug production, as electron conductors in fuel cells, as heat-dissipating agents in lubricants and heat-dissipating applications, and even as a drug delivery system. While this novel material has been used in many different ways and applied in a wide variety of fields ranging from sensors to drug delivery systems, more is yet to come. As research continues, more applications using CNCs will be developed, propelling the use of carbon nanomaterials in advanced technologies.

# References

Chen, Y., Xu, P. & Wu, M., "Colloidal RBC-shaped, hydrophilic, and hollow mesoporous carbon nanocapsules for highly efficient biomedical engineering," *Adv. Mater.* 26 (2014): 4294–4301.

Cheng, W. H., Hung, W. C., Lee, C. H. et al., "Low-cost and low-electromagnetic-interference packaging of optical transceiver modules," *J. Lightwave Technol.* 22 (2004): 2177–2183.

Chu, C. C., Hwang, G. L., Chiou, J. W. et al., "Polymerization of a confined π-system: Chemical synthesis of tetrahedral amorphous carbon nanoballs from graphitic carbon nanocapsules," *Adv. Mater.* 17 (2005): 2707–2710.

Han, Y., Li, P., Xu, Y. et al., "Fluorescent nanosensor for probing histone acetyltransferase activity based on acetylation protection and magnetic graphitic nanocapsules," *Small* 10 (2014): 877–885.

Hu, W., Zhang, H., Zhang, Y., Wang, M. et al., "Biomineralization-induced self-assembly of porous hollow carbon nanocapsule monoliths and their application in Li–S batteries," *Chem. Commun. (Camb.)* 51 (2014): 1085–1088.

Huang, H. C., Hwang, G. L., Chen, H. L. et al., "Immobilization of $TiO_2$ nanoparticles on carbon nanocapsules for photovoltaic applications," *Thin Solid Films* 511–512 (2006a): 203–207.

Huang, H. C., Hwang, G. L., Chen, H. L. et al., "Immobilization of $TiO_2$ nanoparticles on Fe-filled carbon nanocapsules for photocatalytic applications," *Thin Solid Films* 515 (2006b): 1033–1037.

Huang, H. M., Tsai, H. C., Liu, I. C. et al., "Synthesis of polystyrene-grafted carbon nanocapsules," *J. Mater. Res.* 22 (2007): 132–140.

Hwang, G. L., "Preparation of magnetic metal filled carbon nanocapsules," U.S. Patent 6,872,236 (2005).

Hwang, G. L., "Preparation of hollow carbon nanocapsules," U.S. Patent 7,156,958 (2007).

Hwang, K. C., "Recent progress in the preparation and application of carbon nanocapsules," *J. Phys. D Appl. Phys.* 43 (2010): 374001–374014.

Hwang, G. L., "Organically functionalized carbon nanocapsule," U.S. Patent 8,287,835 (2012).

Hwang, G. L. & Abraham, J. K., "Polymer-chain-grafted carbon nanocapsule," U.S. Patent 7,217,748 (2007).

Hwang, G. L. & Chang, C. K., "Carbon nanocapsule supported catalysts," U.S. Patent 6,841,509 (2005).

Hwang, G. L., Tsai, S. J., Hung, S. T. et al., "Carbon nanocapsules as flame retardant and conductive agent for silicone rubber," Paper presented at the annual meeting for the European Polymer Congress, Moscow, Russia, June 28 (2005).

Hwang, G. L., Yeh, S. L., Tsai, S. J. et al., "Using magnetic carbon nanocapsules as support architecture in synthesizing of cisplatin," Paper presented at the Seventh International Conference on the Science and Application of Nanotube, Nagano, Japan, June 18–23 (2006): 242.

Hwang, G. L., Jiang, T. S., Yeh, S. L. et al., "Method for preparing a pharmaceutical compound by way of magnetic carbon nanocapsules," U.S. Patent 8,465,764 (2013a).

Hwang, G. L., Su, T. Y., Tzeng, W. S. et al., "Lube oil compositions," U.S. Patent 8,575,079 (2013b).

Iijima, S., "Helical microtubules of graphitic carbon," *Nature* 354 (1991): 56–58.

Jeng, Y. R., Huang, Y. H., Tsai, P. C. et al., "Tribological properties of carbon nanocapsule particles as lubricant additive," *J. Tribol.* 136 (2014): 0418011–0418019.

Kostarelos, K., "The long and short of carbon nanotube toxicity," *Nat. Biotechnol.* 26 (2008): 774–776.

Lacerda, L., Bianco, A., Prato, M. et al., "Carbon nanotubes as nanomedicines: From toxicology to pharmacology," *Adv. Drug. Deliv. Rev.* 58 (2006): 1460–1470.

Lan, Y. F. & Cheng, S. C., "Dispersion of carbon nanocapsules by using highly aspect-ratio clays," *Appl. Phys. Lett.* 100 (2012): 153109.

Saito, Y. & Matsumoto, T. M., "Hollow and filled rectangular parallelopiped carbon nanocapsules catalyzed by calcium and strontium," *J. Cryst. Growth* 187 (1998): 402–409.

Su, T. T., "Commercialization of nanotechnology—Taiwan experiences," Paper presented at IEEE Conference: Emerging Technologies—Nanoelectronics, Singapore, January 10–13 (2006): 25–28.

Su, F. K., Hong, J. L., Liao, G. F. et al., "Photoluminescent carbon nanocapsule/polycyanate composites with hydrogen bond interactions," *J. Appl. Polym. Sci.* 100 (2006): 3784–3788.

Tang, A. C. L., Hwang, G. L., Tsai, S. J. et al., "Biosafety of non-surface modified carbon nanocapsules as a potential alternative to carbon nanotubes for drug delivery purposes," *PLOS ONE* (2002).

Tang, A. C. L., Chang, M. Y., Tang, Z. C. W. et al., "Treatment of acute thromboembolism in mice using heparin-conjugated carbon nanocapsules," *ACS Nano* 6 (2012): 6099–6107.

Wang, C., Huang, C. L., Chena, Y. C. et al., "Carbon nanocapsules-reinforced syndiotactic polystyrene nanocomposites: Crystallization and morphological features," *Polymer* 49 (2008): 5564–5574.

Wu, C. Y., Wu, P. W., Lin, P. et al., "Silver–carbon nanocapsule electrocatalyst for oxygen reduction reaction," *J. Electrochem. Soc.* 154 (2007): B1059–B1062.

Wu, T. S., Li, S. Y., Weng, S. W. et al., "Synthesis and characterization of poly(*p*-chloromethylstyrene) nanocomposite comprising covalently bonded carbon nanocapsules: Superiority of thermal properties to a physical blend," *Polymer* 53 (2012): 2347–2355.

Yang, X., Ebrahimi, A., Li, J. et al., "Fullerene–biomolecule conjugates and their biomedicinal applications," *Int. J. Nanomed.* 9 (2014): 77–92.

# IV

# Porous Materials

# 20

# Nanoporous Carbon Membranes

Bing Zhang

## 20.1 Introduction

Porous carbon materials are widely used in people's daily lives and industrial sectors because of their polygon morphology, easy fabrication, facile modification, cost-effectiveness, and high surface area and physicochemical properties (Stein et al. 2009, Upare et al. 2011, Inagaki et al. 2014). In the family of porous carbon materials, there are a large number of members featured with various macroscopic morphologies (e.g., colloids, thin films, nanofibrous mats, and monoliths), microscopic pore structure, and promising applications (Stein et al. 2009, Gupta et al. 2013, Roberts et al. 2014). Among them, nanoporous carbon membranes (NCMs) have gained much attention because of their potential in a large variety of application fields, such as in separation of gas mixtures, liquids, and biomaterials; in electrochemistry; and as catalytic materials (Itoh & Haraya 2000, Shah et al. 2007, Ting et al. 2011, Rungta et al. 2012, Wang et al. 2012). Their usage as separation media is the most common application of NCMs since they possess superior separation performance, thermal resistance, chemical inertness, and well-defined, stable pore structures (Saufi & Ismail 2004a, Pietraß 2006, Tin et al. 2006, Ashraf et al. 2012, Paul 2012). NCMs were first made in 1960s by pressing and molding graphitic powder into monolith (Ash & Pope 1963). At that time, the main work was focused on the surface adsorption and diffusion process of gases passing through NCMs (Ash et al. 1963, 1967). However, NCMs have not attracted consistently high levels of research interest until the early 1980s when defect-free NCMs were first successfully fabricated (Koresh & Sofer 1983, Baker 2002). During the last three decades, numerous effort and excellent works have been done on the investigation of NCMs (Ismail & David 2001). Generally, NCMs with pore diameters in the range of 3–10 Å can be produced usually by judicious pyrolysis of polymeric films. On the basis of the effective pore diameters and separation mechanisms, NCMs can be classified into two categories: molecular sieve carbon (MSC) membrane with the pore diameters of 3–5 Å and selective surface flow (SSF) membrane with pore diameters of 5–7 Å. There are four kinds of mechanisms for fluid transportation in NCMs, namely, Knudsen diffusion, selective condensation, selective adsorption, and molecular sieving. The detailed description of their differences has been reviewed in literature (Ismail & David 2001). Actually, the separation mechanisms evolving in MSC and SSF are mainly governed by molecular sieving and adsorption–diffusion, respectively. The former preferentially allows the smaller molecules of a feed gas mixture to enter the carbon pores as adsorbed molecules, followed by their surface diffusion to the lower-pressure side of the membrane. The later preferentially permeates the more polar and/or larger molecules of the feed gas mixture by their selective adsorption and surface diffusion on the pore walls to the lower-pressure side. Generally speaking, MSC is suitable for the separation of permanent gases and SSF is suitable for larger organic vapors in a large variety of promising applications.

The formation and the separation characteristics of NCMs depend on the choice of the precursor polymer and the conditions of pyrolysis and post-treatment, as will be highlighted in the following.

## 20.2 Fabrication Process

### 20.2.1 Precursor Selection

NCMs could be obtained by using a large variety of natural and synthetic products as starting materials. Natural precursors include coal, pitch, and biomass (e.g., lignin). In contrast to synthetic materials, the use of cheap mineral materials (coal, pitch) is favorable to reduce the fabrication cost of NCMs, in particular renewable biomass. Kita et al. prepared NCMs by pyrolysis of lignocresol derived from lignin, of which the selectivities were 50 for $CO_2/N_2$, 8 for $O_2/N_2$, 290 for $H_2/CH_4$, and 87 for $CO_2/CH_4$ (Kita et al. 2002). Yang et al. utilized low-cost common filter paper as the carbon precursor to prepare NCMs with excellent $O_2$ adsorption and transport, electrocatalytic activities toward oxygen reduction reaction (ORR), high tolerance of methanol crossover, and durability in alkaline solution (Yang et al. 2014). In addition, coal (Song et al. 2006a) and its derivatives (Liang et al. 1995) could also be applied in the synthesis of NCMs. However, in most cases NCMs could not be produced to present outstanding gas separation performance because of the complicated components in natural products, other than for the separation of liquid mixture as microfiltration and ultrafiltration.

Synthetic precursors are featured with uniform texture and stable properties, allowing the preferential fabrication of NCMs without defects. There are plenty of synthetic materials to successfully form NCMs, such as polyimide, phenolic resin, polyacrylonitrile (PAN), poly(furfuryl alcohol) (PFA), polyvinylidene chloride, polyphenylene oxide, poly(phthalazinone ether sulfone ketone), and polypyrrolone (Kita et al. 1997, Saufi & Ismail 2004b, Salleh et al. 2011). They can be further subdivided into thermosetting resins and thermoplastic resins. Thermosetting resins are easily transferred to NCMs with high gas separation performance by heat treatment at high temperatures, although heating conditions have to be selected carefully to avoid the formation of cracks during the heating and cooling processes. In controlling the sizes of pores, thermal shrinkage during pyrolysis from organic polymers to carbon must also be taken into consideration. The most popular adopted thermosetting resins are polyimide (Suda & Haraya 1997), PFA (Shiflett & Foley 1999), phenol–formaldehyde resins (Teixeira et al. 2011, 2014), poly(vinylidene fluoride), and so on. In contrast, thermoplastic resins possess good processing ability and good dissolution and are cost-effective. Thermoplastic resins include PAN (David & Ismail 2003), poly(phthalazinone ether sulfone ketone) (Zhang et al. 2006a), poly(2,6-dimethyl-1,4-phenylene oxide) (Lee et al. 2007), polyetherimide (Zhang et al. 2015a), and their derivatives (Wang et al. 2009, Ning & Koros 2014). Thermoplastic resins have to be pretreated to prevent the occurrence of a melting or softening stage during the subsequent pyrolysis, which can collapse the preformed porous structure in membrane matrix and destroy the membrane integrity. After the pretreatment of oxidative heating or chemical agent, the thermoplastic precursors would be cross-linked in molecular structure, along with the enhancement of thermal stability, char yield, and mechanical strength. Consequently, it is favorable for the porosity creation and the molecular sieving ability of NCMs.

Furthermore, even a small difference in the chemical structure of precursors can produce a large divergence of separation property for NCMs. Therefore, modification methods could also be applied to adjust the molecular structure and property of existing precursors in hand. One of the most effective ways is to composite the precursor with organic/inorganic fillers. Organic fillers are usually characteristic with thermal labile segments or moieties, such as PVP (Salleh & Ismail 2011, 2013), PEG (Zhang et al. 2007, Wan Nurul Huda & Ahmad 2012), and F127 (Tanaka et al. 2011, Zhang et al. 2014a, Zhao et al. 2014). They will be thermally degraded and left abundant porosity in the membrane matrix of NCMs. In the case of inorganic fillers, there are two subdivisions: porous and nonporous species (Xiao et al. 2010, Tseng et al. 2011). The feasible inorganic fillers include silica (Tseng et al. 2011), zeolite (Liu et al. 2006), carbon molecular sieves (Zhang et al. 2006b), metal or salt (Teixeira et al. 2012), and ordered mesoporous carbon (Zhang et al. 2014b). The addition of porous inorganic fillers can provide a large amount of additional porosity in membrane matrix, while the nonporous ions, metals, or metal oxides can improve the affinity of NCMs for special gases depending on the function of fillers, the gas separation performance of NCMs could be fine-tuned by tailoring the dosage or types of fillers.

## 20.2.2 Coating Technology

Before pyrolysis, a flawless thin organic film should be fabricated in the form of coating on support or self-standing. The configurations of supported NCMs include plate, tube, hollow fiber, and spiral wound, while that of unsupported (or self-standing) NCMs include plate, hollow fiber, and capillary.

Unsupported plate NCMs are merely significant for academic research in the lab; they are usually simply prepared by the method of casting or solvent evaporation. In contrast, hollow fiber and capillary NCMs have gained much interest owing to their high membrane surface area–to–volume ratio. It is because of this point, as the sole commercial product of unsupported NCMs, hollow fiber NCMs were ever offered by Carbon Membranes Ltd. (Israel) (Lagorsse 2004).

In practice, supported NCMs are more promising for their good mechanics for tolerating harsh environmental conditions in operation. To date, various coating technologies have been developed to form NCMs, including brush coating (Acharya et al. 1997), spray coating (Acharya et al. 1999), dipcoating (Lee et al. 2006a), drop coating (Zhang et al. 2014a), ultrasonic spray coating (Shiflett & Foley 2001), CVD (Hayashi et al. 1997a), and physical vapor deposition (Vick et al. 1999). Among them, it has been proved that ultrasonic vapor deposition can yield the most satisfactory gas separation performance, i.e., 30.4 for $O_2/N_2$, 178 for $He/N_2$, and 331 for $H_2/N_2$, by fine control of the forming conditions (Shiflett & Foley 1999). Supported plate NCMs are another commercial product supplied by Blue Membranes GmbH (Germany) (Lagorsse et al. 2005).

The porous supports used for the preparation of supported NCMs are graphitic plate (Titiloye & Hussain 2008), carbon (Song et al. 2008), ceramic, hollow fiber, stainless steel, polymeric substrate, etc. Moreover, several smooth dense substrates, such as glass or silicon wafer, are temporarily utilized to fabricate unsupported NCMs during the solvent evaporation of casting solution. The selection of support should take the consideration of matching the viscosity and nature of membrane solution used. Support with larger pore apertures tends to trap membrane solution, which is adverse to acquire the surface selective layer with enough little thickness and block the porous passages for fluid passing through. On the other side, if the pore aperture is too little to allow the permeation of fluid without restriction, it would not only make the adhesion of surface layer more difficult owing to the smooth surface, but also diminish the permeability of resultant NCMs.

Among the porous supports, stainless steel and ceramic are most popularly used (Rajagopalan et al. 2006, Tseng et al. 2012, Ma et al. 2013). In addition, it should be noticed that the porous structure and the surface roughness of the support can influence the regularity of polymer chain so as to affect the microstructure of resultant surface of NCMs (Tseng et al. 2012). Nevertheless, the expensive ceramic and stainless steel inevitably boost the total fabrication cost of NCMs. In this regard, some inexpensive materials were attempted to alternate the support, e.g., carbon derived from phenolic resin (Wei et al. 2007, Zhang et al. 2014a) and coal powder (Song et al. 2010a).

## 20.2.3 Carbonization Techniques

Carbonization is the most important step for the evolution from carbonaceous starting materials to NCMs. The carbonization is mainly dependent on the following three ways: pyrolysis, CVD, and ion irradiation.

### 20.2.3.1 Pyrolysis

*Pyrolysis* is commonly mixed up with the term *carbonization* in the preparation since most of NCMs are fabricated through the pyrolysis of polymeric membranes. However, according to the interpretation of the encyclopedia *Wikipedia*, *carbonization* is a term for the conversion of an organic substance into carbon or a carbon-containing residue through pyrolysis or destructive distillation, while pyrolysis is the thermochemical decomposition of organic material at elevated temperatures in the absence of oxygen (or any halogen). Actually, pyrolysis is the most frequently used technique to prepare NCMs for its easy control of process parameters and easy operation with temperature-programmable furnace. The fine control of pyrolysis condition is, of course, very important for the achievement of NCMs with desired structure and performance (Centeno et al. 2004, Song et al. 2006b, Salleh et al. 2011). The essential factors for this technique are pyrolysis temperature, atmosphere, heating rate, soak time, etc. (Centeno et al. 2004, Stell & Koros 2005).

#### 20.2.3.1.1 Pyrolysis Temperature

Usually, the pyrolysis temperature is in the range of 500°C–1000°C. Pyrolysis temperature remarkably determines the pore texture and separation performance of NCMs. The pore size generally increases first then decreases with increasing pyrolysis temperature from 500°C–1000°C, along with the same change trend in porosity of NCMs (Anderson et al. 2008). The temperature location for the highest permeability or porosity of NCMs is mainly subjected to the thermal stability of the precursor. There are two kinds of thermal reactions for precursors, i.e., polycondensation and degradation. These correspondingly dominate the first half and bottom half of the pyrolysis process for precursors, of which the division locates at 600°C–750°C. Sometimes, partial degradation of the precursor is controlled to obtain flexible NCMs at low temperature such as 350°C–500°C (Momose et al. 2004) or remove target groups (e.g., $-SO_3H$ [Islam et al. 2005]).

#### 20.2.3.1.2 Atmosphere

There are two kinds of atmosphere for the pyrolysis, i.e., inert atmosphere and vacuum atmosphere. Inert atmosphere can be reached by the purging of inert gases, including argon, helium, and nitrogen, (Sua & Lua 2007). In contrast, vacuum atmosphere tends to yield more selective but less permeable NCMs than compared to an inert gas pyrolysis system (Yamamoto et al. 1997, Clausi & Koros 2000). Furthermore, the addition of a small amount of oxidative gases (e.g., oxygen) in purge gas with a concentration of 4–50 ppm is preferential to enlarging the porosity and average pore size so as to tune the permeability of NCMs (Kiyono et al. 2010).

#### 20.2.3.1.3 Heating Rate

The heating rate determines the evolution rate of the volatile component from a polymeric membrane during pyrolysis, and consequently affects the formation of pores in NCMs (Suda & Haraya 1997, Fuertes et al. 1999). The common heating rate is in the range of 1°C–9°C/min. Lower heating rates are preferable to produce small pores, to increase carbon crystallinity, and consequently to produce NCMs with higher selectivity. Higher heating rates may lead to the formation of pinholes, microscopic crack, blisters, and distortions, which in extreme cases may render membranes useless (i.e., of low selectivity) for gas separation (Salleh & Ismail 2012).

#### 20.2.3.1.4 Thermal Soak Time

The thermal soak time can be used to fine-tune the transport properties of NCMs using a particular final pyrolysis temperature. Increments in thermal soak time would increase the selectivity and reduce the permeability of NCMs because of the further accomplishment of rearrangement reactions in the membrane matrix during this time (David & Ismail 2003, Zhang et al. 2006a).

In summary, pyrolysis is one of the most controllable and reliable carbonization techniques for the preparation of NCMs. Pyrolysis takes advantage of the good universality of low acquisition facility cost and operation condition. At the same time, the deficiencies of pyrolysis are large energy consumption and long pyrolysis duration ranging from 1 to 3 days (including cooling time). From the viewpoint of future large-scale production, the problems with pyrolysis facility have to be resolved, in particular the fine control of flat-temperature zone because NCMs are very sensitive to pyrolysis temperature.

### 20.2.3.2 Chemical Vapor Deposition

CVD is mainly applied to convert a gaseous organic source into NCMs; sources include methane, ethane, propane, ethylene, benzene, and other hydrocarbons or their mixtures (Liu et al. 2008, Hadi & Rouhollahi 2014, Ito et al. 2014, Lai et al. 2014, Manis-Levy et al. 2014). In a typical CVD, the support (substrate or wafer) is exposed to volatile precursors that react and/or decompose on the substrate surface to quickly produce the desired deposit. Frequently, the derived volatile by-products are removed by gas flow through the reaction chamber. Sometimes, heating is accompanied by the process of CVD to aid the formation of thin carbonaceous film on support. For this thermal CVD, the key operation factors include precursor gas, deposition temperature, working pressure, mass flow rate, substrate size, and the percentage concentration of the precursor (Li et al. 2002, Lai et al. 2014). The crystallinity and the ordering degree of NCMs decrease with increasing deposition temperature (Lai et al. 2014).

To improve the efficiency, plasma-enhanced CVD (Cheng et al. 2014) or remote inductively coupled plasma CVD (Wang & Hong 2005a) can be adopted to prepare NCMs. The high-energy ion bombardment can enhance the gas separation performance, particularly the permeability of NCMs (Wang & Hong 2005b, Cheng et al. 2014).

CVD is now another optional method to form NCMs besides pyrolysis. It is believed that the application of CVD in the preparation of NCMs would become more popular with the fast development of CVD promoted by the advances in the semiconductor industry.

### 20.2.3.3 Ion Irradiation

Apart from the above-mentioned pyrolysis and CVD, ion irradiation could also fulfill the carbonization task of polymeric membranes. Ion irradiation owns the merits of short preparation period and low energy consumption, but has the deficiencies of hard control. For this technique, the key factors are power and the irradiation time adopted, which determine the extent of pyrolysis and result in variations in microstructure and separation performance of resultant NCMs.

Ion irradiation is usually applied in semiconductor device fabrication and in metal finishing, as well as various applications in materials science research. It has also been developed in recent years to prepare NCMs. Through the bombardment of high-energy ions originating from an electrical accelerator, this technique changes the physical, chemical, or electrical properties of the precursor surface and transfers the precursor into NCMs by causing a structural change, in that the crystal structure can be damaged or even destroyed by the energetic collision cascades.

In the preparation of NCMs, the operation factors include ion species (such as krypton, xenon, argon, and helium), energy, irradiation time, and fluence (Hukushima et al. 2010, Koval et al. 2010). Depending on those process parameters, thickness, pore size, density, length, and shape could be controlled, so as to influence the final separation performance of NCMs (Iwase et al. 2004, Laušević et al. 2011).

Under typical circumstances, the function of ion ranges between 10 nm and 1 μm with the variation in typical ion energies ranging from 10 to 500 keV. Thus, ion irradiation is especially useful in cases where the chemical or structural change is desired to be near the surface of the membranes. Ions gradually lose their energy as they travel through the solid, both from occasional collisions with target atoms (which cause abrupt energy transfers) and from a mild drag from overlap of electron orbitals, which is a continuous process. Thus, the application of ion irradiation is restricted to a certain degree.

In addition, other conversion techniques should also be attempted to fabricate NCMs from the standpoint of lowering cost and shortening duration of fabrication, for example, microwave pyrolysis, which has been used in a large variety of applications (Dai et al. 2014, Deng et al. 2014, Undri et al. 2014, Xie et al. 2014, Singh et al. 2015).

## 20.2.4 Structure/Property Modification

### 20.2.4.1 Porous Structure

Generally, the porous structure of NCMs is very complicated like that of activated carbon. The structure of traditional NCMs is an amorphous, microporous architecture, which makes them difficult to characterize compared to crystalline materials. From separation studies of penetrants on these materials, it appears that these materials have both ultramicropores (3–5 Å) and supermicropores (5–7 Å). The ultramicropores are believed to be mainly responsible for molecular sieving, while the supermicropores provide negligible resistance to diffusion but provide high-capacity sorption sites for penetrants.

The porous structure of NCMs is more intuitive from photos of microscopy than its counterpart: carbon structure. Actually, they are the two sides of the same thing. The two kinds of micropores are correspondingly formed by the random stacking of graphitic layers and amorphous carbon sheets. The two types of micropores can compensate for each other and afford NCMs with outstanding permeability and selectivity. If the carbon structure is finely controlled, the void (pore structure) will certainly be determined. For the fact that the carbon is derived from the pyrolysis of polymer, the pyrolysis condition is of significance in tuning the graphitization degree and the pore structure (Steel & Koros 2003).

To acquire a satisfactory separation performance of NCMs, the formed pore texture should meet the requirement of the desired microstucture. In most cases, the porous structure parameters have to be fine-tuned after NCMs are obtained or received as commercial product in future, in addition to preconditioning of pore design and formation during pyrolysis. On the basis of fully understanding the pore structure formation mechanisms, it is therefore essential to adjust the pore structure on purpose to the desired target structure. The porous structure can be adjusted by some post-treatment methods, including CVD, postoxidation (or activation), postpyrolysis, and templating.

*20.2.4.1.1 Postoxidation*

Postoxidation is the popular post-treatment method to enlarge the pore size and alter the pore structure of the NCMs by exposing them to an oxidizing atmosphere at elevated temperature (Suda & Haraya 1997). The oxidation of NCMs can be performed using pure oxygen, oxygen admixed with other gases, air, or other oxidizing agents such as steam, carbon dioxide, nitrogen oxides, ammonia, and chlorine oxides, or solutions of oxidizing agents such as nitric acid, mixtures of nitric and sulfuric acids, chromic acid, and peroxide solutions (Hayashi et al. 1997b, Mahurin et al. 2011, Hirota et al. 2013, Lee et al. 2013). The gas permeability of NCMs usually increases with loss of selectivity with prolonging of the oxidation time, or elevation of the temperature or oxidant concentration (Kusakabe et al. 1998, Lee et al. 2006b,c, Zhang et al. 2012). This way, the separation mechanism of NCMs is shifted from molecular sieving to adsorption–diffusion (Fuertes 2001).

*20.2.4.1.2 Chemical Vapor Deposition*

CVD can not only fabricate NCMs on a substrate, as elaborated in the previous section, but also be applied in the modification of the porous structure of existing NCMs by controlled carbonaceous deposition. For example, Hayashi et al. (1997) modified NCMs with CVD of propylene at 650°C, of which the permselectivities for $O_2/N_2$ and $CO_2/N_2$ were respectively increased from 10 and 47 to 14 and 73, together with a certain loss of permeability.

### 20.2.4.1.3 Postpyrolysis

Through secondary pyrolysis, the microstructure of NCMs could also be modified by further thermal reactions associated with the development or rearrangement of carbon structure. For instance, Acharya and Foley (1999) pyrolyzed PFA-coated membranes in an inert atmosphere to a final temperature of 600°C and modified membrane performance further by postpyrolysis. It was found that postpyrolysis leads to small increases in permeability because of an increase in the porosity of the membrane.

### 20.2.4.1.4 Templating

The templating method is commonly used to create ordered nanopores in NCMs. Different from spontaneous amorphous wormlike pores, tailor-made ordered nanopores are very difficult to be formed in NCMs owing to the shrinkage or collapse of the mesostructure aligned parallel to the substrate during pyrolysis (Gierszal & Jaroniec 2006, Song et al. 2010b, Tao et al. 2011). The creation of ordered nanopores in NCMs can be achieved only through the assembly of the precursor (e.g., resorcinol–formaldehyde, phenolic, or phloroglucinol–formaldehyde resin) with templates of block copolymer, such as Pluronic F127 (Liang et al. 2004, Tanaka et al. 2005, 2009, Simanjuntak et al. 2009). There is a recent review article on NCMs that illustrates many of the points made by the authors (Tao et al. 2011). NCMs with ordered nanopores have super high flux and selectivity for the separation of a liquid mixture rather than a gas mixture because of the big gaps between gas molecular dimensions and the relatively large artificial pores (Sun & Crooks 2000, Holt et al. 2006, Sholl et al. 2006, Lopez-Lorente et al. 2010, Ahn et al. 2012). With the great effort required, only a few works on ordered NCMs have reported separation performance ranging from Knudsen diffusion (Tanaka et al. 2011) and molecular sieving (Zhang et al. 2014a). Besides application as separation media, ordered NCMs have gained more interest in the fields of sensors, membrane electrode, etc.

## 20.2.4.2 Surface Chemical Groups

The modification of surface chemical groups is also essential to the properties of NCMs because the surface structure and property of NCMs can drastically affect the adsorption coefficients and diffusivities of the penetrants. Therefore, the separation process is subject to both the porous structure and the surface properties of NCMs.

The surface properties include rough degree, polarity, and hydrophilicity (or hydrophobicity). The effective modification methods of surface groups are preoxidation, postoxidation, CVD, and postpyrolysis. Preoxidation and postoxidation can introduce oxidative groups into the NCM surface in terms of oxygen-containing groups, such as hydroxyl moieties, or form cross-linking structure of inter- or intramolecules. CVD and postpyrolysis can reduce the surface functional groups by deepening the thermal reactions or carbonaceous deposition (Stein et al. 2009, Mahurin et al. 2011). Mahurin et al. (2011) indicated the introduction of nitrogen-containing functional groups (i.e., pyridinic and nitrile) onto the carbon surface by exposure to ammonia vapor atmosphere in the range of 900°C and 950°C. Hayashi et al. (1997) found that the oxidized pore walls of NCMs provide adsorption sites for water molecules, which enhances the NCM permeablility by surface diffusion mechanism.

## 20.2.4.3 Chemical/Water Resistance

Although NCMs exhibit many outstanding advantages, some deficiencies have to be regarded before realistic applications in terms of chemical/water resistance and mechanical property, as well as separation performance. These topics will be discussed in the following sections.

Although NCMs are believed hydrophobic on the whole, their separation performance is susceptible to steam or some vapors because the adsorption of gas mixture can severely result in the pore clogging of NCMs. It is speculated that the oxygen-containing sites in the NCM structure can act as nucleation points for water droplets to form. Consequently, this brings about an unwanted adverse effect on separation performance and shortens the service life. When NCMs are modified by oxidation, they are preferentially adsorbable and permeable for water vapor, which is attractive for the application of dehydration of air (Sircar & Rao 2000). Nevertheless, NCMs for gas separation are more sensitive to humidity in most cases. Therefore, neither the storage nor the operation condition should be strictly prevented from the effects of water or steam by operation below a certain humidity environment (Jones & Koros 1995, Menendez & Fuertes 2001). The storage environment test of NCMs showed that air can decrease permeability with permselectivity increasing slightly; propylene has a positive effect on permeability, whereas permselectivities decrease slightly (Menendez & Fuertes 2001). Coating with polydimethylsiloxane (Petersen et al. 1997) or Teflon (Jones & Koros 1995) is an effective way to provide a successful protection barrier for NCMs from the influence of water vapor or other impurities such as hydrocarbons.

It is assumed that NCM pores are analogous to the "unrelaxed free volume" in glassy polymers. Over time, these pores tend to shrink to achieve thermodynamically more stable states (Xu et al. 2014).

#### 20.2.4.4 Mechanical Property

For practical application, NCMs should endure certain stress on them by passing fluid or sealing modules. The weak mechanical property is also another issue that prohibits the application of NCMs, although hollow fiber carbon membranes have a much attractive promise because of their large loading area per unit volume; they are commercially provided by Carbon Membranes Ltd. (Israel) on a pilot scale (Lagorsse 2014).

However, during ordinary operation, the membrane performance would absolutely be lost, even though only 1% of the hundreds of fibers leak owing to mechanical deficiency.

There are two approaches to resolving this issue and gaining robust property: one is to composite selective NCMs on the support surface (Wei et al. 2007, Zhang et al. 2014a); the other is to accomplish partial pyrolysis at lower temperature by maintaining some of the flexible nature of the original polymer precursors (Nishiyama et al. 2003, Huertas et al. 2012).

## 20.3 Applications

### 20.3.1 Gas Separation

The applications of NCMs are mainly for gas mixtures, including (a) removal of $SF_6$ from air; (b) separation of $CO_2$ from a mixture with $CH_4$ such as landfill gas; (c) production of $O_2$-enriched air; (d) separation of $C_3H_6$ from $C_3H_8$, which is a difficult distillation problem in petrochemical industry; (e) separation of $H_2$ from $CH_4$; (f) recovery of $H_2$ from refinery waste gases; (g) recovery of $H_2$ from pressure swing adsorption waste gases used in purification of $H_2$ from steam methane reformer off-gas; and (h) separation of bulk $H_2S$ from $H_2$ and $CH_4$.

There are numerous reports on the gas separation performance of NCMs. The detailed data of typical gas pairs not included here have been listed in some special review articles (Saufi & Ismail 2004, Tin et al. 2006, Salleh et al. 2011).

### 20.3.2 Liquid Separation

If the nanopores of NCMs are widened enough by some modification methods such as oxidization (Pugazhenthi et al. 2005), they can be utilized in the separation or purification of liquid mixture (Damle et al. 1994), including decolorization of coke furnace wastewater (Sakoda et al. 1997), pervaporation of azeotropic benzene–cyclohexane mixtures (Sakata et al. 1999), and ultrafiltration of water solution (Pugazhenthi et al. 2005).

In comparison with polymeric membranes, the major advantage of the application of NCMs is their stability in concentrated acid/base solution (Sakata et al. 1999). However, the water flux is limited by the innate hydrophilic nature of NCMs for use in water production.

### 20.3.3 Biomass Purification

The application of NCMs in biomass purification is made effective by maintaining the activity of penetrants. The uses cover protein separation by ultrafiltration (Shah et al. 2007), biofuel separation through pervaporation (Tin et al. 2011), separation of water from bioethanol (Liao et al. 2012), and the treatment of oily wastewater by microfiltration (Liao et al. 2012).

The promising applications offer a viable alternative to current membranes for the separation and purification in advanced fields.

### 20.3.4 Membrane Reactors

The reason behind the application restriction of NCMs is their one to three orders of magnitude higher fabrication cost compared to the cost of polymeric membranes, despite their excellent separation performance. To solve this problem, one has to exploit applications with high added value apart from the optimization of the fabrication process, aiming at lowering cost and improving separation performance. Membrane reactors are developed to enhance the efficiency of chemical reactions by combing the comprehensive advantages of NCMs in terms of thermal stability, chemical inertness, and outstanding separation property (Itoh & Haraya 2000, Strano & Foley 2001, Lapkin et al. 2002, Sheintuch & Efremenko 2004, Sznejer & Sheintuch 2004, Zhang et al. 2006c, Sa et al. 2009). The applications of NCM reactors are included in Table 20.1.

Table 20.1 shows that most of the chemical reactions involve hydrogenation or dehydrogenation for the restriction of the antioxidation property of NCMs (Sa et al. 2009).

### 20.3.5 Other Applications

Researchers have also explored other applications in terms of fuel cell and electrocatalytic chemistry. Wang et al. (2012) prepared NCMs with excellent anticorrosion, hydrophobic, and conductivity, aiming at the modification of the stainless steel electrode of proton exchange membrane fuel cell (PEMFC), to improve the work efficiency of PEMFC. Yang et al. (2014) fabricated NCMs from common filter paper, which exhibited excellent electrocatalytic activities toward the ORR, high tolerance of methanol crossover, and durability in alkaline solution.

On the whole, the key challenges of the application of NCMs lie in reducing the thickness of the membrane without introducing defects and scaling up of fabrication techniques of NCM with large surface

**TABLE 20.1** Examples of Applications of NCM Reactors

| Configuration of Integrated NCMs | Catalyst | Reaction | Refs. |
|---|---|---|---|
| Hollow carbon fibers | $Pt/Al_2O_3$ | Dehydrogenation of cyclohexane | Itoh & Haraya 2000 |
| Catalytic membrane[a] | Pt | Hydrogenation of olefins | Strano & Foley 2001 |
| NCM contactor[a] | Phosphoric acid | Hydration of propylene | Lapkin et al. 2002 |
| Hollow carbon fiber | Chromia/alumina | Dehydrogenation of isobutane | Sheintuch & Efremenko 2004, Sznejer & Sheintuch 2004 |
| Tube | $Cu/ZnO/Al_2O_3$ | Methanol steam reforming | Zhang et al. 2006c |
| Tube | $Cu/ZnO/Al_2O_3$ | Methanol steam reforming | Briceño et al. 2012 |
| Tube | $Pt/Al_2O_3$ | Dehydrogenation of methylcyclohexane | Hirota et al. 2013 |
| Plate catalytic NCMs | $Cu/ZnO/Al_2O_3$ | Methanol steam reforming | Zhang et al. 2015b |

[a] Not disclosed.

areas. Furthermore, for commercial applications, NCMs should be prepared in the form of a honeycomb or hollow fiber module to provide the additional benefits of a low drop in pressure and a high surface-to-volume ratio. It would also be advantageous to shift the thermodynamic equilibrium by improving the porous structure of NCMs.

## 20.4 Summary and Outlook

From the previous elaboration on NCMs, we can see that the research on these promising membrane materials has achieved rich fruits at the academic level. In particular, the underlying rules on the guidance of preparation and application of NCMs would stride largely with the aid of more and more advanced characterization techniques, such as positron annihilation spectroscopy (Anderson et al. 2008, Fu et al. 2011, Liao et al. 2012, 2013, Cheng et al. 2014), scanning AFM (Suk et al. 2012), small angle X-ray scattering (Steriotis et al. 1997, Kimijima et al. 2008, Favvas et al. 2012), and multinuclear pulsed field gradient NMR (Mueller et al. 2012, 2013).

In conjunction with simulation methods, more basic theories will soon be revealed. The simulation methods include nonequilibrium molecular dynamics (Furukawa & Nitta 1997, 2000, Furukawa et al. 1997, MacElroy & Boyle 1999, Kaganov & Sheintuch 2003, Wang et al. 2006a,b,c, Wu et al. 2008), molecular simulation (Sedigh et al. 1998, Ghassemzadeh et al. 2000, Arora & Sandler 2007, He et al. 2009, Müller 2013, Zhai et al. 2013), and Monte Carlo (Seo et al. 2001, 2002, Jia et al. 2007). In addition, some commercial professional software (such as Cerius and ChemOffice) could also be used to simulate the chain properties of precursors for NCMs (Xiao et al. 2005).

Nevertheless, there is still a lot of room for further research on the structure and separation process for NCMs on the basis of proposal of theoretical and practical models, e.g., parallel resistance model with defects span in the entire NCMs (Strano & Foley 2003). Moreover, from the viewpoint of future scale-up of NCMs, the design of membrane module and the study of flow field are required by computational fluid dynamics simulation and particle image velocimetry experimental methods.

## References

Acharya, M. & Foley, H. C., "Spray-coating of nanoporous carbon membranes for air separation," *J. Membrane Sci.* 161 (1999): 1–5.

Acharya, M., Raich, B. A. & Foley, H. C., "Metal-supported carbogenic molecular sieve membranes: Synthesis and applications," *Ind. Eng. Chem. Res.* 36 (1997): 2924–2930.

Ahn, C. H., Baek, Y., Lee, C. et al., "Carbon nanotube-based membranes: Fabrication and application to desalination," *J. Ind. Eng. Chem.* 18 (2012): 1551–1559.

Anderson, C. J., Pas, S. J., Arora, G. et al., "Effect of pyrolysis temperature and operating temperature on the performance of nanoporous carbon membranes," *J. Membrane Sci.* 322 (2008): 19–27.

Arora, G. & Sandler, S. I., "Nanoporous carbon membranes for separation of nitrogen and oxygen: Insight from molecular simulations," *Fluid Phase Equilibr.* 259 (2007): 3–8.

Ash, R. M. B. R. & Pope, C. G., "Flow of adsorbable gases and vapours in a microporous medium: I. Single sorbates," *Proc. R. Soc. Lond. A* 271 (1963): 1–18.

Ash, R., Baker, R. W. R. & Barrer, M., "Sorption and surface flow in graphitized carbon membranes: I. The steady state," *Proc. R. Soc. Lond. A* 299 (1967): 434–454.

Ash, R., Barrer, R. M. & Pope, C. G., "Flow of adsorbable gases and vapours in a microporous medium: II. Binary mixtures," *Proc. R. Soc. Lond. A* 271 (1963): 19–33.

Ashraf, A., Dastgheib, S. A., Mensing, G. A. et al., "Membrane distillation with carbon-based membranes, in: AIChE 2012," 2012 AIChE Annual Meeting, Conference Proceedings (2012).

Baker, R. W., "Future directions of membrane gas separation technology," *Ind. Eng. Chem. Res.* 41 (2002): 1393–1411.

Briceño, K., Iulianelli, A., Montané, D. et al., "Carbon molecular sieve membranes supported on non-modified ceramic tubes for hydrogen separation in membrane reactors," *Int. J. Hydrogen Energy* 37 (2012): 13536–13544.

Centeno, T. A., Vilas, J. L. & Fuertes, A. B., "Effects of phenolic resin pyrolysis conditions on carbon membrane performance for gas separation," *J. Membrane Sci.* 228 (2004): 45–54.

Cheng, L. H., Fu, Y. J., Liao, K. S. et al., "A high-permeance supported carbon molecular sieve membrane fabricated by plasma-enhanced chemical vapor deposition followed by carbonization for $CO_2$ capture," *J. Membrane Sci.* 460 (2014): 1–8.

Clausi, D. T. & Koros, W. J., "Formation of defect-free polyimide hollow fiber membranes for gas separations, "*J. Membrane Sci.* 167 (2000): 79–89.

Dai, Q. J., Jiang, X. G., Jiang, Y. F. et al., "Temperature influence and distribution in three phases of PAHs in wet sewage sludge pyrolysis using conventional and microwave heating," *Energy Fuels* 28 (2014): 3317–3325.

Damle, A. S., Gangwal, S. K. & Venkataraman, V. K., "Carbon membranes for gas separation: Developmental studies," *Gas Separat. Purif.* 8 (1994): 137–147.

David, L. I. B. & Ismail, A. F., "Influence of the thermastabilization process and soak time during pyrolysis process on the polyacrylonitrile carbon membranes for $O_2/N_2$ separation," *J. Membrane Sci.* 213 (2003): 285–291.

Deng, W. Y., Su, Y. X., Liu, S. G. et al., "Microwave-assisted methane decomposition over pyrolysis residue of sewage sludge for hydrogen production," *Int. J. Hydrogen Energy* 39 (2014): 9169–9179.

Favvas, E. P., Stefanopoulos, K. L., Papageorgiou, S. K. et al., "In situ small angle X-ray scattering and benzene adsorption on polymer-based carbon hollow fiber membranes," *Adsorption* 19 (2012): 225–233.

Fu, Y. J., Liao, K. S., Hu, C. C. et al., "Development and characterization of micropores in carbon molecular sieve membrane for gas separation," *Micropor. Mesopor. Mat.* 143 (2011): 78–86.

Fuertes, A. B., "Effect of air oxidation on gas separation properties of adsorption-selective carbon membranes," *Carbon* 39 (2001): 697–706.

Fuertes, A. B., Nevskaia, D. M. & Centeno, T. A., "Carbon composite membranes from Matrimid® and Kapton® polyimides for gas separation," *Micropor. Mesopor. Mat.* 33 (1999): 115–125.

Furukawa, S. & Nitta, T., "Non-equilibrium molecular dynamics simulation studies on gas permeation across carbon membranes with different pore shape composed of micro-graphite crystallites," *J. Membrane Sci.* 178 (2000): 107–119.

Furukawa, S. I. & Nitta, T., "Computer simulation studies on gas permeation through nanoporous carbon membranes by non-equilibrium molecular dynamics," *J. Chem. Eng. Jpn.* 30 (1997): 116–122.

Furukawa, S. I., Hayashi, K. & Nitta, T., "Effects of surface heterogeneity on gas permeation through slit-like carbon membranes by non-equilibrium molecular dynamics simulations," *J. Chem. Eng. Jpn.* 30 (1997): 1107–1112.

Ghassemzadeh, J., Xu, L. F., Tsotsis, T. T. et al., "Statistical mechanics and molecular simulation of adsorption in microporous materials: Pillared clays and carbon molecular sieve membranes," *J. Phys. Chem. B* 104 (2000): 3892–3905.

Gierszal, K. P. & Jaroniec, M., "Carbons with extremely large volume of uniform mesopores synthesized by carbonization of phenolic resin film formed on colloidal silica template," *J. Am. Chem. Soc.* 128 (2006): 10026–10027.

Gupta, V. K. & Saleh, T. A., "Sorption of pollutants by porous carbon, carbon nanotubes and fullerene—An overview," *Env. Science. Pollut. Res. Int.* 20 (2013): 2828–2843.

Hadi, M. & Rouhollahi, A., "Filamentous pyrolytic carbon film and its electroanalytical properties," *J. Electroanal. Chem.* 727 (2014): 13–20.

Hayashi, J.-I., Mizuta, H., Yamamoto, M. et al., "Pore size control of carbonized BPDA-pp' ODA polyimide membrane by chemical vapor deposition of carbon," *J. Membrane Sci.* 124 (1997a): 243–251.

Hayashi, J.-I., Yamamoto, M., Kusakabe, K. et al., "Effect of oxidation on gas permeation of carbon molecular sieving membranes based on BPDA-pp'ODA polyimide," *Ind. Eng. Chem. Res.* 36 (1997b): 2134–2140.

He, X., Lie, J. A., Sheridan, E. et al., "$CO_2$ capture by hollow fibre carbon membranes: Experiments and process simulations," in: *Energy Procedia*, (2009): 261–268.

Hirota, Y., Ishikado, A., Uchida, Y. et al., "Pore size control of microporous carbon membranes by post-synthesis activation and their use in a membrane reactor for dehydrogenation of methylcyclohexane," *J. Membrane Sci.* 440 (2013): 134–139.

Holt, J. K., Park, H. G., Wang, Y. et al., "Fast mass transport through sub-2-nanometer carbon nanotubes," *Science* 312 (2006): 1034–1037.

Huertas, R. M., Doherty, C. M., Hill, A. J. et al., "Preparation and gas separation properties of partially pyrolyzed membranes (PPMs) derived from copolyimides containing polyethylene oxide side chains," *J. Membrane Sci.* 409–410 (2012): 200–211.

Hukushima, S., Sode, K., Yamazaki, K. et al., "Carbon structure in polyimide membrane formed by ion irradiation," *J. Photopolym. Sci. Technol.* 23 (2010): 507–510.

Inagaki, M., Toyoda, M. & Tsumura, T., "Control of crystalline structure of porous carbons," *RSC Adv.* 4 (2014): 41411–41424.

Islam, M. N., Zhou, W. L., Honda, T. et al., "Preparation and gas separation performance of flexible pyrolytic membranes by low-temperature pyrolysis of sulfonated polyimides," *J. Membrane Sci.* 261 (2005): 17–26.

Ismail, A. F. & David, L. I. B., "A review on the latest development of carbon membranes for gas separation," *J. Membrane Sci.* 193 (2001): 1–18.

Ito, Y., Christodoulou, C., Nardi, M. V. et al., "Chemical vapor deposition of N-doped graphene and carbon films: The role of precursors and gas phase," *ACS Nano* 8 (2014): 3337–3346.

Itoh, N. & Haraya, K., "A carbon membrane reactor," *Catal. Today* 56 (2000): 103–111.

Iwase, M., Sannomiya, A., Nagaoka, S. et al., "Gas permeation properties of asymmetric polyimide membranes with partially carbonized skin layer," *Macromolecules* 37 (2004): 6892–6897.

Jia, Y., Wang, M., Wu, L. et al., "Separation of $CO_2/N_2$ gas mixture through carbon membranes: Monte Carlo simulation," *Sep. Sci. Technol.* 42 (2007): 3681–3695.

Jones, C. W. & Koros, W. J., "Carbon composite membranes: A solution to adverse humidity effects," *Ind. Eng. Chem. Res.* 34 (1995): 164–167.

Kaganov, I. V. & Sheintuch, M., "Nonequilibrium molecular dynamics simulation of gas-mixtures transport in carbon-nanopore membranes," *Phys. Rev. E.* 68 (2003): 046701.

Kimijima, K. I., Hayashi, A. & Yagi, I., "Preparation of a self-standing mesoporous carbon membrane with perpendicularly-ordered pore structures," *Chem. Commun.* (2008): 5809–5811.

Kita, H., Nanbu, K., Hamano, T. et al., "Carbon molecular sieving membranes derived from lignin-based materials," *J. Polym. Environ.* 10 (2002): 69–75.

Kita, H., Yoshino, M., Tanaka, K. et al., "Gas permselectivity of carbonized polypyrrolone membrane," *Chem. Commun.* (1997): 1051–1052.

Kiyono, M., Williams, P. J. & Koros, W. J., "Effect of pyrolysis atmosphere on separation performance of carbon molecular sieve membranes," *J. Membrane Sci.* 359 (2010): 2–10.

Koresh, J. E. & Sofer, A., "Molecular sieve carbon permselective membrane—Part I: Presentation of a new device for gas mixture separation," *Sep. Sci. Technol.* 18 (1983): 723–734.

Koval, Y., Geworski, A., Gieb, K. et al., "Fabrication and characterization of glassy carbon membranes," *J. Vac. Sci. Technol. B* 32 (2014): 042001. doi: 10.1116/1.4890008.

Kusakabe, K., Yamamoto, M. & Morooka, S., "Gas permeation and micropore structure of carbon molecular sieving membranes modified by oxidation," *J. Membrane Sci.* 149 (1998): 59–67.

Lagorsse, S., "Carbon molecular sieve membranes: Sorption, kinetic and structural characterization," *J. Membrane Sci.* 241 (2004): 275–287.

Lagorsse, S., Leite, A., Magalhaes, F. D. et al., "Novel carbon molecular sieve honeycomb membrane module: Configuration and membrane characterization," *Carbon* 43 (2005): 809–819.

Lai, L. H., Yang, J. S. & Shiue, S. T., "Characteristics of carbon films prepared by thermal chemical vapor deposition using camphor," *Thin Solid Films* 556 (2014): 544–551.

Lapkin, A. A., Tennison, S. R. & Thomas, W. J., "A porous carbon membrane reactor for the homogeneous catalytic hydration of propene," *Chem. Eng. Sci.* 57 (2002): 2357–2369.

Laušević, Z., Apel, P. Y. & Blonskaya, I. V., "The production of porous glassy carbon membranes from swift heavy ion irradiated Kapton," *Carbon* 49 (2011): 4948–4952.

Lee, H. C., Monji, M., Parsley, D. et al., "Use of steam activation as a post-treatment technique in the preparation of carbon molecular sieve membranes," *Ind. Eng. Chem. Res.* 52 (2013): 1122–1132.

Lee, H. J., Kim, D. P., Suda, H. et al., "Gas permeation properties for the post-oxidized polyphenylene oxide (PPO) derived carbon membranes: Effect of the oxidation temperature," *J. Membrane Sci.* 282 (2006c): 82–88.

Lee, H. J., Suda, H., Haraya, K. et al., "Gas permeation properties of carbon molecular sieving membranes derived from the polymer blend of polyphenylene oxide (PPO)/polyvinylpyrrolidone (PVP)," *J. Membrane Sci.* 296 (2007): 139–146.

Lee, H. J., Yoshimune, M., Suda, H. et al., "Effects of oxidation curing on the permeation performances of polyphenylene oxide-derived carbon membranes," *Desalination* 193 (2006b): 51–57.

Lee, H. J., Yoshimune, M., Suda, H. et al., "Gas permeation properties of poly(2,6-dimethyl-1,4-phenylene oxide) (PPO) derived carbon membranes prepared on a tubular ceramic support," *J. Membrane Sci.* 279 (2006a): 372–379.

Li, Y. Y., Nomura, T., Sakoda, A. et al., "Fabrication of carbon coated ceramic membranes by pyrolysis of methane using a modified chemical vapor deposition apparatus," *J. Membrane Sci.* 197 (2002): 23–35.

Liang, C., Hong, K., Guiochon, G. A. et al., "Synthesis of a large-scale highly ordered porous carbon film by self-assembly of block copolymers," *Angew. Chem.* 43 (2004): 5785–5789.

Liang, C., Sha, G. & Guo, S., "Carbon membrane for gas separation derived from coal tar pitch," *Carbon* 37 (1999): 1391–1397.

Liao, K. S., Fu, Y. J., Hu, C. C. et al., "Microstructure of carbon molecular sieve membranes and their application to separation of aqueous bioethanol," *Carbon* 50 (2012): 4220–4227.

Liao, K.-S., Fu, Y.-J., Hu, C.-C. et al., "Development of the asymmetric microstructure of carbon molecular sieve membranes as probed by positron annihilation spectroscopy," *J. Phys. Chem. C* 117 (2013): 3556–3562.

Liu, B. S., Wang, N., He, F. et al., "Separation performance of nanoporous carbon membranes fabricated by catalytic decomposition of $CH_4$ using Ni/polyamideimide templates," *Ind. Eng. Chem. Res.* 47 (2008): 1896–1902.

Liu, Q. L., Wang, T. H., Liang, C. H. et al., "Zeolite married to carbon: A new family of membrane materials with excellent gas separation performance," *Chem. Mater.* 18 (2006): 6283–6288.

Lopez-Lorente, A. I., Simonet, B. M. & Valcarcel, M., "The potential of carbon nanotube membranes for analytical separations," *Anal. Chem.* 82 (2010): 5399–5407.

Ma, X. L., Lin, B. K., Wei, X. T. et al., "Gamma-alumina supported carbon molecular sieve membrane for propylene/propane separation," *Ind. Eng. Chem. Res.* 52 (2013): 4297–4305.

MacElroy, J. M. D. & Boyle, M. J., "Nonequilibrium molecular dynamics simulation of a model carbon membrane separation of $CH_4$/$H_2$ mixtures," *Chem. Eng. J.* 74 (1999): 85–97.

Mahurin, S. M., Lee, J. S., Wang, X. et al., "Ammonia-activated mesoporous carbon membranes for gas separations," *J. Membrane Sci.* 368 (2011): 41–47.

Manis-Levy, H., Livneh, T., Zukerman, I. et al., "Effect of radio-frequency and low-frequency bias voltage on the formation of amorphous carbon films deposited by plasma enhanced chemical vapor deposition," *Plasma Sci. Technol.* 16 (2014): 954–959.

Menendez, I. & Fuertes, A. B., "Aging of carbon membranes under different environments," *Carbon* 39 (2001): 733–740.

Momose, W., Zheng, T., Nishiyama, N. et al., "Synthesis of partially carbonized polyimide membranes with high resistance to moisture," *J. Chem. Eng. Jpn.* 37 (2004): 1092–1098.

Mueller, R., Kanungo, R., Kiyono-Shimobe, M. et al., "Diffusion of ethane and ethylene in carbon molecular sieve membranes by pulsed field gradient NMR," *Micropor. Mesopor. Mat.* 181 (2013): 228–232.

Mueller, R., Kanungo, R., Kiyono-Shimobe, M. et al., "Diffusion of methane and carbon dioxide in carbon molecular sieve membranes by multinuclear pulsed field gradient NMR," *Langmuir* 28 (2012): 10296–10303.

Müller, E. A., "Purification of water through nanoporous carbon membranes: A molecular simulation viewpoint," *Curr. Opin. Chem. Eng.* 2 (2013): 223–228.

Ning, X. & Koros, W. J., "Carbon molecular sieve membranes derived from Matrimid® polyimide for nitrogen/methane separation," *Carbon* 66 (2014): 511–522.

Nishiyama, N., Momose, W., Egashira, Y. et al., "Partially carbonized polyimide membranes with high permeability for air separation," *J. Chem. Eng. Jpn.* 36 (2003): 603–608.

Paul, D. R., "Materials science: Creating new types of carbon-based membranes," *Science* 335 (2012): 413–414.

Petersen, J., Matsuda, M. & Haraya, K., "Capillary carbon molecular sieve membranes derived from Kapton for high temperature gas separation," *J. Membrane Sci.* 131 (1997): 85–94.

Pietraß, T., "Carbon-based membranes," *MRS Bull.* 31 (2006): 765–769.

Pugazhenthi, G., Sachan, S., Kishore, N. et al., "Separation of chromium (VI) using modified ultrafiltration charged carbon membrane and its mathematical modeling," *J. Membrane Sci.* 254 (2005): 229–239.

Rajagopalan, R., Merritt, A., Tseytlin, A. et al., "Modification of macroporous stainless steel supports with silica nanoparticles for size selective carbon membranes with improved flux," *Carbon* 44 (2006): 2051–2058.

Roberts, A. D., Li, X. & Zhang, H., "Porous carbon spheres and monoliths: Morphology control, pore size tuning and their applications as Li-ion battery anode materials," *Chem. Soc. Rev.* 43 (2014): 4341–4356.

Rungta, M., Xu, L. & Koros, W. J., "Carbon molecular sieve dense film membranes derived from Matrimid® for ethylene/ethane separation," *Carbon* 50 (2012): 1488–1502.

Sa, S., Silva, H., Sousa, J. M. et al., "Hydrogen production by methanol steam reforming in a membrane reactor: Palladium vs carbon molecular sieve membranes," *J. Membrane Sci.* 339 (2009): 160–170.

Sakata, Y., Muto, A., Uddin, M. A. et al., "Preparation of porous carbon membrane plates for pervaporation separation applications," *Sep. Purif. Technol.* 17 (1999): 97–100.

Sakoda, A., Nomura, T. & Suzuki, M., "Activated carbon membrane for water treatments: Application to decolorization of coke furnace wastewater," *Adsorption*, 3 (1997): 93–98.

Salleh, W. N. W. & Ismail, A. F., "Carbon hollow fiber membranes derived from PEI/PVP for gas separation," *Sep. Purif. Technol.* 80 (2011): 541–548.

Salleh, W. N. W. & Ismail, A. F., "Effect of stabilization condition on PEI/PVP-based carbon hollow fiber membranes properties," *Sep. Sci. Technol.* 48 (2013): 1030–1039.

Salleh, W. N. W. & Ismail, A. F., "Effects of carbonization heating rate on $CO_2$ separation of derived carbon membranes," *Sep. Purif. Technol.* 88 (2012): 174–183.

Salleh, W. N. W., Ismail, A. F., Matsuura, T. et al., "Precursor selection and process conditions in the preparation of carbon membrane for gas separation: A review," *Sep. Purif. Rev.* 40 (2011): 261–311.

Saufi, S. M. & Ismail, A. F., "Fabrication of carbon membranes for gas separation—A review," *Carbon* 42 (2004): 241–259.

Sedigh, M. G., Onstot, W. J., Xu, L. F. et al., "Experiments and simulation of transport and separation of gas mixtures in carbon molecular sieve membranes," *J. Phys. Chem. A* 102 (1998): 8580–8589.

Seo, Y. G., Kum, G. H. & Seaton, N. A., "Monte Carlo simulation of transport diffusion in nanoporous carbon membranes," *J. Membrane Sci.* 195 (2001): 65–73.

Seo, Y. G., Kum, G. H. & Seaton, N. A., "Monte Carlo simulation of transport diffusion in nanoporous carbon membranes," *J. Membrane Sci.* 195 (2002): 65–73.

Shah, T. N., Foley, H. C. & Zydney, A. L., "Development and characterization of nanoporous carbon membranes for protein ultrafiltration," *J. Membrane Sci.* 295 (2007): 40–49.

Sheintuch, M. & Efremenko, I., "Analysis of a carbon membrane reactor: From atomistic simulations of single-file diffusion to reactor design," in: *ISCRE18*, Elsevier Ltd., (2004): 4739–4746.

Shiflett, M. B. & Foley, H. C., "Reproducible production of nanoporous carbon membranes," *Carbon* 39 (2001): 1421–1425.

Shiflett, M. B. & Foley, H. C., "Ultrasonic deposition of high-selectivity nanoporous carbon membranes," *Science* 285 (1999): 1902–1905.

Sholl, D. S. & Johnson, J. K., "Materials science: Making high-flux membranes with carbon nanotubes," *Science* 312 (2006): 1003–1004.

Simanjuntak, F. H., Jin, J., Nishiyama, N. et al., "Ordered mesoporous carbon films prepared from 1,5-dihydroxynaphthalene/triblock copolymer composites," *Carbon* 47 (2009): 2531–2533.

Singh, S., Neculaes, V. B., Lissianski, V. et al., "Microwave assisted coal conversion," *Fuel* 140 (2015): 495–501.

Sircar, S. & Rao, M. B., "Nanoporous carbon membranes for gas separation," in: *Recent Advances in Gas Separation by Microporous Ceramic Membrane*, Elsevier Science B.V., (2000): 473–496.

Song, C. W., Wang, T. H., Pan, Y. Q. et al., "Preparation of coal-based microfiltration carbon membrane and application in oily wastewater treatment," *Sep. Purif. Technol.* 51 (2006): 80–84.

Song, C. W., Wang, T. H., Wang, X. Y. et al., "Preparation and gas separation properties of poly(furfuryl alcohol)-based C/CMS composite membranes," *Sep. Purif. Technol.* 58 (2008): 412–418.

Song, C., Wang, T., Jiang, H. et al., "Gas separation performance of C/CMS membranes derived from poly(furfuryl alcohol) (PFA) with different chemical structure," *J. Membrane Sci.* 361 (2010): 22–27.

Song, C., Wang, T., Pan, Y. et al., "Preparation of coal-based microfiltration carbon membrane and application in oily wastewater treatment," *Sep. Purif. Technol.* 51 (2006a): 80–84.

Song, C., Wang, T., Qiu, J. et al., "Effects of carbonization conditions on the properties of coal-based microfiltration carbon membranes," *J. Porous Mat.* 15 (2006b): 1–6.

Song, L., Feng, D., Fredin, N. J. et al., "Challenges in fabrication of mesoporous carbon films with ordered cylindrical pores via phenolic oligomer self-assembly with triblock copolymers," *ACS Nano* 4 (2010b): 189–198.

Steel, K. M. & Koros, W. J., "An investigation of the effects of pyrolysis parameters on gas separation properties of carbon materials," *Carbon* 43 (2005): 1843–1856.

Steel, K. M. & Koros, W. J., "Investigation of porosity of carbon materials and related effects on gas separation properties," *Carbon* 41 (2003): 253–266.

Stein, A., Wang, Z. Y. & Fierke, M. A., "Functionalization of porous carbon materials with designed pore architecture," *Adv. Mater.* 21 (2009): 265–293.

Steriotis, T., Beltsios, K., Mitropoulos, A. C. et al., "On the structure of an asymmetric carbon membrane with a novolac resin precursor," *J. Appl. Polym. Sci.* 64 (1997): 2323–2345.

Strano, M. S. & Foley, H. C., "Modeling ideal selectivity variation in nanoporous membranes," *Chem. Eng. Sci.* 58 (2003): 2745–2758.

Strano, M. S. & Foley, H. C., "Synthesis and characterization of catalytic nanoporous carbon membranes," *Aiche. J.* 47 (2001): 66–78.

Su, J. & Lua, A. C., "Effects of carbonisation atmosphere on the structural characteristics and transport properties of carbon membranes prepared from Kapton® polyimide," *J. Membrane Sci.* 305 (2007): 263–270.

Suda, H. & Haraya, K., "Alkene/alkane permselectivities of a carbon molecular sieve membrane," *Chem. Commun.* (1997): 93–94.

Suda, H. & Haraya, K., "Gas permeation through micropores of carbon molecular sieve membranes derived from Kapton polyimide," *J. Phys. Chem. B* 101 (1997): 3988–3994.

Suk, J. W., Murali, S., An, J. et al., "Mechanical measurements of ultra-thin amorphous carbon membranes using scanning atomic force microscopy," *Carbon* 50 (2012): 2220–2225.

Sun, L. & Crooks, R. M., "Single carbon nanotube membranes: A well-defined model for studying mass transport through nanoporous materials," *J. Am. Chem. Soc.* 122 (2000): 12340–12345.

Sznejer, G. & Sheintuch, M., "Application of a carbon membrane reactor for dehydrogenation reactions," *Chem. Eng. Sci.* 59 (2004): 2013–2021.

Tanaka, S., Doi, A., Nakatani, N. et al., "Synthesis of ordered mesoporous carbon films, powders, and fibers by direct triblock-copolymer-templating method using an ethanol/water system," *Carbon* 47 (2009): 2688–2698.

Tanaka, S., Nakatani, N., Doi, A. et al., "Preparation of ordered mesoporous carbon membranes by a soft-templating method," *Carbon* 49 (2011): 3184–3189.

Tanaka, S., Nishiyama, N., Egashira, Y. et al., "Synthesis of ordered mesoporous carbons with channel structure from an organic–organic nanocomposite," *Chem. Commun. (Camb.)* (2005): 2125–2127.

Tao, Y. S., Endo, M., Inagaki, M. et al., "Recent progress in the synthesis and applications of nanoporous carbon films," *J. Mater. Chem.* 21 (2011): 313–323.

Teixeira, M., Campo, M. C., Tanaka, D. A. P. et al.,"Composite phenolic resin-based carbon molecular sieve membranes for gas separation," *Carbon* 49 (2011): 4348–4358.

Teixeira, M., Campo, M., Tanaka, D. A. et al., "Carbon–Al₂O₃–Ag composite molecular sieve membranes for gas separation," *Chem. Eng. Res. Des.* 90 (2012): 2338–2345.

Teixeira, M., Rodrigues, S. C., Campo, M. et al. "Boehmite–phenolic resin carbon molecular sieve membranes—Permeation and adsorption studies," *Chem. Eng. Res. Des.* 92 (2014): 2668–2680.

Tin, P. S., Lin, H. Y., Ong, R. C. et al., "Carbon molecular sieve membranes for biofuel separation," *Carbon* 49 (2011): 369–375.

Tin, P. S., Xiao, Y. & Chung, T. S., "Polyimide-carbonized membranes for gas separation: Structural, composition, and morphological control of precursors," *Sep. Purif. Rev.* 35 (2006): 285–318.

Titiloye, J. O. & Hussain, I., "Synthesis and characterization of silicalite-1/carbon-graphite membranes," *J. Colloid Interface Sci.* 318 (2008): 50–58.

Tseng, H. H., Shih, K. M., Shiu, P. T. et al., "Influence of support structure on the permeation behavior of polyetherimide-derived carbon molecular sieve composite membrane," *J. Membrane Sci.* 405 (2012): 250–260.

Tseng, H. H., Shiu, P. T. & Lin, Y. S., "Effect of mesoporous silica modification on the structure of hybrid carbon membrane for hydrogen separation," *Int. J. Hydrogen Energy* 36 (2011): 15352–15363.

Undri, A., Rosi, L., Frediani, M. et al., "Fuel from microwave assisted pyrolysis of waste multilayer packaging beverage," *Fuel* 133 (2014): 7–16.

Upare, D. P., Yoon, S. & Lee, C. W., "Nano-structured porous carbon materials for catalysis and energy storage," *Korean J. Chem. Eng.* 28 (2011): 731–743.

Vick, D., Tsui, Y. Y., Brett, M. J. et al., "Production of porous carbon thin films by pulsed laser deposition," *Thin Solid Films* 350 (1999): 49–52.

Wan Nurul Huda, W. Z. & Ahmad, M. A., "A comparison of carbon molecular sieve (CMS) membranes with polymer blend CMS membranes for gas permeation applications," *ASEAN J. Chem. Eng.* 12 (2012): 51–58.

Wang, L. J. & Hong, F. C. N., "Carbon-based molecular sieve membranes for gas separation by inductively-coupled-plasma chemical vapor deposition," *Micropor. Mesopor. Mat.* 77 (2005b): 167–174.

Wang, L. J. & Hong, F. C. N., "Surface structure modification on the gas separation performance of carbon molecular sieve membranes," *Vacuum* 78 (2005a): 1–12.

Wang, S. M., Yu, Y. X. & Gao, G. H., "Effect of temperature on the permeation and separation of oxygen and nitrogen in carbon membranes: A non-equilibrium molecular dynamics simulation study," *Acta Chim. Sin.* 64 (2006b): 1111–1115.

Wang, S. M., Yu, Y. X. & Gao, G. H., "Grand canonical Monte Carlo and non-equilibrium molecular dynamics simulation study on the selective adsorption and fluxes of oxygen/nitrogen gas mixtures through carbon membranes," *J. Membrane Sci.* 271 (2006c): 140–150.

Wang, S. M., Yu, Y. X. & Gao, G. H., "Non-equilibrium molecular dynamics simulation on pure gas permeability through carbon membranes," *Chinese J. Chem. Eng.* 14 (2006a): 164–170.

Wang, T. H., Zhang, B., Qiu, J. S. et al., "Effects of sulfone/ketone in poly(phthalazinone ether sulfone ketone) on the gas permeation of their derived carbon membranes," *J. Membrane Sci.* 330 (2009): 319–325.

Wang, T., Zhang, C. X., Sun, X. et al., "Synthesis of ordered mesoporous boron-containing carbon films and their corrosion behavior in simulated proton exchange membrane fuel cells environment," *J. Power Sources* 212 (2012): 1–12.

Wei, W., Qin, G., Hu, H. et al., "Preparation of supported carbon molecular sieve membrane from novolac phenol–formaldehyde resin," *J. Membrane Sci.* 303 (2007): 80–85.

Wu, Z., Liu, Z., Wang, W. et al., "Non-equilibrium molecular dynamics simulation on permeation and separation of $H_2/CO$ in nanoporous carbon membranes," *Sep. Purif. Technol.* 64 (2008): 71–77.

Xiao, Y. C., Chng, M. L., Chung, T. S. et al., "Asymmetric structure and enhanced gas separation performance induced by in situ growth of silver nanoparticles in carbon membranes," *Carbon* 48 (2010): 408–416.

Xiao, Y. C., Chung, T. S., Chng, M. L. et al., "Structure and properties relationships for aromatic polyimides and their derived carbon membranes: Experimental and simulation approaches," *J. Phys. Chem. B* 109 (2005): 18741–18748.

Xie, Q., Peng, P., Liu, S. et al., "Fast microwave-assisted catalytic pyrolysis of sewage sludge for bio-oil production," *Bioresour. Technol.* 172C (2014): 162–168.

Xu, L., Rungta, M., Hessler, J. et al., "Physical aging in carbon molecular sieve membranes," *Carbon* 80 (2014): 155–166.

Yamamoto, M., Kusakabe, K., Hayashi, J.-I. et al., "Carbon molecular sieve membrane formed by oxidative carbonization of a copolyimide film coated on a porous support tube," *J. Membrane Sci.* 133 (1997): 195–205.

Yang, W., Zhai, Y., Yue, X. et al., "From filter paper to porous carbon composite membrane oxygen reduction catalyst," *Chem. Commun. (Camb.)* 50 (2014): 11151–11153.

Zhai, M. M., Yoshioka, T., Yang, J. H. et al., "Preparation and characterization of amorphous carbon (a-C) membranes by molecular dynamics simulation," *Desalin Water Treat.* 51 (2013): 5231–5236.

Zhang, B. Shi, Y., Wu, Y. H. et al., "Towards the preparation of ordered mesoporous carbon/carbon composite membranes for gas separation," *Sep. Sci. Technol.* 49 (2014b): 171–178.

Zhang, B., Shi, Y., Wu, Y. H. et al., "Preparation and characterization of supported ordered nanoporous carbon membranes for gas separation," *J. Appl. Polym. Sci.* 131 (2014a): 2136–2146.

Zhang, B., Wang, T. H., Zhang, S. H. et al., "Preparation and characterization of carbon membranes made from poly(phthalazinone ether sulfone ketone)," *Carbon* 44 (2006a): 2764–2769.

Zhang, B., Wu, Y., Lu, Y. et al., "Preparation and characterization of carbon and carbon/zeolite membranes from ODPA–ODA type polyetherimide," *J. Membrane Sci.* 474 (2015a): 114–121.

Zhang, B., Zhao, D., Wu, Y. et al., "Fabrication and application of catalytic carbon membranes for hydrogen production from methanol steam reforming," *Ind. Eng. Chem. Res.* 54 (2015b): 623–632.

Zhang, X. Y., Hu, H. Q., Zhu, Y. D. et al., "Carbon molecular sieve membranes derived from phenol formaldehyde novolac resin blended with poly(ethylene glycol)," *J. Membrane Sci.* 289 (2007): 86–91.

Zhang, X. Y., Hu, H. Q., Zhu, Y. D. et al., "Effect of carbon molecular sieve on phenol formaldehyde novolac resin based carbon membranes," *Sep. Purif. Technol.* 52 (2006b): 261–265.

Zhang, X. Y., Liu, R., Hu, H. Q. et al., "Preparation of carbon molecular sieve membranes by KOH activation for gas separation," *New Carbon Mater.* 27 (2012): 61–66.

Zhang, X., Hu, H., Zhu, Y. et al., "Methanol steam reforming to hydrogen in a carbon membrane reactor system," *Ind. Eng. Chem. Res.* 45 (2006c): 7997–8001.

Zhao, X., Li, W. & Liu, S. X., "Ordered mesoporous carbon membrane prepared from liquefied larch by a soft method," *Mater. Lett.* 126 (2014): 174–177.

# 21

# Gas-Adsorbing Nanoporous Carbons

Dolores
Lozano-Castelló

Fabian
Suárez-García

Juan Alcañiz-Monge

Diego
Cazorla-Amorós

Angel
Linares-Solano

## 21.1 Introduction

This chapter presents an overview of synthesis, properties, and gas adsorption applications of nanoporous carbons, including (i) activated carbons (ACs) and activated carbon monoliths (ACMs) prepared by physical or chemical activation of different precursors; (ii) ACs prepared by activation of hydrochars obtained by hydrothermal carbonization (HTC); (iii) templated carbons; and (iv) carbide-derived carbons (CDCs).

Examples of the performance of these nanoporous carbons in gas storage (natural gas [NG], hydrogen, and carbon dioxide) and cryocooler applications are presented, remarking on the importance of carrying out a suitable characterization of the materials to understand and optimize their performance in each application and the relevance of the porosity and the surface chemistry.

## 21.2 Synthesis

### 21.2.1 ACs: Physical and Chemical Activation

ACs are highly porous carbon materials, exhibiting appreciable apparent surface area and micropore volume (MPV), which can present a wide variety of pore size distributions (PSDs) and micropore size distributions (MPSDs) (Bansal et al. 1988, Jankowska et al. 1991, Marsh et al. 1997, Yasuda et al. 2003). ACs are solids which can be prepared in different forms, such as powders, granules, pellets, fibers, and cloths. Because of these features and their special chemical characteristics, ACs can be used for very different applications, e.g., liquid and gas phase treatments and energy storage (Bansal et al. 1988, Jankowska et al. 1991, Marsh et al. 1997, Burchell 1999, Derbyshire et al. 2001, Radovic et al. 2001, Yasuda et al. 2003, Bottani & Tascón 2008).

ACs are not present in nature. To obtain them, the selection of the precursor and the preparation process is necessary. Several precursors, such as wood, coals, pitches, polymers, and residues with a high amount of carbon, and also different preparation methods have been used. These two factors have great importance as they determine the final porous structure of the ACs.

In considering the preparation process, the main stage determining the porous structure is the method of activation. The objective during the activation is both to increase the number of pores and to increase the size of the existing ones so that the AC has a high adsorption capacity. The different activation processes are divided into two different groups: chemical and physical activation (Bansal et al. 1988). The differences between the two are the procedure and the activating agents used.

The preparation of ACs by physical activation (Bansal et al. 1988, Muñoz-Guillena et al. 1992) includes a controlled gasification of the carbonaceous material that has previously been carbonized to increase the carbon content, although, occasionally, the activation of the precursor can be directly done. Thus, the samples are treated to 800°C–1000°C with an oxidant gas (usually $CO_2$ and steam) so that carbon atoms are being removed selectively. The extension of the activation, which must be as selective and controlled as possible, is usually expressed as weight loss percentage, which is named *burn-off percentage*. PSD of the AC depends on the precursor, the preparation conditions (mainly temperature, time, and gas flow), the activating agent used, and the presence of catalysts.

On the other hand, the chemical activation process consists of mixing a carbonaceous precursor with a chemical activating agent, followed by a heat treatment stage and finally by a washing step to remove the chemical agent and the inorganic components (Bansal et al. 1988, Marsh et al. 1997, Derbyshire et al. 2001, Linares-Solano et al. 2007). In the literature the use of several activating agents, such as phosphoric acid (Molina-Sabio et al. 1995, Benaddi et al. 1998, Díaz-Díez et al. 2004, Gonzalez-Serrano et al. 2004, Suárez-García et al. 2004, Bedia et al. 2009, 2010, Puziy & Tascón 2012), zinc chloride (Caturla et al. 1991, Ibarra et al. 1991, Rodríguez-Reinoso & Molina-Sabio 1992, Ahmadpour & Do 1996), alkaline carbonates (Hayashi et al. 2002, Carvalho et al. 2004), KOH (Illan-Gomez et al. 1996, Evans et al. 1999, Lozano-Castelló et al. 2001), and NaOH (Illan-Gomez et al. 1996, Lillo-Ródenas et al. 2001, Perrin et al. 2004), has been reported. In the case of chemical activation, it is better not to use the term *burn-off degree*, as done for physical activation; it is better to talk about the degree of carbon reacted (or degree of activation).

Chemical activation offers well-known advantages (Rodríguez-Reinoso & Molina-Sabio 1992, Marsh et al. 1997, Evans et al. 1999, Lozano-Castelló et al. 2001, Linares-Solano et al. 2007) over the physical one, which can be summarized as follows: (i) it uses lower temperatures and heat treatment times; (ii) it usually consists of one stage; and (iii) the yields obtained are typically higher. On the other hand, chemical activation presents some disadvantages (Marsh et al. 1997, Derbyshire et al. 2001, Lozano-Castelló et al. 2001), such as the need for a washing stage after the heat treatment and the more corrosive behavior of the chemical agents used in comparison with that of $CO_2$ or steam. Traditionally, chemical activation has been carried out using two activating agents, phosphoric acid and zinc chloride.

In the case of chemical activation with phosphoric acid, lignocellulosic materials are preferred as precursors (Molina-Sabio et al. 1995, Derbyshire et al. 2001, Puziy & Tascón 2012). At low degrees of activation, ACs do not have a highly developed area and they are essentially microporous, whereas at higher activation degree, the surface area and the MPV increase but there is also a remarkable increase in the mesopore volume and a widening of the MPSD (Teng et al. 1998, Suárez-García et al. 2002, Puziy & Tascón 2012). Therefore, in the case of activation with phosphoric acid, both high adsorption capacity and narrow MPSD cannot be achieved. However, for ACs that need a well-developed mesoporosity, e.g., for gasoline removal (Burchell 1999), phosphoric acid activation is a very suitable activation method (Baker 1995).

The ACs prepared by chemical activation with $ZnCl_2$ are essentially microporous (Caturla et al. 1991, Rodríguez-Reinoso & Molina-Sabio 1992). The loading of zinc has an important effect on the porosity: samples activated at high zinc loadings present high porosity development and MPV but also a more heterogeneous MPSD (Caturla et al. 1991). Although higher MPVs can be obtained by $ZnCl_2$ activation than by physical activation or by phosphoric acid activation, the increase in porosity development is also concurrent with a widening of the microporosity. The main disadvantage of this activating agent is that the emission of zinc may cause serious environmental problems, which strongly limits its present use.

The purpose of developing ACs with tailored porosity in the whole range of microporosity has motivated continued research toward the use of other activating agents, such as alkaline hydroxides. Since the pioneering patent of Wennerberg and O'Grady (1978), the production of very high–surface area (superactive) ACs using alkaline hydroxide activation started its commercialization, first by Amoco and then by Kansai Coke and Chemical Company (Japan) (Wennerberg & O'Grady 1978, Otowa et al. 1997). In addition, a considerable number of studies have been carried out over the years that focused on chemical activation with hydroxides (Ahmadpour & Do 1996, Illan-Gomez et al. 1996, Lillo-Ródenas et al. 2001, Lozano-Castelló et al. 2001, Linares-Solano et al. 2007, Castro-Muñiz et al. 2013). A systematic study, developed by our research work, has underlined that the reactivity of the carbon precursor, the carbon-to-hydroxide ratio, the heat treatment temperature, and the $N_2$ flow rate are the crucial variables determining the porous character of the ACs (Linares-Solano et al. 2007). In some cases their careful control has allowed the production of highly porous ACs ($S_{BET} > 3500 \text{ m}^2/\text{g}$) with narrow MPSDs.

## 21.2.2 ACs from Hydrothermally Carbonized Precursors

As mentioned in the previous section, the typical approach to converting a raw material into a high–carbon content material is the carbonization process. One of the main efforts in modern materials research is the production of new and effective materials which can be synthesized from cheap natural precursors and also through environment-friendly and efficient processes. In recent years HTC has demonstrated its capability of converting biomass into carbon materials under very mild processing conditions (Titirici & Antonietti 2010, Titirici et al. 2012).

HTC consists of heat treatment of an aqueous solution/dispersion of an organic material, such as saccharides (glucose, cellulose, starch, or sucrose), simpler compounds such as furfural, or more complex substances such as biomass, at temperatures in the range of 150°C–350°C under autogenerated pressure (Titirici & Antonietti 2010). This process has several advantages: (i) the precursors are readily available, cheap, and renewable (i.e., saccharides or biomass); (ii) it is a green and simple process as it involves only water as the solvent and consists of a simple heat treatment in a closed autoclave; and (iii) the resulting solid carbon products exhibit attractive chemical and structural properties. The resulting solid carbon products (called hydrochars) generally exhibit uniform chemical and structural properties as well as a very high content of oxygen-containing functional groups (White et al. 2009). The amount and the type of these groups can be tuned by modifying the operating conditions (i.e., temperature, solution concentration, reaction time, and precursor). Other functionalities (e.g., nitrogen based) can also be introduced into hydrochars by using dopant-containing carbon precursors or additives (Joo et al. 2008, Titirici & Antonietti 2010). Unfortunately, hydrochar materials have the drawback of possessing almost no porosity unless they are synthesized in the presence of a template (Titirici et al. 2007, Joo et al. 2008) or subjected to additional heat treatment at higher temperature (Kim et al. 2006).

To make possible the use of porous materials based on hydrochar in emergent applications, such as hydrogen storage or electrical energy storage (supercapacitors), the synthesis of highly microporous carbon materials based on these materials is required. As explained above, and as reviewed in detail previously (Linares-Solano et al. 2007), chemical activation by KOH and NaOH is a way to successfully prepare highly microporous ACs from very different precursors (coals, lignocellulosic materials, carbon fibers, pitch, and so on), and it was recently demonstrated that controlled chemical activation also constitutes an excellent method to activate these hydrochar products (Sevilla et al. 2011a, Falco et al. 2013).

## 21.2.3 Templated Carbons

Nanocasting or nanotemplating is a powerful tool for creating porous materials that are difficult to synthesize by conventional methods. Nanocasting is based on the use of molecules or ordered porous solids as templates for the production of other materials that are their replicas (Schüth 2003, Lu & Schüth 2006). Thus, by using this technique, many different porous materials can be prepared, including zeolites, silicas, carbons, metal oxides, and other inorganic materials (Corma 1997, Lu & Schüth 2006, Meng et al. 2009, Qisheng 2011, Nishihara & Kyotani 2012). Nanocasting techniques can be classified into two broad categories based on the nature of the template used, namely, endotemplating and exotemplating (Schüth 2003), also called soft-templating and hard-templating, respectively.

The concept of endotemplating refers to the preparation of ordered porous materials by synthesizing them around molecules or supramolecules. The porous network is obtained after the elimination of the template. By using this technique, both microporous solids (e.g., zeolites) and ordered mesoporous silicas, using molecules or molecular aggregates as templates, respectively (Schüth 2003, Qisheng 2011), can be prepared. In the case of carbons, the endotemplating technique has been applied almost for obtaining mesoporous carbons (Ma et al. 2013). This method involves the self-assembly of polymerizable precursors (phenolic resins or resorcinol) and block copolymer templates. The experimental conditions (template type and concentration, carbon precursor, pH, aging temperature, time, etc.) determine the final properties of the carbon (Liang et al. 2004, Liang & Dai 2006, Górka et al. 2009, Xie et al. 2011, Ma et al. 2013). This method is simpler (fewer steps) than exotemplating; however, it has the limitation that the number of organic precursors available is very limited.

The technique of exotemplating consists of the use of rigid materials as a template to create another solid confined in its porous network, the generated solid being an inverse replica of the material used as the template. By using this method, it is possible to prepare microporous (e.g., using zeolites as templates), mesoporous (e.g., using mesoporous silica or other oxides as templates), or macroporous carbon materials (e.g., using opals or colloidal silica as templates) (Lee et al. 2006, Lu & Schüth 2006, Xia et al. 2010b, Nishihara & Kyotani 2012). The exotemplating (hard-templating) technique generally consists of the following steps: (i) synthesis of the template (usually by endotemplating (soft-templating), (ii) infiltration/polymerization of the carbon precursor into the template porosity, (iii) carbonization of the composite, and (iv) elimination of the template (by etching with HF or NaOH).

An important family of ordered carbons in the context of gas adsorption and which are prepared following this technique are the zeolite template carbons (ZTCs). This kind of carbon material was first synthesized by Kyotani et al. (1997) using USY zeolite as the template. In this early work, they filled the zeolite channels by means of CVD of propylene at 700°C–800°C. Later, Kyotani et al. prepared long-range ordered microporous carbons using zeolite Y as the template. The procedure consisted of infiltration with furfuryl alcohol which was polymerized inside the zeolite channels. The composite was heated to 700°C and propylene CVD was performed at this temperature for further carbon deposition to completely fill the zeolite porosity. This ZTC is characterized by a very uniform and ordered pore structure and the highest SSA (about 4000 $m^2/g$) (Ma et al. 2000, 2001, 2002, Matsuoka et al. 2005). Recently, several types of ZTCs with different surface areas, nitrogen-doped ZTC, and ZTC loaded with Pt nanoparticles were prepared (Nishihara et al. 2009, Alam & Mokaya 2010, Xia et al. 2010a, 2011a, Almasoudi & Mokaya 2012).

The nanocasting method for the preparation of mesoporous carbons was first reported in 1999 by Ryoo et al. (1999). In this case, ordered mesoporous silicas are used as templates. Owing to the much larger pore size of the silica templates compared with that of zeolites, many different organic compounds have been used as precursors. Thus, ordered mesoporous carbons (OMCs) have been prepared from different precursors, including saccharides, polymers, resins, pitches, and hydrocarbons (Ryoo et al. 1999, 2001, Lu et al. 2004, Parmentier et al. 2004, Kruk et al. 2005, Lu & Schüth 2006, Gierszal et al. 2008, Enterría et al. 2012b, Sánchez-Sánchez et al. 2014a). In the context of the present chapter (i.e., gas adsorption), the presence of micropores is required, because the adsorption of gases (e.g., $H_2$, $CH_4$, $CO_2$, and He) takes place mainly in the microporosity or narrow microporosity of the adsorbent. Therefore, to make more efficient the OMCs in these applications, the investigation has been focused on developing their microporosity, trying to preserve the ordered mesoporous structure of the starting OMC.

Several activation methods of OMCs have been used, including physical activation (Xia et al. 2007, 2008, Enterría et al. 2012b) and chemical activation with hydroxides (Choi & Ryoo 2007, Enterría et al. 2012a,b), which successfully developed microporosity.

## 21.2.4 Carbide-Derived Carbons

The term *carbide-derived carbons* refers to nanoporous carbons produced by extraction of metals from metal carbides (MeCs). The selective etching of carbides allows the synthesis of carbon with different structures and control of the porosity can be performed by a suitable selection of the synthesis variables (Gogotsi et al. 2001, Presser 2011a). Metals can be leached from carbide by using different processes such as treatments in supercritical water, at high temperature under halogen atmosphere, and by vacuum decomposition.

The reaction of silicon carbide with chlorine gas was reported in 1918 (Hutchins 1918) for the first time as a method for obtaining silicon tetrachloride. This reaction was conducted at 1000°C and it was patented as a method for the production of silicon tetrachloride (Andersen 1956). Thus, the treatment of other MeCs (e.g., Ti, Zr, and Nb) with chlorine gas in the range of temperatures of 450°C–950°C has been used to produce the corresponding metal chlorides (Kirillova et al. 1960, Orekhov et al. 1969). The first time in which authors paid attention to the characteristics of carbon material derived from metal carbides was in 1975, which was done by Boehm and Warnecke (1975), who studied the porous structure and molecular sieve properties of CDCs prepared by chlorination of TaC at 500°C.

A general reaction corresponding to the carbon formation by selective carbide etching from binary carbides and different halogens is shown as follows:

$$x\mathrm{MC} + \frac{y}{2}\mathrm{A}_2 \overset{\Delta}{\rightarrow} \mathrm{M}_x\mathrm{A}_y + \mathrm{C}, \tag{21.1}$$

where A is a gaseous halogen ($F_2$, $Cl_2$, $Br_2$, $I_2$, or mixtures thereof) and $M_xA_y$ is a gaseous reaction product. In the case of ternary carbides, more complex reaction equations must be written.

CDCs with a high bulk porosity (>50 vol%) and specific BET surface areas (>2000 $m^2$/g) can be obtained by this method (Gogotsi et al. 2005, 2009, Dash et al. 2006, Yushin et al. 2006, Kim et al. 2009, Yeon et al. 2010, Presser et al. 2011a,b). The porosity of the CDCs is mainly determined by the structure of the carbide (composition, lattice type, etc.), as the distribution of the carbon atoms in the carbide lattice determines to a large extent the structure of the resulting CDC (Gogotsi et al. 2003, Dash et al. 2004, 2005, Yushin et al. 2005). Besides the carbide structure, the PSD in the final CDC is also influenced by the etching conditions. Thus, the halogenation temperature has a clear effect on the PSD of the CDC and the average pore size increases with increasing halogenation temperature (Gogotsi et al. 2003, Presser 2011a).

The porosity (surface area, pore volume and size, etc.) of the CDCs has also been modified by performing various post-treatments, among which annealing with hydrogen (Dash et al. 2006, Gogotsi et al. 2009) and activation, both chemical and physical (Gogotsi et al. 2009, Osswald et al. 2009, Yeon et al. 2009), were used. The former is important for removal of residual halogen that remains occluded in the CDC. It has been shown that up to 40% residual chlorine can be trapped in the CDC (Yushin et al. 2005). The annealing allows the elimination of halogens and the opening of clogged pores, resulting in an increase in the surface area (Dash et al. 2006, Gogotsi et al. 2009). Post-treatments of activation increase the porosity of the CDC (Gogotsi et al. 2009, Osswald et al. 2009, Yeon et al. 2009) and, in general, porosity is developed as a function of the activation variables in a similar way as with other carbon precursors, as indicated previously in Section 21.2.1.

# 21.3 Properties

## 21.3.1 Porosity

The characterization of the porosity of porous carbons materials is an essential step because the material behavior in the application is mainly determined by its textural parameters (pore size and shape, PSD, pore volumes, surface area, etc.). Pores are classified by the International Union of Pure and Applied Chemistry as a function of their size into micropores (<2 nm), mesopores (2–50 nm), and macropores (>50 nm) (Sing et al. 1985). The complete characterization of the porosity of carbons is complex because of the heterogeneity in the chemistry and structure of these materials. Thus, the selection of the appropriate method can be difficult and, most of the times, a combination of techniques is used. Physical adsorption of gases is, undoubtedly, the most widely used technique (Sing et al. 1985, Rouquerol et al. 1999). Nitrogen at 77 K is the most used, because of its relative experimental simplicity and because at this temperature a wide range of relative pressures can be covered (nitrogen adsorption takes place in both micropores and mesopores) under subatmospheric conditions. However, the main disadvantage of $N_2$ adsorption at 77 K is that when it is used for the characterization of microporous solids, diffusional problems of the molecules inside the narrow microporosity (i.e., pore size below 0.7 nm) may occur. To overcome these problems, the use of $CO_2$ adsorption, at either 273 or 298 K, has been proposed as a useful alternative for the assessment of the narrow microporosity (Rodriguez-Reinoso & Linares-Solano 1988, Cazorla-Amorós et al. 1998, Lozano-Castelló et al. 2004). Although the critical dimension of the $CO_2$ molecule is similar to that of $N_2$, the higher temperature of adsorption used for $CO_2$ results in a larger kinetic energy of the molecules, which are able to enter into the narrow porosity.

The application of different theories and equations to the adsorption/desorption isotherms allows obtaining several porous textural parameters. Numerous specialized works can be found in the literature in this field. Here only some examples are cited (Dubinin 1975, Sing et al. 1985, Patrick 1995, Rouquerol et al. 1999, Lowell et al. 2004, Lozano-Castelló et al. 2009, Tascón & Bottani 2011, Tascón 2012).

## 21.3.2 Density

For those applications using nanoporous materials, where the volume available is a constraint, an essential property is material density. There are several definitions of material density (true density, apparent density, bulk density, tap density, and geometric density). The revision of the terminology and the procedure or equipment available to measure these properties are out of the scope of this chapter and can be found elsewhere (Lowell et al. 2004).

The density of the powder nanoporous carbons is a key property for the gas storage applications presented in the next section, and it can be determined in two different ways: (i) filling a container with the AC and vibrating it (tap density) and (ii) pressing a given amount of AC in a mould at a given pressure (e.g., 550 kg/cm²) and then releasing pressure (packing density) (Lozano-Castelló et al. 2002c).

In general, the material density of porous solids decreases with the development of porosity (Jordá-Beneyto et al. 2008, Alcañiz-Monge et al. 2009), as it can be seen in Figure 21.1 for a series of ACs, where tap and packing density values are plotted versus the MPV. For both types of measurements, a general trend can be observed: the higher the porosity development of the materials, the lower the density, following a linear trend. It should be remarked that the use of the packing density is suitable because this type of porous material (ACs) can be pressed up to 550 kg/cm², and even higher pressure, without changing their porous texture (Alcañiz-Monge et al. 2009, Marco-Lozar et al. 2012a).

## 21.3.3 Surface Chemistry

Carbon materials always have some amount of heteroatoms that have a strong influence on the surface chemistry (especially O, but also N, P, S, etc.) (Figueiredo et al. 1999, Radovic 1999, 2001,

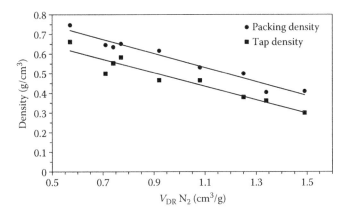

**FIGURE 21.1** Densities of the powder ACs (packing density [●] and tap density [■]) versus the total MPV $V_{DR}$ (N₂) (cm³/g).

Bleda-Martinez et al. 2005, Gorgulho et al. 2008, Tascón & Bottani 2011). These heteroatoms may come from the carbon precursor, or be introduced by post-treatment (i.e., oxidation, amination, and grafting) (Pradhan & Sandle 1999, Cheng & Teng 2003, Shin et al. 2007, Xing et al. 2007, Stein et al. 2009, Peng et al. 2010, Rivera-Utrilla et al. 2011, Sánchez-Sánchez et al. 2014a). Heteratoms usually form different functional surface groups. In the case of oxygen, they can be found as carboxylic acid, hydroxyl, anhydrides, ethers, esters, etc. Nitrogen can be found as pyrroles, pyridone, pyridine groups, quaternary nitrogen, or amines. The surface groups are very important in many applications as they modify the characteristics of carbon material (e.g., increasing its hydrophilicity) and can actively intervene in the application (e.g., establishing specific interactions with the adsorbate).

There are different techniques for the characterization of the surface chemistry of carbons. Because of the difficulty, in many cases the use of different techniques is required. Among them, temperature-programmed desorption (TPD) and XPS can be highlighted, although others such as infrared spectroscopy or potentiometric titration are also useful.

TPD consists of heating the carbon material in an inert atmosphere up to temperatures of about 1000°C and analyzing the evolved gases. Thus, as the temperature is increased, the oxygen-containing groups decompose as $CO_2$ and CO, which are analyzed and quantified (e.g., with a mass spectrometer). There is a relationship between the decomposition temperature and the type of oxygen surface group, allowing quantifying them (Figueiredo et al. 1999, Gorgulho et al. 2008).

XPS is a quantitative spectroscopic technique that measures the elemental composition of a material. XPS spectra are obtained by irradiating the material with a beam of X-rays while measuring the kinetic energy and the number of electrons that escape. XPS gives quantitative information about the chemical groups (Sánchez-Sánchez et al. 2014b). The main drawback of this technique is that it gives information on only the outer layer of the material (<10 nm) and the surface chemistry of the outer layer of the material can be different from the surface chemistry of the pores.

# 21.4 Applications

## 21.4.1 Gas Storage: NG and Hydrogen Storage

NG (mainly methane) or $H_2$ could be used as a substitute for fuel oil in vehicles, but the main drawback of their use is their storage in the vehicular fuel tank. These gases are in a supercritical state at room temperature, so at these standard temperature and pressure conditions, the energy density (defined as the heat of combustion per unit volume) is only 0.038 MJ/L (0.11% of that of gasoline) for NG and

0.0108 MJ/L (0.03% of that of gasoline) for $H_2$, so the mileage per unit volume of the fuel tank is very low. Hence, the suitability of NG and $H_2$ for vehicular application will depend on the ability to store an adequate amount in the onboard fuel tank. Commercially available cars, with a fuel tank of about 50 L, can run a range of around 500 km. The Department of Energy of the United States (DOE) has established different targets for onboard NG and $H_2$ storage systems, including the minimum gravimetric and volumetric capacities and the reversibility of the charging/discharging processes. Thus, to cover the same range as that of commercially available cars, the DOE target for methane has been defined as 150 volumes of methane delivered per volume of storage container (i.e., 150 V/V delivery), when the pressure is reduced from 3.5 MPa to atmospheric pressure at 298 K. For hydrogen, the target is set at 9 wt% of stored hydrogen and a volumetric density of 0.081 kg $H_2$/L, to be achieved by 2015 (Office of Energy Efficiency and Renewable Energy 2010a). These values are referred to the whole system, including the storage medium, the vessel, the refueling infrastructure, any regulators, electronic controllers, and sensors (for more information, see www.eere.energy.gov/hydrogenandfuelcells). The more recently calculated values for light-duty vehicles (Office of Energy Efficiency and Renewable Energy 2010b) reduce the targets down to 5.5 wt% of $H_2$ and 0.040 kg $H_2$/L to be achieved by 2015. In Europe, the targets are less restrictive and in the European Hydrogen and Fuel Cell Strategic Research Agenda and Deployment Strategy, Barrett (2005) points out an energy density value of 1.1 kW h/L, which is equivalent to a volumetric hydrogen storage capacity of about 0.033 kg $H_2$/L.

Several methods have been considered to increase the energy density of NG and $H_2$ and facilitate its use as a road vehicle fuel (Cook et al. 1999, Schlapbach & Zuttel 2001): (a) as a highly compressed gas, (b) as a liquid at very low temperature, and (c) as an adsorbed gas in porous solids. Additionally, in the case of $H_2$, there is also another option, (d) forming metallic hydrides. All of them present advantages and disadvantages.

NG and $H_2$ can be stored as an adsorbed phase in porous materials filling the storage vessel. The search for a suitable porous material, in terms of further improving the storage volumetric energy density and lowering the adsorbent cost to the end use, is currently an active area of research. To obtain high storage capacity of NG and $H_2$, a variety of inorganic solids, such as zeolites, organic–inorganic hybrid porous complexes (MOF), and carbonaceous materials, such as superactivated carbons, AC fibers, carbon monoliths, CNTs, and graphene-based carbons, are being investigated (Lozano-Castelló et al. 2002a, Jordá-Beneyto et al. 2007, Yurum et al. 2009, Froudakis 2011, Orinakova & Orinak 2011, Pumera 2011, Getman et al. 2012, Marco-Lozar et al. 2012b). They predominantly consist of microporous solids (pores of width around 0.8–2 nm). For this application it is very important to optimize not only gravimetric gas uptake but also gas storage in volumetric basis.

At a given pressure, a strong adsorption potential inside the micropores acting on gas molecules significantly increases the density of the adsorbed molecules in relation to the gas phase density. This phenomenon can be exploited for enhancement of gas storage capacity through adsorption. This has been the main reason for the strong interest in using AC as a medium to reduce the pressure required to store gases such as methane and hydrogen. The search for ACs able to store large amounts of NG at a reasonable pressure (3.5–4 MPa), as substitute for NG compressed at much higher pressure (e.g., 21 MPa), has been very intense in the last decades (Parkyns & Quinn 1995, Menon & Komarneni 1998, Cook et al. 1999, Lozano-Castelló et al. 2002a, Marco-Lozar et al. 2012b).

In all previous studies (Parkyns & Quinn 1995, Menon & Komarneni 1998, Cook et al. 1999, Lozano-Castelló et al. 2002a), it was concluded that, in general, the higher the surface area (or MPV), the higher the methane adsorption capacity. However, in a systematic study carried out using KOH-activated carbons, it was concluded that, in addition to surface area and packing density, the MPSD is also important (Lozano-Castelló et al. 2002a). Thus, Figure 21.2 contains the methane uptake versus the apparent BET surface area corresponding to a series of KOH-activated carbons. The linear relationship is seen to reach a maximum for AC with a very high surface area. If only the apparent BET surface area or the MPV was responsible for the methane uptake, sample 5/1 (see Figure 21.2) should have a higher methane capacity than sample 3/1; however, the opposite behavior is observed. These results can be explained by analyzing the porous texture results in detail (results not shown here). Sample 3/1 has a narrower MPSD than

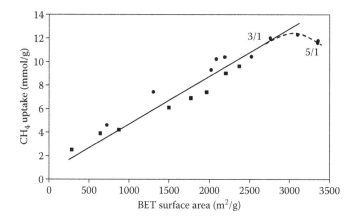

**FIGURE 21.2** $CH_4$ uptake at 298 K and 4 MPa in chemically activated carbons versus the BET surface area.

sample 5/1, which results in a higher methane uptake, consistent with the enhanced adsorption potential argument (Lozano-Castelló et al. 2002a).

From a practical point of view, to increase the methane adsorption capacity, not only do we need to develop the MPV but also we have to control carefully the MPSD. Thus, samples with high surface areas and very narrow MPSD are required for this application.

In addition, another crucial parameter, which has to be taken into account, is the packing density of the adsorbent. This is because in its real application, methane storage would be carried out in an AC-filled vessel. Then, obviously, the higher the amount of AC in the vessel (which is related to the packing density), the higher the volume of methane stored. As mentioned in Section 21.3.2, in general, for porous materials, the higher the MPV, the lower the packing density (Jordá-Beneyto et al. 2008, Alcañiz-Monge et al. 2009).

Because of the enormous importance of the packing density of the ACs for this application, the useless voids (meso-, macro-, and interparticle space), where methane adsorption does not take place in an important amount, must be minimized while maintaining a high MPV. In the case of AC materials, which do not present macroporosity, these voids consist of any space between the particles. By producing the AC as a monolith (ACMs), it is possible to reduce interparticle voids, maximizing the bulk density (Quinn & MacDonald 1992, Parkyns & Quinn 1995). These monoliths consist of cylindrical pieces, which pack uniformly, lessening wasted space in the storage vessel. Another advantage of producing carbons in an efficient space-filling form is that they are also strong and resist attrition. The monoliths can be produced using a binder, which helps to keep the carbon particles in a compressed state. Conventional preparation methods (Quinn & MacDonald 1992, Chen et al. 1997) consist of mixing the AC with a binder, compression and molding using a hydraulic press, and, finally, pyrolysis to improve the binder properties and decrease the weight of binder in the monolith. For methane storage application, the binder must possess certain properties: (i) it must produce monoliths with good mechanical properties using the lowest binder/AC ratio to achieve the maximum amount of AC, which is the adsorbent material, and (ii) its pyrolysis must not produce pore blocking of the AC. This last characteristic of the binder is very important, because the current focus is to increase the packing density while retaining a high MPV, which is responsible for the methane adsorption.

The effect of the type and amount of binder on the MPV, methane adsorption capacity, piece density, and delivery was analyzed in detail in a previous work (Lozano-Castelló et al. 2002b). Different binders were used to prepare ACMs from a powder AC (the powder AC selected had a methane delivery of 145 V/V). These binders included a humic acid-derived sodium salt from Acros Organics, polyvinyl alcohol, a phenolic resin (Georgia-Pacific 5506), a propriety binder from Waterlink Sutcliffe Carbons (WSC), Teflon from DuPont, and a cellulose-based binder. Among all the binders studied, the one which produces monoliths with the best equilibrium between adsorption capacity and piece density, and consequently with the best uptake (140 V/V) and delivery (126 V/V), was the binder WSC.

Analogous to NG storage, in the case of the hydrogen storage systems, they have to achieve the target not only on the gravimetric basis but also on the volumetric one. Thus, by using the values of packing density, hydrogen adsorption capacities at 77 K and up to 4 MPa in volumetric basis were estimated and presented in Figure 21.3 versus the MPV. For comparison purposes, this plot also contains the hydrogen adsorption capacities on a gravimetric basis. It is clearly seen that the hydrogen adsorption capacity on a gravimetric basis (wt%) at 77 K and at pressures of up to 4 MPa follows a good correlation with the total MPV, obtaining the highest value for the sample with the highest MPV. On the other hand, for the data on volumetric basis (g/L), the maximum corresponds to samples with a relatively high porosity and packing density. Thus, according to these results, to get a maximum hydrogen adsorption capacity on volumetric basis, we should choose an AC with a good balance between porosity development and packing density or, in other words, an AC with a high MPV $V_{DR}$ $N_2$ and packing density. Figure 21.3 contains also the values obtained with an ACM, which presents high porosity and also high density. The maximum value obtained for this material is 29.7 g/L.

The results presented so far on both gravimetric and volumetric bases correspond to the adsorption excess of hydrogen on different solids at different temperatures and pressures. However, a more important parameter from an application point of view, which is rarely found in the literature, is the total storage capacity. The storage capacity in a specified volume filled with a physisorption-based hydrogen carrier is the sum of the capacity due to adsorption on the solid surface and the volumetric capacity due to compression in the void space (Zhou et al. 2004). The void space per unit volume (Vs) was assessed using the packing density ($\rho_p$) and the skeleton density ($\rho_s$) of adsorbent (Vs = 1 − $\rho_p/\rho_s$). The packing density of the materials was measured as described in Section 21.3.2. More details about this type of measurements can be found elsewhere (Lozano-Castelló et al. 2002c, Jordá-Beneyto et al. 2007, 2008). The skeleton density of the ACs was determined by the helium expansion method. Figure 21.4 includes the total hydrogen storage capacity of a superactivated carbon at 298 and 77 K on the basis of a 1 L container. This figure also includes the amount of hydrogen stored just by compression and the adsorption excess data, based on the weight of carbon in the 1 L container (the packing density of AC is 0.5 g/cm³). It can be seen that storage of hydrogen at room temperature and 77 K achieves considerable enhancement due to both adsorption and compression. The total storage capacity of a 1 L container filled with carbon powder is 16.7 g $H_2$ at 19.5 MPa and 298 K and 38.8 g $H_2$ at 4 MPa and 77 K. In the case of the best sample prepared in our laboratory in terms of volumetric values (the ACM), the total storage capacity of a 1 L container filled with this material is 39.3 g $H_2$ at 4 MPa and 77 K. Interestingly, this value of volumetric

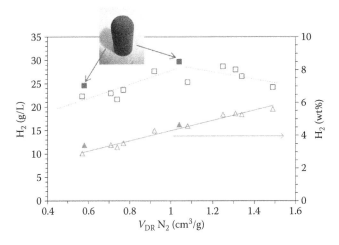

**FIGURE 21.3** Amount of hydrogen adsorbed at 77 K and 4 MPa (on volumetric and gravimetric bases) versus the total MPV. Empty symbols correspond to powder ACs and full symbols to ACMs.

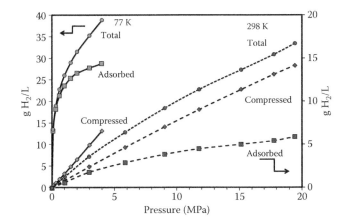

**FIGURE 21.4** Total hydrogen storage capacities of a superactivated carbon (prepared by KOH activation) on the basis of a 1 L container at 298 K (right *y* axis) and 77 K (left *y* axis).

hydrogen storage capacity is almost 1.2 times higher than the European target for onboard hydrogen storage systems (33 g $H_2$/L). Considering the good results obtained with these materials, our effort now focuses on optimizing monolith preparation and properties.

If the weight percentage of total $H_2$ stored on the powder AC at 298 K is assessed from the hydrogen adsorption isotherms measured up to 50 MPa, the values obtained are 3.2 and 6.8 wt% at 20 and 50 MPa, respectively. Very high pressures (50 MPa) are required to approach the DOE target at room temperature with an KOH-activated carbon. On the other hand, at 77 K, this sample gave very interesting values of $H_2$ stored at a quite low pressure (4 MPa), 8.0 wt%.

## 21.4.2 Space Cryocoolers

In some space missions, optical systems require integrated coolers (4.5 K cryocoolers), where any vibration of the optical system cannot be tolerated to guarantee proper mechanical stability. To reach such low temperature, a two-stage vibration-free sorption He/$H_2$ cooler was designed with a suitable AC (Burger et al. 2003). In this section, an example corresponding to the development of such an adsorbent is presented. The detailed results of the study were published elsewhere (Burger et al. 2002, 2007, Lozano-Castelló et al. 2010).

A sorption cooler has two parts: (i) a cold stage and (ii) a sorption compressor. The cold stage consists of a counterflow heat exchanger and a Joule–Thomson expansion valve. This cold stage works as a typical refrigerator. The high-pressure gas needed comes from the sorption compressor. A sorption compressor can be described as a thermodynamic engine that transfers thermal energy to the compressed gas in a system without moving parts. Its operation is based on the principle that large amounts of gas can be adsorbed on certain solids such as highly porous ACs. The amount of gas adsorbed is a function of temperature and pressure. If a pressure container is filled with an adsorbent and gas is adsorbed at low temperature and pressure, then high pressure can be produced inside the closed vessel by an increase in the adsorbent temperature. Subsequently, a controlled gas flow out of the vessel can be maintained at high pressure by a further increase in temperature until most of the gas is desorbed.

ACs are obviously very interesting candidates for this application. They have to satisfy three essential requirements: (i) a large adsorption capacity per mass of adsorbent; (ii) a minimum void volume; and (iii) very good mechanical properties.

To optimize AC properties for this application, studies with samples prepared from different raw materials (anthracite and bituminous coals) and using different activation processes (KOH, NaOH, and

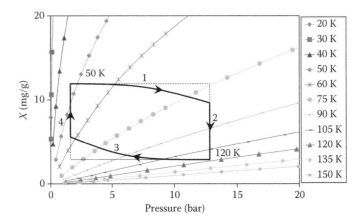

**FIGURE 21.5**   Helium adsorption isotherms at different temperatures corresponding to an AC prepared by KOH activation of anthracite. Each of the numbers 1 to 4 corresponds to a step of a complete cycle of a cell.

$CO_2$ as activating agents) were carried out. To predict their adsorption performance, helium adsorption isotherms at different temperatures (from 20 to 150 K) and high pressures (up to 3.5 MPa) were measured as explained elsewhere (Lozano-Castelló et al. 2010). As an example, Figure 21.5 shows these isotherms corresponding to an AC prepared by anthracite activation. According to the conditions needed in the compressor stage, we are interested in obtaining a material with a maximum helium adsorption capacity at 2 bar (0.2 MPa) and 50 K (adsorption stage) and a minimum adsorption capacity at 13 bar (1.3 MPa) and 120 K (desorption stage). Moreover, the density of the material should be maximized. Then, the selected AC was combined with a binder to prepare an ACM with suitable mechanical properties for machining. The best adsorption capacity was obtained with an ACM with relatively high development of porosity and, most importantly, with high narrow MPV assessed by $CO_2$ adsorption at 273 K. This material also presents quite high density (0.7 g/cm³). A very high activation degree of the material is not desired because the density of the material becomes low. In addition to the porous texture characterization and helium adsorption isotherms for the ACs, the ACM was submitted to a complementary mechanical characterization that included the following tests: maximum compression strength and strain, vibration, pressure drop, and thermal expansion (Lozano-Castelló et al. 2010).

Interestingly, the procedure and binder used to prepare the ACM was very successful because (i) it does not considerably modify the adsorption/desorption properties of the AC used; (ii) it causes a significant increase in the density of the material (0.70 g/cm³ for the ACM versus 0.46 g/cm³ for the AC); (iii) it allows the formation of a dense monolith that is easy to machine, maintains good adsorption characteristics, and has good mechanical properties; and (iv) it presents a low pressure drop and a low thermal expansion (Lozano-Castelló et al. 2010). As a result, the ACM prepared achieved the target performance properties for being used in a sorption compressor.

A sorption cooler was built in the project using the ACM prepared from our study. Such cooler delivered 4.5 mW at 4.5 K with a long-term temperature stability of 1 mK and an input power of 2 W. It operated continuously for 2.5 months and did not show any sign of performance degradation, performing well within all the specifications imposed by the project. A more extensive discussion of tests on the developed breadboard cooler is given elsewhere (Burger et al. 2007).

### 21.4.3  Effect of Surface Chemistry on Gas Adsorption

Surface chemistry of carbons is also a key factor in the adsorption of various compounds, increasing or decreasing their adsorption capacity depending on the nature of the interactions (i.e., favorable or unfavorable) established between the surface groups and the adsorbate. In liquid phase applications (e.g., water as solvent), the importance of surface groups is well known, and in many cases the surface chemistry is the determinant factor of the carbon behavior (Radovic et al. 2001, Rivera-Utrilla et al. 2011). In the case of gases and vapor adsorption, the surface groups are also involved in the process, especially in cases of polar adsorbates, although the effect is usually small compared with the effect of the porosity, as shown in the previous sections.

In the adsorption of volatile organic compounds (VOCs) such as benzene and toluene, the effect of the presence of surface oxygen groups has been studied (García et al. 2004, Lillo-Rodenas et al. 2005). It has been demonstrated that the adsorption of benzene and toluene depends on the narrow microporosity. Thus, the adsorption capacity of both these VOCs increases linearly with the narrow MPV assessed by $CO_2$ adsorption at 273 K (Lillo-Rodenas et al. 2005). However, it has been found that the presence of O-containing surface groups results in a decrease in the adsorption capacity of these VOCs. Thus, the elimination of the oxygen groups (by heating the ACs at high temperature in inert atmosphere) produced an increase in the uptake of these VOCs (Lillo-Rodenas et al. 2005). A similar conclusion has been achieved for the adsorption of phenanthrene vapor on oxidized ACs (García et al. 2004). Thus, the authors observed that the higher the total number of oxygen surface groups, the lower the phenanthrene adsorption capacity. Therefore, it can be concluded that the presence of surface oxygen groups has a negative effect on the adsorption of aromatic VOCs (García et al. 2004, Lillo-Rodenas et al. 2005).

An example of the positive effect of surface groups has been observed in the $SO_2$ adsorption and oxidation on carbon materials (Davini 1990, Raymundo-Pinero et al. 2000, 2003, Mangun et al. 2001). The behavior of carbons in the adsorption and oxidation of this acid adsorbate ($SO_2$) depends to a large extent on the surface chemistry of the material, that is, the nature of the oxygen and mainly of the nitrogen surface groups. Thus, although the adsorption and oxidation of this gas to $SO_3$ or $H_2SO_4$ (in the presence of $O_2$ or $H_2O$, respectively) depend on the pore volume and the PSD, because the porosity strongly influences the catalytic activity of the carbon material, the presence of N-containing groups at the surface increases the catalytic activity. Among the different N functional groups, it has been demonstrated that pyridinic nitrogen is the most active for this reaction (Raymundo-Pinero et al. 2003).

In a similar way, the presence of basic groups (i.e., N-containing groups) has a positive effect on the $CO_2$ adsorption (Zhao et al. 2010, 2012, Plaza et al. 2011, Sevilla et al. 2011b, Xia et al. 2011b, Houshmand et al. 2012, Xing et al. 2012, Zhou et al. 2012, Zhu et al. 2012, Liu et al. 2014, Sánchez-Sánchez et al. 2014c). Thus, a lot of works about $CO_2$ capture and storage technologies can be found in the literature (Metz 2005, D'Alessandro et al. 2010, Olajire 2010), many of them focused on the development of suitable porous carbons for the $CO_2$ capture by adsorption. In this application, it has been shown, again, the importance of the narrow micropores (Presser et al. 2011, Marco-Lozar et al. 2012a,b, Sevilla et al. 2012, 2013, Liu et al. 2014, Marco-Lozar et al. 2014). This porous textural parameter largely determines the $CO_2$ adsorption capacity of the adsorbents. In addition to the porosity, the surface chemistry also affects the $CO_2$ adsorption (Sánchez-Sánchez et al. 2014c). In Figure 21.6, the effect of the concentrations of different heteroatoms (N, O, and P) on the adsorption of $CO_2$ at 1 bar and different temperatures is shown. In that work, the carbon adsorbents were prepared by nanocasting using polyamides as precursor and the SBA-15 mesoporous silica as template. The composites, polyamide/SBA-15, were carbonized in the presence of different amounts of phosphoric acid (Sánchez-Sánchez et al. 2014c).

As shown in Figure 21.6, $CO_2$ adsorption at 273 K depends on the narrow microporosity of the carbon materials, in agreement with previous studies. However, at 298 and 323 K the amount of $CO_2$ adsorbed is also a function of the amount and kind of existing heteroatoms. The amount of $CO_2$ adsorbed increases with the amount of N-containing functional groups (mainly pyrrolic and pyridinic ones), but decreases when the amounts of oxygen and phosphorus increase (Sánchez-Sánchez et al. 2014c). Therefore, the first favors the $CO_2$ adsorption, but the latter two are detrimental.

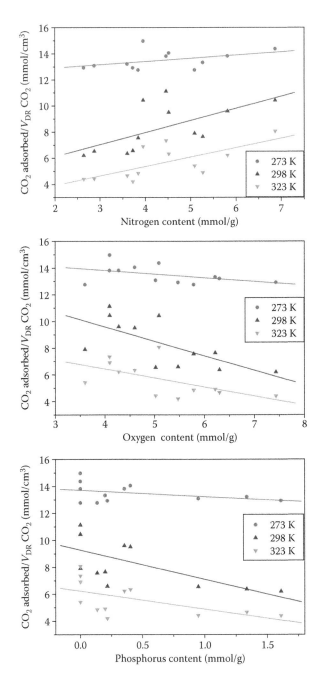

**FIGURE 21.6**  Influence of N, O, and P concentrations on the normalized $CO_2$ adsorption capacity at different temperatures and up to 1 bar.

# Acknowledgments

The authors would like to thank the MINECO (CTQ2012/31762) and GV and FEDER (PROMETEOII/2014/010).

# References

Ahmadpour, A. & Do, D. D., "The preparation of active carbons from coal by chemical and physical activation," *Carbon* 34 (1996): 471–479.

Alam, N. & Mokaya, R., "Evolution of optimal porosity for improved hydrogen storage in templated zeolite-like carbons," *Energy Environ. Sci.* 3 (2010): 1773–1781.

Alcañiz-Monge, J., Trautwein, G., Perez-Cadenas, M. et al., "Effects of compression on the textural properties of porous solids," *Microporous Mesoporous Mater.* 126 (2009): 291–301.

Almasoudi, A. & Mokaya, R., "Preparation and hydrogen storage capacity of templated and activated carbons nanocast from commercially available zeolitic imidazolate framework," *J. Mater. Chem.* 22 (2012): 146–152.

Andersen, J. N., "Silicon tetrachloride manufacture," U.S. Patent 2739041 (1956).

Baker, F. S., "Production of highly microporous activated carbon products," U.S. Patent 5,416,056 (1995).

Bansal, R. C., Donnet, J. & Stoeckli, F., eds., *Active Carbon*, Vol. 10 (Dekker, New York, 1988), Reprint.

Barrett, S., "Patent analysis identifies trends in fuel cell R&D," *Fuel Cells Bull.* 2005 (2005): 12–19.

Bedia, J., Maria Rosas, J., Marquez, J. et al., "Preparation and characterization of carbon based acid catalysts for the dehydration of 2-propanol," *Carbon* 47 (2009): 286–294.

Bedia, J., Rosas, J. M., Rodriguez-Mirasol, J. et al., "Pd supported on mesoporous activated carbons with high oxidation resistance as catalysts for toluene oxidation," *Applied Catal. B Environ.* 94 (2010): 8–18.

Benaddi, H., Legras, D., Rouzaud, J. N. et al., "Influence of the atmosphere in the chemical activation of wood by phosphoric acid," *Carbon* 36 (1998): 306–309.

Bleda-Martinez, M. J., Macia-Agullo, J. A., Lozano-Castelló, D. et al., "Role of surface chemistry on electric double layer capacitance of carbon materials," *Carbon* 43 (2005): 2677–2684.

Boehm, H. P. & Warnecke, H. H., "Structural parameters and molecular-sieve properties of carbons prepared from metal carbides," *Carbon* 13 (1975): 548.

Bottani, E. J. & Tascón, J. M. D., eds., *Adsorption by Carbons* (Elsevier, Amsterdam, Netherlands, 2008).

Burchell, T. D., *Carbon Materials for Advanced Technologies*, Vol. 1 (Pergamon, Amsterdam, Netherlands, 1999), Reprint.

Burger, J. F., ter Brake, H. J. M., Rogalla, H. et al., "Vibration-free 5 K sorption cooler for ESA's Darwin mission," *Cryogenics* 42 (2002): 97–108.

Burger, J. F., Holland, H., ter Brake, M. et al., "Vibration-free 5 K sorption cooler for ESA's Darwin mission," (2003). in Fridlund, M., Henning, T., eds., *Proceedings of the Conference on Towards Other Earths: DARWIN/TPF and the Search for Extrasolar Terrestrial Planets* (ESA Publications Division, Noordwijk, Netherlands, 2003), pp. 379–384.

Burger, J. F., ter Brake, H. J. M., Holland, H. J. et al., "Long-life vibration-free 4.5 K sorption cooler for space applications," *Rev. Sci. Instrum.* 78 (2007): 065102–065110.

Carvalho, A. P., Gomes, M., Mestre, A. S. et al., "Activated carbons from cork waste by chemical activation with $K_2CO_3$: Application to adsorption of natural gas components," *Carbon* 42 (2004): 672–674.

Castro-Muñiz, A., Suárez-García, F., Martínez-Alonso, A. et al., "Energy storage on ultrahigh surface area activated carbon fibers derived from PMIA," *ChemSusChem* 6 (2013): 1406–1413.

Caturla, F., Molina-Sabio, M. & Rodríguez-Reinoso, F., "Preparation of activated carbon by chemical activation with $ZnCl_2$," *Carbon* 29 (1991): 999–1007.

Cazorla-Amorós, D., Alcañiz-Monge, J., Casa-Lillo, M. A. et al., "$CO_2$ as an adsorptive to characterize carbon molecular sieves and activated carbons," *Langmuir* 14 (1998): 4589–4596.

Cook, T. L., Komodromos, C., Quinn, D. F. et al., "Adsorbent storage for natural gas vehicles," in Burchell, T. D., ed., *Carbon Materials for Advanced Technologies* (Pergamon, Amsterdam, Netherlands, etc., 1999), pp. 269–302.

Corma, A., "From microporous to mesoporous molecular sieve materials and their use in catalysis," *Chem. Rev.* 97 (1997): 2373–2420.

Chen, X. S., McEnaney, B., Mays, T. J. et al., "Theoretical and experimental studies of methane adsorption on microporous carbons," *Carbon* 35 (1997): 1251–1258.

Cheng, P.-Z. & Teng, H., "Electrochemical responses from surface oxides present on $HNO_3$-treated carbons," *Carbon* 41 (2003): 2057–2063.

Choi, M. & Ryoo, R., "Mesoporous carbons with KOH activated framework and their hydrogen adsorption," *J. Mater. Chem.* 17 (2007): 4204–4209.

D'Alessandro, D. M., Smit, B. & Long, J. R., "Carbon dioxide capture: Prospects for new materials," *Angew. Chem. Int. Ed.* 49 (2010): 6058–6082.

Dash, R. K., Nikitin, A. & Gogotsi, Y., "Microporous carbon derived from boron carbide," *Microporous Mesoporous Mater.* 72 (2004): 203–208.

Dash, R. K., Yushin, G. & Gogotsi, Y., "Synthesis, structure and porosity analysis of microporous mesoporous carbon derived from zirconium carbide," *Microporous Mesoporous Mater.* 86 (2005): 50–57.

Dash, R., Chmiola, J., Yushin, G. et al., "Titanium carbide derived nanoporous carbon for energy-related applications," *Carbon* 44 (2006): 2489–2497.

Davini, P., "Adsorption and desorption of $SO_2$ on active carbon: The effect of surface basic groups," *Carbon* 28 (1990): 565–571.

Derbyshire, F., Jagtoyen, M., Andrews, R. et al., "Carbon materials in environmental applications," *Chem. Phys. Carbon* 27 (2001): 1–66.

Díaz-Díez, M. A., Gómez-Serrano, V., Fernández González, C. et al., "Porous texture of activated carbons prepared by phosphoric acid activation of woods," *Appl. Surf. Sci.* 238 (2004): 309–313.

Dubinin, M. M., "Physical adsorption of gases and vapors in micropores," in Cadenhead, D. A., Danielli, J. F. & Rosenberg, M. D., ed., *Progress in Surface and Membrane Science* (Academic Press, New York, 1975), pp. 1–70.

Enterría, M., Suárez-García, F., Martínez-Alonso, A. et al., "Avoiding structure degradation during activation of ordered mesoporous carbons," *Carbon* 50 (2012a): 3826–3835.

Enterría, M., Suárez-García, F., Martínez-Alonso, A. et al., "Synthesis of ordered micro–mesoporous carbons by activation of SBA-15 carbon replicas," *Microporous Mesoporous Mater.* 151 (2012b): 390–396.

Evans, M. J. B., Halliop, E. & MacDonald, J. A. F., "The production of chemically-activated carbon," *Carbon* 37 (1999): 269–274.

Falco, C., Marco-Lozar, J. P., Salinas-Torres, D. et al., "Tailoring the porosity of chemically activated hydrothermal carbons: Influence of the precursor and hydrothermal carbonization temperature," *Carbon* 62 (2013): 346–355.

Figueiredo, J. L., Pereira, M. F. R., Freitas, M. M. A. et al., "Modification of the surface chemistry of activated carbons," *Carbon* 37 (1999): 1379–1389.

Froudakis, G. E., "Hydrogen storage in nanotubes & nanostructures," *Mater. Today* 14 (2011): 324–328.

García, T., Murillo, R., Cazorla-Amoros, D. et al., "Role of the activated carbon surface chemistry in the adsorption of phenanthrene," *Carbon* 42 (2004): 1683–1689.

Getman, R. B., Bae, Y.-S., Wilmer, C. E. et al., "Review and analysis of molecular simulations of methane, hydrogen, and acetylene storage in metal–organic frameworks," *Chemical Reviews* 112 (2012): 703–723.

Gierszal, K. P., Jaroniec, M., Kim, T.-W. et al., "High temperature treatment of ordered mesoporous carbons prepared by using various carbon precursors and ordered mesoporous silica templates," *N. J. Chem.* 32 (2008): 981–993.

Gogotsi, Y., Kamyshenko, V., Shevchenko, V. et al., "Nanostructured carbon coatings on silicon carbide: Experimental and theoretical study," in Baraton, M. I. & Uvarova, I., ed., *NATO ASI on Functional Gradient Materials and Surface Layers Prepared by Fine Particles Technology* (Kluwer Academic, Dordrecht, Netherlands, 2001), pp. 239–255.

Gogotsi, Y., Nikitin, A., Ye, H. H. et al., "Nanoporous carbide-derived carbon with tunable pore size," *Nat. Mat.* 2 (2003): 591–594.

Gogotsi, Y., Dash, R. K., Yushin, G. et al., "Tailoring of nanoscale porosity in carbide-derived carbons for hydrogen storage," *J. Am. Chem. Soc.* 127 (2005): 16006–16007.

Gogotsi, Y., Portet, C., Osswald, S. et al., "Importance of pore size in high-pressure hydrogen storage by porous carbons," *Int. J. Hydrog. Energy* 34 (2009): 6314–6319.

Gonzalez-Serrano, E., Cordero, T., Rodriguez-Mirasol, J. et al., "Removal of water pollutants with activated carbons prepared from $H_3PO_4$ activation of lignin from kraft black liquors," *Water Res.* 38 (2004): 3043–3050.

Gorgulho, H. F., Mesquita, J. P., Gonçalves, F. et al., "Characterization of the surface chemistry of carbon materials by potentiometric titrations and temperature-programmed desorption," *Carbon* 46 (2008): 1544–1555.

Górka, J., Fenning, C. & Jaroniec, M., "Influence of temperature, carbon precursor/copolymer ratio and acid concentration on adsorption and structural properties of mesoporous carbons prepared by soft-templating," *Colloids Surf. A* 352 (2009): 113–117.

Hayashi, J. I., Uchibayashi, M., Horikawa, T. et al., "Synthesizing activated carbons from resins by chemical activation with $K_2CO_3$," *Carbon* 40 (2002): 2747–2752.

Houshmand, A., Daud, W. M. A. W., Lee, M.-G. et al., "Carbon dioxide capture with amine-grafted activated carbon," *Water Air Soil Pollut.* 223 (2012): 827–835.

Hutchins, O., "Method for the production of silicon tetrachloride," U.S. Patent 1271713 (1918).

Ibarra, J. V., Moliner, R. & Palacios, J. M., "Catalytic effects of zinc chloride in the pyrolysis of Spanish high sulphur coals," *Fuel* 70 (1991): 727–732.

Illan-Gomez, M. J., Garcia-Garcia, A., Salinas-Martinez de Lecea, C. et al., "Activated carbons from Spanish coals: 2. Chemical activation," *Energy Fuels* 10 (1996): 1108–1114.

Jankowska, H., Swiatkowski, A., Choma, J. et al., *Active Carbon* (Ellis Horwood, London, UK, 1991).

Joo, J. B., Kim, Y. J., Kim, W. et al., "Simple synthesis of graphitic porous carbon by hydrothermal method for use as a catalyst support in methanol electro-oxidation," *Catal. Commun.* 10 (2008): 267–271.

Jordá-Beneyto, M., Suárez-García, F., Lozano-Castelló, D. et al., "Hydrogen storage on chemically activated carbons and carbon nanomaterials at high pressures," *Carbon* 45 (2007): 293–303.

Jordá-Beneyto, M., Lozano-Castelló, D., Suárez-García, F. et al., "Advanced activated carbon monoliths and activated carbons for hydrogen storage," *Microporous Mesoporous Mater.* 112 (2008): 235–242.

Kim, P., Joo, J., Kim, W. et al., "Graphitic spherical carbon as a support for a PtRu-alloy catalyst in the methanol electro-oxidation," *Catal. Lett.* 112 (2006): 213–218.

Kim, H. S., Singer, J. P., Gogotsi, Y. et al., "Molybdenum carbide-derived carbon for hydrogen storage," *Microporous Mesoporous Mater.* 120 (2009): 267–271.

Kirillova, G. F., Meerson, G. A. & Zelikman, A. N., "Kinetics of chlorination of titanium and niobium carbides," *Izvešt. Vuzov Tsvetnaya Metall.* 3 (1960): 90–96.

Kruk, M., Dufour, B., Celer, E. B. et al., "Synthesis of mesoporous carbons using ordered and disordered mesoporous silica templates and polyacrylonitrile as carbon precursor," *J. Phys. Chem. B* 109 (2005): 9216–9225.

Kyotani, T., Nagai, T., Inoue, S. et al., "Formation of new type of porous carbon by carbonization in zeolite nanochannels," *Chem. Mater.* 9 (1997): 609–615.

Lee, J., Kim, J. & Hyeon, T., "Recent progress in the synthesis of porous carbon materials," *Adv. Mater.* 18 (2006): 2073–2094.

Liang, C. & Dai, S., "Synthesis of mesoporous carbon materials via enhanced hydrogen-bonding interaction," *J. Am. Chem. Soc.* 128 (2006): 5316–5317.

Liang, C., Hong, K., Guiochon, G. A. et al., "Synthesis of a large-scale highly ordered porous carbon film by self-assembly of block copolymers," *Angew. Chem. Int. Ed.* 43 (2004): 5785–5789.

Lillo-Ródenas, M. A., Lozano-Castelló, D., Cazorla-Amorós, D. et al., "Preparation of activated carbons from Spanish anthracite: II. Activation by NaOH," *Carbon* 39 (2001): 751–759.

Lillo-Rodenas, M. A., Cazorla-Amoros, D. & Linares-Solano, A., "Behaviour of activated carbons with different pore size distributions and surface oxygen groups for benzene and toluene adsorption at low concentrations," *Carbon* 43 (2005): 1758–1767.

Linares-Solano, A., Lozano-Castelló, D., Lillo-Ródenas, M. A. et al., "Carbon activation by alkaline hydroxides: Preparation and reactions, porosity and performance," in Radovic, L. R., ed., *Chemistry and Physics of Carbon* (CRC Press, Boca Raton, 2007), pp. 1–62.

Liu, Z., Du, Z., Song, H. et al., "The fabrication of porous N-doped carbon from widely available urea formaldehyde resin for carbon dioxide adsorption," *J. Colloid Interface Sci.* 416 (2014): 124–132.

Lowell, S., Shields, J. E., Thomas, M. A. et al., *Characterization of Porous Solids and Powders: Surface Area, Pore Size and Density* (Kluwer Academic Publishers, Dordrecht, Netherlands, 2004).

Lozano-Castelló, D., Lillo-Ródenas, M. A., Cazorla-Amorós, D. et al., "Preparation of activated carbons from Spanish anthracite: I. Activation by KOH," *Carbon* 39 (2001): 741–749.

Lozano-Castelló, D., Alcañiz-Monge, J., Casa-Lillo, M. A. et al., "Advances in the study of methane storage in porous carbonaceous materials," *Fuel* 81 (2002a): 1777–1803.

Lozano-Castelló, D., Cazorla-Amorós, D., Linares-Solano, A. et al., "Activated carbon monoliths for methane storage: Influence of binder," *Carbon* 40 (2002b): 2817–2825.

Lozano-Castelló, D., Cazorla-Amorós, D., Linares-Solano, A. et al., "Influence of pore size distribution on methane storage at relatively low pressure: Preparation of activated carbon with optimum pore size," *Carbon* 40 (2002c): 989–1002.

Lozano-Castelló, D., Cazorla-Amorós, D. & Linares-Solano, A., "Usefulness of $CO_2$ adsorption at 273 K for the characterization of porous carbons," *Carbon* 42 (2004): 1233–1242.

Lozano-Castelló, D., Jordá-Beneyto, M., Cazorla-Amorós, D. et al., "Characteristics of an activated carbon monolith for a helium adsorption compressor," *Carbon* 48 (2010): 123–131.

Lozano-Castelló, D., Suárez-García, F., Cazorla-Amorós, D. et al., "Porous texture of carbons," in *Carbons for Electrochemical Energy Storage and Conversion Systems* (CRC Press, Boca Raton, 2009), pp. 115–162.

Lu, A., Kiefer, A., Schmidt, W. et al., "Synthesis of polyacrylonitrile-based ordered mesoporous carbon with tunable pore structures," *Chem. Mater.* 16 (2004): 100–103.

Lu, A. H. & Schüth, F., "Nanocasting: A versatile strategy for creating nanostructured porous materials," *Adv. Mater.* 18 (2006): 1793–1805.

Ma, Z., Kyotani, T. & Tomita, A., "Preparation of a high surface area microporous carbon having the structural regularity of Y zeolite," *Chem. Commun.* (2000): 2365–2366.

Ma, Z., Kyotani, T., Liu, Z. et al., "Very high surface area microporous carbon with a three-dimensional nano-array structure: Synthesis and its molecular structure," *Chem. Mater.* 13 (2001): 4413–4415.

Ma, Z. X., Kyotani, T. & Tomita, A., "Synthesis methods for preparing microporous carbons with a structural regularity of zeolite Y," *Carbon* 40 (2002): 2367–2374.

Ma, T.-Y., Liu, L. & Yuan, Z.-Y., "Direct synthesis of ordered mesoporous carbons," *Chem. Soc. Rev.* 42 (2013): 3977–4003.

Mangun, C. L., DeBarr, J. A. & Economy, J., "Adsorption of sulfur dioxide on ammonia-treated activated carbon fibers," *Carbon* 39 (2001): 1689–1696.

Marco-Lozar, J. P., Juan-Juan, J., Suárez-García, F. et al., "MOF-5 and activated carbons as adsorbents for gas storage," *Int. J. Hydrog. Energy* 37 (2012a): 2370–2381.

Marco-Lozar, J. P., Kunowsky, M., Suárez-García, F. et al., "Activated carbon monoliths for gas storage at room temperature," *Energy Environ. Sci.* 5 (2012b): 9833–9842.

Marco-Lozar, J. P., Kunowsky, M., Suárez-García, F. et al., "Sorbent design for $CO_2$ capture under different flue gas conditions," *Carbon* 72 (2014): 125–134.

Marsh, H., Heintz, E. A. & Rodríguez Reinoso, F., *Introduction to Carbon Technologies* (Universidad de Alicante, Alicante, Spain, 1997).

Matsuoka, K., Yamagishi, Y., Yamazaki, T. et al., "Extremely high microporosity and sharp pore size distribution of a large surface area carbon prepared in the nanochannels of zeolite Y," *Carbon* 43 (2005): 876–879.

Meng, X., Nawaz, F. & Xiao, F.-S., "Templating route for synthesizing mesoporous zeolites with improved catalytic properties," *Nano Today* 4 (2009): 292–301.

Menon, V. C. & Komarneni, S., "Porous adsorbents for vehicular natural gas storage: A review," *J. Porous Mater.* 5 (1998): 43–58.

Metz, B., *IPCC Special Report on Carbon Dioxide Capture and Storage* (Cambridge University Press for the Intergovernmental Panel on Climate Change, Cambridge, 2005).

Molina-Sabio, M., Rodríguez-Reinoso, F., Caturla, F. et al., "Porosity in granular carbons activated with phosphoric acid," *Carbon* 33 (1995): 1105–1113.

Muñoz-Guillena, M. J., Illán-Gomez, M. J., Martín-Martinez, J. M. et al., "Activated carbons from Spanish coals: 1. Two-stage $CO_2$ activation," *Energy Fuels* 6 (1992): 9–15.

Nishihara, H. & Kyotani, T., "Templated nanocarbons for energy storage," *Adv. Mater.* 24 (2012): 4473–4498.

Nishihara, H., Hou, P. X., Li, L. X. et al., "High-pressure hydrogen storage in zeolite-templated carbon," *J. Phys. Chem. C* 113 (2009): 3189–3196.

Office of Energy Efficiency and Renewable Energy, "DOE targets" (2010a), http://www1.eere.energy.gov /hydrogenandfuelcells/pdfs/freedomcar_targets_explanations.pdf.

Office of Energy Efficiency and Renewable Energy, "DOE targets light-duty vehicles" (2010b), http:// www1.eere.energy.gov/hydrogenandfuelcells/storage/pdfs/targets_onboard_hydro_storage _explanation.pdf.

Olajire, A. A., "$CO_2$ capture and separation technologies for end-of-pipe applications—A review," *Energy* 35 (2010): 2610–2628.

Orekhov, V. P., Seryakov, G. V., Zelikman, A. N. et al., "Kinetics of chlorination of zirconium carbide briquets," *Ž. Prikl. Khim.* 42 (1969): 251–260.

Orinakova, R. & Orinak, A., "Recent applications of carbon nanotubes in hydrogen production and storage," *Fuel* 90 (2011): 3123–3140.

Osswald, S., Portet, C., Gogotsi, Y. et al., "Porosity control in nanoporous carbide-derived carbon by oxidation in air and carbon dioxide," *J. Solid State Chem.* 182 (2009): 1733–1741.

Otowa, T., Nojima, Y. & Miyazaki, T., "Development of KOH activated high surface area carbon and its application to drinking water purification," *Carbon* 35 (1997): 1315–1319.

Parkyns, N. D. & Quinn, D. F., "Natural gas adsorbed on carbon," in Patrick, J. W., ed., *Porosity in Carbons: Characterization and Applications* (Edward Arnold, London, UK, 1995), pp. 293–325.

Parmentier, J., Saadhallah, S., Reda, M. et al., "New carbons with controlled nanoporosity obtained by nanocasting using a SBA-15 mesoporous silica host matrix and different preparation routes," *J. Phys. Chem. Solids* 65 (2004): 139–146.

Patrick, J. W., ed., *Porosity in Carbons: Characterization and Applications* (Edward Arnold, London, UK, 1995).

Peng, L., Philippaerts, A., Ke, X. et al., "Preparation of sulfonated ordered mesoporous carbon and its use for the esterification of fatty acids," *Catal. Today* 150 (2010): 140–146.

Perrin, A., Celzard, A., Albiniak, A. et al., "NaOH activation of anthracites: Effect of temperature on pore textures and methane storage ability," *Carbon* 42 (2004): 2855–2866.

Plaza, M. G., García, S., Rubiera, F. et al., "Evaluation of ammonia modified and conventionally activated biomass based carbons as $CO_2$ adsorbents in postcombustion conditions," *Sep. Purif. Technol.* 80 (2011): 96.

Pradhan, B. K. & Sandle, N. K., "Effect of different oxidizing agent treatments on the surface properties of activated carbons," *Carbon* 37 (1999): 1323–1332.

Presser, V., Heon, M. & Gogotsi, Y., "Carbide-derived carbons—From porous networks to nanotubes and graphene," *Adv. Funct. Mater.* 21 (2011a): 810–833.

Presser, V., McDonough, J., Yeon, S.-H. et al., "Effect of pore size on carbon dioxide sorption by carbide derived carbon," *Energy Environ. Sci.* 4 (2011b): 3059–3066.

Pumera, M., "Graphene-based nanomaterials for energy storage," *Energy Environ. Sci.* 4 (2011): 668–674.

Puziy, A. M. & Tascón, J. M. D., "Adsorption by phosphorus-containing carbons," in Tascon, J. M. D., ed., *Novel Carbon Adsorbents* (Elsevier, Amsterdam, Netherlands, 2012), pp. 245–267.

Qisheng, H., "Chapter 16—Synthetic chemistry of the inorganic ordered porous materials," in *Modern Inorganic Synthetic Chemistry* (Elsevier, Amsterdam, Netherlands, 2011), pp. 339–373.

Quinn, D. F. & MacDonald, J. A., "Natural gas storage," *Carbon* 30 (1992): 1097–1103.

Radovic, L. R., "Surface chemistry of activated carbon materials: State of the art and implications for adsorption," in Schwarz, J. A. & Contescu, C. I., eds., *Surfaces of Nanoparticles and Porous Materials* (Marcel Dekker, New York, 1999).

Radovic, L. R., Moreno-Castilla, C. & Rivera-Utrilla, J., "Carbon materials as adsorbents in aqueous solutions," in Radovic, L. R., ed., *Chemistry and Physics of Carbon* (Marcel Dekker, New York, 2001), pp. 227–405.

Raymundo-Pinero, E., Cazorla-Amoros, D., de Lecea, C. S. M. et al., "Factors controling the $SO_2$ removal by porous carbons: Relevance of the $SO_2$ oxidation step," *Carbon* 38 (2000): 335–344.

Raymundo-Pinero, E., Cazorla-Amoros, D. & Linares-Solano, A., "The role of different nitrogen functional groups on the removal of $SO_2$ from flue gases by N-doped activated carbon powders and fibres," *Carbon* 41 (2003): 1925–1932.

Rivera-Utrilla, J., Sánchez-Polo, M., Gómez-Serrano, V. et al., "Activated carbon modifications to enhance its water treatment applications: An overview," *J. Hazard. Mater.* 187 (2011): 1–23.

Rodriguez-Reinoso, F. & Linares-Solano, A., "Microporous structure of activated carbons as revealed by adsorption methods," in Thrower, P. A., ed., *Chemistry and Physics of Carbon* (CRC Press, Boca Raton, 1988), pp. 1–146.

Rodríguez-Reinoso, F. & Molina-Sabio, M., "Activated carbons from lignocellulosic materials by chemical and/or physical activation: An overview," *Carbon* 30 (1992): 1111–1118.

Rouquerol, F., Rouquerol, J. & Sing, K. S. W., eds., *Adsorption by Powders & Porous Solids: Principles, Methodology and Applications* (Academic Press, New York, 1999), Reprint.

Ryoo, R., Joo, S. H. & Jun, S., "Synthesis of highly ordered carbon molecular sieves via template-mediated structural transformation," *J. Phys. Chem. B* 103 (1999): 7743–7746.

Ryoo, R., Joo, S. H., Kruk, M. et al., "Ordered mesoporous carbons," *Adv. Mater.* 13 (2001): 677–681.

Sánchez-Sánchez, A., Suárez-García, F., Martínez-Alonso, A. et al., "Aromatic polyamides as new precursors of nitrogen and oxygen-doped ordered mesoporous carbons," *Carbon* 70 (2014a): 119–129.

Sánchez-Sánchez, A., Suárez-García, F., Martínez-Alonso, A. et al., "Evolution of the complex surface chemistry in mesoporous carbons obtained from polyaramide precursors," *Appl. Surf. Sci.* 299 (2014b): 19–28.

Sánchez-Sánchez, Á., Suárez-García, F., Martínez-Alonso, A. et al., "Influence of porous texture and surface chemistry on the $CO_2$ adsorption capacity of porous carbons: Acidic and basic site interactions," *ACS Appl. Mater. Interfaces* 6 (2014c): 21237–21247.

Schlapbach, L. & Zuttel, A., "Hydrogen-storage materials for mobile applications," *Nature* 414 (2001): 353–358.

Schüth, F., "Endo- and exotemplating to create high-surface-area inorganic materials," *Angew. Chem. Int. Ed.* 42 (2003): 3604–3622.

Sevilla, M., Fuertes, A. B. & Mokaya, R., "High density hydrogen storage in superactivated carbons from hydrothermally carbonized renewable organic materials," *Energy Environ. Sci.* 4 (2011a): 1400–1410.

Sevilla, M., Valle-Vigón, P. & Fuertes, A. B., "N-doped polypyrrole-based porous carbons for $CO_2$ capture," *Adv. Funct. Mater.* 21 (2011b): 2781–2787.

Sevilla, M., Falco, C., Titirici, M. M. et al., "High-performance $CO_2$ sorbents from algae," *RSC Adv.* 2 (2012): 12792–12797.

Sevilla, M., Parra, J. B. & Fuertes, A. B, "Assessment of the role of micropore size and N-doping in $CO_2$ capture by porous carbons," *ACS Appl. Mater. Interfaces* 5 (2013): 6360–6368.

Shin, Y., Fryxell, G. E., Um, W. et al., "Sulfur-functionalized mesoporous carbon," *Adv. Funct. Mater.* 17 (2007): 2897–2901.

Sing, K. S. W., Everett, D. H., Haul, R. A. W. et al., "Reporting physisorption data for gas solid systems with special reference to the determination of surface-area and porosity (recommendations 1984)," *Pure Appl. Chem.* 57 (1985): 603–619.

Stein, A., Wang, Z. Y. & Fierke, M. A., "Functionalization of porous carbon materials with designed pore architecture," *Adv. Mater.* 21 (2009): 265–293.

Suárez-García, F., Martínez-Alonso, A. & Tascón, J. M. D., "Pyrolysis of apple pulp: Chemical activation with phosphoric acid," *J. Anal. Appl. Pyrolysis* 63 (2002): 283–301.

Suárez-García, F., Martínez-Alonso, A. & Tascón, J. M. D., "Activated carbon fibers from Nomex by chemical activation with phosphoric acid," *Carbon* 42 (2004): 1419–1426.

Tascón, J. M. D., ed., *Novel Carbon Adsorbents* (Elsevier, Amsterdam, Netherlands, 2012).

Tascón, J. M. D. & Bottani, E. J., *Adsorption by Carbons* (Elsevier, Amsterdam, Netherlands, 2011).

Teng, H., Yeh, T.-S. & Hsu, L.-Y., "Preparation of activated carbon from bituminous coal with phosphoric acid activation," *Carbon* 36 (1998): 1387–1395.

Titirici, M. M. & Antonietti, M., "Chemistry and materials options of sustainable carbon materials made by hydrothermal carbonization," *Chem. Soc. Rev.* 39 (2010): 103–116.

Titirici, M. M., Thomas, A. & Antonietti, M., "Replication and coating of silica templates by hydrothermal carbonization," *Adv. Funct. Mater.* 17 (2007): 1010–1018.

Titirici, M.-M., White, R. J., Falco, C. et al., "Black perspectives for a green future: Hydrothermal carbons for environment protection and energy storage," *Energy Environ. Sci.* 5 (2012): 6796–6822.

Wang, J., Senkovska, I., Oschatz, M. et al., "Highly porous nitrogen-doped polyimine-based carbons with adjustable microstructures for $CO_2$ capture," *J. Mater. Chem. A* 1 (2013): 10951–10961.

Wennerberg, A. N. & O'Grady, T. M., "Active carbon process and composition," U.S. Patent 4,082,694 (1978).

White, R. J., Antonietti, M. & Titirici, M.-M., "Naturally inspired nitrogen doped porous carbon," *J. Mater. Chem.* 19 (2009): 8645–8650.

Xia, K. S., Gao, Q. M., Wu, C. D. et al., "Activation, characterization and hydrogen storage properties of the mesoporous carbon CMK-3," *Carbon* 45 (2007): 1989–1996.

Xia, K. S., Gao, Q. M., Song, S. Q. et al., "$CO_2$ activation of ordered porous carbon CMK-1 for hydrogen storage," *Int. J. Hydrog. Energy* 33 (2008): 116–123.

Xia, Y., Yang, Z. & Mokaya, R., "CVD nanocasting routes to zeolite-templated carbons for hydrogen storage," *Chem. Vap. Depos.* 16 (2010a): 322–328.

Xia, Y. D., Yang, Z. X. & Mokaya, R., "Templated nanoscale porous carbons," *Nanoscale* 2 (2010b): 639–659.

Xia, Y. D., Mokaya, R., Grant, D. M. et al., "A simplified synthesis of N-doped zeolite-templated carbons, the control of the level of zeolite-like ordering and its effect on hydrogen storage properties," *Carbon* 49 (2011a): 844–853.

Xia, Y., Mokaya, R., Walker, G. S. et al., "Superior $CO_2$ adsorption capacity on N-doped, high-surface-area, microporous carbons templated from zeolite," *Adv. Energy Mater.* 1 (2011b): 678–683.

Xie, M., Dong, H., Zhang, D. et al., "Simple synthesis of highly ordered mesoporous carbon by self-assembly of phenol–formaldehyde and block copolymers under designed aqueous basic/acidic conditions," *Carbon* 49 (2011): 2459–2464.

Xing, R., Liu, Y., Wang, Y. et al., "Active solid acid catalysts prepared by sulfonation of carbonization-controlled mesoporous carbon materials," *Microporous Mesoporous Mater.* 105 (2007): 41–48.

Xing, W., Qiao, S. Z., Liu, C. et al., "Superior $CO_2$ uptake of N-doped activated carbon through hydrogen-bonding interaction," *Energy Environ. Sci.* 5 (2012): 7323–7327.

Yasuda, E. I., Inagaki, M., Kaneko, K. et al., eds., *Carbon Alloys: Novel Concepts to Develop Carbon Science and Technology*, Vol. 1 (Elsevier, Amsterdam, Netherlands, 2003).

Yeon, S.-H., Osswald, S., Gogotsi, Y. et al., "Enhanced methane storage of chemically and physically activated carbide-derived carbon," *J. Power Sources* 191 (2009): 560–567.

Yeon, S.-H., Knoke, I., Gogotsi, Y. et al., "Enhanced volumetric hydrogen and methane storage capacity of monolithic carbide-derived carbon," *Microporous Mesoporous Mater.* 131 (2010): 423–428.

Yurum, Y., Taralp, A. & Veziroglu, T. N., "Storage of hydrogen in nanostructured carbon materials," *Int. J. Hydrog. Energy* 34 (2009): 3784–3798.

Yushin, G. N., Hoffman, E. N., Nikitin, A. et al., "Synthesis of nanoporous carbide-derived carbon by chlorination of titanium silicon carbide," *Carbon* 43 (2005): 2075–2082.

Yushin, G., Dash, R., Jagiello, J. et al., "Carbide-derived carbons: Effect of pore size on hydrogen uptake and heat of adsorption," *Adv. Funct. Mater.* 16 (2006): 2288–2293.

Zhao, L., Bacsik, Z., Hedin, N. et al., "Carbon dioxide capture on amine-rich carbonaceous materials derived from glucose," *ChemSusChem* 3 (2010): 840–845.

Zhao, Y., Zhao, L., Yao, K. X. et al., "Novel porous carbon materials with ultrahigh nitrogen contents for selective $CO_2$ capture," *J. Mater. Chem.* 22 (2012): 19726–19731.

Zhou, L., Zhou, Y. & Sun, Y., "Enhanced storage of hydrogen at the temperature of liquid nitrogen," *Int. J. Hydrog. Energy* 29 (2004): 319–322.

Zhou, J., Li, W., Zhang, Z. S. et al., "Carbon dioxide adsorption performance of N-doped zeolite Y templated carbons," *RSC Adv.* 2 (2012): 161–167.

Zhu, X., Hillesheim, P. C., Mahurin, S. M. et al., "Efficient $CO_2$ capture by porous, nitrogen-doped carbonaceous adsorbents derived from task-specific ionic liquids," *ChemSusChem* 5 (2012): 1912–1917.

# 22

# Nanoporous Carbon Fibrous Materials

Branka Kaludjerović

## 22.1 Introduction

A nanoporous carbon fibrous material (NCFM) is a fibrous adsorbent material which has been developed by carbonization and activation of organic fibers. NCFM plays an important role in many areas of modern science and technology, such as purification of liquids and gases, separation of mixtures, and catalyst supports, as well as materials for gas storage, electrodes, electric double-layer capacitors, etc. Its main characteristics are high surface area, PSD, and micropore volume (MPV); its surface chemistry depends mostly on the precursor material and the process parameters, and these properties determine its application.

The International Union for Pure and Applied Chemistry classifies pores based on pore size into three classes: macropores (pores with internal width larger than 50 nm), mesopores (internal width size from 2 to 50 nm), and micropores (internal width size less than 2 nm) (Sing et al. 1985). Micropores can be further divided in two groups: ultramicropores (pore size less than 0.7 nm) and supermicropores (pore size is between 0.7 and 2 nm) (Gregg & Sing 1982). Since the diameters of NCFM are small (on the order of micron), pores are directly open to the outer surface and the porosity development process (activation) is followed by pore deepening rather than the creation of new pores (Tascón 2008). Typically, NCFM is a mostly microporous material with narrow PSD, although it can be also prepared with more pronounced mesoporosity. Micropores and mesopores are especially important in the context of adsorption. The sorption capacity of NCFM is 2–15 times higher, whereas the mass transfer coefficient is 10–100 times higher than that of granular carbon (Ermolenko et al. 1990). NCFM has a favorable combination of filtration and sorption properties, but the main obstacle to the widespread use of this material is its cost (Lysenko 2007).

Besides the process parameters, the raw material as well as its textile form significantly affects the NCFM properties. Commonly used raw materials for NCFM production are cellulose fibers, especially rayon fibers; PAN fibers, Saran polymer fibers, pitch-based fibers, resin-based fibers, polysulfone-based fibers, natural fibers, etc. The textile forms of precursor fibers which are retained in the final NCFM are continuous filament, yarn, staple fibers, hollow fibers, tows, cloths, felts, etc. According to the length, fibers are produced in two basic forms: filament (fiber unlimited length) and staple fiber (fiber of a specific length).

## 22.2 Preparation of NCFMs

Carbon fibers have been known since the end of the 19th century, when Thomas Edison first used them as filaments for incandescent in electric lamps. These fibers were prepared by pyrolysis of cellulose-based fibers (Donnet & Bansal 1990).

Union Carbide in the 1950s developed high-modulus and high-strength rayon based carbon fibers (Thornel) by introduction of stress graphitization in the production process. At the same time, Shindo in Japan and Watt and coworkers in United Kingdom, working independently, produced carbon fibers from PAN fibers (Watt 1972, Donnet & Bansal 1990).

The early nanoporous material was made from viscose rayon fiber in 1962 and for the first time porous carbon fibrous material was proposed as a potential material for air filtration (Mays 1999). Soon thereafter in 1966, viscose and acetate cloths were the first materials used to obtain activated carbon cloth (ACC) (Linares-Solano & Cazorla-Amorós 2008).

In 1965, Otani was the first to produce isotropic carbon fibers from molten PVC pitch. Fibers derived from petroleum and coal pitch are now made commercially by Kureha in Japan. Yamada first succeeded in making carbon filament from a complex phenolic resin in 1967. From the commercial Saran fibers, which are in fact polymer polyvinylidene chloride, Adams and coworkers produced carbon fibers in 1970 (Adams et al. 1970, Jenkins & Kawamura 1976).

The preparation of NCFMs is equal to production of activated carbon fibers (ACFs), which are in fact a combination of the technologies for carbon fibers and activated carbons. In principle, it should be very similar to the production of carbon fibers except that the activation step is introduced in the production process. Also, requirement of high mechanical strength is not emphasized in the case of ACFs. Thus, ACF is one of the most important carbon nanoporous fibrous materials to be considered. The fibrous form is also favored because of ease of handling when it is used in felt or fabric forms.

The raw materials used for the preparation of ACFs are chronologically the same as for carbon fibers. Lee et al. (2014) consider ACFs as the hybrid of carbon fiber and activated carbon (AC).

### 22.2.1 Precursors of NCFMs

NCFMs can be made with a wide range of structures, compositions, and properties, depending on the nature of the precursor and subsequent processing and forming methods.

Therefore, it is common to emphasize carbon fiber raw materials in the following way: PAN-based carbon fibers, cellulosic (viscose/rayon)–based carbon fibers, pitch-based carbon fibers, mesophase pitch–based carbon fibers, phenolic-based carbon fibers, etc. This will be discussed later.

Commercial fibers are usually produced using one of the following techniques: melt spinning and wet or dry spinning. In melt spinning, the precursor is merely melted and extruded through a spinneret to get a fiber form that solidifies by cooling. However, when the precursor degrades at or near its melting temperature, either wet or dry spinning technique must be employed. The precursor has to be dissolved to prepare its concentrated solution that could be extruded. If the precursor is extruded through a spinneret into the coagulation bath, where the solvent dissolves in the coagulation liquid, the process is regarded as wet spinning. In dry spinning, the solution is extruded into a drying chamber and the solvent evaporates (Edie 1998). A spinneret contains one to several hundred holes and enables obtaining fibers with different cross-sectional shapes: round, trilobal, irregular, hollow fibers, etc. (see Figures 22.1, 22.2b, and 22.3b). The resulting precursor fibers have diameters of 10–100 µm (Ermolenko et al. 1990), while the hollow fibers have an outer diameter of up to 600 µm with a 300 µm inner diameter (Saufi & Ismail 2002).

AC hollow fibers (see Figure 22.1) compared to ACFs possess higher geometric area–to–volume ratios, which improves the heat and mass transport. Studies of the hollow fibers' activation are, however, quite

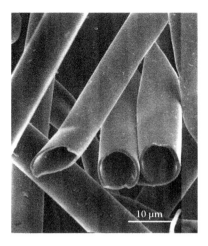

**FIGURE 22.1** SEM micrograph of carbon hollow fibers activated by $H_2O_2$.

**FIGURE 22.2** SEM micrographs of (a) rayon-based ACC and (b) a single fiber.

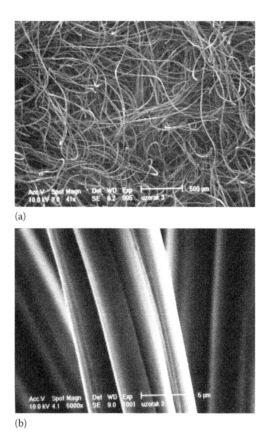

**FIGURE 22.3**　SEM micrographs of (a) rayon-based AC felt and (b) a single fiber.

scarce. Among the precursors applied in the preparation of AC hollow fibers are PAN-based hollow fibers (Sun et al. 2004), cellulose-based hollow fibers, and polysulfone-based hollow fibers (Kaludjerović et al. 2009, Kljajević et al. 2011).

Several thousand precursor fibers are twisted together to form a yarn, which may be used by itself or for preparation of other textile forms (woven, knitted, and needle-punched nonwoven fabrics).

Woven fabrics are produced by using two or more sets of yarn interlaced at right angles to each other, making them more durable. Much variety is produced by weaving, and the basic types of weave patterns are plain, twill, and satin weave. Knitted fabrics (see Figure 22.2) can be made more easily than woven fabrics at comparatively less cost. Knitted fabrics are produced by two general methods: warp knitting and weft knitting. Knitted cloth types are jersey, rib, and double knit. Felt (see Figure 22.3) is also a widely used nonwoven material obtained by the needle punch process.

NCFMs are produced by carbonization and activation of precursor fibers, including its textile forms, although the textile forms of NCFMs could be made from already prepared ACFs.

PAN is currently the most widely used precursor for high-performance engineering in carbon fiber production. Homopolymer polyacrylonitrile, whose molecular formula is $(C_3H_3N)_n$, is produced by addition polymerization of acrylonitrile (Jenkins & Kawamura 1976). Therefore, the starting carbon content in PAN is significant—68%. This initially provides an advantage over other raw materials used for carbon fiber production. Pure PAN is a polar substance with a glass transition temperature of around 120°C, tending to decompose before it melts. Thus, PAN precursor fibers must be produced by either wet- or dry-spinning processes using a highly polar solvent (Edie 1998). Commercial PAN fiber precursor usually contains up to 10% of a copolymer, such as methyl acrylate, itaconic acid, acrylic acid, methacrylic acid, vinyl chloride, vinyl bromide, styrene, and isobutylene (Huang 2009).

After being spun into fibers, the axial orientation within the PAN (fibril orientation) is enhanced by stretching. The production of PAN-based carbon fibers includes two process steps: stabilization and carbonization (Fitzer et al. 1986). The precursor fiber is stabilized under applied tension by heating in an oxidative media (air, oxygen, oxygen-containing mixture, etc.) at temperatures between 200°C and 300°C. Rašković and Marinković (1975) found that rapid cyclization in air atmosphere begins around 240°C, which is followed by partial degradation of the fibers around 320°C. Final degradation of the fibers began around 400°C. This cyclization process is highly exothermic and controlled by oxygen diffusion (Suzuki 1994).

The cyclization process induces formation of the chain ring structures, which increases the thermal stability of the macromolecules. In this way, it is necessary to ensure that both molecular and fibrillar orientation would not be lost during the final heat treatment (Edie 1998).

Huang (2009) reviewed the different ways of improving the stabilization process applied to develop better-quality carbon fibers.

The high degree in order in the structure of PAN-based carbon fibers makes them the carbon fibers most resistant to the activation process. Thus, it is convenient to activate PAN-based fibers during the cyclization process, which, in fact, coincides with the PAN stabilization process. This is called preoxidation process. Subsequent carbonization at relatively low temperature (up to 800°C) followed by physical activation is a common route for ACFs preparation. PAN-based ACFs contain nitrogen in their structure, which provides these materials with specific adsorption properties and makes them suitable for use in catalysis, medicine, and various industries (Ermolenko et al. 1990, Suzuki 1994).

Cellulose is the most common natural polymer which appears as a stiffening component in all plants. In chemical terms, it is a polysaccharide whose chemical formula is $(C_6H_{10}O_5)_n$. Natural cellulose is made of several thousand repeat units of monomer glucose. Because of its large number of hydroxyl groups and high molecular weight, cellulose possesses high rigidity and crystallinity, resulting in a high softening point and insolubility. Also, cellulose decomposes before it begins to soften. To make cellulose appropriate for fiber manufacturing, it is necessary to decrease these intermolecular forces by neutralizing polar hydroxyl groups and/or reducing the molecular weight (Jenkins & Kawamura 1976). Cellulose pulp is changed chemically into another form, which is then changed (regenerated) into cellulose again. This regenerated cellulose consists of several hundred units of glucose. According to that, the obtained regenerated cellulose is a solution, which can be easily converted into fiber form by wet spinning. Textile fibers and filaments composed of regenerated cellulose are defined as rayon. There are three commonly used production methods leading to the formation of distinctly different rayon fibers: viscose rayon, cuprammonium rayon, and saponified cellulose acetate. Rayon is the oldest commercially artificial fiber (Hedge et al. 2004).

Rayon textile forms have been used successfully as a precursor for NCFMs (Babić et al. 1999, Rong et al. 2002, Faur-Brasquet & Le Cloirec 2003, Yuhan et al. 2008). However, NCFMs can be prepared from other cellulose-based precursors such as novel regenerated cellulose such as Tencel (Ramos et al. 2009), lyocell (Ramos et al. 2010), palm waste fiber (Giraldo et al. 2013), plane tree seed (Kaludjerović et al. 2014), and bamboo (Ma et al. 2014). Carbonized fir fibers were successful in grade heavy oil recovery (Toyoda et al. 2008).

Coal and petroleum are the two most important raw materials for natural pitch preparation. Petroleum pitch is preferred for carbon fiber production because coal pitch can contain high solid particle fraction, which causes fiber damage during the extrusion process (Huang 2009).

Otani in 1965 already obtained carbon fiber from molten PVC pitch, which was obtained by heating PVC at 400°C under nitrogen. PVC pitch consists of polynuclear aromatic compounds such as $C_{62}H_{52}$. A preoxidation treatment is needed to prevent not only melting but also crystal growth in the carbonization process (Otani 1965).

Coal tar pitch is a complex mixture made of thousands of polynuclear and aromatics compounds. By controlled heating at temperature between 400°C and 500°C under inert atmosphere, isotropic pitch first transforms into small anisotropic microspheres, called mesophase, which solidifies in anisotropic coke at temperature above 600°C. The formation of mesophase is a spontaneous and homogeneous process within isotropic pitch (Granda et al. 2003).

Isotropic petroleum, and coal tar pitch by low–molecular weight fraction evaporation, can be melt spun into general-purpose carbon fibers. To produce high-performance fibers, it is necessary to apply the hot stretching process. It is more convenient to produce high–performance carbon fibers from an anisotropic pitch—mesophase pitch. The mesophase orients itself along the fiber axis direction during the precursor fiber spinning (Huang 2009).

In 1963, the Carborundum Co. (Economy et al. 1996) showed for the first time that fibers could be prepared from cross-linked phenolic (phenol formaldehyde) resin—novolac type—by melt spinning and then curing with a mixture of hydrogen chloride and formaldehyde. The resulting fibers could be processed into a wide range of textile forms by conventional methods. In 1969, these fibers were commercialized under the trade name Kynol as a potentially low-cost textile for use in highly aggressive environments. Since then, Nippon Kynol, the current manufacturer, has developed a number of niche markets, which today have resulted in a significant industry. These fibers are also known as novoloid fibers since they were obtained from novolac-type resin.

Commercially activated Kynol novoloid fibers and textiles are produced by a one-step process where the carbonization process is followed by the activation process. This process can be carried out using either novoloid tow or finished products such as felts and fabrics. The surface area of this material can approach 3000 m²/g (Kynol 2012).

In Figure 22.4 is the summarized production of NCFMs starting from the precursor material.

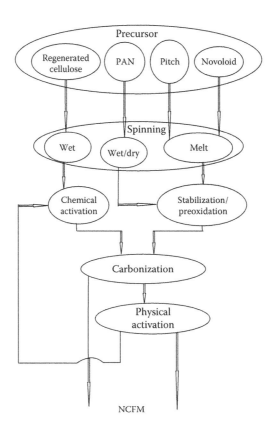

**FIGURE 22.4**  General flow sheet of production of NCFMs starting from the precursor material.

## 22.2.2 Carbonization Process

The carbonization process produces a carbon-rich solid material from liquid or solid carbon containing organic precursor by pyrolysis under an inert atmosphere, usually nitrogen at elevated temperature. For NCFM production this temperature is usually 800°C. However, if there is a need for a material with higher carbon content, the pyrolysis temperature will be higher. Above 1300°C residual solids are pure carbon material. Factors that affect the final product of carbonization are the precursor, the heating rate, the final temperature, and the activation process.

By changing the heating rates and maximum temperature of the carbonization of oxidized textile PAN fiber, Marcuzzo et al. (2013) indicated that it is possible to produce poorly activated carbon by a single step of the production process.

Reactions occurring during transformation of an organic precursor to carbon material are pyrolysis, cyclization, aromatization, polycondensation, and carbonization. However, the term *carbonization process* includes all of these reactions, which are usually consecutive and parallel and are accompanied by an evolution of gaseous products.

Low-volatility compounds such as physically bonded water, aliphatic compounds with low relative mass, and then low–molecular mass aromatics are released at the beginning of the pyrolysis process. The processes of cyclization and aromatization proceed in the residues and are accompanied by release of low–molecular weight hydrocarbons. After this process, polycondensation of aromatic molecules occurs. Mainly heteroatoms such as oxygen and nitrogen in the form of $CO_2$, CO, and $(CN)_2$, together with methane or $CH_4$, are released around 600°C (Inagaki & Feiyu 2006).

With increase in heat treatment temperature, the C/H and C/O ratios progressively increase, but heteroatoms (C, O, H, N, Cl, S, etc.) chemically bonded at the edges of the aromatic macromolecules remain and transform into so-called surface functional groups (Rouquerol et al. 1999). Above 1000°C, mainly $H_2$ is released because of polycondensation of aromatics.

A crucial process for the preparation of nanoporous carbon material is the activation process, which is in mutual dependence with the carbonization process.

## 22.2.3 Activation Process (Physical and Chemical Activation Processes)

There are two types of activation processes: physical activation and chemical activation. Physical activation means that the precursor is first carbonized in an inert atmosphere at 600°C–1000°C and then activated in an oxidizing atmosphere, usually air, $CO_2$, or steam, at a similar or slightly higher temperature. In chemical activation, before the carbonization process, it is necessary to impregnate the precursor with a chemical agent, usually $H_3PO_4$, $Zn_2Cl$, KOH, NaOH, etc. The impregnated precursor is carbonized at temperatures between 400°C and 800°C in an inert atmosphere. After carbonization it is necessary to wash the material to remove the residues of the impregnating agent (Rouquerol et al. 1999, Bandosz et al. 2003, Linares-Solano et al. 2008, Lee et al. 2014). Variations of these sequences of events are usual and necessary to optimize adsorption characteristics. Usually it has been used as a one-step process (Kynol 2012), which combines carbonization and activation processes, such as carbonization followed by physical activation (Shoaib & Al-Swaidan 2015) and chemical activation followed by carbonization (Kaludjerović et al. 2009, 2014, Kljajević et al. 2011) and carbonization of impregnated precursor followed by physical activation (Babić et al. 2002). The possibilities are numerous and can be applied in multistage combinations of various processes (Hu & Vansant 1995, Polovina et al. 1997, 1998, Rouquerol et al. 1999).

Generally, for the adsorption characteristics of an AC obtained by gasification in $CO_2$, steam, etc., of great importance is a careful control of the carbon atom removal, usually termed *burn-off degree* (Linares-Solano et al. 2008).

Activation in steam is often used in industry because of its high gasification kinetics. It also provides chemical reaction control up to very high temperatures and adsorption capacity enlargement because of the pore widening. Adsorption capacity increases with increasing burn-off. However, some applications require only narrow micropores (gas adsorption), while others (liquid adsorption) require wider micropores (Menéndez-Diaz & Martin-Gullón 2006).

The other oxidizing agents are rarely used in industrial purpose, because $CO_2$ may react under diffusion control, which is undesirable, while air and oxygen if independently introduced into the process at higher temperature can damage material (Menéndez-Diaz & Martin-Gullón 2006).

The material activated by different methods as well as by different reagents can have different dominant pores (Ryu et al. 2000).

Common chemicals used today in industry for chemical activation are $H_3PO_4$, KOH, and $ZnCl_2$. Linares-Solano et al. (2008) gave a detailed review of the activation by alkaline hydroxides.

The factors that affect the production of chemically activated ACFs are mass yield of fibrous material after impregnation process (Kaludjerović et al. 2014) and carbonization temperature.

Activation with $H_3PO_4$ is especially proposed for the cellulosic and lignocellulosic-based material, because the acid greatly accelerates the carbonization process (Romero-Anaya et al. 2012).

The NCFMs prepared by chemical activation with KOH or $ZnCl_2$ are essentially nanoporous with high MPV and heterogeneous micropore size distribution (MPSD). However, the main disadvantage of $ZnCl_2$ as a reagent is that the emission of metallic zinc can induce serious environmental contamination problems.

Activation methods, besides developing porous structures, affect the surface chemistry of the adsorbents; i.e., they have a great influence on the formation of surface functional groups.

The above-mentioned activation methods introduce oxygen functional groups on the NCFM surface (see Figure 22.5). Activation with ammonia can induce formation of nitrogen functional groups such as pyridinic or pyrone-type groups (Boudou et al. 2003).

The literature also offers other methods that can affect formation of the ACF porous structure, such as electrochemical oxidation (Pittman, Jr. et al. 1999), ultrasound (Kaludjerović et al. 2014), and laser technique for functional group formation (Kaludjerović et al. 2004).

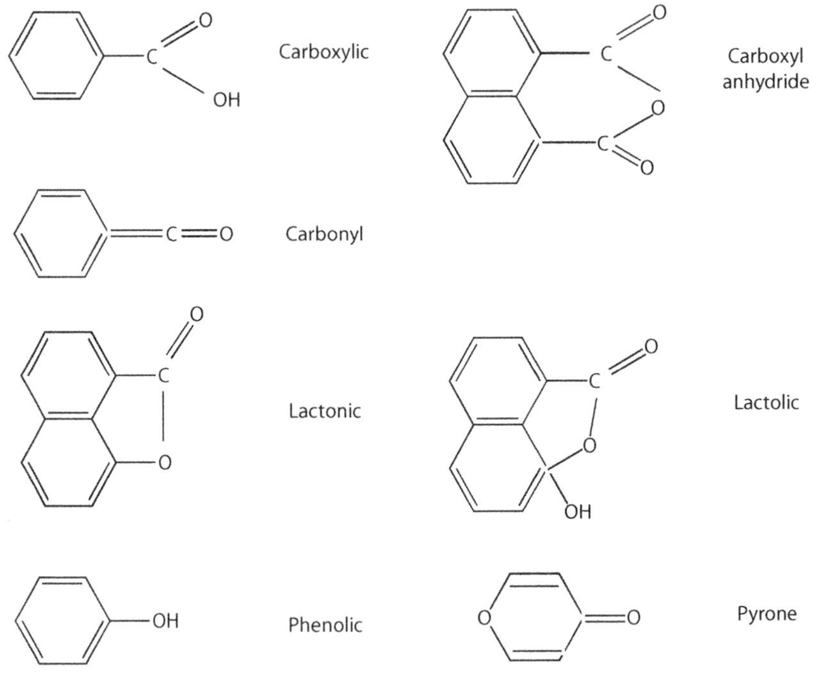

**FIGURE 22.5** Examples of oxygen functional groups on the NCFM surface.

## 22.3 Properties of NCFMs

The advantages of NCFMs over granular or powder ACs are numerous and they are as follows: larger SSA, small and uniform fiber diameter, ease in handling, higher adsorption capacity, and faster adsorption/desorption kinetics owing to small diffusion distance in a fiber. Some commercially available ACFs have a very high SSA, up to 3000 m²/g. Because most nanopores are open to the outer surface of NCFM, they are directly exposed to the adsorbate. Therefore, the adsorption rate, as well as the amount of adsorption of gases into NCFMs, is much higher than those in granular ACs (Inagaki & Tascón 2006).

Other elements attached to the carbon surface in various quantities and composition considerably affect NCFMs' chemical properties. There are two basic types of these components: mineral component or ash and heteroatoms, such as oxygen, hydrogen, nitrogen, sulfur, and halogens. Ash does not chemically react with carbon surface but can decrease porosity by occupying some of the open pores (Mowla et al. 2003).

### 22.3.1 Characterization Methods

To achieve adequate information about structural and surface heterogeneous NCFM structures, the tendency is to use a combination of various techniques. There are a large number of characterization methods and the most commonly used are those which describe their adsorption characteristics. Both the physical and the chemical surface properties of the NCFMs have a significant impact on their adsorption characteristics.

There exist several techniques for the analysis of the structure and porous texture, such as XRD; small-angle X-ray scattering (SAXS); small-angle neutron scattering (SANS); different microscopy techniques, i.e., optical microscopy, SEM, scanning tunneling microscopy (STM), and TEM; physical adsorption of gases; and immersion calorimetry. According to the characterization techniques, various structure parameters have been employed.

The XRD pattern of ACF has features that are common with those of other activated carbon materials; i.e., it has turbostratic structures (Ermolenko et al. 1990, Bandosz et al. 2003). SAXS and SANS techniques can be used to study the pore structure of solids over a range of approximately 0.5–200 nm. These techniques give information on pore sizes as well as surface roughness at the boundaries of pores (Inagaki & Tascón 2006).

Application of optical microscopy and SEM provides insight into the macroscopic texture of the material (see Figure 22.6). Nanopores can be seen only by applying TEM and STM. The PSD, the pore

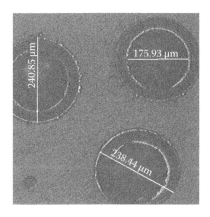

**FIGURE 22.6** Determination of diameters of hollow carbon fibers by light microscope with image analyzer.

volume, and the fractal dimension of the pore wall can be obtained by analysis of ACF TEM image (Inagaki & Feiyu 2006).

Gas adsorption is the most commonly implemented method for characterization of NCFMs because it allows determination of the SSA, the MPV, the pore size, the PSD, and the surface properties of these materials. $N_2$ adsorption at 77 K is the most widely used method for the characterization of the porous materials. However, for better assessment of the ultramicropores, it is recommended to use lower relative pressures for $N_2$ adsorption (i.e., below $10^{-3}$ and up to $10^{-7}$), or to use $CO_2$ as adsorptive at 273 or 298 K.

The BET method is widely used for the evaluation of SSA values of micro- and mesoporous adsorbents. Other methods for calculation of a SSA are based on the comparison of the data from gas adsorption isotherms to the data obtained from standard isotherm, e.g., t-plot and alpha-s plot, which allow one to obtain the external surface area (mesoporous and macroporous). A micropore surface area can also be obtained by the Dubinin–Kaganer method. Over the last two decades tremendous progress has been achieved with the development and application of the DFT of inhomogeneous fluids (e.g., nonlocal density functional theory) or computer simulation methods (e.g., Monte Carlo and molecular dynamics simulations) with regard to the understanding of sorption phenomena in nanoporous carbon materials (Lastoskie et al. 1993).

The Dubinin–Radushkevich (DR) equation is usually used to obtain the MPV and the average pore width of nanoporous carbons. The classical methods which are based on thermodynamic assumptions besides the DR method are the Horvath–Kawazoe method for the analysis of MPSD and Kelvin equation–based methods for analysis of mesopores, such as the Barrett–Joyner–Halenda or Broekhoff–de Boer methods.

A more detailed overview of these methods can be found in the literature (Dubinin 1966, Gregg & Sing 1982, Li et al. 1998, Rouquerol et al. 1999, Ryu et al. 2000, Gauden et al. 2004, Choma & Jaroniec 2006, Inagaki & Tascón 2006, Thommes et al. 2012).

NCFMs have on their surface oxygen functional groups with various acidities, which allows examination of the NCFM surface chemistry by indirect methods of investigation, such as selective neutralization and temperature-programmed desorption. The Boehm titration method is simple and fast and gives good reproducibility (Boehm 1994).

NCFM heating leads to degradation of oxygen-containing surface groups and the release of gaseous products such as CO, $CO_2$, $H_2O$, and $H_2$. The functional groups are determined according to desorption temperature and released gas or gases. The requirement is that the maximum desorption temperature is not higher than those of carbonization (Bandosz et al. 2003).

FTIR provides only qualitative information about surface chemistry.

A very popular method for characterization of surface groups is XPS. The analysis is based on the changes in the intensities of 1 s peaks of carbon, oxygen, and any other heteroatom, which is expected in the material.

To obtain a better insight into the surface chemistry of examined NCFM, a recommended analysis is a combination of several methods (Polovina et al. 1997).

The number of primary adsorption centers is examined by immersion calorimetry (Bandosz et al. 2003).

The point of zero charge for ACC could be determined by the batch equilibrium method (Babić et al. 1999).

Also, there is electrochemical characterization by the cyclic voltammetry method to get information about specific capacitances of the material (Kaludjerović et al. 2014).

## 22.3.2 Physical Properties

The properties of NCFMs greatly depend on the raw materials and process conditions. Activation by $CO_2$ essentially develops ultramicroporosity in isotropic pitch-based ACFs and causes a steady decrease in the tensile strength with burn-off increase, while the fiber diameter is not significantly changed.

However, steam activation results in a wider MPSD and, after the initial stages of the activation, the tensile strength remains nearly constant but the fiber diameter decreases (Linares-Solano & Cazorla-Amorós 2008).

More than 90% of the total surface area of NCFMs corresponds to micropores and the rest is usually mostly mesopores of up to 4 nm. In the micropores a high adsorption potential exists which is responsible for filling of the pore volume by the adsorbate. Thus, most of the adsorption takes place within micropores. At relative pressure below $10^{-3}$, the smaller the micropore width of AC, the larger the adsorbed amount of adsorbate. The relative pressure, at which micropore filling occurs, depends on the pore size and nature of adsorptive molecules, the effective pore width, and the pore shape. However, the pore-filling capacity depends on the accessibility of the pores for the adsorptive molecules, i.e., on the molecular size of adsorptive and experimental conditions. Based on the pioneering work of Polanyi, Dubinin (1966) and coworkers showed the basic understanding of the micropore-filling process.

Woven and knitted forms of ACFs are effective for rapid electrical thermal conditioning. Petkovska et al. (1991) showed that the adsorbent bed from ACC can be directly regenerated as well as cyclic separation could be performed by just passing electrical current through the adsorbent owing to the Joule heat generated inside the ACC. Nowadays it is one of the common ways of recovering of such adsorbents. The conductivity of NCFMs arises from $\pi$ electrons and holes in the $\pi$ band of graphene layers (Leon y Leon & Radovic 1994).

Porous carbons generally show turbostratic structure with less inlayer ordering than graphite and a larger interlayer spacing $d_{002}$. The dimensions of the graphene stacks for ACs determined from the XRD data are a stack height of 1 nm (two to three layers) and a stack width of 1–3 nm. This structural arrangement of the graphene layers is responsible for the physicochemical properties of carbon materials generally.

Analysis of laser interaction with ACC indicates that the larger the surface area of ACC, the higher the laser power density that is needed for causing mechanical damage (Kaludjerović et al. 2011).

## 22.3.3 Chemical Properties

Heteroatoms (C, O, H, N, P, S, etc.) which remained at the edges of the graphene layers were transformed into surface functional groups. The surface of NCFMs can contain a variety of functional groups with Lewis acid/base characteristics. The concentration, the type, and the distribution of the groups present on the carbon surface can vary enormously with the type of carbon material and the pretreatment. Oxygen-containing functional groups (see Figure 22.5) are the most important in influencing the wettability, surface characteristics, and adsorption behavior of NCFMs. The most common acidic surface groups are those with oxygen-containing groups on the edges of crystallite, such as carboxyl, lactole, lactone, phenol, and carbonyl. Functional groups with the oxygen atom bound in the ring, such as pyranes and chromenes, contribute to the basic character of ACFs. Individual functional groups, such as carboxyl groups, will exhibit a spread of their dissociation constants, depending on the neighboring groups, the size of the graphene layers, etc.

The presence of moisture in the air affects the adsorption characteristics of NCFMs. The adsorption of water vapor by ACF is highly specific and depends on the surface chemistry as well as on pore size and shape (Mowla et al. 2003). Because of the relatively hydrophobic nature of the ACF, the adsorption equilibrium is not so much hindered by the coexistence of water vapor in air with low relative humidity. Acidic oxygen functional groups can transform a very hydrophobic carbon material into a very hydrophilic one, while treatment at a carefully chosen temperature in $H_2$ can result in both hydrophobic (basic) and stable carbon surfaces.

In particular, an NCFM immersed into an aqueous solution will develop a surface charge and an electrical double layer will be formed, which depends on both the solution pH and the surface chemistry of the NCFM. Ions and polar molecules in solution will be attracted by opposite charges on the surface.

## 22.4 Application Area

Carbon adsorbents in various fields of applications have proved their superior performance in comparison to mineral ones, such as high adsorption capacity for gas phase and liquid phase adsorption, resistance to chemicals and heat, radiation stability, and comparatively high electrical conductivity that makes it possible to isolate some components upon electrosorption, to carry out electrothermal regeneration of the material, etc. These advantages stem from the physical and chemical properties of AC. AC can act as a base or an acid in aqueous solution, which favors the adsorption of most inorganic substances, but it may be dominant in the adsorption of organic compounds as well (Radovic et al. 2001). Because of these characteristics, these materials are widely used in environmental protection, in various industries, for catalysis, in medicine, for personal protection, and for domestic usage (Mays 1999, Kaludjerović et al. 2002, Figueiredo & Pereira 2012, Goyal 2012). NCFMs are used in many types of adsorbents forms, mainly as fiber, cloth, felt, paper that contains 30%–70% of ACF, and membranes.

NCFMs demonstrate good adsorption capacity at both low and elevated temperatures, which is significant for purification of combustion gases. Thus, ACFs are very attractive for the removal of $SO_2$ (Mochida et al. 1997, Raymundo-Piñero et al. 2001) or $NO_x$ (Lee et al. 2002, Shirahama et al. 2002). When ACF is used for $NO_x$ removal, only a limited amount of NO is physically adsorbed because of the weak interactions between the two of them (Zawadzki & Wiśniewski 2002, Adapa et al. 2006). When using ACF pretreated with salts of transition metals (Ni, Fe, Cu, or Pd) or by electrolytic metal plating (Ag, Cu), the catalytic reduction of NO to $N_2$ and $O_2$ occurred on the ACF surface (Park & Kim 2005, Byeon et al. 2008). Copper introduced on the ACF surface significantly increases ammonia adsorption if the concentration of $NH_3$ is low. Ammonia adsorption at higher concentration is directly influenced by MPV (Polovina et al. 1998). Adsorption of odorous substances such as $H_2S$ onto ACFs depends on the surface structural parameters and surface chemistry (Boudou et al. 2003, Le Leuch et al. 2003).

For environmental pollution control, volatile organic chemicals, such as benzene, toluene, xylene, dichloromethane, and trichloroethylene, have become the focus of considerable attention (Suzuki 1994, Le Cloirec 2012). Gupta and Verma (2002) concluded that adsorption has been found to be more effective than condensation, at low concentration levels, i.e., parts per million. Adsorption of VOC on the ACF or cloth was also studied under dynamic conditions (Kaludjerović et al. 1996, Das et al. 2004).

ACFs have found wide application in the recovery of some solvent vapor. Most effective is rotor-type filter made from ACF cardboard calendered on one side and corrugated on the other. Purified gas flows directly over the filtering layer along the channels formed by goffers. In the area where the clean air leaves the ACFs in the opposite direction, heated air is introduced for regeneration, releasing the desorbed solvent vapor at the end of the filter (Ermolenko et al. 1990, Suzuki 1994).

Also, NCFMs can be used for the storage of natural gases such as methane and hydrogen. ACFs whose SSA was extremely high had shown excellent absorptivity toward supercritical $CH_4$ and $H_2$ (Rejifu et al. 2009).

NCFMs can be very effective in removing both inorganic and organic acids and bases from aqueous solutions (Radovic et al. 2001).

Metal ions are among the important pollutants in wastewater as they are being discharged from various industrial activities. The adsorption of Cd, Zn, Hg, Cu, Ni, and Pb from aqueous solution onto ACC has been widely investigated (Polovina et al. 1995, Gomez-Serrano et al. 1998, Babić et al. 2002, Duman & Ayranci 2010, Leyva-Ramos et al. 2011).

Adsorption of inorganic solutes is of great practical interest because of water purification and metal recovery applications. Its fundamental aspects are very important for the preparation of carbon-supported catalysts because the catalyst precursor is typically dissolved in water before its loading onto the porous carbon as support (Radovic et al. 2001).

It is very interesting that NCFMs can act as both catalyst and catalyst support (Radovic & Rodriguez-Reinoso 1997, Goyal 2012). Also, there is a great potential in loading catalyst such as Pt on the NCFMs as the support. Huang, Chen, and Yuan (2008) obtained excellent Pt dispersion on ACFs. The electrocatalytic

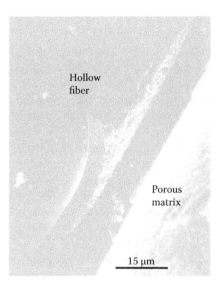

**FIGURE 22.7** SEM micrograph of carbon hollow fibers with electrochemically deposited Pt/carbon cryogel composite.

activity of Pt deposited on a carbon composite with porous hollow fibers and porous matrix is about two times higher than that of Pt deposited on Vulcan XC-72R carbon black (Kaludjerović et al. 2008). Figure 22.7 presents a carbon composite with porous hollow fibers and porous matrix with Pt deposited on the fiber surface.

Wang and Hu (2005) deposited Ru oxides on AC fabrics by electrochemical catalytic oxidation. These modified activated fabrics were demonstrated to be an excellent candidate for the supercapacitor application. ACFs in forms of cloth, felt, or paper found application in electrical double-layer capacitors (Mays 1999).

A very important field of NCFM application is in medicine, especially for wound dressings. ACC dressing is an appropriate wound healing material owing to its biocompatibility and adsorption characteristics. The use of an ACC dressing for the treatment of wound malodor for military purposes was first reported in the United Kingdom in the 1970s. Also, in the former Union of Soviet Socialist Republics, ACC in combined dressings was used to reduce infection, remove toxins, and increase re-epithelialization in burn treatment (Mikhalovsky et al. 2012). Medical cloth produced by Chemviron Carbon, a subsidiary of Calgon Carbon Corporation, is the world's leading manufacturer of ACC for wound dressings and other applications. A number of multicomponent dressings containing this medical cloth are marketed under the trademark names Actisorb (Systagenix Wound Management), CarboFlex (ConvaTec UK), Carbonet (Smith & Nephew), and CliniSorb (CliniMed) (Mikhalovsky et al. 2012). Sekulić et al. (2009) suggest that gamma irradiation is a suitable technique for the sterilization of ACC. To get antibacterial wound dressing, Park and Jang (2003) prepared antibacterial ACFs supporting silver. It was revealed that the obtained ACF/Ag possessed a strong antibacterial activity against and an inhibitory effect on the growths of *Eschecrichia coli* and *Staphylococcus aureus*, respectively.

# References

Adams, L. B., Boucher, E. A. & Everett, D. H., "Adsorption of organic vapours by saran-carbon fibres and powders," *Carbon* 8 (1970): 761–772.

Adapa, S., Gaur, V. & Verma, N., "Catalytic oxidation of NO by activated carbon fiber (ACF)," *Chem. Eng. J.* 116 (2006): 25–37.

Babić, B., Milonjić, S., Polovina, M. et al., "Point of zero charge and intrinsic equilibrium constants of activated carbon cloth," *Carbon* 37 (1999): 477–481.

Babić, B. M., Milonjić, S. K., Polovina, M. J. et al., "Adsorption of zinc, cadmium and mercury ions from agues solutions on activated carbon cloth," *Carbon* 40 (2002): 1109–1115.

Bandosz, T. J., Biggs, M. J., Gubbins, K. E. et al., "Molecular models of porous carbons," in Radovic, L. R., ed., *Chemistry and Physics of Carbon* 28 (Marcel Dekker, New York, 2003), pp. 41–228.

Boehm, H. P., "Some aspects of the surface chemistry of carbon blacks and other carbons," *Carbon* 32 (1994): 759–769.

Boudou, J. P., Chehimi, M., Broniek, E. et al., "Adsorption of $H_2S$ or $SO_2$ on an activated carbon cloth modified by ammonia treatment," *Carbon* 41 (2003): 1999–2007.

Byeon, J. H., Yoon, H. S., Yoon, K. Y. et al., "Electroless copper deposition on a pitch-based activated carbon fiber and an application for NO removal," *Surf. Coat. Technol.* 202 (2008): 3571–3578.

Choma, J. & Jaroniec, M., "Characterization of nanoporous carbons by using gas adsorption isotherms," in Bandosz, T. J., ed., *Activated Carbon Surfaces in Environmental Remediation* (Elsevier, Amsterdam, Netherlands, 2006), pp. 107–158.

Das, D., Gaur, V. & Verma, N., "Removal of volatile organic compound by activated carbon fiber," *Carbon* 42 (2004): 2949–2962.

Donnet, J. B. & Bansal, R. Ch., *Carbon Fibers* (Marcel Dekker, New York, 1990).

Dubinin, M. M., "Porous structure and adsorption properties of active carbon," in Walker, Jr., P., ed., *Chemistry and Physics of Carbon* 2 (Marcel Dekker, New York, 1966), pp. 51–120.

Duman, O. & Ayranci, E., "Attachment of benzo-crown ethers onto activated carbon cloth to enhance the removal of chromium, cobalt and nickel ions from aqueous solutions by adsorption," *J. Hazard. Mater.* 176 (2010): 231–238.

Economy, J., Daley, M. & Mangun, C., "Activated carbon fibers—Past, present and future," *Preprints of Papers, American Chemical Society, Division of Fuel Chemistry* 41, No. CONF-960376 (1996): 321–325.

Edie, D. D., "The effect of processing on the structure and properties of carbon fibers," *Carbon* 36 (1998): 345–362.

Ermolenko, I. N., Lyubliner, I. P. & Gulko, N. V., *Chemically Modified Carbon Fibers* (Wiley-VCH, Weinheim, Germany, 1990).

Faur-Brasquet, C. & Le Cloirec, P., "Modelling of the flow behavior of activated carbon cloths using a neural network approach," *Chem. Eng. Process.* 42 (2003): 645–652.

Figueiredo, J. L. & Pereira, F. M. R., "Porous texture versus surface chemistry in applications of adsorption by carbons," in Tascón, J. M. D., ed., *Novel Carbon Adsorbents* (Elsevier, Amsterdam, Netherlands, 2012), pp. 471–498.

Fitzer, E., Frohs, W. & Heine, M., "Optimization of stabilization and carbonization treatment of PAN fibres and structural characterization of the resulting carbon fibres," *Carbon* 24 (1986): 387–395.

Gauden, P. A., Terzyk, A. P., Rychlicki, G. et al., "Estimating the pore size distribution of activated carbons from adsorption data of different adsorbates by various methods," *J. Colloid Interface Sci.* 273 (2004): 39–63.

Giraldo, L., González-Navarro, M. F. & Moreno-Piraján, J. C., "Activated carbons from African oil palm waste shells and fibre for hydrogen storage," *Carbon Sci. Tech.* 5 (2013): 303–313.

Gomez-Serrano, V., Macias-Garcia, A., Espinosa-Mansilla, A. et al., "Adsorption of mercury, cadmium and lead from aqueous solution on heat-treated and sulphurized activated carbon," *Water Res.* 32 (1998): 1–4.

Goyal, M., "Nonenvironmental industrial applications of activated carbon adsorption," in Tascón, J. M. D., ed., *Novel Carbon Adsorbents* (Elsevier, Amsterdam, Netherlands, 2012), pp. 605–638.

Granda, M., Santamaria, R. & Menéndez, R., "Coal tar pitch: Compositions and pyrolysis behavior," in Radovic, L. R., ed., *Chemistry and Physics of Carbon* 28 (Marcel Dekker, New York, 2003), pp. 263–330.

Gregg, S. J. & Sing, K. S. W., *Adsorption, Surface Area and Porosity* (Academic Press, London, UK, 1982).

Gupta, V. K. & Verma, N., "Removal of volatile organic compounds by cryogenic condensation followed by adsorption," *Chem. Eng. Sci.* 57 (2002): 2679–2696.

Hegde, R. R., Dahiya, A., Kamath, M. G. et al., "Rayon fibers" (2004), http://www.engr.utk.edu/mse /Textiles/Rayon%20fibers.htm, Accessed January 25, 2015.

Hu, Z. & Vansant, E. F., "Carbon molecular sieves produced from walnut shell," *Carbon* 33 (1995): 561–567.

Huang, X., "Fabrication and properties of carbon fibers," *Materials* 2 (2009): 2369–2403.

Huang, H. X., Chen, S. X. & Yuan, C. E., "Platinum nanoparticles supported on activated carbon fiber as catalyst for methanol oxidation," *J. Power Sources* 175 (2008): 166–174.

Inagaki, M. & Feiyu, K. *Carbon Materials Science and Engineering: From Fundamentals to Applications* (Tsinghua University Press, Beijing, China, 2006).

Inagaki, M. & Tascón, J. M. D., "Pore formation and control in carbon materials," in Bandosz, T. J., ed., *Interface Science and Technology* 7 (Elsevier B. V., New York, 2006), pp. 49–105.

Jenkins, G. M. & Kawamura, K., *Polymeric Carbons—Carbon Fibre, Glass and Char* (Cambridge University Press, Cambridge, England, 1976).

Kaludjerović, B., Stojanović, B., Polovina, M. et al., "Kinetics of adsorption of carbon tetrachloride from air mixtures by activated carbon cloth—The application of theoretical models," *J. Serb. Chem. Soc.* 61 (1996): 461–467.

Kaludjerović, B., Milovanović, L. J. & Babić, B., "Carbon fibres and textiles and some of their applications," *Hem. Ind.* 56 (2002): 369–374.

Kaludjerović, B. V., Srećković, M. Z., Trtica, M. S. et al., "A new laser technique for the formation of oxide surface complexes on carbon cloth," *Carbon* 42 (2004): 443–445.

Kaludjerović, B. V., Jovanović, V. M., Babić, B. M. et al., "Characterization of platinum deposited on carbon hollow fibers/carbon cryogel composites," *J. Opt. Adv. Mater.* 10 (2008): 2708–2712.

Kaludjerović, B. V., Kljajević, L., Sekulić, D. et al., "Adsorption characteristics of activated carbon hollow fibers," *Chem. Ind. Chem. Eng. Q.* 15 (2009): 29–31.

Kaludjerović, B. V., Trtica, M. S., Radak, B. B. et al., "Analysis of the interaction of pulsed laser with nanoporous activated carbon cloth," *J. Mater. Sci. Tech.* 27 (2011): 979–984.

Kaludjerović, B. V., Jovanović, V. M., Stevanović, S. I. et al., "Characterization of nanoporous carbon fibrous materials obtained by chemical activation of plane tree seed under ultrasonic irradiation," *Ultrason. Sonochem.* 21 (2014): 782–789.

Kljajević, L. M., Jovanović, V. M., Stevanović, S. I. et al., "Influence of chemical agents on the surface area and porosity of active carbon hollow fibers," *J. Serb. Chem. Soc.* 76 (2011): 1283–1294.

Kynol, "Kynol® activated carbon fibers & textiles" (2012), http://www.kynol.de/pdf/kynol_ac_info_short .pdf, Accessed January 27, 2015.

Lastoskie, C., Gubbins, K. E. & Quirke, N., "Pore size distribution analysis of microporous carbons: A density functional theory approach," *J. Phys. Chem.* 97 (1993): 4786–4796.

Le Cloirec, P., "Adsorption onto activated carbon fiber cloth and electrothermal desorption of volatile organic compound (VOCs): A specific review," *Chin. J. Chem. Eng.* 20 (2012): 461–468.

Le Leuch, L. M., Subrenat, A. & Le Cloirec, P., "Hydrogen sulfide adsorption and oxidation onto activated carbon cloths: Applications to odorous gaseous emission treatments," *Langmuir* 19 (2003): 10869–10877.

Lee, Y. W., Choi, D. K. & Park, J. W., "Performance of fixed-bed KOH impregnated activated carbon adsorber for NO and $NO_2$ removal in the presence of oxygen," *Carbon* 40 (2002): 1409–1417.

Lee, T., Ooi, C.-H., Othman, R. et al., "Activated carbon fiber—The hybrid of carbon fiber and activated carbon," *Rev. Adv. Mater. Sci.* 36 (2014): 118–136.

Leon y Leon, C. A. D. & Radovic, L. R., "Interfacial chemistry and electrochemistry of carbon surfaces," in Thrower, P. A., ed., *Chemistry and Physics of Carbon* 24 (Marcel Dekker, New York, 1994), pp. 213–310.

Leyva-Ramos, R., Berber-Mendoza, M. S., Salazar-Rabago, J. et al., "Adsorption of lead(II) from aqueous solution onto several types of activated carbon fibers," *Adsorption* 17 (2011): 515–526.

Li, Z., Kruk, M., Jaroniec, M. et al., "Characterization of structural and surface properties of activated carbon fibers," *J. Colloid Interface Sci.* 204 (1998): 151–156.

Linares-Solano, A. & Cazorla-Amorós, D., "Adsorption on activated carbon fibers," in Bottani, E. J. & Tascón, J. M. D., eds., *Adsorption by Carbons* (Elsevier, Amsterdam, Netherlands, 2008), pp. 431–454.

Linares-Solano, A., Lozano-Castelló, D., Lillo-Ródenas, M. A. et al., "Carbon activation by alkaline hydroxides," in Radovic, L. R., ed., *Chemistry and Physics of Carbon* 30 (Taylor & Francis, Boca Raton, 2008), pp. 1–62.

Lysenko, A. A., "Prospects for development of research and production of carbon fibre sorbents," *Fibre Chem.* 39 (2007): 93–102.

Ma, X., Yang, H., Yu, L. et al., "Preparation, surface and pore structure of high surface area activated carbon fibers from bamboo by steam activation," *Materials* 7 (2014): 4431–4441.

Marcuzzo, J. S., Otani, C., Polidoro, H. A. et al., "Influence of thermal treatment on porosity formation on carbon fiber from textile PAN," *Mater. Res.* 16 (2013): 137–144.

Mays, T., "Active carbon fibers," in Burchell, T. D., ed., *Carbon Materials for Advanced Technologies* (Pergamon, Amsterdam, Netherlands, 1999), pp. 95–118.

Menéndez-Diaz, J. A. & Martin-Gullón, L., "Types of carbon adsorbents and their production," in Bandosz, T. J., ed., *Activated Carbon Surfaces in Environmental Remediation* (Elsevier, Amsterdam, Netherlands, 2006), pp. 1–47.

Mikhalovsky, S. V., Sandeman, S. R., Howell, C. A. et al., "Biomedical applications of carbon adsorbents," in Tascón, J. M. D., ed., *Novel Carbon Adsorbents* (Elsevier, Amsterdam, Netherlands, 2012), pp. 639–669.

Mochida, I., Kuroda, K., Miyamoto, S. et al., "Remarkable catalytic activity of calcined pitch based activated carbon fiber for oxidative removal of $SO_2$ as aqueous $H_2SO_4$," *Energy Fuels* 11 (1997): 272–276.

Mowla, D., Do, D. D. & Kaneko, K., "Adsorption of water vapor on activated carbon: A brief overview," in Radovic, L. R., ed., *Chemistry and Physics of Carbon* 28 (Marcel Dekker, New York, 2003), pp. 229–262.

Otani, S., "On the carbon fiber from the molten pyrolysis products," *Carbon* 3 (1965): 31–38.

Park, S.-J. & Jang, Y.-S., "Preparation and characterization of activated carbon fibers supported with silver metal for antibacterial behavior," *J. Colloid Interface Sci.* 261 (2003): 238–243.

Park, S. J. & Kim, B. J., "A study on NO removal of activated carbon fibers with deposited silver nanoparticles," *J. Colloid Interface Sci.* 282 (2005): 124–127.

Petkovska, M., Tondeur, D., Grevillot, G. et al., "Temperature-swing gas separation with electrothermal desorption step," *Sep. Sci. Technol.* 26 (1991): 425–444.

Pittman, Jr., C. U., Jiang, W., Yue, Z. R. et al., "Surface area and pore size distribution of microporous carbon fibers prepared by electrochemical oxidation," *Carbon* 37 (1999): 85–96.

Polovina, M., Šurbek, A., Laušević, M. et al., "Adsorption of cadmium ions on activated carbon cloth," *J. Serb. Chem. Soc.* 60 (1995): 43–49.

Polovina, M., Babić, B., Kaludjerović, B. et al., "Surface characterization of oxidized activated carbon cloth," *Carbon* 35 (1997): 1047–1052.

Polovina, M., Kaludjerović, B. & Babić, B., "Ammonia adsorption on chemically modified activated carbon cloth," *J. Serb. Chem. Soc.* 63 (1998): 653–659.

Radovic, L. R. & Rodriguez-Reinoso, F., "Carbon materials in catalysis," in Thrower, P. A., ed., *Chemistry and Physics of Carbon* 25 (Marcel Dekker, New York, 1997), pp. 243–358.

Radovic, L. R., Moreno-Castilla, C. & Rivera-Utrilla, J., "Carbon materials as adsorbents in aqueous solutions," in Radovic, L. R., ed., *Chemistry and Physics of Carbon* 27 (Marcel Dekker, New York, 2001), pp. 41–228.

Ramos, M. E., Bonelli, P. R., Cukierman, A. L. et al., "Influence of thermal treatment conditions on porosity development and mechanical properties of activated carbon cloths from a novel nanofibre-made fabric," *Mater. Chem. Phys.* 116 (2009): 310–34.

Ramos, M. E., Bonelli, P. R., Cukierman, A. L. et al., "Adsorption of volatile organic compounds onto activated carbon cloths derived from a novel regenerated cellulosic precursor," *J. Hazard. Mater.* 177 (2010): 175–182.

Rašković, V. & Marinković, S., "Temperature dependence of processes during oxidation of PAN fibres," *Carbon* 13 (1975): 535–538.

Raymundo-Piñero, E., Cazorla-Amorós, D. & Linares-Solano, A., "Temperature programmed desorption study on the mechanism of $SO_2$ oxidation by activated carbon and activated carbon fibres," *Carbon* 39 (2001): 231–242.

Rejifu, A., Noguchi, H., Ohba, T. et al., "Adsorptivities of extremely high surface area activated carbon fibres for $CH_4$ and $H_2$," *Adsorp. Sci. Technol.* 27 (2009): 877–881.

Romero-Anaya, A. J., Lillo-Ródenas, M. A., Salinas-Martínez de Lecea, C. et al., "Hydrothermal and conventional $H_3PO_4$ activation of two natural bio-fibers," *Carbon* 50 (2012): 3158–3169.

Rong, H., Ryu, Z., Zheng, J. et al., "Effect of air oxidation of rayon-based activated carbon fibers on the adsorption behavior for formaldehyde," *Carbon* 40 (2002): 2291–2300.

Rouquerol, F., Rouquerol, J. & Sing, K., *Adsorption by Powders and Porous Solids* (Academic Press, San Diego, 1999).

Ryu, Z., Zheng, J., Wang, M. et al., "Nitrogen adsorption studies of PAN-based activated carbon fibers prepared by different activation methods," *J. Colloid Interface Sci.* 230 (2000): 312–319.

Saufi, S. M. & Ismail, A. F., "Development and characterization of polyacrylonitrile (PAN) based carbon hollow fiber membrane," *Songklanakarin J. Sci. Technol.* 24 Suppl. (2002): 843–854.

Sekulić, D. R., Babic, B. M., Kljajević, L. M. et al., "The effect of gamma radiation on the properties of activated carbon cloth," *J. Serb. Chem. Soc.* 74 (2009): 1125–1132.

Shirahama, N., Moon, S. H., Choi, K. H. et al., "Mechanistic study on adsorption and reduction of $NO_2$ over activated carbon fibers," *Carbon* 40 (2002): 2605–2611.

Shoaib, M. & Al-Swaidan, H. M., "Optimization and characterization of sliced activated carbon prepared from date palm tree fronds by physical activation," *Biomass Bioenergy* 73 (2015): 124–134.

Sing, K. S. W., Everett, D. H., Haul, R. A. W. et al., "IUPAC: Reporting physisorption data for gas/solid systems; Special reference to the determination of surface area and porosity," *Pure Appl. Chem.* 57 (1985): 603–619.

Sun, J., Wu, G. & Wang, Q., "Adsorption properties of polyacrylonitrile-based activated carbon hollow fiber," *J. Appl. Polym. Sci.* 93 (2004): 602–607.

Suzuki, M., "Activated carbon fiber: Fundamentals and applications," *Carbon* 32 (1994): 577–586.

Tascón, J. M. D., "Overview of carbon materials in relation to adsorption," in Bottani, E. J. & Tascón, J. M. D., eds., *Adsorption by Carbons* (Elsevier, Amsterdam, Netherlands, 2008), pp. 15–49.

Thommes, M., Cychosz, K. A. & Neimark, A. V., "Advanced physical characterization of nanoporous carbon," in Tascón, J. M. D., ed., *Novel Carbon Adsorbents* (Elsevier, Amsterdam, Netherlands, 2012), pp. 107–145.

Toyoda, M., Iwashita, N. & Inagaki, M., "Sorption of heavy oils into carbon materials," in Radovic, L. R., ed., *Chemistry and Physics of Carbon* 30 (Taylor & Francis, Boca Raton, 2008), pp. 177–237.

Wang, C.-C. & Hu, C.-C., "Electrochemical catalytic modification of activated carbon fabrics by ruthenium chloride for supercapacitors," *Carbon* 43 (2005): 1926–1935.

Watt, W., "Carbon work at the Royal Aircraft Establishment," *Carbon* 10 (1972): 121–143.

Yuhan, C., Qilin, W., Ning, P. et al., "Rayon-based activated carbon fibers treated with both alkali metal salt and Lewis acid," *Microporous Mesoporous Mater.* 109 (2008): 138–146.

Zawadzki, J. & Wiśniewski, M., "Adsorption and decomposition of NO on carbon and carbon-supported catalysts," *Carbon* 40 (2002): 119–124.

# Mesoporous Carbon Nanomaterials

Meng Li

Junmin Xue

## 23.1 Introduction

Carbon is the 15th most abundant element in the earth's crust, which has been studied for a long time since ancient time. The emergence of carbon nanomaterials in recent decades well rendered its novel functions and properties for advanced applications in the area of electronics, batteries, and catalysis, as well as drug delivery. In conjunction with the emergence of nanomaterials, there has been a resurgence of interest in the nanostructured porous materials, especially in the hierarchical nanostructured carbons. When carbon nanomaterials have hierarchical porosity, they have enhanced properties compared with single-sized porous carbon materials. For example, they have enhanced mass transport and improved selectivity through macro-/mesopores, and they also have increased SSA and open-pore frameworks on the level of fine pore systems through meso-/micropores. According to the International Union of Pure and Applied Chemistry classification, porous carbon materials can be classified according to their pore diameter as microporous (pore size <2 nm), mesoporous (2 nm < pore size < 50 nm), and macroporous (pore size >50 nm). In particular, mesoporous materials have attracted considerable attention in view of their potential for use in various fields since the researchers from Mobil Corporation first reported the novel family of M41S (mesoporous silica) in 1992 (Beck et al. 1992, Kresge et al. 1992). The mesoporous silica was synthesized from the calcination of aluminosilicate gels in the presence of long-chain cationic surfactants. Mobil Corporation proposed that the formation of these mesoporous materials took place by means of a liquid crystal templating mechanism, in which the silicate material forms inorganic walls between ordered surfactant micelles. MCM-41 is the most extensively researched member of the M41S family of molecular sieves because of its high SSA of 700 $m^2$/g and tailored pore size in the range of 1.6–10 nm. Another important member of the M41S family is MCM-48, which has 3D bicontinuous structure belonging to the cubic *Ia3d* space group (Alfredsson & Anderson 1996, Chen et al. 1997). The typical TEM of MCM-48 is shown in Figure 23.1. Besides mesoporous silica of the M41S family, Santa Barbara Amorphous-15 (SBA-15) with larger pore size of 4.6–30 nm was invented

**FIGURE 23.1** TEM of the uncalcined MCM-48 along [110] recorded at 200 kV. Scale bar: 9.7 nm. (Reprinted with permission from Alfredsson, V. & Anderson, M. W., *Chemistry of Materials* 8, 1141–1146. Copyright 1996 American Chemical Society.)

by the prominent research group of Zhao et al. (1998). SBA-15 with 2D hexagonal *p6mm* structure was first produced in highly acidic media using amphiphilic triblock copolymer of poly(ethylene oxide)–poly(propylene oxide)–poly(ethylene oxide) ($EO_{20}PO_{70}EO_{20}$, P123) as template in 1998. In later years, Zhao's (2005) group further extended the mesoporous silica family such as the FDU-n series.

Besides the typical silica-based mesoporous materials, ordered mesoporous materials with different compositions from pure organic/inorganic frameworks to organic–inorganic hybrid frameworks have also been widely studied in the past decades. For example, mesoporous polymers were fabricated through several routes including controlled foaming, phase separation, molecular imprinting, and hard-template approach by employing colloidal particles or mesoporous silica (Johnson et al. 1999, Jang & Bae 2005). Mesoporous metal oxides and mixed oxides with semicrystalline frameworks, such as $TiO_2$, $ZrO_2$, $Nb_2O_5$, $Ta_2O_5$, $Al_2O_3$, $SiO_2$, $SnO_2$, $WO_3$, $HfO_2$, $SiAlO_y$, $Al_2TiO_y$, $ZrTiO_y$, and $SiTiO_y$, were successfully prepared by Crepaldi et al. (2003a,b), Grosso et al. (2004), and Yang et al. (1998) through the combination of the sol–gel process of metal alkoxides and/or metal salt precursors and the coating techniques. In addition, Bartlett et al. (2002), Whitehead et al. (1999), Bartlett et al. (2001), Nelson et al. (2002), and Attard et al. (1997, 2001) designed several routes to fabricate mesostructured metals and metal alloys, such as Ni, Sn, Co, Pd, Rh, Pt–Ru, and Pt–Pd, by reduction using hydrazine and a hydroboron or by electrodeposition.

Porous carbons including activated carbon (AC) and carbon black have a long history of research and usage in the field of energy storage, adsorption, and catalyst carriers due to their large surface areas, chemical stability, and low cost. However, the practical applications of these traditional porous carbon materials are greatly limited, such as adsorption of large molecules (organic dyes and biomolecules), chromatographic separation, and lithium ion cells, because of the restriction of their micropore size (<2 nm) (Zhao et al. 2013). Thus, mesoporous carbon nanomaterials (MCNs) which possess pore sizes between 2 and 50 nm received tremendous research efforts. More recently, ordered MCNs, as another huge family of porous materials, have been successfully fabricated by a number of research groups. In comparison with mesoporous silica and metal oxide, ordered MCNs possess more fascinating physical and chemical properties, including hydrophobicity of their surfaces, high surface area, enhanced electrical and thermal conductivity, chemical stability, reduced density, and low cost of manufacture. These advantageous properties enable them to be extensively used as guest molecule carriers (Liang et al. 2008, Deng et al. 2010), water and air purifiers, shape-selective catalysts, electrode materials for batteries (Hu et al. 2007), fuel cells (Chang et al. 2007, Fang et al. 2010), and supercapacitors (Li et al. 2011, Wang et al. 2011). Therefore, the synthesis of MCNs with controlled shape, size, pore diameter, and surface property is desirable in the view of fundamental research as well as practical applications.

## 23.2 Synthesis Approach to MCNs

The common approach to ordered MCNs is based on the hard-template strategy, which normally utilizes mesoporous silica as the template. The template mainly serves as molds for the replication of mesoporous carbon materials, and no significant chemical interactions take place between the templates and the carbon precursors (Liang et al. 2008). As shown in Figure 23.2, the general steps in producing MCNs by nanocasting strategy can be summarized as (1) preparation of mesostructured silica/surfactant composites by self-assembly process; (2) removal of surfactant to obtain hard template of mesoporous silica by calcination; (3) incorporation of carbon precursor inside the mesopores; (4) carbonization of carbon precursor at high temperature; and (5) removal of silica scaffold by etching. The synthesis steps are finer to control and the resulting MCNs normally keep the mesostructures and morphologies of the templates well. However, the mesostructures and morphologies of the replicated carbon materials are strongly limited by the mother silica template. Even worse, in order to remove the hard template, the MCNs suffer several complicated steps with aggressive conditions such as strong acid/alkali reactions and high-temperature treatment, which usually result in disordered structure.

Very recently, a supramolecular self-assembly approach has also been developed to directly synthesize ordered MCNs. Generally, this soft-template approach employs cross-linked polymeric materials (e.g., resorcinol–formaldehyde [Dai et al. 2004], phloroglucinol–formaldehyde [Liang & Dai 2006], phenol–formaldehyde [Meng et al. 2006]) as carbon-yielding component and amphiphilic block copolymer (e.g., HTAB [Li et al. 2004], F127 [Liang & Dai 2006, Meng et al. 2006, Fang et al. 2010], P123 [Kim et al. 2005]) as pore-forming components. As shown in Figure 23.2, the synthetic steps for the formation of MCNs by soft-template approach are more concise compared to the nanocasting method: (1) incorporation of carbon precursor and soft template to self-assemble mesostructure; (2) removal of the soft template at the decomposition temperature of the surfactant and leaving open pores and mesostructured polymer; and (3) carbonization of the mesostructured polymer at high temperature and obtaining the final product (or simultaneously conducting this step with the second one).

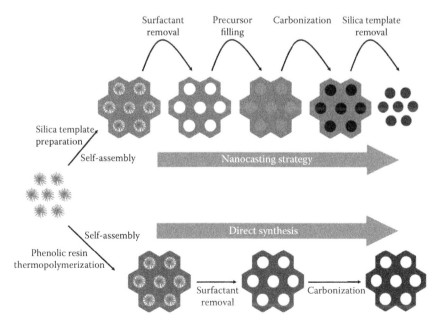

**FIGURE 23.2** Two typical methods for the preparation of MCNs: the nanocasting strategy from mesoporous silica hard templates and the direct synthesis from block copolymer soft templates. (Ma, T.-Y., Liu, L. & Yuan, Z.-Y., *Chemical Society Reviews* 42, 3977–4003, 2013. Reproduced by permission of The Royal Society of Chemistry.)

The focus of the next two sections will be the synthesis approaches to MCNs. Owing to the huge number of publications on MCNs reported over the past decade, the synthesis strategies of hard and soft templates have been proved to be the most successful methods to prepare the MCNs with well-controlled mesostructures. The hard-template method will be reviewed in Section 23.2.1. Development of this methodology will be reviewed and discussed. The soft-template method will be reviewed in Section 23.2.2, with special emphasis on the recent progress and several most reproducible methods. In addition, a quick summary and comparison of hard- and soft-template synthetic approaches will be provided at the end of Section 23.2.2. The merits and demerits of these two methods will be summarized and the summary can hopefully provide beneficial help for researchers who are interested in MCNs.

## 23.2.1 Hard-Template Synthesis of MCNs

In the early 1980s, the pioneering works on fabrication of mesoporous carbon materials through hard-template method were conducted by Knox et al. (1983, 1986). The synthetic steps adopted by Knox et al. were almost the same with the recently developed hard-template approach: (1) Well-bonded silica gel with high porosity was prepared as a template. The spherical silica gel with a mean particle diameter of 7 μm had a pore volume of 1.4 cm³/g with an SSA of 50 m²/g. (2) The silica template was impregnated with a melt of phenol and hexamine (used as carbon source). The impregnated material was heated to 150°C to form the phenol–formaldehyde resin within the silica pores. (3) Such silica–polymer composite was carbonized to 900°C under nitrogen atmosphere. (4) The silica–carbon composite was etched with hot aqueous potassium hydroxide to remove the silica template. The remaining porous glassy carbon had a BET surface area of 450–600 m²/g and a pore volume of 2.0 cm³/g. However, the BET surface area of such carbon decreased to 150 m²/g after further carbonization at 2500°C. The morphology of the obtained porous carbon is shown in Figure 23.3. This porous carbon material was used for high-performance liquid chromatography and offered peak shapes comparable to those from silica gels of the same particle size.

### 23.2.1.1 MCM-41 as Hard Template

Since the researchers from Mobil Corporation first reported the mesoporous silica of M41S series in 1992, people have a better choice of hard template to fabricate ordered MCNs. In 1994, Wu and Bein reported graphite-type carbon wires in the regular 3 nm wide hexagonal channels which were synthesized through the hard template of mesoporous MCM-41. In this method, the encapsulated PAN from

**FIGURE 23.3** SEM image of spherical particle of porous carbon prepared by hard-template method. (Reprinted from *Journal of Chromatography A*, 352, Knox, J. H., Kaur, B. & Millward, G. R., Structure and performance of porous graphitic carbon in liquid chromatography, 3–25, Copyright (1986), with permission from Elsevier.)

the monomer of acrylonitrile was used as a carbon precursor, followed by carbonization and graphitization to increase electronic conductivity at high temperature. The conductivity of PAN-MCM composite was found to increase with pyrolysis temperature and was 10 times that of the bulk PAN, reaching $10^{-1}$ S/cm at 1000°C. However, the template of mesoporous silica was not removed in this study, and the author clearly pointed out that the MCM-41 host played a key role in ordering the carbon structure. Later in 2000, Lee et al. developed a more sophisticated approach to mesoporous carbon by the following steps: (1) preparation of hexagonal mesoporous silica (HMS) aluminosilicate as a template (which has a similar mesostructure as hexagonal mesoporous MCM-41); (2) carbon precursors of phenol and formaldehyde were incorporated into the pore of Al–HMS by heating at 90°C for 12 h under reduced pressure; (3) the composites were reacted at high temperature to polymerize phenol and formaldehyde; (4) carbonization of the resulting Al–HMS/phenolic resin composites at 700°C; and (5) removal of the Al–HMS template by using hydrofluoric acid (HF) to obtain the mesoporous carbon. It is noted that the authors claimed that HMS has several advantages over MCM-48, including higher silica recovery yield than MCM-48, shorter synthesis time, and cost-effectiveness. The obtained MCNs offer a surface area as high as 1056 $m^2$/g, and showed good EDLC performance which was superior to that of the commercially available carbon MSC-25 owing to its improved mesoporosity.

In 2003, Tian et al. synthesized ordered self-supported ultrathin carbon nanowire arrays with small mesopore sizes by using modified MCM-41 as the template. In this published report, MCM-41 was tailored to leaflike in mesoscale texture and macroscale morphology. The carbon precursor used here was sucrose and the silica template was finally removed by 5–10 wt% HF. The diameter of the pore width was around 2–2.4 nm, as indicated in the TEM image in Figure 23.4a. The field emission scanning electron microscopy (FESEM) image in Figure 23.4b shows that the carbon nanowires with the diameter of 4–5 nm were packed side by side. Also, such MCNs exhibited high surface area of 1400 $m^2$/g and large pore volumes above 1.1 $cm^3$/g.

### 23.2.1.2 MCM-48 as Hard Template

MCM-48 as another famous mesoporous silica material has also been extensively used as a hard template to fabricate MCNs since 1999. Ryoo et al. conducted pioneering works in 1999 to first fabricate self-supported highly ordered mesoporous carbon sieves by employing the ordered aluminosilicate MCM-48 as a hard template. The carbon precursor used in this work was sucrose, which was converted to carbon inside the mesochannels of silica through a mild carbonization process using a sulfuric acid catalyst. The ordered MCNs were obtained after the removal of the template by using an aqueous solution of sodium hydroxide. The synthetic process was similar to that of the MCM-41 template. However,

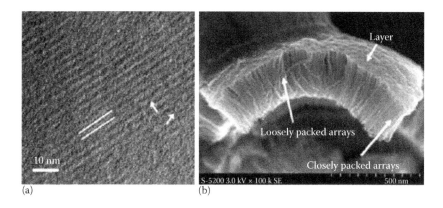

(a)　(b)

**FIGURE 23.4** (a) TEM and (b) FESEM images of MCNs which were obtained by using MCM-41 as hard template. (Tian, B. et al., *Chemical Communications*, 2726–2727, 2003. Reproduced by permission of The Royal Society of Chemistry.)

it is noted that the carbonization process using sucrose and sulfuric acid in this work was suitable for the facile and uniform infiltration of carbon into mesoporous silica.

In 1999, Lee et al. also developed a mesoporous carbon with regular three-dimensionally interconnected 2 nm pore arrays using Al–MCM-48 as a hard template. As listed in Table 23.1, the carbon precursors were phenol and formaldehyde. The template of MCM-48 was modified by aluminum in order to generate strong acid catalytic sites for the polymerization of phenol and formaldehyde (Figure 23.5). The EDLC performance of as-obtained mesoporous carbon SNU-1 was compared to the most popularly applied AC, MSC-25. On the basis of Ryoo et al.'s work in 2005, Li et al. created a series of mesoporous carbon such as C15, CMK-5, and C48 with different mesostructures by using MCM-48 as a hard template. The details of these materials are listed in Table 23.1. Moreover, the surface functionalization and pore size manipulation have also been successfully achieved on these mesoporous carbons of different structures, employing an approach in which diazonium compounds were in situ generated and reacted with the carbon framework.

### 23.2.1.3 SBA-15 as Hard Template

With the emergence of SBA-15 with larger pore size from 4.6 to 30 nm, such mesoporous silica with hexagonal arrays was extensively employed as another hard template candidate to produce ordered MCNs. Joo et al. (2001) reported a general strategy for the synthesis of highly ordered, rigid arrays of nanoporous carbon having uniform but tunable diameters (as shown in Figure 23.6). It was claimed that carbon nanopores were rigidly interconnected into a highly ordered hexagonal array by carbon spacers, and the structural model is provided in Figure 23.6b. The authors also loaded high dispersion of platinum nanoparticles on the as-obtained mesoporous carbon, and the loading amount was quite competitive with the traditional carbons such as carbon black, charcoal, and AC. The high dispersion of this Pt/MCN composite was proved to give rise to promising electrocatalytic activity for oxygen reduction.

**TABLE 23.1** Summary of MCNs Developed by Common Silica-Based Hard Templates

| Silica Template | Carbon Source | Etch Agent | MCN | Space Group | Surface Area ($m^2/g$) | Application | Ref. |
|---|---|---|---|---|---|---|---|
| MCM-41 | Acrylonitrile | – | Nanowires | Hexagonal | – | Improved conductivity | Wu & Bein 1994 |
| | Phenol–formaldehyde | HF | SNU-2 | Hexagonal | 1056 | Electrode | Lee et al. 2000 |
| | Sucrose | HF | Hexagonal nanowires | *P6mm* | 1400 | – | Tian et al. 2003 |
| MCM-48 | Sucrose | NaOH | CMK-1 | *Ia3d* | 1380 | – | Ryoo et al. 1999 |
| | Phenol–formaldehyde | HF | SNU-1 | 3D ordered | 1257 | Electrode | Lee et al. 1999 |
| | Pitch | NaOH | C15 | Hexagonal | 353 | – | Li et al. 2005 |
| | Furfuryl alcohol | NaOH | CMK-5 | Hexagonal | 1576 | – | Li et al. 2005 |
| | Pitch AR | NaOH | C48 | $I4_1/a$ | 510 | – | Li et al. 2005 |
| SBA-15 | Sucrose | NaOH | CMK-3 | 2D Hexagonal | 1160 | – | Chin et al. 2001 |
| | Furfuryl alcohol | NaOH | Ordered MCN | Hexagonal | 2000 | Supporting platform | Chin et al. 2001 |
| | Sucrose | NaOH | CMK-3 (nanorods) | *P6mm* | 1823 | – | Yu et al. 2002 |
| | Furfuryl alcohol | NaOH | CMK-5 | Hexagonal | 2000 | – | Kruk et al. 2003 |
| | Pitch | HF | C15 | – | 353 | – | Li & Dai 2005 |

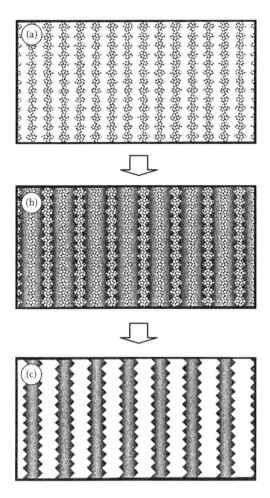

**FIGURE 23.5** Schematic outline of the hard-template synthesis procedure: (a) the mesoporous silica molecular sieve MCM-48; (b) MCM-48 after completing carbonization within pores; and (c) CMK-1 obtained by removing the silica wall after carbonization. (Reprinted with permission from Ryoo, R. et al., *Journal of Physical Chemistry B* 103, 7743–7746. Copyright 1999 American Chemical Society.)

In 2002, Yu et al. modified the synthesis method of SBA-15 and obtained highly condensed meso-porous SBA-15 with high-yield rodlike morphology. They also applied this rodlike SBA-15 as a template and successfully produced hexagonally ordered rodlike mesoporous carbon CMK-3 (shown in Figure 23.7).

## 23.2.2 Soft-Template Synthesis of MCNs

Owing to the drawbacks of hard-template synthesis mentioned previously, a supramolecular self-assembly approach was developed in recent years to directly synthesize ordered MCNs. Generally, this soft-template approach employs cross-linked polymeric materials (e.g., resorcinol–formaldehyde [Dai et al. 2004], phloroglucinol–formaldehyde [Liang & Dai 2006], phenol–formaldehyde [Meng et al. 2006]) as carbon-yielding component and amphiphilic block copolymer (e.g., HTAB [Li et al. 2004], F127 [Liang & Dai 2006, Meng et al. 2006, Fang et al. 2010], P123 [Kim et al. 2005]) as pore-forming component. To the best of our knowledge, Lee et al. (1988) conducted a pioneering work on soft-template synthesis of meso-porous carbon materials. In this work, a first attempt was described to make porous carbon membranes

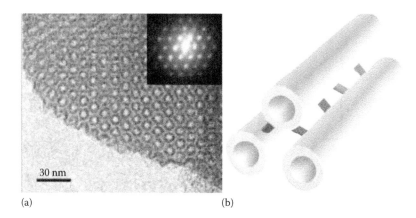

(a)                                                                              (b)

**FIGURE 23.6**  Ordered mesoporous carbon obtained by template synthesis using SBA-15. (a) TEM image of meso-porous carbon and corresponding Fourier diffractogram; and (b) schematic model for the mesoporous carbon structure. (Reprinted by permission from Macmillan Publishers Ltd., Kresge, C. T. et al., *Nature* 359, 710–712, copyright (1992).)

**FIGURE 23.7**  TEM image of ordered mesoporous carbon nanorod obtained by template synthesis using SBA-15. (Yu, C., Fan, J., Tian, B. et al.: High-yield synthesis of periodic mesoporous silica rods and their replication to meso-porous carbon rods. *Advanced Materials*. 2002. 14. 1742–1745. Copyright Wiley-VCH Verlag GmbH & Co. KGaA. Reproduced with permission.)

from the film of block copolymer with well-defined chain structure which is prepared by anionic living polymerization. According to the results of the morphology characterization in Figure 23.8, it is clear to observe the lamellar structure of the segregated microphase on the surface of the block copolymer and the final product. Although the characterizations of this work were not sufficient in today's view, this attempt was a great breakthrough on the preparation of mesoporous MCNs.

### 23.2.2.1  PS-P4VP as Soft Template

In 2004, Liang et al. synthesized large-scale crack-free mesoporous carbon films up to 6 cm² by a stepwise self-assembly approach. This approach included four basic steps as illustrated in Figure 23.9: (1) film casting of PS-P4VP/resorcinol supramolecular assembly; (2) completion of microphase separation by solvent annealing at 80°C in DMF/benzene mixed vapor, wherein one of the carbon precursors of resorcinol was organized in the well-defined P4VP domain; (3) in situ polymerization of resorcinol and formaldehyde by exposing the film to formaldehyde gas, wherein highly cross-linked resorcinol–formaldehyde resin was formed within the template domain; and (4) carbonization of the polymeric film in nitrogen, whrein mesoporous carbon films with a hexagonal array were fabricated by sacrificing the template.

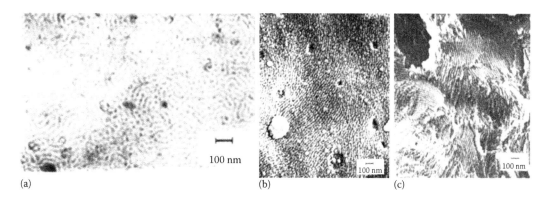

(a)      (b)      (c)

**FIGURE 23.8** TEM image of the thin film of block copolymer: (a) the surface and (b) cross section; (c) SEM image of the ozone-treated and sensed film. (Reprinted with permission from Lee, J. S. et al., *Macromolecules* 21, 274–276. Copyright 1988 American Chemical Society.)

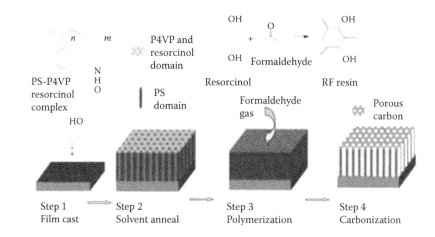

| Step 1 | Step 2 | Step 3 | Step 4 |
| Film cast | Solvent anneal | Polymerization | Carbonization |

**FIGURE 23.9** Scheme illustration of the synthesis protocol used to prepare MCN by using PS-P4VP as a soft template. (Liang, C., Hong, K., Guiochon, G. A. et al.: Synthesis of a large-scale highly ordered porous carbon film by self-assembly of block copolymers. *Angewandte Chemie International Edition*. 2004. 43. 5785–5789. Copyright Wiley-VCH Verlag GmbH & Co. KGaA. Reproduced with permission.)

The morphologies of the obtained mesoporous carbon films are shown in Figure 23.10: the nanopores were oriented perpendicular to the film surface and the Fourier transform of Z-contrast image confirmed that the film had a highly ordered hexagonal pore array. From the high-resolution SEM image, the pore size of the mesoporous carbon film was 33.3 ± 2.5 nm and the wall thickness was 9.0 ± 1.1 nm.

### 23.2.2.2 PEO–PPO–PEO (F127, P123) as Soft Template

The pluronics are a series of nonionic triblock copolymers composed of poly(ethylene oxide)-b-poly(propylene oxide)-b-poly(ethylene oxide) (PEO–PPO–PEO) chains. Owing to their amphiphilic structure, these kinds of polymers have surfactant properties that make them widely used for industrial applications, cosmetics, pharmaceuticals, and drug deliveries (Batrakova & Kabanov 2008, Guan et al. 2011). Figure 23.11 shows the basic structure of pluronic, which is composed of a central hydrophobic chain of propylene oxide flanked by two hydrophilic chains of ethylene oxide. In the field of mesoporous materials, several members of pluronics, such as P123, F127, and F108, are widely applied for soft template due to their amphiphilic properties and easy removal at relatively low temperature.

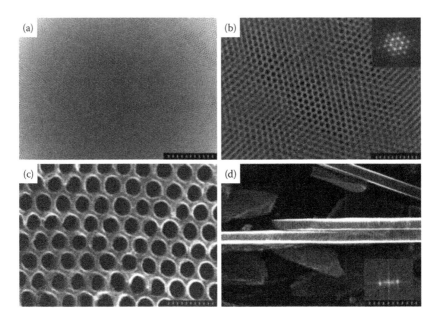

**FIGURE 23.10** Morphologies of mesoporous carbon films synthesized by using PS-P4VP as a soft template. (a) Z-contrast image of the large-scale mesoporous carbon film. The scale bar is 1 mm. (b) Z-contrast image showing details of the highly ordered mesoporous carbon structure. Inset is a Fourier transform of the image. The scale bar is 300 nm. (c) High-resolution SEM image of the surface of the mesoporous carbon film with uniform hexagonal-pore array. The scale bar is 100 nm. (d) Cross-sectional SEM image of mesoporous carbon films. Inset is a Fourier transform of the image. The scale bar is 100 nm. (Liang, C., Hong, K., Guiochon, G. A. et al.: Synthesis of a large-scale highly ordered porous carbon film by self-assembly of block copolymers. *Angewandte Chemie International Edition*. 2004. 43. 5785–5789. Copyright Wiley-VCH Verlag GmbH & Co. KGaA. Reproduced with permission.)

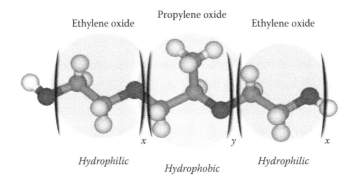

**FIGURE 23.11** Molecule structure of pluronic block copolymer. (Reprinted from *Journal of Controlled Release: Official Journal of the Controlled Release Society*, 130, Batrakova, E. V. & Kabanov, A. V., Pluronic block copolymers: Evolution of drug delivery concept from inert nanocarriers to biological response modifiers, 98–106, Copyright (2008), with permission from Elsevier.)

P123 was first used as a soft template in the synthesis of mesoporous silica in the 1990s (Brinker et al. 1999). In 1999, Lu et al. reported a rapid, aerosol-based process for synthesizing well-ordered spherical particles with stable pore mesostructures of hexagonal and cubic topology. The synthesis started with a homogeneous solution of soluble silica and surfactant of P123 which was prepared in an ethanol/water solvent. The synthetic strategy relied on the evaporation-induced self-assembly (EISA) method.

Such novel EISA process enables the rapid production of patterned porous or nanocomposite materials in the form of films, fibers, or powders. This method was also extensively adopted on the synthesis of mesoporous carbon films. Compared to other synthetic methods such as hydrothermal, EISA has several advantages: (1) EISA processes are conducted in time-effective batch operations without employing hydrothermal processing conditions; (2) the final product can be synthesized in the form of a thin film; and (3) EISA processes are able to readily produce patterns of pores (Brinker et al. 1999). Meng et al (2005, 2006) and Zhang et al (2005, 2006) developed a family of highly ordered mesoporous polymers and carbon frameworks from organic–organic assembly of P123 (soft template) with soluble, low–molecular weight resin precursors by EISA or hydrothermal strategy.

In 2005, Tanaka et al. first reported that pluronic F127 ($PEO_{100}$–$PPO_{65}$–$PEO_{100}$) was used as a soft template to synthesize ordered MCNs. In this work, resorcinol/formaldehyde and triethyl orthoacetate were used as the carbon precursors. The resulting mesoporous carbon (denoted as COU-1 in the paper) (Table 23.2) showed highly ordered mesostructures as illustrated in Figure 23.12. Hexagonally arranged pores were observed in Figure 23.12c and d, and the lattice spacing and the pore diameter were estimated as ca. 7.5 and 6.2 nm, respectively. In late 2005, Meng et al (2005) further accomplished the direct synthesis of highly ordered MCNs through EISA route by using F127 as a template. The photographic image of untreated FDU-15 in Figure 23.13a showed a transparent and soft membrane with a homogeneous organic–organic composite. The corresponding TEM images of FDU-15 in Figure 23.13b through d exhibit well-ordered hexagonal arrays of mesopores with 1D channels. The obtained ordered mesoporous organic polymers and carbon could have many potential applications in catalysis, separation, hydrogen storage, drug delivery, electrode materials, microdevices, and even bioengineering.

### 23.2.2.3 Morphology Control

Since MCNs started to attract extensively researched interest, many efforts have been made to synthesize mesoporous nanomaterials with well-controlled pore structure (Kim et al. 2005), composition (Meng et al. 2006), and porosity (Deng et al. 2010). Despite these achievements, it has come to the notice of the materials scientists that the morphology and the texture of the mesoporous nanomaterials are also important for a variety of practical applications. For example, rod-shaped mesoporous silica nanoparticles are found to be more promising in drug delivery applications than their spherical counterparts (Lu et al. 1999, Huang et al. 2011a). However, to the best of our knowledge, there are few research works reporting the successful fabrication of rod-shaped mesoporous carbon nanoparticles. The hard-template approach has been tried to prepare rod-shaped MCNs (Yu et al. 2002). However, in order to remove the silica template, the mesoporous carbon must withstand several harsh processes such as high-temperature treatment and NaOH/HF reactions, which usually reconstruct and disorder the pore structure. Thus, much attention has been paid to seek the possibility of making rod-shaped MCNs by the soft-template method. Nevertheless, the soft-template method usually produces monolithic or spherical MCNs through the EISA or hydrothermal route (Fang et al. 2010). Fang et al.'s group reported a novel low-concentration hydrothermal route to synthesize MCNs with spherical morphology and uniform size as presented in Figure 23.14. However, the MCNs obtained by this method can offer only spherical morphologies.

Limited works have been made in synthesizing rod-shaped MCNs by using the soft-template method. In the soft-template method of making MCNs, a key step is to form micelles by self-assembly of amphiphilic block copolymer. The micelles are critical for the formation of pore structures. Interestingly, recent studies also pointed out that the morphology of the micelles plays an important role in determining the morphology of the final product of mesoporous carbon nanoparticles, although the exact mechanisms for this phenomenon have yet to be understood. Therefore, it might be possible to fabricate the MCNs with a desired morphology such as rod shaped if the micelle morphology can be carefully designed and controlled.

Recently, our group (2012) reported a facile and reproducible chemical process to synthesize highly ordered MCNs with rod-shaped morphology on the basis of Fang et al.'s (2010) work. In our method,

**TABLE 23.2** Summary of MCNs Developed by Common Soft Templates

| Soft Template | Route | Carbon Source | MCN | Space Group | Surface Area (m²/g) | Application in Article | Ref. |
|---|---|---|---|---|---|---|---|
| Anionic block copolymer | Hydrothermal (acidic) | Polyisoprene | Porous carbon membrane | – | 74 | – | Lee et al. 1988 |
| HTAB[a] | Hydrothermal (basic) | ACM[b] | Carbon vesicles | – | – | – | Li et al. 2004 |
| PS-P4VP[c] (PEO) | EISA (neutral) | Resin | Porous carbon films | *P6mm* | – | – | Liang et al. 2004 |
| | EISA (neutral) | Resin | C-FDU-18 | *Fm3m* | 1510 | – | Deng et al. 2007 |
| PEO-PPO-PEO[d] | EISA (acidic) | Resin | COU-1 | – | 1354 | – | Tanaka et al. 2005 |
| | EISA (basic) | Resin | FDU series | *P6mm/ Im3m/ Ia3d* | 670–1490 | Electrode | Meng et al. 2006 Zhang et al. 2005 |
| | Hydrothermal (basic) | Resin | FDU series | *P6mm/ Im3m/ Ia3d* | 450–1040 | – | Meng et al. 2005 Zhang et al. 2006 |
| | EISA (acidic) | Resin | Mesoporous carbon | *P6mm/ Im3m* | 1549 | – | Lin et al. 2006 |
| | Phase separation (acidic) | Phloroglucinol | Mesoporous carbon | *P6mm* | 288 | Capacitive deionization | Liang & Dai 2006 Mayes et al. 2010 Tsouris et al. 2011 |
| | Hydrothermal (basic) | Resin | MCN | *Im3m* | 894–1131 | Cell permeability | Fang et al. 2010 |
| | Hydrothermal (basic) | Resin | MCN/MCG | *P6mm/ Im3m* | 642–1245 | Electrode | Li & Xue 2012 Li & Xue 2013 |
| | Hydrothermal (basic) | Melamine resin | MCG/NG | – | 588–1056 | Electrode | Li & Xue 2014 |

[a] HTAB: hexadecyltrimethylammonium.
[b] ACM: amphiphilic carbonaceous material.
[c] PS-P4VP: polystyrene-blockpoly(4-vinylpyridine).
[d] PEO-PPO-PEO: poly(ethylene oxide)-b-poly(propylene oxide)-b-poly(ethylene oxide).

low–molecular weight phenolic resol was used as a carbon-yielding component and triblock copolymer pluronic F127 as a pore-forming component. The detailed synthesis strategies are shown in Figure 23.15. In order to obtain the rod-shaped nanoparticles, it is critical to form rodlike F127 micelles in the first place. To serve this purpose, the concentration of F127 was set at ~12 wt%, which was referred to as critical micelle concentration 2 (CMC2) of F127 in aqueous solution (Huo et al. 1994). Moreover, in order to allow the well formation of the rodlike F127 micelles, the carbon-yielding component of resol was added after F127 was completely dissolved. With these carefully controlled parameters, rod-shaped MCNs were successfully fabricated (Figure 23.16d). By decreasing the F127 concentration to 9 wt%, the rod-shaped MCNs evolved to be wormlike (Figure 23.16c). When the F127 concentration was further decreased to CMC1 (~6 wt%) and below (~3 wt%), spherical MCNs with different particle sizes were

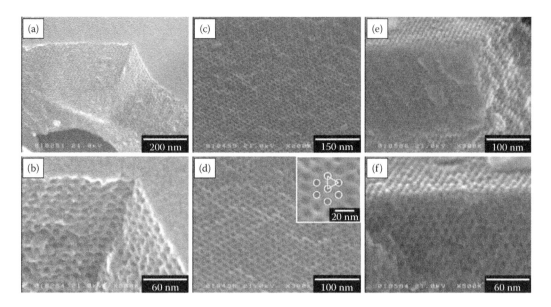

**FIGURE 23.12** SEM images of COU-1 carbonized at different temperatures: (a, b) 400°C; (c, d) 600°C; and (e, f) 800°C. (Tanaka, S. et al., *Chemical Communications*, 2125–2127, 2005. Reproduced by permission of The Royal Society of Chemistry.)

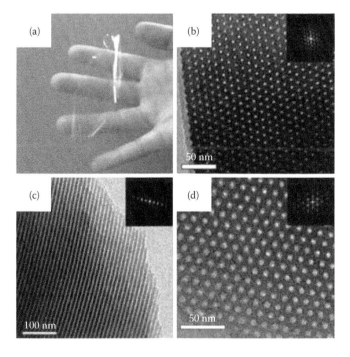

**FIGURE 23.13** (a) Photographic image of untreated FDU-15; (b–d) TEM images of FDU-15 calcined at 350°C viewed at different directions. (Meng, Y. et al.: Ordered mesoporous polymers and homologous carbon frameworks: Amphiphilic surfactant templating and direct transformation. *Angewandte Chemie International Edition*. 2005. 44. 7053–7059. Copyright Wiley-VCH Verlag GmbH & Co. KGaA. Reproduced with permission.)

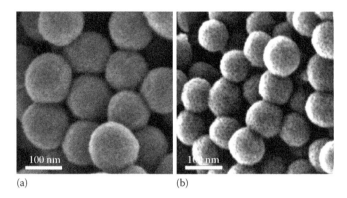

FIGURE 23.14 SEM images of ordered mesoporous nanoparticles with spherical morphology with diameters of (a) 140 and (b) 90 nm. (Fang, Y. et al.: A low-concentration hydrothermal synthesis of biocompatible ordered mesoporous carbon nanospheres with tunable and uniform size. *Angewandte Chemie*. 2010. 49. 7987–7991. Copyright Wiley-VCH Verlag GmbH & Co. KGaA. Reproduced with permission.)

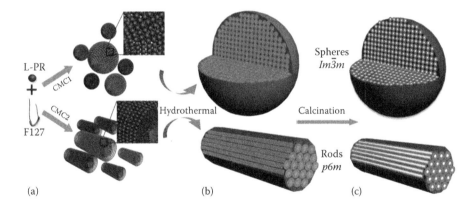

FIGURE 23.15 Schematic illustration of the general strategies to prepare ordered mesoporous polymer and carbon nanoparticles with spherical to rodlike morphologies: (a) composite resol-F127 monomicelles; (b) ordered mesoporous polymer nanoparticles; and (c) ordered mesoporous carbon nanoparticles. (*Journal of Colloid and Interface Science*, 377, Li, M. & Xue, J., Ordered mesoporous carbon nanoparticles with well-controlled morphologies from sphere to rod via a soft-template route, 169–175, Copyright (2012), with permission from Elsevier.)

obtained (Figure 23.16a and b). Moreover, the highly ordered mesostructure can be readily turned from 2D hexagonal (*p6m*) to 3D caged cubic (*Im3m*) along with the tuning of morphologies from rod shaped to spherical. In this work, we also demonstrated that the obtained mesoporous carbon materials showed excellent capacitance in the application as supercapacitors, because of their large surface area, highly ordered mesostructure, and continuous electron transport framework.

Based on the discussion in the aforementioned sections, hard-template synthesis would incur multiple synthesis steps (synthesis of silica template, impregnation of carbon precursor, carbonization, and finally etching of silica template), which were normally time consuming and costly. Moreover, the MCNs prepared by the hard-template route had to withstand harsh process of strong acid/base reactions, and these processes would normally result in the reconstruction and disorder of the mesopores. Thus, the soft-template route was highly recommended for the synthesis of MCNs. Fewer steps were needed in the preparation process and this opens a new way to fabricate mesoporous materials. A detailed comparison in terms of merits and demerits between soft-template and hard-template strategies is summarized in Figure 23.17 based on the review paper by Ma et al. (2013).

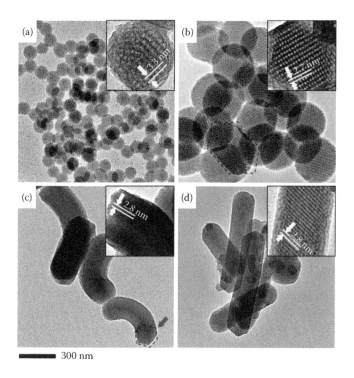

**FIGURE 23.16** TEM images of the ordered MCNs with different morphologies: (a) MCN-S50; (b) MCN-S200; (c) MCN-W wormlike nanoparticles; and (d) MCN-R rodlike nanoparticles; insets are the respective HRTEM images labeled with average pore diameter. (*Journal of Colloid and Interface Science*, 377, Li, M. & Xue, J., Ordered mesoporous carbon nanoparticles with well-controlled morphologies from sphere to rod via a soft-template route, 169–175, Copyright (2012), with permission from Elsevier.)

| Differences | Direct synthesis | Nanocasting |
|---|---|---|
| Procedure | Fewer steps: phenolic resin/surfactant mesophase → surfactant removal → carbonization | Multiple steps: silica/surfactant mesophase → mesoporous silica → carbon precursor filling → carbonization → silica removal |
| Template | Block copolymers | Mesoporous silicas |
| Template removal | Merely heat treatment | Etching by toxic HF |
| Mesophase control | Tunable architectures and surface properties, dependent on pH value, surfactants, temperature, etc. | Fixed and restricted by the preformed structures of hard template |
| Pore size | Highly uniform, narrowly distributed | Relatively wider than silica templates |
| Stability | Mechanically stable, due to the 3D continuous framework | Relatively unstable, due to being composed of nanoarrays connected by small and short nanorods |
| Prospect | Low-cost, convenient, suitable for large-scale industrial applications | Expensive, time-consuming, unsuitable for large-scale industrial production |

**FIGURE 23.17** A comparison between direct synthesis (soft template) and nanocasting (hard template) strategies to prepare MCNs. (Ma, T.-Y. et al., *Chemical Society Reviews* 42, 3977–4003, 2013. Reproduced by permission of The Royal Society of Chemistry.)

# 23.3 Applications of MCNs

MCNs are extensively applied in diverse fields, such as energy storage, catalysis, adsorption, and chromatography, because such materials have an ability to interact with atoms, ion molecules, and foreign nanoparticles not only on their surface, but also throughout the bulk of the materials. Owing to the often-cited desirable physical and chemical properties of MCNs, such materials are the most widely used materials for electrodes or electrode carriers in the energy storage applications. These desirable properties include ease of processability, high surface area, controllable porosity, relatively inert electrochemistry, variety of form (powders, films, composites, monoliths, etc.), and electrocatalytic active sites for a variety of redox reactions (Pandolfo & Hollenkamp 2006, Frackowiak 2007, Zhang & Zhao 2009). Section 23.3.1 will focus on the application of MCNs in energy storage, with special emphasis on the recent progress on EDLCs. Furthermore, the applications in environment-related areas will be reviewed in Section 23.3.2 as well. Information on drug delivery, separation, adsorption, and biology-related field will also be discussed.

## 23.3.1 MCNs for Applications in Energy Storage

To address the increasingly serious environmental problems and excessive consumption of fossil fuels, there has been great urgency in exploiting and refining clean energy technologies as well as more efficient energy storage devices. As one of such energy storage devices, supercapacitors have been widely studied around the world in recent years because they can offer energy density orders of magnitude higher than those of conventional capacitors, lager power density, and better cycling stability than batteries. The electrochemical capacitors (also known as supercapacitors or ultracapacitors) can be divided into two categories according to the charging–discharging mechanism (Simon & Gogotsi 2008). The first group is electrochemical double-layer capacitors (EDLCs, which are based on the physisorption of electrolyte ions onto a porous electrode. The second group is known as pseudocapacitors or redox supercapacitors, which utilize fast and reversible surface or near-surface reactions for charge storage. The typical materials for this group of supercapacitors are transitional metal oxide and conducting polymer. The commercially available supercapacitors are mainly EDLCs, which store energy through the adsorption of both anions and cations at the interface between electrode and electrolyte (Zhang & Zhao 2009). Electrode materials are usually considered to play the most important role in the supercapacitors. Among various electrode materials, MCNs have occupied a special place in development of EDLCs due to their high surface area and moderate cost (Gamby et al. 2001, Babel & Jurewicz 2004, Barbieri et al. 2005, Frackowiak 2007).

The typical construction of EDLC is illustrated in Figure 23.18 (Nishihara & Kyotani 2012). AC is mainly used as electrode material owing to its high surface area and excellent conductivity. When a potential is applied to an electrode, oppositely charged ions are attracted onto the nanopore surface

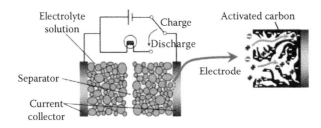

**FIGURE 23.18**  Construction of EDLC. (Nishihara, H. & Kyotani, T.: Templated nanocarbons for energy storage. *Advanced Materials.* 2012. 24. 4473–4498. Copyright Wiley-VCH Verlag GmbH & Co. KGaA. Reproduced with permission.)

of the activated carbon and electric double layers are formed, thereby electricity is being stored (Nishihara & Kyotani 2012). Several critical criteria have been proved for an ideal supercapacitor with high performance: high energy density, high power density, and long cycle life (Conway 1999, Bose et al. 2012, Wang et al. 2012). According to the equation of energy density $E$, $E = \frac{1}{2}CV^2$ (where $C$ is the total capacitance of the cell and $V$ is the cell voltage), the energy density of EDLCs is mainly dependent on specific capacitance, which significantly relies on the surface area of the EDLC electrode materials (Inagaki et al. 2010, Li et al. 2011, Zhu et al. 2011, Zhang et al. 2012). AC has been long used as the commercial electrode material for supercapacitors due to its high surface area and moderate cost (Gamby et al. 2001, Babel & Jurewicz 2004, Barbieri et al. 2005, Frackowiak 2007). However, its application is strongly restricted by its relatively poor energy storage capacity and inferior rate capability. Besides AC, MCNs have received much attention in the past decade due to their high surface area and suitable pore size (Li et al. 2007).

Recently, our group developed a series of MCNs for application as supercapacitor electrodes. The physicochemical and electrochemical properties of the as-obtained MCNs are listed in Table 23.3. The electrochemical measurements of cyclic voltammetry (CV), galvanostatic charge/discharge, and impedance spectrum were used to investigate the electrochemical performance of the as-made mesoporous carbon electrodes. The electrochemical observations also revealed that our carbon electrodes offered reduced equivalent series resistance and excellent cycling stability. Such results demonstrated that the as-obtained mesoporous carbon materials were promising candidates of electrode materials for high-performance energy storage devices.

### 23.3.1.1 Mesoporous Carbon-Decorated Graphene Electrode Materials for Supercapacitor with Improved Conductivity

According to the expression of power density in energy storage devices $P$, $P = V^2/4R_s$ (where $R_s$ is the equivalent series resistance), the high internal resistance will ultimately lead to relatively poor performance in power density (Zhang et al. 2012). In order to fabricate an energy storage device with high energy density and power density, carbon electrode materials not only have the ability of storing more energy but also endow high conductivity. Given that many of the MCNs have complicated microstructure and disordered texture framework, which will significantly limit the efficiency of electron transfer and result in high internal resistance of electrodes, our group fabricated a series of mesoporous carbon-decorated graphene (MCG) electrode materials (Li et al. 2013). The introduction of graphene provided an excellent alternative to the pristine mesoporous carbon electrode materials (Huang et al. 2011b, Huang et al. 2012a). In comparison with the pristine mesoporous carbon electrode materials, graphene possesses very high electrical conductivity, which will greatly contribute to the fast transportation of electrons during the charge/discharge process, especially in the high sweep rate (Yang et al. 2011, Lei et al. 2012, Zhang et al. 2012). Therefore, graphene is a promising candidate for electrode materials to realize high power density. On the basis of the past research work, graphene could be obtained by chemically reducing exfoliated (GO) using reductant such as hydrazine (Stankovich et al. 2007, Gao et al. 2009, Park & Ruoff 2009). However, the loss in negative-charged functional groups of GO after the chemical reduction will dramatically lower the electrostatic repulsion between GO sheets, which will eventually lead to heavy aggregation of the reduced graphene oxide (rGO). As an effort to solve this problem, Fan et al. (2010) reported the preparation of 3D CNT/graphene sandwich structures with CNT pillars grown between the graphene layers by using the CVD method. Similarly, GO/CNT hybrid films were also prepared as an effective electrode material through a film casting process (Huang et al. 2012b). Moreover, rGO sheets intercalated with mesoporous carbon spheres were prepared for addressing the aforementioned issues using a hard-template method (Lei et al. 2011). Although tedious preparation processes were involved in these strategies, the results clearly showed the necessity and possibility of enhancing the performance of EDLC by decorating graphenes with MCNs (Li et al. 2012).

**TABLE 23.3**   Physicochemical and Electrochemical Properties of Various MCNs

| Sample[a] | Morphology | Surface Area (m²/g)[b] | Total Pore Volume (cm³/g) | Average Pore Size (nm)[c] | C (wt%)[d] | N (wt%)[d] | N/C Ratio[d] | Carbonized Temperature (°C)[e] | Specific Capacitance –III (F/g)[f] | Ref. |
|---|---|---|---|---|---|---|---|---|---|---|
| MCN-50 | Spherical | 975 | 0.78 | 3.5 | – | – | – | 420–700 | 112 | Li & Xue 2012 |
| MCN-150 | Spherical | 1056 | 1.14 | 5.5 | – | – | – | 420–800 | 127 | Li & Xue 2013 |
| MCN-200 | Spherical | 1385 | 0.92 | 2.7 | – | – | – | 420–700 | 142 | Li & Xue 2012 |
| MCN-W | Wormlike | 952 | 0.60 | 2.7 | – | – | – | 420–700 | 98 | Li & Xue 2012 |
| MCN-R | Rodlike | 1300 | 0.84 | 2.7 | – | – | – | 420–700 | 130 | Li & Xue 2012 |
| MCG-5 | Nanosheets | 903 | 0.58 | 10.7 | – | – | – | 420–800 | 169 | Li & Xue 2013 |
| MCG-20 | Nanosheets + spherical | 927 | 0.64 | 6.3 | – | – | – | 420–800 | 213 | Li & Xue 2013 |
| MCG-40 | | 1091 | 0.70 | 4.6 | – | – | – | 420–800 | 134 | Li & Xue 2013 |
| N-MCN | Spherical | 962 | 0.84 | 6.9 | 75.00 | 4.12 | 0.055 | 420–600 | 222.9 | Li & Xue 2014 |
| HN-MCN | Nanoparticles | 850 | 0.81 | 8.6 | 78.12 | 5.46 | 0.070 | 420–600 | 238 | Li & Xue 2014 |
| N-G | Nanosheets | 588 | 0.56 | 11.1 | 75.75 | 6.63 | 0.088 | 420–600 | 289 | Li & Xue 2014 |

[a] Abbreviations: MCN: mesoporous carbon nanoparticles; W: wormlike; R: rodlike; MCG: mesoporous carbon decorated graphene; N-MCN: nitrogen-doped mesoporous carbon nanoparticle; HN-MCN: highly nitrogen-doped mesoporous carbon nanoparticle; N-G: nitrogen-doped graphene.

[b] Calculated by the BET model from the adsorption branches of the isotherms.

[c] Calculated by the multipoint DFT model from desorption data.

[d] Data obtained from combustion elemental analysis.

[e] The calcination was carried out in two steps. First, the sample was heated to 420°C in nitrogen environment and kept for 30 min to remove the template. Second, the sample was continuously heated to final temperature to carbonize the product.

[f] Calculated from the galvanostatic discharge at the current density of 0.2 A/g (based on three-electrode system).

We demonstrated a facile and reproducible chemical process to prepare a novel composite of graphene and mesoporous carbon, in which each graphene sheet was fully covered with a layer of mesoporous carbon (Li et al. 2013). Such a composite structure was very promising for application as supercapacitor electrodes. As illustrated in Figure 23.19, the mesoporous carbon layer served as an effective inhibitor of the aggregation between graphene sheets and a place of charge accumulation, while the graphene sheets acted as a highly conductive carbon frame. The porous carbon-decorated graphenes were synthesized by self-assembling polymeric micelles onto GO sheets under the driving of hydrogen bonding between the micelles and the functional groups of GO. After thermal treatment, the polymeric micelles were converted into mesoporous carbons, which were then firmly attached onto the graphene sheets. As a result, the graphene sheets were effectively separated by the mesoporous carbon layers and would not be aggregated easily. With increasing amount of polymeric micelles introduced, mesoporous carbon spheres could be formed between the graphene sheets beside the mesoporous carbon layers attached on the graphene sheets. These mesoporous carbon spheres not only further improved the SSA of the composites but also inhibited the aggregation of the graphene sheets more effectively. Compared to the pristine graphene, the mesoporous carbon-decorated graphene reported in this work had larger surface area and more favorable pore size distribution, which exhibited higher specific capacitance and better cycle stability.

The electrochemical properties of mesoporous carbon-decorated graphene are shown in Figure 23.20. Rectangular CV curves suggested their good double-layer capacitive behaviors. The nearly symmetric charge–discharge profiles suggested the dominant EDLC as well as subordinate pseudocapacitance storage mechanism due to the presence of O-containing functional groups. Figure 23.20c shows the variation in specific capacitance at different current densities. It is noted that the capacitance of MCG electrode was up to 213 F/g at low current density (listed in Table 23.3). A slight decrease in the capacitance of MCG samples was observed from the current density of 0.5 to 20 A/g. Furthermore, the capacitances of the three samples were maintained well under high current densities, suggesting that these electrode materials had good capacitance retention capability. Figure 23.20d shows the cycling stability of the MCG electrode. After 1000 cycles, the capacitance retention remained at 95.4%, revealing that the MCG-5 electrode displayed good stability behavior as supercapacitor electrode materials.

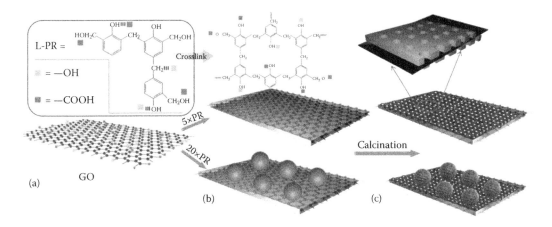

**FIGURE 23.19** Schematic illustration of the general strategies to prepare mesoporous carbon-decorated graphene: (a) GO; (b) GO nanosheets patterned by polymer layer/spheres; and (c) graphene nanosheets decorated by porous carbon (Li, M. et al., *Journal of Materials Chemistry A* 1, 7469, 2013. Reproduced by permission of The Royal Society of Chemistry.)

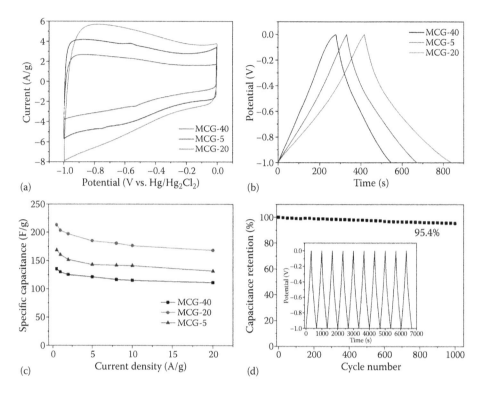

**FIGURE 23.20** Electrochemical properties of mesoporous carbon-decorated graphene electrodes. (a) CV results measured at the scan rate of 25 mV/s; (b) galvanostatic charge–discharge curves at a current density of 0.5 A/g; (c) variation in specific capacitance at different current densities; and (d) cycle performance of MCG-5 at a current density of 0.5 A/g. (Li, M. et al., *Journal of Materials Chemistry A* 1, 7469, 2013. Reproduced by permission of The Royal Society of Chemistry.)

### 23.3.1.2 N-Doped Mesoporous Carbon Electrode Materials for Supercapacitor with Improved Capacitance

In order to further enhance the capacitance of carbonaceous electrode, many efforts have been employed to develop the structural design of porous carbon from pore adjusting to morphology transformation. As a result, 1D, 2D, and 3D structure of carbonaceous materials were developed one after another in recent years. For example, MCNs, a 3D form of carbon atoms with hierarchical porous channels, were considered as one of the most promising electrode materials to realize high-energy EDLCs in the past decades. Besides the structural design, another way to improve the capacitance was to introduce the pseudocapacitance, such as transitional metal oxide and heteroatoms. Transitional metal oxides associated with carbonaceous electrodes normally suffered from poor rate capacity and short cycle life because of their poor conductivity and instability. Comparatively, introducing heteroatoms (e.g., N, O, P, and B), especially nitrogen, enables the improvement of the carbonaceous electrode performance while maintaining the excellent intrinsic characteristics of carbonaceous materials. Posttreatment method was the primary pathway to introduce heteroatoms into carbon materials. In this method, preformed carbonaceous materials such as ACs (Hulicova-Jurcakova et al. 2009a), CNTs, templated carbons (Hulicova-Jurcakova et al. 2009b), and graphenes (Jeong et al. 2011) were treated with N-containing chemicals such as ammonia or urea to attach N atoms. For instance, nitrogen-doped graphene (NG) was fabricated by Li et al. through annealing treatment of GO in the atmosphere of ammonia, and the resulting NG exhibited some unique properties, including improved conductivity and enhanced capacitance value (Li et al. 2009a). However, it is difficult to prepare stable NG with uniform dispersion and tunable N-doping amount by using this method (Wen et al. 2012). In order to get more uniform distribution of the heteroatoms in mesoporous carbon, in situ synthesis method has been developed, of which N-containing polymeric materials such as polyacrylonitrile (Lu et al. 2003, Zhong et al. 2012), polyaniline (Vinu et al. 2007, Li et al. 2010), pyrrole (Yang et al. 2004, Fuertes & Centeno 2005, Chen et al. 2012, Wei et al. 2012), melamine (Kailasam et al. 2009, Lee et al. 2013), or N-containing ionic liquids (Guo et al. 2013) were employed as precursors. For example, Ma et al. (2012) reported a template-free approach for the preparation of N-doped hollow carbon microspheres by direct pyrolysis of solid melamine–formaldehyde resin spheres. Although such hollow carbon spheres possessed an SSA of 753 m$^2$/g, the type I isotherm curve obviously indicated that most of the surface area was contributed by the microporous pores, which was useless for the electrolyte ion adsorption as electrode materials. Moreover, it is hard to tune the nitrogen-doping amount through the direct pyrolysis of N-containing polymer because the N/C ratio is fixed before the polymerization process. Later, Guo et al. (2013) reported the synthesis of N and B codoped porous carbon through the self-assembly of poly(benzoxazine-*co*-resol) with ionic liquid C$_{16}$mimBF$_4$. However, only limited nitrogen content (~1.2 wt%) was doped by using this method, and it is difficult to obtain ordered mesostructure carbon, although the soft template F127 was employed in the synthesis. The ordered mesostructure is a critical factor in order to achieve high surface area and facilitate the ion transportation. Based on the literature referring to the in situ synthesis approach, tunable N-doped carbon with maintaining of ordered mesostructure was seldom synthesized, and this situation will inevitably hinder their large-scale production.

To this end, our group developed an integrated and facile strategy to fabricate N-doped mesoporous carbon (NC) by using melamine resin as the nitrogen source and phenolic resin as the carbon source (Li & Xue 2014). The melamine resin was chosen as the nitrogen source because of its large N content and high reactivity of amino group. Moreover, melamine resin is an inexpensive and commercially available triazine polymer which is normally used in plastic, medicinal, organic coatings, and paper industries (Hulicova et al. 2006, Ma et al. 2012). With the aim of maintaining ordered mesoporous structure and high carbon yield, melamine resin was modified by phenolic resin (serving as carbon source) through the conjunction of –NH– groups with carbon atoms. Two promising aspects of NC include the potential to maintain morphology and ordered mesostructure after introducing nitrogen and the straightforward in situ mechanism for the incorporation of nitrogen atoms into the carbon framework (synthesis strategies illustrated in Figure 23.21). Additionally, the N-doping amount could be easily tuned by controlling the mass ratio of N precursor and C precursor before the polymerization process. The TEM image of NCs is shown in Figure 23.22a. The NCs were in spherical morphology and had an average particle size of 200 nm. Also, the HRTEM image clearly shows their mesostructure with the estimated pore size of 6 nm. By increasing the amount of N source melamine resin introduced, the morphology of NC–H was transformed to larger particles and had more ordered structure (Figure 23.22b). It was supposed that the morphology evolution was related with the interaction between triazine-rich micelles during the self-assembly process.

In order to further enhance the electrochemical performance of such N-doped carbon material, 2D carbon frameworks of graphene sheets with high electronic conductivity were used as a doping matrix. In this NG, each graphene sheet was fully covered with a layer of N-doped mesoporous carbon. This kind of doping could be thought of as an "indirect" doping approach without destroying the excellent conductivity of graphene by maintaining the integrity of graphene sheets. Compared to the pure mesoporous carbon materials, the NC and NG reported in this work exhibited great potential as an efficient electrode material for supercapacitor with higher specific capacitance and excellent rate performance (Table 23.3). Furthermore, all of the precursors used in the experiment are commercially available and suitable for scale-up production.

### 23.3.1.3 Mesoporous Carbon-Based Electrode Materials for Application in LIBs

In conjunction with the emergence of electric vehicles, there has been a resurgence of interest in the development of next-generation LIBs with high energy/power density and cycling stability (Tarascon & Armand 2001, Armand & Tarascon 2008). Mesoporous structure has been proven to be beneficial for LIB application. In previous literature, Yuan et al. (2011a,b) synthesized electrodes consisting of mesoporous

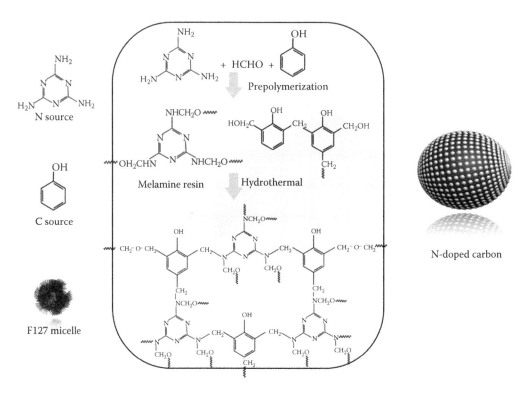

**FIGURE 23.21** Schematic illustration of the general strategies to prepare nitrogen-doped mesoporous carbon. (Reprinted with permission from Li, M. & Xue, J., *Journal of Physical Chemistry C* 118, 2507–2517. Copyright 2014 American Chemical Society.)

metal oxides and carbon coating with superior lithium storage performances. Moreover, mesoporous carbon was considered as an ideal framework to encapsulate electrochemically active particles for LIB applications (Fan et al. 2004, Grigoriants et al. 2005, Hwang et al. 2013). The ordered mesopores provide confined spaces for the accommodation of electrochemically active particles, where the volume expansion of these particles can be buffered by mesopores and carbon matrix; while the interlaced networks of interconnecting pores and carbon framework provide perfect channels for electrolyte penetration and electron conduction, respectively (Shen et al. 2012).

In our recent work, with the aim of reaching high power density and satisfactory cyclic stability, composite anode materials composed of iron oxide ($Fe_3O_4$) nanoparticles with a diameter of around 5 nm confined inside MCNs with uniform size (~70 nm) were synthesized (Chen et al. 2014). $Fe_3O_4$ was chosen as the electrochemical active component in such composite, due to its high theoretical capacity, ease of fabrication into nanosize, and compatibility with the overall synthesis (Poizot et al. 2000, Xiao et al. 2011). To boost the power density of an anode material, nanosized $Fe_3O_4$ particles were preferred due to their significant shorter lithium path length compared with their bulk counterpart (Wang et al. 2010). Moreover, their small size partially buffers the stress and strain related to the particle volume expansion/contraction during the lithium insertion/removal (Guo et al. 2008). Furthermore, the resultant high surface area permits a higher contact area between electrolyte and electrode, thus achieving a higher lithium flux across the interface (Bruce et al. 2008). Being embedded in mesoporous carbon sphere, these nanosized $Fe_3O_4$ particles were in direct contact with both lithium ion and electron transportation media, namely, electrolyte and carbon, which was beneficial for high-rate performances. Each $Fe_3O_4$ nanoparticle was held in position by the carbon matrix, preventing the particles from agglomeration upon cycling. The structure and the formation process of such $Fe_3O_4$ embedded in mesoporous carbon

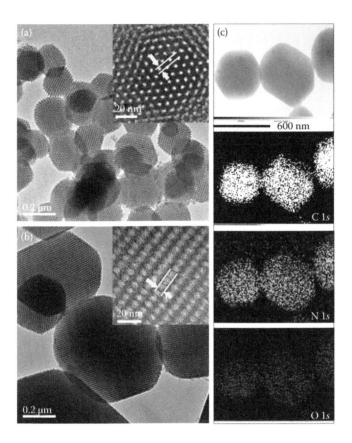

**FIGURE 23.22** TEM images of (a) NC (inset is the respective HRTEM image labeled with average pore diameter); (b) highly N-doped carbon (NC–H); (c) NC–H TEM mapping images with the elements of carbon, nitrogen, and oxygen. (Reprinted with permission from Li, M. & Xue, J., *Journal of Physical Chemistry C* 118, 2507–2517. Copyright 2014 American Chemical Society.)

(IONP@mC) are presented in Figure 23.23. The 3D carbon spheres provided the robustness which was necessary to maintain the structure integrity, as well as the cyclic stability of such anode composite. Compared with the commonly reported graphene in metal oxide composite electrodes, mesoporous carbon had several advantages: relative ease of synthesis via a low-temperature hydrothermal method (Li & Xue 2012), large quantity that can be obtained after every hydrothermal process (Fang et al. 2010), and a much safer fabrication process that does not include highly concentrated acids.

The electrochemical performances of IONP@mC were examined using CR2016 coin cells with lithium metal as counterelectrode and the results are presented in Figure 23.24. The charge/discharge profiles of IONP@mC in Figure 23.24a and b shows a plateau around 0.8 V versus $Li^+/Li^0$ in the first discharge curve, corresponding to the reduction of $Fe_3O_4$ to $Fe^0$ in conversion reaction: $Fe_3O_4 + 8e^- + 8Li^+ \rightarrow Fe^0 + 4Li_2O$ (Kang et al. 2011). In addition, the large surface area of the as-obtained mesoporous structure promoted the formation of solid electrolyte interface, which also contributed to the capacity loss in the first cycle. To demonstrate its advantage for high-rate LIB application, the rate performance of IONP@mC was further tested under higher current densities from 3000 to 10,000 mA $g^{-1}$. As shown in Figure 23.24d, the capacities of IONP@mC-600 (blue spheres) under all current densities were quite stable, showing superior rate capability. The superior cyclic stability of IONP@mC anode materials was further demonstrated by testing the same cell under 2000 mA $g^{-1}$ for 500 cycles. As shown in Figure 23.24e, after being cycled at various high current densities up to 10,000 mA $g^{-1}$, the reversible capacity of

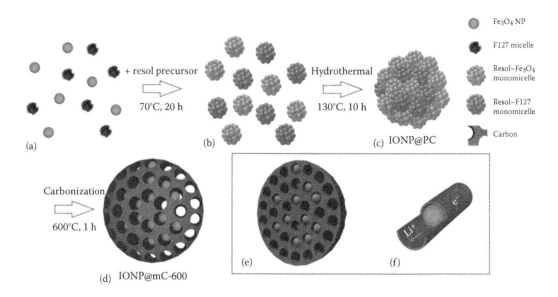

**FIGURE 23.23** Schematic illustration of the formation of $Fe_3O_4$ nanoparticles embedded in MCNs: (a) IONPs and F127 micelles, (b) resol–$Fe_3O_4$ and resol–F127 monomicelles, (c) IONP@PC, (d) IONP@mC, (e) cross section of IONP@mC, and (f) magnified representation of a channel of IONP@mC. (Chen, Y., Song, B., Li, M. et al.: $Fe_3O_4$ nanoparticles embedded in uniform mesoporous carbon spheres for superior high-rate battery applications. *Advanced Functional Materials*. 2014. 24. 319–326. Copyright Wiley-VCH Verlag GmbH & Co. KGaA. Reproduced with permission.)

IONP@mC-600 was restored to 510 mAh $g^{-1}$ at the 1st cycle. Subsequently, the reversible capacity quickly increased to above 530 mAh $g^{-1}$ after 6 cycles and stabilized around 560 mAh $g^{-1}$ during the subsequent cycles. At the 500th cycle, under the current density of 2000 mA $g^{-1}$, which is also the overall 790th cycle, such anode fabricated from IONP@mC-600 was able to deliver a reversible capacity of 547.8 mAh $g^{-1}$. The anode of IONP@mC-450 was also further cycled at 2000 mA $g^{-1}$ for 500 cycles, showing inferior performance with a reversible capacity of 332.5 mAh $g^{-1}$ at the last cycle.

## 23.3.2 MCNs for Applications in Drug Delivery and Environment-Related Fields

The applications of MCNs (or functionalized MCNs) including water purification, propane dehydrogenation, oxygen reduction, $CO_2$ capture, catalysis, magnetism adsorption, and drug delivery have also been widely reported in the past years (Lee et al. 2005, Wang et al. 2006, 2007, Zhang et al. 2008, Frank et al. 2009, Li et al. 2009b, Zhuang et al. 2009). This section will focus on the application of MCNs in drug delivery based on our research works. Moreover, one typical example on the application of MCNs in water purification will be presented in the latter part of this section.

Drug delivery system is defined as "a formulation or a device that enables the introduction of a therapeutic substance in the body and improves its efficacy and safety by controlling the rate, time, and place of release of drugs in the body" (Jain 2008). As a drug delivery system is an interaction between the drug and the patient, its ability to control the drug release in terms of amount, time, and the specificity is very important for an effective treatment. The problems with immediate release (first-order drug release) were the short period of efficacy and the possibility of drug plasma level to reach toxicity level. Thus, attaining a controlled zero-order drug release was desirable due to its stable sustained release kinetic profile which can reduce the problems that may be introduced by first-order drug release. In order to achieve this, a drug delivery system, such as polymeric carriers, could be used. Drug delivery system

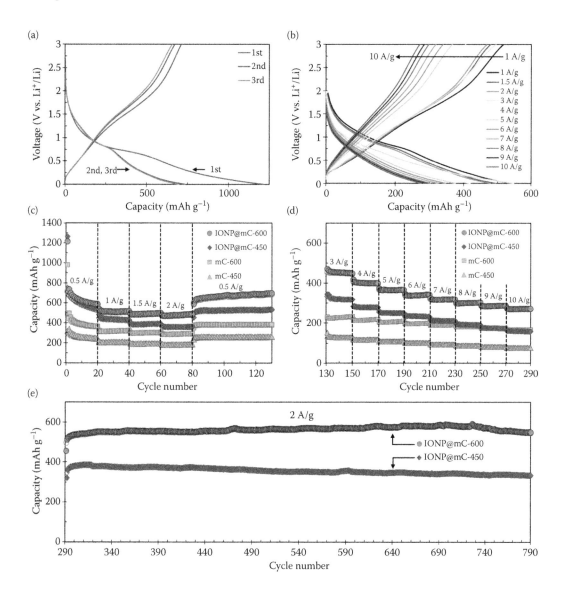

**FIGURE 23.24** Electrochemical performances of IONP@mC-600 (carbonized at 600°C), IONP@mC-450 (carbonized at 450°C), IONP@PC (without carbonization). Charge/discharge profiles of IONP@mC-600 at (a) current densities of 500 mA g⁻¹ and (b) higher rates from 1000 to 10,000 mA g⁻¹. Rate capability tests of IONP@mC-600 (sphere), IONP@mC-450 (diamond), mC-600 (square), and mC-450 (triangle) (c) from 500 to 2000 mA g⁻¹ and (d) from 3000 to 10,000 mA g⁻¹. (e) Subsequent cycling tests of IONP@mC-600 and IONP@mC-450 at 2000 mA g⁻¹ for 500 cycles. (Chen, Y., Song, B., Li, M. et al.: Fe₃O₄ nanoparticles embedded in uniform mesoporous carbon spheres for superior high-rate battery applications. *Advanced Functional Materials*. 2014. 24. 319–326. Copyright Wiley-VCH Verlag GmbH & Co. KGaA. Reproduced with permission.)

relied on carriers, more commonly polymeric carriers, to achieve an ideal zero-order drug release. Some of the conventional carriers used as controlled drug delivery system were polymeric carriers, collagen, liposomes, microsphere, glass-like sugar matrices, resealed red blood cells, and antibody-targeted system (Jain 2008). Among the common drug delivery system carriers, mesoporous carbon nanoparticle has sparked great interest among researchers due to its advantageous properties which made it a promising candidate as a drug delivery system carrier (Kim et al. 2008). Mesoporous carbon nanoparticle is exceptionally useful in drug delivery application. This is because of its high surface area (>1000 m²/g), large pore volume (>1 cm³/g) (Kim et al. 2008), bioinertness and biocompatibility, numerous ways to load the drugs (such as surface adsorption, pore filling, and incorporation into carbon matrix) (Yan et al. 2006), and most importantly low cytotoxicity (Gu et al. 2011). A high surface area and a large pore volume enable larger amounts of drugs to be loaded, while numerous drug loading methods made mesoporous carbon nanoparticles a convenient carrier in drug delivery application. Among these desirable properties, the most crucial criteria when considering mesoporous carbon nanoparticle as a drug delivery system carrier are its biocompatibility and low cytotoxicity as it could reduce the chances of immunogenic response which was potentially harmful to the human body.

Mesoporous carbon nanoparticles could also be modified or unmodified in terms of its surface chemistry. Modified mesoporous carbon nanoparticles allow drugs to be attached through functionalized groups within the pore surfaces and eventually control the drug adsorption/absorption and release rate (Vallet-Regí et al. 2007). Besides surface modification, morphologies of the mesoporous carbon nanoparticles were another critical parameter which can determine penetration or destruction. For example, if mesoporous carbon nanoparticles were shaped like that of a bacteriophage or some parasite, it could invoke an immune response for the human body to eliminate the foreign particle. This elimination or destruction of mesoporous carbon nanoparticles could render its role as drug delivery carrier useless. Thus, to be an effective drug delivery carrier, mesoporous carbon nanoparticles' morphology has to be controlled as we have discussed in Section 23.3.2. Researchers have shown that rodlike mesoporous carbon nanoparticles could penetrate cells better than spherical mesoporous carbon nanoparticles (Vallet-Regí et al. 2007, Fox et al. 2009, Venkataraman et al. 2011). Figure 23.25 shows the

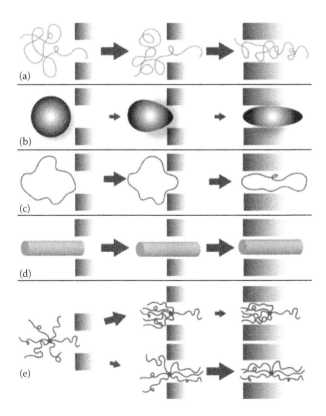

**FIGURE 23.25** Polymer architecture and the passage of a polymer through a pore. (Reprinted with permission from Fox, M. E. et al., *Accounts of Chemical Research* 42, 1141–1151. Copyright 2009 American Chemical Society.)

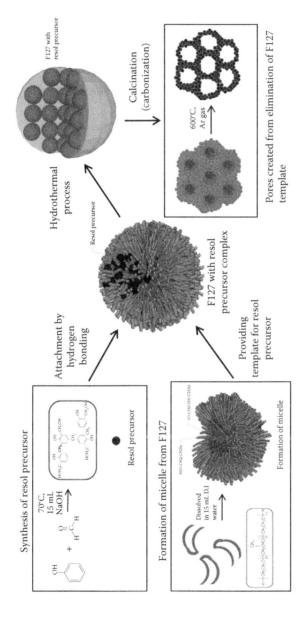

**FIGURE 23.26**  Schematic illustration of the synthesis route to mesoporous carbon nanoparticles.

process of the penetration of different polymeric shapes into a cell pore. For Figure 23.25b, a globular polymer had to undergo substantial deformation to enter and pass through a pore. This required considerable energy, therefore making this process challenging. However, when a rodlike polymer, which is depicted in Figure 23.25c, was used for drug delivery, it could easily enter and pass through the pore. Also, the morphology of the mesoporous carbon nanoparticles can help prevent drugs from entering certain organs. Therefore, the shape of the mesoporous carbon nanoparticles could affect drug delivery application, especially the penetration rate. However, currently, there are limited researches on the morphological control of mesoporous carbon nanoparticles which hinders the purpose of our project: to control drug delivery application. Thus, there is a need to study and understand the morphology control of mesoporous carbon nanoparticles to control the delivery rate.

In order to overcome this challenge, our group recently conducted a project for the synthesis of mesoporous carbon nanoparticles with controlled morphology for the applications in drug delivery. In this project, the soft-template synthesis was used. Several parameters such as pH, temperature, and reaction time were explored to fine-tune the overall size and morphology of the mesoporous carbon nanoparticles. The soft-template synthesis route of mesoporous carbon nanoparticles was achieved with pluronic F127 template. A schematic diagram of the synthesis route used in this experiment is portrayed in Figure 23.26. There are three steps in the synthesis: (1) resol formation (F127 concentration is identified as a potential parameter); (2) hydrothermal process (two potential parameters are identified which are hydrothermal temperature and pH value); and (3) calcination to remove the template.

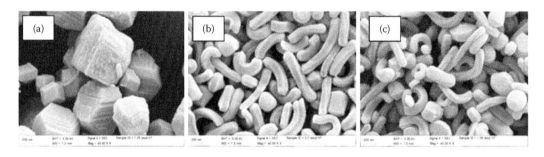

**FIGURE 23.27** SEM images of MCNs with different morphologies via controlling hydrothermal temperature: (a) MCNs at hydrothermal temperature 110°C; (b) MCNs at hydrothermal temperature 130°C; and (c) MCNs at hydrothermal temperature 150°C.

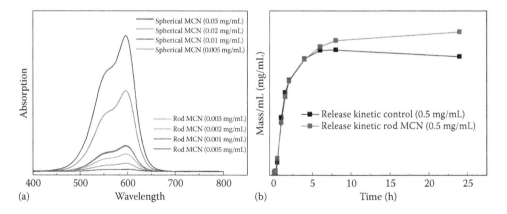

**FIGURE 23.28** (a) UV–Vis results of absorption experiment for MCNs with different morphologies; (b) drug release kinetics of rodlike MCN and control.

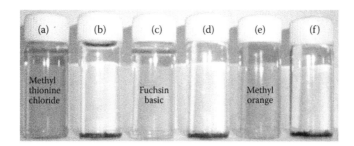

**FIGURE 23.29** Optical photographs of dye-polluted water (a, c, e) before and (b, d, f) after adsorption by the MCNs. (Reprinted with permission from Zhuang, X. et al., *Chemistry of Materials* 21, 706–716. Copyright 2009 American Chemical Society.)

Figure 23.27 shows the transitional morphologies of mesoporous carbon nanoparticles by increasing the hydrothermal temperature. Across Figure 23.27a (110°C) to b (130°C) to c (150°C), the morphologies of mesoporous carbon nanoparticles tend to elongate.

Ultraviolet–visible spectroscopy (UV–Vis) analysis in Figure 23.28a concluded that rodlike mesoporous carbon nanoparticles absorbed a lower dye mass, while spherical mesoporous carbon nanoparticles had a much higher dye mass absorbed in 24 h. Figure 23.28b shows the release kinetics of both the control and rodlike mesoporous carbon nanoparticles over a period of 24 h. Rodlike mesoporous carbon nanoparticles showed an increase in the dye mass in the environment after 24 h. Such properties could be beneficial to the patient as one needs only a lower drug intake frequency with rodlike mesoporous carbon nanoparticles as the drug carrier. Therefore, this could potentially avoid many drug intake-related problems such as overdosing.

As mentioned above, the MCNs also exhibited excellent performance in adsorption for the application in water purification (Wang et al. 2007, Frank et al. 2009, Zhuang et al. 2009). Zhuang et al.'s (2009) group reported that ordered mesoporous carbon derived from phenolic resins were used in the removal of bulky dyes from water. The MCNs with high surface area of 2580 m²/g exhibited twice adsorption capacities for bulky dyes in comparison with commercially activated carbons. The optical photographs in Figure 23.29 were taken before and after dye adsorption by the authors. The polluted water became clear and colorless after adsorption. This phenomenon suggested the efficient adsorption and distinct decoloration for tinctorial wastewater by MCNs' adsorption.

## 23.4 Summary and Outlook

Mesoporous nanomaterials with open-framework structure have received much attention in the past decade due to their ordered mesochannels, large surface areas, and quantum effects in the nanoscale, making them promising for applications in the areas of catalysts, controlled drug delivery, absorption, and energy storage. Among various mesoporous nanomaterials, silica-based mesoporous nanomaterials have been widely studied in the last decade. More recently, ordered MCNs have also been successfully fabricated due to their fascinating physical and chemical properties, such as enhanced electrical and thermal conductivity, chemical stability, and reduced density. These advantageous properties enable them to be extensively used as guest molecules carriers, electrode materials for batteries, fuel cells, and supercapacitors. Seminal studies have been developed in the past decades by many researchers in order to fabricate MCNs. Current synthesis approaches of MCNs were summarized as either hard-template or soft-template methods. Both were elaborated in this chapter along with the potential applications in the field of energy storage, drug delivery, and environmental related fields.

## Acknowledgment

This work was supported by a research grant from the Ministry of Education, Singapore (R-284-000-124-112).

## References

Alfredsson, V. & Anderson, M. W., "Structure of MCM-48 revealed by transmission electron microscopy," *Chem. Mater.* 8 (1996): 1141–1146. doi:10.1021/cm950568k.

Armand, M. & Tarascon, J. M., "Building better batteries," *Nature* 451 (2008): 652–657.

Attard, G. S., Corker, J. M., Göltner, C. G. et al., "Liquid-crystal templates for nanostructured metals," *Angew. Chem. Int. Ed. Eng.* 36 (1997): 1315–1317. doi:10.1002/anie.199713151.

Attard, G. S., Leclerc, S. A. A., Maniguet, S. et al., "Mesoporous Pt/Ru alloy from the hexagonal lyotropic liquid crystalline phase of a nonionic surfactant," *Chem. Mater.* 13 (2001): 1444–1446. doi:10.1021/cm000850d.

Babel, K. & Jurewicz, K., "KOH activated carbon fabrics as supercapacitor material," *J. Phys. Chem. Solids* 65 (2004): 275–280. doi:10.1016/j.jpcs.2003.08.023.

Barbieri, O., Hahn, M., Herzog, A. et al., "Capacitance limits of high surface area activated carbons for double layer capacitors," *Carbon* 43 (2005): 1303–1310. doi:10.1016/j.carbon.2005.01.001.

Bartlett, P. N., Birkin, P. N., Ghanem, M. A. et al., "The electrochemical deposition of nanostructured cobalt films from lyotropic liquid crystalline media," *J. Electrochem. Soc.* 148 (2001): C119–C123. doi:10.1149/1.1342178.

Bartlett, P. N., Gollas, B., Guerin, S. et al., "The preparation and characterisation of $H_1$-e palladium films with a regular hexagonal nanostructure formed by electrochemical deposition from lyotropic liquid crystalline phases," *Phys. Chem. Chem. Phys.* 4 (2002): 3835–3842. doi:10.1039/B201845D.

Batrakova, E. V. & Kabanov, A. V., "Pluronic block copolymers: Evolution of drug delivery concept from inert nanocarriers to biological response modifiers," *J. Control. Release Off. J. Control. Release Soc.* 130 (2008): 98–106. doi:10.1016/j.jconrel.2008.04.013.

Beck, J. S., Vartuli, J. C., Roth, W. J. et al., "A new family of mesoporous molecular sieves prepared with liquid crystal templates," *J. Am. Chem. Soc.* 114 (1992): 10834–10843. doi:10.1021/ja00053a020.

Bose, S., Kuila, T., Mishra, A. K. et al., "Carbon-based nanostructured materials and their composites as supercapacitor electrodes," *J. Mater. Chem.* 22 (2012): 767–784. doi:10.1039/C1JM14468E.

Brinker, C. J., Lu, Y., Sellinger, A. et al., "Evaporation-induced self-assembly: Nanostructures made easy," *Adv. Mater.* 11 (1999): 579–585. doi:10.1002/(SICI)1521-4095(199905)11:7<579::AID-ADMA579>3.0.CO;2-R.

Bruce, P. G., Scrosati, B. & Tarascon, J.-M., "Nanomaterials for rechargeable lithium batteries," *Angew. Chem. Int. Ed.* 47 (2008): 2930–2946. doi:10.1002/anie.200702505.

Chang, H., Joo, S. H. & Pak, C., "Synthesis and characterization of mesoporous carbon for fuel cell applications," *J. Mater. Chem.* 17 (2007): 3078. doi:10.1039/b700389g.

Chen, F., Huang, L. & Li, Q., "Synthesis of MCM-48 using mixed cationic–anionic surfactants as templates," *Chem. Mater.* 9 (1997): 2685–2686. doi:10.1021/cm9703942.

Chen, L.-F., Zhang, X.-D., Liang, H.-W. et al., "Synthesis of nitrogen-doped porous carbon nanofibers as an efficient electrode material for supercapacitors," *ACS Nano* 6 (2012): 7092–7102. doi:10.1021/nn302147s.

Chen, Y., Song, B., Li, M. et al., "$Fe_3O_4$ nanoparticles embedded in uniform mesoporous carbon spheres for superior high-rate battery applications," *Adv. Funct. Mater.* 24 (2014): 319–326. doi:10.1002/adfm.201300872.

Conway, B. E., *Electrochemical Supercapacitors: Scientific Fundamentals and Technological Applications* (Kluwer Academic/Plenum Publisher, New York, 1999).

Crepaldi, E. L., Soler-Illia, G. J. A. A., Grosso, D. et al., "Controlled formation of highly ordered cubic and hexagonal mesoporous nanocrystalline yttria–zirconia and ceria–zirconia thin films exhibiting high thermal stability," *Angew. Chem. Int. Ed.* 42 (2003a): 347–351. doi:10.1002/anie.200390113.

Crepaldi, E. L. et al., "Controlled formation of highly organized mesoporous titania thin films: From mesostructured hybrids to mesoporous nanoanatase $TiO_2$," *J. Am. Chem. Soc.* 125 (2003b): 9770–9786. doi:10.1021/ja030070g.

Dai, S., Guiochon, G. A. & Liang, C., "Robust carbon monolith having hierarchical porosity," US Patent 20050169829 (2004).

Deng, Y., Yu, T., Wan, Y. et al., "Ordered mesoporous silicas and carbons with large accessible pores templated from amphiphilic diblock copolymer poly(ethylene oxide)-b-polystyrene," *J. Am. Chem. Soc.* 129 (2007): 1690–1697. doi:10.1021/ja067379v.

Deng, Y. et al., "Controlled synthesis and functionalization of ordered large-pore mesoporous carbons," *Adv. Funct. Mater.* 20 (2010): 3658–3665. doi:10.1002/adfm.201001202.

Fan, J., Wang, T., Yu, C. et al., "Ordered, nanostructured tin-based oxides/carbon composite as the negative-electrode material for lithium-ion batteries," *Adv. Mater.* 16 (2004): 1432–1436. doi:10.1002/adma.200400106.

Fan, J. et al., "Low-temperature strategy to synthesize highly ordered mesoporous silicas with very large pores," *J. Am. Chem. Soc.* 127 (2005): 10794–10795. doi:10.1021/ja052619c.

Fan, Z. et al., "A three-dimensional carbon nanotube/graphene sandwich and its application as electrode in supercapacitors," *Adv. Mater.* 22 (2010): 3723–3728. doi:10.1002/adma.201001029.

Fang, Y., Gu, D., Zou, Y. et al., "A low-concentration hydrothermal synthesis of biocompatible ordered mesoporous carbon nanospheres with tunable and uniform size," *Angew. Chem. Int. Ed.* 49 (2010): 7987–7991. doi:10.1002/anie.201002849.

Fox, M. E., Szoka, F. C. & Fréchet, J. M. J., "Soluble polymer carriers for the treatment of cancer: The importance of molecular architecture," *Acc. Chem. Res.* 42 (2009): 1141–1151. doi:10.1021/ar900035f.

Frackowiak, E., "Carbon materials for supercapacitor application," *Phys. Chem. Chem. Phys.* 9 (2007): 1774–1785. doi:10.1039/B618139M.

Frank, B., Zhang, J., Blume, R. et al., "Heteroatoms increase the selectivity in oxidative dehydrogenation reactions on nanocarbons," *Angew. Chem. Int. Ed.* 48 (2009): 6913–6917. doi:10.1002/anie.200901826.

Fuertes, A. B. & Centeno, T. A., "Mesoporous carbons with graphitic structures fabricated by using porous silica materials as templates and iron-impregnated polypyrrole as precursor," *J. Mater. Chem.* 15 (2005): 1079–1083. doi:10.1039/B416007J.

Gamby, J., Taberna, P. L., Simon, P. et al., "Studies and characterisations of various activated carbons used for carbon/carbon supercapacitors," *J. Power Sources* 101 (2001): 109–116. doi:10.1016/S0378-7753(01)00707-8.

Gao, W., Alemany, L. B., Ci, L. et al., "New insights into the structure and reduction of graphite oxide," *Nat. Chem.* 1 (2009): 403–408. doi:10.1038/nchem.281.

Grigoriants, I., Sominski, L., Li, H. et al., "The use of tin-decorated mesoporous carbon as an anode material for rechargeable lithium batteries," *Chem. Commun.* (2005): 921–923. doi:10.1039/B414240C.

Grosso, D., Boissiere, C., Smarsly, B. et al., "Periodically ordered nanoscale islands and mesoporous films composed of nanocrystalline multimetallic oxides," *Nat. Mater.* 3 (2004): 787–792.

Gu, J., Su, S., Li, Y. et al., "Hydrophilic mesoporous carbon nanoparticles as carriers for sustained release of hydrophobic anti-cancer drugs," *Chem. Commun.* 47 (2011): 2101–2103. doi:10.1039/C0CC04598E.

Guan, Y., Huang, J., Zuo, L. et al., "Effect of pluronic P123 and F127 block copolymer on P-glycoprotein transport and CYP3A metabolism," *Arch. Pharmacal Res.* 34 (2011): 1719–1728. doi:10.1007/s12272-011-1016-0.

Guo, Y.-G., Hu, J.-S. & Wan, L.-J., "Nanostructured materials for electrochemical energy conversion and storage devices," *Adv. Mater.* 20 (2008): 2878–2887. doi:10.1002/adma.200800627.

Guo, D.-C., Mi, J., Hao, G.-P. et al., "Ionic liquid $C_{16}mimBF_4$ assisted synthesis of poly(benzoxazine-co-resol)-based hierarchically porous carbons with superior performance in supercapacitors," *Energy Environ. Sci.* 6 (2013): 652–659. doi:10.1039/C2EE23127A.

Hu, Y. S., Adelhelm, P., Smarsly, B.M. et al., "Synthesis of hierarchically porous carbon monoliths with highly ordered microstructure and their application in rechargeable lithium batteries with high-rate capability," *Adv. Funct. Mater.* 17 (2007): 1873–1878. doi:10.1002/adfm.200601152.

Huang, X., Yin, Z., Wu, S. et al., "Graphene-based materials: Synthesis, characterization, properties, and applications," *Small* 7 (2011a): 1876–1902. doi:10.1002/smll.201002009.

Huang, X. et al., "The shape effect of mesoporous silica nanoparticles on biodistribution, clearance, and biocompatibility in vivo," *ACS Nano* 5 (2011b): 5390–5399. doi:10.1021/nn200365a.

Huang, Y., Liang, J. & Chen, Y., "An overview of the applications of graphene-based materials in supercapacitors," *Small* 8 (2012a): 1805–1834. doi:10.1002/smll.201102635.

Huang, Z.-D. et al., "Self-assembled reduced graphene oxide/carbon nanotube thin films as electrodes for supercapacitors," *J. Mater. Chem.* 22 (2012b): 3591–3599. doi:10.1039/C2JM15048D.

Hulicova, D., Kodama, M. & Hatori, H., "Electrochemical performance of nitrogen-enriched carbons in aqueous and non-aqueous supercapacitors," *Chem. Mater.* 18 (2006): 2318–2326. doi:10.1021/cm060146i.

Hulicova-Jurcakova, D., Seredych, M., Lu, G. Q. et al., "Combined effect of nitrogen- and oxygen-containing functional groups of microporous activated carbon on its electrochemical performance in supercapacitors," *Adv. Funct. Mater.* 19 (2009a): 438–447. doi:10.1002/adfm.200801236.

Hulicova-Jurcakova, D., Kodama, M., Shiraishi, S. et al., "Nitrogen-enriched nonporous carbon electrodes with extraordinary supercapacitance," *Adv. Funct. Mater.* 19 (2009b): 1800–1809. doi:10.1002/adfm.200801100.

Huo, Q., Margolese, D. I., Ciesla, U. et al., "Organization of organic molecules with inorganic molecular species into nanocomposite biphase arrays," *Chem. Mater.* 6 (1994): 1176–1191. doi:10.1021/cm00044a016.

Hwang, J., Woo, S. H., Shim, J. et al., "One-pot synthesis of tin-embedded carbon/silica nanocomposites for anode materials in lithium-ion batteries," *ACS Nano* 7 (2013): 1036–1044. doi:10.1021/nn303570s.

Inagaki, M., Konno, H. & Tanaike, O., "Carbon materials for electrochemical capacitors," *J. Power Sources* 195 (2010): 7880–7903. doi:10.1016/j.jpowsour.2010.06.036.

Jain, K. K., *Drug Delivery Systems*, Vol. 473 (Humana Press, New York, 2008).

Jang, J. & Bae, J., "Fabrication of mesoporous polymer using soft template method," *Chem. Commun.* (2005): 1200–1202. doi:10.1039/B416518G.

Jeong, H. M., Lee, J. W., Shin, W. H. et al., "Nitrogen-doped graphene for high-performance ultracapacitors and the importance of nitrogen-doped sites at basal planes," *Nano Lett.* 11 (2011): 2472–2477. doi:10.1021/nl2009058.

Johnson, S. A., Ollivier, P. J. & Mallouk, T. E., "Ordered mesoporous polymers of tunable pore size from colloidal silica templates," *Science* 283 (1999): 963–965. doi:10.1126/science.283.5404.963.

Joo, S. H., Choi, S. J., Oh, I. et al., "Ordered nanoporous arrays of carbon supporting high dispersions of platinum nanoparticles," *Nature* 412 (2001): 169–172. doi:10.1038/35084046.

Kailasam, K., Jun, Y.-S., Katekomol, P. et al., "Mesoporous melamine resins by soft templating of block-co-polymer mesophases," *Chem. Mater.* 22 (2009): 428–434. doi:10.1021/cm9029903.

Kang, E., Jung, Y. S., Cavanagh, A. S. et al., "$Fe_3O_4$ nanoparticles confined in mesocellular carbon foam for high performance anode materials for lithium-ion batteries," *Adv. Funct. Mater.* 21 (2011): 2430–2438. doi:10.1002/adfm.201002576.

Kim, T.-W., Kleitz, F., Paul, B. et al., "MCM-48-like large mesoporous silicas with tailored pore structure: Facile synthesis domain in a ternary triblock copolymer–butanol–water system," *J. Am. Chem. Soc.* 127 (2005): 7601–7610. doi:10.1021/ja042601m.

Kim, T.-W., Chung, P.-W., Slowing, I. I. et al., "Structurally ordered mesoporous carbon nanoparticles as trans-membrane delivery vehicle in human cancer cells," *Nano Lett.* 8 (2008): 3724–3727. doi:10.1021/nl801976m.

Knox, J. H., Unger, K. K. & Mueller, H., "Prospects for carbon as packing material in high-performance liquid chromatography," *J. Liq. Chromatogr.* 6 (1983): 1–36. doi:10.1080/01483918308067647.

Knox, J. H., Kaur, B. & Millward, G. R., "Structure and performance of porous graphitic carbon in liquid chromatography," *J. Chromatogr. A* 352 (1986): 3–25. doi:10.1016/S0021-9673(01)83368-9.

Kresge, C. T., Leonowicz, M. E., Roth, W. J. et al., "Ordered mesoporous molecular sieves synthesized by a liquid-crystal template mechanism," *Nature* 359 (1992): 710–712.

Kruk, M., Jaroniec, M., Kim, T.-W. et al., "Synthesis and characterization of hexagonally ordered carbon nanopipes," *Chem. Mater.* 15 (2003): 2815–2823. doi:10.1021/cm034087+.

Lee, J. S., Hirao, A. & Nakahama, S., "Polymerization of monomers containing functional silyl groups: 5. Synthesis of new porous membranes with functional groups," *Macromolecules* 21 (1988): 274–276. doi:10.1021/ma00179a057.

Lee, J., Yoon, S., Hyeon, T. et al., "Synthesis of a new mesoporous carbon and its application to electrochemical double-layer capacitors," *Chem. Commun.* (1999): 2177–2178. doi:10.1039/A906872D.

Lee, J., Yoon, S., Oh, S. M. et al., "Development of a new mesoporous carbon using an HMS aluminosilicate template," *Adv. Mater.* 12 (2000): 359–362. doi:10.1002/(SICI)1521-4095(200003)12:5<359::AID-ADMA359>3.0.CO;2-1.

Lee, J., Jin, S., Hwang, Y. et al., "Simple synthesis of mesoporous carbon with magnetic nanoparticles embedded in carbon rods," *Carbon* 43 (2005): 2536–2543. doi:10.1016/j.carbon.2005.05.005.

Lee, J. H. et al., "Restacking-inhibited 3D reduced graphene oxide for high performance supercapacitor electrodes," *ACS Nano* 7 (2013): 9366–9374. doi:10.1021/nn4040734.

Lei, Z., Christov, N. & Zhao, X. S., "Intercalation of mesoporous carbon spheres between reduced graphene oxide sheets for preparing high-rate supercapacitor electrodes," *Energy Environ. Sci.* 4 (2011): 1866–1873. doi:10.1039/C1EE01094H.

Lei, Z., Lu, L. & Zhao, X. S., "The electrocapacitive properties of graphene oxide reduced by urea," *Energy Environ. Sci.* 5 (2012): 6391–6399. doi:10.1039/C1EE02478G.

Li, Z. & Dai, S., "Surface functionalization and pore size manipulation for carbons of ordered structure," *Chem. Mater.* 17 (2005): 1717–1721. doi:10.1021/cm047954z.

Li, M. & Xue, J., "Ordered mesoporous carbon nanoparticles with well-controlled morphologies from sphere to rod via a soft-template route," *J. Colloid Interface Sci.* 377 (2012): 169–175. doi:10.1016/j.jcis.2012.03.085.

Li, M. & Xue, J., "Integrated synthesis of nitrogen-doped mesoporous carbon from melamine resins with superior performance in supercapacitors," *J. Phys. Chem. C* 118 (2014): 2507–2517. doi:10.1021/jp410198r.

Li, Z., Yan, W. & Dai, S., "A novel vesicular carbon synthesized using amphiphilic carbonaceous material and micelle templating approach," *Carbon* 42 (2004): 767–770. doi:10.1016/j.carbon.2004.01.044.

Li, Z., Yan, W. & Dai, S., "Surface functionalization of ordered mesoporous carbons: A comparative study," *Langmuir* 21 (2005): 11999–12006. doi:10.1021/la051608u.

Li, W., Chen, D., Li, Z. et al., "Nitrogen enriched mesoporous carbon spheres obtained by a facile method and its application for electrochemical capacitor," *Electrochem. Commun.* 9 (2007): 569–573. doi:10.1016/j.elecom.2006.10.027.

Li, X. et al., "Simultaneous nitrogen doping and reduction of graphene oxide," *J. Am. Chem. Soc.* 131 (2009a): 15939–15944. doi:10.1021/ja907098f.

Li, L., Ding, J. & Xue, J., "Macroporous silica hollow microspheres as nanoparticle collectors," *Chem. Mater.* 21 (2009b): 3629–3637. doi:10.1021/cm900874u.

Li, L. et al., "A doped activated carbon prepared from polyaniline for high performance supercapacitors," *J. Power Sources* 195 (2010): 1516–1521. doi:10.1016/j.jpowsour.2009.09.016.

Li, W. et al., "A self-template strategy for the synthesis of mesoporous carbon nanofibers as advanced supercapacitor electrodes," *Adv. Energy Mater.* 1 (2011): 382–386. doi:10.1002/aenm.201000096.

Li, Z. et al., "Carbonized chicken eggshell membranes with 3D architectures as high-performance electrode materials for supercapacitors," *Adv. Energy Mater.* 2 (2012): 431–437. doi:10.1002/aenm.201100548.

Li, M., Ding, J. & Xue, J., "Mesoporous carbon decorated graphene as an efficient electrode material for supercapacitors," *J. Mater. Chem. A* 1 (2013): 7469. doi:10.1039/c3ta10890b.

Liang, C. & Dai, S., "Synthesis of mesoporous carbon materials via enhanced hydrogen-bonding interaction," *J. Am. Chem. Soc.* 128 (2006): 5316–5317. doi:10.1021/ja060242k.

Liang, C., Hong, K., Guiochon, G. A. et al., "Synthesis of a large-scale highly ordered porous carbon film by self-assembly of block copolymers," *Angew. Chem. Int. Ed.* 43 (2004): 5785–5789. doi:10.1002/anie.200461051.

Liang, C., Li, Z. & Dai, S., "Mesoporous carbon materials: Synthesis and modification," *Angew. Chem. Int. Ed.* 47 (2008): 3696–3717. doi:10.1002/anie.200702046.

Lin, H.-P., Chang-Chien, C.-Y., Tang, C.-Y. et al., "Synthesis of *p6mm* hexagonal mesoporous carbons and silicas using Pluronic F127–PF resin polymer blends," *Microporous Mesoporous Mater.* 93 (2006): 344–348. doi:10.1016/j.micromeso.2006.03.011.

Lu, Y., Fan, H., Stump, A. et al., "Aerosol-assisted self-assembly of mesostructured spherical nanoparticles," *Nature* 398 (1999): 223–226.

Lu, A., Kiefer, A., Schmidt, W. et al., "Synthesis of polyacrylonitrile-based ordered mesoporous carbon with tunable pore structures," *Chem. Mater.* 16 (2003): 100–103. doi:10.1021/cm031095h.

Ma, F., Zhao, H., Sun, L. et al., "A facile route for nitrogen-doped hollow graphitic carbon spheres with superior performance in supercapacitors," *J. Mater. Chem.* 22 (2012): 13464–13468. doi:10.1039/C2JM32960C.

Ma, T.-Y., Liu, L. & Yuan, Z.-Y., "Direct synthesis of ordered mesoporous carbons," *Chem. Soc. Rev.* 42 (2013): 3977–4003. doi:10.1039/C2CS35301F.

Mayes, R. T., Tsouris, C., Kiggans, J. O. Jr. et al., "Hierarchical ordered mesoporous carbon from phloroglucinol-glyoxal and its application in capacitive deionization of brackish water," *J. Mater. Chem.* 20 (2010): 8674–8678. doi:10.1039/C0JM01911A.

Meng, Y., Gu, D., Zhang, F. et al., "Ordered mesoporous polymers and homologous carbon frameworks: Amphiphilic surfactant templating and direct transformation," *Angew. Chem. Int. Ed.* 44 (2005): 7053–7059. doi:10.1002/anie.200501561.

Meng, Y. et al., "A family of highly ordered mesoporous polymer resin and carbon structures from organic–organic self-assembly," *Chem. Mater.* 18 (2006): 4447–4464. doi:10.1021/cm060921u.

Nelson, P. A., Elliott, J. M., Attard, G. S. et al., "Mesoporous nickel/nickel oxide—A nanoarchitectured electrode," *Chem. Mater.* 14 (2002): 524–529. doi:10.1021/cm011021a.

Nishihara, H. & Kyotani, T., "Templated nanocarbons for energy storage," *Adv. Mater.* 24 (2012): 4473–4498. doi:10.1002/adma.201201715.

Pandolfo, A. G. & Hollenkamp, A. F., "Carbon properties and their role in supercapacitors," *J. Power Sources* 157 (2006): 11–27. doi:10.1016/j.jpowsour.2006.02.065.

Park, S. & Ruoff, R. S., "Chemical methods for the production of graphenes," *Nat. Nanotechnol.* 4 (2009): 217–224.

Poizot, P., Laruelle, S., Grugeon, S. et al., "Nano-sized transition-metal oxides as negative-electrode materials for lithium-ion batteries," *Nature* 407 (2000): 496–499. doi:10.1038/35035045.

Ryoo, R., Joo, S. H. & Jun, S., "Synthesis of highly ordered carbon molecular sieves via template-mediated structural transformation," *J. Phys. Chem. B* 103 (1999): 7743–7746. doi:10.1021/jp991673a.

Shen, L., Zhang, X., Uchaker, E. et al., "$Li_4Ti_5O_{12}$ nanoparticles embedded in a mesoporous carbon matrix as a superior anode material for high rate lithium ion batteries," *Adv. Energy Mater.* 2 (2012): 691–698. doi:10.1002/aenm.201100720.

Shin, H. J., Ryoo, R., Kruk, M. et al., "Modification of SBA-15 pore connectivity by high-temperature calcination investigated by carbon inverse replication," *Chem. Commun.* (2001): 349–350. doi:10.1039/B009762O.

Simon, P. & Gogotsi, Y., "Materials for electrochemical capacitors," *Nat. Mater.* 7 (2008): 845–854.

Stankovich, S., Dikin, D. A., Piner, R. D. et al., "Synthesis of graphene-based nanosheets via chemical reduction of exfoliated graphite oxide," *Carbon* 45 (2007): 1558–1565. doi:10.1016/j.carbon.2007.02.034.

Tanaka, S., Nishiyama, N., Egashira, Y. et al. "Synthesis of ordered mesoporous carbons with channel structure from an organic–organic nanocomposite," *Chem. Commun.* (2005): 2125–2127. doi:10.1039/B501259G.

Tarascon, J. M. & Armand, M., "Issues and challenges facing rechargeable lithium batteries," *Nature* 414 (2001): 359–367.

Tian, B., Che, S., Liu, Z. et al., "Novel approaches to synthesize self-supported ultrathin carbon nanowire arrays templated by MCM-41," *Chem. Commun.* (2003): 2726–2727. doi:10.1039/B309670J.

Tsouris, C., Mayes, R., Kiggans, J. et al., "Mesoporous carbon for capacitive deionization of saline water," *Environ. Sci. Technol.* 45 (2011): 10243–10249. doi:10.1021/es201551e.

Vallet-Regí, M., Balas, F. & Arcos, D., "Mesoporous materials for drug delivery," *Angew. Chem. Int. Ed.* 46 (2007): 7548–7558. doi:10.1002/anie.200604488.

Venkataraman, S., Hedrick, J. L., Ong, Z. Y. et al., "The effects of polymeric nanostructure shape on drug delivery," *Adv. Drug Deliv. Rev.* 63 (2011): 1228–1246. doi:10.1016/j.addr.2011.06.016.

Vinu, A., Srinivasu, P., Mori, T. et al., "Novel hexagonally ordered nitrogen-doped mesoporous carbon from SBA-15/polyaniline nanocomposite," *Chem. Lett.* 36 (2007): 770–771. doi:10.1246/cl.2007.770.

Wang, L., Zhao, Y., Lin, K. F. et al., "Super-hydrophobic ordered mesoporous carbon monolith," *Carbon* 44 (2006): 1336–1339. doi:10.1016/j.carbon.2005.12.007.

Wang, X., Liu, R., Waje, M. M. et al., "Sulfonated ordered mesoporous carbon as a stable and highly active protonic acid catalyst," *Chem. Mater.* 19 (2007): 2395–2397. doi:10.1021/cm070278r.

Wang, Y., Zhang, H. J., Lu, L. et al., "Designed functional systems from peapod-like Co@carbon to $Co_3O_4$@ carbon nanocomposites," *ACS Nano* 4 (2010): 4753–4761. doi:10.1021/nn1004183.

Wang, J., Xue, C, Lv, Y. et al., "Kilogram-scale synthesis of ordered mesoporous carbons and their electrochemical performance," *Carbon* 49 (2011): 4580–4588. doi:10.1016/j.carbon.2011.06.069.

Wang, G., Zhang, L. & Zhang, J., "A review of electrode materials for electrochemical supercapacitors," *Chem. Soc. Rev.* 41 (2012): 797–828. doi:10.1039/C1CS15060J.

Wei, L., Sevilla, M., Fuertes, A. B. et al., "Polypyrrole-derived activated carbons for high-performance electrical double-layer capacitors with ionic liquid electrolyte," *Adv. Funct. Mater.* 22 (2012): 827–834. doi:10.1002/adfm.201101866.

Wen, Z., Wang, X., Mao, S. et al., "Crumpled nitrogen-doped graphene nanosheets with ultrahigh pore volume for high-performance supercapacitor," *Adv. Mater.* 24 (2012): 5610–5616. doi:10.1002/adma.201201920.

Whitehead, A. H., Elliott, J. M., Owen, J. R. et al., "Electrodeposition of mesoporous tin films," *Chem. Commun.* (1999): 331–332. doi:10.1039/A808775J.

Wu, C.-G. & Bein, T., "Conducting carbon wires in ordered, nanometer-sized channels," *Science* 266 (1994): 1013–1015. doi:10.1126/science.266.5187.1013.

Xiao, L., Li, J., Brougham, D. F. et al., "Water-soluble superparamagnetic magnetite nanoparticles with biocompatible coating for enhanced magnetic resonance imaging," *ACS Nano* 5 (2011): 6315–6324. doi:10.1021/nn201348s.

Yan, A., Lau, B. W., Weissman, B. S. et al., "Biocompatible, hydrophilic, supramolecular carbon nanoparticles for cell delivery," *Adv. Mater.* 18 (2006): 2373–2378. doi:10.1002/adma.200600838.

Yang, P., Zhao, D., Margolese, D. I. et al., "Generalized syntheses of large-pore mesoporous metal oxides with semicrystalline frameworks," *Nature* 396 (1998): 152–155.

Yang, C.-M., Weidenthaler, C., Spliethoff, B. et al., "Facile template synthesis of ordered mesoporous carbon with polypyrrole as carbon precursor," *Chem. Mater.* 17 (2004): 355–358. doi:10.1021/cm049164v.

Yang, X., Zhu, J., Qiu, L. et al., "Bioinspired effective prevention of restacking in multilayered graphene films: Towards the next generation of high-performance supercapacitors," *Adv. Mater.* 23 (2011): 2833–2838. doi:10.1002/adma.201100261.

Yu, C., Fan, J., Tian, B. et al., "High-yield synthesis of periodic mesoporous silica rods and their replication to mesoporous carbon rods," *Adv. Mater.* 14 (2002): 1742–1745. doi:10.1002/1521-4095(20021203)14:23<1742::AID-ADMA1742>3.0.CO;2-3.

Yu, T., Malugin, A. & Ghandehari, H., "Impact of silica nanoparticle design on cellular toxicity and hemolytic activity," *ACS Nano* 5 (2011): 5717–5728. doi:10.1021/nn2013904.

Yuan, S. M., Li, J. X., Yang, L. T. et al., "Preparation and lithium storage performances of mesoporous $Fe_3O_4$@C microcapsules," *ACS Appl. Mater. Interfaces* 3 (2011a): 705–709. doi:10.1021/am1010095.

Yuan, S., Zhou, Z. & Li, G., "Structural evolution from mesoporous α-Fe$_2$O$_3$ to Fe$_3$O$_4$@C and γ-Fe$_2$O$_3$ nanospheres and their lithium storage performances," *CrystEngComm* 13 (2011b): 4709–4713. doi:10.1039/C0CE00902D.

Zhang, L. L. & Zhao, X. S., "Carbon-based materials as supercapacitor electrodes," *Chem. Soc. Rev.* 38 (2009): 2520. doi:10.1039/b813846j.

Zhang, F., Meng, Y., Gu, D. et al., "A facile aqueous route to synthesize highly ordered mesoporous polymers and carbon frameworks with *Ia3d* bicontinuous cubic structure," *J. Am. Chem. Soc.* 127 (2005): 13508–13509. doi:10.1021/ja0545721.

Zhang, F. et al., "An aqueous cooperative assembly route to synthesize ordered mesoporous carbons with controlled structures and morphology," *Chem. Mater.* 18 (2006): 5279–5288. doi:10.1021/cm061400+.

Zhang, J. et al., "Surface-modified carbon nanotubes catalyze oxidative dehydrogenation of *n*-butane," *Science* 322 (2008): 73–77. doi:10.1126/science.1161916.

Zhang, L. L., Zhao, X., Stoller, M. D. et al., "Highly conductive and porous activated reduced graphene oxide films for high-power supercapacitors," *Nano Lett.* 12 (2012): 1806–1812. doi:10.1021/nl203903z.

Zhao, D., Feng, J., Huo, Q. et al., "Triblock copolymer syntheses of mesoporous silica with periodic 50 to 300 angstrom pores," *Science* 279 (1998): 548–552. doi:10.1126/science.279.5350.548.

Zhao, D., Wan, Y. & Zhou, W., "Mesoporous nonsilica materials," in *Ordered Mesoporous Materials* (Wiley-VCH Verlag GmbH & Co. KGaA, Weinheim, Germany, 2013), pp. 293–428.

Zhong, M., Kim, E. K., McGann, J. P. et al., "Electrochemically active nitrogen-enriched nanocarbons with well-defined morphology synthesized by pyrolysis of self-assembled block copolymer," *J. Am. Chem. Soc.* 134 (2012): 14846–14857. doi:10.1021/ja304352n.

Zhu, Y., Murali, S., Stoller, M. D. et al., "Carbon-based supercapacitors produced by activation of graphene," *Science* 332 (2011): 1537–1541. doi:10.1126/science.1200770.

Zhuang, X., Wan, Y., Feng, C. et al., "Highly efficient adsorption of bulky dye molecules in wastewater on ordered mesoporous carbons," *Chem. Mater.* 21 (2009): 706–716. doi:10.1021/cm8028577.

# V

# Hybrids/Composites

24

# Silicon Nanocrystal/Nanocarbon Hybrids

Conor Rocks

Somak Mitra

Manuel
Macias-Montero

Davide Mariotti

Vladimir Svrcek

**Abstract**

In this chapter, we highlight aspects that relate to carbon nanostructures (fullerenes, SWCNTs and/
or MWCNTs, and graphene) and their role in providing advanced functionalization and integration with silicon nanocrystals (Si-NCs). The fabrication of Si-NC/nanocarbon composites can be achieved in some cases with simple solution-chemistry methods; however, optical and, in particular, electronic coupling has suffered from the weak interactions produced with these methods. Therefore, the possibility of forming covalent Si-NC/nanocarbon nanojunctions, for instance, without any metal catalyst, is an exciting scientific opportunity. Synthesis of CNTs on Si-NCs can be achieved through careful control of the Si-NC surface characteristics; in some cases, surfactant-free Si-NCs with oxygen-based terminations have resulted to be one of the main requirements for successful growth. Oxygen coordination on the Si-NC surface can have a strong impact on the chemical bonding with CNTs as well as on the optoelectronic coupling between the two nanostructures.

**Key words:** carbon nanotubes, photocatalyst, photovoltaics, silicon nanocrystals.

## 24.1 Introduction

Silicon (Si) and carbon (C) are two of the most abundant and environment-friendly elements in nature that play a fundamental role in a wide range of very important technologies (e.g., electronics and photovoltaics). At the same time, a wealth of new opportunities has become evident in recent years because of the unique physical and chemical characteristics exhibited by nanoscale silicon and carbon materials. Very attractive also is the possibility of electronically and optically coupling Si and C nanostructures to form advanced nanoscale devices and/or components. Si- and C-based nanostructures can offer the possibility of exploring new nanodevice architectures that might open the way to alternative and improved design approaches. One of the most attractive opportunities is offered, for instance, by silicon nanocrystal (Si-NC)/CNT junctions, which so far have received very limited attention.

## 24.1.1 Properties of Si-NCs

Silicon is one of the most widely studied materials known to people and has remained dominant within the microelectronics industry because of its abundance in resource, its nontoxicity, and also its appropriate optical/electronic properties. However, bulk Si has an indirect bandgap and therefore is a very poor light emitter and absorber. It is for this reason that Si has had restricted use within many optoelectronic devices. For Si, the conduction band minimum does not occur at the same $k$ space location within the Brillouin zone as the valence band maximum. Therefore, interband transitions must involve a momentum transfer typically provided by the additional interaction of a phonon. The reliance on this interaction significantly lowers the probability of absorption/radiative recombination processes when compared to direct bandgap materials such as gallium arsenide. Recent advances in the field of nanotechnology have shown that reducing the size of Si crystal to the nanoscale level brings about new properties and functionalities (Pavesi & Turan 2010). These new phenomena seen in Si-NCs might allow enhancement and improvement of many Si-based applications used today. Si-NCs derived from porous silicon (P-Si) have received extensive interest since the discovery of strong visible room-temperature photoluminescence (PL) by Canham in 1990. Less than a decade previously, Pickering et al. had observed blueshifting PL from P-Si at very low temperatures that indicated an opening of the optical bandgap (Pickering et al. 1984). But it was Canham's argument for the quantum confinement effect (QCE) as the main mechanism responsible that gained significant attention and is still a highly debated area. Si-NCs at sizes comparable to the Bohr radius (4.9 nm) are referred to as quantum confined and can be viewed as a particle-in-a-box analogy, where the particle is confined in all three dimensions and consequently introduces an uncertainty into the momentum owing to Heisenberg's theory. The relaxation in the momentum can allow optical transitions to occur without the interaction of the aforementioned phonon particle, suggesting that the absorption/radiative recombination processes can behave through more "direct-like" transitions. QCE predicts a widening of the indirect bandgap (1.12 eV) in relation to decreasing particle size. When the diameter of a nanoparticle is of the same magnitude as its electron wave function, the charge carriers sense the particle boundaries and respond by adjusting their energy levels (Fox 2001). The complexity of the energy structure at the direct bandgap (3.32 eV) leads to the opposite relationship of shortening the direct bandgap with decreasing particle size (De Boer et al. 2010, Mangolini 2013). There have been numerous experimental and theoretical studies that show that when the size of the Si-NCs becomes comparable to the Bohr radius, QCEs are apparent, causing changes in the electronic band structure (Soni et al. 1999, Altman et al. 2001, Ranjan et al. 2002, Pavesi & Turan 2010). These changes are exhibited through the blueshifting of PL toward higher energies in addition to an increase in intensity. PL external quantum efficiencies (EQEs) exceeding 23% have been obtained for Si-NCs (Gelloz et al. 2005), validating their potential to be incorporated in light-emitting diodes and other optoelectronic devices. One of the most exciting phenomena of the QCE is carrier multiplication (CM) and is perceived by many to be the most promising mechanism to overcome the Shockley and Quiesser efficiency limit of 33% for photovoltaic (PV) devices (Shockley & Queisser 1961). *Carrier multiplication* is a grouped term and refers to the generation of multiple electron–hole pairs or excitons from the absorption of a single high-energy photon either from two neighboring crystals adjacent to one another due to a quantum cutting effect or from inside a single quantum dot. Impact ionization observed in bulk Si is essentially the same process, although high-energy photons many multiples of the bandgap are required, which amount to a small percentage of the solar spectrum (Landsberg et al. 1993). In quantum dots, however, the crystal momentum is relaxed, thus lowering the threshold energy needed for the CM process. Furthermore, because of the increased separation between individual energy levels in quantum dots, the relaxation of hot electrons to the conduction band minimum is much slower when compared to that in bulk Si. This alteration in transition dynamics due to increased carrier lifetime enables CM to compete successfully with the phonon-cooling process (Nozik 2002, Beard et al. 2007). The occurrence of CM was observed in numerous environment-unfriendly nanocrystals (Ellingson et al. 2005, Murphy et al. 2006, Schaller et al. 2007) before Beard et al. (2007) reported CM for Si-NCs, revealing the threshold energy for generation of two electron–hole pairs to be $2.4E_g$, where $E_g$ is the material bandgap, with quantum yield (excitons produced per absorbed photon) reaching 2.6 at $3.4E_g$. This is significantly lower than that for bulk Si, whose threshold energy for impact ionization becomes competitive with the phonon-cooling process only at $\approx 4.4E_g$ (Beard et al. 2007).

Surface characteristics along with the QCE have a strong influence on the exhibited properties and behavior of Si-NCs, specifically in nanocrystals that have a diameter of <3 nm due to the high surface-to-volume ratio that exists. Photon absorption and exciton generation are known to take place exclusively within the nanocrystal core, but because of the exciton's delocalized nature, their wave functions can spread over a large region (several lattice spacings) (Fox 2001). This allows surface states to influence the energy structure or allow localized surface states to facilitate the recombination process of carriers, altering the emission intensity or wavelength (Mariotti et al. 2013), such as in Koch et al.'s (1993) proposed surface state model, where they state that carriers become trapped at these surface defect sites before radiatively recombining at energies lower than the bandgap. This model is in good agreement with the observed Stokes shift that can be frequently seen between absorption band edge and emission spectrum of Si-NC. Electrochemically etching P-Si is a common, low-cost synthetic route to producing hydrogenated Si-NCs; more details on this process will be discussed in Section 24.2. Si-NCs with hydrogen terminations are generally considered as the reference model for understanding the change in optical properties with changing terminations or bond arrangements. Surface characteristics, in particular oxidation, have been the focus of much experimental and theoretical work on investigating the functionality of interface oxide (Deal & Grove 1965, Mott et al. 1989, Okada & Iijima 1991, Scheer et al. 2000). Early stages of oxidation have been shown to passivate dangling bonds at the nanocrystals surface, which allows increased radiative transitions of carriers, resulting in increased EQE (Maruyama & Ohtani 1994). However, while controlled oxide passivation restricted to the outermost layer can improve or stabilize PL, further oxidation is known to result in Si backbonds being attacked, which strains the Si core, creating midband defects as oxidation progresses inward (Ogata et al. 1998). Since the nanocrystal's core dimensions are reduced by the encompassing oxide shell, QCE dominates, blueshifting optical properties because of the further widening of the bandgap (Tischler et al. 1992). The replacement of H terminations with alkyl ligands or other organic terminations as passivating organic monolayers can be beneficial in preventing surface and core oxidation of Si-NCs (Shirahata 2011). However, the inclusion of these capping ligands is detrimental to transport properties of the nanocrystal and limits its use in many optoelectronic applications. Control and understanding of Si-NCs surface characteristics and size effects is of great importance to achieve their uncapped potential within numerous energy applications. We show later in this chapter that control of both nanocrystal size and surface oxidation can be beneficial in the outcome of nanofabrication technique used to produce functional heterojunction nanostructures.

## 24.1.2 Silicon/Carbon Nanostructures

Here, we highlight aspects that relate to carbon nanostructures (fullerenes, SWCNTs and/or MWCNTs, and graphene) and their role in providing advanced functionalization and integration with Si-NCs. The inclusion of nanostructured carbon has previously been shown to benefit already existing PV architectures. Ren et al. (2011) employed SWCNTs in their organic PVs, observing reduced exciton recombination, which is typically a major limiting factor within organic PV devices. Efficiencies of PV devices can be increased by improving the mobility or charge transfer to the electrode, ultimately reducing the overall device series resistance. Dang et al. (2011) incorporated SWCNTs with titanium dioxide ($TiO_2$) quantum dots to produce a dye-sensitized solar cell, increasing short circuit current ($J_{SC}$). However, properties of organic devices are seen to degrade rapidly once exposed to UV light, which affects their feasibility for future third-generation PV devices. With Si's well-established history within the microelectronics industry, carbon/Si nanointerfaces are of great interest for energy and environmental applications. For instance, fullerenes and Si-NCs have been proved to form favorable interactions for bulk heterojunction solar cells. Nevertheless, an Si-NC/CNT system, in principle, would represent an almost ideal fundamental PV nanocomponent (i.e., nanoscale junction between two nanostructures with unique QCE) where the CNT can serve as the acceptor to promote exciton dissociation and as an efficient charge transport channel for flow of carriers to the collecting electrode. Švrček and Mariotti (2012) demonstrated the practicability of an Si-NC/SWCNT bulk heterojunction by increasing the EQE because of carbon's ability to absorb at energies transparent to Si-NCs' sharp band edge.

The control of carrier transport and energy coupling are also key mechanisms in several applications other than PV where Si-NCs and carbon nanostructures could outperform current technologies. For instance, the very large volumetric and gravimetric capacities of Si at room temperature represent attractive properties to be combined with high-conductivity graphene for energy storage. The Si/C nanocomposite could contribute to faster charge/discharge cycles where the mechanical stability of silicon may be enhanced by the rigid CNT structure attached to the nanocrystals. Evanoff et al. (2012) successfully constructed an aligned CNT/silicon composite electrode that showed high reverse capacity of over 3000 Ah g$^{-1}$, with other groups reporting high performance for similar structures (Hu et al. 2011). CNTs are recognized to be an ideal support material for photocatalysts such as $TiO_2$ used for decontamination of organic pollutants in water treatment (Huang & Gao 2003, Xu et al. 2010). The electrons and holes generated through UV light react with water molecules at their surface, producing hydroxyl radicals (OH$^-$) that can oxidize target pollutants. In the past decades, a substantial increase in the level of atmospheric $CO_2$ has caused the ongoing ocean acidification and is predicted to cause a range of harmful consequences. Recent advances have shown Si-NCs to be an ideal photocatalyst for the reduction of $CO_2$ in aqueous environments (Kang et al. 2007). Si-NCs in addition to being eco-friendly and cost-effective have a narrower bandgap when compared to the wide-gap semiconductor $TiO_2$, hence absorbing within a broader range of the solar spectrum. The introduction of an Si-NC/CNT photocatalyst would provide increased surface area, stabilization of charge separation, and enhanced transfer of electrons to facilitate the reduction of $CO_2$.

### 24.1.3 Synthesis of Si-NC/CNT

The most popular method of growing CNT is CVD. Carbon atoms, through decomposed carbon feedstock, are diffused onto the surface of a metal catalyst until the solution (metal–carbon) becomes saturated. When the solution becomes oversaturated with carbon atoms, this induces the precipitation of carbon, which crystallizes and forms cylindrical graphene (Zhang et al. 2013). Strong electrostatic interactions are formed between the CNTs and the metal atoms in contact with them. More details will be discussed about the various growth models that exist for CNT growth in Section 24.4. CVD requires extremely high operating temperatures, restricting the use of many potential substrates, along with high running cost. These elevated temperatures can also lead to stresses and mechanical instabilities in deposited films if the thermal expansion coefficient is dissimilar to that of the substrate material (Creighton & Ho 2011). Fabrication of Si-NC/nanocarbon composites can be achieved in some cases with simple solution-chemistry methods. Svrcek (2010) reviewed low-cost and scalable techniques that successfully encapsulated Si-NCs within the cavity of CNT. Liu et al. (2008) used an electrostatic force process to attach plasma-synthesized Si-NCs to CNT. However, the Si-NC/nanocarbon junctions fabricated using the above techniques do not allow optimization of each partner material unique quantum confinement properties. The reason for this is that their associated optical and electronic coupling suffers from weak interactions. Therefore, the possibility of forming covalently bonded Si-NC/nanocarbon nanojunctions is an exciting scientific opportunity. Synthesis of CNTs from Si-NCs can be achieved through careful control of the Si-NC surface characteristics; in some cases, surfactant-free Si-NCs with oxygen-based terminations have resulted to be one of the main requirements for successful growth. Oxygen coordination on the Si-NC surface can have a strong impact on the chemical bonding with CNTs as well as on the optoelectronic coupling between the two nanostructures. While the variety of oxygen-based surface arrangements represents an initial challenge, it may turn into a range of opportunities. The assessment of the optoelectronic coupling in these chemically bonded nanosystems can also be an intriguing scientific endeavor. In this work we show the synthesis of a covalently bonded Si-NC/CNT structure via microwave plasma-enhanced chemical vapor deposition (MW-PECVD) using a nonmetallic catalyst. Plasma-enhanced chemical vapor deposition (PECVD) is a technique that has eluded many industries, owing to the complex nature of plasma chemical reactions and plasma–surface interactions, although in recent years the PECVD process has gained acceptance, because of applied research into control and understanding of fundamental principles. The versatility of operating conditions (high and low temperatures/pressure) makes it appealing to many industrial uses when compared with the existing technologies of physical vapor deposition and CVD. The high frequencies that PECVD can be operated with enable more efficient dissociation (formation of one ion–electron pair) in precursor gases (Martinu et al. 2010). As a consequence, the ionization and dissociation rates are higher at lower energy, leading to reduced operation costs. This work is of substantial importance in understanding the growth process of CNT from nonmetallic catalysts and influential to the future development and optimization of Si-NCs and carbon-based nanostructures.

# 24.2 Experimental Details

## 24.2.1 Synthesis

Si-NCs were synthesized by electrochemical etching process of a *p*-type boron–doped wafer, ⟨100⟩, 0.1 Ω resistivity, and thickness of around 0.525 mm. Electrolyte of concentration 1:4 (hydrofluoric acid:ethanol) was used along with constant current density of 1.53 mA/cm² maintained for 90 minutes. This process produces a porous layer on the Si surface where after mechanical pulverization, single Si-NCs can be collected in powder form (Dang et al. 2011). Collected Si-NCs (≈3 mg) was dispersed into 4 mL of both ethanol and deionized (DI) water. One hundred microliters of each colloid was then drop casted onto a 1 cm² Si substrate in preparation for the plasma process. Si-NCs then underwent an MW-PECVD process which contained a $CH_4/N_2$ plasma (40/10 standard cm³/min). A pretreatment of our samples with $N_2$ plasma lasted for 4 minutes before introducing $CH_4$ for a further 8 minutes of growth time. Working temperature and pressure were 600°C and 21 mbar, respectively, throughout the process. Initially, no CNT growth was observed from of our samples post-MW-PECVD and this was attributed to large Si-NC aggregates. Ultrasonication of each colloid was performed before the MW-PECVD process at intervals (20/40/60 minutes) to fragment and improve dispersion of drop-casted nanocrystals. Si-NC colloids at each of the highlighted sonication times were then prepared in same manner as previously described and underwent the same MW-PECVD process.

## 24.2.2 Material Characterization

SEM was performed using a Hitachi S-4300 microscope at 20 kV acceleration voltages. TEM was performed for our samples where successful CNT growth had been confirmed from SEM. The grown Si-NC/CNT composites were mechanically removed and collected onto a holey carbon grid. Images were captured in bright-field mode using a Jeol 2100F microscope and 200 kV acceleration voltages. The PL of as-prepared Si-NCs and after sonication 20/40/60 min was carried out at room temperature and measured using Ocean Optics software with excitation wavelength of 365 nm. Absorption was measured using UV–vis spectroscopy with an integrated sphere through a wavelength range of 250–900 nm. Raman spectroscopy was carried out using an ISA LabRAM 300 confocal Raman spectroscope with a 632.8 nm helium–neon 10 mW laser and a 1 μm spot size. A charge-coupled camera and backscattering geometry were used to collect spectra at exposure times of 10 s. XPS to determine the elemental composition of samples was carried out using a Kratos Axis Ultra DLD spectrometer with a monochromatic aluminum X-ray source ($hv$ = 1486.6 eV). Operating pressure of ~$8 \times 10^{-9}$, voltage of 15 KV, and current of 10 mA were used. An electrostatic and magnetic hybrid lens mode was used with a $300 \times 700$ μm² analysis area. High-resolution spectra were obtained using pass energy of 40 cV and 0.05 eV steps. Core-level scans over the Si $2p$ and C $1s$ regions were performed. The peaks were calibrated to the C $1s$ peak at 284.6 eV (Huang & Gao 2003) and normalized to the metallic Si⁰ peak found around 99.7 eV. Data analysis and peak fitting was performed using the data analysis software Origin.

As-prepared and sonicated Si-NCs samples for XPS were prepared by drop casting 100 μL of the Si-NC colloids onto 1 cm² molybdenum (Mo) substrate. Post-MW-PECVD Si-NC/CNT composites were mechanically removed and collected in toluene followed by immediately drop casting 100 μL onto Mo substrate. Because of the expected differential surface charging effects for the samples caused by photoionization, surface charge compensation via neutralizer was carried out for each sample during the scan.

## 24.3 Results

Figure 24.1a shows low- and high-magnification TEM images of our Si-NCs collected from the porous layer formed during electrochemical etching. The associated histogram of particle size distribution is shown in Figure 24.1b and a mean diameter of 1.8 nm has been found for our Si-NCs. Clear lattice fringes can be observed from the high-resolution image of a single nanocrystal, confirming the crystallinity of our particles.

Figure 24.2 shows the evolution of the PL properties from as-prepared Si-NCs in ethanol and DI water to after the ultrasonication process at each interval. It is understood that formation of an oxide layer is a crucial parameter for CNT growth from nonmetallic catalysts. By fragmenting large aggregates, we hope to

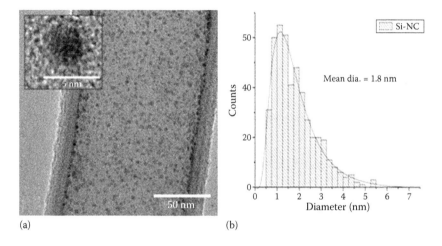

(a)                                                                (b)

**FIGURE 24.1**   (a) TEM images of as-prepared Si-NCs, with the inset showing lattice fringes of single nanocrystal; (b) associated histogram for size distribution of nanocrystals.

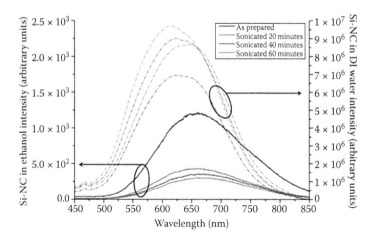

**FIGURE 24.2**   PL spectra for as-prepared Si-NCs in ethanol and DI water as well as after sonication intervals.

accelerate the oxidation process in both ethanol and DI water by exposing previously unexposed Si-NCs to the surrounding environment, thus subjecting cleavage of Si dimers to water molecules. Collected Si-NCs typically contain large agglomerates and almost certainly do not present ideal morphology to induce CNT growth. In addition to enhancing the oxide growth rate, the use of fragmentation process also ensures more dispersed nanocrystals, resulting in increased surface area or active sites for CNT nucleation.

Our results highlight that a blueshift of ≈20 nm is observed for our fragmented Si-NCs in DI water compared to zero shift for our fragmented Si-NCs in ethanol. The blueshift in PL peak wavelength of our nanocrystals after sonication intervals can be attributed to oxidation, consequently reducing the core size and causing a widening of the energy gap in accordance with the QCE (Canham 1990). A progressive increase in PL peak intensity is observed for our Si-NCs in DI water after each sonication interval. Other works have theorized that a low-density oxide shell can passivate dangling bonds at the nanocrystal surface, allowing increased radiative transitions of carriers (Ledoux et al. 2001). The decrease in PL peak intensity for our fragmented Si-NCs in ethanol is attributed to increased defect density after fragmentation. Because of the low water content of ethanol, oxidation occurs at a much slower rate and, therefore, surface defects remain largely unpassivated. These defects present nonradiative channels for carriers to recombine and thus lower the probability of successful radiative transitions (Prokes et al. 1992). Supplementary absorption data displayed in Figure 24.3a and b also highlight blueshifting that can be credited mainly

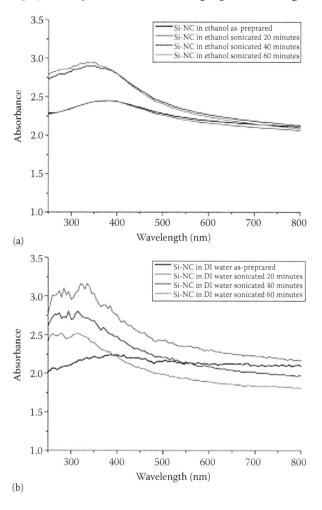

(a)

(b)

**FIGURE 24.3** UV–vis spectra of as-prepared Si-NCs and after sonication intervals in (a) ethanol and (b) DI water.

to oxidation of our nanocrystals. Again, a much larger blueshift in peak absorbance is observed for our nanocrystals in DI water after the fragmentation process, shifting the peak from ≈400 to ≈290 nm. The broad absorption tail observed in our as-prepared samples has previously been attributed to defect states present. The lowering of this tail in both samples, mainly for Si-NCs in DI water, suggests that oxide shell has served to passivate surface defects, therefore allowing absorption from nanocrystal core to dominate.

FTIR analyses of our Si-NCs in ethanol and DI water, displayed in Figure 24.4, show characteristic $SiO_2$ and $SiH_2$ stretching modes at 1100 and 2100 $cm^{-1}$ (Svrcek et al. 2011). Additional peaks are observed for $SiO_2$ and $SiH_2$ bending modes along with $SiH_3$ deformation mode located at 850, 935, and 902 $cm^{-1}$, respectively. It is known that Si-NCs dispersed in DI water are exposed to possible hydroxyl groups or OH ions. This type of radical most likely replaces $H^-$ terminations so that oxidation process progresses by condensation of adjacent $OH^-$ terminations. From accessing the relative increase in ratio of 1100 to 2100 $cm^{-1}$ ($SiO_2$ to $SiH_2$) transmittance peaks between our samples, as-prepared Si-NCs in DI water

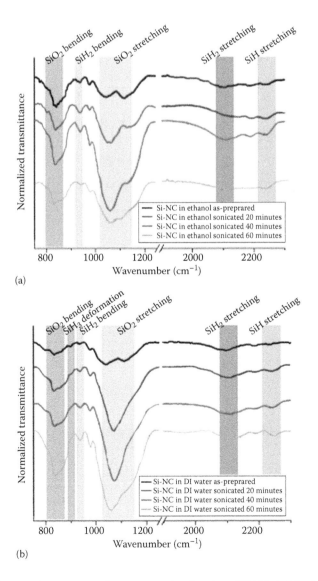

**FIGURE 24.4** FTIR transmittance spectra for as-prepared and sonicated Si-NCs in (a) ethanol and (b) DI water.

increase from 2.4:1 to 8.5:1 after 60-minute sonication, in comparison to 1.7:1 to 5.1:1 for sonicated Si-NCs in ethanol.

Our as-prepared and sonicated Si-NCs in ethanol and DI water were then exposed to an MW-PECVD process as highlighted in Section 24.2. It has to be noted that CNT growth was not observed for our as-prepared Si-NCs in DI water and our as-prepared or sonicated Si-NCs in ethanol. SEM micrographs of our sonicated Si-NCs in DI water after the MW-PECVD process are shown in Figure 24.5a through d. For each sonication time interval, formation of fibrous structures with lengths exceeding 1 μm was observed.

Figure 24.6a shows XPS analysis of the Si $2p$ region for our as-prepared and sonicated Si NCs in DI water (pre MW-PECVD), normalized to the most intense metallic $Si^0$ peak located around 99.7 eV (Sublemontier et al. 2014). Deconvolution of the XPS spectra for as-prepared Si NCs in DI water showed no metallic $Si^0$ peak located around 99.7 eV. A main challenge with the top-down synthesis route of electrochemical etching is that the nanocrystals produced are highly aggregated forming large microsized porous structures. Once exposed to oxidizing species, the aggregate size can play an important role in the packing of absorbed molecules on the surface. The number of water molecules and hydroxyl groups per unit area is found to decrease for smaller particles due to the increasing positive curvature. Meaning for larger particles oxidizing species can be more densely packed, and thus oxidation may occur more rapidly. The absence of a metallic $Si^0$ peak, for as-prepared Si NCs in DI water, is due to the fact that an oxide layer of thickness greater than our XPS penetration depth (10 nm) is present for our large nanocrystal aggregates. After 20 min of sonication, the metallic $Si^0$ peak located around 99.7 eV can be observed. Sonication has induced the fragmentation of heavily oxidized aggregates into smaller arrangements, exposing fresh Si–Si bonds. The intensity of the $Si^0$ state is then seen to decrease with continued sonication time (40 and 60 min). Due to this linear-type regression of the $Si^0$ state with continued sonication time, it can be inferred that the majority of large aggregates have been fragmented and therefore no new Si–Si bonds are uncovered. As a consequence of reducing aggregate size, growth of an oxide shell can now be more accurately controlled as the oxidation process proceeds at a highly reduced rate due to the increased surface curvature. Similarly to the oxidation in water, Figure 24.6b shows the XPS spectra for as-prepared Si NCs in ethanol and after

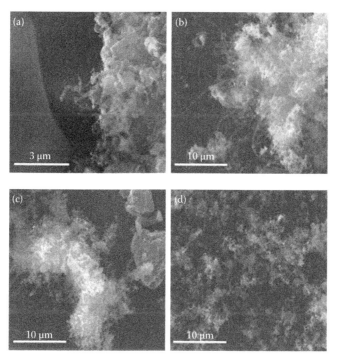

**FIGURE 24.5** Si-NCs in DI water sonicated for (a) 20, (b, c) 40, and (d) 60 minutes.

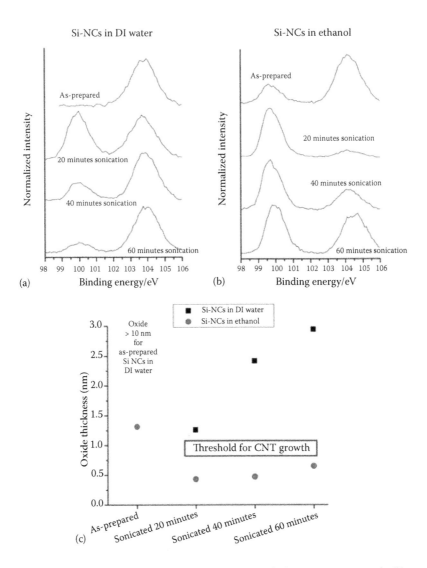

**FIGURE 24.6** (a) XPS spectra for as-prepared Si-NCs in DI water and after sonication intervals; (b) quantification of oxide shell thickness; and (c) shows a plot of oxide thickness that has been quantified from our fitted XPS spectra using a simple calculation. (From Cole, D. A. et al. *Journal of Vacuum Science and Technology B*, 18(1), 440–444, 2000.)

subsequent sonication intervals. As expected, due to the low water content, oxidation of as-prepared Si NCs in ethanol is much slower compared to that of our as-prepared nanocrystals in DI water. As previously stated, top-down synthesis route of electrochemical etching produces large micrometer-sized Si NCs aggregates, which oxidize at a higher rate. Therefore, even in ethanol, we observe that the metallic $Si^0$ peak is less pronounced compared to the $Si^{4+}$ state. After 20 min of sonication, large aggregates are fragmented exposing fresh Si–Si bonds, and an intense $Si^0$ peak appears with minimal oxide states present. Continued sonication (40 and 60 min) shows oxidation progresses at a reduced rate due to smaller Si NCs but also at an overall reduced gradient when compared to sonicated nanocrystals in DI water.

$$T_{ox} = L_{SiO_2} \sin\theta \ln\left[\left(\frac{I_{Si}^\infty}{I_{SiO_2}^\infty} \frac{I_{SiO_2}}{I_{Si}}\right)+1\right]. \tag{24.1}$$

We assume approximation of a constant attenuation length $\left(L_{SiO_2}\right)$ of Si $2p$ photoelectrons in $SiO_2$ equal to 3.485 nm (International Organization for Standardization/Draft International Standard [ISO/DIS] 14701). $I_{SiO_x}$ and $I_{Si}$ are the measured intensities of the oxide and nonoxidized Si, respectively. $I_{SiO_x}$ was calculated by taking into account the relative amount of Si in each oxide subpopulation $I_{Si}^{1+}, I_{Si}^{2+}, I_{Si}^{3+}$, and $I_{Si}^{4+}$, with nearly pure $Si^{4+}$ as a reference:

$$I_{SiO_x} = 0.25 I_{Si^{1+}} + 0.5 I_{Si^{2+}} + 0.75 I_{Si^{3+}} + I_{Si^{4+}}.$$

The ratio $I_{Si}^{\infty}/I_{SiO_2}^{\infty}$ is the ratio of Si $2p$ intensities of "infinitely" thick $Si^0$ and $SiO_2$, respectively, and takes the value 0.9329 (ISO/DIS 14701). The angle $\theta$ is the angle between the surface plane and electron analyzer, which is 58°. This method of calculation assumes the sample to be ideally planar and thus overestimates the oxide thickness of our nanoparticles. This issue has previously been addressed by multiplying the planar thickness by a geometrical correction term. Since our Si-NCs are around a few nanometers in diameter, a correction term of 0.5 is acceptable within a 1% error (Shard 2012). Implementation of this analytical technique allows us to access with relative values the increased rate of oxidation as a consequence of the fragmentation process.

It is understood that during early stages of oxide growth, surface reaction is the rate-limiting factor and oxide thickness varies linearly with time. As the oxide layer becomes thicker, the oxidant must diffuse through the oxide layer to react at the $Si–SiO_2$ interface; therefore, the reaction becomes diffusion limited. Oxide growth is then seen to be proportional to the square root of the oxidizing time that results in parabolic growth rate (Deal & Grove 1965). Our Si-NCs appear to follow this typical growth pattern, although the fragmentation process has assisted to enhance the oxidation gradient. The fact that we observe CNT growth only from our fragmented Si-NCs in DI water indicates a possibility of two things: (a) our as-prepared Si-NCs containing large aggregates do not present ideal size or morphology for CNT growth and/or (b) a certain threshold level of oxidation must be overcome (>0.5 nm) before $SiO_2$ can act as a suitable catalyst.

TEM images of our grown MWCNTs in Figure 24.7a through d show that our nanotubes grow directly from a coalescence of oxidized Si-NCs and produce varying tube diameters. A more detailed TEM analysis (not shown) confirms that our CNTs are represented by MWCNTs of around 15 layers thick. The spacing observed between the walls is ≈0.33 nm, correlating with (002) $d$ spacing of graphite. We also find that single Si-NCs are embedded within the walls of some CNTs, which could additionally represent interesting nanoarchitecture displaying unique optoelectronic properties.

Post-MW-PECVD process Raman spectroscopy was used to confirm the molecular structure of our grown CNTs and also to compare with samples where no CNT growth occurred. Figure 24.8 compares the ratio $I_D/I_G$ of D and G bands for as-prepared Si-NCs in DI water, where no growth was observed, and Si-NCs in DI water sonicated 40 minutes, which showed visible CNT growth. The G band located around 1590 cm$^{-1}$ refers to graphitic carbon and is a result of in-plane vibrations from $sp^2$-bonded carbon. Counter to this, the D band located around 1350 cm$^{-1}$ indicates the presence of some disorder within the graphene structure. The intensity of the disorder or defect band relative to that of the G band is often used as a measure of the quality of nanotubes (Zdrojek et al. 2004). Our sample where no CNT growth was visible had an $I_D/I_G$ ratio of 1.15, which specifies that mainly amorphous carbon was present. The $I_D/I_G$ ratio dropped to 0.92 for our sample that had confirmed CNT growth present. Much lower values would be typically be expected; however, since our CNTs contain multiwalls (≈15 layers), it is likely they contain a high number of defects.

Chemical analysis of our grown Si-NC/CNT composite in Figure 24.9 for (a) C 1$s$ and (b) Si $2p$ confirmed that Si-NCs are covalently bonded to the grown CNTs. Deconvolution of C 1$s$ spectrum showed Si–C, $sp^2$ C, and C–O bonds located at 282.7, 284.4, and 288.3 eV (Bashouti et al. 2012). Additional fitting related to C–C bonds in the amorphous phase were also found at 285.6 eV (Magnuson et al. 2012). Furthermore, the Si $2p$ spectrum contains an intense Si–C bond located around 100.4 eV. Additional Si suboxide states $Si^{2+}$ and $Si^{3+}$ were found at 101.25 and 102.18 eV. Comparison of our Si $2p$ spectra for our sonicated Si-NCs in DI water pre-MW-PECVD process with the Si $2p$ spectra seen below, we notice a

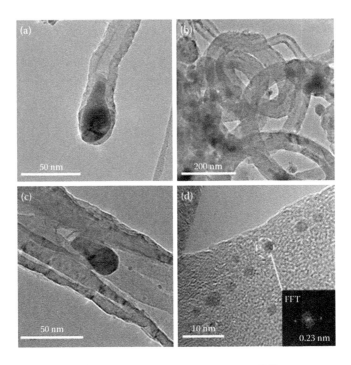

**FIGURE 24.7** (a–d) TEM images of Si-NC/CNT composite post-MW-PECVD process.

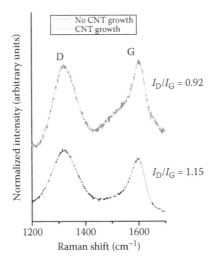

**FIGURE 24.8** Raman spectroscopy showing D and G bands of sample with successful CNT growth compared to sample where no CNT growth occurred after MW-PECVD process.

large reduction for the $Si^{4+}$ state. This is expected as when carbon precursors are absorbed at the surface of the Si-NCs, simulations show that CO is a necessary by-product of the carbothermal reduction of $SiO_2$. Possible growth dynamics will be elaborated on later in Section 24.4.

It is also interesting to note that our Si-NCs maintain their PL properties even after the formation of the Si-NC/CNT nanocomposite. The oxide layer that surrounds the Si-NCs not only provides catalytic functionality for the successful growth of CNTs but we predict that it also aids in protection of the

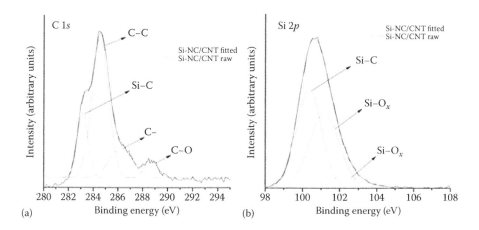

**FIGURE 24.9** XPS of (a) C 1s spectra and (b) Si 2p spectra after CNT growth showing chemically bonded Si–C.

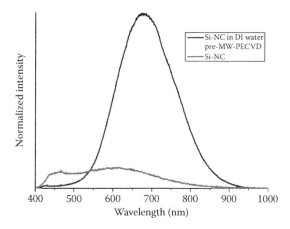

**FIGURE 24.10** PL spectra highlighting that Si-NCs maintain quantum-confined PL properties post-MW-PECVD process.

Si-NCs core. Figure 24.10 shows the PL spectra of Si-NCs in DI water sonicated for 40 min pre-MW-PECVD and Si-NC/CNT nanocomposite post-MW-PECVD process.

Before processing, the PL maximum is located at 677 nm with full width at *half maximum* (FWHM) about 172 nm. This PL is characteristic of our Si-NCs and can be identified as the slow decaying red-orange band, assigned to bound states of electrons and holes (excitons). After CNT formation the "slow" band intensity decreases by a factor of 8, the PL maximum is slightly blueshifted to 620 nm with FWHM increasing to about 285 nm. The appearance of an additional PL peak around 465 nm is seen after the plasma process. This peak is referred to as the fast-decaying blue-green band and is typically associated with the defect sites at the interface between the nanocrystal and its surrounding oxide shell. The changes observed in the positioning and the shape of Si-NC PL properties is almost certainly caused by the bombardment and temperature treatment during the plasma process.

## 24.4 Discussion

### 24.4.1 Typical CNT Growth Mechanism

CNT growth mechanism has been a highly debated topic since its discovery with many contradicting theories proposed. The most widely accepted general mechanism states that when a hydrocarbon gas and a hot metal nanoparticle come into contact, decomposition of the hydrocarbon takes place and the carbon is dissolved into the metal with hydrogen released as a by-product. Once supersaturation (carbon solubility limit) is reached in the metal, the dissolved carbon proceeds to participate outward and crystallizes into a cylindrical network (Kumar & Ando 2010). There is a precise thermal gradient, between the hydrocarbon decomposition (exothermic) and carbon crystallization (endothermic), that needs to be maintained throughout the process. The nanotube can proceed to grow via either the tip-growth model or base-growth model; depending on the contact angle of the metal particle with the substrate, the CNT can precipitate out from the bottom of the particle, pushing away from the substrate (tip growth). Conversely, if the metal particle has a strong interaction with the substrate, then the CNT precipitates outward with the metal particle in a fixed position (base growth). However, in this general model for CNT growth there are several topics that are not yet fully understood such as whether the catalyst is in solid or liquid phase. The vapor–liquid–solid (VLS) model proposed by Baker et al. in the 1970s originally suggested that the role of the metal particle is to form a liquid alloy droplet which absorbs carbon atoms until supersaturation occurs (Baker et al. 1972). This proposition was in agreement with experimental growth activation energies corresponding to bulk diffusion of carbon through metallic catalysts. However, these values were for bulk diffusion in the solid state and not in the liquid state as suggested by the model. This contradiction within the hypothesis created a debate surrounding whether the catalyst particle is liquid or solid phase, giving introduction to a vapor–solid–solid (VSS) mechanism of growth and even vapor–solid surface–solid (VSSS). It has been reported that the resultant properties of the synthesized nanotubes are not seen to be dependent on whether catalysts are in the solid or liquid phase, leaving the specific mechanism of CNT growth from metallic catalysts unknown as both VLS and VSS can occur simultaneously.

## 24.4.2 Nonmetallic Catalyst Growth of CNTs

Recently, many studies have indicated that metals not normally displaying catalytic activity do induce CNT growth when in oxide form (Rümmeli et al. 2007). This suggests that the surrounding oxide layer can act as a suitable catalyst for CNT growth. Similarly, as with CNT growth from metallic catalyst, the mechanisms that allow CNT growth on nonmetallic catalysts such as $SiO_2$ nanoparticles are not yet fully understood. Although some consensus states that because of the generally higher melting temperature of $SiO_2$, the VLS mechanism cannot take place and VSS and VSSS mechanisms have been proposed (Page et al. 2011). Others have speculated that if the catalyst is in solid state, then strain due to the high curvature from the small particle size would dissociate and break bonds between central atoms, thus allowing interaction with hydrocarbons (Saani et al. 2009). Furthermore, it has been suggested that in a semisolid state, distortion in the nanoparticles' overall shape would introduce polarization of Si and O atoms, which could facilitate decomposition of a hydrocarbon.

There are two main differences that exist between nonmetallic catalysts such as our Si-NCs and metallic catalysts. Firstly, the use of nonmetallic catalysts reduces catalytic activity; therefore, carbon precursors must reach the surface of the nanoparticles largely decomposed. This can be achieved through higher processing temperatures or through plasma-induced decomposition. Higher processing temperatures can cause annealing effects that are known to alter or destroy the desired optoelectronic properties of our nanocrystals. Therefore, a low-temperature plasma environment, such as that used in this work, is favorable for the nucleation and formation of a nanocarbon structure (Kumar et al. 2013). Secondly, because a molten layer would be difficult to produce on the nonmetallic $SiO_2$ catalyst, carbon atoms can absorb and diffuse only on the solid surface, i.e., VSS or VSSS. Quantum mechanical–molecular dynamic simulations on $SiO_2$ nanoparticles confirmed the absence of a molten phase in the VLS growth mechanism (Kumar et al. 2013). It was demonstrated that the carbothermal reduction of $SiO_2$ by $CH_x$ was limited to the surface of $SiO_2$ nanoparticles. This localized carbon density results in extended polyyne chains covering the surface and subsurface layers of the $SiO_2$ particles, forming amorphous silicon carbide according to the reaction

$$3C + SiO_2 \rightarrow SiC + 2CO.$$

Silicon carbide is accepted to form through an intermediate SiO stage; thus, overall reaction can be further broken down to (Bachmatiuk et al. 2009)

$$C + SiO_2 \rightarrow SiO + CO,$$

$$2C + SiO \rightarrow SiC + CO.$$

The rate of $sp^2$-hybridized carbon network on the surface of $SiO_2$ nanoparticle is accelerated at higher carbon concentrations and, therefore, it is necessary for supersaturation to occur at the $SiO_2$ surface or subsurface. Page et al. (2011) gave more details on the mechanism of CO formation: the capture of $CH_x$ on the $SiO_2$ nanoparticle surface leads to weakening of an existing Si–O–Si bridge structure, eventually resulting in the formation of Si–O–CH. Simulations show that a neighboring CH carbon then extracts the hydrogen atom from the S–O–CH, followed by the dissociation of CO from $SiO_2$ nanoparticle (Page et al. 2011). Removal of hydrogen in this manner is essential for formation of CO species, a known prerequisite for CNT growth. In addition, it leaves carbon dangling bonds; these surface defects have been shown to play a decisive role in the continuation of CNT nucleation owing to the high reactivity that dangling bonds present (Demuynck et al. 1997). Because CNT growth occurred only on Si-NCs that had been fragmented and heavily oxidized, our assumption is that the role of the oxide can include, but is not limited to, increasing the capture of $CH_x$ radicals.

It was highlighted that if the reduction of $SiO_2$ and production of CO was indeed a limiting factor in CNT growth, then the CNT diameter would decrease during growth because of the reduction of Si–NC particle mass from the reaction

$$SiO + 2C \rightarrow SiC + CO.$$

Because of the high-density network of CNTs grown from our Si-NC sonicated in DI water, we are unable to confirm whether the diameter of our grown nanotubes is uniform throughout. Our TEM images postsynthesis of CNTs show that the silicon carbide particles at the root of the nanotube exhibit smooth curved surfaces. This observation was also made by Bachmatiuk et al. (2009), possibly indicating that during the reaction process the SiC particles were in a liquid-like phase during the process. The melting point for bulk SiC is around 2700°C, which is much greater than our operating temperature of 600°C. Although it is known that particles within the nanoscale dimensions have reduced melting points, coupling them with other species such as hydrogen could further reduce the melting point. This would mean that precipitation of carbon could occur from a liquid-like particle, hence CNT growth occurring through the VLS mechanism as found with metal catalysts. The theory of VLS mechanism from $SiO_2$ nanoparticles was shown to be an unlikely one owing to simulations indicating that carbon atoms absorbed onto the surface of a $SiO_2$ nanoparticle remained and did not proceed to precipitate through a molten state. The diffusion or precipitation of carbon throughout the $SiO_2$ nanoparticles surface/subsurface is also seen to be an energetically unfavorable process, hence concluding that CNT nucleation from $SiO_2$ occurred through a VSS type mechanism and not a VLS mechanism (Page et al. 2011).

$SiO_2$ nanoparticles act as a solid catalyst and in essence become carbon-coated nanoparticles. Following this coating, a graphene island with five-membered rings is formed on the carbon-saturated surface and acts as the nucleus for CNT growth. The nanoparticles high surface-to-volume ratio is beneficial and therefore provides an ideal template for graphene cap formation (Homma et al. 2009, Chen & Zhang 2011). The use of our MW-PECVD process as mentioned assists in the favorable dissociation of carbon precursors, hence substituting the need for a highly functional catalyst. While plasma inter-actions are complementary in this instance, the production of energetic ions leaves our nanocrystals susceptible to radiation damage. Irradiation of a particle with energetic ions can cause damage by dis-placement of atoms, either through direct energetic nuclear collisions or through indirect electronic excitation and ionization (Nastasi et al. 1996). These structural changes can lead to compression or expansion effects that alter the refractive index of the material. In our case for Si-NCs, radiation damage from energetic ions can form bond distortions, ultimately resulting in generation of defects that indi-rectly affects optoelectronic properties. It is therefore important to note that synthesized Si-NC/CNT nanocomposites maintained their quantum-confined PL properties throughout the plasma process. The PL quenching (factor of ≈8) that was experienced in our Si-NCs (slow band) post-CNT synthesis we attribute to a number of factors that act at the same time. Firstly, PL intensity is indirectly proportional to defect density and, therefore, the increased defect density generated due to ion bombardment sig-nificantly lowers the probability of radiative recombination (Tischler et al. 1992). Secondly, the strong absorption properties of the highly dense CNTs can also limit the emission from the Si-NCs. A further consequence of the plasma process is that the "fast" band around 465 nm, related to interface defects, becomes more prominent in the Si-NC/CNT nanocomposite. Nevertheless, our results show that the oxide layer not only provides catalytic functionality for the successful growth of CNTs but also acts as a protective layer from ion bombardment, enabling Si-NCs to maintain their quantum-confined proper-ties throughout energetic plasma interactions. Plasmas have the potential to allow CNT growth at much lower temperatures when compared to CVD, reducing the probability of thermal degradation of our Si-NCs' optical properties.

## 24.5  Conclusions

In this chapter, we discuss the aspects that relate to carbon nanostructures and their integration with Si-NCs. The control of Si-NCs/nanocarbon coupling presents not only an exciting scientific opportunity but key mechanisms where Si-NCs and carbon nanostructures can outperform current technologies. An overview of the Si-NC CNT coupling without any metal catalyst is given. Particularly, we clearly dem-onstrate that CNT growth from quantum-confined Si-NCs is dependent on sufficient surface oxide; we determine critical thickness for our samples to be >0.5 nm and is imperative for successful CNT growth. An additional limiting factor of nonaggregated and thus small nanocrystal assemblies is also vital for successful CNT growth. The presence of large Si-NC aggregates most likely does not present ideal size and morphology to serve as a template/active site for formation of CNT caps. The small size and level of oxidation clearly assists in the unstable physisorption of carbon atoms or for the no–transition metal catalytic decomposition of the hydrocarbons. Importantly, synthesized Si-NC/CNT nanocomposites maintain their quantum-confined properties throughout the plasma process. The oxide layer formed not only provides catalytic functionality for the successful growth of CNTs but also acts as a protective layer against ion bombardment. SEM and TEM imaging showed that CNTs grew directly from a coales-cence of oxidized Si-NCs. Chemical analysis, post-MW-PECVD process, confirmed that the CNTs were covalently bonded with our nanocrystals. Such a Si-NC/CNT junction would be advantageous to many applications by improving dissociation and transport properties. The fundamental results presented within this chapter indicate a promising research direction that could lead to the fabrication of an ideal nanoscale junction between two nanostructures with unique quantum-confined effects.

## Acknowledgments

This work was partially supported by a New Energy and Industrial Technology Development Organization (NEDO) project, the Royal Society International Exchange Scheme (IE120884), the Leverhulme International Network (IN-2012-136,) and the Engineering and Physical Sciences Research Council (EPSRC) (EP/K022237/1).

# References

Altman, I. S., Lee, D., Chung, J. D. et al., "Light absorption of silica nanoparticles," *Phys. Rev. B* 63 (2001): 161402.

Bachmatiuk, A., Börrnert, F., Grobosch, M. et al., "Investigating the graphitization mechanism of $SiO_2$ nanoparticles in chemical vapor deposition," *ACS Nano* 3 (2009): 4098–4104.

Baker, R. T. K., Barber, M. A., Harris, P. S. et al., "Nucleation and growth of carbon deposits from the nickel catalyzed decomposition of acetylene," *J. Catal.* 26 (1972): 51–62.

Bashouti, M. Y., Pietsch, M., Sardashti, K. et al., "Hybrid silicon nanowires: From basic research to applied nanotechnology" in Peng, X., ed., *Nanowires - Recent Advances*. Available from: http://www.intechopen.com/books/nanowires-recent-advances/hybrid-silicon-nanowires-from-basic-research-to-applied-nanotechnology. ISBN: 978-953-51-0898-6, InTech, DOI: 10.5772/54383. (2012).

Beard, M. C., Knutsen, K. P., Yu, P. et al., "Multiple exciton generation in colloidal silicon nanocrystals," *Nano Lett.* 7 (2007): 2506–2512.

Canham, L. T., "Silicon quantum wire array fabrication by electrochemical and chemical dissolution of wafers," *Appl. Phys. Lett.* 57 (1990): 1046–1048.

Chen, Y. & Zhang, J., "Diameter controlled growth of single-walled carbon nanotubes from $SiO_2$ nanoparticle," *Carbon* 49 (2011): 3316–3324.

Cole, D. A., Shallenberger, J. R., Novak, S. W. et al., "$SiO_2$ thickness determination by x-ray photoelectron spectroscopy, Auger electron spectroscopy, secondary ion mass spectrometry, Rutherford backscattering, transmission electron microscopy, and ellipsometry," *J. Vac. Sci. Technol. B* 18 (2000): 440–444.

Creighton, J. R. & Ho, P., "Introduction to chemical vapor deposition (CVD)," (2001): 1–22.

Dang, X., Yi, H., Ham, M. H. et al., "Virus-templated self-assembled single-walled carbon nanotubes for highly efficient electron collection in photovoltaic devices," *Nat. Nanotechnol.* 6 (2011): 377–384.

De Boer, W. D. A. M., Timmerman, D., Dohnalova, K. et al., "Red spectral shift and enhanced quantum efficiency in phonon-free photoluminescence from silicon nanocrystals," *Nat. Nanotechnol.* 5 (2010): 878–884.

Deal, B. E. & Grove, A. S., "General relationship for the thermal oxidation of silicon," *J. Appl. Phys.* 36 (1965): 3770–3778.

Demuynck, L., Arnault, J. C., Polini, R. et al., "CVD diamond nucleation and growth on scratched and virgin Si (100) surfaces investigated by in-situ electron spectroscopy," *Surf. Sci.* 377 (1997): 871–875.

Ellingson, R. J., Beard, M. C., Johnson, J. C. et al., "Highly efficient multiple exciton generation in colloidal PbSe and PbS quantum dots," *Nano Lett.* 5 (2005): 865–871.

Evanoff, K., Khan, J., Balandin, A. A. et al., "Towards ultrathick battery electrodes: Aligned carbon nanotube-enabled architecture," *Adv. Mater.* 24 (2012): 533–537.

Fox, A. M., *Optical Properties of Solids*, Vol. 3 (Oxford University Press, Oxford, England, 2001).

Gelloz, B., Kojima, A. & Koshida, N., "Highly efficient and stable luminescence of nanocrystalline porous silicon treated by high-pressure water vapor annealing," *Appl. Phys. Lett.* 87 (2005): 031107.

Homma, Y., Liu, H., Takagi, D. et al., "Single-walled carbon nanotube growth with non-iron-group 'catalysts' by chemical vapor deposition," *Nano Res.* 2 (2009): 793–799.

Hu, L., Wu, H., Gao, Y. et al., "Silicon–carbon nanotube coaxial sponge as Li-ion anodes with high areal capacity," *Adv. Energy Mater.* 1 (2011): 523–527.

Huang, Q. & Gao, L., "Immobilization of rutile $TiO_2$ on multiwalled carbon nanotubes," *J. Mater. Chem.* 13 (2003): 1517–1519.

Kang, Z., Tsang, C. H. A., Wong, N. B. et al., "Silicon quantum dots: A general photocatalyst for reduction, decomposition, and selective oxidation reactions," *J. Am. Chem. Soc.* 129 (2007): 12090–12091.

Koch, F., Petrova-Koch, V. & Muschik, T., "The luminescence of porous Si: The case for the surface state mechanism," *J. Lumin.* 57 (1993): 271–281.

Kumar, M. & Ando, Y., "Chemical vapor deposition of carbon nanotubes: A review on growth mechanism and mass production," *J. Nanosci. Nanotechnol.* 10 (2010): 3739–3758.

Kumar, S., Mehdipour, H. & Ostrikov, K., "Plasma-enabled graded nanotube biosensing arrays on a Si nanodevice platform: Catalyst-free integration and in situ detection of nucleation events," *Adv. Mater.* 25 (2013): 69–74.

Landsberg, P. T., Nussbaumer, H. & Willeke, G., "Band–band impact ionization and solar cell efficiency," *J. Appl. Phys.* 74 (1993): 1451–1452.

Ledoux, G., Gong, J. & Huisken, F., "Effect of passivation and aging on the photoluminescence of silicon nanocrystals," *Appl. Phys. Lett.* 79 (2001): 4028–4030.

Liu, M., Lu, G. & Chen, J., "Synthesis, assembly, and characterization of Si nanocrystals and Si nanocrystal–carbon nanotube hybrid structures," *Nanotechnology* 19 (2008): 265705.

Magnuson, M., Andersson, M., Lu, J. et al., "Electronic structure and chemical bonding of amorphous chromium carbide thin films," *J. Phys. Cond. Matter* 24 (2012): 225004.

Mangolini, L., "Synthesis, properties, and applications of silicon nanocrystals," *J. Vac. Sci. Technol. B* 31 (2013): 020801.

Mariotti, D., Mitra, S. & Švrček, V., "Surface-engineered silicon nanocrystals," *Nanoscale* 5 (2013): 1385–1398.

Martinu, L., Zabeida, O. & Klemberg-Sapieha, J. E., "Plasma-enhanced chemical vapor deposition of functional coatings," in Martin, P. M., ed., *Handbook of Deposition Technologies for Films and Coatings*, Third edition (Elsevier, Amsterdam, Netherlands, 2010), pp. 394–467.

Maruyama, T. & Ohtani, S., "Photoluminescence of porous silicon exposed to ambient air," *Appl. Phys. Lett.* 65 (1994): 1346–1348.

Mott, N. F., Rigo, S., Rochet, F. et al., "Oxidation of silicon," *Philos. Mag. Part B* 60 (1989): 189–212.

Murphy, J. E., Beard, M. C., Norman, A. G. et al., "PbTe colloidal nanocrystals: Synthesis, characterization, and multiple exciton generation," *J. Am. Chem. Soc.* 128 (2006): 3241–3247.

Nastasi, M., Mayer, J. W. & Hirvonen, J. K., *Ion–Solid Interactions: Fundamentals and Applications* (Cambridge University Press, New York, 1996).

Nozik, A. J., "Quantum dot solar cells," *Phys. E* 14 (2002): 115–120.

Ogata, Y. H., Kato, F., Tsuboi, T. et al., "Changes in the environment of hydrogen in porous silicon with thermal annealing," *J. Electrochem. Soc.* 145 (1998): 2439–2444.

Okada, R. & Iijima, S., "Oxidation property of silicon small particles," *Appl. Phys. Lett.* 58 (1991): 1662–1663.

Page, A. J., Chandrakumar, K. R. S., Irle, S. et al., "SWNT nucleation from carbon-coated $SiO_2$ nanoparticles via a vapor–solid–solid mechanism," *J. Am. Chem. Soc.* 133 (2011): 621–628.

Pavesi, L. & Turan, R., eds., *Silicon Nanocrystals: Fundamentals, Synthesis and Applications* (John Wiley & Sons, Hoboken, 2010).

Pickering, C., Beale, M. I. J., Robbins, D. J. et al., "Optical studies of the structure of porous silicon films formed in p-type degenerate and non-degenerate silicon," *J. Phys. C* 17 (1984): 6535.

Prokes, S. M., Carlos, W. E. & Bermudez, V. M., "Luminescence cycling and defect density measurements in porous silicon: Evidence for hydride based model," *Appl. Phys. Lett.* 61 (1992): 1447–1449.

Ranjan, V., Kapoor, M. & Singh, V. A., "The band gap in silicon nanocrystallites," *J. Phys. Cond. Matter* 14 (2002): 6647.

Ren, S., Bernardi, M., Lunt, R. R. et al., "Toward efficient carbon nanotube/P3HT solar cells: Active layer morphology, electrical, and optical properties," *Nano Lett.* 11 (2011): 5316–5321.

Rümmeli, M. H., Schäffel, F., Kramberger, C. et al., "Oxide-driven carbon nanotube growth in supported catalyst CVD," *J. Am. Chem. Soc.* 129 (2007): 15772–15773.

Saani, M. H., Ghodselahi, T. & Esfarjani, K., "Strain-induced instability of spherical nanodiamond hydrocarbons: Effect of surface $CH_n$ and charging," *Phys. Rev. B* 79 (2009): 125429.

Schaller, R. D., Pietryga, J. M. & Klimov, V. I., "Carrier multiplication in InAs nanocrystal quantum dots with an onset defined by the energy conservation limit," *Nano Lett.* 7 (2007): 3469–3476.

Scheer, K. C., Madhukar, S., Muralidar, R. et al., "Oxidation of silicon nanocrystals," *MRS Proc.* 638 (2000): F6.3.1. doi:10.1557/PROC-638-F6.3.1

Shard, A. G., "A straightforward method for interpreting XPS data from core–shell nanoparticles," *J. Phys. Chem. C* 116 (2012): 16806–16813.

Shirahata, N., "Colloidal Si nanocrystals: A controlled organic–inorganic interface and its implications of color-tuning and chemical design toward sophisticated architectures," *Phys. Chem. Chem. Phys.* 13 (2011): 7284–7294.

Shockley, W. & Queisser, H. J., "Detailed balance limit of efficiency of *p–n* junction solar cells," *J. Appl. Phys.* 32 (1961): 510–519.

Soni, R. K., Fonseca, L. F., Resto, O. et al., "Size-dependent optical properties of silicon nanocrystals," *J. Lumin.* 83 (1999): 187–191.

Sublemontier, O., Nicolas, C., Aureau, D. et al., "X-ray photoelectron spectroscopy of isolated nanoparticles," *J. Phys. Chem. Lett.* 5 (2014): 3399–3403.

Švrček, V., "Fabrication of filled carbon nanotubes with fresh silicon nanocrystals produced in situ by nanosecond pulsed laser processing in environmentally friendly solutions," *J. Phys. Chem. C* 112 (2008): 13181–13186.

Švrček, V. & Mariotti, D., "Electronic interactions of silicon nanocrystals and nanocarbon materials: Hybrid solar cells," *Pure Appl. Chem.* 84 (2012): 2629–2639.

Švrček, V., Mariotti, D., Nagai, T. et al., "Photovoltaic applications of silicon nanocrystal based nanostructures induced by nanosecond laser fragmentation in liquid media," *J. Phys. Chem. C* 115 (2011): 5084–5093.

Tischler, M. A., Collins, R. T., Stathis, J. H. et al., "Luminescence degradation in porous silicon," *Appl. Phys. Lett.* 60 (1992): 639–641.

Xu, Y. J., Zhuang, Y. & Fu, X., "New insight for enhanced photocatalytic activity of $TiO_2$ by doping carbon nanotubes: A case study on degradation of benzene and methyl orange," *J. Phys. Chem. C* 114 (2010): 2669–2676.

Zdrojek, M., Gebicki, W., Jastrzebski, C. et al., "Studies of multiwall carbon nanotubes using Raman spectroscopy and atomic force microscopy," *Solid State Phenom.* 99 (2004): 265–268.

Zhang, Q., Huang, J. Q., Qian, W. Z. et al., "The road for nanomaterials industry: A review of carbon nanotube production, post-treatment, and bulk applications for composites and energy storage," *Small* 9 (2013): 1237–1265.

# 25

# Graphene/Carbon Nanotube Aerogels

Hai M. Duong

Zeng Fan

Son T. Nguyen

## 25.1 Introduction

An aerogel is an open-celled, mesoporous, and solid foam composed of a network of interconnected nanostructures with porosity of over 50%. The term *mesoporous* or *mesoporous material* is commonly defined as a material that contains pores ranging from 2 to 50 nm in diameter. The term *aerogel* does not refer to a specific substance, but rather to a network structure, which a substance can take on. Aerogels can be made of a wide variety of substances, including silica, most of the transition metal oxides, lanthanide and actinide metal oxides, several main group metal oxides, organic polymers, semiconductor nanostructures, metals, and carbon-based nanomaterials (carbon powder, CNTs, and graphene).

Aerogels are well known for their extremely low densities (often ranging from 0.00016 to ~0.5 g/cm$^3$) and have 95%–99% air (or other gases) in their volume (El-Nahhal & El-Ashgar 2007, Du et al. 2013). They are even considered as the lowest-density solid materials that have ever been made (Sun et al. 2012). Because of their extraordinary low densities and the length-scale effects from the nanostructure features, aerogels can exhibit several advanced properties over the nonaerogel forms made of the same substances such as larger surface area and catalytic activity. But the aerogel structure may also have reduced mechanical strength (Fricke & Emmerling 1998, Du et al. 2013).

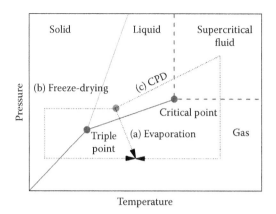

**FIGURE 25.1** Principles of three drying methods (evaporation, freeze-drying, and CPD) to remove liquid from hydrogels on a generic phase diagram. (a) Direct evaporation from liquid phase to gas phase causes collapse of the gel network because of the effect of surface tension. (b) In a freeze-drying process, the hydrogels are fast frozen and then placed in vacuum with temperature increase to allow the ice to sublimate. (c) In the CPD process, the temperature and the pressure of liquid carbon dioxide are increased above the critical point to form a supercritical fluid. By releasing the pressure slowly, the hydrogels are dried to form the aerogels with well-preserved networks.

Based on most gelation routes, hydrogels are usually obtained as direct products, rather than the aerogels. How to properly remove permeated liquids inside the hydrogel pores has been challenging especially for large-scale aerogel fabrication. Figure 25.1 presents three possible drying methods for the liquid removal from the hydrogel structures: (1) evaporation, (2) critical point drying (CPD), and (3) freeze-drying. Because of the effects of capillary force and surface tension, the simple evaporation method normally cannot yield the highly porous structure of aerogels and forms collapsed and condensed solid structures called xerogels (Scherer & Smith 1995, Namatsu et al. 1999). Therefore, the CPD method (García-González et al. 2012) or freeze-drying method (Franks 1998) has been preferred to remove the liquids effectively and still preserve well the gel networks of the carbon-based aerogels. Therefore, the selection between two drying methods in practice strongly depends on the property/cost requirements of the fabricated aerogels.

## 25.1.1 CPD Method

The CPD method, also known as the supercritical drying one, has been widely used to preserve the porous structures in the production of aerogels, biological specimens, and microelectromechanical systems (García-González et al. 2012). As can be seen in Figure 25.1, the CPD process can allow the transformation of liquid to gas passing through a supercritical region instead of crossing the liquid–gas boundary only. Within the supercritical region, the liquid and gas phases mix to form a supercritical fluid. Thus, surface tension causing the collapse of the gel structure can be terminated.

However, only a few materials such as Freon 13 and carbon dioxide can be applied to the CPD method because most permeated liquids/solvents require high temperatures and pressures to reach their critical points (Adams et al. 1981). These operation conditions are critical and may damage many organic gels. The most common solvent is liquid carbon dioxide ($CO_2$), which can be supercritically extracted at a low temperature of 31.1°C and a modest pressure of 72.8 atm (Novak et al. 1994). As liquid $CO_2$ is immiscible with water, an intermediate process called solvent exchange is strongly required. The hydrogels are usually soaked in ethanol for at least several hours, while ethanol is refreshed at least three times (Novak et al. 1994). After this step, the obtained alcogels full of ethanol in their pores are ready to undergo the CPD process to yield their aerogel forms.

## 25.1.2 Freeze-Drying Method

The freeze-drying technique in Figure 25.1 is also another common method to remove the liquid from hydrogels (Franks 1998). In this method, the hydrogels are first frozen and placed in vacuum. After complete transformation to ice, the frozen hydrogels are then heated while still under vacuum, at least below the triple point of water, 0.006 atm. The ice can directly sublimate into vapor and finally the aerogels are formed. A quick freezing should be essential for building better gel structures, as more microcrystals can be simultaneously formed and dominate the ice matrix, compared with slow cooling. Compared to the CPD method, freeze-drying one is more cost-effective, but time consuming. Researchers found that the aerogels from the CPD method can exhibit larger surface areas and superior multifunctional properties over those from the freeze-drying method (Zhang et al. 2011a).

# 25.2 Graphene/CNT Aerogels

CNTs and graphene are two extensively investigated carbon allotropes in recent years. Both of them possess unique multifunctional properties, such as high electrical conductivity up to $10^4$ S/cm (Ijima 1991, Orlita et al. 2008), superior thermal conductivity of 2000–5300 W/m K (Pop et al. 2005, Balandin et al. 2008), and outstanding Young's modulus of ~1 TPa (Yu et al. 2000, Lee et al. 2008). Fabricating these novel carbon-based nanomaterials into their aerogel forms can not only enrich the material family of carbon-based aerogels, but also produce highly porous and interconnected networks with multiple excellent properties. Carbon-based aerogels possess properties almost similar to those of other aerogels such as silica aerogels (Pajonk 1998), but they are also electrically conductive with their density-dependent conductivity. Because of the large surface area, high porosity, and electrical property, carbon-based aerogels such as CNT aerogels, graphene aerogels (GAs) (as shown in Figure 25.2), and CNT–graphene hybrid aerogels are mostly used for industrial applications such as desalination (Wang et al. 2011a), solar energy collection (Biener et al. 2011), and catalyst support (Hrubesh 1998, Meng et al. 2011).

Because of the facile fabrication and dispersion capability of graphene oxide (GO), it has become the most popular reactant among the graphene family such as pristine graphene and epitaxial grown graphene for GA development. GO is normally obtained from the oxidative treatments of graphite and successive exfoliation (Dreyer et al. 2010). GO is the intermediate material of reduced graphene oxide (rGO) before reduction and considered as the oxidized graphene having various reactive oxygen functional groups such as carbonyl, hydroxyl, and epoxide groups on its surface (Nakajima & Matsuo 1994, Sui et al. 2011). Although the exhibited oxygen content of GO is slightly different through different fabrication

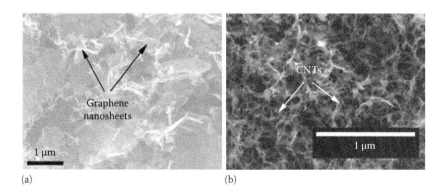

**FIGURE 25.2** SEM images of the highly porous architectures of (a) GA (Reprinted with permission from Worsley et al., 14067–14069. Copyright 2010 American Chemical Socicty.) and (b) CNT aerogel. (Bryning, M. B., Milkie, D. E., Islam, M. F. et al., Carbon Nanotube Aerogels. *Advanced Materials*. 2007. 19. 661–664. Copyright Wiley-VCH Verlag GmbH & Co. KGaA. Reproduced with permission.)

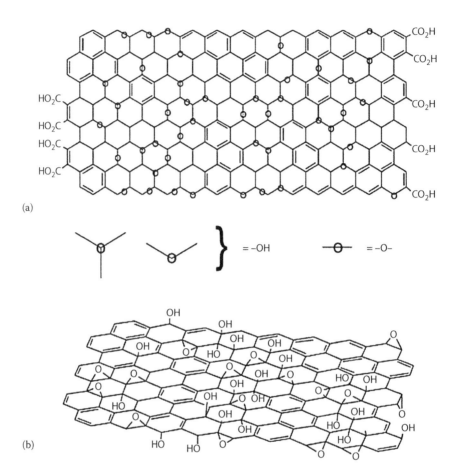

**FIGURE 25.3** The models by Lerf et al. (1998) and He et al. (1998) with carboxylic acid groups (a) (Reprinted with permission from Lerf et al. (1998), 4477–4482. Copyright 1998 American Chemical Society.) and without carboxylic acid groups (b) on the GO (Reprinted from *Chemical Physics Letters*, 287, He, H., Klinowski, J., Forster, M. et al., A new structural model for graphite oxide, 53–56, Copyright (1998), with permission from Elsevier.).

procedures, a C:O ratio of ~2:1 is the upper limit of the oxidation extent, regardless of one-step or multiple oxidation approaches (Brodie 1859, Hummers & Offeman 1958, Dresselhaus & Dresselhaus 2002, Duquesne et al. 2003). Although there are several models of the GO structure (Ruess 1947, Hummers & Offeman 1958, Nakajima et al. 1988, Nakajima & Matsuo 1994), the most well-known model, shown in Figure 25.3, was proposed by Lerf et al. (1998) and He et al. (1998), where solid-state NMR was first employed to characterize GO. The GO structure with and without carboxylic acid groups largely depends on oxidative conditions. However, different raw materials and reaction pathways upon oxidation would probably cause substantial variance to the exact structures and properties of the GO; thus, difficulties still reside in its identification with one unified structural feature (Dreyer et al. 2010).

## 25.2.1 Fabrication of GO

There have been many approaches proposed to prepare GO, but the most popular method is the modified Hummers method (Hummers & Offeman 1958, Stankovich et al. 2006, Xu et al. 2011), owing to the simplicity and cost-effectiveness of its process. Typically, 2 g of sodium nitrate ($NaNO_3$) and 4 g of graphite powder are added to 100 mL of concentrated sulfuric acid ($H_2SO_4$) in an ice bath. After stirring the mixture for 30 min, 14.6 g of potassium permanganate ($KMnO_4$) is then slowly added to the mixture with stirring and cooling to

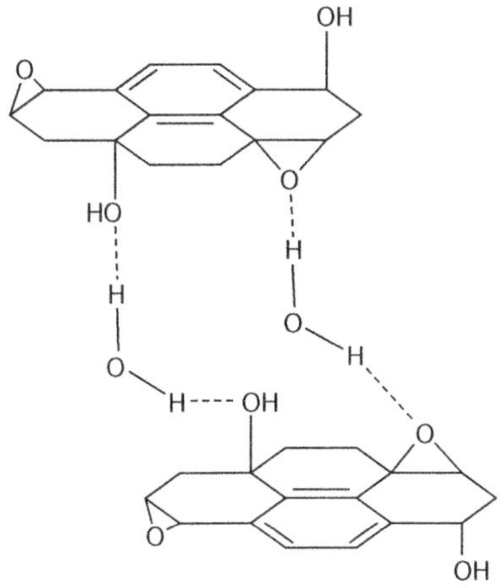

**FIGURE 25.4** Schematic of hydrogen bonding formed between water ($H_2O$) and the oxygen functional groups of the GO. (From Dreyer, D.R. et al., *Chemical Society Reviews*, 39, 228–240, 2010.)

kccp the temperature lower than 20°C for another 2 h. Next, the temperature of the mixture is increased to 35°C with stirring for 12 h to convert graphite to graphite oxide. Then, 180 mL of deionized (DI) water is added to the mixture with stirring for 15 min, and 14 mL of 30% hydrogen peroxide (30% $H_2O_2$) and 110 mL of DI water are gradually added to the reaction mixture. The color of the mixture changes from brown to yellow.

Then the graphite oxide is separated from the mixture by centrifugation and washed with 1 M hydrochloric acid (HCl) and DI water to remove any impurities. The solid of graphite oxide is finally dried at 60°C for 3 days before collection. To convert graphite oxide to GO, the as-prepared graphite oxide solid is first dissolved in DI water to form a 0.1–0.5 wt% aqueous suspension. The suspension is then sonicated for 12 h to exfoliate the interlayers of graphite oxide. GO powder is finally obtained by centrifuging the sonicated suspension and drying the collected solid at 60°C for 3 days.

Because of the abundance of the oxygen functional groups, GO possesses good hydrophilicity in water and aqueous liquids (Left et al. 1997, Lerf et al. 1998). The strong bonding between water ($H_2O$) and the basal lattice of GO is dominated by the oxygen in its epoxides, as illustrated in Figure 25.4. Despite these functional groups forming the good hydrophile GO, they disrupt the $sp^2$ network of the graphitic lattice, consequently resulting in GO with a low electrical conductivity of ~0.021 ± 0.002 S/m (Stankovich et al. 2007). Chemical-reduction (Si & Samulski 2008) or thermal-reduction (Stankovich et al. 2007) processes can be applied to form reduced graphene oxide (rGO) from GO, which can partially recover the conductivity of GO. However, the multifunctional properties of rGO are generally similar to those of pristine graphene but with lower values as the oxidation process itself is detrimental to the graphitic lattice planar (Stankovich et al. 2007, Si & Samulski 2008, Tung et al. 2009a). For GA development, GO obtained through chemical conversion is still the most preferred material, owing to its facile preparation, excellent hydrophobicity, and other good properties.

## 25.2.2 Fabrication of GAs from the Cross-Linking-Induced Method

Based on the synthesis of organic (resorcinol–formaldehyde [RF]) and carbon aerogels (Pekala 1989), Worsley et al. (2009a, 2011) reported the synthesis of ultra-low-density GAs using RF as the organic binders to make carbon cross-links between graphene nanosheets. In this method, the gelation process was the same as the one for RF organic aerogels (Pekala 1989), except the GO was first dispersed in the aqueous RF solutions. After the solvent exchange and supercritical drying (or freeze-drying) processes, dried GO-RF aerogels were obtained. Pyrolysis at 1050°C under inert gases was subsequently applied to carbonize the GO-RF aerogels into carbon-based GAs.

Besides using RF, other polymers such as PVA/poly(styrene sulfonate) (PSS) (Vickery et al. 2009) and Pluronic copolymers (Zu & Han 2009) were also used as the binders for GO stabilization and then GA fabrication. However, the polymer-based binders used for cross-linking the graphene nanosheets were toxic and required extremely high temperature over 1000°C for their carbonization. An ion linkage method to induce cross-links between the graphene nanosheets was developed (Jiang et al. 2010, Tang et al. 2010). Noble particles were in situ present and facilitated the nucleation and assembly of the graphene nanosheets during the hydrothermal process. These embedded particles can further offer the GAs a further potential for various functional applications, such as sensors (Vedala et al. 2011) and catalysis (Scheuermann et al. 2009).

## 25.2.3 Fabrication of GAs from the Hydrothermal Method

In 2011, Xu et al. reported a one-step hydrothermal technique to prepare binderless GAs. A GO aqueous suspension of 2 mg/mL was sealed in a Teflon-lined autoclave and heated at 180°C for 12 h. After this hydrothermal process, graphene hydrogels were formed as a result of hydrothermal reduction and by physical links via $\pi$–$\pi$ interactions. Finally, these hydrogels needed to undergo solvent exchanges and either supercritical/freeze-drying to yield GAs. Based on this hydrothermal method, Nguyen et al. (2012) investigated the effects of GO concentrations and hydrothermal treatment time on the morphology control of as-prepared GAs. The surface area and the total pore volume of the GAs increased with increase in GO concentration, but decreased with increase in treatment time. However, when the GO concentration was lower than 0.5 mg/mL, the GAs could not be formed during the hydrothermal process (Xu et al. 2010). Although the hydrothermal method has been the effective fabrication method for binderless GAs, the GAs still required high temperatures and high pressure during the fabrication, which may limit their industrial scale-up.

## 25.2.4 Fabrication of GAs from the Chemical Reduction Method

Zhang et al. (2011a) and Sui et al. (2011) concluded a facile and environment-friendly method to synthesize GAs. The GO nanosheets were first dispersed in DI water to form aqueous suspensions. A reducing agent of L-ascorbic acid (LAA) was then added to the suspension. After keeping the suspension for several hours at a low temperature below 100°C, graphene hydrogels were then formed through the self-assembly of graphene nanosheets. After the solvent exchange, the graphene hydrogels were dried by the supercritical drying or freeze-drying processes to obtain the GAs. The lowest GO concentration required to form hydrogels was 0.1 mg/mL using oxalic acid and sodium iodide (NaI) (Zhang et al. 2012a).

Fan et al. (2013) investigated morphological effects on the multifunctional properties of the as-prepared GAs by varying GO concentrations, reduction temperatures, and reduction time. The GAs with large GO concentrations exhibited larger surface areas and higher electrical conductivities, while higher reduction temperatures or longer reduction time caused the GAs to have more packed structures with lower surface areas, higher electrical conductivities, and higher thermal stabilities. Chen and Yan (2011) reported the effects of various reducing agents on the gelation time of graphene hydrogels. The gelation took only 10 min for the LAA-reduced and sodium sulfide ($Na_2S$)-reduced hydrogels and 30 min for the sodium hydrogen sulfite ($NaHSO_3$)-reduced ones.

## 25.2.5 Fabrication of CNT Aerogels

Although GA fabrication has been developed well, the development of CNT aerogels are relatively limited. Kohlmeyer et al. (2011) used two cross-linkers of ferrocene-grafted poly($p$-phenyleneethynylene) (Fc-PPE) and ferrocene-grafted poly[($p$-phenyleneethynylene)-*alt*-(2,5-thienyleneethynylene)] (Fc-PPETE) to gelate chlorobenzene and form Fc-PPE–CNT and Fc-PPETE–CNT gels, respectively.

After exchanging in ethanol and supercritically drying with liquid $CO_2$, the as-obtained CNT aerogels were then annealed in air to improve their electrical and mechanical properties, surface areas, and porosities through opening blocked micro- and mesopores in their monoliths. Fc-PPE and Fc-PPETE acted as organogelators for the gelation and also functionalized the CNTs via ferrocenyl groups for better CNT dispersion (Zou et al. 2010, Kohlmeyer et al. 2011).

CNT aerogels can also be synthesized by embedding CNTs into other 3D aerogel structures. Worsley et al. (2008, 2009b) reported the fabrication of SWCNT aerogels (Worsley et al. 2009) and DWCNT aerogels (Worsley et al. 2008) by incorporating CNTs into RF organogels. The obtained gels were then dried and pyrolyzed at 1050°C to carbonize the organic components and form CNT aerogels. Zhang et al. (2011b) demonstrated the fabrication of CNT aerogels by embedding MWCNTs into the frameworks of poly(3,4-ethylenedioxythiophene) (PEDOT)–PSS. Bryning et al. (2007) utilized surfactants to disperse and form binderless CNT aerogels by using sodium dodecylbenzene sulfonate to disperse CNTs in water. The CNT aqueous suspensions were self-gelated into elastic gels overnight. These wet gels could either be directly dried as obtained or be immersed in PVA suspensions before drying to reinforce their structures.

## 25.2.6 Fabrication of Graphene–CNT Hybrid Aerogels

Graphene–CNT hybrid aerogels combined with a 3D network of graphene and CNTs have been successfully developed for various applications such as for LIBs (Yoo et al. 2008), transparent conductors (Tung et al. 2009b), and capacitive deionization (CDI) electrodes for water purification (Sui et al. 2011, Sui et al. 2012). In the graphene–CNT hybrid aerogels shown in Figure 25.5, the CNTs were physically wrapped in the GA networks, without changing the GA structures. The CNTs could prevent the stacking of graphene nanosheets and remarkably enhance the surface area of the GAs (Yoo et al. 2008, Sui et al. 2011).

The fabrication methods for CNT–graphene hybrid aerogels were similar to those for GAs. The chemical reduction method may be preferred because of its facile conditions and high-quality production (Tang et al. 2010). The CNTs were first dispersed in the GO aqueous suspension by sonication, while the mixture was then reduced by reducing agents such as LAA or hydrazine hydrate. The CNTs were well dispersed in the precursors with the GO nanosheets because of the synergistic effect between the CNTs and graphene nanosheets (Kim et al. 2012). An in situ chemical bonding between the graphene nanosheets and the CNTs was formed (Wang et al. 2011b). After the heat treatment, most functional groups and defects of both graphene nanosheets and the functionalized CNTs were removed. Therefore, these two carbon allotropes could possibly be chemically bonded after the simultaneous reduction. Besides the chemical reduction method, there are other

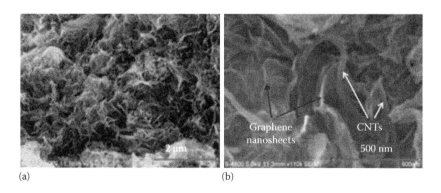

(a)  (b)

**FIGURE 25.5** (a, b) Network structures of the graphene–CNT hybrid aerogels. (From Sui, Z. et al., *Carbon*, 49, 4314–4321, 2011.)

fabrication methods of the CNT–graphene aerogels such as by directly freeze-drying the CNT-GO mixture (Sun et al. 2013), hydrothermal method (Wang et al. 2011b), or two-step CVD method (Dong et al. 2012).

# 25.3 Properties of Graphene/CNT Aerogels

Both graphene and CNT aerogels generally exhibit outstanding multifunctional properties, such as low density, large surface area, high electrical conductivity, and good mechanical stability as discussed in the following.

## 25.3.1 Electrical Properties of Graphene/CNT Aerogels

The electrical conductivities of GAs are reported to be within a large range of 0.1–100 S/m (Worsley et al. 2010, 2011, Xu et al. 2010, Chen et al. 2011). The electrical conductivity of the GAs synthesized from the RF method was up to 87 S/m by a four-probe technique (Worsley et al. 2010), while it was only ~0.1 S/m for the GAs from the hydrothermal method using a two-probe technique (Xu et al. 2010). Such large variations may be due to contact resistance of different characterization methods and the different bonding formed during different fabrication conditions (Nardecchia et al. 2013).

The effects of various drying methods and reducing agents on the electrical properties of GAs were investigated (Chen et al. 2011, Zhang et al. 2011a). The supercritical drying process could yield a more conductive GA than that of the freeze-drying process (Zhang et al. 2011a), and using a more efficient reducing agent would enhance the GA electrical conductivity (Chen et al. 2011). To further enhance electron transfer in GAs, post-treatments such as thermal annealing under inert gases can be performed on the as-prepared GAs. As the functional groups of GO can be effectively removed by thermal treatments above 200°C (Pei & Cheng 2012), annealing the GAs at 400°C in argon could further purify the graphene sheets and increase their electrical conductivities up to ~5 times (Chen et al. 2011).

Compared to GAs, CNT aerogels could achieve slightly higher electrical conductivities because of their relatively more integral graphitic lattice (Nardecchia et al. 2013). This is mainly caused by the different reactants in the processing, i.e., chemically derived rGO for the GAs and CVD-grown CNTs for the CNT aerogels. DWCNT aerogels could achieve an electrical conductivity of up to 8 S/cm measured by a four-probe method (Worsley et al. 2008). The CNT aerogels cross-linked by Fc-PPE and Fc-PPETE exhibited electrical conductivities of ~1–2 S/cm and a large surface area of ~590–680 m²/g (Zuo et al. 2010, Kohlmeyer et al. 2011). The CNT aerogels embedded in PEDOT-PSS had only conductivities of ~$(1.2–6.9) \times 10^{-2}$ S/cm and a surface area of ~280–400 m²/g (Zhang et al. 2011b). For the CNT aerogels prepared with RF organic binders, the pyrolysis of the dried RF could enhance the electrical conductivities of the aerogels. But the PVA reinforcement of the aerogel networks could decrease the electrical conductivities of the pure CNT aerogels from 0.6 to $10^{-5}$ S/cm (Bryning et al. 2007).

## 25.3.2 Mechanical Properties of Graphene/CNT Aerogels

Since graphene/CNT aerogels are constructed by the individual rGO nanosheets or CNTs, their mechanical properties including Young's modulus and strength were far less than those of the graphene monolayer (Lee et al. 2008) or the individual CNT (Krishnan et al. 1998). Therefore, graphene/CNT aerogels are usually considered fragile or brittle. For mechanical property enhancement, polymer binders may be needed to reinforce the networks (Bryning et al. 2007). The GAs developed from LAA and $CO_2$ supercritical drying were reported to be capable of supporting more than 14,000 times their own weights with little deformation (Zhang et al. 2011a). Zhang et al. studied the effects of different drying methods on the GA mechanical performance and found that GAs processed by freeze-drying could support only 3,300 times their own weight, four times smaller than that for supercritical drying. Also,

CNT aerogels could support a thousand times their own weight after the thermal annealing of the binders (Kohlmeyer et al. 2011) and PVA reinforcement (Bryning et al. 2007).

Most mechanical property tests of the graphene/CNT aerogels were compressive tests owing to aerogel brittleness and the simple experimental setup. The compressive stress–strain curves of graphene/CNT aerogels could be divided into three regions: elastic region, yield region, and densification region (Chen et al. 2011, Kim et al. 2012). The elastic region lied at low compressive strains (normally <10%) where the graphene/CNT aerogels behave elastically. From the end of the elastic region to ~60% of compressive strains, it was the yield region (plateau region) where the nanosized pores within the graphene/CNT aerogels started to collapse and stresses slowly increased. The densification region is defined from ~60% of the compressive strains onward, where all the pores collapsed and the graphene/CNT aerogels were completely hardened by further compression. From the compressive tests, the Young's modulus of the GAs varied from 0.26 to 20.0 MPa (Xu et al. 2010, Huang et al. 2013, Zhuo et al. 2013), while those of the CNT aerogels were from 0.12 to 65.0 MPa (Gutiérrez et al. 2011, Kim et al. 2012).

## 25.3.3 Thermal Properties of Graphene/CNT Aerogels

Graphene/CNT aerogels with continuous scaffolds and mesoporous structures may reduce the internal thermal resistance and enhance the thermal efficiency of thermal interface materials (Marconnet et al. 2011, Pettes et al. 2012). However, studies of thermal transport mechanisms and thermal measurement of graphene/CNT aerogels still remain very limited. Zhong et al. (2013) reported that the thermal conductivity of a GA with ~11 vol% of graphene and a relatively low surface area of 43 $m^2/g$ was 2.18 W/m K. Fan et al. (2014) reported a comparative infrared microscopy technique to measure the thermal conductivity of GAs as illustrated in Figure 25.6 using the 1D steady-state heat conduction equation based on Fourier's law. The measured thermal conductivity of GAs was 0.12–0.36 W/m K, much lower than that of pristine graphene because of the high porosity of GAs, the low quality and small size of the chemically derived graphene nanosheets, and the large thermal boundary resistance (TBR) at graphene–graphene and graphene–air contacts (Pettes et al. 2012 et al. Zhong et al. 2013). The post-thermal annealing could be used to increase the thermal conductivity of the GAs with lower than 2.5 vol% of graphene (Fan et al. 2014).

Schiffres et al. (2012) reported that the thermal conductivity of CNT aerogels measured by a modified metal-coated 3ω method was as low as 0.025 ± 0.001 W/m K at room temperature. Similarly as for GAs, the ultralow thermal conductivity of the CNT aerogels could also be attributed to their high porosity, TBR between air–CNT contact, inter-CNT contact, and CNT defects during processing (Schiffres et al. 2012, Kim et al. 2013). An optimization of GO/CNT concentration and other synthesis conditions can possibly enhance the thermal conductivity of graphene/CNT aerogels (Fan et al. 2014).

## 25.3.4 Properties of Graphene–CNT Hybrid Aerogels

The CNTs incorporated in GAs can serve as bridges between the graphene nanosheets, which allows more effective electron transfer, minimizes the CNT/graphene agglomeration, and further increases their surface areas. Figure 25.7 presents that the CNTs could effectively increase the layer distance between the graphene sheets, thus consequently increasing the charge capacity of LIBs up to ~135% (from 540 to 730 mAh/g) (Yoo et al. 2008). The most recently developed graphene–CNT hybrid aerogels had a large surface area of 435 $m^2/g$ and a high conductivity of 7.5 S/m (Sui et al. 2012). For water purification, the graphene–CNT hybrid aerogels surprisingly provided a desalination capacity of 633.3 mg/g with the sodium chloride (NaCl) concentration of 35 g/L, over 15 times higher than that of common CDI materials (Wang et al. 2011a). Hybrid graphene–CNT hybrid aerogels hold great potentials for water purification including CDI of light metal salts, removal of organic dyes, and enrichment of heavy metal ions in water sources (Sui et al. 2012). More potential applications of other carbon/graphene aerogels are discussed in the following.

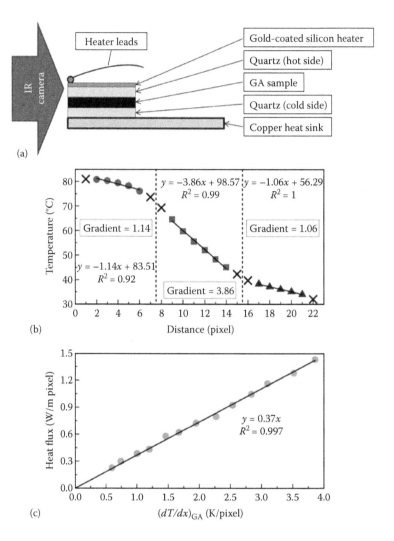

**FIGURE 25.6** (a) Schematic of thermal conductivity measurement using the comparative IR thermography technique. (b) Temperature distribution and linear best-fit curves for a three-layered stack, which consists of a GA sample sandwiched between two amorphous quartz layers. (c) Heat flux as a function of temperature gradient in the GA region at several power levels. During the measurement, the heat flux is typically below 11.25 kW/m². (Reprinted from *International Journal of Heat and Mass Transfer*, 76, Fan, Z., Marconnet, A., Nguyen, S. T. et al., Effects of heat treatment on the thermal properties of highly nanoporous graphene aerogels using the infrared microscopy technique, 122–127, Copyright (2014), with permission from Elsevier.)

# 25.4 Potential Applications of Graphene/CNT Aerogels

## 25.4.1 Graphene/CNT Aerogel Nanocomposites

A nanocomposite consists of a bulk matrix and one or more nanoparticles, nanosheets, and nanofibers. Owing to the high aspect ratio of the nanofillers, nanocomposites can exhibit properties an order of magnitude greater than those of conventional composites. Carbon-based nanofillers such as CNTs and graphene nanosheets dispersed in polymer nanocomposites have been extensively studied in both industrial and academic fields (Gonçalves et al. 2010, Potts et al. 2011, Loomis et al. 2013). However, owing to the hydrophobicity and π–π interaction of CNTs and graphene nanosheets, a homogeneous dispersion

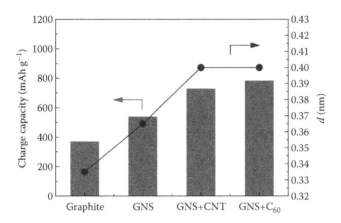

**FIGURE 25.7** Spacing *d* of graphene nanosheets (GNS) families and graphite. (From Yoo et al., *Nano Letters*, 8, 2277–2282, 2008.)

of the carbon-based nanofillers in the polymer matrix has always been a crucial issue (Verdejo et al. 2011, Zhang et al. 2012b). Although many methods of solvent processing, in situ polymerization, and melt processing have been developed (Singh et al. 2011, Du et al. 2012), the properties of these nanocomposites are still much less than expected.

The development of graphene/CNT aerogels, new types of reinforcement materials, could solve the nanofillers' aggregation problems in the nanocomposites. The graphene/CNT aerogel was first prepared and optimized as a scaffold, and then the polymer matrix in a liquid or semiliquid phase could be backfilled into the aerogel scaffold under capillary force. Fan et al. (2015) developed GA-PMMA nanocomposites by backfilling PMMA into the networks of GAs, followed by in situ curing. The uniform distribution of the graphene nanosheets in the PMMA matrix was achieved in Figure 25.8a–d. The GA network was completely infiltrated with PMMA and the rGO also adhered well to PMMA. Furthermore, as there was no interfacial cracking observed, the skeleton of the GAs remained intact in

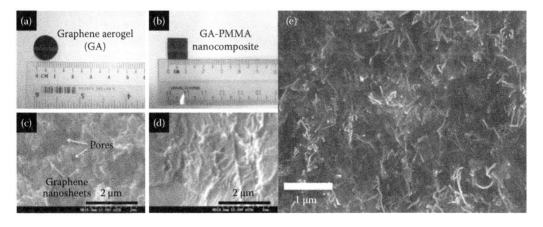

**FIGURE 25.8** (a) A morphology-controlled GA; (b) a poly(methyl methacrylate) (PMMA)–GA composite; (c) SEM image of GA; (d) SEM image of PMMA-GA composite. (Reprinted from *Carbon*, 81, Fan, Z., Gong, F., Nguyen, S. T. et al., Advanced multifunctional graphene aerogel–poly(methyl methacrylate) composites: Experiments and modeling, 396–404, Copyright (2015), with permission from Elsevier.) (e) CNT aerogel–poly(dimethysiloxane) (PDMS) composite. (Worsley, M. A. et al., *Journal of Materials Chemistry* 19, 3370–3372, 2009. Reproduced by permission of The Royal Society of Chemistry.)

the composite. Hence, the GAs not only acted as an ideal uniformly dispersed reinforcement material but also as a porous molding material for the GA-PMMA nanocomposites.

Compared to pure PMMA and graphene-PMMA nanocomposites prepared by the powdery dispersion methods, the developed GA-PMMA nanocomposites exhibited significant enhancements in electrical conductivities (0.160–0.859 S/m), microhardness (303.6–462.5 MPa), and thermal conductivities (0.35–0.70 W/m K) when graphene loading increased from 0.67 to 2.50 vol%. The percolation theory and the Halpin–Tsai theory were further applied and had a good agreement with experimental data on the electrical properties and microhardness of the GA-PMMA nanocomposites, respectively. TBR between graphene and PMMA was further estimated to be $1.906 \times 10^{-8}$ $m^2$ K/W by an off-lattice Monte Carlo algorithm, taking into account the complex morphology, the size distribution, and the dispersion of the rGO nanosheets.

Worsley et al. (2009) applied SWCNT–carbon aerogels for highly conductive PDMS composite development. The CNT–carbon aerogels with 1 vol% of the CNTs were prepared via the organic RF sol–gel method. The composites were then obtained by immersing the as-prepared CNT–carbon aerogels in the PDMS matrix under vacuum and cured at 60°C. Figure 25.8e shows that the CNTs were uniformly dispersed in the PDMS matrix and the infiltration of PDMS caused little degradation in the aerogel networks. The cured CNT–carbon aerogel composites exhibited excellent electrical conductivities of 1 S/cm and increased the Young's modulus up to three times.

### 25.4.2 Energy Storage Devices Using Graphene/CNT Aerogels

Owing to their large surface areas and good electrical conductivities, graphene/CNT aerogels have attracted great interest in energy storage applications such as advanced electrodes of supercapacitors. Meng et al. (2011) prepared alkali-treated GA electrodes in supercapacitor cells, and their specific capacitance was 122 F/g at a discharge currency density of 50 mA/g. Sun et al. (2011) tested the electrochemical performance of GAs in propylene carbonate electrolyte and obtained a specific capacitance of 140 F/g at a current density of 1 A/g. Zhang and Shi (2011) also developed supercapacitors using GA synthesized from the hydrothermal process and obtained a higher specific capacitance of 220 F/g at 1 A/g. The GA-based supercapacitors were also highly stable and could maintain a 92% capacitance after 2000 cycles. In contrast, the electrochemical performance of CNT aerogels was slightly lower than that of GAs at low current densities. Gutiérrez et al. (2011) reported the capacitance of 120 F/g at 1 A/g for the supercapacitor using CNT aerogels tested in 1 M $H_2SO_4$ electrolyte. Extensive research has been conducted to use graphene/CNT aerogels as the advanced electrodes of energy storage devices.

# 25.5 Conclusions and Outlook

Graphene/CNT aerogels with large surface areas of aerogels and their outstanding multifunctional properties have attracted much attention in recent years. GAs may have more advantages than CNT aerogels in terms of their low-cost fabrication, facile preparation of aqueous suspensions using GO, and capabilities of self-assembly upon chemical reduction. The combination of graphene and CNTs in the integrated aerogel monolith is also interesting for exerting synergistic effects between these two allotropes. A fulfillment of graphene/CNT aerogels with enhanced multifunctional properties and applications in diverse fields is still challenging and desired. Several researchers have continuously developed cost-effective fabrication methods and enhanced the multifunctional properties of CNT/graphene aerogels.

# References

Adams, D. O., Edelson, P. J. & Koren, H. S., *Methods for Studying Mononuclear Phagocytes* (Academic Press, New York, 1981).

Balandin, A. A., Ghosh, S., Bao, W. et al., "Superior thermal conductivity of single-layer graphene," *Nano Lett.* 8 (2008): 902–907.

Biener, J., Stadermann, M., Suss, M. et al., "Advanced carbon aerogels for energy applications," *Energy Environ. Sci.* 4 (2011): 656–667.

Brodie, B. C., "On the atomic weight of graphite," *Philos. Trans. R. Soc. Lond.* 149 (1859): 249–259.

Bryning, M. B., Milkie, D. E., Islam, M. F. et al., "Carbon nanotube aerogels," *Adv. Mater.* 19 (2007): 661–664.

Chen, W. & Yan, L., "In situ self-assembly of mild chemical reduction graphene for three-dimensional architectures," *Nanoscale* 3 (2011): 3132–3137.

Dong, X., Chen, J., Ma, Y. et al., "Superhydrophobic and superoleophilic hybrid foam of graphene and carbon nanotube for selective removal of oils or organic solvents from the surface of water," *Chem. Commun.* 48 (2012): 10660–10662.

Dresselhaus, M. S. & Dresselhaus, G., "Intercalation compounds of graphite," *Adv. Phys.* 51 (2002): 1–186.

Dreyer, D. R., Park, S., Bielawski, C. W. et al., "The chemistry of graphene oxide," *Chem. Soc. Rev.* 39 (2010): 228–240.

Du, J. & Cheng, H.-M., "The fabrication, properties, and uses of graphene/polymer composites," *Macromol. Chem. Phys.* 213 (2012): 1060–1077.

Du, A., Zhou, B., Zhang, Z. et al., "A special material or a new state of matter: A review and reconsideration of the aerogel," *Materials* 6 (2013): 941–968.

Duquesne, S., Bras, M. L., Bourbigot, S. et al., "Expandable graphite: A fire retardant additive for polyurethane coatings," *Fire Mater.* 27 (2003): 103–117.

El-Nahhal, I. M. & El-Ashgar, N. M., "A review on polysiloxane-immobilized ligand systems: Synthesis, characterization and applications," *J. Organomet. Chem.* 692 (2007): 2861–2886.

Fan, Z., Tng, D. Z. Y., Nguyen, S. T. et al., "Morphology effects on electrical and thermal properties of binderless graphene aerogels," *Chem. Phys. Lett.* 561–562 (2013): 92–96.

Fan, Z., Marconnet, A., Nguyen, S. T. et al., "Effects of heat treatment on the thermal properties of highly nanoporous graphene aerogels using the infrared microscopy technique," *Int. J. Heat Mass Transf.* 76 (2014): 122–127.

Fan, Z., Gong, F., Nguyen, S. T. et al., "Advanced multifunctional graphene aerogel–poly(methyl methacrylate) composites: Experiments and modeling," *Carbon* 81 (2015): 396–404.

Franks, F., "Freeze-drying of bioproducts: Putting principles into practice," *Eur. J. Pharm. Biopharm.* 45 (1998): 221–229.

Fricke, J. & Emmerling, A., "Aerogels—Recent progress in production techniques and novel applications," *J. Sol–Gel Sci. Technol.* 13 (1998): 299–303.

García-González, C. A., Camino-Rey, M. C., Alnaief, M. et al., "Supercritical drying of aerogels using $CO_2$: Effect of extraction time on the end material textural properties," *J. Supercrit. Fluids* 66 (2012): 297–306.

Gonçalves, G., Marques, P. A. A. P., Barros-Timmons, A. et al., "Graphene oxide modified with PMMA via ATRP as a reinforcement filler," *J. Mater. Chem.* 20 (2010): 9927.

Gutiérrez, M. C., Carriazo, D., Tamayo, A. et al., "Deep-eutectic-solvent-assisted synthesis of hierarchical carbon electrodes exhibiting capacitance retention at high current densities," *Chem. A Eur. J.* 17 (2011): 10533–10537.

He, H., Klinowski, J., Forster, M. et al., "A new structural model for graphite oxide," *Chem. Phys. Lett.* 287 (1998): 53–56.

Hrubesh, L. W., "Aerogel applications," *J. Noncryst. Solids* 225 (1998): 335–342.

Huang, H., Chen, P., Zhang, X. et al., "Edge-to-edge assembled graphene oxide aerogels with outstanding mechanical performance and superhigh chemical activity," *Small* 9 (2013): 1397–1404.

Hummers, W. S. & Offeman, R. E., "Preparation of graphitic oxide," *J. Am. Chem. Soc.* 80 (1958): 1339.

Iijima, S., "Helical microtubules of graphitic carbon," *Nature* 354 (1991): 56–58.

Jiang, X., Ma, Y., Li, J. et al., "Self-assembly of reduced graphene oxide into three-dimensional architecture by divalent ion linkage," *J. Phys. Chem. C* 114 (2010): 22462–22465.

Kim, K. H., Oh, Y. & Islam, M. F., "Graphene coating makes carbon nanotube aerogels superelastic and resistant to fatigue," *Nat. Nanotechnol.* 7 (2012): 562–566.

Kim, K. H., Oh, Y. & Islam, M. F., "Mechanical and thermal management characteristics of ultrahigh surface area single-walled carbon nanotube aerogels," *Adv. Funct. Mater.* 23 (2013): 377–383.

Kohlmeyer, R. R., Lor, M., Deng, J. et al., "Preparation of stable carbon nanotube aerogels with high electrical conductivity and porosity," *Carbon* 49 (2011): 2352–2361.

Krishnan, A., Dujardin, E., Ebbesen, T. et al., "Young's modulus of single-walled nanotubes," *Phys. Rev. B* 58 (1998): 14013–14019.

Lee, C., Wei, X., Kysar, J. W. et al., "Measurement of the elastic properties and intrinsic strength of monolayer graphene," *Science* 321 (2008): 385–388.

Lerf, A., He, H., Riedl, T. et al., "$^{13}$C and $^{1}$H MAS NMR studies of graphite oxide and its chemically modified derivatives," *Solid State Ion.* 101–103, Part 2 (1997): 857–862.

Lerf, A., He, H., Forster, M. et al., "Structure of graphite oxide revisited," *J. Phys. Chem. B* 102 (1998): 4477–4482.

Loomis, J., Fan, X., Khosravi, F. et al., "Graphene/elastomer composite-based photo-thermal nanopositioners," *Sci. Rep.* 3 (2013): 1900.

Marconnet, A. M., Yamamoto, N., Panzer, M. A. et al., "Thermal conductivity in aligned carbon nanotube-polymer nanocomposites with high packing density," *ACS Nano* 5 (2011): 4818–4825.

Meng, F., Zhang, X., Xu, B. et al., "Alkali-treated graphene oxide as a solid base catalyst: Synthesis and electrochemical capacitance of graphene/carbon composite aerogels," *J. Mater. Chem.* 21 (2011): 18537–18539.

Nakajima, T. & Matsuo, Y., "Formation process and structure of graphite oxide," *Carbon* 32 (1994): 469–475.

Nakajima, T., Mabuchi, A. & Hagiwara, R., "A new structure model of graphite oxide," *Carbon* 26 (1988): 357–361.

Namatsu, H., Yamazaki, K. & Kurihara, K., "Supercritical drying for nanostructure fabrication without pattern collapse," *Microelectron. Eng.* 46 (1999): 129–132.

Nardecchia, S., Carriazo, D., Ferrer, M. L. et al., "Three dimensional macroporous architectures and aerogels built of carbon nanotubes and/or graphene: Synthesis and applications," *Chem. Soc. Rev.* 42 (2013): 794–830.

Nguyen, S. T., Nguyen, H. T., Rinaldi, A. et al., "Morphology control and thermal stability of binderless-graphene aerogels from graphite for energy storage applications," *Colloids Surf. A Physicochem. Eng. Aspects* 414 (2012): 352–358.

Novak, B. M., Auerbach, D. & Verrier, C., "Low-density, mutually interpenetrating organic–inorganic composite materials via supercritical drying techniques," *Chem. Mater.* 6 (1994): 282–286.

Orlita, M., Faugeras, C., Plochocka, P. et al., "Approaching the Dirac point in high-mobility multilayer epitaxial graphene," *Phys. Rev. Lett.* 101 (2008): 267601.

Pajonk, G. M., "Transparent silica aerogels," *J. Noncryst. Solids* 225 (1998): 307–314.

Pei, S. & Cheng, H.-M., "The reduction of graphene oxide," *Carbon* 50 (2012): 3210–3228.

Pekala, R., "Organic aerogels from the polycondensation of resorcinol with formaldehyde," *J. Mater. Sci.* 24 (1989): 3221–3227.

Pettes, M. T., Ji, H., Ruoff, R. S. et al., "Thermal transport in three-dimensional foam architectures of few-layer graphene and ultrathin graphite," *Nano Lett.* 12 (2012): 2959–2564.

Pop, E., Mann, D., Wang, Q. et al., "Thermal conductance of an individual single-wall carbon nanotube above room temperature," *Nano Lett.* 6 (2005): 96–100.

Potts, J. R., Dreyer, D. R., Bielawski, C. W. et al., "Graphene-based polymer nanocomposites," *Polymer* 52 (2011): 5–25.

Ruess, G., "Über das graphitoxyhydroxyd (graphitoxyd)," *Monatshefte Chem verwandte Teile anderer Wiss.* 76 (1947): 381–417.

Scherer, G. W. & Smith, D. M., "Cavitation during drying of a gel," *J. Noncryst. Solids* 189 (1995): 197–211.

Scheuermann, G. M., Rumi, L., Steurer, P. et al., "Palladium nanoparticles on graphite oxide and its functionalized graphene derivatives as highly active catalysts for the Suzuki–Miyaura coupling reaction," *J. Am. Chem. Soc.* 131 (2009): 8262–8670.

Schiffres, S. N., Kim, K. H., Hu, L. et al., "Gas diffusion, energy transport, and thermal accommodation in single-walled carbon nanotube aerogels," *Adv. Funct. Mater.* 22 (2012): 5251–5258.

Si, Y. & Samulski, E. T., "Synthesis of water soluble graphene," *Nano Lett.* 8 (2008): 1679–1682.

Singh, V., Joung, D., Zhai, L. et al., "Graphene based materials: Past, present and future," *Prog. Mater. Sci.* 56 (2011): 1178–1271.

Stankovich, S., Piner, R. D., Nguyen, S. T. et al., "Synthesis and exfoliation of isocyanate-treated graphene oxide nanoplatelets," *Carbon* 44 (2006): 3342–3347.

Stankovich, S., Dikin, D. A., Piner, R. D. et al., "Synthesis of graphene-based nanosheets via chemical reduction of exfoliated graphite oxide," *Carbon* 45 (2007): 1558–1565.

Sui, Z., Zhang, X., Lei, Y. et al., "Easy and green synthesis of reduced graphite oxide-based hydrogels," *Carbon* 49 (2011): 4314–4321.

Sui, Z., Meng, Q., Zhang, X. et al., "Green synthesis of carbon nanotube–graphene hybrid aerogels and their use as versatile agents for water purification," *J. Mater. Chem.* 22 (2012): 8767.

Sun, Y., Wu, Q. & Shi, G., "Supercapacitors based on self-assembled graphene organogel," *Phys. Chem. Chem. Phys.* 13 (2011): 17249–17254.

Sun, H., Xu, Z. & Gao, C., "Multifunctional, ultra-flyweight, synergistically assembled carbon aerogels," *Adv. Mater.* 25 (2013): 2554–2560.

Tang, Z., Shen, S., Zhuang, J. et al., "Noble-metal-promoted three-dimensional macroassembly of single-layered graphene oxide," *Angew. Chem. Int. Ed.* 49 (2010): 4603–4607.

Tung, V. C., Allen, M. J., Yang, Y. et al., "High-throughput solution processing of large-scale graphene," *Nat. Nanotechnol.* 4 (2009a): 25–29.

Tung, V. C., Chen, L.-M., Allen, M. J. et al., "Low-temperature solution processing of graphene-carbon nanotube hybrid materials for high-performance transparent conductors," *Nano Lett.* 9 (2009b): 1949–1955.

Vedala, H., Sorescu, D. C., Kotchey, G. P. et al., "Chemical sensitivity of graphene edges decorated with metal nanoparticles," *Nano Lett.* 11 (2011): 2342–2347.

Verdejo, R., Bernal, M. M., Romasanta, L. J. et al., "Graphene filled polymer nanocomposites," *J. Mater. Chem.* 21 (2011): 3301.

Vickery, J. L., Patil, A. J. & Mann, S., "Fabrication of graphene–polymer nanocomposites with higher-order three-dimensional architectures," *Adv. Mater.* 21 (2009): 2180–2184.

Wang, L., Wang, M., Huang, Z. et al., "Capacitive deionization of NaCl solutions using carbon nanotube sponge electrodes," *J. Mater. Chem.* 21 (2011a): 18295–18299.

Wang, Y., Wu, Y., Huang, Y. et al., "Preventing graphene sheets from restacking for high-capacitance performance," *J. Phys. Chem. C* 115 (2011b): 23192–23197.

Worsley, M. A., Satcher, J. H. & Baumann, T. F., "Synthesis and characterization of monolithic carbon aerogel nanocomposites containing double-walled carbon nanotubes," *Langmuir* 24 (2008): 9763–9766.

Worsley, M. A., Pauzauskie, P. J., Kucheyev, S. O. et al., "Properties of single-walled carbon nanotube-based aerogels as a function of nanotube loading," *Acta Mater.* 57 (2009a): 5131–5136.

Worsley, M. A., Kucheyev, S. O., Kuntz, J. D. et al., "Stiff and electrically conductive composites of carbon nanotube aerogels and polymers," *J. Mater. Chem.* 19 (2009b): 3370–3372.

Worsley, M. A., Pauzauskie, P. J., Olson, T. Y. et al., "Synthesis of graphene aerogel with high electrical conductivity," *J. Am. Chem. Soc.* 132 (2010): 14067–14069.

Worsley, M. A., Olson, T. Y., Lee, J. R. I. et al., "High surface area, $sp^2$-cross-linked three-dimensional graphene monoliths," *J. Phys. Chem. Lett.* 2 (2011): 921–925.

Xu, Y., Sheng, K., Li, C. et al., "Self-assembled graphene hydrogel via a one-step hydrothermal process," *ACS Nano* 4 (2010): 4324–4330.

Xu, L. Q., Liu, Y. L., Neoh, K. G. et al., "Reduction of graphene oxide by aniline with its concomitant oxidative polymerization," *Macromol. Rapid Commun.* 32 (2011): 684–688.

Yoo, E., Kim, J. & Hosono, E., "Large reversible Li storage of graphene nanosheet families for use in rechargeable lithium ion batteries," *Nano Lett.* 8 (2008): 2277–2282.

Yu, M., Files, B. S., Arepalli, S. et al., "Tensile loading of ropes of single wall carbon nanotubes and their mechanical properties," *Phys. Rev. Lett.* 84 (2000): 5552–5555.

Zhang, L. & Shi, G., "Preparation of highly conductive graphene hydrogels for fabricating supercapacitors with high rate capability," *J. Phys. Chem. C* 115 (2011): 17206–17212.

Zhang, X., Sui, Z., Xu, B. et al., "Mechanically strong and highly conductive graphene aerogel and its use as electrodes for electrochemical power sources," *J. Mater. Chem.* 21 (2011a): 6494–6497.

Zhang, X., Liu, J., Xu, B. et al., "Ultralight conducting polymer/carbon nanotube composite aerogels," *Carbon* 49 (2011b): 1884–1893.

Zhang, L., Chen, G., Hedhili, M. N. et al., "Three-dimensional assemblies of graphene prepared by a novel chemical reduction-induced self-assembly method," *Nanoscale* 4 (2012a): 7038–7045.

Zhang, K., Zhang, W. L. & Choi, H. J., "Facile fabrication of self-assembled PMMA/graphene oxide composite particles and their electroresponsive properties," *Colloid Polym. Sci.* 291 (2012b): 955–962.

Zhong, Y., Zhou, M., Huang, F. et al., "Effect of graphene aerogel on thermal behavior of phase change materials for thermal management," *Sol. Energy Mater. Sol. Cells* 113 (2013): 195–200.

Zhou, H., Yao, W., Li, G. et al., "Graphene/poly(3,4-ethylenedioxythiophene) hydrogel with excellent mechanical performance and high conductivity," *Carbon* 59 (2013): 495–502.

Zou, J., Liu, J., Karakoti, A. S. et al., "Ultralight multiwalled carbon nanotube aerogel," *ACS Nano* 4 (2010): 7293–7302.

Zu, S. & Han, B., "Aqueous dispersion of graphene sheets stabilized by Pluronic copolymers: Formation of supramolecular hydrogel," *J. Phys. Chem. C* 113 (2009): 13651–13657.

# 26

# Nanotube–Cement Composites

Oscar Aurelio
Mendoza Reales

Romildo Dias
Toledo Filho

## Abstract

This chapter presents a comprehensive review of the properties of CNT–cement composites. A general picture of the advances in the manipulation of cement matrices at the scale of nanometers is provided, discussing the role of CNTs in the development of a new generation of construction materials with superior performance.

A special interest is given to the dispersion of the CNTs in the matrix and how it affects the effectiveness of the nanoparticle and the final properties of the composite. Interactions of CNTs with other nano- and microparticles, such as nanosilica, metakaolin, and fly ash, are also presented.

**Key words:** cement composite, calcium–silicate–hydrate, agglomeration, bridge effect, surface interaction, load distribution, rheology, mechanical strength, electromagnetic properties.

## 26.1 Introduction

Nanotechnology is not usually associated with the cement and construction industries, but today more than ever, there is a growing interest in next-generation construction materials, and nanotechnology has shown a great potential to deliver them. Two factors have influenced the growth of interest of these industries in the use of nanomaterials: first, the industrial scale production of nanoparticles, which has considerably lowered their cost; and second, the growing body of knowledge about the properties of hydrated cement matrices at the nanoscale, which has helped to understand how to nanoengineer construction materials with improved properties, lower cost, and lower polluting emissions associated.

It has been recognized that different nanoparticles can improve the properties of hydrated cement matrices, whether it is due to their chemical affinity and high reactivity with cement or due to their exceptional mechanical properties. Additionally, some of these nanoparticles also have electromagnetic, photocatalytic, bactericide, hydrophilic, or hydrophobic properties, among others. These properties can be exploited to develop smart construction materials, with decontaminating, self-healing, self-cleaning, and self-sensing capabilities. Within this group of nanoparticles, a special interest has been given to the CNTs; their mechanical properties and high aspect ratio put them as a potential nanoreinforcement to generate cement composites with high flexural strength and capable of crack propagation control. Also, their electromagnetic properties can be used to generate composites with novel properties such as self-sensing and electromagnetic shielding.

Several challenges have to be addressed in order to obtain CNT–cement composites suitable for a mass production process that guarantees repeatable properties and homogeneous performance. The effective dispersion and surface interactions of CNTs within the hydrated cement matrix stand out as the key issues. Different chemical treatments and dispersion mechanisms have been studied to maximize the effect of the CNTs over the properties of cement composites; but in most of the cases, it has been found that these mechanisms are too aggressive for the integrity of the nanotubes, and also that the final result is sensitive to the alkaline environment generated by the hydration of cement.

This chapter presents a comprehensive review of the properties of CNT–cement composites. A general picture of the advances in the manipulation of cement matrices at the scale of nanometers is provided, discussing the role of CNTs in the development of new-generation of construction materials with superior performance. The review covers all the modifications induced by CNTs in the properties of the cementitious matrix and how these modifications affect the performance of the composite in both fresh and hardened states. First, a general overview of the properties of the cement matrices at the nanoscale is presented, then it is discussed how the CNTs can modify these properties and how the dispersion and surface interaction issues are related to these modifications. The combinations of CNTs with other nano- and microparticles, such as nanosilica, metakaolin, and fly ash, in cement matrices are highlighted. Finally, the novel electromagnetic properties conferred to cement matrices by the CNTs are presented.

## 26.2 Nanostructures of the Hydrated Cement

Calcium–silicate–hydrate (C-S-H) is the main hydration product of portland cement, comprising approximately 50% to 60% of the total volume of hydrates (Jennings et al. 2002), and is responsible for the cohesion and mechanical response of the hydrated matrix at the micro- and macroscales. TEM images (Richardson 1999), crystallographic models (Chen et al. 2004), molecular simulations (Pellenq et al. 2009), NMR spectra (Beaudoin et al. 2009), and nanomechanical results (Constantinides & Ulm 2007) all agree in the fact that C-S-H can be considered of nanometric nature with a short-range crystalline structure. Therefore, hydrated cement can be considered a naturally nanostructured material, and the nanoscale is considered as the most adequate scale to induce modifications that enhance its performance, and consequently the performance of the whole matrix.

Portland cement is composed of alite ($Ca_3SiO_5$ or $C_3S$), belite ($Ca_2SiO_4$ or $C_2S$), celite ($3CaOAl_2O_3$ or $C_3A$), and ferrite ($4CaOAl_2O_3Fe_2O_3$ or $C_4AF$), which together form the *clinker*. The clinker is produced by the solid solution reactions during the burning of $SiO_2$, $CaO$, $Al_2O_3$, and $Fe_2O_3$ at approximately 1450°C, and rapidly cooled down to ambient temperature to maintain its crystalline structures. After cooling, the clinker is ground together with gypsum ($CaSO_4$) and some inert material such as $CaCO_3$ or $CaMg(CO_3)_2$ (Hewlett 2004). Of all these components, only alite and belite, which are approximately

75% of the mass of cement, are capable of forming C-S-H and Ca(OH)₂ in the presence of water, following the reactions

$$Ca_3SiO_5 + (1.3 + x)H_2O \rightarrow (CaO)_{1.7}SiO_2(H_2O)_x + 1.3Ca(OH)_2, \tag{26.1}$$

$$Ca_2SiO_4 + (0.3 + x)H_2O \rightarrow (CaO)_{1.7}SiO_2(H_2O)_x + 0.3Ca(OH)_2. \tag{26.2}$$

The H₂O, Si, and Ca contents and the crystalline structure of C-S-H are still a matter of discussion; the empirical formula $(CaO)_{1.7}SiO_2(H_2O)_{1.8}$ (Allen et al. 2007) is currently the most accepted one. Regarding its crystalline structure, different models have evolved since it was first proposed that the C-S-H was composed of layers of linear silicate chains or dreierkette chains, constructed of SiO₂ tetrahedrons coordinated with a central Ca–O layer (Bernal et al. 1952), as presented in Figure 26.1. The main issue with this model is the fact that the CaO and SiO₂ contents do not agree with those experimentally found in C-S-H. Currently, the tobermorite/jennite- and tobermorite/calcium hydroxide-based models (Richardson 2004), which include additional Ca octahedrons in the structure, are the ones that agree better with the experimental CaO and SiO₂ contents reported in literature. There are two types of SiO₂ tetrahedrons, depending on how they coordinate with the Ca–O layer. They are of the pairing type if they share O–O edges with the central Ca–O layer, or of the bridging type if they do not share them (Richardson 2004). The continuity of the dreierkette chain is usually interrupted by the absence of some bridging tetrahedrons; this is why the C-S-H crystalline structure is considered short range and difficult to identify in XRD patterns.

The formation of C-S-H is divided into three stages: dissolution, saturation, and precipitation. When in contact with water, the surface of the anhydrous grains suffers protonolysis and dissolution, which releases $Ca^{2+}$ and $SiO_4^{4-}$ into the liquid surrounding the grains. This process occurs until the liquid becomes locally oversaturated and C-S-H starts to precipitate on the surface of the grain (Hewlett 2004). The precipitation process generates individual C-S-H particles of approximately 5 nm and CaO/SiO₂ ratios between 1.2 and 2.3. If the C-S-H precipitates from the original surface to the inside of the grain, it is called *inner product* (Ip), which has a foil-like texture; if it precipitates to the outside of the

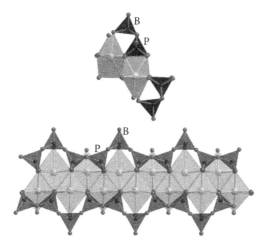

**FIGURE 26.1** Schematic representation of dreierkette chains present in C-S-H. (Reprinted from *Cement and Concrete Research*, 34, Richardson, I. G., Tobermorite/jennite- and tobermorite/calcium hydroxide-based models for the structure of C-S-H: Applicability to hardened pastes of tricalcium silicate, β-dicalcium silicate, Portland cement, and blends of Portland cement with blast-furnace slag, metakaol, or silica fume, Copyright (2004), with permission from Elsevier.)

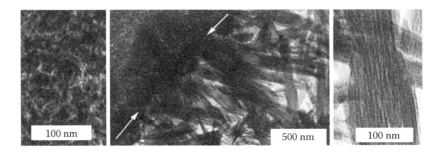

**FIGURE 26.2**　TEM micrograph showing inner and outer products in a hardened C₃S. (Reprinted from *Cement and Concrete Research*, 38, Pellenq, R. J.-M., Lequeux, N. & van Damme, H., Engineering the bonding scheme in C–S–H: The iono-covalent framework, Copyright (2008), with permission from Elsevier.)

grain, it is called *outer product* (Op), which has a fibrillar texture (Richardson 1999). An example of this is presented in Figure 26.2, showing Ip on the left and Op on the right side of the figure in a TEM image. Individual C-S-H particles form agglomerations. It has been found that C-S-H is of a nano-granular nature; this means that its properties depend on the particle-to-particle contact, i.e., its packing density. The more densely packed the C-S-H agglomerations, the better the particle-to-particle contact will be and the more efficiently the load will be distributed through them. Three different kinds of C-S-H agglomerations have been differentiated: low density, high density, and ultrahigh density (Constantinides & Ulm 2007).

As presented in Equations 26.1 and 26.2, crystalline Ca(OH)₂ is precipitated as a by-product of the C-S-H and is the second constituent of the hydrated phase in volume. It is formed as hexagonal platelets of approximately 10 μm in size and is responsible for the elevated pH value of the matrices. Ca(OH)₂ can be easily lixiviated by water, or transformed in CaCO₃ by any CO₂ present in the atmosphere.

## 26.3 Relations among Nano-, Micro-, and Macroscales of Hydrated Cement

The hydrated matrix has intrinsic properties at the nano-, micro-, and macroscales. The processes that modify the intrinsic properties of the matrix at the nanoscale will have an effect on the properties at the microscale, and these in turn will modify the properties at the macroscale. Therefore, all the modifications that occur at the nanoscale affect the behavior of the bulk material with a certain level of proportionality (Sanchez & Sobolev 2010). To better understand these scaling relations, multiscale models ranging from $10^{-8}$ to $10^{1}$ m have been developed. The distinction of scales is made in the function of the size of its fundamental components, rather than using decimal orders of magnitude. As an example, a four-level model (Constantinides & Ulm 2004) starts at Level I, at the scale of the nanometers, including C-S-H particles and gel porosity. Level II follows, at the microscale, including C-S-H agglomerations, Ca(OH)₂ platelets, capillary pores, and air voids. It continues to Level III, at the scale of micrometers and millimeters, including microcracks, sand particles, and the interfacial transition zone (ITZ), which is a zone of higher porosity around the aggregates. And it finally ends at Level IV, at the scale of millimeters to meters, including the coarse aggregates and major aid voids. This multiscale approach can be applied, for example, to the porosity of the hydrated matrix, to show how the scales of the matrix are related to it (Jennings et al. 2002).

Porosity is one of the properties of the matrix that exist at all the scales, from the gel pores of 0.5 to 10 nm, which are considered to be in the structure of the C-S-H, passing through the capillary pores of 10 nm to 10 μm, up to the air voids, ITZ, and microcracks. The size distribution of the pores is continuous throughout the length scale; but each size classification has a different origin and a different functionality, which at the end are correlated. The volume of the pores within the matrix is controlled by the amount of water present in the mixture. The water not chemically bound to the hydrates is called *evaporable water*; gel and capillary pores originate from the evaporable water. Bigger pores are formed by the intentional or unintentional inclusion of air voids in the matrix (Vandamme et al. 2010). Additional porosity is induced by the aggregates at the ITZ, which can later become interconnected by microcracks (Dolado & van Breugel 2011). Using the multiscale approach of four levels applied to the porosity of the matrix, the interlevel relations can be characterized as follows:

Level I—Gel pores: The small size of the gel pores inside the C-S-H generates an electrical unbalance on their inner surface, which causes an orientation of the water molecules inside the pore. This alignment modifies the properties and reduces the mobility of the water. A direct consequence of this decrease in mobility is a modification of the creep behavior of the matrix when a steady load is applied (Jennings 2004).

Level II—Capillary pores: The water inside the capillary pores is not oriented; but due to the high surface area in comparison to their diameter, it is influenced by the surface tension at the interfaces between the water and C-S-H. If the capillary water evaporates or desiccates, differential pressures are generated inside the pore. To relax these internal stresses, the C-S-H goes into compression and causes shrinkage. The shrinkage process is also highly influenced by the gel pores from Level I (Jennings 2004).

Level III—Permeability: The permeability of the matrix is controlled not only by the porosity, but also by the characteristics of the ITZ, and by any possible microcracks induced in the material. Among the other factors, microcracks can be induced by the creep and shrinkage processes, originated in Levels I and II. When the cracks become interconnected with each other, with the ITZ and with the capillary pores, continuous channels are generated and the permeability of the matrix becomes dramatically increased (Hewlett 2004).

Level IV—Strength and durability: The porous structure from Level III, whether it is filled with air or liquid, is not capable of supporting the same load as the solid phase; therefore, porosity is a fundamental factor for the strength of the material. Most of the aggressive agents that penetrate the cement matrix come within the water absorbed through the porosity. Two consequences arise from these degradation processes: the formation of higher volume products that generate internal pressures, which generate microcracks; and the dissolution or lixiviation of hydrated phases, which increases the amount of voids in the matrix. Both processes have one common consequence, the increase in porosity that reduces the strength of the matrix (Jennings et al. 2002).

As with porosity, a similar multiscale model can be applied to the influence of CNTs over the properties of the hydrated cement matrix, taking into account that the fresh and hardened states have to be differentiated. This differentiation is necessary because the mechanisms by which the CNTs influence the properties of the matrix in each state are different. This does not mean that the two states are not correlated; on the contrary, all the changes that occur in the matrix in the fresh state, as a consequence of the presence of CNTs, will have an indirect impact over the properties of the matrix in the hardened state. Sections 26.4 and 26.5 present a multiscale model applied to CNT–cement composites, starting from Level I in the stages of dispersion and mixture with cement, up to Level IV with the mechanical strength of the composite at the macroscale.

## 26.4 Interaction of CNTs with the Cement Matrix in Fresh State

The *fresh state* can be defined as the state at which the matrix is plastic and can be molded. This state has a limited time span, starting from the moment at which cement comes in contact with water, and lasting to approximately 300 min for ordinary portland cement. During this period, the matrix goes through a series of chemical reactions and physical transformations, which are part of the hydration reaction of cement. At the end of this period, the matrix has already gained some of its final strength and is capable of bearing load; it is only at this point when the nanotubes start to work as nanoreinforcement. The reinforcing effectiveness of the CNTs will depend not only on their own properties, but also on how well dispersed they are throughout the matrix. It is of vital importance to understand how the alkaline environment generated by the hydration of cement influences the CNTs, and also how the presence of CNTs influences the physical and chemical processes that lead to the hardening of the matrix. This section presents all the interactions that occur between the CNTs and the cement in the fresh state, focusing on the dispersion as the fundamental parameter to be controlled to obtain a good quality composite.

## 26.4.1 Dispersion of CNTs in the Cement Matrix

The main role of the CNTs, when working as reinforcement in cement composites, is to control the propagation of submicrometric cracks by modifying the load transmission through bridging effects. It has been demonstrated that micro- and macrofibers are ineffective in arresting the propagation of these kind of cracks (Sakulich & Li 2011); therefore, nanotubes or nanofibers are a viable solution to control them. When the composite is exposed to a flexural stress that the cement matrix is not capable of bearing, the cracking process begins; at this stage the CNTs that are in the path of the submicrometric cracks will work as bridges for the load, redistributing it and arresting the propagation of the opening (Metaxa et al. 2013). The formation of submicrometric cracks is a random process that can be predicted only to some extent; therefore, the effective dispersion of the CNTs throughout the entire matrix will ensure that the majority of the microcracks formed will have nanotubes in their path, and that they will act as bridges to redistribute the load.

### 26.4.1.1 Dispersion Methods

The most straightforward method to incorporate CNTs in the cement matrix is using the mixing water. CNTs are dispersed in water using a combination of mechanical waves and chemical-dispersing agents, which together aim to produce a colloid of individual nanotubes that is stable over time and not sensitive to alkaline environments. Some authors have attempted alternative dispersion methods such as dispersing the CNTs together with cement using acetone (Musso et al. 2009), or growing them directly on the anhydrous grains (Sun et al. 2013). This has proven to be effective but not cost effective if not made on an industrial scale. Thus, this section will be focused on the dispersion methods of CNTs in water.

Due to their size, the interaction among CNTs is mainly governed by van der Waals attractions; hence, a tendency to agglomerate is intrinsic to them (Zhang et al. 2011). These agglomerates are usually difficult to separate to obtain individual CNTs; mechanical waves from different sources can be used to overcome the van der Waals forces and disperse the nanotubes. Ultrasonic processors are used to transform a vibration of ultrasonic frequency into a mechanical wave that propagates through water by the generation of cavitation bubbles, which implode and transfer their energy to the CNT bundles, exfoliating individual nanotubes (Konsta-Gdoutos et al. 2010a). High-shear mixers are also used to disperse CNTs in water (Juilland et al. 2012). The high-shear mixer first exerts suction that draws solid and liquid material from the bottom of the container into a rotor–stator pair, where the solids are subjected to milling and hydraulic shear; this expels the material to the sides of the container and draws more material to the rotor–stator pair.

Chemical-dispersing agents are able to adsorb onto the surface of the CNTs due to hydrophobic interactions (Jiang et al. 2003); this effect is used to maintain individual CNTs separated by electrostatic repulsion among the electric charges of the dispersant molecules. Different dispersing agents, such as surfactants (Konsta-Gdoutos et al. 2010a), superplasticizers (Cwirzen et al. 2009), sodium dodecyl sulfate (SDS) (Yu & Kwon 2009), and polyvinylpyrrolidone (PVP) (Chan & Andrawes 2010), have been tested in nanotube–cement composites. Besides the adsorption capacity, the dispersing agent also has to be compatible with the hydration reaction and the highly alkaline conditions generated by it. It has been found that for pH values higher than 8, the adsorption of dispersants onto the surface of the CNTs is not significant (Jiang et al. 2003). It has also been found that for pH values near 12, the dispersion of CNTs in water becomes unstable and tends to reagglomerate (Mendoza et al. 2013b), and also that

(a)                    (b)                    (c)                    (d)

**FIGURE 26.3** Stability of aqueous dispersions of CNTs dispersed using 40,000 J of ultrasonic energy: (a) dispersion without chemical dispersant immediately after sonication; (b) dispersion without chemical dispersant after 24 h of rest; (c) CNT dispersion with chemical dispersant immediately after sonication; (d) CNT dispersion with chemical dispersant after 24 h of rest. (From Mendoza, O. et al., "Influence of super plasticizer and Ca(OH)₂ on the stability of functionalized multi-walled carbon nanotubes dispersions for cement composites applications," *Constr. Build. Mater.* 47 (2013a): 771–778.)

high concentrations of specific surfactants can hamper the cement hydration reaction due to steric hindrance caused by the molecules of surfactant adsorbed onto the surface of the anhydrous cement grains (Ouyang et al. 2008), retarding the heat release during the first hours (Łaźniewska-Piekarczyk 2013).

Chemical-dispersing agents are used in combination with mechanical waves to optimize the dispersion process. The molecules of the dispersant propagate through the gaps in the CNT bundles generated by the mechanical waves; these gaps are free surfaces for the dispersing agents to adsorb onto and prevent the reattachment of the nanotube to the bundle. This process continues through the length of the CNTs until they are completely separated from the bundle. This exfoliation mechanism is called *unzippering* (Strano et al. 2003). If the CNTs are functionalized, the additional electric charge of the functional groups on the surface of the nanotube has an influence on the electrostatic equilibrium of the dispersion. An example of the use of chemical dispersion in combination with mechanical waves is presented in Figure 26.3. Figure 26.3a and c presents aqueous dispersions of CNTs without and with a chemical-dispersing agent, immediately after being exposed to 40,000 J of ultrasonic energy. Figure 26.3b and d presents the same dispersions but after 24 h of rest. It can be seen that a much higher degree of dispersion is achieved with the help of a chemical-dispersing agent, due to the unzippering effect. Additionally, after 24 h of rest, the dispersion with a chemical-dispersing agent remained stable, while the one without a dispersing agent became flocculated.

### 26.4.1.2 Functionalization and Damage Induced by the Dispersion Process

Functional groups are of vital importance to improve the surface interactions between the nanotubes and the matrix. Since plain CNTs have no chemical affinity with the hydrated and nonhydrated phases of the matrix, functional groups work as an anchor between the surfaces to improve the load transmission (Li et al. 2005) and minimize pullout effects of the nanotubes when the matrix is loaded. Different kinds of molecules can be grafted on the surface of the CNTs, via physical or chemical treatments, to be used as functional groups; many tailored functionalization techniques have been developed depending on specific applications (Ma et al. 2010). For cement composite applications, it has been found that the OH– (hydroxylic) functional groups have a chemical affinity to form bonds with C-S-H and Ca(OH)₂ and transmit load (Li et al. 2005). A schematic representation of how the CNTs interact with the hydration products of cement through the OH– functional groups is presented in Figure 26.4.

Hydroxylic groups are generated on the surface of the CNTs using a mixture of sulfuric acid and nitric acid, as an adaptation of a methodology originally proposed for the application of CNTs in polymeric matrices (Abdalla et al. 2008, Kakade & Pillai 2008). Acid treatments generate defects in the organized structure of the CNTs during the functionalization process, affecting their performance (Madni et al. 2010).

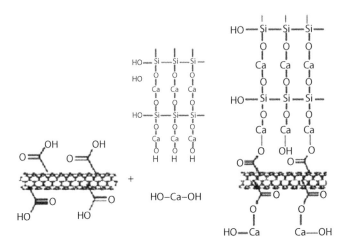

**FIGURE 26.4** Schematic representation of the interaction between OH-functionalized CNTs and hydration products ($Ca(OH)_2$ and C-S-H). (Reprinted from *Carbon*, 6, Li, G. Y., Wang, P. M. & Zhao, X., Mechanical behavior and microstructure of cement composites incorporating surface-treated multi-walled carbon nanotubes, 1239–1245, Copyright (2005), with permission from Elsevier.)

The presence of the defects in the structure, in combination with a highly energetic dispersion process using mechanical waves, can lead to the breakage of the CNTs (Jung et al. 2012), thus decreasing their aspect ratio. Using Raman spectroscopy, a direct relation between the sonication energy and the organization of the structure of the CNTs has been found. Typical Raman spectra of CNTs dispersed in water with different ultrasonic energies are presented in Figure 26.5. The D and G bands, associated with disorganized and organized carbon structures, are clearly identified. A decrease in the intensity of the G band, proportional to the amount of sonication energy, is evidenced. This has been associated with a loss of organization in the structure of the CNTs as a consequence of the mechanical damage caused by the sonication process (Mendoza et al. 2013b). In theory, for applications where the CNTs work as reinforcement, high–aspect ratio CNTs are desirable since they have a higher reinforcing efficiency. But in practice, the longer the CNTs, the higher their tendency to agglomerate (Kim et al. 2010), because their

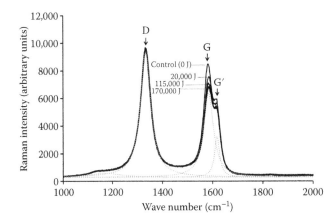

**FIGURE 26.5** Evolution of Raman spectra of CNTs after different sonication energies. (From Mendoza, O. et al., "Influence of super plasticizer and $Ca(OH)_2$ on the stability of functionalized multi-walled carbon nanotubes dispersions for cement composites applications," *Constr. Build. Mater.* 47 (2013a): 771–778.)

surface area is directly proportional to their length, and the attraction phenomenon between the nanotubes that leads to agglomeration is a surface phenomenon. This tendency to agglomerate reduces the reinforcing efficiency of longer CNTs in comparison to shorter CNTs in a cement matrix (Abu Al-Rub et al. 2012a). In contrast, to generate composites with piezoelectric or piezoresistive properties, the CNTs must form an efficient conductive network through the cement matrix, and for this, the highest possible aspect ratio is desirable (Yu & Kwon 2009).

Different solutions have been proposed to minimize the amount of damage induced on the CNTs during the dispersion and functionalization process. It has been proposed that the acid treatments must be controlled not only to obtain functional groups, but also to obtain CNTs of a specific aspect ratio, according to the specific application for which the nanotubes will be used (Kim et al. 2010). Pretreatments with supercritical $CO_2$ on the already functionalized CNTs have been proposed to minimize the amount of mechanical energy necessary to disperse them, therefore decreasing the amount of damage induced during the dispersion process (Jung et al. 2012). Modifications of the functionalization process to eliminate the acid treatments have also been proposed. Alternative treatments with $O_3$ and shortwave UV radiation (Sham & Kim 2006) or hydrogen peroxide and UV radiation (Gong et al. 2011) generate the same carboxylic functional groups and eliminate the need to expose the CNTs to highly aggressive environments.

### 26.4.1.3 Maximum Amount Dispersible

The maximum amount of nanotubes dispersible in the matrix is limited not only by the technical aspects of the dispersion process, but also by the geometry of the space available in between the cement grains. For additions that can be modeled as spheres, the best dispersion scenario is a fully uniform dispersion; this means that all the particles are evenly distributed and equidistant in the matrix. The worst dispersion scenario is a fully nonuniform dispersion, where all the particles are in contact (Yazdanbakhsh et al. 2011). A more realistic approach is a random distribution throughout all the available space, but some of that space is already occupied by the cement grains and other solid constituents of the composite. In the anhydrous state, the matrix is composed mainly of finite size and nonporous particles of angular shapes; this means that the additions are able to occupy only the space left available between cement grains, and that a fully uniform dispersion is impossible to achieve. When the particles of the matrix are much bigger than the additions, a clustering in the shape of the space available appears, causing a low homogeneity in the dispersion of the additions (Yazdanbakhsh et al. 2011).

In the case of fiber-shaped additions, which is the case of the CNTs, the geometry-dependent clustering will also depend on the amount of fibers included in the empty space of the matrix. For low concentrations of fibers, their distribution in the matrix can be considered approximately random, but as the concentration increases, the fibers will start to align following the direction of the borders between cement grains (Yazdanbakhsh & Grasley 2012). This is schematically represented in Figure 26.6 using a ceramic matrix composed of nonporous particles. The distribution of the available space for the CNTs between particles is presented in Figure 26.6a; the distribution of CNTs in low concentration and high concentration are presented in Figure 26.6b and c, respectively. For a low concentration of CNTs, the distribution of the nanotubes can be considered random; while for a high concentration of CNTs, the distribution is strongly influenced by the geometry of the matrix. Cement matrices are usually designed to maximize their packing density in anhydrous state, consequently minimizing the space between particles and making them more sensitive to the geometry-dependent clustering. Therefore, it is important to keep in mind that a higher inclusion of CNTs in the matrix does not necessarily mean an improvement in the properties of the composite.

## 26.4.2 Reagglomeration of CNTs and Its Consequences in the Properties of the Composite

Despite all efforts made to obtain uniform and stable dispersions of CNTs in water, it has been found that this state is hardly maintained when the matrix reaches its hardened state (Yazdanbakhsh et al. 2010). The most

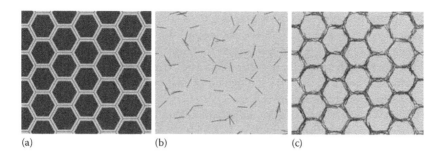

**FIGURE 26.6** Schematic representation of CNTs distributed in a ceramic composite of nonporous particles. (a) The areas between ceramic particles in which CNTs can be distributed are shown with lighter color. (b) When a low dosage of CNT is used to reinforce the ceramic, the distribution of CNTs is relatively uniform. (c) When the CNT dosage is high, the distribution uniformity becomes poor. (Reprinted from *Composites Part A: Applied Science and Manufacturing*, 42, Yazdanbakhsh, A., Grasley, Z., Tyson, B. et al., Dispersion quantification of inclusions in composites, 75–83, Copyright (2011), with permission from Elsevier.)

common result is finding some individual CNTs bridging cracks, and the rest as agglomerations throughout the matrix, whether inside the pores or surrounding specific phases. Two main phenomena have been identified as the cause of reagglomeration of CNT dispersions in cementing matrices: the release of $Ca^{2+}$ cations in the media during the dissolution of cement, and the alkaline environment generated during the precipitation of C-S-H and $Ca(OH)_2$ as a consequence of the hydration reaction of $C_3S$ and $C_2S$. The first phenomenon occurs during the first hours of the hydration reaction, when the cement is undergoing dissolution and $Ca^{2+}$ cations are released into the media; these cations are capable of interacting with the $OH^-$ functional groups generated by the acid treatments, affecting the electrostatic equilibrium of the colloid (Mendoza et al. 2013b). The second phenomenon occurs when the mixing water becomes locally saturated in $Ca^{2+}$ and the pH of the media becomes highly alkaline due to the precipitation of C-S-H and $Ca(OH)_2$. For polymeric dispersing agents with hydrophobic properties, such as surfactants or SDS, it has been found that at elevated pH values, the adsorption of the polymeric molecules on the surface of the CNTs is not significant and the dispersions will have the same stability regardless of the presence of a dispersant molecule (Jiang et al. 2003). Also, the nanotube agglomerates can reach a critical mass where the gravitational forces overcome the electrostatic repulsions and the agglomerations start to precipitate and segregate from the matrix.

After reagglomeration, the CNTs are found in the form of bundles similar to those found prior to dispersion. These bundles can be of micrometric size, with lower surface area and lower aspect ratio than those of the individual nanotubes. A lower surface area implies that there will be a lower load transmission between the matrix and the CNTs, since this interaction is done through the outer wall and surface functional groups of the nanotubes. A lower aspect ratio as a consequence of the twisting and curling of the CNTs will prevent the nanotubes to work as bridges between neighboring C-S-H clusters and across submicrometric cracks (Sobolkina et al. 2012). Thus, it can be said that the reagglomeration process decreases the effectiveness of CNTs as reinforcement because only a portion of the nanotubes works as reinforcement and the other portion remains in the matrix as filler, due to the lack of chemical affinity between the CNT bundles and the components of the matrix.

### 26.4.3 Effect of the CNTs on the Kinetics of the Hydration Reaction

Even though the CNTs do not have chemical affinity with the anhydrous or hydrated phases of the cement matrix, it has been demonstrated that they have an influence on the kinetics of the hydration reaction, acting as extra nucleation spots (Makar & Chan 2009). In a normal hydration reaction, the hydration products are formed by precipitation from nucleation spots, which are located on the surface of the anhydrous grains or on the hydrated phases themselves. The kinetics of this nucleation process depends on the calcium concentration (Thomas 2007), the surface area available for nucleation (Garrault & Nonat 2001), and the temperature (Thomas et al. 2009). Due to their high surface area, CNTs work as nucleation spots, occupying all the relevant nucleation sites and promoting the formation of hydration products, with preference for C-S-H (Makar & Chan 2009). In addition to their surface area, it has been found that the electromagnetic properties of the CNTs also influence the nucleation process. CNTs are capable of attracting metallic ions to their surface (Paillet et al. 2005), and polarize water molecules (Moulin et al. 2005); therefore, the $Ca^{2+}$ cations and water necessary for the precipitation of hydrates are drawn to the surface of the CNTs, enhancing the nucleation process.

When the CNTs are functionalized by acid treatments, there is an additional enhancement of the nucleation process; this is because the OH– functional groups have the ability to adsorb metallic ions (Di et al. 2004), increasing the amount of $Ca^{2+}$ on the surface of the CNTs. It should be noted that the nucleation effect of the CNTs is not limited to the hydration of cement; evidence of nucleation has been identified, for example, in polymers (Ma et al. 2010) and many other materials. Also, the effects of nucleation in cementing matrices is not limited to CNTs; on the contrary, this effect has been found in many different nanoparticles regardless of their pozzolanic activity or chemical affinity with cement (Sanchez & Sobolev 2010).

A direct consequence of the presence of additional nucleation spots is the acceleration of the kinetics of the hydration reaction during the first hours. Therefore, the hydration products are formed faster during the first hours of the reaction, releasing a higher amount of heat in a shorter time. This does not mean that the total amount of hydration products is increased but that its rate of production accelerates, since there is no chemical affinity between the CNTs and the hydration products (Makar & Chan 2009). The accelerating effect takes place only during the first hours of the reaction until the CNTs stop working as nucleation spots, whether it is because all the effective surface area available for nucleation has been used, or because the surface area has decreased due to the reagglomeration phenomenon of the dispersed CNTs. After the accelerating effect of the CNTs has ceased, the hydration reaction takes back its normal course. These phenomena have a direct impact on the initial and final setting times of the cementing matrix.

Not only the CNTs but also the dispersing agent used in combination with the CNTs has an effect on the kinetics of the reaction. A dispersant compatible with the cement should be used to obtain optimal results. Positive results have been found using SDS (Yu & Kwon 2009), PVP (Chan & Andrawes 2010), ether polycarboxylates (Wansom et al. 2006), and surfactants (Han et al. 2010). For the specific case of surfactants, relatively high surfactant-to-CNT ratios have been identified as an effective dispersing agent (Konsta-Gdoutos et al. 2010a); however, it is known that high concentrations of specific surfactants hamper the cement hydration reaction due to steric hindrance caused by the molecules of surfactant adsorbed onto the surface of the anhydrous cement grains (Łaźniewska-Piekarczyk 2013), retarding the heat release during the first hours (Ouyang et al. 2008).

## 26.4.4 Effect of the CNTs on the Rheology and Shrinkage

Some nanoparticles, such as nanosilica and nanoclay, have a direct influence on the rheology of cement matrices (Kawashima et al. 2013). Whether it is due to water adsorption, flocculation phenomena, surface properties or chemical compatibility with the hydration reaction, nanoparticles decrease the workability of pastes, mortars and concretes, increasing the water and superplasticizer demands. In the case of CNTs, the variations in the rheology of the matrices are not caused by the properties of the CNTs themselves (De Paula et al. 2014), but depend mainly on the type and amount of dispersant used (Collins et al. 2012), and on the type of functional groups present on their surface. A valid approach is to use commercial chemical admixtures with proven compatibility with cement as dispersing agents. Some of the most commonly used are sulfonic acid, polycarboxylates, and sulfonates among others; these admixtures have specific functions in the mix, such as air-entraining agents, surface tension modifiers, defoamers, retardants, or accelerants, and also have the potential to work as a dispersing agent for the

CNTs. Usually, high amounts of chemical admixtures are required to obtain an adequate dispersion of CNTs (Konsta-Gdoutos et al. 2010a), and also to maintain its function in the mixture.

The rheological properties of the mixture can be directly measured using a rheometer or viscosimeter, to calculate yield stress and viscosity, or using flow measurements, which give quantitative information about the rheology of the mixture in an easier manner, and work as indicators for viscosity. The most commonly used flow tests are the marsh cone and mini lump for pastes, the flow table for mortars, and the slump cone for concretes. Due to the great variety of chemical admixtures which can be used, a single effect of the CNTs on the rheology of the mix cannot be identified. For example, while CNTs mixed with air-entraining agents decrease the fluidity of the mixture, CNTs mixed with polycarboxylates increase it (Collins et al. 2012); therefore, this has to be studied for each particular case.

Regarding the shrinkage process of the cementing matrix, it has been found that CNTs have a beneficial role (Siddique & Mehta 2014). The magnitude of the shrinkage that a cement matrix undergoes has been found to be proportional to the fraction of the gel and capillary pores. Due to their size, CNTs promote the decrease in this fraction of the pores by two mechanisms: the nucleation of C-S-H and the filling of the pores. This change in the pore structure modifies the transport of water inside the matrix and the internal pressure inside the pores after drying, decreasing the overall amount of shrinkage (Konsta-Gdoutos et al. 2010b). A direct consequence of this decrease in shrinkage is a lower formation of submicrometric and micrometric cracks, which decreases the permeability and increases the durability of the matrix. This effect is not exclusive to CNTs; other nanoparticles exhibit it due to their size and surface area (Sanchez & Sobolev 2010).

## 26.5 Interaction of the CNTs with the Cement Matrix in Hardened State

The *hardened state* can be defined as the state at which the matrix is no longer plastic and cannot be molded. The cement matrix is constantly developing its strength and is capable of bearing the service load for which it was designed. At this point, the matrix behaves as a quasi-brittle material, capable of sustaining high compressive and not so high tensile and flexural loads. It is at this point that the CNTs start to work as nanoreinforcement, enhancing the load transmission throughout the matrix and modifying the cracking behavior of the material at the nano- and microscales. These modifications in the cracking behavior have an effect on the porosity of the matrix, increasing its durability and the overall performance of the material over time. This section is focused on the surface interactions between the CNTs and the hydration products of cement at the hardened state and how these interactions modify the mechanical behavior of the matrix. The dispersion issues presented in Section 26.4.1 have a direct consequence on the reinforcing effectiveness of the CNTs in the hardened state, generating nonuniform reports in literature.

### 26.5.1 Bridge Effect and Cracking Control

It has been previously discussed that pure CNTs do not chemically bond with the components of the cement matrix; this ensures that the CNTs maintain their structure and properties, but at the same time possesses a challenge: the lack of interaction between the surfaces to transmit the tensile load. Due to the high tensile strength of CNTs, it is likely that a CNT–cement composite will fail by fiber pullout, rather than by fracture of CNTs (Chan & Andrawes 2009). The lack of surface interaction between the CNTs and the hydration products causes an easier pullout of CNTs from the matrix; therefore, the reinforcing efficiency of the CNTs is limited by the amount of tensile load the interfaces are capable of transmitting, rather than by the tensile strength of the nanotubes. Acid treatments, which generate OH– functional groups on the surface of the CNTs, are methods to enhance the tensile load transmission and take advantage of the additional reinforcing potential of the CNTs. It has been found that OH– functional groups are capable of bonding with C-S-H and Ca(OH)$_2$ (Li et al. 2005); this interaction enhances the tensile load transmission and increases the efficiency of the CNTs as reinforcement. In the samples with treated CNTs, both the pullout and the breakage of CNTs were observed. The presence of breakage implies that the tensile load transmission between the matrix and the nanotubes significantly improved to the point that a portion of the CNTs was stressed up to their tensile limit (Abu Al-Rub et al. 2012a). The pullout and the breakage of CNTs modify the fracture energy stored in the composite; this can be seen as a multipeak behavior in a stress–strain curve, increasing the total strain capacity and toughness of the material (Abu Al-Rub et al. 2012a).

Another mechanism by which CNTs modify the load transmission within the composite, and enhance its mechanical properties, is the bridge effect. Individual CNTs act as a network of bridges that transmit the tensile load across cracks and pores (Wang et al. 2013). An example of this bridge effect is presented in Figure 26.7, where two agglomerations of C-S-H are bridged by CNTs, some of which suffered pullout and others breakage after the crack opening. Due to their diameter being at the scale of nanometers, CNTs are capable of bridging submicron-sized cracks, for which larger fibers are not effective (Sakulich & Li 2011); and due to their length being at the scale of micrometers, CNTs are capable of bridging micron-sized cracks (Abu Al-Rub et al. 2012b). The aspect ratio of the CNTs also plays an important role, since it has been demonstrated that lower amounts of long CNTs have the same effect on the properties of the matrix as a higher amount of shorter CNTs (Abu Al-Rub et al. 2012a). Finite element analyses of beams reinforced with CNTs have proposed that CNTs bridging the cracks prevented sudden failure by transferring the tensile load across the cracks and controlling the crack propagation, allowing substantial strain of the matrix until complete pullout of the nanotubes is reached. The cracks develop upward, generating sequential pullout of nanotubes until the capacity of the beam abruptly drops at the maximum strain (Chan & Andrawes 2010).

For an effective crack propagation control, CNTs must not only be well dispersed, but also bridge the appropriate phases. C-S-H is usually formed in agglomerations, with a variable spacing among them from nanometers to micrometers; CNTs must reach across these spaces and bridge the C-S-H agglomerations. If the CNTs become agglomerated inside or around individual C-S-H agglomerations, they will not help in the tensile load distribution and will not have any effect on the mechanical properties of the matrix (Sobolkina et al. 2012). This is schematically presented in Figure 26.8 for long and short CNTs.

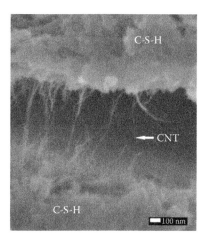

**FIGURE 26.7** SEM of bridge effect of CNTs in a microcrack. (Reprinted from *Construction and Building Materials*, 35, Abu Al-Rub, R. K., Ashour, A. I. & Tyson, B. M., On the aspect ratio effect of multi-walled carbon nanotube reinforcements on the mechanical properties of cementitious nanocomposites, 647–655, Copyright (2012), with permission from Elsevier.)

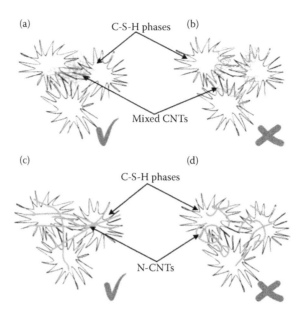

**FIGURE 26.8** Schematic representations of the arrangements of CNTs in a cement matrix: (a, c) advantageous and (b, d) disadvantageous distributions of the mixed CNTs. (Reprinted from *Cement and Concrete Composites*, 34, Sobolkina, A., Mechtcherine, V., Khavrus, V. et al., Dispersion of carbon nanotubes and its influence on the mechanical properties of the cement matrix, 1104–1113, Copyright (2012), with permission from Elsevier.)

Figure 26.8a and c presents short and long CNTs bridging individual agglomerations of C-S-H, while Figure 26.8b and d presents short and long CNTs agglomerated in the individual C-S-H agglomerations.

An extremely high load transmission was found between CNTs and a matrix of geopolymeric cement (Saafi et al. 2013). The experimental results showed that most of the CNTs were found normal to the cracks, with no evidence of pullout from the matrix or breakage of the CNTs. The authors proposed that a slippage occurred between the inner and the outer shells of the nanotubes, allowing the opening of the crack without the full pullout of the nanotubes from the matrix. This slippage had been previously predicted by molecular dynamics simulations and some evidence of it had been found by TEM (Yamamoto et al. 2012). The mechanism proposed for the reinforcing activity of the CNTs in a geopolymeric matrix is based on the fact that the bond strength between the surfaces is high enough to cause a failure of the CNTs instead of a pullout. When the load is applied and a crack opens, a small interfacial debonding between the CNTs and the matrix happens first, followed by the failure of only some of the outer shells of the nanotubes; these shells remain in their original position, allowing the inner shells to slip and be pulled away (Saafi et al. 2013); so far this behavior has not been reported for ordinary portland cement matrices. A representation of this mechanism is presented in Figure 26.9.

## 26.5.2 Modification of Mechanical Properties

The mechanical response of cement composites is usually studied in terms of compressive and flexural strengths, whether it is in pastes, mortars, or concretes. The interactions of CNTs with the cement matrix, in both fresh and hardened states, modify this mechanical response not always in a positive manner. This has been found to depend on many factors, such as amount of water, type, aspect ratio, and amount of CNTs; dispersion and mixing procedures; dispersing agent; and presence of functional groups on the CNTs. Additions of CNTs have been studied in amounts as high as 2.0% or as low as 0.01% by weight of cement, finding that higher amounts of CNTs do not mean better reinforcing efficiency, since they are more sensitive to the reagglomeration phenomenon (Mendoza et al. 2013a). Also, acid

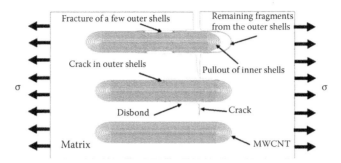

**FIGURE 26.9** Proposed mechanism for crack bridging in geopolymeric composites. (Reprinted from *Construction and Building Materials*, 49, Saafi, M., Andrew, K., Tang, P. L. et al., Multifunctional properties of carbon nanotube/fly ash geopolymeric nanocomposites, 46–55, Copyright (2013), with permission from Elsevier.)

treatments have been found to improve the reinforcing efficiency of the CNTs (Li et al. 2007), but at the same time if the acid treatment is not sufficiently controlled and is combined with sonication, the CNTs will break and an excess of defects will be induced on their structure, degrading their properties (Mendoza et al. 2013b). Finally, the electrostatic interactions among the functional groups on CNTs, the dispersing agents, and the ionic components of cement will also play an important role in the stability of the dispersion of CNTs, determining if the nanotubes will work as reinforcement (Mendoza et al. 2014). All this has led to nonuniform reports in literature and has kept the research community from reaching a general consensus about the beneficial role of CNTs in cement composites.

Regarding compressive strength, mixed results have been reported. While some authors have reported a steady loss of compressive strength with additions of CNTs between 0.5% and 2.0% by weight of cement (Collins et al. 2012), others have found an optimum amount of addition of 0.02%, associated with an increase in compressive strength of 30% (Morsy et al. 2011). At low contents of water, it has been found that variations in the amount of water have a higher impact on the compressive strength than modifications in the amount of CNTs (Cwirzen et al. 2009), and that the hydrophilic behavior of functionalized CNTs is capable of adsorbing enough water to hamper the hydration of cement at early ages, requiring high water contents to reach a full hydration (Musso et al. 2009). In contrast, after 28 days of hydration, composites presented a better compressive strength (Li et al. 2005), and functionalized CNTs generated a higher compressive strength than untreated CNTs (Li et al. 2007).

Regarding flexural strength, there is a higher uniformity of results and a consensus in the fact that CNTs do increase the performance of the composites when exposed to tensional and flexural loads. Small amounts of long CNTs have been found to significantly increase the flexural strength of the composites (Xu et al. 2015), confirming the correlation between the aspect ratio and the reinforcing efficiency (Konsta-Gdoutos et al. 2010b). This has been directly correlated to the dispersion of the CNTs and the bond strength between CNTs and cement. A dispersant-to-CNT ratio of 4.0 was identified as the optimum dispersant amount which generated enough degree of dispersion to substantially increase the fracture load of the composite (Konsta-Gdoutos et al. 2010a). In three-point bending tests, it has been found that the CNT–cement composites are capable of sustaining a higher load and a higher deflection when compared to plain cement samples. Numerical simulations based on these experimental results concluded that it is necessary to drastically increase the bond strength between the CNTs and the cement matrix to reach the full potential of the CNTs as a reinforcing material (Chan & Andrawes 2010). At early ages, specifically during the first hours of hydration, CNTs have been found to improve the early strain capacity of the composites; this is associated with a reduction in capillary stresses, which leads to a smaller drying shrinkage (Konsta-Gdoutos et al. 2010b). Additional evidence that indicates the enhancement of flexural strength of the composites includes increases in the modulus of rupture (Musso et al. 2009), higher fracture energy, and higher fracture toughness (Hu et al. 2014).

Finally, the statistical nanoindentation technique has been used to characterize the nanomechanical response of CNT–cement composites. This technique measures a grid of load versus penetration curves at the scales of millinewtons and nanometers, and from these results, calculates the frequency distribution of elastic moduli of the different phases in the composite. The fundamental principle of the technique is the fact that each phase present in the composite, whether it is hydrated or anhydrous, has an intrinsic modulus of elasticity that does not depend on the proportioning of raw materials used to produce the composite (Constantinides & Ulm 2004). It has been found that the inclusion of CNTs increases the frequency of higher elasticity moduli and decreases the amount of nanoporosity. Higher presence of high-density and ultrahigh-density C-S-H has been associated with better mechanical performance of composites at the macroscale (Sorelli et al. 2008). Some authors have interpreted this as an enhancement in the production of high-density and ultrahigh-density C-S-H (Sanchez & Sobolev 2010), coupled with a decrease in nanoporosity. Since CNTs do not have any chemical affinity with the hydration reaction, and the nanomechanical response of C-S-H has been proposed to be of nanogranular nature, meaning that the particle-to-particle contact has a greater influence on the mechanical behavior than the intrinsic properties of the phase (Constantinides & Ulm 2007), the mechanism by which CNTs increase the elastic moduli of C-S-H has to be of a physical nature, rather than chemical. This is still a matter of discussion.

## 26.6 Hybrids of CNTs with Other Nano- and Microparticles in Cement Composites

CNTs have been combined with other nanoparticles with the goal of maximizing their positive effects and complementing the weak aspects of the other nanoparticles. A complementarity of effects is considered successful when the individual effects of each nanoparticle are enhanced, or when a new effect is generated. For this complementarity of effects to happen, chemical and scale compatibility between CNTs and the other nanoparticles has to be guaranteed. In other areas of knowledge, the potential of the combinations of CNTs with other particles has already been recognized; CNTs have been coated with silicon oxide and gold nanoparticles to fabricate coaxial wires that work as markers for carcinoembryonic biological fluids (Li et al. 2011). Also, silica nanotubes have been produced using CNTs as molds through the decomposition of chlorosilanes (Lin & Shen 2010), to cite only some of the many possibilities that have been studied.

The combination of CNTs with other nanoparticles can be thought from different approaches: combinations with micro- and nanofibers to control the crack propagation at different scales simultaneously; combinations with pozzolanic micro- and nanoparticles to enhance tensile and compressive strengths at the same time; functionalization of the CNTs with other nanoparticles to enhance the bond strength and the load transmission efficiency between CNTs and the matrix; and growth of CNTs on the surface of particles or fibers to modify their surface properties. These combinations, among many others that are possible, can be tailored to solve a specific problem or enhance a specific property of a CNT–cement composite, regarding not only its mechanical response, but also its electromagnetic properties.

One of the most promising combinations is CNTs with silica micro- and nanoparticles. Two approaches, each with a different objective, are used for this combination: a simple combination of CNTs and silica in the mixing water, and a coating of silica particles onto the surface of the CNTs. The goals of the simple combination are to take advantage of the pozzolanic activity of the silica, and to guarantee an increase both in compressive strength as a consequence of the pozzolanic activity, and in flexural and tensile strength as a consequence of the reinforcing effect of the CNTs. As presented before, the effect of CNTs on the compressive strength of the composites is not always beneficial, and this would be a solution for this issue. This combination has been tested with micrometric (Chaipanich et al. 2010) and nanometric (Mendoza et al. 2014) silica particles, and it has been found to be partially effective with high levels of substitution of cement by silica. The main issues of this type of combination are the scale incompatibility between the micrometric silica and the CNTs, and the reagglomeration phenomenon that occurs in the aqueous dispersions of CNTs and nanosilica. These issues cause a loss of surface area of the particles, generating a lower reactivity of the silica and the accumulation of CNTs as agglomerations inside the pores, which limits their reinforcing capability.

The goal of the coating of CNTs with silica is the improvement of the bond strength between the surface of the CNTs and the cement matrix; this is done by inducing the growth of C-S-H from the silica on the surface of the CNTs (Sobolkina et al. 2014). Higher bond strength between the reinforcement and the matrix means higher reinforcing efficiency of the CNTs, since they would be able to withstand more load before suffering pullout from the matrix, even to the point of reaching their maximum tensile strength, a point at which the CNTs reach their maximum reinforcing efficiency. The coating of silica is obtained by sol–gel reaction and individual nanotubes coated with a thin layer of silica on their surface are the result; the silica structure is stable enough to withstand the calcination of the internal nanotube, maintaining its shape and internal structure (Kim et al. 2009). Another method to induce the growth of C-S-H on the surface of CNTs is by pretreating them with $Ca(OH)_2$ and nanosilica particles (Li et al. 2007); the main goal of this treatment is to differentiate the morphology of CNTs from ettringite crystals. A marginal gain in compressive and flexural strength of the composites was obtained with this treatment.

Besides silica, the simple combination approach has also been tested with nanokaolin clay, obtaining similar results (Morsy et al. 2011). Some authors claim that the presence of nanometric clay particles or even nanometric and micrometric silica particles helps to disperse and maintain the stability of CNTs; this is an extrapolation of results from micrometric fibers, where the additional particles help to mechanically separate the individual fibers taking advantage of the gaps between them (Nam et al. 2012). Unfortunately, there are no dispersion studies that directly measure the dispersing effect of these particles on CNT agglomerations; therefore, the affirmation that clay and silica particles work as dispersing agents for CNTs has to be carefully used, taking into account that the most important condition for this to happen is the compatibility of scale (Kim et al. 2014a).

CNTs have also been combined with other fibers with the goal of increasing flexural and tensile strength of composites. As for the combinations with pozzolanic particles, the combinations of CNTs with other fibers can be made by simple mixing or by coating; but in this case, the complementary fibers are the ones coated with nanotubes. The coating of CNTs can be achieved by growing them onto the surface of other fibers via the CVD technique. In this case, CNTs act as elements of interaction between the fiber and the matrix, and their function is to enhance the load transmission and bonding strength between the surfaces. CNTs have been mixed with hemp (Hamzaoui et al. 2014) and bagasse fibers (Kordkheili et al. 2012), finding a significant increase in flexural strength and a modification in the water absorption dynamics of the natural fibers.

## 26.7 Novel Properties of CNT–Cement Composites

Besides modifying the known properties of the cement composites, CNTs can also confer new electromagnetic functionalities to the matrix; these functionalities have opened a new range of potential applications for smart structures. If the CNTs are in an adequate volume fraction and properly dispersed throughout the matrix, they form a conductive network. This network has piezoelectric (Gong et al. 2011) and piezoresistive (Yu & Kwon 2009) properties, which can be used to generate matrices with self-sensing or self-monitoring capabilities, i.e., that are capable of sensing their own state of strain, stress, damage, and temperature (Chung 2012). The self-sensing properties of a CNT–cement composite are widely governed by the amount of nanotubes present in the matrix; if they are present in a volume fraction at which the fibers touch each other, they form a continuous conductive path and the electric response is governed by the effect of strain on the contact between nanotubes (Sun et al. 2014). This volume fraction, at which the matrix exhibits an increase in conductivity, is called the *electrical percolation threshold* (MacDonald et al. 2008). The percolation volume in cement-based composites has been found to be dependent on the conductivity and geometry of the conductive phase, rather than on the composite composition (Xie et al. 1996).

Two pressure-sensitive responses have been found in CNT–cement composites, piezoelectricity and piezoresistivity. The piezoelectric response consists of the generation of a voltage when a load is applied to the matrix (Gong et al. 2011); this voltage originates in the pores of the matrix, which are filled with a conductive electrolytic solution. When a compressive load is applied, additional ions are transported from the matrix to the electrolyte solution, causing a difference in potential (Sun et al. 2000). CNTs enhance the transmission of this voltage throughout the matrix, working as a conductive network. The piezoresistive response consists of the decrease in the electric resistivity of the matrix when a load is applied (Konsta-Gdoutos & Aza 2014); this drop in resistivity has been attributed to two sources: to the intrinsic piezoresistivity of the CNTs and to a higher contact among CNTs due to deformations of the matrix (Yu & Kwon 2009). The electric charges have been found to be conducted mainly by the network of CNTs, and the effect of adsorbed water has been found to be minimum (Kim et al. 2014b). An example of piezoresistive behavior of a CNT–cement composite is presented in Figure 26.10.

The piezoresistive effect of CNT–cement has shown the most potential to generate strain sensors. Not only CNTs but also CNFs have proven to be an effective conductive phase in these composites (Konsta-Gdoutos & Aza 2014), exhibiting not only piezoresistivity but also capacitance characteristics, the latter not sensitive to load (Han et al. 2012). The strain sensors developed with CNT–cement composites can be of small volume and embedded in structural elements, minimizing the consumption of conductive phase and therefore their cost (Sun et al. 2014). Some of the requisites that have to be fulfilled to design an effective fiber-based sensor are that the fibers have to be more conductive than the matrix, their diameters have to be smaller than the crack length, and they must be well dispersed and randomly oriented (Chung 2002b). The water content of the mixture, i.e., the water/cement ratio, is one of the key factors to tailor a strain sensor, since a nonlinear relationship between the water content and the electric resistivity of the matrix has been found (Han et al. 2010). This nonlinear response is caused by a polarization of the water molecules in the matrix, which increases the electrical resistance of the composite during measurement (Konsta-Gdoutos & Aza 2014); this increase is nonlinear and is dependent on the load application rate (Azhari & Banthia 2012). Solutions to this issue go from using dry specimens to measurements with AC or the use of nonlinear calibration curves.

A direct consequence of the addition of CNTs or any other conductive fiber to the composite is the increase in the electric conductivity of the matrix, regardless of the presence of compressive or tension loads. Electrically conductive cement matrices are useful for electrical grounding, lighting protection, resistance heating, static charge dissipation, electromagnetic interference shielding, cathodic protection, and thermoelectric energy generation (Chung 2004). The high electrical conductivity of the CNTs

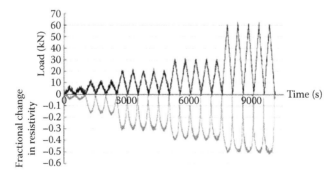

**FIGURE 26.10** Piezoresistive response of a CNT–cement composite in cyclic load. (Reprinted from *Cement and Concrete Composites*, 34, Azhari, F. & Banthia, N., Cement-based sensors with carbon fibers and carbon nanotubes for piezoresistive sensing, 866–873, Copyright (2012), with permission from Elsevier.)

has generated an additional application in cement composites as electromagnetic interference shielding, specifically in the frequency range used for mobile phones communication (Nam et al. 2012). It has been stated that CNT–cement composites have potential as an alternative shielding material; but this property is also limited by the adequate dispersion of the CNTs in the matrix, as is also by the amount of conductive phase in the composite and the thickness of the shielding wall (Micheli et al. 2014). These nonstructural applications have become interesting over time, since they allow a traditionally structural material to be multifunctional, saving costs and increasing its durability (Chung 2002a).

## 26.8 Present Challenges and Future Perspective

Until the recent past, the price and scale of the production of nanoparticles made them prohibitive for industrial-scale applications; but as the price of nanoparticles has been increasingly dropping over time, they have become affordable for general construction applications (Welch et al. 2008). The full development of all the potential applications for CNT–cement composites is not being held back by their price anymore, but by two main technical issues: the effective dispersion of CNTs and their interfacial bond strength with the cement-based matrix. The research community has repeatedly reported that all the improvements obtained in the known and novel properties of the cement matrix are in some way limited by these two issues; therefore, they should become the main focus of the research carried out in the field.

So far, the exceptional properties of individual CNTs are widely underused, making the beneficial effects obtained with them comparable with those obtained from traditional micro- and macrofibers and particles. Novel functionalization methodologies and dispersion methods, specifically designed to target the dispersion and bond strength issues, would provide the leap forward in knowledge necessary. This should be accompanied by detailed studies about the effect of CNT–cement composites on living organisms, since one of the promising applications is their use in housing solutions.

## References

Abdalla, M., Dean, D., Robinson, P. et al., "Cure behavior of epoxy MWCNT nanocomposites: The effect of nanotube surface modification," *Polymer* 49 (2008): 3310–3317.

Abu Al-Rub, R. K., Ashour, A. I. & Tyson, B. M., "On the aspect ratio effect of multi-walled carbon nanotube reinforcements on the mechanical properties of cementitious nanocomposites," *Constr. Build. Mater.* 35 (2012a): 647–655. doi:10.1016/j.conbuildmat.2012.04.086.

Abu Al-Rub, R. K., Tyson, B. M., Yazdanbakhsh, A. et al., "Mechanical properties of nanocomposite cement incorporating surface-treated and untreated carbon nanotubes and carbon nanofibers," *J. Nanomech. Micromech.* 2 (2012b): 3–8. doi:10.1061/(ASCE)NM.2153-5477.0000041.

Allen, A. J., Thomas, J. J. & Jennings, H. M., "Composition and density of nanoscale calcium–silicate–hydrate in cement," *Nat. Mater.* 6 (2007): 311–316. doi:10.1038/nmat1871.

Azhari, F. & Banthia, N., "Cement-based sensors with carbon fibers and carbon nanotubes for piezoresistive sensing," *Cem. Concr. Compos.* 34 (2012): 866–873. doi:10.1016/j.cemconcomp.2012.04.007.

Beaudoin, J. J., Raki, L. & Alizadeh, R., "A $^{29}$Si MAS NMR study of modified C–S–H nanostructures," *Cem. Concr. Compos.* 31 (2009): 585–590. doi:10.1016/j.cemconcomp.2008.11.004.

Bernal, J. D., Jeffrey, J. W. & Taylor, H. F. W., "Crystallographic research on the hydration of Portland cement: A first report on investigations in progress," *Mag. Concr. Res.* 4 (1952): 49–54.

Chaipanich, A., Nochaiya, T., Wongkeo, W. et al., "Compressive strength and microstructure of carbon nanotubes–fly ash cement composites," *Mater. Sci. Eng. A* 527 (2010): 1063–1067. doi:10.1016/j.msea.2009.09.039.

Chan, L. Y. & Andrawes, B., "Characterization of the uncertainties in the constitutive behavior of carbon nanotube/cement composites," *Sci. Technol. Adv. Mater.* 10 (2009): 045007. doi:10.1088/1468-6996/10/4/045007.

Chan, L. Y. & Andrawes, B., "Finite element analysis of carbon nanotube/cement composite with degraded bond strength," *Comput. Mater. Sci.* 47 (2010): 994–1004. doi:10.1016/j.commatsci.2009.11.035.

Chen, J. J., Thomas, J. J., Taylor, H. F. W. et al., "Solubility and structure of calcium silicate hydrate," *Cem. Concr. Res.* 34 (2004): 1499–1519. doi:10.1016/j.cemconres.2004.04.034.

Chung, D. D. L., "Piezoresistive cement-based materials for strain sensing," *J. Intell. Mater. Syst. Struct.* 13 (2002a): 599–609. doi:10.1106/104538902031861.

Chung, D. D. L., "Electrical conduction behavior of cement–matrix composites," *J. Mater. Eng. Perform.* 11 (2002b): 194–204. doi:10.1361/105994902770344268.

Chung, D. D. L., "Electrically conductive cement-based materials," *Adv. Cem. Res.* 4 (2004): 167–176.

Chung, D. D. L., "Carbon materials for structural self-sensing, electromagnetic shielding and thermal interfacing," *Carbon* 50 (2012): 3342–3353. doi:10.1016/j.carbon.2012.01.031.

Collins, F., Lambert, J. & Duan, W. H., "The influences of admixtures on the dispersion, workability, and strength of carbon nanotube–OPC paste mixtures," *Cem. Concr. Compos.* 34 (2012): 201–207. doi:10.1016/j.cemconcomp.2011.09.013.

Constantinides, G. & Ulm, F.-J., "The effect of two types of C-S-H on the elasticity of cement-based materials: Results from nanoindentation and micromechanical modeling," *Cem. Concr. Res.* 34 (2004): 67–80. doi:10.1016/S0008-8846(03)00230-8.

Constantinides, G. & Ulm, F.-J., "The nanogranular nature of C–S–H," *J. Mech. Phys. Solids* 55 (2007): 64–90. doi:10.1016/j.jmps.2006.06.003.

Cwirzen, A., Habermehl-Cwirzen, K., Nasibulin, A. G. et al., "SEM/AFM studies of cementitious binder modified by MWCNT and nano-sized Fe needles," *Mater. Charact.* 60 (2009): 735–740. doi:10.1016/j.matchar.2008.11.001.

De Paula, J. N., Calixto, J. M., Ladeira, L. O. et al., "Mechanical and rheological behavior of oil-well cement slurries produced with clinker containing carbon nanotubes," *J. Petroleum Sci. Eng.* 122 (2014): 274–279. doi:10.1016/j.petrol.2014.07.020.

Di, Z.-C., Li, Y.-H., Luan, Z.-K. et al., "Adsorption of chromium (VI) ions from water by carbon nanotubes," *Adsorpt. Sci. Technol.* 22 (2004): 467–474. doi:10.1260/0263617042879537.

Dolado, J. S. & van Breugel, K., "Recent advances in modeling for cementitious materials," *Cem. Concr. Res.* 41 (2011): 711–726. doi:10.1016/j.cemconres.2011.03.014.

Garrault, S. & Nonat, A., "Hydrated layer formation on tricalcium and dicalcium silicate surfaces: Experimental study and numerical simulations," *Langmuir* 17 (2001): 8131–8138. doi:10.1021/la011201z.

Gong, H., Zhang, Y., Quan, J. et al., "Preparation and properties of cement based piezoelectric composites modified by CNTs," *Curr. Appl. Phys.* 11 (2011): 653–656. doi:10.1016/j.cap.2010.10.021.

Hamzaoui, R., Guessasma, S., Mecheri, B. et al., "Microstructure and mechanical performance of modified mortar using hemp fibres and carbon nanotubes," *Mater. Des.* 56 (2014): 60–68. doi:10.1016/j.matdes.2013.10.084.

Han, B., Yu, X. & Ou, J., "Effect of water content on the piezoresistivity of MWNT/cement composites," *J. Mater. Sci.* 45 (2010): 3714–3719. doi:10.1007/s10853-010-4414-7.

Han, B., Zhang, K., Yu, X. et al., "Electrical characteristics and pressure-sensitive response measurements of carboxyl MWNT/cement composites," *Cem. Concr. Compos.* 34 (2012): 794–800. doi:10.1016/j.cemconcomp.2012.02.012.

Hewlett, P., "Lea's chemistry of cement and concrete," in Hewlett, P. ed., *Science*, Fourth edition, Vol. 58 (Elsevier Science & Technology Books, Oxford, England, 2004), http://www.dbpia.co.kr/view/ar_view.asp?arid=1536305.

Hu, Y., Luo, D., Li, P. et al., "Fracture toughness enhancement of cement paste with multi-walled carbon nanotubes," *Constr. Build. Mater.* 70 (2014): 332–338. doi:10.1016/j.conbuildmat.2014.07.077.

Jennings, H. M., "Colloid model of C-S-H and implications to the problem of creep and shrinkage," *Mater. Struct.* 37 (2004): 59–70.

Jennings, H. M., Thomas, J. J., Chen, J. J. et al., "Cement paste as a porous material," in Schuth, F., Sing. K. & Weitkamp, J., eds., *Handbook of Porous Solids* (Wiley-VCH, Weinheim, Germany, 2002), pp. 2971–3028. doi:10.1002/9783527618286.

Jiang, L., Gao, L. & Sun, J., "Production of aqueous colloidal dispersions of carbon nanotubes," *J. Colloid Interface Sci.* 260 (2003): 89–94. doi:10.1016/S0021-9797(02)00176-5.

Juilland, P., Kumar, A., Gallucci, E. et al., "Effect of mixing on the early hydration of alite and OPC systems," *Cem. Concr. Res.* 42 (2012): 1175–1188. doi:10.1016/j.cemconres.2011.06.011.

Jung, W. R., Choi, J. H., Lee, N. et al., "Reduced damage to carbon nanotubes during ultrasound-assisted dispersion as a result of supercritical-fluid treatment," *Carbon* 50 (2012): 633–636. doi:10.1016/j.carbon.2011.08.075.

Kakade, B. A. & Pillai, V. K., "An efficient route towards the covalent functionalization of single walled carbon nanotubes," *Appl. Surf. Sci.* 254 (2008): 4936–4943. doi:10.1016/j.apsusc.2008.01.148.

Kawashima, S., Hou, P., Corr, D. J. et al., "Cement & concrete composites modification of cement-based materials with nanoparticles," *Cem. Concr. Compos.* 36 (2013): 8–15.

Kim, M., Hong, J., Hong, C. K. et al., "Preparation of silica-layered multi-walled carbon nanotubes activated by grafting of poly(4-vinylpyridine)," *Synth. Met.* 159 (2009): 62–68. doi:10.1016/j.synthmet.2008.07.019.

Kim, D.-Y., Yun, Y. S., Bak, H. et al., "Aspect ratio control of acid modified multiwalled carbon nanotubes," *Curr. Appl. Phys.* 10 (2010): 1046–1052. doi:10.1016/j.cap.2009.12.038.

Kim, H. K., Nam, I. W. & Lee, H. K., "Enhanced effect of carbon nanotube on mechanical and electrical properties of cement composites by incorporation of silica fume," *Compos. Struct.* 107 (2014a): 60–69. doi:10.1016/j.compstruct.2013.07.042.

Kim, H. K., Park, I. S. & Lee, H. K., "Improved piezoresistive sensitivity and stability of CNT/cement mortar composites with low water–binder ratio," *Compos. Struct.* 116 (2014b): 713–719. doi:10.1016/j.compstruct.2014.06.007.

Konsta-Gdoutos, M. S., Metaxa, Z. S. & Shah, S. P., "Multi-scale mechanical and fracture characteristics and early-age strain capacity of high performance carbon nanotube/cement nanocomposites," *Cem. Concr. Compos.* 32 (2010a): 110–115. doi:10.1016/j.cemconcomp.2009.10.007.

Konsta-Gdoutos, M. S., Metaxa, Z. S. & Shah, S. P., "Highly dispersed carbon nanotube reinforced cement based materials," *Cem. Concr. Res.* 40 (2010b): 1052–1059. doi:10.1016/j.cemconres.2010.02.015.

Konsta-Gdoutos, M. S. & Aza, C. A., "Self sensing carbon nanotube (CNT) and nanofiber (CNF) cementitious composites for real time damage assessment in smart structures," *Cem. Concr. Compos.* 53 (2014): 162–169. doi:10.1016/j.cemconcomp.2014.07.003.

Kordkheili, H. Y., Hiziroglu, S. & Farsi, M., "Some of the physical and mechanical properties of cement composites manufactured from carbon nanotubes and bagasse fiber," *Mater. Des.* 33 (2012): 395–398. doi:10.1016/j.matdes.2011.04.027.

Łaźniewska-Piekarczyk, B., "The influence of chemical admixtures on cement hydration and mixture properties of very high performance self-compacting concrete," *Constr. Build. Mater.* 49 (2013): 643–662. doi:10.1016/j.conbuildmat.2013.07.072.

Li, G. Y., Wang, P. M. & Zhao, X., "Mechanical behavior and microstructure of cement composites incorporating surface-treated multi-walled carbon nanotubes," *Carbon* 43 (2005): 1239–1245. doi:10.1016/j.carbon.2004.12.017.

Li, G. Y., Wang, P. M. & Zhao, X., "Pressure-sensitive properties and microstructure of carbon nanotube reinforced cement composites," *Cem. Concr. Compos.* 29 (2007): 377–382. doi:10.1016/j.cemconcomp.2006.12.011.

Li, Q., Tang, D., Tang, J. et al., "Carbon nanotube-based symbiotic coaxial nanocables with nanosilica and nanogold particles as labels for electrochemical immunoassay of carcinoembryonic antigen in biological fluids," *Talanta* 84 (2011): 538–546. doi:10.1016/j.talanta.2011.01.063.

Lin, T.-W. & Shen, H.-H., "The synthesis of silica nanotubes through chlorosilanization of single wall carbon nanotubes," *Nanotechnology* 21 (2010): 365604. doi:10.1088/0957-4484/21/36/365604.

Ma, P.-C., Siddiqui, N. A., Marom, G. et al., "Dispersion and functionalization of carbon nanotubes for polymer-based nanocomposites: A review," *Compos. Part A Appl. Sci. Manuf.* 41 (2010): 1345–1367. doi:10.1016/j.compositesa.2010.07.003.

MacDonald, R. A., Voge, C. M., Kariolis, M. et al., "Carbon nanotubes increase the electrical conductivity of fibroblast-seeded collagen hydrogels," *Acta Biomater.* 4 (2008): 1583–1592. doi:10.1016/j.actbio.2008.07.005.

Madni, I., Hwang, C.-Y., Park, S.-D. et al., "Mixed surfactant system for stable suspension of multiwalled carbon nanotubes," *Colloids Surf. A Physicochem. Eng. Aspects* 358 (2010): 101–107. doi:10.1016/j.colsurfa.2010.01.030.

Makar, J. M. & Chan, G. W., "Growth of cement hydration products on single-walled carbon nanotubes," *J. Am. Ceram. Soc.* 92 (2009): 1303–1310. doi:10.1111/j.1551-2916.2009.03055.x.

Mendoza, O., Sierra, G. & Tobón, J., "Influence of super plasticizer and $Ca(OH)_2$ on the stability of functionalized multi-walled carbon nanotubes dispersions for cement composites applications," *Constr. Build. Mater.* 47 (2013a): 771–778.

Mendoza, O., Sierra, G. & Tobón, J., "Efecto híbrido de los nanotubos de carbono y la nanosilice sobre las propiedades mineralógicas y mecánicas de morteros de cemento Pórtland" (Universidad Nacional de Colombia, Bogotá, Colombia, 2013b).

Mendoza, O., Sierra, G. & Tobón, J., "Effect of the reagglomeration process of multi-walled carbon nanotubes dispersions on the early activity of nanosilica in cement composites," *Constr. Build. Mater.* 54 (2014): 550–557.

Metaxa, Z. S., Konsta-Gdoutos, M. S. & Shah, S. P., "Carbon nanofiber cementitious composites: Effect of debulking procedure on dispersion and reinforcing efficiency," *Cem. Concr. Compos.* 36 (2013): 25–32. doi:10.1016/j.cemconcomp.2012.10.009.

Micheli, D., Pastore, R., Vricella, A. et al., "Electromagnetic characterization and shielding effectiveness of concrete composite reinforced with carbon nanotubes in the mobile phones frequency band," *Mater. Sci. Eng. B* 188 (2014): 119–129. doi:10.1016/j.mseb.2014.07.001.

Morsy, M. S., Alsayed, S. H. & Aqel, M., "Hybrid effect of carbon nanotube and nano-clay on physico-mechanical properties of cement mortar," *Constr. Build. Mater.* 25 (2011): 145–149. doi:10.1016/j.conbuildmat.2010.06.046.

Moulin, F., Devel, M. & Picaud, S., "Molecular dynamics simulations of polarizable nanotubes interacting with water," *Phys. Rev. B* 71 (2005): 1–7. doi:10.1103/PhysRevB.71.165401.

Musso, S., Tulliani, J.-M., Ferro, G. et al., "Influence of carbon nanotubes structure on the mechanical behavior of cement composites," *Compos. Sci. Technol.* 69 (2009): 1985–1990. doi:10.1016/j.compscitech.2009.05.002.

Nam, I. W., Kim, H. K. & Lee, H. K., "Influence of silica fume additions on electromagnetic interference shielding effectiveness of multi-walled carbon nanotube/cement composites," *Constr. Build. Mater.* 30 (2012): 480–487. doi:10.1016/j.conbuildmat.2011.11.025.

Ouyang, X., Guo, Y. & Qiu, X., "The feasibility of synthetic surfactant as an air entraining agent for the cement matrix," *Constr. Build. Mater.* 22 (2008): 1774–1779. doi:10.1016/j.conbuildmat.2007.05.002.

Paillet, M., Poncharal, P. & Zahab, A., "Electrostatics of individual single-walled carbon nanotubes investigated by electrostatic force microscopy," *Phys. Rev. Lett.* 94 (2005): 186801. doi:10.1103/PhysRevLett.94.186801.

Pellenq, R. J.-M., Lequeux, N. & van Damme, H., "Engineering the bonding scheme in C–S–H: The iono-covalent framework," *Cem. Concr. Res.* 38 (2008): 159–174. doi:10.1016/j.cemconres.2007.09.026.

Pellenq, R. J.-M., Kushima, A., Shahsavari, R. et al., "A realistic molecular model of cement hydrates," *Proc. Natl. Acad. Sci. U. S. A.* 106 (2009): 16102–16107. doi:10.1073/pnas.0902180106.

Richardson, I. G., "The nature of C-S-H in hardened cements," *Cem. Concr. Res.* 29 (1999): 1131–1147.

Richardson, I. G., "Tobermorite/jennite- and tobermorite/calcium hydroxide-based models for the structure of C-S-H: Applicability to hardened pastes of tricalcium silicate, β-dicalcium silicate, Portland cement, and blends of Portland cement with blast-furnace slag, metakaol, or silica fume," *Cem. Concr. Res.* 34 (2004): 1733–1777. doi:10.1016/j.cemconres.2004.05.034.

Saafi, M., Andrew, K., Tang, P. L. et al., "Multifunctional properties of carbon nanotube/fly ash geopolymeric nanocomposites," *Constr. Build. Mater.* 49 (2013): 46–55. doi:10.1016/j.conbuildmat.2013.08.007.

Sakulich, A. R. & Li, V. C., "Nanoscale characterization of engineered cementitious composites (ECC)," *Cem. Concr. Res.* 41 (2011): 169–175. doi:10.1016/j.cemconres.2010.11.001.

Sanchez, F. & Sobolev, K., "Nanotechnology in concrete—A review," *Constr. Build. Materials* 24 (2010): 2060–2071. doi:10.1016/j.conbuildmat.2010.03.014.

Sham, M. & Kim, J., "Surface functionalities of multi-wall carbon nanotubes after UV/ozone and TETA treatments," *Carbon* 44 (2006): 768–777. doi:10.1016/j.carbon.2005.09.013.

Siddique, R. & Mehta, A., "Effect of carbon nanotubes on properties of cement mortars," *Constr. Build. Mater.* 50 (2014): 116–129. doi:10.1016/j.conbuildmat.2013.09.019.

Sobolkina, A., Mechtcherine V., Khavrus, V. et al., "Dispersion of carbon nanotubes and its influence on the mechanical properties of the cement matrix," *Cem. Concr. Compos.* 34 (2012): 1104–1113. doi:10.1016/j.cemconcomp.2012.07.008.

Sobolkina, A., Mechtcherine, V., Bellmann, C. et al., "Surface properties of CNTs and their interaction with silica," *J. Colloid Interface Sci.* 413 (2014): 43–53. doi:10.1016/j.jcis.2013.09.033.

Sorelli, L., Constantinides, G., Ulm, F. et al., "The nano-mechanical signature of ultra high performance concrete by statistical nanoindentation techniques," *Cem. Concr. Res.* 38 (2008): 1447–1456. doi:10.1016/j.cemconres.2008.09.002.

Strano, M. S., Moore, V. C., Miller, M. K. et al., "The role of surfactant adsorption during ultrasonication in the dispersion of single walled carbon nanotubes," *J. Nanosci. Nanotechnol.* 3 (2003): 81–86.

Sun, M., Liu, Q., Li, Z. et al., "A study of piezoelectric properties of carbon fiber reinforced concrete and plain cement paste during dynamic loading," *Cem. Concr. Res.* 30 (2000): 1593–1595. doi:10.1016/S0008-8846(00)00338-0.

Sun, S., Yu, X., Han, B. et al., "In situ growth of carbon nanotubes/carbon nanofibers on cement/mineral admixture particles: A review," *Constr. Build. Mater.* 49 (2013): 835–840. doi:10.1016/j.conbuildmat.2013.09.011.

Sun, M., Liew, R. J. Y., Zhang, M.-H. et al., "Development of cement-based strain sensor for health monitoring of ultra high strength concrete," *Constr. Build. Mater.* 65 (2014): 630–637. doi:10.1016/j.conbuildmat.2014.04.105.

Thomas, J. J., "A new approach to modeling the nucleation and growth kinetics of tricalcium silicate hydration," *J. Am. Ceram. Soc.* 90 (2007): 3282–3288. doi:10.1111/j.1551-2916.2007.01858.x.

Thomas, J. J., Jennings, H. M. & Chen, J. J., "Influence of nucleation seeding on the hydration mechanisms of tricalcium silicate and cement," *J. Phys. Chem. C* 113 (2009): 4327–4334. doi:10.1021/jp809811w.

Vandamme, M., Ulm, F.-J. & Fonollosa, P., "Nanogranular packing of C–S–H at substochiometric conditions," *Cem. Concr. Res.* 40 (2010): 14–26. doi:10.1016/j.cemconres.2009.09.017.

Wang, B., Han, Y. & Liu, S., "Effect of highly dispersed carbon nanotubes on the flexural toughness of cement-based composites," *Constr. Build. Mater.* 46 (2013): 8–12. doi:10.1016/j.conbuildmat.2013.04.014.

Wansom, S., Kidner, N., Woo, L. et al., "AC-impedance response of multi-walled carbon nanotube/cement composites," *Cem. Concr. Compos.* 28 (2006): 509–519. doi:10.1016/j.cemconcomp.2006.01.014.

Welch, C., Marcuson, W., Hon, P. E. et al., "Will supermolecules and supercomputers lead to super construction materials?" *Civ. Eng.* (2008): 42–53.

Xie, P., Gu, P. & Beaudoin, J. J., "Electrical percolation phenomena in cement composites containing conductive fibres," *J. Mater. Sci.* 31 (1996): 4093–4097. doi:10.1007/BF00352673.

Xu, S., Liu, J. & Li, Q., "Mechanical properties and microstructure of multi-walled carbon nanotube-reinforced cement paste," *Constr. Build. Mater.* 76 (2015): 16–23. doi:10.1016/j.conbuildmat.2014.11.049.

Yamamoto, G., Liu, S., Hu, N. et al., "Prediction of pull-out force of multi-walled carbon nanotube (MWCNT) in sword-in-sheath mode," *Comput. Mater. Sci.* 60 (2012): 7–12. doi:10.1016/j.commatsci.2012.03.016.

Yazdanbakhsh, A. & Grasley, Z., "The theoretical maximum achievable dispersion of nanoinclusions in cement paste," *Cem. Concr. Res.* 42 (2012): 798–804. doi:10.1016/j.cemconres.2012.03.001.

Yazdanbakhsh, A., Grasley, Z., Tyson, B. et al., "Distribution of carbon nanofibers and nanotubes in cementitious composites," *Transp. Res. Rec. J. Transp. Res. Board* 2142 (2010): 89–95. doi:10.3141/2142-13.

Yazdanbakhsh, A., Grasley, Z., Tyson, B. et al., "Dispersion quantification of inclusions in composites," *Compos. Part A Appl. Sci. Manuf.* 42 (2011): 75–83. doi:10.1016/j.compositesa.2010.10.005.

Yu, X. & Kwon, E., "A carbon nanotube/cement composite with piezoresistive properties," *Smart Mater. Struct.* 18 (2009): 1–5. doi:10.1088/0964-1726/18/5/055010.

Zhang, S., Lu, F. & Zheng, L., "Dispersion of multiwalled carbon nanotubes (MWCNTs) by ionic liquid-based Gemini pyrrolidinium surfactants in aqueous solution," *Colloid Polym. Sci.* 289 (2011): 1815–1819. doi:10.1007/s00396-011-2500-2.

# 27

# Transition Metal/Carbon Nanocomposites

Victor Kuncser

Petru Palade

Gabriel Schinteie

Florin Dumitrache

Claudiu Fleaca

Monica Scarisoreanu

Ion Morjan

George Filoti

## 27.1 Introduction

Being one of the most abundant elements in the universe, carbon plays a special role in both nature- and human-related activities. Three naturally occurring allotropes of carbon are usually mentioned, two of them being recognized to have different properties since antiquity. These are the well-known diamond and graphite. The different electronic and mechanical properties of diamond and graphite are due to the bonding nature between carbon atoms and the involved crystalline structure (King 1998, Terrones & Terrones 2003). Whereas diamond consists of a 3D lattice of tetrahedral coordinated atoms connected via $sp^3$–$sp^3$ σ bonds to form interlocked $C_6$ nonplanar rings, graphite consists of stacked planar layers of hexagons, formed by carbon atoms with trigonal coordination, connected by $sp^2$–$sp^2$ σ bonds and delocalized π bonds between unhybridized $p_z$ orbitals of the same plane atoms. If just one hexagonal layer is considered, it takes the name *graphene*. The strong 3D bonding (leading to a similar C–C distance of about 0.151 nm in all directions) makes diamond a good insulator (large electronic bandgap) and one of the hardest materials ever known. On the other hand, the strong in-plane bonding and relatively soft forces between the hexagonal layers (leading to a separation of about 0.335 nm between carbon successive layers and of only 0.142 nm for the C–C in-plane distance) makes graphite a semimetal with high softness and lubricity (easy relative shifting of the parallel layers). Finally, there is also the possibility of having an amorphous form of carbon (usually noted as a-C), with disordered structure that can be distinguished by XRD or diffusive reflectance IR Fourier transform. The a-C phase may contain different proportions of $sp^2$- and $sp^3$-hybridized carbon atoms. As any type of amorphous phase, a-C might be obtained by vapor condensation or ultrafast quenching of the corresponding liquid phase. There were different experimental and theoretical studies providing evidence that in the amorphous phase, depending on the preparation conditions, average coordination, average bond length, average energy per atom (e.g., relative to diamond), and average density (e.g., relative to diamond) can increase from values closer to those of a graphite-like structure to values closer to those of a diamond-like structure. For example, the average coordination increases from about 3 atoms (specific to the layered graphite) if the a-C is obtained by vapor condensation to an average value of 3.4 atoms (more appropriate to the 3D bonds in diamond) if the a-C is obtained by ultrafast quenching under megabar pressure (Tersoff 1988). Similarly, the average bond length increases from 0.147 to 0.151 nm, and the average density relative to diamond, from 0.62 to 0.86, in the two cases, respectively. However, the large possibilities of shifting the properties of amorphous carbon even outside of the limits of the two well-known structurally ordered phase, as well as the possibilities of obtaining it with different shapes and morphologies, confer to this carbon allotrope a high scientific and technological interest. Besides the disorder degree of their structure, two other important parameters, namely, the degree of $sp^3$ bonding and the relative amount of hydrogenated constituent a-C:H, can introduce differences between the various forms of amorphous carbon. An excellent review of the different forms of a-C, classified according to the mentioned parameters and discussed with respect to preparation, properties, and applications, was provided by Robertson in 2002. For example, among other forms, there is a metastable form of a-C with a significant proportion of $sp^3$ bonding, called diamond-like a-C (usually achieved as a disordered and isotropic thin film with no grain boundaries), and keeping some of the specific properties of diamond (e.g., hardness, elastic modulus, and chemical inertness).

With respect to the ordered species and in direct correlation with the layered structure of graphite, new allotropes of carbon were discovered quite recently.

## 27.2 Nanosized Allotropes of Carbon

Hence, in 1985, finite molecular cage-like structures in which every carbon atom is connected to three neighbors via $sp^2$ bonds (similar to the graphite case yet slightly distorted from trigonal planar coordination) were evidenced (Kroto et al. 1985). These molecular cage-like structures were called *fullerene*, with the most representative example of $C_{60}$, consisting of 60 carbon atoms. Evidently, an additional necessary topological concept should be introduced for this type of new materials, namely, the curvature of the surface. Accordingly, both graphene and graphite belong to the class of materials with zero curvature (obeying the Euclidian geometry), whereas fullerenes belong to the class of materials with positive curvature (obeying the spherical/ellipsoidal geometry). Another form of elemental carbon with each atom connected to three neighbors via $sp^2$ bonds but decorating surfaces of negative curvature was hypothetically assumed and denoted as schwarzite (Mackay & Terrones 1991).

As a matter of fact, since the typical hexagons specific to graphene cannot contribute to the positive curvature of the arrangement in $C_{60}$ (spherical geometry), such a molecule is formed by 12 pentagons with each one surrounded by hexagons (similar to a soccer nanoball with a diameter of 0.7 nm). It was proved in 1990 (Kratschmer et al. 1990, Taylor & Walton 1993) that $C_{60}$ molecules can crystallize in an fcc superarrangement, leading to an fcc supercrystal with a lattice parameter of 1.42 nm, usually known as fullerite. The idea of carbon cages can be generalized to a higher number of atoms (larger is the fullerene specie; lower is its abundance) where the spherical cages imply a specific numbers of rings with four to eight atoms (however, classical fullerenes consist of only pentagons and hexagons). The concentric packing of such larger fullerene gives rise to quasi-spherical nanoparticles known as onion-like nanostructures (Terrones & Terrones 2003). While graphitic onions represent metastable states with very interesting properties, they can have important applications.

Further on, in 1991, another allotrope of carbon was discovered, namely, the CNTs (Iijima 1991). CNTs are cylindrical carbon-based molecules of extremely high aspect ratio (lengths of million times higher than the nanometer-sized diameter) and remarkable structural and electronic properties, offering them a huge potential for applications. They can be formed by rolling up a few stacked graphene sheets at the same time (the rolling is just a virtual and intuitive process having nothing to do with the practical fabrication of the nanotube). According to the number of rolling layers, the CNT can be SWCNT or MWCNT. Structurally, an SWCNT (with the lowest diameter of 0.4 nm) is pictured by cutting out a rectangular region from a graphene sheet and a subsequent rolling into a cylinder (the situation is straightforward extended to MWCNT, with the number of walls corresponding to the number of rolled stacking graphene sheets). The way that the rectangular graphene layer is rolled up, namely, along what direction with respect to the honeycomb hexagonal lattice, defines another important property of SWCNT, namely, the chirality. In fact, the chirality of the CNT determines almost all its properties, including electronic, optical, and mechanical ones, as discussed in the following. Depending on chirality, there are three main categories of CNT, namely, (i) chiral, (ii) zigzag, and (iii) armchair.

It is worth mentioning that many of the properties of all the mentioned systems depend not only on the structural peculiarities of the carbon allotropes but also on an additional parameter, the dimensionality of the nanosystem. While both carbon nanotubes and cage-like nanostructures are obtained by suitable bending/curving a bidimensional graphene-like sheet to obtain tubes (cylindrical geometry) or cages (quasi-spherical geometry) with nanometer range diameters, one may assume all of the following types of dimensionalities specific to nanosized carbon allotropes (see also Figure 27.1): (i) 2D nanosystems consisting of one or a few stacked graphene layers with a total thickness of nanometer order and relative infinite lateral sizes (orders of magnitude larger), (ii) 1D nanosystems consisting of CNTs with a cylindrical diameter of nanometer order and a relative infinite length (orders of magnitude larger), and (iii) 0D nanosystems consisting of spheroidal cage-like nanostructures with nanometer order diameter (fullerenes and onions).

Intensive efforts have been expended in studying the electronic structure of 0D carbon allotropes (Dresselhaus et al. 1996). Concerning the electron transport properties, they are representative models for quantum dots, which currently attract the interest of both scientific and engineering communities. Electron transport through such structures is strongly influenced by single-electron charging and energy level quantization specific to fullerenes (Stone & Wales 1986). For example, single-molecule transistors based on individual $C_{60}$ molecules providing evidence of a novel conduction mechanism leading to nanomechanical oscillations of the $C_{60}$ molecule have been reported (Park et al. 2000).

1D CNTs present unique electric features imposed by the quantum confinement of electrons normal to the nanotube axis (e.g., the electrons propagate can only along the nanotube axis). They can have either metallic or semiconducting character, depending on their chirality. The direction of rolling the graphene sheet is essential because the energy spectrum of the graphene near the Fermi energy is different along different directions in the sheet. For example, zero bandgaps are present along a direction perpendicular to the armchair line while finite, but not too large bandgaps are present along a direction

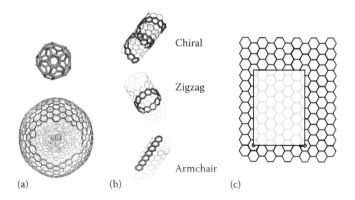

FIGURE 27.1 Carbon allotrope nanostructures of different dimensionalities: (a) 0D fullerene or onion-like, (b) 1D CNT, and (c) 2D graphene.

making 120° with respect to the direction of the armchair line (the chiral vector makes 30° with the armchair line). Thus, the electrons, which propagate only along the tube axis, will "see" a zero bandgap for armchair nanotubes and a finite bandgap for zigzag nanotubes. Accordingly, armchair nanotubes are metallic and the other two types are semiconducting, with typical values for the bandgap between 0.4 and 0.7 eV. On the other hand, the bandgap in semiconducting SWCNT opens inversely proportional to the tube diameter. Hence, SWCNTs of large diameters should have negligible bandgap and a metallic character, independent on their chirality. However, most of the CNTs of large diameter are of multiwall type. It is difficult to establish the conductive character of an MWCNT, but the metallic character seems to prevail also for such cases. The 1D structure of CNT leads in addition to an almost ballistic conduction. Ballistic refers to a conducting system in which there is no scattering of the electrons between the two contacting ends. As a consequence, they can carry a high density of current, without dissipating energy. The remarkable electrical properties of carbon nanotubes make them candidates for a large class of applications, especially electronic devices. There are opportunities for integrating nanotube electronics with other chemical, mechanical, or biological systems. However, not only the electrical properties of CNT are remarkable, but also their mechanical and thermal properties are. For example, it was revealed that MWCNTs possess a Young's modulus 100 times larger than that of steel (Treacy et al. 1996). Such an allotrope structure of carbon can be considered as 100 times stronger than steel and six times lighter. Composite materials containing MWCNTs can be obtained, with greatly enhanced mechanical properties. Owing to their suitable electronic and mechanic properties, CNTs (as well as graphene) can also be used in spintronics (spintronics deals with the manipulation of the electronic spin, as opposed to normal electronics, which relies on the electron charge) (Tombros 2008, Chen & Zhang 2013).

## 27.3 Transition Metal (TM)-Doped Nanosized Allotropes and TM/C Nanocomposites

Computer simulations and experimental works indicate that curvatures in the previously mentioned nanosized allotropes can be modified via different types of defects. Such defects, which can be pentagons, heptagons, rings with more atoms, or even more complex architectures, may drastically modify the electronic, magnetic, optical, and mechanical properties of the systems. Defects can be introduced in a more controllable way by doping the structure with convenient atoms. Accordingly, the doping of curved carbon-based nanostructures is a promising line of research, potentially leading to new carbon-based nanomaterials with remarkable properties to be used in future nanotechnology. For example, by doping fullerites with Cs or Rb, a superconducting behavior was reported below 33 K (Fleming et al. 1991). CNT junctions can be formed around the doping site as well as the functionalization of the nanotubes via different molecules. In case of B-doped CNT, boron atoms act as surfactants, avoiding the tube closure and inducing conducting properties in any type of CNT (Blasé et al. 1999). By doping CNTs with nitrogen atoms, a new local electronic state close to the Fermi level of the nanotube is introduced, hence resulting in a metallic character in both armchair and zigzag tubes (Czerw et al. 2001). There is also a growing interest in using CNTs in biosensor platforms and nanoscale electronic devices owing to the ability of functionalized CNTs to promote electron-transfer reactions with enzymes and other biomolecules (Guiseppi-Elie & Baughman 2002).

Large fullerenes and onion-like structures as well as CNTs have been proposed for encapsulating different types of elements and compounds in their hollow cavities. Radioactive materials can be introduced inside fullerenes or onion-like structures to avoid dangerous leaks in storing nuclear waste (Pasqualini 1997). Heavy metallic elements (Sn, Pb, Bi) (Ajayan et al. 1993, Dujardin et al. 1994) have also been introduced inside nanotubes.

If other nanostructures are introduced or formed in the inner cages of 0D carbon allotropes or 1D CNTs (either open or capped), for example, magnetic nanoparticles/internal cylinders, new nanocomposites which may have important applications (e.g., in the fabrication of magnetic inks and toners for xerography, catalysis, nanosensors, nanoactuators and nanoreactors) can be obtained. Oxides of TMs, for example, Co, Fe, Sn, Au, and Ag (Tsang et al. 1994, Sloan et al. 1997) have been introduced in CNTs, as well as metallic TM nanophases (Grobert et al. 1999). An important difference between such nanocomposites obtained by forming TM-based nanophases inside nanosized allotropes of carbon and those obtained by simple doping process resides in the fact that in the second case, mainly the properties of the carbon allotropes are changed by doping, whereas in the first case mainly the properties of the TM-based nanophases are changed because of their interaction with the carbon cage/tube/matrix. Such situations related to TM–carbon nanocomposites obtained by either bonding/embedding TM nanoparticles to/in nanosized allotropes will be considered in the following with a focus on the properties of the TM which are influenced by different types of interactions with the C constituents.

## 27.4 General Preparation Methods of Nanosized Carbon Allotropes and of TM/C Nanocomposites

There are some general methods for producing all types of nanosized allotropes, some of them used successfully for producing also TM/C nanocomposites. The most important preparation methods in this respect are the arc discharge of graphite electrodes (Ebbesen et al. 1994); electrolysis of graphite electrodes in molten salts (Hsu et al. 1995); sputtering/ablating from graphite targets, plasma- or laser-induced deposition (Logothetidis 1996, Thess 1996) (usually over metallic catalysts) (Ding et al. 2001); and different types of pyrolysis of hydrocarbons (Alexandrescu et al. 2005, Jäger et al. 2006). Characteristic of most of the mentioned methods is the fact that many nanosized allotropes might be obtained at once, the resulting product requiring a further separation or purification. However, even in this case, the contribution of the desired allotrope can be substantially increased by changing or adapting the device parameters or the processing conditions. An alternative way to produce TM/C nanocomposites is to start from TM nanoparticles (e.g., obtained by specific chemical or physical routes) and to connect/embed them to/inside a specific carbon-based nanophase. Some general processing methods will be briefly discussed in the following, to put in evidence the specificities of the main method (laser pyrolysis) used for processing the TM/C nanosystems reported in this work.

In the case of the arc discharge method, an arc is induced between two graphite opposite electrodes (face to face at millimeter distance, less than 10 mm in diameter, hundred amperes, and tenths of volts) placed in an isolated chamber and in an inert atmosphere (e.g., He). If fullerenes are desired, the He pressure should be around 100 mbar, whereas for CNTs, a pressure on the order of 500 mbar is required (CNTs grow into a deposit formed on the cathode). Usually highly crystalline micrometer-length MWCNTs are obtained but with a wide range of diameters and no unique shape, because of the violent arc reaction. The electrolysis method is very similar to the arc discharge, the main differences being related to the fact that the two electrodes are this time inserted inside a molten salt (for example, melted lithium chloride) and the parameters of the electrolysis reaction are very different. While the desired nanosized allotrope can be more homogeneous, the production yield is quite small.

The sputtering of a graphite electrode by an Ar plasma is a quite versatile method (industrially extended), suitable especially for obtaining C-based thin films (e.g., diamond-like a-C is usually obtained). In this case, a DC or radio-frequency plasma discharge is formed between the graphite target and the film substrate facing each other in a low-pressure Ar atmosphere (sometimes mixed with a low amount of methane or hydrogen, depending on the aim). The Ar ions are accelerated toward the target and sputter the C atoms, which are further deposited on the substrate, forming thin films of different morphologies and configurations, as controlled by the discharge parameters. TM/C nanocomposites might be easily obtained by this method via a cosputtering procedure from graphite and TM targets.

Sputtering of carbon atoms from graphite substrates can be induced also by laser irradiation (this specific process is called ablation). While the required power in this case is significant, very short pulses are used and, therefore, the procedure is usually known as pulsed laser deposition method. Pulsed excimer lasers are used to vaporize materials as an intense plasma extending further toward the substrate (the carbon ions incident on the substrate having an analogous energy as in a cathodic arc). The advantage of this method is reflected by its versatility at laboratory scale, including the possibility to control the energy of the carbon ions resulting from the ablation process (and hence the structure and morphology of the obtained carbon allotropes) via the ratio between the laser fluence and the area of the laser spot on the target. Depending on the desired allotrope, the substrate or the target should be kept at specific temperatures. For example, nanotubes are formed if the graphite target is placed inside a furnace at 1200°C (the laser vaporization method). If Ni or Co is added to the graphite target, SWCNTs can be obtained, while in most cases these are grown on metal-based nanoparticles (Thess et al. 1996). In fact, similarly as with the sputtering method, TM/C nanocomposites can be obtained by ablating composite targets.

CVD is the most popular and widely used method to obtain nanosized carbon allotropes (especially CNTs) because of its low cost and high production yield. It is based on the thermal decomposition of hydrocarbon vapors in the presence of a metal catalyst. From here, the specific name *thermal* or *catalytic CVD* is used to distinguish it from other types of CVD methods. It is a story more than 100 years old about obtaining carbon allotropes (starting with threads and fibers) via CVD methods, as excellently reviewed by Kumar and Ando (2010). For example, the process of growing CNTs by the CVD method involves passing a hydrocarbon vapor through a tubular reactor containing the catalyst at a high enough temperature (about 1200°C) to decompose the hydrocarbon. It is worth mentioning that both the hydrocarbon and the catalyst can be initially in solid, liquid, or vapor form, but at the reaction step the hydrocarbon is always in the vapor form, whereas the catalyst is in the form of dispersed very fine nanoparticles (usually metallic). The vapors decompose in contact with the hot nanoparticles in hydrogen (flowing away) and carbon, getting initially dissolved in the nanoparticle and further growing out in cylindrical shapes. Referring to the crystallinity, the arc discharge or sputtering/ablation methods are superior, while the CVD method is net superior with respect to yield, purity, and versatility in controlling the tube architectures (e.g., growing CNTs either as powders or thin films, desired architectures on patterned substrates, or aligned or entangled tubes).

Among the methods based on pyrolysis of hydrocarbons, a powerful method providing either simple carbon allotropes or TM/carbon nanocomposites is laser pyrolysis. While the synthesis mechanism is also based on the condensation of heated gaseous precursors (vapors), its specificity (in contrast to the previously described furnace heated-gas phase technique) is the well-delimited reaction zone induced by the focused beam of a continuous-wave $CO_2$ laser.

A schematic view of a versatile setup providing opportunities to process both C allotropes and complex systems of TM–carbon nanocomposites (as will be considered in the next sections) is presented in Figure 27.2. The system is based on a cross configuration with the reactant flow emerging in the reactor through a specific concentric nozzle system. Here, it is orthogonally intersected by the focused IR radiation from the $CO_2$ laser. Pyrolysis reaction is achieved in the small volume defined by the radiation-gases crossing, where a "flame" usually appears (a visible emission is supposed to result from the hot freshly nucleated particles). The confinement of gas precursors (for both carbon and TM species) toward the flow axis is achieved by a coaxial flow of an additional gas, which can be an oxidizer (e.g., air or $N_2O$) or a protecting one (e.g., Ar), depending on the desired TM/carbon nanocomposites. The nucleated particles are entrained by the gas stream and collected in a removable tank containing a porous filter at the exit of the reaction cell. The main process parameters are the nature of the gas (vapor) precursors, gas flow rates, pressure, laser wavelength, laser power, and specificity of the nozzle system.

The specific features of the $CO_2$ laser pyrolysis are related to the IR vibrational photochemistry. The process is based on the overlapping of the laser emission line at 10.591 μm with an absorption line of one or more gas precursors (Alexandrescu et al. 2005, 2010). The molecular mechanism involves the evolution of the vibrational states through collisional relaxation processes and finally the thermochemical reactions. In case of nonabsorbing gas precursors an additional substance called sensitizer (ethylene [$C_2H_4$] in this case) is used for transferring the energy to the precursors by collisions. As a final effect of the coupling between the laser radiation and the absorbing system, the focalized laser light acts as a well-localized heat source. It is just this

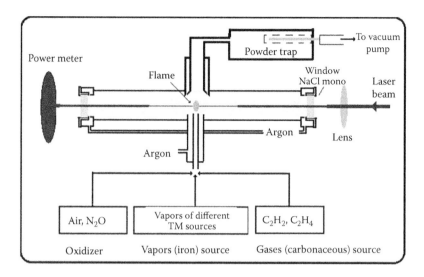

**FIGURE 27.2** Versatile setup for processing complex TM/carbon nanocomposites by laser pyrolysis.

well-defined reaction zone insulated by the cold and nonreactive inert gas coflow against any contact with the reaction chamber walls and avoiding the presence of highly energetic ionic particles which leads to a very clean growing process. Further on, the possibility of controlling the temperature gradients in a very small reaction zone (via the exposure time) allows the control of both the nucleation and growth rates. Owing to all these particularities as well as owing to the diluted nature of reactants in the reaction zone and the very fast cooling rates after exposure to laser beam, the resulting powder consists of very fine and loosely agglomerated nanoparticles (0D systems), mainly spherical and quite monodispersed in size.

If just carbon precursors (hydrocarbons in gas/vapor phase) are used, the obtained nanocarbon forms (0D, as specific to this procedure) present a multiscale organization with different structures and microtextures, in direct relation to the synthesis conditions (Herlin et al. 1998). At high laser powers (e.g., above 200 W), meaning high reaction temperatures, an onion-like structure consisting of a concentric arrangement of polyaromatic basic structural units (BSUs)—meaning a stack of two to three aromatic layers about 1 nm in diameter—will result. Carbon-based nanoparticles with partial structural order (at a distance of a few nanometers) and with diameters of tenths of nanometers are obtained, usually embedded in an amorphous phase or surrounded by a skin of amorphous carbon. In addition to such nanoparticles, amorphous-like nanoparticles (with local order lower than for the polyaromatic BSUs) with diameters ranging from nanometer to tenths of nanometers can also be obtained. Usually, the amorphous character of the samples is enhanced at reduced laser power. Also, by supplementarily introducing controlled amounts of oxidants (such as $N_2O$ or $O_2$) in the laser pyrolysis reactive mixture, carbon nanoparticles containing small fullerenes or nested fullerenic fragments can be obtained.

When precursors of TM are also used, interesting 0D nanocomposite materials can be synthesized. Typically, TM-based nanoparticles (either metallic or oxides) are formed, as encapsulated in mostly disordered or quasi-amorphous carbon matrix. It is worth mentioning that the morphology of the carbon matrix can be also controlled, in the sense that owing to the method particularities, the matrix itself can be composed of fine and spherical carbon-based nanoparticles. Because of the growing mechanism, the initially nucleated TM nanoparticles could have a catalytic role (Jäger et al. 2006) allowing the diffusion of the C atoms inside the nanoparticles and then the growing of either surrounding skin layers of carbon or spherical amorphous-like carbon nanoparticles. The situation is even more complex in case of using an oxidizer, when oxygen species should also diffuse in the system. The TM/carbon nanocomposites obtained by this method open a novel direction of investigation, concerning both fundamental aspects related to nucleation and growing processes, interphase atomic diffusion, new phase compositions obtained in nonequilibrium conditions, etc., as well as technological aspects. For example, if iron–carbon nanocomposites are envisaged, new information can be added to the traditional iron–carbon technologies related to Fe–C relationship in different steels. Iron carbides are also good candidates for applications in magnetic-recording media, whereas Fe or Fe oxide nanoparticles passivated by carbon shells could find important biomedical applications, mainly if they are dispersed in a liquid media, forming a so-called ferrofluid (Alexandrescu et al. 2008, Filoti et al. 2009). On the other hand, $TiO_2$/carbon nanocomposites consisting of $TiO_2$ nanoparticles surrounded by carbon-based shell/matrix of large specific surface can have important applications as catalysts, owing to the possibilities of controlling the phase composition, the particle size distribution, and the specific morphology of the powder-like or supported nanocomposite. Further on, an additional functionality can be added to such a catalyst if magnetic nanoparticles (e.g., iron based) are formed in the $TiO_2$/C composition for the catalyst recovery (even by a weak and nonhomogeneous magnetic field) after the reaction.

As mentioned previously, the main technological applications of such TM/C nanocomposites come mainly from the specific behavior of the TM-based nanoparticles, as influenced by either the interaction with the carbon atoms (through diffusion mechanisms or formation of new mixed phases) or the specific morphology of the surrounding nanosized carbon component. Therefore, except for usual structural and morphological characterization techniques and specific characterization methods related to the carbon phases, specific methods for the extensive characterization of the TM nanosized phases (concerning both microscopic mechanisms and macroscopic properties) are also required.

## 27.5 Characterization Techniques of Carbon Allotropes and TM/C Nanocomposites

Structural and morphological aspects of such complex systems can be directly obtained by diffraction techniques (with the most usual, XRD) and TEM-related techniques. Although XRD provides information on the crystalline structure and structural phase composition in a well-crystallized sample, real difficulties are emphasized in the nanostructured C allotropes, where also the crystallinity might be substantially reduced. For example, in case of TM/C nanocomposites obtained by laser pyrolysis, only the much better-structured TM phases are observed (e.g., in case of not too fine TM-based clusters/nanoparticles), whereas the carbon phases might be separately considered via specificities in the radial distribution function, much weaker in the diffraction pattern. On the other hand, HRTEM is able to provide information on the nanoparticle morphology (including the specific shape of the carbon-based and TM nanoparticles with or without core–shell structure, gradient of crystallinity, and interfacial aspects), whereas SAED provides information on the local structure of the involved nanophases. Electron energy-loss spectroscopy is a preferred method for deriving the $sp^3$ fraction in carbon-based samples as well as the spatial distribution of elements in TM/C composites. Finally, energy-dispersive X-ray (EDX) spectroscopy provides information on the overall sample composition, of high interest in TM/C nanocomposites. XPS provides information on the carbon-involved bonds and on the valence state of the TM, as well as on the sample composition, with high surface sensitivity. However, Raman spectroscopy is the best technique for obtaining the detailed bonding structure of carbon allotropes (Dresselhaus et al. 1996, Baibarac et al. 2013). The hydrogen content and the C–C and C–H bonding in hydrogen-containing allotropes can be analyzed via IR spectroscopy and NMR techniques. ZFC-FC curves, magnetization curves ($M(H)$), and hysteresis loops are used for observing the magnetic behavior of TM/carbon nanocomposites, of high importance in most practical applications.

It is worth mentioning at this point that the magnetic response of any system depends on two main parameters, namely, its spontaneous magnetization and its time-averaged projection along the field direction, related to the so-called magnetic relaxation phenomena. In case of magnetic nanosized systems consisting of magnetic nanoparticles (e.g., TM-based nanoparticles dispersed in the carbon matrix), both parameters are very sensitive to the so-called size effects. On the one hand, the spontaneous magnetization (or similarly, the magnetic moment of the nanoparticle) depends on the particles' composition and the disorder degree of their structure (both influenced by the synthesis conditions) and the diffusion of the carbon atoms at the particle surface. Furthermore, the magnetic relaxation phenomena depend on the particle size and anisotropy, both parameters being also influenced by processing conditions and interdiffusion aspects. Various size effects and specific characterization techniques of magnetic nanocomposites were described in detail by Kuncser et al. (2014). At this point, it is worth mentioning that information about the spontaneous magnetization of the system can be obtained from the experimental magnetization at saturation in the low-temperature range, and the information on the anisotropy energy is derived from the magnetization reversal at low temperature, while the information about the magnetic relaxation phenomena results from the evolution of the ZFC-FC curves (for example, the blocking temperature which is related to the anisotropy energy or magnetic volume is obtained from the maximum of the ZFC curve, whereas the volume dispersion is reflected by the branching process of the ZFC-FC curves) or from the temperature evolution of the coercive field. The information on the nanoparticle interactions is included in the evolution of the blocking temperatures versus the time measurement windows, and finally, the information about the presence of specific magnetic couplings at the interface can be obtained from the shifts of the hysteresis loops or from their increased coercivities. All the mentioned data obtained by magnetic measurements (using field- or temperature-dependent magnetization curves) have a double impact: (i) illustrate the magnetic response of the system with respect to a desired application and (ii) contribute substantially to an indirect characterization of the TM/carbon nanocomposites, being connected to morphologic, structural, magnetic phase composition, interfacial interactions, and atomic interdiffusion phenomena.

Finally, useful information on phase composition, local electronic configurations, and magnetic interactions including relaxation phenomena in TM nanocomposites is provided by Mössbauer spectroscopy (MS) (Greenwood & Gibb 1971, Thomas & Johnson 1986, Kuncser et al. 2013, 2014). The most convenient Mössbauer isotope with respect to the spectral resolution and appearance in magnetic phases of technological interest is $^{57}$Fe. MS is based on the resonant emission/absorption of the γ radiation without loss of energy due to the recoil of the nucleus. Owing to this effect, this spectroscopy becomes similar to any optical emission spectroscopy, with the only technical difference that the involved radiation is much more energetic. In classical configuration, the tuning of the wavelength is realized by the Doppler effect. The spectrum consists of the intensity of the absorbed/reemitted radiation (following Mössbauer events) versus the relative velocity between an oscillatory moving radioactive source and the absorber. A nuclide with identical local configuration as in the source would absorb the incident radiation strictly at zero relative velocity between source and absorber. Different local configurations of Fe in the sample give rise to tiny energy shifts or splitting of the nuclear levels, owing to the hyperfine interactions of the nucleus with the specific electron surroundings. One (singlet), two (doublet), or six (sextet) resonant absorption lines can appear for possible electron configurations of the $^{57}$Fe nuclide in the simplest cases. These specific patterns of the Mössbauer spectrum provide quantitative information on the specific hyperfine parameters: isomer shift (δ), quadrupole splitting (Δ), and hyperfine magnetic field ($B_{hf}$). The first two Mössbauer patterns are specific to nonmagnetic or paramagnetic/superparamagnetic Fe phases/configurations, whereas the third one is specific to Fe configurations presenting magnetic order. The isomer shift contains information about the valence and spin state (oxidation state of Fe, charge delocalization effects, etc.), whereas the quadrupole splitting is related to local electric field asymmetries induced by both the electronic surroundings and specific crystal fields. The hyperfine magnetic field is proportional to the iron net magnetic moment and is antiparallel to it (when provided by spin-only contribution). Hence, the averaged hyperfine magnetic field shows similar evolution with the net magnetization of the observed phase. If this is analyzed versus temperature, reliable information on magnetic relaxation phenomena can be obtained by simply following either the broadening of the sextet lines or the transition from the sextet, specific to the magnetically ordered state, to a doublet/singlet, specific to a paramagnetic/superparamagnetic state (Kuncser et al. 2007). Moreover, in the magnetic ordered state, the relative intensity of the second and fifth Mössbauer lines of the Mössbauer sextet is influenced by the direction of the local magnetic moment (Fe spin) with respect to the direction of the incident γ radiation. For randomly oriented moments (as in the case of powdered samples, without magnetic texture), this ratio is theoretically 2. It is worth mentioning that at variance with the magnetic measurements, all the above extensive information can be evidenced for each Fe configuration or phase in the sample, making $^{57}$Fe MS an extremely powerful local technique for analyzing electron configurations and local magnetic interactions in a wide range of samples containing Fe, as most of the TM/carbon nanocomposite systems to be presented in the following sections are. Lastly, the Fe phase composition in a sample is currently obtained from the relative spectral area of the Mössbauer patterns in the low-temperature spectra (for avoiding thermally induced effects), whereas magnetic relaxation effects are obtained by following the evolution of the spectra with temperature. Each Fe phase is characterized by well-defined hyperfine parameters, giving the possibility of its recognition from the Mössbauer spectra. It is worth mentioning the large meaning of the Fe phase in MS, not only metallurgical phases but also local Fe configurations being distinguishable (e.g., there are spectral differences between an Fe configuration with eight surrounding Fe atoms and that with seven surrounding atoms and an eighth different atom in the same bcc Fe phase, as well as between an $Fe^{3+}$ ion with tetrahedral oxygen coordination and that with octahedral oxygen coordination in the same Fe oxide phase). However, in case of less-ordered or nanosized systems with intrinsic defects (at least with different surroundings and symmetries concerning bulk and surface positions), there may appear a gradual change in some hyperfine parameters, giving rise to broader patterns which can be suitably fitted mainly by a distribution of hyperfine parameters, with possible local maxima corresponding to average specific local configurations. Direct applications of magnetic and MS measurements in the characterization of different TM/C nanocomposite systems will be subsequently presented.

## 27.6 Metallic Nanoparticles in Mesoporous Carbon for Catalytic Applications

Ordered mesoporous carbon (OMC) can be obtained as the carbon replica of the hexagonally ordered silica with large uniform mesopores obtained at the Michigan State University (MSU-H) by sucrose infiltration inside the MSU-H pores and subsequent heat treatment under argon for precursor carbonization, followed by silica elimination using hydrofluoric acid. Further on, TM/C nanocomposites consisting of Fe/Au nanoparticles dispersed in OMC can be obtained by the impregnation of OMC with corresponding metal salt solutions ($FeCl_3$ and $HAuCl_4$ in this case), the process being done under protective argon atmosphere. To analyze the effect of the OMC matrix on the metallic nanoparticles, Mössbauer spectra were collected at different temperatures (Figure 27.3a).

The Mössbauer spectrum collected at the lowest temperature (5 K) consists of three magnetic components. The most external one, with a relative area of about 15%, has specific hyperfine parameters (i.e., a hyperfine magnetic field of 46 T and an isomer shift of about 0.35 mm/s referred to as $\alpha$-Fe at room temperature) attributable to a mixture of Fe oxide phases (magnetite and maghemite). The next inner relatively broad sextet, with an almost equal relative spectral area of about 15%, has hyperfine parameters, including a hyperfine magnetic field of 33.4 T and an isomer shift of about 0.0 mm/s (versus $\alpha$-Fe at room temperature), corresponding to metallic Fe. Finally, the most inner sextet, with a relative spectral area of about 70%, has broad absorption lines indicating a distribution of iron positions and consequently a distribution of hyperfine fields. By the specific shape of the hyperfine magnetic field distribution (shown on the right side of the spectra) with a local well-shaped maximum at about 26 T, it is clearly assigned to a defected cementite-like phase ($Fe_3C$). However, Fe configurations with different C neighbors are also covered by the distribution. All subspectra become increasingly broader at higher temperatures, pointing to magnetic relaxation processes typical to nanosized nanoparticles. The external sextet specific to the Fe oxide phase collapses to a central doublet with quadrupole splitting of about 0.6 mm/s and an isomer shift of 0.3 mm/s at about 200 K, indicating a blocking temperature of 150 K, as deduced along the algorithm described by Kuncser et al. (2007, 2014). The Mössbauer patterns of Fe and cementite phases both collapse in a similar manner (but much more slowly than the sextet belonging to the Fe oxide phase does), indicating a blocking temperature of about 300 K. By corroborating all the Mössbauer information, we may assume that two types of nanoparticles are formed in the OMC, namely, very fine Fe oxide nanoparticles (with an average size of 5–6 nm as resulting from the specific blocking temperature and

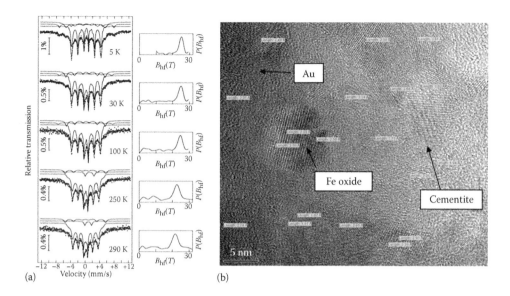

**FIGURE 27.3** (a) Mössbauer spectra obtained at different temperatures for the Au/Fe/OMC sample (on the right side of the spectra are the hyperfine field distributions at the Fe nucleus) and (b) HRTEM image of the same sample.

the algorithm described by Kuncser et al. (2007, 2014) and much larger core–shell nanoparticles with an average size of about 16–20 nm, consisting of a metallic Fe core and a concentric cementite-like shell, as a result of the penetration of the C atoms from the matrix. However, no indication of Fe–Au configurations results from the Mössbauer spectra, pointing clearly to the fact that very fine Au clusters precipitate independently from the Fe-based nanoparticles, as it was also proved by the HRTEM data shown in Figure 27.3b. It is worth mentioning that the metallic nanoparticles dispersed onto carbon-based supports of high surface area are very useful in improving the catalytic reactions and the performances of LIBs and hydrogen storage materials (Wang et al. 2011, Palade et al. 2012, Blanita et al. 2015) and, therefore, studies concerning information on the TM phase content, particle size, and morphology, as well as on atomic interdiffusion and impurification at the particle surface, become of significant importance.

## 27.7 $Fe_xO_y/C$ Nanocomposites Obtained by Laser Pyrolysis

Fe-based oxides/C nanocomposites can be obtained by laser pyrolysis if vapors of the Fe precursors are mixed with an oxidizing agent (e.g., air, in the simplest case). A typical pyrolysis setup, as schematically described in Figure 27.2, was used. Accordingly, the pyrolysis process was initiated by a continuous-wave $CO_2$ laser radiation (80 W maximum output power, $\lambda = 10.6$ μm) crossing orthogonally the gas flows emerging through two concentric nozzles. The gas mixture which contains iron pentacarbonyl vapors and air, carried by the $C_2H_4$ gas sensitizer, was admitted through the central inner tube. The reactive gas flow was confined to the flow axis by a coaxial Ar stream (the outer tube). Two nanocomposite samples will be reported, as obtained by decreasing the laser power from 55 W (sample S1) to 35 W (sample S2). All the other parameters, namely, the reactor pressure, the relative flow of air, and the flow of ethylene/precursor, were maintained at 300 mbar, 70 sccm, and 145 sccm, respectively.

XRD investigations of the considered nanopowders (Figure 27.4) evidence broad diffraction maxima which may be indexed in the cubic system and can be attributed to either the maghemite ($\gamma$-$Fe_2O_3$) or the magnetite ($Fe_3O_4$) phase, with slightly lower lattice parameters relative to the standard values ($a = 0.8351$ nm for maghemite and $a = 0.8396$ nm for magnetite). The broadness of the diffraction lines are clearly associated not only to the structural/crystallographic disorder but also to the very small average size of the Fe oxide nanoparticles (rough estimations of the crystalline coherence length indicates average

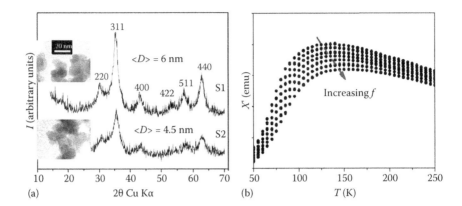

**FIGURE 27.4** XRD data of samples S1 and S2 with corresponding TEM images (a) and temperature-dependent AC magnetic susceptibilities at different frequencies *f* of the exciting magnetic field for sample S1 (b).

sizes in the range of 4–10 nm, with the particles obtained at higher laser power being slightly larger). There is no evidence for diffraction picks coming from the C allotropes, suggesting the presence of an amorphous-like carbon matrix.

The TEM analysis reveals very slightly aspherical nanoparticles (form factor $k = 0.86$), enough to induce uniaxial magnetic anisotropy but to consider an almost spherical morphology. The histograms of the Fe oxide particles' size distribution point for average diameters of 6 nm in S1 and 4.5 nm in S2 (see also Figure 27.4a). On Figure 27.4b are shown temperature-dependent magnetic susceptibility data obtained for sample S1 under excitation with a small alternative magnetic field, the in-phase magnetization being represented at different frequencies *f* from 11 Hz (topmost curve) to 8999 Hz (lowest curve). According to the methodology specific to systems of nanoparticles described by Kuncser et al. (2014), the blocking temperature can be obtained from the maximum of the AC susceptibility data following a ZFC procedure. While the blocking temperature depends on the specific time window of the investigation method (inverse of the AC frequency in the present case), there is an expected corresponding temperature shift of the blocking temperature versus the frequency *f*, which is clearly observed in the figure. Based on the Néel relation and the definition of the blocking temperature applied for measuring time windows inversely proportional to the AC frequency in systems of noninteracting nanoparticles (Kuncser et al. 2014), an average magnetic size in the range of 6–9 nm is obtained for the magnetic nanoparticles (the specific anisotropy constant of bulk magnetite was assumed). This magnetic result supports the assumption for the formation of fine Fe–O nanoparticles, with nonsignificant carbon inclusions, but with an enough thick shell of amorphous carbon to cut the magnetic dipolar interactions specific to simple magnetic nanopowders.

The Mössbauer spectra obtained for both samples at two different temperatures are shown in Figure 27.5. The 80 K spectra consist of very broad magnetic patterns which are still broadening at higher temperatures, suggesting enhanced magnetic relaxation regimes existing already at 80 K, specific to Fe–O nanoparticles of quite low magnetic anisotropy energy. It is worthy to mention that the spectra collected in the magnetic frozen regime at 5 K (not shown) consist of only one magnetic pattern, with still broad lines owing to the defective structure and the consistent presence of Fe positions at the nanoparticle surface. The average hyperfine magnetic field corresponding to the 5 K magnetic sextet is about 52 T, namely, between specific values for magnetite and maghemite (hence supporting a gradient-like structure of magnetite/maghemite per nanoparticle). Accordingly, both the broad external sextet and the central pattern observed in each spectrum collected at higher temperature were fitted by hyperfine field distributions being assigned to a bilobar distribution of iron oxide nanoparticles. The Mössbauer-derived blocking temperature of about 40 K (the corresponding pattern is already collapsed at 80 K) assigned to the finest nanoparticles infers an average size of 2–3 nm. The blocking temperatures of

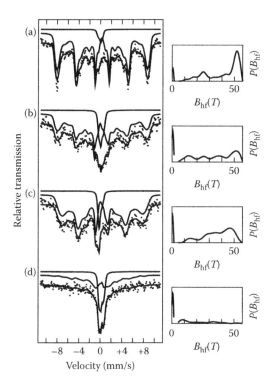

**FIGURE 27.5** Mössbauer spectra obtained for sample S1 at (a) 80 and (b) 160 K and for sample S2 at (c) 80 and (d) 160 K. The corresponding hyperfine magnetic field distributions are shown on the right side of each spectrum.

the largest nanoparticles as deduced from the collapsing behavior of the most intense external sextet are about 200 K for sample S1 and 120 K for sample S2. Hence, average magnetic sizes in the range of 6–8 nm for sample S1 and 4–5 nm for sample S2, in good agreement with TEM measurements, were obtained (however, the TEM data were not able to evidence the contribution of the finest nanoparticles). It is worth mentioning that the MS data gave evidence of quite pure Fe–O nanoparticles (without C inclusions), which corroborated the TEM and AC magnetic susceptibility data, inferring quite sharp interfaces between the magnetic core and the rather thick carbon shell, leading to negligible magnetic dipolar interactions. Such specific core (iron oxide)–shell (amorphous carbon) nanocomposite systems could find different applications in biomedicine (MRI drug delivery, AC magnetic field–assisted cancer therapy, etc.). Useful ferrofluids can be also obtained by their dispersion in different liquid carriers (Alexandrescu et al. 2008, Filoti et al. 2009).

## 27.8 Fe/C Nanocomposites Obtained by Laser Pyrolysis

Fe-based nanocomposites obtained by dispersing metallic Fe nanoparticles in different matrices are of peculiar interest; compared to Fe oxide, metallic Fe presents more convenient magnetic properties for certain applications (e.g., a much higher spontaneous magnetization and lower coercive and saturation field). However, to prevent the easy oxidation of Fe nanoparticles, most of the applications involve core–shell systems in which the metallic core is covered by thin shells of corrosion-resistant compounds/noble metals (Crisan et al. 2006) or carbides (Morjan et al. 2012). Laser pyrolysis is a convenient method of producing such Fe nanoparticles encapsulated in nonmagnetic carbon shell preventing the oxidation and reducing the interparticle interactions if suitable cautions are taken to avoid the Fe oxidation process during the synthesis (Schinteie et al. 2013). In this respect, the same general preparation procedure as previously reported was used, with the difference of using a triple–concentric nozzle system. The vapors of $Fe(CO)_5$ are carried by an ethylene flux pass through the central nozzle while a mixture of ethylene/argon (50 vol%) goes through the middle nozzle with both sensitizing and oxygen-protecting aims. Finally, an additional Ar flux flows through the outer nozzle for ensuring the lamellar flow of the reactive mixture and for additional protecting purposes. Two samples, P1 and P2, are reported here, as obtained with ethylene/Fe pentacarbonyl ratios of 10 and 33 vol%, both central and middle nozzle fluxes of 65 and 43 sccm, and power laser of 75 W, respectively. The pressure in the reaction chamber was maintained at 650 mbar and the temperature in the flame was about 690°C.

The elemental composition of samples, as evaluated by EDX spectroscopy, are 39 and 39 at% of Fe, 54 and 46 at% of C, and 7 and 15 at% of oxygen for samples P1 and P2, respectively. XRD data (Figure 27.6a) show the presence of both metallic Fe and Fe–C ($Fe_3C$ and $Fe_7C_3$) species and give some hints of the presence of a more ordered type of diamond-like carbon. There is no evidence of iron oxide phases, meaning that either Fe oxide clusters are very fine or of low crystallinity, or that the oxygen reported by EDX is localized on the carbon phase. HRTEM pictures (Figure 27.6b) give evidence of the formation of metallic Fe and $Fe_3C$ nanoparticles of average size of less than 10 nm, surrounded by slightly ordered carbon shells of nanometer order thickness.

Hysteresis loops collected at different temperatures for the two samples (Figure 27.7a and b, for samples P1 and P2, respectively) show a decrease in the coercive field versus temperature. The dependence of the coercive field on $T^{1/2}$, as shown in Figure 27.7c, is quite linear, suggesting the typical magnetic relaxation behavior toward the superparamagnetic regime of noninteracting magnetic nanoparticles, respecting the low $H_C = H_0[1 - (T/T_B)]^{1/2}$ (e.g., Kuncser et al. 2014). Accordingly, maximum values for the coercive fields $H_0$, in the magnetic frozen regime, of 1100 Oe in P1 and 660 Oe in P2 are deduced as well as blocking temperatures of more than 600 K in P1 and 400 K in P2. While the blocking temperature deduced this way is not conclusively related to a specific measurement time window, it has to be considered just qualitatively, to underline a noticeably higher average anisotropy energy per particle in sample P1 compared to P2. Taking into account also the much higher average anisotropy constant $K$ in P1 ($K$ being proportional to $H_0$), it might be expected that the corresponding higher anisotropy energy (expressed as the product $KV$, with $V$ as the average magnetic volume) leads to not very different average volumes of the magnetic nanoparticles in the two samples. Concerning the values of the spontaneous magnetization (deduced from the magnetization at saturation at 5 K), these are in the range of 100–110 emu/g per magnetic component, being intermediate between the bulk values for metallic Fe, cementite, and magnetite. In conclusion, the magnetic measurements stand for the presence of both Fe oxide and metallic Fe/cementite nanoparticles in a strong magnetic relaxation regime at room temperature, for both samples.

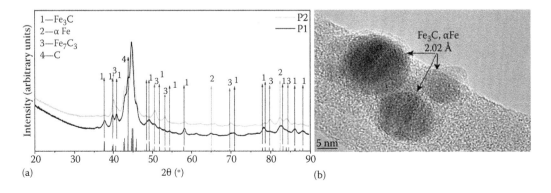

**FIGURE 27.6** XRD data of the two samples P1 and P2 (a) and HRTEM picture of sample P1 (b).

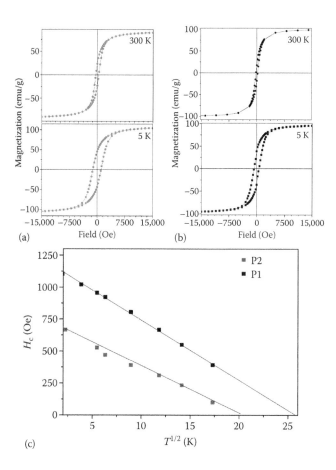

**FIGURE 27.7** Hysteresis loops collected at different temperatures for samples (a) P1 and (b) P2 and (c) the peculiar dependence of the corresponding coercive fields versus $T^{1/2}$.

Finally, the Mössbauer spectra collected at different temperatures for the two samples (as presented in Figure 27.8) provide the most comprehensive (structural, compositional, and magnetic) information. The spectra obtained at the lowest temperature of 4.5 K (in the magnetic frozen regime) evidence three magnetic patterns in sample P1 and four magnetic patterns in sample P2. By their specific hyperfine parameters, these correspond to the following Fe phases: (i) the most external sextet, with an average hyperfine field of about 48 T, to magnetite; (ii) the next inner one, with a hyperfine field of 34.5 T, to metallic Fe; (iii) the only internal one in sample P1, with an average hyperfine magnetic field of 25.2 T, to $Fe_3C$; and (iv) the most internal one, appearing only in sample P2, with an average hyperfine magnetic field of 18.2 T, to $Fe_7C_3$. It is worth mentioning that both Fe carbide phases can be accounted also by only one hyperfine magnetic field distribution, which will be predominantly monolobar (local maxima at about 25 T) in sample P1 and bilobar (with local maxima at about 25 T and 18 T) in sample P2. The relative contents of the three Fe phases (looking at the Fe carbide species as a unique Fe–C phase), as deduced from the spectral relative areas, are 76% and 65% Fe carbide, 11% and 27% Fe oxide, and 13% and 8% metallic Fe, for samples P1 and P2, respectively.

The linewidths of the magnetic components become broader at increasing temperatures and start to collapse in a manner specific to magnetic relaxation phenomena for nanoparticulate systems. From plotting the evolution of the hyperfine magnetic fields specific to the three Fe phases (see Figure 27.9a and b), what result straightforwardly are blocking temperatures of about 30 K for the Fe oxide nanoparticles in P1 and 50 K for those in P2 (Kuncser et al. 2007, 2014), in line with nanoparticle average sizes in the range

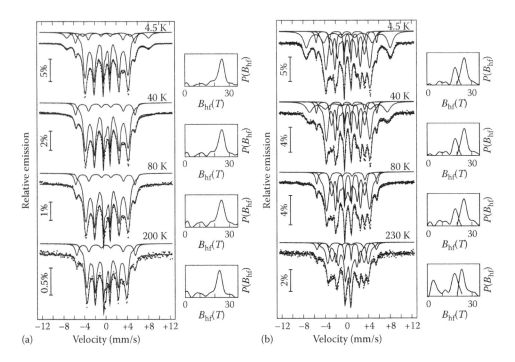

**FIGURE 27.8** Mössbauer spectra collected at different temperatures for samples P1 (a) and P2 (b). The hyperfine magnetic field distribution is presented on the right side of each spectrum.

of 3–4 nm. The relaxation regime for the metallic nanoparticles is much reduced, suggesting evidently larger nanoparticles. In addition, the linear decreases versus temperature for the reduced hyperfine field in the two metallic phases, with much higher slope for the Fe carbide nanoparticles, clearly stand for (i) the collective excitation regime up to room temperature and (ii) the lower average sizes of the iron carbide nanoparticles compared to the metallic Fe nanoparticles. A comprehensive analysis of such systems providing information on relative and absolute contents of each phase including nonreacting carbon (in the shell of all nanoparticles), anisotropy energy, anisotropy constant, and average size for each type of nanoparticle was presented by Schinteie et al. (2013). As a general conclusion, by using the triple–concentric nozzle system and adequate protecting gas fluxes, a significant amount of metallic core/carbon shell nanoparticles can be obtained. However, the oxidation process cannot be completely removed and a small amount of Fe oxide nanoparticles (always the finest ones with average size smaller than 3–4 nm) is also present. By optimizing the synthesis parameters, the relative amount of Fe oxide phases can decrease down to a few percent. In these conditions, the main products of the laser pyrolysis process are Fe carbide nanoparticles (6–7 nm average diameter) covered by thin shells of less-ordered (diamond-like) carbon as well as larger nanoparticles (14–16 nm) consisting of an Fe core surrounded by the corresponding carbon shell. The mechanism of their formation consists of the penetration of C atoms into the metallic clusters within a skin depth of about 3 nm, leading to the full penetration for nanoparticles of about 6 nm (with the formation of Fe carbide nanoparticles with an increased content of C if they are thinner) and just a partial penetration in case of larger nanoparticles. The layer of penetrating carbon (called reacted carbon) is different from the shell layer of nonreacted carbon, most probably each particle possessing a gradient-like carbon composition. The most desired nanoparticles are the ones with metallic Fe core, meaning further improvements of the synthesis parameters to reduce the carbon penetration skin and to shift the size of most magnetic nanoparticles toward needed values of 10–12 nm. The type of the nonreacted carbon shell can be also modified by synthesis parameters or subsequent processing.

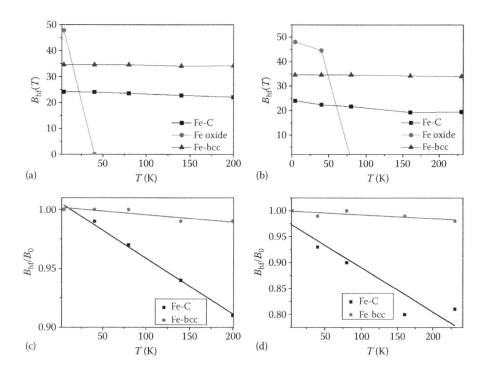

**FIGURE 27.9** The temperature-dependent evolution of the hyperfine magnetic fields assigned to the three Fe configurations for samples (a) P1 and (b) P2 and the detailed evolution of the reduced hyperfine magnetic field corresponding to the metallic Fe and Fe–C phases for samples (c) P1 and (d) P2.

## 27.9 Fe–Ti–O/C Nanocomposites

There is currently a growing interest related to oxide semiconductor photocatalysts, with respect to controllable synthesis, reaction mechanisms, and applications. Among them, $TiO_2$ nanoparticulate systems present many advantages related to a peculiar electronic structure leading to suitable bandgap (3.2 eV), high photoactivity, and chemical inertness (O'Regan & Gratzel 1991, Palmisano et al. 1994, Alexandrescu et al. 2008). While the photocatalytic activity of such systems depends on both the crystalline structure and the specific morphology (including shape, size distribution, dispersion, surface penetrability, and hydrolysis) and in case of supported catalyst on the specific interaction with the support, special synthesis procedures assuring a large variety of the above parameters in an "as much as" controlled way are desired. Among them, laser pyrolysis can be extremely useful because it promotes the controlled formation of nanoparticulate systems with different structures and morphologies. In this respect, $TiO_2$/carbon nanocomposites can be easily obtained by starting from $TiCl_4$ liquid precursor in the presence of a carbon donor. An additional idea, along with the previously presented results, was to promote the formation of Fe-based magnetic nanoparticles, directly/indirectly connected to the $TiO_2$ ones, to recover the catalyst via nonhomogeneous applied fields (Fleaca et al. 2013). That might be also simply obtained by the simultaneous use of the $Fe(CO)_5$ and $TiCl_4$ precursors. Hence, two samples will be further considered as obtained via the usual pyrolysis setup with three concentric nozzles and the following specificities: $Fe(CO)_5/C_2H_4$ vapors are introduced in the reaction chamber through the central nozzle, whereas $FeCl_4/C_2H_4$/air gas mixture (345 sccm) is introduced by the next nozzle. A strong Ar flux (2000 sccm) is introduced through the third (external) nozzle for assuring the laminar flow and the shielding of the reactive mixture. The laser power was 450 W and the pressure in the reactor chamber was 400 mbar (Fleaca et al. 2014). The first sample, denoted as TO, was obtained without the central flux containing Fe pentacarbonyl, whereas the second one, denoted as TOF, was obtained with a central flux of 30 sccm. The EDX analysis of the two samples showed 0 and 6 at% Fe, 30 and 20 at% Ti, 60 and 44 at% O, and 10 and 30 at% C. The XRD analysis pointed to the lack of oxidic Fe phase in TO and a low amount of metallic Fe and cementite in sample TOF.

Concerning the two main crystallizing structures for the $TiO_2$ compounds (anatase and rutile), the XRD data of sample TO (Figure 27.10) have shown predominantly the photocatalytic active phase of anatase. This phase is also dominant in sample TOF, but in a lower relative amount (90% in TO and 80% in TOF). The structural coherence length approaching the mean crystalline size was in the range of 25–30 nm for anatase and 30–40 nm for rutile.

Different HRTEM images obtained for the two samples clearly prove the formation of nanoparticles of average size of about 30 nm of anatase, with thin coating layers (2–3 nm) of carbon (most probable turbostratic). The TEM images of the TOF sample evidence besides the larger $TiO_2$ nanoparticles a few nanometer-sized nodules which were assigned to metallic/carbidic Fe.

To evidence better the Fe phase composition in the TOF sample, Mössbauer spectra were acquired at different temperatures (Figure 27.11). It can be observed that the low-temperature spectrum consists of three magnetic sextet patterns and two paramagnetic central ones. The three magnetic sextets were assigned by their specific hyperfine parameters as follows: (i) the most external sextet, of relative spectral area of 22%, to a Fe oxide phase (magnetite/maghemite); (ii) the next inner one, with a spectral area of 16%, to metallic Fe; and (iii) the most inner one, with a spectral area of 18%, to iron carbide. Finally, the two central doublets with an overall spectral contribution of about 35% are assigned to very small and poor crystallized clusters (less than 1–2 nm in size) of Fe oxide or Fe carbide or even to dissolved Fe in the carbon shell. The magnetic patterns of the metallic phases evolve very slightly with temperature, inferring quite large sizes (some 20–25 nm, except for Fe oxide, with average size of 3–4 nm). It is worth noticing that the mentioned percentages refer to the total Fe configurations, which are in turn just 6% of the total atoms in the sample. Therefore, the contribution of the relatively large Fe-based nanoparticles to the X-ray patterns is negligible, as well as the probability to catch them in the TEM pictures.

Magnetic hysteresis loops obtained for both the TOF and TO samples are shown in Figure 27.12. While the magnetization and the coercive field (including its evolution with temperature) in sample TOF are expectable owing to a low amount (a few atomic percent and about 10 wt%) of metallic Fe and Fe carbide phase, the presence of the magnetic signal of finite coercive field in sample TO is quite unexpected and can be interpreted only in terms of the specific magnetism appearing in diluted magnetic oxides/semiconductors without TM doping (Nistor et al. 2013).

It is worth mentioning that the elemental difference between the two samples (due to the absence/presence of Fe) is the lower amount of anatase phase and the higher amount of C in the Fe-containing

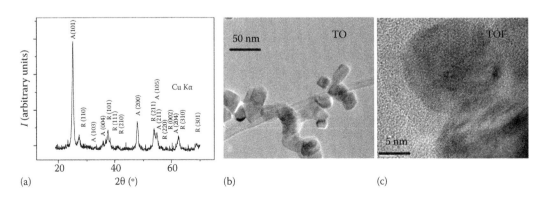

**FIGURE 27.10**    XRD data for sample TO (a) and HRTEM data for samples TO (b) and TOF (c). A: anatase; R: rutile.

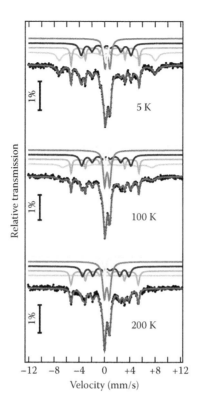

**FIGURE 27.11** Mössbauer spectra acquired at different temperatures for sample TOF.

sample, both ingredients having a negative impact on the photocatalytic activity of the product. Indeed, preliminary experiments concerning the photocatalytic activity of such samples (Fleaca et al. 2014) confirmed the higher activity of sample TO (low C content and higher content of anatase phase) with respect to TOF. Hence, at a first view it seems that the Fe addition aimed at a possible recovery of the catalyst under the magnetic field would not be recommended with respect to the activity of the catalyst. In this condition, a new direction remains to be exploited in $TiO_2/C$ nanocomposites (without Fe addition) based on the weak intrinsic magnetisms of diluted magnetic oxide/semiconductor systems.

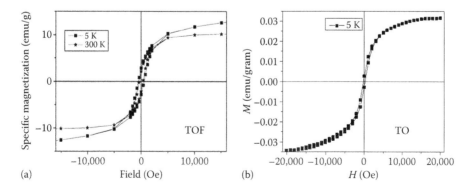

**FIGURE 27.12** Magnetic hysteresis loops collected (a) at 5 K and 300 K for sample TOF and (b) at 5 K for sample TO.

# Acknowledgments

Research was supported by Executive Agency for Higher Education, Research, Development and Innovation Funding (UEFISCDI), Romania, under Grants PN-II-ID-PCE-2011 (contract No. 75/2011), PN-II-PT-PCCA-2013 (Contract No. 275/2014) and PN-II-Capacities (Contract No. E11/30.06.2014) at National Institute of Materials Physics (NIMP) and Grants PN-II-ID-PCE-2011 (Contract No. 63/2011 and Contract No. 80/2011) at National Institute for Laser, Plasma and Radiation Physics (NILPRP). VK acknowledges the kind assistance of the Alexander von Humboldt Foundation.

# References

Ajayan, P. M., Ebbesen, T. W., Ichihashi, T. et al., "Opening nanotubes with oxygen and implications for filling," *Nature* 362 (1993): 522.

Alexandrescu, R., Morjan, I., Voicu, I. et al., "Combining resonant/non-resonant processes: Nano-meter-scale iron based material preparation via the $CO_2$ laser pyrolysis," *Appl. Surf. Sci.* 248 (2005): 138–146.

Alexandrescu, R., Morjan, I., Dumitrache, F. et al., "Photochemistry aspects of the laser pyrolysis addressing preparation of oxide semiconductor photocatalysts," *Int. J. Photoenergy* (2008): 1–11, article ID 604181.

Alexandrescu, R., Morjan, I., Scarisoreanu, M. et al., "Development of the IR laser pyrolysis for the synthesis of iron-doped $TiO_2$ nanoparticles: Structural properties and photoactivity," *Infrared Phys. Technol.* 53 (2010): 94–102.

Baibarac, M., Baltog, I., Gomez-Romero, P. et al., "Hybrid organic–inorganic materials based on carbon nanotubes and conducting polymers with applications in energy storage," in Balasoiu, M. & Arzumanyan, G. M., eds., *Modern Trends in Nanoscience* (Editura Academiei Romane, Bucharest, Romania, 2013), pp. 157–179.

Blanita, M., Mihet, M., Borodi, G. et al., "Ball milling and compression effects on hydrogen absorption by MOF:Pt/carbon mixtures," *Microporous Mesoporous Mater.* 203 (2015): 195–201.

Blasé, X., Charlier, J. C., De Vita, A. et al., "Boron mediated growth of long helicity-selected carbon nanotubes," *Phys. Rev. Lett.* 83 (1999): 5078.

Chen, P. & Zhang, G. Y., "Carbon-based spintronics," *Sci. China* 56 (2013): 207–221.

Crisan, O., Angelakeris, M., Kehagias, Th. et al., "Structure effects on the magnetism of AgCo nanoparticles," *Acta Mater.* 54 (2006): 5251.

Czerw, R., Terrones, M., Charlier, J. C. et al., "Identification of electron donor states in N-doped carbon nanotubes," *Nano Lett.* 1 (2001): 457–460.

Ding, R. G., Lu, G. Q., Yan, Z. F. et al., "Recent advances in the preparation and utilization of carbon nanotubes for hydrogen storage," *J. Nanosci. Nanotechnol.* 1 (2001): 7.

Dresselhaus, M. S., Dresselhaus, G. & Ecklund, P. C., *Science of Fullerenes and Carbon Nanotubes* (Academic Press, Waltham, 1996).

Dujardin, E., Ebbesen, T. W., Hiura, H. et al., "Capillarity and wetting of carbon nanotubes," *Science* 265 (1994): 1850.

Ebbesen, T. W., Ajayan, P. M., Hiura, H. et al., "Purification of nanotubes," *Nature* 367 (1994): 519.

Filoti, G., Kuncser, V., Schinteie, G. et al., "Characterization of magnetic nano-fluids via Mössbauer spectroscopy," *Hyperfine Interact.* 191 (2009): 55–60.

Fleaca, C. T., Dumitrache, F., Morjan, I. et al., "Novel Fe@C–$TiO_2$ and Fe@C–$SiO_2$ water-dispersible magnetic nanocomposites," *Appl. Surf. Sci.* 278 (2013): 284–288.

Fleaca, C. T., Scarisoreanua, M., Morjan, I. et al., "Recent progress in the synthesis of magnetic titania/iron-based composite nanoparticles manufactured by laser pyrolysis," *Appl. Surf. Sci.* 302 (2014): 198–204.

Fleming, R. M., Ramirez, A. P., Rosseinsky, M. J. et al., "Relation of structure and superconducting transition temperatures in $A_3C_{60}$," *Nature* 1352 (1991): 787.

Greenwood, N. N. & Gibb, T. G., *Mössbauer Spectroscopy* (Chapman and Hall Ltd., London, UK, 1971).

Grobert, N., Hsu, W. K., Zhu, Y. Q. et al., "Enhanced magnetic coercivities in Fe nanowires," *Appl. Phys. Lett.* 75 (1999): 3363.

Guiseppi-Elie, A., Lei, C. & Baughman, R. H., "Direct electron transfer of glucose oxidize on carbon nanotubes," *Nanotechnology* 13 (2002): 559.

Herlin, N., Bohn, I., Reynaoud, C. et al.,"Nanoparticles produced by laser pyrolysis of hydrocarbons: Analogy with carbon cosmic dust," *Astron. Astrophys.* 330 (1998): 1127–1135.

Hsu, W. K., Hare, J. P., Terrones, M. et al., "Condensed-phase nanotubes," *Nature* 377 (1995): 687.

Iijima, S., "Helical microtubules of graphitic carbon," *Nature* 354 (1991): 56.

Jäger, C., Mutschke, H., Huisken, F. et al., "Iron–carbon nanoparticles prepared by $CO_2$ laser pyrolysis of toluene and iron pentacarbonyl," *Appl. Phys. A* 85 (2006): 53–62.

King, R. B., "Some aspects of the symmetry and topology of possible carbon allotrope structures," *J. Math. Chem.* 23 (1998): 197–227.

Kratschmer, W., Lamb, L. D., Fostiropoulos, K. et al., "Solid $C_{60}$: A new form of carbon," *Nature* 347 (1990): 354.

Kroto, H. W., Heath, J. R., O'Brien, S. C. et al., "$C_{60}$: Buckminsterfullerene," *Nature* 318 (1985): 162.

Kumar, M. & Ando, Y., "Chemical vapor deposition of carbon nanotubes: A review on growth mechanism and mass production," *J. Nanosci. Nanotechnol.* 10 (2010): 3739.

Kuncser, V., Schinteie, G., Sahoo, B. et al., "Magnetic interactions in water based ferrofluids studied by Mössbauer spectroscopy," *J. Phys. Condens. Matter* 19 (2007): 016205.

Kuncser, V., Crisan, O., Schinteie, G. et al., "Magnetic nanophases: From exchange coupled multilayers to nanopowders and nanocomposites," in Balasoiu, M. & Arzumanyan, G. M., eds., *Modern Trends in Nanoscience* (Editura Academiei Romane, Bucharest, Romania, 2013), pp. 197–222.

Kuncser, V., Palade, P., Kuncser, A. et al., "Engineering magnetic properties of nanostructures via size effects and interparticle interactions," in Kuncser, V. & Miu, L., eds., *Size Effects in Nanostructures* (Springer-Verlag, Berlin, Germany, 2014), 297–314.

Logothetidis, S., "Hydrogen-free amorphous carbon films approaching diamond prepared by magnetron sputtering," *Appl. Phys. Lett.* 69 (1996): 158.

Mackay, A. L. & Terrones, H., "Diamond from graphite," *Nature* 352 (1991): 762.

Morjan, I., Dumitrache, F., Alexandrescu, R. et al., "Laser synthesis of magnetic iron–carbon nanocomposites with size dependent properties," *Adv. Powder Technol.* 23 (2012): 88–96.

Nistor, L. C., Ghica, C., Kuncser, V. et al., "Microstructure-related magnetic properties in Co-implanted ZnO thin films," *J. Phys. D Appl. Phys.* 46 (2013): 065003–065013.

O'Regan, B. & Gratzel, M., "A low-cost, high-efficiency solar cell based on dye-sensitized colloidal $TiO_2$ films," *Nature* 353 (1991): 737–740.

Palade, P., Comanescu, C. & Mercioniu, I., "Improvements of hydrogen desorption of lithium borohydride by impregnation onto MSH-H carbon replica," *J. Ovonic Res.* 8 (6) (2012): 155.

Palmisano, L., Schiavello, M., Sclafani, A. et al., "Surface properties of iron–titania photocatalysts employed for 4-nitrophenol photodegradation in aqueous $TiO_2$ dispersion," *Catal. Lett.* 24 (1994): 303–315.

Park, H., Park, J., Lim, A. K. L. et al., "Nano-mechanical oscillations in a single-$C_{60}$ transistor," *Nature* 407 (2000): 57–60.

Pasqualini, E., "Nuclear nanocapsules and curved carbon structures," *Carbon* 35 (1997): 783.

Robertson, J., "Diamond-like amorphous carbon," *Mater. Sci. Eng. R Rep.* (2002): 129–281.

Schinteie, G., Kuncser, V., Palade, P. et al., "Magnetic properties of iron–carbon nanocomposites obtained by laser pyrolysis in specific configurations," *J. Alloys Compd.* 564 (2013): 27–34.

Sloan, J., Cook, J., Heesom, J. R. et al., "The encapsulation and in situ rearrangement of polycrystalline SnO inside carbon nanotubes," *J. Cryst. Growth* 173 (1997): 81.

Stone, A. J. & Wales, D. J., "Theoretical studies of icosahedral $C_{60}$ and some related species," *Chem. Phys. Lett.* 128 (1986): 501.

Taylor, R. & Walton, D. R. M., "The chemistry of fullerenes," *Nature* 363 (1993): 685–693.

Terrones, H. & Terrones, M., "Curved nanostructured materials," *New J. Phys.* 5 (2003): 126.1–126.7.

Tersoff, J., "Empirical interatomic potential for carbon, with applications to amorphous carbon," *Phys. Rev. Lett.* 61 (1988): 2879.

Thess, A., Lee, R., Nikolaev, P. et al., "Crystalline ropes of metallic carbon nanotubes," *Science* 273 (1996): 483.

Thomas, M. F. & Johnson, C. E., *Mössbauer Spectroscopy* (Cambridge University Press, Cambridge, England, 1986).

Tombros, N., "Electron spin transport in graphene and carbon nanotubes," PhD thesis, University of Groningen, Zernike Institute, Groningen, Netherlands (2008).

Treacy, M. M. J., Ebbesen, T. W. & Gibson, J. M., "Exceptionally high Young's modulus observed for individual carbon nanotubes," *Nature* 381 (1996): 678.

Tsang, S. C., Chen, Y. K., Harris, P. J. F. et al., "A simple chemical method of opening and filling carbon nanotubes," *Nature* 372 (1994): 15.

Wang, G., Luo, Y., Zhao, Y. et al., "Preparation of $Fe_2O_3$/graphene composite and its electrochemical performance as an anode material for lithium ion batteries," *J. Alloys Compd.* 509 (24) (2011): L216–L220.

# 28

# Nanocarbon Hybrid Materials

Alexey S. Cherevan

Paul Gebhardt

Cameron J. Shearer

Dominik Eder

## 28.1 Introduction to Nanocarbon Hybrids

Excellent physicochemical properties of various nanocarbons along with mechanical characteristics have attracted enormous attention of both applied and fundamental research. As a consequence, bare nanocarbons have already found their niche in a variety of fields such as transistors, photodetectors, transparent conducting films, supercapacitors, and field emission displays. In order to widen the application range of the nanocarbon materials, they have been further combined with other functional components, leading to the appearance of two new classes of multifunctional materials: nanocarbon composites and nanocarbon hybrids.

Research on nanocarbon composites has always been in a close connection to the nanocarbon research progress and was initially inspired by the fiber-reinforced polymer industry, which introduced this new class of strong yet lightweight materials for various applications. Both CNTs (Ajayan et al. 1995) and graphene (Stankovich et al. 2006) were incorporated into composite materials soon after the first reports on their excellent strength and electric conductivity. A few years ago, another member of the composite family—nanocarbon hybrids—has rapidly appeared and already received tremendous attention due to its high potential in environmental and sustainable energy applications (Eder 2010, Eder & Schlögl 2014). Despite early attempts to distinguish between the two classes, the terms *hybrid* and *composite* have been used interchangeably for a long time and only recently have been differentiated in the research community. It is of great importance for this chapter to summarize the main differences between these two types of materials.

**TABLE 28.1**    Main Differences between Nanocarbon Composites and Nanocarbon Hybrids

|  | Composite | Hybrid |
| --- | --- | --- |
| Synthesis | Nanocarbon as filler in inorganic matrix | Ex situ linkage of premade components or in situ deposition of a functional compound onto the CNT/graphene surface |
| Structure | Poor control over morphology and interface, low nanocarbon content (0.01–5 wt%) | Comparable volume fractions of the components, nanocarbon affects size and morphology, ready access to a large interior surface area |
| Properties | Combination of individual properties | Additional synergistic properties due to interfacial charge/energy transfer |
| Application | Lightweight devices with increased mechanical and electrical properties | Energy applications: catalysis, batteries, photovoltaics, supercapacitors, and sensors |

## 28.1.1  Differences between Hybrids and Composites

In nanocarbon composites, the nanocarbon part (e.g., CNTs or graphene) is used as filler and is typically dispersed within an inorganic or polymeric matrix, rarely reaching concentrations of a few weight percent (Figure 28.1a). The main purpose of a nanocomposite material is to combine beneficial properties of its components. For instance, electrical conductivity of CNTs with the optical properties of a polymer to create a transparent polymeric conductor (Huang & Terentjev 2010) or mechanical properties of CNTs to reinforce metals and ceramics (Ci et al. 2008). As a result, nanocarbon composites are ideal candidates for applications in which lightweight materials with excellent mechanical or electronic properties are required. Commercial CNT composites are already available in the market and are implemented

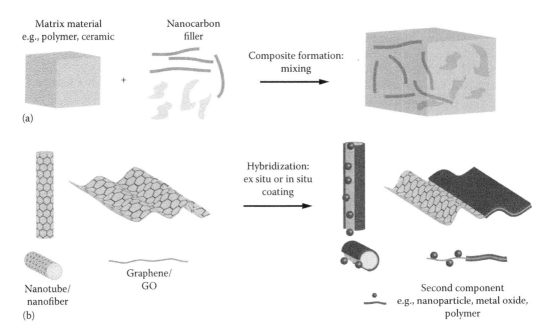

**FIGURE 28.1**    Scheme showing the main difference between (a) nanocarbon composites prepared by mixing of nanocarbons into a matrix material and (b) nanocarbon hybrids prepared by coating of a second component onto the surface of the nanocarbon. (Shearer, C. J., Cherevan, A. & Eder, D.: Application and Future Challenges of Functional Nanocarbon Hybrids. *Advanced Materials.* 2014. 26. 2295–2318. Copyright Wiley-VCH Verlag GmbH & Co. KGaA. Reproduced with permission.)

in a wide range of consumer products such as LIBs, automotive/nautical/aerospace materials, sports equipment, conducive pastes and flame retardants (De Volder et al. 2013). However, the properties of the composites are still far from ideal. This originates from poor interfacial adhesion and inhomogeneous mixing of the composite's components. Such scarce control over the morphology and interface on the nanoscale strongly limits further development and application of this material class. Finally, due to comparatively low accessible surface area and limited access to solid–liquid and solid–gas interfaces, nanocarbon composites are generally not suited for applications such as photocatalysis, thermal cataly-sis, energy storage and conversion, and chemical sensors.

In nanocarbon hybrids, the nanocarbon part is used as an active substrate, while the second compo-nent is typically deposited on the nanocarbon surface in a form of thin layers or nanoparticles (Figure 28.1b). This second component in hybrids constitutes an active, often functional compound and may include a range of options from molecules and polymers to ceramics and semiconductors, opening up possibilities for many new applications compared to composites. Another important property of the hybrids originates from similar volume fractions of the hybrid's components and, thus, a large interface between them. This allows designing functional hybrid materials with enhanced charge and energy transfer between the components. The resulting synergistic effects (see Section 28.4) can improve the properties of the individual components and even create new properties that commend these nano-carbon hybrids as ideal multifunctional materials for various applications that require gas–surface or liquid–surface interactions as well as solid–solid interaction central to energy conversion and storage applications (Vilatela & Eder 2012, Shearer et al. 2014).

## 28.2 Preparation of Nanocarbons for Further Hybridization

The morphology, structure, and properties of a hybrid material are directly linked to the type and struc-ture of the nanocarbon chosen for hybridization. Depending on the targeted application, nanocarbons usually have to be modified or processed in order to harness the beneficial characteristics of the nano-carbons (e.g., electric and mechanical properties) in an optimal way or to allow hybridization with an inorganic material in the first place. This section describes the most useful postprocessing methods for as-synthesized nanocarbons, including purification, functionalization, and noncarbon doping.

### 28.2.1 Purification of Nanocarbons

The amount of impurities in nanocarbons obtained after synthesis depends on the synthesis method. For example, the CVD process produces metal catalyst particles (often encapsulated within the nano-carbon), the arc-discharge yields relatively large amounts of amorphous carbon by-products, and Hummer's method often leads to chemical residues (e.g., manganates). Only the removal of these impu-rities ensures the reproducible and controlled high performance of nanocarbons when implemented in hybrids, independent of the synthesis method. The specific purification method has to be chosen care-fully to trade the additional cost against possible unwanted modifications of the material. Table 28.2 summarizes the various possibilities and respective effects on nanocarbons.

Treatment with hydrochloric acid removes nonencapsulated metal catalyst particles and, if assisted by microwave irradiation, also amorphous carbon residues. This simple method does not, however, remove Fe particles encapsulated within CNTs. This would require opening the CNT tips by reflux in strongly oxidizing acids such as $HNO_3$ and $H_2SO_4$, which subsequently dissolves the metal residues. An additional effect of acids is the surface corrosion, which introduces defects and oxygen-containing func-tional groups on the surface of the nanocarbons. This is often undesired, since it considerably impairs the chemical and physical properties of the nanocarbons.

Gas-phase methods are milder than wet chemical acid treatments. Carefully varying the applied tem-perature and gas atmosphere allows control over the selective removal. Oxidative heat treatments in wet air selectively remove the amorphous carbon residues, if the temperature is low enough to leave the

**TABLE 28.2** Overview of Postsynthesis Treatments of Nanocarbons

| Treatment | Metals[a] | Carbon[b] | Functional[c] | CNT Structure |
|---|---|---|---|---|
| Ultrasonic | | | | Opens tips, shortens CNTs |
| HCl$_{conc}$ | +[d] | | | |
| HNO$_3$/H$_2$SO$_4$ | + | + | + | Opens tips, shortens CNTs |
| Heating in air/O$_2$ | | + | + | Opens tips |
| Heating in wet O$_2$ | | + | | |
| Microwave | | +[e] | | |
| Ar @ 1000°C | | + | | |
| Ar @ 2000°C | + | + | | Induces graphitization and defect annealing |

[a] Treatment can remove metal catalyst residues.
[b] Treatment can remove carbon residues (e.g., amorphous or organic aromatic debris).
[c] Purification introduces covalently bonded functional groups.
[d] Only if not covered with carbon or encapsulated within CNT.
[e] Only amorphous carbon around metal particles.

nanocarbon unharmed. This typically means keeping the temperature below 600°C; however, hybridization with metal oxides can decrease the nanocarbon combustion temperature considerably (see Section 28.4.1), which can require adjustment. Higher temperatures can be utilized in inert atmosphere, such as argon, without harming the nanocarbon. Between 900°C and 1000°C, amorphous carbon residues are removed; at 1600°C, the tubes are opened so that the metal particles from within the CNTs can be removed; and at even higher temperatures (>2000°C), structural defects are annealed and the overall level of graphitization increases (Eder 2010).

## 28.2.2 Functionalization

Since high-purity and defect-free nanocarbons consist only of *sp*$^2$ carbon, they generally exhibit a high degree of hydrophobicity. This can be disadvantageous when trying to disperse them in aqueous solvents and implement them into hybrids. Functionalization on the surface not only increases hydrophilicity, but also offers reaction sites for functional inorganic compounds' deposition (see Section 28.3). We can distinguish covalent and noncovalent functionalization.

Covalent functionalization implies formation of a covalent bond that connects the functional group and the nanocarbon surface—most simply achieved by an oxidative acid treatment at elevated temperatures (Figure 28.2). However, this method is harsh and lacks control over the type, location, and amount of functional groups. Instead of selectively producing one kind of functionality, a mixture of carboxyl, hydroxyl, and epoxy groups is obtained. Furthermore, the functionalization is not evenly distributed,

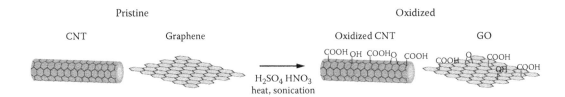

**FIGURE 28.2** Scheme demonstrating the simplest covalent functionalization route of pristine nanocarbons used—an oxidative acid treatment at elevated temperatures—where a number of oxygen-containing groups are introduced on the nanocarbon surface. (Shearer, C. J., Cherevan, A. & Eder, D.: Application and Future Challenges of Functional Nanocarbon Hybrids. *Advanced Materials*. 2014. 26. 2295–2318. Copyright Wiley-VCH Verlag GmbH & Co. KGaA. Reproduced with permission.)

but mainly located on the CNT tips, graphene edge atoms, and defects because all these sites exhibit an intrinsically higher activity compared to the in-plane $sp^2$ carbon atoms. This also explains why this process works better with SWCNTs: the curvature renders the carbon atoms more reactive compared with more graphitic MWCNTs.

The advantage of covalent functionalization in hybrids is the strong interaction between the two components, which can withstand chemical and mechanical stress. However, the additional bond on the carbon atom causes a shift in hybridization, which, in turn, downgrades the nanocarbon's properties. This applies for graphene as well as CNTs with different layer counts, since the outermost layer in MWCNTs contributes most to the exceptional material characteristics (Delaney et al. 1999), which are dramatically influenced by structural defects.

A milder alternative is noncovalent functionalization, which relies on $\pi$–$\pi$ or van der Waals interactions and thus preserves the material characteristics. If the aim is simply to improve the dispersability, a wide range of surfactants, aromatics, biomolecules, polyelectrolytes, or polymers can be combined with nanocarbons in aqueous solvents. A more advanced molecule for noncovalent functionalization is benzyl alcohol. While the aromatic ring interacts with the graphitic plane via $\pi$–$\pi$ interactions (Cooke et al. 2010), it offers hydroxyl groups as reaction centers for inorganic compounds, which are uniformly distributed over the entire carbon surface. Examples with other aromatic systems (e.g., naphthalene or pyrene) and other functional groups (e.g., carboxylic) exist. A drawback of this approach, however, is the lower interfacial interaction: the exact bond strength between nanocarbon and inorganic material varies and a gap could be introduced at later sintering or crystallization steps.

## 28.2.3 Doping and Defects

Various defect types in nanocarbons, such as dislocations, terminations, and carbon vacancies, can be produced during synthesis, processing, or by a physical impact (e.g., by the electron beam in electron microscopy).

One of them, so-called topological defects, describes the deviation from six-membered rings in the carbon layers. The so-called Stone–Wales defects constitute an energetically favored version, consisting of adjacent five- and seven-membered rings. Defects generally alter the mechanical, optical, electronic, and charge transfer properties of nanocarbons, which can be crucial for the implementation in hybrids. For example, just one carbon vacancy in the outer layer of an MWCNT reduces its tensile strength by 30% (Pugno 2006). Furthermore, induced $sp^3$ hybridization increases the reactivity on that defect site, which is then more prone to oxidation, but can also offer sites for crystallization or nucleation, affecting the crystal structure, size, and morphology of inorganic coating materials (Gebhardt et al. 2014).

Another type of defect is the complete replacement of a carbon atom with another element. This is usually carried out to influence the electronic properties, e.g., the so-called on-wall doping, where a carbon atom is replaced with N or B to achieve *n*- or *p*-type doping, respectively. In some synthesis approaches, this can be achieved in situ, e.g., using either a nitrogen-containing carbon source or electrode in the CVD or arc-discharge process, respectively. In the case of nitrogen doping, there exists a special taxonomy:

- *Graphite-like*: The N atom occupies the position of the former C atom, i.e., with three direct neighbors. The additional *p* electron of N leads to the (usually desired) *n*-type doping.
- *Pyridine type*: The induced defect is stabilized by an adjacent vacancy. In this case, either *n*- or *p*-type doping can be obtained, depending on the concentration of dopants and defect sites.

## 28.2.4 Separation of Metallic and Semiconducting SWCNTs

SWCNTs can exhibit either a metallic or semiconducting band structure depending on their chirality. Since none of the standard synthesis methods for SWCNTs can sufficiently control the band structure until now, separation methods are required in order to separate each of the CNT fractions. Specific use of metallic or semiconducting CNTs can be beneficial for many applications—such as in electric devices, field emission, photovoltaics, and photocatalysis—depending on the particular task of the nanocarbon.

Chemical separation methods rely on surfactant molecules that preferentially adsorb on SWCNTs of metallic (e.g., diazonium reagents, DNA) or semiconducting (octadecylamine, agarose gel) type. The aggregated CNTs can then be separated by standard techniques: ion-exchange chromatography or microfiltration. Alternating current dielectrophoresis is a physical separation approach that relies on the difference in dielectric constant between the two types of SWCNTs. When an electric field gradient is applied, metallic SWCNTs deposit on the microelectrode array, while semiconducting SWCNTs remain in the solution. This approach requires relatively small batch sizes of well-dispersed CNTs to avoid aggregation of CNTs with mixed band structure.

## 28.3 Synthesis of Nanocarbon Hybrids

Once the nanocarbon is synthesized, purified, and suitably modified (i.e., functionalized or annealed), it can be used as a template to prepare a hybrid material following one of the synthetic procedures summarized in Figure 28.3. In general, they can be classified into two routes, namely, ex situ and in situ

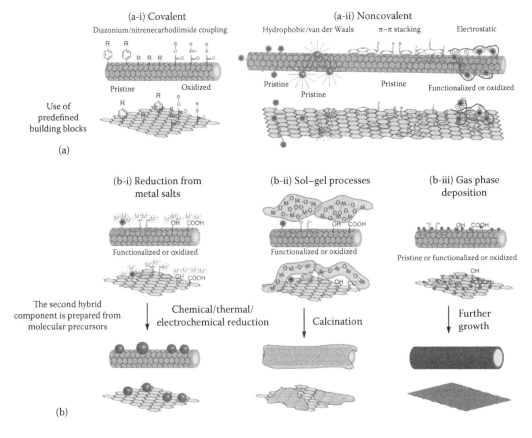

**FIGURE 28.3** Scheme illustrating the different hybridization strategies of nanocarbon hybrids: (a) ex situ synthesis from presynthesized building blocks (i) via covalent and (ii) noncovalent techniques and (b) in situ synthesis (i) via reduction, (ii) sol–gel processes, (iii) and gas phase techniques. (Shearer, C. J., Cherevan, A. & Eder, D.: Application and Future Challenges of Functional Nanocarbon Hybrids. *Advanced Materials*. 2014. 26. 2295–2318. Copyright Wiley-VCH Verlag GmbH & Co. KGaA. Reproduced with permission.)

methods; however, the implementation of a particular synthetic technique strongly depends on the type of nanocarbon and the degree of its functionalization (see Section 28.2.2). This section offers a brief overview of the various synthetic routes, while more detailed information can be found in the following references: Eder (2010); Eder and Schlögl (2014); Shearer et al. (2014); and Vilatela and Eder (2012).

## 28.3.1 Ex Situ Route

In the ex situ route (known as the building block approach), both hybrid components are first synthesized separately with predefined dimensions, type, and morphology; then modified with suitable functional groups or linker molecules to become attachable; and finally combined by means of covalent, noncovalent, or electrostatic interactions (Figure 28.3a).

The most effective route is to induce the formation of amide and ester bonds by linking functionalized nanocarbons (e.g., acid treated) with molecules or nanoparticles (NPs) premodified with alcohol or amines (Figure 28.3a-i) (Shearer et al. 2014). A wide range of quantum dots, inorganic and metal NPs, polymers, porphyrins, and proteins have been connected this way (Eder 2010).

The noncovalent route relies upon electrostatic and van der Waals interactions, thus offering a nondestructive path toward hybridization (Figure 28.3a-ii). Aromatic linker molecules are widely used, as they utilize comparatively strong $\pi$–$\pi$ interactions between the delocalized $\pi$ electrons of nanocarbons and those in aromatic organic compounds, such as derivatives of pyrene, perylene, porphyrins, benzyl alcohol, triphenylphosphine, and phthalocyanines. These molecules are often modified with long alkyl chains that are terminated with thiol, amine, acid, or charged groups which can then be used for further hybridization.

While the aliphatic chains in surfactants or polymers prefer the graphitic surface of annealed nanocarbons, the attachment through electrostatic interactions typically requires functionalized nanocarbons. In a typical example, Yang et al. (2008) have modified CNTs with poly(sodium 4-styrenesulfonate) to facilitate the adsorption of Pt nanocubes capped with charged cetyltrimethylammonium bromide. This has resulted in the attachment of uniformly sized metal NPs with unique morphology (Figure 28.4a).

In summary, the ex situ methods allow utilizing presynthesized NPs with defined structure, size, and shape and therefore enable a good structure–property control over the hybrid material. Moreover, the ex situ route remains the only method that enables coupling of biomolecules with nanocarbons. The drawback of this approach, however, is its complexity and the necessity to chemically modify either the nanocarbon and/or the hybridizing component.

## 28.3.2 In Situ Route

In the in situ method (often referred to as bottom–up approach), the hybridizing component is deposited directly on the nanocarbon surface using molecular precursors, such as metal salts or metal–organic compounds (Figure 28.3b). The nanocarbon here acts as both a substrate and an active template, thus affecting the size, the crystalline structure, and especially the morphology of the second compound during its deposition and postprocessing (e.g., calcination). As a result, depending on the synthetic conditions (e.g., temperature, deposition time, precursor type, nanocarbon type), the produced coating may consist of nanoparticles, nanowires, nanoplatelets, or thin films, while being either amorphous or crystalline. In contrast to the ex situ approach, the functionalization of the nanocarbon is not necessarily required, although the presence of functional groups or the use of linking agents often facilitate initial stages of hybrid formation (e.g., hydrolysis, formation of chemical bonds between the components).

The simplest way to obtain nanocarbon–metal hybrids is via coprecipitation reaction using metal salts (Figure 28.3b-i). These are rather cheap, generally soluble in water containing solutions and can be used to form metal NPs with tunable size and morphology (Walsh & Herron 1991). This, however, requires a reduction of the corresponding precursor metal complex and can be achieved by means of

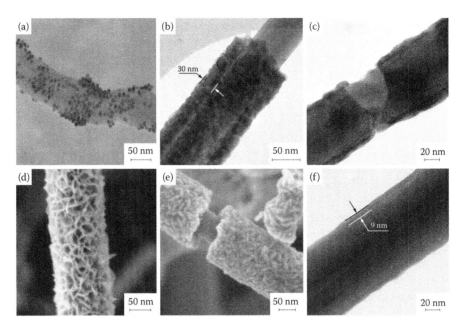

**FIGURE 28.4** SEM and TEM images of exemplary nanocarbon hybrids. (a) A hybrid composed of CNTs and Pt nanocubes prepared via ex situ approach. (Yang, W., Wang, X., Yang, F. et al.: Carbon Nanotubes Decorated with Pt Nanocubes by a Noncovalent Functionalization Method and Their Role in Oxygen Reduction. *Advanced Materials.* 2008. 20. 2579–2587. Copyright Wiley-VCH Verlag GmbH & Co. KGaA. Reproduced with permission.) Hybrids via in situ sol–gel method: (b) CNTs-TiO$_2$ NPs, (c) CNTs–Ta$_2$O$_5$ layers, and (d) CNTs–V$_2$O$_5$ platelets. Hybrids via in situ ALD: (e) CNTs–ZnO NPs (Kemnade, N. et al., *Nanoscale* 7 (7): 3028–3034, 2015. Reproduced by permission of The Royal Society of Chemistry.) and (f) CNTs–Al$_2$O$_3$ layers.

chemical agents (chemical reduction/oxidation), light (photoreduction), heat (thermal reduction), or electrons (electrodeposition). Common reducing agents include NaBH$_4$, sodium citrate, formic acid, ethylene glycol, and thiourea. An interesting route is offered for metal NP-decorated graphene hybrids' synthesis: they can be obtained via concurrent reduction of GO impregnated with metal salts.

The sol–gel method is another conventional process, especially useful for metal oxide deposition in the form of NPs, nanowires, or thin films (Figure 28.3b-ii). The reaction utilizes metal–organic precursors able to undergo hydrolysis and condensation reactions and involves the transition of a liquid, colloidal, or polymeric "sol" into a solid "gel" phase. During the transition, the sol forms a continuous network with oxo or hydroxo bridges, leading to a gel. The attachment of metal oxides typically requires a hydrophilic nanocarbon surface (such as in GO or acid-treated CNTs), or alternatively the addition of linking agents with aromatic system (i.e., via noncovalent functionalization [Eder and Windle 2008a]). Benzyl alcohol (Cooke et al. 2010) or pyrenecarboxylic acid (Kemnade et al. 2015) are an excellent choice for such a linker as it can adsorb onto the surface of the CNT due to π–π interactions, therefore providing a large number of polar hydroxyl or acidic groups for the reaction with metal–organic molecules. This method has already been used to deposit more than 20 different metal oxides in the form of NPs (e.g., in CNT–TiO$_2$ hybrids in Figure 28.4b [Aksel & Eder 2010]), layers (e.g., in CNT–Ta$_2$O$_5$ hybrids in Figure 28.4c [Cherevan et al. 2014]), and platelets (e.g., in CNT–V$_2$O$_5$ hybrids in Figure 28.4d).

Gas phase deposition techniques provide an excellent control over the morphology, thickness, and uniformity of the nanocarbon–inorganic hybrids (Figure 28.3b-iii). They include CVD, physical vapor deposition (PVD), and atomic layer deposition (ALD) methods as well as various sputtering, plasma, and evaporation processes. In particular, ALD offers many advantages being a low-temperature, surface-terminated process that allows conformal deposition of inorganic layers with precise control over the

**TABLE 28.3**  Main Characteristics of Both Ex Situ and In Situ Techniques for Nanocarbon Hybrid Synthesis

| Characteristic | Ex Situ | In Situ Wet Chemical | In Situ Gas Phase |
| --- | --- | --- | --- |
| Morphology | As predefined | Affected by the nanocarbon (substrate/heat sink), i.e., nanoparticles, nanowires/rods, nanoplatelets, thin films | |
| Structure | As predefined | Amorphous or crystalline depending on reaction conditions and postprocessing | |
| Size control | Narrow size distribution (e.g., using capping agents) | Less defined size distribution Depends on concentration and reaction time | Good control in film thickness Depends on reaction time and temperature |
| Coating | Limited to monolayer | Multilayer Conformal Directed (electrodeposition) | Multilayer Conformal (chemical methods) Directed (physical methods) |
| Interface | Needs anchoring molecules or functionalization of both compounds | Pristine or functionalized CNTs possible | Good coverage on pristine CNTs |

coating thickness. The process has been applied to deposit various metal oxides and metal nitrides, e.g., CNT–ZnO hybrids (Kemnade et al. 2015) in Figure 28.4e or CNT–Al$_2$O$_3$ hybrids in Figure 28.4f. An important advantage of the gas phase deposition techniques is that the methods can be realized for immobilized nanocarbons and freestanding architectures such as CNT fibers or bucky papers.

### 28.3.3 Comparison between Techniques

The summary and the main features of both techniques are presented in Table 28.3. In comparison to ex situ hybrid synthesis, in situ methods often require fewer reaction and purification steps. The ex situ approach results in more uniform coatings with tunable thickness, while the in situ methods also take advantage of the nanocarbon as an active substrate and heat sink, which can facilitate nucleation and crystallization (resulting in new interesting morphologies) and prevent particle growth upon heat transfer.

## 28.4 Synergistic Properties

The hybridization of nanocarbon with an active compound offers a variety of beneficial synergistic effects that can affect both structure and properties of the final material. Here we discuss the main characteristics of the effects and provide some vivid experimental results.

### 28.4.1 Structural Effects

As was mentioned in the previous sections, nanocarbons can affect nucleation, growth, crystallization, and phase transformation of the inorganic component when the hybrid is prepared via the in situ approach. For example, this can result in considerably smaller particles on the nanocarbon when compared to the bulk synthesis, directed growth in the form of nanowires or nanosheets (i.e., perpendicular or parallel to the carbon surface), epitaxial growth of single crystals, and stabilization of alternative crystal phases (Shearer et al. 2014).

One particular example of it is a so-called heat-sink effect, in which nanocarbons suppress the growth of inorganic nanoparticles during thermal treatments and therefore stabilize unusually small particles/thin layers. This growth often happens during crystallization, which is exothermic in nature, when the excess heat is being consumed in the grain growth. However, if crystallization happens on the nanocarbon surface, this excess heat can be effectively transferred to the nanocarbon and therefore dissipated due to their exceptionally high thermal conductivity. As such, this heat-sink effect has enabled the synthesis of very small rutile $TiO_2$ NPs (ca. 8 nm) onto CNT surface (Eder et al. 2006). In a similar example, it was possible to grow and stabilize titanium silicalite (TS-1 zeolite) on CNTs and graphene with unprecedented small sizes of 8–10 nm, which is about two orders of magnitude smaller than those grown in the absence of the nanocarbon (Ren et al. 2012). In general, close proximity between the two components is beneficial for these structural effects. However, the heat-sink effect, which is based on interfacial heat transfer, has been observed even up to distances of about 50–60 nm (Eder & Windle 2008b).

Another interesting effect originating from the close interface between the hybrid's components is a strong influence of the deposited metal oxide type on the thermal stability of the hybrid. It has been observed that different metal oxides combined with CNTs in a form of hybrid can either increase or reduce onset temperature for the oxidation of the nanocarbon (see Table 28.4). This phenomenon has been directly correlated with the oxide reducibility (i.e., the ability to create oxygen vacancies) and further enabled the prediction of the chemical stability of a wide range of CNT–metal oxide hybrids (Aksel & Eder 2010).

## 28.4.2 Property Effects

Another synergistic effect that may alter properties of the hybrid is based on interfacial charge transfer processes. In principle, nanocarbon materials can be both electron donors and acceptors depending on their chemical nature (i.e., type of nanocarbon, possible doping, work function value, etc.). The hybridization of a nanocarbon (either semiconducting or metallic; see Section 28.2.4) with another component such as dye molecules or inorganic compound (either semiconductor or metal) creates an interfacial

**TABLE 28.4**　A List of Onset Oxidation Temperatures for Various CNTs–Metal Oxide Hybrids

| Metal Oxide | Oxidation Temperature (°C) |
| --- | --- |
| $Al_2O_3$ | 760 |
| $SiO_2$ | 690 |
| MgO | 650 |
| $ZrO_2$ | 640 |
| $WO_3$ | 630 |
| ZnO | 625 |
| $Fe_2O_3$ | 600 |
| $SnO_2$ | 590 |
| $Cr_2O_3$ | 575 |
| $TiO_2$ anatase | 565 |
| $MoO_3$ | 555 |
| CuO | 540 |
| $TiO_2$ rutile | 530 |
| $V_2O_5$ | 490 |
| NiO | 460 |
| $La_2O_3$ | 460 |
| $Mn_2O_3$ | 440 |
| $CeO_2$ | 425 |
| $Co_3O_4$ | 395 |
| $\alpha$-PbO | 345 |
| $\beta$-$Bi_2O_3$ | 330 |

*Source:* Aksel, S. & Eder, D., *Journal of Materials Chemistry* 20 (41): 9149–9154, 2010. Reproduced by permission of The Royal Society of Chemistry.

heterojunction and thus an internal electric field between the components, which stimulates the transfer of photoexcited electrons at the interface. The position of the respective Fermi levels generally defines the final state of the junction and direction of the charge transfer in hybrids.

Current research aims at understanding the nature and extent of charge/energy transfer processes by monitoring them with a number of complex methods including transient fluorescence spectroscopy (Valeur & Berberan-Santos 2012), time-resolved transient absorption spectroscopy (Berera et al. 2009), electrochemical impedance spectroscopy (Lvovich 2012), and intensity modulated photocurrent/photovoltage spectroscopy. For the simplest hybrid scenario (fluorophore molecules hybridized with nanocarbons), exciton transfer can be governed by Förster resonance energy transfer or Dexter electron transfer both resulting in acceptor excitation. Experimental evidence of the electron transfer in such systems has been already reported between dye molecules and SWCNTs, MWCNTs, and GO, as well as graphene.

The mechanism of the charge/energy transfer in nanocarbon–inorganic hybrids is more complex due to the presence of two solid components with continuous electronic band structures. However, there are a few examples of such preliminary studies. For example, a decrease in ZnO photoluminescence intensity with increased nanocarbon concentration was observed in both CNT–ZnO (Vietmeyer et al. 2007) and GO–ZnO (Williams & Kamat 2009) hybrids. The authors have therefore concluded the presence of photoexcited electrons being transferred from the ZnO to the nanocarbon, supporting the presence of charge transfer (Figure 28.5).

Despite some promising results, more studies with carefully designed model systems are required to fully understand and eventually be able to tune these interfacial charge and energy processes. This will further allow purposeful design of hybrid materials to fulfill the requirements for every specific application.

## 28.5 Applications of Nanocarbon Hybrids

The aforementioned synergistic effects along with a high specific surface area and accessibility of the functional surface in nanocarbon hybrids commend them for diverse energy-related applications such as (photo)catalysis, supercapacitors, batteries, fuel cell, and chemical sensors (Figure 28.6; Table 28.5) (Eder & Schlögl 2014, Shearer et al. 2014). This section discusses each of the applications, provides some important experimental results, and explains how and why the use of the nanocarbon hybrids can be beneficial for every particular case.

### 28.5.1 Batteries

Some of the most advanced research fields are now focused on solving the world's energy problem. The development of electrochemical energy storage devices with excellent properties is of great importance

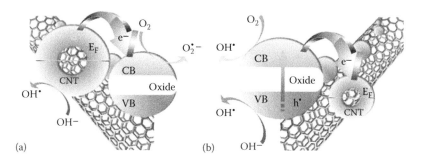

**FIGURE 28.5** Scheme showing the proposed mechanism of charge transfer (a) from CNT to a metal oxide (photosensitization) and (b) from metal oxide to CNT (photoexcited electron transfer). (Vilatela, J. J. & Eder, D.: Nanocarbon Composites and Hybrids in Sustainability: A Review. *ChemSusChem.* 2012. 5. 456–478. Copyright Wiley-VCH Verlag GmbH & Co. KGaA. Reproduced with permission.)

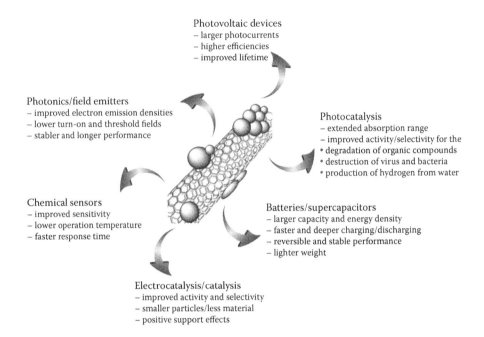

Photovoltaic devices
– larger photocurrents
– higher efficiencies
– improved lifetime

Photonics/field emitters
– improved electron emission densities
– lower turn-on and threshold fields
– stabler and longer performance

Photocatalysis
– extended absorption range
– improved activity/selectivity for the
  * degradation of organic compounds
  * destruction of virus and bacteria
  * production of hydrogen from water

Chemical sensors
– improved sensitivity
– lower operation temperature
– faster response time

Batteries/supercapacitors
– larger capacity and energy density
– faster and deeper charging/discharging
– reversible and stable performance
– lighter weight

Electrocatalysis/catalysis
– improved activity and selectivity
– smaller particles/less material
– positive support effects

**FIGURE 28.6** Overview of the various applications where nanocarbon hybrids have already been tested, resulting in enhanced performance. (Vilatela, J. J. & Eder, D.: Nanocarbon Composites and Hybrids in Sustainability: A Review. *ChemSusChem.* 2012. 5. 456–478. Copyright Wiley-VCH Verlag GmbH & Co. KGaA. Reproduced with permission.)

if one aims to utilize energy derived from renewable energy sources and therefore ensures sustainable and constant economy growth. Rechargeable batteries constitute a broad research field where improved charging/discharging rates, power density, and energy density are needed in order to meet the requirements of future devices such as electric vehicles and electric grid power storage stations. Currently, lithium ion batteries (LIBs) are by far the most widely used type of rechargeable batteries; however, many more types of batteries are under development. A typical LIB consists of three components: the $Li^+$ intercalation anode (e.g., graphite able to accommodate the ions), the $Li^+$ intercalation cathode (e.g., a metal oxide able to undergo redox reaction), and the electrolyte able to transport the ions (Marom et al. 2011). When the battery is being charged, $Li^+$ ions diffuse from the cathode to the anode, and back when being discharged (Figure 28.7). Graphite is currently the most intensively used anode material due to its low redox potential (~0.2 V versus $Li/Li^+$), good capacity (theoretical maximum: 372 mAh $g^{-1}$), and good recyclability. Metal oxides used as cathodes can offer a higher redox potential (~4 V versus $Li/Li^+$) and higher theoretical intercalation capacity. However, their low conductivity dramatically limits rates of charge and discharge. In addition, $Li^+$ intercalation often results in the metal oxide expansion, which causes cracking of the cathode and, possibly, leakage of the electrolyte.

Pure nanocarbon materials (e.g., CNTs or graphene) used instead of graphite offer high capacity values; however, a large portion of the intercalated ions forms a passivation layer, which reduces subsequent cycling performance. One way to minimize the issue and increase the capacity is to use nanocarbon hybrids containing both CNTs and graphene, which results in larger surface area with better access. Nanocarbons can also provide conductive, mechanically stable, and high surface area supports for metal oxides (Wu et al. 2012) and conducting polymers (Song et al. 2012). Depending on their redox potential, nanocarbon hybrids with metal oxides can be used as both anodes and cathodes. The main advantages shown by the incorporation of nanocarbon hybrids in LIBs are improved charge transfer kinetics and the capability of the hybrids to withstand severe volume charges upon charging/discharging cycles.

**TABLE 28.5** A List of Inorganic Compounds Used in Hybrid Materials Including Applied Synthesis Techniques and Tested Applications

| Inorganic Compound | Ex Situ | In Situ | Applications |
|---|---|---|---|
| $Al_2O_3$ | Noncovalent | Hydrothermal | Field emission |
| | | Sol–gel | Oxidation resistance |
| | | CVD, ALD, PVD | |
| $BaSrO_3$ | | PVD | Field emission |
| $CeO_2$ | | Sol–gel | Heterogeneous catalysis |
| | | Hydrothermal | Gas sensors |
| $Co_3O_4$ | | Sol–gel | Magnetics |
| | | CVD | Batteries |
| $Cu_2O$ | | Hydrothermal | Photocatalysis |
| $Eu_2O_3$ | Noncovalent | Hydrothermal | Diodes, lasers |
| $Fe_xO_y$ | | Covalent | Magnetics |
| | | Hydrothermal | Biosensors |
| | | | Heterogeneous catalysis |
| | | | Photoelectrochemical water splitting |
| $HfO_2$ | | PVD | Oxidation resistance |
| $MgO$ | | PVD | Field emission |
| | | | Electrocatalysis |
| $MnO_2$ | | Electrochemical | Electrocatalysis |
| | | Electrodeposition | Heterogeneous catalysis |
| | | Hydrolysis | Supercapacitors |
| | | CVD | Oxidation resistance |
| $MoO_2$ | | Hydrolysis | Electrocatalysis |
| $NiO$ | | Sol–gel | Supercapacitors |
| | | CVD | |
| $RuO_2$ | | Electrodeposition | Supercapacitors |
| | | Sol–gel | Biosensors |
| | | Hydrothermal | Heterogeneous catalysis |
| | | PVD | |
| | | CVD | |
| $SiO_2$ | Covalent | Sol–gel | Field emission |
| | | PVD | Oxidation resistance |
| $SnO_2$ | | Sol–gel | Gas sensors |
| | | Hydrothermal | Electrocatalyis |
| | | CVD | Nanofluids |
| | | | Batteries |
| $TiO_2$ | Covalent | Electrodeposition | Photocatalysis |
| | Noncovalent | Microemulsion | Optoelectronics |
| | | Sol–gel | Biosensors |
| | | Hydrothermal | Electrocatalysis |
| | | CVD | Supercapacitors |
| | | ALD | Batteries |
| | | | Oxidation resistance |
| $VO_2, V_2O_5$ | | Electrochemical | Batteries |
| | | | Heterogeneous catalysis |
| $WO_3$ | | Electrochemical | Gas sensors |
| | | | Heterogeneous catalysis |
| | | | Selective oxidation |

*(Continued)*

**TABLE 28.5 (CONTINUED)**   A List of Inorganic Compounds Used in Hybrid Materials Including Applied Synthesis Techniques and Tested Applications

| Inorganic Compound | Ex Situ | In Situ | Applications |
| --- | --- | --- | --- |
| ZnO | | Electrochemical | Photocatalysis |
| | | Microemulsion | Optoelectronics |
| | | Sol–gel | Diodes, lasers |
| | | Hydrothermal | Field emission |
| | | PVD, ALD, CVD | |
| $ZrO_2$ | Noncovalent | Hydrothermal | Oxidation resistance |
| | | PVD, ALD, CVD | Dielectric devices |
| | | | Heterogeneous catalysis |
| | | | Chemical sensors |
| Carbides (Fe, W, Ta, Ti) | | Electrochemical | Heterogeneous catalysis |
| | | Sol–gel | Electrochemistry |
| | | CVD | Field emission |
| Chalcogenides (Zn, Cd, Hg; $X$ = S, Se, Te) | Covalent | Electrochemical | Optoelectronics |
| | Noncovalent | Hydrothermal | Photocatalytic water splitting |
| | | Sol–gel | |
| Nitrides (Ti, Fe, Ta, Nb) | | Sol–gel | Field emission |
| | | CVD, ALD, PVD | Heterogeneous catalysis |

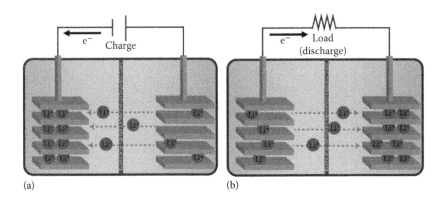

**FIGURE 28.7**   Scheme of the basic function of LIB showing the direction of the current and Li ion flow during both (a) charge and (b) discharge. (Shearer, C. J., Cherevan, A. & Eder, D.: Application and Future Challenges of Functional Nanocarbon Hybrids. *Advanced Materials*. 2014. 26. 2295–2318. Copyright Wiley-VCH Verlag GmbH & Co. KGaA. Reproduced with permission.)

## 28.5.2 Supercapacitor Devices

Supercapacitors are another type of electrochemical storage devices that have been extensively developed in the last years. Recent reports have confirmed that the hybridization of electroactive compounds with various nanocarbons improves their electrochemical performance due to the high surface area of the nanocarbons, their excellent electrical properties, tunable surface chemistry, and mechanical stability.

Supercapacitors are used for quick charge, fast-release applications with lifetimes of over a million cycles (Miller & Simon 2008) and can be categorized into two classes depending on their energy storage mechanism: (a) EDLCs, based on charge separation at the electrode/electrolyte interface, and (b) pseudocapacitors, based on various faradaic processes in active electrode materials (e.g., redox reactions or intercalation). Nanocarbon hybrids have demonstrated enhanced performance based on both mechanisms of energy storage. EDLC supercapacitors can be improved by introducing CNTs and graphene as active substrates due to their large surface area, high conductivity, and stability. Most studies have investigated hybrids of CNTs and graphene with polymers such as polyaniline, polypyrrole, and polythiophene derivatives, which have resulted in specific capacity values as high as 100–300 F g⁻¹. The major synthetic advantages of the conducting polymers are their flexibility and ease of processing. However, conducting polymers generally exhibit relatively low specific capacity and can suffer from low recycling rates due to degradation of the organic polymers upon charge/discharge cycling.

In contrast to this, metal oxides offer higher intrinsic specific capacitance and better electrochemical stability, but are generally semiconducting and therefore require a conductive support. A wide range of oxides (Table 28.5) has already been incorporated into nanocarbon hybrids. Reported results highlight both the potential of nanocarbon hybrids in energy storage devices and the importance of the interface between the components toward maximizing performance.

### 28.5.3 Heterogeneous Catalysis

Heterogeneous catalysis is another important field of application where nanocarbon hybrids have recently attracted much interest. For industrial processes, high–surface area substrates, e.g., ACs or carbon black, are used as supports for catalytic NPs due to their low cost and chemical inertness (Rodríguez-reinoso 1998). So far, only a few studies have reported on using nanocarbons as supports for noble metals, such as Pd and Pt, for hydrogenation reactions and Suzuki and Heck couplings (Su et al. 2013). However, nanocarbons can potentially offer a number of advantages compared to other carbon-based supports. For instance, the heat-sink effect (see Section 28.4.1) allows nanocarbon stabilization of comparatively smaller metal and metal oxide NPs on their surface. This results in higher surface area of the catalyst and enhances the overall catalytic activity, while the mechanically and thermally stable nanocarbon support enhances the lifetime and recyclability of the catalyst. Interfacial energy transfer is another process that can change selectivity of partial oxidation reactions due to redistribution of redox sites. Additionally, some recent report demonstrated that the defects in the nanocarbons can selectively drive a number of catalytic oxidative dehydrogenation reactions (Zhang et al. 2008). Another potential use of nanocarbon-based hybrids originates from the tubular structure of CNTs, whose inner channels with defined dimensions can provide a confined environment for catalysis. NPs of restricted size trapped inside the CNTs can affect reaction pathways due to the limited diffusion toward the active sites (Pan & Bao 2011). Despite the fact that research in this area has only just started, the first results commend nanocarbon hybrids as promising candidates for heterogeneous catalysis with the potential to improve catalyst recyclability, tailor mechanisms, and affect both activity and selectivity of a target process.

### 28.5.4 Photocatalysis

Photocatalysis has recently emerged as a highly attractive research field that addresses environmental (e.g., $CO_2$ fixation, water and air purification), health (e.g., disinfection), and sustainable energy (e.g., water splitting and chemical synthesis) issues. In photocatalysis, light photons are converted into reactive species (e.g., electron/hole pairs, radicals) that in turn catalyze the desired reaction. In particular, when a semiconductor is illuminated by light with energy greater than its bandgap ($hv > E_g$), electrons are excited from the valence band (VB) to the conduction band (CB), thereby leaving positively charged holes in the VB. These charge carriers may then diffuse to the particle surface provided that this diffusion happens faster than the competitive charge recombination/relaxation. As illustrated in Figure 28.8, the separated holes and electrons are consequently able to drive redox processes on the particle surface, either via direct charge transfer to the adsorbed reactants or via formation of reactive intermediates, such as $OOH^{\bullet}$, $OH^{\bullet}$, and $O_2^{\bullet-}$.

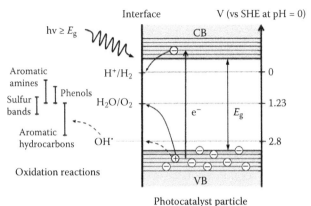

**FIGURE 28.8** Energy-space diagram of the photocatalytic processes that occur on a semiconductor surface: excitation of an electron from the VB to the CB by a light photon, and reduction and oxidation of corresponding species (depends on a particular application) by electrons and holes. (Shearer, C. J., Cherevan, A. & Eder, D.: Application and Future Challenges of Functional Nanocarbon Hybrids. *Advanced Materials*. 2014. 26. 2295–2318. Copyright Wiley-VCH Verlag GmbH & Co. KGaA. Reproduced with permission.)

When CNTs or graphene (or GO and rGO) are incorporated into a photocatalyst (i.e., a semiconducting oxide), several processes can be activated while leading to an increased photocatalytic performance of the hybrid material:

1. The nanocarbon acts as a chemically stable photosensitizer and extends the absorption range of a semiconductor through charge injection (Figure 28.5a) (Wang et al. 2005).
2. The nanocarbon extracts photogenerated electrons from the semiconductor, leading to charge separation and, thus, reducing electron–hole pair recombination, the major drawback of semiconductor photocatalysts (Figure 28.5b) (Leary & Westwood 2011).
3. The nanocarbon acts a cocatalyst and can effectively reduce adsorption and reaction overpotentials and facilitate the red/ox processes on its surface (Yeh et al. 2010).
4. The nanocarbon acts as an active substrate and a heat sink (Section 28.4.1), allowing to stabilize small NPs (i.e., increases surface area and thus number of active sites) and prevents particle aggregation (i.e., reduces diffusion limitations [Eder et al. 2006, Ren et al. 2012]).

Nanocarbon–inorganic hybrids have already demonstrated their excellent performance in different photocatalytic applications, including water splitting (e.g., for $H_2$ generation), selective oxidation reactions, and air/water purification. The photocatalytic enhancement mainly depends on the type of the nanocarbon used (e.g., graphene or CNTs; as-prepared or functionalized), morphology (e.g., particles or layers), structure (thickness, crystallinity) of the inorganic coating, and the choice of test reaction (reduction or oxidation, concentration of the target compound, etc.).

An average improvement of hybrid materials toward oxidation (e.g., dye degradation tested for $TiO_2$-, ZnO-, $WO_3$-, CdS-, and ZnS-based hybrids) is typically in the range of two to five times compared to bare photocatalyst without the nanocarbon and can be fully attributed to the aforementioned mechanisms. In contrast, the reduction of water to hydrogen is a more challenging task as it requires the use of photoexcited electrons, which also limits the choice of semiconductor with respect to the band positions (Figure 28.8).

$TiO_2$ is a suitable photocatalyst for both water oxidation and reduction, and has consequently become the most investigated material by far. However, its CB lies just about 0.1 eV above the reduction potential of water and the reduction potential of the electrons is often further lowered by the presence of defects, thus offering only a small thermodynamic driving force for proton reduction. Furthermore, with the bandgap of 3.2 eV (in anatase), it is not well suited for visible light irradiation, as it can absorb only about 4% of sunlight. Most nanocarbon–TiO$_2$ hybrids indeed showed a significant improvement over bare TiO$_2$, typically being associated with interfacial charge separation. As such, Fan et al. (2011) demonstrated that hybrids with rGO were considerably more active (10.9 times) than the corresponding CNT-based hybrid (4.7 times).

Many attempts have been made where visible light–active semiconductors with a smaller bandgap (e.g., various sulfides and nitrides) were implemented into nanocarbon hybrids. As such, in several works CNTs–CdS and graphene-based CdS hybrids were prepared and examined. Interestingly, photocatalytic results of Ye et al. (2012) revealed a similar trend: the enhancement of graphene-based CdS hybrid was found to be stronger compared to the CNTs–CdS hybrid (4.8 and 3.6 times compared to bare CdS, respectively). In contrast to this, when Ta$_2$O$_5$ was hybridized with both CNTs and GO (Cherevan et al. 2014), an opposite result was recorded (Figure 28.9a). This work also demonstrated that the photocatalytic properties of the hybrids could be further improved by purposefully engineering the interfaces and morphology of the active layer. In particular, the authors have grown ultrathin single crystalline layers of Ta$_2$O$_5$ over CNTs, therefore creating a tight junction (Figure 28.9d through f) between the components. The resulting hybrid (H2 type) possessed an excellent activity which was approximately three times higher than that of a CNTs–Ta$_2$O$_5$ hybrid of H1 type (see Figure 28.9c) where the oxide layer was composed of small NPs, and a poor interface to the nanocarbon was formed in the hybrid.

In conclusion, comparisons between graphene and CNT hybrids are often complicated by the apparent differences in size and morphology of the inorganic layer, electronic and mechanical properties of the nanocarbon (e.g., stiff MWCNTs versus flexible SWCNTs and graphene), light absorption properties, different carbon contents and synthetic protocols used, and nature and extent of the interface. More

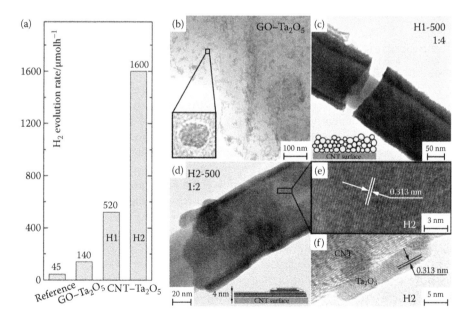

**FIGURE 28.9** (a) Hydrogen evolution rates in water splitting for 50 mg of a photocatalyst loaded with a Pt cocatalyst for a reference sample (Ta$_2$O$_5$ nanotubes), GO–Ta$_2$O$_5$ hybrid, and CNTs–Ta$_2$O$_5$ hybrids of H1 and H2 morphologies. HRTEM images of the (b) GO–Ta$_2$O$_5$ hybrid consisting of well-distributed crystalline oxide particles on the GO surface and of CNTs–Ta$_2$O$_5$ hybrid with morphology types (c) H1 and (d) H2; insets schematically represent the corresponding coating morphology. HRTEM images of the H2 hybrid showing (e) (101) lattice fringes of the Ta$_2$O$_5$ layer and (f) the tight interface between CNTs and Ta$_2$O$_5$. (Cherevan, A. S. et al., *Energy & Environmental Science* 7 (2): 791–796, 2014. Reproduced by permission of The Royal Society of Chemistry.)

efforts have to be made in order to maximize synergistic effects offered by the hybrids by means of a rational design of the morphology and the interface for any nanocarbon type.

## 28.5.5 Electrocatalysis and Fuel Cells

Electrocatalytic electrodes are able to increase the performance and reduce the energy requirements of a wide range of fuel cells (Kunze & Stimming 2009). ORR—a bottleneck of fuel cell technology—is commonly catalyzed by noble metal catalysts supported by AC. The application demands that the support material is highly resistant toward corrosion and has a high surface area and electronic conductivity. The major drawbacks of contemporary direct alcohol fuel cells are unavoidable poisoning of the catalyst electrode by CO and insufficiently fast kinetics of the electro-oxidation of alcohols. In this perspective, CNTs and graphene can reduce the poisoning of Ru, Pt, and other NPs during CO oxidation while increasing the reaction kinetics. In addition, they are able to enhance power densities at higher backpressures, lower the onset potentials, and allow higher anodic currents compared to AC supports. It has also been shown that depositing Pt NPs on CNTs–metal oxide hybrids improves the lifetime of the electrocatalyst. In most reported cases, the enhanced performance is due to smaller and more uniformly distributed noble metal particles and their stronger interaction with the nanocarbon. This results in better interface and enhances the electron transfer between the components, at the same time improving the stability of the NPs.

Contemporary research aims at designing fuel cells that do not require the presence of expensive precious metal NPs. Preliminary results have already shown that N-doping of CNTs and graphene-based nanocarbons can turn them into efficient ORR catalysts without the use of any noble metal (Lai et al. 2012). Further research, however, is required to determine which functional groups or defect sites are responsible for this performance. This knowledge will allow us to produce efficient and reliable fuel cells from cheap and available materials to fulfill the requirements of future devices.

## 28.5.6 Photovoltaic Devices

Solar energy is the most abundant and the largest source of renewable energy available for humankind (Lewis 2007). A number of approaches have been developed in order to convert sunlight into electricity, among which the most famous is the solar cell technology based on photovoltaic effect. All photovoltaic cells share four common components: they require a transparent conducting electrode (TCE), a photo-excitable active layer, an electron-transporting layer, and a hole-transporting layer. So far, nanocarbons have been incorporated into a number of photovoltaic cells both as additives and as individual components (Figure 28.10).

TCEs (Figure 28.10) are currently based on thin films of indium-doped (or fluorine-doped) tin oxide (ITO) deposited onto a glass substrate. These have low resistivity and high transparency, but are brittle and prone to cracking at high temperatures. TCEs based on or involving nanocarbons have recently become very popular alternatives. Here, the nanocarbon is used either as an additive (e.g., with transparent polymers) in order to enhance overall conductivity or alone to provide a freestanding, thin, and porous network. In order to obtain conductive electrodes with sufficient transparency, the nanocarbon film has to be very thin (containing only a few layers of SWCNTs or graphene) (Tune & Shapter 2013). This has been achieved via a number of techniques such as spin coating, drop casting, or direct transfer for both pristine CNTs and graphene. Despite the fact that sheet resistance recorded for such films is still relatively high due to the charge percolation between individual CNTs or graphene flakes, it can be further optimized by hybridizing the nanocarbons with both metal NPs and conducting polymers. Such hybrids combine high flexibility of the nanocarbons with a low sheet resistance.

The incorporation of nanocarbon hybrids into more complex photovoltaic components (e.g., photoactive or charge-transporting layers) requires more efforts and fundamental studies. However, the first results demonstrate their superior performance and commend the hybrids as promising candidates for flexible electronic devices (Gomez De Arco et al. 2010).

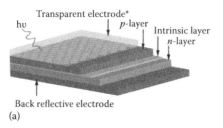

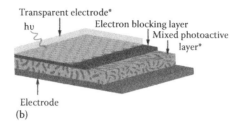

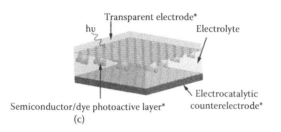

**FIGURE 28.10** Nanocarbon hybrid-based photovoltaics. Schematics of (a) an inorganic, (b) an organic, and (c) a dye-sensitized solar cell showing major components of each system. Components for which nanocarbon hybrids can be incorporated are marked with an asterisk. (Shearer, C. J., Cherevan, A. & Eder, D.: Application and Future Challenges of Functional Nanocarbon Hybrids. *Advanced Materials*. 2014. 26. 2295–2318. Copyright Wiley-VCH Verlag GmbH & Co. KGaA. Reproduced with permission.)

## 28.5.7 Chemical Sensors and Biosensors

Chemical sensors (and biosensors) are devices that allow the detection of chemical (and biological) compounds with both high sensitivity and selectivity. This application is important for a range of tasks including air/water/soil quality monitoring, medical diagnosis, and chemical synthesis (He et al. 2012). Nanocarbon materials (e.g., CNTs and graphene) hold great promise for the development of such devices due to their high surface area (resulting in large interaction zone between the analyte and the sensor), and excellent electrical (resulting in faster response at lower operation temperature) and mechanical properties (making it possible to fabricate a flexible device). The selectivity and sensitivity of nanocarbon-based sensors can be further enhanced by hybridization with a sensing element, such as metal oxide and metal NPs, organic fluorophores, and biomolecules, depending on a particular application. For example, excellent electronic properties of nanocarbons define their use in electronic and electrochemical sensors, while their ability to quench fluorescence can be used in optical sensors.

The most common configuration of nanocarbon hybrid electronic sensors is based on field-effect transistors (FETs, or so-called chemresistors). In this setup, two metal electrodes (the source and the drain) are connected via a (semi)conducting channel (i.e., CNT or graphene) on an insulating substrate (Figure 28.11a). When a molecule adsorbs on the nanocarbon surface, the device records changes in the FET conductance. Nanocarbon hybrid FET-based sensors are capable of detecting gases, chemicals, and biomolecules with very high sensitivity, and also work in solutions (Figure 28.11b); however, further

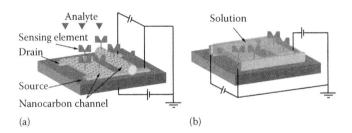

**FIGURE 28.11**  Nanocarbon FET-based nanocarbon-hybrid sensors showing schematics of (a) a gas sensor and (b) an aqueous sensor. (Shearer, C. J., Cherevan, A. & Eder, D.: Application and Future Challenges of Functional Nanocarbon Hybrids. *Advanced Materials.* 2014. 26. 2295–2318. Copyright Wiley-VCH Verlag GmbH & Co. KGaA. Reproduced with permission.)

progress is required for commercialization. The major drawbacks of such systems are their high cost and poor reproducibility upon fabrication.

Nanocarbon hybrids can also be used as optical sensors. The mechanism is based on the ability of nanocarbons to quench the fluorescence of a sensing element which is hybridized with the nanocarbon. When the analyte is present, the sensing element is separated from the nanocarbon and its fluorescence returns. From the amount of fluorescent light detected, one can determine the concentration of the analyte.

## 28.6  Conclusion

Nanocarbon hybrids are a new class of multifunctional materials that have already proved to be an excellent choice for a variety of fields, especially concerning environmental and sustainable energy applications (Table 28.5). Apart from the extended surface area and the combination of unprecedented mechanical and electrical properties of nanocarbons with functional properties of the second component, the hybrids offer a number of synergistic effects that originate from the extended interface between the components. Processes such as charge, energy, and heat transfer are therefore greatly facilitated and can enhance or even create new properties of the system leading to a number of potential applications. We would further like to highlight the importance of the hybrids' rational design and engineering on the nanoscale for maximizing those functions and mastering the structure–property relation.

## References

Ajayan, P. M., Stephan, O., Redlich, Ph. et al., "Carbon nanotubes as removable templates for metal oxide nanocomposites and nanostructures," *Nature* 375 (1995): 564–567. doi:10.1038/375564a0.

Aksel, S. & Eder. D., "Catalytic effect of metal oxides on the oxidation resistance in carbon nanotube–inorganic hybrids," *J. Mater. Chem.* 20 (2010): 9149–9154. doi:10.1039/C0JM01129K.

Berera, R., van Grondelle, R. & Kennis, J. T. M., "Ultrafast transient absorption spectroscopy: Principles and application to photosynthetic systems," *Photosynth. Res.* 101 (2009): 105–118. doi:10.1007/s11120-009-9454-y.

Cherevan, A. S., Gebhardt, P., Shearer, C. J. et al., "Interface engineering in nanocarbon–Ta$_2$O$_5$ hybrid photocatalysts," *Energy Environ. Sci.* 7 (2014): 791–796. doi:10.1039/C3EE42558D.

Ci, L., Suhr, J., Pushparaj, V. et al. "Continuous carbon nanotube reinforced composites," *Nano Lett.* 8 (2008): 2762–2766. doi:10.1021/nl8012715.

Cooke, D. J., Eder, D. & Elliott, J. A., "Role of benzyl alcohol in controlling the growth of TiO$_2$ on carbon nanotubes," *J. Phys. Chem. C* 114 (2010): 2462–2470. doi:10.1021/jp909117x.

Delaney, P., Di Ventra, M. & Pantelides, S. T., "Quantized conductance of multiwalled carbon nanotubes," *Appl. Phys. Lett.* 75 (1999): 3787–3789. doi:10.1063/1.125456.

De Volder, M. F. L., Tawfick, S. H., Baughman, R. H. et al., "Carbon nanotubes: Present and future commercial applications," *Science* 339 (2013): 535–539. doi:10.1126/science.1222453.

Eder, D., "Carbon nanotube–inorganic hybrids," *Chem. Rev.* 110 (2010): 1348–1385. doi:10.1021/cr800433k.

Eder, D. & Schlögl, R., *Nanocarbon–Inorganic Hybrids: Next Generation Composites for Sustainable Energy Applications* (De Gruyter, Berlin, Boston, 2014), http://www.degruyter.com/view/product/179736.

Eder, D. & Windle, A. H., "Carbon–inorganic hybrid materials: The carbon-nanotube/TiO$_2$ interface," *Adv. Mater.* 20 (2008a): 1787–1793. doi:10.1002/adma.200702835.

Eder, D. & Windle, A. H., "Morphology control of CNT–TiO$_2$ hybrid materials and rutile nanotubes," *J. Mater. Chem.* 18 (2008b): 2036–2043. doi:10.1039/B800499D.

Eder, D., Kinloch, I. A. & Windle, A. H., "Pure rutile nanotubes," *Chem. Commun.* 13 (2006): 1448–1450. doi:10.1039/B517260H.

Fan, W., Lai, Q., Zhang, Q. et al., "Nanocomposites of TiO$_2$ and reduced graphene oxide as efficient photocatalysts for hydrogen evolution," *J. Phys. Chem. C* 115 (2011): 10694–10701. doi:10.1021/jp2008804.

Gebhardt, P., Pattinson, S. W., Ren, Z. et al., "Crystal engineering of zeolites with graphene," *Nanoscale* 6 (2014): 7319–7324. doi:10.1039/C4NR00320A.

Gomez De Arco, L., Zhang, Y., Schlenker, C. W. et al., "Continuous, highly flexible, and transparent graphene films by chemical vapor deposition for organic photovoltaics," *ACS Nano* 4 (2010): 2865–2873. doi:10.1021/nn901587x.

He, Q., Wu, S., Yin, Z. et al., "Graphene-based electronic sensors," *Chem. Sci.* 3 (2012): 1764–1772.

Huang, Y. Y. & Terentjev, E. M., "Tailoring the electrical properties of carbon nanotube–polymer composites," *Adv. Funct. Mater.* 20 (2010): 4062–4068. doi:10.1002/adfm.201000861.

Kemnade, N., Shearer, C. J., Dieterle, D. J. et al., "Non-destructive functionalisation for atomic layer deposition of metal oxides on carbon nanotubes: Effect of linking agents and defects," *Nanoscale* 7 (2015): 3028–3034. doi:10.1039/C4NR04615C.

Kunze, J. & Stimming, U., "Electrochemical versus heat-engine energy technology: A tribute to Wilhelm Ostwald's visionary statements," *Angew. Chem. Int. Ed.* 48 (2009): 9230–9237. doi:10.1002/anie.200903603.

Lai, L., Potts, J. R., Zhan, D. et al., "Exploration of the active center structure of nitrogen-doped graphene-based catalysts for oxygen reduction reaction," *Energy Environ. Sci.* 5 (2012): 7936–7942. doi:10.1039/C2EE21802J.

Leary, R. & Westwood, A., "Carbonaceous nanomaterials for the enhancement of TiO$_2$ photocatalysis," *Carbon* 49 (2011): 741–772. doi:10.1016/j.carbon.2010.10.010.

Lewis, N. S., "Toward cost-effective solar energy use," *Science* 315 (2007): 798–801. doi:10.1126/science.1137014.

Lvovich, V. F., *Impedance Spectroscopy: Applications to Electrochemical and Dielectric Phenomena* (John Wiley & Sons, Hoboken, 2012).

Marom, R., Amalraj, S. F., Leifer, N. et al., "A review of advanced and practical lithium battery materials," *J. Mater. Chem.* 21 (2011): 9938–9954.

Miller, J. R. & Simon, P., "Electrochemical capacitors for energy management," *Science* 321 (2008): 651–652. doi:10.1126/science.1158736.

Pan, X. & Bao, X., "The effects of confinement inside carbon nanotubes on catalysis," *Acc. Chem. Res.* 44 (2011): 553–562. doi:10.1021/ar100160t.

Pugno, N. M., "On the strength of the carbon nanotube-based space elevator cable: From nanomechanics to megamechanics," *J. Phys. Cond. Matter* 18 (2006): S1971. doi:10.1088/0953-8984/18/33/S14.

Ren, Z., Kim, E., Pattinson, S. W. et al., "Hybridizing photoactive zeolites with graphene: A powerful strategy towards superior photocatalytic properties," *Chem. Sci.* 3 (2012): 209–216.

Rodríguez-reinoso, F., "The role of carbon materials in heterogeneous catalysis," *Carbon* 36 (1998): 159–175. doi:10.1016/S0008-6223(97)00173-5.

Shearer, C. J., Cherevan, A. & Eder, D., "Application and future challenges of functional nanocarbon hybrids," *Adv. Mater.* 26 (2014.): 2295–2318. doi:10.1002/adma.201305254.

Song, Z., Xu, T., Gordin, M. L. et al., "Polymer–graphene nanocomposites as ultrafast-charge and -discharge cathodes for rechargeable lithium batteries," *Nano Lett.* 12 (2012): 2205–2211. doi:10.1021/nl2039666.

Stankovich, S., Dikin, D. A., Dommett, G. H. B. et al., "Graphene-based composite materials," *Nature* 442 (2006): 282–286. doi:10.1038/nature04969.

Su, D. S., Perathoner, S. & Centi, G., "Nanocarbons for the development of advanced catalysts," *Chem. Rev.* 113 (2013): 5782–5816. doi:10.1021/cr300367d.

Tune, D. D. & Shapter, J. G., "The potential sunlight harvesting efficiency of carbon nanotube solar cells," *Energy Environ. Sci.* 6 (2013): 2572–2577. doi:10.1039/C3EE41731J.

Valeur, B. & Berberan-Santos, M. N., *Molecular Fluorescence: Principles and Applications* (John Wiley & Sons, Hoboken, 2012).

Vietmeyer, F., Seger, B. & Kamat, P. V., "Anchoring ZnO particles on functionalized single wall carbon nanotubes: Excited state interactions and charge collection," *Adv. Mater.* 19 (2007): 2935–2940. doi:10.1002/adma.200602773.

Vilatela, J. J. & Eder, D., "Nanocarbon composites and hybrids in sustainability: A review," *ChemSusChem* 5 (2012): 456–478. doi:10.1002/cssc.201100536.

Walsh, F. C. & Herron, M. E., "Electrocrystallization and electrochemical control of crystal growth: Fundamental considerations and electrodeposition of metals," *J. Phys. D Appl. Phys.* 24 (1991): 217. doi:10.1088/0022-3727/24/2/019.

Wang, W., Serp, P., Kalck, P. et al., "Visible light photodegradation of phenol on MWNT–$TiO_2$ composite catalysts prepared by a modified sol–gel method," *J. Mol. Catal. A Chem.* 235 (2005): 194–199. doi:10.1016/j.molcata.2005.02.027.

Williams, G. & Kamat, P. V., "Graphene–semiconductor nanocomposites: Excited-state interactions between ZnO nanoparticles and graphene oxide," *Langmuir* 25 (2009): 13869–13873. doi:10.1021/la900905h.

Wu, Z. S., Zhou, G., Yin, L. C. et al., "Graphene/metal oxide composite electrode materials for energy storage," *Nano Energy* 1 (2012): 107–131.

Yang, W., Wang, X., Yang, F. et al., "Carbon nanotubes decorated with Pt nanocubes by a noncovalent functionalization method and their role in oxygen reduction," *Adv. Mater.* 20 (2008): 2579–2587. doi:10.1002/adma.200702949.

Ye, A., Fan, W., Zhang, Q. et al., "CdS–graphene and CdS–CNT nanocomposites as visible-light photocatalysts for hydrogen evolution and organic dye degradation," *Catal. Sci. Technol.* 2 (2012): 969–978. doi:10.1039/C2CY20027A.

Yeh, T.-F., Syu, J.-M., Cheng, C. et al., "Graphite oxide as a photocatalyst for hydrogen production from water," *Adv. Funct. Mater.* 20 (2010): 2255–2262. doi:10.1002/adfm.201000274.

Zhang, J., Liu, X., Blume, R. et al., "Surface-modified carbon nanotubes catalyze oxidative dehydrogenation of N-butane," *Science* 322 (2008): 73–77. doi:10.1126/science.1161916.

# 29
# Nanographite–Polymer Composites

Georgios Chr.
Psarras

**Abstract**

Nanographite–polymer composites are extensively studied worldwide. Research efforts are guided by the very high values of modulus of elasticity, mechanical strength, and electrical conductivity, exhibited by the nanocarbon filler. Embedding carbonaceous nanofiller within a polymer matrix has been proved to be beneficial to the mechanical and physical properties of carbon–polymer nanocomposites, even at a low loading level. An overview of the fabrication procedures is presented, since the state of the achieved dispersion of nanoinclusions and the interfacial effects between matrix and reinforcing phase are crucial for the overall performance of the nanocomposites. The morphology, mechanical properties, thermal stability and conductivity, and electrical behavior of carbon nanocomposites are also presented and discussed. Graphitic nanofillers, such as CNTs, graphene, and graphite nanoplatelets, when suitably dispersed in a polymer matrix, provide a strong impetus for both thermomechanical and electrical performance. However, the effect on the electrical properties appears to be more pronounced. Electrical properties include the transition from the insulating to the conductive behavior at a low critical concentration of the conductive phase, the dielectric response, and the energy storage efficiency of the nanocomposites. Finally, current and potential applications of nanographite–polymer composites are discussed.

## 29.1 Introduction

Polymer nanocomposites represent a well-promising class of engineering materials. Intensive research efforts worldwide have been concentrated on the development and exploitation of novel polymer-based nanocomposites. The scientific, technological, and social impacts of nanotechnology and nanomaterials are well established nowadays, since nanotechnology and nanomaterials opened novel capabilities in designing and developing new structures and devices starting from the molecular scale level. The predicted influence of the "nano era" on human civilization and everyday life is sometimes compared to the explosion of informatics a few decades ago.

The large variety of different reinforcing nanoentities and the great number of the polymers available as matrices make the number of the possible polymer nanocomposites enormously high. Polymer nanocomposites should keep the polymers' advantages, such as good processability, light weight, thermomechanical performance, and low cost, and at the same time significantly improve their mechanical and physical properties, tending to approach the properties of the employed nanofiller. In contrast to conventional microcomposites, the optimum performance can be achieved at a low, or even very low, reinforcing phase content.

Carbon-reinforced polymer composites (CRPs) were at the front line of technological applications for decades. Aerospace and automotive industries, sports and leisure goods, tire industry, and microelectronic and electrochemical manufacturing are only a few examples. The industrial production of high-modulus and high-strength carbon fibers gave the real impetus in CRP applications. The next breakthrough in CRPs is related to the advent of nanofillers, such as CNTs, CNFs, and graphene nanosheets. The extremely high values of Young's modulus, mechanical strength, and electrical conductivity of CNTs and graphene created great expectations for the development of nano-CRPs with a dramatic improvement in their properties. A lot of efforts, time, and money are invested in this field by the scientific community and the society, targeting superb materials with tailored mechanical, electrical, thermal, optical, and other properties. Besides nanofiller properties and geometrical characteristics, key factors for the performance of carbon–polymer nanocomposites are the state of the distribution of nanoinclusions, and the interface between polymer matrix and nanofiller. The latter appears to have a crucial role on the overall behavior of nanocomposites and, thus, is the focus of scientific research worldwide.

## 29.2 Types and Forms of Nanographite Fillers

Carbon is the sixth element of the periodic table of elements, lying in the IV column. Its substantial difference from the properties of the rest of the column's elements is related to its ability to form $sp^2$ bonding.

Graphite is one of the three forms of carbon which can be found in nature. The other two are coal and diamonds, with the latter existing in very limited quantities. In the last three decades, new allotropic forms of carbon were discovered, starting in 1985 with fullerene (Kroto et al. 1985), followed in 1991 by CNTs (Iijima 1991), and more recently graphene (Geim & Novoselov 2007). While another form of elemental carbon is carbyne (Dresselhaus et al. 1996). The name *graphite* is related to its ability to leave marks and letters on a paper sheet, and its etymological origin is the Greek word *graphē* (γραφή in both ancient and modern Greek), which means "writing."

Graphite is the oldest known allotrope of crystalline carbon. Its structure consists of parallel stacks of 2D sheets, where hexagonals of elemental carbon are organized in a honeycomb lattice. The graphite lattice is anisotropic because of the different type of bonding between in-plane and out-of-plane atomic bonds. As a result, the modulus of elasticity is much higher in the direction parallel to the sheets than perpendicular to them. A single graphitic layer is called *graphene layer* or *graphene sheet*. Weak van der Waals forces are responsible for keeping the parallel graphene layers together in the graphitic structure. The distance of adjacent layers is 0.335 nm; and because of their mutual weak bonding, the layers are able to slide with respect to each other.

Graphene is a 2D monoatomic sheet of carbon, which can be considered as the building block for different carbon nanostructures. 3D structures, like CNTs or vapor-grown CNF, can be prepared by enfolding single graphene sheets.

Graphite oxide is produced by the oxidation of graphite by employing strong acids and oxidizing agents. Graphene oxide (GO) is an oxygenated graphene, with the carbon atoms connected to the epoxide and hydroxyl groups, while the atoms at the edges are connected to the carbonyl and carboxyl groups. Methods for the synthesis of GO have been reported elsewhere (Kovtyukhova et al. 1999, Tantis et al. 2012).

Graphite nanoplatelets (GNPs) are produced by exfoliating flat graphite sheets to platelets of few graphene sheets with varying thickness (Han et al. 2012). GNPs are stacks of few graphitic monolayers and constitute a low-cost alternative type of reinforcing phase to graphene, with high mechanical and electrical properties in tandem with its light weight. The variety of carbon molecular structures and properties is connected with the hybridization states of the carbon–carbon bonds. Carbon hybridizes in $sp$, $sp^2$, and $sp^3$ configurations, corresponding to chain, planar, and tetrahedral structures, respectively.

# 29.3 Fabrication of Graphite–Polymer Nanocomposites

The fabrication of graphite–polymer nanocomposites follows the methods employed for the preparation of all types of polymer matrix nanocomposites. There are three main methods used for the fabrication of polymer nanocomposites: (a) solution blending or solvent-assisted methods; (b) melt blending; and (c) in situ polymerization (Karger-Kocsis & Wu 2004, Moniruzzaman & Winey 2006, Galpaya et al. 2012, Saravanan et al. 2014).

## 29.3.1 Solution Blending

Solution blending is a widely used method for the preparation of polymer matrix nanocomposites and is considered as an easy and straightforward technique. The method is based on the dispersion of the carbon nanofiller within a suitable solvent. Dispersion can be achieved by mechanical stirring, although in most cases, a successful dispersion of nanoinclusions requires the ultrasonication of the mixture. The next step involves the addition of the polymer in order to form a filler–solvent–polymer blend. In the case of epoxy matrices, oligomers and curing agents can be added in two successive steps (Yang et al. 2011). At this stage, it is essential to avoid the formation of clusters and agglomerates, the presence of which reduces the beneficial influence of nanoreinforcement and subsequently affects the mechanical, electrical, and other properties of the produced nanocomposites. The last step involves the removal of the solvent-mediator by evaporation or distillation and casting of the remaining nanocarbon–polymer blend. Choosing the appropriate solvent and removing the solvent from the final blend are crucial factors for the quality of the produced nanocomposites. A possible incompatibility of the solution constituents could negatively affect the quality of the achieved dispersion. Carbon nanocomposites with various polymer matrices have been prepared using the solution blending method, indicatively could be referred to the cases of poly(vinyl alcohol) (PVA), polyvinyl fluoride, polyethylene (PE), poly(methylmethacrylate), rubbers, polyurethane, epoxy resins, and others (Moniruzzaman & Winey 2006, Jiang et al. 2010b, Kim et al. 2010, Zhao et al. 2010, Jinhong et al. 2011, Kim et al. 2011, Chen et al. 2012, Layek et al. 2012, Li & McKenna 2012, Tantis et al. 2012, Zhang et al. 2012).

The chemical functionalization of nanocarbon inclusions improves their solubility and therefore their dispersion within the polymer matrix. The reduction of agglomeration allows the usage of higher filler loadings in the nanocomposites. Chemical functionalization can be done either by oxidation or by physical adsorption/grafting procedures. Attached functional groups could influence the interfacial interactions between carbon nanoinclusions and polymer matrix. In the case of graphene nanofiller, modification of its surface via physical adsorption has the advantage of keeping the structural integrity of the conjugated network (Tantis et al. 2012). On the other hand, the formation of defects on the graphitic nanostructures has been shown, upon treatment in oxidative conditions, grafting reactions, and by overexposure to high power sonication (Singh et al. 2011, Tantis et al. 2012). Induced defects on the graphitic nanostructures have a detrimental effect on the integrity and performance of the nanocomposites.

Graphene–polymer nanocomposites can be also fabricated by using noncovalently modified pristine graphene nanoplatelets. Incorporating amphiphilic block copolymer-modified graphenes within PVA, using water as the compounding solvent, constitute a simple and environment-friendly fabrication method (Tantis et al. 2012).

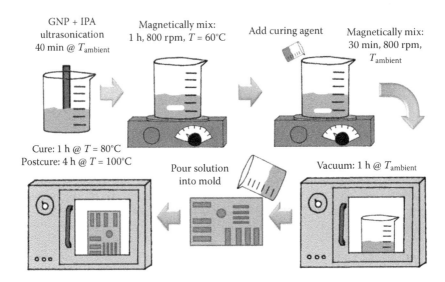

**FIGURE 29.1** Schematic representation of GNP/epoxy resin nanocomposites manufacturing procedure. (From Patsidis, A. C., Hybrid nanodielectrics of polymer matrix/functional inclusions: Development, characterization and functionality, PhD thesis, University of Patras, Patras, Greece, 2015.)

The conjugated structure of GO sheets could be broken by functional groups that contain oxygen, causing the localization of π-electrons, with the subsequent decrease in charge carriers' mobility and concentration. Attached groups act as scattering points, thus obstructing the charge migration and reducing the values of conductivity in the range of insulators. The reduction of GO has been used for recovering both the conjugated network of graphene layers and the initial high values of electrical conductivity (Wei et al. 2009, An et al. 2010, Ansari et al. 2010, Traina & Pegoretti 2012). However, since most of the polymers cannot withstand high temperatures, chemical reduction is the preferable route. Reduction agents should be carefully selected in order to avoid matrix degradation. A schematic representation of GNP/epoxy resin nanocomposite manufacturing procedure, via solution blending, is depicted in Figure 29.1.

## 29.3.2 Melt Blending

Melt blending is a feasible and convenient technique for preparing carbon–polymer nanocomposites. This technique is more suitable for thermoplastic polymers, since it involves high temperatures and shear forces for the dispersion of nanoinclusions. The polymer matrix is heated up to a temperature at which it becomes soft or semiliquid, and nanoinclusions are dispersed under the influence of high shear forces applied by a rotating mixer. The absence of any solvent, and particularly of toxic solvents, is the main advantage of the method, although the dispersion of nanoinclusions becomes difficult at high filler loadings due to the increase in the mixture's viscosity (Singh et al. 2011). Additional problems of buckling, rolling, or even cutting of the graphene sheets during the shear mixing procedure have been reported (An et al. 2010).

The masterbatch approach is a common way to fabricate series of carbon–polymer nanocomposites with varying concentration of the reinforcing phase. The polymer–carbon (CNT or graphene) mixture is placed in a stirring device under vacuum. Applied shear forces reduce the particle agglomeration, while vacuum removes air inclusions. Reduction of the mixture's viscosity can be achieved by increasing the temperature. The dispersion of nanoinclusion can be further improved by applying a sonication procedure to the produced mixture. Precalculated values of composite constituents are used in order to obtain the desired concentration of the reinforcing phase (Vavouliotis et al. 2010).

### 29.3.3 In Situ Polymerization

Another widely used technique for the preparation of carbon–polymer nanocomposites is in situ polymerization. A variety of polymers, both thermosetting and thermoplastics, can be employed as matrices (Potts et al. 2011, Teng et al. 2011, Wang et al. 2011a, Zhang et al. 2011, Chatterjee et al. 2012, Fabbri et al. 2012, Huang & Lin 2012, Mo et al. 2012, Zaman et al. 2012, Fim et al. 2013). Initially, the reinforcing phase is dispersed within the monomer, which is in liquid form. Afterward, the appropriate chemical agent is added, and polymerization is triggered by heating or by exerting radiation. In the case of 2D fillers, such as graphene sheets or graphene-modified sheets, molecules intercalate inside their galleries, thus increasing the separation space between the adjacent layers. Moreover, exfoliated graphene layers are also formed, leading to nanocomposites with fine filler dispersion. During in situ polymerization, the covalent bonds between modified carbon nanofiller and macromolecules could be developed, providing composites with increased durability. Two serious disadvantages of this fabrication technique are the progressive increase in mixture viscosity with the polymerization level and the usage in some case of solvents for the better initial distribution of nanofiller. The first one makes the manipulation of the mixture difficult, shortens the available time for the composite preparation, and also limits the amount of reinforcing phase which can be loaded. The problems of the second one resemble those of solution blending and are related to the efficient removal of the employed solvent.

Rubbers are an important family of engineering materials, which are used in various fields, such as mechanical engineering applications, automotive industry, construction/buildings, and agriculture. The wide range of rubber applications is related to their low modulus of elasticity, their ability to recover from very high strains, and their high internal damping (Wrana et al. 2001, Felhös et al. 2008, Wang et al. 2008). Rubbers exhibit low glass-to-rubber transition temperature; thus, the majority of their applications are used in the elastomeric phase. Because of the broad range of applications, the performance of rubbers has to be improved in order to address several problems arising in service. These problems include thermal and chemical resistance, environmental stability, conservation of elasticity at low temperatures, and problems arising from the accumulation of electrostatic charges on the surface of moving or rotating rubber-made parts. The latter could cause an early failure of the component via the appearance of unexpected sparks (Psarras et al. 2014). The incorporation of carbon nanoinclusions, such as MWCNTs, in rubber matrices could improve the thermomechanical and electrical performance of the composites. Since most rubbers are available in latex, the latex technique can be employed for the preparation of carbon–rubber nanocomposites. Latex can be considered as a stable aqueous dispersion of fine rubber particles (Karger-Kocsis & Wu 2004). The aqueous latex solution can be mixed with the dispersion of MWCNT in an SDS water solution. Prior to mixing, ultrasonication of the MWCNT dispersion minimizes the formation of clusters and agglomerates. Furthermore, care should be taken in order to limit the conductivity reduction caused by SDS (Yu et al. 2007, Jiang & Drzal 2010a, Jiang et al. 2012). The MWCNT/rubber nanocomposites are fabricated by freezing the mixture, at low temperature, and removing the aqueous solvent in a freeze-dryer (Yu et al. 2007, Jiang & Drzal 2010a, Jiang et al. 2012).

## 29.4 Properties

### 29.4.1 Morphology

Morphology is related to both the preparation techniques and the properties of the nanocomposites, since it is the result, the outcome, of the composite fabrication and provides the structure from where properties are derived. Morphology can be investigated in various types of graphite–polymer nanocomposites by means of XRD spectra and electron microscopy images. X-ray spectra verify the presence of graphitic inclusions and give information about their dimensions, which are strong indications of the intercalation of inclusions by polymer macromolecules, by evaluating the distance between successive graphitic layers and valuable information with respect to the crystallinity of the composites. However, in order to confirm the intercalation of inclusions, TEM images should also be examined. Differential scanning calorimetry (DSC) is able to provide additional information concerning the systems' morphology. Crystallinity level, melting enthalpy, crystallization, melting, and glass transition temperatures can be deduced from DSC thermographs. The crystallization process, the achieved crystallinity level, and the portion of the amorphous part of the polymer matrix are of great importance because of their influence on the macroscopic properties of nanocomposites. The effect of graphitic nanofillers on the crystallization behavior and crystalline parameters is consequently of great importance. Examples of filler dispersion in nanographite–polymer nanocomposites are given in Figure 29.2. Figure 29.2a corresponds to a modified MWCNT/PEO, and Figure 29.2b to GNP/epoxy resin nanocomposite. Nanodispersions and small clusters can be detected. The increase in filler content favors the formation of clusters, because of the strong interactions between the nanoinclusions.

Polymer nanocomposites reinforced with 2D graphitic layers can attain three different configuration types: (a) microphase-separated composites, where the polymer matrix and graphitic stacks of layers remain immiscible; (b) intercalated structures, where polymer molecules are inserted between the graphitic layers; and (c) exfoliated structures, where individual graphitic layers are dispersed in the polymer matrix. In real systems, more than one or even all three configurations might coexist.

Carbonaceous nanofillers, incorporated within a semicrystalline polymer, act as nucleating agents directly affecting the crystallization process. Nucleation activation energy diminishes, induction period becomes shorter, and the whole crystallization process is accelerated. SWCNTs and MWCNTs, graphene, and GNPs have been reported to act as nucleating agents in various polymers, including PE, polypropylene, poly(butylene terephthalate), polyethyleneterephthalate, PVA, poly(vinylidene fluoride) (PVDF), hydrogenated acrylonitrile butadiene rubber (HNBR), and others (Huang et al. 2009, Li et al. 2009, Salavagione et al. 2009, Sridhar et al. 2013, Psarras et al. 2014, Xu et al. 2014). The crystallinity level can be examined by means of XRD spectra and DSC thermographs; however, the crystallization process and the nucleating ability of carbon nanofillers are investigated via DSC studies.

The XRD spectra of GNP-reinforced epoxy resin nanocomposites are depicted in Figure 29.3. The sharp peak at $2\theta = 26°$ is the characteristic diffraction pattern from the graphitic interlayer (Cao et al. 2001, Li et al. 2003, Psarras 2007a). This peak, in comparison to the corresponding one of the micrographitic structures (Psarras 2007a), becomes broader in the case of GNP/epoxy nanocomposites due to the reduction of the particle sizes.

HNBR is a semicrystalline rubber giving a broad peak at $2\theta = 19°$, which is attributed to the long-range stereoregularity of the amorphous state of acrylonitrile butadiene rubber (Psarras et al. 2014).

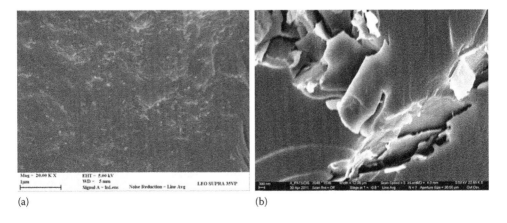

(a)                                                                                            (b)

**FIGURE 29.2** SEM images of (a) 5.0% w/w MWCNT/PEO nanocomposite (From Pontikopoulos P., Electric response of poly(ethylene oxide)/modified multiwall carbon nanotubes nanocomposites, MSc thesis, University of Patras, Patras, Greece, 2009.) and (b) 5 phr GNP/epoxy resin nanocomposite (Reprinted from Patsidis, A. C., Kalaitzidou, K. & Psarras, G. C., *Journal of Advanced Physics*, 2, 7–12, 2013. With permission.).

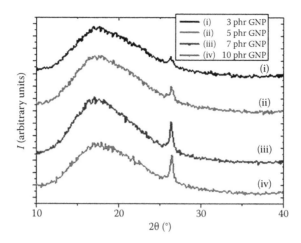

**FIGURE 29.3** XRD spectra of GNP/epoxy resin nanocomposites. (From Patsidis, A. C., Hybrid nanodielectrics of polymer matrix/functional inclusions: Development, characterization and functionality, PhD thesis, University of Patras, Patras, Greece, 2015.)

Recording of additional sharp peaks in the HNBR diffractograms is related to the presence of inorganic ingredients, such as ZnO or MgO, in the composition of HNBR. Incorporating MWCNTs in the HNBR matrix gives rise to an extra, rather wide and shallow, peak located at ~26°. This is the (002) graphite peak, which is also present in the MWCNT spectra, and in the spectra of other allotropic forms of carbon (Cao et al. 2001, Li et al. 2003, Psarras 2007a). The level of crystallinity of pure HNBR and HNBR-based nanocomposites can be calculated as the ratio of the area under the 19° peak to the area under the whole XRD spectrum (Psarras et al. 2014). The initial degree of crystallinity for the unfilled HNBR was found to be 30.0%, and increases with the MWCNT content. The crystallinity level of HNBR + 10 parts per hundred rubber per weight (phr) MWCNT system was found to be equal to 35.8%, while the corresponding level for the HNBR + 20 phr MWCNT system was calculated as 47.6%. DSC studies in the same systems revealed a tendency of $T_g$ to increase with MWCNT content. This trend is in accordance with the increase in crystallinity, since the growth of the hard and rigid crystalline regions exerts restrictions on the motions of the amorphous parts. The segmental mobility of constrained amorphous polymer chains becomes more difficult and requires enhanced thermal agitation, leading to increased values of glass transition temperature.

PVDF is a ferroelectric semicrystalline polymer. It exhibits polymorphism of several crystalline phases. The most common one is α-phase, while the polar β-phase is of great importance in applications due to its high dielectric permittivity and piezoelectric properties. Recent studies on CNF-, CNT-, and GO-loaded PVDF have shown that the presence of carbon nanofillers promotes the transformation of α- to β-phase (Huang et al. 2009, Sun et al. 2010, Yu et al. 2011). The effect of the geometrical characteristics of the carbon nanofiller on the phases of PVDF was studied by employing long and short MWCNTs (Huang et al. 2009). Long MWCNTs had a diameter of 10–20 nm and a length of 5–15 μm, while the corresponding values for short MWCNTs were 10–20 nm and 1–2 μm, respectively. It is obvious that the long ones exhibit a higher value of aspect ratio. TEM images revealed a homogeneous and fine dispersion in the PVDF matrix with no preferred alignment. XRD spectra of the high–aspect ratio CNT specimens provided evidence for the presence of a mixture of both phases, for filler loadings up to 1.5% w/w. However, the trace of the XRD spectrum of the 2.0% w/w long CNT composite indicated that only the β-phase is present. XRD spectra from composites with low CNT aspect ratio include peaks corresponding to both phases at all filler loadings, and the portion of β-phase increases with the CNT content at the expense of the amount of α-phase. DSC studies carried out on the same two sets of composites revealed that pure PVDF and low-content, high–aspect ratio CNT composites present only one melting peak, corresponding to α-phase. With increasing concentration, a second melting peak is recorded, at a higher temperature, indicating the existence of β-phase. Finally, in the thermograph of the nanocomposite with 2.0% w/w of high–aspect ratio CNT, only the second melting peak is recorded, in accordance with the XRD spectra (Huang et al. 2009). The existence of both phases in the low–aspect ratio CNT–PVDF nanocomposites was confirmed by the relative DSC thermographs (Huang et al. 2009). Both types of MWCNTs act as nucleating agents; however, the influence of the high–aspect ratio ones is more effective, and their efficiency could be assigned to the larger specific interactions at the CNT–PVDF interface. The acceleration of the crystallization process of the β-phase, in the high–aspect ratio CNT composites, could be attributed to the large number of zigzag carbon atoms on the nanotubes' surface (Huang et al. 2009).

Increase in crystallinity and crystallites' size was detected by XRD and DSC studies on graphene-reinforced poly(3-hydroxybutyrate-co-4-hydroxybutyrate) nanocomposites, due to the addition of graphene (Sridhar et al. 2013).

Although it is generally accepted that the presence of carbon nanoinclusions inside semicrystalline polymers facilitates the crystallization process and alters crystallinity (Huang et al. 2009, Li et al. 2009, Sridhar et al. 2013, Psarras 2014, Xu et al. 2014), the effect of the filler content increase does not monotonically enhance the crystallinity level. XRD studies on reduced graphite oxide/PVA nanocomposites show that the intensity of the main peaks of pure PVA (namely, 11.3°, 19.4°, and 22.4°) reduces and the peaks become broader with reduced graphite oxide content, indicating a reduction in crystallinity, probably due to interactions between the polymer macromolecules and the employed filler (Salavagione et al. 2009). The glass transition temperature was found to decrease with the filler content, while melting temperature, crystallization temperature, melting enthalpy, and crystallinity level systematically decrease with the reduced graphite oxide loading. The recorded performance was explained by suggesting interactions between the filler and the polymer due to the remaining oxygenated groups in the graphene (Salavagione et al. 2009).

It seems that the incorporation of carbon nanofillers within semicrystalline polymers affects the crystallization process by two different mutual competing ways. Initially, the nanofillers act as nuclei for heterogeneous crystallization, while afterward they exert constraints on the polymer chains, limiting their mobility and obstructing the crystal growth. Nanocomposites with high filler content and fine dispersion of carbon nanoinclusions are a prerequisite for the improvement of the mechanical or electrical performance. However, as the filler content increases, restrictions to polymer chain mobility are imposed and a constrained network, via geometrical contacts, is formed reducing the rate of crystal growth (Xu et al. 2014). The crystallinity and crystallization parameters do not linearly depend on the carbon nanofiller content, and approach a saturation level at a filler concentration, which varies with the type of the employed nanofiller, the specific surface area of the constituents' interface, and the matrix (Grady et al. 2002, Miltner et al. 2008, Deng et al. 2010).

## 29.4.2 Thermomechanical Properties

Polymers are used as engineering materials in numerous applications; however, their low mechanical strength and stiffness restrict their potential use as structural materials. Thus, from the early steps of composite materials, it was clear that considerable benefit could be gained by reinforcing polymers. The significant impetus in the CRPs was achieved with carbon fiber–polymer composites. The incorporation of carbon nanoinclusions in polymer matrices created a breakthrough in CRPs, resulting in improved mechanical, tribological, and thermal performance, at significantly lower filler loading (Psarras 2014). Suitable mechanical reinforcement should be characterized by high stiffness (Young's modulus), high strength, and if possible, large elongation at break. Additional important issues are the geometrical characteristics, achieved level of dispersion, and adhesion and bonding between the matrix and the reinforcing phase.

The advent of CNTs created great expectations of superb mechanical performance of CNT–polymer composites, because of the extremely high values of CNT stiffness and strength. Various methods have been used for the estimation or measurement of the modulus of elasticity and mechanical strength of SWCNT and MWCNT (Wong et al. 1997, Krishnan et al. 1998, Salvetat et al. 1999, Yu et al. 2000a, Yu et al. 2000b). These methods include measurements of the vibration amplitude of freestanding SWCNTs in a TEM, employment of an MWCNT in a single cantilever-like configuration on an AFM, and stress–strain curves on individual MWCNTs obtained at a nanostressing stage within a SEM. The outcomes of all these studies were values for Young's modulus ranging from 0.95 to 1.25 TPa and for strength ranging from 30 to 63 GPa. These tremendously high values attracted the interest of the composite research community, almost exclusively for many years. Simple calculations via the rule of mixing lead to very high values of stiffness and strength of CNT–polymer nanocomposites, even at low filler loading. However, high predictions and expectations have not been fulfilled yet. The strong tendency of CNTs to form agglomerations could be considered as the main reason for the discrepancies between predicted and experimentally observed mechanical performance in CNT–polymer systems. Many efforts have been carried out in order to optimize the dispersion of CNTs or their alignment within the polymer matrix (Xie et al. 2005, Papagelis et al. 2007, Tasis et al. 2007, Sengupta et al. 2011, Roy et al. 2012). Additional problems related to the dispersion of CNTs arise from the weak interactions between the CNTs and the matrix, as well as the mismatch in the respective coefficients of thermal expansions. Surface modification of CNTs is employed for the improvement of their adhesion with the polymer matrix. Besides the influence on the distribution of the CNTs within the polymer matrix, a stronger bonding level between the constituents of the composite results in an increase in the glass transition temperature of the matrix.

Embedding carbon nanoinclusions in polymer matrices is beneficial for both static and dynamic mechanical properties, although the expected tremendous alteration has not been achieved. In fluorinated SWCNT-PEO nanocomposites, increases of 145% and 300% in Young's modulus and yield strength, respectively, have been reported (Geng et al. 2002) at 1% w/w loading. In carboxylated SWCNT/nylon-6 nanocomposites, the Young's modulus increases by 153% and tensile strength by 103% at 1% w/w filler content (Gao et al. 2005). Elastic modulus of MWCNT/phenylethyl-terminated polyimide has been found to systematically increase with the MWCNT content from 2.84 GPa for pure matrix to 3.90 GPa at 14.3% w/w filler loading (Ogasawara et al. 2004). In the same study, dynamic mechanical analysis (DMA) results showed that the storage modulus and the damping coefficient are not significantly altered with the MWCNT content, while $T_g$ monotonously increases with the reinforcing phase. The latter has been attributed to the formation of a secondary network by the MWCNTs in addition to the cross-linking structure of the polymer, or to the development of covalent bonds between the polymer chains and the functional groups on the surface of MWCNTs (Ogasawara et al. 2004, Thostenson et al. 2005). The price of the modulus and strength enhancement is a considerable reduction of ductility and fracture toughness. In many cases, elongation at break diminishes by a factor almost equal to 4 (Ogasawara et al. 2004, Gao et al. 2005, Thostenson et al. 2005, Moniruzzaman et al. 2006).

Graphene nanosheets exhibit Young's modulus higher than 1 TPa and strength of ~125 GPa (Singh et al. 2011). Similar to previous referred cases, the mechanical behavior of nanocomposites is strongly related to reinforcing phase content, distribution in the polymer matrix, interface bonding, and inclusions' aspect ratio. The thermal response and mechanical properties of graphene–polystyrene (PS) nanocomposites, prepared via covalently grafting PS chains on the graphene surface, improved significantly (Fang et al. 2009). DSC studies revealed a 15°C increase in $T_g$ in the composite specimen containing 12% w/w graphene sheets, with respect to the neat PS. Since glass-to-rubber transition is related to cooperative segmental motions of the polymer chains, the recorded enhancement was attributed to confinement effects exerted by graphene on the polymer chains (Fang et al. 2009). Mechanical properties were determined via standard tensile tests. Modulus of elasticity and tensile strength were found to systematically increase with the graphene content, reaching 2.28 GPa and 41.42 MPa, respectively, for the 0.9% w/w graphene–PS system. The corresponding values for the neat PS were 1.45 GPa and 24.44 MPa, respectively (Fang et al. 2009). At the same time, plastic deformation and elongation at break reduced significantly. The improvement of the mechanical properties was connected to the effective load transfer between the matrix and the filler (Fang et al. 2009).

Pristine graphene sheets are considered to exhibit higher mechanical strength with respect to functionalized GO sheets (Singh et al. 2011). However, functional groups on the GO sheets provide benefits regarding the level of dispersion and interfacial interactions, which could lead to better mechanical performance.

Glucose-reduced GO–PVA nanocomposites were fabricated by blending reduced graphene oxide (rGO) with aqueous solution of PVA (Ma et al. 2013). rGO suspensions in water were stabilized by adding surfactants, such as sodium dodecyl benzene sulfonate and poly(*N*-vinyl-2-pyrrolidone). The presence of the surfactants improves the dispersion of graphene in PVA (Ma et al. 2013). Glass transition temperature and crystallinity level were examined by DSC tests. Moreover, the filler effect on the mechanical properties of the nanocomposites was investigated by employing tensile stress–strain curves (Ma et al. 2013). $T_g$ increased from 73.5°C for the neat PVA to 81.5°C for the 0.7% w/w PVP-stabilized rGO (P-rGO)–PVA nanocomposite. This behavior was ascribed to the restrictions on the polymer chain mobility, imposed by hydrogen bonds between P-rGO nanosheets and PVA. The level of crystallinity appeared to be not affected by the presence of rGO, although the addition of filler in PVA could act as a nucleating agent for the formation of crystal structure (Ma et al. 2013). Young's modulus and tensile strength increase with the rGO content. The corresponding values for the neat PVA are 3.3 GPa and 105 MPa, while for the 0.7% w/w P-rGO–PVA nanocomposite, the values are 4.9 GPa and 154 MPa, respectively, indicating a 48% increase in modulus and a 47% increase in strength. Strong interactions between the filler and the matrix were considered responsible for the mechanical behavior. A significant decrease in ductility in the nanocomposites was also reported (Ma et al. 2013).

Noncovalently functionalized graphene sheets (with aromatic amino acid, tryptophan) were dispersed in PVA at very low filler loading (0.2% w/w). The nanocomposite produced was characterized by enhanced thermal stability (determined via TG analysis tests), a small increase in modulus of elasticity, and a 23% increase in tensile strength, compared to pure matrix (Guo et al. 2011).

Flexural modulus derived from a three-point bending test in GNP/epoxy resin specimens increases with the filler content, and optimum performance is exhibited by the nanocomposite with the highest GNP concentration (Figure 29.4) (Patsidis et al. 2014). On the other hand, the flexural strength decreases with the filler content, indicating that nanoinclusions could act as stress raisers within the polymer matrix (Patsidis et al. 2013b).

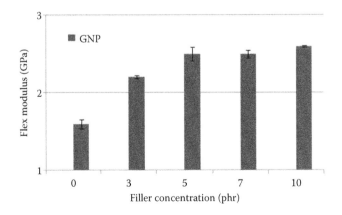

**FIGURE 29.4** Flexural modulus as a function of filler content for neat epoxy and GNP nanocomposites. (With kind permission from Springer Science+Business Media: *Journal of Thermal Analysis and Calorimetry*, Graphite nanoplatelets/polymer nanocomposites: Thermomechanical, dielectric, and functional behavior, 116, 2014, 41–49, Patsidis, A. C., Kalaitzidou, K. & Psarras, G. C.)

DMA studies on the same nanocomposite systems revealed an increase in storage modulus with the GNP content from ambient temperature up to 100°C (Patsidis et al. 2014). The storage modulus increases with the filler content and optimum performance is observed at the higher GNP loading (10 phr) (Figure 29.5). At temperatures higher than 110°C, the storage modulus exhibits an abrupt decrease, which is related to the glass-to-rubber transition of the polymer matrix. The transition from the rigid glassy state to the soft rubbery one is accompanied by a significant decrease in the composite stiffness. Mechanical loss, expressed via the variation of $\tan\delta$ with temperature, represents the ratio of dissipated energy to the stored one. The intensity of $\tan\delta$ peaks diminishes with the GNP content, while peaks become narrower. These observations lead to the conclusions that the nanocomposite energy damping capability decreases and the elastic behavior increases, while macromolecular segments between cross-links become constrained and immobilized (Patsidis et al. 2014).

Fracture toughness, impact strength, and fatigue have also been studied in graphene–polymer nanocomposites. A comparative study of the mechanical performance of GNP/epoxy, SWCNT/epoxy, and MWCNT/epoxy nanocomposites, at constant nanofiller content (0.2% w/w), denoted that the GNPs are significantly more beneficial to mechanical properties than CNTs are (Rafiee et al. 2009). In particular, the modulus of elasticity was found to be 31% greater than pure epoxy, in the case of the GNP/epoxy

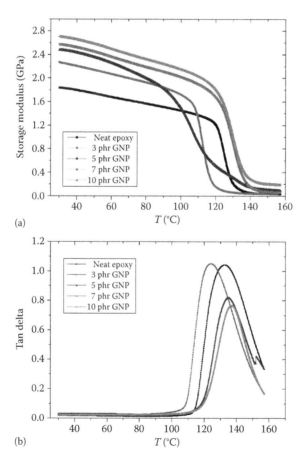

**FIGURE 29.5** (a) Storage modulus and (b) loss tangent delta as a function of temperature, at $f = 1$ Hz, GNP/epoxy resin nanocomposites. (With kind permission from Springer Science+Business Media: *Journal of Thermal Analysis and Calorimetry*, Graphite nanoplatelets/polymer nanocomposites: Thermomechanical, dielectric, and functional behavior, 116, 2014, 41–49, Patsidis, A. C., Kalaitzidou, K. & Psarras, G. C.)

system, while the relative increase in SWCNT/epoxy system was limited to 3%. The increase in tensile strength was almost 40% in GNP/epoxy composites and 14% in MWCNT/epoxy composites. The corresponding increases in fracture toughness were 53% and 20% for the GNP- and MWCNT-reinforced composites, respectively. GNP/epoxy nanocomposites also show superior performance under fatigue testing (Rafiee et al. 2009, Rafiee et al. 2010).

GNP/epoxy nanocomposites with hierarchically constructed interphase through locally rich environment in amine demonstrated improved mechanical response, including a 93.8% increase in fracture toughness and a 91.5% increase in flexural strength, at 0.6% w/w GNP reinforcement (Fang et al. 2010). The superior reinforcing efficiency of GNPs on CNTs has been attributed to their high specific area, increased adhesion/bonding of polymer matrix on their rough surface, and their 2D geometry (Rafiee et al. 2009, Rafiee et al. 2010).

Analogous performance has been observed in composite systems reinforced by different types of carbon nanoinclusions. For example, the incorporation of CNFs in polyoxymethylene has been found to be beneficial for stiffness, mechanical strength, resistance to creep, and stress relaxation. On the other hand, nanocomposite ductility decreases with CNF incorporation (Siengchin et al. 2010).

## 29.4.3 Electrical Properties

Polymers are basically electrical insulators, while carbon inclusions are conductive. Mixing these two components leads to nanocomposites with very interesting dielectric behavior and variable conductivity. By controlling the type and the amount of nanoinclusions, the dielectric properties can be tuned according to the applications' demands. Dielectric behavior can be studied using broadband dielectric spectroscopy in a wide frequency and temperature range. Additionally, the dependence of conductivity on temperature, frequency, and intensity of the applied field allows the determination of the physical origin of the occurring charge transport mechanisms (Psarras 2006, Psarras 2007a, Psarras 2009a, Psarras 2010). The content of the conductive phase, carbon nanoinclusions in our case, governs the overall conductivity of the polymer nanocomposites; and at a critical concentration (also known as percolation threshold), a transition from the insulating to the conductive state occurs (Zallen 2004, Tjong 2014).

### 29.4.3.1 Dielectric Behavior

As previously mentioned, the majority of the polymers are insulators, because of their low concentration of free charge carriers. Moreover, in polymer nanocomposites, the polymeric matrix represents the major part of the composite system, and thus the dielectric response of the nanocomposites is reasonable to be significantly affected by the dielectric properties of the matrix. Dielectric data can be analyzed by means of different formalisms, namely, dielectric permittivity, electric modulus, and AC conductivity. All three formalisms describe the same physical effects; however, under certain conditions, one of them could be proven more helpful in deriving information for the studied physical mechanisms (Psarras 2010). Thermosetting polymers are amorphous, and their dielectric spectra include relaxation processes arising from the cooperative segmental motion of the macromolecules, related to the glass-to-rubber transition of the polymer, local rearrangement of polar side groups of the main polymer chain, and locally restricted motions of small and soft parts of the main chains. These processes are typically denoted with the small letters of the Greek alphabet, as $\alpha$-, $\beta$-, and $\gamma$-*relaxations*, respectively. On the other hand, semicrystalline polymers—like thermoplastics or rubbers—exhibit all three previously mentioned processes plus interfacial polarization (IP), which occurs at the interface of crystalline and amorphous regions. In some cases, $\delta$-relaxation process is present in the dielectric spectra of semicrystalline polymers. $\delta$-relaxation is not recorded very often, and its origin has been attributed to defect dipoles in the crystalline phase (Karahaliou et al. 2014).

Figure 29.6a depicts the variation of the real part of dielectric permittivity with temperature, at constant frequency ($f$ = 1 Hz), for a thermosetting epoxy resin/GNP-reinforced nanocomposite system, at various filler loadings. Values of $\varepsilon'$ systematically increase, in the whole temperature range,

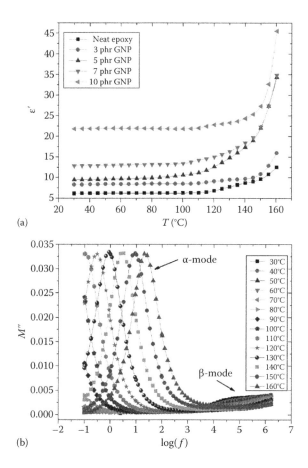

**FIGURE 29.6** (a) Real part of dielectric permittivity as a function of temperature for the GNP/epoxy resin nanocomposites, at 1 Hz; (b) imaginary part of electric modulus as a function of frequency, at various temperatures for the 7 phr GNP/epoxy nanocomposite. (With kind permission from Springer Science+Business Media: *Journal of Thermal Analysis and Calorimetry*, Graphite nanoplatelets/polymer nanocomposites: Thermomechanical, dielectric, and functional behavior, 116, 2014, 41–49, Patsidis, A. C., Kalaitzidou, K. & Psarras, G. C.)

with filler content (Patsidis et al. 2014). Mainly responsible for this behavior is the enhanced conductivity of the GNPs. The electrical properties of this binary composite system differ significantly, and thus the examined nanocomposite system is characterized by high electrical heterogeneity. The IP or Maxwell–Wagner–Sillars effect is detected in electrically heterogeneous systems. Unbounded charges, which are present from the manufacturing stage of the composites, are gathered at the constituents' interface, forming large dipoles (Tsangaris et al. 1998, Kalini et al. 2010). These dipoles are characterized by increased inertia in following the alternation of the applied electric field. By providing sufficient time, low-frequency range, and thermal agitation, the induced dipoles are oriented parallel to the field, increasing specimen's polarization and permittivity. Thus, IP is a slow relaxation process, recorded at high temperatures and low frequencies. The increase in the real part of dielectric permittivity, at temperatures higher than 130°C, shown in Figure 29.6a, expresses the influence of IP. In the vicinity of 120°C, close to the glass transition temperature of the specific resin, a moderate rise in ε′ values is observed (Patsidis et al. 2014). In this temperature area, large parts of the polymer chains acquire enhanced mobility and follow the exerted orientation of the applied field, leading to higher polarization and permittivity values. Loss modulus index ($M''$) as a function of frequency, at various temperatures, for the 7 phr GNP/epoxy nanocomposite is depicted in Figure 29.6b. The dielectric loss spectra of the examined GNP/epoxy nanocomposites include two peaks, also present in the spectrum of pure matrix. The relaxation processes are assigned as follows: (i) the more intensive one to the glass-to-rubber transition (α-relaxation) of the amorphous matrix and (ii) the weaker one in the high-frequency range to local motions of polar side groups (β-relaxation) (Patsidis et al. 2014). Nanoinclusions and especially 2D nanofillers could be considered as a dispersed network of nanocapacitors inside the polymer matrix, acting as nanodevices where electric energy can be stored and harvested (Patsidis & Psarras 2013a, Patsidis et al. 2014). The efficiency of energy storing can be expressed via the energy density of the material, which is determined by

$$U = \frac{1}{2}\varepsilon_0\varepsilon'E^2, \tag{29.1}$$

where $\varepsilon_0$ is the permittivity of free space, $\varepsilon'$ is the real part of dielectric permittivity, and $E$ is the intensity of electric field. The only material parameter in Equation 29.1 is the dielectric permittivity, while the maximum value of energy density is restricted by the material's dielectric strength. Considering a constant applied field, below the dielectric strength, the variation of energy density with frequency and temperature follows the corresponding variations in permittivity. Figure 29.7 presents the variation in the relative energy density of GNP/epoxy resin nanocomposites, normalized on the energy density of

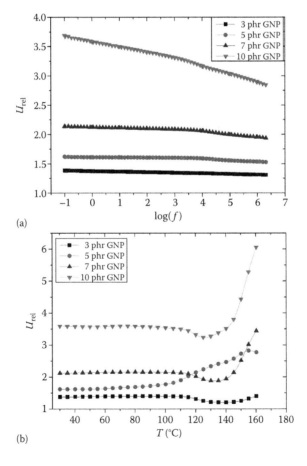

(a)

(b)

**FIGURE 29.7** Relative energy density (values normalized upon energy density of neat resin) as a function (a) of frequency at 30°C and (b) of temperature at $f = 1$ Hz, for the GNP/epoxy resin nanocomposites.

the neat matrix (Patsidis et al. 2014). Embedding GNPs in epoxy resin is beneficial for energy storage in the whole frequency and temperature range, and at 10 phr loading, the storing energy is multiplied by a factor close to 4 at 30°C, or 6 at high temperatures (Patsidis et al. 2014).

PEO is a semicrystalline polymer with low glass transition temperature (~−28°C), used in many applications in batteries, fuel cells, and in other solid polymer electrolyte technologies. MWCNT/PEO nanocomposites exhibit improved mechanical and electrical performances. Dielectric relaxations arising from both the polymer matrix and the filler, for the 0.50% w/w MWCNT/PEO nanocomposite, are depicted in Figure 29.8. It should be noted that the employed CNTs have been chemically modified targeting a fine distribution of PEO (Ratna et al. 2009). The loss modulus index spectra as a function of frequency, at various temperatures, include peaks corresponding to IP, glass-to-rubber transition (α-mode), while the relaxation arising from the polar side groups' rearrangement (β-mode) has been shifted to higher frequencies out of the experimental window (Pontikopoulos 2009, Psarras 2010, Pontikopoulos & Psarras 2013).

HNBR is an elastomer used by the automotive industry in power transmission belts, timing belts, vibration dampers, hoses, seals, etc. Embedding CNTs within HNBR considerably increases its mechanical properties as well as its wear resistance (Dang et al. 2007, Jiang et al. 2008). Many moving or rotating insulating parts are manufactured from HNBR; during their service life, electrostatic charges are developed on their surface, leading to sparks and undesirable early failure of the component. The existence of conductive inclusions, like CNTs, inside HNBR forms conductive paths through which leakage current can flow. Dielectric relaxations detected in the spectra of the imaginary part of electric modulus in Figure 29.9 are assigned to α-mode and IP. HNBR is a polar semicrystalline rubber with $T_g$ close to −30°C; the recorded peaks in the same temperature region correspond to the glass-to-rubber process (Psarras et al. 2014). The loss peak of interfacial polarization is located at higher temperatures. IP in the neat HNBR is related to the accumulation of charges at the interfaces between amorphous and crystalline parts. The presence of MWCNTs alters the heterogeneity of the system and introduces an additional interface between CNTs and the matrix. The recorded peak shifts to higher temperature and becomes broader, reflecting the increased interactions between the constituents of the system (Psarras et al. 2014). The lower values of $M''$, according to the definition of electric modulus as the inverse quantity of complex permittivity, reflect the enhancement to both $\varepsilon'$ and $\varepsilon''$, which is a characteristic of increased electrical heterogeneity (Tsangaris et al. 1998, Psarras et al. 2007b). The

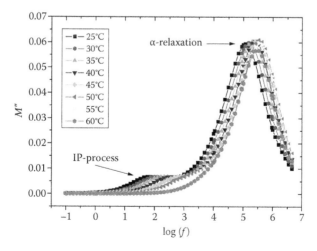

**FIGURE 29.8** Electric modulus loss index as a function of frequency, at various temperatures, of the 0.5% w/w MWCNT/PEO nanocomposite. (From Pontikopoulos, P., Electric response of poly(ethylene oxide)/modified multiwall carbon nanotubes nanocomposites, MSc thesis, University of Patras, Patras, Greece, 2009.)

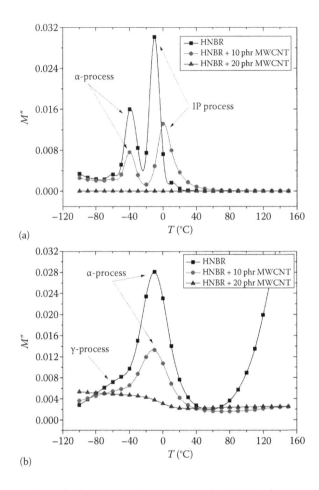

**FIGURE 29.9** Variation of modulus loss index with temperature for HNBR and MWCNT/HNBR nanocomposites at (a) $10^{-1}$ and (b) $10^5$ Hz.

spectrum of the 20 phr MWCNT/HNBR specimen is suppressed to lower values, expressing the additional increase in heterogeneity, while no relaxation process is evident (Figure 29.9a). The absence of IP loss peak has been attributed to the formation of conductive paths inside the nanocomposite through which the accumulated charges at the interfaces can flow, diminishing or even annulling the IP process (Pontikopoulos & Psarras 2013, Psarras et al. 2014). Furthermore, the presence of MWCNTs increases the crystalline regions of the system, reducing the remaining amorphous areas. The latter becomes constrained between the rigid crystalline parts; their relaxation dynamics is delayed and obstructed (Psarras et al. 2014). Fast relaxation processes, characterized by small relaxation times, are depicted in Figure 29.9b. These processes are recorded at higher frequencies and lower temperatures; the loss peak of α-relaxation is formed in the vicinity of 0°C, while γ-relaxation appears as hump at even lower temperatures. The superposition of both processes leads to a plateau in the case of the 20 phr MWCNT/HNBR specimen (Psarras et al. 2014).

The dielectric response of chemically modified graphene/ PVA nanocomposites revealed that, in GO/PVA systems, dielectric permittivity systematically grows with the filler content, while it decreases in graphene/copolymer/PVA system, due to the formation of an insulating coating between the nanographite inclusions and the PVA, because of the presence of the copolymer (Tantis et al. 2012).

### 29.4.3.2 Conductivity

The property of polymer nanocomposites, which is most remarkably affected by the presence of carbon nanoinclusions, is electrical conductivity. Insulating matrices reinforced with conductive inclusions gradually alter their conductivity up to a certain filler content, where an abrupt increase of several orders of magnitude for conductivity values occurs. This significant increase corresponds to a small variation in conductive phase content, and can be described in terms of percolation theory (Lux 1993, Zallen 2004, Psarras 2010). Percolation theory describes the transition from the state of limited connections between the conductive sites to the state where a conductive path is formed within the polymer matrix, allowing charges to percolate the whole composite. The transition from the insulating to the conductive behavior can be mathematically expressed by

$$\sigma \approx \sigma_0 \, (P - P_c)^t, \tag{29.2}$$

where $\sigma_0$ is the preexponential factor, $P$ is the content of conductive phase, $P_c$ is the critical concentration or percolation threshold, and $t$ is the critical exponent, a constant related to the dimensionality (1D, 2D, 3D) of the conduction process. According to the classical percolation theory, the increase in conductive phase content brings the conductive inclusions closer, and at the percolation threshold, a conductive path is formed via the geometrical contacts of the inclusions (Figure 29.10a). Charges are now able to flow through the whole material and conductivity dramatically increases. Further increase in conductive filler content results to the formation of a 3D network, and at the same time conductivity remains unaffected by the reinforcing phase concentration (Psarras 2006, Psarras 2007a, Psarras 2009a). In the same approach, conductive inclusions are considered as hard spheres, randomly distributed in the polymer matrix, with no mutual interactions. Percolation threshold and critical exponent are considered dependent only on the geometrical characteristics of the inclusions and the dimensionality of the conductive path (Psarras 2010, Vavouliotis et al. 2010, Pontikopoulos & Psarras 2013). It is evident that in this approach, the dynamic nature of nanocomposites, the influence of the preparation procedure, the interactions between filler and macromolecules, and the interactions between adjacent charged nanoinclusions are omitted (Vavouliotis et al. 2010, Pontikopoulos & Psarras 2013). Critical concentration

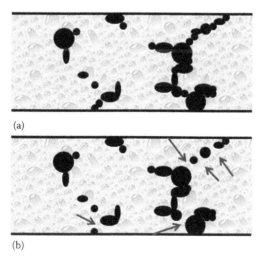

(a)

(b)

**FIGURE 29.10** Schematic representations of conductive percolating path in an insulating matrix: (a) via totally geometrical contacts and (b) in the absence of full geometrical contacts.

has been found to be lower than the predicted value from the statistical percolation theory. Moreover, the underlying conduction mechanism in the vicinity of the sharp increase in conductivity has been assigned not only to the physical contacts of the conductive sites, but also to the hopping and tunneling conduction (Psarras 2006, Psarras 2007a, Psarras 2009a, Psarras 2010, Pontikopoulos & Psarras 2013). In that case, not all conductive sites are in geometrical contact, and charges can hop from site to site if they possess the appropriate activation energy. A schematic representation of this process is shown in Figure 29.10b.

Comparing the variation of conductivity with filler content between nanocarbon–polymer composites and microcarbon–polymer composites, it is clear that the first category achieves the same or even higher values at much lower values of critical concentrations (Balberg 2002, Barrau et al. 2003, Vavouliotis et al. 2010, Pontikopoulos & Psarras 2013, Tjong 2014). Conductivity measurements can be conducted under DC and AC conditions. AC conductivity, as stated by von Hippel, sums all dissipative effects including an actual ohmic conductivity, caused by migrating charge carriers on isolated conductive clusters, as well as a frequency dielectric dispersion (von Hippel 1995). Figure 29.11 presents the variation of conductivity versus the filler content in modified MWCNT/PEO composites at various temperatures. The abrupt increase in conductivity values in a narrow range of filler concentration (less than 2%) signifies the accordance of data with the percolation theory. The effect of temperature appears to be more intensive in nanocomposites with MWCNT content lower than the critical one. The fitting experimental data via Equation 29.2 allows the determination of percolation threshold and critical exponent. Critical concentration and exponent have been found to deviate from the values predicted by classical or statistical percolation theory (Connor et al. 1998, Balberg 2002, Barrau et al. 2003, Vavouliotis et al. 2010, Pontikopoulos & Psarras 2013). These deviations reflect the differences between an ideal network of conductive and noninteracting spherical inclusions and a real polymer composite. Polymer composites are disordered solids; their kinetics is influenced by temperature, and interactions appear between the matrix and the filler and between the conductive inclusions. Other factors inducing differences are the inclusions' aspect ratio, the functionalization, the distribution of the filler in the matrix, and the employed manufacturing method (Singh et al. 2011). The dynamic nature of polymer composites is considered to be responsible for the discrepancies from the theoretical values of percolation theory (Connor et al. 1998, Balberg 2002, Barrau et al. 2003, Vavouliotis et al. 2010, Pontikopoulos & Psarras 2013). For SWCNT- and MWCNT-reinforced polymers, low values for percolation threshold,

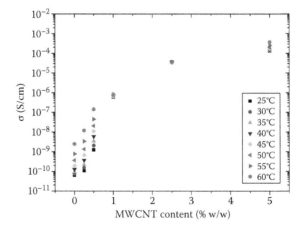

**FIGURE 29.11** Conductivity as a function of modified MWCNT content, at 1 Hz, for various temperatures. (From Pontikopoulos, P., Electric response of poly(ethylene oxide)/modified multiwall carbon nanotubes nanocomposites, MSc thesis, University of Patras, Patras, Greece, 2009.)

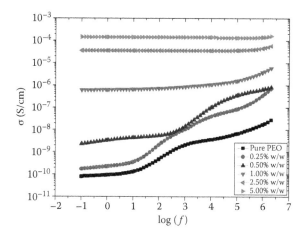

**FIGURE 29.12** AC conductivity as a function of frequency for modified MWCNT/PEO nanocomposites at 35°C. (From Pontikopoulos, P., Electric response of poly(ethylene oxide)/modified multiwall carbon nanotubes nano-composites, MSc thesis, University of Patras, Patras, Greece, 2009.)

such as 0.005 and 0.002 vol%, have been reported (Sandler et al. 2003, Bryning et al. 2005); while for graphene-reinforced systems, the value of 0.1 vol% has been reported (Stankovich et al. 2006). The very low values of percolation threshold are a strong indication that current flows through the material prior to the formation of a conductive path via physical–geometrical contacts (Psarras 2007a, Psarras 2009a, Pontikopoulos & Psarras 2013).

The type of the conduction mechanism, prior, in the vicinity, and after the percolation threshold, can be investigated by examining the variation of conductivity with temperature ($d\sigma/dT$), by examining the AC conductivity as a function of frequency at various temperatures, and by employing suitable theoretical models. Figure 29.12 depicts the AC conductivity as a function of frequency, at various temperatures, for the modified MWCNT/PEO systems. AC conductivity data tend to attain constant values at low frequencies and vary as an exponent of frequency after a critical one. AC curves are in accordance with the so-called AC universality law (Jonscher 1992):

$$\sigma(\omega) = \sigma_{DC} + A\omega^s, \tag{29.3}$$

where $\sigma_{DC}$ is the DC value of conductivity and $A$, $s$ are constant depending on filler content and temperature. The form of conductivity variation with frequency in Figure 29.12 is considered as indicative of hopping conductivity (Psarras 2006, Psarras 2010, Pontikopoulos & Psarras 2013). The variable range hopping model (Mott & Davis 1979) and the symmetric hopping model (Dyre & Schröder 2002) applied on the conductivity data of MWCNT/PEO system implied that the main conduction mechanism prior and close to percolation threshold is hopping conductivity. Conductivity increases with modified MWCNT content. Moreover, at filler concentrations higher than the critical one, another conduction mechanism, via physical contacts, contributes to the recorded conductivity. Finally, the above percolation threshold conductivity appears to be not dependent on the frequency of the applied electric field (Pontikopoulos & Psarras 2013).

## 29.4.4 Other Properties

The thermal conductivities of CNT and graphene attain very high values, like $10^3$ and $3 \times 10^3$ W/m K, respectively (Moniruzzaman & Winey 2006, Balandin et al. 2008, Singh et al. 2011). These high values created great expectation for improving thermal conductivity of nanocarbon–polymer composites.

Nanocomposites with enhanced thermal conductivity could be exploited in electronic circuit boards, heat sinks, connectors, and many others (Moniruzzaman & Winey 2006). However, the reinforcing efficiency of CNTs and graphene sheets in the case of thermal conductivity proved to be significantly poorer than the corresponding electrical conductivity. Two main factors determine this behavior: (i) the ratio of CNT or graphene thermal conductivity to the polymer thermal conductivity is on the order of $10^4$, while the relative ratio for electrical conductivities ranges from $10^{15}$ to $10^{19}$, and (ii) the interfacial area exhibits high thermal resistance between the matrix and the filler (Moniruzzaman & Winey 2006, Singh et al. 2011). Despite these problems, a 125% and a 300% increase in thermal conductivity, at room temperature, in 1% and 3% w/w SWCNT/epoxy composite, respectively, have been observed (Biercuk et al. 2002, Choi et al. 2003). A synergistic effect of GNP and MWCNTs in hybrid epoxy composites reduces the interfacial phonon scattering, allowing an increase of 147% in thermal conductivity at 1% w/w filler loading (MWCNT:GNP content ratio, 1:9) (Yang et al. 2011).

Several studies via TG analysis show improved thermal stability in CNT and graphene–polymer composites (Huxtable et al. 2003, Ge et al. 2004) at loadings ranging from 1% to 3% w/w. Improved thermal stability has been attributed to the increase in thermal conductivity due to the presence of nanocarbon filler and to hindering the flux of degradation products (Huxtable et al. 2003). Moreover, the thermal stability of carbon–polymer nanocomposites effectively alters their flame retardancy (Beyer 2002, Kashiwagi et al. 2005).

Finally, the incorporation of nanographitic inclusions in polymer matrices reduces the gas permeability, since they act as gas barriers, forcing diffusing gas molecules to follow a spiral path within the nanocomposite (Compton et al. 2010, Jiang et al. 2010a).

## 29.5 Applications

Applications of nanographite–polymer composites are based on their properties. From the discussion previously conducted, it is apparent that the property with the most impressive improvement is electrical conductivity. High electrical conductivity, at a low percolation threshold, enhanced modulus of elasticity and mechanical strength, increased thermal conductivity and thermal stability, and reduced gas permeability; these properties and the composites' light weight and flexibility constitute a set of properties for engineering materials which can be exploited in many industrial fields. Important drawbacks are the cost of high-quality nanographite inclusions, dispersion state of the nanocarbon filler, adhesion between the filler and the matrix, repeatability of properties, and mass production.

Carbon–polymer microcomposites have been used for decades in the fields of automotive and aerospace industries. However, because of the superior mechanical performance of carbon–polymer nanocomposites, it is expected that many structural components of automotive and aerospace industries will be constructed with nanocomposites in the near future, thus replacing conventional composites. Electrical properties of graphite–polymer nanocomposites could be exploited in the field of electronic industry as circuit boards, as conductive adhesives, and for thermal management in electronic devices.

Current applications of polymer matrix carbonaceous filler nanocomposites include (i) devices for energy storage, based on their increased permittivity and thus capacitance (Wang et al. 2009, Wang et al. 2010, Wu et al. 2010, Yan et al. 2010, Zhang et al. 2010a, Gómez et al. 2011); (ii) electrically conductive polymers in electronics (Stankovich et al. 2006, Eda & Chhowalla 2009); (iii) antistatic coatings in machines, devices, and apparatus, where moving and rotating insulating members exist, for the protection from detrimental sparks; and (iv) electromagnetic interference shielding (Liang et al. 2009).

Capacitance is the property of dielectric materials that determines energy storage, via dielectric permittivity. The capacitance increases as the capacitor's width decreases. In the case of finely dispersed nanoinclusions, especially in the case of 2D inclusions (like graphene), an inherent network of nanocapacitors is defined. Charging and discharging this network constitute a system of materials where energy can be stored and be harvested, and at the same time, this material can be used for the fabrication of a structural component. This type of reinforced polymers can be characterized as structural batteries (Psarras 2008, Psarras 2009b). Additionally, electrical conductivity is governed by a small amount of carbon nanofillers. Controlling the percolation threshold results in nanocomposites with tunable conductivity and finally in polymer-based materials which could have insulating, semiconductive, or even conductive behavior. This could prove very helpful in new technologies like printing circuits in nanoelectronics and transmission lines. A single-polymer system (also referred as *all-polymer cable*) with gradient conductivity across its cross section could transport electrical power or signals and at the same time protect and insulate the caring line (Psarras 2008, 2009b). Further, potential applications of the electrical properties of carbon–polymer nanocomposites refer to self-current regulators, self-heating systems, transparent conductive electrode for dye-sensitized solar cells, field-emitting diodes, forthcoming touch screens, polymer solar cells, etc. (Hong et al. 2008, Xu et al. 2009, Zhao et al. 2009, Rahman et al. 2011, Wang et al. 2011b).

An important issue related to the exploitation of nanographite–polymer nanocomposites into commercially available products is their influence on human or animal health and on the environment. This is part of a more general aspect concerning the biosafety of nanomaterials and nanotechnology. The importance of this aspect forced the scientific community, international organizations, and governments to pay attention to the effects of long-term exposure to nanoparticles and nanomaterials. Up to now, the scientific community reported controversial results on the toxicity and other health hazards of carbon–nanocomposites (Dreher 2004, Zhang et al. 2010b, Wang et al. 2011b, Singh et al. 2011).

## 29.6 Concluding Remarks

It is expected that nanomaterials and nanocomposites will provide unique solutions and advantages to many emerging technological problems in areas like construction, transportation, energy storage, power transmission, safety, and others. Nanographite–polymer nanocomposites offer a very promising improvement in mechanical, electrical, thermal, and gas permeation properties, thus giving the opportunity to explore their performance as structural or functional materials. Lightweight polymer nanocomposites with high modulus and strength are suitable for aerospace and automotive industries, as well as for the construction field. Polymer composites with enhanced mechanical behavior are also useful in biomedical and petrochemical applications. Polymer nanocomposites with controllable conductivity can be exploited as antistatic coatings, electrodes for solar cells, sensors, electromagnetic interference shieldings, and conductive adhesives. Furthermore, research results provide evidence of the improvement of thermal stability, thermal conductivity, optical, and other properties, leading to a general impression that nanographite–polymer composites constitute the appropriate solution for all technological problems and issues, being the contemporary panacea. However, the achieved improvement in all kinds of properties, and especially in mechanical properties, is not as high as it was predicted, concerning the relative properties of the nanographite filler. Up to today, scientific results signify that, like in ancient times, panacea (a single solution to all kind of problems) does not exist.

In order to approach the state where many products of everyday life will be composed of carbon–polymer nanocomposites, there are important challenges to overcome or drawbacks to address. Manufacturing routes for mass production and repeatability of properties is still a question. The dispersion level of the fine nanoinclusions within the polymer matrix should be controlled. Strong interactions between nanoinclusions and interactions between the filler and the polymer's macromolecules could prove detrimental to the overall performance of the nanocomposite and could cancel the effectiveness of the reinforcing phase. Functionalization of inclusions and tuning the polymer–nanofiller interactions will allow tailoring of the nanocomposite properties. Finally, the commercialization of products made with carbon–polymer nanocomposites should prove their cost-effectiveness and be economically competitive with conventional solutions.

# References

An, X., Simmons, T., Shah, R. et al., "Stable aqueous dispersions of noncovalently functionalized graphene from graphite and their multifunctional high-performance applications," *Nano Lett.* 10 (2010): 4295–4301.

Ansari, S., Kelarakis, A., Estevez, L. et al., "Oriented arrays of graphene in a polymer matrix by in situ reduction of graphite oxide nanosheets," *Small* 6 (2010): 205–209.

Balandin, A. A., Ghosh, S., Bao, W. et al., "Superior thermal conductivity of single-layer graphene," *Nano Lett.* 8 (2008): 902–907.

Balberg, I., "A comprehensive picture of the electrical phenomena in carbon black–polymer composites," *Carbon* 40 (2002): 139–143.

Barrau, S., Demont, P., Peigney, A. et al., "DC and AC conductivity of carbon nanotubes–polyepoxy composites," *Macromolecules* 35 (2003): 5187–5194.

Beyer, G., "Carbon nanotubes as flame retardants for polymers," *Fire Mater.* 26 (2002): 291–293.

Biercuk, M. J., Llaguno, M. C., Radosavljevic, M. et al., "Carbon nanotube composites for thermal management," *Appl. Phys. Lett.* 80 (2002): 2767–2769.

Bryning, M. B., Islam, M. F., Kikkawa, J. M. et al., "Very low conductivity threshold in bulk isotropic single-walled carbon nanotube–epoxy composites," *Adv. Mater.* 17 (2005): 1186–1191.

Cao, A., Xu, C., Liang, J. et al., "X-ray diffraction characterization on the alignment degree of carbon nanotubes," *Chem. Phys. Lett.* 344 (2001): 13–17.

Chatterjee, S., Wang, J. W., Kuo, W. S. et al., "Mechanical reinforcement and thermal conductivity in expanded graphene nanoplatelets reinforced epoxy composites," *Chem. Phys. Lett.* 531 (2012): 6–10.

Chen, Y., Qi, Y., Tai, Z. et al., "Preparation, mechanical properties and biocompatibility of graphene oxide/ ultrahigh molecular weight polyethylene composites," *Eur. Polym. J.* 48 (2012): 1026–1033.

Choi, E. S., Brooks, J. S., Eaton, D. L. et al., "Enhancement of thermal and electrical properties of carbon nanotube polymer composites by magnetic field processing," *J. Appl. Phys.* 94 (2003): 6034–6039.

Compton, O. C., Kim, S., Pierre, C. et al., "Crumpled graphene nanosheets as highly effective barrier, property enhancers," *Adv. Mater.* 22 (2010): 4759–4763.

Connor, M. T., Roy, S., Ezquerra, T. A. et al., "Broadband ac conductivity of conductor–polymer composites," *Phys. Rev. B* 57 (1998): 2286–2294.

Dang, Z.-M., Yao, S.-H. & Yu, H.-P., "Effect of tensile strain on morphology and dielectric property in nanotube/polymer nanocomposites," *Appl. Phys. Lett.* 90 (2007): 012907.

Deng, H., Bilotti, E., Zhang, R. et al., "Effective reinforcement of carbon nanotubes in polypropylene matrices," *J. Appl. Polym. Sci.* 118 (2010): 30–41.

Dreher, K. L., "Health and environmental impact of nanotechnology: Toxicological assessment of manufactured, nanoparticles," *Toxicol. Sci.* 77 (2004): 3–5.

Dresselhaus, M. S., Dresselhaus, G. & Eklund, P. C., *Science of Fullerenes and Carbon Nanotubes* (Academic Press, San Diego, 1996).

Dyre, J. C. & Schrøder, T. B., "Hopping models and ac universality," *Phys. Status Solidi B* 230 (2002): 5–13.

Eda, G. & Chhowalla, M., "Graphene-based composite thin films for electronics," *Nano Lett.* 9 (2009): 814–818.

Fabbri, P., Bassoli, E., Bon, S. B. et al., "Preparation and characterization of poly (butylene terephthalate)/ graphene composites by in situ polymerization of cyclic butylene terephthalate," *Polymer* 53 (2012): 897–902.

Fang, M., Wang, K., Lu, H. et al., "Covalent polymer functionalization of graphene nanosheets and mechanical properties of composites," *J. Mater. Chem.* 19 (2009): 7098–7105.

Fang, M., Zhang, Z., Li, J. et al., "Constructing hierarchically structured interphases for strong and tough epoxy nanocomposites by amine-rich graphene surfaces," *J. Mater. Chem.* 20 (2010): 9635–9643.

Felhös, D., Karger-Kocsis, J. & Xu, D., "Tribological testing of peroxide cured HNBR with different MWCNT and silica contents under dry sliding and rolling conditions against steel," *J. Appl. Polym. Sci.* 108 (2008): 2840–2851.

Fim, F. D. C., Basso, N. R. S., Graebin, A. P. et al., "Thermal, electrical, and mechanical properties of polyethylene–graphene nanocomposites obtained by in situ polymerization," *J. Appl. Polym. Sci.* 128 (2013): 2630–2637.

Galpaya, D., Wang, M., Liu, M. et al., "Recent advances in fabrication and characterization of graphene-polymer nanocomposites," *Graphene* 1 (2012): 30–49.

Gao, J., Itkis, M. E., Yu, A. et al., "Continuous spinning of a single-walled carbon nanotube–nylon composite fiber," *J. Am. Chem. Soc.* 127 (2005): 3847–3854.

Ge, J. J., Hou, H., Li, Q. et al., "Assembly of well-aligned multiwalled carbon nanotubes in confined poly-acrylonitrile environments: Electrospun composite nanofiber sheets," *J. Am. Chem. Soc.* 126 (2004): 15754–15761.

Geim, A. K. & Novoselov, K. S., "The rise of graphene," *Nat. Mater.* 6 (2007): 183–191.

Geng, H., Rosen, R., Zheng, B. et al., "Fabrication and properties of composites of poly(ethylene oxide) and functionalized carbon nanotubes," *Adv. Mater.* 14 (2002): 1387–1390.

Gómez, H., Ram, M. K., Alvi, F. et al., "Graphene-conducting polymer nanocomposite as novel electrode for supercapacitors," *J. Power Sources* 196 (2011): 4102–4108.

Grady, B. P., Pompeo, F., Shambaugh, R. L. et al., "Nucleation of polypropylene crystallization by single-walled carbon nanotubes," *J. Phys. Chem. B* 106 (2002): 5852–5858.

Guo, J., Ren, L., Wang, R. et al., "Water dispersible graphene noncovalently functionalized with trypto-phan and its poly(vinyl alcohol) nanocomposite," *Compos. Part B* 42 (2011): 2130–2135.

Han, S. O., Karevan, M., Bhuiyan, M. A. et al., "Effect of exfoliated graphite nanoplatelets on the mechanical and viscoelastic properties of poly(lactic acid) biocomposites reinforced with kenaf fibers," *J. Mater. Sci.* 47 (2012): 3535–3543.

Hong, W., Xu, Y., Lu, G. et al., "Transparent graphene/PEDOT-PSS composite films as counter electrodes of dye-sensitized solar cells," *Electrochem. Commun.* 10 (2008): 1555–1558.

Huang, X., Jiang, P., Kim, C. et al., "Influence of aspect ratio of carbon nanotubes on crystalline phases and dielectric properties of poly(vinylidene fluoride)," *Eur. Polym. J.* 45 (2009): 377–386.

Huang, Y. F. & Lin, C. W., "Facile synthesis and morphology control of graphene oxide/polyaniline nanocomposites via in situ polymerization process," *Polymer* 53 (2012): 2574–2582.

Huxtable, S. T., Cahill, D. G., Shenogin, S. et al., "Interfacial heat flow in carbon nanotube suspensions," *Nat. Mater.* 2 (2003): 731–734.

Iijima, S., "Helical microtubules of graphitic carbon," *Nature* 354 (1991): 56–58.

Jiang, M.-J., Dang, Z.-M., Yao, S.-H. et al., "Effects of surface modification of carbon nanotubes on the microstructure and electrical properties of carbon nanotubes/rubber nanocomposites," *Chem. Phys. Lett.* 457 (2008): 352–356.

Jiang, X. & Drzal, L. T., "Multifunctional high density polyethylene nanocomposites produced by incorporation of exfoliated, graphite nanoplatelets—1: Morphology and mechanical properties," *Polym. Compos.* 31 (2010a): 1091–1098.

Jiang, L., Shen, X. P., Wu, J. L. et al., "Preparation and characterization of graphene/poly (vinyl alcohol) nanocomposites," *J. Appl. Polym. Sci.* 118 (2010b): 275–279.

Jiang, H., Ni, Q., Wang, H. et al., "Fabrication and characterization of NBR/MWCNT composites by latex technology," *Polym. Compos.* 33 (2012): 1586–1592.

Jinhong, Y., Huang, X., Wu, C. et al., "Permittivity, thermal conductivity and thermal stability of poly (vinylidene fluoride)/graphene nanocomposites," *IEEE Trans. Dielectr. Electr. Insul.* 18 (2011): 478–484.

Jonscher, A. K., *Universal Relaxation Law* (Chelsea Dielectrics Press, London, UK, 1992).

Kalini, A., Gatos, K. G., Karahaliou, P. K. et al., "Probing the dielectric response of polyurethane/alumina nanocomposites," *J. Polym. Sci. Part B Polym. Phys.* 48 (2010): 2346–2354.

Karahaliou, P. K., Kerasidou, A. P., Georga, S. N. et al., "Dielectric relaxations in polyoxymethylene and in related nanocomposites: Identification and molecular dynamics," *Polymer* 55 (2014): 6819–6826.

Karger-Kocsis, J. & Wu, C.-M., "Thermoset rubber/layered silicate nanocomposites: Status and future trends," *Polym. Eng. Sci.* 44 (2004): 1083–1093.

Kashiwagi, T., Du, F., Douglas, J. F. et al., "Nanoparticle networks reduce the flammability of polymer nanocomposites," *Nat. Mater.* 4 (2005): 928–933.

Kim, H., Miura, Y. & Macosko, C. W., "Graphene/polyurethane nanocomposites for improved gas barrier and electrical conductivity," *Chem. Mater.* 22 (2010): 3441–3450.

Kim, H., Kobayashi, S., AbdurRahim, M. A. et al., "Graphene/polyethylene nanocomposites: Effect of polyethylene functionalization and blending methods," *Polymer* 52 (2011): 1837–1846.

Kovtyukhova, N. I., Ollivier, P. J., Martin, B. R. et al., "Layer-by-layer assembly of ultrathin composite films from micron-sized graphite oxide sheets and polycations," *Chem. Mater.* 11 (1999): 771–778.

Krishnan, A., Dujardin, E., Ebbesen, T. et al., "Young's modulus of single-walled nanotubes," *Phys. Rev. B* 58 (1998): 14013–14019.

Kroto, H. W., Heath, J. R., O'Brien, S. C. et al., "$C_{60}$: Buckminsterfullerene," *Nature* 318 (1985): 162–163.

Layek, R. K., Samanta, S. & Nandi, A. K., "The physical properties of sulfonated graphene/poly (vinyl alcohol) composites," *Carbon* 50 (2012): 815–827.

Li, X. & McKenna, G. B., "Considering viscoelastic micromechanics for the reinforcement of graphene polymer nanocomposites," *ACS Macro Lett.* 1 (2012): 388–391.

Li, W., Liang, C., Zhou, W. et al., "Preparation and characterization of multiwalled carbon nanotube-supported platinum for cathode catalysts of direct methanol fuel cells," *J. Phys. Chem. B* 107 (2003): 6292–6299.

Li, Q., Xue, Q. Z., Gao, X. L. et al., "Temperature dependence of the electrical properties of the carbon nanotube/polymer composites," *Express Polym. Lett.* 3 (2009): 769–777.

Liang, J., Wang, Y., Huang, Y. et al., "Electromagnetic interference shielding of graphene/epoxy composites," *Carbon* 47 (2009): 922–925.

Lux, F., "Models proposed to explain the electrical conductivity of mixtures made of conductive and insulating materials," *J. Mater. Sci.* 28 (1993): 285–301.

Ma, H.-L., Zhang, Y., Hu, Q.-H. et al., "Enhanced mechanical properties of poly(vinyl alcohol) nanocomposites with glucose-reduced graphene oxide," *Mater. Lett.* 102–103 (2013): 15–18.

Miltner, H. E., Grossiord, N., Lu, K. B. et al., "Isotactic polypropylene/carbon nanotube composites prepared by latex technology: Thermal analysis of carbon nanotube-induced nucleation," *Macromolecules* 41 (2008): 5753–5762.

Mo, Z.-L., Xie, T.-T., Zhang, J.-X. et al., "Synthesis and characterization of nanoGs-PPy/epoxy nanocomposites by *in situ* polymerization," *Synth. React. Inorg. Metal-Organic Nano-Metal Chemistry* 42 (2012): 1172–1176.

Moniruzzaman, M. & Winey, K. I., "Review: Polymer nanocomposites containing carbon nanotubes," *Macromolecules* 39 (2006): 5194–5205.

Mott, N. F. & Davis, E. A., *Electronic Conduction in Non-Crystalline Materials* (Clarenton Press, Oxford, England, 1979).

Ogasawara, T., Ishida, Y., Ishikawa, T. et al., "Characterization of multi-walled carbon nanotube/phenyl-ethynyl terminated polyimide composites," *Compos. Part A* 35 (2004): 67–74.

Papagelis, K., Kalyva, M., Tasis, D. et al., "Covalently functionalized carbon nanotubes as macroinitiators for radical environmental," *Phys. Status Solidi B* 244 (2007): 4046–4050.

Patsidis, A. C. & Psarras, G. C., "Structural transition, dielectric properties and functionality in epoxy resin–barium titanate nanocomposites," *Smart Mater. Struct.* 22 (2013a): 115006.

Patsidis, A. C., Kalaitzidou, K. & Psarras, G. C., "Carbon or barium titanate reinforced epoxy resin nanocomposites: Dielectric, thermomechanical and functional behavior," *J. Adv. Phys.* 2 (2013b): 7–12.

Patsidis, A. C., Kalaitzidou, K. & Psarras, G. C., "Graphite nanoplatelets/polymer nanocomposites: Thermomechanical, dielectric, and functional behavior," *J. Thermal Anal. Calorim.* 116 (2014): 41–49.

Patsidis, A. C., "Hybrid nanodielectrics of polymer matrix/functional inclusions: Development, characterization and functionality," PhD thesis, Polymer Science and Technology Interdepartmental Program, University of Patras, Patras, Greece (2015).

Pontikopoulos, P., "Electric response of poly(ethylene oxide)/modified multiwall carbon nanotubes nanocomposites," MSc thesis, Department of Materials Science, University of Patras, Patras, Greece (2009).

Pontikopoulos, P. L. & Psarras, G. C., "Dynamic percolation and dielectric response in multiwall carbon nanotubes/poly(ethylene oxide) composites," *Sci. Adv. Mater.* 5 (2013): 1–7.

Potts, J. R., Lee, S. H., Alam, T. M. et al., "Thermomechanical properties of chemically modified graphene/poly(methyl methacrylate) composites made by *in situ* polymerization," *Carbon,* 49 (2011): 2615–2623.

Psarras, G. C., "Hopping conductivity in polymer matrix–metal particles composites," *Compos. Part A* 37 (2006): 1545–1553.

Psarras, G. C., "Charge transport properties in carbon black/polymer composites," *J. Polym. Sci. Part B Polym. Phys.* 45 (2007a): 2535–2545.

Psarras, G. C., Gatos, K. G., Karahaliou, P. K. et al., "Relaxation phenomena in rubber/layered silicate nanocomposites," *Express Polym. Lett.* 1 (2007b): 837–845.

Psarras, G. C., "Nanodielectrics: An emerging sector of polymer nanocomposites," *Express Polym. Lett.* 2 (2008): 460.

Psarras, G. C., "Conduction processes in percolative epoxy resin/silver particles composites," *Sci. Adv. Mater.* 1 (2009a): 101–106.

Psarras, G. C., "Electrical properties of polymer matrix composites: Current impact and future trends," *Express Polym. Lett.* 3 (2009b): 533.

Psarras, G. C., "Conductivity and dielectric characterization of polymer nanocomposites," in Tjong, S. C. & Mai, Y.-M., eds., *Polymer Nanocomposites: Physical Properties and Applications* (Woodhead Publishing Limited, Cambridge, England, 2010), pp. 31–69.

Psarras, G. C., "The ongoing impact of carbon (allotropic forms)/polymer composites," *Express Polym. Lett.* 8 (2014): 73.

Psarras, G. C., Sofos, G. A., Vradis, A. et al., "HNBR and its MWCNT reinforced nanocomposites: Crystalline morphology and electrical response," *Eur. Polym. J.* 54 (2014): 190–199.

Rafiee, M. A., Rafiee, J., Wang, Z. et al., "Enhanced mechanical properties of nanocomposites at low graphene content," *ACS Nano* 3 (2009): 3884–3890.

Rafiee, M. A., Rafiee, J., Srivastava, I. et al., "Fracture and fatigue in graphene nanocomposites," *Small* 6 (2010): 179–183.

Rahman, A., Ali, I., Al Zahrani, S. M. et al., "A review of the applications of nanocarbon polymer composites," *NANO Brief Rep. Rev.* 6 (2011): 185–203.

Ratna, D., Abraham, T. N., Siengchin, S. et al., "Novel method for dispersion of multiwall carbon nanotubes in poly(ethylene oxide) matrix using dicarboxylic acid salts," *J. Polym. Sci. B Polym. Phys.* 47 (2009): 1156–1165.

Roy, N., Sengupta, N. & Bhowmick, A. K., "Modifications of carbon for polymer composites and nanocomposites," *Progr. Polym. Sci.* 37 (2012): 781–819.

Salavagione, H. J., Martínez, G. & Gómez, M. A., "Synthesis of poly(vinyl alcohol)/reduced graphite oxide nanocomposites with improved thermal and electrical properties," *J. Mater. Chem.* 19 (2009): 5027–5032.

Salvetat, J. P., Briggs, G. A. D., Bonard, J. M. et al., "Elastic and shear moduli of single-walled carbon nanotube ropes," *Phys. Rev. Lett.* 82 (1999): 944–947.

Sandler, J. K. W., Kirk, J. E., Kinloch, I. A. et al., "Ultra-low electrical percolation threshold in carbon-nanotube-epoxy composites," *Polymer* 44 (2003): 5893–5899.

Saravanan, N., Rajasekar, R., Mahalakshmi, S. et al., "Graphene and modified graphene-based polymer nanocomposites—A review," *J. Reinf. Plastics Compos.* 33 (2014): 1158–1170.

Sengupta, R., Bhattacharya, M., Bandyopadhyay, S. et al., "A review on the mechanical and electrical properties of graphite and modified graphite reinforced polymer composites," *Prog. Polym. Sci.* 36 (2011): 638–670.

Siengchin, S., Psarras, G. C. & Karger-Kocsis, J., "POM/PU/Carbon nanofiber composites produced by water-mediated melt compounding: Structure, thermomechanical and dielectrical properties," *J. Appl. Polym. Sci.* 117 (2010): 1804–1812.

Singh, V., Joung, D., Zhai, L. et al., "Graphene based materials: Past, present and future," *Prog. Mater. Sci.* 56 (2011): 1178–1271.

Sridhar, V., Lee, I., Chum, H. H. et al., "Graphene reinforced biodegradable poly(3-hydroxybutyrate-co-4-hydroxybutyrate) nano-composites," *Express Polym. Lett.* 7 (2013): 320–32.

Stankovich, S., Dikin, D. A., Dommett, G. H. B. et al., "Graphene-based composite materials," *Nature* 442 (2006): 282–286.

Sun, L. L., Li, B., Zhao, Y. et al., "Suppression of AC conductivity by crystalline transformation in poly(vinylidene fluoride)/carbon nanofiber composites," *Polymer* 51 (2010): 3230–3237.

Tantis, I., Psarras, G. C. & Tasis, D., "Functionalized graphene–poly(vinyl alcohol) nanocomposites: Physical and dielectric properties," *Express Polym. Lett.* 6 (2012): 283–292.

Tasis, D., Pispas, S., Galiotis, C. et al., "Growth of calcium carbonate on non-covalently modified carbon nanotubes," *Mater. Lett.* 261 (2007): 5044–5046.

Teng, C.-C., Ma, C.-C. M., Lu, C.-H. et al., "Thermal conductivity and structure of non-covalent functionalized graphene/epoxy composites," *Carbon* 49 (2011): 5107–5116.

Thostenson, E. T., Li, C. & Chou, T.-W., "Nanocomposites in context," *Compos. Sci. Technol.* 65 (2005): 491–516.

Tjong, S. C., "Polymer composites with graphene nanofillers: Electrical properties and applications," *J. Nanosci. Nanotechnol.* 1 (2014): 1154–1168.

Traina, M., & Pegoretti, A., "In situ reduction of graphene oxide dispersed in a polymer matrix," *J. Nanoparticle Res.* 14 (2012): 1–6.

Tsangaris, G. M., Psarras, G. C. & Kouloumbi, N., "Electric modulus and interfacial polarization in composite polymeric systems," *J. Mater. Sci.* 33 (1998): 2027–2037.

Vavouliotis, A., Fiamegou, E., Karapappas, P. et al., "DC and AC conductivity in epoxy resin/multiwall carbon nanotubes percolative system," *Polym. Compos.* 31 (2010): 1874–1880.

von Hippel, A. R., *Dielectrics and Waves* (Artech, Boston, 1995).

Wang, X. P., Huang, A.-M., Jia, D.-M. et al., "From exfoliation to intercalation—Changes in morphology of HNBR/organoclay nanocomposites," *Eur. Polym. J.* 44 (2008): 2784–2789.

Wang, D. W., Li, F., Zhao, J. P. et al., "Fabrication of graphene/polyaniline composite paper via in situ anodic electropolymerization for high-performance flexible electrode," *ACS Nano* 3 (2009):1745–1752.

Wang, H., Hao, Q., Yang, X. et al., "A nanostructured graphene/polyaniline hybrid material for supercapacitors," *Nanoscale* 2 (2010): 2164–2170.

Wang, X., Hu, Y., Song, L. et al., "In situ polymerization of graphene nanosheets and polyurethane with enhanced mechanical and thermal properties," *J. Mater. Chem.* 21 (2011a): 4222–4227.

Wang, K., Ruan, J., Song, H. et al., "Biocompatibility of graphene oxide," *Nanoscale Res. Lett.* 6 (2011b): 1–8.

Wei, T., Luo, G., Fan, Z. et al., "Preparation of graphene nanosheet/polymer composites using in situ reduction-extractive dispersion," *Carbon* 47 (2009): 2296–2299.

Wong, E. W., Sheehan, P. E. & Lieber, C. M., "Nanobeam mechanics: Elasticity, strength, and toughness of nanorods and nanotubes," *Science* 277 (1997): 1971–1975.

Wrana, C., Reinartz, K. & Winkelbach, H. R., "Therban®—The high performance elastomer for the new millennium," *Macromol. Mater. Eng.* 286 (2001): 657–662.

Wu, Q., Xu, Y., Yao, Z. et al., "Supercapacitors based on flexible graphene/polyaniline nanofiber composite films," *ACS Nano* 4 (2010): 1963–1970.

Xie, X. L., Mai, Y. W. & Zhou, X. P., "Dispersion and alignment of carbon nanotubes in polymer matrix: A review," *Mater. Sci. Eng. R Rep.* 49 (2005): 89–112.

Xu, Y., Wang, Y., Liang, J. et al., "A hybrid material of graphene and poly (3,4-ethyldioxythiophene) with high conductivity, flexibility, and transparency," *Nano Res.* 2 (2009): 343–348.

Xu, J.-Z., Zhong, G.-J., Hsiao, B. S. et al., "Low-dimensional carbonaceous nanofiller induced polymer crystallization," *Prog. Polym. Sci.* 39 (2014): 555–593.

Yan, J., Wei, T., Shao, B. et al., "Preparation of a graphene nanosheet/polyaniline composite with high specific capacitance," *Carbon* 48 (2010): 487–493.

Yang, S.-Y., Lin, W.-N., Huang, Y.-L. et al., "Synergetic effects of graphene platelets and carbon, nanotubes on the mechanical and thermal properties of epoxy composites," *Carbon* 49 (2011): 793–803.

Yu, M. F., Files, B. S., Arepalli, S. et al., "Tensile loading of ropes of singlewall carbon nanotubes and their mechanical properties," *Phys. Rev. Lett.* 84 (2000a): 5552–5555.

Yu, M. F., Lourie, O., Dyer, M. J. et al., "Strength and breaking mechanism of multiwalled carbon nanotubes under tensile load," *Science* 287 (2000b): 637–640.

Yu, J., Lu, K., Sourty, E. et al., "Characterization of conductive composites prepared by latex technology," *Carbon* 45 (2007): 2897– 2903.

Yu, J., Jiang, P., Wu, C. et al., "Graphene nanocomposites based on poly(vinylidene fluoride): Structure and properties," *Polym. Compos.* 32 (2011): 1483–1491.

Zallen, R., *The Physics of Amorphous Solids* (Wiley, New York, 2004).

Zaman, I., Kuan, H. C., Meng, Q. et al., "A facile approach to chemically modified graphene and its polymer nanocomposites," *Adv. Funct. Mater.* 22 (2012): 2735–2743.

Zhang, K., Zhang, L. L., Zhao, X. S. et al., "Graphene/polyaniline nanofiber composites as supercapacitor electrodes," *Chem. Mater.* 22 (2010a): 1392–1401.

Zhang, Y., Ali, S. F., Dervishi, E. et al., "Cytotoxicity effects of graphene and single-wall carbon nanotubes, in neural phaeochromocytoma-derived PC12 cells," *ACS Nano* 4 (2010b): 3181–3186.

Zhang, F., Peng, X., Yan, W. et al., "Nonisothermal crystallization kinetics of in situ nylon 6/graphene composites by differential scanning calorimetry," *J. Polym. Sci. Part B Polym. Phys.* 49 (2011): 1381–1388.

Zhang, H.-B., Zhang, W.-G., Yan, Q. et al., "The effect of surface chemistry of graphene on rheological and electrical properties of polymethyl–methacrylate composites," *Carbon* 50 (2012): 5117–5125.

Zhao, L., Zhao, L., Xu, Y. et al., "Polyaniline electrochromic devices with transparent graphene electrodes," *Electrochim. Acta* 55 (2009): 491–497.

Zhao, X., Zhang, Q. & Chen, D., "Enhanced mechanical properties of graphene-based poly(vinyl alcohol) composites," *Macromolecules* 43 (2010): 2357–2363.

# 30

# Graphite- and Graphene-Based Nanocomposites

Luigi Sorrentino

Marco Aurilia

## 30.1 Introduction

Graphite and graphene are materials based on carbon atoms and are strictly related, since graphene can be considered as the constitutive building block of graphite particles. They are allotropic forms of carbon, alongside other natural species such as coal and, in very small amounts, diamonds. Recently, some nanosized species were discovered, namely, fullerenes (discovered in 1985), CNFs, and CNTs (discovered in 1991). From then on, the study of carbon nanostructures grew rapidly and several other synthetic forms of carbon-based structures have been created such as synthetic graphite and synthetic diamonds, adsorbent carbon, cokes, carbon black, carbon and graphitic fibers, and glassy carbons. Carbon-based particles are used for different applications such as gas adsorption, electrodes, lubricants, toner for photocopying machines, cutting wheels, catalytic support, gas barrier, tire and elastomer reinforcement, inks, aircraft and spacecraft composites, and heat sinks for ultrafast semiconductors. All allotropic forms of carbon depend on the carbon's unique orbital hybridization properties. The ground-state orbital configuration of carbon is $1s^2\,2s^2\,2p^2$. The narrow energy gap between the $2s$ and $2p$ orbitals facilitates the promotion of one $2s$ electron to a vacant higher-energy $2p$ orbital. This electron promotion allows carbon to hybridize into $sp$, $sp^2$, and $sp^3$ configurations, leading to fascinating and diverse molecular structures. The $sp$ bonding gives rise to chain structures; $sp^2$ bonding, to planar structures; and $sp^3$ bonding, to tetrahedral structures. The hybridization states of some typical carbon nanomaterials are summarized in Figure 30.1 (Sengupta et al. 2011).

Graphite, as well as graphite oxides, is known since the 18th century and its name comes from the Greek word γράφω (*grafo*), which means "I write." The earliest images showing its nanoscale structures are TEM images of few-layer graphite (FLG) published by G. Ruess and F. Vogt in 1948 (Geim 2012).

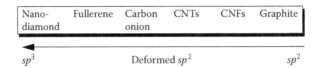

FIGURE 30.1  Hybridization states of some typical carbon-based nanomaterials. (Reprinted from *Progress in Polymer Science*, 36, Sengupta, R., Bhattacharya, M., Bandyopadhyay, S. et al., A review on the mechanical and electrical properties of graphite and modified graphite reinforced polymer composites, 638–670, Copyright (2011), with permission from Elsevier.)

Later, single graphene layers were also directly observed by electron microscopy (Geim & Novoselov 2007). Graphite compounds, after being treated with special processes aimed at increasing the inter-layer spacing (graphite intercalated compounds [GICs]), were studied under a TEM and researchers occasionally observed thin graphitic flakes (FLG) and also individual layers. Such individual layers were subsequently identified as graphene.

The term *graphene* first appeared in 1987 (Mouras et al. 2015) to describe single sheets of graphite as a constituent of GICs, which can be considered as a crystalline salt of intercalant and graphene. The term was also initially used in early descriptions of CNTs, epitaxial graphene, and polycyclic aromatic hydro-carbons. Even if graphene is sometimes referred to as graphite layer, it is actually considered incorrect, according to IUPAC (Fitzer et al. 1995, p. 491), to use for the single layer a term that includes the term *graphite*, which, although nanometric in size, implies a 3D structure. In fact, "graphene is a single atomic plane of graphite, which—and this is essential—is sufficiently isolated from its environment to be considered free-standing" (Geim 2009). Initial attempts to make atomically thin graphitic films employed exfoliation techniques similar to the drawing method. Multilayer samples down to 10 nm in thickness were obtained (Geim & Novoselov 2007).

Graphene was first reliably produced in a laboratory in 2004 by Andre Geim and Kostya Novoselov at the University of Manchester through a very simple process. In fact, single-atom-thick layers were obtained from graphite by peeling with an adhesive tape a single graphene layer and transferring them onto a silicon wafer in a process called either *micromechanical cleavage* or *the Scotch tape technique*. This very basic process was able to give a 2D carbon-based structure with very peculiar and extreme proper-ties, which granted Geim and Novoselov the Nobel Prize in Physics in 2010.

Graphite is one of the most versatile nonmetallic minerals. It is widely available in nature but can also be manufactured synthetically, primarily via the Acheson process, which utilizes lower-purity carbon-bearing raw materials blended with tar pitch. Natural graphite is mined and is available in three com-mercial forms, depending on the carbon content and purity: amorphous (60%–85% of carbon atoms [C]; the most abundant type), flake (>85% C; most used in higher-value applications), and vein (>90% C; used in high-end applications since it is the most pure and needs low processing). Vein graphite is being mined only in Sri Lanka. Graphite's main markets are refractories (commonly in the form of high temperature–resistant bricks and linings utilized in metal production, ceramics, petrochemicals, and cement industries) and batteries. Since it has an extremely low friction coefficient, it is also used as a solid lubricant. Natural graphite production was 1.1 million tons worldwide in 2013 (Figure 30.2). Of this total, flake accounted for 55%, amorphous 44%, and vein 1%. China is the major producer of natural graphite, accounting for more than 70% of the total world production. Other significant producers are India, Brazil, North Korea, Canada, and, to a minor extent, Norway, Zimbabwe, Madagascar, Russia, Ukraine, and Germany.

Graphene nanoparticles, available as powder or dispersion (in a polymeric matrix, adhesive, oil, aqueous, or nonaqueous solutions), as of 2015 are not largely employed in commercial applications, but their huge structural and functional performance potentials are pushing toward new developments and many solutions are arising in industrial fields such as lightweight/strong composite materials, photovol-taic and energy storage, electronics, biological engineering, and filtration.

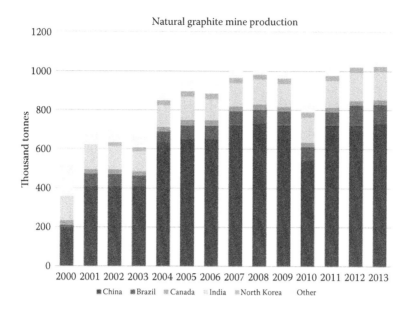

FIGURE 30.2 Global graphite supplier countries. (From Collerson, K., Graphite Commodity Update, August 6, 2014, HDR International, Inc. With permission.)

## 30.2 Graphite and Graphene Nanoparticles

### 30.2.1 Graphite

Graphite has a layered, planar structure formed by stacks of parallel 2D graphene sheets (a single carbon layer in the crystalline honeycomb graphite lattice is known as a graphene layer or graphene sheet) with $sp^2$-hybridized carbon atoms bonded in hexagonal rings. In each layer, the carbon atoms are arranged in a honeycomb lattice with a separation of 0.142 nm, and the distance between the planes is 0.335 nm (Figure 30.3). The atoms on the plane are bonded covalently, with only three of the four potential bonding sites satisfied. Bonding between layers is achieved via weak van der Waals bonds, which allows

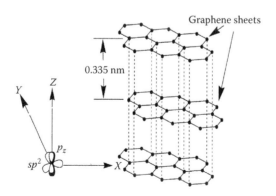

FIGURE 30.3 Layered structure of graphite showing the $sp^2$-hybridized carbon atoms tightly bonded in hexagonal rings. (Reprinted from *Progress in Polymer Science*, 36, Sengupta, R., Bhattacharya, M., Bandyopadhyay, S. et al., A review on the mechanical and electrical properties of graphite and modified graphite reinforced polymer composites, 638–670, Copyright (2011), with permission from Elsevier.)

**TABLE 30.1**  Main Properties of Graphite Nanoparticles and Graphene and Comparison with CNTs

| Property | Unit | Graphite (GNP) | Graphene | CNT |
|---|---|---|---|---|
| Density | g cm$^{-3}$ | 2.26 | | 0.8 for SWCNT<br>1.8 for MWCNT (theoretical) |
| Thickness | nm | 2–150 | 0.34 | |
| SSA | m$^2$/g | 15–2630 | 2630 | |
| Elastic modulus | TPa | 1 (in-plane) | 1 (in-plane) | ~1 for SWCNT<br>~0.3–1 for MWCNT |
| Strength | GPa | 130 | 130 | 50–500 for SWCNT<br>10–60 for MWCNT |
| Electrical conductivity | S/m | $5.98 \times 10^4$ | $7.2 \times 10^3$ | ~5–50 |
| Thermal conductivity | W m$^{-1}$ K$^{-1}$ | 3000 (in-plane) 6 (z-axis) | 5300 (in-plane) | 3000 (theoretical) |
| Thermal expansion | K$^{-1}$ | ~$1 \times 10^{-6}$ (in-plane)<br>$29 \times 10^{-6}$ (z-axis) | | Negligible (theoretical) |

*Source:* *Progress in Polymer Science*, 36, Sengupta, R., Bhattacharya, M., Bandyopadhyay, S. et al., A review on the mechanical and electrical properties of graphite and modified graphite reinforced polymer composites, 638–670, Copyright (2011), with permission from Elsevier.

layers of graphite to be, at least in principle, easily separated or to slide past each other. The fourth electron is free to migrate on the plane and renders graphite capable of conducting electricity thanks to the vast electron delocalization within the carbon layers (a phenomenon called *aromaticity*). However, electricity is primarily conducted in directions parallel to the layers, inducing a strong anisotropy of all molecular structure properties.

Thus, graphite has an anisotropic molecular structure due to the difference between in-plane and out-of-plane bonding of carbon atoms. All properties exhibit higher values in the $sp^2$-hybridized planes, such as mechanical, acoustic, thermal, and elastic properties. The elastic modulus parallel to the graphene plane is so high (Table 30.1) that it results to be stronger than the elastic modulus of diamond. The acoustic and thermal properties of graphite are also highly anisotropic, since phonons propagate quickly along the tightly bound planes, but are slower to travel from one plane to another.

## 30.2.2  Modifications of Graphite

In order to achieve graphite particles nanometric in size, special exfoliating treatments have to be applied to natural flaky graphite. The resulting nanoparticles, GNPs, can be obtained with thickness ranging from 0.34 to 100 nm. Moreover, graphite can be easily intercalated, hosting various types of atoms, molecules, metal complexes, and salts between the expanded graphene sheets to form GICs. Chemical intercalation is a viable route to modify graphite and create GO, GICs, and expanded graphite (EG).

### 30.2.2.1  GO

GO is also known as graphite oxide, graphitic oxide, or graphitic acid, and is usually prepared by treating graphite flakes with oxidizing agents so that the polar groups, introduced on the graphite surface, widen the interlayer spacing of the graphene planes. GO, prepared for the first time by Brodie (1859), typically involves the reaction of graphite flakes with potassium chlorate and fuming nitric acid. Hummers and Offeman (1958) developed a faster and safer method for the preparation of GO by reacting anhydrous sulfuric acid, sodium nitrate, and potassium permanganate, a method which is widely followed even today. Unlike graphite, GO has low electrical conductivity since surface electrons are involved in bonds with oxide compounds and are not available as charge carriers. GO is employed as an important intermediate for the preparation of GNPs.

### 30.2.2.2 GICs

GICs are formed by the insertion of atomic or molecular layers of different chemical species between the layers of the graphite host lattice (Dresselhaus & Dresselhaus 1981, Boehm et al. 1994). In GICs, the graphene layers either accept electrons from or donate electrons to the intercalated species. Graphite intercalated by electron donors like alkali metals (e.g., potassium and sodium) are known as donor-type GICs, whereas compounds formed by the intercalation of molecular species acting as electron acceptors like halogens, halide ions, and acids are known as acceptor-type GICs.

### 30.2.2.3 EG

When intercalated graphite is heated past a critical temperature or exposed to microwave radiation, a large expansion (up to hundreds of times) of the graphite flakes occurs along the out-of-plane direction and forms vermicular structures (Figure 30.4) with low density and high temperature resistance. This product is known as exfoliated graphite or EG. EG is composed of stacks of nanosheets that may vary from 100 to 400 nm, but the use of pressure wave–dispersing techniques (ultrasonic sonication or ultrasonication) in solvents is usually performed to obtain thinner GNPs (~30–80 nm in thickness). These nanoparticles are easier to disperse by high-speed shearing in a polymeric matrix.

## 30.2.3 Graphene

Graphene has the same hexagonal arrangement of carbon atoms as graphite, and differs only in thickness, being only one atom thick. Graphene single layers are remarkably strong (about 100 times stronger than steel) and can conduct heat and electricity with great efficiency (Table 30.1). The scientific community was aware of the graphene structure and its properties since they were all theoretically calculated

**FIGURE 30.4** SEM micrograph of EG. (Chen, G., Weng, W., Wu, D. et al.: Preparation of Polymer/Graphite Conducting Nanocomposite by Intercalation Polymerization. *Journal of Applied Polymer Science.* 2001. 82. 2506–2513. Copyright Wiley-VCH Verlag GmbH & Co. KGaA. Reproduced with permission.)

decades earlier than its first production in the lab in 2004. Graphene can be considered the basic structural element of other carbon allotrope nanoparticles besides graphite, such as CNTs and fullerenes.

Different routes are available to produce graphene nanolayers. The most significant routes are as follows:

- CVD of monolayers on transition metal surfaces.
- Micromechanical exfoliation of graphite, by peeling of graphene monolayers from graphite using adhesive tape. The tape is then dipped in a solvent, usually acetone, to release the graphene nanolayers and subsequently allow nanoparticle capture on a silicon wafer with a $SiO_2$ layer on top.
- Epitaxial growth of graphene on electrically insulating substrates like silicon carbide.
- Specific treatments of intermediate graphite-based nanoparticles, such as GO and GICs.

The last two are the only routes to prepare bulk quantities of graphene and chemically modified graphene. In fact, the availability of hydrophilic functional groups in GO (hydroxyl and epoxide groups on the basal planes, and carboxyl and carbonyl groups on the edges) renders the graphene precursors water soluble. GO layers easily intercalate in water and then exfoliate in a colloidal suspension through the use of opportune dispersing aid techniques, such as ultrasonic sonication or microwave irradiation.

Chemical reduction of GO with reducing agents (such as hydrazine or hydrazine derivatives) is necessary to detach and expel all functional groups bonded to the surface plane. This step allows the recovery of the pristine conductive electronic structure. In this way, the low-conducting graphite oxide turns into the extremely thermal and electrical conducting graphene. Usually, the chemical reduction approach is not very effective in completely exfoliating the nanoparticles since it results in GNPs essentially made of thin FLG stacks.

The most effective route to prepare large amounts of graphene layers is to use thermal treatments by means of heat or microwave irradiation of GO (Figure 30.5). This technique involves rapidly heating GO by thermal energy in inert gas (argon or nitrogen) or microwave irradiation in a microwave oven to free the graphene surface from all attached compounds and produce the thermally reduced GO (TRGO). The heating is also responsible for the exfoliation of graphene sheets. TRGO is also known as functionalized graphene sheet (FGS) and can have a wrinkled sheet structure due to residual epoxy groups forming chains across the graphene surface. Due to their wrinkled nature, FGSs do not collapse back to GO but remain highly agglomerated.

Some modifying agents and solvents for the chemical modification of graphene starting from graphite oxides prepared from naturally occurring graphite are shown in Table 30.2. After oxidation, several chemical methods to obtain soluble graphene have been explored, including the reduction of GO in

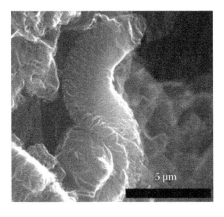

**FIGURE 30.5** Graphite expanded with microwave irradiation. (Reprinted with permission from Potts et al. 2011b, 6488–6495. Copyright 2011 American Chemical Society.)

**TABLE 30.2** Modifying Agents for Graphite Oxide and Solvents Used for the Dispersion

| Method of Modification | Modifying Agent | Dispersing Medium |
| --- | --- | --- |
| Reduction in stabilization medium | KOH | Water |
| | Octadecyl amine | THF/CCl$_4$/1,2-dichloroethane |
| | Alkyl lithium | THF |
| Covalent modification | Organic isocyanate | DMF/NMP/DMSO/HMPA |
| | Organic diisocyanate | DMF |
| | PVA | Water/DMSO |
| | Porphyrin | DMF |
| | Poly-l-lysine | Water |
| Non-covalent modification | Poly(sodium 4-styrenesulfonate) | Water |
| | TCNQ | Water/DMF/DMSO |
| | PBA | Water |
| | SPANI | Water |
| Nucleophilic substitution | Alkyl amine/amino acid | CHCl$_3$/THF/toluene/DCM |
| Diazonium salt coupling | Aryl diazonium salt | DMF/DMAc/NMP |
| Electrochemical modification | Imidazolium based ionic liquids | DMF |
| Thermal treatment | – | NMP |
| π–π interaction | PNIPAAm | Water |

*Source: Progress in Polymer Science*, 35, Kuilla, T., Bhadra, S., Yao, D. et al., Recent advances in graphene based polymer composites, 1350–1375, Copyright (2010), with permission from Elsevier.

a stabilization medium, covalent modification by the amidation of the carboxylic groups, noncovalent functionalization of rGO, nucleophilic substitution to epoxy groups, and diazonium salt coupling (Kuilla et al. 2010).

# 30.3 Nanocomposite Processing

Processing methods used for the preparation of graphite and graphene-based polymer nanocomposites are similar to those used for other platelets like nanoparticles such as organo-modified clays but peculiar treatments and approaches in improving their interfacial and chemical properties should be performed on particles or on the host polymer (Figure 30.6). In particular, the compatibility between the polymer and the nanoparticle must be specifically designed or their dispersion must be optimized due to the different nature of graphite and graphene (organic materials) with respect to clays (inorganic materials).

Structural and functional properties of nanocomposites are strictly related to the extent of the regular dispersion of nanoparticles. For this reason, the main goal of each processing method is to achieve both the highest feasible degree of particle dispersion and the retention of the particle aspect ratio to exploit the huge potential of nanoparticles. Consequently, the selection of the most appropriate process should be carefully assessed according to costs, productivity, and final nanocomposite quality. A random, homogeneous, and aggregate-free nanoparticle dispersion allows for the best structural and/or functional properties at the lowest filler content unless a percolation path is required, such as for electrical conductivity.

The most interesting form of graphite nanoparticles for the production of well-dispersed nanocomposites is characterized by very thin sheet thickness (equal to a few graphene sheets) and high aspect ratio, as shown in Figure 30.7. In this form, graphite can compete with the most diffuse nanoreinforcements (layered silicate nanoclays and CNTs). Graphite nanosheets and graphene are actually the most investigated nanoscale fillers for research and industrial projects. Exfoliated GNPs can combine the lower price and layered structure of clays with the superior thermal and electrical properties of CNTs

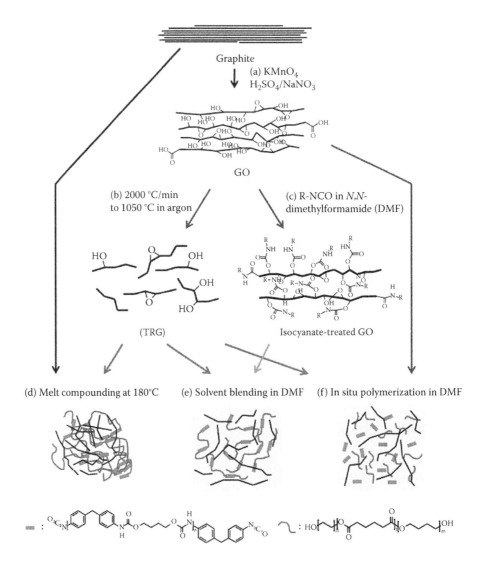

**FIGURE 30.6** Different routes for the dispersion of graphene or graphite nanoparticles into polymers. (Reprinted with permission from Kim et al. 2010a, 3441–3450. Copyright 2010 American Chemical Society.)

(Kalaitzidou et al. 2007b). This unique combination of characteristics renders graphite a serious competitor of nanoclays and nanotubes for both structural and functional applications.

Thin nanosheets of graphite can be obtained from EG through different processing routes. The first patents on the use of EG as reinforcement for polymers date back to the beginning of the 20th century (Aylsworth 1915, 1916). The use of conventional processing techniques, on the other hand, is more recent (Lincoln & Claude 1983). Since then, exfoliated graphite nanocomposites based on a wide array of polymers have been investigated, in all cases starting from particles with increased interplatelet gap. Huge efforts have been made to improve the starting morphology of graphite nanoparticles, because the effective dispersion of nanoplatelets is difficult to obtain (Chen et al. 2001a) and different procedures have been developed to prepare separate graphite nanosheets from EG (Weng et al. 2003). Graphite nanosheets prepared according to such a procedure were successfully incorporated into polymethylmethacrylate (PMMA) (Chen et al. 2003a) and polystyrene (PS) (Chen et al. 2003b).

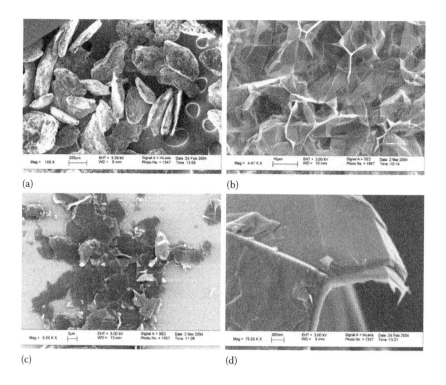

**FIGURE 30.7** SEM images of (a) intercalated, (b) expanded, (c) sonicated (exfoliated), and (d) graphite nanosheets. (Reprinted from *Composites Science and Technology*, 66, Yasmin, A., Luo, J.-J. & Daniel, I. M., Processing of expanded graphite reinforced polymer nanocomposites, 1182–1189, Copyright (2006), with permission from Elsevier.)

Graphene sheets, whichever technique is used to obtain their single layers, present more challenging issues than graphite nanosheets. In fact, graphite nanosheets possess an inherent stiffness due to the thickness of the particles, while graphene is very easily foldable. Its planar shape is difficult to keep intact and so its outstanding properties too. The main goal in preparing graphene-based nanocomposites is to get a satisfactory dispersion of nanolayers without folding a significant amount of them. As a consequence, the proper selection of the blending method is of paramount importance and gentle treatments are preferred. In all cases, the choice of the dispersing technique comes, again, from a balance between costs, productivity, and target performance requirements.

The main processing routes to disperse graphitic nanoparticles in polymers can be summarized as follows:

- In situ polymerization
- Polymerization filling technique
- Solution intercalation
- Direct mixing
- Melt blending
- Combined approaches

It is worth noting that, in general, in situ processing techniques allow for the best results in terms of nanoparticle dispersion and retention of the starting shape morphology and aspect ratios, because the low stresses applied to the particles do not break them and hinder their reaggregation. Sheet breaking and reduction of the aspect ratio, in fact, strongly increase when high shear rates are applied, such as in extruders or internal mixers. As a result, nanocomposites from in situ methods show better structural and functional properties compared to nanocomposites made by melt mixing processes. Another advantage of in situ methods is the lower processing temperature. In fact, a temperature higher than 220°C is usually required to exfoliate EG particles in a polymeric melt; thus, thermal sensitive polymers cannot be processed. Conversely, melt blending techniques allow higher productivity, lower environmental concerns, and are characterized by a lower technological complexity.

### 30.3.1 In Situ Polymerization

In situ polymerization is a method based on the use of one or more monomers that may be in situ linearly polymerized or cross-linked within nanoparticles (Shen et al. 2002) and was the first method used to synthesize nanocomposites, such as polymer-layered silicate nanocomposites based on polyamide 6 (Fukushima et al. 1988, Yano et al. 1993). It is still widely used in many studies due to the very good results it allows to reach. It has been typically used in nanocomposite systems based on thermosetting polymers but applications to thermoplastic precursors have also been investigated. Nylon 6/graphite nanocomposites, via intercalation polymerization of ε-caprolactam in the presence of EG (Pan et al. 2000), graphite–polystyrene, starting from a styrene–graphite–benzoyl peroxide mixture (Chen et al. 2001a,b, Zheng et al. 2002), and graphite–PAN nanocomposites (Mack et al. 2005) have been successfully prepared.

Nanoparticles are mixed with monomeric or oligomeric precursor of the polymeric matrix (Figure 30.8). Such components are intimately dispersed by means of functionalized nanoparticles or through the use of a specific compatibilizing agent. The low viscosity of the precursors eases the dispersion and intercalation processes without compromising the starting aspect ratio of the nanofiller. After the homogeneous dispersion of particles, precursors are polymerized to get a high–molecular weight polymer. In situ polymerization technique needs some aids for improving and/or speeding the dispersion process, such as ultrasound or other intensive local mixing processes. In particular, the nonpolar nature of graphite and graphene nanolayers requires special treatments for polar polymers to reach their complete exfoliation. In situ polymerization has the advantage of bypassing the thermodynamic or kinetic limitations associated with direct polymer intercalation, because monomers can penetrate into interlayer gaps more easily than large molecules (Huang et al. 2010).

In situ polymerization is usually performed by mixing the components in a stainless steel autoclave reactor equipped with a mechanical stirrer. In a typical reaction, the polymer precursor is poured

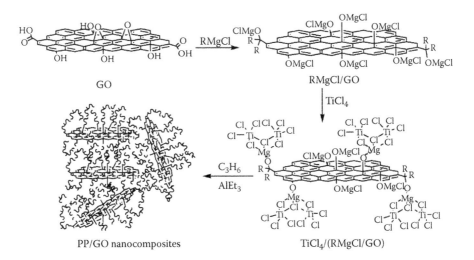

**FIGURE 30.8**  GO layers dispersed in an olefinic matrix before in situ polymerization takes place. (Reprinted with permission from Huang et al. 2010, 4096–4102. Copyright 2010 American Chemical Society.)

into the reactor with a solvent, and the compatibilizing agent is added if necessary. The reactor is then heated to the desired temperature. Nanoparticles and catalyst are added and vigorously stirred with the liquid mixture for an adequate time to reach the complete dispersion (exfoliation) of nanoparticles. The polymerization reaction is then initiated and a time interval is awaited, depending on the specific polymerization reaction, before the nanocomposite is obtained.

The dispersion of nanoparticles with the in situ polymerization technique is very good, and the aspect ratio is preserved even if stirring takes place because shear stresses are applied when the viscosity is very low (before polymerization). In graphite nanocomposites, this translates to no or very low particle ruptures. Due to the good compatibilization of polymer and particles, graphene nanosheets are usually not folded and their full potential can be exploited.

The main disadvantages of this technique are the high costs, long process, and very low productivity, since polymers have to be polymerized in small quantities. Furthermore, the presence of low–molecular weight species and solvents results in environmental issues of managing and recycling of wastes.

## 30.3.2 Solution Intercalation (or Solution Blending)

Solution-based methods for dispersing graphite or graphene nanoparticles are generally based on the mixing of colloidal suspensions of platelets with the desired polymer. This processing approach uses a solvent to dissolve the polymer, which being in solution has an increased mobility and facilitates the mutual interpenetration of all species by simple stirring or gentle shear mixing.

The process starts with the polymer dissolution and, when this step is complete, the required amount of nanoparticles is added. The solution is stirred for an adequate time depending on the peculiar polymer/nanoparticle system. This approach can be used with thermoplastic polymers and is similar to the in situ polymerization because the low viscosity is exploited to intercalate/exfoliate graphite or graphene sheets and to preserve, at the same time, the aspect ratio of particles. After the selected mixing time has ended, the solution is then cast into a mold and the solvent is removed. Special precautions should be taken to avoid the formation of particle aggregates during the solvent evaporation because this can reduce the properties of the nanocomposite (Moniruzzaman & Winey 2006). Another way to obtain the nanocomposite from the solution is through solute precipitation using a nonsolvent for the polymer. The precipitated nanocomposite has to be extracted and dried to ensure the complete evacuation of the solvent. Different solvents, from water to organic species such as chloroform or xylene, can be used depending on the polymer hydrophilicity.

Solution intercalation has been widely used because graphite nanoparticles or graphene can be processed in either water or organic solvents. Different thermoplastic polymers have been investigated to prepare graphite or graphene-based nanocomposites such as PMMA (Zheng et al. 2002, Zheng & Wong 2003), maleic anhydride-compatibilized polypropylene (Shen et al. 2003), polystyrene (Fang et al. 2009), polycarbonate (Higginbotham et al. 2009), polyacrylamide, and polyimides (Potts et al. 2011a). Water-soluble polymers are easily processed by means of this processing technique, such as PVA (Liang et al. 2009a) and poly(allylamine) (Xu et al. 2009, Satti et al. 2010).

Although the aggregation of nanoparticles can occur during the solvent evaporation, the dispersion of platelets in the nanocomposite is mainly governed by the level of nanoparticle exfoliation achieved during mixing. Due to the frequent need to functionalize the nanoparticle surface to obtain complete exfoliation, it should be considered that these treatments may affect the mechanical or functional (such as electric or thermal conductivity) properties of the final material. This processing method results in nanocomposites with higher mechanical performance and exhibits lower percolation thresholds compared to nanocomposites made from the exact same materials using the melt mixing technique (Pan et al. 2000).

## 30.3.3 Direct Mixing

The direct mixing method is often used in the case of low-viscosity thermosetting polymers (Celzard et al. 1996). It consists of the direct dispersion of graphite nanoparticles or graphene within a precursor of the polymer. A continuous and long stirring is applied to reach an adequate degree of particle intercalation or exfoliation. After this time interval, the other reactants are added to the blend and again mixed for the time needed to obtain the complete homogenization of the mixture. The final solution is then cast in a mold to give the final shape and cured at the required curing temperature.

Direct mixing differs from in situ polymerization. In the former, in fact, the polymer precursors are mixed with nanoparticles and then cross-linking is activated in a separate mold. In the in situ polymerization process, the polymer precursors are monomers, and the type of reaction is different and is performed usually in the same container where nanoparticles were mixed.

Direct mixing, such as all dispersion processes working on low- or very low–molecular weight polymers, may need some aids for improving and/or speeding the dispersion process, such as ultrasound or other intensive local shear mixing processes.

## 30.3.4 High-Shear Mixing (Melt Compounding) Techniques

In high-shear melt mixing, the polymer and nanoparticles are mixed in the melt state under high-shear conditions applied by means of internal mixers, single- or twin-screw extruders, or three-roll mills (Kalaitzidou et al. 2007b). Melt mixing techniques are more economical compared to all other dispersing approaches due to the higher productivity, the absence of solvents to be removed (and recycled), and the simpler process that can be easily implemented through standard industrial practices (Paul & Robeson 2008). However, this approach proved to be not able to achieve the same level of dispersion (intercalation or exfoliation) of the fillers as in the solvent mixing or in the in situ polymerization methods (Kim et al. 2010a).

In Figure 30.9, SEM micrographs from graphene nanocomposites obtained by means of the three main processing techniques are shown. It is evident that the best morphologies are obtained with in situ polymerization and solution mixing, while melt blending results in more compact, less intercalated/ exfoliated nanoparticles. The usual high temperatures and the high shear rates applied to the polymer can degrade nanoparticles or their surface treatment (Jeong et al. 2009) and foster an undesired aggregation process that lowers the performances of the final nanocomposite. Furthermore, it is very unlikely to reach an effective and broad dispersion of graphene or nanosized graphite particles, and the aspect ratio of the particles is reduced by either breaking or folding. For this reason, melt mixing is sometimes preceded by pretreatments of nanoparticles or predispersion in high-concentration masterbatches prepared with other techniques (Steurer et al. 2009). Several studies report melt mixing using TRGO (Zhang et al. 2010) and GNPs (Kalaitzidou et al. 2007a, Kim & Jeong 2010, Kim et al. 2010b) as fillers. These materials could be fed directly into the extruder and dispersed into the polymer without the use of any solvents or surfactants (Potts et al. 2011a). Another technical issue related to the melt mixing approach is due to the very low bulk density of nanoparticles (approximately 0.004 g/cm$^3$ based on a volumetric expansion of 500 for TRGO [Schniepp et al. 2006]). This renders the handling of the dry

| Melt compounding | Solution mixing | In situ polymerization |
| --- | --- | --- |

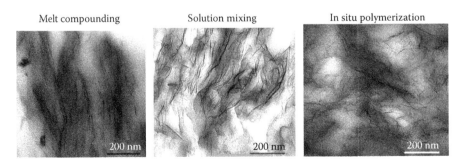

**FIGURE 30.9** Comparison of graphene nanoparticle dispersion obtained by using different dispersion techniques. (Reprinted with permission from Kim et al. 2010a, 3441–3450. Copyright 2010 American Chemical Society.)

nanoparticles very difficult and the feeding into the processing equipment, such as the hopper of the extruder, very challenging.

Despite all these issues, many investigations have been carried out on the melt mixing approach with the aim to solve such difficulties and obtain a more industrial-ready process. Melt mixing has been applied to prepare nanographite-based high-density polyethylene (HDPE) (Wang et al. 1986), PE and PS using a Brabender mixer (Krupa & Chodák 2001), HDPE using a HAAKE mixer and twin-screw extruder (Zheng et al. 2004), poly(ethylene naphthalate) (PEN) using a twin-screw extruder (Sorrentino et al. 2012, Sorrentino et al. 2014), nylon 6,6 and polycarbonate using a twin-screw extruder (Clingerman et al. 2002), and HDPE using a two-roll mill (Weng et al. 2004).

### 30.3.5 Three-Roll Milling

The three-roll mill, also known as triple-roll mill (Pötschke et al. 2013, Chandrasekaran et al. 2014), is a processing technique characterized by the use of three horizontal rolls, internally cooled due to the development of large amounts of heat, rotating at different relative speeds in opposite directions. The three-roll mill is typically used for thermoplastics and elastomers but thermosetting polymers are also treated. The mixing principle is to apply shear forces between the rolls to the blend in order to homogenize the viscous material. The three adjacent rolls are called *feed roll*, *center roll*, and *apron roll*, and rotate at progressively higher speeds. The recipe components are fed between the feed roll and the center roll. Due to the friction with rolls, the polymer undergoes very strong shear stresses that increase the temperature up to melting. When plasticized (thermoplastics) or reduced in paste (thermosettings), the polymer and all other mixture components flow between the rolls and are blended. Upon exiting from the first roll gap, some material adheres to the center roll and moves through the second roll gap between the center and the apron rolls. In this stage the blend undergoes even higher shear stresses to the higher speed of the apron roll and, typically, to the smaller gap than that between the feed and center rolls. A knife blade then scrapes the blend off the apron roll to pick it up. The milling process is repeated several times on the same blend to get the highest level of dispersion, especially if nanoparticles have to be dispersed. During each subsequent milling cycle, the gaps between the rolls are usually reduced, by means of mechanical or hydraulic actuators, to induce higher shear stresses and promote a finer nanoparticle dispersion. It is worth noting that the roll gap at each cycle has to be greater than the particle size, and after each cycle, a minimum gap limit is defined according to the particle size obtained during the previous milling step.

### 30.3.6 Pressure Wave (Ultrasound)–Assisted Dispersion Aids

Dispersing aids are very helpful in enhancing the nanoparticle dispersion and their use is highly recommended in all nanoparticle mixing techniques. Dispersing aids are able to concentrate high specific energies on nanoparticles for short times. The most diffuse approach lies on ultrasound (pressure waves). The ultrasound technique is increasingly used in the generation of nanoparticles using a top–down approach from microsized starting particles. The formation of nanometer-thick layers is achieved through the deagglomeration of graphite by means of ultrasound waves. This process relies on the acoustic cavitation phenomenon, which consists of the formation, growth, and implosive collapse of bubbles in a liquid such as a solvent or a low viscosity polymer precursor (Figure 30.10) (Unalan et al. 2014). The sonotrode in the ultrasonicator device generates high-intensity ultrasound waves, characterized by alternating expansive and compressive cycles. This causes the formation of yields intense shockwaves that promote interparticle collisions and, eventually, the exfoliation of the layered filler.

Cavitation induces particle intercalation or exfoliation because the interlayer distance increases as a consequence of different effects: temperature increase (>4500°C), high pressures (>100 MPa), heating–cooling cycles rates (>10⁹°C s⁻¹), and liquid jet streams (>10⁵ m s⁻¹).

Ultrasonic sonication (ultrasonication) can also lead to the detaching of a few graphene layers from graphite flakes in proper conditions, in either water-based or organic-based solutions. Examples of direct

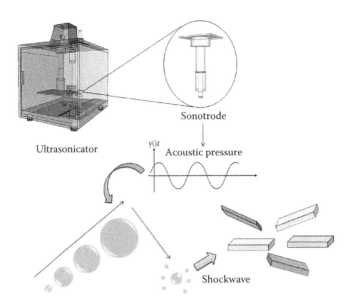

**FIGURE 30.10** Schematic illustration of the ultrasonication method based on the acoustic cavitation. (Unalan, I., Cerri, G. & Marcuzzo, E., *RSC Advances*: 4; 29393–29428, 2014. Reproduced by permission of The Royal Society of Chemistry.)

exfoliation of graphite into FLG sheets by (bio)polymer-assisted ultrasonication have been reported for PVP and PVA, a pyrene-functionalized amphiphilic block copolymer, pyrene polymers, gum arabic, and acrylate polymers (Zheng et al. 2012, Liu et al. 2013, Sun et al. 2013, Unalan et al. 2014).

### 30.3.7 Solid-State Shear Pulverization

Solid-state shear pulverization (SSSP) is another dispersion technique that allows the production of polymer–graphite nanocomposites without undergoing the thermodynamic/kinetic limitations associated with the conventional high–shear rate mixing processes (Wakabayashi et al. 2008). As described by Lebovitz et al. (2003), a twin-screw extruder has been modified to apply both shear and compressive stresses to materials (polymer and nanoparticles) in the solid state (Figure 30.11). The SSSP has been

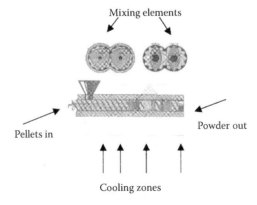

**FIGURE 30.11** Schematics of the SSSP dispersion technique. (Reprinted from *Polymer*, 44, Lebovitz, A. H., Khait, K. & Torkelson, J. M., Sub-micron dispersed-phase particle size in polymer blends: Overcoming the Taylor limit via solid-state shear pulverization, 199–206, Copyright (2003), with permission from Elsevier.)

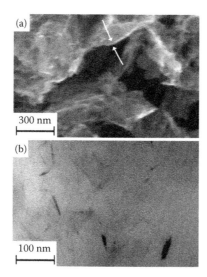

**FIGURE 30.12** Distribution of graphite nanoparticles (a) before and (b) after SSSP process. (Reprinted with permission from Wakabayashi et al. 2008, 1905–1908. Copyright 2008 American Chemical Society.)

applied to polypropylene (PP) and unmodified graphite. The SSSP process resulted in well-dispersed unmodified graphite in PP. Even though particle pretreatments are not needed and the extent of particle dispersion is very good, this technique showed a huge reduction in the aspect ratio of nanocomposite particles compared to other techniques (Figure 30.12), leading to improvements of structural properties lower than expected. In fact, the authors failed to find completely isolated graphene sheets or nanoplatelets in a fully extended form and only wavy or wrinkled shapes were detected.

## 30.3.8 Combined Approaches

Since none of the mentioned processes is very effective in dispersing nanoparticles alone, the most promising approach is to combine different techniques to exploit the best characteristics of each one (see, for example, the schematics in Figure 30.13). Since it is widely assessed that ultrasonic treatments during mixing in low-viscosity polymers help in lowering the mixing time and in improving the dispersing efficiency, several efforts are being made to extend their use to melt blending processes. This would reduce the intensity of shear stresses and mitigate the aspect ratio decrease and the complexity of the extruders' design.

A remarkable example of a combined approach has been proposed by Kalaitzidou et al. (2007b), where the authors used a solution-based pretreatment to coat a polymer in powder form with GNP and then used conventional melt blending techniques to produce the final nanocomposite. The coating was obtained in an ultrasonicated solution by stirring a mixture of PP powder and GNP for an adequate time interval. After the coating process stopped, the solvent was removed. Coated PP powders were thus ready to be processed with conventional meld blending techniques, such as compression or injection molding. In this way, a very good exfoliation of GNP has been achieved and particles have been evenly dispersed in the polymer as a very thin coating. In this approach, the melt blending process only served to shape the nanocomposite and not to disperse the filler. The comparison of such an approach with the conventional melt blending process showed that the coating method was more effective than the polymer solution method in terms of lowering the percolation threshold of thermoplastic nanocomposites and enhancing the probability to preserve the large platelet morphology of EG nanoparticles in the final composite.

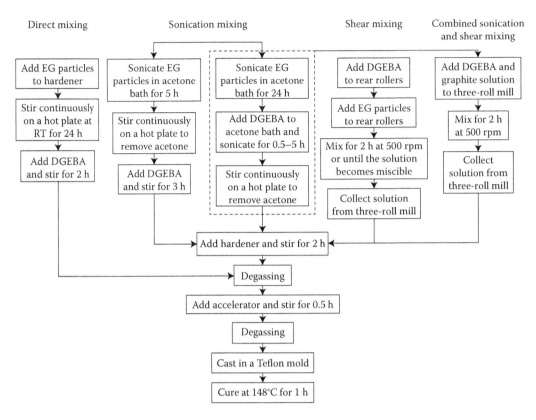

**FIGURE 30.13** Flowchart of combined processing techniques used by Yasmin, Luo, and Daniel (2006). (Reprinted from *Composites Science and Technology*, 66, Yasmin, A., Luo, J.-J. & Daniel, I. M., Processing of expanded graphite reinforced polymer nanocomposites, 1182–1189, Copyright (2006), with permission from Elsevier.)

Another example of the combined approach has been proposed by Yasmin et al. (2006) for a thermo-setting matrix. The authors prepared nanocomposites based on an epoxy resin and EG by using different processing techniques (Figure 30.13) and compared the dispersion efficiencies of direct mixing, sonication mixing, high-shear mixing, and a combined method (sonication and shear mixing). The results of the mechanical characterization showed that nanocomposites prepared with the aid of shear mixing provided the best exfoliation and dispersion of graphite nanosheets and consequently the best matrix stiffness enhancement over other processing techniques (Schilling et al. 2005).

## 30.4 Properties and Applications

The dispersion of graphite and graphene nanoparticles into polymeric matrices may result in the improvement of several mechanical (tensile and flexural modulus, tensile and flexural strength, strain at break, fracture toughness) and functional properties (electrical and thermal conductivity, electromagnetic shielding, thermal stability, and flame retardancy) (Huang et al. 2012, Wang et al. 2013, 2014).

Graphite- and graphene-based polymer nanocomposites can be produced with different processing techniques, several particle sizes, and surface functionalizations as shown in Sections 30.2 and 30.3. The selection of the dispersing process and of the nanoparticle types depends on the hosting matrix (thermoplastic, thermosetting, or elastomeric), on the required application, namely, the required improvement in mechanical and/or functional properties, and on the required amount of material to produce. For example, if the target requirement is to obtain a thermosetting system with high electrical conductivity, a nonfunctionalized graphene should be dispersed in the low-viscosity precursors via a solution mixing process aided by ultrasonication. Such a mixing could be followed by further processing with a three-roll milling machine. Conversely, if improved mechanical properties are the main target together with a large production of material, a functionalized graphite-based nanocomposite manufactured through the melt blending dispersing approach (e.g., by means of a twin-screw extruder) could be a viable route. It is worth highlighting that in the design process of nanoreinforced systems, the availability and the quality of the graphite/graphene nanoparticles strongly affect the cost of raw materials and can also affect the selection of the manufacturing process. Several graphite-/graphene-based nanoparticles are currently available on the market (see http://www.nanowerk.com/ for an example), but their cost is variable and usually strictly related to their purity and/or surface functionalization.

The polymer type commonly drives the selection of the dispersion technique; in situ polymerization, direct mixing, and three-roll milling are methods used for thermosetting polymers, whereas hot blending mixing at high temperatures, by means of internal mixers as well as single- or twin-screw extruders, or SSSP is adopted for thermoplastic polymers. Nevertheless, intensive mixing facilities, such as extruders, could be employed to produce thermosetting nanocomposites and the three-roll milling approach has been used to manufacture nanoreinforced thermoplastic polymers (Pötschke et al. 2013).

The performances of the nanocomposite are influenced by several factors such as the dispersing/manufacturing process, the matrix type and nanoparticle characteristics (amount, mean particle size), and the surface functionalization, which can heavily affect both the dispersion of nanoparticles and the interaction with the hosting matrix. The particle–polymer interaction is responsible for the structural properties' increase, such as strength or toughness, but may also be used to control the degree of crystallinity if the matrix is a semicrystalline thermoplastic polymer (Aurilia et al. 2011).

Since the dispersing/manufacturing process and the matrix and nanoparticle types are parameters with mutual influence, graphite-based or graphene-based nanocomposites may largely differ in properties as a consequence of small variation in those parameters. Relevant properties and their correlations with nanocomposite composition and production process are reported in the following. In particular, the effect of the graphene functionalization and dispersion on mechanical behavior, electrical and electromagnetic shielding properties, gas barrier properties, and thermal conductivity has been addressed. A brief mention has been made on the role that graphitic nanoparticles have on the main characteristics of porous lightweight materials (foams).

Some reviews in the scientific literature report on specific performances of graphene- and graphite-based nanocomposites based on thermoplastics and thermosetting polymers (Kuilla et al. 2010), elastomers (Wan et al. 2014a), and foamed polymers (Sorrentino et al. 2011, Antunes & Velasco 2014). In fact, it is worth pointing out that the research and development of either graphite/graphene particles' synthesis or their nanocomposite preparation is very dynamic, and new valuable papers are issued almost monthly; therefore, a scientific research survey on the latest articles is strongly recommended to be up to date on this topic.

## 30.4.1 Mechanical Properties

One of the main reasons why nanoparticles based on graphite or graphene are used lies on their capability to impart a strong mechanical reinforcement (Tables 30.3 and 30.4). In particular, due to the very wide ratio between the surface and the mass, just a few thousandths in volume of particles can significantly raise the stiffness of the nanocomposite. In some cases, a particle content as low as 0.01 wt% can almost double the elastic modulus of the pristine polymer (Figure 30.14). However, to get such an increase and minimize the amount of particles needed, the dispersion in the matrix has to be very effective and layers must be isolated. The latter requirement allows to maximize the stiffness increase since any aggregation of particles can reduce the reinforcing effect.

The strength of a nanocomposite, on the other hand, is influenced by the compatibility between the particle and the polymers. Particle functionalization, useful in promoting the dispersion of graphite nanoparticles or graphene nanolayers into the polymer, can also play a major role in improving the

**TABLE 30.3** Mechanical Performance Increase for Elastomeric Nanocomposites Reinforced with Graphitic Nanoparticles

| Elastomer | Filler | Compatibilizer | Content | Processing | Modulus (MPa) (Composite/Pure Matrix) | Tensile Strength (MPa) (Composite/Pure Matrix) | Elongation at Break (%) (Composite/Pure Matrix) |
|---|---|---|---|---|---|---|---|
| NR latex | CRG in situ reduction | – | 2 wt% | Solution | M 300 6.6/2.4 | 25.20/17.10 | 564/579 |
| | | | | Melt | 2.47/2.4 | 18.80/17.10 | 600/579 |
| NR | TRG | – | 1.2 (vol%) | Melt/Solution | 1.3/– | – | – |
| SBR | FGS | | 2 wt% | – | | 11.00/2.00 | – |
| SR | FGS | – | 0.05 wt% | – | 1.42/1.33 | 0.52/0.57 | 66/74 |
| | | | 3 wt% | – | 4.8/1.33 | 3.43/0.57 | 112/74 |
| HXNBR | GO | – | 0.44 (vol%) | Solution | M 200 3.4/1.7 | 22.40/14.80 | 419/534 |
| | | | 1.3 (vol%) | Solution | 6.5/1.7 | 10.30/14.80 | 248/534 |
| IIR | MG | CTAB | 1 wt% | Solution | ~0.5/0.17 | ~0.22/0.20 | ~320/600 |
| | | | 10 wt% | Solution | ~3.5/0.17 | ~0.50/0.20 | ~300/600 |
| SBR | FGS | – | 2 phr | – | 5/1 | 11.00/1.80 | 600/400 |
| SBR | AG | – | 5 phr | Melt | M 300 22.6/21.9 | 29.00/25.70 | 397/342 |
| | | bis-(3-triethoxy silylpropyl)tetra sulfane | 5 phr | Melt | 25.3/21.9 | 30.00/25.70 | 395/342 |
| SBR | Graphite:HAF | – | 30:20 phr | Melt | M 100 2.70/– | 13.40/– | 511/– |
| SBR | Graphite (<53 pm) | | 140 phr | Melt | | 4.35/1.03 | 80/235 |
| | Graphite (125 pm) | | 140 phr | Melt | | 3.25/1.03 | 67/235 |
| TPU | TRG | – | 1.6 (vol%) | Solution | 6.1–7.1/– | – | – |
| TPU | GNPs | – | 2.7 (vol%) | Solution | 35/10 | 25.00/28.00 | – |

*(Continued)*

**TABLE 30.3 (CONTINUED)**  Mechanical Performance Increase for Elastomeric Nanocomposites Reinforced with Graphitic Nanoparticles

| Elastomer | Filler | Compatibilizer | Content | Processing | Modulus (MPa) (Composite/Pure Matrix) | Tensile Strength (MPa) (Composite/Pure Matrix) | Elongation at Break (%) (Composite/Pure Matrix) |
|---|---|---|---|---|---|---|---|
| NBR latex | EG | – | 10 phr | Solution | 11.5/1.1 | 11.80/4.00 | 110/410 |
| | | | | Melt | 1.8/1.1 | 5.80/4.00 | 610/410 |
| NBR | Graphite (diameter < 2 pm thickness ~130 nm) | – | 60 phr | Melt | M100 4.3/1.0 | 10.20/3.00 | 563/480 |
| XNBR | EG | SDS | 5 phr | Solution | M100 2.4/1.3 | 12.20/7.40 | 590/590 |
| NBR-PVC (60:40) | Graphite (<30 pm) | – | 70 phr | Melt | 25/5 | 4.50/1.80 | 53/56 |
| SR | Nickel-coated graphite 40:60 | Vinyltriethoxy silane | 200 phr | Melt | M100 3.26/1.03 | 3.87/4.56 | 142/475 |
| PDMS | EG | – | 15 wt% | Solution | 1/– (220% increase) M100 | – | – |
| PDMS | FGS | – | 2 wt% | Melt | 0.99 ± 0.03/0.33 ± 0.05 M100 | – | 528 ± 32/842 ± 23 |
| | CNT | – | 2 wt% | Melt | 0.69 ± 0.06/0.33 ± 0.05 M100 | – | 583 ± 14/842 ± 23 |
| NBR | EG (micro) | – | 10 phr | Melt | – | 6.00/5.80 | 250/300 |
| | EG (nano) | – | 10 phr | Melt | – | 10.00/5.80 | 220/300 |
| NBR | EG/CB (micro) | – | 5/40 phr | Melt | 4.30/1.70 | 19.80/2.80 | 381/300 |
| | EG/CB (nano) | – | 5/40 phr | Melt | 7.60/1.70 | 18.80/2.80 | 218/300 |

*Source: Progress in Polymer Science*, 39, Sadasivuni, K. K., Ponnamma, D., Thomas, S. et al., Evolution from graphite to graphene elastomer composites, 749–80, Copyright (2014), with permission from Elsevier.

**TABLE 30.4** Mechanical Properties of Graphene/Graphite-Based Polymer Nanocomposites

| Matrix | Filler Type | Filler Loading (wt.%[a], vol.%[b]) | | Process | % Increase E | % Increase TS | % Increase Flexural Strength |
|---|---|---|---|---|---|---|---|
| Epoxy | EG | 1.00 | a | Sonication | 8 | −20 | |
| | EG | 1.00 | a | Shear | 11 | −7 | |
| | EG | 1.80 | a | Sonication and shear | 15 | −6 | |
| | EG | 0.10 | a | Solution | | | 87 |
| PMMA | EG | 21.00 | a | Solution | 21 | | |
| | GNP | 5.00 | a | Solution | 133 | | |
| PP | EG | 3.00 | b | Melt | | | 8 |
| | xGnP-1 | 3.00 | b | Melt | | | 26 |
| | xGnP-15 | 3.00 | b | Melt | | | 8 |
| | Graphite | 2.50 | b | SSSP | | 60 | |
| LLDPE | xGnP | 15.00 | a | Solution | | 200 | |
| | Parrafin-coated xGnP | 30.00 | | Solution | | 22 | |
| HDPE | EG | 3.00 | a | Melt | 100 | 4 | |
| | UG | 3.00 | a | Melt | 33 | | |
| PPS | EG | 4.00 | a | Melt | | | −20 |
| | S–EG | 4.00 | a | Melt | | | −33 |
| PVA | GO | 0.70 | a | Solution | | 76 | |
| | Graphene | 1.80 | b | Solution | | 150 | |
| TPU | Graphene | 5.10 | b | Solution | 200 | | |
| | Sulfonated Graphene | 1.00 | a | Solution | | 75 | |
| PETI | EG | 5.00 | a | In situ | 39 | | |
| | | 10.00 | a | In situ | 42 | | |

*Source: Progress in Polymer Science,* 35, Kuilla, T., Bhadra, S., Yao, D. et al., Recent advances in graphene based polymer composites, 1350–1375, Copyright (2010), with permission from Elsevier.

physical interactions between polymer and nanoreinforcement in the solid-state and can be a viable route to foster the increase of the strength (tensile, compressive, flexural, etc.) of the final material. Since from a mechanical point of view, the stronger the interaction, the stronger the interface strength, the potential of graphite and graphene addition is huge because they are among the strongest materials available to humans. In an example by Wan et al. (2014b), the effect of GO functionalization (GO with a diglycidyl ether of bisphenol-A) on nanoparticle dispersion and mechanical properties has been presented. The surface functionalization of the diglycidyl ether of bisphenol-A layer was found to effectively improve the compatibility and dispersion of GO sheets in epoxy matrix. The tensile and flexural properties of the systems reinforced with both types of GO presented an increment compared to the neat resin but the epoxy-containing functionalized GO showed the largest improvements (Figure 30.14).

Another important effect of incorporating graphite or graphene nanoparticles into the polymeric matrix can be the increase in the fracture toughness. This property is particularly important for thermosetting matrices, which are typically prone to damaging after weak impacts. Occasional microcracks, originated from the manufacturing of the object or induced during its service life, grow very quickly in thermosetting matrices, a condition that can end with a premature failure. The use of graphite or graphene nanoparticles can help in both retarding their appearance and blocking their propagation through the matrix and is more effective than other carbon-based nanofillers (Figure 30.15). Also,

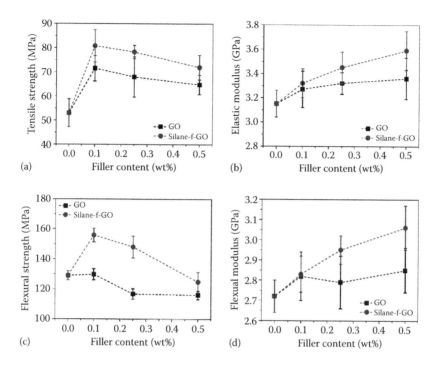

**FIGURE 30.14** Tensile and flexural properties of neat epoxy and its nanocomposites: (a) elastic modulus, (b) tensile strength, (c) flexural modulus, and (d) flexural strength. (Reprinted from *Composites Part A: Applied Science and Manufacturing*, 64, Wan, Y.-J., Gong, L.-X., Tang, L.-C. et al., Mechanical properties of epoxy composites filled with silane-functionalized graphene oxide, 79–89, Copyright (2014), with permission from Elsevier.)

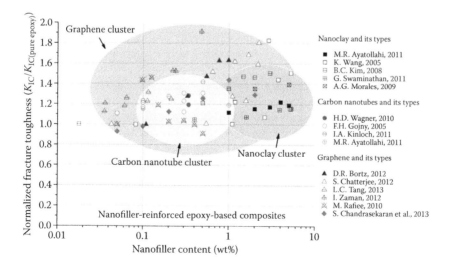

**FIGURE 30.15** Normalized fracture toughness as a function of filler content for different nanofiller-reinforced epoxies. (Reprinted from *Composites Science and Technology*, 97, Chandrasekaran, S., Sato, N., Tölle, F. et al., Fracture toughness and failure mechanism of graphene based epoxy composites, 90–99, Copyright (2014), with permission from Elsevier.)

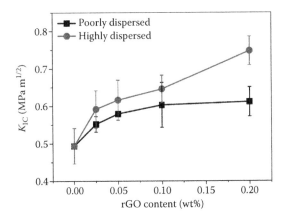

**FIGURE 30.16** Fracture toughness of cured epoxy composites as a function of rGO content. (Reprinted from *Carbon*, 60, Tang, L.-C., Wan, Y.-J., Yan, D. et al., The effect of graphene dispersion on the mechanical properties of graphene/epoxy composites, 16–27, Copyright (2013), with permission from Elsevier.)

for fracture toughness, the quality of the nanoparticle dispersion positively affects the performance increase, as shown in Figure 30.16 (Tang et al. 2013).

## 30.4.2 Electric and Electromagnetic Properties

Electrical conductivity of graphite/graphene nanocomposites varies over a very wide range, as shown in Tables 30.5 and 30.6. Such variability depends on the hosting matrix type, the processing technique used to prepare the nanocomposite, and the nanoparticle used, in particular its size and surface functionalization. In this regard, it is worth noting that particle percolation should be achieved to obtain electrical conduction through graphite nanoparticles or graphene in a nanocomposite. For this reason, the goal is to shift the percolation threshold to the lowest particle content. Unfunctionalized graphene nanoparticles dispersed with ultrasonication technique, due to the availability of all free electrons, yield the largest electrical conductivity improvement with the lowest percolation threshold (Table 30.5). For electron transport, as well as for structural properties, the particle dispersion plays a major role; but in this case, the surface functionalization is detrimental to electrical conductivity because it introduces defects on the particle surface or reduces the availability of free conducting electrons (Kuilla et al. 2010, Sadasivuni et al. 2014).

Electromagnetic interference (EMI) shielding of radio frequency radiations is an important requirement in different industrial fields such as electronics, communications, and the military. Polymer-based shielding materials are lightweight, are resistant to corrosion, and can be easily processed in the desired shapes. They are, as a consequence, very promising and are replacing conventional EMI shielding systems based on metals. The EMI shielding effectiveness (SE) of a nanocomposite mainly depends on the filler's intrinsic conductivity, dielectric constant, and aspect ratio. The large aspect ratio and the high conductivity render graphene and graphite nanolayers an easy choice for imparting EMI shielding properties to polymers. In fact, such as suggested by results shown in Figure 30.17, the addition of graphene into a polymer is able to strongly improve the EMI SE. In the 8–12 GHz microwaves band, the graphene/epoxy nanocomposite exhibited EMI SE equal to −21 dB at 8.2 GHz with 15 wt% (8.8 vol%) loading of functionalized graphene sheets (Liang et al. 2009b).

## 30.4.3 Gas Transport Properties

Gas permeation can be strongly reduced by the presence of exfoliated graphene sheets. The reduction of oxygen or carbon dioxide permeability in films for packaging and the reduction of gas diffusion

**TABLE 30.5** Electrical Percolation Threshold of Elastomeric Nanocomposites Based on Graphitic Nanoparticles

| Elastomer | Filler | Compatibilizer | Processing | Electrical Percolation Threshold |
|---|---|---|---|---|
| NR latex | TRG | – | Melt | >2.00 (vol%) |
| | | – | Solution | 0.80 (vol%) |
| | | – | Solution | 1.00 (vol%) |
| TPU | TRG | | In situ polymerization | 1.60 (vol%) |
| SBR latex | MLGS | HTAB | Solution | 0.50–1.00 wt% |
| NBR | GNP | – | Melt | 0.50 phr |
| SR | FGS | – | – | 0.10–0.20 wt% |
| SR | Graphene | | Solution | 2 wt% |
| | GNP | – | Solution | 0.009 (v/v) |
| SR | 8000 mesh graphite | – | Solution | 0.053 (v/v) |
| | 2000 mesh graphite | – | Solution | 0.07 (v/v) |
| PDMS | Exfoliated graphite | – | Solution | 3.00 wt% |
| NBR:PVC (60:40) | Graphite | – | Melt | 40–50 phr |
| | GNP | – | Solution | 3.00 wt% |
| RTVSR | | BYK–9076 | | |
| | GNP | (30 wt%) | Solution | 1.00 wt% |
| SR | Nickel-coated graphite (40:60) | Vinyltriethoxy silane | Melt | ~15 to 16 (vol%) |

*Source: Progress in Polymer Science*, 39, Sadasivuni, K. K., Ponnamma, D., Thomas, S. et al., Evolution from graphite to graphene elastomer composites, 749–80, Copyright (2014), with permission from Elsevier.

through elastomeric materials are some applications of such a concept. The use of fillers having sheet morphology, such as graphene, is more effective with respect to 1D fillers, such as CNTs, in lowering gas transport properties.

The permeability of a polymer to a specific gas is directly influenced by the diffusion coefficients and the solubility. The sheet morphology of nanoparticles affects both factors (Sadasivuni et al. 2014). In particular, diffusion coefficients are reduced because the tortuosity of the gas migration path is increased by the presence of nonpermeable species (Aurilia et al. 2010), while solubility is reduced due to the filling of the free volume into the polymer. Barrier properties are influenced by the filler amount, its shape factor, and the orientation of the carbon sheets. A perpendicular alignment and higher aspect ratios of filler platelets can provide highly tortuous paths (Figure 30.18), thus giving an increased barrier effect (Figure 30.19).

The influence of EG, GNP, and graphene on the barrier properties of rubber nanocomposites prepared by different processing methods is exemplified in Table 30.7. EG nanoplatelets are able to induce barrier properties similar to those exhibited by layered silicates. The dispersion of EG in the matrix and the interaction between carbon nanoparticles and the polymeric matrix (through the surface functionalization) influence the nanocomposite performance. Other applications on thermoplastics confirmed such behavior (Figure 30.20).

## 30.4.4 Thermal Properties

The presence of graphitic nanoparticles in polymers can affect all thermal phenomena, such as transition temperatures and thermal conductivity. A large scientific literature reports on changes in the glass transition temperature after graphene addition. Such effect is very important because it can increase

**TABLE 30.6**   Electrical Properties of Graphene/Graphite-Based Polymer Nanocomposites

| Matrix | Filler | Filler Loading (wt.%[a], vol.%[b]) | | Process | $\sigma$ (S m$^{-1}$) of Matrix | $\sigma$ (S m$^{-1}$) of Composite |
|---|---|---|---|---|---|---|
| Epoxy | EG | 3.00 | a | Sonication | $10^{-13}$ | $10^{-4}$ |
| | EG | 2.50 | b | Solution | $10^{-15}$ | $10^{-2}$ |
| | Graphene | 0.52 | b | Solution | $10^{-10}$ | $10^{-2}$ |
| PMMA | NanoG | 0.68 | b | In situ | $10^{-13}$ | $10^{-3}$ |
| | EG | 1.00 | a | Solution | $10^{-15}$ | $10^{-3}$ |
| | EG | 10.00 | a | In situ | – | 77.65 |
| PS | NanoG | 1.00 | a | In situ | $10^{-14}$ | $10^{-4}$ |
| | Graphene | 0.10 | b | Solution | $10^{-16}$ | $10^{-5}$ |
| | GNSC4P | 0.40 | b | Solution | $10^{-14}$ | $10^{-5}$ |
| | GNSC4P | 0.10 | b | Solution | $10^{-14}$ | 4 |
| | GNS8B | 0.20 | b | Solution | $10^{-14}$ | $10^{-5}$ |
| | GNS5D | 0.30 | b | Solution | $10^{-14}$ | $10^{-5}$ |
| | Graphene | – | – | Solution | $10^{-16}$ | 24 |
| | Graphene | 2.00 | a | In situ | $10^{-10}$ | $10^{-2}$ |
| | EG | 1.50 | b | In situ | $10^{-16}$ | $10^{-4}$ |
| | K-GIC | 8.20 | a | Solution | NA | – |
| Nylon-6 | EG | 1.50 | b | In situ | $10^{-15}$ | 0.1 |
| | FG | 0.75 | b | In situ | $10^{-15}$ | $10^{-5}$ |
| PP | xGnP–1 | 3.00 | b | Coating | $10^{-12}$ | 0.1 |
| | xGnP–1 | 3.00 | b | Solution | $10^{-12}$ | $10^{-2}$ |
| | xGnP–15 | 7.00 | b | Melt | $10^{-12}$ | $10^{-3}$ |
| | xGnP–15 | 5.00 | b | Coating | $10^{-12}$ | 0.1 |
| | EG | 0.67 | b | Solution | $10^{-16}$ | 0.1 |
| HDPE | EG | 3.00 | a | Melt | $10^{-16}$ | $10^{-8}$ |
| | UG | 5.00 | a | Melt | $10^{-16}$ | $10^{-10}$ |
| PPS | EG | 4.00 | a | Melt | $10s^{-12}$ | $10^{-3}$ |
| | S-EG | 4.00 | a | Melt | $10^{-12}$ | $10^{-2}$ |
| PANI | Graphite | 1.50 | a | In situ | 5 | 3300.3 |
| | GO | – | – | In situ | 2 | 1000 |
| PVDF | FGS | 2.00 | a | Solution | $10^{-11}$ | $10^{-2}$ |
| | EG | 5.00 | a | Solution | $10^{-11}$ | $10^{-3}$ |
| PVA-S | NanoG | 0.20 | a | Solution | $10^{-13}$ | $10^{-3}$ |
| PET | Graphene | 0.47 | b | Melt | $10^{-14}$ | $7.4 \times 10^{-2}$ |
| Polycarbonate | FGS | 2.00 | a | Melt | $10^{-14}$ | $10^{-9}$ |
| | Graphite | 12.00 | | Melt | $10^{-14}$ | $6.6 \times 10^{-11}$ |

*Source: Progress in Polymer Science*, 35, Kuilla, T., Bhadra, S., Yao, D. et al., Recent advances in graphene based polymer composites, 1350–1375, Copyright (2010), with permission from Elsevier.

the service temperature of the polymer, in particular in epoxy resin-based composites since gains in thermal stability can be critical for many high-performance applications. The use of reduced graphite nanoparticles (RGNs), such as, for example, that reported by Tang et al. (2013), can significantly increase the glass transition temperature of RGN/epoxy nanocomposites. Highly dispersed nanoparticles can improve the $T_g$ of the nanocomposite to more than 10°C, whereas a poor dispersion increases it to only a maximum of 2.5°C. This result is very interesting since comparable amounts of CNTs are not able to reach such a performance increase (Tang et al. 2011).

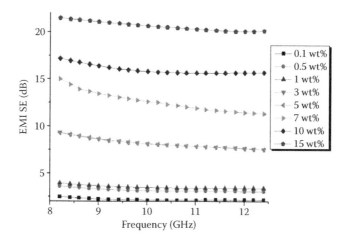

**FIGURE 30.17**  EMI SE of epoxy/graphene composites with various functionalized graphene sheet loadings as a function of the frequency in the X-band. (Reprinted from *Carbon*, 47, Liang, J., Wang, Y., Huang, Y. et al., Electromagnetic interference shielding of graphene/epoxy composites, 922–925, Copyright (2009), with permission from Elsevier.)

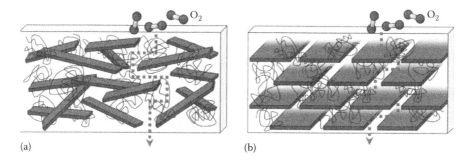

**FIGURE 30.18**  Formation of a tortuous path of platelets inhibiting diffusion of gases through an elastomer composite (a) without alignment and (b) with alignment. (Compton, O. C., Kim, S., Pierre, C. et al., Crumpled Graphene Nanosheets as Highly Effective Barrier Property Enhancers. *Advanced Materials*. 2010. 22. 4759–4763. Copyright Wiley-VCH Verlag GmbH & Co. KGaA. Reproduced with permission.)

Due to the highest thermal conductivity among all known materials (5000 W/m K), 2D platelet-like GNPs and graphene have a clear advantage over 1D carbon-based nanofillers, such as CNTs (3000 W/m K) or CNFs. This is due to their higher capability to transfer thermal energy through lattice vibrations (phonons). A poor coupling at the filler–polymer and filler–filler interfaces reduces the thermal conductivity, and a significant thermal resistance appears. To get a high thermal conductivity, the interfacial thermal resistance must be kept low and the presence of interfaces should be minimized. Unlike electrical conductivity, no rapid increase was observed in the thermal conductivity percolation threshold with the use of nanoparticles.

Several factors affect the thermal conductivity of the polymer nanocomposite: the geometry of the filler, the degree of exfoliation, and the orientation and interfacial interaction between the filler particles and the matrix. To minimize the presence of nonconducting interfaces, the coupling of 1D (SWCNTs) and 2D (GNPs) carbon-based nanoparticles proved to be an effective approach. Their synergistic effect improved the thermal conductivity by establishing an extended network of fillers in direct contact. In Table 30.8, the thermal conductivity values of graphene- and graphite-based elastomer nanocomposites

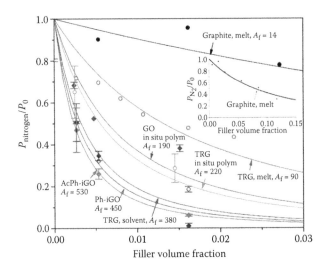

**FIGURE 30.19** Nitrogen permeability $P_{nitrogen}$ of thermoplastic polyurethane (TPU) nanocomposites based on graphite or graphene nanoparticles. (Reprinted with permission from Kim et al. 2010a, 3441–3450. Copyright 2010 American Chemical Society.)

**TABLE 30.7** Gas Permeabilities of Graphene, Graphite Derivative/Elastomer Nanocomposites

| Elastomer | Filler | Content | Processing | Permeant | Relative Reduction |
|---|---|---|---|---|---|
| NR | TRG | 1.70 vol% | Solution/melt | Air | 60 |
| TPU | TRG | 1.60 vol% | Melt | $N_2$ | 52 |
| | | 1.60 vol% | Solvent | $N_2$ | 82 |
| | | 1.50 vol% | In situ polymerization | $N_2$ | 71 |
| | IGO | 1.60 vol% | Solvent | $N_2$ | 94–99 |
| | GO | 1.50 vol% | In situ polymerization | $N_2$ | 62 |
| SR | FGS | 0.43 vol% | – | $N_2$ | 7.3 |
| | | 1.31 vol% | – | $N_2$ | 49 |
| NBR | EG | 10 phr | Melt | $N_2$ | 43 |
| | | 10 phr | Solution | $N_2$ | 62 |
| | | 5 phr | Solution | $N_2$ | 38 |
| XNBR | EG | 5 phr | Surfactant Aq-solution | $N_2$ | 52 |

*Source:* Kim et al. 2010a, 3441–3450, Copyright 2010 American Chemical Society; Prud'homme, R. K. et al., WIPO; Shuyang, Aksay, and Prud'homme 2011; Yang et al. 2007; Yang et al. 2010. With permission.

are reported. The thermal conductivity of nanocomposites is improved, in general, with respect to that of unfilled polymers. This can be attributed to the increased interfacial interactions. The proper functionalization of nanoparticles can enhance the dispersion and the physical–chemical interactions with the matrix. In some cases, the functionalization of EG improved the thermal conductivity by up to 20% over unmodified EG at the same loading (Ganguli et al. 2008).

## 30.4.5 Lightweight Porous Structures (Foams)

Polymeric foams are a class of materials which can benefit from the dispersion of graphene/graphite nanoparticles (Antunes & Velasco 2014). They are lightweight porous structures obtained by inducing

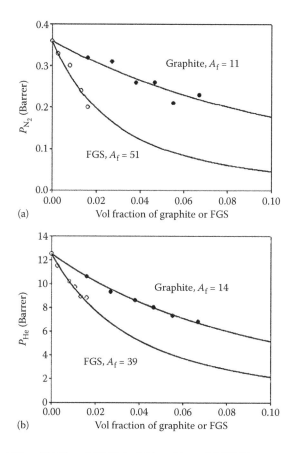

**FIGURE 30.20** Permeability of (a) $N_2$ and (b) He in PC/graphite and PC/FGS nanocomposites at 35°C. The permeability of the polycarbonate (PC)/FGS nanocomposites is significantly smaller than that of the PC/graphite composites. (Reprinted with permission from Kim and Macosko, 3797–3809, 2009. Copyright 2010 American Chemical Society.)

the internal formation of bubbles through blowing agents (physical or chemical in nature depending on the specific foaming process and on the final cellular structure expected). Due to the capability of graphite nanoparticles and graphene to reinforce the polymer and to act as a barrier to gas diffusion, such nanoparticles are very promising for the production of lightweight materials with high specific mechanical properties. In fact, the platelet-like nanoparticle shape is aligned in the polymer during the foaming process and can strongly reinforce the cell walls in the porous structure in the same way as fiber fabrics can reinforce laminated composites in the in-plane directions.

Graphitic nanoparticles can not only tune mechanical properties but also impart functional characteristics (EMI shielding, electrical conduction, barrier properties, and so on) to foams and improve their morphological parameters. In particular, their addition generally has the effect of increasing the number of cells per unit volume compared to foams obtained from neat polymers (Sorrentino et al. 2012) because they improve the nucleation and growth process of cells. As shown in Figure 30.21, it is evident that cells in EG/PEN nanocomposite have a mean size five times lower than in the unfilled polymer. This peculiar capability is responsible for the performance gains such as the mechanical behavior increase (tensile and compressive elastic moduli and strengths, impact toughness) and lowering of the thermal conductivity.

**TABLE 30.8** Thermal Conductivities of Graphene, Graphite Derivative/Elastomer Nanocomposites

| Elastomer | Filler | Compatibilization | Content | Processing | Thermal Conductivity (W/mK) (Composite/ Pure Matrix) |
|---|---|---|---|---|---|
| NR latex | CRG in situ reduction | – | 2.00 wt% | Solution | 0.19/0.17 |
| NR | Nanographene platelets (thickness <0.8 nm) | | 2.00 wt% | | 21.30/0.13 |
| | Nanographene platelets (thickness <6 nm) | | 2.00 wt% | | 2.11/0.13 |
| | Graphite nanoparticles (thickness = 350 nm) | | 2.00 wt% | | 0.85/0.13 |
| SBR | Graphite | – | 4.76 wt% | Melt | 0.34/0.30 |
| | Acid graphite | – | 4.76 wt% | Melt | 0.34/0.30 |
| | | bis-(3-triethoxysilylpropyl) tetrasulfane (0.8 phr) | 4.76 wt% | Melt | 0.36/0.30 |
| SBR | NG | – | 4.76 wt% | Melt | 0.33/0.30 |
| | AG | – | 4.76 wt% | Melt | 0.34/0.30 |
| | NGST | – | 4.76 wt% | Melt | 0.36/0.30 |
| | NGST | – | 4.76 wt% | Melt | 0.36/0.30 |
| NBR latex | EG | – | 9.09 wt% | Solution | 0.30/0.19 |
| | | | | Melt | 0.23/0.19 |
| SR | EG | – | 8.25 wt% | Solution | 0.24/0.17 |
| | | | | Melt | 0.32/0.17 |

*Source: Progress in Polymer Science*, 39, Sadasivuni, K. K., Ponnamma, D., Thomas, S. et al., Evolution from graphite to graphene elastomer composites, 749–80, Copyright (2014), with permission from Elsevier.

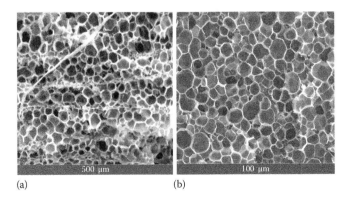

(a)                                                       (b)

**FIGURE 30.21** Effect of EG addition on the foam morphology of PEN nanocomposite foams: (a) unfilled PEN, (b) 1.0 wt% EG-filled PEN. (Reprinted from Sorrentino, L. et al., *Journal of Cellular Plastics*, 48, 355–268, 2012.)

# References

Antunes, M. & Velasco, J. I., "Multifunctional polymer foams with carbon nanoparticles," *Prog. Polym. Sci.* 39 (2014): 486–509. doi:10.1016/j.progpolymsci.2013.11.002.

Aurilia, M., Sorrentino, L., Sanguigno, L. et al., "Nanofilled polyethersulfone as matrix for continuous glass fibers composites: Mechanical properties and solvent resistance," *Adv. Polym. Technol.* 29 (2010): 146–160. doi:10.1002/adv.20187.

Aurilia, M., Piscitelli, F., Sorrentino, L. et al., "Detailed analysis of dynamic mechanical properties of TPU nanocomposite: The role of the interfaces," *Eur. Polym. J.* 47 (2011): 925–936. doi:10.1016/j.eurpolymj.2011.01.005.

Aylsworth, J. W., "Expanded graphite and composition thereof," US Patent 1,137,373 (1915).

Aylsworth, J. W., "Expanded graphite," US Patent 1,191,383 (1916).

Boehm, H.-P., Setton, R. & Stumpp, E., "Nomenclature and terminology of graphite intercalation compounds (IUPAC Recommendations 1994)," *Pure Appl. Chem.* 66 (1994): 1893–1901. doi:10.1351/pac199466091893.

Brodie, B. C., "On the atomic weight of graphite," *Philos. Trans. R. Soc. Lond.* 149 (1859): 249–259.

Celzard, A., McRae, E., Marêché, J. F. et al., "Composites based on micron-sized exfoliated graphite particles: Electrical conduction, critical exponents and anisotropy," *J. Phys. Chem. Solids* 57 (1996): 715–718. doi:10.1016/0022-3697(95)00337-1.

Chandrasekaran, S., Sato, N., Tölle, F. et al., "Fracture toughness and failure mechanism of graphene based epoxy composites," *Compos. Sci. Technol.* 97 (2014): 90–99. doi:10.1016/j.compscitech.2014.03.014.

Chen, G.-H., Wu, D.-J., Weng, W.-G. et al., "Preparation of polymer/graphite conducting nanocomposite by intercalation polymerization," *J. Appl. Polym. Sci.* 82 (2001a): 2506–2513. doi:10.1002/app.2101.

Chen, G.-H., Wu, D.-J., Weng, W.-G. et al., "Preparation of polystyrene–graphite conducting nanocomposites via intercalation polymerization," *Polym. Int.* 50 (2001b): 980–985. doi:10.1002/pi.729.

Chen, G., Weng, W., Wu, D. et al., "PMMA/graphite nanosheets composite and its conducting properties," *Eur. Polym. J.* 39 (2003a): 2329–2335. doi:10.1016/j.eurpolymj.2003.08.005.

Chen, G., Wu, C., Weng, W. et al., "Preparation of polystyrene/graphite nanosheet composite," *Polymer* 44 (2003b): 1781–1784. doi:10.1016/S0032-3861(03)00050-8.

Clingerman, M. L., King, J. A., Schulz, K. H. et al., "Evaluation of electrical conductivity models for conductive polymer composites," *J. Appl. Polym. Sci.* 83 (2002): 1341–1356. doi:10.1002/app.10014.

Collerson, K., "Graphite commodity update," HDR International, Inc. (2014).

Compton, O. C., Kim, S., Pierre, C. et al., "Crumpled graphene nanosheets as highly effective barrier property enhancers," *Adv. Mater.* 22 (2010): 4759–4763. doi:10.1002/adma.201000960.

Dresselhaus, M. S. & Dresselhaus, G., "Intercalation compounds of graphite," *Adv. Phys.* 30 (1981): 139–326. doi:10.1080/00018738100101367.

Fang, M., Wang, K., Lu, H. et al., "Covalent polymer functionalization of graphene nanosheets and mechanical properties of composites," *J. Mater. Chem.* 19 (2009): 7098–7105. doi:10.1039/b908220d.

Fitzer, E., Köchling, K. H., Boehm, H. P. et al., "Recommended terminology for the description of carbon as a solid (IUPAC Recommendations 1995)," *Pure Appl. Chem.* 67 (1995): 473–506.

Fukushima, Y., Okada, A., Kawasumi, M. et al., "Swelling poly-6-amide by poly-6-amide," *Clay Minerals* 23 (1988): 27–34.

Ganguli, S., Roy, A. K. & Anderson, D. P., "Improved thermal conductivity for chemically functionalized exfoliated graphite/epoxy composites," *Carbon* 46 (2008): 806–817. doi:10.1016/j.carbon.2008.02.008.

Geim, A. K., "Graphene: Status and prospects," *Science (New York, N. Y.)* 324 (2009): 1530–1534. doi:10.1126/science.1158877.

Geim, A. K., "Graphene prehistory," *Phys. Scripta* T146 (2012): 014003. doi:10.1088/0031-8949/2012/T146/014003.

Geim, A. K. & Novoselov, K. S., "The rise of graphene," *Nat. Mater.* 6 (2007): 183–191. doi:10.1038/nmat1849.

Higginbotham, A. L., Lomeda, J. R., Morgan, A. B. et al., "Graphite oxide flame-retardant polymer nano-composites," *ACS Appl. Mater. Interfaces* 1 (2009): 2256–2261. doi:10.1021/am900419m.

Huang, Y., Qin, Y., Zhou, Y. et al., "Polypropylene/graphene oxide nanocomposites prepared by in situ Ziegler–Natta polymerization," *Chem. Mater.* 22 (2010): 4096–4102. doi:10.1021/cm100998e.

Huang, G., Gao, J., Wang, X. et al., "How can graphene reduce the flammability of polymer nanocomposites?" *Mater. Lett.* 66 (2012): 187–189. doi:10.1016/j.matlet.2011.08.063.

Hummers, Jr., W. S. & Offeman, R. E., "Preparation of graphitic oxide," *J. Am. Chem. Soc.* 80 (1958): 1339. doi:10.1021/ja01539a017.

Jeong, H.-K., Lee, Y. P., Jin, M. H. et al., "Thermal stability of graphite oxide," *Chem. Phys. Lett.* 470 (2009): 255–258. doi:10.1016/j.cplett.2009.01.050.

Kalaitzidou, K., Fukushima, H. & Drzal, L. T., "Mechanical properties and morphological characterization of exfoliated graphite–polypropylene nanocomposites," *Compos. Part A Appl. Sci. Manuf.* 38 (2007a): 1675–1682. doi:10.1016/j.compositesa.2007.02.003.

Kalaitzidou, K., Fukushima, H. & Drzal, L. T., "A new compounding method for exfoliated graphite-polypropylene nanocomposites with enhanced flexural properties and lower percolation threshold," *Compos. Sci. Technol.* 67 (2007b): 2045–2051. doi:10.1016/j.compscitech.2006.11.014.

Kim, I.-H. & Jeong, Y. G., "Polylactide/exfoliated graphite nanocomposites with enhanced thermal stability, mechanical modulus, and electrical conductivity," *J. Polym. Sci. Part B Polym. Phys.* 48 (2010): 850–858. doi:10.1002/polb.21956.

Kim, H., Miura, Y. & Macosko, C. W., "Graphene/polyurethane nanocomposites for improved gas barrier and electrical conductivity," *Chem. Mater.* 22 (2010a): 3441–3450. doi:10.1021/cm100477v.

Kim, S., Do, I. & Drzal, L. T., "Thermal stability and dynamic mechanical behavior of exfoliated graphite nanoplatelets–LLDPE nanocomposites," *Polym. Compos.* 31 (2010b): 755–761. doi:10.1002/pc.20781.

Krupa, I. & Chodák, I., "Physical properties of thermoplastic/graphite composites," *Eur. Polym. J.* 37 (2001): 2159–2168. doi:10.1016/S0014-3057(01)00115-X.

Kuilla, T., Bhadra, S., Yao, D. et al., "Recent advances in graphene based polymer composites," *Prog. Polym. Sci.* 35 (2010): 1350–1375. doi:10.1016/j.progpolymsci.2010.07.005.

Lebovitz, A. H., Khait, K. & Torkelson, J. M., "Sub-micron dispersed-phase particle size in polymer blends: Overcoming the Taylor limit via solid-state shear pulverization," *Polymer* 44 (2003): 199–206. doi:10.1016/S0032-3861(02)00717-6.

Liang, J., Huang, Y., Zhang, L. et al., "Molecular-level dispersion of graphene into poly(vinyl alcohol) and effective reinforcement of their nanocomposites," *Adv. Funct. Mater.* 19 (2009a): 2297–2302. doi:10.1002/adfm.200801776.

Liang, J., Wang, Y., Huang, Y. et al., "Electromagnetic interference shielding of graphene/epoxy composites," *Carbon* 47 (2009b): 922–925. doi:10.1016/j.carbon.2008.12.038.

Lincoln, V. F. & Claude, Z., "Organic matrix composites reinforced with intercalated graphite," US Patent 4,414,142 (1983).

Liu, Z., Liu, J., Cui, L. et al., "Preparation of graphene/polymer composites by direct exfoliation of graphite in functionalised block copolymer matrix," *Carbon* 51 (2013): 148–155. doi:10.1016/j.carbon.2012.08.023.

Mack, J. J., Viculis, L. M., Ali, A. et al., "Graphite nanoplatelet reinforcement of electrospun polyacrylonitrile nanofibers," *Adv. Mater.* 17 (2005): 77–80. doi:10.1002/adma.200400133.

Moniruzzaman, M. & Winey, K. I., "Polymer nanocomposites containing carbon nanotubes," *Macromolecules* 39 (2006): 5194–5205. doi:10.1021/ma060733p.

Mouras, S., Hamm, A., Djurado, D. et al., "Synthesis of first stage graphite intercalation compounds with fluorides," *Rev. Chim. Minérale* 24 (1987): 572–582, http://cat.inist.fr/?aModele=afficheN&cps idt=7578318, Accessed April 17, 2015.

Pan, Y. X., Yu, Z. Z., Ou, Y. Z. et al., "New process of fabricating electrically conducting nylon 6/graphite nanocomposites via intercalation polymerization," *J. Polym. Sci. Part B Polym. Phys.* 38 (2000): 1626–1633. doi:10.1002/(SICI)1099-0488(20000615)38:12<1626::AID-POLB80>3.0.CO;2-R.

Paul, D. R. & Robeson, L. M., "Polymer nanotechnology: Nanocomposites," *Polymer* 49 (2008): 3187–3204. doi:10.1016/j.polymer.2008.04.017.

Pötschke, P., Krause, B., Buschhorn, S. T. et al., "Improvement of carbon nanotube dispersion in thermoplastic composites using a three roll mill at elevated temperatures," *Compos. Sci. Technol.* 74 (2013): 78–84. doi:10.1016/j.compscitech.2012.10.010.

Potts, J. R., Dreyer, D. R., Bielawski, C. W. et al., "Graphene-based polymer nanocomposites," *Polymer* 52 (2011a): 5–25. doi:10.1016/j.polymer.2010.11.042.

Potts, J. R., Murali, S., Zhu, Y. et al., "Microwave-exfoliated graphite oxide/polycarbonate composites," *Macromolecules* 44 (2011b): 6488–6495. doi:10.1021/ma2007317.

Prud'homme, R. K., Ozbas, B., Aksay, I. A. et al., "Functionalized graphene sheets having high carbon to oxygen ratios," WIPO 2009/134492 A2 2009 (2009).

Sadasivuni, K. K., Ponnamma, D., Thomas, S. et al., "Evolution from graphite to graphene elastomer composites," *Prog. Polym. Sci.* 39 (2014): 749–80. doi:10.1016/j.progpolymsci.2013.08.003.

Satti, A., Larpent, P. & Gun'ko, Y., "Improvement of mechanical properties of graphene oxide/poly(allylamine) composites by chemical crosslinking," *Carbon* 48 (2010): 3376–3381. doi:10.1016/j.carbon.2010.05.030.

Schniepp, H. C., Li, J. L., McAllister, M. J. et al., "Functionalized single graphene sheets derived from splitting graphite oxide," *J. Phys. Chem. B* 110 (2006): 8535–8539. doi:10.1021/jp060936f.

Sengupta, R., Bhattacharya, M., Bandyopadhyay, S. et al., "A review on the mechanical and electrical properties of graphite and modified graphite reinforced polymer composites," *Prog. Polym. Sci.* 36 (2011): 638–670. doi:10.1016/j.progpolymsci.2010.11.003.

Shen, Z., Simon, G. P. & Cheng, Y.-B., "Comparison of solution intercalation and melt intercalation of polymer–clay nanocomposites," *Polymer* 43 (2002): 4251–4260. doi:10.1016/S0032-3861(02)00230-6.

Shen, J. W., Chen, X. M. & Huang, W. Y., "Structure and electrical properties of grafted polypropylene/graphite nanocomposites prepared by solution intercalation," *J. Appl. Polym. Sci.* 88 (2003): 1864–1869. doi:10.1002/app.11892.

Sorrentino, L., Aurilia, M., Cafiero, L. et al., "Nanocomposite foams from high-performance thermoplastics," *J. Appl. Polym. Sci.* 122 (2011): 3701–3710. doi:10.1002/app.34784.

Sorrentino, L., Aurilia, M., Cafiero, L. et al., "Mechanical behavior of solid and foamed polyester/expanded graphite nanocomposites," *J. Cell. Plastics* 48 (2012): 355–268. doi:10.1177/0021955X12449641.

Sorrentino, L., Cafiero, L. & Iannace, S., "Foams from high performance thermoplastic PEN/PES blends with expanded graphite," in D'Amore, A., Acierno, D. & Grassia, L., eds., *Proceedings of PPS-29*, 1593 (2014): 358–361 (AIP Publishing LLC, Melville). doi:10.1063/1.4873800.

Steurer, P., Wissert, R., Thomann, R. et al., "Functionalized graphenes and thermoplastic nanocomposites based upon expanded graphite oxide," *Macromol. Rapid Commun.* 30 (2009): 316–327. doi:10.1002/marc.200800754.

Sun, Z., Pöller, S., Huang, S. et al., "High-yield exfoliation of graphite in acrylate polymers: A stable few-layer graphene nanofluid with enhanced thermal conductivity," *Carbon* 64 (2013): 288–294. doi:10.1016/j.carbon.2013.07.063.

Tang, L.-c., Zhang, H., Han, J.-h. et al., "Fracture mechanisms of epoxy filled with ozone functionalized multi-wall carbon nanotubes," *Compos. Sci. Technol.* 72 (2011): 7–13. doi:10.1016/j.compscitech.2011.07.016.

Tang, L.-C., Wan, Y.-J., Yan, D. et al., "The effect of graphene dispersion on the mechanical properties of graphene/epoxy composites," *Carbon* 60 (2013): 16–27. doi:10.1016/j.carbon.2013.03.050.

Unalan, I., Cerri, G. & Marcuzzo, E., "Nanocomposite films and coatings using inorganic nanobuilding blocks (NBB): Current applications and future opportunities in the food packaging sector," *RSC Adv.* (2014): 29393–29428. doi:10.1039/c4ra01778a.

Wakabayashi, K., Pierre, C., Diking, D. A. et al., "Polymer–graphite nanocomposites: Effective dispersion and major property enhancement via solid-state shear pulverization," *Macromolecules* 41 (2008): 1905–1908. doi:10.1021/ma071687b.

Wan, Y.-J., Tang, L.-C., Gong, L.-X. et al., "Grafting of epoxy chains onto graphene oxide for epoxy composites with improved mechanical and thermal properties," *Carbon* 69 (2014a): 467–480. doi:10.1016/j.carbon.2013.12.050.

Wan, Y.-J., Gong, L.-X., Tang, L.-C. et al., "Mechanical properties of epoxy composites filled with silane-functionalized graphene oxide," *Compos. Part A Appl. Sci. Manuf.* 64 (2014b): 79–89. doi:10.1016/j.compositesa.2014.04.023.

Wang, Y. S., O'Gurkis, M. A. & Lindt, J. T., "Electrical properties of exfoliated-graphite filled polyethylene composites," *Polym. Compos.* 7 (1986): 349–354. doi:10.1002/pc.750070512.

Wang, X., Song, L., Pornwannchai, W. et al., "The effect of graphene presence in flame retarded epoxy resin matrix on the mechanical and flammability properties of glass fiber-reinforced composites," *Compos. Part A Appl. Sci. Manuf.* 53 (2013): 88–96. doi:10.1016/j.compositesa.2013.05.017.

Wang, Z., Wei, P., Qian, Y. et al., "The synthesis of a novel graphene-based inorganic–organic hybrid flame retardant and its application in epoxy resin," *Compos. Part B Eng.* 60 (2014): 341–349. doi:10.1016/j.compositesb.2013.12.033.

Weng, W., Chen, G. & Wu, D., "Crystallization kinetics and melting behaviors of nylon 6/foliated graphite nanocomposites," *Polymer* 44 (2003): 8119–8132. doi:10.1016/j.polymer.2003.10.028.

Weng, W., Chen, G., Wu, D. et al., "HDPE/expanded graphite electrically conducting composite," *Compos. Interfaces* 11 (2004): 131–143. doi:10.1163/156855404322971404.

Xu, Y., Hong, W., Bai, H. et al., "Strong and ductile poly(vinyl alcohol)/graphene oxide composite films with a layered structure," *Carbon* 47 (2009): 3538–3543. doi:10.1016/j.carbon.2009.08.022.

Yano, K., Usuki, A., Okada, A. et al., "Synthesis and properties of polymide–clay hybrid," *J. Polym. Sci. Part A Polym. Chem.* 31 (1993): 2493–2498. doi:10.1002/pola.1993.080311009.

Yasmin, A., Luo, J.-J. & Daniel, I. M., "Processing of expanded graphite reinforced polymer nanocomposites," *Compos. Sci. Technol.* 66 (2006): 1182–1189. doi:10.1016/j.compscitech.2005.10.014.

Zhang, H.-B., Zheng, W.-G., Yan, Q. et al., "Electrically conductive polyethylene terephthalate/graphene nanocomposites prepared by melt compounding," *Polymer* 51 (2010): 1191–1196. doi:10.1016/j.polymer.2010.01.027.

Zheng, W., Wong, S.-C. & Sue, H.-J., "Transport behavior of PMMA/expanded graphite nanocomposites," *Polymer* 43 (2002): 6767–6773. doi:10.1016/S0032-3861(02)00599-2.

Zheng, W. & Wong, S.-C., "Electrical conductivity and dielectric properties of PMMA/expanded graphite composites," *Compos. Sci. Technol.* 63 (2003): 225–235. doi:10.1016/S0266-3538(02)00201-4.

Zheng, W., Lu, X. & Wong, S-c., "Electrical and mechanical properties of expanded graphite-reinforced high-density polyethylene," *J. Appl. Polym. Sci.* 91 (2004): 2781–2788. doi:10.1002/app.13460.

Zheng, X., Xu, Q., Li, J. et al., "High-throughput, direct exfoliation of graphite to graphene via a cooperation of supercritical $CO_2$ and pyrene-polymers," *RSC Adv.* 2 (2012): 10632–10638. doi:10.1039/c2ra21316h.

# Index

Page numbers followed by f and t indicate figures and tables, respectively.

Printed and bound by CPI Group (UK) Ltd, Croydon, CR0 4YY

01/11/2024

01782604-0017